ILLUSTRATED HANDBOOK OF PHYSICAL-CHEMICAL PROPERTIES AND ENVIRONMENTAL FATE FOR ORGANIC CHEMICALS

Volume IV

Oxygen, Nitrogen, and Sulfur Containing Compounds

Donald Mackay
Wan-Ying Shiu
Kuo-Ching Ma

LEWIS PUBLISHERS

Boca Raton New York London Tokyo

Library of Congress Cataloging-in-Publication Data

Mackay, Donald, Ph.D.

Illustrated handbook of physical-chemical properties and environmental fate for organic chemicals.

Includes bibliographical references and index.

Contents: v. 1. Monoaromatic hydrocarbons, chlorobenzens, and PCBs — v. 2. Polynuclear aromatic hydrocarbons, polychlorinated dioxins, and dibenzofurans — v. 3. Volatile organic chemicals — v. 4. Oxygen, nitrogen, and sulfur containing compounds.

1. Organic compounds — Environmental aspects — Handbooks, manuals, etc. 2. Environmental chemistry — Handbooks, manuals, etc. I. Shiu, Wan Ying II. Ma, Kuo Ching. III. Title.

TD196.073M32 1992 628.5'2 91-33888

International Standard Book Number 0-87371-513-6 (v. 1)
International Standard Book Number 0-87371-583-7 (v. 2)
International Standard Book Number 0-87371-973-5 (v. 3)
International Standard Book Number 1-56670-035-3 (v. 4)
Library of Congress Card Number 91-33888
Printed in the United States of America 1 2 3 4 5 6 7 8 9 0
Printed on acid-free paper

Illustrated Handbook of Physical-Chemical Properties and Environmental Fate for Organic Chemicals

Volume I
Monoaromatic Hydrocarbons, Chlorobenzenes, and PCBs

Monoaromatic Hydrocarbons
Chlorobenzenes
Polychlorinated Biphenyls (PCBs)

Volume II
Polynuclear Aromatic Hydrocarbons, Polychlorinated Dioxins, Dibenzofurans

Polynuclear Aromatic Hydrocarbons (PAHs)
Chlorinated Dibenzo-*p*-Dioxins
Chlorinated Dibenzofurans

Volume III
Volatile Organic Chemicals

Hydrocarbons
Halogenated Hydrocarbons
Ethers

Volume IV
Oxygen, Nitrogen, and Sulfur Containing Compounds

Alcohols
Aldehydes and Ketones
Phenolic Compounds
Carboxylic Acids
Esters
Nitrogen and Sulfur Compounds

PREFACE

This series of Handbooks brings together physical-chemical data for similarly structured groups of chemical substances which influence their fate in the multimedia environment of air, water, soils, sediments and their resident biota. The task of assessing chemical fate locally, regionally and globally is complicated by the large (and increasing) number of chemicals of potential concern, by uncertainties in their physical-chemical properties, and by lack of knowledge of prevailing environmental conditions such as temperature, pH and deposition rates of solid matter from the atmosphere to water, or from water to bottom sediments. Further, reported values of properties such as solubility are often in conflict. Some are measured accurately, some approximately and some are estimated by various correlation schemes from molecular structure. In some cases, units or chemical identity are wrongly reported. The user of such data thus has the difficult task of selecting the "best" or "right" values. There is justifiable concern that the resulting deductions of environmental fate may be in substantial error. For example, the potential for evaporation may be greatly underestimated if an erroneously low vapor pressure is selected.

To assist the environmental scientist and engineer in such assessments, this Handbook contains compilations of physical-chemical property data for series of chemicals such as, in this case, the oxygen, nitrogen and sulfur containing organic chemicals. It has long been recognized that within such series, properties vary systematically with molecular size, thus providing guidance about the properties of one substance from those of its homologs. Plots of these systematic property variations are provided to check the reported data and provide an opportunity for interpolation and even modest extrapolation to estimate unreported properties of other homologs. Most handbooks treat chemicals only on an individual basis, and do not contain this feature of chemical-to-chemical comparison which can be valuable for identifying errors and estimating properties.

The data are taken a stage further and used to estimate likely environmental partitioning tendencies, i.e., how the chemical is likely to become distributed between the various media which comprise our biosphere. The results are presented numerically and pictorially to provide a visual impression of likely environmental behavior. This will be of interest to those assessing environmental fate by confirming the general fate characteristics or behavior profile. It is, of course, only possible here to assess fate in a "typical" or "generic" or "evaluative" environment; thus, no claim is made that a chemical will behave in this manner in all situations, but this assessment should reveal the broad characteristics of behavior. These evaluative fate assessments are generated using simple fugacity models which flow naturally from the compilations of data on physical-chemical properties of relevant chemicals.

It is hoped that this series of Handbooks will be of value to environmental scientists and engineers and to students and teachers of environmental science. Its aim is to contribute to better assessments of chemical fate in our multimedia environment by serving as a reference source for environmentally relevant physical-chemical property data of classes of chemicals and by illustrating the likely behavior of these chemicals as they migrate throughout our biosphere.

Donald Mackay, born and educated in Scotland, received his degrees in Chemical Engineering from The University of Glasgow. After a period of time in the petrochemical industry he joined The University of Toronto, where he is now a Professor in the Department of Chemical Engineering and Applied Chemistry, and in the Institute for Environmental Studies. Professor Mackay's primary research is the study of organic environmental contaminants, their sources, fates, effects, and control, and particularly in understanding and modeling their behavior with the aid of the fugacity concept. His work has focused especially on the Great Lakes Basin and on cold northern climates.

Wan-Ying Shiu is a Research Associate in the Department of Chemical Engineering and Applied Chemistry, and the Institute for Environmental Studies, University of Toronto. She received her Ph.D. in Physical Chemistry from the Department of Chemistry, University of Toronto, M.Sc. in Physical Chemistry from St. Francis Xavier University and B.Sc. in Chemistry from Hong Kong Baptist College. Her research interest is in the area of physical-chemical properties and thermodynamics for organic chemicals of environmental concern.

Kuo-Ching Ma obtained his Ph.D. from The Florida State University, M.Sc. from The University of Saskatchewan and B.Sc. from The National Taiwan University; all in Physical Chemistry. After working many years in the Aerospace, Battery Research, Fine Chemicals and Metal Finishing industries in Canada as Research Scientist, Technical Supervisor/Director, he is now dedicating his time and interests to environmental research.

Quick Reference: A Guide to Readers

The reader requiring information on the properties of a specific chemical such as pentachlorophenol should first consult the list of chemicals in the appropriate class (in this case phenols in Chapter 4, Section 4.1). The page number for the detailed data compilation is given in that section.

An index of compiled chemicals (using both their chemical names and IUPAC names) in alphabetical order is included at the end of this volume.

Summary tables giving selected values are given in Section 4.2, as are QSPR plots which may be used to estimate properties if the specific compound is not listed. To improve the QSPR correlations some compounds are included in the QSPR plots and in the summary tables of selected values which are not listed in the data compilations.

Illustrations of estimated environmental fate are given in Section 4.3, which in this case includes pentachlorophenol, but not all chemicals are treated. Guidance about likely environmental fate can be obtained by examining the illustrations for a similar compound. Reference can be made to the Commentary in Section 4.4.

It is recommended that the reader should consult the original references for verification, and determine if a more recent determination has been reported. Attempts have been made to make the compilation complete up to the end of 1993.

A diskette is provided with this volume which contains the programs used to calculate environmental fate. Three programs are provided which should yield identical results. One is a program written in simple BASIC or GWBASIC which regrettably are obsolete and are not available in the newer MS-DOS systems (MS-DOS 5.0 and up), but are still widely available. The program is saved in ASCII code and can be loaded and run in QBASIC. The other two are written for a spreadsheet program, in this case Lotus 123® (for both DOS and Window versions) which not only performs fugacity calculations of the chemical of interest but is also able to produce distribution pie-charts. A "readme" file is also included.

TABLE OF CONTENTS

ILLUSTRATED HANDBOOK
OF
PHYSICAL-CHEMICAL PROPERTIES
AND
ENVIRONMENTAL FATE
FOR
ORGANIC CHEMICALS

Volume IV

Oxygen, Nitrogen, and
Sulfur Containing Compounds

1. INTRODUCTION

1.1 THE INCENTIVE

It is alleged that there are some 60,000 chemicals in current commercial production, with approximately 1000 being added each year. Most are organic chemicals. Of these, perhaps 500 are of environmental concern because of their presence in detectable quantities in various components of the environment, their toxicity, their tendency to bioaccumulate, or their persistence. A view is emerging that some of these chemicals are of such extreme environmental concern that all production and use should be ceased, i.e., as a global society we should elect not to synthesize or use these chemicals. They should be "sunsetted." PCBs, "dioxins" and Freons are examples. A second group consists of chemicals which are of concern because they are used or discharged in large quantities, or they are toxic or persistent. They are, however, of sufficient value to society that their continued use is justified, but only under conditions in which we fully understand their sources, fate and effects. This understanding is essential if society is to be assured that there are no adverse ecological or human health effects. Other groups of increasingly benign chemicals can presumably be treated with less rigor.

A key feature of this "cradle-to-grave" approach is that society must improve its skills in assessing chemical fate in the environment. We must better understand where chemicals originate, how they migrate in, and between, the various media of air, water, soils, sediments and their biota which comprise our biosphere. We must understand how these chemicals are transformed by chemical and biochemical processes and thus how long they will persist in the environment. We must seek a fuller understanding of the effects which they will have on the multitude of interacting organisms which occupy these media, including ourselves.

It is now clear that the fate of chemicals in the environment is controlled by a combination of two groups of factors. First are the prevailing environmental conditions such as temperatures, flows and accumulations of air, water and solid matter and the composition of these media. Second are the properties of the chemicals which influence partitioning and reaction tendencies, i.e., whether the chemical evaporates or associates with sediments, and how the chemical is eventually destroyed by conversion to other chemical species.

In recent decades there has emerged a discipline within environmental science concerned with increasing our understanding of how chemicals behave in our multimedia environment. It has been termed "chemodynamics." Practitioners of this discipline include scientists and engineers, students and teachers who attempt to measure, assess and predict how this large number of chemicals will behave in laboratory, local, regional and global environments. These individuals need data on physical-chemical and reactivity properties, as well as information on how these properties translate into environmental fate. This Handbook provides a compilation of such data and uses them to estimate the broad features of environmental fate. It does so for classes or groups of chemicals, instead of the usual approach of treating chemicals on an individual basis. This has the advantage that systematic variations in properties with molecular size can be revealed and used to check reported values, interpolate and even extrapolate to other chemicals of similar structure.

With the advent of inexpensive and rapid computation there has been a remarkable growth of interest in this general area of Quantitative Structure-Property Relationships (QSPRs). The ultimate goal is to use information about chemical structure to deduce physical-chemical properties, environmental partitioning and reaction tendencies, and even uptake and effects on biota. The goal is far from being realized, but considerable progress has been made, as is briefly reviewed in a following section. In this Handbook we adopt a simple, and well tried, approach of using molecular structure to deduce a molar volume, which in turn is related to physical-chemical properties. Undoubtedly, other molecular descriptors such as surface area or topological indices have the potential to give more accurate correlations and will be increasingly used in the future, but at this stage we believe that the improvements in accuracy obtained by using these more complex descriptors do not justify the computational effort of generating them, at least for the purposes of routine, general assessments. In some cases, the fundamental causes of the relationships remain obscure.

A major benefit of this simple QSPR analysis is that it reveals likely errors in reported data. Regrettably, the scientific literature contains a great deal of conflicting data with reported values often varying over several orders of magnitude. There are some good, but more not-so-good reasons for this lack of accuracy. Many of these properties are difficult to measure because they involve analyzing very low concentrations of 1 part in 10^9 or 10^{12}. For many purposes an approximate value, for example, that a solubility is less than 1 mg/L, is adequate. There has been a mistaken impression that if a vapor pressure is low, as is the case with DDT, it is not important. DDT evaporates appreciably from solution in water, despite its low vapor pressure, because of its low solubility in water. In some cases the units are reported incorrectly or there are uncertainties about temperature or pH. In other cases the chemical is wrongly identified. One aim of this Handbook is to assist the user to identify such problems and provide guidance when selecting appropriate values.

The final aspect of chemical fate treated in this Handbook is the depiction or illustration of likely chemical fate. This is done using a series of multimedia "fugacity" models as is described in a later section. The authors' aim is to convey an impression of likely environmental partitioning and transformation characteristics, i.e., we seek to generate a "behavior profile." A fascinating feature of chemodynamics is that chemicals differ so greatly in their behavior. Some, such as chloroform, evaporate rapidly and are dissipated in the atmosphere. Others, such as DDT, partition into the organic matter of soils and sediments and the lipids of fish, birds and mammals. Phenols tend to remain in water subject to fairly rapid transformation processes such as biodegradation and photolysis. By entering the physical-chemical data into a model of chemical fate in a generic or evaluative environment, it is possible to estimate the likely general features of the chemical's behavior and fate. The output of these calculations is presented numerically and pictorially.

In total, the aim of this series of Handbooks is to provide a useful reference work for those concerned with the assessment of the fate of existing and new chemicals in the environment.

1.2 PHYSICAL-CHEMICAL PROPERTIES

1.2.1 The Key Physical-Chemical Properties

The major differences between behavior profiles of organic chemicals in the environment are attributable to physical-chemical properties. The key properties are believed to be solubility in water, vapor pressure, octanol-water partition coefficient, dissociation constant in water (when relevant) and susceptibility to degrading or transformation reactions. Other essential molecular descriptors are molecular mass and molar volume, with properties such as critical temperature and pressure and molecular area being occasionally useful for specific purposes.

Chemical identity may appear to present a trivial problem, but many chemicals have several names, and subtle differences between isomers (e.g., cis and trans) may be ignored. The most commonly accepted identifiers are the IUPAC name and the Chemical Abstracts System (CAS) number. More recently, methods have been sought of expressing the structure in line notation form so that computer entry of a series of symbols can be used to define a three-dimensional structure. The Wiswesser Line Notation is quite widely used, but it appears that for environmental purposes it will be superseded by the SMILES (Simplified Molecular Identification and Line Entry System, Anderson et al. 1987).

Molecular mass is readily obtained from structure. Also of interest are molecular volume and area, which may be estimated by a variety of methods.

Solubility in water and vapor pressure are both "saturation" properties, i.e., they are measurements of the maximum capacity which a phase has for dissolved chemical. Vapor pressure P (Pa) can be viewed as a "solubility in air," the corresponding concentration C (mol/m^3) being P/RT where R is the ideal gas constant (8.314 J/mol • K) and T is absolute temperature (K). Although most chemicals are present in the environment at concentrations well below saturation, these concentrations are useful for estimating air-water partition coefficients as ratios of saturation values. It is usually assumed that the same partition coefficient applies at lower sub-saturation concentrations. Vapor pressure and solubility thus provide estimates of air-water partition coefficients K_{AW} or Henry's law constants H (Pa·m^3/mol), and thus the relative air-water partitioning tendency.

The octanol-water partition coefficient K_{OW} provides a direct estimate of hydrophobicity or of partitioning tendency from water to organic media such as lipids, waxes and natural organic matter such as humin or humic acid. It is invaluable as a method of estimating K_{OC}, the organic carbon-water partition coefficient, the usual correlation invoked being that of Karickhoff (1981)

$$K_{OC} = 0.41\ K_{OW}$$

It is also used to estimate fish-water bioconcentration factors K_B or BCF using a correlation similar to that of Mackay (1982)

$$K_B = 0.05\ K_{OW}$$

where the term 0.05 corresponds to a 5% lipid content of the fish.

For dissociating chemicals it is essential to quantify the extent of dissociation as a function of pH using the dissociation constant pKa. The parent and ionic forms behave and partition quite differently, thus pH and the presence of other ions may profoundly affect chemical fate.

Characterization of chemical reactivity presents a severe problem in Handbooks. Whereas radioisotopes have fixed half-lives, the half-life of a chemical in the environment depends not only on the intrinsic properties of the chemical, but also on the nature of the surrounding environment. Factors such as sunlight intensity, hydroxyl radical concentration and the nature of the microbial community, as well as temperature, affect the chemical's half-life so it is impossible (and misleading) to document a single reliable half-life. The compilation by Howard et al. (1991) provides an excellent review of the existing literature for a large number of chemicals. It is widely used as a source document in this work. The best that can be done is to suggest a semi-quantitative classification of half-lives into groups, assuming average environmental conditions to apply. Obviously, a different class will generally apply in air and bottom sediment. In this compilation we use the following class ranges for chemical reactivity in a single medium such as water.

Class	Mean half-life (hours)	Range (hours)
1	5	< 10
2	17 (~ 1 day)	10-30
3	55 (~ 2 days)	30-100
4	170 (~ 1 week)	100-300
5	550 (~ 3 weeks)	300-1,000
6	1700 (~ 2 months)	1,000-3,000
7	5500 (~ 8 months)	3,000-10,000
8	17000 (~ 2 years)	10,000-30,000
9	55000 (~ 6 years)	> 30,000

These times are divided logarithmically with a factor of approximately 3 between adjacent classes. With the present state of knowledge it is probably misleading to divide the classes into finer groupings; indeed, a single chemical may experience half-lives ranging over three classes, depending on season.

When compiling suggested reactivity classes the authors have examined the available information as reaction rates of the chemical in each medium, by all relevant processes. These were expressed as an overall half-life for transformation. For example, a chemical may be subject to biodegradation with a half-life of 20 days (rate constant 0.0014 h^{-1}) and a photolysis with a rate constant of 0.0011 h^{-1} (half-life 630 hours). The overall rate constant is thus 0.0025 h^{-1} and the half-life 277 hours or 12 days. Data for homologous chemicals were also compiled, and insights into the reactivity of various functional groups considered. In most cases a single reaction class was assigned to the series; in this case, class 4 with a mean half-life of 170 hours was chosen. It must be appreciated that this chemical could fall into class 3 in summer and class 5 in winter. These half-lives must be used with caution and it is wise to test the implications of selecting longer and shorter half-lives.

The chemicals treated in this volume are generallly more polar and soluble in water than the more hydrophobic chemicals treated in Volumes I, II, and III. In many cases the substances are subject to dissociation, thus complicating the estimation of physical-chemical properties and environmental fate since both ionic and neutral species must be considered. There is thus a tendency for these chemicals to be more associated with water, to evaporate less and to be less sorptive, and less bioaccumulative than the more common environmental contaminants. The esters are the exception in this regard because of the often sparing solubilities in water.

Extensive research has been conducted into the atmospheric chemistry of the volatile and semi-volatile organic chemicals because of air quality concerns. Recently, Atkinson and coworkers (1984, 1985, 1987, 1989, 1990, 1991), Altshuller (1980, 1991) and Sabljic and Güsten (1990) have reviewed the photochemistry of many organic chemicals of environmental interest for their gas phase reactions with hydroxyl radicals (OH), ozone (O_3) and nitrate radicals (NO_3) and have provided detailed information on reaction rate constants and experimental conditions, which allowed the estimation of atmospheric lifetimes. Klöpffer (1991) has estimated the atmospheric lifetimes for the reaction with OH radicals to range from 1 hour to 130 years, based on these reaction rate constants and an assumed constant concentration of OH radicals in air. As Atkinson (1985) has pointed out, the gas-phase reactions with OH radicals are the major tropospheric loss process for the alkanes, haloalkanes, the lower alkenes, the aromatic hydrocarbons, and a majority of the oxygen-containing organics. In addition, photooxidation reactions with O_3 and NO_3 radicals can result in transformation of these compounds. The night-time reaction with NO_3 radicals may also be important (Atkinson and Carter 1984, Sabljic and Güsten 1990).

There are fewer studies on direct or indirect photochemical degradation in the water phase; however, Klöpffer (1991) had pointed out that the rate constant or lifetimes derived from these studies "is valid only for the top layer or surface waters." Mill (1982, 1989 1993) and Mill and Mabey (1985) have estimated half-lives of various chemicals in aqueous solutions from their reaction rate constants with singlet oxygen, as well as photooxidation with OH and peroxy radicals. Buxton et al. (1988) gave a critical review of rate constants for reactions with hydrated electrons, hydrogen atoms and hydroxyl radicals in aqueous solutions. Mabey and Mill (1978) also reviewed the hydrolysis of organic chemicals in water under environmental conditions. Recently, Ellington and coworkers (1987a,b, 1988, 1989) also reported the hydrolysis rate constants in aqueous solutions for a variety of organic chemicals.

Other processes, such as biodegradation, may be important, especially in the subsurface environment where there is widespread concern about groundwater contamination.

In most cases a review of the literature suggested that reaction rates in water by chemical processes are 1 to 2 orders of magnitude slower than in air, but with biodegradation often being significant, especially for hydrocarbons and oxygen-containing chemicals. Generally, the water half-life class is three more than that in air, i.e., a factor of about 30 slower. Chemicals in soils tend to be shielded from photolytic processes, and they are less bioavailable, thus the authors have frequently assigned a reactivity class to soil of one more than that for water. Bottom sediments are assigned an additional class to that of soils largely on the basis that there is little or no photolysis, there may be lack of oxygen, and the intimate sorption to sediments renders the chemicals less bioavailable.

The chemical reactivity of these substances is a topic which continues to be the subject of extensive research; thus there is often detailed, more recent information about the fate of chemical species which is of particular relevance to air or water quality. The reader is thus urged to consult the original and recent references because it is impossible in a volume such as this, which considers the entire multimedia picture, to treat this subject in the detail it deserves.

When selecting physical-chemical properties or reactivity classes the authors have been guided by:

(1) the age of the data and acknowledgment of previous conflicting or supporting values,
(2) the method of determination,
(3) the perception of the objectives of the authors, not necessarily as an indication of competence, but often as an indication of the need of the authors for accurate values, and
(4) the reported values for structurally similar, or homologous compounds.

In this Handbook we have used these considerations as well as information derived from the QSPR analyses.

It is appropriate, therefore, to review briefly the experimental methods which are commonly used for property determinations and comment on their accuracy.

1.2.2. Experimental Methods

Solubility in Water

Most conventional organic contaminants are fairly hydrophobic and thus exhibit a measurable solubility in water. Many of the chemicals treated in this volume are hydrophilic, are miscible with water in all proportions and no solubility is measurable. An obvious example is ethanol. Solubility is often used to estimate the air-water partition coefficient or Henry's law constant, but this is not possible for miscible chemicals, indeed the method is suspect for chemicals of appreciable solubility in water, i.e., 1 g/100 g or more. Direct measurement of the Henry's law constant is thus required.

The conventional method of preparing saturated solutions for the determination of solubility is batch equilibration. An excess amount of solute chemical is added to water and equilibrium is achieved by shaking gently (generally referred as the "shake flask method") or slow stirring with a magnetic stirrer. The aim is to prevent formation of emulsions or suspensions and thus avoid extra experimental procedures such as filtration or centrifuging which may be required to ensure that a true solution is obtained. Experimental difficulties can still occur because of the formation of emulsion or microcrystal suspensions with the sparingly soluble chemicals such as higher normal alkanes and polycyclic aromatic hydrocarbons (PAHs). An alternative approach is to coat a thin layer of the chemical on the surface of the equilibration flask before water is added. An accurate "generator column" method has also been developed (Weil et al. 1974, May et al. 1978a,b) in which a column is packed with an inert solid support, such as glass beads or Chromosorb, and then coated with the solute chemical. Water is pumped through the column at a controlled, known flow rate to achieve saturation.

The method of concentration measurement of the saturated solution depends on the solute solubility and its chemical properties. Some common methods used for solubility measurement are listed below.

1. Gravimetric or volumetric methods (Booth and Everson 1948)
   An excess amount of solid compound is added to a flask containing water to achieve saturation solution by shaking, stirring, centrifuging until the water is saturated with solute and undissolved solid or liquid residue appears, often as a cloudy phase. For liquids, successive known amounts of solute may be added to water and allowed to reach equilibrium, and the volume of excess undissolved solute is measured.
2. Instrumental methods
   a. UV spectrometry (Andrews and Keffer 1950, Bohon and Claussen 1951, Yalkowsky and Valvani 1976);
   b. Gas chromatographic analysis with FID, ECD or other detectors (McAuliffe 1966, Mackay et al. 1975, Chiou et al. 1982);
   c. Fluorescence spectrophotometry (Mackay and Shiu 1977);
   d. Interferometry (Gross and Saylor 1931);
   e. High-pressure liquid chromatography (HPLC) with R.I., UV or fluorescence detection (May et al. 1978a,b, Wasik et al. 1983, Shiu et al. 1988, Doucette and Andren 1988a);
   f. Liquid phase elution chromatography (Schwarz 1980, Schwarz and Miller 1980);
   g. Nephelometric methods (Davis and Parke 1942, Davis et al. 1942, Hollifield 1979);
   h. Radiotracer or liquid scintillation counting (LSC) method (Banerjee et al. 1980, Lo et al. 1986).

For most organic chemicals the solubility is reported at a defined temperature in distilled water from completely miscible chemicals such as lower alcohols and amines to sparingly soluble hydrophobic chemicals such as PAHs and dioxins. For substances which dissociate (e.g. phenols, carboxylic acids and amines) it is essential to report the pH of the determination because the extent of dissociation affects the solubility. It is common to maintain the desired pH by buffering with an appropriate electrolyte mixure. This raises the complication that the presence of electrolytes modifies the water structure and changes the solubility. The effect is usually "salting-out." For example, many hydrocarbons have solubilities in seawater about 75% of their solubilities in distilled water. Care must thus be taken to interpret and use reported data properly when electrolytes are present.

The most common problem encountered with reported data is inaccuracy associated with very low solubilities, i.e., those less than 1.0 mg/L. Such solutions are difficult to prepare, handle and analyze, and reported data often contain appreciable errors.

Octanol-Water Partition Coefficient K_{OW}

The experimental approaches are similar to those for solubility, i.e., employing shake flask or generator-column techniques. Concentrations in both the water and octanol phases may be determined after equilibration. Both phases can then be analyzed by the instrumental methods discussed above and the partition coefficient is calculated from the concentration ratio C_O/C_W.

This is actually the ratio of solute concentration in octanol saturated with water to that in water saturated with octanol.

As with solubility, K_{OW} is a function of the presence of electrolytes and for dissociating chemicals it is a function of pH. Accurate values can generally be measured up to about 10^6, but accurate measurement beyond this requires meticulous technique. A common problem is that the presence of small quantities of emulsified octanol in the water phase could create a high concentration of chemical in that emulsion which would cause an erroneously high apparent water phase concentration.

Considerable success has been achieved by calculating K_{OW} from molecular structure; thus, there has been a tendency to calculate K_{OW} rather than measure it, especially for "difficult" hydrophobic chemicals. These calculations are, in some cases, extrapolations and can be in serious error. Any calculated log K_{OW} value above 7 should be regarded as suspect, and any experimental or calculated value above 8 should be treated with extreme caution.

For many hydrophilic compounds such as the alcohols, K_{OW} is low and can indeed be less than 1.0, resulting in negative values of log K_{OW}. In such cases, care should be taken when using correlations developed for more hydrophobic chemicals since partitioning into biota or organic carbon phases may be primarily into aqueous rather than organic media.

Details of experimental methods are described by Fujita et al. (1964), Leo et al. (1971); Hansch and Leo (1979), Rekker (1977), Chiou et al. (1977), Miller et al. (1984), Bowman and Sans (1983), Woodburn et al. (1984), Doucette and Andren (1987), and De Bruijn et al. (1989).

Vapor Pressure

In principle, the determination of vapor pressure involves the measurement of the saturation concentration or pressure of the solute in a gas phase. The most reliable methods involve direct determination of these concentrations, but convenient indirect methods are also available based on evaporation rate measurement or chromatographic retention times. Some methods and approaches are listed below.

a. Direct measurement by use of pressure gauges: diaphragm gauge (Ambrose et al. 1975), Rodebush gauge (Sears and Hopke 1947), inclined-piston gauge (Osborn and Douslin 1975);
b. Comparative ebulliometry (Ambrose 1981);
c. Effusion methods, torsion and weight-loss (Balson 1947, Bradley and Cleasby 1953, Hamaker and Kerlinger 1969, De Kruif 1980);
d. Gas saturation or transpiration methods (Spencer and Cliath 1970, 1972, Sinke 1974, Macknick and Prausnitz 1979, Westcott et al. 1981, Rordorf 1985a,b, 1986);
e. Dynamic coupled-column liquid chromatographic method - a gas saturation method (Sonnefeld et al. 1983);
f. Calculation from evaporation rates and vapor pressures of reference compound (Gückel et al. 1974, 1982, Dobbs and Grant 1980, Dobbs and Cull 1982);

g. Calculation from GC retention time data (Hamilton 1980, Westcott and Bidleman 1982, Bidleman 1984, Kim et al. 1984, Foreman and Bidleman 1985, Burkhard et al. 1985a, Hinckley et al. 1990).

The greatest difficulty and uncertainty arises when determining the vapor pressure of chemicals of low volatility, i.e., those with vapor pressures below 1.0 Pa. Vapor pressures are strongly dependent on temperature, thus accurate temperature control is essential. Data are often regressed against temperature and reported as Antoine or Clapeyron constants. Care must be taken when using the Antoine or other equations to extrapolate data beyond the temperature range specified. It must be clear if the data apply to the solid or liquid phase of the chemical.

Henry's Law Constant

The Henry's law constant is essentially an air-water partition coefficient which can be determined by measurement of solute concentrations in both phases. This raises the difficulty of accurate analytical determination in two very different media which require different techniques. Accordingly, some effort has been devoted to devising techniques in which concentrations are measured in only one phase and the other concentration is deduced by a mass balance. These methods are generally more accurate. The principal difficulty arises with hydrophobic, low-volatility chemicals which can establish only very small concentrations in both phases.

Henry's law constant can be regarded as a ratio of vapor pressure to solubility, thus it is subject to the same effects which electrolytes have on solubility and temperature has on both properties. Some methods are as follows:

a. Volatility measurement of dilute aqueous solutions (Butler et al. 1935, Burnett 1963, Buttery et al. 1969);
b. Multiple equilibration method (McAuliffe 1971, Munz and Roberts 1987);
c. Equilibrium batch stripping (Mackay et al. 1979, Dunnivant et al. 1988, Betterton and Hoffmann 1988; Zhou and Moppper 1990);
d. GC-determined distribution coefficients (Leighton and Calo 1981);
e. GC analysis of both air/water phases (Vejrosta et al. 1982, Jönsson et al. 1982);
f. EPICS (Equilibrium Partioning In Closed Systems) method (Lincoff and Gossett 1984; Gossett 1987, Ashworth et al. 1988);
g. Wetted-wall column (Fendinger and Glotfelty 1988, 1990);
h. Headspace analyses (Hussam and Carr 1985);
i. Calculation from vapor pressure and solubility (Mackay and Shiu 1981);
j. GC retention volume/time determined activity coefficient at infinite dilution γ^∞ (Karger et al. 1971a,b, Sugiyama et al. 1975, Tse et al. 1992).

When using vapor pressure and solubility data, it is essential to ensure that both properties apply to the same chemical phase, i.e., both are of the liquid, or of the solid. Occasionally, a solubility is of a solid while a vapor pressure is extrapolated from higher temperature liquid phase data.

As was discussed earlier under solubility, for miscible chemicals it is necessary to determine the Henry's law constant directly, since solubilities are not measurable.

1.3 QUANTITATIVE STRUCTURE-PROPERTY RELATIONSHIPS (QSPRs)

1.3.1 Objectives

Because of the large number of chemicals of actual and potential concern, the difficulties and cost of experimental determinations, and scientific interest in elucidating the fundamental molecular determinants of physical-chemical properties, a considerable effort has been devoted to generating quantitative structure-property relationships (QSPRs). This concept of structure-property relationships or structure-activity relationships (QSARs) is based on observations of linear free-energy relationships, and usually takes the form of a plot or regression of the property of interest as a function of an appropriate molecular descriptor which can be obtained from merely a knowledge of molecular structure.

Such relationships have been applied to solubility, vapor pressure, K_{OW}, Henry's law constant, reactivities, bioconcentration data and several other environmentally relevant partition coefficients. Of particular value are relationships involving various manifestations of toxicity, but these are beyond the scope of this handbook. These relationships are valuable because they permit values to be checked for "reasonableness" and (with some caution) interpolation is possible to estimate undetermined values. They may be used (with extreme caution!) for extrapolation.

A large number of descriptors have been, and are being, proposed and tested. Dearden (1990) and the compilation by Karcher and Devillers (1990) and Hermens and Opperhuizen (1991) give comprehensive accounts of descriptors and their applications.

Among the most commonly used molecular descriptors are molecular weight and volume, the number of specific atoms (e.g., carbon or chlorine), surface areas (which may be defined in various ways), refractivity, parachor, steric parameters, connectivities and various topological parameters. Several quantum chemical parameters can be calculated from molecular orbital calculations including charge, electron density and superdelocalizability.

It is likely that existing and new descriptors will continue to be tested, and that eventually a generally preferred set of readily accessible parameters will be adopted of routine use for correlating purposes. From the viewpoint of developing quantitative correlations it is very desirable to seek a linear relationship between descriptor and property, but a nonlinear or curvilinear relationship is quite adequate for illustrating relationships and interpolating purposes. In this handbook we have elected to use the simple descriptor of molar volume at the normal boiling point as estimated by the LeBas method (Reid et al. 1987). This parameter is very easily calculated and proves to be adequate for the present purposes of plotting property versus relationship without seeking linearity.

The LeBas method is based on a summation of atomic volumes with adjustment for the volume decrease arising from ring formation. The full method is described by Reid et al. (1987), but for the purposes of this compilation, the volumes and rules as listed in Table 1.1 are used.

Table 1.1 LeBas Molar Volume

	increment, cm^3/mol
carbon	14.8
hydrogen	3.7
oxygen	7.4
in methyl esters and ethers	9.1
in ethyl esters and ethers	9.9
join to S, P, or N	8.3
nitrogen	
doubly bonded	15.6
in primary amines	10.5
in secondary amines	12.0
bromine	27.0
chlorine	24.6
fluorine	8.7
iodine	37.0
sulfur	25.6
ring	
three-membered	-6.0
four-membered	-8.5
five-membered	-11.5
six-membered	-15.0
naphthalene	-30.0
anthracene	-47.5

Example: The experimental molar volume of chlorobenzene 115 cm^3/mol (Reid et al. 1987). From the above rules, the LeBas molar volume for chlorobenzene (C_6H_5Cl) is:

$$V = 6 \times 14.8 + 5 \times 3.7 + 24.6 - 15 = 117 \text{ cm}^3/\text{mol}$$

Accordingly, plots are presented at the end of each chapter for solubility, vapor pressure, K_{OW}, and Henry's law constant versus LeBas molar volume.

A complication arises in that two of these properties (solubility and vapor pressure) are dependent on whether the solute is in the liquid or solid state. Solid solutes have lower solubilities and vapor pressures than they would have if they had been liquids. The ratio of the (actual) solid to the (hypothetical supercooled) liquid solubility or vapor pressure is termed the fugacity ratio F and can be estimated from the melting point and the entropy of fusion ΔS_{fus}, as discussed by Mackay and Shiu (1981). For solid solutes, the correct property to plot is the calculated or extrapolated subcooled liquid solubility. This is calculated in this handbook using the relationship suggested by Yalkowsky (1979) which implies an entropy of fusion of 56 J/mol K or 13.5 cal/mol·K

$$F = C^S_S/C^S_L = P^S_S/P^S_L = \exp\{6.79(1 - T_M/T)\}$$

where C^S is solubility, P^S is vapor pressure, subscripts S and L refer to solid and liquid phases, T_M is melting point and T is the system temperature, both in absolute (K) units. The fugacity ratio is given in the data tables at 25 °C, the usual temperature at which physical-chemical property data are reported. For liquids, the fugacity ratio is 1.0.

1.3.2 Examples

Recently, there have been efforts to extend the long-established concept of Quantitative Structure-Activity Relationships (QSARs) to Quantitative Structure-Property Relationships (QSPRs) to compute all relevant environmental physical-chemical properties (such as aqueous solubility, vapor pressure, octanol-water partition coefficient, Henry's law constant, bioconcentration factor (BCF), sorption coefficient and envrionmental reaction rate constants from molecular structure. Examples are Burkhard (1984) and Burkhard et al. (1985a) who calculated solubility, vapor pressure, Henry's law constant, K_{OW} and K_{OC} for all PCB congeners. Hawker and Connell (1988) also calculated log K_{OW}; Abramowitz and Yalkowsky (1990) calculated melting point and solubility for all PCB congeners based on the correlation with total surface area (planar TSAs). Doucette and Andren (1988b) used six molecular descriptors to compute the K_{OW} of some chlorobenzenes, PCBs and PCDDs. Mailhot and Peters (1988) employed seven molecular descriptors to compute physical-chemical properties of some 300 compounds. Isnard and Lambert (1988, 1989) correlated solubility, K_{OW} and BCF for a large number of organic chemicals. Nirmalakhandan and Speece (1988a,b, 1989) used molecular connectivity indices to predict aqueous solubility and Henry's law constants for 300 compounds over 12 logarithmic units in solubility. Kamlet and co-workers (1987, 1988) have developed the solvatochromic parameters with the intrinsic molar volume to predict solubility, log K_{OW} and toxicity of organic chemicals. Warne et al. (1990) correlated solubility and K_{OW} for lipophilic organic compound with 39 molecular descriptors and physical-chemical properties. Atkinson (1987, 1988) has used the structure-activity relationship (SAR) to estimated gas-phase reaction rate constants of hydroxyl radicals for organic chemicals. Mabey et al. (1984) have reviewed the estimation methods from SAR correlation for reaction rate constants and physical-chemical properties in environmental fate assessment. Other correlations are reviewed by Lyman et al. (1982) and Yalkowsky and Banerjee (1992). As Dearden (1990) has pointed out, "new parameters are continually being devised and tested, although the necessity of that may be questioned, given the vast number already available." It must be emphasized, however, that regardless of how accurate these predicted or estimated properties are claimed to be, utimately they have to be confirmed or verified by experimental measurement.

A fundamental problem encountered in these correlations is that the molecular descriptors can be calculated with relatively high precision, usually within a few percent. The accuracy may not always be high, but for empirical correlation purposes precision is more important than accuracy. The precision and accuracy of the experimental data are often poor, frequently ranging over a factor of two or more. Certain isomers may yield identical descriptors, but have different properties. There is thus an inherent limit to the applicability of QSPRs imposed by the quality of the experimental data, and further efforts to improve descriptors, while interesting and potentially useful, are unlikely to yield demonstrably improved QSPRs.

For correlation of **solubility**, the correct thermodynamic quantities for correlation are the activity coefficient γ, or the excess Gibbs free energy ΔG, as discussed by Pierotti et al. (1959) and Tsonopoulos and Prausnitz (1971). Examples of such correlations are given below.

1. Carbon number or carbon plus chlorine number (Tsonopoulos and Prausnitz 1971, Mackay and Shiu 1977);
2. Molar volume cm^3/mol
   a. Liquid molar volume - from density (McAuliffe 1966, Lande and Banerjee 1981, Chiou et al. 1982, Abernethy et al. 1988, Wang et al. 1992);
   b. Molar volume by additive group contribution method, e.g., LeBas method, Schroeder method (Reid et al. 1987, Miller et al. 1985);
   c. Intrinsic molar volume, V_I, cm^3/mol - from van der Waals radius with solvatochromic parameters α and β (Leahy 1986, Kamlet et al. 1987, 1988);
   d. Characteristic molecular volume, m^3/mol (McGowan and Mellors 1986);
3. Group contribution method (Irmann 1965, Korenman et al. 1971, Polak and Lu 1973);
4. Molecular volume - $Å^3/mol$ (cubic Angstrom per mole)
   a. van der Waals volume (Bondi 1964);
   b. Total Molecular Volume (TMV) (Pearlman et al. 1984, Pearlman 1986);
5. Total Surface Area (TSA) - $Å^2/mol$ (Hermann 1971, Amidon et al. 1975, Yalkowsky and Valvani 1976, Yalkowsky et al. 1979, Pearlman 1986, Andren et al. 1987, Hawker and Connell 1988, Dunnivant et al. 1992);
6. Molecular Connectivity indices (MCI) or χ (Kier and Hall 1976, Andren et al. 1987, Nirmalakhandan and Speece 1988b, 1989);
7. Boiling point (Almgren et al. 1979);
8. Melting point (Amidon and Williams 1982);
9. Melting point and TSA (Abramowitz and Yalkowsky 1990);
10. High Pressure Liquid Chromatography (HPLC) - retention data (Locke 1974, Whitehouse and Cooke 1982, Brodsky and Ballschmiter 1988);
11. Adsorbability index (AI) (Okouchi et al. 1992);
12. Fragment solubility constants (Wakita et al. 1986).

Several workers have explored the linear relationship between octanol-water partition coefficient and solubility as a means of estimating solubility.

Hansch et al. (1968) established the linear free-energy relationship between aqueous and octanol-water partition of organic liquid. Others, such as Tulp and Hutzinger (1978), Yalkowsky et al. (1979), Mackay et al. (1980), Banerjee et al. (1980), Chiou et al. (1982), Bowman and Sans (1983), Miller et al. (1985), Andren et al. (1987) and Doucette and Andren (1988b) have all presented similar but modified relationships.

The UNIFAC (UNIQUAC Functional Group Activity Coefficient) group contribution (Fredenslund et al. 1975, Kikic et al. 1980, Magnussen et al. 1981, Gmehling et al. 1982 and Hansen et al. 1991) is widely used for predicting the activity coefficient in nonelectrolyte liquid mixtures by using group-interaction parameters. This method has been used by Kabadi and Danner (1979), Banerjee (1985), Arbuckle (1983, 1986), Banerjee and Howard (1988) and Al-

Sahhaf (1989) for predicting solubility (as a function of the infinite dilution acitivity coefficient, γ^{∞}) in aqueous systems. Its performance is reviewed by Yalkowsky and Banerjee (1992).

HPLC retention time data have been used as a psuedo-molecular descriptor by Whitehouse and Cooke (1982), Hafkenscheid and Tomlinson (1981), Tomlinson and Hafkenscheid (1986) and Swann et al. (1983).

The **octanol-water partition coefficient** K_{OW} is widely used as a descriptor of hydrophobicity. Variation in K_{OW} is primarily attributable to variation in activity coefficient in the aqueous phase (Miller et al. 1985); thus, the same correlations used for solubility in water are applicable to K_{OW}. Most widely used is the Hansch-Leo compilation of data (Leo et al. 1971, Hansch and Leo 1979) and related predictive methods. Examples of K_{OW} correlations are

I. Molecular descriptors
  1. Molar volumes: LeBas method; from density; intrinsic molar volume; characteristic molecular volume (Abernethy et al. 1988, Chiou 1985, Kamlet et al. 1988, McGowan and Mellors 1986);
  2. TMV (De Bruijn and Hermens 1990);
  3. TSA (Yalkowsky et al. 1979, Yalkowsky et al. 1983, Hawker and Connell 1988);
  4. Molecular connectivity indices (Doucette and Andren 1988b);
  5. Molecular weight (Doucette and Andren 1988b).

II. Group contribution methods
  1. π-constant or hydrophobic substituent method (Hansch et al. 1968, Hansch and Leo 1979, Doucette and Andren 1988b);
  2. Fragmental constants or f-constant (Rekker 1977, Yalkowsky et al. 1983);
  3. Hansch and Leo's f-constant (Hansch and Leo 1979; Doucette and Andren 1988b).

III. From solubility - K_{OW} relationship

IV. HPLC retention data
  1. HPLC-k' capacity factor (Könemann et al. 1979, McDuffie 1981);
  2. HPLC-RT retention time (Veith et al. 1979, Rapaport and Eisenreich 1984, Doucette and Andren 1988b);
  3. HPLC-RV retention volume (Garst 1984);
  4. HPLC-RT/MS HPLC retention time with mass spectrometry (Burkhard et al. 1985c).

V. Reversed-phase thin-layer chromatography (TLC) (Bruggeman et al. 1982).

VI. Molar refractivity (Yoshida et al. 1983).

As with solubility and octanol-water partition coefficient, **vapor pressure** can be estimated with a variety of correlations as discussed in detail by Burkhard et al. (1985a) and summarized as follows:

1. Interpolation or extrapolation from equation for correlating temperature relationships, e.g., the Clausius-Clapeyron, Antoine equations (Burkhard et al. 1985a);
2. Carbon or chlorine numbers (Mackay et al. 1980, Shiu and Mackay 1986);
3. LeBas molar volume (Shiu et al. 1987, 1988);
4. Boiling point T_B and heat of vaporization ΔH_v (Mackay et al. 1982);
5. Group contribution method (Macknick and Prausnitz 1979);

6. UNIFAC group contribution method (Jenson et al. 1981, Yair and Fredenslund 1983, Burkhard et al. 1985a, Banerjee et al. 1990);
7. Molecular weight and Gibbs' free energy of vaporization ΔG_v (Burkhard et al. 1985a);
8. TSA and ΔG_v (Amidon and Anik 1981, Burkhard et al. 1985a, Hawker 1989);
9. Molecular connectivity indices (Kier and Hall 1976, 1986, Burkhard et al. 1985a);
10. Melting point T_M and GC retention index (Bidleman 1984, Burkhard et al. 1985a);
11. Solvatochromic parameters and intrinsic molar volume (Banerjee et al. 1990).

As described earlier, **Henry's law constants** can be calculated from the ratio of vapor pressure and aqueous solubility. Henry's law constants do not show a simple linear pattern as solubility, K_{OW} or vapor pressure when plotted against simple molecular descriptors, such as numbers of chlorine or LeBas molar volume, e.g. PCBs (Burkhard et al. 1985b), pesticides (Suntio et al. 1988), and chlorinated dioxins (Shiu et al. 1988). Henry's law constants can be estimated from:

1. UNIFAC-derived infinite dilution activity coefficients (Arbuckle 1983);
2. Group contribution and bond contribution method (Hine and Mookerjee 1975, Meylan and Howard 1991);
3. Molecular connectivity indices (Nirmalakhandan and Speece 1988b, Sabljic and Güsten 1989, Dunnivant et al. 1992);
4. Total surface area - planar TSA (Hawker 1989);

For water-miscible compounds the use of aqueous solubility data is obviously impossible.

Bioconcentration Factors:

1. Correlation with K_{OW} (Neely et al. 1974, Könemann and van Leeuwen 1980, Veith et al. 1980, Chiou et al. 1977, Mackay 1982, Briggs 1981, Garten and Trabalka 1983, Davies and Dobbs 1984, Oliver and Niimi 1988, Isnard and Lambert 1988);
2. Correlation with solubility (Kenaga 1980, Kenaga and Goring 1980, Briggs 1981, Garten and Trabalka 1983, Davies and Dobbs 1984, Isnard and Lambert 1988);
3. Correlation with K_{OC} (Kenaga 1980, Kenaga and Goring 1980, Briggs 1981);
4. Calculation with HPLC retention data (Swann et al. 1983);
5. Calculation with solvatochromic parameters (Hawker 1989a, 1990).

Sorption Coefficients:

1. Correlation with K_{OW} (Karickhoff et al. 1979, Schwarzenbach and Westall 1981, Mackay 1982, Oliver 1984);
2. Correlation with solubility (Karickhoff et al. 1979);
3. Molecular connectivity indices (Sabljic 1984, 1987, Sabljic et al. 1989, Meylan et al. 1992);
4. Estimation from molecular connectivity index/fragment contribution method (Meylan et al. 1992);
5. From HPLC retention data (Swann et al. 1983, Szabo et al. 1990).

1.4 FATE MODELS

1.4.1 Evaluative Environmental Calculations

The nature of these calculations has been described in a series of papers, notably Mackay (1979), Paterson and Mackay (1985), Mackay and Paterson (1990, 1991), and a recent text (Mackay 1991). Only the salient features are presented here. Three calculations are completed for each chemical, namely the Level I, II and III fugacity calculations. The results of these calculations are presented in illustrated form for representative chemicals throughout this Volume. No calculations are, however, done for the nitrogen and sulfur containing compounds because of the lack of available data. This is discussed in more detail in Chapter 7 Section 7.3.

1.4.2 Treatment of Dissociating Compounds

Corrections for Dissociation

In the case of dissociating or ionizing organic chemicals such as organic acids and bases, e.g., phenols, carboxylic acids and amines, it is desirable to calculate the concentrations of ionic and non-ionic species, and correct for this effect. Recently, Westall et al. (1985), Schwarzenbach et al. (1988), Jafvert et al. (1990), Johnson and Westall (1990) and the text by Schwarzenbach, Gschwend and Imboden (1993) have discussed and reviewed the effect of pH and ionic strength to the distribution of these types of chemicals in the environment.

In the following section, an approach is suggested for estimating the effect of pH on properties and environmental fate using the phenols as an example. A similar approach can be used for bases. The extent of dissociation is characterized by the acid dissociation constant, K_a, expressed as its negative logarithm, pK_a, which for most chloro- phenolic compounds range between 4.75 for pentachlorophenol and 10.2 to phenol, and between 10.0 and 10.6 for the alkylphenols. The dissolved concentration in water is thus the sum of the undissociated, parent or protonated compound and the dissociated phenolate ionic form. When the pK_a exceeds pH by 2 or more units, dissociation is 1 % or less and for most purposes is negligible. The ratio of ionic to non-ionic or dissociated to undissociated species is given by,

$$\text{ionic/non-ionic} = 10^{(pH-pKa)} = I$$

The fraction ionic X_I is $I/(1 + I)$. The fraction non-ionic X_N is $1/(1 + I)$. For compounds such as pentachlorophenol in which pH generally exceeds pK_a, I and X_I can be appreciable and there is an apparently enhanced solubility (IUPAC 1985, NRCC 1982, Yoshida et al. 1987, Acrand et al. 1993). There are other reports of pH effects on octanol-water partition coefficient (Kaiser and Valdmanis 1982, Westall et al. 1985, Lee et al. 1990, Smejtek and Wang 1993), soil sorption behaviour (Choi and Amoine 1974, Lee et al. 1990, Schellenberg et al. 1984, Yoshida et al. 1987, Lee et al. 1990), bioconcentration and uptake kinetics to goldfish (Stehly and Hayton 1990) and toxicity to algae (Smith et al. 1987, Shigeoka et al. 1988).

The following treatment has been suggested by Shiu et al. (1994) and is reproduced briefly below. It is suggested that the simplest, "first order" approach is to take in to account the effect of dissociation by deducing the ratio of ionic to non-ionic species I, the fraction ionic X_I and fraction non-ionic X_N for the chemical at both the pH of experimental data determination (I_D, X_{ID}, X_{ND}) and at the pH of the desired environmental simulation (I_E, X_{IE}, X_{NE}). It is assumed that dissociation takes place only in aqueous solution, not in air, organic carbon, octanol or lipid phases. Some ions and ion pairs are known to exist in the latter two phases, but there are insufficient data to give a general procedure for estimating the quantities. No correction is made for the effect of cations other than H^+. This approach must be regarded as merely a first correction for the dissociation effect, and not a full evaluation. This should preferably be based on direct experimental determinations. The reported solubility S mol/m^3 and K_{OW} presumably refer to the total of ionic and non-ionic forms, i.e., S_T and $K_{OW,T}$, at the pH of experimental determination, i.e.,

$$S_T = S_N + S_I$$

The solubility and K_{OW} of the non-ionic forms can be estimated as

$$S_N = S_T \cdot X_{ND}; \quad K_{OW,N} = K_{OW,T}/X_{ND}$$

Vapor pressure P^S is not affected, an apparent Henry's law constant H_T, must also be adjusted to H_T/X_N, being P^S/S_N or $P^S/(S_T \cdot X_N)$.

The Z values are calculated using the conventional equations at the pH of the experimental data (i.e., the system pH). The total Z value in water is then separated into its ionic and non-ionic contributions, i.e., fractions of I/(I + 1) and 1/(I + 1). The Z value for the non-ionic form in water is assumed to apply at all pHs i.e., including the environmental pH, but an additional and possibly different ionic Z value in water is deduced at the environmental pH using I calculated at that pH. The total Z values in water are then calculated. Z values in other media are unaffected.

The calculation is illustrated for pentachlorophenol. The experimental aqueous solubility is 14.0 g/m^3 at a pH of 5.1. The environmental pH is 7. Higher environmental pH increases the extent of dissociation thus increasing the Z value in water, increasing the apparent solubility, decreasing the apparent K_{OW} and Henry's law constant and the air-water partition coefficient, and decreasing the soil-water partition coefficient.

Table 1.2 Calculated Z values at different experimental and environmental pHs of pentachlorophenol. Z values at 25 °C, log K_{OW} is 5.05, pK_a 4.74, at data pH of 5.1 and environmental pH of 7.0 for air, water and soil of fraction organic carbon 0.02 and density of soil 2.4 kg/L.

	at data pH of 5.1 (I = 2.29)			*at environ. pH of 7 (I = 182)*		
	Neutral	Ionic	Total	Neutral	Ionic	Total
Air	4.03×10^{-4}	0	4.03×10^{-4}	4.03×10^{-4}	0	4.03×10^{-4}
Water	3.85	8.82	12.67	3.85	700.4	704.2
Soil solids	27970	0	27970	27970	0	27970

Note: at pH of 5.1, K_{OW} is 112200 and is the ratio of concentration in octanol to total concentration in water comprising fractions 1/(1 + I) or 1/(1 + 2.29) or 0.304 of neutral and 0.696 of ionic species. K_{OW} is thus 112200/0.304 or 369000 for the neutral species and zero for the ionic species. For the neutral species K_{OC} is assumed to be $0.41 \cdot K_{OW}$ or 151300, thus K_P is 151300×0.02 L/kg, i.e., 3027 for a soil of 2% organic carbon. K_{SW} is thus 3027×2.4 where 2.4 is the solid density (kg/L) or 7265. Z_S for the neutral species is thus $7265 \times Z_W$ or 27970. At pH of 7, the neutral species Z values are unaffected, but the Z value for water increases to 704 because of the greater extent of dissociation. K_{SW} thus decreases to 27970/704 or 39.72.

This is further demonstrated in Table 1.3 which shows the effects of environmental pH on the partitioning behaviour of 2,4-dichlorophenol (pK_a = 7.90, solubility of 6000 g/m^3 at pH of 5.1 and log K_{OW} = 3.20), 2,4,6-trichlorophenol (pK_a = 6.10, solubility of 430 g/m^3 at pH of 5.1 and log K_{OW} = 3.69), pentachlorophenol (pK_a = 4.74, solubility of 14.0 g/m^3 at pH of 5.1 and log K_{OW} = 5.05) and *p*-cresol (pK_a = 10.26, a solubility of 22000 g/m^3 and log Kow = 2.0) in the multimedia environment at 25 °C. For environmental pH from 4 to 7, there is no significant effect for *p*-cresol (or for chemicals for which pK_a > > pH), very little effect for 2,4-dichlorophenol (and chemicals with pK_a ranging between 7-10). There is some effect on 2,4,6-trichlorophenol (and chemicals with pK_a of 6-7) and a large effect for pentachlorophenol.

Table 1.3. Calculated Z_W values and some partition coefficients at different environmental pHs for pentachlorophenol (PCP), 2,4-dichlorophenol (2,4-DCP), 2,4,6-trichlorophenol (2,4,6-TCP) and *p*-cresol at 25 °C. K_{AW} is the air-water partition coefficient and K_{SW} is the soil-water partition coefficient calculated as in Table 1.2.

	Z values in water:				**Partitioning properties:**			
at pH	Neutral	Ionic	Total Z_W	Fraction X_N	S_T g/m^3	H_T Pa·m^3/mol	K_{AW}	K_{SW}
PCP								
4	3.849	0.7004	4.549	0.846	16.55	0.224	8.9×10^{-5}	6147
6	3.849	70.04	73.89	0.052	268.8	0.0135	5.46×10^{-6}	378.5
7	3.849	700.4	704.2	0.0055	2562	0.00142	5.73×10^{-7}	39.7
2,4,6-TCP								
4	1.7677	0.0140	1.7817	0.992	434	0.5612	2.26×10^{-4}	105.2
6	1.7677	1.4041	3.1718	0.557	772	0.315	1.272×10^{-4}	59.09
7	1.7677	14.041	15.8088	0.118	3644	0.172	6.945×10^{-5}	11.86
2,4-DCP								
4	3.063	0.00038	3.063	1.0	6000	0.326	0.000132	31.24
6	3.063	0.0386	3.101	0.988	6073	0.322	0.000130	30.85
7	3.063	0.386	3.448	0.888	6760	0.290	0.000117	27.75
p-cresol								
4	11.97	0	11.948	1.0	22000	0.0836	3.37×10^{-5}	1.968
7	11.97	0.0066	11.975	1.0	22000	0.0836	3.35×10^{-5}	1.968

Similar treatment can be applied to other dissociating compounds such as the carboxylic acids, nitrophenols. For bases such as amines the pK_b is defined and the extent of dissociation I estimated as $10^{-(pH - pKb)}$.

1.4.3 Treatment of Water-Miscible Compounds

In the multimedia models used in this series of volumes, an air-water partition coefficient K_{AW} or Henry's law constant (HLC) is required, and is calculated from the ratio of the pure substance vapor pressure and aqueous solubility. This method is widely used for hydrophobic chemicals but is inappropriate for water-miscible chemicals for which no solubility can be measured (although there are some reported "calculated solubilities" which have been derived from correlations with molecular descriptors for alcohols, aldehydes and amines by Leahy 1986; Kamlet et al. 1987, 1988 and Nirmalakhandan and Speece 1988). The obvious option is to input the HLC directly either as such or as the air-water partition coefficient. If the chemical's activity coefficient γ in water is known, then H can be estimated as $v_W \gamma P^S_L$ where v_W is the molar volume of water and P^S_L is the liquid vapor pressure. Since H can be regarded as P^S_L/C^S_L where C^S_L is the solubility, it is apparent that $(1/v_W\gamma)$ is a "pseudo-solubility." Correlations for γ are available as has been discussed earlier. For example, if γ is 5.0, the pseudo-solubility is 11100 mol/m^3 since v_W is 18×10^{-6} m^3/mol or 18 cm^3/mol. Chemicals with γ less than about 20 are usually miscible in water. In this case if the liquid vapor pressure is 1000 Pa, H will be 1000/11100 or 0.090 Pa·m^3/mol and K_{AW} will be H/RT or 3.6×10^{-5} at 25°C. Alternatively, if H or K_{AW} is known, C^S_L can be calculated. It is possible to apply existing models to hydrophilic chemicals if this pseudo-solubility is calculated from the activity coefficient or from a known H, i.e., C^S_L is P^S_L/H or $P^S_L/(K_{AW}\cdot RT)$. This is the approach used in this volume. In the fugacity illustrations all pseudo-solubilities are so designated and should not be regarded as real experimentally accessible quantities.

1.4.4 Level I Fugacity Calculation

The Level I calculation describes how a given amount of chemical partitions at equilibrium between six media: air, water, soil, bottom sediment, suspended sediment and fish. No account is taken of reactivity. Whereas most early evaluative environments have treated a one square kilometer region with about 70% water surface (simulating the global proportion of ocean surface), it has become apparent that a more useful approach is to treat a larger, principally terrestrial area similar to a jurisdictional region such as a U.S. state. The area selected is 100,000 km^2 or 10^{11} m^2, which is about the area of Ohio, Greece or England.

The atmospheric height is selected as a fairly arbitrary 1000 m reflecting that region of the troposphere which is most affected by local air emissions. A water surface area of 10% or 10,000 km^2 is used, with a water depth of 20 m. The water volume is thus 2×10^{11} m^3. The soil is viewed as being well mixed to a depth of 10 cm and is considered to be 2% organic carbon. It has a volume of 9×10^9 m^3. The bottom sediment has the same area as the water, a depth of 1 cm and an organic carbon content of 4%. It thus has a volume of 10^8 m^3.

For the Level I calculation both the soil and sediment are treated as simple solid phases with the above volumes, i.e., the presence of air or water in the pores of these phases is ignored.

Two other phases are included for interest. Suspended matter in water is often an important medium when compared in sorbing capacity to that of water. It is treated as having 20% organic carbon and being present at a volume fraction in the water of 5×10^{-6}, i.e., it is about 5 mg/L. Fishes are also included at an entirely arbitrary volume fraction of 10^{-6} and are assumed to contain 5% lipid, equivalent in sorbing capacity to octanol. These two phases are small in volume and rarely contain an appreciable fraction of the chemical present, but it is in these phases that the highest concentration of chemical often exists.

Another phase which is introduced later in the Level III model is aerosol particles with a volume fraction in air of 2×10^{-11}, i.e., approximately 30 $\mu g/m^3$. Although negligible in volume, an appreciable fraction of the chemical present in the air phase may be associated with aerosols. Aerosols are not treated in Level I or II calculations because their capacity for chemical is usually negligible when compared with soil.

These dimensions and properties are summarized in Table 1.4. The user is encouraged to modify these dimensions to reflect conditions in a specific area of interest.

Table 1.4*a* Compartment Dimensions and Properties for Level I and II Calculations

Compartment	Air	Water	Soil	Sediment	Suspended Sediment	Fish
Volume, V (m^3)	10^{14}	2×10^{11}	9×10^{9}	10^{8}	10^{6}	2×10^{5}
Depth, h (m)	1000	20	0.1	0.01	-	-
Area, A (m^2)	100×10^{9}	10×10^{9}	90×10^{9}	10×10^{9}	-	-
Org. Fraction (ϕ_{OC})	-	-	0.02	0.04	0.2	-
Density, ρ (kg/m^3)	1.2	1000	2400	2400	1500	1000
Adv. Residence Time, t (hours)	100	1000	-	50,000	-	-
Adv. flow, G (m^3/h)	10^{12}	2×10^{8}	-	2000	-	-

Table 1.4*b* Bulk Compartment Dimensions and Volume Fraction (v) for Level III Calculations

Air	Total volume	10^{14} m^3	(as above)
	Air phase	10^{14} m^3	
	Aerosol phase	2000 m^3	($v = 2\times10^{-11}$)
Water	Total volume	2×10^{11} m^3	
	Water phase	2×10^{11} m^3	(as above)
	Suspended sediment phase	10^{6} m^3	($v = 5\times10^{-6}$)
	Fish phase	2×10^{5} m^3	($v = 1\times10^{-6}$)
Soil	Total volume	18×10^{9} m^3	
	Air phase	3.6×10^{9} m^3	($v = 0.2$)
	Water phase	5.4×10^{9} m^3	($v = 0.3$)
	Solid phase	9.0×10^{9} m^3	($v = 0.5$) (as above)
Sediment	Total volume	500×10^{6} m^3	
	Water phase	400×10^{6} m^3	($v = 0.8$)
	Solid phase	100×10^{6} m^3	($v = 0.2$) (as above)

The amount of chemical introduced in the Level I calculation is an arbitrary 100,000 kg or 100 tonnes. If dispersed entirely in the air, this amount yields a concentration of 1 $\mu g/m^3$ which is not unusual for ubiquitous contaminants such as hydrocarbons. If dispersed entirely in the water, the concentration is a higher 500 $\mu g/m^3$ or 500 ng/L, which again is reasonable for a well-used chemical of commerce. The corresponding value in soil is about 0.0046 $\mu g/g$. It is believed that this amount is a reasonable common value for evaluative purposes. Clearly for restricted chemicals such as PCBs, this amount is too large, but it is preferable to adopt a common evaluative amount for all substances. No significance should, of course, be attached to the absolute values of the concentrations which are deduced from this arbitrary amount. Only the relative values have significance.

The Level I calculation proceeds by deducing the fugacity capacities, Z values for each medium (see Table 1.5), following the procedures described by Mackay (1991). These working equations show the necessity of having data on molecular mass, water solubility, vapor pressure, and octanol-water partition coefficient. The fugacity f (Pa) common to all media is deduced as

$$f = M/\sum V_i Z_i$$

where M is the total amount of chemical (mol), V_i is the medium volume (m^3) and Z_i is the corresponding fugacity capacity for the chemical in each medium.

The molar concentration C (mol/m^3) can then be deduced as Zf mol/m^3 or as WZf g/m^3 or 1000 WZf/ρ $\mu g/g$ where ρ is the phase density (kg/m^3) and W is the molecular mass (g/mol). The amount m_i in each medium is C_iV_i mol, and the total in all media is M mol. The **BASIC** computer program for undertaking this calculation is appended. For those who prefer a spreadsheet format, an identical Lotus 123® program is also provided.

The information obtained from this calculation includes the concentrations, amounts and distribution. In the figures, a pie chart illustrates the distribution between the four primary compartments of air, water, soil and sediment, the amount in fish and suspended sediment being ignored. This information is useful as an indication of the relative concentrations.

Note that this simple treatment assumes that the soil and sediment phases are entirely solid, i.e., there are no air or water phases present to "dilute" the solids. Later in the Level III calculation these phases and aerosols are included.

Table 1.5*a* Equations for Phase Z Values Used in Levels I and II and the Bulk Phase Values Used in Level III

Air	$Z_1 = 1/RT$
Water	$Z_2 = 1/H = C^S/P^S$
Soil	$Z_3 = Z_2 \cdot \rho_3 \cdot \phi_3 \cdot K_{OC}/1000$
Sediment	$Z_4 = Z_2 \cdot \rho_4 \cdot \phi_4 \cdot K_{OC}/1000$
Suspended Sediment	$Z_5 = Z_2 \cdot \rho_5 \cdot \phi_5 \cdot K_{OC}/1000$
Fish	$Z_6 = Z_2 \cdot \rho_6 \cdot L \cdot K_{OW}/1000$
Aerosol	$Z_7 = Z_1 \cdot 6x10^6/P^S_L$
where	R = gas constant (8.314 J/mol K)
	T = absolute temperature (K)
	C^S = solubility in water (mol/m^3)
	P^S = vapor pressure (Pa)
	H = Henry's law constant ($Pa \cdot m^3/mol$)
	P^S_L = liquid vapor pressure (Pa)
	K_{OW} = octanol-water partition coefficient
	K_{OC} = organic-carbon partition coefficient (= 0.41 K_{OW})
	ρ_i = density of phase i (kg/m^3)
	ϕ_i = mass fraction organic-carbon in phase i (g/g)
	L = lipid content of fish

Note for solids $P^S_L = P^S_S/\exp\{6.79(1 - T_M/T)\}$ where T_M is melting point (K) of the solute.

Table 1.5*b* Bulk Phase Z Values, Z_{Bi} Deduced as $\Sigma\, v_iZ_i$, in Which the Coefficients, e.g., 2×10^{-11}, are the Volume Fractions v_i of Each Pure Phase as Specified in Table 1.4*b*

Air	$Z_{B1} = Z_1 + 2\times10^{-11}\, Z_7$	(approximately 30 $\mu g/m^3$ aerosols)
Water	$Z_{B2} = Z_2 + 5\times10^{-6}\, Z_5 + 1\times10^{-6}\, Z_6$	(5 ppm solids, 1 ppm fish by volume)
Soil	$Z_{B3} = 0.2\, Z_1 + 0.3\, Z_2 + 0.5\, Z_3$	(20% air, 30% water, 50% solids)
Sediment	$Z_{B4} = 0.8\, Z_2 + 0.2\, Z_4$	(80% water, 20% solids)

1.4.5 Level II Fugacity Calculation

The Level II calculation simulates a situation in which a chemical is continuously discharged into the multimedia environment and achieves a steady-state equilibrium condition at which input and output rates are equal. The task is to deduce the rates of loss by reaction and advection.

The reaction rate data developed for each chemical in the tables are used to select a reactivity class as described earlier, and hence a first-order rate constant for each medium. Often these rates are in considerable doubt, thus the quantities selected should be used with extreme caution because they may not be widely applicable. The rate constants k_i h^{-1} are used to calculate reaction D values for each medium D_{Ri} as $V_iZ_ik_i$. The rate of reactive loss is then $D_{Ri}f$ mol/h.

For advection, it is necessary to select flow rates. This is conveniently done in the form of advective residence times, t in hour(h), thus the advection rate G_i is V_i/t m^3/h for each medium. For air, a residence time of 100 hours is used (approximately 4 days), which is probably too long for the geographic area considered, but shorter residence times tend to cause air advective loss to be a dominant mechanism. For water, a figure of 1000 hours (42 days) is used, reflecting a mixture of rivers and lakes. For sediment burial (which is treated as an advective loss), a time of 50,000 hours or 5.7 years is used. Only for very persistent, hydrophobic chemicals is this process important. No advective loss from soil is included. The D value for loss by advection D_{Ai} is G_iZ_i and the rates are $D_{Ai}f$ mol/h. These rates are listed in Table 1.6.

There may thus be losses caused by both reaction and advection D values for the four primary media. These loss processes are not included for fish or suspended matter. At steady state, equilibrium conditions the input rate E mol/h can be equated to the sum of the output rates, from which the common fugacity can be calculated as follows

$$E = f\cdot\sum D_{Ai} + f\cdot\sum D_{Ri}$$

thus,

$$f = E/(\sum D_{Ai} + \sum D_{Ri})$$

The common assumed emission rate is 1000 kg/h or 1 tonne/h. To achieve an amount equivalent to the 100 tonnes in the Level I calculation requires an overall residence time of 100 hours. Again, the concentrations and amounts m_i and $\sum m_i$ or M can be deduced, as well as the reaction and advection rates. These rates obviously total to give the input rate E. Of particular interest are the relative rates of these loss processes, and the overall persistence or residence time which is calculated as

$$t_O = M/E$$

where M is the total amount present. It is also useful to calculate a reaction and an advection persistence t_R and t_A as

$$t_R = M/\sum D_{Ri}f \qquad t_A = M/\sum D_{Ai}f$$

Obviously,

$$1/t_O = 1/t_R + 1/t_A$$

These persistences indicate the likelihood of the chemical being lost by reaction as distinct from advection. The percentage distribution of chemical between phases is identical to that in Level I. A pie-chart depicting the distribution of losses is presented.

1.4.6 Level III Fugacity Calculation

Whereas the Level I and II calculations assume equilibrium to prevail between all media, this is recognized as being excessively simplistic and even misleading. In the interests of algebraic simplicity only the four primary media are treated for this level. The task is to develop expressions for intermedia transport rates by the various diffusive and nondiffusive processes as described by Mackay (1991). This is done by selecting values for 12 intermedia transport velocity parameters which have dimensions of velocity (m/h or m/year), are designated as U_i m/h and are applied to all chemicals. These parameters are used to calculate seven intermedia transport D values.

It is desirable to calculate new "bulk phase" Z values for the four primary media which include the contribution of dispersed phases within each medium as described by Mackay and Paterson (1991) and as listed in Tables 1.4 and 1.5. The air is now treated as an air-aerosol mixture, water as water plus suspended particles and fish, soil as solids, air and water, and sediment as solids and porewater. The Z values thus differ from the Level I and Level II "pure phase" values. The necessity for introducing this complication arises from the fact that much of the intermedia transport of the chemicals occurs in association with the movement of chemical in these dispersed phases. To accommodate this change the same volumes of the soil solids and sediment solids are retained, but the total phase volumes are increasd. These Level III volumes are also given in Table 1.4. The reaction and advection D values employ the generally smaller bulk phase Z values but the same residence times, thus the G values are increased and the D values are generally larger.

Intermedia D Values

The justisfication for each intermedia D value follows. It is noteworthy that, for example, air-to-water and water-to-air values differ because of the presence of one-way nondiffusive processes. A fuller description of the background to these calculations is given by Mackay (1991).

1. Air to Water (D_{12})

Four processes are considered: diffusion (absorption), dissolution in rain of gaseous chemical, and wet and dry deposition of particle-associated chemical.

For diffusion, the conventional two-film approach is taken with water-side (k_W) and air-side (k_A) mass transfer coefficients (m/h) being defined. Values of 0.05, 5 m/h for k_W and k_A are used. The absorption D value is then

$$D_{VW} = 1/(1/(k_A A_W Z_1) + 1/(k_W A_W Z_2))$$

where A_W is the air-water area (m^2) and Z_1 and Z_2 are the pure air and water Z values. The velocities k_A and k_W are designated as U_1 and U_2.

For rain dissolution, a rainfall rate of 0.876 m/year is used, i.e., U_R or U_3 is 10^{-4} m/h. The D value for dissolution D_{RW} is then

$$D_{RW} = U_R A_W Z_2 = U_3 A_W Z_2$$

For wet deposition, it is assumed that the rain scavenges Q (scavenging ratio) or about 200,000 times its volume of air. Using a particle concentration (volume fraction) v_Q of 2×10^{-11}, this corresponds to the removal of Qv_Q or 4×10^{-6} volumes of aerosol per volume of rain. The total rate of particle removal by wet deposition is then $Qv_QU_RA_W$ m^3/h, thus the wet "transport velocity" Qv_QU_R is 4×10^{-10} m/h.

For dry deposition, a typical deposition velocity U_Q of 10 m/h is selected yielding a rate of particle removal of $U_Qv_QA_W$ or $2\times10^{-10}A_W$ m^3/h corresponding to a transport velocity of 2×10^{-10} m/h. Thus,

$$U_4 = Qv_QU_R + U_Qv_Q = v_Q(QU_R + U_Q)$$

The total particle transport velocity U_4 for wet and dry deposition is thus $6\text{x}10^{-10}$ m/h and the total D value D_{QW} is

$$D_{QW} = U_4A_WZ_7$$

where Z_7 is the aerosol Z value.

The overall D value is given by

$$D_{12} = D_{VW} + D_{RW} + D_{QW}$$

2. Water to Air (D_{21})

Evaporation is treated as the reverse of absorption; thus D_{21} is simply D_{VW} as before.

3. Air to Soil (D_{13})

A similar approach is adopted as for air to water transfer. Four processes are considered with rain dissolution (D_{RS}) and wet and dry deposition (D_{QS}) being treated identically except that the area term is now the air-soil area A_S.

For diffusion, the approach of Jury et al. (1983, 1984) is used as described by Mackay and Stiver (1991) and Mackay (1991) in which three diffusive processes are treated. The air boundary layer is characterized by a mass transfer coefficient k_S or U_7 of 5 m/h, equal to that of the air-water mass transfer coefficient k_A used in D_{12}.

For diffusion in the soil air-pores, a molecular diffusivity of 0.02 m^2/h is reduced to an effective diffusivity using a Millington-Quirk type of relationship by a factor of about 20 to 10^{-3} m^2/h. Combining this with a path length of 0.05 m gives an effective air to soil mass transfer coefficient k_{SA} of 0.02 m/h which is designated as U_5.

Similarly, for diffusion in water a molecular diffusivity of 2×10^{-6} m^2/h is reduced by a factor of 20 to an effective diffusivity of 10^{-7} m^2/h, which is combined with a path length of 0.05 m to give an effective soil to water mass transfer coefficient of k_{SW} 2×10^{-6} m/h.

It is probable that capillary flow of water contributes to transport in the soil. For example, a rate of 7 cm/year would yield an equivalent water velocity of 8×10^{-6} m/h which exceeds the water diffusion rate by a factor of four. For illustrative purposes we thus select a water transport velocity or coefficient U_6 in the soil of 10×10^{-6} m/h, recognizing that this may be in error by a substantial amount, and will vary with rainfall characteristics and soil type.

The soil processes are in parallel with boundary layer diffusion in series, so the final equation is

$$D_{VS} = 1/[1/D_S + 1/(D_{SW} + D_{SA})]$$

where

$$D_S = U_7 A_S Z_1 \qquad (U_7 = 5 \text{ m/h})$$
$$D_{SW} = U_6 A_S Z_2 \qquad (U_6 = 10 \times 10^{-6} \text{ m/h})$$
$$D_{SA} = U_5 A_S Z_1 \qquad (U_5 = 0.02 \text{ m/h})$$

where A_S is the soil horizontal area.

Air-soil diffusion thus appears to be much slower than air-water diffusion because of the slow migration in the soil matrix. In practice, the result will be a nonuniform composition in the soil with the surface soil (which is much more accessible to the air than the deeper soil) being closer in fugacity to the atmosphere.

The overall D value is given as

$$D_{13} = D_{VS} + D_{QS} + D_{RS}$$

4. Soil to Air (D_{31})

Evaporation is treated as the reverse of absorption, thus the D value is simply D_{VS}.

5. Water to Sediment (D_{24})

Two processes are treated, diffusion and deposition.

Diffusion is characterized by a mass transfer coefficient U_8 of 10^{-4} m/h which can be regarded as a molecular diffusivity of 2×10^{-6} m^2/h divided by a path length of 0.02 m. In practice, bioturbation may contribute substantially to this exchange process, and in shallow water current-induced turbulence may also increase the rate of transport. Diffusion in association with organic colloids is not included. The D value is thus given as $U_8 A_W Z_2$.

Deposition is assumed to occur at a rate of 5000 m^3/h which corresponds to addition of a depth of solids of 0.438 cm/year; thus 43.8% of the solids resident in the accessible bottom sediment is added each year. This rate is about 12 cm^3/m^2 • day which is high compared to values observed in large lakes. The velocity U_9, corresponding to the addition of 5000 m^3/h over the area of 10^{10} m^2, is thus 5×10^{-7} m/h.

It is assumed that of this 5000 m^3/h deposited, 2000 m^3/h or 40% is buried (yielding the advective flow rate in Table 1.2), 2000 m^3/h or 40% is resuspended (as discussed later) and the remaining 20% is mineralized organic matter. The organic carbon balance is thus only approximate.

The transport velocities are thus:

deposition U_9 — 5.0×10^{-7} m/h or 0.438 cm/years

resuspension U_{10} — 2.0×10^{-7} m/h or 0.175 cm/year

burial U_B — 2.0×10^{-7} m/h or 0.175 cm/year
(included as an advective residence time of 50,000 h)

The water to sediment D value is thus

$$D_{24} = U_8A_WZ_2 + U_9A_WZ_5$$

where Z_5 is the Z value of the particles in the water column.

6. Sediment to Water (D_{42})

This is treated similarly to D_{24} giving:

$$D_{42} = U_8A_WZ_2 + U_{10}A_WZ_4$$

where U_{10} is the sediment resuspension velocity of 2.0×10^{-7} m/h and Z_4 is the Z value of the sediment solids.

7. Sediment Advection (D_{A4})

This D value is $U_BA_WZ_4$ where U_B, the sediment burial rate, is 2.0×10^{-7} m/h. It can be viewed as G_BZ_{B4} where G_B is the total burial rate specified as V_S/t_B where t_B (residence time) is 50,000 h, and V_S (the sediment volume) is the product of sediment depth (0.01 cm) and area A_W. Z_4, Z_{B4} are the Z values of the sediment solids and of the bulk sediment respectively. Since there are 20% solids, Z_{B4} is about 0.2 Z_4. There is a slight difference between these approaches because in the advection approach (which is used here) there is burial of water as well as solids.

8. Soil to Water (D_{32})

It is assumed that there is run-off of water at a rate of 50% of the rain rate, i.e., the D value is

$$D = 0.5\ U_3A_SZ_2 = U_{11}A_SZ_2$$

thus the transport velocity term U_{11} is $0.5U_3$ or 5×10^{-5} m/h.

For solids run-off it is assumed that this run-off water contains 200 parts per million by volume of solids; thus the corresponding velocity term U_{12} is $200\times10^{-6}U_{11}$, i.e., 10^{-8} m/h. This corresponds to the loss of soil at a rate of about 0.1 mm per year. If these solids were completely deposited in the aquatic environment (which is about 1/10th the soil area), they would accumulate at about 0.1 cm per year, which is about a factor of four less than the deposition rate to sediments. The implication is that most of this deposition is of naturally generated organic carbon and from sources such as bank erosion.

Summary

The twelve intermedia transport parameters are listed in Table 1.6 and the equations are summarized in Table 1.7.

Table 1.6 Intermedia Transport Parameters

U		m/h	m/year
1	Air side, air-water MTC*, k_A	5	43800
2	Water side, air-water MTC, k_W	0.05	438
3	Rain rate, U_R	10^{-4}	0.876
4	Aerosol deposition	6×10^{-10}	5.256×10^{-6}
5	Soil-air phase diffusion MTC, k_{SA}	0.02	175.2
6	Soil-water phase diffusion MTC	10×10^{-6}	0.0876
7	Soil-air boundary layer MTC, k_S	5	43800
8	Sediment-water MTC	10^{-4}	0.876
9	Sediment deposition	5.0×10^{-7}	0.00438
10	Sediment resuspension	2.0×10^{-7}	0.00175
11	Soil-water run-off	5.0×10^{-5}	0.438
12	Soil-solids run-off	10^{-8}	0.0000876

* Mass transfer coefficient
with,
Scavenging ratio $Q = 2\times10^5$
Dry deposition velocity $U_Q = 10$ m/h
Sediment burial rate $U_B = 2.0\times10^{-7}$ m/h

Table 1.7 Intermedia Transport D Value Equations

Air-Water	
	$D_{12} = D_{VW} + D_{RW} + D_{QW}$
	$D_{VW} = A_W/(1/U_1Z_1 + 1/U_2Z_2)$
	$D_{RW} = U_3A_WZ_2$
	$D_{QW} = U_4A_WZ_7$
Water-Air	
	$D_{21} = D_{VW}$
Air-Soil	
	$D_{13} = D_{VS} + D_{RS} + D_{QS}$
	$D_{VS} = 1/(1/D_S + 1/(D_W + D_A))$
	$D_S = U_7A_SZ_1$
	$D_{SA} = U_5A_SZ_1$
	$D_{SW} = U_6A_SZ_2$
	$D_{RS} = U_3A_SZ_2$
	$D_{QS} = U_4A_SZ_7$
Soil-Air	
	$D_{31} = D_{VS}$
Water-Sediment	
	$D_{24} = U_8A_WZ_2 + U_9A_WZ_5$
Sediment-Water	
	$D_{42} = U_8A_WZ_2 + U_{10}A_WZ_4$
Soil-Water	
	$D_{32} = U_{11}A_SZ_2 + U_{12}A_SZ_3$

Algebraic Solution

Four mass balance equations can be written, one for each medium, resulting in a total of four unknown fugacities, enabling simple algebraic solution as shown in Table 1.8. From the four fugacities, the concentration, amounts and rates of all transport and transformation processes can be deduced, yielding a complete mass balance.

Table 1.8 Level III Solutions to Mass Balance Equations

Mass balance equations:

Air $E_1 + f_2D_{21} + f_3D_{31} = f_1D_{T1}$

Water $E_2 + f_1D_{12} + f_3D_{32} + f_4D_{42} = f_2 D_{T2}$

Soil $E_3 + f_1D_{13} = f_3D_{T3}$

Sediment $E_4 + f_2D_{24} = f_4D_{T4}$

where E_i is discharge rate, E_4 usually being zero.

$$D_{T1} = D_{R1} + D_{A1} + D_{12} + D_{13}$$

$$D_{T2} = D_{R2} + D_{A2} + D_{21} + D_{23} + D_{24}, \quad (D_{23} = 0)$$

$$D_{T3} = D_{R3} + D_{A3} + D_{31} + D_{32}, \quad (D_{A3} = 0)$$

$$D_{T4} = D_{R4} + D_{A4} + D_{42}$$

Solution:

$$f_2 = [E_2 + J_1J_4/J_3 + E_3D_{32}/D_{T3} + E_4D_{42}/D_{T4}]/(D_{T2} - J_2J_4/J_3 - D_{24}\cdot D_{42}/D_{T4})$$

$$f_1 = (J_1 + f_2J_2)/J_3$$

$$f_3 = (E_3 + f_1D_{13})/D_{T3}$$

$$f_4 = (E_4 + f_2D_{24})/D_{T4}$$

where

$$J_1 = E_1/D_{T1} + E_3D_{31}/(D_{T3}\cdot D_{T1})$$

$$J_2 = D_{21}/D_{T1}$$

$$J_3 = 1 - D_{31}\cdot D_{13}/(D_{T1}\cdot D_{T3})$$

$$J_4 = D_{12} + D_{32}\cdot D_{13}/D_{T3}$$

The new information from the Level III calculations are the intermedia transport data, i.e., the extent to which chemical discharged into one medium tends to migrate into another. This migration pattern depends strongly on the proportions of the chemical discharged into each medium; indeed, the relative amounts in each medium are largely a reflection of the locations of discharge. It is difficult to interpret these mass balance diagrams because, for example, chemical depositing from air to water may have been discharged to air, or to soil from which it evaporated, or even to water from which it is cycling to and from air.

To simplify this interpretation, it is best to conduct three separate Level III calculations in which unit amounts (1000 kg/h) are introduced individually into air, soil and water. Direct discharges to sediment are unlikely and are not considered here. These calculations show clearly the extent to which intermedia transport occurs. If, for example, the intermedia D values are small compared to the reaction and advection values, the discharged chemical will tend to remain in the discharge or "source" medium with only a small proportion migrating to other media. Conversely, if the intermedia D values are relatively large the chemical becomes very susceptible to intermedia transport. This behavior is observed for persistent substances such as PCBs, which have very low rates of reaction.

A direct assessment of multimedia behavior is thus possible by examining the proportions of chemical found at steady state in the "source" medium and in other media. For example, when discharged to water, an appreciable fraction of the benzene is found in air, whereas for atrazine, only a negligible fraction of atrazine reaches air.

Linear Additivity

Because these equations are entirely linear, the solutions can be scaled linearly. The concentrations resulting from a discharge of 2000 kg/h are simply twice those of 1000 kg/h. Further, if discharge of 1000 kg/h to air causes 500 kg in water and discharge of 1000 kg/h to soil causes 100 kg in water, then if both discharges occur simultaneously, there will be 600 kg in water. If the discharge to soil is increased to 3000 kg/h, the total amount in the water will rise to (500 + 300) or 800 kg. It is thus possible to deduce the amount in any medium arising from any combination of discharge rates by scaling and adding the responses from the unit inputs. This "linear additivity principle" is more fully discussed by Stiver and Mackay (1989).

In the diagrams presented later, these three-unit (1000 kg/h) responses are given. Also, an illustrative "three discharge" mass balance is given in which a total of 1000 kg/h is discharged, but in proportions judged to be typical of chemical use and discharge to the environment. For example, benzene is believed to be mostly discharged to air with minor amounts to soil and water.

Also given in the tables are the rates of reaction, advection and intermedia transport for each case.

The reader can deduce the fate of any desired discharge pattern by appropriate scaling and addition. It is important to re-emphasize that because the values of transport velocity parameters are only illustrative, actual environmental conditions may be quite different; thus, simulation of conditions in a specific region requires determination of appropriate parameter values as well as the site specific dimensions, reaction rate constants and the physical-chemical properties which prevail at the desired temperature.

In total, the aim is to convey an impression of the likely environmental behavior of the chemical in a readily assimilable form.

1.5. DATA SOURCES AND PRESENTATIONS

1.5.1 Data Sources

Most physical properties such as molecular weight (MW, g/mol), melting point (M.P., °C), boiling point (B.P., °C), and density have been obtained from commonly used handbooks such as the CRC Handbook of Physics and Chemistry (Weast 1972, 1984), Lange's Handbook of Chemistry (Dean 1979, 1985), Dreisbach's Physical Properties of Chemical Compounds, Vol. I, II and III (1955, 1959, 1961), Organic Solvents, Physical Properties and Methods of Purification (Riddick et al. 1986) and the Merck Index (1983, 1987). Other physical-chemical properties such as aqueous solubility, vapor pressure, octanol-water partition coefficient, Henry's law constant, bioconcentration factor and sorption coefficient have been obtained from scientific journals or other environmental handbooks, notably Verschueren's Handbook of Environmental Data on Organic Chemicals (1977, 1983) and Howard et al.'s Handbook of Environmental Fate and Exposure Data, Vol. I, II, III and IV (1989, 1990, 1991 and 1993). Other important sources of vapor pressure are the CRC Handbook of Physics and Chemistry (Weast 1972, 1982), Lange's Handbook of Chemistry (Dean 1985), the Handbook of Vapor Pressures and Heats of Vaporization of Hydrocarbons and Related Compounds (Zwolinski and Wilhoit 1971), the Vapor Pressure of Pure Substances (Boublik et al. 1973, 1984), the Handbook of the Thermodynamics of Organic Compounds (Stephenson and Malanowski 1987). For aqueous solubilities, valuable sources include the IUPAC Solubility Data Series (1984, 1985, 1989a,b) and Horvath's Halogenated Hydrocarbons, Solubility-Miscibility with Water (Horvath 1982). Octanol-water partition coefficients are conveniently obtained from the compilation by Leo et al. (1971) and Hansch and Leo (1979), or can be calculated from molecular structure by the methods of Hansch and Leo (1979) or Rekker (1977). Lyman et al. (1982) also outline methods of estimating solubility, K_{OW}, vapor pressure, and bioconcentration factor for organic chemicals. The recent Handbook of Environmental Degradation Rates by Howard et al. (1991) is a valuable source of rate constants and half-lives for inclusion in subsequent fugacity calculations.

The most reliable sources of data are the original citations in the reviewed scientific literature. Particularly reliable are those papers which contain a critical review of data from a number of sources as well as independent experimental determinations. Calculated or correlated values are reported in the tables but are viewed as being less reliable. A recurring problem is that a value is frequently quoted, then requoted and the original paper may not be cited. The aim in this work has been to gather and list the citations, interpret them and select a "best" or "most likely" value. To assist in this process, plots are prepared of properties as a function of molar volume as the molecular descriptors. These are discussed at the end of each chapter.

1.5.2 Data Format

Each data sheet lists the following properties, although not all quantities are included for all chemicals. In all cases citations are provided.

Common Name:
Synonym:
Chemical Name:
CAS Registry No:
Molecular Formula:
Molecular Weight (g/mol):

Melting Point (°C):
Boiling Point (°C):
Density (g/cm^3 at 20 °C):
Molar Volume (cm^3/mol):
Molecular Volume, TMV ($Å^3$):
Dissociation constant: pK_a or pK_b:
Total Surface Area, TSA ($Å^2$):
Heat of Fusion, ΔH_{fus}, kcal/mol:
Entropy of Fusion, ΔS_{fus}, cal/mol K (or e.u.):
Fugacity Ratio at 25 °C:
Water Solubility (g/m^3 or mg/L at 25°C):
Vapor Pressure (Pa at 25°C):
Henry's Law Constant (Pa m^3/mol) or Air-Water Partition Coefficient:
Octanol-Water Partition Coefficient K_{OW} or log K_{OW}:
Bioconcentration Factor K_B or BCF (or log K_B):
Sorption Partition Coefficient to Organic Carbon K_{OC} or to Organic Matter K_{OM}:
Half-lives in the Environment:

- Air:
- Surface water:
- Groundwater:
- Soil:
- Sediment:
- Biota:

Environmental Fate Rate Constants or Half-Lives:

- Volatilization/Evaporation:
- Photolysis:
- Oxidation or Photooxidation:
- Hydrolysis:
- Biotransformation/Biodegradation:
- Bioconcentration, Uptake (k_1) and Elimination (k_2) Rate Constants:

1.5.3 Explanation of Data Presentations

Example: Pentachlorophenol (data sheets presented in Chapter 4)

1. Chemical Properties.

The names, formula, melting and boiling point and density data are self-explanatory.

The molar volumes are in some cases at the stated temperature and in others at the normal boiling point. Certain calculated molecular volumes are also used; thus the reader is cautioned to ensure that when using a molar volume in any correlation, it is correctly selected. In the case of polynuclear aromatic hydrocarbons, the LeBas molar volume is regarded as suspect because of the compact nature of the multi-ring compounds. It should thus be regarded as merely an indication of relative volume, not absolute volume.

The total surface areas (TSAs) are calculated in various ways and may contain the hydration shell, thus giving a much larger area. Again, the reader is cautioned to ensure that values are consistent.

Heats of fusion, ΔH_{fus}, are generally expressed in kcal/mol or kJ/mol and entropies of fusion, ΔS_{fus} in cal/mol·K (e.u. or entropy unit) or J/mol·K. The fugacity ratio F, defined below, is used to calculate the supercooled liquid vapor pressure or solubility for correlation purposes. In the case of liquids such as benzene, it is 1.0. For solids it is a fraction representing the ratio of solid to liquid solubility or vapor pressure. It is generally assumed that for a rigid organic molecule, the entropy of fusion is 13.5 e.u. or 56 J/mol·K, which is an average value of a number of organic compounds (Yalkowsky 1979, Miller et al. 1984). The fugacity ratio, F, given is calculated using ΔS_{fus} = 56 J/mol·K in the following expression

$$F = \exp\{(\Delta H_{fus}/RT)(1 - T_M/T)\} = \exp\{(\Delta S_{fus}/R)(1 - T_M/T)\}$$

where R is the ideal gas constant (8.314 J/mol·K or 1.987 cal/mol·K) and T_M is the melting point and T is the system temperature (K).

As is apparent, a wide variety of solubilities (in units of g/m^3 or the equivalent mg/L) have been reported. Experimental data have the method of determination indicated. In other compilations of data the reported value has merely been quoted from another secondary source. In some cases the value has been calculated. The abbreviations are generally self-explanatory and usually include two entries, the method of equilibration followed by the method of determination. From these values a single value is selected for inclusion in the summary data table. In the case of pentachlorophenol (PCP), the selected solubility value is 14.0 g/m^3 at 25°C and pH of 5.1. From an examination of the data it is judged that the true value almost certainly lies between 10.0 and 20.0 g/m^3.

The vapor pressure data are treated similarly with a value of 0.00415 Pa being selected. The true value is judged to lie between 0.003 and 0.005 Pa. Vapor pressures are, of course very temperature-dependent.

The Henry's law constant data are not available experimentally but are reported to lie in the range from 0.013 to 0.0025 estimated by calculated from the ratio of vapor pressure and solubility (in units of mol/m^3) and the bond contribution method. In this case a value of 0.080 Pa·m^3/mol is selected based on the selected vapor pressure and solubility, the actual value probably lying between 0.010 and 0.10 Pa·m^3/mol at low pH. Care must be exercised when dissociation occurs, since the Henry's law constant has to be corrected from the "apparent Henry's law constant" as discussed earlier.

The octanol-water partition coefficient data are similarly a combination of calculated and experimental values. Again, due to low pK_a, PCP will dissociate significantly in water at pH 7, therefore, a correction has to be made for the "apparent K_{OW}." A value of 5.05 is selected as being the most likely value of log K_{OW}, i.e., K_{OW} is 112200. A range of log K_{OW} value of about 3.80 to 5.86 is reported, but the authors believe that the actual value lies between 5.0 to 5.25. Again, consideration of the pH of the solution must be taken. In situations in which two or more species partition between water and octanol, it is possible to assign individual (and generally different) K_{OW} values to each species and deduce the overall partition coefficient from a knowledge of the proportions of the species in the aqueous phase. These proportions can be estimated from the pH. The reader is referred to the more detailed discussion of this issue by Kaiser and Valdmanis 1982, Westall et al. 1985, Lee et al. 1990 and Smejtek and Wang 1993.

A number of (log) bioconcentration factors are listed. Most lie in the range from 2.0 to 3.20 (with the exception of the Killifish data 0.9-1.70 by Trejilo et al. 1982), which corresponds to BCFs of 100 to 1585. Due to the complexity of pH effect on various physical-chemical properties this range could not be straightly interpreted as lipid contents as in the case of hydrophobic chemicals, as BCF is often estimated from K_{OW}, if only the neutral species is considered then it is possible that giving lower-than-equilibrium BCFs are expected.

The (log) organic-carbon partition coefficients listed range mostly from 2.20 to 4.50. It is expected that K_{OC} usually lies in the range of 20 to 80% of K_{OW}, i.e., K_{OC} will range from 158 to 31620. This is clustering of experimental values around 10000, i.e., log K_{OC} is 4.0. Organic matter partition coefficients are also reported. Since organic carbon accounts for some 50 to 60% of the content of organic matter, K_{OM} is expected to be about half K_{OC}.

The reader is advised to consult the original reference when using these values of BCF, K_{OC} and K_{OM}, to ensure that conditions are as close as possible to those of specific interest.

The "Half-life in the Environment" data reflect observations of the rate of disappearance of the chemical from a medium, without necessarily identifying the cause of mechanism of loss. For example, loss from water may be a combination of evaporation, biodegradation and photolysis. Clearly these times are highly variable and depend on factors such as temperature, meteorology and the nature of the media. Again, the reader is urged to consult the original reference.

The "Environmental Fate Rate Constants" refer to specific degradation processes rather than media. As far as possible the original numerical quantities are given and thus there is a variety of time units with some expressions being rate constants and others half-lives.

The conversion is

$$k = 0.693/t_{1/2}$$

where k is the first-order rate constant (h^{-1}) and $t_{1/2}$ is the half-life (h).

From these data a set of medium-specific degradation reaction half-lives was selected for use in Level II and III calculations. Emphasis was based on the fastest and the most plausible degradation process for each of the environmental compartments considered. Instead of assuming an equal half-life for both the water and soil compartment, as suggested by Howard et al. (1991), a slower active class (in the reactivity table described earlier) was assigned for soil and sediment compared to that of the water compartment. This is in part because the major degradation processes are often photolysis (or photooxidation) and biodegradation. There is an element of judgement in this selection and it may be desirable to explore the implications of selecting other values. The selected values of the phenolic compounds are given in Table 4.4 at the end of Chapter 4.

In summary, the physical-chemical and environmental fate data listed result in the selection of values of solubility, vapor pressure, K_{OW} and reaction half-lives which are used in the evaluative environmental calculations.

These QSPR (quantitative structure-property relationship) plots display the usual approximately linear relationships which are reported for the series of chlorinated aromatic hydrocarbons (Mackay et al. 1992a, b). We do not develop linear regressions of these data since we suggest that the plots be used directly for data estimation purposes. This enables the user to assess into account the values of similarly structured compounds and it gives a direct impression of likely error. We discuss, below, the general nature of the relationships and in particular the slopes of the QSPR plots.

Most acid dissociation constants pK_a exceed environmental pH values, the exceptions being the highly chlorinated phenols. As a result, these substances tend to have a higher apparent solubility in water because of dissociation. The structure-property relationships apply to the unionized or protonated species.

Figure 1.1 shows the relationship between various molecular descriptors and LeBas molar volume for the phenols.

Figure 1.2 shows that the chlorophenols behaved similarly to other chemical series with slopes of about 0.62 log units per 20.9 cm^3/mol which is the volume difference of one chlorine. The result is a factor of $10^{0.62}$ or 4.2 drop in solubility per chlorine. The alkyl phenols have a lower slope of about 0.5 per CH_2 and usually have higher solubilities at the same molar volume. The two sets of data are, however, generally similar.

The vapor pressure data in Figure 1.3 show a slope of about 0.60 log units per 20.9 cm^3/mol (i.e., a factor or 4.0) per chlorine. There is a lower slope for the alkylphenols and they usually have higher vapor pressures especially for the larger molecules.

The K_{OW} data in Figure 1.4 show that the chlorophenols and alkylphenols are different in properties. There is uncertainty about the K_{OW} of the longer-chain phenols. The chlorophenols tend to partition more into octanol at the same molar volume and are thus expected to be more bioaccumulative. The slope of the chlorophenol line is about 0.78 log units per chlorine or a factor of 6.0. The alkylphenol slope is lower and about 0.36 log units per CH_2, i.e., a factor of 2.3.

The Henry's law constant data calculated as the ratio of vapour pressure to solubility in Figure 1.5 are quite scattered. There is little systematic variation with molar volume. Most values of log H lie between -0.1 to -1.10, i.e., H lies between 0.8 and 0.08 and the resulting air-water partition coefficient K_{AW} or H/RT thus lies between 3×10^{-4} and 3×10^{-5}.

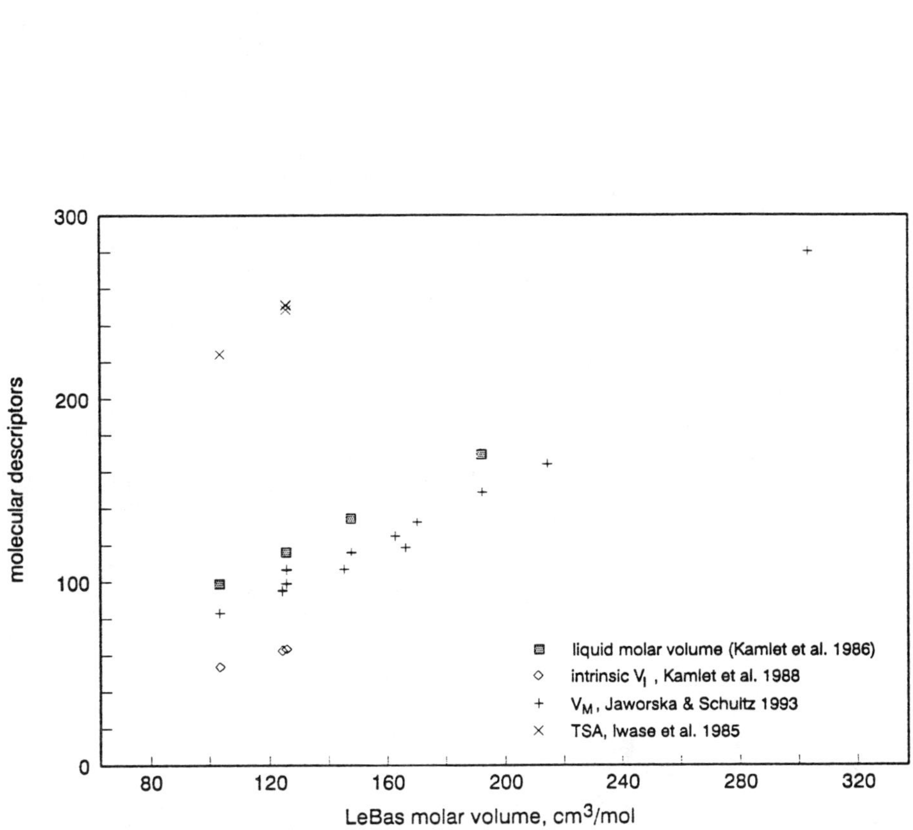

Figure 1.1 Plot of molecular descriptors versus LeBas molar volume.

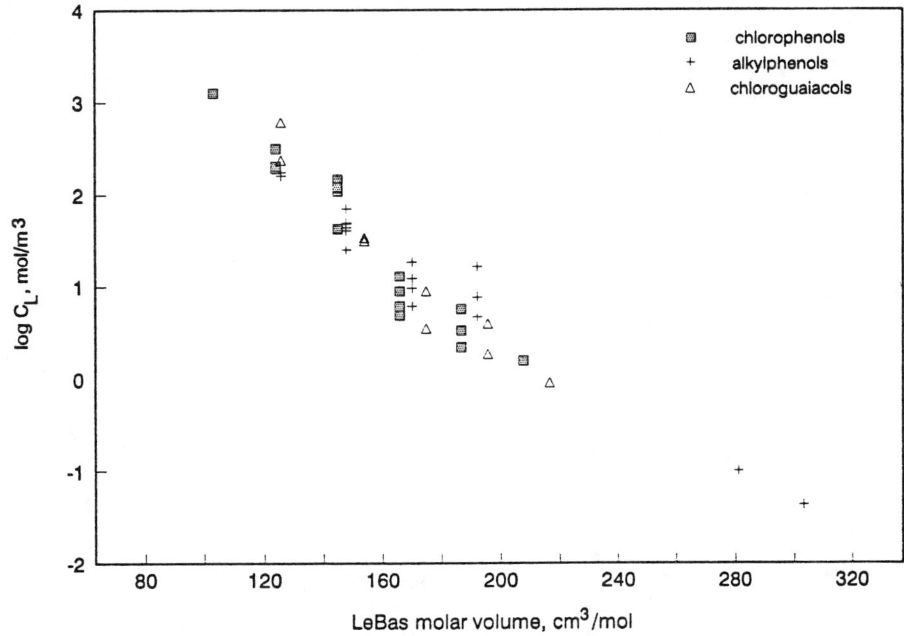

Figure 1.2 Plot of solubility (liquid or subcooled liquid) vs LeBas molar volume

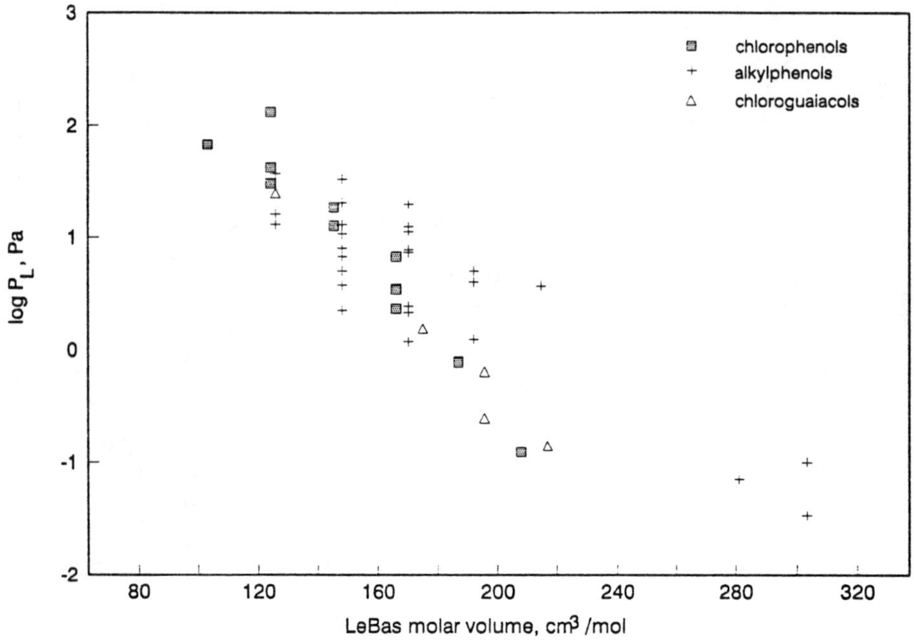

Figure 1.3 Plot of log vapour pressure (liquid or subcooled liquid) vs LeBas molar volume

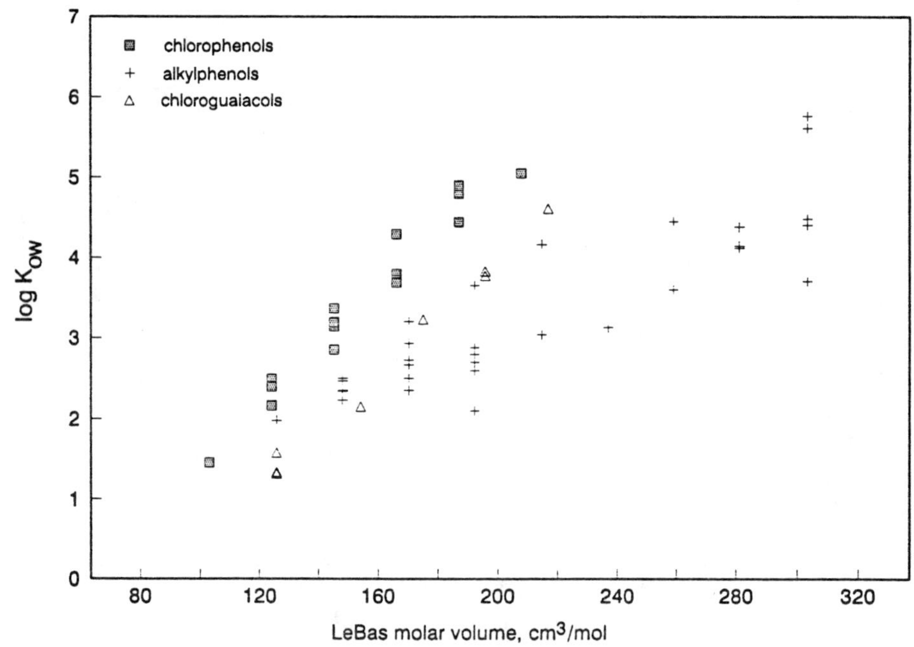

Figure 1.4 Plot of log K_{OW} versus LeBas molar volume

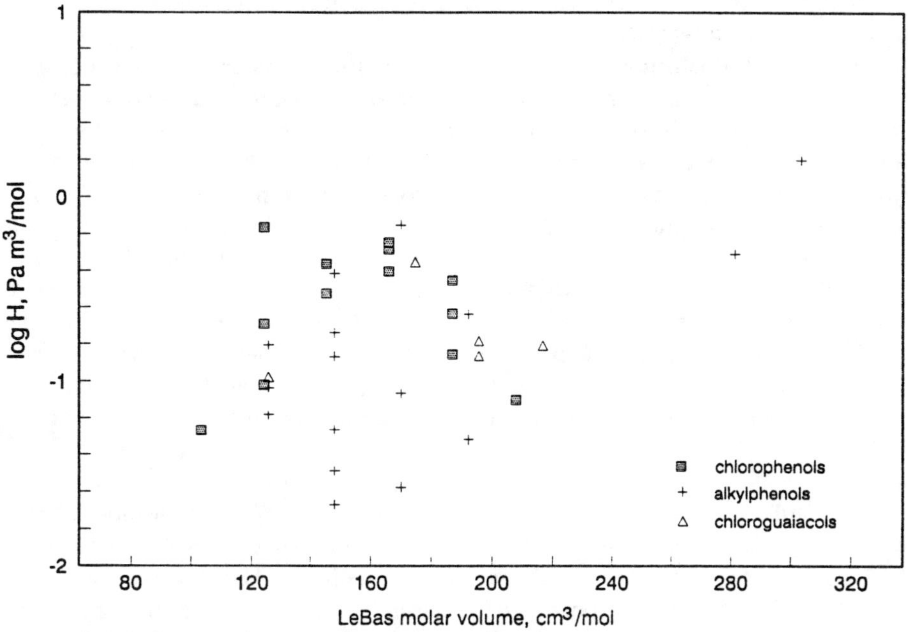

Figure 1.5 Plot of log Henry's law constant vs LeBas molar volume

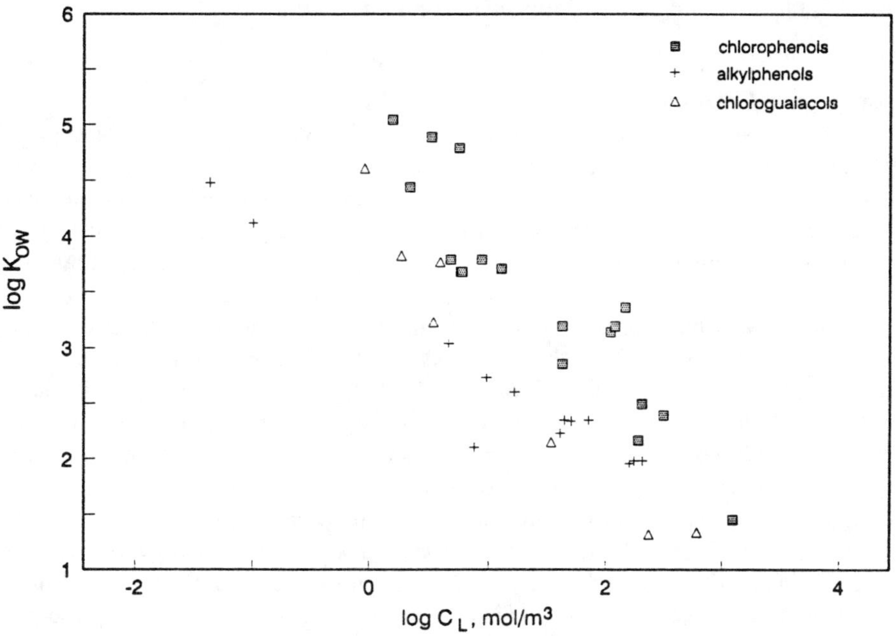

Figure 1.6 Plot of log K_{OW} versus log solubility (liquid or subcooled liquid)

Figure 1.6, the plot of log K_{OW} versus log solubility shows a relatively high slope of 1.25 for the chlorophenols and a lower slope of 0.70 for the alkylphenols.

Addition of a chlorine causes a drop in solubility in water by about 0.62 log units and thus K_{OW} increases by about 0.78 log units. For the alklylphenols addition of a methylene causes about a 0.5 log unit drop in solubility in water, and a drop in solubility in octanol of about 0.1 log units, thus K_{OW} increases by only about 0.36 log units. The slope of the log K_{OW} versus solubility lines are thus about 0.78/0.62 or 1.25 for the chlorophenols and 0.36/0.5 or 0.72 for the alkylphenols. An implication is that, since K_{OW} can be viewed as a ratio of "solubility" in octanol and solubility in water, the solubility of the chlorophenols in octanol increases by (0.78 - 0.62) or 0.16 log unit per chlorine while for the alkyl phenols the corresponding change is (0.36 - 0.50) or -0.14 log unit or a decrease of a factor of 1.4. The reasons for this difference are not known. The chlorophenols thus appear to have an unusually strong tendency to partition into octanol. Whether or not this tendency applies to lipid phases in biota or to organic carbon is not certain, but is obviously of considerable interest when interpreting the toxicity and fate of these chemicals.

These data show clearly that the structure-property relationships which apply to hydrophobic organic chemicals such as the chloro- and alkyl-aromatics also apply to the phenols, but the relationships are more scattered and less well defined. The absolute values of properties differ greatly. This scatter is probably attributable in part to insufficient experimental data or errors in experimental measurements, to dissociation and to the greater polar character of these chemicals. It is not recommended that correlations developed for non-polar organic chemicals be applied to the phenols. Separate treatment is required.

1.5.4. Evaluative Calculations

The illustrative evaluative environmental calculations discussed here and presented later for a number of chemicals have been modified from those in Volume I of this series to give more data in a more spacious layout. Level I and II diagrams are assigned to separate pages and the physical-chemical properties are included in the Level I diagram. The Level III diagram is identical to that in Volume I, but additional pie-chart diagrams are included to show how the mass of chemical is distributed between compartments, and how loss processes are distributed as a function of the compartment of discharge. For dissociating compounds the environmental pH is specified and the calculation of Z values has been modified to include ionic species as discussed in Section 1.4.2. For water-miscible compounds, a "pseudo-solubility" is calculated from the measured Henry's law constant. Generally, if discharge is into a compartment such as water, most chemical will be found in that compartment, and will react there but a quantity does migrate to other compartments and is lost from these media. Three pie-charts corresponding to discharges of 1000 kg/h into air, water and soil are included. The percentage emission in each medium is different from the previous Volumes, i.e., 5, 25 and 70% will be discharged to air, water and soil respectively (instead of 60, 30 and 10% to air, water and soil compartments). A fourth pie-chart with discharges to all three compartments is also given. This latter chart is in principle the linear sum of the first three, but since the overall residence times differ, the diagram with the longer residence time, and greater resident mass, tends to dominate.

Figures 7 and 8 show the mass distributions obtained in Level I calculations and the removal distribution from Level II fugacity calculation of pentachlorophenol (PCP) at different environmental pHs for the generic environment. Fig. 1.9*a* - Fig. 1.12*a* show the Level III fugacity calculation. Both mass and removal distributions are shown in Fig. 1.9*b* - Fig. 1.12*b* for the four scenarios of discharges into air, water, soil, and mixed compartments.

Level I

The Level I calculations illustrated in Fig. 1.7*a* (for system pH of 5.1), and Figures 1.7*b* to 1.7*d* (for environmental pH of 4, 6, and 7) suggest that if 100,000 kg (100 tonnes) of pentachlorophenol (PCP) are introduced into the 100,000 km^2 environment, most PCP will tend to be associated with soil. This is especially the case at low pH when the protonated form dominates. Only 0.015% will partition into air at a concentration of 1.55×10^{-10} g/m^3 or about 155 pg/m^3. The water will contain 0.97% at a low concentration of 4.9 $\mu g/m^3$ or equivalently 4.9 ng/L. Soil would contain 96.8% of the PCP at 4.48×10^{-3} $\mu g/g$ and sediments about 2.15% at 8.96×10^{-3} $\mu g/g$. The fugacity is calculated to be 1.44×10^{-9} Pa. The dimensionless soil-water and sediment-water partition coefficients or ratios of Z values are 2207 and 4415 as a result of a K_{OC} of about 46,000 and a few percent organic carbon in these media. There is evidence of bioconcentration with a rather high fish concentration of 2.73×10^{-2} $\mu g/g$ corresponding to a BCF of 5610. The pie chart in Figure 1.7*a* (which is the same as the Level I diagram for PCP in Chapter 4) clearly shows that soil is the primary medium of accumulation. Note that only four media (air, water, soil and bottom sediment) are depicted in the pie chart, therefore the sum of the percent distribution figures is slightly less than 100%. The air-water partition coefficient is very low at about 3.18×10^{-5}. As pH increases, dissociation increases and there is a tendency for partitioning to water become more important. At pH 7 (Fig. 1.7*d*) over one-third of the PCP will be dissolved in water. Essentially the capacity of water for the chemical increases. Partitioning to air is always negligible.

Level II

The Level II calculations illustrated in Fig. 1.8*a* at pH 5.1 include the reaction half-lives of 550 h in air, 550 h in water, 1700 h in soil and 5500 h in sediment. No reaction is included for suspended sediment or fish. The steady-state input of 1000 kg/h results in an overall fugacity of 3.43×10^{-8} Pa which is about 24 times the Level I value. The concentrations and amounts in each medium are thus about 24 times the Level I values. The relative mass distribution is identical to Level I. The primary loss mechanism is reaction in soil which accounts for 936 kg/h or 94% of the input. Most of the remainder is lost by reaction and advection in water. The air and sediment loss processes are unimportant largely because so little of the PCP is present in these media. The overall residence time is 2373 h; thus, there is an inventory of PCP in the system of 2373 × 1000 or 2,373,000 kg. The pie-chart in Figure 1.8*a* shows the dominance of soil reaction as a removal process.

The primary loss mechanism of soil reaction has a D value of 1.03×10^{11}, thus, for any other process to compete with this would require a D value of at least 10^{10} mol/Pa·h. The next largest D values are 3.19×10^9 and 2.53×10^9 for reaction and advection in water, which are

Chemical name: Pentachlorophenol

Level I calculation: (six-compartment model)

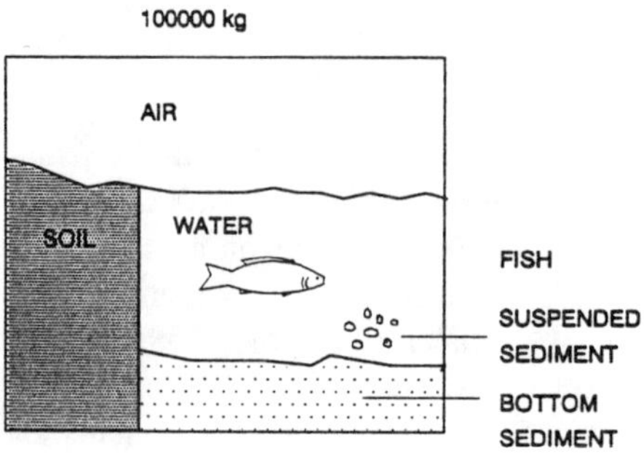

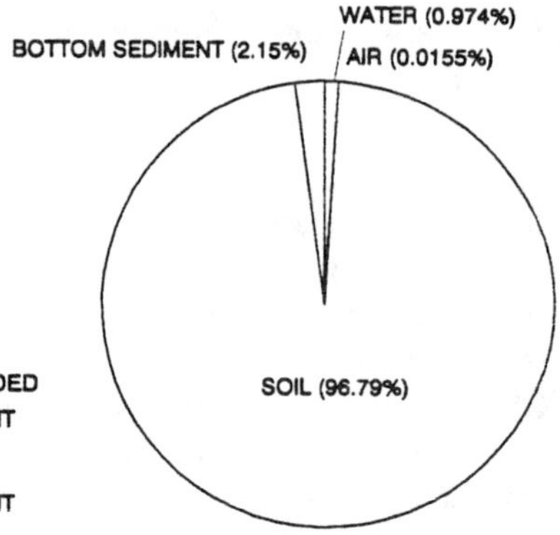

Distribution of mass

Level I calculation:

Physical-chemical properties:		Partition coeff.:		Z values in water:	
molecular wt., g/mol	266.34	HLC	0.07895	at data pH	5.1
melting point, C	174	Kaw	3.19E-05	neutral	3.849
solubility, g/m3	14	Koc	46002.76	ionic	8.817
vapor pressure, Pa	4.15E-03	BCF	5610.09	total	12.666
log Kow	5.05	Ksw	2208.13	at environ. pH	5.1
fugacity ratio, F	0.0336	Ksd/w	4416.26	neutral	3.849
dissoc. const. pKa	4.74	Kssd/w	13800.83	ionic	8.817
		Kar/w	4.86E+07	total	12.666

Compartment	Z	Concentration			Amount	Amount
	mol/m3 Pa	mol/m3	mg/L (or g/m3)	ug/g	kg	%
Air	4.03E-04	5.82E-13	1.55E-10	1.31E-07	15.512	0.0155
Water	1.27E+01	1.83E-08	4.87E-06	4.87E-06	974.05	0.9740
Soil	2.80E+04	4.04E-05	1.08E-02	4.48E-03	96786.94	96.7869
Biota (fish)	7.11E+04	1.03E-04	2.73E-02	2.73E-02	5.4645	5.46E-03
Suspended sediment	1.75E+05	2.52E-04	6.72E-02	4.48E-02	67.2132	6.72E-02
Bottom sediment	5.59E+04	8.08E-05	2.15E-02	8.96E-03	2150.821	2.1508
	Total				100000	100

f = 1.444E-09 Pa

Figure 1.7a Level I fugacity calculation of PCP at data determination pH of 5.1

Chemical name: Pentachlorophenol

Level II calculation: (six-compartment model)

at data pH of 5.1

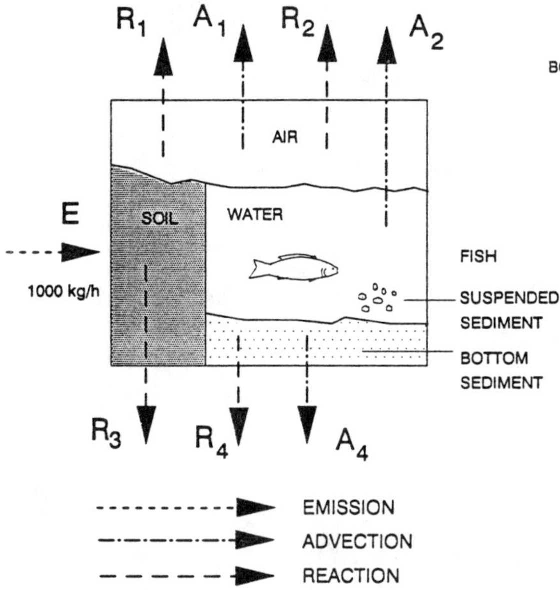

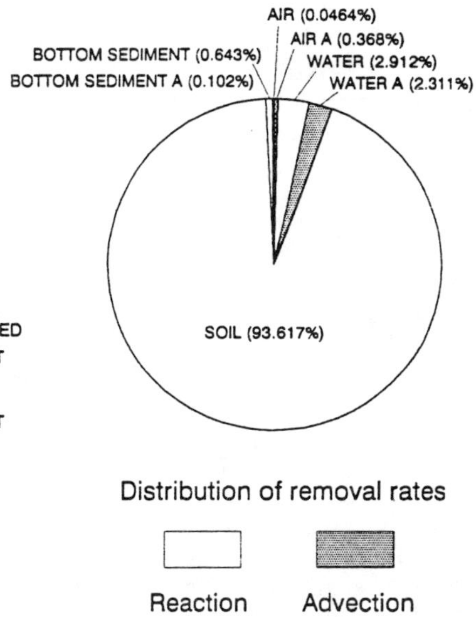

Level II Calculation:

Compartment	Half-life h	D Values Reaction mol/Pa.h	D Values Advection mol/Pa.h	Conc'n mol/m3	Loss Reaction kg/h	Loss Advection kg/h	Removal %
Air	550	5.08E+07	4.03E+08	1.38E-11	0.464	3.681	0.4144
Water	550	3.19E+09	2.53E+09	4.34E-07	29.121	23.112	5.223
Soil	1700	1.03E+11		9.58E-04	936.172		93.617
Biota (fish)				2.43E-03			
Suspended sediment				5.99E-03			
Bottom sediment	5500	7.05E+08	1.12E+08	1.92E-03	6.430	1.021	0.7451
	Total	1.06E+11	2.94E+09		972.19	27.813	100
	R + A		1.09E+11			1000	

f = 3.425E-08 Pa
Total amt = 2372764 kg

Overall residence time = 2372.76 h
Reaction time = 2440.65 h
Advection time = 85311.18 h

Figure 1.8a Level II fugacity calculation for PCP at data determination pH 5.1

Chemial Name: Pentachlorophenol

Fugacity Level I calculation:

(six-compartment model)

at environmental pH 4.0

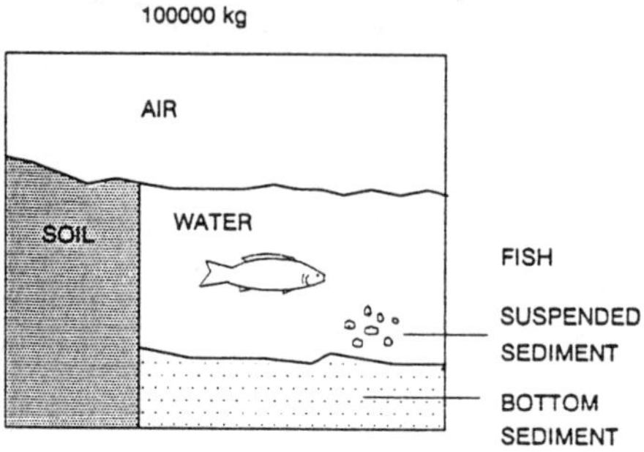

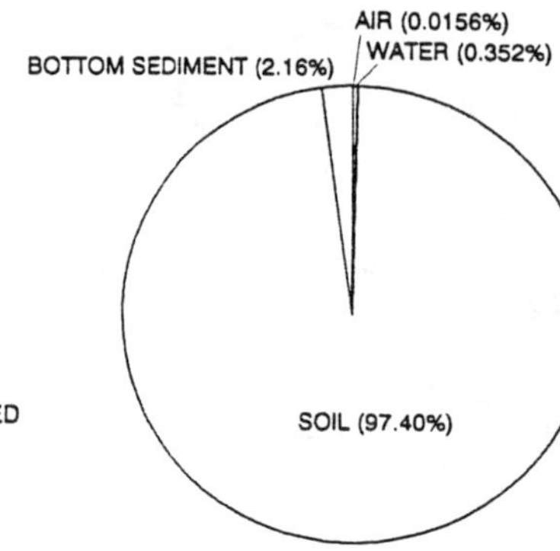

Distribution of mass

Level I calculation:

Physical-chemical properties:		Partition coeff.:		Z values in water:	
molecular wt., g/mol	266.34	HLC	0.21982	at data pH	5.1
melting point, C	174	Kaw	8.87E-05	neutral	3.849
solubility, g/m3	14	Koc	46002.76	ionic	8.817
vapor pressure, Pa	4.15E-03	BCF	5610.09	total	12.666
log Kow	5.05	Ksw	6147.93	at environ. pH	4
fugacity ratio, F	0.0336	Ksd/w	12295.86	neutral	3.849
dissoc. const. pKa	4.74	Kssd/w	38424.57	ionic	0.700
		Kar/w	4.86E+07	total	4.549

Compartment	Z	Concentration			Amount	Amount
	mol/m3 Pa	mol/m3	mg/L (or g/m3)	ug/g	kg	%
Air	4.03E-04	5.86E-13	1.56E-10	1.32E-07	15.609	0.0156
Water	4.55E+00	6.61E-09	1.76E-06	1.76E-06	352.04	0.3520
Soil	2.80E+04	4.06E-05	1.08E-02	4.51E-03	97394.88	97.3949
Biota (fish)	7.11E+04	1.03E-04	2.75E-02	2.75E-02	5.4988	5.50E-03
Suspended sediment	1.75E+05	2.54E-04	6.76E-02	4.51E-02	67.6353	6.76E-02
Bottom sediment	5.59E+04	8.13E-05	2.16E-02	9.02E-03	2164.331	2.1643
	Total				100000	100

f = 1.453E-09 Pa

Figure 1.7b Level I fugacity calculation of PCP at pH 4

Chemical name: Pentachlorophenol

Level II calculation: (six-compartment model) at environmental pH 4.0

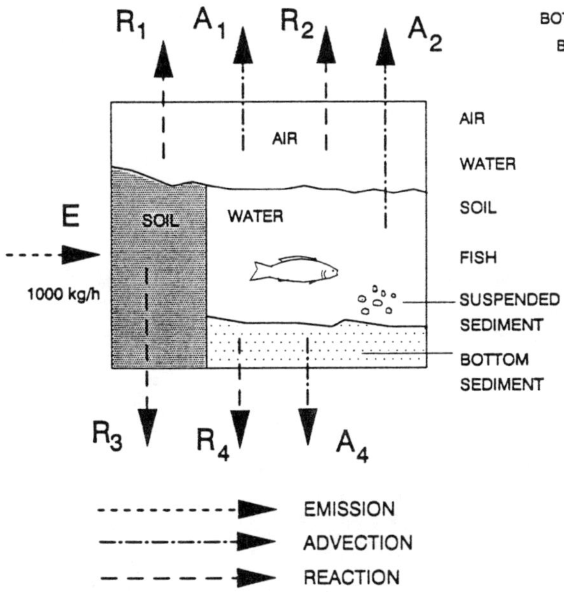

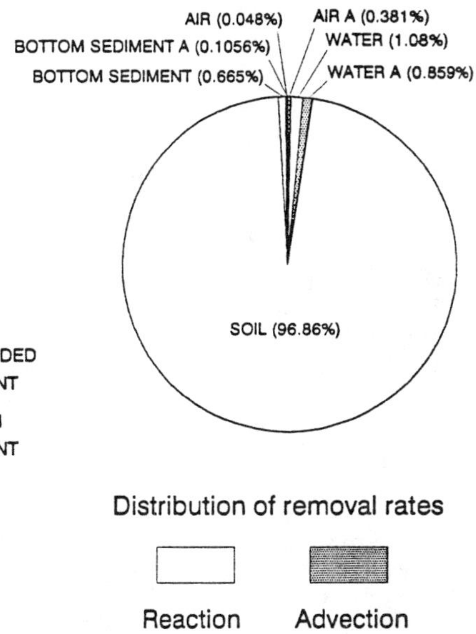

Level II Calculation:

Compartment	Half-life h	D Values Reaction mol/Pa.h	D Values Advection mol/Pa.h	Conc'n mol/m3	Loss Reaction kg/h	Loss Advection kg/h	Removal %
Air	550	5.08E+07	4.03E+08	1.43E-11	0.480	3.808	0.4288
Water	550	1.15E+09	9.10E+08	1.61E-07	10.821	8.588	1.941
Soil	1700	1.03E+11		9.91E-04	968.593		96.859
Biota (fish)				2.52E-03			
Suspended sediment				6.20E-03			
Bottom sediment	5500	7.05E+08	1.12E+08	1.98E-03	6.653	1.056	0.7709
	Total	1.04E+11	1.31E+09		986.55	13.453	100
	R + A		1.05E+11			1000	

f = 3.544E-08 Pa

Total amt = 2439613 kg

Overall residence time = 2439.61 h

Reaction time = 2472.88 h

Advection time = 181349.79 h

Figure 1.8b Level II fugacity calculation of PCP at environmental pH of 4

Chemical Name: Pentachlorophenol

Fugacity Level I calculation: at environmental pH 6.0

(six-compartment model)

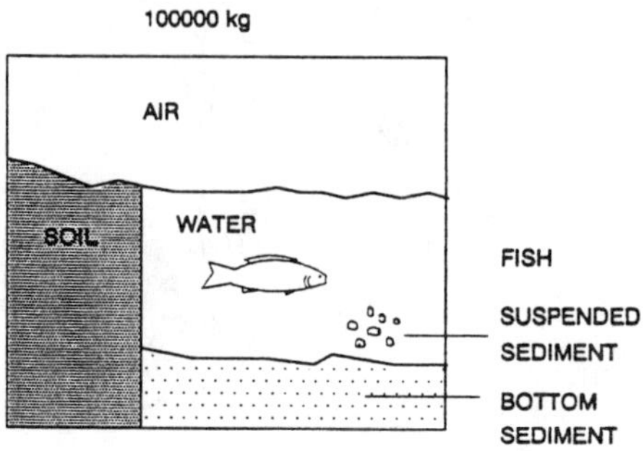

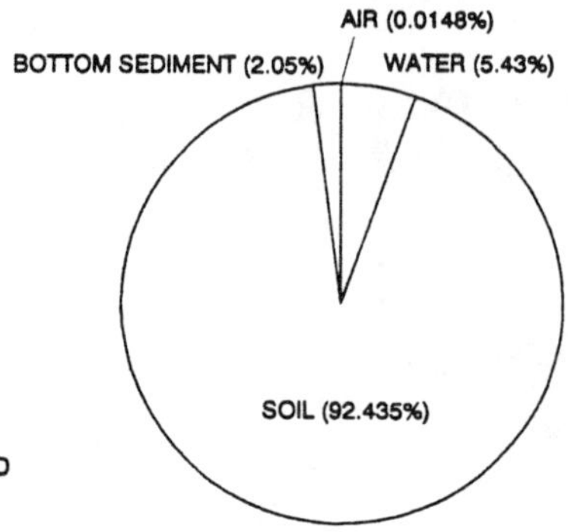

Distribution of mass

Level I calculation:

Physical-chemical properties:		Partition coeff.:		Z values in water:	
molecular wt., g/mol	266.34	HLC	0.01353	at data pH	5.1
melting point, C	174	Kaw	5.46E-06	neutral	3.849
solubility, g/m3	14	Koc	46002.76	ionic	8.817
vapor pressure, Pa	4.15E-03	BCF	5610.09	total	12.666
log Kow	5.05	Ksw	378.531	at environ. pH	6
fugacity ratio, F	0.0336	Ksd/w	757.063	neutral	3.849
dissoc. const. pKa	4.74	Kssd/w	2365.821	ionic	70.038
		Kar/w	4.86E+07	total	73.887

Compartment	Z	Concentration			Amount	Amount
	mol/m3 Pa	mol/m3	mg/L (or g/m3)	ug/g	kg	%
Air	4.03E-04	5.56E-13	1.48E-10	1.25E-07	14.814	0.0148
Water	7.39E+01	1.02E-07	2.71E-05	2.71E-05	5426.53	5.4265
Soil	2.80E+04	3.86E-05	1.03E-02	4.28E-03	92435	92.4351
Biota (fish)	7.11E+04	9.80E-05	2.61E-02	2.61E-02	5.2188	5.22E-03
Suspended sediment	1.75E+05	2.41E-04	6.42E-02	4.28E-02	64.191	6.42E-02
Bottom sediment	5.59E+04	7.71E-05	2.05E-02	8.56E-03	2054.11	2.0541
	Total				100000	100

f = 1.379E-09 Pa

Figure 1.7c Level I fugacity calculation of PCP at pH 6

Chemical name: Pentachlorophenol

Level II calculation: (six-compartment model)

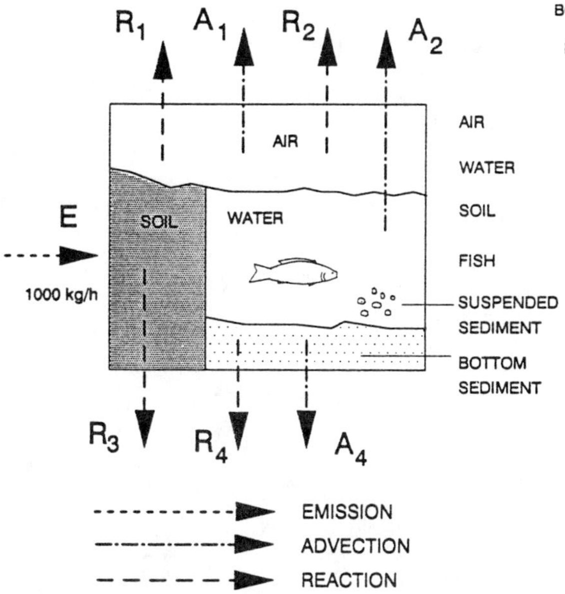

at environmental pH 6.0

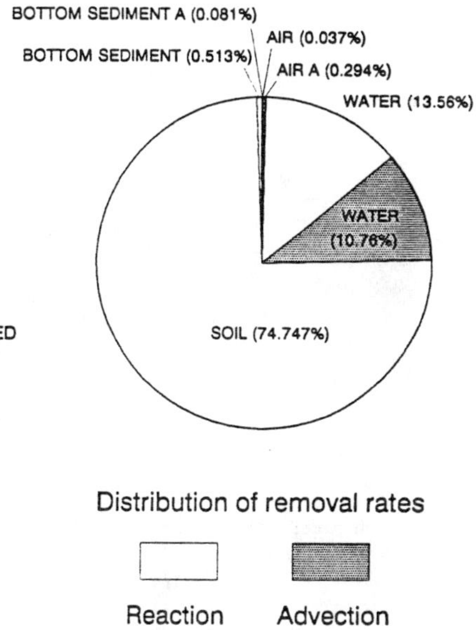

Level II Calculation:

Compartment	Half-life h	D Values Reaction mol/Pa.h	D Values Advection mol/Pa.h	Conc'n mol/m3	Loss Reaction kg/h	Loss Advection kg/h	Removal %
Air	550	5.08E+07	4.03E+08	1.10E-11	0.370	2.939	0.3309
Water	550	1.86E+10	1.48E+10	2.02E-06	135.632	107.645	24.328
Soil	1700	1.03E+11		7.65E-04	747.465		74.747
Biota (fish)				1.94E-03			
Suspended sediment				4.78E-03			
Bottom sediment	5500	7.05E+08	1.12E+08	1.53E-03	5.134	0.815	0.5949
	Total	1.21E+11	1.52E+10		888.60	111.398	100
	R + A		1.36E+11			1000	

f = 2.735E-08 Pa
Total amt = 1983671 kg

Overall residence time = 1983.67 h
Reaction time = 2232.35 h
Advection time = 17807.03 h

Figure 1.8c Level II fugacity calculation of PCP at environmental pH of 6

Chemical Name: Pentachlorophenol
Fugacity Level I calculation:
(six-compartment model)

at environmental pH 7.0

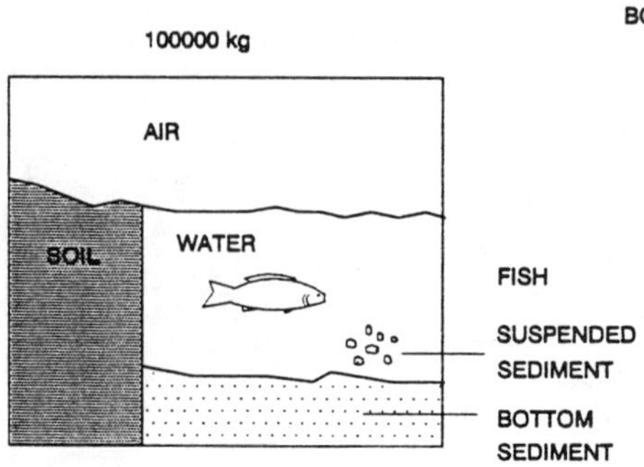

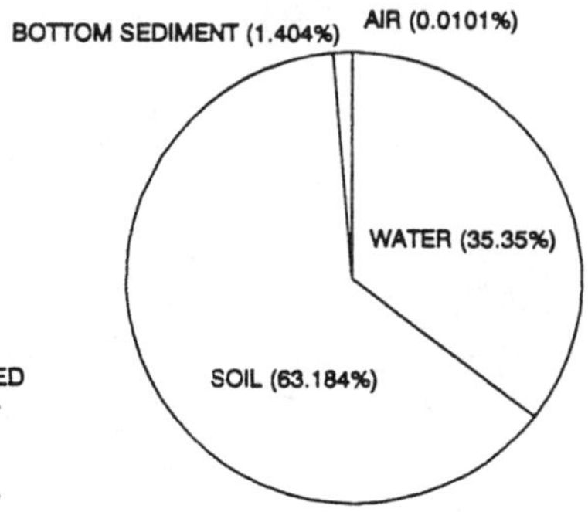

Distribution of mass

Level I calculation:

Physical-chemical properties:		Partition coeff.:		Z values in water:	
molecular wt., g/mol	266.34	HLC	0.00142	at data pH	5.1
melting point, C	174	Kaw	5.73E-07	neutral	3.849
solubility, g/m3	14	Koc	46002.76	ionic	8.817
vapor pressure, Pa	4.15E-03	BCF	5610.09	total	12.666
log Kow	5.05	Ksw	39.715	at environ. pH	7
fugacity ratio, F	0.0336	Ksd/w	79.430	neutral	3.849
dissoc. const. pKa	4.74	Kssd/w	248.219	ionic	700.379
		Kar/w	4.86E+07	total	704.228

Compartment	Z	Concentration			Amount	Amount
	mol/m3 Pa	mol/m3	mg/L (or g/m3)	ug/g	kg	%
Air	4.03E-04	3.80E-13	1.01E-10	8.54E-08	10.126	0.0101
Water	7.04E+02	6.64E-07	1.77E-04	1.77E-04	35354	35.3542
Soil	2.80E+04	2.64E-05	7.02E-03	2.93E-03	63184	63.1842
Biota (fish)	7.11E+04	6.70E-05	1.78E-02	1.78E-02	3.5673	3.57E-03
Suspended sediment	1.75E+05	1.65E-04	4.39E-02	2.93E-02	43.878	4.39E-02
Bottom sediment	5.59E+04	5.27E-05	1.40E-02	5.85E-03	1404.09	1.4041
	Total				100000	100

f = 9.425E-10 Pa

Figure 1.7d Level I fugacity calculation of PCP at pH 7

Chemical name: Pentachlorophenol

Level II calculation: (six-compartment model)

at environmental pH 7.0

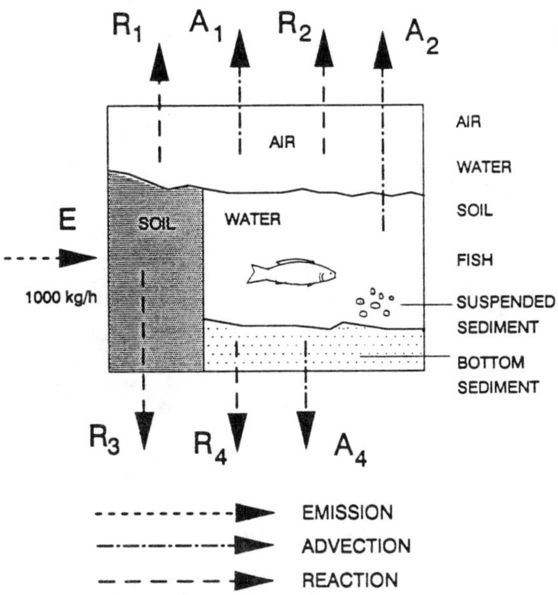

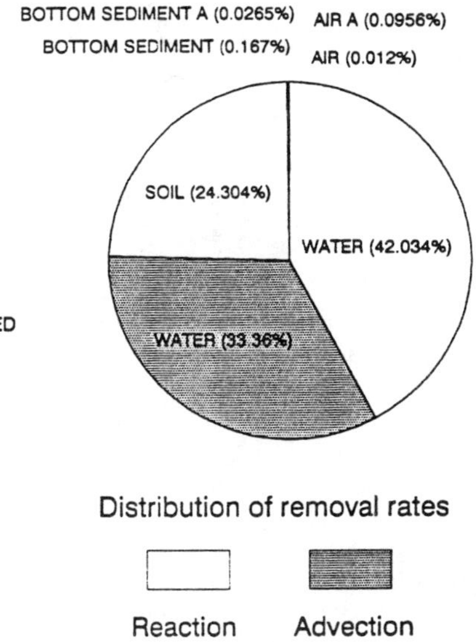

Level II Calculation: Compartment	Half-life h	D Values Reaction mol/Pa.h	D Values Advection mol/Pa.h	Conc'n mol/m3	Loss Reaction kg/h	Loss Advection kg/h	Removal %
Air	550	5.08E+07	4.03E+08	3.59E-12	0.120	0.956	0.1076
Water	550	1.77E+11	1.41E+11	6.26E-06	420.342	333.604	75.395
Soil	1700	1.03E+11		2.49E-04	243.044		24.304
Biota (fish)				6.32E-04			
Suspended sediment				1.55E-03			
Bottom sediment	5500	7.05E+08	1.12E+08	4.97E-04	1.669	0.265	0.1934
	Total	2.80E+11	1.41E+11		665.18	334.825	100
	R + A		4.21E+11			1000	

f = 8.893E-09 Pa

Total amt = 943608 kg

Overall residence time = 943.61 h

Reaction time = 1418.59 h

Advection time = 2818.21 h

Figure 1.8d Level II fugacity calculation of PCP at environmental pH of 7

Fugacity Level III calculations: (four-compartment model)

Chemical name: Pentachlorophenol at environmental pH of 4.0

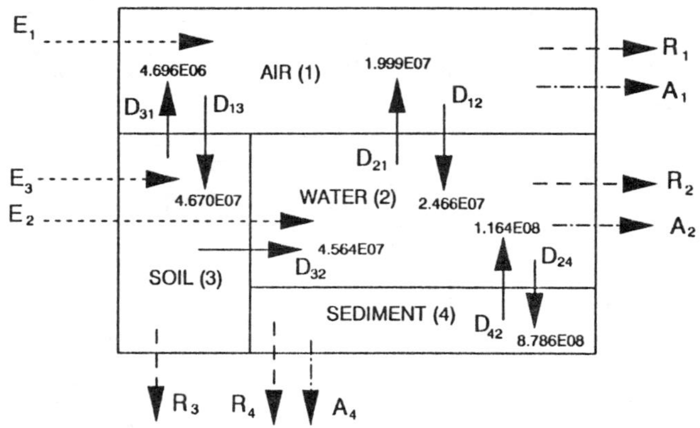

Phase Properties and Rates:

Compartment	Bulk Z mol/m3 Pa	Half-life h	D Values Reaction mol/Pa h	Advection mol/Pa h
Air (1)	4.038E-04	550	5.09E+07	4.04E+08
Water (2)	5.494E+00	550	1.38E+09	1.10E+09
Soil (3)	1.399E+04	1700	1.03E+11	
Sediment (4)	1.119E+04	5500	7.05E+08	1.12E+08

	E(1)=1000	E(2)=1000	E(3)=1000	E(1,2,3)	
Overall residence time =	385.80	1947.66	2452.78	2223.15	h
Reaction time =	1795.45	3105.25	2453.28	2562.99	h
Advection time =	491.39	5224.64	12164680	16766.70	h

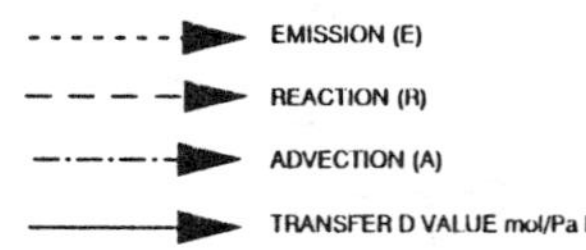

Phase Properties, Compositions, Transport and Transformation Rates:

Emission, kg/h			Fugacity, Pa				Concentration, g/m3				Amounts, kg				Total Amount, kg
E(1)	E(2)	E(3)	f(1)	f(2)	f(3)	f(4)	C(1)	C(2)	C(3)	C(4)	m(1)	m(2)	m(3)	m(4)	
1000	0	0	7.139E-06	5.385E-08	3.247E-09	5.069E-08	7.679E-07	7.880E-05	1.210E-02	1.511E-01	7.679E+04	1.576E+04	2.177E+05	7.554E+04	3.858E+05
0	1000	0	4.362E-08	1.148E-06	1.984E-11	1.080E-06	4.691E-09	1.679E-03	7.390E-05	3.220E+00	4.691E+02	3.359E+05	1.330E+03	1.610E+06	1.948E+06
0	0	1000	3.459E-10	5.127E-10	3.657E-08	4.826E-10	3.720E-11	7.502E-07	1.362E-01	1.438E-03	3.720E+00	1.500E+02	2.452E+06	7.192E+02	2.453E+06
50	250	700	3.681E-07	2.900E-07	2.577E-08	2.729E-07	3.959E-08	4.243E-04	9.598E-02	8.135E-01	3.959E+03	8.487E+04	1.728E+06	4.068E+05	2.223E+06

Emission, kg/h			Loss, Reaction, kg/h				Loss, Advection, kg/h			Intermedia Rate of Transport, kg/h T12	T21	T13	T31	T32	T24	T42
E(1)	E(2)	E(3)	R(1)	R(2)	R(3)	R(4)	A(1)	A(2)	A(4)	air-water	water-air	air-soil	soil-air	soil-water	water-sed	sed-water
1000	0	0	9.675E+01	1.986E+01	8.88E+01	9.518E+00	7.679E+02	1.576E+01	1.511E+00	4.689E+01	2.867E-01	8.880E+01	4.061E-03	3.947E-02	1.260E+01	1.572E+00
0	1000	0	5.911E-01	4.232E+02	5.42E-01	2.029E+02	4.691E+00	3.359E+02	3.220E+01	2.865E-01	6.112E+00	5.425E-01	2.481E-05	2.412E-04	2.686E+02	3.350E+01
0	0	1000	4.688E-03	1.891E-01	1.00E+03	9.062E-02	3.720E-02	1.500E-01	1.438E-02	2.272E-03	2.730E-03	4.302E-03	4.574E-02	4.446E-01	1.200E-01	1.496E-02
50	250	700	4.989E+00	1.069E+02	7.04E+02	5.125E+01	3.959E+01	8.487E+01	8.135E+00	2.418E+00	1.544E+00	4.578E+00	3.222E-02	3.132E-01	6.785E+01	8.464E+00

Figure 1.9a Level III calculation for PCP in a generic environment at pH of 4.0

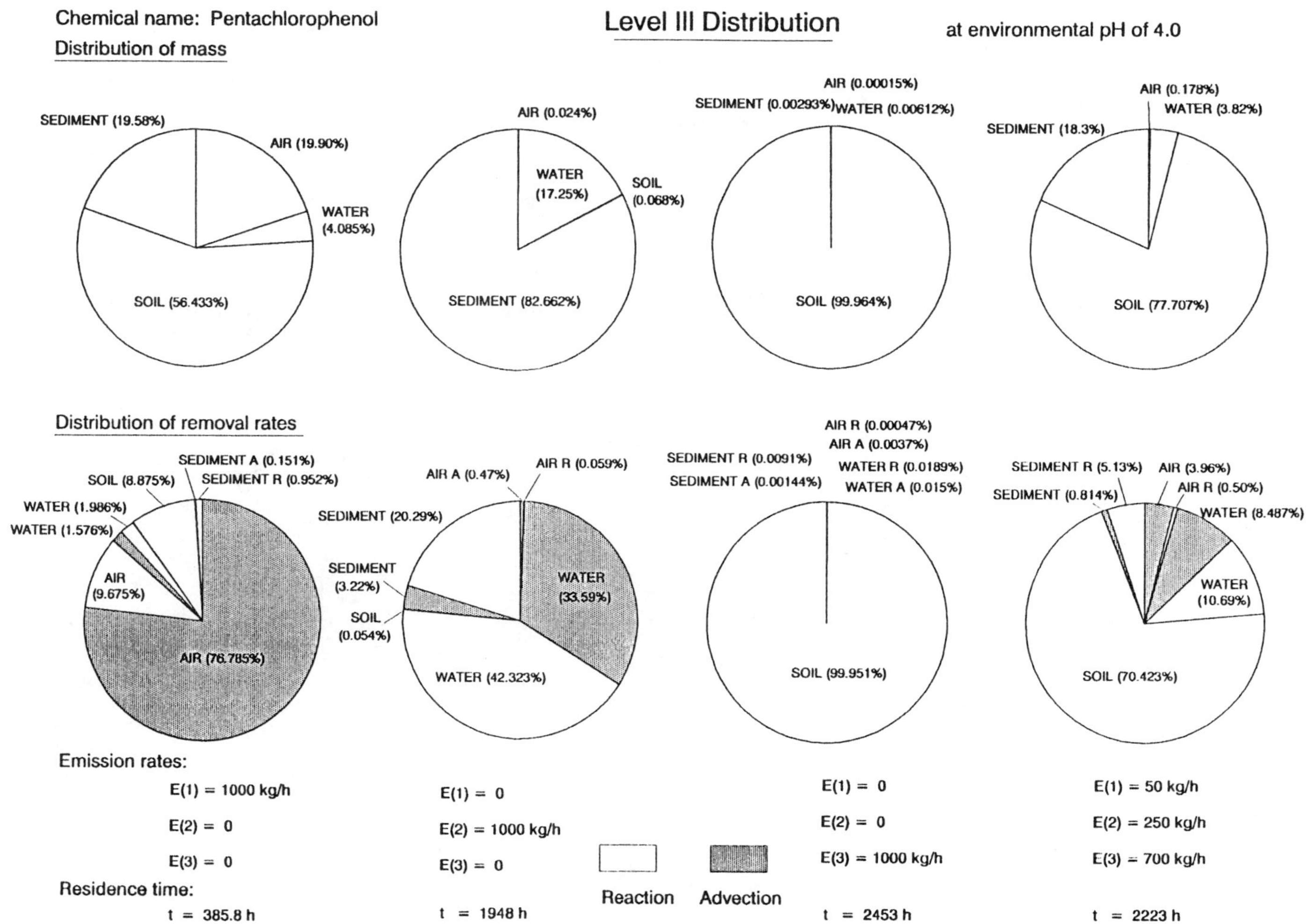

Figure 1.9b Level III distributions of PCP for four emission scenarios at pH 4.0

Fugacity Level III calculations: (four-compartment model)

Chemical name: Pentachlorophenol   at data determination pH of 5.1

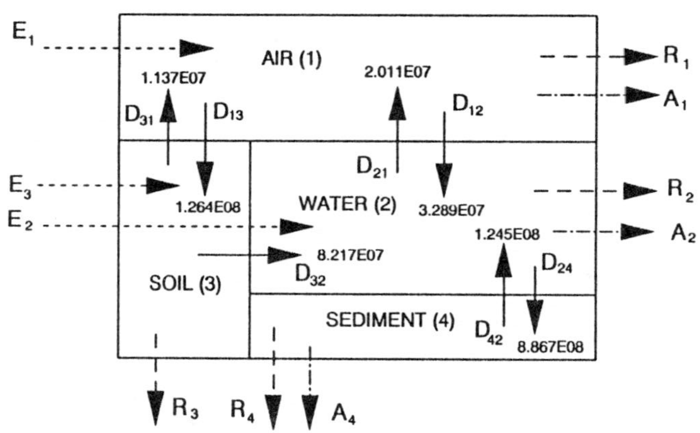

Phase Properties and Rates:

Compartment	Bulk Z mol/m3 Pa	Half-life h	D Values Reaction mol/Pa h	Advection mol/Pa h
Air (1)	4.038E-04	550	5.09E+07	4.04E+08
Water (2)	1.361E+01	550	3.43E+09	2.72E+09
Soil (3)	1.399E+04	1700	1.03E+11	
Sediment (4)	1.120E+04	5500	7.05E+08	1.12E+08

	E(1)=1000	E(2)=1000	E(3)=1000	E(1,2,3)	
Overall residence time =	632.37	1153.18	2451.86	2036.22	h
Reaction time =	1974.17	1952.24	2452.85	2358.35	h
Advection time =	930.39	2817.42	6090446	14907.33	h

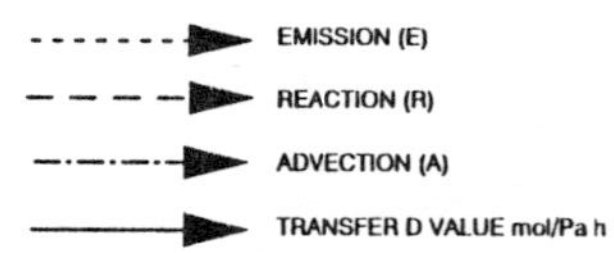

Phase Properties, Compositions, Transport and Transformation Rates:

Emission, kg/h			Fugacity, Pa				Concentration, g/m3				Amounts, kg				Total Amount, kg
E(1)	E(2)	E(3)	f(1)	f(2)	f(3)	f(4)	C(1)	C(2)	C(3)	C(4)	m(1)	m(2)	m(3)	m(4)	
1000	0	0	6.116E-06	2.907E-08	7.526E-09	2.736E-08	6.578E-07	1.054E-04	2.804E-02	8.160E-02	6.578E+04	2.107E+04	5.047E+05	4.080E+04	6.324E+05
0	1000	0	1.772E-08	5.410E-07	2.180E-11	5.092E-07	1.905E-09	1.961E-03	8.122E-05	1.519E+00	1.905E+02	3.922E+05	1.462E+03	7.593E+05	1.153E+06
0	0	1000	6.909E-10	4.359E-10	3.655E-08	4.103E-10	7.430E-11	1.580E-06	1.362E-01	1.224E-03	7.430E+00	3.160E+02	2.451E+06	6.118E+02	2.452E+06
50	250	700	3.107E-07	1.370E-07	2.597E-08	1.290E-07	3.342E-08	4.966E-04	9.674E-02	3.846E-01	3.342E+03	9.933E+04	1.741E+06	1.923E+05	2.036E+06

Emission, kg/h			Loss, Reaction, kg/h				Loss, Advection, kg/h			Intermedia Rate of Transport, kg/h T12	T21	T13	T31	T32	T24	T42
E(1)	E(2)	E(3)	R(1)	R(2)	R(3)	R(4)	A(1)	A(2)	A(4)	air-water	water-air	air-soil	soil-air	soil-water	water-sed	sed-water
1000	0	0	8.288E+01	2.655E+01	2.06E+02	5.141E+00	6.578E+02	2.107E+01	8.160E-01	5.358E+01	1.557E-01	2.059E+02	2.278E-02	1.647E-01	6.864E+00	9.076E-01
0	1000	0	2.401E-01	4.942E+02	5.96E-01	9.567E+01	1.905E+00	3.922E+02	1.519E+01	1.552E-01	2.897E+00	5.965E-01	6.599E-05	4.771E-04	1.278E+02	1.689E+01
0	0	1000	9.362E-03	3.982E-01	9.99E+02	7.709E-02	7.430E-02	3.160E-01	1.224E-02	6.052E-03	2.334E-03	2.326E-02	1.106E-01	7.999E-01	1.029E-01	1.361E-02
50	250	700	4.211E+00	1.252E+02	7.10E+02	2.423E+01	3.342E+01	9.933E+01	3.846E+00	2.722E+00	7.337E-01	1.046E+01	7.861E-02	5.683E-01	3.235E+01	4.277E+00

Figure 1.10a Fugacity Level III calculation for PCP in a generic environment

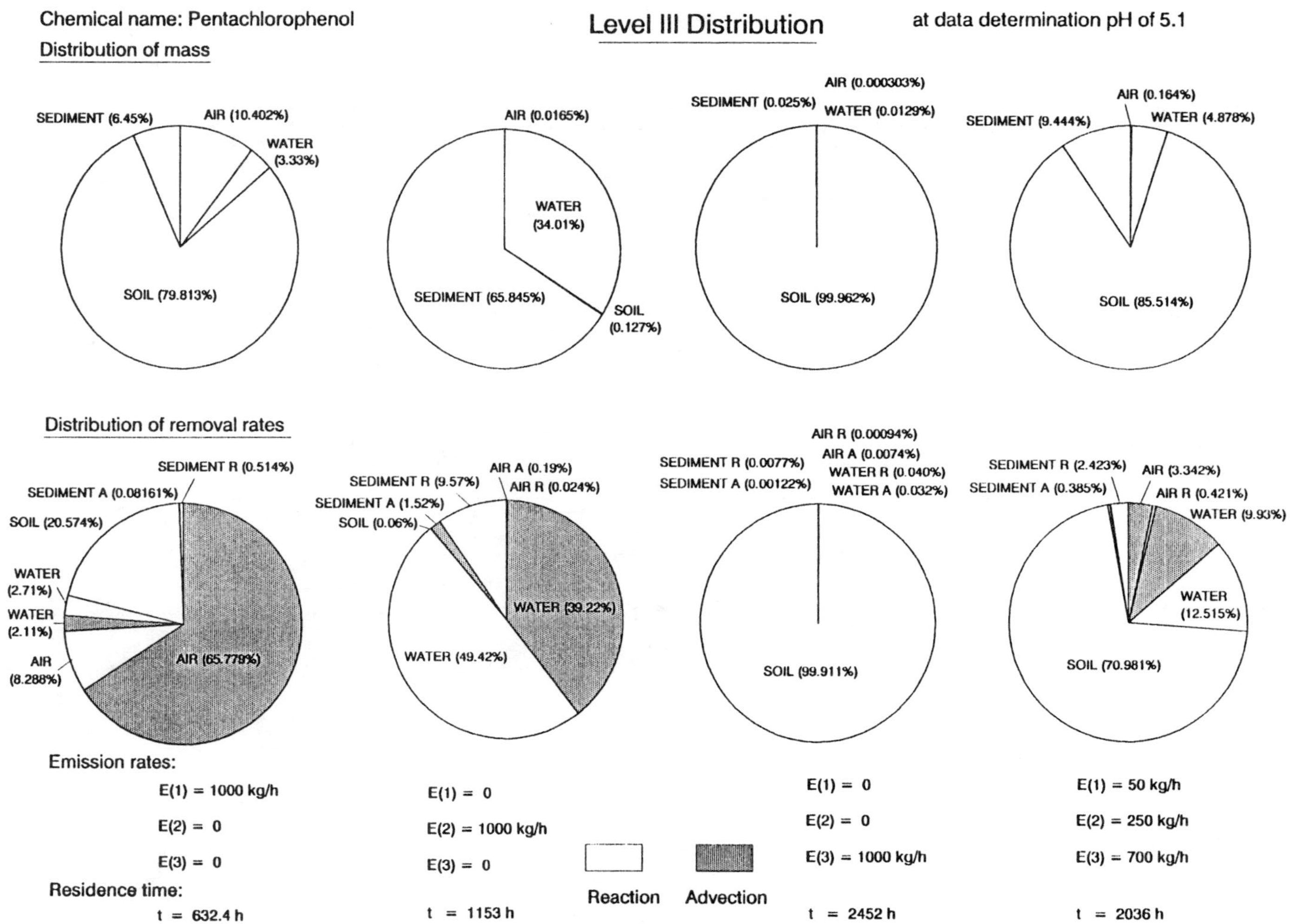

Figure 1.10b Level III distributions of PCP for four emission scenarios at pH 5.1

Fugacity Level III calculations: (four-compartment model)

Chemical name: Pentachlorophenol at environmental pH of 6.0

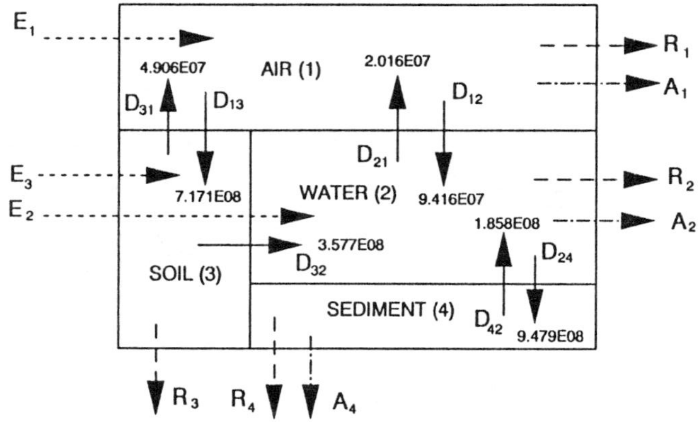

Phase Properties and Rates:

Compartment	Bulk Z mol/m3 Pa	Half-life h	D Values Reaction mol/Pa h	Advection mol/Pa h
Air (1)	4.038E-04	550	5.09E+07	4.04E+08
Water (2)	7.483E+01	550	1.89E+10	1.50E+10
Soil (3)	1.401E+04	1700	1.03E+11	
Sediment (4)	1.125E+04	5500	7.09E+08	1.12E+08

	E(1)=1000	E(2)=1000	E(3)=1000	E(1,2,3)	
Overall residence time =	1459.56	586.13	2446.16	1931.82	h
Reaction time =	2255.49	1038.52	2450.27	2214.68	h
Advection time =	4136.10	1345.57	1458005	15125.50	h

- - - - - -▶ EMISSION (E)
— — — ▶ REACTION (R)
—·—·—▶ ADVECTION (A)
———▶ TRANSFER D VALUE mol/Pa h

Phase Properties, Compositions, Transport and Transformation Rates:

Emission, kg/h			Fugacity, Pa				Concentration, g/m3				Amounts, kg				Total Amount, kg
E(1)	E(2)	E(3)	f(1)	f(2)	f(3)	f(4)	C(1)	C(2)	C(3)	C(4)	m(1)	m(2)	m(3)	m(4)	
1000	0	0	2.971E-06	8.296E-09	2.059E-08	7.811E-09	3.196E-07	1.653E-04	7.682E-02	2.340E-02	3.196E+04	3.307E+04	1.383E+06	1.170E+04	1.460E+06
0	1000	0	1.730E-09	1.085E-07	1.199E-11	1.021E-07	1.861E-10	2.162E-03	4.474E-05	3.059E-01	1.861E+01	4.324E+05	8.053E+02	1.530E+05	5.861E+05
0	0	1000	1.419E-09	3.799E-10	3.640E-08	3.577E-10	1.526E-10	7.572E-06	1.358E-01	1.071E-03	1.526E+01	1.514E+03	2.444E+06	5.357E+02	2.446E+06
50	250	700	1.500E-07	2.780E-08	2.651E-08	2.617E-08	1.613E-08	5.540E-04	9.890E-02	7.840E-02	1.613E+03	1.108E+05	1.780E+06	3.920E+04	1.932E+06

Emission, kg/h			Loss, Reaction, kg/h				Loss, Advection, kg/h			Intermedia Rate of Transport, kg/h T12	T21	T13	T31	T32	T24	T42
E(1)	E(2)	E(3)	R(1)	R(2)	R(3)	R(4)	A(1)	A(2)	A(4)	air-water	water-air	air-soil	soil-air	soil-water	water-sed	sed-water
1000	0	0	4.027E+01	4.167E+01	5.64E+02	1.474E+00	3.196E+02	3.307E+01	2.340E-01	7.452E+01	4.454E-02	5.659E+02	2.691E-01	1.962E+00	2.094E+00	3.864E-01
0	1000	0	2.345E-02	5.448E+02	3.28E-01	1.927E+01	1.861E-01	4.324E+02	3.059E+00	4.340E-02	5.824E-01	3.296E-01	1.567E-04	1.142E-03	2.738E+01	5.053E+00
0	0	1000	1.923E-02	1.908E+00	9.96E+02	6.750E-02	1.526E-01	1.514E+00	1.071E-02	3.558E-02	2.040E-03	2.702E-01	4.756E-01	3.467E+00	9.592E-02	1.770E-02
50	250	700	2.033E+00	1.396E+02	7.26E+02	4.939E+00	1.613E+01	1.108E+02	7.840E-01	3.762E+00	1.493E-01	2.857E+01	3.464E-01	2.525E+00	7.018E+00	1.295E+00

Figure 1.11a Fugacity Level III calculation for PCP in a generic environment.

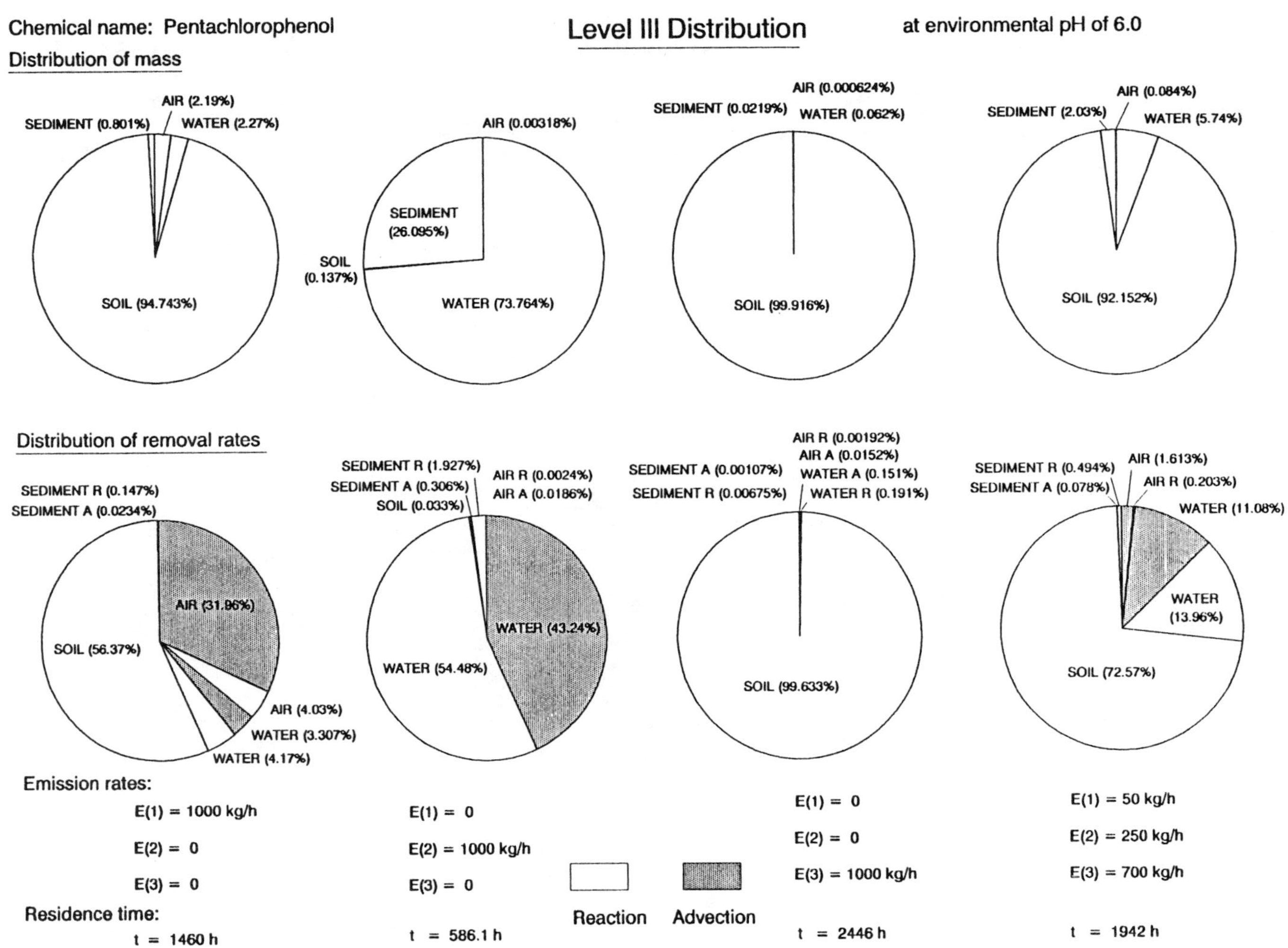

Figure 1.11b Level III distributions of PCP for four emission scenarios at pH 6

Fugacity Level III calculations: (four-compartment model)

Chemical name: Pentachlorophenol at environmental pH of 7.0

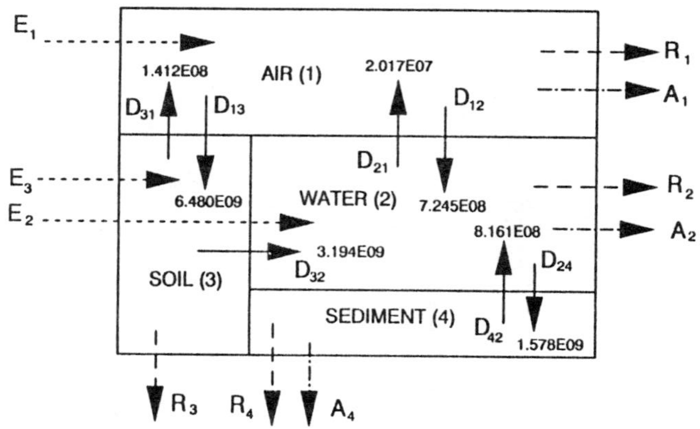

Phase Properties and Rates:

Compartment	Bulk Z mol/m3 Pa	Half-life h	D Values Reaction mol/Pa h	Advection mol/Pa h
Air (1)	4.038E-04	550	5.09E+07	4.04E+08
Water (2)	7.052E+02	550	1.78E+11	1.41E+11
Soil (3)	1.420E+04	1700	1.04E+11	
Sediment (4)	1.175E+04	5500	7.40E+08	1.18E+08

	E(1)=1000	E(2)=1000	E(3)=1000	E(1,2,3)	
Overall residence time =	2073.55	458.80	2393.34	1893.72	h
Reaction time =	2318.68	821.75	2425.51	2164.23	h
Advection time =	19613.46	1038.74	180451	15150.82	h

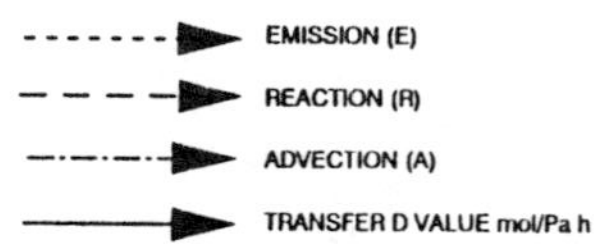

Phase Properties, Compositions, Transport and Transformation Rates:

Emission, kg/h			Fugacity, Pa				Concentration, g/m3				Amounts, kg				Total Amount, kg
E(1)	E(2)	E(3)	f(1)	f(2)	f(3)	f(4)	C(1)	C(2)	C(3)	C(4)	m(1)	m(2)	m(3)	m(4)	
1000	0	0	4.907E-07	1.408E-09	2.958E-08	1.328E-09	5.278E-08	2.645E-04	1.118E-01	4.156E-03	5.278E+03	5.290E+04	2.013E+06	2.078E+03	2.074E+06
0	1000	0	3.097E-11	1.175E-08	1.867E-12	1.108E-08	3.331E-12	2.207E-03	7.060E-06	3.467E-02	3.331E-01	4.413E+05	1.271E+02	1.733E+04	4.588E+05
0	0	1000	6.453E-10	3.510E-10	3.497E-08	3.309E-10	6.940E-11	6.592E-05	1.322E-01	1.036E-03	6.940E+00	1.318E+04	2.380E+06	5.178E+02	2.393E+06
50	250	700	2.500E-08	3.253E-09	2.596E-08	3.067E-09	2.688E-09	6.110E-04	9.814E-02	9.600E-03	2.688E+02	1.222E+05	1.766E+06	4.800E+03	1.894E+06

Emission, kg/h			Loss, Reaction, kg/h				Loss, Advection, kg/h			Intermedia Rate of Transport, kg/h T12	T21	T13	T31	T32	T24	T42
E(1)	E(2)	E(3)	R(1)	R(2)	R(3)	R(4)	A(1)	A(2)	A(4)	air-water	water-air	air-soil	soil-air	soil-water	water-sed	sed-water
1000	0	0	6.650E+00	6.665E+01	8.21E+02	2.618E-01	5.278E+01	5.290E+01	4.156E-02	9.470E+01	7.565E-03	8.470E+02	1.112E+00	2.517E+01	5.920E-01	2.886E-01
0	1000	0	4.197E-04	5.561E+02	5.18E-02	2.184E+00	3.331E-03	4.413E+02	3.467E-01	5.977E-03	6.312E-02	5.346E-02	7.020E-05	1.589E-03	4.939E+00	2.408E+00
0	0	1000	8.745E-03	1.661E+01	9.70E+02	6.525E-02	6.940E-02	1.318E+01	1.036E-02	1.245E-01	1.885E-03	1.114E+00	1.315E+00	2.975E+01	1.475E-01	7.193E-02
50	250	700	3.387E-01	1.540E+02	7.20E+02	6.048E-01	2.688E+00	1.222E+02	9.600E-02	4.823E+00	1.748E-02	4.314E+01	9.758E-01	2.208E+01	1.368E+00	6.667E-01

Figure 1.12a Fugacity Level III calculation for PCP in a generic environment.

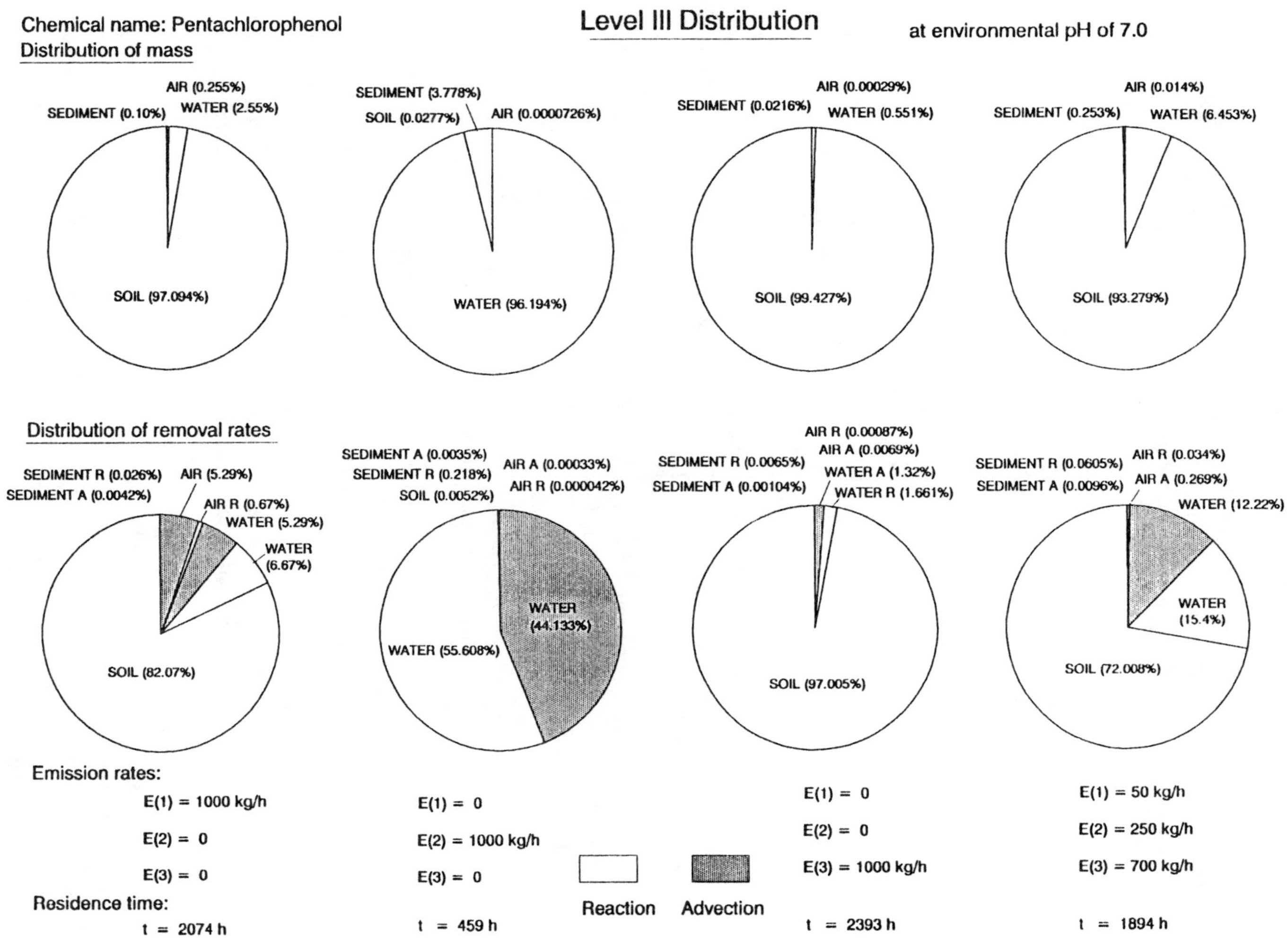

Figure 1.12b Level III distributions of PCP for four emission scenarios at pH 7.

about a factor of 30 smaller. Only if the water advection or reaction rates are increased by about this factor will these processes become significant.

Figures 1.8*a* to 1.8*d*, the Level II diagrams at selected pH values, suggest that soil degradation (corresponding to a half-life of 1700 h) is most important at low pH. As pH increases, reaction in, and advection from, water increase in importance. The overall residence times are 2440 h at pH of 4, decreasing to 1984 h at pH of 6 and to 944 h at pH of 7.

Level III

The Level III diagrams (Fig. 1.9 to Fig. 1.12 for selected pHs) are regarded as the most realistic depictions of chemical fate.

This calculation includes an estimation of intermedia transport. Examination of the magnitude of the intermedia D values given in the fate diagrams (Fig. 1.9 - Fig. 1.12, which are the same as the Level III calculations and distribution graphs for PCP in Chapter 4) suggest that water-sediment and air-soil transport are most important, with soil-water, and air-water exchange being slower. The magnitude of these larger intermedia transport D values (approximately 1-9 $\times$ 10^8 mol/Pa·h) compared to the soil reaction value of 10^{11} and the water reaction and advection values of some 3 $\times$ 10^9 suggests that reaction in soil and water and advection from water will be very fast relative to transport from water to sediment or from air to water and soil. This chemical tends to be fairly immobile in terms of intermedia transport.

The bulk Z values are similar for air and water to the values for the "pure" phases in Level I and II, but they are lower for soil and sediment because of the "dilution" of the solid soil and sediment phases with air or water.

The complete discussion of PCP fate as deduced in these calculations is beyond our scope, but to assist the reader we describe the behavior at a pH of 5.1 in some detail.

These tabulated data are given in illustrated or pictorial form in Figures 1.9*a* to Fig. 1.12*a*. The first row of figures at the foot of Fig. 1.10*a* describes the condition if 1000 kg/h is emitted into the air. The result is similar to the Level II calculation with 65780 kg in air, 21070 kg in water, 504700 kg in soil and only 40800 kg in sediment. It can be concluded that PCP discharged to the atmosphere has fairly high potential to enter other media. The rate of transfer from air to water (T_{12}) is about 54 kg/h and that from air to soil (T_{13}) 206 kg/h. The reason for this is the value of the mass transfer coefficients which control this transport process. The overall residence time is 632 hours.

If 1000 kg/h of PCP is discharged to water, as in the second row, there is, as expected a much higher concentration in water. There is reaction of 494 kg/h in water, advective outflow of 392 kg/h and transfer to air (T_{21}) of 2.90 kg/h with substantial loss of 128 kg/h to sediment. The amount in the water is 392,200 kg, thus the residence time in the water is 392 hours and the overall environmental residence time is a longer 1153 hours. The key processes are thus reaction in water (half-life 550 h) and advective outflow (residence time 1000 h). The evaporation half-life can be calculated as (0.693 × mass in water)/rate of transfer, i.e., (0.693 × 392,200)/2.90 = 93700 h. Clearly competition between advection and reaction in the water determines the overall fate. Thirty-four percent of the PCP discharged is now found in the water and the concentration is fairly high, namely 1.96 × 10^{-3} g/m^3 or 1.96 μg/L.

The third row shows the fate if PCP is discharged into soil. The amount in soil is 245100 kg, with only 7.43 kg in air. The overall residence time is 2452 hours, which is largely controlled by the reaction rate in soil. The rate of reaction in soil is 999 kg/h, there is no advection, thus the other loss mechanism is transfer to air (T_{31}) at a rate of 0.11 kg/h, with a relatively minor 0.8 kg/h to water by run-off. The soil concentration of 0.136 g/m^3 is controlled almost entirely by the rate at which the PCP reacts.

The net result is that PCP behaves entirely differently when discharged to the three media. If discharged to air, it advects rapidly and reacts with a residence time of 632 h or about 26.3 days with substantial transport to soil or water. If discharged to water, it reacts and evaporates to air with a residence time of 1153 h or 48 days. If discharged to soil, it mostly reacts with an overall residence time of about 2452 h or 102 days.

The final scenario is a combination of discharges, 50 kg/h to air, 250 kg/h to water, and 700 kg/h to soil (which are different from the previous conditions in Volume I, II, and III as 600 kg/h to air, 200 kg/h to water and 100 kg/h to soil). The concentrations, amounts and transport and transformation rates are merely linearly combined versions of the three initial scenarios. For example, the rate of reaction in air is now 4.21 kg/h. This is 0.05 of the first (air emission) rate of 82.9 kg/h, i.e., 4.14 kg/h, plus 0.25 of the second (water emission) rate of 0.24 kg/h, i.e., 0.06 kg/h and 0.7 of the third (soil emission) rate of 0.0094 kg/h, i.e., 0.0066 kg/h yielding a total of (4.14 + 0.06 + 0.0066) or 4.21 kg/h. It is also apparent that the amount in the air of 3342 kg causing a concentration of 3.342 × 10^{-8} g/m^3 or 33 ng/m^3 is attributable to emissions to air (0.05 × 658 or 33 ng/m^3), emissions to water (0.25 × 1.9 or 0.5 ng/m^3) and emissions to soil (0.7 × 0.0743 or 0.052 μg/m^3). The concentration in water of 4.97 × 10^{-4} g/m^3 or 497 ng/L is largely attributable to the discharges to water which alone cause 0.25 × 1.96 × 10^{-3} g/m^3 or 4.9 × 10^{-4} g/m^3 or 490 μg/m^3 or 490 ng/L. Although more is emitted to soil it contributes only about 1.1 μg/m^3 to the water with air emissions accounting for about 5.27 μg/m^3. Similarly, the prevailing soil concentration is controlled by the rate of discharge to the soil.

In this multimedia discharge scenario the overall residence time is 2036 hours, which can be viewed as 5% of the air emission residence time of 632 h, 25% of the water emission residence time of 1153 h and 70% of the soil emission residence time of 2452 h. The overall amount in the environment of 2.04 × 10^6 kg is thus largely controlled by the discharges to soil and water.

Finally, it is interesting to note that the fugacities in this final case (in units of μPa) are for the four media 0.31 (air), 0.137 (water), 0.026 (soil) and 0.129 (sediment). The media are fairly close to equilibrium, i.e., within a factor of about 5 of the average value.

At pH 7, Fig. 1.12*a*, the capacity of water for PCP increases, thus the water compartment becomes more important as do intermedia transport processes involving water such as wet deposition in dissolved form and run-off from soil to water. The net effect is that if discharged to air, the amounts transferred to soil and water increase as does the overall residence time. If discharged to water, there is less water to sediment transfer because of the reduced apparent hydrophobicity, and the residence time decreases. If discharged to soil there is little effect of the pH increase because the PCP tends to remain there.

Diagrams similar to Figures 1.7 to 1.12 could be prepared for other phenolic compounds at a range of pH values. The results suggest that the same broad patterns of behavior apply as for PCP but the residence times are generally shorter because of reduced hydrophobicity and more rapid reactions. The lower chlorinated phenols are relatively short lived and are not subject to appreciable intermedia transport, i.e., when discharged to a medium they tend to remain there until degraded or advected. The longest persistence occurs when the chemical is present in soils.

Such simulations suggest that, because of their relatively high water solubility which in combination with low vapor pressure causes low air-water partition coefficients, the phenols tend to remain in water or in soil and show little tendency to evaporate. Their environmental fate tends to be dominated by reaction in soil and water, and for the more sorptive species, in sediments. Their half-lives are relatively short, because of their susceptibility to degradation.

It is believed that examining these three behavior profiles, and their combination in the fourth, illustrates and explains the environmental fate characteristics of this and other chemicals. Important intermedia transport processes and levels in various media arise from discharges into other media became clear. It is believed that the broad fate characteristics of environmental fate as described in the generic environment are applicable to other environments.

1.6 REFERENCES

Abernethy, S., Mackay, D., McCarty, L.S. (1988) "Volume fraction" correlation for narcosis in aquatic organisms: The key role or partitioning. *Environ. Toxicol. Chem.* 7, 469-481.

Abramowitz, R., Yalkowsky, S.H. (1990) Estimation of aqueous solubility and melting point of PCB congeners. *Chemosphere* 21, 1221-1229.

Almgren, M., Grieser, F., Powell, J.R., Thomas, J.K. (1979) A correlation between the solubility of aromatic hydrocarbons in water and micellar solutions, with their normal boiling points. *J. Chem. Eng. Data* 24, 285-287.

Al-Sahhaf, T.A. (1989) Prediction of the solubility of hydrocarbons in water using UNIFAC. *J. Environ. Sci. Health* A24, 49-56.

Altshuller, A.P. (1980) Lifetimes of organic molecules in the troposphere and lower stratosphere. *Adv. Environ. Sci. Technol.* 10, 181-219.

Altshuller, A.P. (1991) Chemical reactions and transport of alkanes and their products in the troposphere. *J. Atmos. Chem.* 12, 19-61.

Ambrose, D. (1981) Reference value of vapor pressure. The vapor pressures of benzene and hexafluorobenzene. *J. Chem. Thermodyn.* 13, 1161-1167.

Ambrose, D., Lawrenson, L.J., Sprake, C.H.S. (1975) The vapour pressure of naphthalene. *J. Chem. Thermodyn.*7, 1173-1176.

Amidon, G.L., Yalkowsky, S.H., Anik, S.T., Leung, S. (1975) Solubility of nonelectrolytes in polar solvents. V. Estimation of the solubility of aliphatic monofunctional compounds in water using a molecular surface area approach. *J. Phys. Chem.* 9, 2239-2245.

Amidon, G.L., Anik, S.T. (1981) Application of the surface area approach to the correlation and estimation of aqueous solubility and vapor pressure. Alkyl aromatic hydrocarbons. *J. Chem. Eng. Data* 26, 28-33.

Amidon, G.L., Williams, N.A. (1982) A solubility equation for non-electrolytes in water. *Intl. J. Pharm.* 11, 249-156.

Anderson, E., Veith, G.D., Weininger, D. (1987) *SMILES: A Line Notation and Computerized Interpreter for Chemical Structures.* US EPA Environmental Research Brief, EPA/600/M-87/021.

Andren, A.W., Doucette, W.J., Dickhut, R.M. (1987) Methods for estimating solubilities of hydrophobic organic compounds: Environmental modeling efforts. In: *Sources and Fates of Aquatic Pollutants.* Hites, R.A., Eisenreich, S.J., Eds., pp. 3-26, Advances in Chemistry Series 216, American Chemical Society, Washington, D.C.

Andrews, L.J., Keffer, R.M. (1950a) Cation complexes of compounds containing carbon-carbon double bonds. IV. The argentation of aromatic hydrocarbons. *J. Am. Chem. Soc.* 72, 3644-3647.

Andrews, L.J., Keefer, R.M. (1950b) Cation complexes of compounds containing carbon-carbon double bonds. VII. Further studies on the argentation of substituted benzenes. *J. Am. Chem. Soc.* 72, 5034-5037.

Arbuckle, W.B. (1983) Estimating activity coefficients for use in calculating environmental parameters. *Environ. Sci. Technol.* 17, 537-542.

Arbuckle, W.B. (1986) Using UNIFAC to calculate aqueous solubilities. *Environ. Sci. Technol.* 20, 1060-1064.

Arcand, Y., Hawari, J., Guiot, S.R. (1994) Solubility of pentachlorophenol in aqueous solutions: the pH effect. *Water Res.* in press.

Ashworth, R.A., Howe, G.B., Mullins, M.E., Roger, T.N. (1988) Air-water partitioning coefficients of organics in dilute aqueous solutions. *J. Hazard. Materials* 18, 25-36.

Atkinson, R. (1985) Kinetics and mechanisms of the gas phase reaction of hydroxyl radicals with organic compounds under atmospheric conditions. *Chem. Rev.* 85, 69-201.

Atkinson, R. (1987) A structure-activity relationship for the estimation of the rate constants for the gas phase reactions of OH radicals with organic compounds. *Int. J. Chem. Kinetics* 19, 790-828.

Atkinson, R. (1988) Estimation of gas-phase hydroxyl radical rate constants for organic chemicals. *Environ. Toxicol. Chem.* 7, 435-442.

Atkinson, R. (1989) Kinetics and mechanisms of the gas-phase reactions of the hydroxyl radical with organic compounds. *J. Phys. Chem. Ref. Data Monograph No. 1.* 1-246.

Atkinson, R. (1990) Gas-phase tropospheric chemistry of organic compounds, a review. *Atmos. Environ.* 24A, 1-41.

Atkinson, R. (1991) Kinetics and mechanisms of the gas-phase reactions of the NO_3 radicals with organic compounds. *J. Phys. Chem. Data* 20, 450-507.

Atkinson, R., Carter, W.L. (1984) Kinetics and mechanisms of the gas-phase reactions of ozone with organic compounds under atmospheric conditions. *Chem. Rev.* 84, 437-470.

Balson, E.W. (1947) Studies in vapour pressure measurement. Part III. An effusion manometer sensitive to $5x10^{-6}$ millimetres of mercury: vapour pressure of D.D.T. and other slightly volatile substances. *Trans. Farad. Soc.* 43, 54-60.

Banerjee, S. (1985) Calculation of water solubility of organic compounds with UNIFAC-derived parameters. *Environ. Sci. Technol.* 19, 369-370.

Banerjee, S., Howard, P.H. (1988) Improved estimation of solubility and partitioning through correction of UNIFAC-derived activity coefficients. *Environ. Sci. Technol.* 22, 839-841.

Banerjee, S., Howard, P.H., Lande, S.S. (1990) General structure-vapor pressure relationships for organics. *Chemosphere* 21, 1173-1180.

Banerjee, S., Yalkowsky, S.H., Valvani, S.C. (1980) Water solubiltiy and octanol/water partition coefficients of organics. Limitations of the solubility-partition coefficient correlation. *Environ. Sci. Technol.* 14, 1227-1229.

Betterton, E.A., Hoffmann, M.R. (1988) Henry's law constants of some enviromentally inportant aldehydes. *Environ. Sci. Technol.* 22, 1415-1418.

Bidleman, T.F. (1984) Estimation of vapor pressures for nonpolar organic compounds by capillary gas chromatography. *Anal. Chem.* 56, 2490-2496.

Bohon, R.L., Claussen, W.F. (1951) The solubility of aromatic hydrocarbons in water. *J. Am. Chem. Soc.* 73, 1571-1576.

Bondi, A. (1964) van der Waals volumes and radii. *J. Phys. Chem.* 68, 441-451.

Booth, H.S., Everson, H.E. (1948) Hydrotropic soublities: soublities in 40 percent sodium xylenesulfonate. *Ind. Eng. Chem.* 40, 1491-1493.

Boublik, T., Fried, V., Hala, E. (1973) *The Vapor Pressure of Pure Substances*, Elsevier, Amsterdam.

Boublik, T., Fried, V., Hala, E. (1984) *The Vapor Pressure of Pure Substances*, 2nd revised ed., Elsevier, Amsterdam.

Bowman, B.T., Sans, W.W. (1983) Determination of octanol-water partitioning coefficient (K_{OW}) of 61 organophosphorus and carbamate insecticides and their relationship to respective water solubility (S) values. *J. Environ. Sci. Health* B18, 667-683.

Bradley, R.S., Cleasby, T.G. (1953) The vapour pressure and lattice energy of some aromatic ring compounds. *J. Chem. Soc.* 1953, 1690-1692.

Briggs, G.G. (1981) Theoretical and experimental relationships between soil adsorption, octanol-water partition coefficients, water solubilities, bioconcentration factors, and the parachor. *J. Agric. Food Chem.* 29, 1050-1059.

Brodsky, J., Ballschmiter, K. (1988) Reversed phase liquid chromatography of PCBs as a basis for the calculation of water solubility and log K_{OW} for polychlorobiphenyls.

Bruggeman, W.A., van der Steen, J., Hutzinger, O. (1982) Reversed-phase thin-layer chromatography of polynuclear aromatic hydrocarbons and chlorinated biphenyls. Relationship with hydrophobicity as measured by aqueous solubility and octanol-water partition coefficient. *J. Chromatogr.* 238, 335-346.

Budavari, S., Ed. (1989) *The Merck Index. An Encyclopedia of Chemicals, Drugs and Biologicals.* 11th ed., Merck & Co. Inc., Rahway, New Jersey.

Burkhard, L.P. (1984) *Physical-Chemical Properties of the Polychlorinated Biphenyls: Measurement, Estimation, and Application to Environmental Systems.* Ph.D. Thesis, University of Wisconsin-Madison, Wisconsin.

Burkhard, L.P., Andren, A.W., Armstrong, D.E. (1985a) Estimation of vapor pressures for polychlorinated biphenyls: A comparison of eleven predictive methods. *Environ. Sci. Technol.* 19, 500-507.

Burkhard, L.P., Armstrong, D.E., Andren, A.W. (1985b) Henry's law constants for polychlorinated biphenyls. *Environ. Sci. Technol.* 590-595.

Burkhard, L.P., Kuehl, D.W., Veith G.D. (1985c) Evaluation of reversed phase liquid chromatograph/mass spectrometry for estimation of n-octanol/water partition coefficients of organic chemicals. *Chemosphere* 14, 1551-1560.

Burnett, M.G. (1963) Determination of partition coefficients in infinite dilution by the gas chromatographic analysis of the vapor above dilute solutions. *Anal. Chem.* 35, 1567-1570.

Butler, J.A.V., Ramchandani, C.N., Thomson, D.W. (1935) The solubility of non-electrolytes. Part I. The free energy of hydration of some aliphatic alcohols. *J. Chem. Soc.* 280-285.

Buttery, R.B., Ling, L.C., Guadagni, D.G. (1969) Volatilities of aldehydes, ketones, and esters in dilute water solution. *J. Agric. Food Chem.* 17, 385-389.

Buxton, G.V., Greenstock, G.L., Helman, W.P., Ross, A.B. (1988) Critical review of rate constants for reactions of hydrated electrons, hydrogen atoms and hydroxyl radicals in aqueous solutions. *J. Phys. Chem. Data* 17, 513-886.

Chiou, C.T. (1981) Partition coefficient and water solubility in environmental chemistry. In: *Hazard Assessment of Chemicals Current Developments.* Vol. 1, pp. 117-153. Academic Press, New York.

Chiou, C.T. (1985) Partition coefficients of organic compounds in lipid-water systems and correlations with fish bioconcentration factors. *Environ. Sci. Technol.* 19, 57-62.

Chiou, C.T., Freed, V.H., Schmedding, D.W. (1977) Partition coefficient and bioaccumulation of selected organic chemicals. *Environ. Sci. Technol.* 11, 475-478.

Chiou, C.T., Schmedding, D.W., Manes, M. (1982) Partitioning of organic compounds in octanol-water system. *Environ. Sci. Technol.* 16, 4-10.

Choi, J., Aomine, S. (1974) Adsorption of pentachlorophenol by soils. *Soil Sci. Plant Nutr.* 20, 135-144.

Coates, M., Connell, D.W., Barron, D.M. (1985) Aqueous solubility and octan-1-ol to water partition coefficients of aliphatic hydrocarbons. *Environ. Sci. Technol.* 19, 628-632.

Davies, R.P., Dobbs, A.J. (1984) The prediction of bioconcentration in fish. *Water Res.* 18, 1253-1262.

Davis, W.W., Krahl, M.E., Clowes, G.H. (1942) Solubility of carcinogenic and related hydrocarbons in water *J. Am. Chem. Soc.* 64, 108-110.

Davis, W.W., Parke, Jr, T.V. (1942) A nephelometric method for determination of solubilities of extremely low order. *J. Am. Chem. Soc.* 64, 101-107.

Dean, J.D., Ed. (1979) *Lange's Handbook of Chemistry*. 12th ed., McGraw-Hill, New York, N.Y.

Dean, J.D., Ed. (1985) *Lange's Handbook of Chemistry*. 13th ed., McGraw-Hill, New York, N.Y.

Dearden, J.C. (1990) Physico-chemical descriptors. In: *Practical Applications of Quantitative Structure-Activity Relationships (QSAR) in Environmental Chemistry and Toxicology*. Karcher, W. and Devillers, J., Eds., pp. 25-60. Kluwer Academic Publisher, Dordrecht, Netherlands.

De Bruijn, J., Busser, G., Seinen, W., Hermens, J. (1989) Determination of octanol/water partition coefficient for hydrophobic organic chemicals with the "slow-stirring" method. *Environ. Toxicol. Chem.* 8, 499-512.

De Bruijn, J., Hermens, J. (1990) Relationships between octanol/water partition coefficients and total molecular surface area and total molecular volume of hydrophobic organic chemicals. *Quant. Struct.-Act. Relat.* 9, 11-21.

De Kruif, C.G. (1980) Enthalpies of sublimation and vapour pressures of 11 polycyclic hydrocarbons. *J. Chem. Thermodyn.* 12, 243-248.

Dobbs, A.J., Grant, C. (1980) Pesticide volatilization rate: a new measurement of the vapor pressure of pentachlorophenol at room temperature. *Pestic. Sci.* 11, 29-32.

Dobbs, A.J., Cull, M.R. (1982) Volatilization of chemical relative loss rates and the estimation of vapor pressures. *Environ. Pollut. (Ser. B)* 3, 289-298.

Doucette, W.J., Andren, A.W. (1987) Correlation of octanol/water partition coefficients and total molecular surface area for highly hydrophobic aromatic compounds. *Environ. Sci. Technol.* 21, 521-524.

Doucette, W.J., Andren, A.W. (1988a) Aqueous solubility of selected biphenyl, furan, and dioxin congeners. *Chemosphere* 17, 243-252.

Doucette, W.J., Andren, A.W. (1988b) Estimation of octanol/water partition coefficients: Evaluation of six methods for highly hydrophobic aromatic hydrocarbons. *Chemosphere* 17, 345-359.

Dreisbach, R.R. (1955) *Physical Properties of Chemical Compounds*. No. 15 of the Adv. in Chemistry Series, American Chemical Society, Washington, D.C.

Dreisbach, R.R. (1959) *Physical Properties of Chemical Compounds-II*. No. 22, Adv. in Chemistry Series, American Chemical Society, Washington, D.C.

Dreisbach, R.R. (1961) *Physical Properties of Chemical Compounds-III*. No. 29, Adv. in Chemistry Series, American Chemical Society, Washington, D.C.

Dunnivant, F.M., Coate, J.T., Elzerman, A.W. (1988) Experimentally determined Henry's law constants for 17 polychlorobiphenyl congeners. *Environ. Sci. Technol.* 22, 448-453.

Dunnivant, F.M., Elzerman, A.W., Jurs, P.C., Hansen, M.N. (1992) Quantitative structure-property relationships for aqueous solubilities and Henry's law constants of polychlorinated biphenyls. *Environ. Sci. Technol.* 26, 1567-1573.

Ellington, J.J. (1989) *Hydrolysis Rate Constants for Enhancing Property-Reactivity Relationships*. USEPA, EPA/600/3-89/063, Athens, GA.

Ellington, J.J., Stancil, Jr., F.E., Payne, W.D. (1987a) *Measurements of Hydrolysis Rate Constant for Evaluation of Hazardous Land Disposal: Volume I. Data on 32 Chemicals.* USEPA, EPA/600/3-86/043, Athens, GA.

Ellington, J.J., Stancil, Jr., F.E., Payne, W.D., Trusty, C.D. (1987b) *Measurements of Hydrolysis Rate Constant for Evaluation of Hazardous Land Disposal: Volume II. Data on 54 Chemicals.* USEPA, EPA/600/3-88/028, Athens, GA.

Ellington, J.J., Stancil, Jr., F.E., Payne, W.D., Trusty, C.D. (1988) *Interim Protocol for Measurement Hydrolysis Rate Constants in Aqueous Solutions.* USEPA, EPA/600/3-88/014, Athens, GA.

Fendinger, N.J., Glotfelty, D.E. (1988) A laboratory method for the experimental determination of air-water Henry's law constants for several pesticides. *Environ. Sci. Technol.* 22, 1289-1293.

Fendinger, N.J., Glotfelty, D.E. (1990) Henry's law constants for selected pesticides, PAHs and PCBs. *Environ. Toxicol. Chem.* 9, 731-735.

Foreman, W.T., Bidleman, T.F. (1985) Vapor pressure estimates of individual polychlorinated biphenyls and commercial fluids using gas chromatographic retention data. *J. Chromatogr.* 330, 203-216.

Fredenslund, A., Jones, R.L., Prausnitz, J.M. (1975) Group-contribution estimation of activity coefficients in nonideal liquid mixtures. *AIChE J.* 21, 1086-1099.

Fujita, T., Iwasa, J., Hansch, C. (1964) A new substituent constant, "pi" derived from partition coefficients. *J. Am. Chem. Soc.* 86, 5175-5180.

Garst, J.E. (1984) Accurate, wide-range, automated, high-performance chromatographic method for the estimation of octanol/water partition coefficients. II: Equilibrium in partition coefficient measurements, additivity of substituent constants, and correlation of biological data. *J. Pharm. Sci.* 73, 1623-1629.

Garten, C.T., Trabalka, J.R. (1983) Evaluation of models for predicting terrestrial food chain behavior of xenobiotics. *Environ. Sci. Technol.* 17, 590-595.

Gmehling, J., Rasmussen, P., Fredenslund, A. (1982) Vapor-liquid equilibria by UNIFAC group contribution. Revision and extension. 2 *Ind. Eng. Chem. Process Des. Dev.* 21, 118-127.

Gossett, R. (1987) Measurement of Henry's law constants for C_1 and C_2 chlorinated hydrocarbons. *Environ. Sci. Technol.* 21, 202-208.

Gross, P.M., Saylor, J.H. (1931) The solubilities of certain slightly soluble organic compounds in water. *J. Am. Chem. Soc.* 1931, 1744-1751.

Gückel, W., Rittig, R., Synnatschke, G. (1974) A method for determining the volatility of active ingredients used in plant protection. II. Application to formulated products. *Pestic. Sci.* 5, 393-400.

Gückel, W., Kästel, R., Lawerenz, J., Synnatschke, G. (1982) A method for determining the volatility of active ingredients used in plant protection. Part III: The temperature relationship between vapour pressure and evaporation rate. *Pestic. Sci.* 13, 161-168.

Hafkenscheid, T.L., Tomlinson, E. (1981) Estimation of aqueous solubilities of organic non-electrolytes using liquid chromatographic retention data. *J. Chromatogr.* 218, 409-425.

Hamaker, J.W. Kerlinger, H.O. (1969) Vapor pressures of pesticides. *Adv. Chem. Ser.* 86, 39-54.

Hamilton, D.J. (1980) Gas chromatographic measurement of volatility of herbicide esters. *J. Chromatogr.* 195, 75-83.

Hansch, C., Leo, A. (1979) *Substituent Constants for Correlation Analysis in Chemistry and Biology.* Wiley-Interscience, New York, N.Y.

Hansch, C., Quinlan, J.E., Lawrence, G.L. (1968) The linear-free energy relationship between partition coefficient and aqueous solubility of organic liquids. *J. Org. Chem.* 33, 347-350.

Hansen, H.K., Schiller, M., Gmehling, J. (1991) Vapor-liquid equlibria by UNIFAC group contribution. 5. Revision and extension. *Ind. Eng. Chem. Res.* 30, 2362-2356.

Hawker, D.W. (1989) The relationship between octan-1-ol/water partition coefficient and aqueous solubility in terms of solvatochromic parameters. *Chemosphere* 19, 1586-1593.

Hawker, D.W. (1990a) Vapor pressures and Henry's law constants of polychlorinated biphenyls. *Environ. Sci. Technol.* 23, 1250-1253.

Hawker, D.W. (1990b) Description of fish bioconcentration factors in terms of solvatochromic parameters. *Chemosphere* 20, 267-477.

Hawker, D.W., Connell, D.W. (1988) Octanol-water partition coefficients of polychlorinated biphenyl congeners. *Environ. Sci. Technol.* 22, 382-387.

Hermann, R.B. (1971) Theory of hydrophobic bonding. II. The correlation of hydrocarbon solubility in water with solvent cavity surface area. *J. Phys. Chem.* 76, 2754-2758.

Hermens, J.L.M., Opperhuizen, A., Eds. (1991) *QSAR in Environmental Toxicology IV.* Elsevier, Amsterdam, The Netherlands. Also published as Volumes 109/110 in *Science for the Total Environment* 1991.

Hinckley, D.A., Bidleman, T.F., Foreman, W.T. (1990) Determination of vapor pressures for nonpolar and semipolar organic compounds from gas chromatographic retention data. *J. Chem. Eng. Data* 35, 232-237.

Hine, J., Mookerjee, P.K. (1975) The intrinsic hydrophilic character of organic compounds. Correlations in terms of structural contributions. *J. Org. Chem.* 40, 292-298.

Hollifield, H.C. (1979) Rapid nephelometric estimate of water solubility of highly insoluble organic chemicals of environmental interest. *Bull. Environ. Contam. Toxicol.* 23, 579-586.

Horvath, A.L. (1982) *Halogenated Hydrocarbons, Solubility - Miscibility with Water.* Marcel Dekker, Inc., New York, N.Y.

Howard, P.H., Ed. (1989) *Handbook of Fate and Exposure Data for Organic Chemicals. Vol. I. Large Production and Priority Pollutants.* Lewis Publishers, Chelsea, Michigan.

Howard, P.H., Ed. (1990) *Handbook of Fate and Exposure Data for Organic Chemicals. Vol. - II - Solvents.* Lewis Publishers, Inc., Chelsea, Michigan.

Howard, P.H., Ed. (1991) *Handbook of Fate and Exposure Data for Organic Chemicals. Vol. III. Pesticides.* Lewis Publishers, Inc., Chelsea, Michigan.

Howard, P.H., Ed. (1993) *Handbook of Fate and Exposure Data for Organic Chemicals. Vol. IV. Solvent 2.* Lewis Publishers, Inc., Chelsea, Michigan.

Howard, P.H., Boethling, R.S., Jarvis, W.F., Meylan, W.M., Michalenko, E.M. (1991) *Handbook of Environmental Degradation Rates.* Lewis Publishers, Inc., Chelsea, Michigan.

Hussam, A., Carr, P.W. (1985) A study of a rapid and precise methodology for the measurement of vapor liquid equilibria by headspace gas chromatography. *Anal. Chem.* 57, 793-801.

Irmann, F. (1965) Eine einfache korrelation zwishen wasserlöslichkeit und strucktur von kohlenwasserstoffen und halogenkohlenwasserstoffen. *Chemie-Ing.-Techn.* 37, 789-798.

Isnard, P., Lambert, S. (1988) Estimating bioconcentration factors for octanol-water partition coefficient and aqueous solubility. *Chemosphere* 17, 21-34.

Isnard, P., Lambert, S. (1989) Aqueous solubility/n-octanol-water partition coefficient correlations. *Chemosphere* 18, 1837-1853.

IUPAC Solubility Data Series (1984) Vol. 15: *Alcohols with Water.* Barton, A.F.M., Ed., Pergamon Press, Oxford, England.

IUPAC Solubility Data Series (1985) Vol. 20: *Halogenated Benzenes, Toluenes and Phenols with Water.* Horvath, A.L., Getzen, F.W., Eds., Pergamon Press, Oxford, England.

IUPAC Solubility Data Series (1989a) Vol. 37: *Hydrocarbons (C_5 - C_7) with Water and Seawater.* Shaw, D.G., Ed., Pergamon Press, Oxford, England.

IUPAC Solubility Data Series (1989b) Vol. 38: *Hydrocarbons (C_8 -C_{36}) with Water and Seawater.* Shaw, D.G., Ed., Pergamon Press, Oxford, England.

Iwase, K., Komatsu, K., Kirono, Nakagawa, S., Moriguchi, I. (1985) Estimation of hydrophobicity based on the solvent-accessible surface area of molecules. *Chem Pharm. Bull.* 33, 2114-2121.

Jafvert, C.T., Westall, J.C., Grieder, E., Schwarzenbach, P. (1990) Distribution of hydrophobic ionogenic organic compounds between octanol and water: organic acids. *Environ. Sci. Technol.* 24, 1795-1803.

Jaworska, J.S., Schultz, T.W. (1992) Quantitive relationships of structure-activity and volume fraction for selected nonpolar and polar narcotic chemicals. *SAR. & QSAR in Environ. Res.* 1, 3-19.

Jensen, T., Fredenslund, A., Rasmussen, P. (1981) Pure-compound vapor pressures using UNIFAC group contribution. *Ind. Eng. Chem. Fundam.* 20, 239-246.

Johnson, C.A., Westall, J.C. (1990) Effect of pH and KCl concentration on the octanol-water distribution of methylanilines. *Environ. Sci. Technol.* 24, 1869-1875.

Jönsson, J.A., Vejrosta, J., Novak, J. (1982) Air/water partition coefficients for normal alkanes (*n*-pentane to *n*-nonane). *Fluid Phase Equil.* 9, 279-286.

Jury, W.A., Spencer, W.F., Farmer, W.J. (1983) Behavior assessment model for trace organics in soil: I. Model description. *J. Environ. Qual.* 12, 558-566.

Jury, W.A., Farmer, W.J., Spencer, W.F. (1984a) Behavior assessment model for trace organics in soil: II. Chemical classification and parameter sensitivity. *J. Environ. Qual.* 13, 567-572.

Jury, W.A., Farmer, W.J., Spencer, W.F. (1984b) Behavior assessment model for trace organics in soil: III. Application of screening model. *J. Environ. Qual.* 13, 573-579.

Jury, W.A., Spencer, W.F., Farmer, W.J. (1984) Behavior assessment model for trace organics in soil: IV. Review of experimental evidence. *J. Environ. Qual.* 13, 580-587.

Kabadi, V.N., Danner, R.P. (1979) Nomograph solves for solubilities of hydrocarbons in water. *Hydrocarbon Processing* 68, 245-246.

Kaiser, K.L.E., Valdmanis, I. (1982) Apparent octanol/water partition coefficients of pentachlorophenol as a function of pH. *Can. J. Chem.* 61, 2104-2106.

Kamlet, M.J., Doherty, R.M., Veith, G.D., Taft, R.W., Abraham, M.H. (1986) Solubility properties in polymers and biological media. 7. An analysis toxicant properties that influence inhibition of bioluminescence in *Photobacterium phosphoreum* (the Microtox test). *Environ. Sci. Technol.* 20, 690-695.

Kamlet, M.J., Doherty, R.M., Abraham, M.H., Carr, P.W., Doherty, R.F., Raft, R.W. (1987) Linear solvation energy relationships. Important differences between aqueous solublity relationships for aliphatic and aromatic solutes. *J. Phy. Chem.* 91, 1996.

Kamlet, M.J., Doherty, R.M., Carr, P.W., Mackay, D., Abraham, M.H., Taft, R.W. (1988) Linear solvation energy relationships. 44. Parameter estimation rules that allow accurate prediction of octanol/water partition coefficients and other solubility and toxicity properties of polychlorinated biphenyls and polycyclic aromatic hydrocarbons. *Environ. Sci. Technol.* 22, 503-509.

Karcher, W., Devillers, J., Eds., (1990) *Practical Applications of Quantitative-Structure-Activity Relationships (QSAR) in Environmental Chemistry and Toxicology.* Kluwer Academic Publisher, Dordrecht, The Netherlands.

Karger, B.L., Castells, R.C., Sewell, P.A., Hartkopf, A. (1971a) Study of the adsorption of insoluble and sparingly soluble vapors at the gas-liquid interface of water by gas chromatography. *J. Phys. Chem.* 75, 3870-3879.

Karger, B.L., Sewell, P.A., Castells, R.C., Hartkopf, A. (1971b) Gas chromatographic study of the adsorption of insoluble vapors on water. *J. Colloid Interface Sci.* 35(2), 328-339.

Karickhoff, S.W. (1981) Semiempirical estimation of sorption of hydrophobic pollutants on natural sediments and soil. *Chemosphere* 10, 833-846.

Karickhoff, S.W., Brown, D.S., Scott, T.A. (1979) Sorption of hydrophobic pollutants on natural water sediments. *Water Res.* 13, 241-248.

Kenaga, E.E. (1980) Predicted bioconcentration factors and soil sorption coefficients of pesticides and other chemicals. *Ecotox. Environ. Saf.* 4, 26-38.

Kenaga, E.E., Goring, C.A.I. (1980) Relationship between water solubility, soil sorption, octanol-water partitioning, and concentration of chemicals in biota. In: *Aquatic Toxicology.* Eaton, J.G., Parrish, P.R., Hendrick, A.C., Eds., pp. 78-115, Am. Soc. for Testing and Materials, STP 707, Philadelphia, Pa.

Kier, L.B., Hall, L.H. (1976) Molar properties and molecular connectivity. In: *Molecular Connectivity in Chemistry and Drug Design.* Medicinal Chem. Vol. 14, pp. 123-167, Academic Press, New York.

Kier, L.B., Hall, L.H. (1986) *Molecular Connectivity in Structure-Activity Analysis.* Wiley, New York.

Kikic, I., Alesse, P., Rasmussen, P., Fredenslunds, A. (1980) On the combinatorial part of the UNIFAC and UNIQUAC models. *Can. J. Chem. Eng.* 58, 253-258.

Kim, Y.-H., Woodrow, J.E., Seiber, J.N. (1984) Evaluation of a gas chromatographic method for calculating vapor pressures with organophosphorus pesticides. *J. Chromotagr.* 314, 37-53.

Klöpffer, W. (1991) Photochemistry in environmental research: Its role in abiotic degradation and exposure analysis. *EPA Newsletter* 41, 24-39.

Könemann, H., van Leeuewen, K. (1980) Toxicokinetics in fish: accumulation of six chlorobenzenes by guppies. *Chemosphere* 9, 3-19.

Könemann, H., Zelle, R., Busser, F. (1979) Determination of log P_{oct} values of chloro-substituted benzenes, toluenes and anilines by high-performance liquid chromatography on ODS-silica. *J. Chromatogr.* 178, 559-565.

Korenman, I.M., Gur'ev, I.A., Gur'eva, Z.M. (1971) Solubility of liquid aliphatic compounds in water. *Russ. J. Phys. Chem.* 45, 1065-1066.

Lande, S.S., Banerjee, S. (1981) Predicting aqueous solubility of organic nonelectrolytes from molar volume. *Chemosphere* 10, 751-759.

Leahy, D.E. (1986) Intrinsic molecular volume as a measure of the cavity term in linear solvation energy relationships: octanol-water partition coefficients and aqueous solubilities. *J. Pharm. Sci.* 75, 629-636.

Lee, L.S., Rao, P.S.C., Nkedi-Kizza, P., Delfino, J. (1990) Influence of solvent and sorbent characteristics on distribution of pentachlorophenol in octanol-water and soil-water systems. *Environ. Sci. Technol.* 24, 654-661.

Leighton, D.T.,Jr., Calo, J.M. (1981) Distribution coefficients of chlorinated hydrocarbons in dilute air-water systems for groundwater contamination applications. *J. Chem. Eng. Data* 26, 382-385.

Leo, A., Hansch, C., Elkins, D. (1971) Partition coefficients and their uses. *Chem. Rev.* 71, 525-616.

Lincoff, A.H., Gossett, J.M. (1984) The determination of Henry's law constants for volatile organics by equilibrium partitioning in closed systems. In: *Gas Transfer at Water Surfaces*. Brutsaert, W., Jirka, G.H., Eds., pp. 17-26, D. Reidel Publishing Co., Dordrecht, The Netherlands.

Lo, J.M., Tseng, C.L., Yang, J.Y. (1986) Radiometric method for determining solubility of organic solvents in water. *Anal. Chem.* 58, 1596-1597.

Locke, D. (1974) Selectivity in reversed-phase liquid chromatography using chemically bonded stationary phases. *J. Chromatogr. Sci.* 12, 433-437.

Lyman, W.J., Reehl, W.F., Rosenblatt, D.H. (1982) *Handbook of Chemical Property Estimation Methods*. McGraw-Hill, New York.

Mabey, W., Mill, T. (1978) Critical review of hydrolysis of organic compounds in water under environmental conditions. *J. Phys. Chem. Ref. Data* 7, 383-414.

Mabey, W.J., Mill, T., Podoll, R.T. (1984) *Estimation Methods for Process Constants and Properties used in Fate Assessment*. USEPA, EPA-600/3-84-035, Athens, GA.

Mackay, D. (1979) Finding fugacity feasible. *Environ. Sci. Technol.* 13, 1218-1223.

Mackay, D. (1982) Correlation of bioconcentration factors. *Environ. Sci. Technol.* 16, 274-278.

Mackay, D. (1991) *Multimedia Environmental Models. The Fugacity Approach.* Lewis Publishers, Inc., Chelsea, Michigan.

Mackay, D., Bobra, A.M., Shiu, W.Y., Yalkowsky, S.H. (1980) Relationships between aqueous solubility and octanol-water partition coefficient. *Chemosphere* 9, 701-711.

Mackay, D., Bobra, A.M., Chan, D.W., Shiu, W.Y. (1982) Vapor pressure correlation for low-volatility environmental chemicals. *Environ. Sci. Technol.* 16, 645-649.

Mackay, D., Paterson, S. (1990) Fugacity models. In: *Practical Applications of Quantitative Structure-Activity Relationships (QSAR) in Environmental Chemistry and Toxicology.* Karcher, W., Devillers, J., Eds., pp. 433-460, Kluwer Academic Publishers, Dordrecht, Netherlands.

Mackay, D., Paterson, S. (1991) Evaluating the multimedia fate of organic chemicals: A Level III fugacity model. *Environ. Sci. Technol.* 25, 427-436.

Mackay, D., Shiu, W.Y. (1977) Aqueous solubility of polynuclear aromatic hydrocarbons. *J. Chem. Eng. Data* 22, 339-402.

Mackay, D., Shiu, W.Y. (1981) A critical review of Henry's law constants for chemicals of environmental interest. *J. Phys. Chem. Ref. Data* 11, 1175-1199.

Mackay, D., Shiu, W.Y., Sutherland, R.P. (1979) Determination of air-water Henry's law constants for hydrophobic pollutants. *Envrion. Sci. Technol.* 13, 333-337.

Mackay, D., Shiu, W.Y., Wolkoff, A.W. (1975) Gas chromatographic determination of low concentration of hydrocarbons in water by vapor phase extraction. *ASTM STP 573*, pp. 251-258, American Society for Testing and Materials, Philadelphia, Pa.

Mackay, D., Stiver, W.H. (1991) Predictability and environmental chemistry. In: *Environmental Chemistry of Herbicides*. Vol. II. Grover, R., Lessna, A.J., Eds., pp. 281-297. CRC Press, Boca Raton, FL.

Macknick, A.B., Prausnitz, J.M. (1979) Vapor pressure of high-molecular weight hydrocarbons. *J. Chem. Eng. Data* 24, 175-178.

Magnussen, T., Rasmussen, P., Fredenslund, A. (1981) UNIFAC parameter table for prediction of liquid-liquid equilibria. *Ind. Eng. Chem. Process Des. Dev.* 20, 331-339.

Mailhot, H., Peters, R.H. (1988) Empirical relationships between the 1-octanol/water partition coefficient and nine physicochemical properties. *Environ. Sci. Technol.* 22, 1479-1488.

May, W.E., Wasik, S.P., Freeman, D.H. (1978a) Determination of the aqueous solubility of polynuclear aromatic hydrocarbons by a coupled-column liquid chromatographic technique. *Anal. Chem.* 50, 175-179.

May, W.E., Wasik, S.P., Freeman, D.H. (1978b) Determination of the solubility behavior of some polycyclic aromatic hydrocarbons in water. *Anal. Chem.* 50, 997-1000.

McAuliffe, C. (1966) Solubility in water of paraffin, cycloparaffin, olefin, acetylene, cycloolefin and aromatic hydrocarbons. *J. Phys. Chem.* 76, 1267-1275.

McAuliffe, C. (1971) GC determination of solutes by multiple phase equilibration. *Chem. Tech.* 1, 46-51.

McDuffie, B. (1981) Estimation of octanol/water partition coefficient for organic pollutants using reversed phase HPLC. *Chemosphere* 10, 73-83.

McGowan, J.C., Mellors, A. (1986) *Molecular Volumes in Chemistry and Biology-Applications including Partitioning and Toxicity.* Ellis Horwood Limited, Chichester, England.

The Merck Index (1989) *An Encyclopedia of Chemicals, Drugs and Biologicals.* 11th ed., Budavari, S., Ed., Merck and Co., Inc., Rahway, N.J.

Meylan, W.M., Howard, P.H. (1991) Bond contribution method for estimating Henry's law constants. *Environ. Toxicol. Chem.* 10, 1283-1293.

Meylan, W.M., Howard, P.H., Boethling, R.S. (1992) Molecular topology/fragment contribution for predicting soil sorption coefficient. *Environ. Sci. Technol.* 26, 1560-1567.

Mill, T. (1982) Hydrolysis and oxidation processes in the environment. *Environ. Toxicol. Chem.* 1, 135-141.

Mill, T. (1989) Structure-activity relationships for photooxidation processes in the environment. *Environ. Toxicol. Chem.* 8, 31-45.

Mill, T. (1993) Environmental chemistry. In: *Ecological Risk Assessment*. Suter, II, G.W., Ed., pp. 91-127.

Mill, T., Mabey, W. (1985) Photodegradation in water. In: *Environmental Exposure from Chemicals.* Vol. 1. Neely, W.B., Blau, G.E., Eds., pp. 175-216, CRC Press, Boca Raton, FL.

Miller, M.M., Ghodbane, S., Wasik, S.P., Tewari, Y.B., Martire, D.E. (1984) Aqueous solubilities, octanol/water partition coefficients and entropies of melting of chlorinated benzenes and biphenyls. *J. Chem. Eng. Data* 29, 184-190.

Miller, M.M., Wasik, S.P., Huang, G.L., Shiu, W.Y., Mackay, D. (1985) Relationships between octanol-water partition coefficient and aqueous solublity. *Environ. Sci. Technol.* 19, 522-529.

Munz, C., Roberts, P.V. (1987) Air-water phase equilibria of volatile organic solutes. *J. Am. Wat. Works Assoc.* 79, 62-69.

NRCC (1982) *Chlorinated Phenols: Criteria for Environmental Quality.* National Research Council Canada, Publication NRCC No. 18578, Ottawa, Canada.

Neely, W.B., Branson, D.R., Blau, G.E. (1974) Partition coefficient to measure bioconcentration potential of organic chemicals in fish. *Environ. Sci. Technol.* 8, 1113-1115.

Nirmalakhandan, N.N., Speece, R.E. (1988a) Prediction of aqueous solubility of organic chemicals based on molecular structure. *Environ. Sci. Technol.* 22, 328-338.

Nirmalakhandan, N.N., Speece, R.E. (1988b) QSAR model for predicting Henry's law constant. *Environ. Sci. Technol.* 22, 1349-1357.

Nirmalakhandan, N.N., Speece, R.E. (1989) Prediction of aqueous solubility of organic chemicals based on molecular structure. 2. Application to PNAs, PCBs, PCDDs, etc. *Environ. Sci. Technol.* 23, 708-713.

Okouchi, H., Saegusa, H., Nojima, O. (1992) Prediction of environmental parameters by adsorbability index: water solubilities of hydrophobic organic pollutants. *Environ. Intl.* 18, 249-261.

Oliver, B.G. (1984) The relationship between bioconcentration factor in rainbow trout and physical-chemical properties for some halogenated compounds. In: *QSAR in Environmental Toxicology.* Kaiser, K.L.E., Ed., pp. 300-317, D. Reidel Publishing Co., Dordrecht, The Netherlands.

Oliver, B.G., Niimi, A.J. (1988) Trophodynamic analysis of polychlorinated biphenyl congeners and other chlorinated hydrocarbons in the Lake Ontario ecosystem. *Environ. Sci. Technol.* 22, 388-397.

Osborn, A.G., Douslin, D.R. (1975) Vapor pressures and derived enthalpies of vaporization of some condensed-ring hydrocarbons. *J. Chem. Eng. Data* 20, 229-231.

Paterson, S., Mackay, D. (1985) The fugacity concept in environmental modelling. In: *The Handbook of Environmental Chemistry.* Vol. 2/Part C, Hutzinger, O., Ed., pp. 121-140. Springer-Verlag, Heidelberg, Germany.

Pearlman, R.S. (1980) Molecular surface areas and volumes and their use in structure/activity relationships. In: *Physical Chemical Properties of Drugs.* Yalkowsky, S.H., Sinkula, A.A., Valvani, S.C., Eds., Medicinal Research Series, Vol. 10., pp. 321-317, Marcel Dekker, Inc., New York.

Pearlman, R.S. (1986) Molecular surface area and volume: Their calculation and use in predicting solubilities and free energies of desolvation. In: *Partition coefficient, Determination and Estimation.* Dunn III, W.J., Block, J.H., Pearlman R.S., Eds., pp. 3-20, Pergamon Press, New York.

Pearlman, R.S., Yalkowsky, S.H., Banerjee, S. (1984) Water solubilities of polynuclear aromatic and heteroaromatic compounds. *J. Phys. Chem. Ref. Data* 13, 555-562.

Pierotti, C., Deal, C., Derr, E. (1959) Activity coefficient and molecular structure. *Ind. Eng. Chem. Fundam.* 51, 95-101.

Polak, J., Lu, B.C.Y. (1973) Mutual solubilities of hydrocarbons and water at 0° and 25°C. *Can. J. Chem.* 51, 4018-4023.

Rapaport, R.A., Eisenreich, S.J. (1984) Chromatographic determination of octanol-water partition coefficients (K_{OW}'s) for 58 polychlorinated biphenyl congeners. *Environ. Sci. Technol.* 18, 163-170.

Reid, R.C., Prausnitz, J.M., Polling, B.E. (1987) *The Properties of Gases and Liquids.* 4th ed., McGraw-Hill, New York, N.Y.

Rekker, R.F. (1977) *The Hydrophobic Fragmental Constant.* Elsevier, Amsterdam/New York N.Y.

Riddick, J.A., Bunger, W.B., Sakano, T.K. (1986) *Organic Solvents, Physical Properties and Methods of Purification.* 4th ed., Wiley-Science Publication, John Wiley & Sons, New York, N.Y.

Rordorf, B.F. (1985a) Thermodynamic and thermal properties of polychlorinated compounds: the vapor pressures and flow tube kinetic of ten dibenzo-*para*-dioxins. *Chemosphere* 14, 885-892.

Rordorf, B.F. (1985b) Thermodynamic properties of polychlorinated compounds: the vapor pressures and enthalpies of sublimation of ten dibenzo-p-dioxins. *Thermochimica Acta*, 85, 435-438.

Rordorf, B.F. (1986) Thermal properties of dioxins, furans and related compounds, *Chemosphere* 15, 1325-1332.

Sabljic, A. (1984) Predictions of the nature and strength of soil sorption of organic pollutants by molecular topology. *J. Agric. Food Chem.* 32, 243-246.

Sabljic, A. (1987) On the prediction of soil sorption coefficients of organic pollutants from molecular structure: Application of molecular topology model. *Environ. Sci. Technol.* 21, 358-366.

Sabljic, A., Lara, R., Ernst, W. (1989) Modelling association of highly chlorinated biphenyls with marine humic substances. *Chemosphere* 19, 1665-1676.

Sabljic, A., Güsten, H. (1989) Predicting Henry's law constants for polychlorinated biphenyls, *Chemosphere* 19, 1503-1511.

Sabljic, A., Güsten, H. (1990) Predicting the night-time NO_3 radical reactivity in the troposphere. *Atmos. Environ.* 24A, 73-78.

Schellenberg, K., Leuenberger, C., Schwarzenbach, R.P. (1984) Sorption of chlorinated phenols by natural sediments and aquifer materials. *Environ. Sci. Technol.* 18, 652-657.

Schwarz, F.P. (1980) Measurement of the solubilities of slightly soluble organic liquids in water by elution chromatography. *Anal. Chem.* 52, 10-15.

Schwarz, F.P., Miller, J. (1980) Determination of the aqueous solubilities of organic liquids at 10.0, 20.0, 30.0 °C by elution chromatography. *Anal. Chem.* 52, 2162-2164.

Schwarzenbach, R.P., Gschwend, P.M., Imboden, D.M. (1993) *Environmental Organic Chemistry.* John Wiley & Sons, Inc., New York, N.Y.

Schwarzenbach, R.P., Stierli, R., Folsom, B.R., Zeyer, J. (1988) Compound properties relevant for assessing the environmental partitioning of nitrophenols. *Environ. Sci. Technol.* 22, 83-92.

Schwarzenbach, R.P., Westall, J. (1981) Transport of nonpolar compounds from surface water to groundwater. Laboratory sorption studies. *Environ. Sci. Technol.* 11, 1360-1367.

Sears, G.W., Hopke, E.R. (1947) Vapor pressures of naphthalene, anthracene and hexachlorobenzene in a low pressure region. *J. Am. Chem. Soc.* 71, 1632-1634.

Shigeoka, T., Sato, Y., Takeda, Y. (1988) Acute toxicity of chlorophenols to green algae, *selenastrum capricornutum* and *chlorella vulgaris*, and quantitative structure-activity relationships. *Environ. Toxicol. Chem.* 7, 847-854.

Shiu, W.Y., Ma, K.C., Verhanikova, D., Mackay, D. (1994) Chlorophenols and alkylphenols: A review and correlation of environmentally relevant properties and fate in an evaluative environment. *Chemosphere* 29(6), 1155-1224.

Shiu, W.Y., Mackay, D. (1986) A critical review of aqueous solubilities, vapor pressures, Henry's law constants, and octanol-water partition coefficients of the polychlorinated biphenyls. *J. Phys. Chem. Ref. Data* 15, 911-929.

Shiu, W.Y., Gobas, F.A.P.C., Mackay, D. (1987) Physical-chemical properties of three congeneric series of chlorinated aromatic hydrocarbons. In: *QSAR in Environmental Toxicology -II*. Kaiser, K.L.E., Ed., pp. 347-362. D. Reidel Publishing Co., Dordrecht, The Netherlands.

Shiu, W.Y., Doucette, W., Gobas, F.A.P.C., Mackay, D., Andren, A.W. (1988) Physical-chemical properties of chlorinated dibenzo-*p*-dioxins. *Environ. Sci. Technol.* 22, 651-658.

Sinke, G.C. (1974) A method for measurement of vapor pressures of organic compounds below 0.1 torr. Naphthalene as reference substance. *J. Chem. Thermodyn.* 6, 311-316.

Smejtek, P., Wang, S. (1993) Distribution of hydrophobic ionizable xenobiotics between water and lipid membranes: pentachlorophenol and pentachlorophenate. A comparison with octanol-water partition. *Arch. Environ. Contam. Toxicol.* 25, 394-404.

Smith, P.D., Brockwaty, D.L., Stancil, Jr., F.E. (1987) Effects of hardness, alkalinity and pH on the toxicity of pentachlorophenol to *senastrum capricornutum* (printz). *Environ. Toxicol. Chem.* 6, 891-990.

Sonnefeld, W.J., Zoller, W.H., May, W.E. (1983) Dynamic coupled-column liquid chromatographic determination of ambient temperature vapor pressures of polynuclear aromatic hydrocarbons. *Anal. Chem.* 55, 275-280.

Spencer, W.F., Cliath, M.M. (1969) Vapor density of dieldrin. *Environ. Sci. Technol.* 3, 670-674.

Spencer, W.F., Cliath, M.M. (1970) Vapor density and apparent vapor pressure of lindane (γ-BHC). *J. Agric. Food Chem.* 18, 529-530.

Spencer, W.F., Cliath, M.M. (1972) Volatility of DDT and related compounds. *J. Agric. Food Chem.* 20, 645-649.

Stehly, G.R., Hayton, W.L. (1990) Effect of pH on the accumulation kinetics of pentachlorophenol in goldfish . *Arch. Environ. Contam. Toxicol.* 19, 464-470.

Stephenson, R.M., Malanowski, A. (1987) *Handbook of the Thermodynamics of Organic Compounds*. Elsevier, New York.

Stiver, W., Mackay, D. (1989) The linear additivity principle in environmental modelling: Application to chemical behaviour in soil. *Chemosphere* 19, 1187-1198.

Sugiyama, T., Takeuchi, T., Suzuki, Y. (1975) Thermodynamic properties of solute molecules at infinite dilution determined by gas-liquid chromatography. I. Intermolecular energies of *n*-alkane solutes in C_{28} - C_{36} *n*-alkane solvents. *J. Chromatogr.* 105, 265-272.

Suntio, L.R., Shiu, W.Y., Mackay, D. (1988) Critical review of Henry's law constants for pesticides. *Rev. Environ. Contam. Toxicol.* 103, 1-59.

Swann, R.L., Laskowski, D.A., McCall, P.J., Vander Kuy, K., Dishburger, H.J. (1983) A rapid method for the estimation of the environmental parameters octanol/water partition coefficient, soil sorption constant, water to air ratio, and water solubility. *Residue Rev.* 85, 17-28.

Szabo, G., Prosser, S., Bulman, R.A. (1990) Determination of the adsorption coefficient (K_{OC}) of some aromatics for soil by RP-HPLC on two immobilized humic acid phases. *Chemosphere* 21, 777-788.

Tomlinson, E., Hafkenscheid, T.L. (1986) Aqueous solution and partition coefficient estimation from HPLC data. In: *Partition Coefficient, Determination and Estimation*. Dunn III, W.J., Block, J.H., Pearlman, R.S., Eds., pp. 101-141, Pergamon Press, New York.

Trujilo, D.A., Ray, L.E., Murray, H.E., Giam, C.S. (1982) Bioaccumulation of pentachlorophenol by killifish (*Fundulus similus*). *Chemosphere* 11, 25-31.

Tse, G., Orbey, H., Sandler, S.I. (1992) Infinite dilution activity coefficients and Henry's law coefficients for some priority water pollutants determined by a relative gas chromatographic method. *Environ. Sci. Technol.* 26, 2017-2022.

Tsonopoulos, C., Prausnitz, J.M. (1971) Activity coefficients of aromatic solutes in dilute aqueous solutions. *Ind. Eng. Chem. Fundam.* 10, 593-600.

Tulp, M.T.M., Hutzinger, O. (1978) Some thoughts on the aqueous solubilities and partition coefficients of PCB, and the mathematical correlation between bioaccumulation and physico-chemical properties. *Chemosphere* 7, 849-760.

Veith, G.D., Austin, N.M., Morris, R.T. (1979) A rapid method for estimating log P for organic chemicals. *Water Res.* 13, 43-47.

Veith, G.D., Macek, K.J., Petrocelli, S.R., Caroll, J. (1980) An evaluation of using partition coefficients and water solubilities to estimate bioconcentration factors for organic chemicals in fish. In: *Aquatic Toxicology.* Eaton, J.G., Parrish, P.R., Hendrick, A.C., Eds, pp. 116-129, ASTM ATP 707, Am. Soc. for Testing and Materials, Philadelphia, Pa.

Vejrosta, J., Novak, J., Jönsson, J. (1982) A method for measuring infinite-dilution partition coefficients of volatile compounds between the gas and liquid phases of aqueous systems. *Fluid Phase Equil.* 8, 25-35.

Verschueren, K. (1977) *Handbook of Environmental Data on Organic Chemicals.* Van Nostrand Reinhold, New York, N.Y.

Verschueren, K. (1983) *Handbook of Environmental Data on Organic Chemicals.* Van Nostrand Reinhold, New York, N.Y.

Wakita, K., Yoshimoto, M., Miyamoto, S., Watsnabe, H. (1986) A method for calculations of the aqueous solubility of organic compounds by using new fragment solubility constants. *Chem. Pharm. Bull.* 34, 4663-4681.

Wang, L., Zhao, Y., Hong, G. (1992) Predicting aqueous solubility and octanol/water partition coefficients of organic chemicals from molar volume. *Environ. Chem.* 11, 55-70.

Warne, M., St. J., Connell, D.W., Hawker, D.W. (1990) Prediction of aqueous solubility and the octanol-water partition coefficient for lipophilic organic compounds using molecular descriptors and physicochemical properties. *Chemosphere* 16, 109-116.

Wasik, S.P., Miller, M.M., Tewari, Y.B., May, W.E., Sonnefeld, W.J., DeVoe, H., Zoller, W.H. (1983) Determination of the vapor pressure, aqueous solubility, and octanol/water partition coefficient of hydrophobic substances by coupled generator column/liquid chromatographic methods. *Res. Rev.* 85, 29-42.

Weast, R. (1972-73) *Handbook of Chemistry and Physics.* 53th ed., CRC Press, Cleveland, OH.

Weast, R. (1982-83) *Handbook of Chemistry and Physics.* 64th ed., CRC Press, Boca Raton, FL.

Weil, L., Dure, G., Quentin, K.L. (1974) Solubility in water of insecticide, chlorinated hydrocarbons and polychlorinated biphenyls in view of water pollution. *Z. Wasser Abwasser Forsch.* 7, 169-175.

Westall, J.C., Leuenberger, C., Schwarzenbach, R.P. (1985) Influence of pH and ionic strength on the aqueous-nonaqueous distribution of chlorinated phenols. *Environ. Sci. Technol.* 19, 193-198.

Westcott, J.W., Bidleman, T.F. (1982) Determination of polychlorinated biphenyl vapor pressures by capillary gas chromatography. *J. Chromatogr.* 210, 331-336.

Westcott, J.W., Simon, J.J., Bidleman, T.F. (1981) Determination of polychlorinated biphenyl vapor pressures by a semimicro gas saturation method. *Environ. Sci. Technol.* 15, 1375-1378.

Whitehouse, B.G., Cooke, R.C. (1982) Estimating the aqueous solubility of aromatic hydrocarbons by high performance liquid chromatography. *Chemosphere* 11, 689-699.

Windholz, M., Editor (1983) *The Merck Index, An Encyclopedia of Chemicals, Drugs and Biologicals.* 10th ed., Merck & Co. Inc. Rahway, New Jersey.

Woodburn, K.B., Doucette, W.J., Andren, A.W. (1984) Generator column determination of octanol/water partition coefficients for selected polychlorinated biphenyl congeners. *Environ. Sci. Technol.* 18, 457-459.

Yair, O.B., Fredenslund, A. (1983) Extension of the UNIFAC group-contribution method for the prediction of pure-component vapor pressure. *Ind. Eng. Chem. Fundam. Des. Dev.* 22, 433-436.

Yalkowsky, S.H. (1979) Estimation of entropies of fusion of organic compounds. *Ind. Eng. Chem. Fundam.* 18, 108-111.

Yalkowsky, S.H., Banerjee, S. (1992) *Aqueous Solubility, Methods of Estimation for Organic Compounds.* Marcel Dekker, Inc., New York, N.Y.

Yalkowsky, S.H., Valvani, S.C. (1976) Partition coefficients and surface areas of some alkylbenzenes. *J. Med. Chem.* 19, 727-728.

Yalkowsky, S.H., Valvani, S.C. (1979) Solubility and partitioning. I: Solubility of nonelectrolytes in water. *J. Pharm. Sci.* 69, 912-922.

Yalkowsky, S.H., Orr, R.J., Valvani, S.C. (1979) Solubility and partitioning. 3. The solubility of halobenzenes in water. *I&EC Fundam.* 18, 351-353.

Yalkowsky, S.H., Valvani, S.S., Mackay, D. (1983) Estimation of the aqueous solubility of some aromatic compounds. *Res. Rev.* 85, 43-55.

Yoshida, K., Shigeoka, T., Yamauchi, F. (1983) Relationship between molar refraction and n-octanol/water partition coefficient. *Ecotox. Environ. Saf.* 7, 558-565.

Yoshida, K., Shigeoka, T., Yamauchi, F. (1987) Evaluation of aquatic environmental fate of 2,4,6-trichlorophenol with a mathematical model. *Chemosphere* 16, 2531-2544.

Zhou, X., Mopper, K. (1990) Apparent partition coefficients of 15 carbonyl compounds between air and seawater and between air and freshwater: Implications for air-sea exchange. *Environ. Sci. Technol.* 24, 1864-1869.

Zwolinski, B.J. Wilhoit, R.C. (1971) *Handbook of Vapor Pressures and Heats of Vaporization of Hydrocarbons and Related Compounds.* API-44, TRC Publication No. 101, Texas A&M University, College Station, TX.

Chapter 2 Alcohols

2.1 List of Chemicals and Data Compilations

Common Name: Methanol
Synonym: methyl alcohol, carbinol, wood alcohol, wood spirit
CAS Registry No: 67-56-1
Molecular Formula: CH_3OH
Molecular Weight: 32.04
Melting Point (°C):
-97.8 (Stull 1947; Howard 1990)
-93.9 (Weast 1982-83)
-98.0 (Verschueren 1983)
Boiling Point (°C):
64.7 (Stull 1947; Howard 1990)
65.0 (Weast 1982-83; Verschueren 1983; Banerjee et al. 1990)
Density (g/cm^3 at 20°C):
0.7914 (Weast 1982-83)
0.7960 (15°C, Verschueren 1983)
Molar Volume (cm^3/mol):
42.5 (exptl. at normal B.P., Lee et al. 1972; quoted, Reid et al. 1977)
40.6 (calculated-density, Rohrschneider 1973)
37.0 (calculated-LeBas method, Reid et al. 1977)
40.5 (Taft et al. 1985; Kamlet et al. 1986; Leahy 1986)
20.5 (intrinsic volume, calculated, Leahy 1986; Kamlet et al. 1987)
33.7 (calculated-density, Jaworska & Schultz 1993)
Molecular Volume (Å^3):
Total Surface Area, TSA (Å^2):
136.2 (Iwase et al. 1985)
Heat of Fusion ΔH_{fus}, kcal/mol:
Entropy of Fusion ΔS_{fus}, cal/mol·K (e.u.):
Fugacity Ratio at 25°C (assuming ΔS_{fus} = 13.5 e.u.), F: 1.0

Water Solubility (g/m^3 or mg/L at 25°C):
miscible (Dean 1985; Howard 1990; Pinal et al. 1991)
1163300 (calculated-solvatochromic p., Taft et al. 1985)
1163300, 882455 (quoted, calculated-solvatochromic p., Leahy 1986)
1163300, 967592 (quoted, calculated-solvatochromic p., Kamlet et al. 1987)
1196740, 3556313 (quoted, calculated- χ , Nirmalakhandan & Speece 1988)
945180 (calculated as per Micro QSAR 2.0, Schultz et al. 1990)
miscible (Yaws et al. 1990)

Vapor Pressure (Pa at 25°C):
16210 (interpolated-regression of tabulated data, Stull 1947)
13330 (21.2°C, Stull 1947)
12264 (Weber et al. 1981; quoted, Howard 1990)
12264 (20°C, quoted, Verschueren 1983)
21328 (30°C, quoted, Verschueren 1983)

16942 (quoted, Abraham 1984)
16958, 16958 (quoted exptl., calculated-Antoine eqn., Boublik et al. 1984)
16669 (calculated-Antoine eqn., Dean 1985)
16930 (Howard et al. 1986; quoted, Banerjee et al. 1990)
16930, 169290 (quoted, calculated-solvatochromic p., Banerjee et al. 1990)

Henry's Law Constant (Pa·m³/mol):
0.446 (partial pressure-isotensiscope, Butler et al. 1935; quoted, Buttery et al. 1971)
0.444 (entrainment method-GC, Burnett 1963)
0.472 (quoted exptl., Hine & Mookerjee 1975)
0.367, 0.319 (calculated-group contribution, bond contribution, Hine & Mookerjee 1975)
0.450 (headspace-GC, Snider & Dawson 1985)
13.68 (quoted from Snider & Dawson 1985; Howard 1990)
0.461 (quoted, Gaffney et al. 1987)
0.472, 0.555 (quoted, calculated- χ , Nirmalakhandan & Speece 1988)
0.704 (computed, Yaws et al. 1991)

Octanol/Water Partition Coefficient, log K_{OW}:
-0.82 (shake flask-CR, Collander 1951; quoted, Leo et al. 1971; Rekker 1977; Sangster 1989)
-1.27 (calculated, Iwasa et al. 1965)
-0.66 (shake flask-GC, Hansch & Anderson 1967; quoted, Leo et al. 1969,71; Hansch & Dunn III 1972; Rekker 1977; Veith et al. 1983)
-0.66 (Tute 1971; quoted, Fujisawa & Masuhara 1981)
-0.77 ± 0.02 (shake flask-GC, Leo et al. 1975)
-0.74, -0.73 (calculated-f const., Rekker 1977)
-0.66 (calculated-π const., Rekker 1977)
-0.77 (Hansch & Leo 1979; quoted, Cornford 1982)
-0.52 (shake flask-RC, Cornford 1982; quoted, Sangster 1989)
-0.79 (calculated-f const., Hermens & Leeuwangh 1982)
-0.82 (Verschueren 1983; quoted, Pinal et al. 1991)
-0.66 (quoted, Verschueren 1983)
-0.70 (shake flask, OECD 1981 Guidelines, Geyer et al. 1984)
-0.77 (Hansch & Leo 1985; quoted, Howard 1990)
-0.64, -0.63 (quoted, calculated-TSA, Iwase et al. 1985)
-0.25 (computed, Klopman et al. 1985; quoted, Pangrekar et al. 1994)
-0.65 -0.49 (quoted, calculated-V_I and solvatochromic p., Taft et al. 1985)
-0.70 -0.49 (quoted, calculated-V_I and solvatochromic p., Leahy 1986)
-0.70 -0.61 (quoted, calculated-V_I and solvatochromic p., Kamlet et al. 1986)
-0.71 (shake flask-GC at pH 7.0, Riebesehl & Tomlinson 1986; quoted, Sangster 1989)
-0.73 (quoted, Gerstl & Helling 1987)
-0.77 (measured value, GEMS 1986; quoted, Vaishnav et al. 1987)
-0.71 (quoted from Hansch & Leo 1979, Bodor et al. 1989)
-0.69 (calculated-M.O. indices, Bodor et al. 1989)

-0.74 (quoted from CLOGP, Bodor et al. 1989)
-0.764 (calculated-CLOGP, Bodor et al. 1989)
-0.77 (quoted from Pomona College unpublished data, Sangster 1989)
-0.74 (recommended, Sangster 1989)
-0.77 (quoted measured value, Schultz et al. 1990)
-0.77 (quoted-CLOGP 3.15, Jawoska & Schultz 1993)
-0.764 (calculated-QSAR, Kollig 1993)

Bioconcentration Factor, log BCF:
-0.70 (calculated-K_{OW}, Lyman et al. 1982; quoted, Howard 1990)
4.45 (alga *chlorella fusca*, wet wt. basis, Geyer et al. 1984)
4.45 (alga *chlorella fusca*, calculated-K_{OW}, Geyer et al. 1984)
<1.0 (golden ide, after 3 days, Freitag et al. 1985; quoted, Howard 1990)
4.46 (algae, after 1 day, Freitag et al. 1985)
2.67 (activated sludge, after 5 days, Freitag et al. 1985)
-2.09 (calculated as per Mackay 1982, Schultz 1990)

Sorption Partition Coefficient, log K_{OC}:
0.44, -0.23 (quoted, calculated- χ , Gerstl & Helling 1987)
0.44 (soil, quoted exptl., Meylan et al. 1992)
-0.36 (soil, calculated- χ & fragment contribution, Meylan et al. 1992)
-1.08 (calculated-K_{OW}, Kollig 1993)

Half-Lives in the Environment:
Air: half-life of >9.9 days for the gas-phase reaction with hydroxyl radicals in air, based on the rate of disappearance of hydrocarbon due to reaction with hydroxyl radicals (Darnall et al. 1976); estimated half-life of 17.8 days in ambient atmosphere (Howard 1990); atmospheric transformation lifetime was estimated to be 1 to 5 days (Kelly et al. 1994).
Surface water:
Groundwater:
Sediment:
Soil:
Biota:

Environmental Fate Rate Constants or Half-Lives:
Volatilization: estimated half-lives of 5.3 hours and 2.6 days for a model river 1 m deep and an environmental pond (Lyman et al. 1982; selected, Howard 1990).
Photolysis:
Oxidation: photooxidation half-life of >9.9 days for the gas-phase reaction with hydroxyl radicals in air, based on the rate of disapperance of hydrocarbon due to reaction with hydroxyl radicals (Darnall et al. 1976); atmospheric second-

order rate constant of $(5.7 \pm 0.6) \times 10^8$ $L \cdot mol^{-1} \cdot sec^{-1}$ for reaction with hydroxyl radicals at 292 K (Campbell et al. 1976); rate constant of $(1.06 \pm 0.10) \times 10^{-12}$ $cm^3 \cdot molecule^{-1} \cdot sec^{-1}$ for the gas-phase reaction with OH radicals at 296 $\pm$ 2 K in air (Overend & Paraskevopoulos 1978; quoted, Atkinson 1985); rate constant 5.7×10^{-11} $cm^3 \cdot mol^{-1} \cdot sec^{-1}$ for reaction with OH radical at 300 K (Lyman 1982); rate constants of $(1.00 \pm 0.10) \times 10^{-12}$ $cm^3 \cdot molecule^{-1} \cdot sec^{-1}$ (Ravishankara & Davis 1978) and 0.76×10^{-12} $cm^3 \cdot molecule^{-1} \cdot sec^{-1}$ for the gas-phase reaction with OH radicals at 296 K in air (Meier et al. 1985; quoted, Atkinson 1985); rate constant of 0.80×10^{-12} $cm^3 \cdot molecule^{-1} \cdot sec^{-1}$ for the gas-phase reaction with OH radicals at 298 K in the atmosphere (Zellner & Lorenz 1984; quoted, Carlier et al. 1986); second-order rate constant of $< 6.0 \times 10^{16}$ $cm^3 \cdot molecule^{-1} \cdot sec^{-1}$ for the gas-phase reaction with NO_3 radicals by flash photolysis-absortion technique at 298 $\pm$ 2 K in air (Wallington et al. 1987; selected, Atkinson 1991); exptl. rate constant of 0.9×10^{-12} $cm^3 \cdot molecule^{-1} \cdot sec^{-1}$ for reactions with OH radicals in air at room temperature (Atkinson 1987); exptl rate constant of 8.61×10^{-13} $cm^3 \cdot molecule^{-1} \cdot sec^{-1}$ for the gas-phase reaction with OH radicals in air (Wallington et al. 1988a,b) and 1.60×10^{-12} $cm^3 \cdot molecule^{-1} \cdot sec^{-1}$ for the solution-phase reaction with OH radicals in aqueous solution (Wallington & Kurylo 1987; quoted, Wallington et al. 1988b); 932×10^{-12} $cm^3 \cdot molecule^{-1} \cdot sec^{-1}$, average reviewed value of measured rate constant for the gas-phase reaction with OH radicals at 298 K in air (Atkinson 1990).

Hydrolysis:

Biodegradation:

Biotransformation:

Bioconcentration, Uptake (k_1) and Elimination (k_2) Rate Constants:

Common Name: Ethanol
Synonym: ethyl alcohol, methylcarbinol
Chemical Name: ethanol, ethyl alcohol
CAS Registry No: 64-17-5
Molecular Formula: CH_3CH_2OH
Molecular Weight: 46.07
Melting Point (°C):
-112.0 (Stull 1947)
-117.3 (Weast 1982-83)
-114/-117 (Verschueren 1983)
-114.0 (Dean 1985)
-114.5 (Riddick et al. 1986)
-114.1 (Howard 1990)
Boiling Point (°C):
78.4 (Stull 1947; Verschueren 1983)
78.5 (Weast 1982-83; Howard 1990)
78.3 (Dean 1985; Riddick et al. 1986)
78.0 (Banerjee et al. 1990)
Density (g/cm^3 at 20°C):
0.7893 (Weast 1982-83)
0.7890 (Verschueren 1983)
0.7894 (Dean 1985)
0.7892 (Riddick et al. 1986)
Molar Volume (cm^3/mol):
58.6 (calculated-density, Rohrschneider 1973)
58.4 (Taft et al. 1985; Kamlet et al. 1986; Leahy 1986; Taft et al. 1986)
30.5 (intrinsic volume, calculated, Leahy 1986; Kamlet et al. 1987)
50.2 (calculated-density, Jawoska & Schultz 1993)
59.2 (calculated-LeBas method)
Molecular Volume ($Å^3$):
Dissociation Constant, pK_a:
15.9 (Riddick et al. 1986)
Total Surface Area, TSA ($Å^2$):
163.8 (Iwase et al. 1985)
136.0 (Funasaki et al. 1986)
Heat of Fusion ΔH_{fus}, kcal/mol:
1.20 (Riddick et al. 1986)
Entropy of Fusion ΔS_{fus}, cal/mol·K (e.u.):
Fugacity Ratio at 25°C (assuming ΔS_{fus} = 13.5 e.u.), F: 1.0

Water Solubility (g/m^3 or mg/L at 25°C):
miscible (Dean 1985)
580000, 593500 (quoted, calculated-solvatochromic p., Kamlet et al. 1986)
580000, 529000 (quoted, calculated-solvatochromic p., Leahy 1986)
infinite (Riddick et al. 1986; quoted, Howard 1990)
580000, 528850 (quoted, calculated-solvatochromic p., Kamlet et al. 1987)

576700, 1127200 (quoted, calculated- χ , Nirmalakhandan & Speece 1988)
341400 (calculated as per Micro QSAR 2.0, Schultz et al. 1990)
miscible (Yaws et al. 1990)

Vapor Pressure (Pa at 25°C):
7538 (interpolated-regression tabulated data, Stull 1947)
7998 (26°C, Stull 1947)
6665 (quoted, Verschueren 1983)
5852 (20°C, quoted, Verschueren 1983)
9998 (30°C, quoted, Verschueren 1983)
7863 (quoted, Abraham 1984)
7870, 8060 (quoted exptl., calculated-Antoine eqn., Boublik et al. 1984)
7968 (calculated-Antoine eqn., Dean 1985)
7905 (Howard et al. 1986; quoted, Banerjee et al. 1990)
7870 (Riddick et al. 1986; quoted, Howard 1990)
8010 (calculated-solvatochromic p., Banerjee et al. 1990)

Henry's Law Constant (Pa·m^3/mol):
0.520 (partial pressure-isotensiscope, Butler et al. 1935; quoted, Buttery et al. 1971)
0.0118 (partial vapor pressure, concn.-GC, Burnett & Swobada 1962)
0.466 (entrainment method-GC, Burnett 1963)
0.637 (exptl., Hine & Mookerjee 1975; quoted, Howard 1990)
0.495 (calculated-group contribution, Hine & Mookerjee 1975)
0.472 (calculated-bond contribution, Hine & Mookerjee 1975)
0.527 (headspace-GC, Snider & Dawson 1985)
0.507 (quoted, Gaffney et al. 1987)
0.637, 0.637 (quoted, calculated-χ , Nirmalakhandan & Speece 1988)
0.823 (computed, Yaws et al. 1991)
0.637, 0.70 (quoted, correlated-molecular structure, Russell et al. 1992)

Octanol/Water Partition Coefficient, log K_{OW}:
-0.32 (shake flask-CR, Collander 1951; quoted, Leo et al. 1971; Rekker 1977; Funasaki et al. 1986; Sangster 1989)
-0.16 (Leo et al. 1969)
-0.16 (Hansch & Dunn III 1972)
-0.30 (shake flask-GC, Dillingham et al. 1973; quoted, Sangster 1989)
-0.75 (calculated, Iwasa et al. 1965)
-0.31 $\pm$ 0.02 (shake flask-GC, Leo et al. 1975)
-0.22, -0.20 (calculated-f const., Rekker 1977)
-0.16 (calculated-π const., Rekker 1977)
-0.31, -0.21 (quoted, calculated-π const., Chou & Jurs 1979)
-0.31 (Hansch & Leo 1979; quoted, Cornford 1982)
-0.26 (calculated-f const., Könemann 1981; quoted, Verhaar et al. 1992)
-0.18 (shake flask-RC, Cornford 1982; quoted, Sangster 1989)
-0.16 (quoted, Veith et al. 1983)
-0.32 (quoted, Verschueren 1983)
-0.31 (Hansch & Leo 1985; quoted, Howard 1990)

-0.30, -0.11 (quoted, calculated-TSA, Iwase et al. 1985)
0.07 (computed, Klopman et al. 1985; quoted, Pangrekar et al. 1994)
-0.30, -0.16 (quoted, calculated-solvatochromic p., Taft et al. 1985)
-0.25, -0.25 (quoted, calculated-solvatochromic p., Kamlet et al. 1986)
-0.25, -0.12 (quoted, calculated-solvatochromic p., Leahy 1986)
-0.20 (shake flask-GC at pH 7, Riebesehl & Tomlinson 1986; quoted, Sangster 1989)
-0.31 (measured, GEMS 1986; quoted, Vaishnav et al. 1987)
-0.31 (quoted, Gerstl & Helling 1987)
-0.28, -0.31, -0.235, -0.22 (quoted from Hansch & Leo 1979, quoted-CLOGP, calculated-CLOGP, calculated-M.O. indices, Bodor et al. 1989)
-0.31 (quoted from Pomona College unpublished data, Sangster 1989)
-0.30 (recommended, Sangster 1989)
-0.31 (quoted measured value, Schultz et al. 1990)
-0.31 (quoted, Verhaar et al. 1992)
-0.21 (calculated as per Hansch & Leo 1979, Dallos et al. 1993)
-0.29 (calculated- γ from UNIFAC, Dallos et al. 1993; quoted, Sangster 1993)
-0.31 (quoted from CLOGP 3.15, Jawoska & Schultz 1993)

Bioconcentration Factor, log BCF:
-1.63 (calculated as per Mackay 1982, Schultz et al. 1990)

Sorption Partition Coefficient, log K_{OC}:
0.20, 0.09 (quoted, calculated- χ , Gerstl & Helling 1987)
0.20 (soil, quoted exptl., Meylan et al. 1992)
-0.14 (soil, calculated- χ and fragment contribution, Meylan et al. 1992)

Half-Lives in the Environment:

Air: half-life of 0.24-2.4 hours in air for the gas-phase reaction with hydroxyl radicals, based on the rate of disappearance of hydrocarbon due to reaction with hydroxyl radicals (Darnall et al. 1976); estimated half-life in the atmosphere ranges from 5.9 days (Graedel 1978; quoted, Howard 1990) to 4.0 days, based on the reaction with a hydroxyl radical concentration of 8×10^5 molecules/cm^3 (Campbell 1976; Lyman et al. 1982; quoted, Howard 1990); photooxidation half-life of 12.2-122 hours, based on measured rate constant for the reaction with hydroxyl radicals in air (Atkinson 1987; selected, Howard et al. 1991).

Surface water: photooxidation half-life of 334 days-36.6 years, based on measured rate constant for the reaction with hydroxyl radicals in water (Anbar & Neta 1967; selected, Howard et al. 1991).

Groundwater: 48-336 hours, based on estimated unacclimated aqueous aerobic biodegradation half-life (Howard et al. 1991).

Sediment:

Soil: 24-168 hours, based on estimated unacclimated aqueous aerobic biodegradation half-life (Howard et al. 1991).

Biota:

Environmental Fate Rate Constants or Half-Lives:

Volatilization: estimated half-life of 6 days from water (Howard 1990).

Photolysis:

Oxidation: photooxidation half-life of 334 days to 36.6 years in water, based on measured rate constant for the reaction with hydroxyl radicals in water (Anbar & Neta 1967; selected, Howard et al. 1991); rate constant of 0.01 M^{-1} sec^{-1} for oxidation by RO_2 radicals at 30°C in aquatic systems with half-life of $8x10^5$ days (Howard 1972; Hendry et al. 1974; quoted, Mill 1982); photooxidation half-life of 0.24-2.4 hours in air for the gas-phase reaction with hydroxyl radicals, based on the rate of disappearance of hydrocarbon due to reaction with hydroxyl radicals (Darnall et al. 1976); atmospheric second-order rate constant of $(1.8\pm0.2)x10^9$ $L\cdot mol^{-1}\cdot sec^{-1}$ for reaction with OH radicals at 292 K (Campbell et al. 1976); rate constant of $<2x10^2$ M^{-1} sec^{-1} for oxidation by singlet oxygen in aquatic systems at 25°C with half-life >100 years (Foote 1976; Mill 1979; quoted, Mill 1982) and rate constant of $2x10^{-9}$ M^{-1} sec^{-1} for oxidation by OH radicals at 25°C in the atmosphere with half-life of 2.8 days (Hendry & Kenley 1979; quoted, Mill 1982); rate constants of $(3.74 \pm 0.37)x10^{-12}$ cm^3 $molecule^{-1}$ sec^{-1} at 296 ± 2 K (Overend & Paraskevopoulos 1978); $1.8x10^{12}$ $cm^3\cdot mol^{-1}\cdot sec^{-1}$ for reaction with OH radical at 300 K (Lyman et al. 1982); $(2.62 \pm 0.36)x10^{-12}$ cm^3 $molecule^{-1}$ sec^{-1} (Ravishankara & Davis 1978) and $1.75x10^{-12}$ cm^3 $molecule^{-1}$ sec^{-1} at 298 K (Meier et al. 1985) for the gas-phase reaction with OH radicals in air (reviewed & quoted, Atkinson 1985); rate constant of $2.9x10^{-12}$ cm^3 $molecule^{-1}$ sec^{-1} at 298 K for the gas-phase reaction with OH radicals in the atmosphere (Zellner & Lorenz 1984; quoted, Carlier et al. 1986); rate constants of $2.9x10^{-12}$ cm^3 $molecule^{-1}$ sec^{-1} (exptl.) and $3.1x10^{-12}$ cm^3 $molecule^{-1}$ sec^{-1} (calculated) for reaction with OH radicals in the gas phase at room temperature (Atkinson 1987); rate constants of $2.0x10^{-12}$ rate constant of $<9.0x10^{-16}$ cm^3 $molecule^{-1}$ sec^{-1} for the gas-phase reaction with NO_3 radicals by flash photolysis-absorption technique at 298 ± 2 K in air (Wallington et al. 1987; quoted, Atkinson 1991); experimental rate constant of $3.33x10^{-12}$ cm^3 $molecule^{-1}$ sec^{-1} for the gas-phase reaction with hydroxyl radicals in air (Wallington et al. 1988a,b) and $3.20x10^{-12}$ cm^3 $molecule^{-1}$ sec^{-1} for the solution-phase reaction with hydroxyl radicals in aqueous solution (Wallington et al. 1988b); rate constant of an average value $3.27x10^{-12}$ cm^3 $molecule^{-1}$ sec^{-1} for the gas-phase reaction with hydroxyl radicals in air at 298 K (reviewed by Atkinson 1990); photooxidation half-life of 12.2-122 hours in air, based on measured rate constant for the reaction with hydroxyl radicals in air (Howard et al. 1991).

Hydrolysis:

Biodegradation: aqueous aerobic half-life of 24-168 hours, based on estimated unacclimated aqueous aerobic biodegradation screening test data (Heukelekian & Rand 1955; Malaney & Gerhold 1969; selected, Howard et al. 1991); aqueous anaerobic half-life of 96-672 hours, based on estimated unacclimated aqueous aerobic biodegradation half-life (Howard et al. 1991); biodegradation rate constants of 0.0.43-0.055 h^{-1} in 30 mg/L activated sludge after a time lag of 5-10 hours (Urano & Kato 1986).

Biotransformation:

Bioconcentration, Uptake (k_1) and Elimination (k_2) Rate Constants:

Common Name: Propanol
Synonym: propyl alcohol, 1-propanol, *n*-propyl alcohol
Chemical Name: propanol, propyl alcohol, *n*-propyl alcohol
CAS Registry No: 71-23-8
Molecular Formula: $CH_3CH_2CH_2OH$
Molecular Weight: 60.11
Melting Point (°C):
- -127.0 (Stull 1947; Verschueren 1983)
- -126.5 (Weast 1982-83)
- -126.2 (Dean 1985; Riddick et al. 1986)

Boiling Point (°C):
- 97.8 (Stull 1947; Verschueren 1983)
- 97.3 (Kemme & Kreps 1969)
- 97.4 (Weast 1982-83; Dean 1985)
- 97.2 (Boublik et al. 1984; Dean 1985; Riddick et al. 1986)

Density (g/cm^3 at 20°C):
- 0.8035 (Weast 1982-83)
- 0.8040 (Verschueren 1983)
- 0.8037 (Dean 1985)
- 0.8036 (Riddick et al. 1986)

Molar Volume (cm^3/mol):
- 81.8 (exptl. at normal B.P., Lee et al. 1972; quoted, Reid et al. 1977)
- 75.1 (calculated-density, Rohrschneider 1973)
- 77.7 (calculated-LeBas method, Reid et al. 1977)
- 74.8 (Taft et al. 1985; Kamlet et al. 1986; Leahy 1986)
- 40.2 (intrinsic volume, calculated, Leahy 1986; Kamlet et al. 1986,87)
- 75.7 (Kamlet et al. 1987)
- 81.0 (LeBas method, Abernethy et al. 1988)
- 40.0 (intrinsic volume, Abernethy et al. 1988)
- 67.3 (calculated-density, Jawoska & Schultz 1993)

Molecular Volume (Å^3):
Dissociation Constant, pK:
- 19.4 (pK_s, Riddick et al. 1986)

Total Surface Area, TSA (Å^2):
- 190.0 (Iwase et al. 1985)
- 160.0 (Funasaki et al. 1986)

Heat of Fusion ΔH_{fus}, kcal/mol:
- 1.284 (Riddick et al. 1986)

Entropy of Fusion ΔS_{fus}, cal/mol·K (e.u.):
Fugacity Ratio at 25°C (assuming ΔS_{fus} = 13.5 e.u.), F: 1.0

Water Solubility (g/m^3 or mg/L at 25°C):
- $>3.1 \times 10^6$ (Booth & Everson 1948)
- miscible (Dean 1985)
- 250580, 190085 (quoted, calculated-solvatochromic p., Leahy 1986)
- infinite (Riddick et al. 1986)

250580, 185758 (quoted, calculated-solvatochromic p., Kamlet et al. 1987)
122624 (calculated as per Micro QSAR 2.0, Schultz et al. 1990)
miscible (Yaws et al. 1990)

Vapor Pressure (Pa at 25°C):
2594 (interpolated-regression of tabulated data, Stull 1947)
2666 (25.3°C, Stull 1947)
1986 (19.8°C, DTA-thermocouple, Kemme & Kreps 1969)
2720 (24.5°C, DTA-thermocouple, Kemme & Kreps 1969)
3920 (30.3°C, DTA-thermocouple, Kemme & Kreps 1969)
2773 (quoted, Verschueren 1983)
2675 (quoted, Abraham 1984)
2744 (extrapolated-Antoine eqn., Boublik et al. 1984)
2780 (calculated-Antoine eqn., Dean 1985)
2798 (Riddick et al. 1986; quoted, Howard 1990)

Henry's Law Constant (Pa·m^3/mol):
0.694 (partial pressure-isotensiscope, Butler et al. 1935; quoted, Buttery et al. 1971)
0.620 (entrainment method-GC, Burnett 1963)
0.683 (exptl., Hine & Mookerjee 1975; quoted, Howard 1990)
0.699 (calculated-group contribution, Hine & Mookerjee 1975)
0.710 (calculated-bond contribution, Hine & Mookerjee 1975)
0.751 (headspace-GC, Snider & Dawson 1985)
0.925 (computed, Yaws et al. 1991)
0.683, 0.942 (quoted, correlated-molecular structure, Russell et al. 1992)

Octanol/Water Partition Coefficient, log K_{OW}:
-0.23 (calculated, Iwasa et al. 1965)
0.34 (shake flask-GC, Hansch & Anderson 1967; quoted, Leo et al. 1969,1971; Hansch & Dunn III 1972; Rekker 1977; Funasaki et al. 1986)
0.34 (Tute 1971; quoted, Fujisawa & Masuhara 1981)
0.30 (shake flask-GC, Dillingham et al. 1973; quoted, Sangster 1989)
0.25 $\pm$ 0.01 (shake flask-GC, Leo et al. 1975)
0.31, 0.33 (calculated-f const., Rekker 1977)
0.34 (calculated-π, Rekker 1977)
0.27 (calculated by Rekker's method, Hanai et al. 1981)
0.34 (quoted, Verschueren 1983)
0.30, 0.39 (quoted, calculated-TSA, Iwase et al. 1985)
0.30 (measured value, GEMS 1986; quoted, Vaishnav et al. 1987)
0.28, 0.30 (quoted, calculated-solvatochromic p., Taft 1985)
0.28, 0.21 (quoted, calculated-solvatochromic p., Kamlet et al. 1986)
0.28, 0.38 (quoted, calculated-solvatochromic p., Leahy 1986)
0.32 (shake flask-GC at pH 7, Riebesehl & Tomlinson 1986; quoted, Sangster 1989)
0.28 (quoted, Gerstl & Helling 1987)

0.30 (quoted, Abernethy et al. 1988)
0.28 (quoted from Hansch & Leo 1979, Bodor et al. 1989)
0.25 (quoted from CLOGP, Bodor et al. 1989)
0.29 (calculated-MO, Bodor et al. 1989)
0.294 (calculated-CLOGP, Bodor et al. 1989)
0.25 (quoted from Pomona College unpublished data, Sangster 1989)
0.25 (recommended, Sangster 1989)
0.25 (quoted measured value, Schultz et al. 1990)
0.25 (quoted from CLOGP 3.15, Jaworska & Schultz 1993)

Bioconcentration Factor, log BCF:
-1.069 (calculated as per Mackay 1982, Schultz et al. 1990)

Sorption Partition Coefficient, log K_{OC}:
0.48, 0.37 (quoted, calculated- χ , Gerstl & Helling 1987)
0.48 (soil, quoted exptl., Meylan et al. 1992)
0.12 (soil, calculated- χ & fragment contribution, Meylan et al. 1992)

Half-Lives in the Environment:
Air: half-life of 0.24-2.4 hours for the gas-phase reaction with hydroxyl radicals in air, based on the rate of disappearance of hydrocarbon due to reaction with hydroxyl radicals (Darnall et al. 1976).
Surface water:
Groundwater:
Sediment:
Soil:
Biota:

Environmental Fate Rate Constants or Half-Lives:
Volatilization:
Photolysis:
Hydrolysis:
Oxidation: photooxidation half-life of 0.24-2.4 hours for the gas-phase reaction with hydroxyl radicals in air, based on the rate of disappearance of hydrocarbon due to reaction with hydroxyl radicals (Darnall et al. 1976); atmospheric second-order rate constant of $(2.3 \pm 0.2)x10^9$ $L \cdot mol^{-1} \cdot sec^{-1}$ for reaction with OH radicals at 292 K (Campbell et al. 1976); rate constant of $<2x10^2$ M^{-1} sec^{-1} for the oxidation by singlet oxygen at 25°C in aquatic systems with half-life >100 years (Foote 1976; Mill 1979; quoted, Mill 1982) and $2x10^9$ M^{-1} sec^{-1} for the oxidation by OH radicals at 25°C in the atmosphere with half-life of 2.8 days (Hendry & Kenley 1979; quoted, Mill 1982); measured rate constant of $(5.33 \pm 0.53)x10^{-12}$ $cm^3 \cdot molecule^{-1} \cdot sec^{-1}$ for the gas phase reaction with OH radicals at 296 ± 2 K in air (Overend & Paraskevopoulos 1978; quoted, Atkinson 1985);

experimental rate constant of 0.9×10^{-12} $cm^3 \cdot molecule^{-1} \cdot sec^{-1}$ and the calculated rate constant of 5.0 $cm^3 \cdot molecule^{-1} \cdot sec^{-1}$ for reactions with OH radicals at room temperature (Atkinson 1987); experimental rate constant of 5.34×10^{-12} $cm^3 \cdot molecule^{-1}$ sec^{-1} for the gas-phase reaction with OH radicals in air and 4.70×10^{-12} $cm^3 \cdot molecule^{-1} \cdot sec^{-1}$ for the solution-phase reaction with OH radicals in aqueous solution (Wallington & Kurylo 1987; quoted, Wallington et al. 1988); experimental rate constant of 0.9×10^{-12} $cm^3 \cdot molecule^{-1} \cdot sec^{-1}$ for the gas phase reaction with OH radicals at 298 K in air (Atkinson 1990).

Biodegradation: average rate of biodegradation 71.0 mg COD $g^{-1} \cdot h^{-1}$ based on measurements of COD decrease using activated sludge inoculum with 20 days of adaptation to the substrate (Pitter 1976; quoted, Scow 1982).

Biotransformation:

Bioconcentration, Uptake (k_1) and Elimination (k_2) Rate Constants:

Common Name: Isopropanol
Synonym: isopropyl alcohol, 2-propanol, *i*-propyl alcohol, dimethylcarbinol, *sec*-propylalcohol, perspirit, petrohol, avantine, IPA
Chemical Name: isopropanol, isopropyl alcohol, *i*-propyl alcohol
CAS Registry No: 67-63-0
Molecular Formula: $CH_3(CH_3)CHOH$
Molecular Weight: 60.11
Melting Point (°C):
-85.8 (Stull 1947)
-89.5 (Weast 1982-83; Dean 1985)
-86/-89 (Verschueren 1983)
-88.5 (Howard 1990)
Boiling Point (°C):
82.4 (Butler et al. 1935; Weast 1982-83; Verschueren 1983; Dean 1985)
82.5 (Stull 1947; Howard 1990)
Density (g/cm^3 at 20°C):
0.7812 (25°C, Butler et al. 1935)
0.7850 (Verschueren 1983)
0.7855 (Weast 1982-83; Dean 1985)
Molar Volume (cm^3/mol):
76.8 (calculated-density, Rohrschneider 1973)
76.5 (Taft et al. 1985; Kamlet et al. 1986)
81.0 (LeBas method, Abernethy & Mackay 1988)
40.1 (intrinsic volume, Kamlet et al. 1987)
40.0 (intrinsic volume, Abernethy et al. 1988)
67.5 (calculated-density, Jawoska & Schultz 1993)
Molecular Volume (Å^3):
Dissociation Constant, pK_a:
17.1 (Serjeant & Dempsey 1979; Howard 1990)
Total Surface Area, TSA (Å^2):
188.4 (Iwase et al. 1985)
156.0 (Funasaki et al. 1986)
Heat of Fusion ΔH_{fus}, kcal/mol:
1.292 (Riddick et al. 1986)
Entropy of Fusion ΔS_{fus}, cal/mol·K (e.u.):
Fugacity Ratio at 25°C (assuming ΔS_{fus} = 13.5 e.u.), F: 1.0

Water Solubility (g/m^3 or mg/L at 25°C):
$>3.1\times10^6$ (Booth & Everson 1948)
miscible (Dean 1985; Howard 1990; Pinal et al. 1991)
445600 (calculated-solvatochromic p., Taft et al. 1985)
infinite (Riddick et al. 1986)
406393, 388103 (quoted, calculated-solvatochromic p., Kamlet et al. 1987)

Vapor Pressure (Pa at 25°C):

5699 (interpolated-regression of tabulated data, Stull 1947)
5332 (23.8°C, Stull 1947)
4266 (20°C, quoted, Verschueren 1983)
7598 (30°C, quoted, Verschueren 1983)
5070 (extrapolated-Antoine eqn., Boublik et al. 1984)
5700 (extrapolated-Antoine eqn., Boublik et al. 1984)
6021 (calculated-Antoine eqn., Dean 1985)
5775 (Riddick et al. 1986; quoted, Howard 1990)

Henry's Law Constant (Pa·m^3/mol):

0.815 (partial pressure-isotensiscope, Butler et al. 1935)
0.820 (quoted exptl., Hine & Mookerjee 1975; quoted, Howard 1990)
1.159 (calculated-group contribution, Hine & Mookerjee 1975)
0.710 (calculated-bond contribution, Hine & Mookerjee 1975)
1.131 (computed, Yaws et al. 1991)
0.820, 0.715 (quoted, correlated-molecular structure, Russell et al. 1992)

Octanol/Water Partition Coefficient, log K_{OW}:

-0.36 (calculated, Iwasa et al. 1965)
0.14 (Leo et al. 1969)
0.14 (Hansch & Dunn III 1972)
0.05 (shake flask-GC, Dillingham et al. 1973; quoted, Sangster 1989)
0.05, 0.11 (quoted, calculated-f const., Chou & Jurs 1979)
0.15 (calculated-f const., Könemann 1981)
0.14 (quoted, Veith et al. 1983)
-0.16 (Verschueren 1983; quoted, Pinal et al. 1991)
0.28 (calculated, Verschueren 1983)
0.05 (Hansch & Leo 1985; quoted, Howard 1990)
0.05, 0.36 (quoted, calculated-TSA, Iwase et al. 1985)
0.05, 0.13 (quoted, calculated-solvatochromic p., Taft et al. 1985)
0.14 (HPLC-k', Funasaki et al. 1986; quoted, Sangster 1989)
0.05, 0.09 (quoted, calculated-solvatochromic p., Kamlet et al. 1986)
0.14 (quoted, Purdy 1987; Abernethy et al. 1988)
0.05 (measured, GEMS 1986; quoted, Vaishnav et al. 1987)
0.05, 0.55 (quoted, calculated-γ from UNIFAC, Banerjee & Howard 1988)
0.05 (recommended, Sangster 1989)
0.05 (quoted, Verhaar et al. 1992)
0.05, 0.4463 (quoted exptl., calculated-original UNIFAC, Chen et al. 1993)
0.05 (quoted from CLOGP 3.15, Jawoska & Schultz 1993)

Bioconcentration Factor, log BCF:

Sorption Partition Coefficient, log K_{OC}:

Half-Lives in the Environment:

Air: photooxidation half-life of 0.24-2.4 hours in air for the gas-phase reaction with hydroxyl radicals, based on the rate of disappearance of hydrocarbon due to reaction with hydroxyl radicals (Darnall et al. 1976); photooxidation half-life of 6.2-72 hours, based on measured rate constant for the reaction with hydroxyl radicals in air (Atkinson 1985, 1987; quoted, Howard 1990; selected, Howard et al. 1991).

Surface water: photooxidation half-life of 197 days to 22 years, based on measured rate constant for the reaction with hydroxyl radicals in water (Dorfman & Adams 1973; selected, Howard et al. 1991); 26-168 hours, based on estimated unacclimated aerobic aqueous biodegradation half-life (Howard et al. 1991).

Groundwater: 48-336 hours, based on estimated unacclimated aerobic aqueous biodegradation half-life (Howard et al. 1991).

Sediment:

Soil: 24-168 hours, based on estimated unacclimated aerobic aqueous biodegradation half-life (Howard et al. 1991).

Biota:

Environmental Fate Rate Constants or Half-Lives:

Volatilization: the estimated half-life for evaporation from water 1 m deep with 1 m/sec current and 3 m/sec wind is 3.6 days (Lyman et al. 1982; quoted, Howard 1990).

Photolysis:

Oxidation: photooxidation half-life of 197 days to 22 years in water, based on measured rate constant for the reaction with hydroxyl radicals in water (Dorfman & Adams 1973; selected, Howard et al. 1991); photooxidation half-life of 0.24-2.4 hours in air for the gas-phase reaction with hydroxyl radicals, based on the rate of disappearance of hydrocarbon due to reaction with hydroxyl radicals (Darnall et al. 1976); rate constant of $<2\times10^{2}$ M^{-1} sec^{-1} for oxidation by singlet oxygen at 25°C in aquatic systems with half-life >100 years (Foote 1976; Mill 1979; quoted, Mill 1982) and 4.1×10^{9} M^{-1} sec^{-1} for oxidation by OH radicals at 25°C in the atmosphere with half-life of 1.3 day (Hendry & Kenley 1979; quoted, Mill 1982); measured rate constant of $(5.48 \pm 0.55)\times10^{-12}$ cm^3 $molecule^{-1}$ sec^{-1} for the gas-phase reaction with OH radicals at 296 ± 2 K (Overend & Paraskevopoulos 1978; quoted, Atkinson 1985); rate constant of 4.3×10^{12} $cm^3 \cdot mole^{-1} \cdot sec^{-1}$ for reaction with OH radical at 300 K (Lyman et al. 1982); photooxidation half-life of 6.2-72 hours, based on measured rate constant for the reaction with hydroxyl radicals in air (Atkinson 1985, 1987; quoted, Howard et al. 1991); experimental rate constant of 5.3×10^{-12} $cm^3 \cdot molecule^{-1} \cdot sec^{-1}$ and calculated rate constant of 5.0×10^{-12} $cm^3 \cdot molecule^{-1} \cdot sec^{-1}$ for reaction with OH radicals in air at room temperature (Atkinson 1987); rate constant of $<2.3\times10^{-15}$ cm^3 $molecule^{-1}$ sec^{-1} by flash photolysis-absorption technique for the gas-phase reaction with NO_3 radicals at 298 ± 2 K (Wallington et al. 1987; quoted, Atkinson 1991); rate constant of 5.81×10^{-12} cm^3 $molecule^{-1}$ sec^{-1} for the gas-phase reaction with OH radicals in air (Wallington et al. 1988a,b) and 3.20×10^{-12} $cm^3 \cdot molecule^{-1}$ sec^{-1} for the solution-phase reaction with OH radicals in aqueous solution (Wallington & Kurylo 1987; quoted, Wallington et al. 1988b); rate constant of 5.21×10^{-12} cm^3 $molecule^{-1}$ sec^{-1} for the gas-phase reaction with OH radicals at 298 (Atkinson 1991).

Hydrolysis:

Biodegradation: average rate of biodegradation 52.0 mg COD $g^{-1} \cdot h^{-1}$ based on measurements of COD decrease using activated sludge inoculum with 20 days of adaptation to the substrate (Pitter 1976; quoted, Scow 1982); aqueous aerobic half-life of 24-168 hours, based on unacclimated aerobic aqueous screening test data (Gellman & Heukelekian 1955; Heukelekian & Rand 1955; Price et al. 1974; Takemoto et al. 1981; Wagner 1976; selected, Howard et al. 1991); aqueous anaerobic half-life of 96-672 hours, based on estimated aerobic aqueous biodegradation half-life and unacclimated anaerobic aqueous screening test data (Hou et al. 1983; Sonoda & Seiko 1968; Speece 1983; selected, Howard et al. 1991).

Biotransformation:

Bioconcentration, Uptake (k_1) and Elimination (k_2) Rate Constants:

Common Name: 1-Butanol
Synonym: 1-butanol, *n*-butyl alcohol, *n*-propylcarbinol
Chemical Name: *n*-butyl alcohol, 1-butanol
CAS Registry No: 71-36-3
Molecular Formula: $CH_3CH_2CH_2CH_2OH$
Molecular Weight: 74.122
Melting Point (°C):
-79.9 (Stull 1947)
-89.5 (Weast 1982-83; Howard 1990)
-89.9 (Verschueren 1983)
-88.6 (Dean 1985; Riddick et al. 1986)
Boiling Point (°C):
117.6 (Kahlbaum 1898)
117.7 (Butler et al. 1935; Verschueren 1983; Dean 1985; Riddick et al. 1986)
117.5 (Stull 1947)
117.8 (Kemme & Kreps 1969)
117.2 (Weast 1982-83; Howard 1990)
118.0 (Banerjee et al. 1990)
Density (g/cm^3 at 20°C):
0.80978 (Kahlbaum 1898)
0.80970 (Stockhardt & Hull 1931; Dean 1985)
0.80593 (25°C, Butler et al. 1935)
0.80980 (Weast 1982-83)
0.8100 (Verschueren 1983)
0.80956, 0.80575 (20°C, 25°C, Riddick et al. 1986)
Molar Volume (cm^3/mol):
91.7 (calculated-density, Rohrschneider 1973)
91.0 (calculated-density, Lande & Banerjee 1981)
91.5 (Taft et al. 1985; Kamlet et al. 1986, 1987; Leahy 1986)
49.9 (intrinsic volume, calculated, Leahy 1986; Kamlet et al. 1987)
104, 50 (LeBas method, intrinsic molar volume V_I, Abernethy et al. 1988)
83.6 (calculated-density, Jawoska & Schultz 1993)
88.27 (Yalkowsky & Pinal 1993)
Molecular Volume ($Å^3$):
Dissociation Constant, pK_a:
20.89 (Riddick et al. 1986)
Total Surface Area, TSA ($Å^2$):
272.1 (Amidon et al. 1974; Horvath 1982)
264.5, 272.1, 115.8 (Valvani et al. 1976)
216.3 (Iwase et al. 1985)
184.0 (Funasaki et al. 1986)
Heat of Fusion ΔH_{fus}, kcal/mol:
2.240 (Riddick et al. 1986)
Entropy of Fusion ΔS_{fus}, cal/mol·K (e.u.):
13.5 (estimated, Yalkowsky & Valvani 1980)
Fugacity Ratio at 25°C (assuming ΔS_{fus} = 13.5 e.u.), F: 1.0

Water Solubility (g/m^3 or mg/L at 25°C):

68118 (18°C, shake flask-turbidity, Fühner 1924)
79000 (20°C, shake flask-turbidity, Fühner 1924; quoted, Palit 1947)
79730 (volumetric method, Hill & Malisoff 1926)
65000 (20°C, synthetic method, Jones 1929; quoted, Palit 1947)
74500 (gravimetric, Stockhart & Hull 1931)
73180 (24.85°C, shake flask-interferometer, Butler et al. 1933)
73100 (shake flask-cloud point, Butler et al. 1933)
70000 (26°C, synthetic method, Othmer et al. 1945)
73320 (shake flask-residue volume, Booth & Everson 1948)
74100 (shake flask-interferometry, Hansen et al. 1949)
73000 (shake flask-interferometry, quoted from Butler et al. 1933, Donahue & Bartell 1952)
75849 (estimated, McGowan 1954)
78700 (26.7°C, shake flask-turbidity, Skrzec & Murphy 1954)
74000 (surface tension, Kinoshita et al. 1958)
70000 (titration, Petriris & Geankopolis 1959)
77800 (shake flask-GC, Korenman et al. 1974, 1975)
78690, 50230 (quoted, calculated-K_{OW}, Hansch et al. 1968)
74600, 81530; 60855 (quoted values; calculated-TSA, Amidon et al. 1974)
21209 (quoted, Amidon et al. 1975)
74000 (shake flask-colorimetric analysis, De Santis et al. 1976)
72435, 83166 (quoted, estimated, Yalkowsky & Valvani 1980)
74122 (quoted, Lande & Banerjee 1981)
72435 (quoted average, Valvani et al. 1981)
30900 (predicted-K_{OW}, Valvani et al. 1981)
63300 (gen. col.-GC, Wasik et al. 1981; Tewari et al. 1982; quoted, Li & Andren 1994)
77000 (quoted, Verschueren 1983)
76960 (20°C, Mackay & Yuen 1983)
74000 (recommended best value, IUPAC 1984)
74000 (quoted, Dean 1985)
74636 (quoted from Amidon et al. 1974, Funasaki et al. 1986)
74500 (quoted, Riddick et al. 1986)
72435, 63090 (quoted, calculated-solvatochromic p., Leahy 1986)
72435, 75850 (quoted, calculated-solvatochromic p., Kamlet et al. 1986)
72435, 61652 (quoted, calculated-solvatochromic p., Kamlet et al. 1987)
78100 (shake flask-GC, Doucette & Andren 1988; quoted, Li et al. 1992)
72280, 62950 (quoted, predicted- χ , Nirmalakhandan & Speece 1988)
77000 (quoted, Howard 1990)
74600 (selected, Yaws et al. 1990)
41660 (calculated as per Micro QSAR 2.0, Schultz et al. 1990)
65720 (shake flask-GC, Li et al. 1992)
78570 (quoted, Müller & Klein 1992)
70786 (quoted, Yalkowsky & Pinal 1993)
74600, 71230 (quoted, shake flask-GC, Li & Andren 1994)

Vapor Pressure (Pa at 25°C):

- 800 (24.0°C, Kahlbaum 1898)
- 933 (26.1°C, Kahlbaum 1898)
- 879 (interpolated-regression of tabulated data, Kahlbaum 1898)
- 904 (modified isotensiscope method, Butler et al. 1935)
- 660 (20°C, Butler et al. 1935; quoted, Gückel et al. 1973)
- 939 (interpolated-regression of tabulated data, Stull 1947)
- 667 (20°C, Stull 1947)
- 820 (calculated-Antoine eqn., Reid & Sherwood 1966; quoted, Djerki & Laub 1988)
- 733 (22.6°C, DTA-thermocouple, Kemme & Kreps 1969)
- 1373 (30.9°C, DTA-thermocouple, Kemme & Kreps 1969)
- 864 (Hoy 1970)
- 586, 1333 (20°C, 30°C, quoted, Verschueren 1983)
- 866 (quoted, Verschueren 1983)
- 598 (20°C, quoted, Mackay & Yuen 1983)
- 904 (quoted, Abraham 1984)
- 733 (22°C, quoted exptl., Boublik et al. 1984)
- 890 (calculated-Antoine eqn., Boublik et al. 1984)
- 936 (Daubert & Danner 1985; quoted, Howard 1990)
- 824 (calculated-Antoine eqn., Dean 1985)
- 893 (Howard et al. 1986; quoted, Banerjee et al. 1990)
- 910 (quoted, Riddick et al. 1986)
- 998 (calculated-solvatochromic p., Banerjee et al. 1990)

Henry's Law Constant (Pa·m^3/mol):

- 0.866 (partial pressure-isotensiscope, Bulter et al. 1935; quoted, Buttery et al. 1971)
- 0.731 (entrainment method-GC, Burnett 1963)
- 0.892 (shake flask, partial vapor pressure-GC, Buttery et al. 1969)
- 0.868 (quoted, calculated K_{AW}s, Buttery et al. 1969)
- 0.860 (exptl., Hine & Mookerjee 1975)
- 0.964 (calculated-group contribution, Hine & Mookerjee 1975)
- 1.057 (calculated-bond contribution, Hine & Mookerjee 1975)
- 0.564 (Mackay & Yuan 1983; quoted, Howard 1990)
- 0.80 (headspace-GC, Snider & Dawson 1985)
- 0.860, 1.010 (quoted, calculated- χ Nirmalakhandan & Speece 1988b)
- 0.860, 1.160 (quoted, correlated-molecular structure, Russell et al. 1992)

Octanol/Water Partition Coefficient, log K_{OW}:

- 0.83 (shake flask-CR, Collander 1951; quoted, Leo et al. 1971)
- 0.29 (calculated, Iwasa et al. 1965)
- 0.84 (calculated-π constant, Hansch et al. 1968; quoted, Leo et al. 1969; Wasik et al. 1981; Tewari et al. 1982)
- 0.88 (quoted Hansch & Dunn III unpublished result, Leo et al. 1971; quoted, Iwase et al. 1985; Funasaki et al. 1986)
- 0.84 (Tute 1971; quoted, Fujisawa & Masuhara 1981)
- 0.88 (Hansch & Dunn III 1972; quoted, Rekker 1977)

0.84, 0.86 (calculated-f const., Rekker 1977)
0.84 (calculated-π constant, Rekker 1977)
0.89 (Hansch & Leo 1979; quoted, Valvani et al. 1981; Cornford 1982)
1.02, 0.955 (observed, calculated-HPLC-RV/RT, Yonezawa & Urushigawa 1979)
1.02 (HPLC-RV, Yonezawa & Urushigawa 1979; quoted, Sangster 1989)
0.84 (estimated, Yalkowsky & Valvani 1980)
0.80 (calculated by Rekker's method, Hanai et al. 1981)
0.79 (gen. col.-GC, Wasik et al. 1981; Tewari et al. 1982; quoted, Sangster 1989)
0.76 (calculated from measured γ, Wasik et al. 1981)
0.93 (shake flask-RC, Cornford 1982; quoted, Sangster 1989)
0.80 (RPHPLC-k', D'Amboise & Hanai 1982)
0.88 (quoted, Veith et al. 1983; Verschueren 1983)
0.88 (Hansch & Leo 1985; quoted, Howard 1990)
0.88, 0.89 (quoted, calculated-TSA, Iwase et al. 1985)
0.99, 0.77 (quoted, calculated-solvatochromic p., Taft 1985)
0.75, 0.88 (quoted, calculated-solvatochromic p., Leahy 1986)
0.75, 0.69 (quoted, calculated-solvatochromic p., Kamlet et al. 1986)
0.87 (shake flask-GC at pH 7, Riebesehl & Tomlinson 1986; quoted, Sangster 1989)
0.88 (quoted, Gerstl & Helling 1987)
0.29 (quoted, Purdy 1987)
0.79 (gen. col.-GC, Schantz & Martire 1987; quoted, Sangster 1989)
0.80 (calculated from measured γ, Schantz & Martire 1987; quoted, Sangster 1989)
0.88 (measured, GEMS 1986; quoted, Vaishnav et al. 1987)
0.88 (quoted, Abernethy et al. 1988)
0.88, 0.87 (quoted, calculated-γ from UNIFAC, Banerjee & Howard 1988)
0.88, 0.88, 0.823, 0.84 (quoted from Hansch & Leo 1979, quoted from CLOGP, calculated-CLOGP, calculated-M.O. indices, Bodor et al. 1989)
0.88, 0.84 (quoted, recommended, Sangster 1989)
0.88 (quoted measured value, Schultz et al. 1990)
0.82 (calculated-CLOGP, Müller & Klein 1992)
0.87 (calculated as per Hansch & Leo 1979, Dallos et al. 1993)
0.89 (calculated-UNIFAC, Dallos et al. 1993)
0.88 (quoted from CLOGP 3.15, Jawoska & Schultz 1993)
0.823 (calculated-QSAR, Kollig 1993)

Bioconcentration Factor, log BCF:
-0.439 (calculated as per Mackay 1982, Schultz et al. 1990)

Sorption Partition Coefficient, log K_{OC}:
1.85 (calculated-K_{OW}, Lyman et al. 1982; quoted, Howard 1990)
0.50, 0.64 (quoted, calculated- χ , Gerstl & Helling 1987)
0.50 (soil, quoted exptl., Meylan et al. 1992)
0.39 (soil, calculated- χ & fragment contribution, Meylan et al. 1992)
0.503 (calculated-K_{OW}, Kollig 1993)

Half-Lives in the Environment:

Air: half-life of 0.24-2.4 hours for the gas-phase reaction with hydroxyl radicals in air, based on the rate of disappearance of hydrocarbon due to reaction with hydroxyl radicals (Darnall et al. 1976); photooxidation half-life of 8.8-87.7 hours, based on measured rate constant for the vapor-phase reaction with hydroxyl radicals in air (Atkinson 1985; selected Howard et al. 1991).

Surface water: 24-168 hours, based on unacclimated freshwater grab sample data (Hammerton 1955; selected, Howard et al. 1991) and aqueous screening test data (Bridie et al. 1979; selected, Howard et al. 1991); photoxidation half-life of 2602-104000 hours, based on measured rate constant for reaction with hydroxyl radicals in aqueous solution (Dorfman & Adams 1973; selected, Howard et al. 1991); volatilization half-life estimated to be 2.4 hours in streams, 3.9 hours in rivers and 125.9 hours in lakes (Lyman et al. 1982; quoted, Howard 1990).

Groundwater: 48-1296 hours, based on estimated aqueous anaerobic biodegradation half-life (Howard et al. 1991).

Sediment:

Soil: 24-168 hours, based on estimated aqueous aerobic biodegradation half-life (Howard et al. 1991).

Biota:

Environmental Fate Rate Constants or Half-Lives:

Volatilization: evaporation rate of 2.538×10^{-4} $mol \cdot m^{-2} \cdot h^{-1}$ was determined by gravimetric method with an air flow rate of (50 ± 1) $L \cdot h^{-1}$ at (20 ± 0.1)°C (Gückel et al. 1973); half-life estimated to be 2.4 hours in streams, 3.9 hours in rivers and 125.9 hours in lakes (Lyman et al. 1982; quoted, Howard 1990).

Photolysis:

Oxidation: rate constant of 2.2×10^{9} $L \cdot mol^{-1} \cdot sec^{-1}$ for the reaction with 1×10^{-17} M hydroxyl radicals in water with a half-life about a year (Anbar & Neta 1967; quoted, Howard 1990); photooxidation half-life of 2602-104000 hours in water, based on measured rate constant for reaction with hydroxyl radicals in aqueous solution (Dorfman & Adams 1973; quoted, Howard et al. 1991); rate constant of 6.8×10^{-10} $cm^3 \cdot molecule^{-1} \cdot sec^{-1}$ for the atmospheric reaction with 1×10^{-6} molecules/cm^3 of hydroxyl radicals with a calculated half-life of 1.2 days (Campbell et al. 1976; quoted, Howard 1990); photooxidation half-life of 0.24-2.4 hours for the gas-phase reaction with hydroxyl radicals in air, based on the rate of disappearance of hydrocarbon due to reaction with hydroxyl radicals (Darnall et al. 1976); half-life of 6.5 hours for the vapor phase reaction with photochemically produced NO radicals in the atmosphere (Dilling et al. 1976; quoted, Howard 1990); rate constant of 4.1×10^{-12} $cm^3 \cdot molecule^{-1} \cdot sec^{-1}$ for reaction with OH radical at 300 K (Lyman et al. 1982); photooxidation half-life of 8.8-87.7 hours in air, based on measured rate constant for the vapor-phase reaction with hydroxyl radicals in air (Atkinson 1985; selected, Howard et al. 1991); estimated half-life of 2.3 days for the vapor phase reaction with photochemically produced hydroxyl radicals in the atmosphere (GEMS 1986; quoted, Howard 1990); experimental rate constant 7.3×10^{-12} $cm^3 \cdot molecule^{-1} \cdot sec^{-1}$ compared with the calculated rate constant of 6.4x10 $cm^3 \cdot molecule^{-1} \cdot sec^{-1}$ for reactions of OH radicals at room temperature (Atkinson 1987); experimental rate constant of 8.31×10^{-12} $cm^3 \cdot molecule^{-1} \cdot sec^{-1}$ for the gas-phase reaction with hydroxyl radicals in air and 7.00×10^{-12} $cm^3 \cdot molecule^{-1} \cdot sec^{-1}$ for the solution-phase reaction with hydroxyl radicals in aqueous solution (Wallington & Kurylo 1987; quoted, Wallington et al. 1988); atmospheric second-order rate constant of 8.3×10^{-12} $cm^3 \cdot molecule^{-1} \cdot sec^{-1}$ for the gas phase reaction with OH radicals at 298 K in air (Atkinson 1990).

Hydrolysis:

Biodegradation: aqueous aerobic half-life of 24-168 hours, based on unacclimated freshwater grab sample data (Hammerton 1955; selected, Howard et al. 1991) and aqueous screening test data (Bridie et al. 1979; selected, Howard et al. 1991); average rate of biodegradation 84.0 mg COD $g^{-1} \cdot h^{-1}$ based on measurements of COD decrease using activated sludge inoculum with 20 days of adaptation to the substrate (Pitter 1976; quoted, Scow 1982); aqueous anaerobic half-life of 96-1296 hours, based on acclimated screening test data (Chou et al. 1979; selected, Howard et al. 1991) and aqueous aerobic biodegradation half-life (Howard et al. 1991); calculated rate constant of $(0.959 \pm 0.063) \times 10^2$ h^{-1} by activated sludge (Yonezawa & Urushigawa 1979); biodegradation rate constants of 0.035-0.046 h^{-1} in 30 mg/L activated sludge after a time lag of 5-10 hours (Urano & Kato 1986).

Biotransformation:

Bioconcentration, Uptake (k_1) and Elimination (k_2) Rate Constants:

Common Name: Isobutanol
Synonym: isobutyl alcohol, 2-methyl-1-propanol
Chemical Name: isobutanol, isobutyl alcohol, 2-methyl-1-propanol
CAS Registry No: 78-83-1
Molecular Formula: $(CH_3)_2CHCH_2OH$
Molecular Weight: 74.12
Melting Point (°C):
-108.0 (Stull 1947; Verschueren 1983; Riddick et al. 1986)
Boiling Point (°C):
107.0-107.2 (Stockhardt & Hull 1931)
107.9 (Butler et al. 1935; Verschueren 1983; Dean 1985; Riddick et al. 1986)
108.0 (Stull 1947; Weast 1982-83; Banerjee et al. 1990)
Density (g/cm³ at 20°C):
0.8020 (Stockhardt & Hull 1931)
0.7883 (25°C, Butler et al. 1935)
0.8018 (Weast 1982-83)
0.7980 (25°C, Verschueren 1983)
0.8016 (Dean 1985; Riddick et al. 1986)
0.7978 (25°C, Riddick et al. 1986)
Molar Volume (cm³/mol):
92.0 (Taft et al. 1985; Kamlet et al. 1986)
49.9 (intrinsic volume, Kamlet et al. 1987,88)
82.7 (calculated, Jaworska & Schultz 1993)
88.31 (Yalkowsky & Pinal 1993)
103.6 (calculated-LeBas method)
92.7 (calculated-density)
Molecular Volume (Å³):
Total Surface Area, TSA (Å²):
263.8 (Amidon et al. 1974)
263.8, 257.7, 113.8 (with and without hydration cell, Valvani et al. 1976)
213.2 (Iwase et al. 1985)
180.0 (Funasaki et al. 1986)
Heat of Fusion ΔH_{fus}, kcal/mol:
1.511 (Riddick et al. 1986)
Entropy of Fusion ΔS_{fus}, cal/mol·K (e.u.):
13.5 (estimated, Yalkowsky & Valvani 1980)
Fugacity Ratio at 25°C (assuming ΔS_{fus} = 13.5 e.u.), F: 1.0

Water Solubility (g/m³ or mg/L at 25°C):
100140 (18°C, shake flask-turbidity, Führer 1924)
85000 (20°C, Seidell 1941; quoted, Palit 1947)
92885 (Palit 1947; quoted, Hansch et al. 1968)
75600 (shake flask-interferometry, Donahue & Bartell 1952)
95490 (calculated-K_{OW}, Hansch et al. 1968; quoted, Müller & Klein 1992)
75825, 96280 (quoted, calculated-TSA, Amidon et al. 1974)
94000 (shake flask-colorimetric, De Santis et al. 1976)

85104, 107140 (quoted, estimated, Yalkowsky & Valvani 1980)
85100, 41680 (quoted average, calculated-K_{OW}, Valvani et al. 1981)
95000 (18°C, quoted, Verschueren 1983)
94874 (20°C, quoted, Mackay & Yuen 1983)
76270 (Barton 1984; quoted, Howard 1990)
100000 (Dean 1985; Riddick et al. 1986)
75850 (quoted from Amidon et al. 1974, Funasaki et al. 1986)
85103, 72435 (quoted, calculated-V_I and solvatochromic p., Kamlet et al. 1986)
84920, 79250 (quoted, predicted- χ , Nirmalakhandan & Speece 1988a)
89110 (selected, Yalkowsky & Pinal 1993)

Vapor Pressure (Pa at 25°C):
1541 (modified isotensiscope method, Butler et al. 1935)
1648 (interpolated-regression of tabulated data, Stull 1947)
1333 (21.7°C, Stull 1947)
1333 (quoted, Verschueren 1983)
1337 (20°C, quoted, Mackay & Yuen 1983)
1388 (Daubert & Danner 1985; quoted, Howard 1990)
1680 (Howard et al. 1986; quoted, Banerjee et al. 1990)
1527 (Riddick et al. 1986)
1008 (calculated-solvatochromic p., Banerjee et al. 1990)

Henry's Law Constant (Pa·m^3/mol):
1.20 (partial pressure-isotensiscope, Butler et al. 1935)
1.214 (quoted exptl., Hine & Mookerjee 1975; quoted Nirmalakhandan & Speece 1988)
1.159 (calculated-group contribution, Hine & Mookerjee 1975)
1.057 (calculated-bond contribution, Hine & Mookerjee 1975)
104 (calculated-P/C, Mackay & Yuen 1983)
0.992 (headspace-GC, Snider & Dawson 1985)
40.52 (quoted K_{WA} as K_{AW} from Snider & Dawson 1985; Howard 1990)
1.214, 1.186 (quoted, calculated- χ , Nirmalakhandan & Speece 1988b)

Octanol/Water Partition Coefficient, log K_{OW}:
0.83 (shake flask-CR, Collander 1951; quoted, Rekker 1977; Sangster 1989)
0.80 (selected, McGowan 1954)
0.16 (calculated-π constant, Iwasa et al. 1965)
0.61 (calculated-π constant, Hansch et al. 1968)
0.64 (Leo et al. 1969)
0.65 (quoted from Hansch & Dunn III unpublished result, Leo et al. 1971; quoted, Rekker 1977; Funasaki et al. 1986)
0.76 (shake flask-GC, Dillingham et al. 1973; quoted, Sangster 1989)
0.74 (calculated-f const., Rekker 1977)
0.64 (calculated-π constant, Rekker 1977)

0.73 (estimated, Yalkowsky & Valvani 1980)
0.75 (quoted, Valvani et al. 1981)
0.74 (calculated-f const., as per Hansch & Leo 1979, Veith et al. 1983)
0.65, 0.83 (quoted, Verschueren 1983)
0.76 (Hansch & Leo 1985; quoted, Howard 1990)
0.76, 0.83 (quoted, calculated-TSA, Iwase et al. 1985)
0.76, 0.78 (quoted, calculated-V_I and solvatochromic p., Taft et al. 1985)
0.75, 0.72 (quoted, calculated-V_I and solvatochromic p., Kamlet et al. 1986)
0.76 (measured, GEMS 1986; quoted, Vaishnav et al. 1987)
0.65 (quoted from Pomona College unpublished data, Sangster 1989)
0.76 (recommended, Sangster 1989)
0.69 (calculated-CLOGP, Müller & Klein 1992)
0.35, 0.8602 (quoted exptl., calculated-UNIFAC, Chen et al. 1993)
0.76 (EPA CLOGP data base, Hulzebos et al. 1993)
0.76 (quoted from CLOGP 3.15, Jaworska & Schultz 1993)

Bioconcentration Factor, log BCF:

Sorption Partition Coefficient, log K_{OC}:

Half-Lives in the Environment:

Air: photooxidation half-life of 0.24-2.4 hours in air for the gas-phase reaction with hydroxyl radicals, based on the rate of disappearance of hydrocarbon due to reaction with hydroxyl radicals (Darnall et al. 1976); photooxidation half-life of 9.96-99.6 hours, based on estimated rate constant for the reaction with hydroxyl radicals in air (Atkinson 1987; quoted, Howard et al. 1991).

Surface water: photooxidation half-life of 201 days to 22 years, based on measured rate for the reaction with hydroxyl radicals in water (Anbar & Neta 1967; quoted, Howard et al. 1991); 43-173 hours, based on estimated unacclimated aqueous aerobic biodegradation half-life (Howard et al. 1991).

Groundwater: 86-346 hours, based on estimated unacclimated aqueous aerobic biodegradation half-life (Howard et al. 1991).

Sediment:

Soil: 43-173 hours, based on estimated unacclimated aqueous aerobic biodegradation half-life (Howard et al. 1991).

biota:

Environmental Fate Rate Constants or Half-Lives:

Volatilization: using Henry's law constant, a half-life of 79.7 hours was calculated for evaporation from a model river of 1 m deep with a current of 3 m/sec and with a wind velocity of 3 m/sec (Lyman et al. 1982; selected, Howard 1990).

Photolysis:

Oxidation: photooxidation half-life of 201 days to 22 years in water, based on measured rate for the reaction with hydroxyl radicals in water (Anbar & Neta 1967;

selected, Howard et al. 1991); photooxidation half-life of 0.24-2.4 hours in air for the gas-phase reaction with hydroxyl radicals, based on the rate of disappearance of hydrocarbon due to reaction with hydroxyl radicals (Darnall et al. 1976); photooxidation half-life of 9.96-99.6 hours in air, based on estimated rate constant for the reaction with hydroxyl radicals in air (Atkinson 1987; selected, Howard et al. 1991).

Hydrolysis:

Biodegradation: biodegradation rate constants of 0.015-0.020 h^{-1} in 30 mg/L activated sludge after a time lag of 5-10 hours (Urano & Kato 1986); aqueous aerobic half-life of 43-173 hours, based on river die-away data for one sample of water from one river (Hammerton 1955; selected, Howard et al. 1991); aqueous anaerobic half-life of 172-692 hours, based on estimated unacclimated aqueous aerobic biodegradation half-life (Howard et al. 1991).

Biotransformation:

Bioconcentration, Uptake (k_1) and Elimination (k_2) Rate Constants:

Common Name: *sec*-Butyl Alcohol
Synonym: 2-butyl alcohol, 2-butanol, methylethylcarbinol
Chemical Name: 2-butanol, *sec*-butyl alcohol
CAS Registry No: 78-92-2
Molecular Formula: $CH_3CH_2CHOHCH_3$
Molecular Weight: 74.12
Melting Point (°C):
-114.7 (Stull 1947; Dean 1985; Riddick et al. 1986)
Boiling Point (°C):
99.95 (Butler et al. 1935)
99.50 (Stull 1947; Weast 1982-83; Dean 1985)
99.51 (Riddick et al. 1986)
99.0 (Banerjee et al. 1990)
Density (g/cm^3 at 20°C):
0.8029 (25°C, Butler et al. 1935)
0.8063 (Weast 1982-83)
0.8080 (Verschueren 1983)
0.8069 (Dean 1985)
0.8065 (Riddick et al. 1986)
0.8024 (25°C, Riddick et al. 1986)
Molar Volume (cm^3/mol):
92.4 (Hoy 1970; Amidon & Williams 1982)
91.7 (Taft et al. 1985; Kamlet et al. 1986; Leahy 1986)
50.0 (intrinsic volume V_I, Leahy 1986; Abernethy et al. 1988)
49.9 (intrinsic volume V_I, Kamlet et al. 1987)
104.0 (LeBas method, Abernethy et al. 1988)
83.8 (Jaworska & Schultz 1993)
88.18 (Yalkowsky & Pinal 1993)
Molecular Volume ($Å^3$):
Total Surface Area, TSA ($Å^2$):
264.1 (Amidon et al. 1974)
178.0 (Funasaki et al. 1986)
Heat of Fusion ΔH_{fus}, kcal/mol:
Entropy of Fusion ΔS_{fus}, cal/mol·K (e.u.):
13.5 (estimated, Yalkowsky & Valvani 1980)
Fugacity Ratio at 25°C (assuming ΔS_{fus} = 13.5 e.u.), F: 1.0

Water Solubility (g/m^3 or mg/L at 25°C):
200000 (20°C, synthetic method, Jones 1929; quoted, Palit 1947)
142870 (Palit 1947)
142890, 95490 (quoted, calculated-K_{OW}, Hansch et al. 1968)
79162, 110886 (quoted, predicted-TSA, Amidon et al. 1974)
181950, 10714 (quoted, estimated, Yalkowsky & Valvani 1980)
181950 (quoted average, Valvani et al. 1981; quoted, Banerjee 1985)
58880 (predicted-K_{OW}, Valvani et al. 1981)
181950, 104700 (quoted, calculated-K_{OW} and M.P., Amidon & Williams 1982)
125000/201000 (20°C, quoted, Verschueren 1983)

120000 (Dean 1985)
79240 (quoted from Amidon et al. 1974, Funasaki et al. 1986)
441000 (Riddick et al. 1986)
181950, 151340 (quoted, calculated-V_I and solvatochromic p., Kamlet et al. 1986)
181950, 123010 (quoted, calculated-V_I and solvatochromic p., Leahy 1986)
85100, 61650 (quoted, calculated-V_I and solvatochromic p., Kamlet et al. 1987)
181550, 148940 (quoted, predicted- χ , Nirmalakhandan & Speece 1988)
85100 (selected, Yalkowsky & Pinal 1993)

Vapor Pressure (Pa at 25°C):
2286 (modified isotensiscope method, Butler et al. 1935)
2266 (interpolated-regression of tabulated data, Stull 1947)
2666 (27.3°C, Stull 1947)
2357 (Hoy 1970)
1600 (20°C, quoted, Verschueren 1983)
3200 (30°C, quoted, Verschueren 1983)
2438 (calculated-Antoine eqn., Dean 1985)
2440 (Howard et al. 1986; quoted, Banerjee et al. 1990)
2317 (Riddick et al. 1986)
2440, 997 (measured, calculated-solvatochromic p., Banerjee et al. 1990)

Henry's Law Constant (Pa·m^3/mol):
1.040 (partial pressure-isotensiscope, Butler 1935)
1.033 (quoted exptl., Hine & Mookerjee 1975; quoted, Nirmalakhandan & Speece 1988b)
1.600 (calculated-group contribution, Hine & Mookerjee 1975)
1.057 (calculated-bond contribution, Hine & Mookerjee 1975)
0.80 (headspace-GC, Snider & Dawson 1985)
1.033, 1.107 (quoted, calculated- χ , Nirmalakhandan & Speece 1988b)

Octanol/Water Partition Coefficient, log K_{OW}:
0.61 (shake flask-GC, Hansch & Anderson 1967; Hansch et al. 1968; quoted, Leo et al. 1971; Hansch & Dunn III 1972; Rekker 1977; Valvani et al. 1981; Amidon & Williams 1982; Funasaki et al. 1986; Sangster 1989)
0.74 (calculated-f const., Rekker 1977)
0.64 (calculated-π const., Rekker 1977)
0.81 (HPLC, Yonezawa & Urushigawa 1979; quoted, Sangster 1989)
0.73 (estimated, Yalkowsky & Valvani 1980)
0.61 (quoted, Verschueren 1983)
0.61, 0.56 (quoted, calculated-V_I and solvatochromic p., Taft et al. 1985)
0.61, 0.50 (quoted, calculated-V_I and solvatochromic p., Kamlet et al. 1986)
0.67 (calculated-V_I & solvatochromic p., Leahy 1986)
0.88 (selected, Abernethy et al. 1988)
0.61, 0.87 (quoted, calculated-γ from UNIFAC, Banerjee & Howard 1988)
0.65 (recommended, Sangster 1989)
0.61, 0.8932 (quoted, calculated-original UNIFAC, Chen et al. 1993)
0.61 (quoted from CLOGP 3.15, Jaworska & Schultz 1993)

Bioconcentration Factor, log BCF:

Sorption Partition Coefficient, log K_{OC}:

Half-Lives in the Environment:

Air: photooxidation half-life of 0.24-2.4 hours for the gas-phase reaction with hydroxyl radicals in air, based on the rate of disappearance of hydrocarbon due to reaction with hydroxyl radicals (Darnall et al. 1976); photooxidation half-life of 7.2-72 hours, based on measured rate constant for the reaction with hydroxyl radicals in air (Edney & Corse 1986; selected, Howard et al. 1991).

Surface water: 24-168 hours, based on estimated aqueous aerobic biodegradation half-life (Howard et al. 1991).

Groundwater: 48-336 hours, based on estimated aqueous aerobic biodegradation half-life (Howard et al. 1991).

Sediment:

Soil: 24-168 hours, based on estimated aqueous aerobic biodegradation half-life (Howard et al. 1991).

Biota:

Environmental Fate Rate Constants or Half-Lives:

Volatilization:

Photolysis:

Hydrolysis:

Oxidation: photooxidation half-life of 129 days to 23 years, based on measured rate for the reaction with hydroxyl radicals in aqueous solution (Anbar & Neta 1967; Dorfman & Adams 1973; selected, Howard et al. 1991); photooxidation half-life of 0.24-2.4 hours for the gas-phase reaction with hydroxyl radicals in air, based on the rate of disappearance of hydrocarbon due to reaction with hydroxyl radicals (Darnall et al. 1976); rate constant of $<2\times10^{2}$ M^{-1} sec^{-1} for oxidation by singlet oxygen at 25°C in aquatic systems with half-life >100 years (Foote 1976; Mill 1979; quoted, Mill 1982) and 4.1×10^{9} M^{-1} sec^{-1} for oxidation by OH radicals at 25°C in the atmosphere with half-life of 1.3 day (Hendry & Kenley 1979; quoted, Mill 1982); photooxidation half-life of 7.2-72 hours, based on measured rate constant for the reaction with hydroxyl radicals in air (Edney & Corse 1986; selected, Howard et al. 1991).

Biodegradation: aqueous aerobic half-life of 24-168 hours, based on river die-away studies (Hammerton 1955; selected, Howard et al. 1991) and aqueous anaerobic half-life of 96-672 hours, based on estimated aqueous aerobic biodegradation half-life (selected, Howard et al. 1991); average rate of biodegradation 30.0 mg COD $g^{-1}\cdot h^{-1}$ based on measurements of COD decrease using activated sludge inoculum with 20 days of adaptation to the substrate (Pitter 1976; quoted, Scow 1982);

Biotransformation:

Bioconcentration, Uptake (k_1) and Elimination (k_2) Rate Constants:

Common Name: *tert*-Butyl Alcohol
Synonym: 3-butanol, *t*-butyl alcohol, 2-methyl-2-propanol, trimethylcarbinol,
Chemical Name: *t*-butanol, *t*-butyl alcohol, 2-methyl-2-propanol
CAS Registry No: 75-65-0
Molecular Formula: $(CH_3)_3COH$
Molecular Weight: 74.12
Melting Point (°C):
  25.3 (Stull 1947)
  25.4 (Yalkowsky & Valvani 1980)
  25.5 (Weast 1982-83)
  25.0 (Verschueren 1983)
  25.8 (Dean 1985)
  25.62 (Riddick et al. 1986)
  25.7 (Howard 1990)
Boiling Point (°C):
  82.75 (Butler et al. 1935)
  82.90 (Stull 1947; Verschueren 1983; Banerjee et al. 1990)
  82.30 (Weast 1982-83)
  82.40 (Dean 1985; Howard 1990)
  82.35 (Riddick et al. 1986)
Density (g/cm^3 at 20°C):
  0.7883 (Weast 1982-83)
  0.7880 (Verschueren 1983)
  0.7858 (Dean 1985)
  0.7812 (25°C, Riddick et al. 1986)
Molar Volume (cm^3/mol):
  94.2 (calculated-density, Rohrschneider 1973)
  93.9 (Taft et al. 1985)
  94.0 (Kamlet et al. 1986; Leahy 1986)
  49.9 (intrinsic volume, Leahy 1986)
  49.8 (intrinsic volume, Kamlet et al. 1987)
  87.28 (Yalkowsky & Pinal 1993)
  103.6 (calculated-LeBas method)
Molecular Volume ($Å^3$):
Total Surface Area, TSA ($Å^2$):
  176.0 (Funasaki et al. 1986)
Heat of Fusion, ΔH_{fus}, kcal/mol:
  1.587 (Riddick et al. 1986)
Entropy of Fusion, ΔS_{fus}, cal/mol·K (e.u.):
  5.44 (observed, Yalkowsky & Valvani 1980)
  13.5 (estimated, Yalkowsky & Valvani 1980)
Fugacity Ratio at 25°C (assuming ΔS = 13.5 e.u.), F: 1.0

Water Solubility (g/m^3 or mg/L at 25°C):
  infinite (Palit 1947; Riddick et al. 1986)
  41373 (calculated-ΔS, Yalkowsky & Valvani 1980)
  41563 (estimated-M.P., Yalkowsky & Valvani 1980)

miscible (Barton 1984; Dean 1985; quoted, Howard 1990)
316190, 257008 (quoted, calculated-solvatochromic p., Kamlet et al. 1986)
316190, 245440 (quoted, calculated-solvatochromic p., Leahy 1986)
316190, 269113 (quoted, calculated-solvatochromic p., Kamlet et al. 1987)
miscible (Yaws et al. 1990)
316180 (quoted, Yalkowsky & Pinal 1993)

Vapor Pressure (Pa at 25°C):
5653.3 (modified isotensiscope method, Butler et al. 1935)
5019 (interpolated-regression of tabulated data, Stull 1947)
5332 (24.5°C, Stull 1947)
5554.6 (Wilhoit & Zwolinski 1973; quoted, Howard 1990)
5599 (quoted, Verschueren 1983)
5599 (Howard et al. 1986; quoted, Banerjee et al. 1990)
5637 (Riddick et al. 1986)
1008 (calculated-solvatochromic p., Banerjee et al. 1990)

Henry's Law Constant (Pa·m^3/mol):
1.21 (partial pressure-isotensiscope, Butler et al. 1935)
1.214 (quoted exptl., Hine & Mookerjee 1975; quoted, Nirmalakhandan & Speece 1988b; Howard 1990)
1.057 (calculated-group contribution, Hine & Mookerjee 1975)
1.057 (calculated-bond contribution, Hine & Mookerjee 1975)
1.46 (headspace-GC, Snider & Dawson 1985)
1.214, 1.426 (quoted, calculated- χ , Nirmalakhandan & Speece 1988b)

Octanol/Water Partition Coefficient, log K_{OW}:
0.070 (calculated, Iwasa et al. 1965)
0.37 (shake flask-GC, Hansch & Anderson 1967; quoted, Leo et al. 1969; Sangster 1989)
0.59 (HPLC, Yonezawa & Urushigawa 1979; quoted, Sangster 1989)
0.37 (quoted, Verschueren 1983)
0.35 (Hansch & Leo 1985; quoted, Howard 1990)
0.36, 0.40 (quoted, calculated-solvatochromic p., Taft et al. 1985)
0.39 (HPLC-k', Funasaki et al. 1986; quoted, Sangster 1989)
0.36, 0.42 (quoted, calculated-solvatochromic p., Kamlet et al. 1986)
0.36, 0.45 (quoted, calculated-solvatochromic p., Leahy 1986)
0.35 (quoted from Pomona College unpublished data, Sangster 1989)
0.35 (recommended, Sangster 1989)
0.35, 0.8602 (quoted, calculated-original UNIFAC, Chen et al. 1993)

Bioconcentration Factor, log BCF:
-1.46 (calculated-K_{OW}, Lyman et al. 1982; quoted, Howard 1990)

Sorption Partition Coefficient, log K_{OC}:
1.57 (soil, calculated-K_{OW}, Lyman et al. 1982; quoted, Howard 1990)

Half-Lives in the Environment:

Air: photooxidation half-life of >9.9 days for the gas-phase reaction with hydroxyl radicals in air, based on the rate of disappearance of hydrocarbon due to reaction with hydroxyl radicals (Darnall et al. 1976); for 10 ppm in the air reacted with 5 ppm NO would give a half-life of 34.5 hours in air (Dilling et al. 1976; quoted, Howard 1990); photooxidation half-life of 59-590 hours, based on measured rate for the reaction with hydroxyl radicals in air (Atkinson 1985; selected, Howard et al. 1991); estimated half-life of 1.09 month for the vapor phase reaction with photochemically produced hydroxyl radicals in air (GEMS 1986; quoted, Howard 1990).

Surface water: about 8.8 years, based on rate constant of $2.5x10^8$ liter/mol-sec for the reaction with hydroxyl radicals in water (Anbar & Neta 1967; quoted, Howard 1990); 672-4320 hours, based on unacclimated aqueous aerobic biodegradation half-life (Howard et al. 1991).

Groundwater: 1344-8640 hours, based on unacclimated aqueous aerobic biodegradation half-life (Howard et al. 1991).

Sediment:

Soil: 360-4800 hours, based on soil microcosm studies (Novak et al. 1985; Wilson et al. 1986; selected, Howard et al. 1991).

Biota:

Environmental Fate Rate Constants or Half-Lives:

Volatilization: estimated half-lives of 51.6 hours in streams, 65.37 hours in rivers and 3104.3 hours in lakes were estimated by using Henry's law constant and assuming the wind velocity to be 3 m/sec, the current velocities of the streams, rivers, and lakes 2,1, and 0.01 m/sec with depths of lakes 50 m and that of the streams and lakes 1m (Lyman et al. 1982; quoted, Howard 1990).

Photolysis:

Oxidation: photooxidation half-life of 771 days to 64500 years, based on measured rate for the reaction with hydroxyl radicals in water (Anbar & Neta 1967; Dorfman & Adams 1973; selected, Howard et al. 1991); photooxidation half-life of >9.9 days for the gas-phase reaction with hydroxyl radicals in air, based on the rate of disappearance of hydrocarbon due to reaction with hydroxyl radicals (Darnall et al. 1976); photooxidation half-life of 59-590 hours, based on measured rate for the reaction with hydroxyl radicals in air (Atkinson 1985; selected, Howard et al. 1991); experimental rate constant of $0.59x10^{-12}$ $cm^3 \cdot molecule^{-1} \cdot sec^{-1}$ and calculated rate constant of $1.09x10^{-12}$ $cm^3 \cdot molecule^{-1} \cdot sec^{-1}$ for reactions with OH radicals in gas phase at room temperature (Atkinson 1987).

Hydrolysis:

Biodegradation: average rate of biodegradation 30.0 mg COD $g^{-1} \cdot h^{-1}$ based on measurements of COD decrease using activated sludge inoculum with 20 days of adaptation to the substrate (Pitter 1976; quoted, Scow 1982); aqueous aerobic half-life of 677-4320 hours, based on river die-away studies (Hammerton 1955; Sokolova & Kaplin 1969; selected, Howard et al. 1991); aqueous anaerobic half-life of 2400-12000 hours, based on degradation rates in microcosm studies simulating anaerobic aquifers (Novak et al. 1985; selected, Howard et al. 1991).

Bioconcentration, Uptake (k_1) and Elimination (k_2) Rate Constants:

Common Name: 1-Pentanol
Synonym: amyl alcohol, *n*-butylcarbinol, 1-pentanol, *n*-pentyl alcohol, pentyl alcohol
Chemical Name: *n*-amyl alcohol, *n*-pentyl alcohol, 1-pentanol
CAS Registry No: 71-41-0
Molecular Formula: $CH_3CH_2CH_2CH_2CH_2OH$
Molecular Weight: 88.15
Melting Point (°C):

-79.0 (Weast 1982-83; Mackay et al. 1982)
-78.9 (Verschueren 1983; Dean 1985)
-78.2 (Riddick et al. 1986)

Boiling Point (°C):

137.75 (Butler et al. 1935)
137.6-138.3 (Ginnings & Baum 1937)
137.8 (Stull 1947; Mackay et al. 1982; Dean 1985)
138.0 (Kemme & Kreps 1969; Verschueren 1983)
137.3 (at 748 mm Hg, Weast 1982-83)
137.98 (Riddick et al. 1986)

Density (g/cm^3 at 20°C):

0.81253 (25°C, Butler et al. 1935)
0.8110 (25°C, Ginnings & Baum 1937)
0.81440 (Weast 1982-83)
0.8240 (Verschueren 1983)
0.81480 (Dean 1985)
0.81445, 0.81080 (20°C, 25°C, Riddick et al. 1986)

Molar Volume (cm^3/mol):

108.7 (Hoy 1970; Amidon & Williams 1982)
108 (calculated-density, Lande & Banerjee 1981)
108.2 (Taft et al. 1985; Leahy 1986)
59.3 (intrinsic volume, calculated, Leahy 1986; Kamlet et al. 1987)
108.6 (Kamlet et al. 1987)
126, 59.0 (LeBas method, intrinsic volume V_I, Abernethy et al. 1988)
100.1 (calculated-density, Jaworska & Schultz 1993)
103.64 (Yalkowsky & Pinal 1993)

Molecular Volume (Å^3):
Acid Dissociation Constant, pK_a:

20.81 (Riddick et al. 1986)

Total Surface Area, TSA (Å^2):

303.9 (with hydration cell, Amidon et al. 1974)
303.9, 294.4, 134.0 (Valvani et al. 1976)
303.9, 303.754 (calculated, correlation, Horvath 1982)
244.9 (Iwase et al. 1985)
207.0 (Funasaki et al. 1986)

Heat of Fusion ΔH_{fus}, kcal/mol:

2.510 (Riddick et al. 1986)

Entropy of Fusion ΔS_{fus}, cal/mol·K (e.u.):

13.5 (estimated, Yalkowsky & Valvani 1980)

Fugacity Ratio at 25°C (assuming ΔS_{fus} = 13.5 e.u.), F: 1.0

Water Solubility (g/m^3 or mg/L at 25°C):

22080 (shake flask-interferometer, Butler et al. 1933; quoted, Ginnings & Baum 1937)
21900 (volumetric method, Ginnings & Baum 1937)
22100 (20°C, surface tension, Addison 1945)
23600 (20°C, Seidell 1941; quoted, Palit 1947)
27570 (shake flask-residue volume, Booth & Everson 1948)
15230 (20°C, shake flask-turbidity, Laddha & Smith 1948)
25400 (shake flask-interferometry, Hansen et al. 1949)
22100 (quoted from Butler et al. 1933, Donahue & Bartell 1952)
22000 (shake flask-titration, Crittenden & Hixon 1954)
22142 (estimated, McGowan 1954)
22000 (surface tension, Kinoshita et al. 1958)
23800 (shake flask-GC, Korenman 1974, 1975)
22550, 14765 (quoted, calculated-K_{OW}, Hansch et al. 1968)
21950 (Riddick & Bunger 1955; quoted, Wasik et al. 1981; Tewari et al. 1982)
22920 (Amidon et al. 1974; quoted, Funasaki et al. 1986; Li & Andren 1994)
22920, 22038; 18423 (quoted; predicted-TSA, Amidon et al. 1974)
13860 (quoted, Amidon et al. 1975)
18800 (shake flask-GC, Evans et al. 1978)
21640, 30565 (quoted, estimated, Yalkowsky & Valvani 1980)
22660 (quoted, Lande & Banerjee 1981)
22190, 9020 (quoted average, predicted-K_{OW}, Valvani et al. 1981)
11720 (gen. col.-GC, Wasik et al. 1981; Tewari et al. 1982; quoted, Li & Andren 1994)
22140, 16410 (quoted, calculated-K_{OW} and M.P., Amidon & Williams 1982)
27000 (22°C, quoted, Verschueren 1983)
26450 (20°C, quoted, Mackay & Yuen 1983)
27000 (Windholz 1983; quoted, Chessells et al. 1992)
22000 (recommended best value, IUPAC 1984)
27000 (quoted, Dean 1985)
21900 (quoted, Riddick et al. 1986)
22500 (20.2°C, shake flask-GC/TC, Stephenson et al. 1984)
20200 (30.6°C, shake flask-GC/TC, Stephenson et al. 1984)
21640, 21640 (quoted, calculated-solvatochromic p., Leahy 1986)
21640, 24840 (quoted, calculated-solvatochromic p., Kamlet et al. 1986)
21640, 20660 (quoted, calculated-solvatochromic p., Kamlet et al. 1987)
21580, 19450 (quoted, predicted- χ , Nirmalakhandan & Speece 1988)
22000 (quoted, Yaws et al. 1990)
13310 (calculated as per Micro QSAR 2.0, Schultz et al. 1990)
22570 (quoted, Müller & Klein 1992)
20190 (shake flask-GC, Li & Andren 1994)

Vapor Pressure (Pa at 25°C):

333 (modified isotensiscope method, Butler et al. 1935)
220 (20°C, Butler et al. 1935, quoted, Gückel et al. 1973)
349 (interpolated-regression of tabulated data, Stull 1947)
286 (DTA-thermocouple, Kemme & Kreps 1969)

313 (calculated-Antoine eqn., Riddick & Bunger 1970; quoted, Djerki & Laub 1988)
375, 892 (quoted, calculated-B.P., Mackay et al. 1982)
373 (20°C, quoted, Verschueren 1983)
375 (quoted, Mackay & Yuen 1983)
289 (quoted, Abraham 1984)
212 (extrapolated-Antoine eqn., Boublik et al. 1984)
313 (extrapolated-Antoine eqn., Dean 1985)
293 (Riddick et al. 1986)

Henry's Law Constant (Pa m^3/mol):
1.324 (partial pressure-isotensiscope, Butler et al. 1937; quoted, Buttery et al. 1971)
1.271 (exptl., Hine & Mookerjee 1975; quoted, Nirmalakhandan & Speece 1988)
1.426 (calculated-group contribution, Hine & Mookerjee 1975)
1.600 (calculated-bond contribution, Hine & Mookerjee 1975)
1.04 (calculated-P/C, Mackay & Yuen 1983)
1.271, 1.271 (quoted, calculated- χ , Nirmalakhandan & Speece 1988)
1.236 (computed, Yaws et al. 1991)
1.271, 1.600 (quoted, correlated-molecular structure, Russell et al. 1992)

Octanol/Water Partition Coefficient, log K_{OW}:
0.81 (calculated, Iwasa et al. 1965)
1.34 (calculated-π constant, Hansch et al. 1968)
1.40 (quoted from Hansch & Dunn III unpublished result, Leo et al. 1971; Hansch & Dunn III 1972; quoted, Rekker 1977; Funasaki et al. 1986)
1.37, 1.38 (calculated-f const., Rekker 1977)
1.34 (calculated-π constant, Rekker 1977)
1.34 (calculated, Hansch & Leo 1979; quoted, Wasik et al. 1981; Tewari et al. 1982)
1.48, 1.44 (observed, calculated-HPLC-RV/RT, Yonezawa & Urushigawa 1979)
1.48 (HPLC-RV, Yonezawa & Urushigawa 1979; quoted, Sangster 1989)
1.37 (estimated, Yalkowsky & Valvani 1980)
1.33 (calculated by Rekker's method, Hanai et al. 1981)
1.53 (gen. col.-GC, Wasik et al. 1981; Tewari et al. 1982; quoted, Sangster 1989)
1.48 (quoted from Valvani et al. 1981, Amidon & Williams 1982)
1.33 (HPLC, D'Amboise & Hanai 1982; quoted, Sangster 1989)
1.40 (quoted, Verschueren 1983)
1.40, 1.43 (quoted, calculated-TSA, Iwase et al. 1985)
1.00 (computed, Klopman et al. 1985; quoted, Pangrekar et al. 1994)
1.48, 1.23 (quoted, calculated-solvatochromic p., Taft et al. 1985)
1.40, 1.17 (quoted, calculated-solvatochromic p., Kamlet et al. 1986)
1.40, 1.37 (quoted, calculated-solvatochromic p., Leahy 1986)
1.49 (shake flask-GC at pH 7, Riebesehl & Tomlinson 1986; quoted, Sangster 1989)
1.48 (quoted, Gerstl & Helling 1987)
1.53 (gen. col.-GC, Schantz & Martire 1987; quoted, Sangster 1989)
1.55 (calculated from measured γ, Schantz & Martire 1987; quoted, Sangster 1989)
1.40 (measured, GEMS 1986; quoted, Vaishnav et al. 1987)
1.40 (quoted, Abernethy et al. 1988)
1.56, 1.40 (quoted unpublished data, Sangster 1989)

1.51 (recommended, Sangster 1989)
1.56 (quoted measured value, Schultz et al. 1990)
1.56 (quoted from Medchem software value, Chessells et al. 1992)
1.35 (calculated-CLOGP, Müller & Klein 1992)
1.56 (quoted from CLOGP 3.15, Jaworska & Schultz 1993)

Bioconcentration Factor, log BCF:
0.241 (calculated as per Mackay 1982, Schultz et al. 1990)

Sorption Partition Coefficient, log K_{OC}:
0.70, 0.93 (quoted, calculated- χ , Gerstl & Helling 1987)
0.70 (soil, quoted exptl., Meylan et al. 1992)
0.65 (soil, calculated- χ & fragment contribution, Meylan et al. 1992)

Half-Lives in the Environment:
Air: half-life of 0.24-2.4 hours for the gas-phase reaction with hydroxyl radicals in air, based on the rate of disappearance of hydrocarbon due to reaction with hydroxyl radicals (Darnall et al. 1976).
Surface water:
Groundwater:
Sediment:
Soil:
Biota:

Environmental Fate Rate Constants or Half-Lives:
Volatilization: evaporation rate of 6.92×10^{-5} mol cm^{-2} h^{-1} was determined by gravimetric method with an air flow rate of (50 $\pm$ 1) L h^{-1} at (20 $\pm$ 0.1)°C (Gückel et al. 1973).
Photolysis:
Oxidation: photooxidation half-life of 0.24-2.4 hours for the gas-phase reaction with hydroxyl radicals in air, based on the rate of disappearance of hydrocarbon due to reaction with hydroxyl radicals (Darnall et al. 1976); experimental rate constant of 1.08×10^{-11} cm^3 $molecule^{-1}$ sec^{-1} for the gas-phase reaction with hydroxyl radicals in air and 6.50×10^{-12} cm^3 $molecule^{-1}$ sec^{-1} for the solution-phase reaction with hydroxyl radicals in aqueous solution (Wallington & Kurylo 1987; quoted, Wallington et al. 1988).
Hydrolysis:
Biodegradation:
Biotransformation:
Bioconcentration, Uptake (k_1) and Elimination (k_2) Rate Constants:

Common Name: 1-Hexanol
Synonym: amylcarbinol, 1-hexanol, *n*-hexyl alcohol
Chemical Name: *n*-hexyl alcohol, 1-hexanol
CAS Registry No: 111-27-3
Molecular Formula: $CH_3CH_2CH_2CH_2CH_2CH_2OH$
Molecular Weight: 102.176
Melting Point (°C):
- -51.6 (Stull 1947; Dean 1985)
- -46.7 (Weast 1982-83)
- -51/-44 (Verschueren 1983)
- -44.6 (Riddick et al. 1986)

Boiling Point (°C):
- 155.7 (Butler et al. 1935)
- 157.0 (Stull 1947; Riddick et al. 1986)
- 157.3 (Kemme & Kreps 1969)
- 158.0 (Weast 1982-83; Verschueren 1983)
- 157.5 (Dean 1985)

Density (g/cm^3 at 20°C):
- 0.81835 (25°C, Butler et al. 1935)
- 0.81360 (Weast 1982-83)
- 0.8200 (Verschueren 1983)
- 0.81860 (Dean 1985)
- 0.81875, 0.81534 (20°C, 25°C, Riddick et al. 1986)

Molar Volume (cm^3/mol):
- 126.0 (calculated-density, Lande & Banerjee 1981)
- 125.6 (Taft et al. 1985; Kamlet et al. 1986)
- 148.0 (LeBas method, Abernethy et al. 1988)
- 69.0 (intrinsic volumeV_I, Kamlet et al. 1987; Abernethy et al. 1988)
- 117.3 (calculated-density, Jaworska & Schultz 1993)
- 118.45 (Yalkowsky & Pinal 1993)

Molecular Volume ($Å^3$):
Acid Dissociation Constant, pK_a:
Total Surface Area, TSA ($Å^2$):
- 335.7 (with hydration cell, Amidon et al. 1975)
- 335.7, 335.672 (calculated, correlated, Horvath 1982)
- 269.1 (Iwase et al. 1985)
- 231.0 (Funasaki et al. 1986)

Heat of Fusion ΔH_{fus}, kcal/mol:
- 3.680 (Riddick et al. 1986)

Entropy of Fusion ΔS_{fus}, cal/mol K (e.u.):
- 13.5 (estimated, Yalkowsky & Valvani 1980)

Fugacity Ratio at 25°C (assuming ΔS_{fus} = 13.5 e.u.), F: 1.0

Water Solubility (g/m^3 or mg/L at 25°C):
- 5000 (20°C, shake flask-turbidity, Fühner 1924; quoted, Palit 1947)
- 6240 (shake flask-interferometer, Butler et al. 1933; quoted, Ginnings & Webb 1938)

5025 (20°C, shake flask-turbidity, Laddha & Smith 1948)
6200 (quoted from Butler et al. 1933, Donahue & Bartell 1952)
6000 (shake flask-titration, Crittenden & Hixon 1954)
6300 (estimated, McGowan 1954)
6000 (surface tension, Kinoshita et al. 1958)
6270, 4230 (quoted, calculated-K_{OW}, Hansch et al. 1968)
6300 (Hansch et al. 1968; quoted, Hine & Mookerjee 1975; Müller & Klein 1992)
6300, 6230, 6744; 5436 (quoted, values; calculated-TSA, Amidon et al. 1974)
5838 (shake flask-refractive index, Hill & White 1974)
3590 (shake flask-GC, Korenman et al. 1974)
2203 (quoted, Amidon et al. 1975)
5880, 11203 (observed, estimated, Yalkowsky & Valvani 1980)
5880, 2880 (quoted average, predicted-K_{OW}, Valvani et al. 1981)
6300 (quoted, Lande & Banerjee 1981; Banerjee 1985; Li & Andren 1994)
4230 (gen. col.-GC, Wasik et al. 1981; Tewari et al. 1982; quoted, Li & Andren 1994)
5900 (20°C, quoted, Verschueren 1983)
6000 (recommended best value, IUPAC 1984)
8000 (quoted, Dean 1985)
6270 (quoted from Amidon et al. 1974, Funasaki et al. 1986)
7060 (20°C, Riddick et al. 1986)
6660 (20°C, shake flask-GC/TC, Stephenson et al. 1984)
5580 (29.7°C, shake flask-GC/TC, Stephenson et al. 1984)
5880, 7750 (quoted, calculated-solvatochromic p., Kamlet et al. 1986)
5880, 6750 (quoted, calculated-solvatochromic p., Leahy 1986)
5880, 6600 (quoted, calculated-solvatochromic p., Kamlet et al. 1987)
5875, 6000 (quoted, predicted- χ , Nirmalakhandan & Speece 1988)
4260 (calculated as per Micro QSAR 2.0, Schultz et al. 1990)
5870 (quoted, Yaws et al. 1990)
6016 (selected, Yalkowsky & Pinal 1993)
5354 (shake flask-GC, Li & Andren 1994)

Vapor Pressure (Pa at 25°C):
95.84 (modified isotensiscope method, Butler et al. 1935)
94.38 (20°C, Butler et al. 1935; quoted, Gückel et al. 1973)
151.1 (interpolated-regression of tabulated data, Stull 1947)
133.3 (24.4°C, Stull 1947)
88.25 (DTA-thermocouple, Kemme & Kreps 1969)
109.6 (Riddick & Bunger 1970; quoted, Djerki & Laub 1988)
95.84 (quoted, Hine & Mookerjee 1975)
130.6 (20°C, quoted, Verschueren 1983)
95.95 (quoted, Abraham 1984)
79.65 (extrapolated-Antoine eqn., Boublik et al. 1984)
109.5 (extrapolated-Antoine eqn., Dean 1985)
110 (quoted, Riddick et al. 1986)

Henry's Law Constant (Pa m^3/mol):

1.562 (partial pressure-isotensiscope, Butler et al. 1935; quoted, Buttery et al. 1971)
1.735 (shake flask, partial vapor pressure-GC, Buttery et al. 1969)
1.562 (quoted lit. calculated K_{AW}s, Buttery et al. 1969)
1.562 (quoted, exptl., Hine & Mookerjee 1975)
1.880 (calculated-group contribution, Hine & Mookerjee 1975)
2.367 (calculated-bond contribution, Hine & Mookerjee 1975)
1.564, 1.600 (quoted, calculated- χ , Nirmalakhandan & Speece 1988b)
1.896 (computed, Yaws et al. 1991)
1.564, 2.64 (quoted, correlated-molecular structure, Russell et al. 1992)

Octanol/Water Partition Coefficient, log K_{OW}:

2.03 (shake flask-GC, Hansch & Anderson 1967; quoted, Leo et al. 1971; Hansch & Dunn III 1972; Rekker 1977; Iwase et al. 1985)
1.84 (calculated-S, Hansch et al. 1968; quoted, Wasik et al. 1981; Tewari et al. 1982)
1.84 (Tute 1971; quoted, Fujisawa & Masuhara 1981)
1.84 (calculated-π constant, Rekker 1977)
1.90, 1.91 (calculated-f const., Rekker 1977)
1.99, 1.95 (observed, calculated-HPLC-RV/RT, Yonezawa & Urushigawa 1979)
1.99 (HPLC-RV, Yonezawa & Urushigawa 1979; quoted, Sangster 1989)
1.90 (estimated, Yalkowsky & Valvani 1980)
1.86 (calculated by Rekker's method, Hanai et al. 1981)
2.03 (quoted, Valvani et al. 1981; Veith et al. 1983)
2.03 (gen. col.-GC, Wasik et al. 1981; Tewari et al. 1982; quoted, Sangster 1989)
1.86 (RPHPLC-k', D'Amboise & Hanai 1982; quoted, Sangster 1989)
2.03 (quoted, Verschueren 1983)
2.03, 1.89 (quoted, calculated-TSA, Iwase et al. 1985)
1.32 (computed, Klopman et al. 1985; quoted, Pangrekar et al. 1994)
2.03, 1.73 (quoted, calculated-solvatochromic p., Taft et al. 1985)
2.03 (HPLC-k', Funasaki et al. 1986; quoted, Sangster 1989)
2.03, 1.64 (quoted, calculated-solvatochromic p., Kamlet et al. 1986)
2.03, 1.87 (quoted, calculated-solvatochromic p., Leahy 1986)
2.04 (shake flask-GC at pH 7, Riebesehl & Tomlinson 1986; quoted, Sangster 1989)
2.03 (quoted, Gerstl & Helling 1987; Purdy 1987)
2.03 (gen. col.-GC, Schantz & Martire 1987; quoted, Sangster 1989)
2.05 (calculated from measured γ, Schantz & Martire 1987; quoted, Sangster 1989)
2.03 (measured, GEMS 1986; quoted, Vaishnav et al. 1987)
2.03, 1.50 (quoted, calculated-γ from UNIFAC, Banerjee & Howard 1988)
2.03, 2.03, 1.881, 1.88 (quoted from Hansch & Leo 1979, quoted from CLOGP, calculated-CLOGP, calculated-M.O. indices, Bodor et al. 1989)
2.03 (quoted and recommended, Sangster 1989)
2.03 (quoted measured value, Schultz et al. 1990)
1.88 (calculated-CLOGP, Müller & Klein 1992)
2.03, 1.7863 (quoted exptl., calculated-UNIFAC, Chen et al. 1993)
2.03 (quoted from CLOGP 3.15, Jaworska & Schultz 1993)

Bioconcentration Factor, log BCF:

0.711 (calculated as per Mackay 1982, Schultz et al. 1990)

Sorption Partition Coefficient, log K_{OC}:

1.01 (soil, quoted exptl., Meylan et al. 1992)

0.92 (soil, calculated- χ and fragment contribution, Meylan et al. 1992)

Half-Lives in the Environment:

Air: half-life of 0.24-2.4 hours in air for the gas-phase reaction with hydroxyl radicals, based on the rate of disappearance of hydrocarbon due to reaction with hydroxyl radicals (Darnall et al. 1976).

Surface water:

Groundwater:

Sediment:

Soil:

Biota:

Environmental Fate Rate Constants or Half-Lives:

Volatilization: evaporation rate of 1.85×10^{-5} mol·cm^{-2}·h^{-1} was determined by gravimetric method with an air flow rate of (50 ± 1) L·h^{-1} at (20 ± 0.1)°C (Gückel et al. 1973).

Photolysis:

Oxidation: photooxidation half-life of 0.24-2.4 hours in air for the gas-phase reaction with hydroxyl radicals, based on the rate of disappearance of hydrocarbon due to reaction with hydroxyl radicals (Darnall et al. 1976); measured rate constant of $(12.4 \pm 0.7) \times 10^{-12}$ cm^3·molecule^{-1}·sec^{-1}, and 10.0×10^{-12} cm^3·molecule^{-1}·sec^{-1} (predicted) for the gas phase reaction with OH radicals at 298 K (Wallington et al. 1988a); experimental rate constant of 1.24×10^{-11} cm^3·molecule^{-1}·sec^{-1} for the gas-phase reaction with OH radicals in air and 1.20×10^{-11} cm^3·molecule^{-1}·sec^{-1} for the solution-phase reaction with OH radicals in aqueous solution (Wallington et al. 1988b).

Hydrolysis:

Biodegradation:

Biotransformation:

Bioconcentration, Uptake (k_1) and Elimination (k_2) Rate Constants:

Common Name: 1-Heptanol
Synonym: 1-heptanol, *n*-heptyl alcohol
Chemical Name: *n*-heptyl alcohol, 1-heptanol
CAS Registry No: 111-70-6
Molecular Formula: $CH_3CH_2CH_2CH_2CH_2CH_2CH_2OH$
Molecular Weight: 116.21
Melting Point (°C):
-34.1 (Weast 1982-83)
-34.6 (Dean 1985)
Boiling Point (°C):
175.6 (Butler et al. 1935)
175.8 (Stull 1947; Dean 1985)
176.4 (Kemme & Kreps 1969)
176.0 (Weast 1982-83; Verschueren 1983)
Density (g/cm^3 at 20°C):
0.82053 (25°C, Butler et al. 1935)
0.8219 (Weast 1982-83; Dean 1985)
Molar Volume (cm^3/mol):
142.0 (calculated-density, Lande & Banerjee 1981)
141.4 (Kamlet et al. 1986; Leahy 1986)
78.9 (intrinsic volume V_I, Leahy 1986; Kamlet et al. 1987)
79.0, 170 (intrinsic volume V_I, LeBas method, Abernethy et al. 1988)
132.1 (calculated-density, Jaworska & Schultz 1993)
Molecular Volume (Å^3):
Acid Dissociation Constant, pK_a:
Total Surface Area, TSA (Å^2):
367.5 (with hydration cell, Amidon et al. 1974)
367.5, 354.2, 170.3 (Valvani et al. 1976)
3.67.5, 367.59 (calculated, correlation, Horvath 1982)
255 (Funasaki et al. 1986)
Heat of Fusion ΔH_{fus}, kcal/mol:
Entropy of Fusion ΔS_{fus}, cal/mol·K (e.u.): 13.5 (estimated, Yalkowsky & Valvani 1980)
Fugacity Ratio at 25°C (assuming ΔS_{fus} = 13.5 e.u.), F: 1.0

Water Solubility (g/m^3 or mg/L at 25°C):
930 (18°C, shake flask-turbidity, Führer 1924)
1804 (shake flask-interferometer, Butler et al. 1933)
1720 (20°C, shake flask-titration, Addison 1943)
3288 (shake flask-centrifuge, Booth & Everson 1948)
1200 (shake flask-turbidimetric method, Harkins & Oppenheimer 1949)
1800 (quoted from Butler et al. 1933, Donahue & Bartell 1952)
1800 (estimated, McGowan 1954)
1700 (surface tension, Kinoshita et al. 1958)
1904, 1190 (quoted, calculated-K_{OW}, Hansch et al. 1968)
1800, 1685, 1976; 1580 (quoted values; calculated-TSA, Amidon et al. 1974)
1676 (shake flask-interferometric method, Hill & White 1974)
1804 (quoted, Hine & Mookerjee 1975)

1720, 4030 (observed, estimated, Yalkowsky & Valvani 1980)
1800 (quoted, Lande & Banerjee 1981; Banerjee 1985; Li & Andren 1994)
1313 (gen. col.-GC, Wasik et al. 1981; Tewari et al. 1982; Li & Andren 1994)
900, 2000 (18°C, 20°C, quoted, Verschueren 1983)
1740 (best value, IUPAC 1984)
1840, 1540 (20.2°C, 30.6°C, shake flask-GC/TC, Stephenson et al. 1984)
1000 (colloidal, Dean 1985)
1800 (quoted from Amidon et al. 1974, Funasaki et al. 1986)
1720, 2602 (quoted, calculated-solvatochromic p., Kamlet et al. 1986)
1720, 2066 (quoted, calculated-solvatochromic p., Leahy 1986)
1720, 1973 (quoted, calculated-solvatochromic p., Kamlet et al. 1987)
1714, 1854 (quoted, predicted- χ , Nirmalakhandan & Speece 1988)
1740 (selected, Yaws et al. 1990)
1336 (calculated as per Micro QSAR 2.0, Schultz et al. 1990)
1800 (quoted, Müller & Klein 1992)
1510 (shake flask-GC, Li & Andren 1994)

Vapor Pressure (Pa at 25°C):
29.86 (modified isotensiscope method, Butler et al. 1935)
23.19 (20°C, Butler et al. 1935; quoted, Gückel et al. 1973)
44.04 (extrapolated-regression of tabulated data, Stull 1947)
22.2 (DTA-thermocouple, Kemme & Kreps 1969)
31.3 (calculated-Antoine eqn., Jordan 1970; quoted, Djerki & Laub 1988)
29.86 (quoted, Hine & Mookerjee 1975)
133.3 (42°C, quoted, Verschueren 1983)
24.0 (extrapolated-Antoine eqn., Boublik et al. 1984)
17.64 (extrapolated-Antoine eqn., Dean 1985)

Henry's Law Constant (Pa m^3/mol):
1.909 (partial pressure-isotensiscope, Butler et al. 1935; quoted, Buttery et al. 1971)
1.880 (quoted exptl., Hine & Mookerjee 1975)
2.656, 3.583 (calc-group contribution, bond constribution, Hine & Mookerjee 1975)
3.583 (calculated-bond contribution, Hine & Mookerjee 1975)
1.880, 2.015 (quoted, calculated- χ , Nirmalakhandan & Speece 1988)
1.880, 5.55 (quoted, calculated-molecular structure, Russell et al. 1992)

Octanol/Water Partition Coefficient, log K_{OW}:
2.34 (calculated-π constant, Hansch et al. 1968; quoted, Wasik et al. 1981; Tewari et al. 1982)
2.53 (Hansch & Dunn III 1972)
2.41, 2.47 (observed, calculated-HPLC-RV/RT, Yonezawa & Urushigawa 1979)
2.41 (HPLC-RV, Yonezawa & Urushigawa 1979; quoted, Sangster 1989)
2.42 (estimated, Yalkowsky & Valvani 1980)
2.39 (calculated by Rekker's method, Hanai et al. 1981)
2.57 (gen. col.-GC, Wasik et al. 1981; Tewari et al. 1982; quoted, Sangster 1989)
2.39 (RPHPLC-k', D'Amboise & Hanai 1982; quoted, Sangster 1989)
2.41 (calculated, Verschueren 1983; quoted, Gerstl & Helling 1987)

2.60 (HPLC-k', Funasaki et al. 1986)
2.38 (calculated-solvatochromic p., Leahy 1986)
2.65 (shake flask-GC at pH 7, Riebesehl & Tomlinson 1986; quoted, Sangster 1989)
2.57 (gen. col.-GC, Schantz & Martire 1987; quoted, Sangster 1989)
2.57 (calculated- γ , Schantz & Martire 1987)
2.34 (estimated-CLOGP 3, Vaishnav et al. 1987)
2.38 (selected, Abernethy et al. 1988)
2.57, 1.83 (quoted, calculated-γ from UNIFAC, Banerjee & Howard 1988)
2.62 (recommended, Sangster 1989)
2.41 (quoted measured value, Schultz et al. 1990)
2.41 (calculated-CLOGP, Müller & Klein 1992; Jaworska & Schultz 1993)
2.57, 2.232 (quoted exptl., calculated-original UNIFAC, Chen et al. 1993)

Bioconcentration Factor, log BCF:
1.09 (calculated as per Mackay 1982, Schultz et al. 1990)

Sorption Partition Coefficient, log K_{OC}:
1.14, 1.48 (quoted, calculated- χ , Gerstl & Helling 1987)
1.14 (soil, quoted exptl., Meylan et al. 1992)
1.19 (soil, calculated- χ & fragment contribution, Meylan et al. 1992)

Half-Lives in the Environment:
Air: half-life of 0.24-2.4 hours in air for the gas-phase reaction with hydroxyl radicals, based on the rate of disappearance of hydrocarbon due to reaction with hydroxyl radicals in air (Darnall et al. 1976).
Surface water:
Sediment:
Soil:
Biota:

Environmental Fate Rate Constants or Half-Lives:
Volatilization: evaporation rate of 6.804×10^{-6} mol·cm^{-2}·h^{-1} was determined by gravimetric method with an air flow rate of (50 ± 1) L·h^{-1} at 20°C (Gückel et al. 1973).
Photolysis:
Oxidation: photooxidation half-life of 0.24-2.4 hours in air for the gas-phase reaction with hydroxyl radicals, based on the rate of disappearance of hydrocarbon due to reaction with hydroxyl radicals in air (Darnall et al. 1976); second-order rate constant of $(13.6 \pm 1.3) \times 10^{-12}$ cm^{-1}·molecule^{-1}·sec^{-1} (measured) and 10.5 x 10^{-12} cm^{-1}·molecule^{-1}·sec^{-1} (predicted) for gas phase reaction reaction with OH radicals at 298 K (Wallington et al. 1988a); experimental rate constant of 1.36 x 10^{-11} cm^{-1}·molecule^{-1}·sec^{-1} for the gas-phase reaction reaction with OH radicals in air and 1.20×10^{-11} cm^{-1}·molecule^{-1}·sec^{-1} for the solution-phase reaction with OH radicals in aqueous solution (Wallington et al. 1988b).
Hydrolysis:
Biodegradation:
Biotransformation:
Bioconcentration, Uptake (k_1) and Elimination (k_2) Rate Constants:

Common Name: 1-Octanol
Synonym: 1-octanol, *n*-octyl alcohol, capryl alcohol, heptylcarbinol
Chemical Name: *n*-octyl alcohol, 1-octanol
CAS Registry No: 111-87-5
Molecular Formula: $CH_3CH_2CH_2CH_2CH_2CH_2CH_2CH_2OH$
Molecular Weight: 130.23
Melting Point (°C):
-15.4 (Stull 1947)
-17.2 (Mackay et al. 1982)
-16.7 (Weast 1982-83)
-17.0 (Verschueren 1983)
-15.0 (Dean 1985; Riddick et al. 1986)
Boiling Point (°C):
194.5 (Butler et al. 1935)
195.2 (Stull 1947; Dean 1985; Riddick et al. 1986)
195.3 (Kemme & Kreps 1969)
194.9 (Mackay et al. 1982)
194.4 (Weast 1982-83)
195.0 (Verschueren 1983)
Density (g/cm³ at 20°C):
0.82322 (25°C, Butler et al. 1935)
0.8270 (Weast 1982-83)
0.8240 (Verschueren 1983)
0.8258 (Dean 1985)
0.82499, 82157 (20°C, 25°C, Riddick et al. 1986)
Molar Volume (cm³/mol):
158.0 (calculated-density, Rohrschneider 1973; Lande & Banerjee 1981)
157.5 (Kamlet et al. 1986; Leahy 1986)
88.2 (intrinsic volume V_I, Leahy 1986; Kamlet et al. 1987)
88.0 (intrinsic volume V_I, Abernethy et al. 1988)
192.0 (LeBas method, Abernethy et al. 1988)
148.7 (calculated-density, Jaworska & Schultz 1993)
151.74 (Yalkowsky & Pinal 1993)
Molecular Volume (Å³):
Acid Dissociation Constant, pK_a:
Total Surface Area, TSA (Å²):
399.4 (with hydration cell, Amidon et al. 1974)
399.4, 384.1, 188.4 (Valvani et al. 1976)
399.4, 399.486 (calculated, correlated, Horvath 1982)
321.2 (Iwase et al. 1985)
279.0 (Funasaki et al. 1986)
Heat of Fusion ΔH_{fus}, kcal/mol:
10.10 (Riddick et al. 1986)
Entropy of Fusion ΔS_{fus}, cal/mol·K (e.u.):
13.5 (estimated, Yalkowsky & Valvani 1980)
Fugacity Ratio at 25°C (assuming ΔS_{fus} = 13.5 e.u.), F: 1.0

Water Solubility (g/m^3 or mg/L at 25°C):

586 (shake flask-interferometer, Butler et al. 1933; quoted, Ginnings & Coltrane 1939)
590 (shake flask-turbidimeter, McBain & Richard 1946)
500 (shake flask-titration, Crittenden & Hixon 1954)
582 (estimated, McGowan 1954)
490 (shake flask-surface tension, Kinoshita et al. 1958)
495 (shake flask-turbidity, Shinoda et al. 1959)
586, 330 (quoted, calculated-K_{OW}, Hansch et al. 1968)
590 (Hansch et al. 1968; quoted, Hine & Mookerjee 1975; Müller & Klein 1992)
590, 480, 534; 450 (quoted values; calculated-TSA, Amidon et al. 1974)
556, 1400 (observed, estimated, Yalkowsky & Valvani 1980)
530, 272 (quoted, predicted-K_{OW}, Valvani et al. 1981)
582 (Lande & Banerjee 1981; Banerjee 1985; Li & Andren 1994)
300 (20°C, quoted, Verschueren 1983)
540 (best value, IUPAC 1984; quoted, Howard 1993)
590 (quoted, Dean 1985)
590 (quoted from Amidon et al. 1974, Funasaki et al. 1986)
586 (Munz & Roberts 1986; quoted, Li et al. 1992)
538 (quoted, Riddick et al. 1986)
490 (20.5°C, shake flask-GC/TC, Stephenson et al. 1984)
640 (30.6°C, shake flask-GC/TC, Stephenson et al. 1984)
582 (quoted, Kamlet et al. 1986)
582, 668 (quoted, calculated-V_I and solvatochromic p., Leahy 1986)
582, 623 (quoted, calculated-V_I and solvatochromic p., Kamlet et al. 1987)
555, 571 (quoted, predicted- χ , Nirmalakhandan & Speece 1988)
540 (selected, Yaws et al. 1990)
402 (calculated as per Micro QSAR 2.0, Schultz et al. 1990)
517 (shake flask-GC, Li et al. 1992)
1108 (selected, Yalkowsky & Pinal 1993)
417 (shake flask-GC, Li & Andren 1994)

Vapor Pressure (Pa at 25°C):

11.06 (modified isotensiscope method, Butler et al. 1935)
9.22 (20°C, Butler et al. 1935; quoted, Gückel et al. 1973)
18.8 (extrapolated-regression of tabulated data, Stull 1947)
5.28 (DTA-thermocouple, Kemme & Kreps 1969)
11.06 (quoted, Hine & Mookerjee 1975)
132, 578 (quoted, calculated-B.P., Mackay et al. 1982)
8.15 (average, extrapolated-Antoine eqn., Boublik et al. 1984)
10.84 (quoted, Abraham 1984)
13.43 (calculated-Antoine eqn., Dean 1985)
14.50 (static measurement, Berti et al. 1986)
10 (quoted, Riddick et al. 1986)
13.5 (extrapolated-Antoine eqn., Stephenson & Malanowski 1987)
10.58 (Daubert & Danner 1989; quoted, Howard 1993)

Henry's Law Constant (Pa·m³/mol):

2.454 (partial pressure-isotensiscope, Butler et al. 1935; quoted, Buttery et al. 1971)
2.479 (shake flask, partial vapor pressure-GC, Buttery et al. 1969)
2.54 (calculated-P/C, Buttery et al. 1969)
2.422 (exptl., Hine & Mookerjee 1975; quoted, Nirmalakhandan & Speece 1988)
3.344 (calculated-group contribution, Hine & Mookerjee 1975)
6.085 (calculated-bond contribution, Hine & Mookerjee 1975)
2.422, 2.537 (quoted, calculated- χ , Nirmalakhandan & Speece 1988)
1.609 (computed, Yaws et al. 1991)

Octanol/Water Partition Coefficient, log K_{OW}:

3.15 (shake flask-CR, Collander 1951; quoted, Leo et al. 1971; Rekker 1977; Hansch & Leo 1979; Verschueren 1983; Iwase et al. 1985; Funasaki et al. 1986; Bodor et al. 1989; Sangster 1989)
2.37 (calculated-π constant, Iwasa et al. 1965)
2.84 (calculated-π constant, Hansch et al. 1968; quoted, Leo et al. 1969)
3.03 (Hansch & Dunn III 1972)
2.95, 2.97 (calculated-f const., Rekker 1977)
2.84 (calculated-π constant, Rekker 1977)
2.97 (HPLC-k', Könemann et al. 1979; quoted, Sangster 1989)
2.95 (calculated-f const., Könemann et al. 1979)
2.80, 2.83 (observed, calculated-HPLC-RV/RT, Yonezawa & Urushigawa 1979)
2.80 (HPLC-RV, Yonezawa & Urushigawa 1979; quoted, Sangster 1989)
2.95 (estimated, Yalkowsky & Valvani 1980)
2.92 (calculated by Rekker's method, Hanai et al. 1981)
3.15 (quoted, Valvani et al. 1981)
2.92 (RPHPLC-k', D'Amboise & Hanai 1982; quoted, Sangster 1989)
3.16 (shake flask, Platford 1983; quoted, Sangster 1989)
3.03 (quoted, Veith et al. 1983)
2.97 (Hansch & Leo 1985; quoted, Howard 1993)
3.15, 2.88 (quoted, calculated-TSA, Iwase et al. 1985)
1.95 (computed, Klopman et al. 1985; quoted, Pangrekar et al. 1994)
2.85 (calculated-solvatochromic p., Leahy 1986)
3.27 (shake flask-GC at pH 7, Riebesehl & Tomlinson 1986; quoted, Sangster 1989)
2.84 (quoted, Gerstl & Helling 1987)
3.15 (measured, GEMS 1986; quoted, Vaishnav et al. 1987)
3.03 (selected, Abernethy et al. 1988)
2.97, 2.14 (quoted, calculated-γ from UNIFAC, Banerjee & Howard 1988)
2.97 (quoted from CLOGP, Bodor et al. 1989; Jaworska & Schultz 1993)
2.03 (calculated from CLOGP, Bodor et al. 1989)
2.80 (calculated-M.O. indices, Bodor et al. 1989)
3.07 (recommended, Sangster 1989)
2.97 (calculated-CLOGP, Schultz et al. 1990)
2.94 (calculated-CLOGP, Müller & Klein 1992)
2.97, 2.6802 (quoted exptl., calculated-original UNIFAC, Chen et al. 1993)

Bioconcentration Factor, log BCF:

0.307 (calculated-K_{OW}, Lyman et al. 1982; quoted, Howard 1993)
0.097 (calculated-S, Lyman et al. 1982; quoted, Howard 1993)
1.651 (calculated as per Mackay 1982, Schultz et al. 1990)

Sorption Partition Coefficient, log K_{OC}:

2.99 (soil, calculated-K_{OW}, Lyman et al. 1982; quoted, Howard 1993)
2.14 (soil, calculated-S, Lyman et al. 1982; quoted, Howard 1993)
1.56, 1.76 (quoted, calculated- χ , Gerstl & Helling 1987)
1.56 (soil, quoted exptl., Meylan et al. 1992)
1.45 (soil, calculated- χ and fragment contribution, Meylan et al. 1992)

Half-Lives in the Environment:

Air: half-life of 0.24-2.4 hours in air for the gas-phase reaction with hydroxyl radicals, based on the rate of disappearance of hydrocarbon due to reaction with hydroxyl radicals (Darnall et al. 1976); photooxidation half-life of 1.3 day in air, based on measured rate constant of 1.195×10^{-11} cm^3 $molecule^{-1}$ sec^{-1} for the vapor-phase reaction with photochemically produced hydroxyl radicals of $5\times10^5/cm^3$ at 25°C in air (Atkinson 1987; quoted, Howard 1993).

Surface water:

Groundwater:

Sediment:

Soil:

Biota:

Environmental Fate Rate Constants or Half-Lives:

Volatilization: evaporation rate of 1.752×10^{-6} $mol\cdot cm^{-2}\cdot h^{-1}$ was determined by gravimetric method with an air flow rate of (50 ± 1) $L\cdot h^{-1}$ at (20 ± 0.1)°C (Gückel et al. 1973); based on calculated Henry's law constant, half-life from a model river of one meter deep flowing at 1 m/sec with a wind speed of 3 m/sec has been estimated to be 1.8 day (Lyman et al. 1982; quoted, Howard 1993); half-life from a model pond with the consideration of adsorption, has been estimated to be about 82 days (USEPA 1987; quoted, Howard 1993).

Photolysis:

Oxidation: photooxidation half-life of 0.24-2.4 hours in air for the gas-phase reaction with hydroxyl radicals, based on the rate of disappearance of hydrocarbon due to reaction with hydroxyl radicals (Darnall et al. 1976); measured rate constant of 1.195×10^{-11} $cm^3\cdot molecule^{-1}\cdot sec^{-1}$ for the vapor-phase reaction with photochemically produced hydroxyl radicals of $5\times10^5/cm^3$ at 25°C in air corresponds to an atmospheric half-life of 1.3 day (Atkinson 1987; quoted, Howard 1993).

Hydrolysis:

Biodegradation:

Biotransformation:

Bioconcentration, Uptake (k_1) and Elimination (k_2) Rate Constants:

Common Name: Ethylene Glycol
Synonym: 1,2-ethanediol, 1,2-dihydroxyethane, MEG
Chemical Name: 1,2-ethanediol, ethylene glycol
CAS Registry No: 107-21-1
Molecular Formula: $HOCH_2CH_2OH$
Molecular Weight: 62.068
Melting Point (°C):
-11.5 (Weast 1982-83)
-17/-12.6 (Verschueren 1983)
-13.0 (Dean 1985; Howard 1990)
Boiling Point (°C):
197.40 (Gallaugher & Hibbert 1937)
198.93 (at 13 mm Hg, Weast 1982-83)
198.0 (Verschueren 1983)
197.60 (Dean 1985; Howard 1990)
197.54 (Riddick et al. 1986)
197.0 (Banerjee et al. 1990)
Density (g/cm^3 at 20°C):
1.1088 (Weast 1982-83)
1.1130 (Verschueren 1983)
1.1154 (Dean 1985)
1.1135 (Riddick et al. 1986)
Molar Volume (cm^3/mol):
55.8 (Rohrschneider 1973)
66.6 (LeBas method)
Molecular Volume ($Å^3$):
Dissociation Constant, pK_a:
14.22 (Dean 1985)
14.24 (Riddick et al. 1986)
Total Surface Area, TSA ($Å^2$):
177.1 (Iwase et al. 1985)
Heat of Fusion ΔH_{fus}, kcal/mol:
2.380 (Riddick et al. 1986)
Entropy of Fusion ΔS_{fus}, cal/mol K (e.u.):
Fugacity Ratio at 25°C (assuming ΔS_{fus} = 13.5 e.u.), F: 1.0

Water Solubility (g/m^3 or mg/L at 25°C):
miscible (Dean 1985; Howard 1990)
infinite (Riddick et al. 1986)
miscible (Yaws et al. 1990)

Vapor Pressure (Pa at 25°C):
22.39 (Gallaugher & Hibbert 1937)
11.86 (20°C, Gallaugher & Hibbert 1937; quoted, Gückel et al. 1973)
11.70 (ebulliometry, extrapolated-Antoine eqn., Ambrose & Hall 1981)

14.90 (20°C, evaporation method, Gückel et al. 1982)
6.67 (20°C, quoted, Verschueren 1983)
26.66 (30°C, quoted, Verschueren 1983)
9.299 (extrapolated-Antoine eqn., Boublik et al. 1984)
11.86 (extrapolated-Antoine eqn., Dean 1985)
12.26 (Howard et al. 1986; quoted, Banerjee et al. 1990)
11.70 (Riddick et al. 1986; quoted, Howard 1990)
0.0786 (calculated-solvatochromic p., Banerjee et al. 1990)

Henry's Law Constant (Pa m^3/mol):
0.006 (quoted exptl., Hine & Mookerjee 1975)
5.81×10^{-6} (calculated-group contribution, Hine & Mookerjee 1975)
2.37×10^{-5} (calculated-bond contribution, Hine & Mookerjee 1975)

Octanol/Water Partition Coefficient, log K_{OW}:
-0.54 (shake flask-CR, Collander 1951)
-1.93 (Hansch & Leo 1979; quoted, Cornford 1982; Iwase et al. 1985)
-1.93 (quoted, Verschueren 1983)
-1.36 (Hansch & Leo 1985; quoted, Howard 1990)
-2.02 (calculated-TSA, Iwase et al. 1985)

Bioconcentration Factor, log BCF:
1.0 (golden ide, after 3 days, Freitag et al. 1985; quoted, Howard 1990)
2.28 (algae, after 1 day, Freitag et al. 1985; quoted, Howard 1990)
2.30 (activated sludge, after 5 days, Freitag et al. 1985)

Sorption Partition Coefficient, log K_{OC}:

Half-Lives in the Environment:

Air: half-life of 0.24-2.4 hours in air for the gas-phase reaction with hydroxyl radicals, based on the rate of disappearance of hydrocarbon due to reaction with hydroxyl radicals (Darnall et al. 1976); photooxidation half-life of 8.3-83 hours, based on measured rate constant for the reaction with hydroxyl radicals in air (Atkinson 1985; selected, Howard et al. 1991); atmospheric transformation lifetime was estimated to be 1 to 5 days (Kelly et al. 1994).

Surface water: photooxidation half-life of 267 days-64.6 years, based on measured rate constant for the reaction with hydroxyl radicals in water (Anbar & Neta 1967; Dorfman & Adams 1973; quoted, Howard et al. 1991); 48-288 hours, based on aqueous aerobic biodegradation half-life (Howard et al. 1991).

Groundwater: 96-576 hours, based on aqueous aerobic biodegradation half-life (Howard et al. 1991).

Sediment:

Soil: 48-288 hours, based on aqueous aerobic biodegradation half-life (Howard et al. 1991).

Biota:

Environmental Fate Rate Constants or Half-Lives:

Volatilization: evaporation rate of 2.915×10^{-6} mol·cm^{-2}·h^{-1} was determined by gravimetric method with an air flow rate of (50 ± 1) L·h^{-1} at (20 ± 0.1)°C (Gückel et al. 1973); evaporation rate of 2.97×10^{-8} mol·cm^{-2}·h^{-1} was determined at 20°C (Gückel et al. 1982).

Photolysis:

Oxidation: photooxidation half-life of 267 days-64.6 years in water, based on measured rate constant for the reaction with hydroxyl radicals in water (Anbar & Neta 1967; Dorfman & Adams 1973; selected, Howard et al. 1991); photooxidation half-life of 0.24-2.4 hours in air for the gas-phase reaction with hydroxyl radicals, based on the rate of disappearance of hydrocarbon due to reaction with hydroxyl radicals (Darnall et al. 1976); measured rate constant of $(7.7 \pm 1.1) \times 10^{-12}$ cm^3 molecule^{-1} sec^{-1} for the gas-phase reaction with OH radicals at 295 K in air (Wiedermann & Zetzsch 1982; quoted, Atkinson 1985); photooxidation half-life of 8.3-83 hours in air, based on measured rate constant for the reaction with hydroxyl radicals in air (Atkinson 1985; selected, Howard et al. 1991).

Hydrolysis:

Biodegradation: average rate of biodegradation 41.7 mg COD g^{-1}·h^{-1} based on measurements of COD decrease using activated sludge inoculum with 20 days of adaptation to the substrate (Pitter 1976; quoted, Scow 1982); biodegradation rate constants of 0.026-0.035 h^{-1} in 30 mg/L activataed sludge after a time lag of 10-15 hours (Urano & Kato 1986b); aqueous aerobic half-life of 48-288 hours, based on grab sample from river die-away studies (Evans & David 1974; selected, Howard et al. 1991); aqueous anaerobic half-life of 192-1152 hours, based on aqueous aerobic biodegradation half-life (Howard et al. 1991).

Biotransformation:

Bioconcentration, Uptake (k_1) and Elimination (k_2) Rate Constants:

Common Name: Allyl Alcohol
Synonym: propenylalcohol, 2-propen-1-ol, 1-propene-3-ol, vinylcarbinol
Chemical Name: allyl alcohol, 1-propene-3-ol, 2-propen-1-ol
CAS Registry No: 107-18-6
Molecular Formula: $CH_2=CHCH_2OH$
Molecular Weight: 58.08
Melting Point (°C):
-129 (Stull 1947; Weast 1982-83; Verschueren 1983; Riddick et al. 1986)
Boiling Point (°C):
96.6 (Stull 1947)
97.1 (Weast 1982-83; Dean 1985; Riddick et al. 1986)
96.9 (Verschueren 1983)
97.0 (Banerjee et al. 1990)
Density (g/cm^3 at 20°C):
0.85400 (Weast 1982-83; Dean 1985)
0.8250 (Verschueren 1983)
0.85511 (15°C, Riddick et al. 1986)
Molar Volume (cm^3/mol):
64.8 (Kamlet et al. 1986; Leahy 1986)
37.2 (intrinsic volume, Leahy 1986)
74.0 (calculated-LeBas method)
Molecular Volume (Å^3):
Dissociation Constant, pK_a:
15.5 (Riddick et al. 1986)
Total Surface Area, TSA (Å^2):
Heat of Fusion, kcal/mol:
Entropy of Fusion, cal/mol K (e.u.):
Fugacity Ratio at 25°C (assuming ΔS = 13.5 e.u.), F: 1.0

Water Solubility (g/m^3 or mg/L at 25°C):
miscible (Dean 1985)
infinite (Riddick et al. 1986)
366460 (quoted from Banerjee et al. 1980, Kamlet et al. 1986)
375000, 383730 (quoted, calculated-solvatochromic p., Leahy 1986)

Vapor Pressure (Pa at 25°C):
3287 (interpolated-regression of tabulated data, Stull 1947)
2666 (21.7°C, Stull 1947)
3140 (Hoy 1970)
2666 (20°C, quoted, Verschueren 1983)
4266 (30°C, quoted, Verschueren 1983)
3746 (Howard et al. 1986; quoted, Banerjee et al. 1990)
3750 (Riddick et al. 1986)
3840 (calculated-solvatochromic p., Banerjee et al. 1990)

Henry's Law Constant (Pa·m^3/mol):
0.506 (exptl., Hine & Mookerjee 1975)

0.510 (calculated-bond contribution, Hine & Mookerjee 1975)
0.564 (computed, Yaws et al. 1991)

Octanol/Water Partition Coefficient, log K_{OW}:
-0.03 (calculated-f const., Hermens & Leeuwangh 1982; quoted, Verhaar et al. 1992)
0.17 (quoted, Verschueren 1983)
0.17 (quoted from Hansch & Leo 1979, Haky & Young 1984)
0.17, 0.11 (quoted, calculated-solvatochromic p., Kamlet et al. 1986)
0.17, 0.03 (quoted, calculated-solvatochromic p., Leahy 1986)
0.17 (quoted and recommended, Sangster 1989)

Bioconcentration Factor, log BCF:

Sorption Partition Coefficient, log K_{OC}:

Half-Lives in the Environment:

Air: photooxidation half-life of 2.2-22 hours, based on estimated rate constant for the reaction with hydroxyl radicals in air (Atkinson 1987; quoted, Howard et al. 1991).

Surface water: photooxidation half-life of 334 days-37 years, based on measured rate constant for the reaction with hydroxyl radicals in water (Anbar & Neta 1967; quoted, Howard et al. 1991).

Groundwater: 48-336 hours, based on estimated unacclimated aqueous aerobic biodegradation half-life (Howard et al. 1991).

Sediment:

Soil: 24-168 hours, based on estimated unacclimated aqueous aerobic biodegradation half-life (Howard et al. 1991).

Environmental Fate Rate Constants or Half-Lives:

Volatilization:

Photolysis:

Hydrolysis:

Oxidation: photooxidation half-life of 334 days-37 years, based on measured rate constant for the reaction with hydroxyl radicals in water (Anbar & Neta 1967; quoted, Howard et al. 1991); photooxidation half-life of 2.2-22 hours, based on estimated rate constant for the reaction with hydroxyl radicals in air (Atkinson 1987; quoted, Howard et al. 1991).

Biodegradation: aqueous aerobic half-life of 24-168 hours, based on estimated unacclimated aqueous aerobic biodegradation screening test data (Sasaki 1978; quoted, Howard et al. 1991); aqueous anaerobic half-life of 96-672 hours, based on estimated unacclimated aqueous aerobic biodegradation half-life (Howard et al. 1991).

Biotransformation:

Bioconcentration, Uptake (k_1) and Elimination (k_2) Rate Constants:

Common Name: Cyclohexanol
Synonym: adronol, anol, cyclohexylalcohol, hexahydrophenol, hexalin, hydralin, hydrophenol
Chemical Name: cyclohaxanol
CAS Registry No: 108-93-0
Molecular Formula: $C_6H_{11}OH$
Molecular Weight: 100.16
Melting Point (°C):
23.9 (Stull 1947)
Boiling Point (°C):
161.0 (Stull 1947)
Density (g/cm^3): 0.9684
Acid Dissociation Constants, pKa:
Molar Volume (cm^3/mol):
104 (calculated-density, Lande & Banerjee 1981)
125.6 (calculated-LeBas method)
103.4 (calculated-density)
0.636 (intrisic $V_I/100$, Leahy 1986; Kamlet et al. 1988)
114.1 (liquid molar volume, Leahy 1986)
Molecular Volume, TMV ($Å^3$):
Total Surface Area, TSA ($Å^2$):
290.5 (with hydration cell, Amidon et al. 1974)
Heat of Fusion, ΔH_{fus}, kcal/mol: 0.46 (Riddick et al. 1986)
Entropy of Fusion, ΔS_{fus}, kcal/mol·K:
Fugacity Ratio at 25°C (assuming ΔS_{fus} = 13.5 kcal/mol·K), F: 0.997

Water Solubility (g/m^3 or mg/L at 25°C):
37000 (shake flask-turbidity, Sidgwick & Sutton 1930)
39200 (shake flask-interferometry, Hansen et al. 1949)
32920 (residue volume, Booth & Everson 1942)
48960 (shake flask-interferometry, Donahue & Bartell 1952)
37300 (26.7°C, shake flask-turbidity, Skrzec & Murphy 1954)
38340, 22840 (quoted, calculated-K_{OW}, Hansch et al. 1968)
38360, 43069 (quoted, calculated-TSA, Amidon et al. 1974)
38080 (quoted, Hine & Mookerjee 1975)
36000 (20°C, quoted, Verschueren 1983)
38000 (recommended "best" value, IUPAC 1984; Howard 1993)
38020, 20920 (quoted, calculated-UNIFAC, Banerjee 1985)
35540, 33940 (quoted, calculated-V_I and solvatochromic p., Leahy 1986)
37500 (selected, Riddick et al. 1986)
38020 (quoted, Isnard & Lambert 1989)
38200 (selected, Yaws et al. 1990)
38200 (selected, Müller & Klein 1992)

Vapor Pressure (Pa at 25°C):

193.7 (interpolated-regression of tabulated data, Stull 1947)
133.3 (21°C, Stull 1947)
221 (quoted, Hine & Mookerjee 1975)
174.8 (calculated-Cox eqn., Chao et al. 1983)
133.3 (20°C, quoted, Verschueren 1983)
37.53 (extrapolated-Antoine eqn., Dean 1985)
14.66 (Howard et al. 1986; quoted, Banerjee et al. 1990)
1300 (56°C, Riddick et al. 1986)
97.36 (interpolated, solid, Antoine eqn., Stephenson & Malanowski 1987)
80.0 (extrapolated, liquid, Antoine eqn., Stephenson & Malanowski 1987)
84.85 (extrapolated, liquid, Antoine eqn., Stephenson & Malanowski 1987)
106.6 (Daubert & Danner 1989; quoted, Howard 1993)
14.66, 196.0 (quoted, calculated-solvatochromic p., Banerjee et al. 1990)

Henry's Law Constant (Pa·m^3/mol):

0.581 (quoted, exptl., Hine & Mookerjee 1975)
2.48, 2.37 (calculated-group contribution, bond contribution, Hine & Mookerjee 1975)
0.281 (calculated-P/C, Howard 1993)
0.581, 0.278 (quoted, correlated-molecular structure, Russell et al. 1992)

Octanol/Water Partition Coefficient, log K_{OW}:

1.23 (shake flask-AS, Hansch & Anderson 1967, Hansch et al. 1968)
1.23 (quoted, Leo et al. 1971; Hansch & Leo 1979; Veith et al. 1979; Yalkowsky & Morozowich 1980; Verschueren 1983; Isnard & Lambert 1988, 1989)
1.23, 1.42 (quoted, calculated-π constant, Chou & Jurs 1979)
1.55, 1.45, 1.30 (calculated values, Yalkowsky & Morozowich 1980)
1.23 (Hansch & Leo 1985; quoted, Howard 1993)
1.23, 1.11 (quoted, calculated-V_I and solvatochromic p., Leahy 1986)
1.23, 1.36 (quoted, calculated- γ from UNIFAC, Banerjee & Howard 1988)
1.23, 1.33 (quoted, calculated-V_I and solvatochromic p., Kamlet et al. 1988)
1.23 (quoted and recommended, Sangster 1989)
1.27 (selected, Müller & Klein 1992)
1.23, 1.589 (quoted exptl., calculated-original UNIFAC, Chen et al. 1993)

Bioconcentration Factor, log BCF:

0.708 (estimated-K_{OW}, Lyman et al. 1990; quoted, Howard 1993)
0.176 (estimated-S, Lyman et al. 1990; quoted, Howard 1993)

Sorption Partition Coefficient, log K_{OC}:

2.045 (soil, estimated-K_{OW}, Lyman et al. 1990; quoted, Howard 1993)
1.114 (soil, estimated-S, Lyman et al. 1990; quoted, Howard 1993)

Half-Lives in the Environment:

Air: atmosphere photooxidation half-life of 22 hours, based on estimated rate constant of 17.4×10^{-12} cm^3 $molecule^{-1}$ sec^{-1} for the vapor phase reaction with 5×10^5 hydroxyl radicals per cm^3 in air at 25°C (Atkinson 1987; quoted, Howard 1993).

Surface water:

Ground water:

Sediment:

Soil:

Biota:

Environmental Fate Rate Constants and Half-Lives:

Volatilization: half-life of 145 days has been estimated from a model environmental pond with the consideration of the effect of adsorption (USEPA 1987; quoted, Howard 1993); based on the Henry's law constant, half-life from a model river of one meter deep flowing at 1 m/sec with a wind velocity of 3 m/sec is estimated to be 13.3 days at 25°C (Lyman et al. 1990; quoted, Howard 1993).

Photolysis:

Oxidation: estimated rate constant of 17.4×10^{-12} cm^3 $molecule^{-1}$ sec^{-1} for the vapor phase reaction with 5×10^5 hydroxyl radicals per cm^3 in air at 25°C corresponds to an atmosphere half-life of 22 hours (Atkinson 1987; quoted, Howard 1993).

Hydrolysis:

Biodegradation: average rate of biodegradation 28.0 mg COD $g^{-1} \cdot h^{-1}$ based on measurements of COD decrease using activated sludge inoculum with 20 days of adaptation to the substrate (Pitter 1976; quoted, Scow 1982).

Biotransformation:

Bioconcentration Uptake (k_1) and Elimination (k_2) Rate Constants:

Common Name: Benzyl Alcohol
Synonym: (hydroxymethyl)benzene, *o*-hydroxy toluene, phenylcarbinol, phenylmethanol
Chemical Name: benzyl alcohol
CAS Registry No: 100-51-6
Molecular Formula: $C_6H_5CH_2OH$
Molecular Weight: 108.15
Melting Point (°C):
 -15.3 (Stull 1947; Weast 1982-83; Riddick et al. 1986)
 -15.0 (Verschueren 1983; Stephenson & Malanowski 1987)
 -15.2 (Howard 1993)
Boiling Point (°C):
 204.7 (Stull 1947; Howard 1993)
 205.0 (Urano et al. 1982; Stephenson & Malanowski 1987)
 205.3 (Weast 1982-83; Ambrose & Ghiassee 1990)
 206.0 (Verschueren 1983; Riddick et al. 1986)
Density (g/cm^3 at 20°C):
 1.04127 (25°C, Riddick et al. 1986)
Molar Volume (cm^3/mol):
 103.8 (Hoy 1970; Amidon & Williams 1982)
 103.5 (calculated-density, Rohrschneider 1973)
 116.9 (Taft et al. 1985)
 125.6 (calculated-LeBas method)
Molecular Volume ($Å^3$):
Total Surface Area, TSA ($Å^2$):
 207.0 (Funasaki et al. 1986)
Heat of Fusion ΔH_{fus}, kcal/mol:
 4.074 (Riddick et al. 1986)
Entropy of Fusion ΔS_{fus}, cal/mol·K (e.u.):
Fugacity Ratio at 25°C (assuming ΔS_{fus} = 13.5 e.u.), F: 1.0

Water Solubility (g/m^3 or mg/L at 25°C):
 38020 (Seidell 1941; quoted, Müller & Klein 1992)
 34200 (estimated, McGowan 1954)
 38020, 35400 (quoted, calculated-K_{OW}, Hansch et al. 1968)
 38370 (quoted average, Valvani et al. 1981)
 35400 (predicted-K_{OW}, Valvani et al. 1981)
 38370, 38370 (quoted, calculated-K_{OW} & M.P., Amidon & Williams 1982)
 38000 (20-25°C, shake flask-GC, Urano et al. 1982)
 35000 (20°C, quoted, Verschueren 1983)
 42900 (shake flask-LSC, Banerjee 1984; quoted, Banerjee 1985; Li et al. 1992)
 800 (20°C, Riddick et al. 1986)

43000 (20.1°C, shake flask-GC/TC, Stephenson & Stuart 1986)
43000 (29.6°C, shake flask-GC/TC, Stephenson & Stuart 1986)
38370, 38370 (quoted, calculated-fragment const., Wakita et al. 1986)
39810 (quoted, Isnard & Lambert 1989)
46070 (shake flask-GC, Li et al. 1992)
37960 (quoted, Müller & Klein 1992)
40000 (quoted, Howard 1993)

Vapor Pressure (Pa at 25°C):
15.4 (extrapolated-regression of tabulated data, Stull 1947)
12.0 (Hoy 1970)
10.06 (gas-saturation method, Grayson & Fosbraey 1982)
4.75 (extrapolated-Cox eqn., Chao et al. 1983)
10.88 (extrapolated-Antoine eqn., Boublik et al. 1984)
11.72 (extrapolated-Antoine eqn., Dean 1985)
15 (Riddick et al. 1986; quoted, Howard 1993)
15.25 (calculated-Antoine eqn., Stephenson & Malanowski 1987)
7.395 (ebulliometry, fitted to Antoine eqn., Ambrose & Ghiassee 1990)
12.0 (calculated-Wagner eqn., Ambrose & Ghiassee 1990)

Henry's Law Constant (Pa·m^3/mol):
0.0231 (quoted estimated value of Hine & Mookerjee 1975, Howard 1993)
0.0396 (calculated-P/C, Howard 1993)

Octanol/Water Partition Coefficient, log K_{OW}:
1.10 (shake flask-AS, Fujita et al. 1964; Hansch et al. 1968; quoted, Leo et al. 1969,1971; McCall 1975; Rekker 1977; Hansch & Leo 1979; Cornford 1982; Verschueren 1983; Funasaki et al. 1986; Miyake et al. 1987; Isnard & Lambert 1989, Sangster 1989)
1.10 ± 0.02 (shake flask-AS, Iwasa et al. 1965)
1.10 (shake flask-UV, Hansch et al. 1968)
1.00, 0.99 (calculated-f const., Rekker 1977)
1.47 (calculated-π const., Rekker 1977)
1.02 (quoted, Valvani et al. 1981; quoted, Amidon & Williams 1982)
1.10 (shake flask-RC, Cornford 1982; quoted, Sangster 1989)
1.00 (shake flask-AS, Mayer et al. 1982; quoted, Sangster 1989)
1.10 (Hansch & Leo 1983,85; quoted, Tayar et al. 1991; Howard 1993)
1.10 (selected best lit. value, Garst & Wilson 1984)
1.16 ± 0.02 (exptl.-ALPM, Garst & Wilson 1984)
1.16 (quoted from Hansch & Leo 1979, Haky & Young 1984)
1.63 (computed, Klopman et al. 1985; quoted, Pangrekar et al. 1994)
1.10, 1.10 (quoted, calculated-solvatochromic p., Taft et al. 1985)
1.06 (HPLC-k', Eadsforth 1986; quoted, Sangster 1989)

1.10, 1.11 (quoted: shake flask, calculated, Eadsforth 1986)
1.10, 1.58 (quoted, calculated-γ from UNIFAC, Banerjee & Howard 1988)
1.05 (recommended, Sangster 1989)
1.22 (CPC-RV, Tayar et al. 1991)
1.10 (calculated-CLOGP, Müller & Klein 1992)
1.13 (calculated as per Hansch & Leo 1979, Dallos et al. 1993)
0.96 (calculated-UNIFAC, Dallos et al. 1993)
1.10 (calculated-QSAR, Kollig 1993)

Bioconcentration Factor, log BCF:
0.602 (calculated-K_{OW}, Lyman et al. 1982; quoted, Howard 1993)

Sorption Partition Coefficient, log K_{OC}:
1.193 (red-brown Australian soil with 1.09% organic carbon, Briggs 1981; quoted, Howard 1993)
0.79 (calculated-K_{OW}, Kollig 1993)

Half-Lives in the Environment:
Air: the estimated half-life for the vapor-phase reaction with photochemically produced hydroxyl radicals in the atmosphere is 2 days (Atkinson 1985; quoted, Howard 1993).
Surface water:
Groundwater:
Sediment:
Soil:
Biota:

Environmental Fate Rate Constants or Half-Lives:
Volatilization: estimated half-life of 97 days for a model river of 1 meter deep, flowing at 1 m/sec with wind velocity of 3 m/sec (Lyman et al. 1982; quoted, Howard 1993).
Photolysis:
Oxidation: Photooxidation reaction rate constants were 7.99×10^{-12} cm^3/molecule·sec (estimated from Atmospheric Oxidation Program), 22.9×10^{-12} cm^3/molecule·sec (exptl.) and 7.98×10^{-12} cm^3/molecule·sec (estimated from Fate of Atmospheric Pollutants Program) with OH radicals in the gas phase (Meylan & Howard 1993).
Hydrolysis:
Biodegradation: biodegradation rate constant of 0.042-0.062 h^{-1} in 30 mg/L activated sludge after a lag time of 5-15 hours (Urano & Kato 1986b).
Biotransformation:
Bioconcentration, Uptake (k_1) and Elimination (k_2) Rate Constants:

2.2 Summary Tables and QSPR Plots

Table 2.1 Physical-chemical properties of alcohols at 25 °C.

compound	CAS no.	formula	MW	mp, °C	bp, °C	fugacity ratio, F at 25°C	molar volume, cm^3/mol V_M from ρ	V_M (a)	V_I (a)	V_M LeBas	V_M (b)	surface area, $Å^2$ TSA (c)	TSA (c)	TSA (d)
methanol	67-56-1	CH_3OH	32.04	-98	64.6	1	40.6			37	33.7	136.2		136.2
ethanol	64-17-5	C_2H_5OH	46.069	-114.69	78.29	1	58.692			55.5	50.2	163.8		163.8
propanol	73-21-8	C_3H_7OH	60.10	-123.2	97.3	1	75.1	75.7	40.2	81	67.3			190
isopropanol	67-63-0	iC_3H_7OH	60.096	-88	82.24	1	76.922		40.1	81	67.5	188.4		
1-butanol	71-36-3	C_4H_9OH	74.122	-88.42	117.725	1	91.991		50	103.6	83.6		272.1	216.3
isobutanol	78-83-1	iC_4H_9OH	74.122		108	1	92	92	49.9	103.6	82.7	213.2		213.2
sec-butyl alcohol	78-92-2	sC_4H_9OH	74.122	-114.7	99.512	1	92.37	91.7	50	103.6	83.8		264.1	
tert-butyl alcohol	75-65-0	tC_4H_9OH	74.122	25.62	82.347	1	94.2		49.9	103.6				
1-pentanol	71-41-0	$C_5H_{11}OH$	88.149	-78.2	137.98	1	108.72		59.3	126	100.1	244.9	303.9	
2-pentanol	6032-29-7	$C_5H_{11}OH$	88.149		119	1	109.45		59.3	126			295.9	238
1-hexanol	111-26-3	$C_6H_{13}OH$	102.176	-34.1	157	1	125.32		69	148	117.3	269.1	335.7	
1-heptanol	111-70-6	$C_7H_{15}OH$	116.2	-36	176	1	142.00		79	170	132.1		367.5	
1-octanol	111-87-5	$C_8H_{17}OH$	130.23	-14.97	195.2	1	158.51		88	192	148.7	321.2	399.4	
allyl alcohol	107-18-6	C_3H_5OH	58.09	-129	96-97	1	68.01		37.2	103.6				
benzyl alcohol	100-51-6	C_7H_8O	108.14	-15.3	205.45	1	103.06							
cyclopentanol	96-41-3	$C_5H_{10}O$	86.13	-19	139	1	96.7	100.9	56	118				
cyclohexanol	108-93-0	$C_6H_{12}O$	100.16	25.15	161.1	1	103.43			125.6				232.3
ethylene glycol	107-21-1	$(CH_2OH)_2$	62.07	-13	195-198	1	114			122				177.1

(a) V_M liquid molar volume, V_I intrinsic molar volume, Kamlet et al. 1987, 1988
(b) Jaworska & Schultz 1993
(c) TSA with and without hydration cell, Amidon et al. 1975
(d) Iwase et al. 1985

Table 2.2 Summary and selected physical-chemical properties of alcohols at 25 °C.

compound	selected properties:				Henry's law constants, Pa·m³/mol			
	S, g/m³	C_L, mol/m³	P_L, Pa	log K_{OW}	calc, P/C	exptl.(a)	exptl.(b)	exptl.(c)
methanol	miscible	miscible	16210	-0.77		0.45		0.45
ethanol	miscible	miscible	7800	-0.31		0.53		0.527
propanol	miscible	miscible	2780	0.25				0.751
isopropanol	miscible	miscible	5700	0.05				0.80
1-butanol	74000	998.35	900	0.82	1.373	0.80	0.892	0.80
isobutanol	81000	1092.79	1500	0.76	1.373	0.99		0.99
sec-butyl alcohol	181000	2441.92	2300	0.61	0.942	0.80		0.918
tert-butyl alcohol	miscible	miscible	5500	0.35		1.46		1.46
1-pentanol	22000	249.58	300	1.50	1.202	1.314		
2-pentanol	45000	510.50	777	1.14	1.522			
1-hexanol	6000	58.72	110	2.03	1.873	1.562	1.735	
1-heptanol	1740	14.97	24	2.62	1.603	1.909		
1-octanol	340	2.611	11	3.07	4.213	2.454	2.48	
1-nonanol	106	0.735		3.77				
allyl alcohol	miscible	miscible	3750					
benzyl alcohol	80	0.740	12	1.10	16.221			
cyclopentanol			292					
cyclohexanol	38000	379.39	85	1.23	0.224			
ethylene glycol	miscible	miscible	12	-1.36		0.006		

(a) Butler et al. 1935
(b) Buttery et al. 1969
(c) Snider & Dawson 1985

Table 2.3 Suggested half-life classes of alcohols at 25 °C in various environmental compartments

Compounds	Air class	Water class	Soil class	Sediment class
methanol	4	3	3	4
ethanol	3	3	3	4
propanol	3	3	3	4
isopropanol	3	3	3	4
1-butanol	3	3	3	4
2-methyl-1-propanol	3	3	3	4
tert-butyl alcohol	4	4	4	5
1-pentanol	3	3	3	4
1-hexanol	3	3	3	4
1-octanol	3	3	3	4
allyl alcohol	3	3	3	4
benzyl alcohol	3	3	3	4
cyclohexanol	3	3	3	4
ethylene glycol	3	3	3	4

where,

Class	Mean half-life (hours)	Range (hours)
1	5	< 10
2	17 (~ 1 day)	10-30
3	55 (~ 2 days)	30-100
4	170 (~ 1 week)	100-300
5	550 (~ 3 weeks)	300-1,000
6	1700 (~ 2 months)	1,000-3,000
7	5500 (~ 8 months)	3,000-10,000
8	17000 (~ 2 years)	10,000-30,000
9	~ 5 years	> 30,000

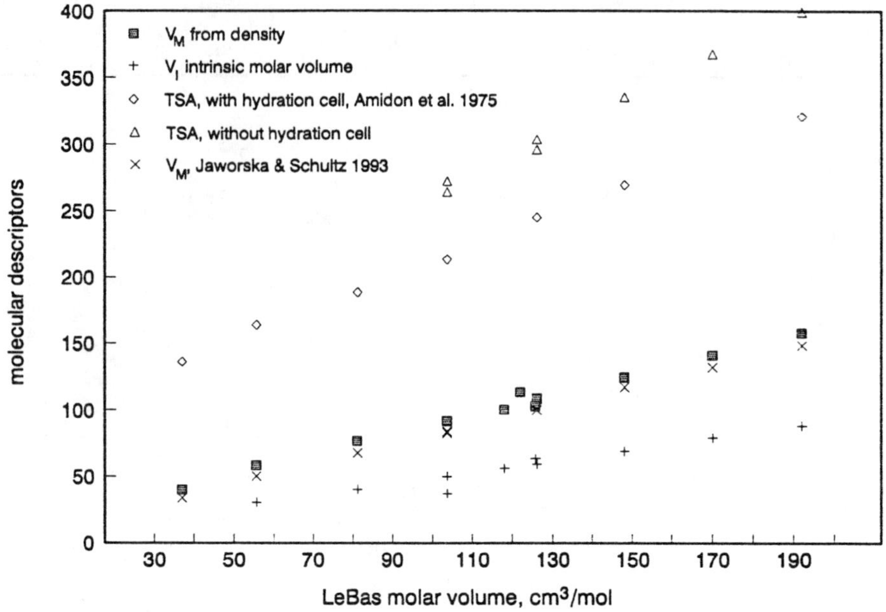

Figure 2.1 Plot of molecular descriptors vs LeBas molar volume

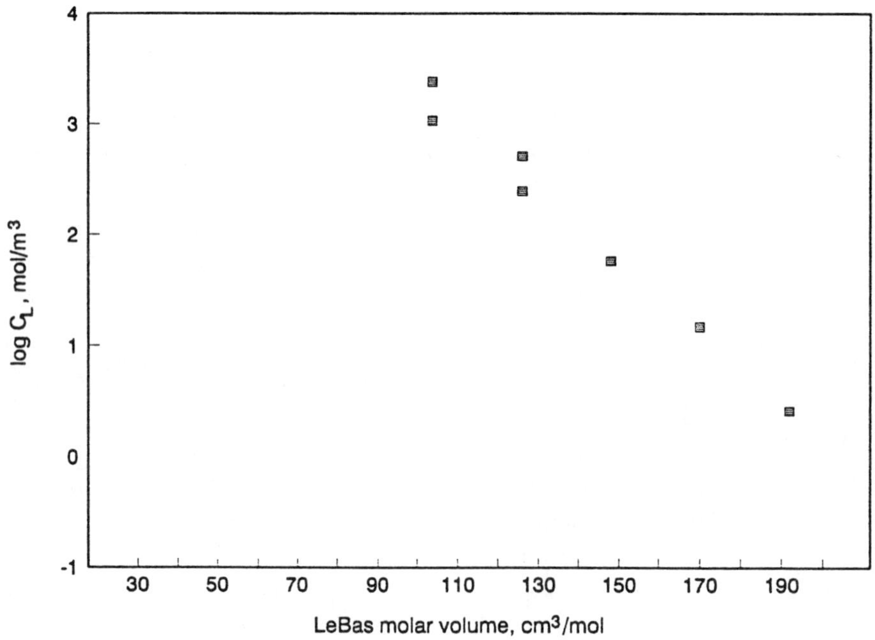

Figure 2.2 Plot of log liquid solubility vs LeBas molar volume

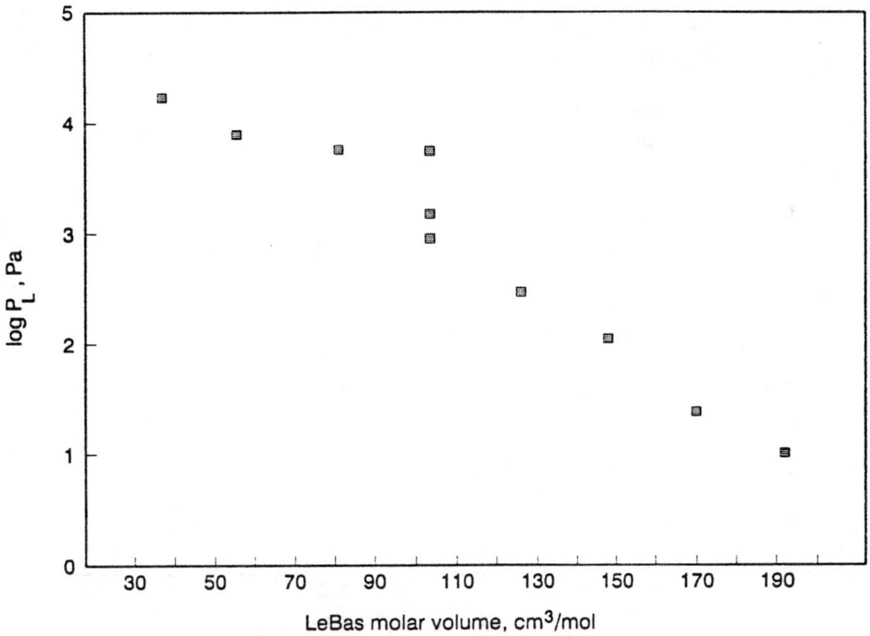

Figure 2.3 Plot of log liquid vapor pressure vs LeBas molar volume

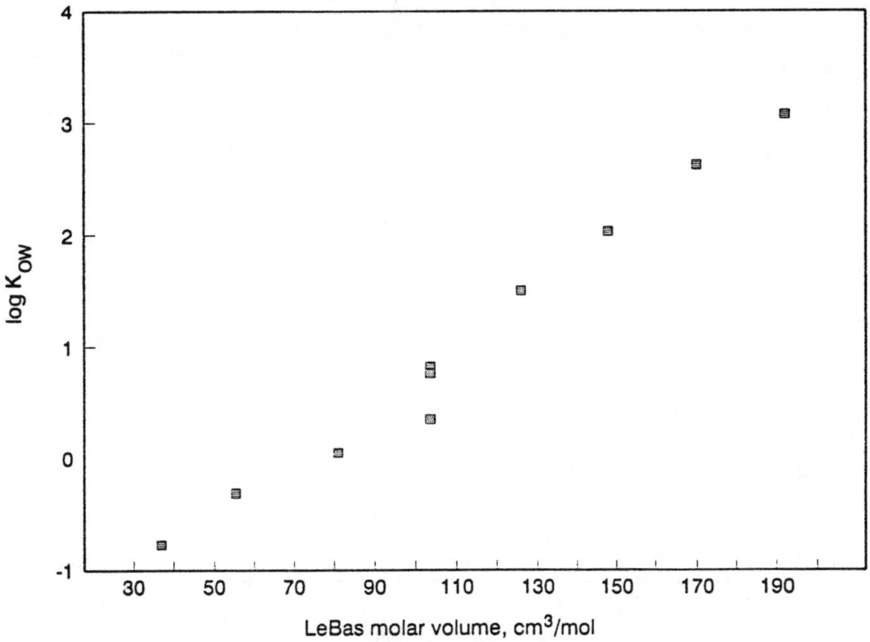

Figure 2.4 Plot of log K_{OW} versus LeBas molar volume

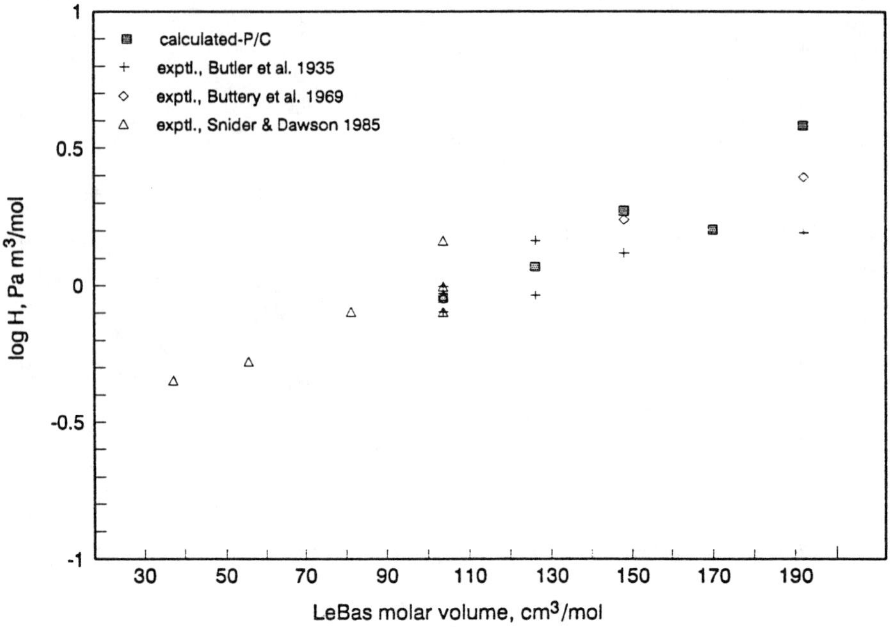

Figure 2.5 Plot of log H vs LeBas molar volume

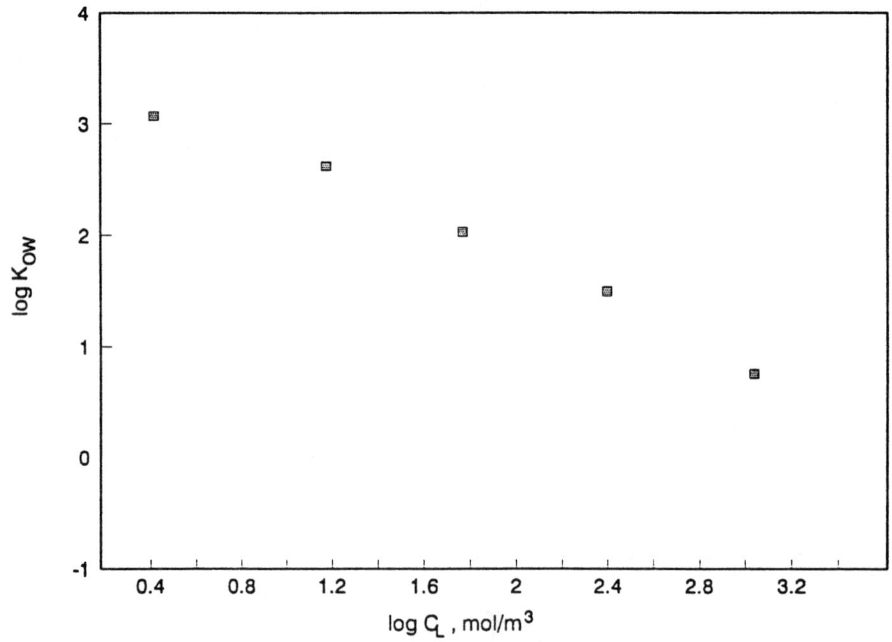

Figure 2.6 Plot of log K_{OW} versus log liquid solubility

2.3 Illustrative Fugacity Calculations: Levels I, II, III

Chemical name: Ethanol

Level I calculation: (six-compartment model)

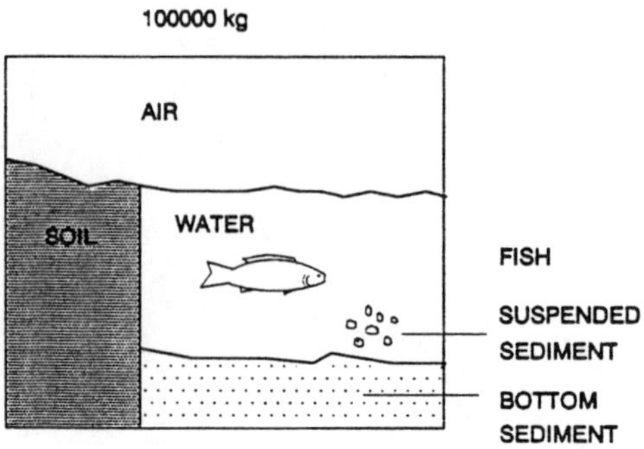

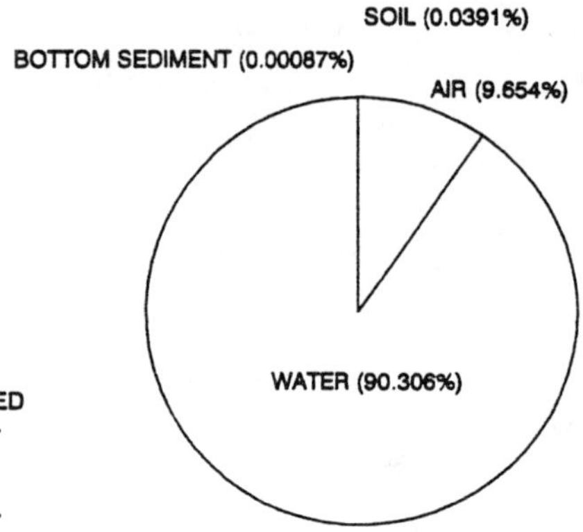

Distribution of mass

Level I calculation:

Physical-chemical properties:		Partition coeff.:	
molecular wt., g/mol	46.07	HLC	0.53001
melting point, C	-115	Kaw	2.14E-04
solubility, g/m3	678000	Koc	0.2008
vapor pressure, Pa	7800	BCF	0.0245
log Kow	-0.31	Ksw	9.64E-03
fugacity ratio, F	1.0000	Ksd/w	0.0193
dissoc. const. pKa		Kssd/w	0.0602
		Kar/w	769.231

Compartment	Z	Concentration			Amount	Amount
	mol/m3 Pa	mol/m3	mg/L (or g/m3)	ug/g	kg	%
Air	4.03E-04	2.10E-09	9.65E-08	8.14E-05	9654.35	9.6543
Water	1.89E+00	9.80E-06	4.52E-04	4.52E-04	90306	90.306
Soil	1.82E-02	9.45E-08	4.35E-06	1.81E-06	39.170	0.0392
Biota (fish)	4.62E-02	2.40E-07	1.11E-05	1.11E-05	0.00221	2.21E-06
Suspended sediment	1.14E-01	5.90E-07	2.72E-05	1.81E-05	0.0272	2.72E-05
Bottom sediment	3.64E-02	1.89E-07	8.70E-06	3.63E-06	0.8704	8.70E-04
	Total				100000	100

f = 5.195E-06 Pa

Chemical name: Ethanol

Level II calculation: (six-compartment model)

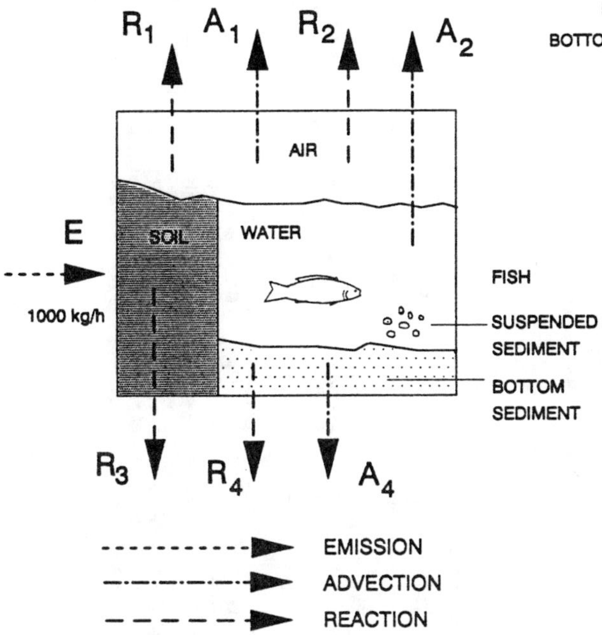

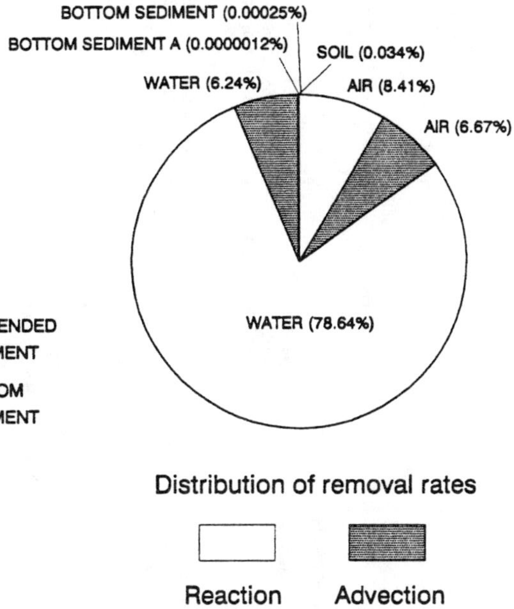

Distribution of removal rates

Level II Calculation:

Compartment	Half-life h	D Values Reaction mol/Pa.h	D Values Advection mol/Pa.h	Conc'n mol/m3	Loss Reaction kg/h	Loss Advection kg/h	Removal %
Air	55	5.08E+08	4.03E+08	1.45E-09	84.076	66.727	15.080
Water	55	4.75E+09	3.77E+08	6.77E-06	786.438	62.416	84.885
Soil	55	2.06E+06		6.53E-08	0.341		0.0341
Biota (fish)				1.66E-07			
Suspended sediment				4.08E-07			
Bottom sediment	170	1.48E+04	7.27E+01	1.31E-07	2.45E-03	1.20E-05	2.46E-04
	Total	5.27E+09	7.81E+08		870.86	129.143	100
	R + A		6.05E+09			1000	

f = 3.590E-06 Pa

Total amt = 69116 kg

Overall residence time = 69.12 h

Reaction time = 79.37 h

Advection time = 535.19 h

Fugacity Level III calculations: (four-compartment model)

Chemical name: Ethanol

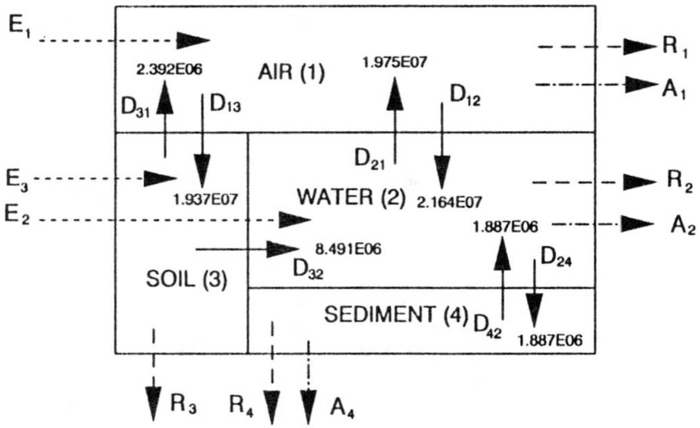

Phase Properties and Rates:

Compartment	Bulk Z mol/m3 Pa	Half-life h	D Values Reaction mol/Pa h	D Values Advection mol/Pa h
Air (1)	4.034E-04	55	5.08E+08	4.03E+08
Water (2)	1.887E+00	55	4.75E+09	3.77E+08
Soil (3)	5.752E-01	55	1.30E+08	
Sediment (4)	1.517E+00	170	3.09E+06	1.52E+04

	E(1)=1000	E(2)=1000	E(3)=1000	E(1,2,3)
Overall residence time =	45.61	73.46	78.44	75.55 h
Reaction time =	79.37	79.41	79.37	79.38 h
Advection time =	107.22	981.29	6706	1568.40 h

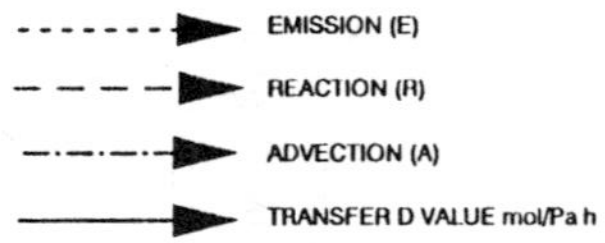

Phase Properties, Compositions, Transport and Transformation Rates:

Emission, kg/h			Fugacity, Pa				Concentration, g/m3				Amounts, kg				Total Amount, kg
E(1)	E(2)	E(3)	f(1)	f(2)	f(3)	f(4)	C(1)	C(2)	C(3)	C(4)	m(1)	m(2)	m(3)	m(4)	
1000	0	0	2.279E-05	1.008E-07	3.124E-06	3.812E-08	4.236E-07	8.766E-06	8.279E-05	2.663E-06	4.236E+04	1.753E+03	1.490E+03	1.332E+00	4.561E+04
0	1000	0	8.735E-08	4.213E-06	1.197E-08	1.592E-06	1.624E-09	3.662E-04	3.173E-07	1.113E-04	1.624E+02	7.324E+04	5.711E+00	5.563E+01	7.346E+04
0	0	1000	3.910E-07	2.548E-07	1.536E-04	9.630E-08	7.268E-09	2.215E-05	4.071E-03	6.729E-06	7.268E+02	4.429E+03	7.328E+04	3.364E+00	7.844E+04
50	250	700	1.435E-06	1.237E-06	1.077E-04	4.674E-07	2.667E-08	1.075E-04	2.854E-03	3.266E-05	2.667E+03	2.150E+04	5.137E+04	1.633E+01	7.555E+04

Emission, kg/h			Loss, Reaction, kg/h				Loss, Advection, kg/h			Intermedia Rate of Transport, kg/h T12	T21	T13	T31	T32	T24	T42
E(1)	E(2)	E(3)	R(1)	R(2)	R(3)	R(4)	A(1)	A(2)	A(4)	air-water	water-air	air-soil	soil-air	soil-water	water-sed	sed-water
1000	0	0	5.338E+02	2.209E+01	1.88E+01	5.429E-03	4.236E+02	1.753E+00	2.663E-05	2.272E+01	9.175E-02	2.034E+01	3.443E-01	1.222E+00	8.769E-03	3.313E-03
0	1000	0	2.046E+00	9.228E+02	7.20E-02	2.268E-01	1.624E+00	7.324E+01	1.113E-03	8.707E-02	3.833E+00	7.797E-02	1.320E-03	4.684E-03	3.663E-01	1.384E-01
0	0	1000	9.157E+00	5.581E+01	9.23E+02	1.371E-02	7.268E+00	4.429E+00	6.729E-05	3.898E-01	2.318E-01	3.490E-01	1.693E+01	6.009E+01	2.215E-02	8.371E-03
50	250	700	3.361E+01	2.709E+02	6.47E+02	6.656E-02	2.667E+01	2.150E+01	3.266E-04	1.431E+00	1.125E+00	1.281E+00	1.187E+01	4.213E+01	1.075E-01	4.063E-02

Chemical name: Ethanol

Level III Distribution

Distribution of mass

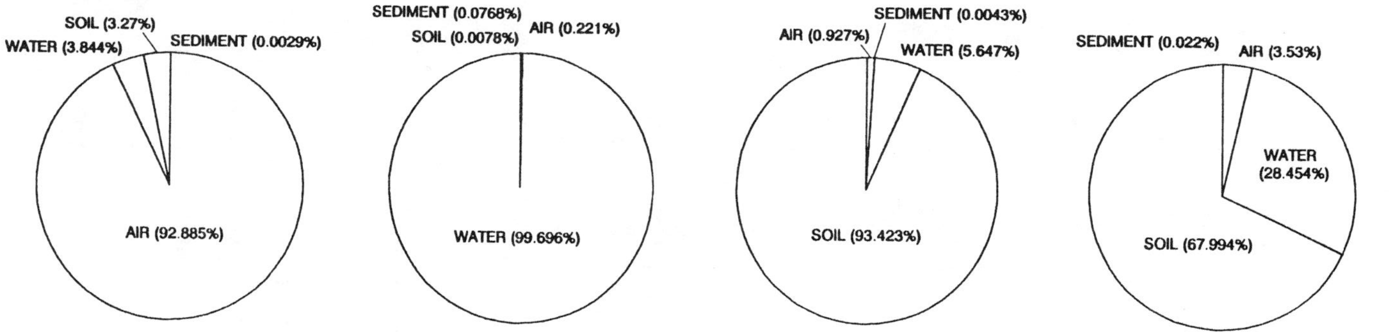

Distribution of removal rates

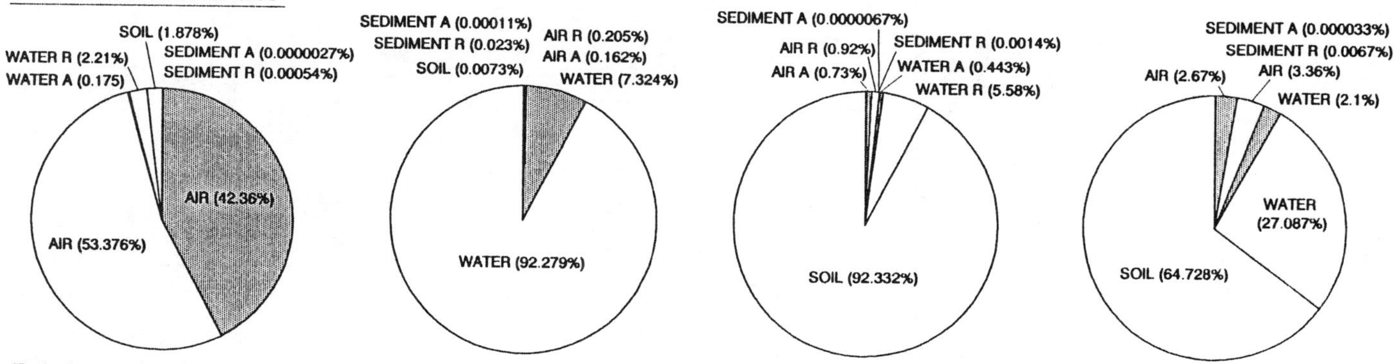

Reaction (unshaded) Advection (shaded)

Emission rates:	E(1) = 1000 kg/h	E(1) = 0	E(1) = 0	E(1) = 50 kg/h
	E(2) = 0	E(2) = 1000 kg/h	E(2) = 0	E(2) = 250 kg/h
	E(3) = 0	E(3) = 0	E(3) = 1000 kg/h	E(3) = 700 kg/h
Residence time:	t = 45.61 h	t = 73.46 h	t = 78.44 h	t = 75.55 h

Chemical name: t-Butyl alcohol

Level I calculation: (six-compartment model)

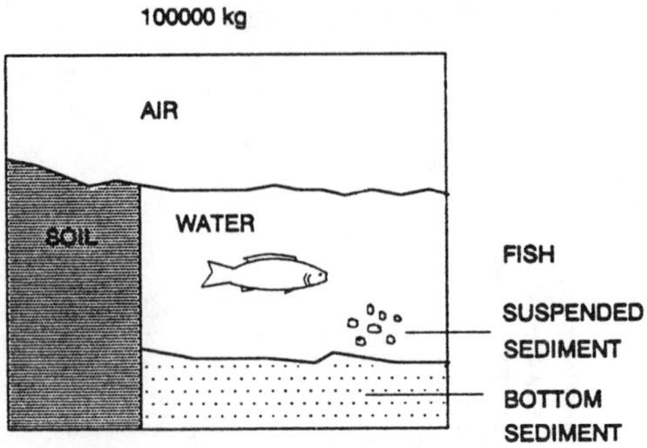

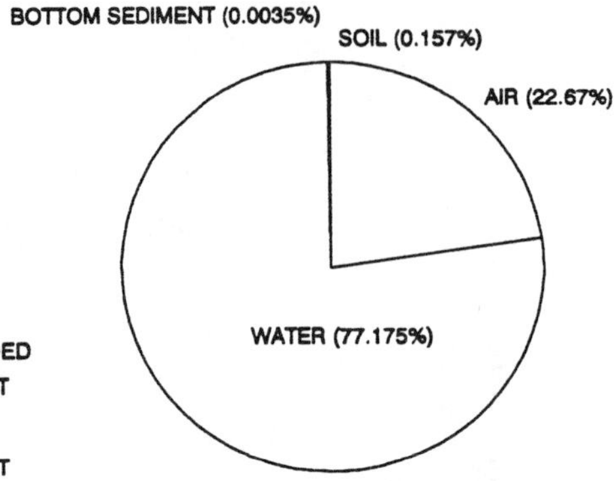

Distribution of mass

Level I calculation:

Physical-chemical properties:	
molecular wt., g/mol	74.12
melting point, C	-115
solubility, g/m3	280000
vapor pressure, Pa	5500
log Kow	0.36
fugacity ratio, F	1.0000
dissoc. const. pKa	

Partition coeff.:	
HLC	1.45593
Kaw	5.87E-04
Koc	0.9393
BCF	0.1145
Ksw	4.51E-02
Ksd/w	0.0902
Kssd/w	0.2818
Kar/w	1090.909

Compartment	Z	Concentration			Amount	Amount
	mol/m3 Pa	mol/m3	mg/L (or g/m3)	ug/g	kg	%
Air	4.03E-04	3.06E-09	2.27E-07	1.91E-04	22664.40	22.6644
Water	6.87E-01	5.21E-06	3.86E-04	3.86E-04	77175	77.175
Soil	3.10E-02	2.35E-07	1.74E-05	7.25E-06	156.573	0.1566
Biota (fish)	7.87E-02	5.96E-07	4.42E-05	4.42E-05	0.00884	8.84E-06
Suspended sediment	1.94E-01	1.47E-06	1.09E-04	7.25E-05	0.1087	1.09E-04
Bottom sediment	6.19E-02	4.69E-07	3.48E-05	1.45E-05	3.4794	3.48E-03
	Total				100000	100

f = 7.580E-06 Pa

Chemical name: t-Butyl alcohol

Level II calculation: (six-compartment model)

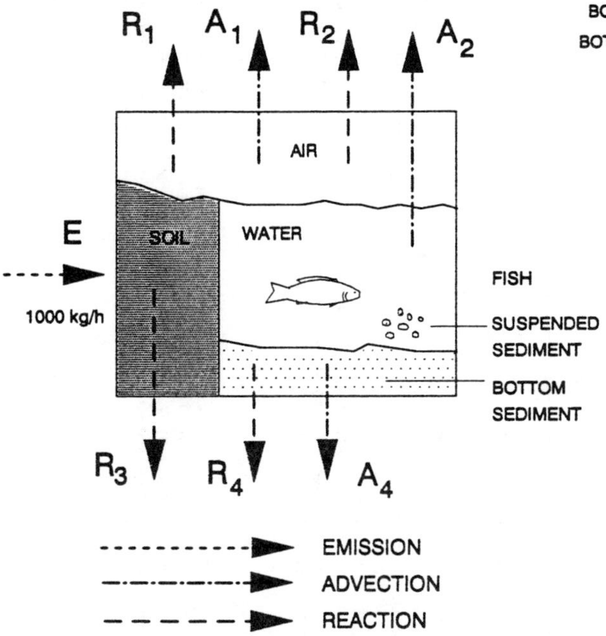

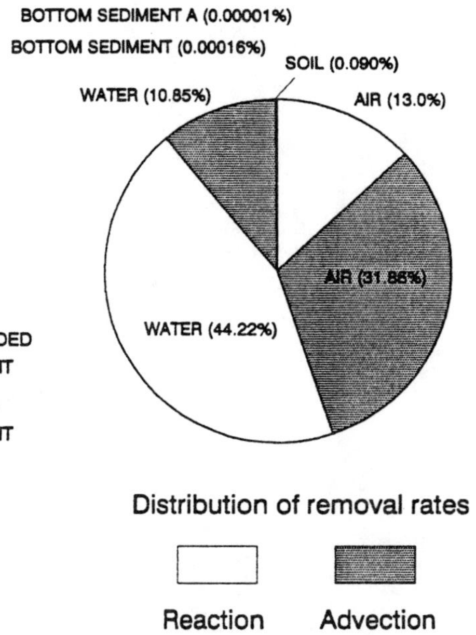

Level II Calculation:

Compartment	Half-life h	D Values Reaction mol/Pa.h	D Values Advection mol/Pa.h	Conc'n mol/m3	Loss Reaction kg/h	Loss Advection kg/h	Removal %
Air	170	1.64E+08	4.03E+08	4.30E-09	129.861	318.564	44.843
Water	170	5.60E+08	1.37E+08	7.32E-06	442.196	108.475	55.067
Soil	170	1.14E+06		3.30E-07	0.8971		0.0897
Biota (fish)				8.38E-07			
Suspended sediment				2.06E-06			
Bottom sediment	550	7.80E+03	1.24E+02	6.60E-07	6.16E-03	9.78E-05	6.26E-04
	Total	7.26E+08	5.41E+08		572.96	427.039	100
	R + A		1.27E+09			1000	

f = 1.065E-05 Pa

Total amt = 140557 kg

Overall residence time = 140.56 h

Reaction time = 245.32 h

Advection time = 329.14 h

Fugacity Level III calculations: (four-compartment model)

Chemical name: t-Butyl alcohol

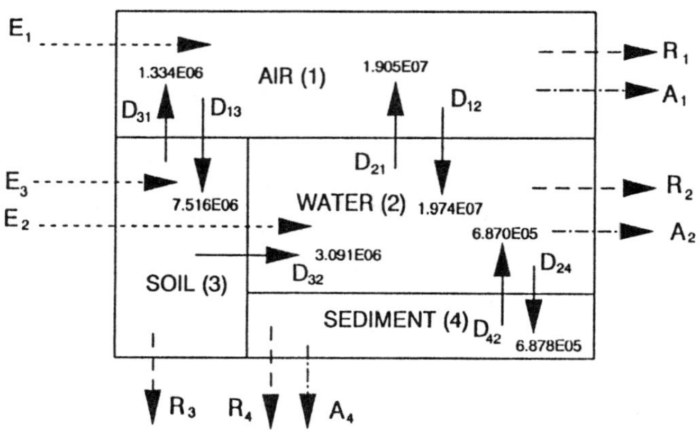

Phase Properties and Rates:

Compartment	Bulk Z mol/m3 Pa	Half-life h	D Values Reaction mol/Pa h	Advection mol/Pa h
Air (1)	4.034E-04	170	1.64E+08	4.03E+08
Water (2)	6.868E-01	170	5.60E+08	1.37E+08
Soil (3)	2.216E-01	170	1.63E+07	
Sediment (4)	5.619E-01	550	3.54E+05	5.62E+03

	E(1)=1000	E(2)=1000	E(3)=1000	E(1,2,3)	
Overall residence time =	77.09	193.99	226.79	211.11	h
Reaction time =	245.33	245.54	245.34	245.38	h
Advection time =	112.40	924.12	3000	1511.23	h

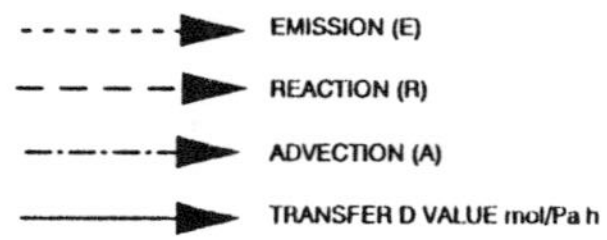

Phase Properties, Compositions, Transport and Transformation Rates:

Emission, kg/h			Fugacity, Pa				Concentration, g/m3				Amounts, kg				Total Amount, kg
E(1)	E(2)	E(3)	f(1)	f(2)	f(3)	f(4)	C(1)	C(2)	C(3)	C(4)	m(1)	m(2)	m(3)	m(4)	
1000	0	0	2.271E-05	6.611E-07	8.251E-06	4.345E-07	6.791E-07	3.366E-05	1.355E-04	1.809E-05	6.791E+04	6.731E+03	2.440E+03	9.047E+00	7.709E+04
0	1000	0	6.037E-07	1.884E-05	2.194E-07	1.238E-05	1.805E-08	9.593E-04	3.603E-06	5.158E-04	1.805E+03	1.919E+05	6.486E+01	2.579E+02	1.940E+05
0	0	1000	1.555E-06	2.858E-06	6.528E-04	1.878E-06	4.650E-08	1.455E-04	1.072E-02	7.823E-05	4.650E+03	2.910E+04	1.930E+05	3.911E+01	2.268E+05
50	250	700	2.375E-06	6.745E-06	4.574E-04	4.433E-06	7.102E-08	3.434E-04	7.513E-03	1.846E-04	7.102E+03	6.867E+04	1.352E+05	9.230E+01	2.111E+05

Emission, kg/h			Loss, Reaction, kg/h				Loss, Advection, kg/h			Intermedia Rate of Transport, kg/h T12	T21	T13	T31	T32	T24	T42
E(1)	E(2)	E(3)	R(1)	R(2)	R(3)	R(4)	A(1)	A(2)	A(4)	air-water	water-air	air-soil	soil-air	soil-water	water-sed	sed-water
1000	0	0	2.768E+02	2.744E+01	9.95E+00	1.140E-02	6.791E+02	6.731E+00	1.809E-04	3.323E+01	9.336E-01	1.265E+01	8.161E-01	1.890E+00	3.370E-02	2.212E-02
0	1000	0	7.359E+00	7.821E+02	2.64E-01	3.249E-01	1.805E+01	1.919E+02	5.158E-03	8.833E-01	2.661E+01	3.363E-01	2.170E-02	5.025E-02	9.607E-01	6.306E-01
0	0	1000	1.896E+01	1.186E+02	7.87E+02	4.928E-02	4.650E+01	2.910E+01	7.823E-04	2.275E+00	4.036E+00	8.664E-01	6.456E+01	1.495E+02	1.457E-01	9.564E-02
50	250	700	2.895E+01	2.799E+02	5.51E+02	1.163E-01	7.102E+01	6.867E+01	1.846E-03	3.475E+00	9.524E+00	1.323E+00	4.524E+01	1.048E+02	3.438E-01	2.257E-01

Chemical name: t-Butyl alcohol

Level III Distribution

Distribution of mass

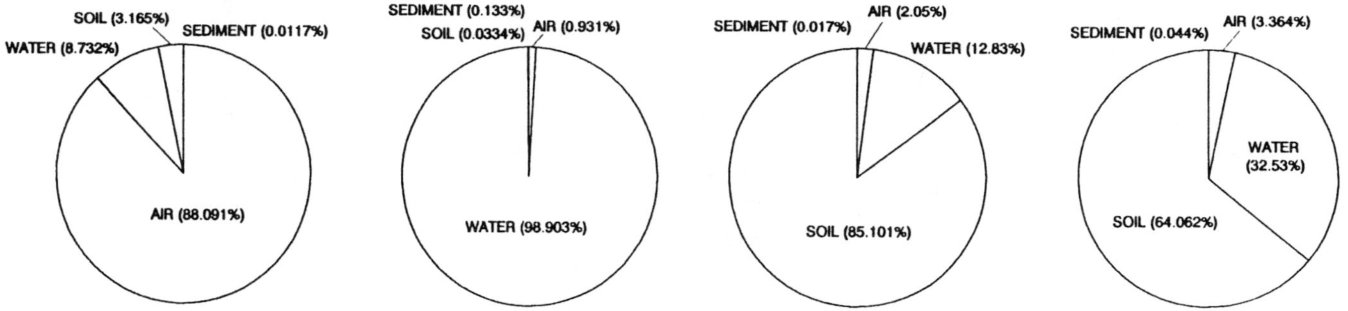

Distribution of removal rates

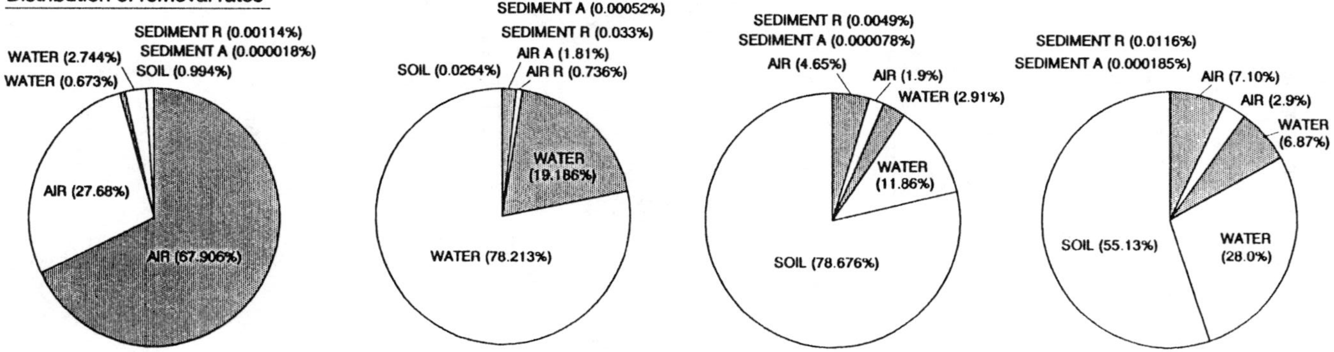

Reaction Advection

Emission rates:			
E(1) = 1000 kg/h	E(1) = 0	E(1) = 0	E(1) = 50 kg/h
E(2) = 0	E(2) = 1000 kg/h	E(2) = 0	E(2) = 250 kg/h
E(3) = 0	E(3) = 0	E(3) = 1000 kg/h	E(3) = 700 kg/h
Residence time:			
t = 77.09 h	t = 194 h	t = 227 h	t = 211 h

Chemical name: 1-Hexanol

Level I calculation: (six-compartment model)

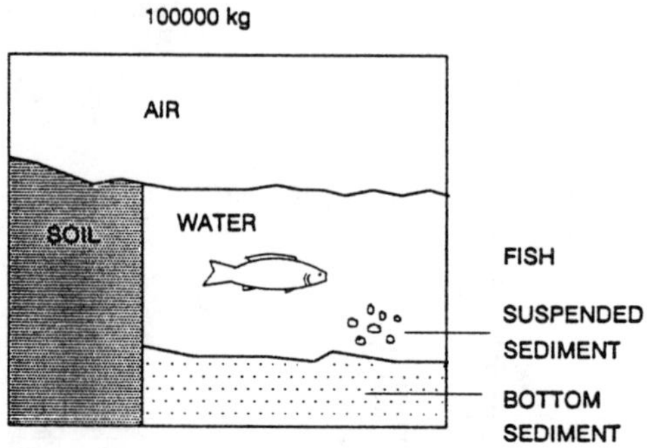

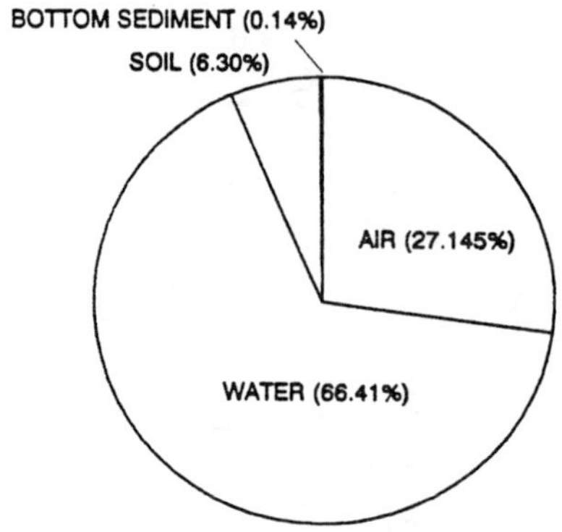

Distribution of mass

Level I calculation:

Physical-chemical properties:		Partition coeff.:	
molecular wt., g/mol	102.18	HLC	2.02657
melting point, C	-44.6	Kaw	8.18E-04
solubility, g/m3	6000	Koc	43.9323
vapor pressure, Pa	119	BCF	5.3576
log Kow	2.03	Ksw	2.1087
fugacity ratio, F	1.0000	Ksd/w	4.2175
dissoc. const. pKa		Kssd/w	13.1797
		Kar/w	50420.17

Compartment	Z	Concentration			Amount	Amount
	mol/m3 Pa	mol/m3	mg/L (or g/m3)	ug/g	kg	%
Air	4.03E-04	2.66E-09	2.71E-07	2.29E-04	27145.93	27.1459
Water	4.93E-01	3.25E-06	3.32E-04	3.32E-04	66408	66.408
Soil	1.04E+00	6.85E-06	7.00E-04	2.92E-04	6301.67	6.3017
Biota (fish)	2.64E+00	1.74E-05	1.78E-03	1.78E-03	0.35579	3.56E-04
Suspended sediment	6.50E+00	4.28E-05	4.38E-03	2.92E-03	4.3762	4.38E-03
Bottom sediment	2.08E+00	1.37E-05	1.40E-03	5.83E-04	140.04	0.1400
	Total				100000	100

f = 6.585E-06 Pa

Chemical name: 1-Hexanol

Level II calculation: (six-compartment model)

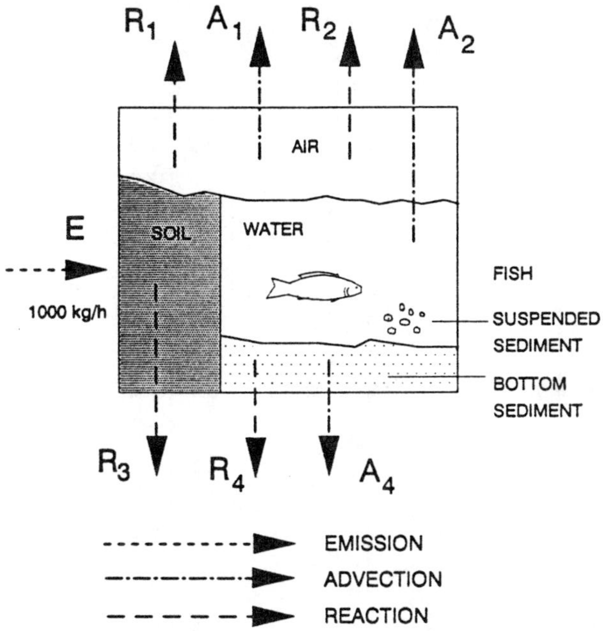

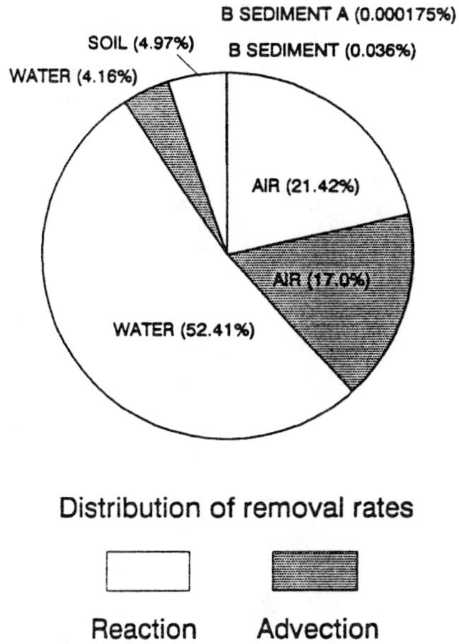

Distribution of removal rates

Reaction Advection

Level II Calculation:

Compartment	Half-life h	D Values Reaction mol/Pa.h	D Values Advection mol/Pa.h	Conc'n mol/m3	Loss Reaction kg/h	Loss Advection kg/h	Removal %
Air	55	5.08E+08	4.03E+08	1.66E-09	214.227	170.022	38.425
Water	55	1.24E+09	9.87E+07	2.04E-06	524.068	41.593	56.566
Soil	55	1.18E+08		4.29E-06	49.731		4.9731
Biota (fish)				1.09E-05			
Suspended sediment				2.68E-05			
Bottom sediment	170	8.48E+05	4.16E+03	8.58E-06	3.58E-01	1.75E-03	3.59E-02
	Total	1.87E+09	5.02E+08		788.38	211.616	100
	R + A		2.37E+09			1000	

f = 4.125E-06 Pa
Total amt = 62632 kg

Overall residence time = 62.63 h
Reaction time = 79.44 h
Advection time = 295.97 h

Fugacity Level III calculations: (four-compartment model)

Chemical name: 1-Hexanol

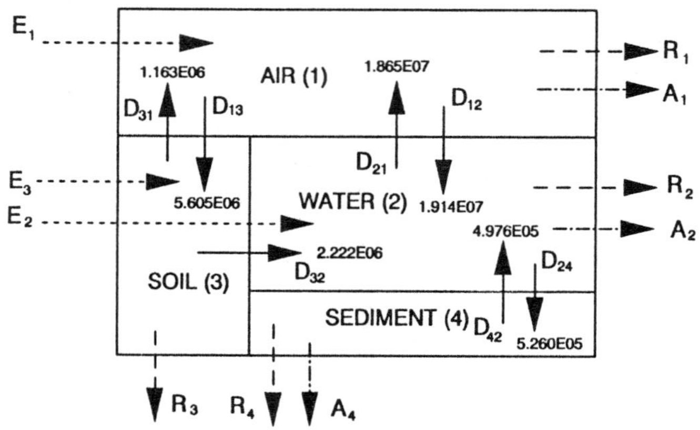

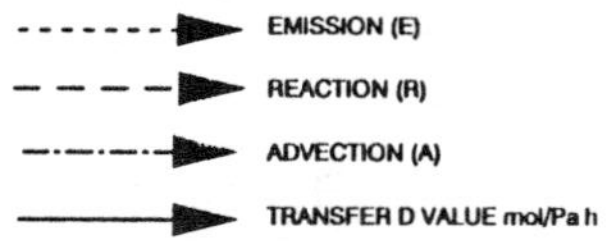

Phase Properties and Rates:

Compartment	Bulk Z mol/m3 Pa	Half-life h	D Values Reaction mol/Pa h	D Values Advection mol/Pa h
Air (1)	4.034E-04	55	5.08E+08	4.03E+08
Water (2)	4.935E-01	55	1.24E+09	9.87E+07
Soil (3)	6.684E-01	55	1.52E+08	
Sediment (4)	8.110E-01	170	1.65E+06	8.11E+03

	E(1)=1000	E(2)=1000	E(3)=1000	E(1,2,3)	
Overall residence time =	45.05	73.19	79.02	75.86	h
Reaction time =	79.37	79.42	79.37	79.38	h
Advection time =	104.18	933.25	18088	1713.07	h

Phase Properties, Compositions, Transport and Transformation Rates:

Emission, kg/h			Fugacity, Pa				Concentration, g/m3				Amounts, kg				Total Amount, kg
E(1)	E(2)	E(3)	f(1)	f(2)	f(3)	f(4)	C(1)	C(2)	C(3)	C(4)	m(1)	m(2)	m(3)	m(4)	
1000	0	0	1.045E-05	1.476E-07	3.781E-07	3.596E-08	4.309E-07	7.443E-06	2.582E-05	2.980E-06	4.309E+04	1.489E+03	4.648E+02	1.490E+00	4.505E+04
0	1000	0	1.432E-07	7.191E-06	5.179E-09	1.752E-06	5.903E-09	3.626E-04	3.537E-07	1.452E-04	5.903E+02	7.252E+04	6.366E+00	7.260E+01	7.319E+04
0	0	1000	8.049E-08	1.042E-07	6.315E-05	2.539E-08	3.318E-09	5.253E-06	4.313E-03	2.104E-06	3.318E+02	1.051E+03	7.764E+04	1.052E+00	7.902E+04
50	250	700	6.148E-07	1.878E-06	4.423E-05	4.576E-07	2.534E-08	9.470E-05	3.021E-03	3.792E-05	2.534E+03	1.894E+04	5.437E+04	1.896E+01	7.586E+04

Emission, kg/h			Loss, Reaction, kg/h				Loss, Advection, kg/h			Intermedia Rate of Transport, kg/h						
										T12	T21	T13	T31	T32	T24	T42
E(1)	E(2)	E(3)	R(1)	R(2)	R(3)	R(4)	A(1)	A(2)	A(4)	air-water	water-air	air-soil	soil-air	soil-water	water-sed	sed-water
1000	0	0	5.430E+02	1.876E+01	5.86E+00	6.074E-03	4.309E+02	1.489E+00	2.980E-05	2.045E+01	2.812E-01	5.987E+00	4.492E-02	8.582E-03	7.932E-03	1.829E-03
0	1000	0	7.437E+00	9.138E+02	8.02E-02	2.959E-01	5.903E+00	7.252E+01	1.452E-03	2.800E-01	1.370E+01	8.201E-02	6.153E-04	1.175E-03	3.865E-01	8.909E-02
0	0	1000	4.180E+00	1.324E+01	9.78E+02	4.288E-03	3.318E+00	1.051E+00	2.104E-05	1.574E-01	1.985E-01	4.610E-02	7.503E+00	1.433E+01	5.599E-03	1.291E-03
50	250	700	3.193E+01	2.386E+02	6.85E+02	7.729E-02	2.534E+01	1.894E+01	3.792E-04	1.202E+00	3.578E+00	3.521E-01	5.255E+00	1.004E+01	1.009E-01	2.327E-02

Level III Distribution

Chemical name: 1-Hexanol

Distribution of mass

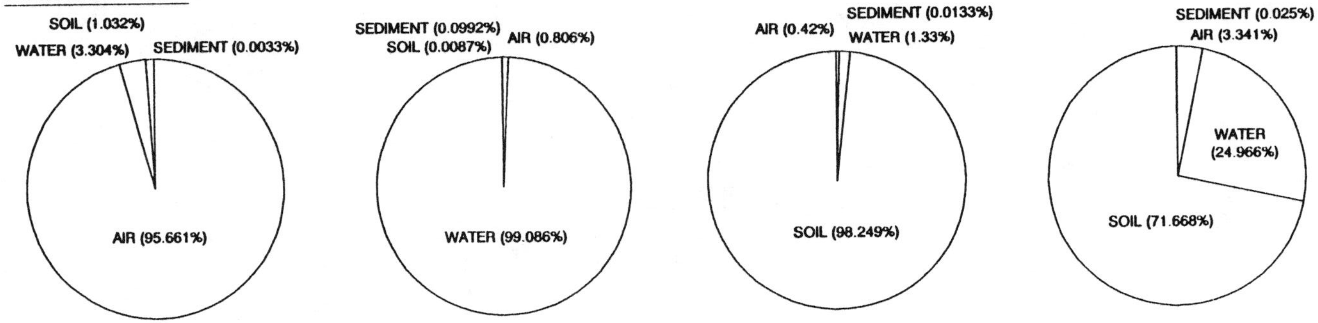

Distribution of removal rates

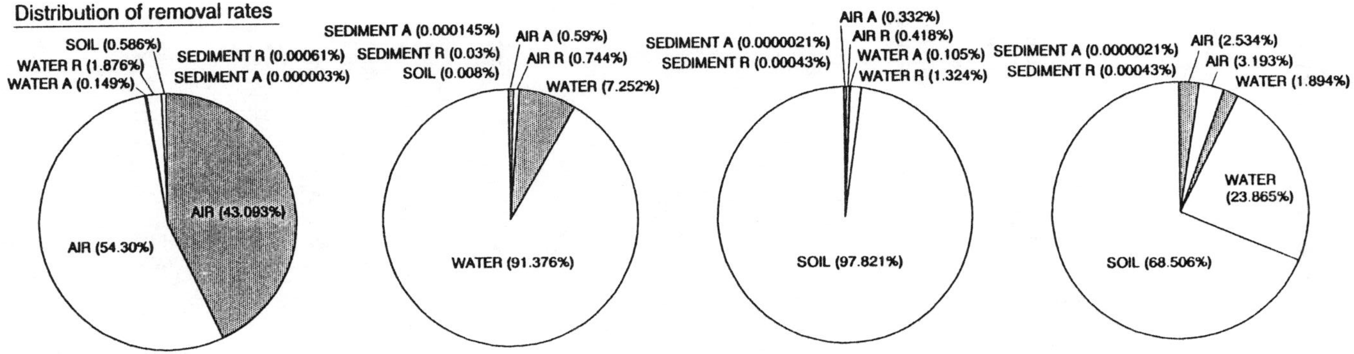

Reaction   Advection

Emission rates:			
E(1) = 1000 kg/h	E(1) = 0	E(1) = 0	E(1) = 50 kg/h
E(2) = 0	E(2) = 1000 kg/h	E(2) = 0	E(2) = 250 kg/h
E(3) = 0	E(3) = 0	E(3) = 1000 kg/h	E(3) = 700 kg/h
Residence time:			
t = 45.05 h	t = 73.19 h	t = 79.02 h	t = 75.86 h

Chemical name: 1-Octanol

Level I calculation: (six-compartment model)

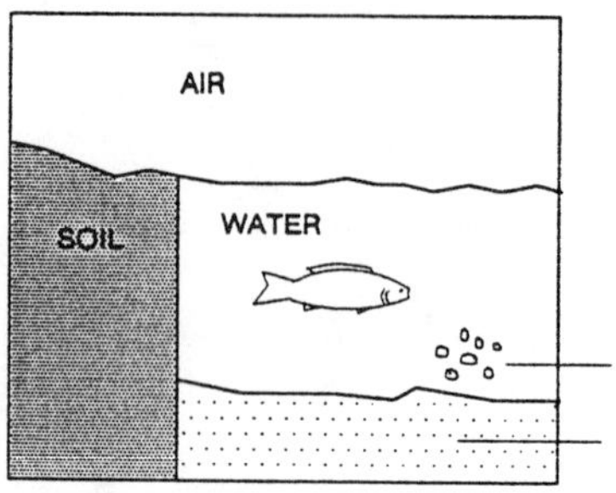

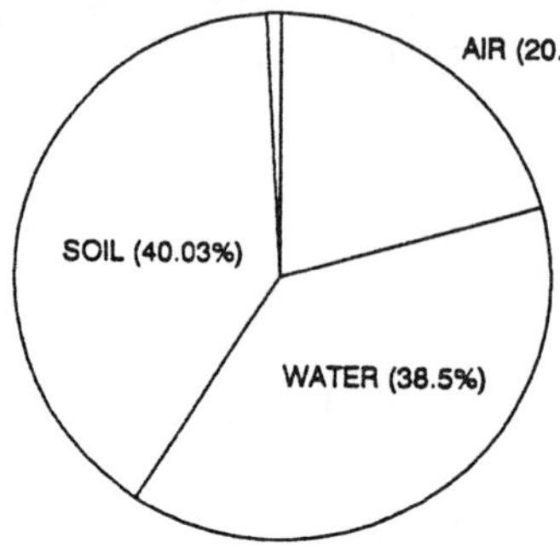

Distribution of mass

Level I calculation:

Physical-chemical properties:		Partition coeff.:	
molecular wt., g/mol	130.23	HLC	2.65283
melting point, C	-14.97	Kaw	1.07E-03
solubility, g/m3	540	Koc	481.7080
vapor pressure, Pa	11	BCF	58.7449
log Kow	3.07	Ksw	23.1220
fugacity ratio, F	1.0000	Ksd/w	46.2440
dissoc. const. pKa		Kssd/w	144.5124
		Kar/w	545454.55

Compartment	Z	Concentration			Amount	Amount
	mol/m3 Pa	mol/m3	mg/L (or g/m3)	ug/g	kg	%
Air	4.03E-04	1.58E-09	2.06E-07	1.74E-04	20584.79	20.5848
Water	3.77E-01	1.48E-06	1.92E-04	1.92E-04	38469	38.469
Soil	8.72E+00	3.42E-05	4.45E-03	1.85E-03	40026.63	40.0266
Biota (fish)	2.21E+01	8.68E-05	1.13E-02	1.13E-02	2.25986	2.26E-03
Suspended sediment	5.45E+01	2.13E-04	2.78E-02	1.85E-02	27.7963	2.78E-02
Bottom sediment	1.74E+01	6.83E-05	8.89E-03	3.71E-03	889.48	0.8895
	Total				100000	100

f = 3.918E-06 Pa

Chemical name: 1-Octanol

Level II calculation: (six-compartment model)

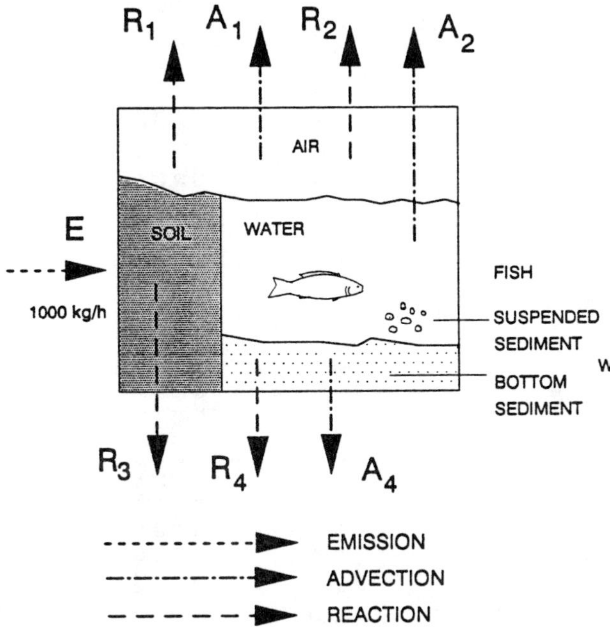

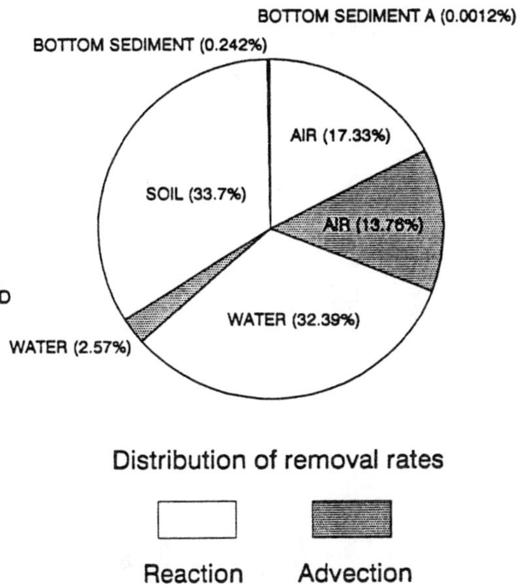

Level II Calculation:

Compartment	Half-life h	D Values Reaction mol/Pa.h	D Values Advection mol/Pa.h	Conc'n mol/m3	Loss Reaction kg/h	Loss Advection kg/h	Removal %
Air	55	5.08E+08	4.03E+08	1.06E-09	173.331	137.564	31.090
Water	55	9.50E+08	7.54E+07	9.87E-07	323.923	25.708	34.963
Soil	55	9.88E+08		2.28E-05	337.038		33.7038
Biota (fish)				5.80E-05			
Suspended sediment				1.43E-04			
Bottom sediment	170	7.11E+06	3.49E+04	4.56E-05	2.423	1.19E-02	0.2435
	Total	2.45E+09	4.79E+08		836.72	163.284	100
	R + A		2.93E+09			1000	

f = 2.618E-06 Pa

Total amt = 66828 kg

Overall residence time = 66.83 h

Reaction time = 79.87 h

Advection time = 409.27 h

Fugacity Level III calculations: (four-compartment model)

Chemical name: 1-Octanol

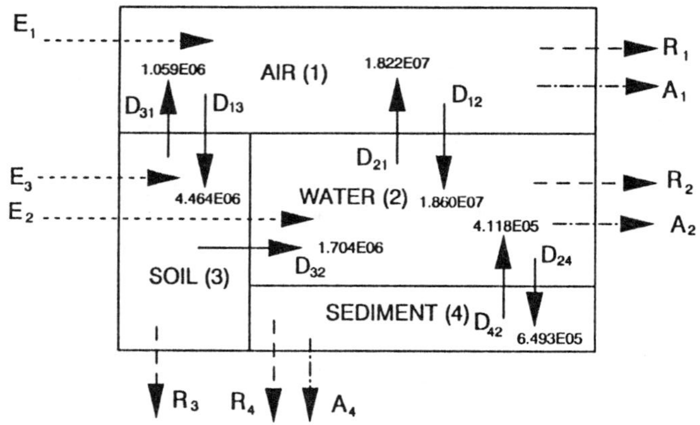

Phase Properties and Rates:

Compartment	Bulk Z mol/m3 Pa	Half-life h	D Values Reaction mol/Pa h	D Values Advection mol/Pa h
Air (1)	4.034E-04	55	5.08E+08	4.03E+08
Water (2)	3.772E-01	55	9.51E+08	7.54E+07
Soil (3)	4.471E+00	55	1.01E+09	
Sediment (4)	3.788E+00	170	7.72E+06	3.79E+04

	E(1)=1000	E(2)=1000	E(3)=1000	E(1,2,3)
Overall residence time =	44.99	73.13	79.32	76.06 h
Reaction time =	79.37	79.47	79.37	79.39 h
Advection time =	103.87	916.91	135616	1810.56 h

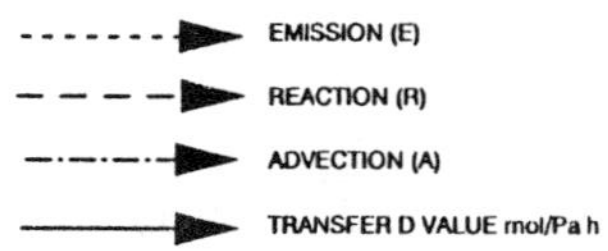

Phase Properties, Compositions, Transport and Transformation Rates:

Emission, kg/h			Fugacity, Pa				Concentration, g/m3				Amounts, kg				Total Amount, kg
E(1)	E(2)	E(3)	f(1)	f(2)	f(3)	f(4)	C(1)	C(2)	C(3)	C(4)	m(1)	m(2)	m(3)	m(4)	
1000	0	0	8.217E-06	1.463E-07	3.607E-08	1.163E-08	4.317E-07	7.188E-06	2.100E-05	5.736E-06	4.317E+04	1.438E+03	3.781E+02	2.868E+00	4.499E+04
0	1000	0	1.433E-07	7.351E-06	6.290E-10	5.842E-07	7.528E-09	3.611E-04	3.662E-07	2.882E-04	7.528E+02	7.223E+04	6.592E+00	1.441E+02	7.313E+04
0	0	1000	8.800E-09	1.247E-08	7.552E-06	9.912E-10	4.623E-10	6.128E-07	4.397E-03	4.890E-07	4.623E+01	1.226E+02	7.915E+04	2.445E-01	7.932E+04
50	250	700	4.528E-07	1.854E-06	5.288E-06	1.473E-07	2.379E-08	9.107E-05	3.079E-03	7.268E-05	2.379E+03	1.821E+04	5.543E+04	3.634E+01	7.606E+04

Emission, kg/h			Loss, Reaction, kg/h				Loss, Advection, kg/h			Intermedia Rate of Transport, kg/h T12	T21	T13	T31	T32	T24	T42
E(1)	E(2)	E(3)	R(1)	R(2)	R(3)	R(4)	A(1)	A(2)	A(4)	air-water	water-air	air-soil	soil-air	soil-water	water-sed	sed-water
1000	0	0	5.440E+02	1.811E+01	4.76E+00	1.169E-02	4.317E+02	1.438E+00	5.736E-05	1.990E+01	3.472E-01	4.777E+00	4.976E-03	8.006E-03	1.237E-02	6.236E-04
0	1000	0	9.485E+00	9.101E+02	8.31E-02	5.874E-01	7.528E+00	7.223E+01	2.882E-03	3.471E-01	1.744E+01	8.329E-02	8.676E-05	1.396E-04	6.216E-01	3.133E-02
0	0	1000	5.825E-01	1.544E+00	9.97E+02	9.966E-04	4.623E-01	1.226E-01	4.890E-06	2.131E-02	2.960E-02	5.115E-03	1.042E+00	1.676E+00	1.055E-03	5.316E-05
50	250	700	2.998E+01	2.295E+02	6.98E+02	1.481E-01	2.379E+01	1.821E+01	7.268E-04	1.097E+00	4.399E+00	2.632E-01	7.294E-01	1.174E+00	1.568E-01	7.901E-03

Level III Distribution

Chemical name: 1-Octanol

Distribution of mass

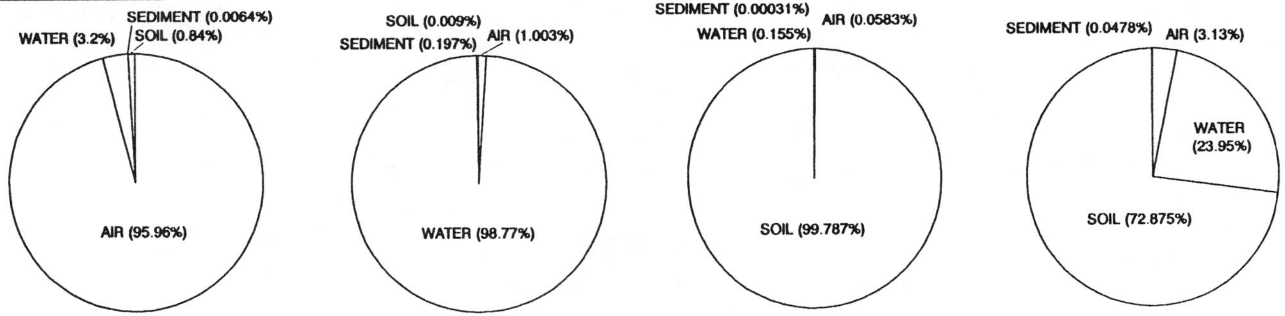

Distribution of removal rates

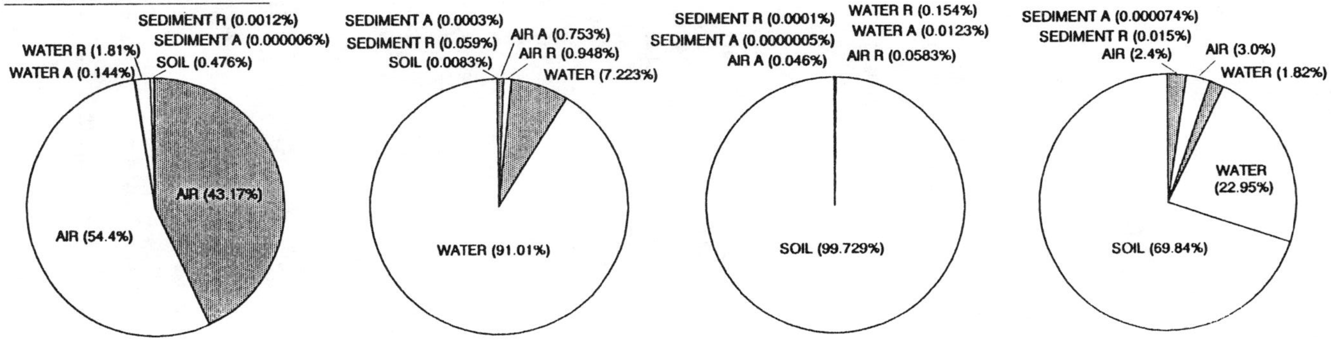

Emission rates:			
E(1) = 1000 kg/h	E(1) = 0	E(1) = 0	E(1) = 50 kg/h
E(2) = 0	E(2) = 1000 kg/h	E(2) = 0	E(2) = 250 kg/h
E(3) = 0	E(3) = 0	E(3) = 1000 kg/h	E(3) = 700 kg/h
Residence time:			
t = 45 h	t = 73.13 h	t = 79.32 h	t = 76.06 h

Reaction (unshaded) — Advection (shaded)

Chemical name: Benzyl alcohol

Level I calculation: (six-compartment model)

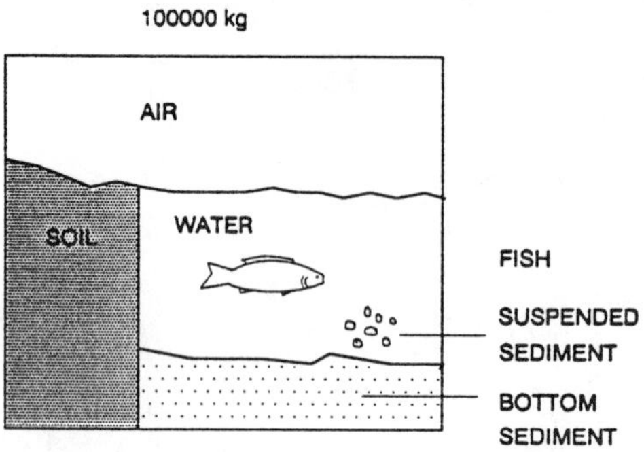

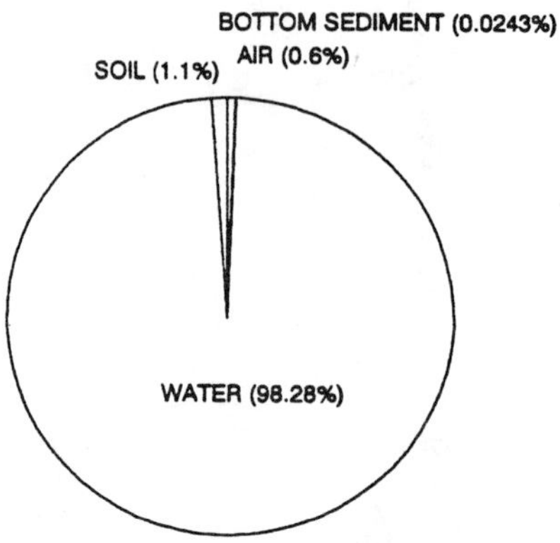

Distribution of mass

Level I calculation:

Physical-chemical properties:	
molecular wt., g/mol	108.14
melting point, C	-15.3
solubility, g/m3	43000
vapor pressure, Pa	12
log Kow	1.10
fugacity ratio, F	1.0000
dissoc. const. pKa	

Partition coeff.:	
HLC	0.03018
Kaw	1.22E-05
Koc	5.1616
BCF	0.6295
Ksw	0.2478
Ksd/w	0.4955
Kssd/w	1.5485
Kar/w	5.00E+05

Compartment	Z	Concentration			Amount	Amount
	mol/m3 Pa	mol/m3	mg/L (or g/m3)	ug/g	kg	%
Air	4.03E-04	5.53E-11	5.98E-09	5.05E-06	598.26	0.5983
Water	3.31E+01	4.54E-06	4.91E-04	4.91E-04	98281	98.281
Soil	8.21E+00	1.13E-06	1.22E-04	5.07E-05	1095.74	1.0957
Biota (fish)	2.09E+01	2.86E-06	3.09E-04	3.09E-04	0.06186	6.19E-05
Suspended sediment	5.13E+01	7.04E-06	7.61E-04	5.07E-04	0.7609	7.61E-04
Bottom sediment	1.64E+01	2.25E-06	2.43E-04	1.01E-04	24.35	0.0243
	Total				100000	100

f = 1.371E-07 Pa

Chemical name: Benzyl alcohol

Level II calculation: (six-compartment model)

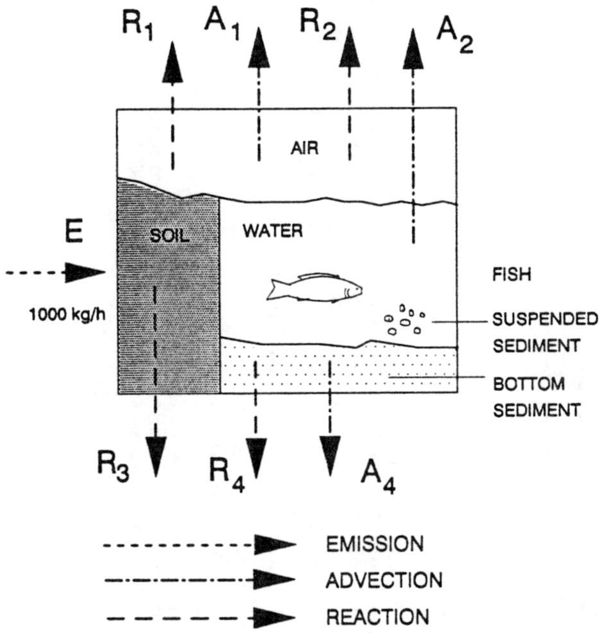

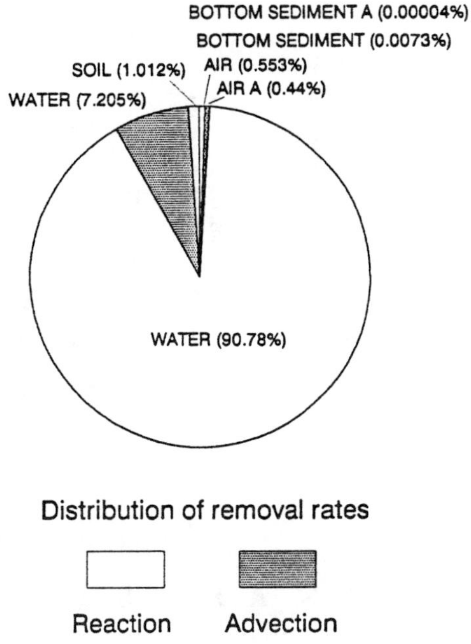

Distribution of removal rates

Level II Calculation:

Compartment	Half-life h	D Values Reaction mol/Pa.h	D Values Advection mol/Pa.h	Conc'n mol/m3	Loss Reaction kg/h	Loss Advection kg/h	Removal %
Air	55	5.08E+08	4.03E+08	4.06E-11	5.526	4.386	0.991
Water	55	8.35E+10	6.63E+09	3.33E-06	907.842	72.051	97.989
Soil	55	9.31E+08		8.25E-07	10.122		1.0122
Biota (fish)				2.10E-06			
Suspended sediment				5.16E-06			
Bottom sediment	170	6.69E+06	3.28E+04	1.65E-06	0.0728	3.57E-04	0.00731
	Total	8.49E+10	7.03E+09		923.56	76.437	100
	R + A		9.20E+10			1000	

f = 1.005E-07 Pa

Total amt = 73311 kg

Overall residence time = 73.31 h

Reaction time = 79.38 h

Advection time = 959.10 h

Fugacity Level III calculations: (four-compartment model)

Chemical name: Benzyl alcohol

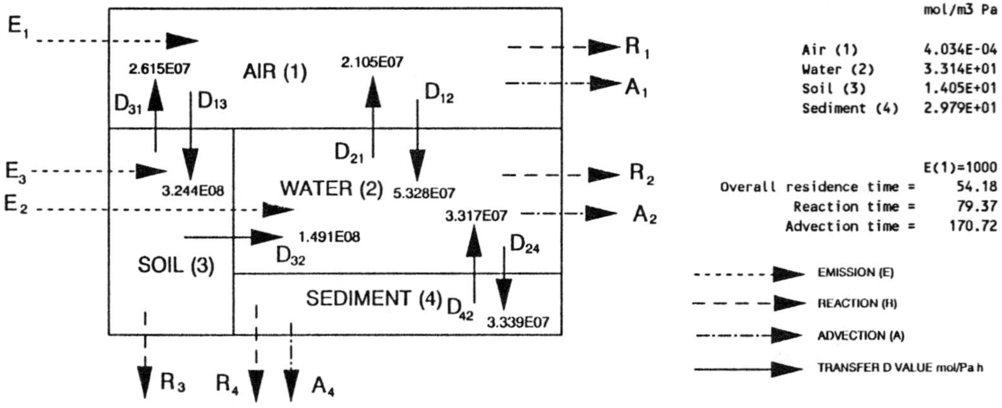

Phase Properties and Rates:

Compartment	Bulk Z mol/m3 Pa	Half-life h	D Values Reaction mol/Pa h	D Values Advection mol/Pa h
Air (1)	4.034E-04	55	5.08E+08	4.03E+08
Water (2)	3.314E+01	55	8.35E+10	6.63E+09
Soil (3)	1.405E+01	55	3.19E+09	
Sediment (4)	2.979E+01	170	6.07E+07	2.98E+05

	E(1)=1000	E(2)=1000	E(3)=1000	E(1,2,3)
Overall residence time =	54.18	73.57	78.91	76.34 h
Reaction time =	79.37	79.41	79.37	79.38 h
Advection time =	170.72	999.98	13764	1994.58 h

Phase Properties, Compositions, Transport and Transformation Rates:

Emission, kg/h			Fugacity, Pa				Concentration, g/m3				Amounts, kg				Total
E(1)	E(2)	E(3)	f(1)	f(2)	f(3)	f(4)	C(1)	C(2)	C(3)	C(4)	m(1)	m(2)	m(3)	m(4)	Amount, kg
1000	0	0	7.186E-06	5.393E-09	6.936E-07	1.912E-09	3.135E-07	1.933E-05	1.053E-03	6.160E-06	3.135E+04	3.865E+03	1.896E+04	3.080E+00	5.418E+04
0	1000	0	1.605E-09	1.026E-07	1.550E-10	3.636E-08	7.004E-11	3.675E-04	2.354E-07	1.171E-04	7.004E+00	7.350E+04	4.237E+00	5.857E+01	7.357E+04
0	0	1000	5.598E-08	4.592E-09	2.757E-06	1.628E-09	2.442E-09	1.646E-05	4.187E-03	5.245E-06	2.442E+02	3.291E+03	7.537E+04	2.623E+00	7.891E+04
50	250	700	3.989E-07	2.912E-08	1.965E-06	1.032E-08	1.740E-08	1.044E-04	2.984E-03	3.326E-05	1.740E+03	2.087E+04	5.371E+04	1.663E+01	7.634E+04

Emission, kg/h			Loss, Reaction, kg/h				Loss, Advection, kg/h			Intermedia Rate of Transport, kg/h T12	T21	T13	T31	T32	T24	T42
E(1)	E(2)	E(3)	R(1)	R(2)	R(3)	R(4)	A(1)	A(2)	A(4)	air-water	water-air	air-soil	soil-air	soil-water	water-sed	sed-water
1000	0	0	3.950E+02	4.870E+01	2.39E+02	1.256E-02	3.135E+02	3.865E+00	6.160E-05	4.141E+01	1.175E-02	2.521E+02	1.961E+00	1.118E+01	1.948E-02	6.858E-03
0	1000	0	8.825E-02	9.261E+02	5.34E-02	2.387E-01	7.004E-02	7.350E+01	1.171E-03	9.251E-03	2.234E-01	5.632E-02	4.382E-04	2.499E-03	3.703E-01	1.304E-01
0	0	1000	3.077E+00	4.147E+01	9.50E+02	1.069E-02	2.442E+00	3.291E+00	5.245E-05	3.226E-01	1.000E-02	1.964E+00	7.796E+00	4.446E+01	1.658E-02	5.839E-03
50	250	700	2.193E+01	2.630E+02	6.77E+02	6.780E-02	1.740E+01	2.087E+01	3.326E-04	2.298E+00	6.345E-02	1.399E+01	5.555E+00	3.168E+01	1.052E-01	3.703E-02

Level III Distribution

Chemical name: Benzyl alcohol

Distribution of mass

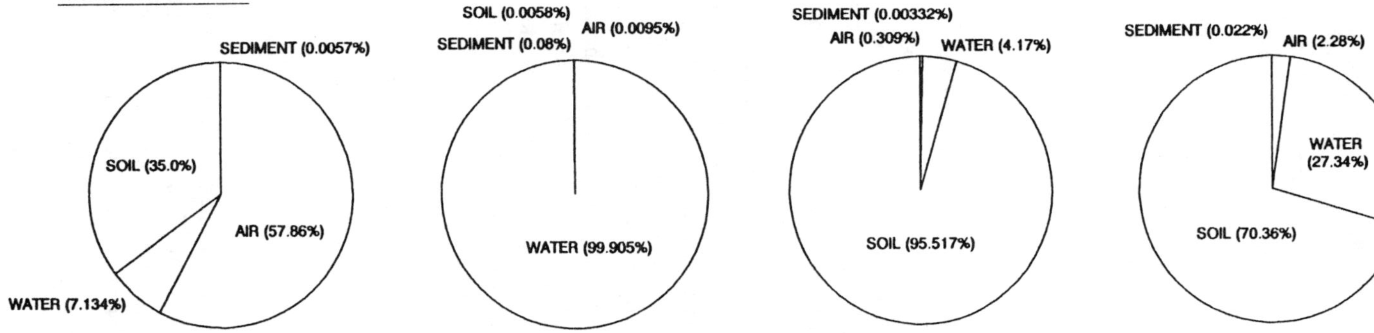

Distribution of removal rates

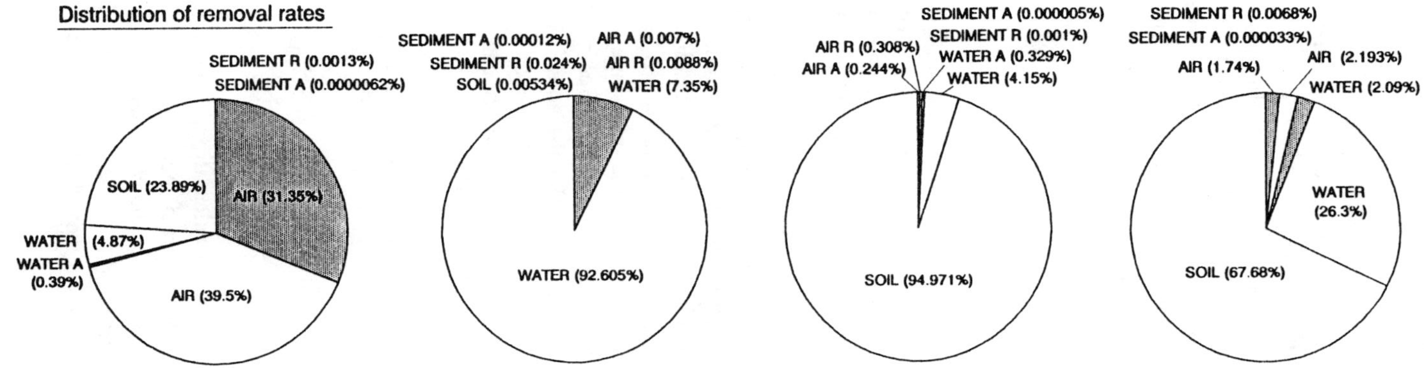

Reaction   Advection

Emission rates:			
E(1) = 1000 kg/h	E(1) = 0	E(1) = 0	E(1) = 50 kg/h
E(2) = 0	E(2) = 1000 kg/h	E(2) = 0	E(2) = 250 kg/h
E(3) = 0	E(3) = 0	E(3) = 1000 kg/h	E(3) = 700 kg/h
Residence time:			
t = 54.18 h	t = 73.57 h	t = 78.91 h	t = 76.34 h

2.4 Commentary on the Physical-Chemical Properties and Environmental Fate

Properties and Reactivity

The selected physical-chemical properties of the alcohols are given in Tables 2.1 and 2.2. Table 2.3 gives the estimated half-lives. These half-lives should be regarded as only tentative, average values or ranges. From reviews of the literature it is suggested that air, water and soil half-lives are typically Class 3, i.e., 30 to 100 hours with an average of 55 hours, although in some cases a longer half-life is possible. It is expected that sediment half-lives will be somewhat longer than those in soils. Only *t*-butyl alcohol displays slower reactions presumably because of the stability of the *t*-butyl group. Generally, the alcohols are fairly reactive and short lived in the environment, being susceptible to hydroxyl radical attack and to biodegradation which is facilitated by the solubility in water.

QSPR Plots

Figures 2.1 to 2.6 give the QSPR plots. The various molecular descriptors in Fig. 2.1 are clearly closely related to molar volume and to each other. The solubility data in Fig. 2.2 are well behaved, showing the expected decrease in solubility with increased molar volume. Although pseudo-solubilities are calculated for the lower miscible alcohols, these data are not included in Fig. 2.2. Addition of each methylene group causes an approximate drop in solubility by 0.70 log units, i.e., a factor of 5.0. The vapor pressure data in Fig. 2.3 are similarly behaved, displaying a steady drop of 0.46 log units per methylene or a factor of 2.9. The log K_{OW} data in Fig. 2.4 suggest an increase of 0.55 log units per methylene or a factor of 3.5. An implication of the K_{OW} and solubility data is that alcohols display a steadily decreasing solubility or pseudo-solubility in octanol with increased molar volume of about (0.70 - 0.55) or 0.16 log units or a factor of 1.4 per methylene group. This is reflected in a slope of about 0.77, a typical slope for such a plot. In contrast with other homologous series the Henry's law constant plot in Fig. 2.5 shows a steady increase with molar volume caused by the solubility decreasing faster than the vapor pressure. The slope is about 0.11 log units or a factor of 1.3 per methylene group. It is likely that H calculated from vapor pressure and solubility can be in substantial error for the more soluble alcohols, thus the experimental data should be assigned greater reliability.

These plots can be used with appropriate caution to deduce the properties of other alcohols.

Evaluative Calculations

The evaluative fugacity calculations show that most alcohols partition primarily into water with a slight tendency to evaporate, because of the relatively low Henry's law constant (HLC) and little tendency to sorb to organic matter (because of the low K_{OW}). The higher alcohols partition more into air and soil because of the larger HLCs and K_{OW}s. The Level II calculations show that they are primarily removed by reaction in water.

The Level III calculations suggest that the alcohols are not susceptible to appreciable intermedia transport; they tend to remain in the medium of dischage and are removed primarily by reaction. Taking 1-hexanol as an example, the reaction D values vary from 0.15 to $1.2x10^9$ mol/Pa·h (ignoring sediment) while the intermedia values are generally in the range 10^6 to $2x10^7$, a factor of about 100 slower. Intermedia transport is thus not generally competitive with reaction. In other and equivalent terms, the half-lives for intermedia transport greatly exceed those for reaction. For example, when 1000 kg/h of hexanol is discharged to water, 914 kg/h reacts, 72 kg/h advects and only 14 kg/h evaporates. Sedimentation is negligible.

The key determinants of environmental concentrations are thus the reaction half-lives.

2.5 REFERENCES

Abernethy, S.G., Mackay, D., McCarty, L.S. (1988) "Volume fraction" correlation for narcosis in aquatic organisms: The key role of partitioning. *Environ. Toxicol. Chem.* 7, 469-481.

Abraham, M.H. (1984) Thermodynamics of solution of homologous series of solutes in water. *J. Chem. Soc., Farad. Trans. 1* 80, 153-181.

Addison, C.C. (1943) The properties of freshly formed surfaces. Part I. The application of the vibrating-jet technique to surface-tension measurements of mobile liquids. *J. Chem. Soc.* 535-541.

Ambrose, D., Ghiassee, N.B. (1990) Vapour pressures, critical temperatures, and critical pressures of benzyl alcohol, octan-2-ol, and 2-ethylhexan-1-ol. *J. Chem. Thermodyn.* 22, 307-311.

Ambrose, D., Hall, D.J. (1981) Thermodynamic properties of organic oxygen compounds. L. The vapour pressures of 1,2-ethanediol (ethylene glycol) and bis(2-hydroxyethyl) ether (diethylene glycol). *J. Chem. Thermodyn.* 13, 61-66.

Amidon, G.L., Williams, N.A. (1982) A solubility equation for non-electrolytes in water. *Int'l J. Pharm.* 11, 249-256.

Amidon, G.L., Yalkowsky, S.H., Anik, S.T., Valvani, S.C. (1975) Solubility of nonelectrolytes in polar solvents. V. Estimation of the solubility of aliphatic monofunctional compounds in water using a molecular surface area approach. *J. Phys. Chem.* 79(21), 2239-2246.

Amidon, G.L., Yalkowsky, S.H., Leung, S.J. (1974) Solubility of nonelectrolytes in polar solvents. II: Solubility of aliphatic alcohols in water. *J. Pharm. Sci.* 63, 1858-1866.

Anbar, M., Neta, P. (1967) A compilation of specific bimolecular rate hydroxyl radical with inorganic and organic compounds in aqueous solution. *Int'l J. Appl. Radiation Isotopes* 18, 493-523.

Atkinson, R. (1985) Kinetics and mechanisms of the gas-phase reactions of hydroxyl radical with organic compounds under atmospheric conditions. *Chem. Rev.* 85, 69-201.

Atkinson, R. (1987) Structure-activity relationship for the estimation of rate constants for the gas-phase reactions of OH radicals with organic compounds. *Int'l. J. Chem. Kinetics* 19, 799-828.

Atkinson, R. (1987) Estimation of OH radical rate constants and atmospheric lifetimes for polychlorobiphenyls, dibenzo-*p*-dioxins. *Environ. Sci. Technol.* 21, 305-307.

Atkinson, R. (1990) Gas-phase tropospheric chemistry of organic chemistry: A review. *Atmos. Environ.* 24A(1), 1-41.

Atkinson, R. (1991) Kinetics and mechanisms of the gas-phase reactions of the NO_3 radical with organic compounds. *J. Phys. Chem. Ref. Data* 20(3), 459-507.

Banerjee, S. (1985) Calculation of water solubility of organic compounds with UNIFAC-derived parameters. *Environ. Sci. Technol.* 19, 369-370.

Banerjee, S., Howard, P.H. (1988) Improved estimation of solubility and partitioning through correction of UNIFAC-derived activity coefficients. *Environ. Sci. Technol.* 22, 839-841.

Banerjee, S., Howard, P.H., Lande, S.S. (1990) General structure-vapor pressure relationships for organics. *Chemosphere* 21(10/11), 1173-1180.

Barton, A.F.M. (1984) Alcohols with water. In: *IUPAC Solubility Data Series Vol. 15.* Pergamon Press Inc., New York. p.438.

Berti, P., Cabani, S., Conti, G., Mollica, V. (1986) Thermodynamic study of organic compounds in octan-1-ol. *J. Chem. Soc. Faraday Trans.* 1, 82, 2547-2556.

Bodor, N., Gabanyi, Z., Wong, C.K. (1989) A new method for the estimation of partition coefficient. *J. Am. Chem. Soc.* 111, 3783-3786.

Booth, H.S., Everson, H.E. (1948) Hydrotropic solubilities. Solubilities in 40 per cent sodium xylenesulfonate. *Ind. & Eng. Chem.* 40(8), 1491-1493.

Boublik, T., Fried, V., Hala, E. (1973) *The Vapour Pressures of Pure Substances*. Elsevier, Amsterdam, The Netherlands.

Boublik, T., Fried, V., Hala, E. (1984) *The Vapour Pressures of Pure Substances*, 2nd ed., Elsevier, Amsterdam, The Netherlands.

Bridie, A.L., Wolff, C.J.M., Winter, M. (1979) BOD and COD of some petrochemicals. *Water Res.* 13, 627-630.

Briggs, G.G. (1981) Theoretical and experimental relationships between soil adsorption, octanol-water partition coefficients, water solubilities, bioconcentration factors, and the parachor. *J. Agric. Food & Chem.*29, 1050-1059.

Burnett, M.G. (1963) Determination of partition coefficients at infinite dilution by the gas chromatographic analysis of the vapor above dilute solutions. *Anal. Chem.* 35, 1567-1570.

Burnett, M.G., Swoboda, P.A.T. (1962) A simple method for the calibration of sensitive gas chromatographic detectors. *Anal. Chem.* 34, 1162-1163.

Butler, J.A.V., Ramchandani, C.N., Thomson, D.W. (1935) The solubility of nonelectrolytes. Part I. The free energy of hydration of some aliphatic alcohols. *J. Chem. Soc.* 280-285.

Butler, J.A.V., Thomson, D.W., Maclennan, W.H. (1933) The free energy of the normal aliphatic alcohols in aqueous solution. Part I. The partial vapour pressures of aqueous solutions of methyl, *n*-propyl, and *n*-butyl alcohols. Part II. The solubilities of some normal aliphatic alcohols in water. Part III. The theory of binary solutions, and its application in aqueous-alcoholic solutions. *J. Chem. Soc.* Part 1, 674-686.

Buttery, R.G., Bomben, J.L., Guadagni, D.G., Ling, L.C. (1971) Some considerations of the volatilities of organic flavor compounds in foods. *J. Agric. Food Chem.* 19, 1045-1048.

Buttery, R.G., Ling, L.C., Guadagni, D.G. (1969) Food volatiles. Volatilities of aldehydes, ketones, and esters in dilute water solution. *J. Agric. Food Chem.* 17(2), 385-389.

Campbell, I.M., McLaughlin, D.F., Handy, B.J. (1976) Rate constants for reactions of hydroxyl radicals with alcohol vapours at 292 K. *Chem. Phys. Lett.* 38, 362-364.

Carlier, P., Hannachi, H., Mouvier, G. (1986) The chemistry of carbonyl compounds in the atmosphere-A review. *Atmos. Environ.* 20(11), 2079-2099.

Chao, J., Lin, C.T., Chung, T.H. (1983) Vapor pressure of coal chemicals. *J. Phys. Chem. Ref. Data* 12(4), 1034-1063.

Chen, F., Holten-Andersen, J., Tyle, H. (1993) New developments of the UNIFAC model for environmental application. *Chemosphere* 26, 1325-1354.

Chessells, M., Hawker, D.W., Connell, D.W. (1992) Influence of solubility in lipid on bioconcentration of hydrophobic compounds. *Ecotoxicol. & Environ. Safety* 23, 260-273.

Chou, J.T., Jurs, P.C. (1979) Chapter 5, Computation of partition coefficients from molecular structures by a fragment addition method. In: *Physical and Chemical Properties of Drugs*. Medical Research Series, Vol. 10. Yalkowsky, S.H., Sinkula, A.A., Valvani, S.C., Editors.

Chou, W.L., Speece, R.E., Siddiqi, R.H. (1979) Acclimation and degradation of petrochemical wastewater components by methane fermentation. *Biotechnol. Bioeng. Symp.* 8, 391-414.

Collander, R. (1951) Partition of organic compounds between higher alcohols and water. *Acta. Chem. Scand.* 5, 774-780.

Cornford, E.M. (1982) Correlation between liquid partition coefficients and surface permeation in *schistosoma japonicum*. *J. Membr. Biol.* 64, 217.

CRC (1976-77) *Handbook of Chemistry and Physics.* 57th Edition, Weast, R.C., Editor, CRC Press, Boca Raton, Florida.

Crittenden, Jr. E.D., Hixon, A.W. (1954) Extraction of hydrogen chloride from aqueous solutions. *Ind. Eng. Chem.* 46, 265-168.

Dallos, A., Wienke, G., Ilchman, A., Gmehlong, J. (1993) Vorausberechnung von octanol/wasser-verteilungskoeffizienten mit hilfe der UNIFAC-method. *Chem.-Ing.-Tech.* 65(2), 201-203.

D'Amboise, M., Hanai, T. (1982) Hydrophobicity and retention in reverse phase liquid chromatography. *J. Chromatogr.* 5, 229-244.

Darnall, K.R., Lloyd, A.C., Winer, A.M., Pitts Jr., J.N. (1976) Reactivity scale for the atmospheric hydrocarbons based on reaction with hydroxyl radical. *Environ. Sci. Technol.* 10(7), 692-696.

Daubert, T.E., Danner, R.P. (1985) *Data Compilation Tables of Properties of Pure Compounds.* Am. Institute of Chemical Engineers. p.450.

Daubert, T.E., Danner, R.P. (1989) *Physical & Thermodynamic Properties of Pure Compounds.* Am. Institute of Chemical Engineers.

Dean, J. (1985) *Lange's Handbook of Chemistry.* 13th ed., McGraw-Hill, New York.

De Santis, R., Marrelli, L., Muscetta, P.N. (1976) Liquid-liquid equilibria in water-aliphatic alcohol systems in the presence of sodium chloride. *Chem. Eng. J.* 11, 207-214.

Dilling, W.L., Bredweg, C.J., Terfertiller, N.B. (1976) Organic photochemistry. Simulated atmospheric rates of methylene chloride, 1,1,1-trichloroethane, trichloroethylene, tetrachloroethylene, and other compounds. *Environ. Sci. Technol.* 10(4), 351-356.

Dillingham, E.O., Mast, R.W., Bass, G.E., Autian, J. (1973) Toxicity of methyl- and halogen-substituted alcohols in tissue culture relative to structure-activity models and acute toxicity in mice. *J. Pharm. Sci.* 62, 22.

Djerki, R., Laub, R.J. (1988) Solute retention in column liquid chromatography. X. Determination of solute infinite-dilution activity coefficients in methanol, water, and their mixtures, by combined gas-liquid and liquid-liquid chromatography. *J. Liq. Chromatogr.* 11(3), 585-612.

Donahue, D.J., Bartell, F.E. (1952) The boundary tension at water-organic liquid interfaces. *J. Phys. Chem.* 56, 480.

Dorfman, L.M., Adams, G.E. (1973) *Reactivity of the Hydroxyl Radical in Aqueous Solution.* NSRD-NDB-46. NTIS COM-73-50623. National Bureau of Standards, Washington D.C. 51 pp.

Eadsforth, C.V. (1986) Application of reverse-phase HPLC for the determination of partition coefficients. *Pest. Sci.* 17(3), 311-325.

Edney, E.O., Corse, E.W. (1986) *Validation of OH Radical Reaction Rate Constant Test Protocol.* NTIS PB86-166 758/as. U.S. Environmental Protection Agency, Washington, DC.

Evans, B.K., James, K.C., Luscombe, D.K. (1978) Quantitative structure-activity relationships and carminative activity. *J. Pharm. Sci.* 67, 277-278.

Evans, W.H., David, E.J. (1974) Biodegradation of mono-, di- and triethylene glycols in river water under controlled laboratory conditions. *Water Res.* 8, 97-100.

Foote, C.S. (1976) *Free Radicals in Biology.* Pryor, W.A., Editor, Academic Press, Inc., New York.

Freitag, D., Ballhorn, L., Geyer, H., Körte, F. (1985) Environmental hazard profile of organic chemicals. An experimental method for the assessment of the behaviour of chemicals in the ecosphere by simple laboratory tests with C-14 labelled chemicals. *Chemosphere* 14, 1589-1616.

Führer, H. (1924) Water solubilities in homologous series. *Ber. Dtsch. Chem. Gse.* 57, 510-515.

Fujisawa, S., Masuhara, E. (1981) Determination of partition coefficients of acrylates, methacrylates, and vinyl monomers using high performance liquid chromatography (HPLC). *J. Biomed. Mat. Res.* 15, 787-793.

Fujita, T., Iwasa, J., Hansch, C. (1964) A new substituent constant derived from partition coefficients. *J. Am. Chem. Soc.* 86(23), 5175-5180.

Funasaki, N., Hada, S., Neya, S. (1986) Prediction of retention times in reversed-phase high-performance liquid chromatography from the chemical structure. *J. Chromatogr.* 361, 33-45.

Gaffney, J.S., Streit, G.E., Spall, W.D., Hall, J.H. (1987) Beyond acid rain. Do soluble oxidants and organic toxins interact with SO_2 and NO_2 to increase ecosystem effects? *Environ. Sci. Technol.* 21(6), 519-524.

Gallaugher, A.F., Hibbert, H.J. (1937) Reactions relating to carbohydrates and polysaccharides. LIV. The surface-tension constants of the polyethylene glycols and their derivatives. LV. Vapor pressures of the polyethylene glycols and their derivatives. *J. Am. Chem. Soc.* 59, 2521-2514.

Garst, J.E., Wilson, W.C. (1984) Accurate, wide range, automated, high performance liquid chromatographic method for the estimation of octanol/water partition coefficients. II: Effects of chromatographic method and procedure variables on accuracy and reproducibility of the method. *J. Pharm. Sci.* 73(11), 1616-11623.

Gellman, I., Heukelekian, H. (1955) Studies of biochemical oxidation by direct methods. V. Effects of various seed materials in rates of oxidation of industrial wastes and organic compounds. *Sew. Indust. Wastes* 27, 793-801.

GEMS (1986) Graphic Exposure Modeling System. FAP, Fate of Atmospheric Pollutants.

Gerstl, Z., Helling, C.S. (1987) Evaluation of molecular connectivity as a predictive method for the adsorption of pesticides in soil. *J. Environ. Sci. & Health* B22(1), 55-69.

Geyer, H., Politzki, G., Freitag, D. (1984) Prediction of ecotoxicological behaviour of chemicals: Relationship between *n*-octanol/water partition coefficient and bioaccumulation of organic chemicals by alga *chlorella*. *Chemosphere* 13(2), 269-284.

Ginnings, P.M., Baum, R. (1937) Aqueous solubilities of the isomeric pentanols. *J. Am. Chem. Soc.* 59, 1111-1113.

Ginnings, P.M., Coltrane, D. (1939) Aqueous solubility of 2,2,3-trimethylpentanol-3. *J. Am. Chem. Soc.* 61-525-526.

Ginnings, P.M., Webb, R. (1938) Aqueous solubilities of some isomeric hexanols. *J. Am. Chem. Soc.* 60, 1388-1389.

Graedel, T.E. (1978) *Chemical Compounds in the Atmosphere.* Academic Press, New York.

Grayson, B.T., Fosbraey, L.A. (1982) Determination of vapour pressure of pesticides. *Pestic. Sci.* 13, 269-278.

Gückel, W., Kästel, R., Lewerenz, J., Synnatschke, G. (1982) A method for determining the volatility of active ingredients used in plant protection. Part III: The temperature relationship between vapour pressure and evaporation rate. *Pestic. Sci.* 13, 161-168.

Gückel, W., Synnatschke, G., Rittig, R. (1973) A method for determining the volatility of active ingredients used in plant protection. *Pestic. Sci.* 4, 137-147.

Haky, J.E., Young, A.M. (1984) Evaluation of a simple HPLC correlation method for the estimation of the octanol-water partition coefficients of organic compounds. *J. Liq. Chromatogr.* 7, 675-689.

Hammers, W.E., Meurs, G.J., DeLigny, C.L. (1982) Correlations betweeen liquid chromatographic capacity factors ratio data on Lichrosorb RP-18 and partition coefficients in the octanol-water system. *J. Chromatogr.* 247, 1-13.

Hammerton, C. (1955) Observations of the decay of synthetic anionic detergents in natural water. *J. Appl. Chem.* 5, 517-524.

Hanai, T., Tran, C., Hubert, J. (1981) An approach to the prediction of retention times in liquid chromatography. *J. HRC & CC* 4, 454-460.

Hansch, C., Anderson, S. (1967) The effect of intermolecular hydrophobic bonding on partition coefficients. *J. Org. Chem.* 32, 2583.

Hansch, C., Dunn, III, W.J. (1972) Linear relationships between lipophilic character and biological activity of drugs. *J. Pharm. Sci.* 61(1), 1-19.

Hansch, C., Leo, A. (1979) *Substituent Constants for Correlation Analysis in Chemistry and Biology.* John-Wiley & Sons, New York.

Hansch, C., Leo, A. (1983) *Medchem. Project*, Pomona College, Claremont, California.

Hansch, C., Leo, A. (1985) *Medchem. Project*, Pomona College, Claremont, California.

Hansch, C., Quinlan, J.E., Lawrence, G.L. (1968) The linear free-energy relationship between partition coefficients and aqueous solubility of organic liquids. *J. Org. Chem.* 33, 347-350.

Hansen, R.S., Fu, Y., Bartell, F.E. (1949) Multimolecular adsorption from binary liquid solutions. *J. Phys. & Colloid Chem.* 53, 769-785.

Harkins, W.D., Oppenheimer, H. (1949) Solubilization of polar-nonpolar substances in solutions of long chain electrolytes. *J. Am. Chem. Soc.* 71, 808-811.

Hendry, D.G., Kenley, R.A. (1979) *Atmospheric Reaction Products of Organic Compounds.* EPA Final Report, EPA560/12-79-001.

Hendry, D.G., Mill, T., Piszkiewicz, L., Howard, J.A., Eigenmann, H.K. (1974) Critical review of hydrogen-atom transfer in the liquid phase. Chlorine atom, alkyltrichloromethyl, alkoxy, and alkyl peroxy radicals. *J. Phys. Chem. Ref. Data* 3, 937-978.

Hermens, J., Leeuwangh, P. (1982) Joint toxicity of mixtures of 8 and 24 chemicals to the guppy *(poecilia reticulata). Ecotoxicol. & Environ. Safety* 6, 302-310.

Heukelekian, H., Rand, M.C. (1955) Biochemical oxygen demand of pure organic compounds. *J. Water Pollut. Control Assoc.* 29, 1040-1053.

Hill, A.E., Malisoff, W.M. (1926) The mutual solubility of liquids. III. The mutual solubility of phenol and water. IV. The mutual solubility of normal butyl alcohol and water. *J. Am. Chem. Soc.* 48, 918-927.

Hill, D.J.T., White, L. R. (1974) Enthalpies of solution of 1-hexanol and 1-heptanol in water. *Aust. J. Chem.* 27, 1905-1916.

Hine, J., Mookerjee, P.K. (1975) The intrinsic hydrophilic character of organic compounds. Correlations in terms of structural contributions. *J. Org. Chem.* 40, 292-298.

Horvath, A.L. (1982) *Halogenated Hydrocarbons: Solubility-Miscibility with Water.* Marcel Dekker, New York.

Hou, C.T., Patel, R., Laskin, A.I., Barnabe, N., Barist, I. (1983) Production of methyl ketones from secondary alcohols by cell suspensions of C_2 to C_4 *n*-alkane-grown bacteria. *Appl. Environ. Microbiol.* 46, 178-184.

Howard, J.A. (1972) Absolute rate constants for reactions of oxy radicals. *Adv. Free Radical Chem.* 4, 49-173.

Howard, P.H., Ed. (1990) *Handbook of Fate and Exposure Data for Organic Chemicals. Vol. II, Solvents.* Lewis Publishers, Inc., Chelsea, Michigan.

Howard, P.H., Ed. (1993) *Handbook of Fate and Exposure Data for Organic Chemicals. Vol. IV, Solvents 2.* Lewis Publishers, Inc., Chelsea, Michigan.

Howard, P.H., Boethling, R.S., Jarvis, W.F., Meylan, W.M., Michalenko, E.M., Editors (1991) *Handbook of Environmental Degradation Rates.* Lewis Publishers Inc., Chelsea, Michigan.

Howard, P.H., Hueber, A.E., Mulesky, B.C., Crisman, J.S., Meylan, W., Crosbie, E., Gray, D.A., Sage, G.W., Howard, K.P., LaMacchia, A., Boethling, R.S., Troast, R. (1986) BIOLOG, BIODEG, and FATE/EXPOS: New files on microbial degradation and toxicity as well as environmental fate/exposure of chemicals. *Environ. Toxicol. & Chem.* 5, 977-988.

Hoy, K.L. (1970) New values of solubility parameters from vapor pressure data. *J. Paint Technol.* 42(541), 76-118.

Hulzebos, E.M., Adema, D.M.M., D.-van Breemen, E.M., Henzen, L., van Dis, W.A., Herbold, H.A., Hoekstra, J.A., Baerselman, R., van Gestel, C.A.M. (1993) Phytotoxicity studies with *lactuca saltiva* in soil and nutrient solution. *Environ. Toxicol. Chem.* 12, 1079-1094.

Isnard, P., Lambert, S. (1988) Estimating bioconcentration factors from octanol partition coefficient and aqueous solubility. *Chemosphere* 17, 21-34.

Isnard, P., Lambert, S. (1989) Aqueous solubility/*n*-octanol water partition coefficient correlations. *Chemosphere 18, 1837-1853.*

IUPAC (1984) *Alcohols with Water. Solubility Data Series.* Vol. 15, Barton, A.F.M., Editor, Pergamon Press, Oxford.

Iwasa, J., Fujita, T., Hansch, C. (1965) Substituent constants for aliphatic functions obtained from partition coefficients. *J. Med. Chem.* 8, 150-153.

Iwase, K., Komatsu, K., Hirono, S., Nakagawa, S., Moriguchi, I. (1985) Estimation of hydrophobicity based on the solvent accessible surface area of molecules. *Chem. Pharm. Bull.* 33(5), 2114-2121.

Jaworska, J.S., Schultz, T.W. (1993) Quantitative relationships of structure-activity and volume fraction for selected nonpolar and polar narcotic chemicals. *SAR & QSAR in Environ. Res.* 1, 3-19.

Jones, D.C. (1929) The systems *n*-butyl alcohol-water and *n*-butyl alcohol-acetone-water. *J. Chem. Soc.* 799-816.

Jordan, T.E. (1970) *Vapor Pressure of Organic Compounds.* Interscience, New York.

Joshi, S.B., Dodge, M.C., Bufalini, J.J. (1982) Reactivities of organic compounds and contamination effects. *Atmos. Environ.* 16(6), 1301-1310.

Kahlbaum, G.W.A. (1898) Studien über dampfspannkraftmessungen. II. *Z. Phys. Chem.* 26, 577-658.

Kamlet, M.J., Abraham, D.J., Doherty, R.M., Taft, R.W., Abraham, M.H. (1986) Solubility properties in polymers and biological media: 6. An equation for correlation and prediction of solubilities of liquid organic nonelectrolytes in blood. *J. Pharm. Sci.* 75(4), 350-355.

Kamlet, M.J., Doherty, R.M., Abboud, J.L.M., Abraham, M.H., Taft, R.W. (1986) Linear solvation energy relationships. 36. Molecular properties governing solubilities of organic nonelectrolytes in water. *J. Pharm. Sci.* 75(4), 338-349.

Kamlet, M.J., Doherty, R.M., Abraham, M.H., Carr, P.W., Doherty, R.F., Taft, R.W. (1987) Linear solvation energy relationships. 42. Important differences between aqueous solubility relationships for aliphatic and aromatic solutes. *J. Phys. Chem.* 91, 1996-2004.

Kamlet, M.J., Doherty, R.M., Taft, R.W., Abraham, M.H., Veith, G.D., Abraham, D.J. (1987) Solubility properties in polymers and biological media. 8. An analysis of the factors that influence toxicities of organic nonelectrolytes to the golden orfe fish (*leuciscus idus melanotus*). *Environ. Sci. Technol.* 21(2), 149-155.

Kelly, T.J., Mukund, R., Spicer, C.W., Pollack, A.J. (1994) Concentration and transformations of hazardous air pollutants. *Envriron. Sci. Technol.* 28, 378A-387A.

Kemme, H.R., Kreps, S.I. (1969) Vapor pressure of primary *n*-alkyl chlorides and alcohols. *J. Chem. Eng. Data* 14(1), 98-102.

Kenaga, E.E., Goring, C.A.I. (1980) In: *Aquatic Toxicology.* Eaton, J.G., Parrish, P.R., Hendricks, A.C., Editors, Am. Soc. for Testing and Materials, STP 707, Philadelphia, Pa. pp.78-115.

Kinoshita, K., Ishikawa, H., Shinoda, K. (1958) Solubility of alcohols in water determined by the surface tension measurements. *Bull. Chem. Soc. Japan* 31, 1081-1082.

Klopman, G., Namboodiri, K., Schochet, M. (1985) Simple method of computing the partition coefficient. *J. Comput. Chem.* 6, 28-38.

Kollig, H.P., Ed. (1993) *Environmental Fate Constants for Organic Chemicals Under Consideration for EPA's Hazardous Waste Identification Projects.* EPA Report EPA/600/R-93/132, Environmental Research Lab., Athens, GA.

Könemann, H. (1981) Quantitative structure-activity relationships in fish toxicity studies. Part 1. Relationship for 50 industrial pollutants. *Toxicology* 19, 209-221.

Könemann, H., Zelle, R., Busser, F., Hammers, W.E. (1979) Determination of log P_{OCT} values of chloro-substituted benzenes, toluenes and anilines by high-performance liquid chromatography on ODS-silica. *J. Chromatogr.* 178, 559-565.

Korenman, I.M., Gorokhov, A.A., Polozenkom G.N. (1974) Salting out in a two-phase systems of higher alcohols-alkali metal chlorides-water. *Zh. Fiz. Khim.* (7), 1810-1812.

Korenman, I.M., Gorokhov, A.A., Polozenkom G.N. (1975) Salting out in a two-phase system. II. Higher alcohols-potassium halides water. *Russ. J. Phys. Chem.* 49, 1490-1493.

Laddha, G.S., Smith, J.M. (1948) The systems: Glycol-*n*-amyl alcohol-water and glycol-*n*-hexyl alcohol-water. *Ing. Eng. Chem.* 40, 494-496.

Lande, S.S., Banerjee, S. (1981) Predicting aqueous solubility of organic nonelectrolytes from molar volume. *Chemosphere* 10(7), 751-759.

Leahy, D.E. (1986) Intrinsic molecular volume as a measure of the cavity term in linear solvation energy relatioships: octanol-water partition coefficients and aqueous solubilities. *J. Pharm. Sci.* 75(7), 629-636.

Lee, B.I., Erbar, J.H., Edmister, W.C. (1972) Thermodynamic properties at low temperatures. *Chem. Eng. Prog.* 68(9), 83-84.

Leo, A., Hansch, C., Church, C. (1969) Comparison of parameters currently used in the study of structure-activity relationships. *J. Med. Chem.* 12, 766-771.

Leo, A., Hansch, C., Elkins, D. (1971) Partition coefficients and their uses. *Chem. Rev.* 71, 525-616.

Leo, A., Jow, P.Y.C., Silipo, C., Hansch, C. (1975) Calculation of hydrophobic constant (Log *P*) from π and f constants. *J. Med. Chem.* 18(9), 865-868.

Li, A., Andren, A.W. (1994) Solubility of polychlorinated biphenyls in water/alcohol mixtures. 1. Experimental Data. *Environ. Sci. Technol.* 28(1), 47-52.

Li, A., Doucette, W.J., Andren, A.W. (1992) Solubility of polychlorinated biphenyls in binary water/organic solvent systems. *Chemosphere* 24(9), 1347-1360.

Lyman, W.J., Reehl, W.F., Rosenblatt, D.H. (1982) *Handbook on Chemical Property Estimation Methods. Environmental Behaviour of Organic Compounds.* McGraw-Hill, New York.

Lyman, W.J., Reehl, W.F., Rosenblatt, D.H. (1990) *Handbook on Chemical Property Estimation Methods.* pp. 4-9, 5-4, 5-10, 7-14, 15-16 to 15-29., American Chemical Society, Washington DC.

Mackay, D. (1982) Correlation of bioconcentration factors. *Environ. Sci. Technol.* 16, 274-278.

Mackay, D., Bobra, A., Chan, D.W., Shiu, W.Y. (1982) Vapor pressure correlations for low-volatility environmental chemicals. *Environ. Sci. Technol.* 16, 645-649.

Mackay, D., Yuen, T.K. (1983) Mass transfer coefficients correlations for volatilization of organic solutes from water. *Environ. Sci. Technol.* 17, 211-217.

Malaney, G.W., Gerhold, R.W. (1969) Structural determinants in the oxidation of aliphatic compounds by activated sludge. *J. Water Pollut. Control Fed.* 41, R18-R33.

Mayer, J.M., Testa, B., van de Waterbeemd, H., Bornand-Crausaz, A. (1982) Nonadditivity and intramolecular effects in the partition coefficients of ω-functionalized alkylpyridines. *Eur. J. Med. Chem.* 17(5), 453-459.

McBain, J.W., Richard, P.H. (1946) Solubilization of insoluble organic liquids by detergents. *Ind. Eng. Chem.* 38, 642-646.

McCall, J.M. (1975) Liquid-liquid partition coefficients by high-pressure liquid chromatography. *J. Med. Chem.* 18(6), 549-552.

McGowan, J.C. (1954) The physical toxicity of chemicals. IV. Solubilities, partition coefficients and physical toxicities. *J. Appl. Chem.* 4, 41-47.

Meier, U., Grotheer, H.H., Riekert, G., Just, T. (1985) Temperature dependence and branching ratio of the C_2H_5OH + OH reaction. *Chem. Phys. Lett.* 115, 221-225.

Meylan, W., Howard, P.H., Boethling, R.S. (1992) Molecular topology/fragment contribution method for predicting soil sorption coefficients. *Environ. Sci. Technol.* 26(8), 1560-1567.

Meylan, W., Howard, P.H. (1993) Computer estimation of the atmospheric gas-phase reaction rate of organic compounds with hydroxyl radicals and ozone. *Chemosphere* 26, 2293-2299.

Mill, T. (1979) *Structure Reactivity Correlations for Environmental Reactions.* EPA Final Report, EPA-560/11-79-012.

Mill, T. (1982) Hydrolysis and oxidation processes in the environment. *Environ. Toxicol. & Chem.* 1, 135-141.

Miyake, K., Kitaura, F., Mizuno, N., Tereda, H. (1987) Determination of partition coefficient and acid dissociation constant by high-performance liquid chromatography on porous polymer gel as a stationary phase. *Chem. Pharm. Bull. 35(1)*, 377-388.

Müller, M., Klein, W. (1991) Estimating atmospheric degradation processes by SARs. *Total Sci. Environ.* 109/110, 261-273.

Müller, M., Klein, W. (1992) Comparative evaluation of methods predicting water solubility for organic compounds. *Chemosphere* 25, 769-782.

Munz, C., Roberts, P.V. (1986) Effects of solute concentration and cosolvents on the aqueous activity coefficients of halogenated hydrocarbons. *Environ. Sci. Technol.* 20, 830-836.

Nirmalakhandan, N.N., Speece, R.E. (1988a) Prediction of aqueous solubility of organic chemicals based on molecular structure. *Environ. Sci. Technol.* 22(3), 328-338.

Nirmalakhandan, N.N., Speece, R.E. (1988b) QSAR model for predicting Henry's constant. *Environ. Sci. Technol.* 22(11), 1349-1357.

Novak, J.T., Goldsmith, C.D., Benoit, R.E., O'Brien, J.H. (1985) Biodegradation of methanol and tertiary butyl alcohol in subsurface systems. *Water Sci. Technol.* 17, 71-85.

Othmer, D.F., Bergen, W.S., Shlechter, N., Bruins, P.F. (1945) Liquid-liquid extraction data. Systems used in butadiene manufacture from butylene glycol. *Ind. Eng. Chem.* 37. 890-894.

Overend, R., Paraskevopoulos, G. (1978) Rates of OH radical reactions. 4. Reactions with methanol, ethanol, 1-propanol, and 2-propanol at 296 K. *J. Phys. Chem.* 82, 1329-1333.

Palit, S.R. (1947) Electronic interpretations of organic chemistry. II. Interpretation of the solubility of organic compounds. *J. Phys. Chem.* 51, 837-857.

Pangrekar, J., Klopman, G., Rosenkranz, H.S. (1994) Expert-system comparison of structural determinants of chemical toxicity to environmental bacteria. *Environ. Toxicol. & Chem.* 13(6), 979-1001.

Petriris, V.E., Geankopolis, C.J. (1959) Phase equilibrium in 1-butanol-water-lactic acid system. *J. Chem. Eng. Data* 4, 197-198.

Pinal, R., Lee, L.S., Rao, P.S.C. (1991) Prediction of the solubility of hydrophobic compounds in nonideal solvent mixtures. *Chemosphere* 22(9/10), 939-951.

Pitter, P. (1976) Determination of biological degradability of organic substances. *Water Res.* 10, 231.

Platford, R.F. (1983) The octanol-water partitioning of some hydrophobic and hydrophilic compounds. *Chemosphere* 12, 1107-1111.

Price, K.S., Waggy, G.T., Conway, R.A. (1974) Brine shrimp bioassay and seawater BOD of of petrochemicals. *J. Water Poll. Control Fed.* 46, 63-77.

Purdy, R. (1987) Taft sigma and $sigma_I$ constants improve log octanol/water partition coefficient based QSAR for fathead minnow toxicity. In: *QSAR in Experimental Toxicology-II.* Kaiser, K.L.E., Editor, pp.271-280, D. Reidel Publishing Co., Dordrecht, The Netherlands.

Ravishankara, A.R., Davis, D.D. (1978) Kinetic rate constants for the reaction of OH with methanol, ethanol, and tetrahydrofuran at 298 K. *J. Phys. Chem.* 82, 2852-2853.

Reid, R.C., Sherwood, T.K. (1966) *The Properties of Gases and Liquids-Their Estimation and Correlation.* 2nd. Edition, McGraw-Hill, New York.

Reid, R.C., Prausnitz, J.M., Sherwood, T.K. (1977) *The Properties of Gases and Liquids.* 3rd Edition, McGraw-Hill Book Company, New York.

Rekker, R.F. (1977) *The Hydrophobic Fragmental Constant.* Its derivation and application. A means of characterizing membrane systems. Vol. 1, Pharmacochemistry Library, Nauta, W.T., Rekker, R.F., Editor. Elsevier Scientific Publishing Company, Oxford.

Riddick, J., Bunger, W.B. (1955) *Techniques of Organic Chemistry.* Wiessberger, A., Editor, Vol. 2. Wiley-Interscience, New York.

Riddick, J., Bunger, W.B. (1970) *Organic Solvents: Physical Properties and Method of Purification.* 3rd. Edition, *Techniques of Organic Chemistry.* Wiessberger, A., Editor, Vol. 2. John Wiley & Sons, New York, New York.

Riddick, J., Bunger, W.B., Sakano, T.K. (1986) *Organic Solvents: Physical Properties and Method of Purification.* 4th Edition, John Wiley & Sons, New York, New York.

Riebesehl, , W., Tomlinson, E. (1986) Thermodynamics of non-electrolyte transfer between octanol and water. *J. Soln. Chem.* 15, 141-150.

Rogers, K.S., Cammarata, A. (1969) Molecular orbital description of the partitioning aromatic compounds between polar and nonpolar phases. *Biochim. Biophys. Acta* 193(1), 22-29.

Rohrschneider, L. (1973) Solvent characterization by gas-liquid partition coefficients of selected solutes. *Anal. Chem.* 45(7), 1241-1247.

Russell, C.J., Dixon, S.L., Jurs, P.C. (1992) Computer-assisted study of the relationship between molecular structure and Henry's law constant. *Anal. Chem.* 64, 1350-1355.

Sangster, J. (1989) Octanol-water partition coefficients of simple organic compounds. *J. Phys. Chem. Ref. Data* 18, 1111-1230.

Sangster, J. (1993) LOGKOW databank, Sangster Research Labs., Montreal, Canada.

Sasaki, S. (1978) The scientific aspects of the chemical substance control law in Japan. In: *Aquatic Pollutants: Transformation and Biological Effects.* Hutzinger, O., et al., Editors, Pergamon Press, Oxford, U.K. pp. 283-298.

Schantz, M.M., Martire, D.E. (1987) Determination of hydrocarbon-water partition coefficients from chromatographic data and based on solution thermodynamics and theory. *J. Chromatogr.* 391, 35-51.

Schild, R., Donkin, P., Cotsifis, P.A., Donkin, M.E. (1993) A quantitative structure-activity relationship for the effect of alcohols on the neutral red retention by the marine marcoralga *enteromorpha intestinalis*. *Chemosphere* 27, 1777-1788.

Schultz, T.W., Arnold, L.M., Wilke, T.S., Moulton, M.P. (1990) Relationships of quantitative structure-activity to normal aliphatic alcohols. *Ecotoxicol. & Environ. Safety* 19, 243-253.

Scow, K.M. (1982) Chapter 9. Rate of biodegradation. In: *Handbook on Chemical Property Estimation Methods.* pp.9-1 to 9-85. Lyman, W.J., Reehl, W.F., Rosenblatt, D.H., Editors, McGraw-Hill, New York.

Seidell, A. (1941) *Solubilities of Organic Compounds.* Vol. 2, 3rd. Edition, D. Van Nostrand Co., New York, New York.

Serjeant, E.P., Dempsey, B. (1979) *Ionization Constants of Organic Acids in Aqueous Solution.* Pergamon Press, Oxford, U.K.

Shinoda, K., Yamanaka, T., Kinoshita, K. (1959) Surface chemical properties in aqueous solutions of non-ionic surfactants: Octyl glycol ether, α-octyl glyceryl ether and octyl glycoside. *J. Phys. Chem.* 63, 648-650.

Sidgwick, N.V., Sutton, L.E. (1930) CLXX.-The system cyclohexanol and water. *J. Chem. Soc.* 1323-1326.

Skrzec, A.E., Murphy, N.F. (1954) Liquid-liquid equilibria. Acetic acid in water, with 1-butanol, methyl ethyl ketone, furfural, cyclohexanol, and nitromethane. *Ind. Eng. Chem.* 46, 2245-2247.

Snider, J.R., Dawson, G.A. (1985) Tropospheric light alcohols, carbonyls, and acetonitrile: concentrations in the southwestern United States and Henry's law data. *J. Geophys. Res.* 90, 3797-3805.

Sokolova, L.P., Kaplin, V.T. (1969) Self-cleaning of natural waters. Elimination of alcohols of various structure under simulated conditions. *Gidrokhim Mater.* 51, 186-191.

Sonoda, Y., Seiko, Y. (1968) Degradation and toxicity of various compounds in anaerobic digestion. (I). Carbohydrates and alcohols. *J. Ferment. Technol.* 46, 796-801.

Speece, R.E. (1983) Anaerobic biotechnology for industrial wastewater treatment. *Environ. Sci. Technol.* 17, 416A-427A.

Stephen, H., Stephen, Y. (1963) *Solubilities of Inorganic and Organic Compounds.* Vol. 1 and 2. Pergamon Press, Oxford.

Stephenson, R., Stuart, J. (1986) Mutual solubilities: Water-alcohols and water-esters. *J. Chem. Eng. Data* 31, 56-70.

Stephenson, R., Stuart, J., Tabak, M. (1984) Mutual solubility of water and aliphatic alcohols. *J. Chem. Eng. Data* 29, 287-290.

Stephenson, R.M., Malanowski, A. (1987) *Handbook of the Thermodynamics of Organic Compounds*, Elsevier, New York.

Stockhardt, J.S., Hull, C.M. (1931) Vapor-liquid equilibria and boiling-point composition relations for systems *n*-butanol-water and isobutanol-water. *Ind. Eng. Chem.* 33, 1348.

Stull, D.R. (1947) Vapor pressure of pure substances: Organic compounds. *Ind. Eng. Chem.* 39(4), 517-560.

Suntio, L.R., Shiu, W.Y., Mackay, D. (1988) A review of the nature and properties of chemicals present in pulp mill effluents. *Chemosphere* 17, 1249-1290.

Taft, R.W., Abraham, M.H., Doherty, R.M., Kamlet, M.J. (1985) The molecular properties governing solubilities of organic nonelectrolytes in water. *Nature* 313, 384-386.

Taft, R.W., Abraham, M.H., Famini, G.R., Doherty, R.M., Abboud, J.-L.M., Kamlet, M.J. (1985) Solubility properties in polymers and biological media 5: An analysis of the physicochemical properties which influence octanol-water partition coefficients of aliphatic and aromatic solutes. *J. Pharm. Sci.* 74(8), 807-814.

Takemoto, S., Kuge, Y., Nakamoto, M. (1981) The measurement of BOD in seawater. *Suishitsu Odaku Kenkyu.* 4, 80-90.

Tayar, N. El, Tsai, R.-S., Vallat, P., Altomare, C., Testa, B. (1991) Measurement of partition coefficients by various centrifugal partition chromatographic techniques. *J. Chromatogr.* 556, 181-194.

Tewari, Y.B., Miller, M.M., Wasik, S.P., Martire, D.E. (1982) Aqueous solubility and octanol/water partition coefficient of organic compounds at 25.0°C. *J. Chem. Eng. Data* 27, 451-454.

Tute, M.S. (1971) Principle and practice of Hansch analysis: A guide structure-activity correlation for the medical chemist. *Adv. Drug Res.* 6, 1-77.

Urano, K., Maeda, H., Ogura, K., Wada, H. (1982) Direct analytical method for aromatic compounds in water by steam carrier gas chromatography. *Water Res.* 16, 323-327.

Urano, K., Kato, Z. (1986) Evaluation of biodegradation ranks of priority organic compounds. *J. Hazard. Mat.* 13, 147-159.

USEPA (1987) EXAMS II Computer Simulation. U.S. Environmental Protection Agency.

Vaishnav, D.A., Boethling, R.S., Babeu, L. (1987) Quantitative structure-biodegradability relationships for alcohols, ketones and alicyclic compounds. *Chemosphere* 16(4), 695-703.

Valvani, S.C., Yalkowsky, S.H., Amidon, G.L. (1976) Solubility of nonelectrolytes on polar solvents. VI. Refinements in molecular surface area computations. *J. Phys. Chem.* 80, 829-835.

Valvani, S.C., Yalkowsky, S.H., Roseman, T.J. (1981) Solubility and partitioning. IV. Aqueous solubility and octanol/water partition coefficient of liquid electrolytes. *J. Pharm. Sci.* 70, 502-507.

Veith, G.D., Call, D.J., Brooke, L.T. (1983) Structure-toxicity relationships for the fathead minnow, *pimephales promelas*: Narcotic industrial chemicals. *Can. J. Fish Aquat. Sci.* 40, 743-748.

Verhaar, H.J.M., van Leeuwen, C.J., Hermens, J.L.M. (1992) Classifying environmental pollutants. 1: Structure-activity relationships for prediction of aquatic toxicity. *Chemosphere* 25(4), 471-491.

Verschueren, K. (1977) *Handbook of Environmental Data on Organic Chemicals*. Van Nostrand Reinhold Co., New York.

Verschueren, K. (1983) *Handbook of Environmental Data on Organic Chemicals*, 2nd ed., Van Nostrand Reinhold Co., New York.

Wagner, R. (1976) Investigation of the biodegradation of organic substances using the dilution bottle method. II. The degradation of test substances. *Vom. Wasser* 47, 241-265.

Wakita, K., Yoshimoto, M., Miyamoto, S. (1986) A method for calculation of the aqueous solubility of organic compounds by using fragment solubility constants. *Chem. Pharm. Bull.* 34, 4663-4681.

Wallington, T.J., Atkinson, R., Winer, A.M., Pitts, J.N., Jr. (1987) A study of reaction $NO_3 + NO_2 + M \rightarrow N_2O_5 + M$ ($M = N_2, O_2$). *Intl. J. Chem. Kinet.* 19, 243-249.

Wallington, T.J., Dagaut, P., Liu, R., Kurylo, M.J. (1988a) Rate constants for the gas phase reactions of OH with C_5 through C_7 aliphatic alcohols and ethers: predicted and experimental values. *Intl. J. Chem. Kinet.* 20, 541-547.

Wallington, T.J., Dagaut, P., Kurylo, M.J. (1988b) Correlation between gas-phase and solution-phase reactivities of hydroxyl radicals toward saturated organic compounds. *J. Phys. Chem.* 92, 5024-5028.

Wallington, T.J., Kurylo, M.J. (1987) Flash photolysis resonance fluorescence investigation of the gas-phase reactions of OH radicals with a series of aliphatic ketones over the temperature range 240-440 K. *J. Phys. Chem.* 91, 5050-5054.

Wang, L., Zhao, Y., Hong, G. (1992) Predicting aqueous solubility and octanol/water partition coefficients of organic chemicals from molar volume. *Environ. Chem.* 11, 55-70.

Wasik, S.P., Tewari, Y.B., Miller, M.M., Martire, D.E. (1981) *Octanol/Water Partition Coefficients and Aqueous Solubilities of Organic Compounds*. NBSIR No.81-2406. U.S. Dept. of Commerce, Washington.

Weast, R.C., Editor (1972-73) *Handbook of Chemistry and Physics*. 53rd Edition, CRC Press, Cleveland, Ohio.

Weast, R.C., Editor (1973-74) *Handbook of Chemistry and Physics*. 54th ed., CRC Press, Cleveland Ohio.

Weast, R.C., Editor (1976-77) *Handbook of Chemistry and Physics*. 57th Edition, CRC Press, Cleveland, Ohio.

Weast, R.C., Editor (1978) *Handbook of Chemistry and Physics*. 58th Edition, CRC Press, Inc., West Palm Beach, Florida.

Weast, R.C., Editor (1982-83) *Handbook of Chemistry and Physics*. 63th ed., CRC Press, Boca Raton, Florida.

Weast, R.C., Editor (1983-84) *Handbook of Chemistry and Physics*. 63th ed., CRC Press, Boca Raton, Florida.

Weber, R.C. et al. (1981) *Vapor Pressure Distribution of Selected Organic Chemicals.* US-EPA-600/2-81-021. p.24.

Wiedermann, A., Zetzsch, C. (1982) presented at Bunsentagung Ulm and Neu-Ulm, May 20-22, 1982.

Wilhoit, R.C., Zwolinski, B.J. (1971) *Handbook of Vapor Pressures and Heats of Vaporization of Hydrocarbons and Other Related Compounds.* API 44-TRC Publications in Science and Engineering.

Wilson, W.G., Novak, J.T., White, K.D. (1986) Enhancement of degradation of alcohols in groundwater systems. In: *Toxic Hazard. Wastes, Proc. of Mid-Atlantic Industrial Waste Conference* 18, 421-430.

Windholz, M., Editor (1983) *The Merck Index.* Merck Co., Rahway, New Jersey.

Yalkowsky, S.H., Morozowich, W. (1980) Chapter 3, A physical chemical basis for the design of orally active prodrugs. In: *Drug Design.* Vol. IX. Academic Press, Inc., New York.

Yalkowsky, S.H., Pinal, R. (1993) Estimation of the aqueous solubility of complex organic compounds. *Chemosphere* 26, 1239-1261.

Yalkowsky, S.H., Valvani, S.C. (1980) Solubility and partitioning I: Solubility of nonelectrolytes in water. *J. Pharma. Sci.*, 69(8), 912-922.

Yaws, C., Yang, H.C., Hopper, J.R., Hansen, K.C. (1990) Organic chemicals: water solubility data. *Chem. Eng.* July, 115-118.

Yaws, C., Yang, H.C., Pan, X. (1991) Henry's law constants for 362 organic compounds in water. *Chem. Eng.* 98(11), 179-185.

Yonezawa, Y., Urushigawa, Y. (1979) Chemico-biological interactions in biological purification systems. V. Relation between biodegradation rate constants of aliphatic alcohols by activated sludge and their partition coefficients in a 1-octanol-water system. *Chemosphere* 8(3), 139-142.

Zellner, R., Lorenz, K. (1984) Laser-photolysis/resonance fluorescence study of the rate constants for the reactions of OH radicals with ethane and propane. *J. Phys. Chem.* 88, 984-989.

Zhou, X., Mopper, K. (1990) Apparent partition coefficients of 15 carbonyl compounds between air and seawater and between air and freshwater; implications for air-sea exchange. *Environ. Sci. Technol.* 24, 1864-1869.

Chapter 3 Aldehydes and Ketones

3.1 List of Chemicals and Data Compilations

Common Name: Formaldehyde
Synonym: formalin, methanal, oxomethane
Chemical Name: formaldehyde
CAS Registry No: 50-00-0
Molecular Formula: HCHO
Molecular Weight: 30.03
Melting Point (°C):
-92.0 (Weast 1982-83; Dean 1985)
-118/-92 (Verschueren 1983)
Boiling Point (°C):
-21.0 (Weast 1982-83)
-21/-19 (Verschueren 1983)
-19.5 (Dean 1985)
Density (g/cm^3):
0.815 (Weast 1982-83)
0.815 (-20°C, Verschueren 1983; Dean 1985)
Dissociation Constants, pK_a:
Molar Volume (cm^3/mol):
29.6 (calculated-LeBas method)
Molecular Volume, TMV ($Å^3$):
Total Surface Area, TSA ($Å^2$):
Heat of Fusion, ΔH_{fus}, kcal/mol:
Entropy of Fusion, ΔS_{fus}, cal/mol·K (e.u.):
Fugacity Ratio at 25°C (assuming ΔS_{fus} = 13.5 e.u.), F: 1.0

Water Solubility (g/m^3 or mg/L at 25°C):
1220000 (Dean 1985)
very soluble, up to 55% (Howard 1989)

Vapor Pressure (Pa at 25°C):
1333 (-88°C, Verschueren 1983)
5.17×10^5 (extrapolated-Antoine eqn., Dean 1985)
451030 (> 1 atm., Howard 1989)

Henry's Law Constant ($Pa \cdot m^3/mol$):
0.0331 (Dong & Dasgupta 1986, quoted, Howard 1989)
0.0169 (quoted, Gaffney et al. 1987)

Octanol/Water Partition Coefficient, log K_{OW}:
-0.75 (calculated-f const. per Rekker 1977, Deneer et al. 1988, quoted, Verhaar et al. 1992)
0.00 (calculated, Verschueren 1983)
0.35 (quoted, Howard 1989)
0.35 (recommended, Sangster 1989)

Bioconcentration Factor, log BCF:
no bioconcn. in fish and shrimp observed (Howard 1989)

Sorption Partition Coefficient, log K_{OC}:

Half-Lives in the Environmental Compartments:

Air: photooxidation half-life of 7.13-71.3 hours, based on measured rate constant for the vapor-phase reaction with hydroxyl radicals in air (Atkinson 1985; quoted, Howard et al. 1991); half-life of 1.26-6.0 hours, based on photolysis half-life in air (Howard et al. 1991); atmospheric transformation lifetime was estimated to be 1 to 5 days (Kelly et al. 1994).

Surface water: 24-168 hours, based on unacclimated aqueous aerobic biodegradation half-life (Howard et al. 1991).

Ground water: 48-336 hours, based on unacclimated aqueous aerobic biodegradation half-life (Howard et al. 1991).

Sediment:

Soil: 24-168 hours, based on unacclimated aqueous aerobic biodegradation half-life (Howard et al. 1991).

Environmental Fate Rate Constants and Half-Lives:

Volatilization:

Photolysis: sunlight photolysis half-life of 1.25-6.0 hours, based on measured gas-phase photolysis by simulated sunlight (Calvert et al. 1972; Su et al. 1979; quoted, Howard et al. 1991).

Oxidation: rate constant of $1.4x10^{-11}$ cm^3 $molecule^{-1}$ sec^{-1} for the vapor-phase reaction with OH radical was determined by the discharge-flow of mass spectrometric method (Morris & Niki 1971; quoted, Niki et al. 1978) and rate constant of $1.5x10^{-11}$ cm^3 $molecule^{-1}$ sec^{-1} for the vapor-phase reaction with OH radical was determined by the long-path Fourier transform IR spectroscopic method (Niki et al. 1978) both were studied with reference to that for C_2H_4 or C_2D_4; rate constant of $3.2x10^{-16}$ cm^3 $molecule^{-1}$ sec^{-1} for the vapor-phase reaction with NO_3 radicals in the atmosphere at 298 ± 1 K (Atkinson & Lloyd 1984; quoted, Carlier et al. 1986); rate constant of $4.5x10^{-14}$ cm^3 $molecule^{-1}$ sec^{-1} for the vapor-phase reaction with HO_2 radicals in the atmosphere at 298 K (Baulch et al. 1984; quoted, Carlier et al. 1986); rate constant of $111x10^{-11}$ cm^3 $molecule^{-1}$ sec^{-1} for the vapor-phase reaction with OH radicals in the atmosphere at 298 K (Baulch et al. 1984; quoted, Carlier et al. 1986); atmospheric photooxidation half-life of 7.13-71.3 hours, based on measured rate constant for the vapor-phase reaction with OH radicals in air (Atkinson 1985; quoted, Howard et al. 1991); aqueous photooxidation half-life of 4813-190000 hours, based on measured rate constant for the reaction with OH radicals in water (Dorfman & Adams 1973; quoted, Howard et al. 1991).

Hydrolysis: no hydrolyzable group (Howard et al. 1991).

Biodegradation: aqueous aerobic half-life of 24-168 hours, based on unacclimated aqueous aerobic biodegradation screening test data (Gellman & Heukelekian 1950; Heukelekian & Rand 1955; quoted, Howard et al. 1991); aqueous anaerobic half-life of 96-672 hours, based on unacclimated aqueous aerobic biodegradation half-life (Howard et al. 1991).

Biotransformation:

Bioconcentration Uptake (k_1) and Elimination (k_2) Rate Constants:

Common Name: Acetaldehyde
Synonym: acetic aldehyde, aldehyde, ethanal, ethylaldehyde
Chemical Name: acetaldehyde, ethanal
CAS Registry No: 75-07-0
Molecular Formula: CH_3CHO
Molecular Weight: 44.05
Melting Point (°C):
-123.5 (Stull 1947; Verschueren 1983; Dean 1985)
-121 (Weast 1982-83)
-123 (Riddick et al. 1986)
Boiling Point (°C):
20.2 (Stull 1947; Verschueren 1983; Dean 1985)
20.8 (Weast 1982-83)
20.4 (Boublik et al. 1984; Riddick et al. 1986)
Density (g/cm^3 at 20°C): 0.566
Molar Volume (cm^3/mol):
56.6 (Kamlet et al. 1986,87; Leahy 1986)
28.3 (intrinsic volume V_I, Leahy 1986; Kamlet et al. 1987)
56.2 (Kamlet et al. 1987)
51.8 (calculated-LeBas method)
Molecular Volume ($Å^3$):
Dissociation Constant, pK_b:
-10.2 (Riddick et al. 1986)
Total Surface Area, TSA ($Å^2$):
Heat of Fusion ΔH_{fus}, kcal/mol:
0.775 (Riddick et al. 1986)
Entropy of Fusion ΔS_{fus}, cal/mol·K (e.u.):
Fugacity Ratio at 25°C (assuming ΔS_{fus} = 13.5 e.u.), F: 1.0

Water Solubility (g/m^3 or mg/L at 25°C):
infinite (Palit 1947; Riddick et al. 1986)
miscible (Verschueren 1983; Dean 1985)
682254, 899386 (quoted, calculated-V_I and solvatochromic p., Leahy 1986)
682254, 920335 (quoted, calculated-V_I and solvatochromic p., Kamlet et al. 1987)
680770, 147570 (quoted, calculated- χ , Nirmalakhandan & Speece 1988)
miscible (Yaws et al. 1990)

Vapor Pressure (Pa at 25°C):
134018 (extrapolated-regression of tabulated data, Stull 1947)
101308 (20.2°C, Stull 1947)
120060 (Hoy 1970)
98642 (20°C, Verschueren 1983)
102125 (20.7°C, quoted exptl., Boublik et al. 1984)
120223 (calculated-Antoine eqn., Boublik et al. 1984)
120229 (calculated-Antoine eqn., Dean 1985)
121300 (Riddick et al. 1986)

Henry's Law Constant (Pa·m^3/mol):

6.692 (shake flask, partial vapor pressure-GC, Buttery et al. 1969)
8.924 (quoted lit. calculated K_{AW}s, Buttery et al. 1969)
6.672 (quoted exptl., Hine & Mookerjee 1975)
5.946 (calculated-group contribution, Hine & Mookerjee 1975)
5.423 (calculated-bond contribution, Hine & Mookerjee 1975)
6.75 (quoted, Gaffney et al. 1987)
8.90 (gas stripping-GC, Betterton & Hoffmann 1988)
6.80 (gas stripping-HPLC, Zhou & Mopper 1990)
10.18 (computed, Yaws et al. 1991)

Octanol/Water Partition Coefficient, log K_{OW}:

0.52 (gen. col.-HPLC, Wasik et al. 1981; quoted, Sangster 1989)
0.36 (gen. col.-GC, Tewari et al. 1982)
0.43 (calculated, Verschueren 1983)
-0.47 (calculated-V_I & solvatochromic p., Leahy 1986)
0.06 (selected, Yoshida et al. 1986)
-0.53 (calculated-f const. per Rekker 1977, Deneer et al. 1988; quoted, Verhaar et al. 1992)
0.45 (recommended, Sangster 1989)

Bioconcentration Factor, log BCF:

Sorption Partition Coefficient, log K_{OC}:

Half-Lives in the Environment:

Air: half-life of less than 0.24 hour for the gas-phase reaction with hydroxyl radicals in air, based on the rate of disappearance of hydrocarbon due to reaction with hydroxyl radicals (Darnall et al. 1976); atmospheric transformation lifetime was estimated to be < 1 day (Kelly et al. 1994).
Surface water:
Groundwater:
Sediment:
Soil:
Biota:

Environmental Fate Rate Constants or Half-Lives:

Volatilization:
Photolysis:
Oxidation: rate constant of 1.5×10^{-11} cm^3 molecule^{-1} sec^{-1} for the vapor-phase reaction with OH radical was determined by the discharge-flow of mass spectrometric method (Morris et al. 1971; quoted, Niki et al. 1978) and rate constant of 1.6×10^{-11} cm^3 molecule^{-1} sec^{-1} for the vapor-phase reaction with OH radical was determined by the long-path Fourier transform IR spectroscopic method (Niki

et al. 1978) both were studied with reference to that for C_2H_4 or C_2D_4; second-order rate constants of $(15 \pm 3.8)x10^{-12}$ $cm^3 \cdot molecule^{-1} \cdot sec^{-1}$ for the gas-phase reaction with hydroxyl radicals at 300 K in air (Morris & Niki 1971; quoted, Atkinson 1985); rate constant of 0.1 M^{-1} sec^{-1} for oxidation by RO_2 radicals at 30°C in aquatic systems with half-life of $8x10^4$ days (Howard 1972; Hendry et al. 1974; quoted, Mill 1982); $<2x10^2$ M^{-1} sec^{-1} for oxidation by singlet oxygen at 25°C in aquatic systems with half-life >100 years (Foote 1976; Mill 1979; quoted, Mill 1982); $(16.2 \pm 1.7)x10^{-12}$ $cm^3 \cdot molecule^{-1} \cdot sec^{-1}$ (Niki et al. 1978; quoted, Atkinson 1985); rate constant of $(12.8 \pm 4.3)x10^{-12}$ $cm^3 \cdot molecule^{-1} \cdot sec^{-1}$ (Kerr & Sheppard 1981; quoted, Atkinson 1985), $(12.2 \pm 2.7)x10^{-12}$ $cm^3 \cdot molecule^{-1} \cdot sec^{-1}$ (Semmes et al. 1985; quoted, Atkinson 1985) all at 298 K; rate constant of $(3.4 \pm 0.5)x10^{-20}$ $cm^3 \cdot molecule^{-1} \cdot sec^{-1}$ for the gas-phase reaction with ozone at 298 $\pm$ 2 K (Stedman & Niki 1973; quoted, Atkinson & Carter 1984); rate constant of $9.6x10^{12}$ $cm^3 \cdot mol^{-1} \cdot sec^{-1}$ for reaction with OH radical at 300 K (Lyman 1982); rate constant of $\leq 6x10^{-21}$ $cm^3 \cdot molecule^{-1} \cdot sec^{-1}$ for the gas-phase reaction with ozone at 296 $\pm$ 2 K (Atkinson et al. 1981, 1982; quoted, Atkinson & Carter 1984; Atkinson 1985); rate constant of $1.0x10^{-17}$ $cm^3 \cdot molecule^{-1} \cdot sec^{-1}$ for the gas-phase reaction with HO_2 radicals at 293 K in the atmosphere (Barnes et al. 1982; quoted, Carlier et al. 1986); second-order rate constant of $1.62x10^{-11}$ $cm^3 \cdot molecule^{-1} \cdot sec^{-1}$ for the gas-phase reaction with OH radicals and $3.02x10^{-15}$ $cm^3 \cdot molecule^{-1} \cdot sec^{-1}$ for the gas-phase reaction with NO_3 radicals both at 298 K in air (Atkinson et al. 1984; quoted, Atkinson 1985); rate constant of $1.40x10^{-15}$ $cm^3 \cdot molecule^{-1} \cdot sec^{-1}$ for the gas-phase reaction with NO_3 radicals at 298 $\pm$ 1 K in the atmosphere (Atkinson & Lloyd 1984; quoted, Carlier et al. 1986); $1.60x10^{-11} \cdot cm^3 \cdot molecule^{-1} \cdot sec^{-1}$ for the gas-phase reaction with hydroxyl radicals at 298 K in the atmosphere (Baulch et al. 1984; quoted, Carlier et al. 1986); rate constant of $(14.7 \pm 2.8)x10^{-12}$ $cm^3 \cdot molecule^{-1} \cdot sec^{-1}$ for gas-phase reaction with OH radicals at 298 K (Michael et al. 1985; quoted, Atkinson 1985); room temperature gas-phase rate constants of $1.6x10^{11}$ $cm^3 \cdot molecule^{-1} \cdot sec^{-1}$ with a loss rate of 0.7 day^{-1} for reaction with OH radicals (concn. of $5x10^5$ $molecule \cdot cm^{-3}$), $\leq 6.0x10^{-21}$ $cm^3 \cdot molecule^{-1} \cdot sec^{-1}$ with a loss rate of ≤ 0.0004 $days^{-1}$ for reactions with O_3 (concn of $7.2x10^{11}$ $molecule \cdot cm^{-3}$), $2.5x10^{-15}$ $cm^3 \cdot molecule^{-1} \cdot sec^{-1}$ with a loss rate of 0.05 day^{-1} with reactions with NO_3 radicals (concn. of $2.4x10^8$ $molecule \cdot cm^{-3}$) (Atkinson 1985); atmospheric rate constants of $16.22x10^{-12}$ $cm^3 \cdot molecule^{-1} \cdot sec^{-1}$ for reactions with OH radicals and $3.02x10^{-15}$ $cm^3 \cdot molecule^{-1} \cdot sec^{-1}$ for reaction with NO_3 radicals (quoted, Sabljic & Güsten 1990; Müller & Klein 1991); rate constants of $(2.44 \pm 0.52)x10^{-15}$ $cm^3 \cdot molecule^{-1} \cdot sec^{-1}$; $(2.74 \pm 0.33)x10^{-15}$ $cm^3 \cdot molecule^{-1} \cdot sec^{-1}$, at 298K, $(3.15 \pm 0.60)x10^{-15}$ $cm^3 \cdot molecule^{-1} \cdot sec^{-1}$ at 299K and $(2.54 \pm 0.64)x10^{-15}$ $cm^3 \cdot molecule^{-1} \cdot sec^{-1}$ at 300 K for the gas-phase reactions with NO_3 radicals (Atkinson 1991).

Hydrolysis:

Biodegradation:

Biotransformation:

Bioconcentration, Uptake (k_1) and Elimination (k_2) Rate Constants:

Common Name: Propionaldehyde
Synonym: propanal
Chemical Name: propionaldehyde, propanal
CAS Registry No: 123-38-6
Molecular Formula: CH_3CH_2CHO
Molecular Weight: 58.08
Melting Point (°C):
-81 (Weast 1982-83; Verschueren 1983; Dean 1985)
-80 (Riddick et al. 1986)
Boiling Point (°C):
48.8 (Weast 1982-83)
48.0 (Boublik et al. 1984; Riddick et al. 1986; Banerjee et al. 1990)
49.0 (Verschueren 1983)
48.9 (Dean 1985)
Density (g/cm^3 at 20°C):
0.8058 (Weast 1982-83)
0.8070 (Verschueren 1983)
0.8071 (Dean 1985)
0.7970 (Riddick et al. 1986)
Molar Volume (cm^3/mol):
72.0 (Taft et al. 1985; Leahy 1986)
38.1 (intrinsic volume V_I, Leahy 1986)
74.0 (calculated-LeBas method)
72.6 (calculated-density)
Molecular Volume ($Å^3$):
Dissociation Constant, pK_a:
Total Surface Area, TSA ($Å^2$):
224.0 (Amidon et al. 1975)
Heat of Fusion ΔH_{fus}, kcal/mol:
Entropy of Fusion ΔS_{fus}, cal/mol·K (e.u.):
Fugacity Ratio at 25°C (assuming ΔS_{fus} = 13.5 e.u.), F: 1.0

Water Solubility (g/m^3 or mg/L at 25°C):
160000 (20°C, Seidell 1941; quoted, Palit 1947)
10650 (quoted, Amidon et al. 1975)
160000 (quoted, Urano et al. 1981)
200000 (20°C, quoted, Verschueren 1983)
306000 (Dean 1985; Riddick et al. 1986)
136150 (calculated-V_I and solvatochromic p., Leahy 1986)
405000 (selected, Yaws et al. 1990)
310000 (20°C, shake flask-GC/TC, Stephenson 1993)
269000 (30°C, shake flask-GC/TC, Stephenson 1993)

Vapor Pressure (Pa at 25°C):

31330 (20°C, quoted, Verschueren 1983)
42300 (quoted exptl., Boublik et al. 1984; Riddick et al. 1986)
42340 (calculated-Antoine eqn., Boublik et al. 1984)
42260 (Howard et al. 1986; quoted, Banerjee et al. 1990)
23190 (calculated-solvatochromic p., Banerjee et al. 1990)

Henry's Law Constant (Pa·m^3/mol):

7.436 (shake flask, partial vapor pressure-GC, Buttery et al. 1969)
10.91 (quoted lit. calculated K_{AW}s, Buttery et al. 1969)
7.486 (quoted exptl., Hine & Mookerjee 1975)
8.399 (calculated-group contribution, Hine & Mookerjee 1975)
8.208 (calculated-bond contribution, Hine & Mookerjee 1975)
13.77 (headspace-GC, Snider & Dawson 1985)
8.31 (gas stripping-HPLC, Zhou & Mopper 1990)
7.486, 13.62 (quoted, correlated-molecular structure, Russell et al. 1992)

Octanol/Water Partition Coefficient, log K_{OW}:

0.83 (calculated, Verschueren 1983)
0.59, 0.40 (quoted, calculated-V_I & solvatochromic p., Leahy 1986)
0.0 (calcd.-f const. per Rekker 1977, Deneer et al. 1988; quoted, Verhaar et al. 1992)
0.59 (quoted unpublished data, Sangster 1989)
0.59 (recommended, Sangster 1989)

Bioconcentration Factor, log BCF:

Sorption Partition Coefficient, log K_{OC}:

Half-Lives in the Environment:

Air: half-life of <0.24 hour for the gas-phase reaction with hydroxyl radicals in air, based on the rate of disappearance of hydrocarbon due to reaction with hydroxyl radicals (Darnall et al. 1976); photooxidation half-life of 3.3-33 hours in air, based on measured reaction rate constant for the vapor-phase reaction with hydroxyl radicals in air (Atkinson 1987; selected, Howard et al. 1991).

Surface water: 24-168 hours, based on estimated aqueous aerobic biodegradation half-lives (Howard et al. 1991).

Groundwater: 48-336 hours, based on estimated aqueous aerobic biodegradation half-lives (Howard et al. 1991).

Sediment:

Soil: 24-168 hours, based on estimated aqueous aerobic biodegradation half-lives (Howard et al. 1991).

Environmental Fate Rate Constants or Half-Lives:

Volatilization:

Photolysis:

Oxidation: rate constant of 3.0×10^{-11} cm^3 $molecule^{-1}$ sec^{-1} for the vapor-phase reaction with hydroxyl radicals was determined by the discharge-flow of mass spectrometric method (Morris & Niki 1971; quoted, Niki et al. 1978) and rate constant of 2.1×10^{-11} cm^3 $molecule^{-1}$ sec^{-1} for the vapor-phase reaction with OH radicals was determined by the long-path Fourier transform IR spectroscopic method (Niki et al. 1978) both were studied with reference to that for C_2H_4 or C_2D_4; photooxidation half-life of <0.24 hour for the gas-phase reaction with OH radicals in air, based on the rate of disappearance of hydrocarbon due to reaction with hydroxyl radicals (Darnall et al. 1976); measured rate constants of 30.6×10^{-12} cm^3 $molecule^{-1}$ sec^{-1} (Morris & Niki 1971; quoted, Atkinson 1985), $(18.5 \pm 2.1) \times 10^{-12}$ cm^3 $molecule^{-1}$ sec^{-1} (Audley et al. 1981; quoted, Atkinson 1985), $(19.5 \pm 1.5) \times 10^{-12}$ cm^3 $molecule^{-1}$ sec^{-1} (Kerr & Sheppard 1981; quoted, Atkinson 1985), 18×10^{-12} cm^3 $molecule^{-1}$ sec^{-1} (Atkinson & Lloyd 1984; quoted, Carlier et al. 1986) and $(17.1 \pm 2.4) \times 10^{-12}$ cm^3 $molecule^{-1}$ sec^{-1} for the gas-phase reaction with OH radicals at 298 K (Semmes et al. 1985; quoted, Atkinson 1985); photooxidation half-life of 3.3-33 hours in air, based on measured reaction rate constant for the vapor-phase reaction with hydroxyl radicals in air (Atkinson 1987; selected, Howard et al. 1991).

Hydrolysis: no hydrolyzable groups (Howard et al. 1991).

Biodegradation: biodegradation rate constants of 0.046-0.063 h^{-1} in 30 mg/L activated sludge after a time lag of ~5 hours (Urano & Kato 1986b); aqueous aerobic half-life of 24-168 hours, based on aerobic biological screening test data (Gerhold & Malaney 1966; Dore et al. 1975; Urano & Kato 1986; selected, Howard et al. 1991); aqueous anaerobic half-life of 96-672 hours, based on estimated aqueous aerobic biodegradation half-lives (Howard et al. 1991).

Biotransformation:

Bioconcentration, Uptake (k_1) and Elimination (k_2) Rate Constants:

Common Name: Butyraldehyde
Synonym: 1-butanal, butylaldehyde, butyric aldehyde
Chemical Name: butyraldehyde, butanal
CAS Registry No: 123-72-8
Molecular Formula: $CH_3CH_2CH_2CHO$
Molecular Weight: 72.11
Melting Point (°C):
- -99.0 (Weast 1982-83; Dean 1985; Verschueren 1983)
- -96.4 (Dean 1985; Riddick et al. 1986)

Boiling Point (°C):
- 75.7 (Weast 1982-83)
- 75-76 (Verschueren 1983)
- 75.2 (Boublik et al. 1984)
- 74.8 (Dean 1985; Riddick et al. 1986)

Density (g/cm^3 at 20°C):
- 0.8170 (Weast 1982-83; Verschueren 1983)
- 0.8016 (Dean 1985; Riddick et al. 1986)

Molar Volume (cm^3/mol):
- 88.3 (Taft et al. 1985; Leahy 1986)
- 48.0 (intrinsic volume V_I, Leahy 1986)
- 88.2 (Kamlet et al. 1987)
- 96.2 (calculated-LeBas method)
- 89.0 (calculated-density)

Molecular Volume ($Å^3$):
Dissociation Constant, pK_a:
Total Surface Area, TSA ($Å^2$):
- 255.8 (Amidon et al. 1975)

Heat of Fusion ΔH_{fus}, kcal/mol:
- 2.65 (Riddick et al. 1986)

Entropy of Fusion ΔS_{fus}, cal/mol·K (e.u.):
Fugacity Ratio at 25°C (assuming ΔS_{fus} = 13.5 e.u.), F: 1.0

Water Solubility (g/m^3 or mg/L at 25°C):
- 36000 (20°C, quoted, Palit 1947)
- 37842 (quoted, Deno & Berkheimer 1960)
- 44070 (Amidon et al. 1975)
- 36000 (quoted, Urano et al. 1981)
- 37000, 71000 (quoted, Verschueren 1983)
- 51048, 86692 (quoted, calculated-V_I and solvatochromic p., Leahy 1986)
- 71000 (quoted, Dean 1985; Riddick et al. 1986)
- 50933 (quoted, Nirmalakhandan & Speece 1988)
- 16070 (calculated- χ , Nirmalakhandan & Speece 1988)
- 83700 (selected, Yaws et al. 1990)
- 74400 (20°C, shake flask-GC/TC, Stephenson 1993)
- 54800 (30°C, shake flask-GC/TC, Stephenson 1993)

Vapor Pressure (Pa at 25°C):

9464 (20°C, Verschueren 1983)
19372 (30.7°C, quoted exptl., Boublik et al. 1984)
14798 (extrapolated-Antoine eqn., Boublik et al. 1984)
14785 (extrapolated-Antoine eqn., Dean 1985)
15700 (quoted, Riddick et al. 1986)

Henry's Law Constant (Pa·m³/mol):

11.650 (shake flask, partial vapor pressure-GC, Buttery et al. 1969)
12.890 (quoted, lit. calculated K_{AW}s, Buttery et al. 1969)
11.590 (quoted exptl., Hine & Mookerjee 1975)
11.594 (calculated-group contribution, Hine & Mookerjee 1975)
12.140 (calculated-bond contribution, Hine & Mookerjee 1975)
11.65 (gas stripping-HPLC, Zhou & Mopper 1990)
11.59, 11.59 (quoted, correlated-molecular structure, Russell et al. 1992)

Octanol/Water Partition Coefficient, log K_{OW}:

1.20 (Hansch & Leo 1979)
1.18 (quoted, Verschueren 1983)
0.88, 0.75 (quoted, calculated-V_I & solvatochromic p., Taft et al. 1985)
0.88 (quoted, Leahy 1986; Kamlet et al. 1988)
0.63 (calculated-V_I & solvatochromic p., Leahy 1986)
0.53 (calculated-f const. per Rekker 1977, Deneer et al. 1988; quoted, Verhaar et al. 1992)
0.63 (calculated-solvatochromic p., Kamlet et al. 1988)
0.88 (quoted and recommended, Sangster 1989)

Bioconcentration Factor, log BCF:

Sorption Partition Coefficient, log K_{OC}:

Half-Lives in the Environment:

Air: photooxidation half-life of 2.8-28 hours, based on measured rate constant for the gas-phase reaction with OH radicals in air (Atkinson 1987; selected, Howard et al. 1991).

Surface water: photooxidation half-life of 114-days to 19 years, based on measured rate constant for the reaction with OH radicals in water (Anbar & Neta 1967; Dorfman & Adams 1973; selected, Howard et al. 1991); 24-168 hours, based on estimated aqueous aerobic biodegradation half-life (Howard et al. 1991).

Groundwater: 48-336 hours, based on estimated aqueous aerobic biodegradation half-life (Howard et al. 1991).

Sediment:

Soil: 24-168 hours, based on estimated aqueous aerobic biodegradation half-life (Howard et al. 1991).

Biota:

Environmental Fate Rate Constants or Half-Lives:

Volatilization:

Photolysis:

Oxidation: photooxidation half-life of 114-days to 19 years in water, based on measured rate constant for the reaction with OH radicals in water (Anbar & Neta 1967; Dorfman & Adams 1973; selected, Howard et al. 1991); measured rate constants of $(26.2 \pm 3.3)\times 10^{-12}$ $cm^3\cdot molecule^{-1}\cdot s^{-1}$, based on relative rate technique to acetaldehyde (Audley et al. 1981; quoted, Atkinson 1985) and $(25.3 \pm 0.6)\times 10^{-12}$ $cm^3\cdot molecule^{-1}\cdot s^{-1}$, based on relative rate technique to ethene (Kerr & Sheppard 1981; quoted, Atkinson 1985) for the reaction with OH radicals at 298 K in air; rate constant of 25×10^{-12} $cm^3\cdot molecule^{-1}\cdot sec^{-1}$ for the gas-phase reaction with hydroxyl radicals at 298 K in the atmosphere (Atkinson & Lloyd 1984; quoted, Carlier et al. 1986); measured rate constant of $(30.8 \pm 4.2)\times 10^{-12}$ $cm^3\cdot molecule^{-1}\cdot s^{-1}$ by flash photolysis-resonance fluorescence technique for the gas-phase reaction with hydroxyl radicals at 298 K in air (Semmes et al. 1985; quoted, Atkinson 1985); photooxidation half-life of 2.8-28 hours in air, based on measured rate constant for the gas-phase reaction with OH radicals in air (Atkinson 1987; selected, Howard et al. 1991).

Hydrolysis:

Biodegradation: rate constants of 0.044-0.069 h^{-1} in 30 mg/L activated sludge after a time lag of ~5 hours (Urano & Kato 1986b); aqueous aerobic half-life of 24-168 hours, based on aerobic biological screening test data (Lamb & Jenkins 1952; Heukelekian & Rand 1955; Dore et al. 1975; Urano & Kato 1986b; selected, Howard et al. 1991); aqueous anaerobic half-life of 96-672 hours, based on estimated aqueous aerobic biodegradation half-life (Howard et al. 1991).

Biotransformation:

Bioconcentration, Uptake (k_1) and Elimination (k_2) Rate Constants:

Common Name: Acrolein
Synonym: 2-propenal, acraldehyde, acrylic aldehyde, allyaldehyde, acrylaldehyde, aqualin
Chemical Name: 2-propenal
CAS Registry No: 107-02-8
Molecular Formula: $CH_2=CHCHO$
Molecular Weight: 56.064
Melting Point (°C):

-86.95 (Riddick et al. 1986)
-87.7 (Verschueren 1983)

Boiling Point (°C):

52.69 (Riddick et al. 1986)
52.5 (Verschueren 1983)

Density (g/cm^3 at 20°C):

0.8389 (Riddick et al. 1986)
0.8427 (Verschueren 1983)

Dissociation Constants, pKa:
Molar Volume (cm^3/mol):

66.6 (calculated-LeBas method)

Molecular Volume, TMV (Å^3):
Total Surface Area, TSA (Å^2):
Heat of Fusion, ΔH_{fus}, kcal/mol:
Entropy of Fusion, ΔS_{fus}, cal/mol·K:
Fugacity Ratio at 25°C (assuming ΔS_{fus} = 13.5 kcal/mol·K), F: 1.0

Water Solubility (g/m^3 or mg/L at 25°C):

20800 (20°C, quoted, Günther et al. 1968)
20800 (quoted, Callahan et al. 1979)
102020 (shake flask-radioactive analysis, Veith et al. 1980)
32261 (calculated-K_{OW} as per Hansch et al. 1968, Veith et al. 1980)
20800 (20°C, quoted, Verschueren 1983)
20800 (Riddick et al. 1986, quoted, Howard 1989)
20893 (quoted, Isnard & Lambert 1989)
20800 (selected, Müller & Klein 1992)

Vapor Pressure (Pa at 25°C):

29330 (20°C, Verschueren 1977, 1983; quoted, Callahan et al. 1979)
44000 (30°C, Verschueren 1977, 1983; quoted, Callahan et al. 1979)
35864 (Smith 1962, quoted, Callahan et al. 1979)
35300 (quoted, Riddick et al. 1986, Howard 1989)

Henry's Law Constant (Pa·m^3/mol):

12.36 (quoted, Gaffney et al. 1987)
0.446 (quoted, Howard 1989)

Octanol/Water Partition Coefficient, log K_{OW}:

-0.090 (quoted, Callahan et al. 1979)
0.90 (measured value, Veith et al. 1980)
0.90 (quoted, Isnard & Lambert 1989)
-0.10 (quoted, Howard 1989)
0.10 (quoted, Müller & Klein 1992)
0.101 (calculated-QSAR, Kollig 1993)

Bioconcentration Factor, log BCF:

2.54 (bluegill sunfish, Barrows et al. 1980)
2.54 (bluegill sunfish, Veith et al. 1980)
0.06 (calculated-K_{OW}, Veith et al. 1980)
-0.22 (estimated-K_{OW}, Howard 1989)

Sorption Partition Coefficient, log K_{OC}:

1.38 (estimated-K_{OW}, Howard et al. 1989)
-0.219 (calculated-K_{OW}, Kollig 1993)

Half-Lives in the Environmental Compartments:

Air: half-lives 10-13 h for reaction with photochemically generated hydroxyl radicals, 18 days for reaction with ozone and 3.5 days for photodissociation in the atmosphere (Howard 1989); half-lives of 3.4-33.7 hours, based on photooxidation half-life in air (Howard et al. 1991); atmospheric transformation lifetime was estimated to be < 1 day (Kelly et al. 1994).

Surface water: removal half-lives of 2.0 to 2.5 days from water (Bowmer & Higgins 1976, quoted, Howard 1989); half-life for reacting with singlet oxygen and alkylperoxy radicals in natural sunlit water was estimated to be 8 and 23 days (Howard 1989), half-lives of 168-672 hours, based on aqueous aerobic biodegradation half-life (Howard et al. 1991).

Ground water: half-lives of 336-1344 hours, based on aqueous aerobic biodegradation half-life (Howard et al. 1991).

Sediment:

Soil: half-lives of 168-672 hours, based on aqueous aerobic biodegradation half-life (Howard et al. 1991).

Biota:

Environmental Fate Rate Constants and Half-Lives:

Volatilization: half-life 10 days from a model river (Howard 1989);

Photolysis: half-life of 3.5 days, based on measured quantum yields (Howard 1989).

Oxidation: photooxidation rate constant 6.4×10^{-19} $cm^3 \cdot molecule^{-1} \cdot sec^{-1}$ for reaction with ozone in the gas phase (Atkinson & Carter 1984); rate constant of (1.90 - 2.53) x 10^{-13} $cm^3 \cdot molecule^{-1} \cdot sec^{-1}$ for reaction with OH radicals in air (Atkinson 1985); rate constant of 19.9×10^{-12} $cm^3 \cdot molecule^{-1} \cdot sec^{-1}$ for the gas-phase reactions of the

OH radicals (Atkinson 1990); atmospheric half-lives of 3.4-33.7 hours, based on rate constant for reaction with OH radicals (Howard et al. 1991); atmospheric second-order rate constants of 22.9×10^{-12} $cm^3 \cdot molecule^{-1} \cdot sec^{-1}$ for reaction with OH radicals and 1.15×10^{-15} $cm^3 \cdot molecule^{-1} \cdot sec^{-1}$ for reaction with NO_3 radicals (Sabljic & Güsten 1990; Müller & Klein 1991).

Hydrolysis: no hydrolyzable groups (Callahan et al. 1979).

Biodegradation: half-life of 29 hours (natural unsterilized water) and 43 hours in sterilized (thymol-treated) water (Bowmer & Higgins 1976, quoted, Howard 1989); aqueous aerobic half-lives of 168-672 hours, based on acclimated aqueous screening test data; and aqueous anaerobic half-lives of 672-2880 hours, based on aqueous aerobic biodegradation half-life (Howard et al. 1991).

Biotransformation: half-life < 4 days (Callahan et al. 1979)

Bioconcentration Uptake (k_1) and Elimination (k_2) Rate Constants:

Common Name: Furfural
Synonym: 2-furaldehyde, 2-furancarboxaldehyde; furfurole, 2-furancarbonal, fural, furfuraldehyde, furole
Chemical Name: furfural
CAS Registry No: 98-01-1
Molecular Formula: $C_5H_4O_2$
Molecular Weight: 96.08
Melting Point (°C):
-36.5 (Stull 1947)
-38.0 (Verschueren 1983)
-38.7 (Stephenson & Malanowski 1987)
Boiling Point (°C):
161.8 (Stull 1947; Verschueren 1983)
Density (g/cm³ at 20°C):
1.16 (Verschueren 1983)
Acid Dissociation Constants, pK_a:
Molar Volume (cm³/mol):
91.1 (calculated-LeBas method)
82.9 (calculated-density, Stephenson & Malanowski 1987)
Molecular Volume, TMV (Å³):
Total Surface Area, TSA (Å²):
Heat of Fusion, ΔH_{fus}, kcal/mol:
Entropy of Fusion, ΔS_{fus}, cal/mol·K:
Fugacity Ratio at 25°C (assuming ΔS_{fus} = 13.5 kcal/mol·K), F: 1.0

Water Solubility (g/m³ or mg/L at 25°C):
82000 (40°C, synthetic method, Jones 1929)
91593 (Booth & Everson 1948)
92900 (26.7°C, shake flask-turbidity, Skrzec & Murphy 1954)
81680 (Hansch et al. 1968)
77830 (gen. col.-HPLC/UV, Tewari et al. 1982)
83000 (20°C, quoted, Verschueren 1983)
86000 (quoted, Howard 1993)
79400 (20°C, shake flask-GC/TC, Stephenson 1993)
84000 (30°C, shake flask-GC/TC, Stephenson 1993)

Vapor Pressure (Pa at 25°C):
220.4 (interpolated-regression of tabulated data, Stull 1947)
133.3 (18.5°C, Stull 1947)
133.3 (20°C, quoted, Verschueren 1983)
399.9 (30°C, quoted, Verschueren 1983)
208.0 (extrapolated-Antoine eqn., Dean 1985)
333.3 (Riddick et al. 1986; quoted, Howard 1993)
313.2 (extrapolated-Antoine eqn., Stephenson & Malanowski 1987)

Henry's Law Constant (Pa·m^3/mol):

0.375 (calculated-P/C, Howard 1993)

Octanol/Water Partition Coefficient, log K_{OW}:

0.52 (gen. col.-HPLC/UV, Tewari et al. 1982)

0.81 (calculated-f const. per Rekker 1977, Deneer et al. 1988; quoted, Verhaar et al. 1992)

0.41 (Hansch & Leo 1985; quoted, Howard 1993)

0.96 (computed, Klopman et al. 1985; quoted, Pangrekar et al. 1994)

Bioconcentration Factor, log BCF:

0.079 (estimated-K_{OW}, Lyman et al. 1982; quoted, Howard 1993)

-2.097 (estimated-S, Lyman et al. 1982; quoted, Howard 1993)

Sorption Partition Coefficient, log K_{OC}:

1.602 (soil, estimated-K_{OW}, Lyman et al. 1982; quoted, Howard 1993)

0.00 (soil, estimated-S, Lyman et al. 1982; quoted, Howard 1993)

Half-Lives in the Environmental Compartments:

Air: estimated photooxidation half-life of 0.44 day for the vapor-phase reaction with hydroxyl radicals in air (Atkinson 1985; quoted, Howard 1993).

Surface water:

Ground water:

Sediment:

Soil:

Biota:

Environmental Fate Rate Constants and Half-Lives:

Volatilization: based on calculated Henry's law constant, the estimated half-life from a model river of one meter deep flowing at 1 m/s with a wind velocity of 3 m/s is 9.9 days (Lyman et al. 1982; quoted, Howard 1993).

Photolysis:

Photooxidation:

Hydrolysis:

Biodegradation: average rate of biodegradation 41.0 mgCOD·g^{-1}·h^{-1} based on measurements of COD decrease using activated sludge inoculum with 20 days of adaptation to the substrate (Pitter 1976; quoted, Scow 1982).

Biotransformation:

Bioconcentration Uptake (k_1) and Elimination (k_2) Rate Constants:

Common Name: Benzaldehyde
Synonym: benzenecarbonal, oil of bitter almonds
Chemical Name: benzaldehyde
CAS Registry No: 100-52-7
Molecular Formula: C_6H_5CHO
Molecular Weight: 106.12
Melting Point (°C):
-26.0 (Stull 1947; Weast 1982-83; Verschueren 1983; Dean 1985)
-55.6 (Riddick et al. 1986)
Boiling Point (°C):
179.0 (Stull 1947; Urano et al. 1982; Verschueren 1983)
178.62 (at 10 mm Hg, Weast 1982-83)
178.75 (Boublik et al. 1984; Riddick et al. 1986)
178.90 (Dean 1985)
Density (g/cm^3 at 20°C):
1.0401 (24.94°C, measured, Ambrose et al. 1975)
1.0446 (calculated-Riedel's eqn., Ambrose et al. 1975)
1.050 (15.0°C, Verschueren 1983)
1.0447 (Dean 1985)
1.0446 (Riddick et al. 1986)
Molar Volume (cm^3/mol):
102.0 (calculated-density, Chiou 1985)
111.9 (Taft et al. 1985; Leahy 1986; Kamlet et al. 1987)
60.6 (intrinsic volume V_I, Leahy 1986; Kamlet et al. 1987)
103.2 (calculated-LeBas method)
Molecular Volume (Å^3):
Dissociation Constant, pK_a:
Total Surface Area, TSA (Å^2):
243.6 (Iwase et al. 1985)
Heat of Fusion ΔH_{fus}, kcal/mol:
2.228 (Riddick et al. 1986)
Entropy of Fusion ΔS_{fus}, cal/mol K (e.u.):
Fugacity Ratio at 25°C (assuming ΔS_{fus} = 13.5 e.u.), F: 1.0

Water Solubility (g/m^3 or mg/L at 25°C):
2991 (quoted, Deno & Berkheimer 1960)
6696 (Hansch et al. 1968; quoted, Hine & Mookerjee 1975)
4527 (quoted average, Valvani et al. 1981)
10860 (predicted-K_{OW}, Valvani et al. 1981)
7200 (20°C, shake flask-GC, Tewari et al. 1982)
3000 (20-25°C, shake flask-GC, Urano et al. 1982)
3300 (quoted, Verschueren 1983)
3279 (shake flask-GC, Chiou 1985; quoted, Chessells et al. 1992)
3000 (quoted, Dean 1985)
3000 (20°C, Riddick et al. 1986)
6543, 6543 (quoted, calculated-fragment const., Wakita et al. 1986)

4527, 25457 (quoted, calculated-V_I and solvatochromic p., Leahy 1986)
6543, 34340 (quoted, calculated-solvatochromic p., Kamlet et al. 1987)
6543, 8238 (quoted, calculated-V_I and M.P., Kamlet et al. 1987)
4170 (quoted, Isnard & Lambert 1988)
398, 914 (quoted, calculated- χ , Nirmalakhandan & Speece 1988)
7200 (20°C, shake flask-GC/TC, Stephenson 1993)
7400 (30°C, shake flask-GC/TC, Stephenson 1993)
5700 (selected, Yalkowsky et al. 1993)

Vapor Pressure (Pa at 25°C):
145.2 (extrapolated-regression of tabulated data, Stull 1947)
133.3 (26.2°C, Stull 1947)
137 (ebulliometry-extrapolated from Antoine eqn., Ambrose et al. 1975)
169 (calculated from different eqns., Ambrose et al. 1975)
174 (quoted, Hine & Mookerjee 1975)
160.5 (extrapolated-Antoine eqn., Ambrose et al. 1979)
133.3 (26.0°C, quoted, Verschueren 1983)
160.5 (extrapolated-Antoine eqn., Boublik et al. 1984)
169 (Riddick et al. 1986)

Henry's Law Constant ($Pa \cdot m^3/mol$):
2.781 (quoted exptl., Hine & Mookerjee 1975)
1.494 (calculated-bond contribution, Hine & Mookerjee 1975)
2.815 (quoted, Gaffney et al. 1987)
2.71 (gas stripping-GC, Betterton & Hoffmann 1988)
2.28 (gas stripping-HPLC, Zhou & Mopper 1990)

Octanol/Water Partition Coefficient, log K_{OW}:
1.48 (Leo et al. 1971; quoted, Hansch et al. 1972; McCall 1975; Veith et al. 1979; Chiou 1985; Iwase et al. 1985)
1.45 (shake flask-AS at pH 5.62, Umeyama et al. 1971; quoted, Sangster 1989)
1.43 (shake flask-AS, Holmes & Lough 1976; quoted, Sangster 1989)
1.45 $\pm$ 0.03 (shake flask at pH 7, Unger et al. 1978)
2.33 (estimated-HPLC-RT, Veith et al. 1979; quoted, Sangster 1989)
1.48 (Valvani et al. 1981)
1.48 (quoted, Verschueren 1983)
1.64 (calculated-TSA, Iwase et al. 1985)
1.56 (HPLC-k', Eadsforth 1986; quoted, Sangster 1989)
1.48, 1.49 (quoted: shake flask, calculated, Eadsforth 1986)
1.48, 1.33 (quoted, calculated-V_I & solvatochromic p., Taft et al. 1985)
1.48, 1.31 (quoted, calculated-V_I & solvatochromic p., Leahy 1986)
1.45 (quoted from Hansch & Leo 1979, Brooke et al. 1986; Bodor et al. 1989)
1.44 (HPLC average, Ge et al. 1987; quoted, Sangster 1989)
1.49 (calculated-f const. as per Rekker 1977, Deneer et al. 1988)

1.48 (quoted from CLOGP, Bodor et al. 1989)
1.495 (calculated-CLOGP, Bodor et al. 1989)
1.46 (calculated-MO, Bodor et al. 1989)
1.48 (quoted from unpublished data, Sangster 1989)
1.48 (recommended, Sangster 1989)
1.72 (shake flask-AS, Kramer & Henze 1990)
1.48 (quoted from Hansch & Leo 1983, Tayar et al. 1991)
1.48 (quoted from Medchem software value, Chessells et al. 1992)
1.48 (EPA CLOGP data base, Hulzebos et al. 1993)

Bioconcentration Factor, log BCF:

Sorption Partition Coefficient, log K_{OC}:

Half-Lives in the Environment:

Air: half-life of > 9.9 days for the gas-phase reaction with hydroxyl radacals in air, based on the rate of disappearance of hydrocarbon due to reaction with hydroxyl radicals (Darnall et al. 1976); half-life of 0.74 d based on rate constants for oxidation by OH radicals in the atmosphere at 25°C (Mill 1982); loss rate of $(3.4 \pm 1.7)\times10^{-4}$ min^{-1} in outdoor Teflon chambers in the dark (Grosjean 1985).

Surface water:

Groundwater:

Sediment:

Soil:

Biota:

Environmental Fate Rate Constants or Half-Lives:

Volatilization:

Photolysis: loss rate of 3.4×10^{-4} min^{-1} in outdoor Teflon chambers in dark (Grosjean 1985).

Oxidation: rate constant of $<2.0\times10^{2}$ $M^{-1}\cdot sec^{-1}$ for oxidation by singlet oxygen in the aquatic systems at 25°C with half-life >100 years (Foote 1976; Mill 1979; quoted, Mill 1982); rate constant of 1.3×10^{-11} cm^3 $molecule^{-1}$ sec^{-1} for the vapor-phase reaction with OH radicals was determined by the long-path Fourier transform IR spectroscopic method with reference to that for C_2H_4 or C_2D_4 (Niki et al. 1978); rate constant of 7.6×10^{9} $M^{-1}\cdot sec^{-1}$ for oxidation by OH radicals in the atmosphere at 25°C with a half-life of 0.74 day (Hendry & Kenley 1979; quoted, Mill 1982); rate constant of 7.8×10^{12} $cm^3\cdot mol^{-1}\cdot sec^{-1}$ for reaction with OH radical at 300 K (Lyman 1982); (2.5 ± 0.5) $M^{-1}\cdot sec^{-1}$ for 2-10 mM to react with ozone in water at pH 1.7 and 20-23°C (Hoigné & Bader 1983); using relative rate technique to propene, rate constant for the gas-phase reaction with NO_3 radicals at 300 ± 1 K in air was determined to be $<9.6\times10^{-15}$ $cm^3\cdot molecule^{-1}\cdot sec^{-1}$ (Carter et al. 1981; quoted, Atkinson 1991) as compared to

$(2.55 \pm 0.08)\times10^{-15}$ $cm^3\cdot molecule^{-1}\cdot sec^{-1}$ at 294 K (Atkinson et al. 1984; quoted, Atkinson 1991); using relative rate technique, rate constant for the gas-phase reaction with NO_3 radicals at 296 ± 2 K in air was determined to be $(1.13 \pm 0.25)\times10^{-15}$ $cm^3\cdot molecule^{-1}\cdot sec^{-1}$ compared to a previous measured value of $<5.2\times10^{-15}$ $cm^3\cdot molecule^{-1}\cdot sec^{-1}$ (Atkinson et al. 1984); rate constant of 1.2×10^{-11} $cm^3\cdot molecule^{-1}\cdot sec^{-1}$ for the gas-phase reaction with OH radicals at 298 K in the atmosphere (Atkinson & Lloyd 1984; quoted, Carlier et al. 1986); reevaluated rate constant of 2.0×10^{-15} $cm^3\cdot molecule^{-1}\cdot sec^{-1}$ for the gas-phase reaction with NO_3 radicals at 298 ± 2K in air (Atkinson et al. 1987); observed rate constant of 1.30×10^{-11} $cm^3\cdot molecule^{-1}\cdot sec^{-1}$ for the gas-phase reaction with hydroxyl radicals at 298 K in air compared with a calculated value of 1.698×10^{-11} $cm^3\cdot molecule^{-1}\cdot sec^{-1}$ (Atkinson 1987, 1990; quoted, Sabljic & Güsten 1990; Müller & Klein 1991); rate constant of $2.54\times10^{-15}\cdot cm^3\cdot molecule^{-1}\cdot sec^{-1}$ for the gas-phase reaction with NO_3 radicals at 298 K in air (Atkinson et al. 1988; quoted, Sabljic & Güesten 1990; Müller & Klein 1991).

Hydrolysis:

Biodegradation: average rate of biodegradation 119.0 mg COD $g^{-1}\cdot h^{-1}$ based on measurements of COD decrease using activated sludge inoculum with 20 days of adaptation to the substrate (Pitter 1976; quoted, Scow 1982), biodegradation rate constants of 0.065-0.074 h^{-1} in 30 mg/L activated sludge after a lag time of 5-10 hours (Urano & Kato 1986b).

Biotransformation:

Bioconcentration, Uptake (k_1) and Elimination (k_2) Rate Constants:

Common Name: Acetone
Synonym: 2-propanone, dimethylketone, DMK
Chemical Name: acetone, 2-propanone
CAS Registry No: 67-64-1
Molecular Formula: CH_3COCH_3
Molecular Weight: 58.08
Melting Point (°C):
-94.6 (Stull 1947; Verschueren 1983)
-95.35 (Weast 1982-83; Dean 1985; Howard 1990)
-94.7 (Riddick et al. 1986)
Boiling Point (°C):
56.5 (Stull 1947)
56.20 (Weast 1982-83; Verschueren 1983; Howard 1990)
56.06 (Boublik et al. 1984; Riddick et al. 1986)
56.24 (Dean 1985)
56.00 (Banerjee et al. 1990)
Density (g/cm^3 at 20°C):
0.7899 (Weast 1982-83)
0.7910 (Verschueren 1983)
0.7908 (Dean 1985)
0.78998 (Riddick et al. 1986)
Molar Volume (cm^3/mol):
77.5 (exptl. at normal B.P., Lee et al. 1972; quoted, Reid et al. 1977)
73.8 (Rohrschneider 1973)
74.0 (LeBas method, Reid et al. 1977; Abernethy et al. 1988)
73.4 (Taft et al. 1985; Kamlet et al. 1986,87; Leahy 1986)
38.0 (intrinsic volume V_I, Leahy 1986; Kamlet et al. 1987; Abernethy et al. 1988)
59.3 (calculated-density, Jarworska & Schultz 1993)
Molecular Volume (Å^3):
Total Surface Area, TSA (Å^2):
Heat of Fusion ΔH_{fus}, kcal/mol:
1.360 (Riddick et al. 1986)
Entropy of Fusion ΔS_{fus}, cal/mol·K (e.u.):
Fugacity Ratio at 25°C (assuming ΔS_{fus} = 13.5 e.u.), F: 1.0

Water Solubility (g/m^3 or mg/L at 25°C):
infinite (20°C, Palit 1947)
485610 (calculated-activity coefficient of Perry 1950, Rathbun et al. 1982)
miscible (Dean 1985; Pinal et al. 1991)
infinite (Riddick et al. 1986; Howard 1990)
483090, 483090 (quoted, calculated-V_I and solvatochromic p., Taft et al. 1986)
440580, 698275 (quoted, calculated-V_I and solvatochromic p., Leahy 1986)
440580, 748215 (quoted, calculated-V_I and solvatochromic p., Kamlet et al. 1987)
miscible (Yaws et al. 1990)

Vapor Pressure (Pa at 25°C):

29613 (interpolated-regression of tabulated data, Stull 1947)
26660 (22.7°C, Stull 1947)
30494 (Perry 1950; quoted, Rathbun et al. 1982)
30792 (Buttery et al. 1969; quoted, Howard 1990)
30812 (Hoy 1970)
30806 (ebulliometry, fitted to Antoine eqn., Ambrose et al. 1975)
30867, 31522 (quoted exptl., calculated-Antoine eqn., Boublik et al. 1984)
30780 (calculated-Antoine eqn., Dean 1985)
30926 (Howard et al. 1986; quoted, Banerjee et al. 1990)
30806 (Riddick et al. 1986)
19330 (calculated-solvatochromic p., Banerjee et al. 1990)

Henry's Law Constant ($Pa \cdot m^3/mol$):

3.342 (partial pressure, Butler & Ramchandani 1935)
3.960 (shake flask, partial vapor pressure-GC, Burnett 1963; quoted, Betterton 1991)
3.25 (28°C, concn. ratio-GC, Nelson & Hoff 1968)
3.97 (shake flask, partial vapor pressure-GC, Buttery et al. 1969; quoted, Betterton 1991)
4.140 (quoted, calculated-K_{AW}, Buttery et al. 1969)
4.020 (quoted, exptl., Hine & Mookerjee 1975)
3.929 (calculated-group contribution, Hine & Mookerjee 1975)
2.912 (calculated-bond contribution, Hine & Mookerjee 1975)
4.104 (calculated-$\gamma \cdot P$, Rathbun & Tai 1982)
3.334 (calculated-MW, Rathbun & Tai 1982)
3.648 (calculated-P/C from Perry 1950 data, Rathbun et al. 1982)
3.93 (headspace-GC, Snider & Dawson 1985; quoted, Howard 1990)
3.378 (quoted, Gaffney et al. 1987)
2.93 (gas stripping-HPLC, Zhou & Mopper 1990)
3.070 (gas stripping-GC, Betterton 1991)
3.27, 3.90, 3.75, 3.96 (quoted literature values, Betterton)
4.331 (computed, Yaws et al. 1991)

Octanol/Water Partition Coefficient, log K_{OW}:

-0.24 (shake flask-CR, Collander 1951; quoted, Leo et al. 1971; Sangster 1989)
-0.73 (calculated, Iwasa et al. 1965)
-0.21 (Leo et al. 1969)
-0.24 (quoted, shake flask at pH 7, Unger et al. 1978)
-0.30 (calculated-f const., Könemann 1981)
-0.24 (calculated-f const. as per Hansch & Leo 1979, Veith et al. 1983)
-0.24 (Verschueren 1983; quoted, Pinal et al. 1991)
-0.24 (Hansch & Leo 1985; quoted, Howard 1990)
0.38 (computed, Klopman et al. 1985; quoted, Pangrekar et al. 1994)
-0.24, -0.13 (quoted, calculated-V_I and solvatochromic p., Taft et al. 1985)
-0.24 (measured, GEMS 1986; quoted, Vaishnav et al. 1987)

-0.24, -0.24 (quoted, calculated-V_I and solvatochromic p., Leahy 1986)
-0.48 (shake flask-GC, Tanii et al. 1986; quoted, Sangster 1989)
-0.24 (quoted, Abernethy et al. 1988)
-0.24, -0.24 (quoted from Hansch & Leo 1979, quoted from CLOGP, Bodor et al. 1989)
-0.027, -0.07 (calculated from CLOGP, calculated-MO, Bodor et al. 1989)
-0.24 (recommended, Sangster 1989)
-0.24 (quoted, Verhaar et al. 1992)
-0.24 (calculated as per Hansch & Leo 1979, Dallos et al. 1993)
-0.37 (calculated-UNIFAC, Dallos et al. 1993)
-0.24 (quoted from CLOGP 3.15, Jaworska & Schultz 1993)
-0.266 (calculated-QSAR, Kollig 1993)

Bioconcentration Factor, log BCF:

Sorption Partition Coefficient, log K_{OC}:
-0.586 (calculated-K_{OW}, Kollig 1993)

Half-Lives in the Environment:
Air: half-life of >9.9 days for the gas-phase reaction with hydroxyl radicals in air, based on the rate of disappearance of hydrocarbon due to reaction with hydroxyl radicals (Darnall et al. 1976); photooxidation half-life of 279-2790 hours, based on measured data for the vapor-phase reaction with hydroxyl radicals in air (Howard et al. 1991).
Surface water: photooxidation half-life of 11.3-453 years, based on measured data for the reaction with hydroxyl radicals in aqueous solution (Dorfman & Adams 1973; quoted, Howard et al. 1991); 24-168 hours, based on estimated unacclimated aqueous aerobic biodegradation half-life (Howard et al. 1991).
Groundwater: 48-336 hours, based on estimated unacclimated aqueous aerobic biodegradation half-life (Howard et al. 1991).
Sediment:
Soil: 24-168 hours, based on estimated unacclimated aqueous aerobic biodegradation half-life (Howard et al. 1991).
Biota:

Environmental Fate Rate Constants or Half-Lives:
Volatilization: using Henry's law constant, a half-life of 20 hours was estimated for a model river 1 m deep flowing at 1 m/sec with a wind velocity of 3 m/sec (Lyman et al. 1982; quoted, Howard 1990).
Photolysis:
Oxidation: photooxidation half-life of 11.3-453 years, based on measured data for the reaction with hydroxyl radicals in aqueous solution (Dorfman & Adams 1973; selected, Howard et al. 1991); photooxidation half-life of >9.9 days for the gas-phase reaction with hydroxyl radicals in air, based on the rate of disappearance

of hydrocarbon due to reaction with hydroxyl radicals (Darnall et al. 1976); rate constant of $(0.23 \pm 0.03)\times10^{-12}$ $cm^3 \cdot molecule^{-1} \cdot sec^{-1}$ by flash photolysis-resonance fluorescence for the gas-phase reaction with OH radicals at 300 K in air (Zetzsch 1982; quoted, Atkinson 1985); rate constant of $(0.62 \pm 0.09)\times10^{-12}$ $cm^3 \cdot molecule^{-1} \cdot sec^{-1}$ by relative rate technique to *n*-hexane for the gas-phase reaction with OH radicals at 298 K in air (Chiorboli et al. 1983; quoted, Atkinson 1985); photooxidation half-life of 279-2790 hours, based on measured data for the vapor-phase reaction with hydroxyl radicals in air (Atkinson 1985; selected, Howard et al. 1991); rate constant of 2.16×10^{13} $cm^3 \cdot molecule^{-1} \cdot sec^{-1}$ for the gas-phase reaction with hydroxyl radicals in air and 1.80×10^{-13} $cm^3 \cdot molecule^{-1} \cdot sec^{-1}$ for the aqueous-phase reaction with hydroxyl radicals in solution (Wallington & Kurylo 1987; quoted, Wallington et al. 1988).

Hydrolysis:

Biodegradation: biodegradation rate constants of 0.016-0.020 h^{-1} in 30 mg/L activated sludge after a time lag of 20-25 hours (Urano & Kato 1986b); aqueous aerobic half-life of 24-168 hours, based on unacclimated aqueous screening test data (Bridie et al. 1979; Dore et al. 1975; selected, Howard et al. 1991); aqueous anaerobic half-life of 96-672 hours, based on estimated unacclimated aqueous aerobic biodegradation half-life (Howard et al. 1991); experimental rate constant of 0.0440 h^{-1} compared to group contribution method predicted rate constants of 0.0433 h^{-1} (nonlinear) and 0.043 h^{-1} (linear) (Tabak & Govind 1993).

Biotransformation:

Bioconcentration, Uptake (k_1) and Elimination (k_2) Rate Constants:

Common Name: Methyl Ethyl Ketone
Synonym: 2-butanone, butan-2-one, MEK
Chemical Name: 2-butanone, methyl ethyl ketone
CAS Registry No: 78-93-3
Molecular Formula: $CH_3CH_2COCH_3$
Molecular Weight: 72.107
Melting Point (°C):
-85.90 (Stull 1947)
-86.30 (Weast 1982-83)
-86.40 (Verschueren 1983)
-86.69 (Dean 1985; Riddick et al. 1986)
-86.35 (Howard 1990)
Boiling Point (°C):
80.75 (Ginnings et al. 1940)
79.60 (Stull 1947; Weast 1982-83; Verschueren 1983; Dean 1985; Howard 1990)
79.58 (Boublik et al. 1984; Riddick et al. 1986)
80.0 (Banerjee et al. 1990)
Density (g/cm^3 at 20°C):
0.8007 (Ginnings et al. 1940)
0.8054 (Weast 1982-83)
0.8049 (Verschueren 1983; Dean 1985; Riddick et al. 1986)
0.7997 (25°C, Riddick et al. 1986)
Molar Volume (cm^3/mol):
89.9 (calculated-density, Rohrschneider 1973)
89.5 (Taft et al. 1985; Kamlet et al. 1986; Leahy 1986)
47.7 (intrinsic volume V_I, Leahy 1986; Kamlet et al. 1987)
48.0, 96.0 (intrinsic volume V_I, LeBas method, Abernethy et al. 1988)
75.7 (calculated-density, Jaworska & Schultz 1993)
Molecular Volume ($Å^3$):
Dissociation Constants:
14.7 (pK_a, Riddick et al. 1986)
-7.2 (pK_{BH}^+, Riddick et al. 1986)
Total Surface Area, TSA ($Å^2$):
258.0 (Amidon et al. 1975)
211.8 (Iwase et al. 1985)
Heat of Fusion ΔH_{fus}, kcal/mol:
2.017 (Riddick et al. 1986)
Entropy of Fusion ΔS_{fus}, cal/mol·K (e.u.):
Fugacity Ratio at 25°C (assuming ΔS_{fus} = 13.5 e.u.), F: 1.0

Water Solubility (g/m^3 or mg/L at 25°C):
240000 (20°C, synthetic method, Jones 1929; quoted, Palit 1947)
273300 (20°C, shake flask-volumetric, Ginnings et al. 1940)
343550 (shake flask-volumetric, Ginnings et al. 1940; quoted, Hansch et al. 1968)
228022 (estimated, McGowan 1954)
226980 (calculated, Hansch et al. 1968)

12420 (20°C, quoted, Amidon et al. 1975)
238770 (quoted average, Valvani et al. 1981; quoted, Banerjee 1985; Howard 1990; Müller & Klein 1992)
104230 (predicted-K_{OW}, Valvani et al. 1981)
136280 (gen. col.-GC, Wasik et al. 1981; Tewari et al. 1982)
353000 (20°C, Verschueren 1983; quoted, Pinal et al. 1991)
240000 (quoted, Dean 1985)
240000 (20°C, Riddick et al. 1986)
222840, 165190 (quoted, calculated-V_I and solvatochromic p., Taft et al. 1985)
222840, 217770 (quoted, calculated-V_I and solvatochromic p., Leahy 1986)
222840, 161430 (quoted, calculated-V_I and solvatochromic p., Kamlet et al. 1986)
222840, 228020 (quoted, calculated-V_I and solvatochromic p., Kamlet et al. 1987)
249000 (selected, Yaws et al. 1990)
276000 (19.3°C, shake flask-GC, Stephenson 1992)
245000 (29.7°C, shake flask-GC, Stephenson 1992)

Vapor Pressure (Pa at 25°C):
13330 (Stull 1947)
12954 (interpolated-regression of tabulated data, Stull 1947)
12080 (ebulliometry, Ambrose et al. 1975; quoted, Howard 1990)
12230 (extrapolated-Antoine eqn., Boublik et al. 1984)
12060 (quoted, Abraham 1984)
12023 (extrapolated-Antoine eqn., Dean 1985)
12700 (Howard et al. 1986; quoted, Banerjee et al. 1990)
11790 (quoted, Kamlet et al. 1986)
12080 (quoted, Riddick et al. 1986)
5425 (calculated-solvatochromic p., Banerjee et al. 1990)

Henry's Law Constant (Pa·m^3/mol):
4.71 (shake flask, partial vapor pressure-GC, Buttery et al. 1971)
4.723 (quoted, exptl., Hine & Mookerjee 1975)
5.549 (calculated-group contribution, Hine & Mookerjee 1975)
4.408 (calculated-bond contribution, Hine & Mookerjee 1975)
6.191 (calculated-$\gamma \cdot P$, Rathbun & Tai 1982)
4.215 (calculated-MW, Rathbun & Tai 1982)
4.356 (gas-stripping, Hawthorne 1984)
5.76 (headspace-GC, Snider & Dawson 1985)
1.064 (quoted K_{WA} as K_{AW} from Snider & Dawson 1985; Howard 1990)
13.17 (EPICS-GC, Ashworth et al. 1988)
5.21 (gas stripping-HPLC/UV, Zhou & Mopper 1990)

Octanol/Water Partition Coefficient, log K_{OW}:

0.26 (shake flask-CR, Collander 1957; quoted, Yalkowsky & Morozowich 1980; Sangster 1989)
0.28 (shake flask-AS, Fujita et al. 1964)
0.32 ± 0.01, -0.27 (shake flask-AS, calculated, Iwasa et al. 1965)
0.29 (shake flask-AS,GC, Hansch & Anderson 1967; Hansch et al. 1968; quoted, Leo et al. 1969, 1971; Yalkowsky & Morozowich 1980; Wasik et al. 1981; Tewari et al. 1982; Iwase et al. 1985; Sangster 1989; Kollig 1993)
0.28 ± 0.02 (quoted, shake flaks at pH 7, Unger et al. 1978)
0.29, 0.24, 0.18 (calculated values, Yalkowsky & Morozowich 1980)
0.35 (exptl., Valvani et al. 1981)
0.69 (gen. col.-GC, Wasik et al. 1981; Tewari et al. 1982)
0.28 (quoted from Hansch & Leo 1979, Veith et al. 1983)
0.26 (Verschueren 1983; quoted, Pinal et al. 1991)
0.29 (Hansch & Leo 1985; quoted, Howard 1990)
0.29, 0.37 (quoted, calculated-TSA, Iwase et al. 1985)
0.29, 0.36 (quoted, calculated-V_I & solvatochromic p., Taft et al. 1985)
0.26 (GEMS 1986; quoted, Vaishnav et al. 1987)
0.29, 0.31 (quoted, calculated-V_I & solvatochromic p., Leahy 1986)
0.26 (shake flask-GC, Tanii et al. 1986; quoted, Sangster 1989)
0.28 (selected, Abernethy et al. 1988)
0.29, 0.62 (quoted, calculated-γ from UNIFAC, Banerjee & Howard 1988)
0.35, 0.29 (quoted from Hansch & Leo 1979, quoted from CLOGP, Bodor et al. 1989)
0.261, 0.45 (calculated-CLOGP, calculated-M.O. indices,Bodor et al. 1989)
0.50 (quoted from Pomona College unpublished data, Sangster 1989)
0.29 (recommended, Sangster 1989)
0.26 (calculated-CLOGP, Müller & Klein 1992)
0.28 (quoted average of Lu et al. 1986, Walton et al. 1992)
0.29, 0.785 (quoted exptl., calculated-UNIFAC, Chen et al. 1993)
0.29 (quoted from CLOGP 3.15, Jaworska & Schultz 1993)

Bioconcentration Factor, log BCF:

0.0 (estimated-K_{OW}, Lyman et al. 1982; quoted, Howard 1990)

Sorption Partition Coefficient, log K_{OC}:

1.53 (soil, estimated-K_{OW}, Lyman et al. 1982; quoted, Howard 1990)
1.47 ± 0.55 (Captina silt loam, Walton et al. 1992)
1.53 ± 0.88 (McLaurin sandy loam, Walton et al. 1992)
0.070 (predicted-K_{OW}, Walton et al. 1992)
-0.03 (calculated-K_{OW}, Kollig 1993)

Half-Lives in the Environment:

Air: half-life of 2.4-24 hours in air for the gas-phase reaction with hydroxyl radicals, based on the rate of disappearance of hydrocarbon due to reaction with hydroxyl radicals (Darnall et al. 1976); half-life of 2.3 days for the atmospheric reaction with photochemically produced hydroxyl radicals (Cox et al. 1981; quoted, Howard 1990); photooxidation half-life of 64.2-642 hours, based on measured rate data for the vapor phase reaction with hydroxyl radicals (Atkinson 1985; selected, Howard et al. 1991); atmospheric transformation lifetime was estimated to be < 1 day (Kelly et al. 1994).

Surface water: photooxidation half-life of 48.8-81.4 years, based on measured rate data for the reaction with hydroxyl radicals in aqueous solution (Anbar & Neta 1967; quoted, Howard et al. 1991); 24-168 hours, based on unacclimated grab sample of aerobic freshwater (Dojlido 1979; selected, Howard et al. 1991) and aerobic aqueous screening test data (Takemoto et al. 1981; selected, Howard et al. 1991).

Groundwater: 48-336 hours, based on estimated unacclimated aqueous aerobic biodegradation half-life (Howard et al. 1991).

Sediment:

Soil: calculated half-life of 4.9 days from first-order kinetic of degradation under both sterile and nonsterile conditions (Anderson et al. 1991); 24-168 hours, based on estimated unacclimated aqueous aerobic biodegradation half-life (Howard et al. 1991).

Biota:

Environmental Fate Rate Constants or Half-Lives:

Volatilization: based on lab. data for evaporation relative to the reaeration rate of 0.27 (Mackay et al. 1982; Rathbun & Tai 1982; quoted, Howard 1990) and typical reaeration rates of rivers and lakes (Mill et al. 1982; quoted, Howard), the half-life for evaporation from a river and lake will be 3 and 12 days, respectively (Howard 1990).

Photolysis:

Oxidation: photooxidation half-life of 48.8-81.4 years, based on measured rate data for the reaction with hydroxyl radicals in aqueous solution (Anbar & Neta 1967; selected, Howard et al. 1991); photooxidation half-life of 2.4-24 hours in air for the gas-phase reaction with hydroxyl radicals, based on the rate of disappearance of hydrocarbon due to reaction with hydroxyl radicals (Darnall et al. 1976); rate constant of $<2.0\times10^{2}$ M^{-1} sec^{-1} for oxidation with singlet oxygen at 25°C in aquatic systems with half-life >100 years (Foote 1976; Mill 1979; quoted, Mill 1982); rate constant of 1.9×10^{9} M^{-1} sec^{-1} for oxidation with OH radicals at 25°C in the atmosphere with a half-life of 2.9 days (Hendry & Kenley 1979; quoted, Mill 1982); second-order rate constant of $(0.95 \pm 0.09)\times10^{12}$ cm^3 $molecule^{-1}$ sec^{-1} by relative rate technique to ethene for the gas-phase reaction with OH radicals at 295 ± 2 K in air (Cox et al. 1981; quoted, Atkinson 1985); rate constant of $(1.20 \pm 0.20)x10^{-12}$ cm^3 $molecule^{-1}$ sec^{-1} by flash photolysis-resonance fluorescence for the gas-phase reaction with OH radicals at 300 K in air (Zetzsch 1982; quoted, Atkinson 1985); rate constant of $1.15x10^{-13}$ cm^3 $molecule^{-1}$ sec^{-1} for the gas-phase reaction with hydroxyl radicals in air and $1.50x10^{-13}$ cm^3 $molecule^{-1}$ sec^{-1} for the aqueous-phase reaction with hydroxyl radicals in solution (Wallington & Kurylo 1987; quoted, Wallington et al. 1988); photooxidation half-life of 64.2-642 hours, based on measured rate data for the vapor phase reaction with hydroxyl radicals (Howard et al. 1991).

Hydrolysis: first-order hydrolysis half-life of more than 50 years, based on nonreactive hydrolysis from pH 5 to 9 at 15°C (Kollig et al. 1987; selected, Howard et al. 1991).

Biodegradation: biodegradation rate constants of 0.021-0.025 h^{-1} in 30 mg/L activated sludge after a time lag of ~5 hours (Urano & Kato 1986b); aqueous aerobic half-life of 24-168 hours, based on unacclimated grab sample of aerobic freshwater (Dojilido 1979; selected, Howard et al. 1991) and aerobic aqueous screening test data (Takemoto et al. 1981; selected, Howard et al. 1991); aqueous anaerobic half-life of 96-672 hours, based on estimated unacclimated aqueous aerobic biodegradation half-life (Howard et al. 1991).

Biotransformation:

Bioconcentration, Uptake (k_1) and Elimination (k_2) Rate Constants:

Common Name: 2-Pentanone
Synonym: 2-pentanone, methyl propyl ketone, methyl *n*-propyl ketone, ethylacetone
Chemical Name: 2-pentanone, methyl propyl ketone
CAS Registry No: 107-87-9
Molecular Formula: $CH_3CH_2CH_2COCH_3$
Molecular Weight: 86.14
Melting Point (°C):
-77.80 (Stull 1947; Weast 1982-83; Dean 1985; Howard 1990)
-78/-83 (Verschueren 1983)
-76.86 (Riddick et al. 1986)
-78.10 (Stephenson & Malanowski 1987)
Boiling Point (°C):
102.3 (Ginnings et al. 1940; Boublik et al. 1984; Riddick et al. 1986)
103.3 (Stull 1947)
102.0 (Weast 1982-83; Verschueren 1983)
101.7 (Dean 1985; Howard 1990)
102.4 (Stephenson & Malanowski 1987)
Density (g/cm^3 at 20°C):
0.8018 (Ginnings et al. 1940)
0.8089 (Weast 1982-83)
0.8095 (Dean 1985)
0.8064 (Riddick et al. 1986)
0.8015 (25°C, Riddick et al. 1986)
Molar Volume (cm^3/mol):
106.5 (Taft et al. 1985; Kamlet et al. 1986; Leahy 1986)
57.4 (intrinsic volume V_I, Leahy 1986)
91.20 (calculated-density, Jaworska & Schultz 1993)
118.4 (calculated-LeBas method)
Molecular Volume ($Å^3$):
Total Surface Area, TSA ($Å^2$):
289.4 (Amidon et al. 1975)
Heat of Fusion ΔH_{fus}, kcal/mol:
2.541 (Riddick et al. 1986)
Entropy of Fusion ΔS_{fus}, cal/mol K (e.u.):
Fugacity Ratio at 25°C (assuming ΔS_{fus} = 13.5 e.u.), F: 1.0

Water Solubility (g/m^3 or mg/L at 25°C):
76406 (10°C, shake flask-interferometer, Gross et al. 1939)
54268 (30°C, shake flask-interferometer, Gross et al. 1939)
30150 (30°C, Gross et al. 1939; quoted, Palit et al. 1947)
55100 (shake flask-volumetric, Ginnings et al. 1940; quoted, Hansch et al. 1968)
59500 (20°C, shake flask-volumetric, Ginnings et al. 1940)
64014 (shake flask-interferometer, Donahue & Bartell 1952)
54350 (estimated, McGowan 1954)
54348 (quoted, Deno & Berkheimer 1960)
67016 (calculated-K_{OW}, Hansch et al. 1968)

5265 (20°C, quoted, Amidon et al. 1975)
43000 (quoted, Verschueren 1077, 1983)
43070 (quoted, Mackay & Yuen 1983)
59500 (20°C, Riddick et al. 1986; quoted, Howard 1990)
55400 (selected, Yaws et al. 1990)
56910, 70013 (quoted, calculated-V_I and solvatochromic p., Leahy 1986)
59000 (19.7°C, shake flask-GC, Stephenson 1992)

Vapor Pressure (Pa at 25°C):
2088 (interpolated-regression of tabulated data, Stull 1947)
2666 (28.5°C, Stull 1947)
4720 (ebulliometry-fitted to Antoine eqn., Ambrose et al. 1975)
2133 (quoted, Verschueren 1983)
1621 (quoted, Mackay & Yuen 1983)
4648 (extrapolated-Antoine eqn., Boublik et al. 1984)
4722 (quoted, Abraham 1984)
4720 (Riddick et al. 1986; quoted, Howard 1990)
4702 (extrapolated-Antoine eqn., Stephenson & Malanowski 1987)

Henry's Law Constant (Pa·m³/mol):
10.91 (28°C, concn. ratio-GC, Nelson & Hoff 1968)
6.44 (partial vapor pressure-GC, Buttery et al. 1969)
6.520 (quoted, exptl., Hine & Mookerjee 1975; quoted, Howard 1990)
7.660 (calculated-group contribution, Hine & Mookerjee 1975)
6.520 (calculated-bond contribution, Hine & Mookerjee 1975)
3.830 (calculated-$\gamma \cdot P$, Rathbun & Tai 1982)
5.340 (calculated-MW, Rathbun & Tai 1982)
3.20 (calculated-P/C, Mackay & Yuen 1983)
5.876 (gas-stripping, Hawthorne 1984; quoted, Hawthorne et al. 1985)
10.13 (modified gas-stripping, Hawthorne et al. 1985)
6.52, 8.021 (quoted, correlated-molecular structure, Russell et al. 1992)

Octanol/Water Partition Coefficient, log K_{OW}:
0.31 (calculated-π const., Iwasa et al. 1965)
0.79 (calculated-S, Hansch et al. 1968; quoted, Leo et al. 1969)
0.91 $\pm$ 0.03 (shake flask-UV at pH 7, Unger et al. 1978; quoted, Sangster 1989)
0.91 (Hansch & Leo 1985; quoted, Howard 1990)
0.91, 0.76 (quoted, calculated-V_I and solvatochromic p., Taft et al. 1985)
0.91 (quoted, Leahy 1986; Kamlet et al. 1988)
0.91, 0.81 (quoted, calculated-V_I and solvatochromic p., Leahy 1986)
0.78 (shake flask-GC, Tanii et al. 1986; quoted, Sangster 1989)
0.87 (calculated-V_I and solvatochromic p., Kamlet et al. 1988)
0.91, 0.95 (quoted, calculated- γ from UNIFAC, Banerjee & Howard 1988)
0.84 (recommended, Sangster 1989)
0.91 (quoted from CLOGP 3.15, Jaworska & Schultz 1993)

Bioconcentration Factor, log BCF:

0.477 (estimated-K_{OW}, Lyman et al. 1982; quoted, Howard 1990)

Sorption Partition Coefficient, log K_{OC}:

1.869 (soil, estimated-K_{OW}, Lyman et al. 1982; quoted, Howard 1990)

Half-Lives in the Environment:

Air: half-life of 2.4-24 hours for the gas-phase reaction with hydroxyl radicals in air, based on the rate of disappearance of hydrocarbon due to reaction with hydroxyl radicals (Darnall et al. 1976); half-life of 3.5 days for reactions with photochemically produced OH radicals (Howard 1990).

Surface water: volatilization half-life of 11-17 hours from a model river (Howard 1990).

Groundwater:

Sediment:

Soil:

Biota:

Environmental Fate Rate Constants or Half-Lives:

Volatilization: using Henry's law constant, a half-life of 15.5 hours was estimated for a model river of 1 m deep flowing at 1 m/sec with a wind velocity of 3 m/sec, and 14.5 hours in a wind-wave tank with a 6m/sec wind speed (Howard 1990).

Photolysis:

Oxidation: photooxidation half-life of 2.4-24 hours for the gas-phase reaction with hydroxyl radicals in air, based on the rate of disappearance of hydrocarbon due to reaction with hydroxyl radicals (Darnall et al. 1976); rate constant of $(4.64 \pm 0.14)\times10^{-12}$ cm^3 $molecule^{-1}$ sec^{-1} by relative rate technique to cyclohexane for the reaction with OH radicals at 299 ± 2 K (Atkinson et al. 1982; quoted, Atkinson 1985); rate constant of 4.00×10^{-12} cm^3 $molecule^{-1}$ sec^{-1} for the gas-phase reaction with OH radicals in air and 3.20×10^{-12} cm^3 $molecule^{-1}$ sec^{-1} for the aqueous-phase reaction with OH radicals in solution (Wallington & Kurylo 1987; quoted, Wallington et al. 1988).

Hydrolysis:

Biodegradation:

Biotransformation:

Bioconcentration, Uptake (k_1) and Elimination (k_2) Rate Constants:

Common Name: 3-Pentanone
Synonym: diethyl ketone, ethylketone, propione, *sym*-dimethylacetone
Chemical Name: 3-pentanone, diethyl ketone
CAS Registry No: 96-22-0
Molecular Formula: $CH_3CH_2COCH_2CH_3$
Molecular Weight: 86.14
Melting Point (°C):
-42.0 (Stull 1947; Verschueren 1983)
-39.80 (Weast 1982-83)
-38.97 (Dean 1985; Riddick et al. 1986; Howard 1993)
Boiling Point (°C):
101.7 (Gross et al. 1933; Weast 1982-83)
101.6-101.9 (Ginnings et al. 1940)
102.7 (Stull 1947)
102.0 (Verschueren 1983; Dean 1985)
102.06 (Boublik et al. 1984)
101.96 (Riddick et al. 1986; Howard 1993)
Density (g/cm^3 at 20°C):
0.8116 (Ginnings et al. 1940)
0.8138 (Weast 1982-83)
0.8160 (19°C, Verschueren 1983)
0.8143 (Dean 1985; Riddick et al. 1986)
0.80945 (25°C, Riddick et al. 1986)
Molar Volume (cm^3/mol):
92.60 (calculated-density, Jaworska & Schultz 1993)
118.4 (calculated-LeBas method)
Molecular Volume ($Å^3$):
Total Surface Area, TSA ($Å^2$):
Dissociation Constant:
27.1 (pK_a, Riddick et al. 1986; Howard 1993)
Heat of Fusion ΔH_{fus}, kcal/mol:
2.771 (Riddick et al. 1986)
Entropy of Fusion ΔS_{fus}, cal/mol·K (e.u.):
Fugacity Ratio at 25°C (assuming ΔS_{fus} = 13.5 e.u.), F: 1.0

Water Solubility (g/m^3 or mg/L at 25°C):
67275 (10°C, shake flask-interferometer, Gross et al. 1933)
49617 (30°C, shake flask-interferometer, Gross et al. 1933; quoted, Palit 1947)
50800 (20°C, shake flask-volumetric, Ginnings et al. 1940)
48100 (shake flask-volumetric, Ginnings et al. 1940)
44520 (shake flask-centrifuge, Booth & Everson 1948)
43172 (estimated, McGowan 1954)
47334 (quoted, Deno & Berkheimer 1960)
49330, 65480 (quoted, calculated-K_{OW}, Hansch et al. 1968)
41350 (quoted, Wasik et al. 1981; Tewari et al. 1982)
46000 (quoted, Urano et al. 1981)

45654 (gen. col.-GC, Wasik et al. 1981; Tewari et al. 1982)
47000 (20°C, quoted, Verschueren 1983)
38000 (100°C, quoted, Verschueren 1983)
34000 (20°C, Riddick et al. 1986)
48100 (Yalkowsky et al. 1987; quoted, Howard 1993)
47610 (quoted, Isnard & Lambert 1989)
67020 (quoted, Müller & Klein 1992)
53000 (19.3°C, shake flask-GC, Stephenson 1992)

Vapor Pressure (Pa at 25°C):
2169 (interpolated-regression of tabulated data, Stull 1947)
2666 (27.9°C, Stull 1947)
4723 (ebulliometry-fitted to Antoine eqn., Ambrose et al. 1975)
1733 (20°C, quoted, Verschueren 1983)
5316 (extrapolated-Antoine eqn., Boublik et al. 1984)
4714 (extrapolated-Antoine eqn., Boublik et al. 1984)
4693 (extrapolated-Antoine eqn., Dean 1985)
4723 (Riddick et al. 1986)
4932 (Daubert & Danner 1989; quoted, Howard 1993)

Henry's Law Constant ($Pa \cdot m^3/mol$):
3.617 (calculated-$\gamma \cdot P$, Rathbun & Tai 1982)
5.340 (calculated-MW, Rathbun & Tai 1982)
8.834 (calculated-P/C, Howard 1993)
8.132 (calculated-P/C, this work)

Octanol/Water Partition Coefficient, log K_{OW}:
0.31 (calculated-π constant, Iwasa et al. 1965)
0.79 (calculated-π constant, Hansch et al. 1968; quoted, Leo et al. 1969; Wasik et al. 1981; Tewari et al. 1982)
0.99 (gen. col.-GC, Wasik et al. 1981; Tewari et al. 1982; quoted, Sangster 1989; Howard 1993)
0.75 (calculated-γ, Wasik et al. 1981; quoted, Sangster 1989)
0.84 (calculated-f const. as per Hansch & Leo 1979, Veith et al. 1983)
0.67 (calculated-γ, Berti et al. 1986; quoted, Sangster 1989)
0.79 (estimated-CLOGP 3, Vaishnav et al. 1987)
0.99, 1.15 (quoted, calculated- γ from UNIFAC, Banerjee & Howard 1988)
0.82 (recommended, Sangster 1989)
0.79 (calculated-CLOGP, Müller & Klein 1992)
0.99, 1.2319 (quoted exptl., calculated-original UNIFAC, Chen et al. 1993)
0.79 (quoted from CLOGP 3.15, Jaworska & Schultz 1993)

Bioconcentration Factor, log BCF:

0.519 (estimated-K_{OW}, Lyman et al. 1990; quoted, Howard 1993)

0.146 (estimated-S, Lyman et al. 1990; quoted, Howard 1993)

Sorption Partition Coefficient, log K_{OC}:

1.914 (soil, estimated-K_{OW}, Lyman et al. 1990; quoted, Howard 1993)

1.08 (soil, estimated-S, Lyman et al. 1990; quoted, Howard 1993)

Half-Lives in the Environment:

Air: half-life of 2.4-24 hours for the gas-phase reaction with hydroxyl radicals in air, based on the rate of disappearance of hydrocarbon due to reaction with hydroxyl radicals (Darnall et al. 1976); photooxidation half-life of 5.9 days, based on an experimentally determined rate constant for the vapor-phase reaction with hydroxyl radicals in air (Wallington & Kurylo 1987; quoted, Howard 1993).

Surface water: estimated volatilization half-lives of 12 hours in a model river (1 m deep flowing at 1 m/sec with a 3m/sec wind) and 5.6 days in a model environmental pond (2 m deep) (Howard 1993).

Groundwater:

Sediment:

Soil:

Biota:

Environmental Fate Rate Constants or Half-Lives:

Volatilization: estimated half-life from a model pond of two meter deep is 5.6 days (USEPA 1987; quoted, Howard 1993); based on calculated Henry's law constant, estimated half-life from a model river of one meter deep flowing at 1 m/sec with a wind velocity of 3 m/sec is 12 hours (Lyman et al. 1990; quoted, Howard 1993).

Photolysis:

Oxidation: photooxidation half-life of 2.4-24 hours for the gas-phase reaction with hydroxyl radicals in air, based on the rate of disappearance of hydrocarbon due to reaction with hydroxyl radicals (Darnall et al. 1976); rate constant of $(1.82 \pm 0.33) \times 10^{-12}$ cm^3 $molecule^{-1}$ sec^{-1} by relative rate technique to cyclohexane for the gas-phase reaction with OH radicals at 299 ± 2 K in air (Atkinson et al. 1982; quoted, Atkinson 1985); experimentally determined rate constant of 2.74×10^{-12} cm^3 $molecule^{-1}$ sec^{-1} for the vapor-phase reaction with hydroxyl radicals in air (Wallington & Kurylo 1987; quoted, Wallington et al. 1988; Howard 1993); experimentally determined rate constant of 2.30×10^{-12} cm^3 $molecule^{-1}$ sec^{-1} for the solution-phase reaction with hydroxyl radicals in aqueous solution (Wallington & Kurylo 1987; quoted, Wallington et al. 1988).

Hydrolysis: not expected to be important (Howard 1993).

Biodegradation: half-lives approximately 5 to 10 days in acclimated cultures during screening tests (Howard 1993).

Biotransformation:

Bioconcentration, Uptake (k_1) and Elimination (k_2) Rate Constants:

Common Name: Methyl Isobutyl Ketone
Synonym: hexone, hexanone, 4-methyl-2-pentanone, MIBK
Chemical Name: methyl isobutyl ketone, 4-methyl-2-pentanone
CAS Registry No: 108-10-1
Molecular Formula: $(CH_3)_2CHCH_2COCH_3$
Molecular Weight: 100.16
Melting Point (°C):
-84.70 (Weast 1982-83; Howard 1990)
-85/-80 (Verschueren 1983)
-83.5 (Dean 1985)
-84 (Riddick et al. 1986)
Boiling Point (°C):
115.7 (Ginnings et al. 1940; Dean 1985)
116.8 (Weast 1982-83; Howard 1990)
116/119 (Verschueren 1983)
116.2 (Boublik et al. 1984)
117.40 (Riddick et al. 1986)
117.0 (Banerjee et al. 1990)
Density (g/cm^3 at 20°C):
0.7969 (Ginnings et al. 1940)
0.7978 (Weast 1982-83)
0.8017 (Verschueren 1983)
0.8006 (Dean 1985)
0.8010 (Riddick et al. 1986)
Molar Volume (cm^3/mol):
140.6 (calculated-LeBas method)
108.40 (calculated-density, Jaworska & Schultz 1993)
Molecular Volume ($Å^3$):
Total Surface Area, TSA ($Å^2$):
Heat of Fusion ΔH_{fus}, kcal/mol:
Entropy of Fusion ΔS_{fus}, cal/mol·K (e.u.):
Fugacity Ratio at 25°C (assuming ΔS_{fus} = 13.5 e.u.), F: 1.0

Water Solubility (g/m^3 or mg/L at 25°C):
23137 (10°C, shake flask-interferometer, Gross et al. 1939)
16627 (30°C, shake flask-interferometer, Gross et al. 1939)
20400 (20°C, shake flask-volumetric, Ginnings et al. 1940; Howard 1990)
19100 (shake flask-volumetric, Ginnings et al. 1940; quoted, Hansch et al. 1968)
18200 (shake flask-turbidimeter, McBain & Richards 1946)
19085 (quoted, Deno & Berkheimer 1960)
21430, 33705 (quoted, calculated-K_{OW}, Hansch et al. 1968)
18000 (20°C, quoted average, Verschueren 1983)
17000 (Dean 1985; Riddick et al. 1986)

Vapor Pressure (Pa at 25°C):

799.8 (20°C, quoted, Verschueren 1983)
1333 (30°C, quoted, Verschueren 1983)
2200 (21.7°C, quoted exptl., Boublik et al. 1984; Howard 1990)
2587 (calculated-Antoine eqn., Boublik et al. 1984)
2575 (calculated-Antoine eqn., Dean 1985)
2653 (Howard et al. 1986; quoted, Banerjee et al. 1990)
2510 (Riddick et al. 1986)
2653, 605.18 (quoted, calculated-solvatochromic p., Banerjee et al. 1990)
2686 (measured & fitted to Antoine eqn., Ambrose et al. 1988)

Henry's Law Constant (Pa·m^3/mol):

39.52 (EPICS-GC, Ashworth et al. 1988)
9.523 (calculated-P/C, Howard 1990)

Octanol/Water Partition Coefficient, log K_{OW}:

1.09 (calculated-π constant, Hansch et al. 1968)
1.19 (Hansch & Leo 1985; quoted, Howard 1990)
1.31 (shake flask-GC, Tanii et al. 1986; quoted, Sangster 1989)
1.31 (recommended, Sangster 1989)
1.19 (quoted from CLOGP 3.15, Jaworska & Schultz 1993)
1.19 (calculated-QSAR, Kollig 1993)

Bioconcentration Factor, log BCF:

0.301-0.699 (estimated-K_{OW} & S, Lyman et al. 1982; quoted, Howard 1990)

Sorption Partition Coefficient, log K_{OC}:

1.279-2.025 (soil, estimated-K_{OW} & S, Lyman et al. 1982; quoted, Howard 1990)
0.87 (calculated-K_{OW}, Kollig 1993)

Half-Lives in the Environment:

Air: half-life of 0.24-2.4 hours for the gas-phase reaction with hydroxyl radicals in air, based on the disappearance rate of hydrocarbon due to reaction with hydroxyl radicals (Darnall et al. 1976); direct photolysis half-life of 15 hours and photooxidation half-life of 16-17 hours for reactions with hydroxyl radicals in air (Howard 1990); photooxidation half-life of 4.6-45.5 hours, based on measured rate data for the vapor phase reaction with hydroxyl radicals (Atkinson 1985; selected, Howard et al. 1991); atmospheric transformation lifetime was estimated to be 1 to 5 days (Kelly et al. 1994).

Surface water: estimated volatilization half-lives of 15-33 hours (Howard 1990); 24-168 hours, based on unacclimated aerobic aqueous biodegradation half-life (Howard et al. 1991).

Groundwater: 48-336 hours, based on estimated unacclimated aqueous aerobic biodegradation half-life (Howard et al. 1991).

Sediment:

Soil: 24-168 hours, based on estimated unacclimated aqueous aerobic biodegradation half-life (Howard et al. 1991).

Biota:

Environmental Fate Rate Constants or Half-Lives:

Volatilization: 15-33 hours from water of 1 m deep, based on lab. mass-transfer coefficient for volatilization from a stirred 557-2020 rpm water bath at 25°C (Rathbun & Tai 1982; quoted, Howard 1990).

Photolysis: direct photolysis half-life of 15 hours in air (Howard 1990).

Oxidation: photooxidation half-life of 0.24-2.4 hours for the gas-phase reaction with hydroxyl radicals in air, based on the disappearance rate of hydrocarbon due to reaction with hydroxyl radicals (Darnall et al. 1976); second-order rate constants of 14.0×10^{-11} $cm^3\cdot molecule^{-1}\cdot sec^{-1}$ at 295K, 14.2×10^{-11} $cm^3\cdot molecule^{-1}\cdot sec^{-1}$ at 299K and 13.2×10^{-11} $cm^3\cdot molecule^{-1}\cdot sec^{-1}$ at 300K for gas-phase reactions with hydroxyl radicals (Atkinson 1985); rate constants of $(1.31 - 1.45)\times10^{-11}$ $cm^3\cdot molecule^{-1}\cdot sec^{-1}$ for the gas-phase reaction with OH radicals (concn. of 8.0×10^5 $molecule\cdot cm^{-3}$) at 22-27°C giving an estimated half-life in air of 16-17 hours (Howard 1990); photooxidation half-life of 4.6-45.5 hours, based on measured rate data for the vapor phase reaction with hydroxyl radicals (Atkinson 1985; selected, Howard et al. 1991).

Hydrolysis: not expected to undergo chemical hydrolysis (Howard 1990).

Biodegradation: aqueous aerobic half-life of 24-168 hours, based on unacclimated aerobic aqueous screening test data (Bridie et al. 1979; Takemoto et al. 1981; selected, Howard et al. 1991); aqueous anaerobic half-life of 96-672 hours, based on estimated unacclimated aqueous aerobic biodegradation half-life (Howard et al. 1991).

Biotransformation:

Bioconcentration, Uptake (k_1) and Elimination (k_2) Rate Constants:

Common Name: Methyl butyl ketone
Synonym: 2-hexanone, methyl *n*-butyl ketone
Chemical Name: 2-hexanone, methylbutyl ketone
CAS Registry No: 591-78-6
Molecular Formula: $CH_3COCH_2CH_2CH_2CH_3$
Molecular Weight: 100.16
Melting Point (°C):
-56.9 (Stull 1947)
-57.0 (Weast 1982-83; Verschueren 1983; Dean 1985)
-55.8 (Riddick et al. 1986; Howard 1993)
Boiling Point (°C):
127.6 (Ginnings et al. 1940; Riddick et al. 1986; Howard 1993)
127.5 (Stull 1947)
128.0 (Weast 1982-83; Verschueren 1983)
127.2 (Dean 1985)
127.0 (Banerjee et al. 1990)
Density (g/cm^3 at 20°C):
0.8072 (Ginnings et al. 1940)
0.8113 (Weast 1982-83; Riddick et al. 1986)
0.8209 (Dean 1985)
0.8067 (25°C, Riddick et al. 1986)
Molar Volume (cm^3/mol):
124.2 (Hoy 1970; Amidon & Williams 1982)
123.5 (Taft et al. 1985; Kamlet et al. 1986; Leahy 1986)
67.0 (intrinsic volume V_I, Leahy 1986)
108.1 (calculated-density, Jaworska & Schultz 1993)
140.6 (calculated-LeBas method)
Molecular Volume (Å^3):
Total Surface Area, TSA (Å^2):
265.0 (Iwase et al. 1985)
Dissociation Constnat:
-8.30 (pK_a, Riddick et al. 1986)
25.30 (pK_s, Riddick et al. 1986)
Heat of Fusion ΔH_{fus}, kcal/mol:
3.56 (Riddick et al. 1986)
Entropy of Fusion ΔS_{fus}, cal/mol·K (e.u.):
Fugacity Ratio at 25°C (assuming ΔS_{fus} = 13.5 e.u.), F: 1.0

Water Solubility (g/m^3 or mg/L at 25°C):
20430 (10°C, shake flask-interferometer, Gross et al. 1939)
14520 (30°C, shake flask-interferometer, Gross et al. 1939)
17500 (20°C, shake flask-volumetric, Ginnings et al. 1940)
16400 (shake flask-volumetric, Ginnings et al. 1940; quoted Hansch et al. 1968)
16000 (Erichsen 1952; quoted, Howard 1993)
23930 (shake flask-interferometer, Donahue & Bartell 1952)
15870 (esimated, McGowan 1954)

16620 (quoted, Deno & Berkheimer 1960)
16660; 19260 (quoted, calculated-K_{OW}, Hansch et al. 1968)
14815 (quoted exptl. average, Valvani et al. 1981)
12900 (predicted-K_{OW}, Valvani et al. 1981)
14815, 14480 (quoted, calculated-K_{OW} and M.P., Amidon & Williams 1982)
35000 (20°C, quoted, Verschueren 1983)
16620, 22945 (quoted, calculated-V_I and solvatochromic p., Leahy 1986)
17500 (20°C, Riddick et al. 1986)
15140 (quoted, Isnard & Lambert 1988)
19260 (quoted, Müller & Klein 1992)
15100 (19.8°C, shake flask-GC, Stephenson 1992)
13700 (29.7°C, shake flask-GC, Stephenson 1992)

Vapor Pressure (Pa at 25°C):
514.4 (interpolated-regression of tabulated data, Stull 1947)
666.5 (28.8°C, Stull 1947)
1550 (ebulliometry-fitted to Antoine eqn., Ambrose et al. 1975)
509.8 (calculated-Cox eqn., Chao et al. 1983)
266.6 (20°C, quoted, Verschueren 1983)
1550 (quoted, Abraham 1984)
1546 (Howard et al. 1986; quoted, Banerjee et al. 1990)
1550 (Riddick et al. 1986; quoted, Howard 1993)
1546, 611 (measured, calculated-solvatochromic p., Banerjee et al. 1990)

Henry's Law Constant (Pa·m^3/mol):
9.689 (calculated-P/C, Meylan & Howard 1991)
11.76 (calculated-bond contribution, Meylan & Howard 1991)
9.695 (calculated-P/C, Howard 1993)

Octanol/Water Partition Coefficient, log K_{OW}:
1.38 (shake flask-AS, Iwasa et al. 1965; quoted, Leo et al. 1971; Valvani et al. 1981; quoted, Amidon & Williams 1982; Verschueren 1983; Iwase et al. 1985)
1.29 (calculated-π constant, Hansch et al. 1968)
1.38 (quoted from Hansch & Leo 1979, Haky & Young 1984)
1.39 (HPLC-k', Haky & Young 1984; quoted, Sangster 1989)
1.38 (Hansch & Leo 1985; quoted, Howard 1993)
1.38, 1.39 (quoted, calculated-TSA, Iwase et al. 1985)
1.38 (measured value, GEMS 1986; quoted, Vaishnav et al. 1987)
1.38, 1.26 (quoted, calculated-V_I and solvatochromic p., Taft et al. 1985)
1.38, 1.31 (quoted, calculated-V_I and solvatochromic p., Leahy 1986)
1.19 (shake flask-GC, Tanii et al. 1986; quoted, Sangster 1989)
1.40 (calculated-V_I and solvatochromic p., Kamlet et al. 1988)
1.78, 1.38, 1,319, 1.55 (quoted from Hansch & Leo 1979, quoted-CLOGP, calculated-CLOGP, calculated-M.O. indiecs, Bodor et al. 1989)
1.38 (quoted and recommended, Sangster 1989)
1.32 (calculated-CLOGP, Müller & Klein 1992)
1.38 (quoted from CLOGP 3.15, Jaworska & Schultz 1993)

Bioconcentration Factor, log BCF:

0.778 (estimated-K_{OW}, Lyman et al. 1990; quoted, Howard 1993)

Sorption Partition Coefficient, log K_{OC}:

1.322 (soil, estimated-K_{OW}, Lyman et al. 1990; quoted, Howard 1993)

2.127 (soil, estimated-S, Lyman et al. 1990; quoted, Howard 1993)

Half-Lives in the Environment:

Air: half-life of 2.4-24 hours for the gas-phase reaction with hydroxyl radicals in air, based on the rate of disappearance of hydrocarbon due to reaction with hydroxyl radicals (Darnall et al. 1976); estimated half-life for the vapor-phase reaction with OH radicals is 42 hours (Atkinson 1985; quoted, Howard 1993).

Surface water: volatilization half-lives of 12.1 hours from a model river, 135 hours from a model environmental pond (Howard 1993).

Groundwater:

Sediment:

Soil:

Biota:

Environmental Fate Rate Constants or Half-Lives:

Volatilization: half-life from a model environmental pond of 2 meter deep was estimated to be about 135 hours (USEPA 1987; quoted, Howard 1993); using Henry's law constant, half-life from a model river of one meter deep flowing at 1 m/sec with a wind speed of 3 m/sec was estimated to be 12.1 hours (Lyman et al. 1990; quoted, Howard 1993).

Photolysis:

Oxidation: photooxidation half-life of 2.4-24 hours for the gas-phase reaction with hydroxyl radicals in air, based on the rate of disappearance of hydrocarbon due to reaction with hydroxyl radicals (Darnall et al. 1976); rate constant of $(6.81 \pm 0.29)\times10^{-12}$ cm^3 $molecule^{-1}$ sec^{-1} by relative rate technique to cyclohexane for the reaction with OH radicals at 299 ± 2 K in air (Atkinson et al. 1982; quoted, Atkinson 1985); second-order rate constant of 8.97×10^{-12} $cm^3\cdot molecule^{-1}\cdot sec^{-1}$ for gas-phase reaction with hydroxyl radicals at 299 K (Atkinson 1985); rate constant of 9.1×10^{-12} $cm^3\cdot molecule^{-1}\cdot sec^{-1}$ for reactions with OH radicals at 25°C, which corresponds to a half-life of 42 hours at an atmospheric OH• concentration of 5×10^5 $molecule\cdot cm^{-3}$ (quoted, Howard 1993)

Hydrolysis:

Biodegradation: half-life of approximately 5 days in acclimated mixed microbial culture (Howard 193).

Biotransformation:

Bioconcentration, Uptake (k_1) and Elimination (k_2) Rate Constants:

Common Name: 2-Heptanone
Synonym: methyl *n*-amyl ketone
Chemical Name: 2-heptanone
CAS Registry No: 110-43-0
Molecular Formula: $CH_3(CH_2)_4COCH_3$
Molecular Weight: 114.187
Melting Point (°C): -35.0
Boiling Point (°C): 151.058
Density (g/cm^3):
0.81537, 0.81123 (20, 25°C, Riddick et al. 1986)
Acid Dissociation Constants, pKa:
Molar Volume (cm^3/mol):
125.30 (calculated-density, Jaworska & Schultz 1993)
162.8 (LeBas method)
Molecular Volume, TMV (Å^3):
Total Surface Area, TSA (Å^2):
Heat of Fusion, ΔH_{fus}, kcal/mol:
Entropy of Fusion, ΔS_{fus}, cal/mol·K:
Fugacity Ratio at 25°C (assuming ΔS_{fus} = 13.5 cal/mol·K), F: 1.0

Water Solubility (g/m^3 or mg/L at 25°C):
4546 (estimated, McGowan 1954)
4320, 5428 (quoted, calculated-K_{OW}, Hansch et al. 1968)
4340 (quoted, Mackay & Yuen 1983)
4300 (Riddick et al. 1986)

Vapor Pressure (Pa at 25°C):
345 (quoted, Mackay & Yuen 1983)
514 (quoted, Abraham 1984; Riddick et al. 1986)

Henry's Law Constant (Pa·m^3/mol):
14.63 (partial vapor pressure-GC, Buttery et al. 1969)
14.6; 14.94, 14.6 (quoted exptl.; calculated-group contribution, bond contribution, Hine & Mookerjee 1975)
5.715; 8.552 (calculated-γ·P, calculated-MW, Rathbun & Tai 1982)
9.12 (calculated-P/C, Mackay & Yuen 1983)
14.60, 16.0 (quoted, correlated-molecular structure, Russell et al. 1992)

Octanol/Water Partition Coefficient, log K_{OW}:
1.79 (calculated-π constant, Hansch et al. 1968)
2.03 (shake flask-GC, Tanii et al. 1986)
1.98 (quoted from CLOGP 3.15, Jaworska & Schultz 1993)

Bioconcentration Factor, log BCF: no data
Sorption Partition Coefficient, log K_{OC}: no data
Half-Lives in the Environment/Environmental Fate Rate Constants and Half-Lives: no data

Common Name: Cyclohexanone
Synonym:
Chemical Name: cyclohexanone
CAS Registry No: 108-94-1
Molecular Formula: $C_6H_{10}O$
Molecular Weight: 98.144
Melting Point (°C):

-16.4 (Weast 1982-83)
-32.1 (Riddick et al. 1986; Howard 1990)

Boiling Point (°C):

155.60 (Weast 1982-83; Howard 1990)
155.42 (Boublik et al. 1984)
155.70 (Dean 1985)
155.65 (Riddick et al. 1986)
155.0 (Banerjee et al. 1990)

Density (g/cm^3 at 20°C):

0.9478 (Weast 1982-83; Dean 1985)
0.9452 (Riddick et al. 1986)

Molar Volume (cm^3/mol):

104.2 (Hoy 1970; Amidon & Williams 1982)
113.6 (Taft et al. 1985; Leahy 1986; Kamlet et al. 1986,87)
61.9 (intrinsic volume V_I, Leahy 1986; Kamlet et al. 1987)
62.0 (intrinsic volume V_I, Abernethy et al. 1988)
118 (LeBas method, Abernethy et al. 1988)
103.3 (calculated-density)
103.2 (calculated-LeBas method)

Molecular Volume ($Å^3$):
Total Surface Area, TSA ($Å^2$):

229.4 (Iwase et al. 1985)

Heat of Fusion ΔH_{fus}, kcal/mol:
Entropy of Fusion ΔS_{fus}, cal/mol·K (e.u.):
Fugacity Ratio at 25°C (assuming ΔS_{fus} = 13.5 e.u.), F: 1.0

Water Solubility (g/m^3 or mg/L at 25°C):

45905 (quoted average, Valvani et al. 1981)
49189 (predicted-K_{OW}, Valvani et al. 1981)
87470, 95910 (quoted, calculated-K_{OW} and M.P., Amidon & Williams 1982)
23000 (Riddick et al. 1986; quoted, Howard 1990)
100430, 95910 (quoted, calculated-V_I and solvatochromic p., Leahy 1986)
100430, 105163 (quoted, calculated-V_I and solvatochromic p., Kamlet et al. 1987)
22910 (quoted, Isnard & Lambert 1988)
93200 (selected, Yaws et al. 1990)
22966 (quoted from Howard 1990, Müller & Klein 1992)
97000 (19.5°C, shake flask-GC, Stephenson 1992)
82000 (29.8°C, shake flask-GC, Stephenson 1992)

Vapor Pressure (Pa at 25°C):

533 (ebulliometry-fitted to Antoine eqn., Meyer & Hotz 1973)
614 (calculated-Cox eqn., Chao et al. 1983)
533 (extrapolated-Antoine eqn., Boublik et al. 1984)
600 (Howard et al. 1986; quoted, Banerjee et al. 1990)
640 (Riddick et al. 1986; quoted, Howard 1990)
600, 837 (measured, calculated-solvatochromic p., Banerjee et al. 1990)

Henry's Law Constant (Pa m^3/mol):

1.216 (Hawthorne et al. 1985; quoted, Howard 1990)
2.266 (calculated-P/C, Meylan & Howard 1991)
5.179 (estimated-bond contribution, Meylan & Howard 1991)

Octanol/Water Partition Coefficient, log K_{OW}:

0.81 (Valvani et al. 1981; quoted, Amidon & Williams 1982)
0.81, 0.71 (quoted, calculated-TSA, Iwase et al. 1985)
0.81, 0.77 (quoted, calculated-V_I and solvatochromic p., Taft et al. 1985)
0.81, 0.75 (quoted, calculated-V_I and solvatochromic p., Leahy 1986)
0.81 (quoted from Hansch & Leo 1979, Veith et al. 1983; Bodor et al. 1989)
0.81 (quoted from CLOGP, Bodor et al. 1989)
0.805 (calculated from CLOGP, Bodor et al. 1989)
1.32 (calculated-M.O. indices, Bodor et al. 1989)
0.81 (quoted and recommended, Sangster 1989)
0.81 (quoted from Hansch & Leo 1985, Howard 1990)
0.81 (calculated-CLOGP, Müller & Klein 1992)

Bioconcentration Factor, log BCF:

0.39 (calculated-K_{OW} with regression eqns., Lyman et al. 1982; quoted, Howard 1990)

Sorption Partition Coefficient, log K_{OC}:

1.00 (calculated-S, K_{OW} with regression eqns., Roy & Griffin 1985; quoted, Howard 1990)

Half-Lives in the Environment:

Air: half-life of 24 hours to 9.9 days in air for the gas-phase reaction with hydroxyl radicals, based on the rate of disappearance of hydrocarbon due to reaction with hydroxyl radicals (Darnall et al. 1976); degrade rapidly by reaction with sunlight produced hydroxyl radicals with a half-life of about one day and by direct sunlight photolysis with half-life about 4.3 days (Howard 1990).
Surface water: estimated half-life of 0.003 year at Noordwijk (Zoeteman et al. 1981).
Groundwater:
Sediment:

Soil:
Biota:

Environmental Fate Rate Constants or Half-Lives:

Volatilization: using Henry's law constant, half-life from a model river 1 m deep flowing 1 m/sec with wind velocity of 3 m/sec has been estimated to be 3.1 days (Lyman et al. 1982; quoted, Howard 1990).

Photolysis: direct sunlight photolysis rate constant of about 0.16 day^{-1} corresponding to a half-life of 4.3 days (Mill & Davenport 1986; quoted, Howard 1990).

Oxidation: photooxidation half-life of 24 hours to 9.9 days in air for the gas-phase reaction with hydroxyl radicals, based on the rate of disappearance of hydrocarbon due to reaction with hydroxyl radicals (Darnall et al. 1976); estimated rate constant of 1.56×10^{-11} cm^3 $molecule^{-1}$ sec^{-1} for the vapor phase reaction with hydroxyl radicals in air at 25°C with a half-life of about one day (Atkinson 1985; quoted, Howard 1990); photooxidation reaction rate constants were estimated to be 12.55×10^{-12} cm^3/molecule·sec (from Atmospheric Oxidation Program), 6.39×10^{-12} cm^3/molecule·sec (exptl.) and 2.92×10^{-12} cm^3/molecule • s (from Fate of Atmospheric Pollutants Program) with OH radicals in gas phase (Meylan & Howard 1993).

Hydrolysis:

Biodegradation: average rate of biodegradation 51.5 mg COD $g^{-1} \cdot h^{-1}$ based on measurements of COD decrease using activated sludge inoculum with 20 days of adaptation to the substrate (Pitter 1976; quoted, Scow 1982).

Biotransformation:

Bioconcentration, Uptake (k_1) and Elimination (k_2) Rate Constants:

Common Name: Acetophenone
Synonym: 1-phenylethanone, methyl phenyl ketone, acetylbenzene, phenylmethylketone, hypnone
Chemical Name: acetophenone, methyl phenyl ketone
CAS Registry No: 98-86-2
Molecular Formula: $C_6H_5COCH_3$
Molecular Weight: 120.16
Melting Point (°C):
20.50 (Stull 1947; Weast 1982-83; Howard 1993)
19.0 (Verschueren 1983)
19.62 (Dean 1985; Riddick et al. 1986)
Boiling Point (°C):
202.4 (Stull 1947)
202.60 (Weast 1982-83)
202.08 (Dean 1985)
202.0 (Verschueren 1983; Riddick et al. 1986; Banerjee et al. 1990; Howard 1993)
Density (g/cm^3 at 20°C):
1.0281 (Weast 1982-83; Riddick et al. 1986)
1.030 (Verschueren 1983)
1.0238 (25°C, Riddick et al. 1986)
Molar Volume (cm^3/mol):
117.1 (calculated-density, Rohrschneider 1973)
126.9 (Taft et al. 1985; Leahy 1986; Kamlet et al. 1987)
69.0 (intrinsic volume V_I, Leahy 1986; Kamlet et al. 1987)
140.4 (calculated-LeBas method)
Molecular Volume ($Å^3$):
Total Surface Area, TSA ($Å^2$):
265.9 (Iwase et al. 1985)
Heat of Fusion ΔH_{fus}, kcal/mol:
3.98 (Tsonopoulos & Prausnitz 1971)
Entropy of Fusion ΔS_{fus}, cal/mol·K (e.u.):
10.1 (Tsonopoulos & Prausnitz 1971)
Fugacity Ratio at 25°C (assuming ΔS_{fus} = 13.5 e.u.), F: 1.0

Water Solubility (g/m^3 or mg/L at 25°C):
5529 (shake flask-centrifuge, Booth & Everson 1948)
6300 (shake flask-AS, Andrews & Keefer 1950)
6022 (estimated, McGowan 1954)
5492 (quoted, Deno & Berkheimer 1960)
5934 (selected, Tsonopoulos & Prausnitz 1971)
3465 (Hansch et al. 1968; quoted, Hine & Mookerjee 1975)
5540 (24°C, shake flask-LSC, Means et al. 1980)
5885, 8124 (quoted, predicted-K_{OW}, Valvani et al. 1981)
5500 (quoted, Verschueren 1983)
5500 (quoted, Dean 1985)
5885, 14116 (quoted, calculated-V_I and solvatochromic p., Leahy 1986)
6130 (Southworth & Keller 1986; quoted, Howard 1993)

5492, 5492 (quoted, calculated-fragment const., Wakita et al. 1986)
5492 (quoted, Gerstl & Helling 1987)
5492, 20881 (quoted, calculated-V_I and solvatochromic p., Kamlet et al. 1987)
5492, 5010 (quoted, calculated-V_I and M.P., Kamlet et al. 1987)

Vapor Pressure (Pa at 25°C):
66.21 (extrapolated-regression of tabulated data, Stull 1947)
41.3 (Hoy 1970)
50 (quoted, Hine & Mookerjee 1975)
57.48 (extrapolated-Cox eqn., Chao et al. 1983)
52.92 (Daubert & Danner 1985; quoted, Howard 1993)
40.23 (extrapolated-Antoine eqn., Dean 1985)
49.32 (Howard et al. 1986; quoted, Banerjee et al. 1990)
49 (Riddick et al. 1986)
75.98 (calculated-solvatochromic p., Banerjee et al. 1990)

Henry's Law Constant (Pa m^3/mol):
1.082 (calculated-1/K_{AW}, C_W/C_A, reported as exptl., Hine & Mookerjee 1975)
0.784 (calculated-bond contribution, Hine & Mookerjee 1975)
1.084 (Mackay et al. 1982; quoted, Howard 1993)
0.921 (gas stripping-GC, Betterton 1991)

Octanol/Water Partition Coefficient, log K_{OW}:
1.58 (shake flask-AS, Fujita et al. 1964; quoted, Leo et al. 1969,71; Hansch et al. 1972; McCall 1975; Briggs 1981; Verschueren 1983; Sangster 1989)
1.58 (calculated-π const., Iwasa et al. 1965)
1.73 (shake flask-AS, Chapman et al. 1972; quoted, Sangster 1989)
1.58 (HPLC, Lazare et al. 1974; quoted, Sangster 1989)
1.68 (shake flask-AS, Holmes & Lough 1976; quoted, Sangster 1989)
1.66 ± 0.08 (shake flask at pH 7, Unger et al. 1978)
1.66 (Hansch & Leo 1979; quoted, Nahum & Horvath 1980; Veith et al. 1983)
1.59 (shake flask-RC, Khan et al 1979; quoted, Sangster 1989; Bintein & Devillers 1994)
1.59 (shake flask-LSC, Means et al. 1980)
1.58 (shake flask-HPLC, Nahum & Horvath 1980; quoted, Sangster 1989)
1.75 (calculated by Rekker's method, Hanai et al. 1981)
1.58; 1.77, 1.56 (quoted, correlated-k', McDuffie 1981)
1.65 (quoted, Valvani et al. 1981)
1.80 (HPLC, Hammers et al. 1982; quoted, Sangster 1989)
1.58, 1.70, 1.73 (quoted from Rekker 1977, Hanai & Hubert 1982)
1.83 (RP-LC, Hanai & Hubert 1982)
1.75 (calculated-f const. as per Rekker 1977, Hanai & Hubert 1982)
1.73 (quoted from Hansch & Leo 1979, Miyake & Terada 1982)
1.56 (RP-HPLC-k', Miyake & Terada 1982)

1.58 (Hansch & Leo 1983; quoted, Tayar et al. 1991)
1.63 (inter-lab. shake flask average, Eadsforth & Moser 1983)
1.65 (inter-lab. HPLC average, Eadsforth & Moser 1983; quoted, Sangster 1989)
1.58 (quoted, Verschueren 1983)
1.65 (selected best lit. value, Garst & Wilson 1984)
1.68 ± 0.02 (exptl.-ALPM, Garst & Wilson 1984)
1.58 (Hansch & Leo 1985; quoted, Howard 1993)
1.68, 1.65 (quoted, calculated-TSA, Iwase et al. 1985)
1.72 (HPLC average, Eadsforth 1986; quoted, Sangster 1989)
1.57, 1.59 (quoted, shake flask, calculated, Eadsforth 1986)
1.58, 1.63 (quoted, calculated-V_I and solvatochromic p., Taft et al. 1985)
1.58, 1.63 (quoted, calculated-V_I and solvatochromic p., Leahy 1986)
1.71 (HPLC, Ge et al. 1987; quoted, Sangster 1989)
1.59 (quoted, Gerstl & Helling 1987)
1.58 (estimated-CLOGP 3, Vaishnav et al. 1987)
1.58, 1.65 (quoted, calculated-γ from UNIFAC, Banerjee & Howard 1988)
1.58, 1.60 (quoted from Leo et al. 1971, shake flask-CPC, Berthod et al. 1988)
1.58 (RP-HPLC-capacity ratio, Minick et al. 1988)
1.66, 1.58, 1.581, 1.87 (quoted, quoted-CLOGP, calculated-CLOGP, calculated-M.O., Bodor et al. 1989)
1.63 (recommended, Sangster 1989)
1.58 (calculated-QSAR, Kollig 1993)

Bioconcentration Factor, log BCF:
0.699-0.954 (estimated-K_{OW} and S, Lyman et al. 1982; quoted, Howard 1993)

Sorption Partition Coefficient, log K_{OC}:
1.544 (average of sediments and soil samples, Means et al. 1980)
1.380 (calculated, Means et al. 1980)
1.54, 1.89 (quoted, calculated- χ , Gerstl & Helling 1987)
1.630 (soil, quoted, Sabljic 1987)
1.34-2.43 (soils and sediments, quoted exptl. and estimated values, Howard 1993)
1.26 (calculated-K_{OW}, Kollig 1993)

Half-Lives in the Environment:
Air: estimated photooxidation half-life of 22 days in atmosphere, based on the vapor-phase reaction with photochemically produced hydroxyl radicals in air (Atkinson 1988; quoted, Howard 1993); atmospheric transformation lifetime was estimated to be 1 to 5 days (Kelly et al. 11994).
Surface water: biodegradation half-lives in river water of 8 days and in lake water of 4.5 days (Vaishnav & Babeu 1986, 1987; quoted, Howard 1993) and 4 days in Superior harbor waters (Vashinav & Babeu 1987).

Groundwater: biodegradation half-life of 32 days (Vaishnav & Babeu 1986, 1987; quoted, Howard 1993); estimated half-life of 0.01 year at Noordwijk (Zoeteman et al. 1981).

Sediment:

Soil:

Biota:

Environmental Fate Rate Constants or Half-Lives:

Volatilization: based on calculated Henry's law constant, half-life from a model river of 1 meter deep flowing at 1 m/sec with a wind velocity of 3 m/sec was estimated to be 3.8 days (Lyman et al. 1982; quoted, Howard 1993).

Photolysis:

Oxidation: reaction rate constants 1.61×10^{-12} $cm^3/molecule \cdot sec$ (estimated from Atmospheric Oxidation Program), 2.74×10^{-12} $cm^3/molecule \cdot sec$ (quoted exptl.) and 2.74×10^{-12} $cm^3/molecule \cdot sec$ (estimated from Fate of Atmospheric Pollutants Program) with OH radicals in vapor phase (Meylan & Howard 1993).

Hydrolysis:

Biodegradation: biodegradation rate constant of 0.029-0.042 h^{-1} in 30 mg/L activated sludge after a time lag of 15-20 hours (Urano & Kato 1986b); rate constants: 0.022 day^{-1} in groundwater, 0.083 day^{-1} in river waters and 0.155 day^{-1} in Superior harbor waters (Vaishnav & Babeu 1987).

Biotransformation:

Bioconcentration, Uptake (k_1) and Elimination (k_2) Rate Constants:

Common Name: Benzophenone
Synonym: diphenyl ketone
Chemical Name: benzophenone, diphenyl ketone
CAS Registry No: 119-61-9
Molecular Formula: $C_6H_5COC_6H_5$
Molecular Weight: 182.22
Melting Point (°C): 48.1, 26.0 (α, β, Weast 1982-83; Dean 1985)
Boiling Point (°C):
305.4 (Stull 1947)
305.9 (Weast 1982-83)
Density (g/cm^3 at 20°C):
1.6077; 1.6059 (α, 19°C; β, 23°C, Weast 1982-83)
Molar Volume (cm^3/mol):
184.4 (Taft et al. 1985; Leahy 1986)
99.9 (intrinsic volume V_I, Leahy 1986)
206.8 (calculated-LeBas method)
Molecular Volume ($Å^3$):
Total Surface Area, TSA ($Å^2$): 342.8 (Iwase et al. 1985)
Dissociation Constant, pK_a:
Heat of Fusion ΔH_{fus}, kcal/mol: 13.36 (Yalkowsky & Valvani 1980)
Entropy of Fusion ΔS_{fus}, cal/mol·K (e.u.):
Fugacity Ratio at 25°C (assuming ΔS_{fus} = 13.5 e.u.), F: 0.591 at M.P. 48.1 °C

Water Solubility (g/m^3 or mg/L at 25°C):
20720, 21000 (calculated-ΔS_{fus} and M.P., estimated, Yalkowsky & Valvani 1980)
276 (calculated-V_I and solvatochromic p., Leahy 1986)

Vapor Pressure (Pa at 25°C):
0.617 (extrapolated-regression tabulated data, Stull 1947)
0.08 (extrapolated-Antoine eqn., Dean 1985)

Henry's Law Constant ($Pa \cdot m^3/mol$):

Octanol/Water Partition Coefficient, log K_{OW}:
3.18 (shake flask-AS, Leo et al. 1971; quoted, Unger et al. 1978)
3.58 (shake flask-AS, Holmes & Lough 1976; quoted, Sangster 1989)
3.10 (shake flask-AS, Unger & Chiang 1981)
3.18; 3.02, 2.86 (quoted, correlated HPLC-k', McDuffie 1981)
3.38 (quoted from Hansch & Leo 1979, Veith et al. 1983)
3.18, 3.11 (quoted, calculated-TSA, Iwase et al. 1985)
3.38, 3.36 (quoted, calculated-V_I and solvatochromic p., Taft et al. 1985)
3.38, 3.24 (quoted, calculated-V_I and solvatochromic p., Leahy 1986)
3.12, 3.18; 3.18 (quoted values; recommended, Sangster 1989)

Bioconcentration Factor, log BCF: no data
Sorption Partition Coefficient, log K_{OC}: no data
Half-Lives in the Environment/Environmental Fate Rate Constants or Half-Lives: no data

3.2 Summary Tables and QSPR Plots

Table 3.1 Physical-chemical properties of aldehydes and ketones at 25 °C.

Compound	CAS no.	formula	MW	mp, °C	bp, °C	fugacity ratio F at 25	molar volume, cm^3/mol V_M from ρ	V_M (a)	V_I (a)	V_M LeBas	V_M (b)	surface area, $Å^2$ TSA (c)	TSA (d)
Aldehydes:													
methanal	50-00-0	HCHO	30.03	-92	-21	1				29.6			
ethanal	75-07-0	CH_3CHO	44.053	-123	20.8	1	56.233	56.2	28.3	51.8			
propanal	123-38-6	C_2H_5CHO	58.08	-80	48	1	73.41	72	38.1	74		224	224
butanal	123-72-8	C_3H_7CHO	72.107	-96.4	74.8	1	88.26	88.2	48	96.2		255.8	
isobutraldehyde	78-84-2	iC_3H_7CHO	72.107	-65	64.1	1	92.02			96.2			255.8
n-valeraldehyde	110-62-3	C_4H_9CHO	86.14	-93.5	103	1	106.41			118.4			
hexanal	66-25-1	$C_5H_{11}CHO$	100.16	-56	129	1	123.06	123.1	67.4	140.6			
heptanal	111-71-7	$C_6H_{13}CHO$	114.19	-43.3	152.8	1	134.42			162.8			
octanal	124-13-0	$C_8H_{15}CHO$	128.22		171.72	1	156.16			185			
acrolein	107-02-8	$H_2C=CHCHO$	56.064	-87	52.69	1		66.6		66.6			
2-butenal	123-73-9	$CH_3CH=CHCHO$	70.091	-76.5	104.1	1	82.51			88.6			
furfural	98-1-1	$C_5H_4O_2$	96.09	-38	162	1	82.9			103.6			
benzyldehyde	100-52-7	C_6H_5CHO	106.12	-55.6	178.8	1	102.04	111.9	60.6	118.2			243.6
Ketones:													
acetone	67-64-1	CH_3COCH_3	58.08	-94.7	56.06	1	73.8	73.4	38	74	59.3		
methyl ethyl ketone	78-93-3	$C_2H_5COCH_3$	72.11	-87	79.58	1	89.9	89.5	47.7	96.2	75.7	258	211.8
4-methyl-2-pentanone	108-10-1	$(CH_3)_2CHCH_2COCH_3$	100.16	-84	117.4	1	125.78					308.7	
2-pentanone	107-87-9	$C_3H_7COCH_3$	86.135	-76.9	102.26	1	107.47	106.5	57.4	118.4	91.2		
3-pentanone	96-22-0	$CH_3CH_2COCH_2CH_3$	86.135	-38.97	101.96	1	107		57	118.4	92.6	289.4	
2-hexanone	591-78-6	$C_4H_9COCH_3$	100.16	-55.8	128.58	1	124.16	123.5	67	140.6	108.1		265
2-heptanone	110-43-0	$CH_3(CH_2)_4COCH_3$	114.18	-35	151.06	1	140.76	140.8	76.7	162.8	125.3		
2-octanone	111-13-7	$C_7H_{16}CO$	128.21	-20.29	173	1	158		87	185	141.1		
cyclopentanone	120-92-3	C_4H_8CO	84.118	-51.3	130.7	1	99	98.6	52	111			
cyclohexanone	108-94-1	$C_5H_{10}CO$	98.144	-32.1	155.6	1	114	113.6	61.9	118.2			229.4
acetophenone	98-86-2	$C_6H_5COCH_3$	120.15	19.62	202	1	127	126.9	69	140.4			265.9
benzophenone	119-61-9	$C_6H_5COC_6H_5$	182.22	48	305.4	0.592	184	184.4	100	206.8			342.8

(a) Kamlet et al., (b) Jaworska & Schultz 1992, (c) Amidon et al. 1975, (d) Iwase et al. 1985

Table 3.2 Summary and selected physical-chemical properties of aldehydes and ketones at 25 °C.

Compound	selected properties:					Henry's law constant, Pa·m³/mol				
	S, g/m³	C_L, Pa	P_S, Pa	P_L, Pa	log K_{OW}	calc-P/C	exptl.(a)	exptl.(b)	exptl.(c)	exptl.(c)
Aldehydes:										
methanal	miscible	miscible			0.35			0.0298		0.0341
ethanal	miscible	miscible	121300	121300	0.45		6.69	6.80	8.00	8.90
propanal	31000	533.75	42400	42400	0.59	79.438	7.44	8.40	7.51	
butanal	71000	984.65	15200	15200	0.88	15.437	11.65	11.65		
n-valeraldehyde			2180	2180			14.87	15.59		
hexanal	5020	50.12			1.78		21.57	19.49		
acrolein	208000	371.00	36500	36500	-0.01	98.382	13.37		13.17	
2-butenal	15600	222.57	5100	5100		22.914				
furfural	79400	826.31	310	310	0.41	0.375				
benzyldehyde	3000	28.27	174	174	1.48	6.155	2.28			2.71
Ketones:										
acetone	miscible	miscible	30806	30800	-0.24		3.97		3.93	3.07
methyl ethyl ketone	240000	3328.25	12100	12100	0.29	3.636	9.71		5.76	
4-methyl-2-pentanone		169.73	2600	2600		15.319				
2-pentanone	59500	690.78	4720	4720	0.84	6.833	6.44			
3-pentanone	34000	394.73	4700	4700	0.82	11.907				
2-hexanone	17500	174.72	1600	1600	1.38	9.157				
2-heptanone	4300	37.66	500	500	2.08	13.278	14.73			
2-octanone	1130	8.81	180	180	2.37	20.423	19.09			
cyclopentanone			1540	1540						
cyclohexanone	23000	234.35	620	620	0.61	2.646				
acetophenone	5500	45.78	45	45	1.63	0.983				0.92
benzophenone	276	1.51	0.09	0.1520	3.18					

(a) Butler et al. 1935, Buttery et al. 1969; (b) Zhou & Mopper 1990; (c) Snider & Dawson 1985; (d) Betterton & Hoffmann 1988.

Table 3.3 Suggested half-life classes for aldehydes and ketones at 25°C at various environmental compartments.

Compound	Air class	Water class	Soil class	Sediment class
Aldehydes:				
methanal	1	3	3	4
ethanal	1	3	3	4
propanal	1	3	3	4
butanal	1	3	3	4
acrolein	1	3	3	4
2-butenal	1	3	3	4
furfural	1	3	3	4
benzaldehyde	1	3	3	4
Ketones:				
acetone	6	4	4	5
methyl ethyl ketone	4	3	3	5
4-methyl-2-pentanone	3	4	4	5
2-pentanone	3	4	4	5
3-pentanone	3	4	4	5
2-hexanone	3	4	4	5
2-heptanone	3	4	4	5
cyclohexanone	3	4	4	5
acetophenone	5	4	4	5
benzophenone	5	4	4	5

where,

Class	Mean half-life (hours)	Range (hours)
1	5	< 10
2	17 (~ 1 day)	10-30
3	55 (~ 2 days)	30-100
4	170 (~ 1 week)	100-300
5	550 (~ 3 weeks)	300-1,000
6	1700 (~ 2 months)	1,000-3,000
7	5500 (~ 8 months)	3,000-10,000
8	17000 (~ 2 years)	10,000-30,000
9	~ 5 years	$> 30,000$

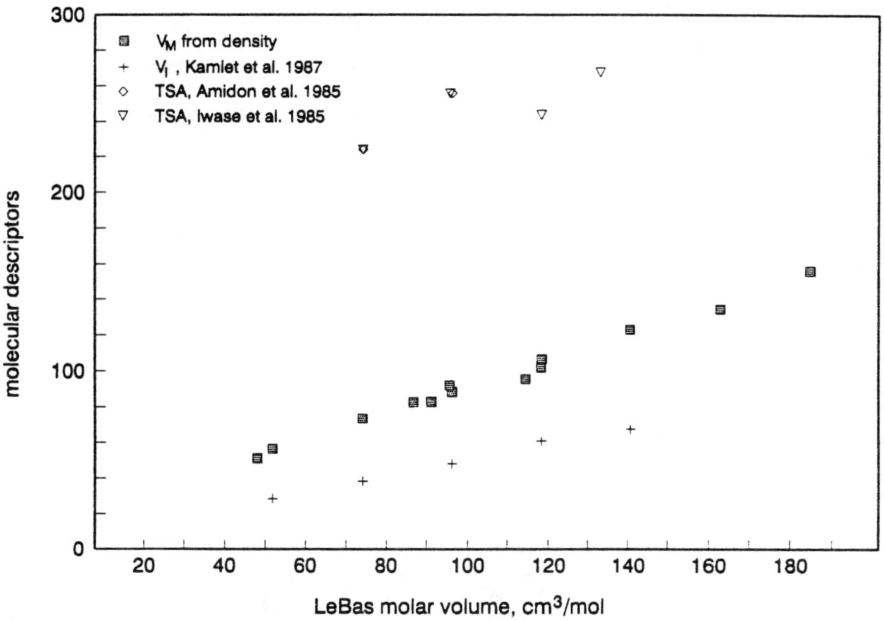

Figure 3.1a Plot of molecular descriptors vs LeBas molar volume

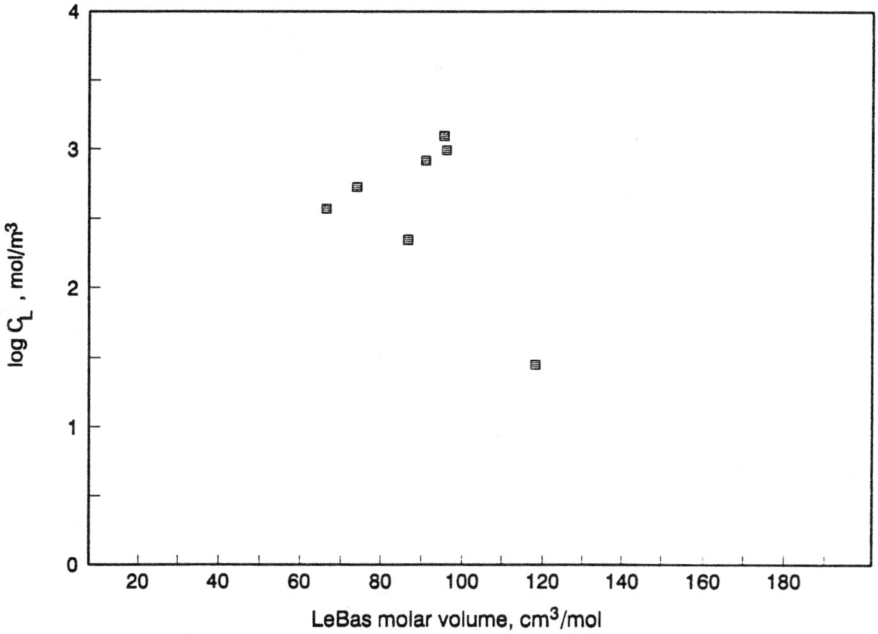

Figure 3.2a Plot of log liquid solubility versus LeBas molar volume

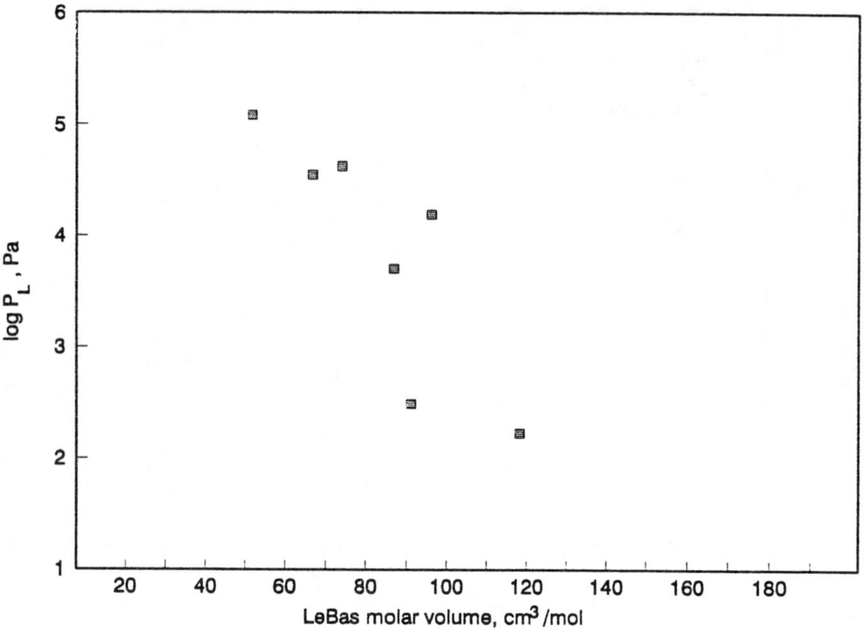

Figure 3.3a Plot of log vapor pressure versus LeBas molar volume

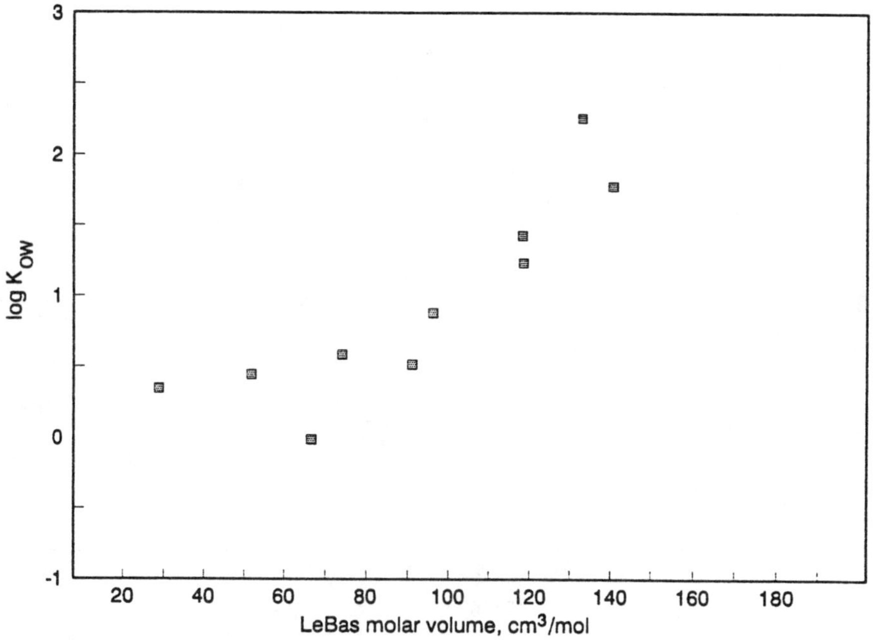

Figure 3.4a Plot of log K_{OW} versus LeBas molar volume

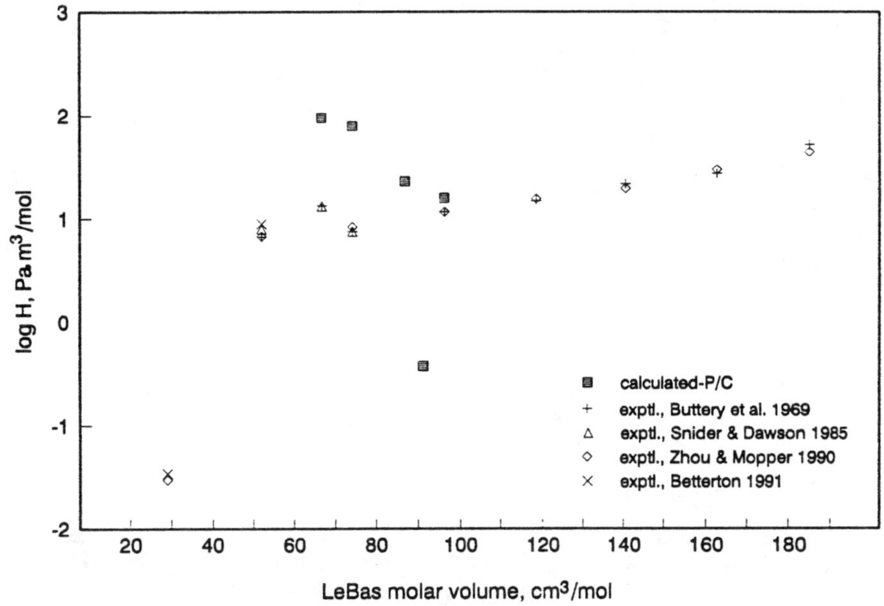

Figure 3.5a Plot of log H versus LeBas molar volume

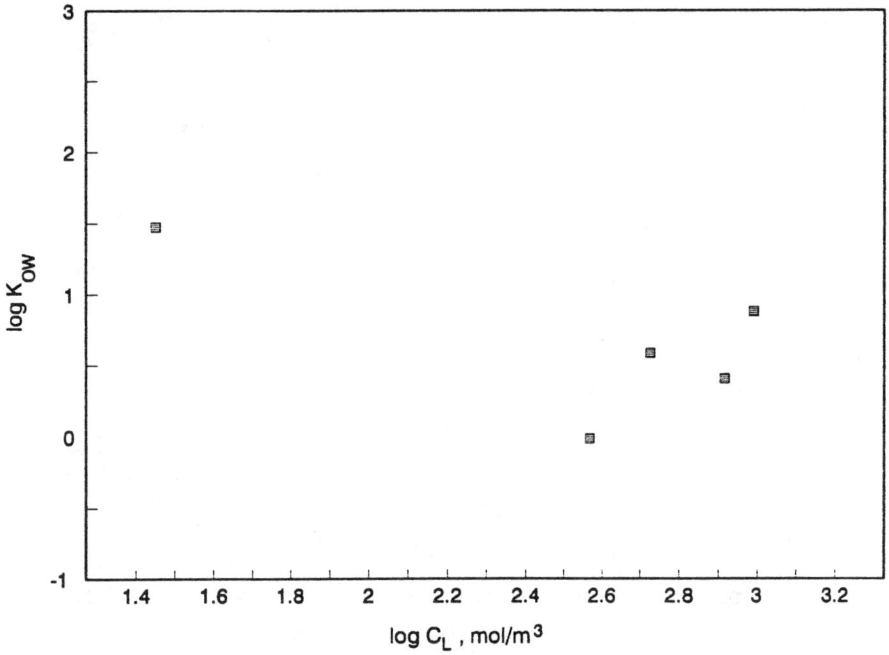

Figure 3.6a Plot of log K_{OW} versus log liquid solubility

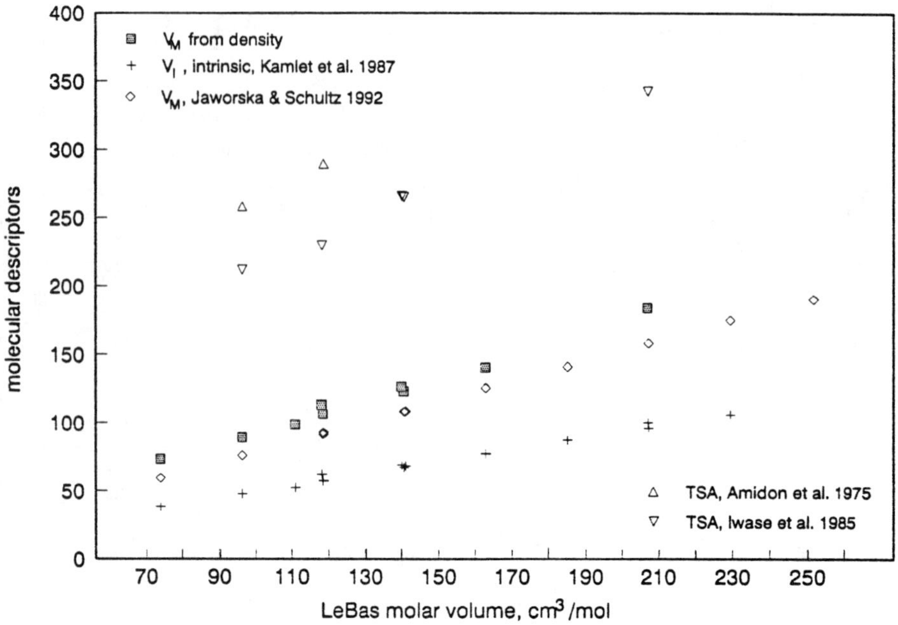

Figure 3.1b Plot of molecular descriptors vs LeBas molar volume

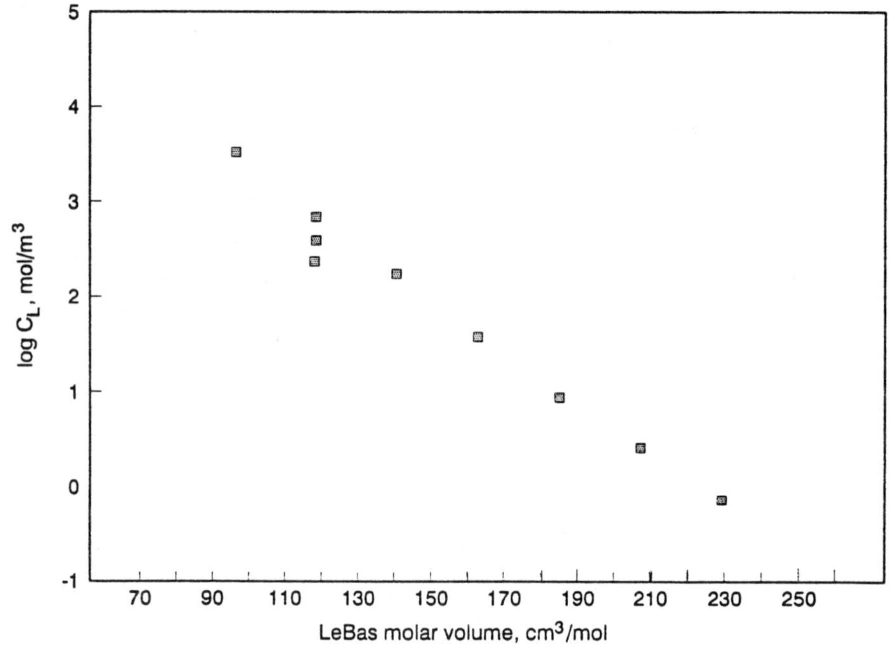

Figure 3.2b Plot of log liquid solubility vs LeBas molar volume

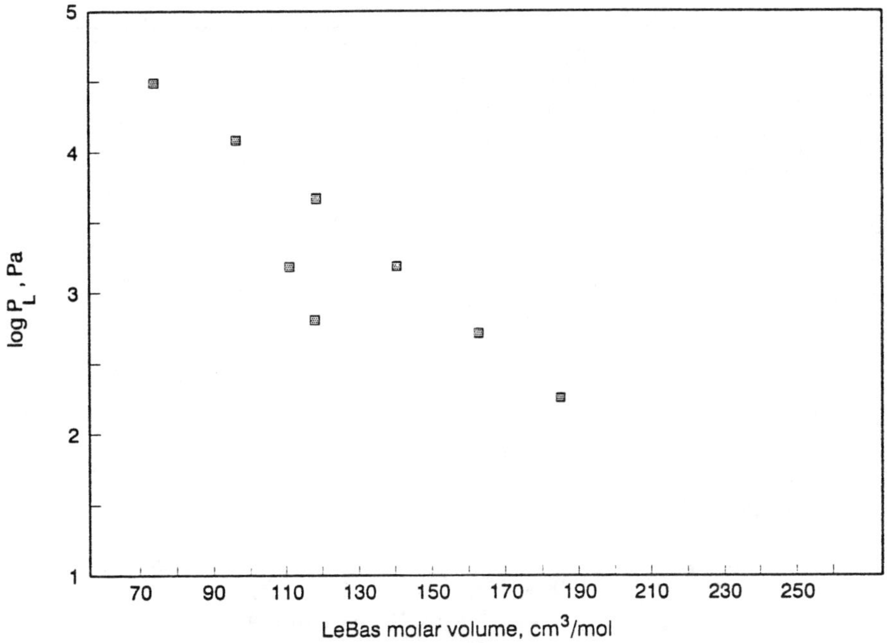

Figure 3.3b Plot of log vapor pressure vs LeBas molar volume

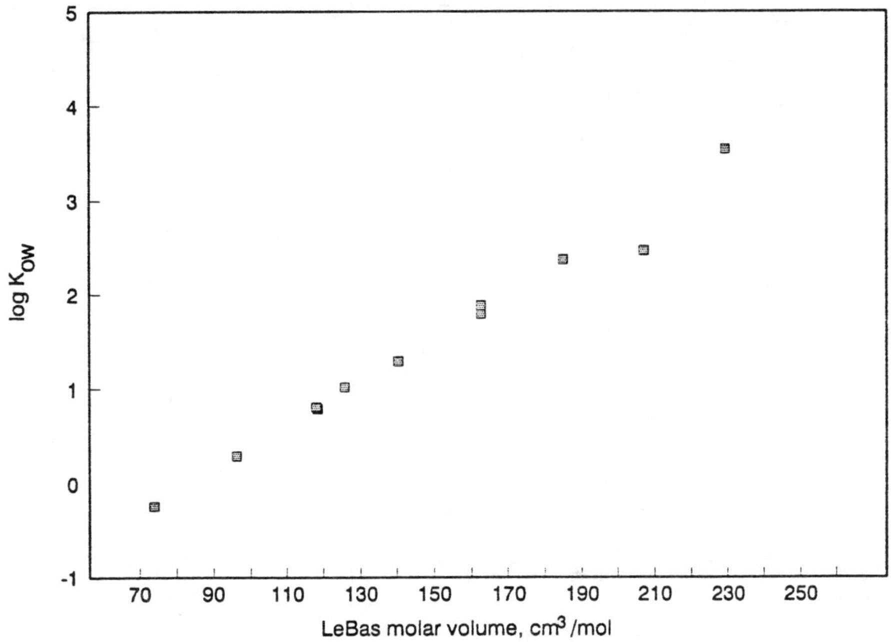

Figure 3.4b Plot of log K_{OW} versus LeBas molar volume

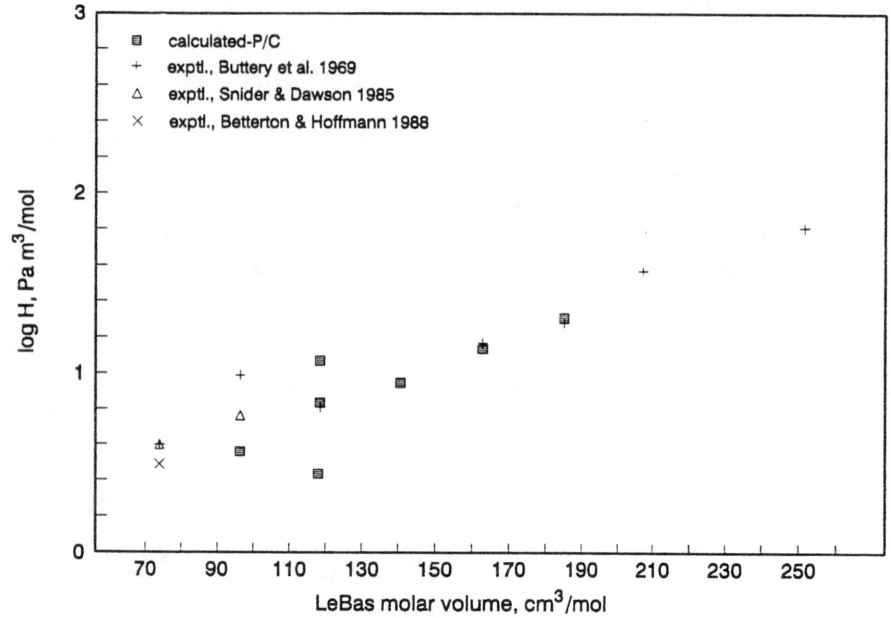

Figure 3.5b Plot of log H versus LeBas molar volume

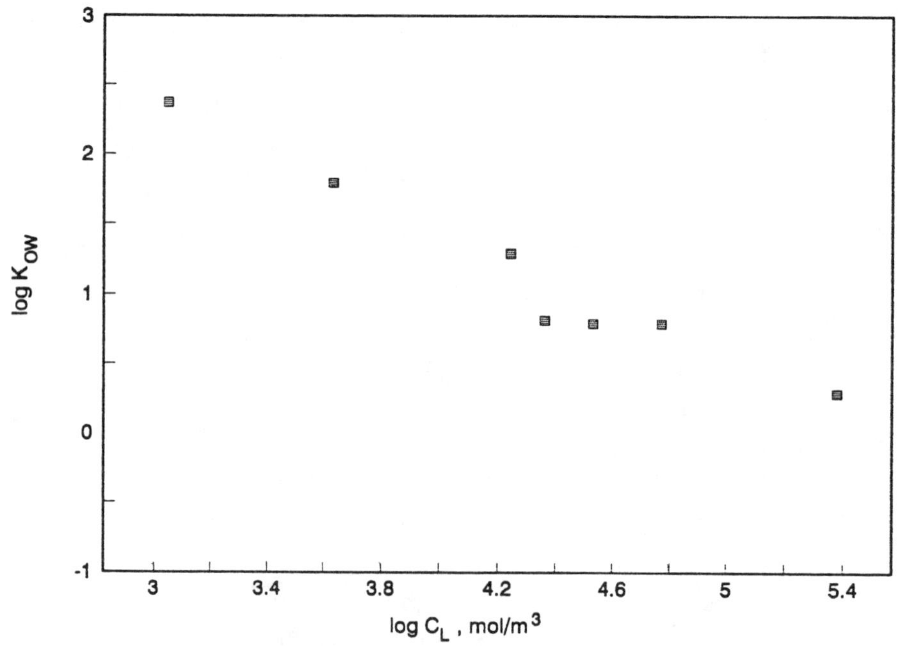

Figure 3.6b Plot of log K_{OW} versus log liquid solubility

3.3 Illustrative Fugacity Calculations: Levels I, II, III

Chemical name: Acetaldehyde

Level I calculation: (six-compartment model)

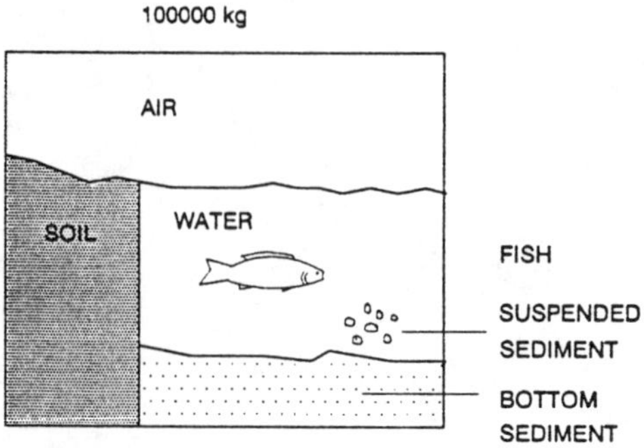

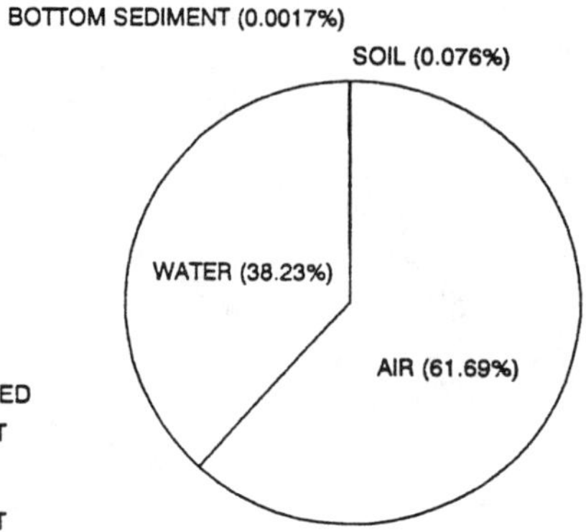

Distribution of mass

Level I calculation:

Physical-chemical properties:			Partition coeff.:	
molecular wt., g/mol	44.05		HLC	7.99890
melting point, C	-123		Kaw	3.23E-03
solubility, g/m3	668000 *		Koc	0.9179
vapor pressure, Pa	121300		BCF	0.1119
log Kow	0.35		Ksw	0.0441
fugacity ratio, F	1.0000		Ksd/w	0.0881
dissoc. const. pKa			Kssd/w	0.2754
			Kar/w	49.464

Compartment	Z	Concentration			Amount	Amount
	mol/m3 Pa	mol/m3	mg/L (or g/m3)	ug/g	kg	%
Air	4.03E-04	1.40E-08	6.17E-07	5.20E-04	61688.52	61.6885
Water	1.25E-01	4.34E-06	1.91E-04	1.91E-04	38233.93	38.2339
Soil	5.51E-03	1.91E-07	8.42E-06	3.51E-06	75.803	0.0758
Biota (fish)	1.40E-02	4.86E-07	2.14E-05	2.14E-05	0.00428	4.28E-06
Suspended sediment	3.44E-02	1.20E-06	5.26E-05	3.51E-05	0.0526	5.26E-05
Bottom sediment	1.10E-02	3.82E-07	1.68E-05	7.02E-06	1.6845	0.00168
	Total				100000	100

f = 3.471E-05 Pa

* pseudo-solubility of a miscible compound

Chemical name: Acetaldehyde

Level II calculation: (six-compartment model)

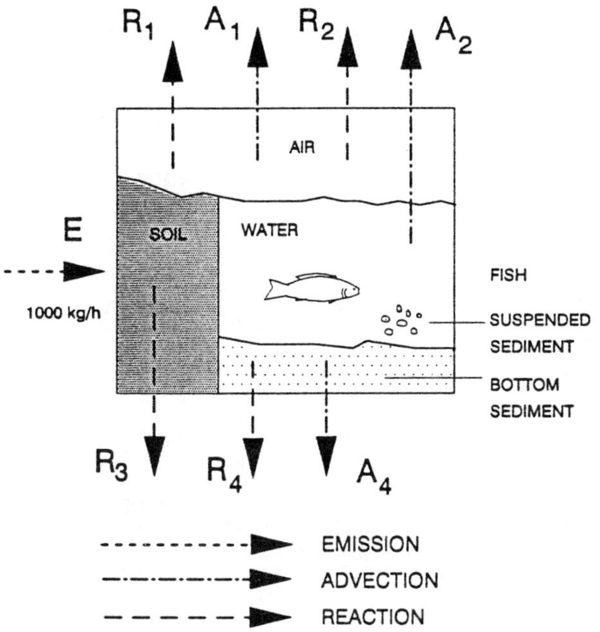

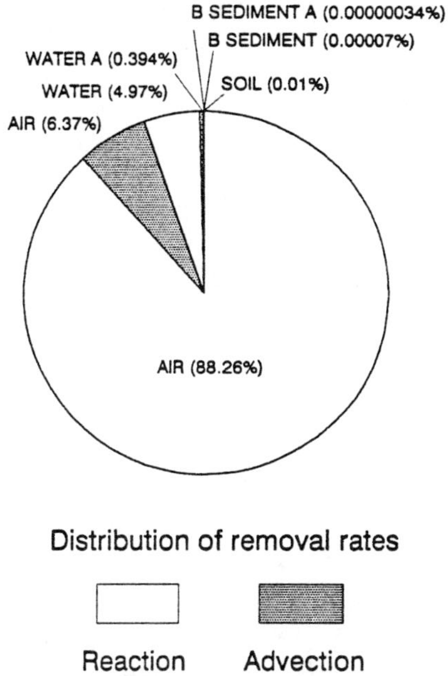

Level II Calculation:

Compartment	Half-life h	D Values Reaction mol/Pa.h	D Values Advection mol/Pa.h	Conc'n mol/m3	Loss Reaction kg/h	Loss Advection kg/h	Removal %
Air	5	5.59E+09	4.03E+08	1.45E-09	882.551	63.676	94.623
Water	55	3.15E+08	2.50E+07	4.48E-07	49.727	3.947	5.367
Soil	55	6.25E+05		1.97E-08	0.0986		0.0099
Biota (fish)				5.01E-08			
Suspended sediment				1.23E-07			
Bottom sediment	170	4.49E+03	2.20E+01	3.95E-08	7.09E-04	3.48E-06	7.12E-05
	Total	5.91E+09	4.28E+08		932.38	67.623	100
	R + A		6.34E+09			1000	

f = 3.583E-06 Pa

Total amt = 10322 kg

Overall residence time = 10.32 h

Reaction time = 11.07 h

Advection time = 152.64 h

Fugacity Level III calculations: (four-compartment model)

Chemical name: Acetaldehyde

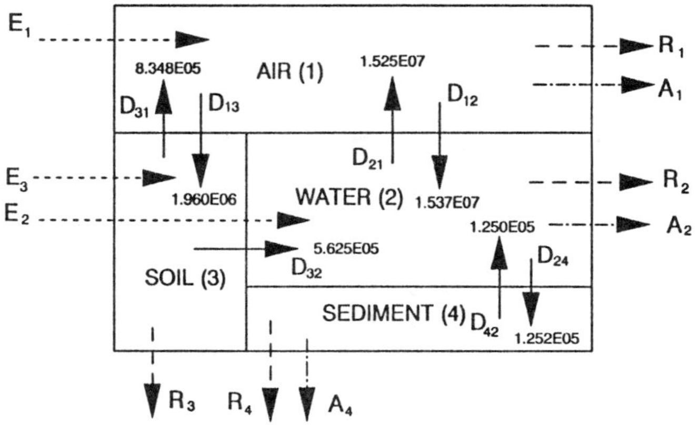

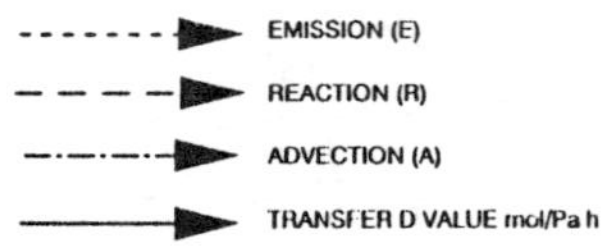

Phase Properties and Rates:

Compartment	Bulk Z mol/m3 Pa	Half-life h	D Values Reaction mol/Pa h	D Values Advection mol/Pa h
Air (1)	4.034E-04	5	5.59E+09	4.03E+08
Water (2)	1.250E-01	55	3.15E+08	2.50E+07
Soil (3)	4.034E-02	55	9.15E+06	
Sediment (4)	1.022E-01	170	2.08E+05	1.02E+03

	E(1)=1000	E(2)=1000	E(3)=1000	E(1,2,3)
Overall residence time =	6.91	70.71	73.17	69.24 h
Reaction time =	7.41	76.30	73.85	71.25 h
Advection time =	102.76	965.35	7924	2460.60 h

Phase Properties, Compositions, Transport and Transformation Rates:

Emission, kg/h			Fugacity, Pa				Concentration, g/m3				Amounts, kg				Total Amount, kg
E(1)	E(2)	E(3)	f(1)	f(2)	f(3)	f(4)	C(1)	C(2)	C(3)	C(4)	m(1)	m(2)	m(3)	m(4)	
1000	0	0	3.776E-06	1.645E-07	7.018E-07	6.158E-08	6.711E-08	9.059E-07	1.247E-06	2.773E-07	6.711E+03	1.812E+02	2.245E+01	1.386E-01	6.915E+03
0	1000	0	1.621E-07	6.389E-05	3.012E-08	2.392E-05	2.880E-09	3.518E-04	5.352E-08	1.077E-04	2.880E+02	7.037E+04	9.633E-01	5.385E+01	7.071E+04
0	0	1000	3.076E-07	3.421E-06	2.153E-03	1.281E-06	5.466E-09	1.884E-05	3.825E-03	5.767E-06	5.466E+02	3.768E+03	6.885E+04	2.883E+00	7.317E+04
50	250	700	4.446E-07	1.837E-05	1.507E-03	6.879E-06	7.901E-09	1.012E-04	2.678E-03	3.097E-05	7.901E+02	2.024E+04	4.820E+04	1.549E+01	6.924E+04

Emission, kg/h			Loss, Reaction, kg/h				Loss, Advection, kg/h			Intermedia Rate of Transport, kg/h T12	T21	T13	T31	T32	T24	T42
E(1)	E(2)	E(3)	R(1)	R(2)	R(3)	R(4)	A(1)	A(2)	A(4)	air-water	water-air	air-soil	soil-air	soil-water	water-sed	sed-water
1000	0	0	9.301E+02	2.283E+00	2.83E-01	5.652E-04	6.711E+01	1.812E-01	2.773E-06	2.558E+00	1.105E-01	3.260E-01	2.581E-02	1.739E-02	9.071E-04	3.392E-04
0	1000	0	3.991E+01	8.866E+02	1.21E-02	2.195E-01	2.880E+00	7.037E+01	1.077E-03	1.098E-01	4.292E+01	1.399E-02	1.107E-03	7.463E-04	3.523E-01	1.317E-01
0	0	1000	7.576E+01	4.748E+01	8.68E+02	1.175E-02	5.466E+00	3.768E+00	5.767E-05	2.083E-01	2.298E+00	2.655E-02	7.916E+01	5.334E+01	1.887E-02	7.054E-03
50	250	700	1.095E+02	2.550E+02	6.07E+02	6.313E-02	7.901E+00	2.024E+01	3.097E-04	3.011E-01	1.234E+01	3.839E-02	5.541E+01	3.734E+01	1.013E-01	3.789E-02

Level III Distribution

Chemical name: Acetaldehyde

Distribution of mass

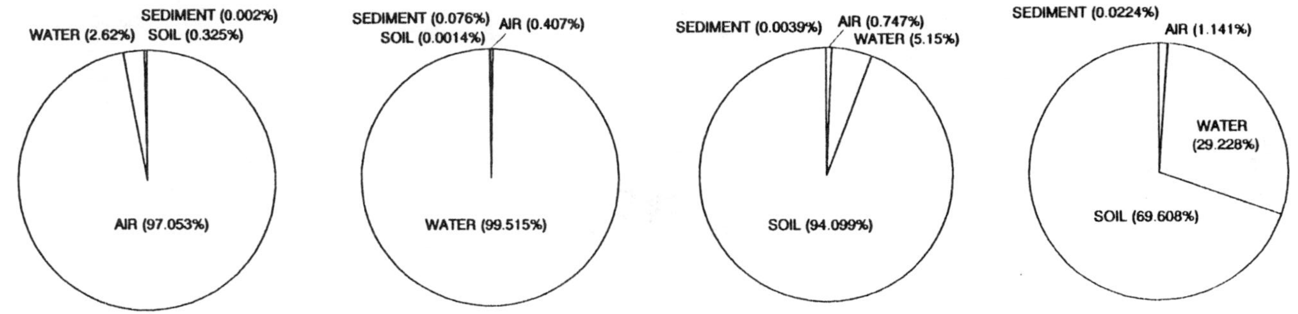

Distribution of removal rates

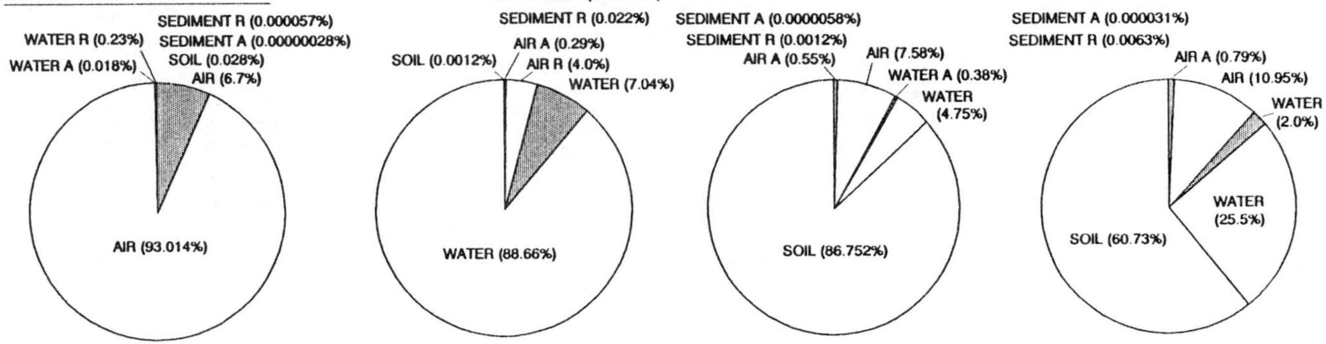

Emission rates:

E(1) = 1000 kg/h	E(1) = 0	E(1) = 0	E(1) = 50 kg/h
E(2) = 0	E(2) = 1000 kg/h	E(2) = 0	E(2) = 250 kg/h
E(3) = 0	E(3) = 0	E(3) = 1000 kg/h	E(3) = 700 kg/h

Reaction Advection

Residence time:

t = 6.91 h	t = 70.71 h	t = 73.17 h	t = 69.24 h

Chemical name: Butyraldehyde

Level I calculation: (six-compartment model)

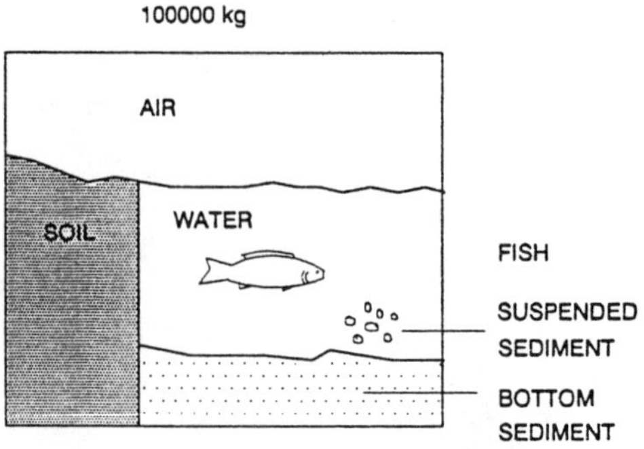

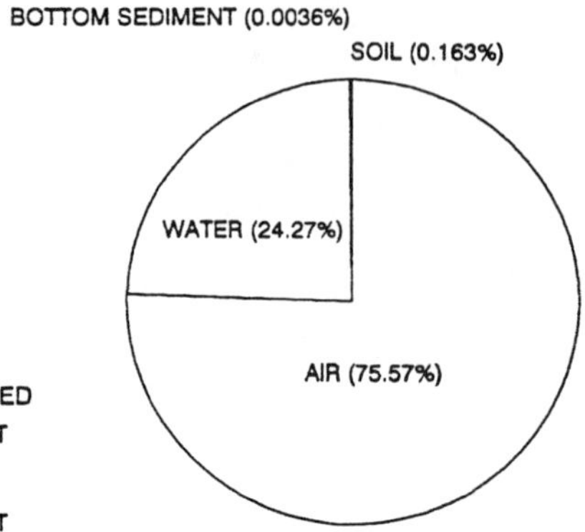

Distribution of mass

Level I calculation:

Physical-chemical properties:		Partition coeff.:	
molecular wt., g/mol	72.11	HLC	15.43763
melting point, C	-96	Kaw	6.23E-03
solubility, g/m3	71000	Koc	3.1102
vapor pressure, Pa	15200	BCF	0.3793
log Kow	0.88	Ksw	0.1493
fugacity ratio, F	1.0000	Ksd/w	0.2986
dissoc. const. pKa		Kssd/w	0.9331
		Kar/w	394.737

Compartment	Z	Concentration			Amount	Amount
	mol/m3 Pa	mol/m3	mg/L (or g/m3)	ug/g	kg	%
Air	4.03E-04	1.05E-08	7.56E-07	6.37E-04	75565.99	75.5660
Water	6.48E-02	1.68E-06	1.21E-04	1.21E-04	24267.24	24.2672
Soil	9.67E-03	2.51E-07	1.81E-05	7.55E-06	163.026	0.1630
Biota (fish)	2.46E-02	6.38E-07	4.60E-05	4.60E-05	0.00920	9.20E-06
Suspended sediment	6.04E-02	1.57E-06	1.13E-04	7.55E-05	0.1132	1.13E-04
Bottom sediment	1.93E-02	5.02E-07	3.62E-05	1.51E-05	3.6228	0.00362
	Total				100000	100

f = 2.598E-05 Pa

Chemical name: Butyraldehyde

Level II calculation: (six-compartment model)

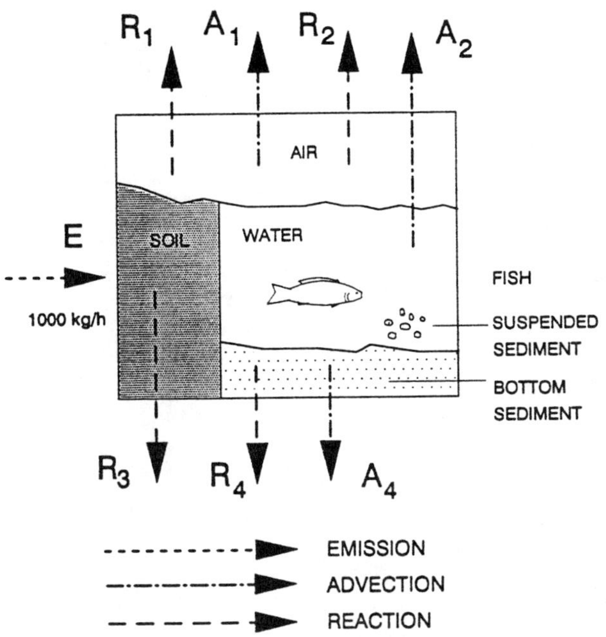

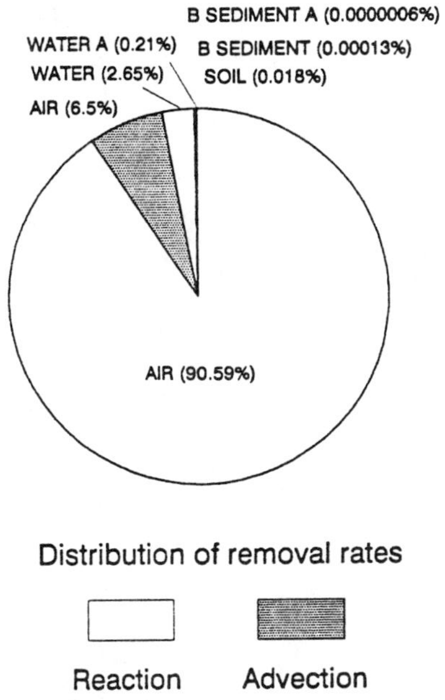

Distribution of removal rates

Level II Calculation:

Compartment	Half-life h	D Values Reaction mol/Pa.h	D Values Advection mol/Pa.h	Conc'n mol/m3	Loss Reaction kg/h	Loss Advection kg/h	Removal %
Air	5	5.59E+09	4.03E+08	9.06E-10	905.913	65.362	97.127
Water	55	1.63E+08	1.30E+07	1.46E-07	26.448	2.099	2.855
Soil	55	1.10E+06		2.17E-08	0.1777		0.0178
Biota (fish)				5.52E-08			
Suspended sediment				1.36E-07			
Bottom sediment	170	7.88E+03	3.87E+01	4.35E-08	1.28E-03	6.27E-06	1.28E-04
	Total	5.76E+09	4.16E+08		932.54	67.461	100
	R + A		6.17E+09			1000	

f = 2.247E-06 Pa

Total amt = 8650 kg

Overall residence time = 8.65 h

Reaction time = 9.28 h

Advection time = 128.22 h

Fugacity Level III calculations: (four-compartment model)

Chemical name: Butyraldehyde

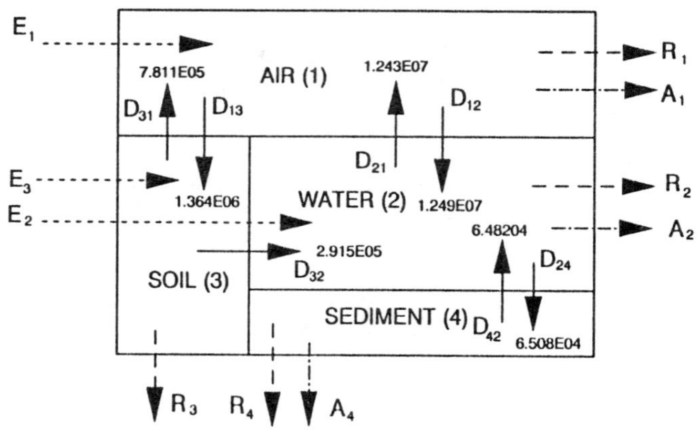

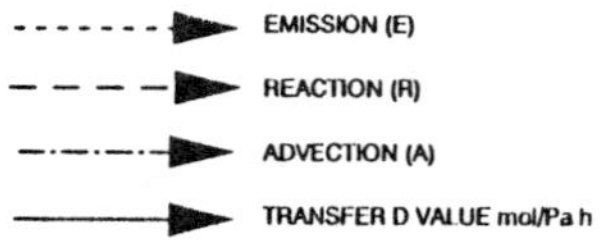

Phase Properties and Rates:

Compartment	Bulk Z mol/m3 Pa	Half-life h	D Values Reaction mol/Pa h	D Values Advection mol/Pa h
Air (1)	4.034E-04	5	5.59E+09	4.03E+08
Water (2)	6.478E-02	55	1.63E+08	1.30E+07
Soil (3)	2.435E-02	55	5.52E+06	
Sediment (4)	5.569E-02	170	1.14E+05	5.57E+02

	E(1)=1000	E(2)=1000	E(3)=1000	E(1,2,3)	
Overall residence time =	6.87	69.18	70.33	66.87	h
Reaction time =	7.37	74.63	71.13	68.90	h
Advection time =	102.15	946.27	6279	2268.13	h

Phase Properties, Compositions, Transport and Transformation Rates:

Emission, kg/h			Fugacity, Pa				Concentration, g/m3				Amounts, kg				Total Amount, kg
E(1)	E(2)	E(3)	f(1)	f(2)	f(3)	f(4)	C(1)	C(2)	C(3)	C(4)	m(1)	m(2)	m(3)	m(4)	
1000	0	0	2.308E-06	1.536E-07	4.775E-07	5.589E-08	6.715E-08	7.175E-07	8.383E-07	2.244E-07	6.715E+03	1.435E+02	1.509E+01	1.122E-01	6.874E+03
0	1000	0	1.521E-07	7.351E-05	3.146E-08	2.675E-05	4.424E-09	3.434E-04	5.523E-08	1.074E-04	4.424E+02	6.868E+04	9.941E-01	5.370E+01	6.918E+04
0	0	1000	2.801E-07	3.268E-06	2.103E-03	1.189E-06	8.149E-09	1.526E-05	3.692E-03	4.774E-06	8.149E+02	3.053E+03	6.646E+04	2.387E+00	7.033E+04
50	250	700	3.495E-07	2.067E-05	1.472E-03	7.521E-06	1.017E-08	9.657E-05	2.585E-03	3.020E-05	1.017E+03	1.931E+04	4.652E+04	1.510E+01	6.687E+04

Emission, kg/h			Loss, Reaction, kg/h				Loss, Advection, kg/h			Intermedia Rate of Transport, kg/h T12	T21	T13	T31	T32	T24	T42
E(1)	E(2)	E(3)	R(1)	R(2)	R(3)	R(4)	A(1)	A(2)	A(4)	air-water	water-air	air-soil	soil-air	soil-water	water-sed	sed-water
1000	0	0	9.307E+02	1.808E+00	1.90E-01	4.574E-04	6.715E+01	1.435E-01	2.244E-06	2.080E+00	1.377E-01	2.271E-01	2.689E-02	1.004E-02	7.209E-04	2.612E-04
0	1000	0	6.132E+01	8.653E+02	1.25E-02	2.189E-01	4.424E+00	6.868E+01	1.074E-03	1.370E-01	6.589E+01	1.496E-02	1.772E-03	6.612E-04	3.450E-01	1.250E-01
0	0	1000	1.129E+02	3.846E+01	8.37E+02	9.730E-03	8.149E+00	3.053E+00	4.774E-05	2.524E-01	2.929E+00	2.755E-02	1.184E+02	4.420E+01	1.533E-02	5.556E-03
50	250	700	1.409E+02	2.434E+02	5.86E+02	6.156E-02	1.017E+01	1.931E+01	3.020E-04	3.149E-01	1.853E+01	3.438E-02	8.291E+01	3.094E+01	9.702E-02	3.515E-02

Level III Distribution

Chemical name: Butyraldehyde

Distribution of mass

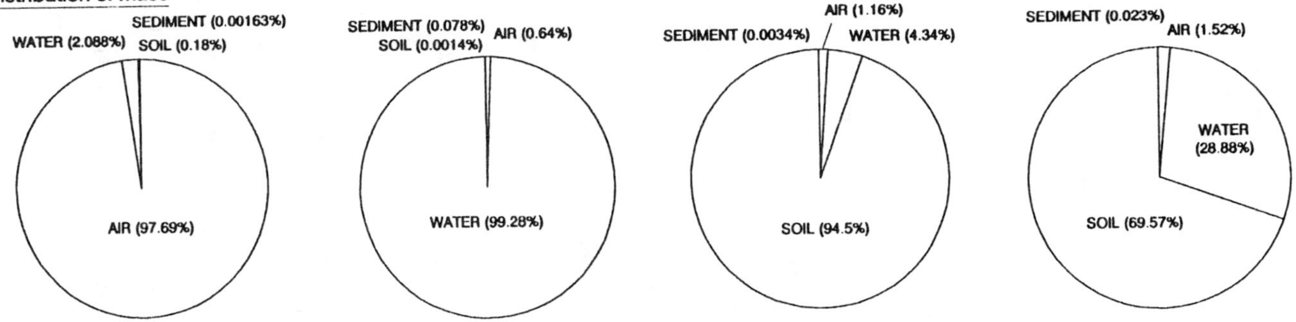

Distribution of removal rates

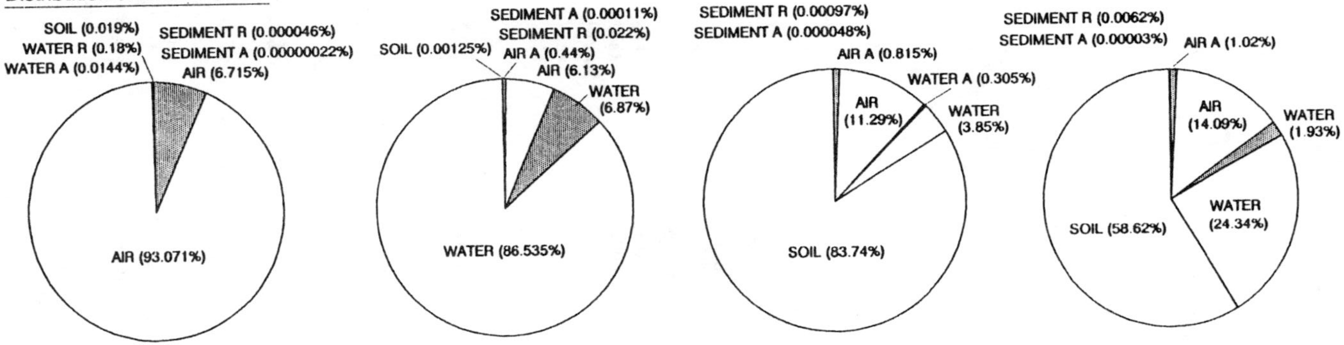

Emission rates:

E(1) = 1000 kg/h	E(1) = 0	E(1) = 0	E(1) = 50 kg/h
E(2) = 0	E(2) = 1000 kg/h	E(2) = 0	E(2) = 250 kg/h
E(3) = 0	E(3) = 0	E(3) = 1000 kg/h	E(3) = 700 kg/h

Residence time:

t = 6.87 h	t = 69.18 h	t = 70.33 h	t = 66.87 h

Reaction (open) Advection (shaded)

Chemical name: Acrolein

Level I calculation: (six-compartment model)

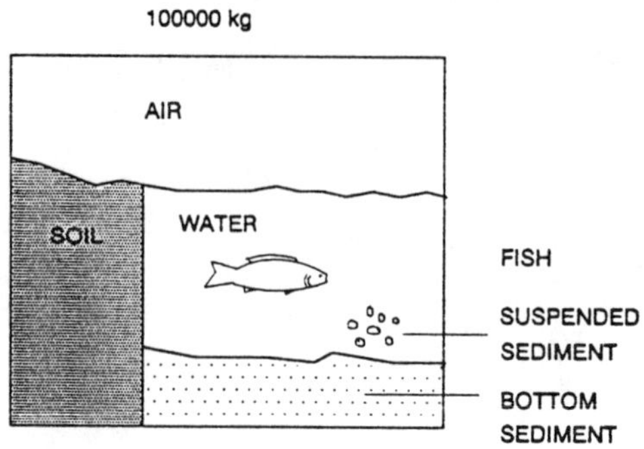

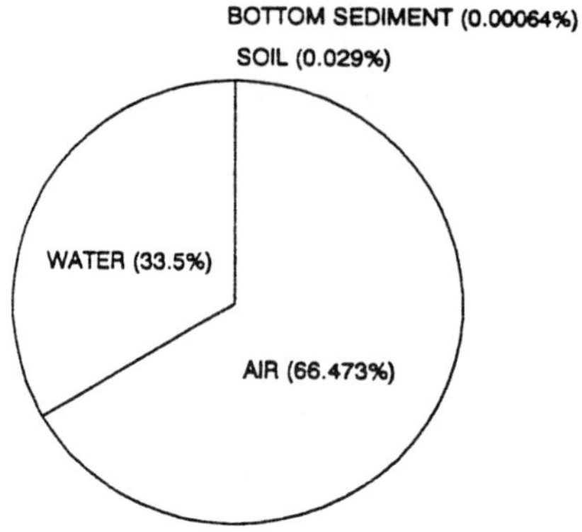

Distribution of mass

Level I calculation:

Physical-chemical properties:		Partition coeff.:	
molecular wt., g/mol	56.06	HLC	9.83745
melting point, C	-86.95	Kaw	3.97E-03
solubility, g/m3	208000	Koc	0.4007
vapor pressure, Pa	36500	BCF	0.0489
log Kow	-0.01	Ksw	0.0192
fugacity ratio, F	1.0000	Ksd/w	0.0385
dissoc. const. pKa		Kssd/w	0.1202
		Kar/w	164.384

Compartment	Z	Concentration			Amount	Amount
	mol/m3 Pa	mol/m3	mg/L (or g/m3)	ug/g	kg	%
Air	4.03E-04	1.19E-08	6.65E-07	5.61E-04	66471.61	66.4716
Water	1.02E-01	2.99E-06	1.67E-04	1.67E-04	33498.73	33.4987
Soil	1.95E-03	5.75E-08	3.22E-06	1.34E-06	28.991	0.0290
Biota (fish)	4.97E-03	1.46E-07	8.18E-06	8.18E-06	0.00164	1.64E-06
Suspended sediment	1.22E-02	3.59E-07	2.01E-05	1.34E-05	0.0201	2.01E-05
Bottom sediment	3.91E-03	1.15E-07	6.44E-06	2.68E-06	0.6442	6.44E-04
	Total				100000	100

f = 2.939E-05 Pa

Chemical name: Acrolein

Level II calculation: (six-compartment model)

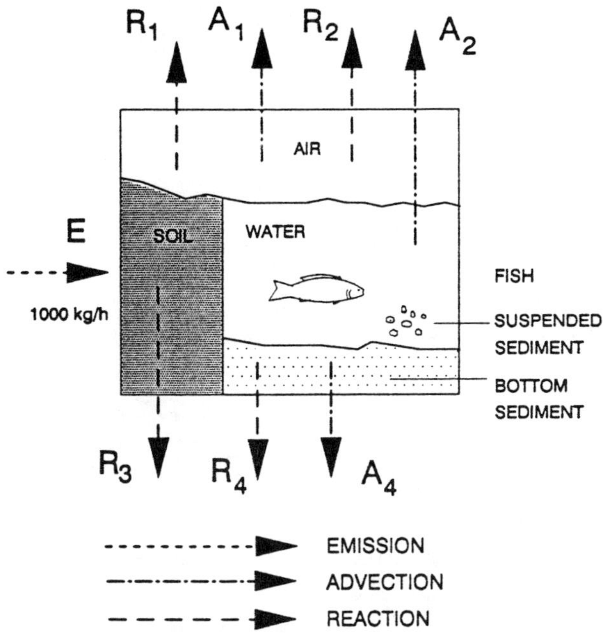

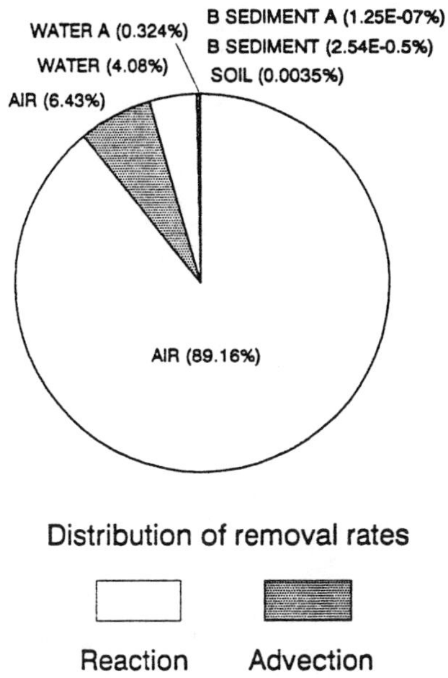

Distribution of removal rates

Level II Calculation:

Compartment	Half-life h	D Values Reaction mol/Pa.h	D Values Advection mol/Pa.h	Conc'n mol/m3	Loss Reaction kg/h	Loss Advection kg/h	Removal %
Air	5	5.59E+09	4.03E+08	1.15E-09	891.552	64.326	95.588
Water	55	2.56E+08	2.03E+07	2.89E-07	40.846	3.242	4.409
Soil	55	2.22E+05		5.56E-09	0.0353		0.00353
Biota (fish)				1.41E-08			
Suspended sediment				3.48E-08			
Bottom sediment	170	1.59E+03	7.82E+00	1.11E-08	2.54E-04	1.25E-06	2.55E-05
	Total	5.85E+09	4.24E+08		932.43	67.567	100
	R + A		6.27E+09			1000	

f = 2.844E-06 Pa

Total amt = 9677 kg

Overall residence time = 9.68 h

Reaction time = 10.38 h

Advection time = 143.22 h

Fugacity Level III calculations: (four-compartment model)

Chemical name: Acrolein

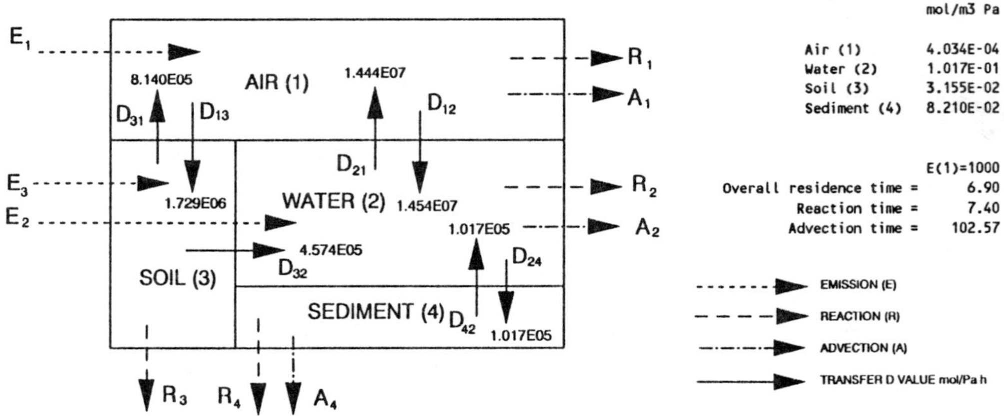

Phase Properties and Rates:

Compartment	Bulk Z mol/m3 Pa	Half-life h	D Values Reaction mol/Pa h	D Values Advection mol/Pa h
Air (1)	4.034E-04	5	5.59E+09	4.03E+08
Water (2)	1.017E-01	55	2.56E+08	2.03E+07
Soil (3)	3.155E-02	55	7.16E+06	
Sediment (4)	8.210E-02	170	1.67E+05	8.21E+02

	E(1)=1000	E(2)=1000	E(3)=1000	E(1,2,3)
Overall residence time =	6.90	70.26	71.87	68.22 h
Reaction time =	7.40	75.81	72.63	70.26 h
Advection time =	102.57	959.78	6863	2352.71 h

Phase Properties, Compositions, Transport and Transformation Rates:

Emission, kg/h			Fugacity, Pa				Concentration, g/m3				Amounts, kg				Total
E(1)	E(2)	E(3)	f(1)	f(2)	f(3)	f(4)	C(1)	C(2)	C(3)	C(4)	m(1)	m(2)	m(3)	m(4)	Amount, kg
1000	0	0	2.968E-06	1.493E-07	6.088E-07	5.627E-08	6.712E-08	8.507E-07	1.077E-06	2.590E-07	6.712E+03	1.701E+02	1.939E+01	1.295E-01	6.902E+03
0	1000	0	1.473E-07	6.131E-05	3.021E-08	2.311E-05	3.331E-09	3.494E-04	5.344E-08	1.064E-04	3.331E+02	6.987E+04	9.620E-01	5.318E+01	7.026E+04
0	0	1000	2.946E-07	3.342E-06	2.117E-03	1.260E-06	6.664E-09	1.904E-05	3.744E-03	5.798E-06	6.664E+02	3.809E+03	6.739E+04	2.899E+00	7.187E+04
50	250	700	3.915E-07	1.767E-05	1.482E-03	6.662E-06	8.853E-09	1.007E-04	2.621E-03	3.066E-05	8.853E+02	2.014E+04	4.718E+04	1.533E+01	6.822E+04

Emission, kg/h			Loss, Reaction, kg/h				Loss, Advection, kg/h			Intermedia Rate of Transport, kg/h T12	T21	T13	T31	T32	T24	T42
E(1)	E(2)	E(3)	R(1)	R(2)	R(3)	R(4)	A(1)	A(2)	A(4)	air-water	water-air	air-soil	soil-air	soil-water	water-sed	sed-water
1000	0	0	9.303E+02	2.144E+00	2.44E-01	5.279E-04	6.712E+01	1.701E-01	2.590E-06	2.420E+00	1.208E-01	2.877E-01	2.778E-02	1.561E-02	8.512E-04	3.207E-04
0	1000	0	4.617E+01	8.804E+02	1.21E-02	2.168E-01	3.331E+00	6.987E+01	1.064E-03	1.201E-01	4.963E+01	1.427E-02	1.379E-03	7.748E-04	3.496E-01	1.317E-01
0	0	1000	9.236E+01	4.799E+01	8.49E+02	1.182E-02	6.664E+00	3.809E+00	5.798E-05	2.402E-01	2.705E+00	2.856E-02	9.658E+01	5.428E+01	1.906E-02	7.180E-03
50	250	700	1.227E+02	2.538E+02	5.94E+02	6.250E-02	8.853E+00	2.014E+01	3.066E-04	3.191E-01	1.431E+01	3.794E-02	6.761E+01	3.800E+01	1.008E-01	3.797E-02

Chemical name: Acrolein

Level III Distribution

Distribution of mass

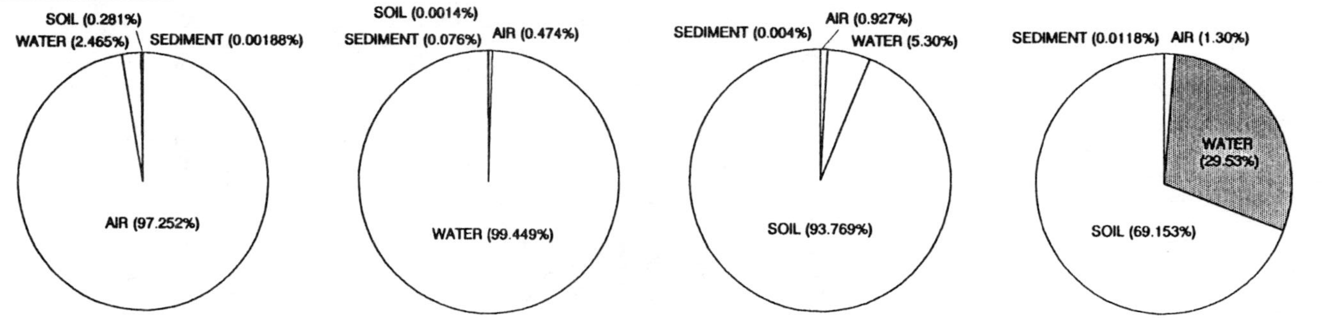

Distribution of removal rates

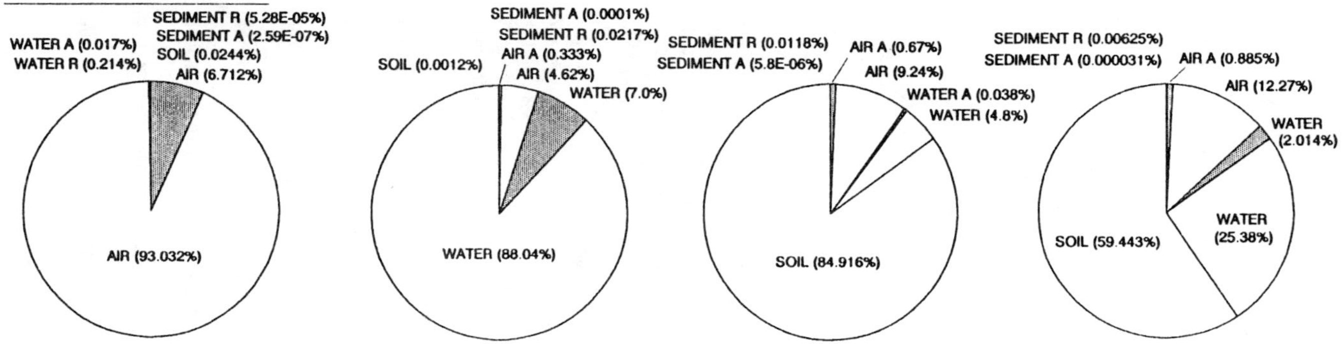

Reaction (white) | Advection (shaded)

Emission rates:

E(1) = 1000 kg/h	E(1) = 0	E(1) = 0	E(1) = 50 kg/h
E(2) = 0	E(2) = 1000 kg/h	E(2) = 0	E(2) = 250 kg/h
E(3) = 0	E(3) = 0	E(3) = 1000 kg/h	E(3) = 700 kg/h

Residence time:

t = 6.90 h	t = 70.26 h	t = 71.87 h	t = 68.22 h

Chemical name: Acetone

Level I calculation: (six-compartment model)

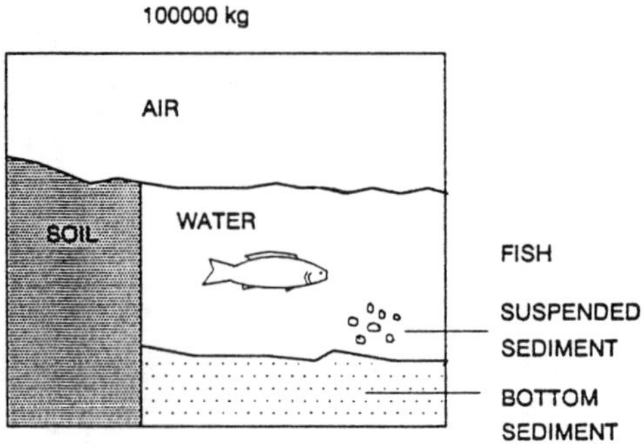

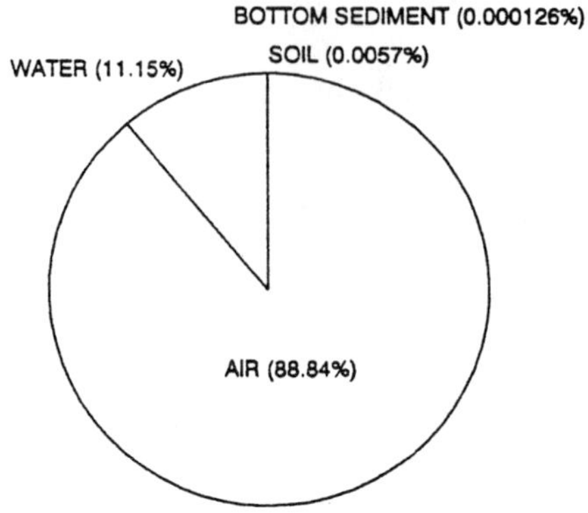

Distribution of mass

Level I calculation:

Physical-chemical properties:		Partition coeff.:	
molecular wt., g/mol	58.08	HLC	3.9500
melting point, C	-94.7	Kaw	1.59E-03
solubility, g/m3	452880 *	Koc	0.2359
vapor pressure, Pa	30800	BCF	0.0288
log Kow	-0.24	Ksw	0.0113
fugacity ratio, F	1.0000	Ksd/w	0.0226
dissoc. const. pKa		Kssd/w	0.0708
		Kar/w	194.805

Compartment	Z	Concentration			Amount	Amount
	mol/m3 Pa	mol/m3	mg/L (or g/m3)	ug/g	kg	%
Air	4.03E-04	7.63E-09	4.43E-07	3.74E-04	44330.94	44.3309
Water	2.53E-01	4.79E-06	2.78E-04	2.78E-04	55640.06	55.6401
Soil	2.87E-03	5.42E-08	3.15E-06	1.31E-06	28.355	0.0284
Biota (fish)	7.28E-03	1.38E-07	8.00E-06	8.00E-06	0.00160	1.60E-06
Suspended sediment	1.79E-02	3.39E-07	1.97E-05	1.31E-05	0.0197	1.97E-05
Bottom sediment	5.73E-03	1.08E-07	6.30E-06	2.63E-06	0.6301	6.30E-04
	Total				100000	100

f = 1.892E-05 Pa

* pseudo-solubility of a miscible compound

Chemical name: Acetone

Level II calculation: (six-compartment model)

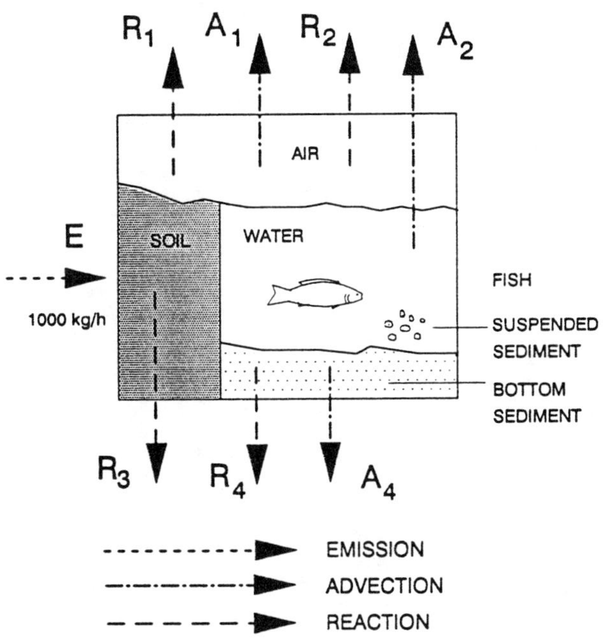

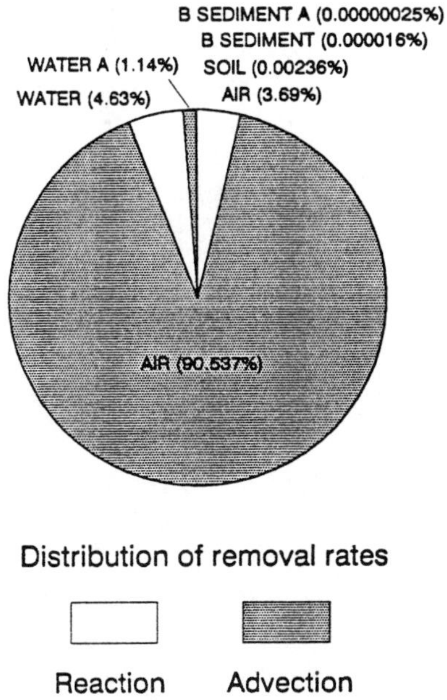

Level II Calculation:

Compartment	Half-life h	D Values Reaction mol/Pa.h	D Values Advection mol/Pa.h	Conc'n mol/m3	Loss Reaction kg/h	Loss Advection kg/h	Removal %
Air	1700	1.64E+07	4.03E+08	1.03E-08	24.291	595.884	62.018
Water	170	2.06E+08	5.06E+07	6.44E-06	304.879	74.790	37.967
Soil	170	1.05E+05		7.29E-08	0.1554		0.01554
Biota (fish)				1.85E-07			
Suspended sediment				4.56E-07			
Bottom sediment	550	7.22E+02	1.15E+01	1.46E-07	1.07E-03	1.69E-05	1.08E-04
	Total	2.23E+08	4.54E+08		329.33	670.674	100
	R + A		6.77E+08			1000	

f = 2.543E-05 Pa

Total amt = 134417 kg

Overall residence time = 134.42 h

Reaction time = 408.16 h

Advection time = 200.42 h

Fugacity Level III calculations: (four-compartment model)

Chemical name: Acetone

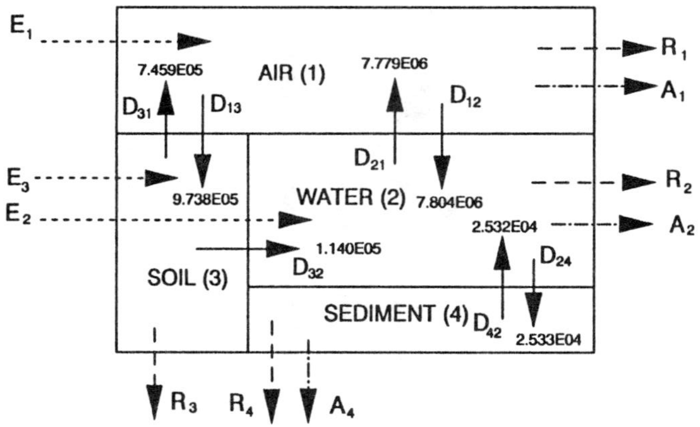

Phase Properties and Rates:

Compartment	Bulk Z mol/m3 Pa	Half-life h	D Values Reaction mol/Pa h	D Values Advection mol/Pa h
Air (1)	4.034E-04	1700	1.64E+07	4.03E+08
Water (2)	2.532E-01	170	2.06E+08	5.06E+07
Soil (3)	7.746E-02	170	5.68E+06	
Sediment (4)	2.037E-01	550	1.28E+05	2.04E+03

	E(1)=1000	E(2)=1000	E(3)=1000	E(1,2,3)
Overall residence time =	100.79	191.07	219.71	206.61 h
Reaction time =	1364.68	252.46	258.10	261.93 h
Advection time =	108.83	785.85	1477	978.26 h

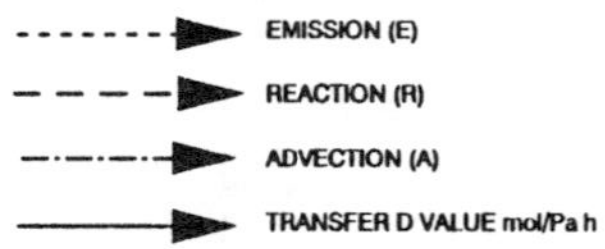

Phase Properties, Compositions, Transport and Transformation Rates:

Emission, kg/h			Fugacity, Pa				Concentration, g/m3				Amounts, kg				Total Amount, kg
E(1)	E(2)	E(3)	f(1)	f(2)	f(3)	f(4)	C(1)	C(2)	C(3)	C(4)	m(1)	m(2)	m(3)	m(4)	
1000	0	0	3.920E-05	2.588E-06	1.628E-05	1.709E-06	9.185E-07	3.806E-05	7.324E-05	2.022E-05	9.185E+04	7.611E+03	1.318E+03	1.011E+01	1.008E+05
0	1000	0	2.485E-06	6.288E-05	1.032E-06	4.152E-05	5.821E-08	9.246E-04	4.642E-06	4.912E-04	5.821E+03	1.849E+05	8.355E+01	2.456E+02	1.911E+05
0	0	1000	5.151E-06	9.533E-06	2.217E-03	6.295E-06	1.207E-07	1.402E-04	9.976E-03	7.447E-05	1.207E+04	2.804E+04	1.796E+05	3.723E+01	2.197E+05
50	250	700	6.187E-06	2.252E-05	1.553E-03	1.487E-05	1.450E-07	3.312E-04	6.988E-03	1.759E-04	1.450E+04	6.624E+04	1.258E+05	8.797E+01	2.066E+05

Emission, kg/h			Loss, Reaction, kg/h				Loss, Advection, kg/h			Intermedia Rate of Transport, kg/h T12	T21	T13	T31	T32	T24	T42
E(1)	E(2)	E(3)	R(1)	R(2)	R(3)	R(4)	A(1)	A(2)	A(4)	air-water	water-air	air-soil	soil-air	soil-water	water-sed	sed-water
1000	0	0	3.744E+01	3.103E+01	5.37E+00	1.274E-02	9.185E+02	7.611E+00	2.022E-04	4.019E+01	2.615E+00	7.349E+00	8.973E-01	1.077E+00	3.807E-02	2.513E-02
0	1000	0	2.373E+00	7.538E+02	3.41E-01	3.095E-01	5.821E+01	1.849E+02	4.912E-03	2.547E+00	6.354E+01	4.657E-01	5.687E-02	6.827E-02	9.250E-01	6.106E-01
0	0	1000	4.920E+00	1.143E+02	7.32E+02	4.692E-02	1.207E+02	2.804E+01	7.447E-04	5.281E+00	9.633E+00	9.655E-01	1.222E+02	1.467E+02	1.402E-01	9.257E-02
50	250	700	5.909E+00	2.700E+02	5.13E+02	1.108E-01	1.450E+02	6.624E+01	1.759E-03	6.343E+00	2.276E+01	1.160E+00	8.561E+01	1.028E+02	3.313E-01	2.187E-01

Level III Distribution

Chemical name: Acetone

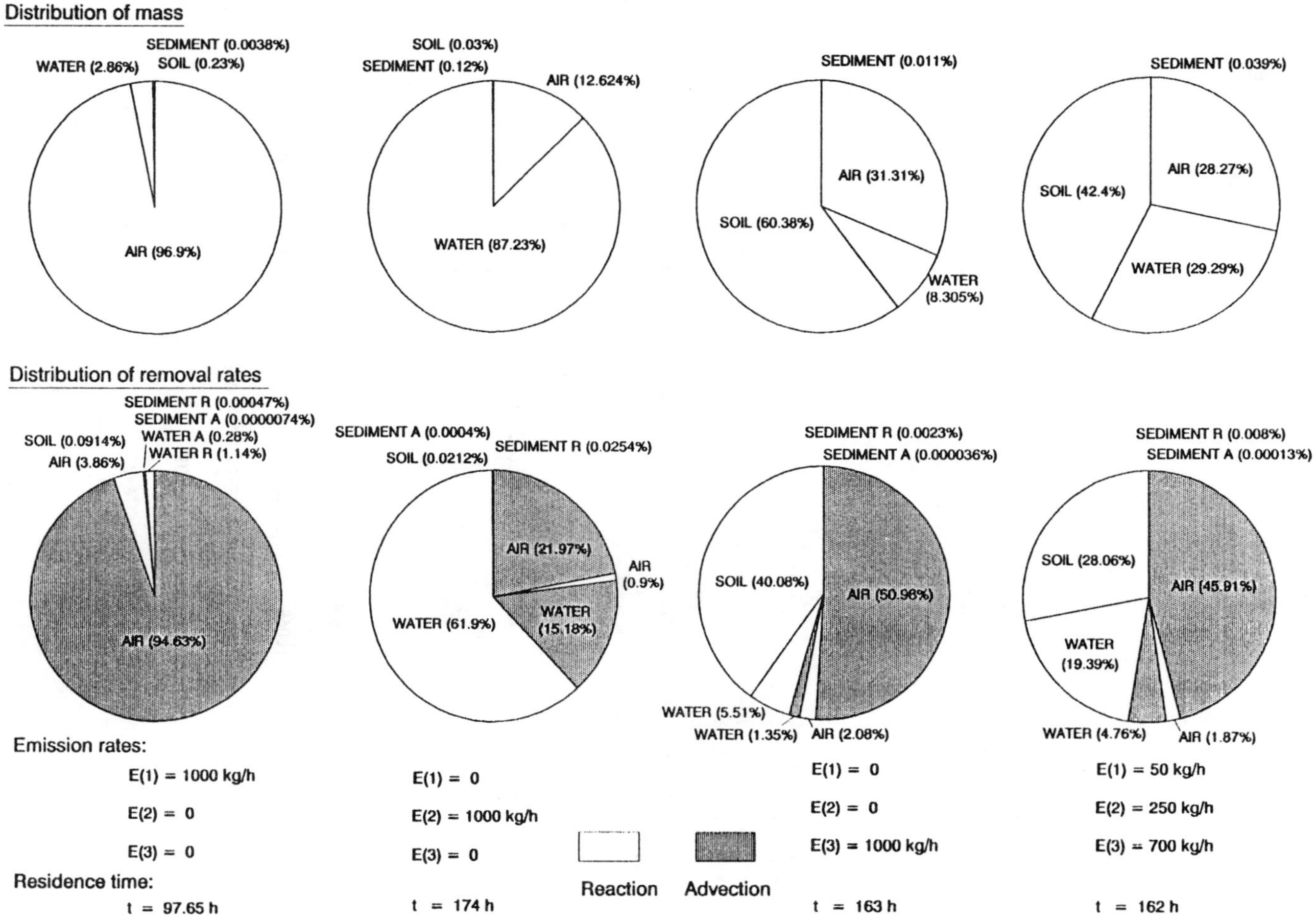

Chemical name: Methyl ethyl ketone

Level I calculation: (six-compartment model)

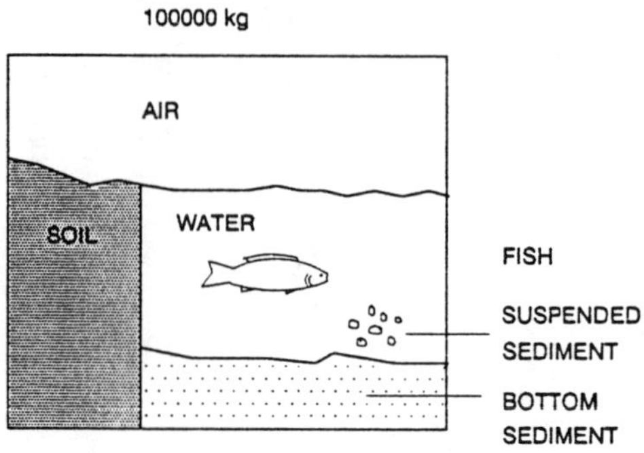

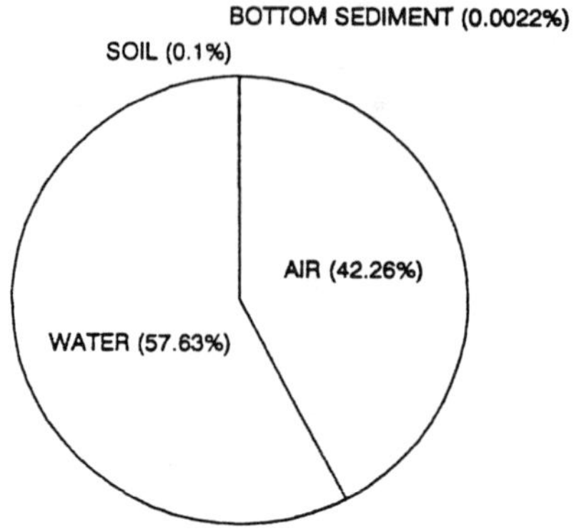

Distribution of mass

Level I calculation:

Physical-chemical properties:		Partition coeff.:	
molecular wt., g/mol	72.11	HLC	3.6355
melting point, C	-87	Kaw	1.47E-03
solubility, g/m3	240000	Koc	0.7994
vapor pressure, Pa	12100	BCF	0.0975
log Kow	0.29	Ksw	0.0384
fugacity ratio, F	1.0000	Ksd/w	0.0767
dissoc. const. pKa		Kssd/w	0.2398
		Kar/w	495.868

Compartment	Z	Concentration			Amount	Amount
	mol/m3 Pa	mol/m3	mg/L (or g/m3)	ug/g	kg	%
Air	4.03E-04	5.86E-09	4.23E-07	3.57E-04	42264.25	42.2642
Water	2.75E-01	4.00E-06	2.88E-04	2.88E-04	57633.94	57.6339
Soil	1.06E-02	1.53E-07	1.11E-05	4.61E-06	99.521	0.0995
Biota (fish)	2.68E-02	3.90E-07	2.81E-05	2.81E-05	0.00562	5.62E-06
Suspended sediment	6.60E-02	9.58E-07	6.91E-05	4.61E-05	0.0691	6.91E-05
Bottom sediment	2.11E-02	3.07E-07	2.21E-05	9.21E-06	2.2116	2.21E-03
	Total				100000	100

f = 1.453E-05 Pa

Chemical name: Methyl ethyl ketone

Level II calculation: (six-compartment model)

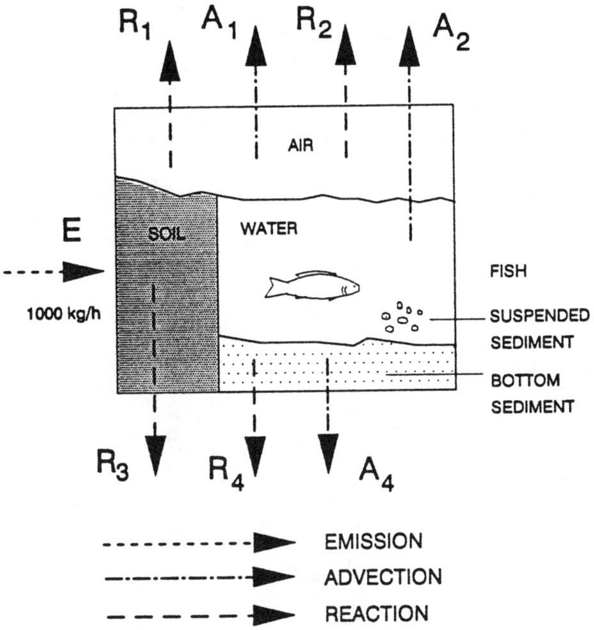

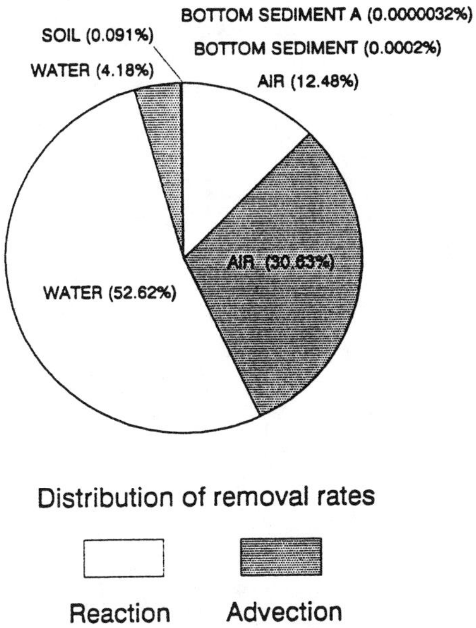

Distribution of removal rates

Level II Calculation:

Compartment	Half-life h	D Values Reaction mol/Pa.h	D Values Advection mol/Pa.h	Conc'n mol/m3	Loss Reaction kg/h	Loss Advection kg/h	Removal %
Air	170	1.64E+08	4.03E+08	4.25E-09	124.846	306.260	43.111
Water	55	6.93E+08	5.50E+07	2.90E-06	526.219	41.763	56.798
Soil	55	1.20E+06		1.11E-07	0.9087		0.09087
Biota (fish)				2.82E-07			
Suspended sediment				6.95E-07			
Bottom sediment	550	2.66E+03	4.22E+01	2.22E-07	2.02E-03	3.21E-05	2.05E-04
	Total	8.59E+08	4.58E+08		651.98	348.024	100
	R + A		1.32E+09			1000	

f = 1.053E-05 Pa

Total amt = 72463 kg

Overall residence time = 72.46 h

Reaction time = 111.14 h

Advection time = 208.21 h

Fugacity Level III calculations: (four-compartment model)

Chemical name: Methyl ethyl ketone

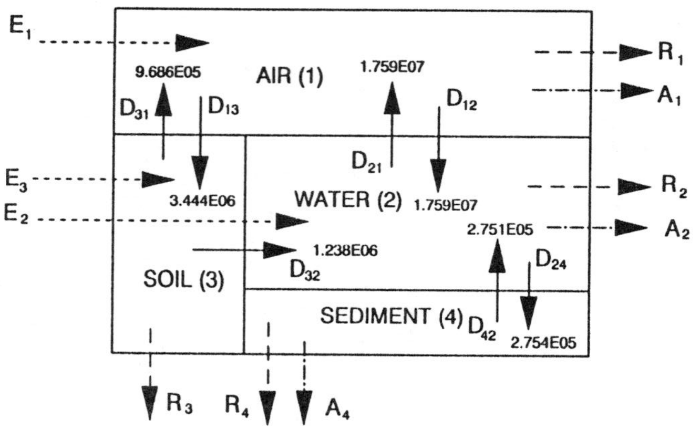

Phase Properties and Rates:

Compartment	Bulk Z mol/m3 Pa	Half-life h	D Values Reaction mol/Pa h	D Values Advection mol/Pa h
Air (1)	4.034E-04	170	1.64E+08	4.03E+08
Water (2)	2.751E-01	55	6.93E+08	5.50E+07
Soil (3)	8.788E-02	55	1.99E+07	
Sediment (4)	2.243E-01	550	1.41E+05	2.24E+03

	E(1)=1000	E(2)=1000	E(3)=1000	E(1,2,3)
Overall residence time =	71.16	73.56	78.68	77.03 h
Reaction time =	227.77	80.63	81.53	83.79 h
Advection time =	103.50	839.50	2249	953.61 h

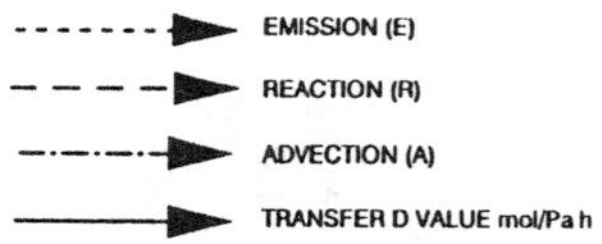

Phase Properties, Compositions, Transport and Transformation Rates:

Emission, kg/h			Fugacity, Pa				Concentration, g/m3				Amounts, kg				Total
E(1)	E(2)	E(3)	f(1)	f(2)	f(3)	f(4)	C(1)	C(2)	C(3)	C(4)	m(1)	m(2)	m(3)	m(4)	Amount, kg
1000	0	0	2.356E-05	5.555E-07	3.665E-06	3.654E-07	6.854E-07	1.102E-05	2.323E-05	5.910E-06	6.854E+04	2.204E+03	4.181E+02	2.955E+00	7.116E+04
0	1000	0	5.412E-07	1.812E-05	8.419E-08	1.192E-05	1.574E-08	3.594E-04	5.335E-07	1.928E-04	1.574E+03	7.188E+04	9.603E+00	9.639E+01	7.356E+04
0	0	1000	1.061E-06	1.038E-06	6.266E-04	6.825E-07	3.087E-08	2.058E-05	3.971E-03	1.104E-05	3.087E+03	4.116E+03	7.147E+04	5.519E+00	7.868E+04
50	250	700	2.056E-06	5.284E-06	4.388E-04	3.476E-06	5.981E-08	1.048E-04	2.781E-03	5.622E-05	5.981E+03	2.096E+04	5.005E+04	2.811E+01	7.703E+04

Emission, kg/h			Loss, Reaction, kg/h				Loss, Advection, kg/h			Intermedia Rate of Transport, kg/h T12	T21	T13	T31	T32	T24	T42
E(1)	E(2)	E(3)	R(1)	R(2)	R(3)	R(4)	A(1)	A(2)	A(4)	air-water	water-air	air-soil	soil-air	soil-water	water-sed	sed-water
1000	0	0	2.794E+02	2.777E+01	5.27E+00	3.723E-03	6.854E+02	2.204E+00	5.910E-05	3.035E+01	7.047E-01	5.851E+00	2.560E-01	3.272E-01	1.103E-02	7.250E-03
0	1000	0	6.417E+00	9.057E+02	1.21E-01	1.214E-01	1.574E+01	7.188E+01	1.928E-03	6.972E-01	2.299E+01	1.344E-01	5.880E-03	7.515E-03	3.598E-01	2.365E-01
0	0	1000	1.258E+01	5.186E+01	9.01E+02	6.954E-03	3.087E+01	4.116E+00	1.104E-04	1.367E+00	1.316E+00	2.635E-01	4.376E+01	5.593E+01	2.060E-02	1.354E-02
50	250	700	2.438E+01	2.641E+02	6.31E+02	3.542E-02	5.981E+01	2.096E+01	5.622E-04	2.649E+00	6.703E+00	5.106E-01	3.065E+01	3.917E+01	1.049E-01	6.896E-02

Level III Distribution

Chemical name: Methyl ethyl ketone

Distribution of mass

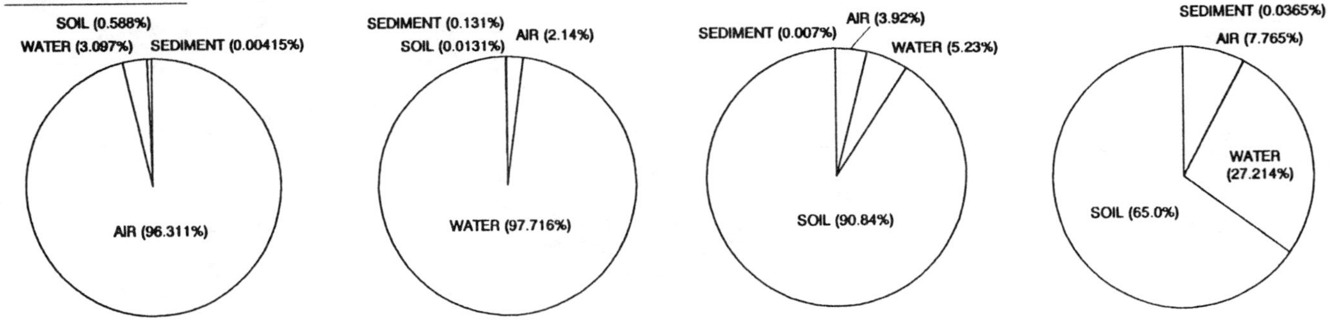

Distribution of removal rates

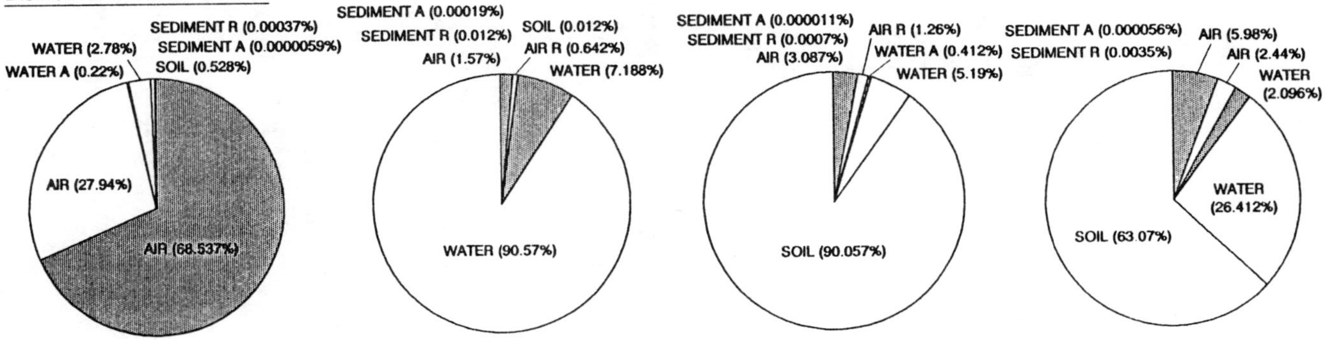

Reaction (white) Advection (shaded)

Emission rates:

E(1) = 1000 kg/h	E(1) = 0	E(1) = 0	E(1) = 50 kg/h
E(2) = 0	E(2) = 1000 kg/h	E(2) = 0	E(2) = 250 kg/h
E(3) = 0	E(3) = 0	E(3) = 1000 kg/h	E(3) = 700 kg/h

Residence time:

t = 71.16 h	t = 73.56 h	t = 76.68 h	t = 77.03 h

Chemical name: 2-Hexanone

Level I calculation: (six-compartment model)

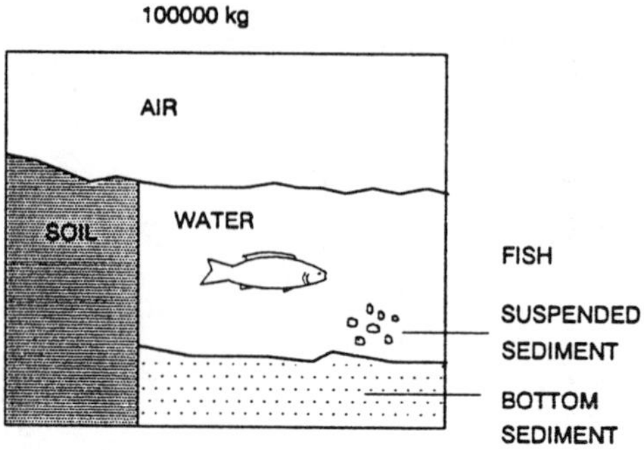

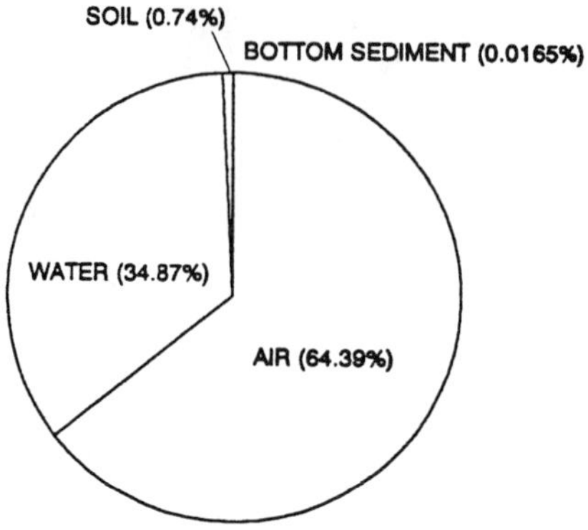

Distribution of mass

Level I calculation:

Physical-chemical properties:		Partition coeff.:	
molecular wt., g/mol	100.16	HLC	9.1575
melting point, C	-55.8	Kaw	3.69E-03
solubility, g/m3	17500	Koc	9.8352
vapor pressure, Pa	1600	BCF	1.1994
log Kow	1.38	Ksw	0.4721
fugacity ratio, F	1.0000	Ksd/w	0.9442
dissoc. const. pKa		Kssd/w	2.9506
		Kar/w	3750.000

Compartment	Z	Concentration			Amount	Amount
	mol/m3 Pa	mol/m3	mg/L (or g/m3)	ug/g	kg	%
Air	4.03E-04	6.43E-09	6.44E-07	5.43E-04	64385.67	64.3857
Water	1.09E-01	1.74E-06	1.74E-04	1.74E-04	34856.82	34.8568
Soil	5.16E-02	8.21E-07	8.23E-05	3.43E-05	740.501	0.7405
Biota (fish)	1.31E-01	2.09E-06	2.09E-04	2.09E-04	0.04181	4.18E-05
Suspended sediment	3.22E-01	5.13E-06	5.14E-04	3.43E-04	0.5142	5.14E-04
Bottom sediment	1.03E-01	1.64E-06	1.65E-04	6.86E-05	16.4556	1.65E-02
	Total				100000	100

f = 1.593E-05 Pa

Chemical name: 2-Hexanone

Level II calculation: (six-compartment model)

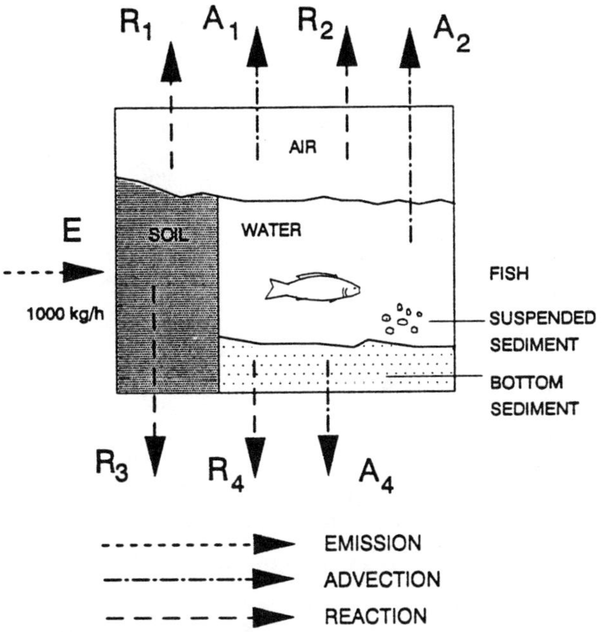

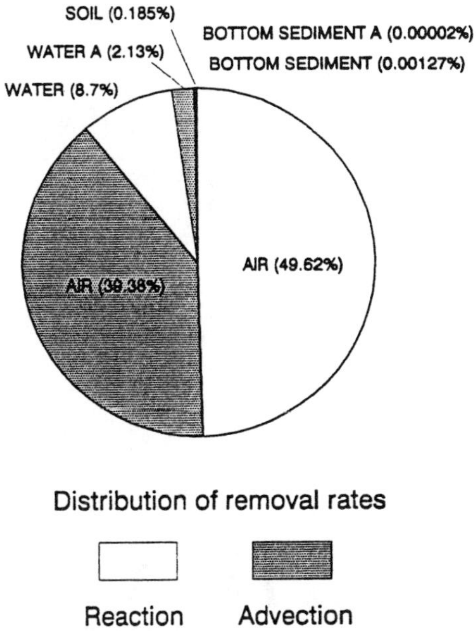

Distribution of removal rates

Level II Calculation:

Compartment	Half-life h	D Values Reaction mol/Pa.h	D Values Advection mol/Pa.h	Conc'n mol/m3	Loss Reaction kg/h	Loss Advection kg/h	Removal %
Air	55	5.08E+08	4.03E+08	3.93E-09	496.151	393.771	88.992
Water	170	8.90E+07	2.18E+07	1.06E-06	86.901	21.318	10.822
Soil	170	1.89E+06		5.02E-07	1.8461		0.18461
Biota (fish)				1.28E-06			
Suspended sediment				3.14E-06			
Bottom sediment	550	1.30E+04	2.06E+02	1.00E-06	1.27E-02	2.01E-04	1.29E-03
	Total	5.99E+08	4.25E+08		584.91	415.089	100
	R + A		1.02E+09			1000	

f = 9.745E-06 Pa

Total amt = 61158 kg

Overall residence time = 61.16 h

Reaction time = 104.56 h

Advection time = 147.34 h

Fugacity Level III calculations: (four-compartment model)

Chemical name: 2-Hexanone

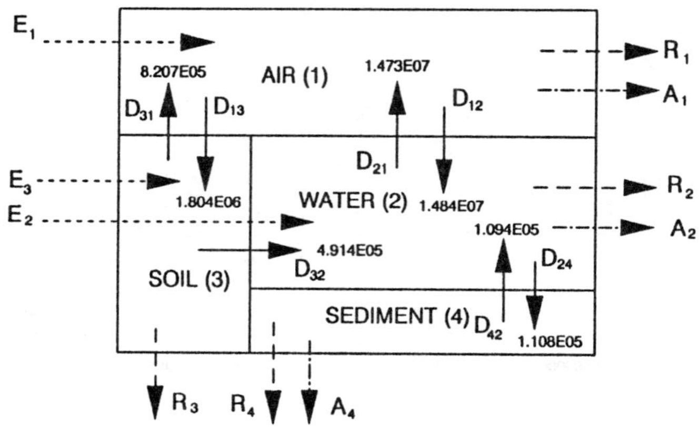

Phase Properties and Rates:

Compartment	Bulk Z mol/m3 Pa	Half-life h	D Values Reaction mol/Pa h	Advection mol/Pa h
Air (1)	4.034E-04	55	5.08E+08	4.03E+08
Water (2)	1.092E-01	170	8.90E+07	2.18E+07
Soil (3)	5.862E-02	170	4.30E+06	
Sediment (4)	1.080E-01	550	6.80E+04	1.08E+03

	E(1)=1000	E(2)=1000	E(3)=1000	E(1,2,3)	
Overall residence time =	46.73	179.57	210.52	194.59	h
Reaction time =	83.20	231.77	229.78	225.46	h
Advection time =	106.62	797.34	2512	1421.63	h

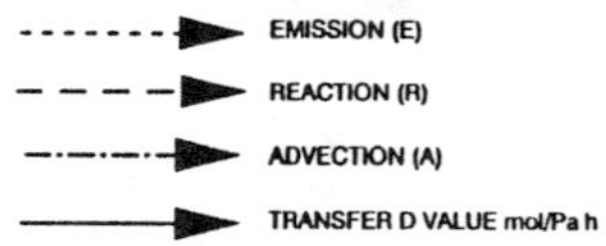

Phase Properties, Compositions, Transport and Transformation Rates:

Emission, kg/h			Fugacity, Pa				Concentration, g/m3				Amounts, kg				Total
E(1)	E(2)	E(3)	f(1)	f(2)	f(3)	f(4)	C(1)	C(2)	C(3)	C(4)	m(1)	m(2)	m(3)	m(4)	Amount, kg
1000	0	0	1.078E-05	1.286E-06	3.463E-06	7.985E-07	4.355E-07	1.407E-05	2.033E-05	8.636E-06	4.355E+04	2.814E+03	3.660E+02	4.318E+00	4.673E+04
0	1000	0	1.264E-06	7.961E-05	4.060E-07	4.942E-05	5.105E-08	8.708E-04	2.384E-06	5.345E-04	5.105E+03	1.742E+05	4.290E+01	2.672E+02	1.796E+05
0	0	1000	1.686E-06	7.158E-06	1.779E-03	4.443E-06	6.814E-08	7.830E-05	1.045E-02	4.806E-05	6.814E+03	1.566E+04	1.880E+05	2.403E+01	2.105E+05
50	250	700	2.035E-06	2.498E-05	1.246E-03	1.551E-05	8.224E-08	2.732E-04	7.314E-03	1.677E-04	8.224E+03	5.464E+04	1.316E+05	8.385E+01	1.946E+05

Emission, kg/h			Loss, Reaction, kg/h				Loss, Advection, kg/h			Intermedia Rate of Transport, kg/h T12	T21	T13	T31	T32	T24	T42
E(1)	E(2)	E(3)	R(1)	R(2)	R(3)	R(4)	A(1)	A(2)	A(4)	air-water	water-air	air-soil	soil-air	soil-water	water-sed	sed-water
1000	0	0	5.487E+02	1.147E+01	1.49E+00	5.441E-03	4.355E+02	2.814E+00	8.636E-05	1.602E+01	1.898E+00	1.947E+00	2.847E-01	1.705E-01	1.428E-02	8.750E-03
0	1000	0	6.433E+01	7.100E+02	1.75E-01	3.367E-01	5.105E+01	1.742E+02	5.345E-03	1.878E+00	1.175E+02	2.283E-01	3.337E-02	1.998E-02	8.836E-01	5.415E-01
0	0	1000	8.586E+01	6.383E+01	7.66E+02	3.028E-02	6.814E+01	1.566E+01	4.806E-04	2.506E+00	1.056E+01	3.047E-01	1.463E+02	8.758E+01	7.945E-02	4.869E-02
50	250	700	1.036E+02	2.227E+02	5.37E+02	1.056E-01	8.224E+01	5.464E+01	1.677E-03	3.025E+00	3.685E+01	3.677E-01	1.024E+02	6.132E+01	2.772E-01	1.699E-01

Level III Distribution

Chemical name: 2-Hexanone

Distribution of mass

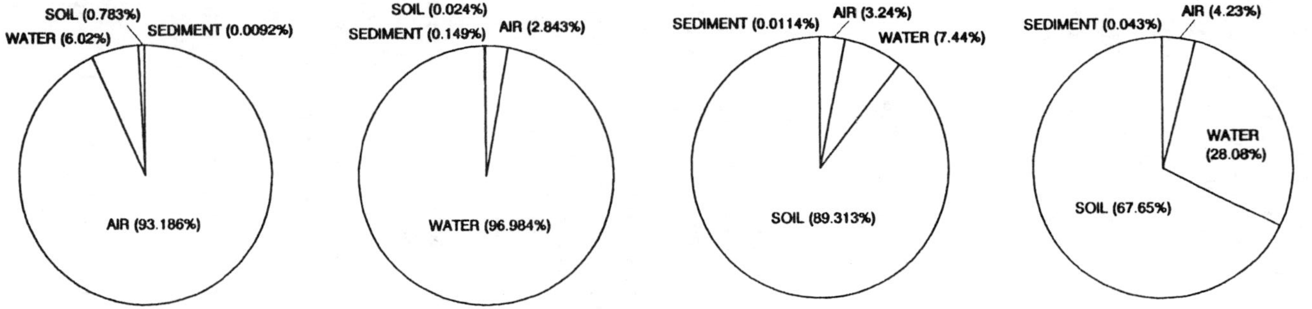

Distribution of removal rates

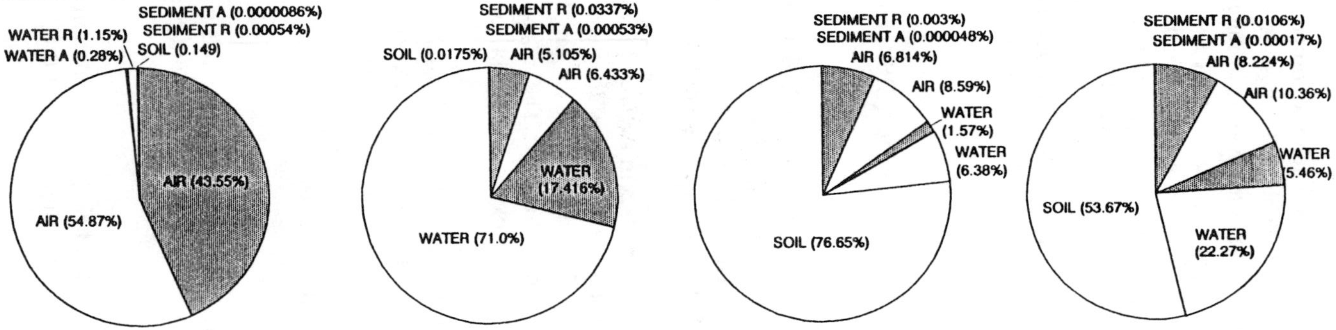

Emission rates:

E(1) = 1000 kg/h	E(1) = 0	E(1) = 0	E(1) = 50 kg/h
E(2) = 0	E(2) = 1000 kg/h	E(2) = 0	E(2) = 250 kg/h
E(3) = 0	E(3) = 0	E(3) = 1000 kg/h	E(3) = 700 kg/h

Residence time:

t = 46.73 h	t = 180 h	t = 210 h	t = 195 h

Reaction (white) Advection (shaded)

3.4 Commentary on the Physical-Chemical Properties and Environmental Fate

Properties and Reactivity

The selected physical-chemical properties of the aldehydes and ketones are given in Tables 3.1 and 3.2, while Table 3.3 gives the estimated half-lives. These half-lives should be regarded as only tentative, average values or ranges. From reviews of the literature it is suggested that air half-lives are typically class 1, i.e., less than 10 hours with an average of 5 hours. It is expected that the water, soil and sediment half-lives will be somewhat longer, i.e., Classes 3 or 4. The ketones are believed to be generally more stable, especially acetone. These carbonyl compounds are generally fairly reactive in the environment especially under oxidizing conditions, and they are susceptible to biodegradation. There are no reactivity data for many of these substances, but half-lives can probably be inferred from reported reactivity of similar compounds.

QSPR Plots

Figures 3.1*a* to 3.6*a* give the QSPR plots for aldehydes while Fig. 3.1*b* to 3.6*b* give the QSPR plots for ketones.

The solubility data for the aldehydes shown in Figure 3.2*a* show no systematic trend, partly because there are relatively few data and there is hydration of the aldehyde leading to variable non-ideality. The ketones in Figure 3.2*b* are better behaved showing a slope of about 0.61 log units or a factor of 4.1 per methylene group, a typical slope for a homologous series.

The vapor pressure data in Figures 3.3*a* and 3.3*b* show the expected drop with increasing molar volume, but again the aldehydes are more scattered. The K_{OW} values in Figure 3.4*a* and 3.4*b* show a systematic increase with chain length of typically of 0.54 log units or a factor of 3.5 per methylene for the ketones. The K_{OW} values are fairly low, thus relatively little bioconcentration in lipids is expected, except for the longer chain ketones.

The Henry's law constants in Figure 3.5*a* and 3.5*b* are quite scattered in the case of the aldehydes, probably because of variable interactions in the aqueous phase. The ketones show a steady increase in H with chain length, but the slope is low and comparable to that of the alcohols, i.e., about 0.15 log units of a factor of 1.4 per methylene group. The H values of the ketones are about a factor of 10 larger than those of the corresponding alcohols primarily because of the higher vapor pressures.

As with any fairly soluble compounds it is best to use experimental data for H since the ratio of vapor pressure to solubility may not accurately express the real air-water partitioning characteristics.

Finally, the plots of K_{OW} versus solubility in Figure 3.6a for aldehydes is scattered while for ketones in Fig. 3.6b the slope is about 0.87.

These plots can be used to estimate the properties of other carbonyl compounds, but such estimation can be in considerable error in the case of aldehydes which are fundamentally less amenable to QSPR analysis.

Evaluative Calculations

The aldehydes, having relatively low K_{OW} values will partition primarily into air and water with relatively little in soil and sediment. Most reaction is in these media, especially in air for which the half-life is relatively short. Half-lives in soil and sediment are, in the context of Level II calculations relatively unimportant.

The Level III diagrams suggest that the aldehydes are not very susceptible to intermedia transport. They tend to remain in, and react in the medium of discharge. Rates of evaporation, absorption or run-off are fairly slow. This is largely because the intermedia transport D values (as controlled by quantities such as mass transfer coefficients) are small compared to the reaction D values. Advection is usually relatively slow compared to reaction. The implication is that aldehydes will tend to present only relatively localized contamination problems of fairly short duration. They are rarely susceptible to long range transport.

The ketones also tend to associate with air and water in proportions controlled by Henry's law constant. Slower reactivity tends to increase the rate of loss by advection thus contamination may extend over a larger area than is the case for aldehydes. A consequence of slower reactions is that they are more susceptible to intermedia transport expecially when discharged to soil from which evaporation and run-off may be appreciable. Air-water exchange can also be significant. The ketones are thus more mobile in the environment than the aldehydes.

The relatively fast reaction with half-lives, generally of some days or weeks, results in relatively short persistence of some hundreds of hours, thus they are not persistent compounds in the same class as chlorinated hydrocarbons. Their effects will tend to be quite localized.

3.5 REFERENCES

Abernethy, S.G., Mackay, D., McCarty, L.S. (1988) "Volume fraction" correlation for narcosis in aquatic organisms: The key role of partitioning. *Environ. Toxicol. Chem.* 7, 469-481.

Abraham, M.H. (1984) Thermodynamics of solution of homologous series of solutes in water. *J. Chem. Soc., Farad. Trans. 1* 80, 153-181.

Ambrose, D., Connett, J.E., Green, J.H.S., Hales, J.L., Head, A.J., Martin, J.F. (1975) Thermodynamic properties of organic oxygen compounds. 42. Physical and thermodynamic properties of benzaldehyde. *J. Chem. Thermodyn.* 7, 1143-1157.

Ambrose, D., Ellender, J.H., Lees, E.B., Sprake, C.H.S., Townsend, R. (1975) Thermodynamic properties of organic oxygen compounds. XXXVIII. Vapour pressures of some aliphatic ketones. *J. Chem. Thermodyn.* 7, 453-472.

Ambrose, D., Ghiassee, N.B., Tuckerman, R. (1988) Vapour pressure and critical temperature and critical pressure of 4-methylpentan-2-one. *J. Chem. Thermodyn.* 20, 767-768.

Amidon, G.L., Williams, N.A. (1982) A solubility equation for non-electrolytes in water. *Int'l J. Pharm.* 11, 249-256.

Amidon, G.L., Yalkowsky, S.H., Anik, S.T., Valvani, S.C. (1975) Solubility of nonelectrolytes in polar solvents. V. Estimation of the solubility of aliphatic monofunctional compounds in water using a molecular surface area approach. *J. Phys. Chem.* 79(21), 2239-2246.

Anbar, M., Neta, P. (1967) A compilation of specific bimolecular rate and hydroxyl radical with inorganic and organic compounds in aqueous solution. *Int'l. J. Appl. Radiation Isotopes* 18, 493-523.

Anderson, T.A., Beauchamp, J.J., Walton, B.T. (1991) Organic chemicals in the environment. *J. Environ. Qual.* 20, 420-424.

Andrews, L.J., Keefer, R.M. (1950) Cation complexes of compounds containing carbon-carbon double bonds. VI. The argentation of substituted benzenes. *J. Am. Chem. Soc.* 72, 3113-3116.

Ashworth, R.A., Howe, G.B., Mullins, M.E., Rogers, T.N. (1988) Air-water partition coefficients of organics in dilute solutions. *J. Haz. Materials* 18, 25-36.

Atkinson, R. (1985) Kinetics and mechanisms of the gas-phase reactions of hydroxyl radical with organic compounds under atmospheric conditions. *Chem. Rev.* 85, 69-201.

Atkinson, R. (1987) A structure-activity relationship for the estimation of rate constants for the gas-phase reactions of OH radicals with organic compounds. *Int'l J. Chem. Kinetics* 19, 799-828.

Atkinson, R. (1988) Estimation of gas-phase hydroxyl radical rate constants for organic chemicals. *Environ. Toxicol. Chem.* 7, 435-442.

Atkinson, R. (1990) Gas-phase tropospheric chemistry of organic compounds: A review. *Atmos. Environ.* 24A(1), 1-41.

Atkinson, R. (1991) Kinetics and mechanisms of the gas-phase reactions of NO_3 radical with organic compounds. *J. Phys. Chem. Ref. Data* 20(3), 459-507.

Atkinson, R., Aschmann, S.M., Carter, W.P.L., Pitts, J.N. Jr. (1982) Rate constants for the gas-phase reaction of hydroxyl radicals with a series of ketones at 299 $\pm$ 2 K. *Int'l J. Chem. Kinet.* 14, 839-847.

Atkinson, R., Aschmann, S.M., Fitz, D.R., Winer, A.M., Pitts, J.N. Jr. (1982) Rate constants for the gas-phase reactions of O_3 with selected organics at 296 K. *Int'l J. Chem. Kinet.* 14, 13-18.

Atkinson, R., Aschmann, S.M., Pitts, J.N. Jr. (1988) Rate constants for the gas-phase reactions of NO_3 radicals with a series of organic compounds at 296 $\pm$ 2 K. *J. Phys. Chem.* 92, 3454-3457.

Atkinson, R., Aschmann, S.M., Winer, A.M. (1987) Kinetics of the reactions of NO_3 radicals with a series of aromatic compounds. *Environ. Sci. Technol.* 21(11), 1123-1126.

Atkinson, R., Aschmann, S.M., Winer, A.M., Pitts, J.N. Jr. (1981) Rate constants for the gas-phase reactions of ozone with a series of carbonyls at 296 K. *Int'l J. Chem. Kinet.* 13(11), 1133-1142.

Atkinson, R., Aschmann, S.M., Winer, A.M., Pitts, J.N. Jr. (1982) Rate constants for the reactions of hydroxyl radicals with a series of alkanes and alkenes at 299 $\pm$ 2 K. *Int'l J. Chem. Kinet.* 14, 507-516.

Atkinson, R., Carter, W.P.L. (1984) Kinetics and mechanisms of the gas-phase reactions of ozone with organic compounds under atmospheric conditions. *Chem. Rev.* 84, 437-470.

Atkinson, R., Carter, W.P.L., Plum Ch. N., Winer, A.M., Pitts, J.N. Jr. (1984) Kinetics of the gas-phase reactions of the NO_3 radicals with a series of aromatics at 296 $\pm$ 2 K. *Int'l J. Chem. Kinet.* 16, 887-898.

Atkinson, R., Lloyd, A.C. (1984) Evaluation of kinetic and mechanistic data for modeling of photochemical smog. *J. Phys. Chem. Ref. Data* 13, 315-444.

Audley, G.J., Baulch, D.L., Campbell, I.M. (1981) Gas-phase reactions of hydroxyl radicals with aldehydes in flowing hydrogen peroxide + nitrogen dioxide + carbon monoxide mixtures. *J. Chem. Soc. Faraday Trans* 1,77, 2541-2549.

Banerjee, S. (1985) Calculation of water solubility of organic compounds with UNIFAC-derived parameters. *Environ. Sci. Technol.* 19, 369-370.

Banerjee, S., Howard, P.H. (1988) Improved estimation of solubility and partitioning through correction of UNIFAC-derived activity coefficients. *Environ. Sci. Technol.* 22, 839-841.

Banerjee, S., Howard, P.H., Lande, S.S. (1990) General structure-vapor pressure relationships for organics. *Chemosphere* 21(10/11), 1173-1180.

Barnes, I., Becker, K.H., Fink, E.H., Zabel, F. (1982) The reaction of HO_2 radicals with aldehydes and some sulfur containing molecules. In: *Physical-Chemical Behaviour of Atmospheric Pollutants.* Report on Discussion Meeting held at University College, Dublin, October 5-6, 1982. pp.20-26.

Barrows, M.E., Petrocelli, S.R., Macek, K.J. (1980) Bioconcentration and elimination of selected water pollutants by bluegill sunfish *(lepomis machrochirus).* In: *Dynamic, Exposure, Hazard Assessment Toxic Chemicals.* Haque, R., Editor, pp.379-392, Ann Arbor Science Publishers, Inc., Ann Arbor, Michigan.

Baulch, D.L., Cox, R.A., Hampson, R.F., Jr., Kerr, J.A., Troe, J., Watson, R.T. (1984) Evaluated kinetic and photochemical data for atmospheric chemistry (Supplement II). *J. Phys. Chem. Ref. Data* 13, 1255-1380.

Berthod, A., Han, Y.I., Armstrong, D.W. (1988) Centrifugal partition chromatography. V. Octanol-water partition coefficients, direct and indirect determination. *J. Liq. Chromatogr.* 11(7), 1441-1456.

Berti, P., Cabani, S., Conti, G., Mollica, V. (1986) Thermodynamic study of organic compounds in octan-1-ol. *J. Chem. Soc., Faraday Trans.* 1 82, 2547.

Betterton, E.A. (1991) The partitioning of ketones between the gas and aqueous phases. *Atmos. Environ.* 25A, 1473-1477.

Betterton, E.A., Hoffmann, M.R. (1988) Henry's law constants of some environmental important aldehydes. *Environ. Sci. Technol.* 22, 1415-1418.

Bintein, S., Devillers, J. (1994) QSAR for organic chemical sorption in soils and sediments. *Chemosphere* 28(6), 1171-1188.

Bodor, N., Gabanyi, Z., Wong, C.K. (1989) A new method for the estimation of partition coefficient. *J. Am. Chem. Soc.* 111, 3783-3786.

Booth, H.S., Everson, H.E. (1948) Hydrotropic solubilities. Solubilities in 40 per cent sodium xylenesulfonate. *Ind. & Eng. Chem.* 40(8), 1491-1493.

Boublik, T., Fried, V., Hala, E. (1973) *The Vapour Pressures of Pure Substances. Elsevier, Amsterdam, The Netherlands.*

Boublik, T., Fried, V., Hala, E. (1984) The Vapour Pressures of Pure Substances, 2nd ed., Elsevier, Amsterdam, The Netherlands.

Bowmer, K.H., Higgins, M.L. (1976) Some aspects of the persistence and fate of acrolein from water by volatilization and degradation. *Arch. Environ. Contam. Toxicol.* 5, 87-96.

Brazo, J.F., Coca, C.M. (1989) Relation between the partition coefficient (Log P_{OW}) and capacity factor (Log k') derived from reversed-phase HPLC of a series of benzaldehyde derivatives. *Cienc. Ind. Farm.* 8(10-12), 223-226.

Bridie, A.L., Wolff, C.J.M., Winter, M. (1979) BOD and COD of some petrochemicals. *Water Res.* 13, 627-630.

Briggs, G.G. (!981) Theoretical and experimental relationships between soil adsorption, octanol-water partition coefficients, water solubilities, bioconcentration factors, and the parachor. *J. Agric. Food & Chem.* 29, 1050-1059.

Brooke, D.N., Doobs, A.J., Williams, N. (1986) Octanol/water partition coefficients P: measurement, estimation, and interpretation particularly for chemicals with $P > 10^6$. *Ecotoxicol. Environ. Safety* 11, 251-260.

Burnett, M.G. (1963) Determination of partition coefficients at infinite dilution by gas chromatographic analysis of vapor above dilute solutions. *Anal. Chem.* 35, 1567-1569.

Burnett, M.G., Swodoba, P.A.T. (1962) A simple method for the calibration of sensitive gas chromatographic detectors. *Anal. Chem.* 34, 1162-1163.

Butler, J.A.V., Ramchandani, C.N. (1935) The solubilities of nonelectrolytes. Part II. The influence of the polar group on the free energy of hydration of aliphatic compounds. *J. Chem. Soc.* 952-955.

Buttery, R.G., Bomben, J.L., Guadagni, D.G., Ling, L.C. (1971) Some considerations of the volatilities of organic flavor compounds in foods. *J. Agric. Food Chem.* 19, 1045-1048.

Buttery, R.G., Ling, L.C., Guadagni, D.G. (1969) Food volatiles. Volatilities of aldehydes, ketones, and esters in dilute water solution. *J. Agric. Food & Chem.* 17, 385-389.

Callahan, M.A., Slimak, M.W., Gabel, N.W., May, I.P., Fowler, C.F., Freed, J.R., Jennings, P., Durfee, R.L., Whitemore, F.C., Maestri, B., Mabey, W.R., Holt, B.R., Gould, C. (1979) *Water Related Environmental Fate of 129 Priority Pollutants.* U.S. EPA-440-4-79-029a,b. Versar, Inc., Springfield, Virginia.

Calvert, J.G., Demeyan, K.L., Kerr, J.A., McQuigg, R.D. (1972) Photolysis of formaldehyde as a hydrogen atom source in the lower atmosphere. *Science* 175, 751-752.

Carlier, P., Hannachi, H., Mouvier, G. (1986) The chemistry of carbonyl compounds in the atmosphere-A review. *Atmos. Environ.* 20(11), 2079-2099.

Carter, W.P.L., Winer, A.M., Pitts, J.N. Jr. (1981) Major atmospheric sink for phenol and the cresols. Reaction with nitrate radical. *Environ. Sci. Technol.* 15, 829-831.

Chao, J., Lin, C.T., Chung, T.H. (1983) Vapor pressure of coal chemicals. *J. Phys. Chem. Ref. Data* 12(40), 1034-1063.

Chapman, J.D., Raleigh, J.A., Borsa, J., Webb, R.G., Whitehorse, R. (1972) Radiosensitization of mammalin cells by *p*-nitroacetophenone. II. Effectiveness of analogs. *Int'l J. Radiat. Biol.* 21, 475-482.

Chen, F., Holten-Andersen, J., Tyle, H. (1993) New developments of the UNIFAC model for environmental application. *Chemosphere* 26(7), 1325-1354.

Chessells, M., Hawker, D.W., Connell, D.W. (1992) Influence of solubility in lipid on bioconcentration of hydrophobic compounds. *Ecotoxicol. Environ. Safety* 23, 260-273.

Chiorboli, C., Bignozzi, C.A., Maldotti, A., Giardini, P.F., Rossi, A., Carassiti, V. (1983) Rate constants for the gas-phase reactions of hydroxyl radicals with β,β-dimethyl styrene and acetone. Mechanism of β,β-dimethyl styrene nitrogen oxide (NO_x)-air photooxidation. *Int'l J. Chem. Kinet.* 15, 579-586.

Chiou, C.T. (1985) Partition coefficients of organic compounds in lipid water system and correlations with fish bioconcentration factors. *Environ. Sci. Technol.* 19, 57-62.

Collander, R. (1951) Partition of organic compounds between higher alcohols and water. *Acta. Chem. Scand.* 5, 774-780.

Cox, R.A., Patrick, K.F., Chant, S.A. (1981) Mechanism of atmospheric photooxidation of organic compounds. Reactions of alkoxy radicals in oxidation of *n*-butanol and simple ketones. *Environ. Sci. Technol.* 15, 587-592.

Dallos, A., Wienke, G., Ilchman, A., Gmehlong, J. (1993) Vauausberechnung von octanol/wasser-vertilungskoeffizienten mit hilfe der UNIFAC-method. *Chem.-Ing.-Tech.* 65(2), 201-203.

Darnall, K.R., Lloyd, A.C., Winer, A.M., Pitts, Jr., J.N. (1976) Reactivity scale for the atmospheric hydrocarbons based on reaction with hydroxyl radical. *Environ. Sci. Technol.* 10(7), 692-696.

Daubert, T.E., Danner, R.P. (1989) *Physical and Thermodynamic Properties of Pure Chemicals: Data Compilation*. Hemisphere Publication Corporation, New York.

Dean, J. (1985) *Lange's Handbook of Chemistry.* 13th ed., McGraw-Hill, New York.

Deneer, J.W., Seinen, W., Hermens, J.L.M. (1988) The acute toxicity of aldehydes to the guppy. *Aquatic Toxicol.* 12, 185-192.

Deno, N.C., Berkheimer, H.E. (1960) Phase equilibria molecular transport thermodynamics: activity coefficients as a function of structure and media. *J. Chem. Eng. Data* 5, 1-5.

Djerki, R.A., Laub, R.J. (1988) Solute retention in column liquid chromatography. X. Determination of solute infinite-dilution activity coefficients in methanol, water, and their mixtures, by combined gas-liquid chromatography and liquid-liquid chromatography. *J. Liq. Chromatogr.* 11(3), 585-612.

Dojlido, J., Stodja, A., Gantz, E., Kowaslski, J. (1984) Biodegradation of methyl ethyl ketone, cyclohexanone, and cyclohexanol in sewage and surface water. *Arch. Ochr. Srodowiska.* 1, 115-123.

Donahue, D.J., Bartell, F.E. (1952) The boundary tension at water-organic liquid interfaces. *J. Phys. Chem.* 56, 480-484.

Dong, S., Dasgupta, P.K. (1986) Solubility of gaseous formaldehyde in liquid water and generation of trace standard gaseous formaldehyde. *Environ. Sci. Technol.* 6(20), 637-640.

Dore, M., Brunet, N., Legube, B. (1975) Participation of various organic compounds in the evaluation of global pollution critiria. *Trib. Cebedeau.* 28, 3-11.

Dorfman, L.M., Adams, G.E. (1973) *Reactivity of the Hydroxyl radical in Aqueous Solution.* NSRD-NDB-46. NTIS COM-73-50623. National Bureau of Standards, Washington D.C. 51 pp.

Eadsforth, C.V. (1986) Application of reverse-phase HPLC for the determination of partition coefficients. *Pest. Sci.* 17(3), 311-325.

Eadsforth, C.V., Moser, P. (1983) Assessment of reversed phase chromatographic methods for determining partition coefficients. *Chemosphere* 12, 1459-1475.

El Tayar, N., Tsai, R.-S., Vallat, P., Altomare, C., Testa, B. (1991) Measurement of partition coefficients by various centrifugal partition chromatographic techniques. *J. Chromatogr.* 556, 181-194.

Erichsen, R. (1952) *Naturwissenschaften* 39, 41-42.

Foote, C.S. (1976) *Free Radicals in Biology.* Academic Press, Inc., New York.

Fujita, T., Iwasa, J., Hansch, C. (1964) A new substituent constant derived from partition coefficients. *J. Am. Chem. Soc.* 86(23), 5175-5180.

Gaffney, J.S., Streit, G.E., Spall, W.D., Hall, J.H. (1987) Beyond acid rain. Do soluble oxidants and organic toxins interact with SO_2 and NO_2 to increase ecosystem effects? *Environ. Sci. Technol.* 21(6), 519-524.

Garst, J.E., Wilson, W.C. (1984) Accurate, wide-range, automated, high-performance liquid chromatographic method for the estimation of octanol/water partition coefficients. I: Effects of chromatographic conditions and procedure variables on accuracy and reproducibility of the method. *J. Pharm. Sci.* 73, 1616-1622.

Ge, J., Liu, W., Dong, S. (1987) Determination of partition coefficient with chemically bonded omega-hydroxysilica as HPLC column packing. *Sepu* 5, 182.

Gellman, I., Heukelekian, H. (1950) Biological oxidation of formaldehyde. *Sew. Indust. Waste* 22, 13-21.

Gerhold, R.M., Malaney, G.W. (1966) Structural determinants in the oxidation of aliphatic compounds by activated sludge. *J. Water Pollut. Control Fed.* 38, 562-579.

Gerstl, Z., Helling, C.S. (1987) Evaluation of molecular connectivity as a predictive method for the adsorption of pesticides in soil. *J. Environ. Sci. Health* B22(1), 55-69.

Ginnings, P.M., Plonk, D., Carter, E. (1940) Aqueous solubilities of some aliphatic ketones. *J. Am. Chem. Soc.* 62, 1923-1924.

Grosjean, D. (1985) Wall loss of gaseous pollutants in outdoor Teflon chambers. *Environ. Sci. Technol.* 19, 1059-1065.

Gross, P., Rintelen, J.C., Saylor, J.H. (1939) Energy and volume relations in the solubilities of some ketones in water. *J. Phys. Chem.* 43, 197-205.

Gross, P.M., Saylor, J.H., Gorman, M.A. (1933) Solubilities studies. IV. The solubilities of certain slightly soluble organic compounds in water. *J. Am. Chem. Soc.* 55, 650-652.

Günther, F.A., Westlake, W.E., Jaglan, P.S. (1968) Reported solubilities of 738 pesticide chemicals in water. *Residue Rev.* 20, 1-148.

Haky, J.E., Young, A.M. (1984) Evaluation of a simple correlation method for the estimation of octanol-water partition coefficients of organic compounds. *J. Liq. Chromatogr.* 7, 675-689.

Hammers, W.E., Meurs, G.J., DeLigny, C.L. (1982) Correlations betweeen liquid chromatographic capacity factors ratio data on Lichrosorb RP-18 and partition coefficients in the octanol-water system. *J. Chromatogr.* 247, 1-13.

Hanai, T., Hubert, J. (1982) Hydrophobicity and chromatographic behaviour of aromatic acids found in urine. *J. Chromatogr.* 239, 527-536.

Hanai, T., Tran, C., Hubert, J. (1981) An approach to the prediction of retention times in liquid chromatography. *J. HRC & CC* 4, 454-460.

Hansch, C., Anderson, S. (1967) The effect of intermolecular hydrophobic bonding on partition coefficients. *J. Org. Chem.* 32, 2583.

Hansch, C., Leo, A. (1979) *Substituent Constants for Correlation Analysis in Chemistry and Biology.* John-Wiley & Sons, New York.

Hansch, C., Leo, A. (1983) *Medchem. Project*, Pomona College, Claremont, California.

Hansch, C., Leo, A. (1985) *Medchem. Project*, Pomona College, Claremont, California.

Hansch, C., Leo, A., Nikaitani, D. (1972) On the additive-constitutive character of partition coefficients. *J. Org. Chem.* 37, 3090-3092.

Hansch, C., Quinlan, J.E., Lawrence, G.L. (1968) The linear free-energy relationship between partition coefficients and aqueous solubility of organic liquids. *J. Org. Chem.* 33, 347-350.

Hawthorne, S.B. (1984) *Ph.D. Thesis.* University of Colorado, Boulder, Colorado.

Hawthorne, S.B., Sievers, R.E., Barkley, R.M. (1985) Organic emissions from shale oil wastewaters and their implications for air quality. *Environ. Sci. Technol.* 19, 992-997.

Hendry, D.G., Kenley, R.A. (1979) *Atmospheric Reaction Products of Organic Compounds.* EPA Final Report, EPA 560/12-79-001.

Hendry, D.G., Mill, T., Piszkiewicz, L., Howard, J.A., Eigenmann, H.K. (1974) Critical review of hydrogen-atom transfer in the liquid phase. Chlorine atom, alkylchloromethyl, alkoxy, and alkyl peroxy radicals. *J. Phys. Chem. Ref. Data* 3, 937-978.

Heukelekian, H., Rand, M.C. (1955) Biochemical oxygen demand for pure organic compounds. *J. Water Pollut. Control Assoc.* 29, 1040-1053.

Hine, J., Mookerjee, P.K. (1975) The intrinsic hydrophilic character of organic compounds. Correlations in term of structural contributions. *J. Org. Chem.*40, 292-298.

Hoigné, J., Bader, H. (1983) Rate constants of reactions of ozone with organic and inorganic compounds in water-I. Non-dissociating organic compounds. *Water Res.* 17, 173-183.

Holmes, H.L., Lough, C.E. (1976) *Effect of Intramolecular Hydrogen bonding on Partition Coefficients.* Suffield Technical Note No. DRES-TN-365. Defence Research Establishment Suffield-Information Canada. U.S. NTIS Report No. AD-A030683.

Howard, J.A. (1972) Absolute rate constants for reactions of oxy radicals. *Adv. Free Radical Chem.* 4, 49-173.

Howard, P.H., Ed. (1989) *Handbook of Fate and Exposure Data for Organic Chemicals. Vol. I-Large Production and Priority Pollutants.* Lewis Publishers, Inc., Chelsea, Michigan.

Howard, P.H., Ed. (1990) *Handbook of Fate and Exposure Data for Organic Chemicals. Vol. II, Solvents.* Lewis Publishers, Inc., Chelsea, Michigan.

Howard, P.H., Ed. (1993) *Handbook of Fate and Exposure Data for Organic Chemicals. Vol. IV, Solvents 2.* Lewis Publishers, Inc., Chelsea, Michigan.

Howard, P.H., Boethling, R.S., Jarvis, W.F., Meylan, W.M., Michalenko, E.M., Editors (1991) *Handbook of Environmental Degradation Rates.* Lewis Publishers Inc., Chelsea, Michigan.

Howard, P.H., Hueber, A.E., Mulesky, B.C., Crisman, J.S., Meylan, W., Crosby, E., Gray, D.A., Sage, G.W., Howard, K.P., LaMacchia, A., Boethling, R.S., Troast, R. (1986) BIOLOG, BIODEG, and FATE/EXPOS: New files on microbial degradation and toxicity as well as environmental fate/exposure of chemicals. *Environ. Toxicol. & Chem.* 5, 977-988.

Hoy, K.L. (1970) New values of the solubility parameters from vapor pressure data. *J. Paint Technol.* 42(541) 76-118.

Hulzebos, E.M., Adema, D.M.M., D.-van Breemen, E.M., Henzen, L., van Dis, W.A., Herbold, H.A., Hoekstra, J.A., Baerselman, E.R., van Gestel, C.A.M. (1993) Phytotoxicity studies with *lactuca sativa* in soil and nutrient solution. *Environ. Toxicol. Chem.* 12, 1079-1094.

Isnard, P., Lambert, S. (1989) Aqueous solubility/*n*-octanol water partition coefficient correlations. *Chemosphere* 18, 1837-1853.

Iwasa, J., Fujita, T., Hansch, C. (1965) Substituent constants for aliphatic functions obtained from partition coefficients. *J. Med. Chem.* 8, 150-153.

Iwase, K., Komatsu, K., Hirono, S., Nakagawa, S., Moriguchi, I. (1985) Estimation of hydrophobicity based on the solvent-accessible surface area of molecules. *Chem. Pharm. Bull.* 33(5), 2114-2121.

Jaworska, J.S., Schultz, T.W. (1993) Quantitative relationships of structure-activity and volume fraction for selected nonpolar and polar narcotic chemicals. *SAR & QSAR in Environ. Res.* 1, 3-19.

Jones, D.C. (1929) The systems *n*-butyl alcohol-water and *n*-butyl alcohol-acetone-water. *J. Chem. Soc.* (London) 799-816.

Kahlbaum, G.W.A. (1898) Studien über dampfspannkraftmessungen. II. *Z. Phys. Chem.* 26, 577-658.

Kamlet, M.J., Abraham, D.J., Doherty, R.M., Taft, R.W., Abraham, M.H. (1986) Solubilities in polymers and biological media: 6. An equation for correlation and prediction of solubilities of liquid organic nonelectrolytes in blood. *J. Pharm. Sci.* 75(4), 350-355.

Kamlet, M.J., Doherty, R.M., Abboud, J.-L.M., Abraham, M.H., Taft, R.W. (1986) Linear solvation energy relationships. 36. Molecular properties governing solubilities of organic nonelectrolytes in water. *J. Pharm. Sci.* 75(4), 338-349.

Kamlet, M.J., Doherty, R.M., Abraham, M.H., Carr, P.W., Doherty, R.F., Taft, R.W. (1987) Linear solvation energy relationships. 41. Important differences between aqueous solubility relatioships for aliphatic and aromatic solutes. *J. Phys. Chem.* 91, 1996-2004.

Kamlet, M.J., Doherty, R.M., Abraham, M.H., Marcus, Y., Taft, R.W. (1988) Linear solvation energy relationships. 46. An improved equation for correlation and prediction of octanol/water partition coefficients of organic nonelectrolytes (including strong hydrogen bond donor solutes). *J. Phys. Chem.* 92, 5244-5255.

Kamlet, M.J., Doherty, R.M., Taft, R.W., Abraham, M.H., Veith, G.D., Abraham, D.J. (1986) Solubilities in polymers and biological media: 8. An analysis of the factors that influence toxicities of organic nonelectrolytes to the golden orfe fish *(leuciscus idus melanotus) Environ. Sci. Technol.* 21(2), 149-155.

Kelly, T.J., Mukund, R., Spicer, C.W., Pollack, A.J. (1994) Concentrations and transformations of hazardous air pollutants. *Environ. Sci. Technol.* 28, 378A-387A.

Kerr, J.A., Sheppard, D.W. (1981) Kinetics of the reactions of hydroxyl radicals with aldehydes studied under atmospheric conditions. *Environ. Sci. Technol.* 15, 960-962.

Khan, A., Hassett, J.J., Banwart, W.L., Means, J.C., Wood, S.G. (1979) Sorption of acetophenone by sediments and soils. *Soil Sci.* 128, 297-302.

Klopman, G., Namboodiri, K., Scochet, M. (1985) Simple method of computing partition coefficient. *J. Comput. Chem.* 6, 28-38.

Kollig, H.P., Ed. (1993) *Environmental Fate Constants for Organic Chemicals Under Consideration for EPA's Hazardous Waste Identification Projects.* EPA Report EPA/600/R-93/132, Environmental Research Lab., USEPA, Athens, GA.

Kollig, H.P., Ellington, J.J., Hamrick, K.J., Jafverts, C.T., Weber, E.J., Wolfe, N.L. (1987) Hydrolysis rate constants, partition coefficients, and water solubilities for 129 chemicals. A summary of fate constants provided for the concentration-based Listing Program of 1987. USEPA Environ. Res. Lab., Off. Res. Development Prepublication. Athens, GA.

Könemann, H. (1981) Quantitative structure-activity relationships in fish toxicity studies. Part 1: Relationship for 50 industrial pollutants. *Toxicology* 19, 209-221.

Kramer, C.-R., Henze, U. (1990) Partitioning properties of benzene derivatives. I. Temperature dependence of the partitioning of monosubstituted benzenes and nitrobenzenes in the *n*-octanol/water system. *Z. Phys. Chem.* (Leipzig), 271(3), 503-513.

Lamb, C.B., Jenkins, G.F. (1952) BOD of synthetic organic chemicals. *Proc. 8th Indust. Waste Conf.*, pp.326-329. Purdue University.

Lazare, A., Lazare, C., Bastide, J., Couquelet, J. (1974) Hypnotic activity and physicochemical parameters. Application to some acetophenone derivatives. *Ann. Pharm. France* 32, 677-684.

Leahy, D.E. (1986) Intrinsic molecular volume as a measure of the cavity term in linear solvation energy relationships: octanol-water partition coefficients and aqueous solubilities. *J. Pharm. Sci.* 75(7), 629-636.

Lee, B.I., Erbar, J.H., Edmister, W.C. (1972) Thermodynamic properties at low temperatures. *Chem. Eng. Prog.* 68(9), 83-84.

Leo, A., Hansch, C., Church, C. (1969) Comparison of parameters currently used in the study of structure-activity relationships. *J. Med. Chem.* 12, 766-771.

Leo, A., Hansch, C., Elkins, D. (1971) Partition coefficients and their uses. *Chem. Rev.* 71(6), 525-616.

Lu, P.Y., Walton, B.T., Lewis, E.B., Kine, B.W., Scott, J.H., Groover, G.E. (1986) *Chemical Information Profile of Selected Resource Conservation and Recovery Act Chemicals.* ORNL Internal Report to the Robert S. Kerr Res. Laboratory, U.S. Environmental Protection Agency, Oak Ridge, TN.

Lyman, W.J. (1982) Chapter 4. Adsorption coefficient for soils and sediments. In: *Handbook of Chemical Property Estimation Methods*. Lyman, W.J., Reehl, W.F., Rosenblatt, D.H., Editors, McGraw-Hill, New York.

Lyman, W.J., Reehl, W.F., Rosenblatt, D.H. (1982) *Handbook of Chemical Property Estimation Methods. Environmental Behaviour of Organic Compounds.* Lyman, W.J., Reehl, W.F., Rosenblatt, D.H., Editors, McGraw-Hill, New York.

Mackay, D., et al. (1982) *Volatilization of Organic Pollutants from Water.* USEPA-600/53-82-019, NTIS PB 82-230 939. U.S. Environmental Protection Agency, Athens, Georgia.

Mackay, D., Bobra, A.M., Chan, D.W., Shiu, W.Y. (1982) Vapor pressure correlation for low volatility environmental chemicals. *Environ. Sci. Technol.* 16, 645-649.

Mackay, D., Yuen, A.T.K. (1983) Mass transfer coefficient correlations for volatilization of organic solutes from water. *Environ. Sci. Technol.* 17, 211-217.

Mailhot, H., Peters, R.H. (1988) Empirical relationships between the 1-octanol/water partition coefficient and nine physicochemical properties. *Environ. Sci. Technol.* 22, 1479-1488.

McBain, J.W., Richards, P.H. (1946) Solubilization of insoluble organic liquids by detergents. *Ind. Eng. Chem.* 38(6), 642-644.

McCall, J.M. (1975) Liquid-liquid partition coefficients by high-performance liquid chromatography. *J. Med. Chem.* 18(6), 549-552.

McDuffie, D. (1981) Estimation of octanol/water partition coefficients for organic pollutants using reversed-phase HPLC. *Chemosphere* 10, 73-83.

McGowan, J.C. (1954) The physical toxicity of chemicals. IV. Solubilities, partition coefficients and physical toxicities. *J. Appl. Chem.* 4, 41-47.

Means, J.C., Hassett, J.J., Wood, S.G., Banwart, W.L., Ali, S., Khan, A. (1980) Sorption properties of polynuclear aromatic hydrocarbons and sediments: Heterocyclic and substituted compounds. In: *Polynuclear Hydrocarbons: Chemistry and Biological Effects.* Bjorseth, A., Dennis, A.J., Editors, pp.395-404, Ann Arbor Sci. Publ., Ann Arbor, Michigan.

Meyer, E.F., Hotz, R.D. (1973) High-precision vapor-pressure data for eight organic compounds. *J. Chem. Eng. Data* 18(4), 359-362.

Meylan, W.M., Howard, P.H. (1991) Bond contribution method for estimating Henry's law constants. *Environ. Toxicol. & Chem.* 10, 1283-1293.

Meylan, W.M., Howard, P.H. (1993) Computer estimation of the atmospheric gas-phase reaction rate of organic compounds with hydroxyl radicals and ozone. *Chemosphere* 26, 2293-2299.

Mill, T., Davenport, J. (1986) *ACS Environ. Chem. Div. 192nd Nat'l Meeting* 26, 59-63.

Mill, T. (1979) *Structure Reactivity Correlations for Environmental Reactions.* EPA Final Report, EPA 560/11-79-012.

Mill, T. (1982) Hydrolysis and oxidation processes in the environment. *Environ. Toxicol. & Chem.* 1, 135-141.

Mill, T., et al. (1982) *Laboratory Protocols for Evaluating the Fate of Organic Chemicals in Air and Water.* p.255. U.S. EPA-600/3-82-022.

Minick, D.J., Frenz, J.H., Patrick, M.A., Brent, D.A. (1988) A comprehensive method for determining hydrophobicity constants by reversed-phase high-performance liquid chromatography. *J. Med. Chem.* 31, 1923-1933.

Miyake, K., Terada, H. (1982) Determination of partition coefficients of very hydrophobic compounds by high-performance liquid chromatography on glyceryl-coated controlled-pore glass. *J. Chromatogr.* 240, 9-20.

Morris, E.D., Jr., Stedman, D.H., Niki, H. (1971) Mass spectrometric study of the reactions of the hydroxyl radical with ethylene, propylene, and acetaldehyde in a discharge-flow system. *J. Am. Chem. Soc.* 93, 3570-3572.

Morris, E.D., Jr., Niki, H. (1971) Mass spectrometric study of the reaction of hydroxyl radical with formaldehyde. *J. Chem. Phys.* 55, 1991-1992.

Morris, E.D., Jr., Niki, H. (1971) Reactivity of hydroxyl radicals with olefins. *J. Phys. Chem.* 75, 3640-3641.

Müller, M., Klein, W. (1991) Estimating atmospheric degradation processes by SARs. *Sci. Total Environ.* 109/110, 261-273.

Müller, M., Klein, W. (1992) Comparative evaluation of methods predicting water solubility for organic compounds. *Chemosphere* 25(6), 769-782.

Nahum, A. Horvath, C. (1980) Evaluation of octanol-water partition coefficients by using high-performance liquid chromatography. *J. Chromatogr.* 192, 315-322.

Nelson, P.E., Hoff, J.E. (1968) Food volatiles: gas chromatographic determination of partition coefficient in water-lipid systems. *J. Food Sci.* 33, 479-482.

Niki, H., Maker, P.D., Savage, C.M., Breitenbach, L.P. (1978) Relative rate constants for the reaction of hydroxyl radical with aldehydes. *J. Phys. Chem.* 82, 132-134.

Nirmalakhandan, N.N., Speece, R.E. (1988) Prediction of aqueous solubility of organic chemicals based on molecular structure. *Environ. Sci. Technol.* 22(3), 328-338.

Norrington, F.E., Hyde, R.M., Williams, S.G., Wootton, R. (1979) Physicochemical-activity relations in practice. I. A rational and self-consistent data bank. *J. Med. Chem.* 18, 604-607.

Palit, S.R. (1947) Electronic interpretations of organic chemistry. II. Interpretation of the solubility of organic compounds. *J. Phys. Chem.* 51, 837-857.

Pangrekar, J., Klopman, G., Rosenkranz, H.S. (1994) Expert-system comparison of structural determinants of chemical toxicity to environmental bacteria. *Environ. Toxicol. & Chem.* 13(6), 979-1001.

Perry, J.H., Editor (1950) *Chemical Engineer's Handbook.* 3rd. Edition, McGraw-Hill, New York.

Pinal, R., Lee, L.S., Rao, P.S.C. (1991) Prediction of the solubility of hydrophobic compounds in nonideal solvent mixtures. *Chemosphere* 22(9/10), 939-951.

Pitter, P. (1976) Determination of biological degradability of organic substances. *Water Res.* 10, 231.

Rathbun, R.E., Stephens, D.W., Shultz, D.J., Tai, D.Y. (1982) Fate of acetone in water. *Chemosphere* 11(11), 1097-1114.

Rathbun, R.E., Tai, D.Y. (1982) Volatilization of ketones from water. *Water Air Soil Pollut.* 17, 281-293.

Reid, R.C., Prausnitz, J.M., Sherwood, T.K. (1977) *The Properties of Gases and Liquids.* 3rd. Edition, McGraw-Hill Book Company, New York.

Rekker, R.F. (1977) *The Hydrophobic Fragmental Constants.* Its derivation and application. A means of characterizing membrane systems. Vol. 1, Pharmacochemistry Library. Nauta, W.T., Rekker, R.F., Editors. Elsevier Scientific Publishing Company, Oxford.

Riddick, J., Bunger, W.B., Sakano, T.K. (1986) *Organic Solvents: Physical Properties and Method of Purification.* 4th Edition, John Wiley & Sons, New York, New York.

Rohrschneider, L. (1973) Solvent characterization of gas-liquid partition coefficients of selected solutes. *Anal. Chem.* 45(7), 1241-1247.

Roy, W.R., Griffin, R.A. (1985) Mobility of organic solvents in water-saturated soil materials. *Environ. Geol. Water Sci.* 7, 241-247.

Russell, C.J., Dixon, S.L., Jurs, P.C. (1992) Computer-assisted study of the relationship between molecular structure and Henry's law constant. *Anal. Chem.* 64, 1350-1355.

Sabljic, A. (1987) On the prediction of soil sorption coefficients of organic pollutants from molecular structure: Application of molecular topology model. *Environ. Sci. Technol.* 21, 358-366.

Sabljic, A., Güesten, H. (1990) Predicting the night-time NO_3 radical reactivity in the troposphere. *Atmos. Environ.* 24A(1), 73-78.

Sangster, J. (1989) Octanol-water partition coefficients of simple organic compounds. *J. Phys. Chem. Ref. Data* 18, 1111-1230.

Scow, K.M. (1982) Chapter 9, Rate of biodegradation. In: *Handbook of Chemical Property Estimation Methods*. Lyman, W.J., Reehl, W.F., Rosenblatt, D.H., Editors, McGraw-Hill, New York. pp.9-1 to 9-85.

Seidell, A. (1941) *Solubility of Organic Compounds.* Vol. 2, Van Nostrand, New York.

Semmes, D.H., Ravishankara, A.R., Gump-Perkins, C.A., Wine, P.H. (1985) Kinetics of the reactions of hydroxyl radical with aliphatic aldehydes. *Int'l J. Chem. Kinetics* 17, 303-313.

Skrzec, A.E., Murphy, N.F. (1954) Liquid-Liquid Equilibria. Acetic acid in water, with 1-butanol, methyl ethyl ketone, furfural, cyclohexanol, and nitromethane. *Ind. & Eng. Chem.* 46(10), 2245-2247.

Smith, C.W., Editor (1962) *Acrolein.* John Wiley & Sons, Inc., New York.

Snider, J.R., Dawson, G.A. (1985) Tropospheric light alcohols, carbonyls, and acetonitrile: concentrations in the southwestern United States and Henry' law data. *J. Geophys. Res.* 90, 3797-3805.

Southworth, G.R., Keller, J.L. (1986) Hydrophobic sorption of polar organics by low organic carbon soils. *Water Air Soil Pollut.* 28, 239-248.

Stedman, D.H., Niki, H. (1973) Ozonolysis rates of some atmosphere gases. *Environ. Lett.* 4, 303-310.

Stephen, H., Stephen, Y. (1963) *Solubilities of Inorganic and Organic Compounds.* Vol. 1 and 2. Pergamon Press, Oxford.

Stephenson, R.M. (1992) Mutual solubilities: water-ketones, water-ethers, and water-gasoline-alcohols. *J. Chem. Eng. Data* 37, 80-95.

Stephenson, R.M. (1993) Mutual solubility of water and aldehydes. *J. Chem. Eng. Data* 38, 630-633.

Stephenson, R.M., Malanowski, A. (1987) *Handbook of the Thermodynamics of Organic Compounds*, Elsevier, New York.

Stull, D.R. (1947) Vapor pressure of pure substances: Organic compounds. *Ind. Eng. Chem.* 39(4), 517-560.

Su, F., Calvert, J.G., Shaw, J.H. (1979) Mechanism of the photooxidation of gaseous formaldehyde. *J. Phys. Chem.* 83, 3185-3191.

Tabak, H.H., Govind, R. (1993) Prediction of biodegradation kinetics using a nonlinear group contribution method. *Environ. Toxicol. Chem.* 12, 251-260.

Takemoto, S., Kuge, Y., Nakamoto, M. (1981) The measurement of BOD in seawater. *Suishitsu Odaku Kenkyu.* 4, 80-90.

Taft, R.W., Abraham, M.H., Doherty, R.M., Kamlet, M.J. (1985) The molecular properties governing solubilities of organic nonelectrolytes in water. *Nature* 313, 384-386.

Taft, R.W., Abraham, M.H., Famini, G.R., Doherty, R.M., Abboud, J.-L.M., Kamlet, M.J. (1985) Solubility properties in polymers and biological media 5: An analysis of the physicochemical properties which influence octanol-water partition coefficients of aliphatic and aromatic solutes. *J. Pharm. Sci.* 74(8), 807-814.

Tanii, H., Tsuji, H., Hashimoto, K. (1986) Structure-toxicity relationship of monoketones. *Toxicol. Letters* 30, 13-17.

Tewari, Y.B., Miller, M.M., Wasik, S.P., Martire, D.E. (1982) Aqueous solubility and octanol/water partition coefficient of organic compounds at 25.0°C. *J. Chem. Eng. Data* 27, 451-454.

Tsonopoulos, C., Prausnitz, J.M. (1971) Activity coefficients of aromatic solutes in dilute aqueous solutions. *Ind. & Eng. Chem. Fundamentals* 10, 593-600.

Umeyama, H., Nagai, T., Nogami, H. (1971) Mechanism of adsorption of phenols by carbon black from aqueous solution. *Chem. Pharm. Bull.* 19(8), 1714-1721.

Unger, S.H., Chiang, G.H. (1981) Octanol-physiological buffer distribution coefficients of lipophilic amines by reversed-phase high-performance liquid chromatography and their correlation with biological activity. *J. Med. Chem.* 24(3), 262-270.

Unger, S.H., Cook, J.R., Hollenberg, J.S. (1978) Simple procedure for determining octanol-aqueous partition, distribution, and ionization coefficients by reversed-phase high-pressure liquid chromatography. *J. Pharm. Sci.* 67, 1364-1367.

Urano, K., Kato, Z. (1986) Evaluation of biodegradation ranks of priority organic compounds. *J. Hazardous Materials* 13, 135-145.

Urano, K., Maeda, H., Ogura, K., Wada, H. (1982) Direct analytical method for aromatic compounds in water by steam carrier gas chromatography. *Water, Res.* 16, 323-327.

Urano, K., Ogura, K., Wada, H. (1981) Direct analytical method for aliphatic compounds in water by steam carrier gas chromatography. *Water, Res.* 15, 225-231.

USEPA (1987) EXAMS II Computer Simulation.

Vaishnav, D.D., Babeu, L. (1986) Occurence and rates of chemical biodegradation in Superior harbor water. *J. Great Lakes Res.* 12, 184-191.

Vaishnav, D.D., Babeu, L. (1987) Comparison of occurrence and rates of chemical biodegradation in natural waters. *Bull. Environ. Contam. Toxicol.* 39, 237-244.

Vaishnav, D.D., Boethling, R.S., Babeu, L. (1987) Quantitative structure-biodegradability relationships for alcohols, ketones and alicyclic compounds. *Chemosphere* 16(4), 695-703.

Valvani, S.C., Yalkowsky, S.H., Roseman, T.J. (1981) Solubility and partitioning. VI. Aqueous solubility and octanol/water partition coefficient of liquid electrolytes. *J. Pharm. Sci.* 70, 502-507.

Veith, G.D., Call, D.J., Brooke, L.T. (1983) Structure-toxicity relationships for the fathead minnow, *pimephales promelas*: Narcotic industrial chemicals. *Can. J. Aquat. Sci.* 40, 743-748.

Veith, G.D., DeFoe, D.L., Bergstedt, B.V. (1979) Measuring and estimating the bioconcentration factors of chemicals in fish. *J. Fish. Res. Board. Can.* 36, 1040-1048.

Veith, G.D., Macek, K.J., Petrocelli, S.R., Carroll, J. (1980) An evaluation of using partition coefficients and solubility to estimate bioconcentration factors for organic chemicals in fish. In: *Aquatic Toxicology.* pp.116-129. ASTM STP 707.

Verhaar, H.J.M., van Leeuwen, C.J., Hermens, J.L.M. (1992) Classifying environmental pollutants. 1: Structure-activity relationships for prediction of aquatic toxicity. *Chemosphere* 25(4), 471-491.

Verschueren, K. (1977) *Handbook of Environmental Data on Organic Chemicals*. Van Nostrand Reinhold Co., New York.

Verschueren, K. (1983) *Handbook of Environmental Data on Organic Chemiocals,* 2nd ed., Van Nostrand Reinhold Co., New York.

Wakita, K., Yoshimoto, M., Miyamoto, S. (1986) A method for calculation of the aqueous solubility of organic compounds by using new fragment constants. *Chem. Pharm. Bull.* 34, 4663-4681.

Wallington, T.J., Daganet, P., Kurylo, M.J. (1988) Correlation between gas-phase and solution phase reactivities of hydroxyl radicals toward saturated organic compounds. *J. Phys. Chem.* 92, 5024-5028.

Wallington, T.J., Kurylo, M.J. (1987) Flash photolysis resonance fluorescence investigation of the gas-phase reactions of OH radicals with a series of aliphatic ketones over the temperature range 240-440 K. *J. Phys. Chem.* 91, 5050-5054.

Walton, B.T., Hendricks, M.S., Anderson, T.A., Griest, W.H., Merriweather, R., Beauchamp, J.J., Francis, C.W. (1992) Soil sorption and semivolatile organic compounds in a mixture. *J. Environ. Qual.* 21, 552-558.

Wasik, S.P., Tewari, Y.B., Miller, M.M., Martire, D.E. (1981) *Octanol/Water Partition Coefficients and Aqueous Solubilities of Organic Compounds.* NBSIR No.81-2406. U.S. Dept. of Commerce, Washington.

Weast, R.C., Editor (1972-73) *Handbook of Chemistry and Physics.* 53rd Edition, CRC Press, Cleveland, Ohio.

Weast, R.C., Editor (1973-74) *Handbook of Chemistry and Physics.* 54th ed., CRC Press, Cleveland Ohio.

Weast, R.C., Editor (1976-77) *Handbook of Chemistry and Physics.* 57th Edition, CRC Press, Cleveland, Ohio.

Weast, R.C., Editor (1978) *Handbook of Chemistry and Physics.* 58th Edition, CRC Press, Inc., West Palm Beach, Florida.

Weast, R.C., Editor (1983-84) *Handbook of Chemistry and Physics.* 63th ed., CRC Press, Boca Raton, Florida.

Yalkowsky, S.H., Morozowich, W. (1980) Chapter 3, A physical chemical basis for the design of orally active prodrugs. In: *Drug Design.* Vol. IX. Academic Press, Inc., New York.

Yalkowsky, S.H., Pinal, R. (1993) Estimation of the aqueous solubility of complex organic compounds. *Chemosphere* 26, 1239-1261.

Yaws, C.L., Yang, H.-C., Hopper, J.R., Hansen, K.C. (1990) Organic chemicals: water solubility data. *Chem. Eng.* Jul, 115-118.

Yaws, C., Yang, H.C., Pan, X. (1991) Henry's law constants for 362 organic compounds in water. *Chem. Eng.* 98(11), 179-185.

Yoshida, Y., Mizuno, T., Ose, Y., Sato, T. (1986) The estimation for toxicity of chemicals on fish by physico-chemical properties. *Chemosphere* 15(2), 195-203.

Zetzsch, C., (1982) 7th International Symposium on Gas Kinetics, August 23-28, 1982. University of Göttingen, Göttingen, Germany.

Zhou, X., Mopper, K. (1990) Apparent partition coefficients of 15 carbonyl compounds between air and seawater and between air and freshwater; implications for air-sea exchange. *Environ. Sci. Technol.* 24, 1864-1869.

Zoeteman, B.C.J., De Greef, E., Brinkmann, F.J.J. (1981) Persistency of organic contaminants in groundwater, lessons from soil pollution incidents in the Netherland. *Chemosphere* 21, 187-202.

Chapter 4 Phenolic Compounds

4.1 List of Chemicals and Data Compilations:

4.1 List of Chemicals and Data Compilations

Common Name: Phenol
Synonym: carbolic acid, phenic acid, phenylic acid, phenyl hydrate, phenyl hydroxide, hydroxybenzene, oxybenzene
Chemical Name: phenol
CAS Registry No: 108-95-2
Molecular Formula: C_6H_5OH
Molecular Weight: 94.113
Melting Point (°C):
40.6 (Stull 1947)
42.0 (Ericksen & Dobbert 1955)
40.9 (Dreisbach 1955; Andon et al. 1960; Yalkowsky & Valvani 1980; Mackay et al. 1982; Mackay 1991; Guinee & Heijungs 1993)
41.5 (Weast 1978; Kaiser et al. 1984)
43.0 (Weast 1982; Valvani & Yalkowsky 1980; Howard 1989)
41.0 (Verschueren 1983; Yalkowsky et al. 1983; Banerjee et al. 1990)
Boiling Point (°C):
181.9 (Stull 1947)
181.60 (Ericksen & Dobbert 1955)
181.75 (Dreisbach 1955, Boublik et al. 1973; Weast 1977)
181.84 (Biddiscombe & Martin 1958)
181.85 (Mackay et al. 1982)
182.0 (Verschueren; Banerjee et al. 1990)
181.65 (Boublik et al. 1984)
Density (g/cm^3 at 20°C):
1.5479 (supercooled liq., Ericksen & Dobbert 1955)
1.0576 (Weast 1982)
1.070 (Verschueren 1983)
Dissociation Constant, pK_a :
9.90 (Blackman et al. 1955, McLeese et al. 1979)
10.02 (Herington & Kynaston 1957; Callahan et al. 1979)
9.89 (Doedens 1967; Weast 1982-83)
9.82 (Sillén & Martell 1971; Kaiser et al. 1984; Shigeoka et al. 1988)
10.0 (Serjeant & Dempsey 1979; Paris et al. 1982; Miyake et al. 1987; Tratnyek & Hoigné 1991)
9.92 (Könemann 1981; Könemann & Musch 1981; Ugland et al. 1981)
10.05 (Saarikoski 1982; Saarikoski & Viluksela 1982)
10.0 (25°C, Windholz 1983)
9.90 (UV absorption, Hoigné & Bader 1983; Scully & Hoigné 1987)
9.99 (Dean 1985; Schultz & Cajina-Quezada 1987; Hersey et al. 1989)
10.93 (Miyake et al. 1987)
10.12 (UV spectrophotometry, Nendza & Seydel 1988)
9.994 (Howard 1989)
Molar Volume (cm^3/mol):
83.14 (calculated-density)
103.4 (calculated-LeBas method)
105 (Andon et al. 1960)
0.989 ($V_M/100$, Kamlet et al. 1987)

0.536 (intrinsic $V_I/100$, Kamlet et al. 1988; Hickey & Passino-Reader 1991)
118.0, 54.0 (LeBas method, instrinsic volume V_I, Abernethy et al. 1988)
83.0 (calculated-density, Jaworska & Schultz 1993)
94.11 (Yalkowsky & Pinal 1993)

Molecular Volume (A^3):

Total Surface Area, TSA (A^2):
224.2 (Iwase et al. 1985)

Heat of Fusion, ΔH_{fus}, kcal/mol: 2.752 (Riddick et al. 1986)

Entropy of Fusion ΔS_{fus}, cal/mol·K (e.u.):
8.755 (Tsonopoulos & Prausnitz 1971)
8.60, 13.5 (observed, estimated, Yalkowsky & Valvani 1980)

Fugacity Ratio at 25°C, (assuming ΔS_{fus} = 13.5 e.u.) F: 0.6946

Water Solubility (g/m³ or mg/L at 25 °C):
83000 (20°C, synthetic method, Jones 1929)
83000 (Seidell 1941; Urano et al. 1982)
88360 (shake flask-UV at pH 5.1, Blackman et al. 1955)
79000 (20°C, shake flask-optical, Ericksen & Dobbert 1955)
82000 (quoted, Dreisbach 1955; Mackay 1991; Guinee & Heijungs 1993)
86600 (Stephen & Stephen 1963)
80000 (20°C, Mulley & Metcalf 1966)
93000 (Morrison & Boyd 1973; quoted, Callahan et al. 1979)
78000 (shake flask-AS, Roberts et al. 1977)
79750 (shake flask-GC, Kraij & Sincic 1980)
83878, 36614 (quoted, calculated-K_{OW}, Yalkowsky & Morozowich 1980)
78280, 42039 (quoted, calculated-K_{OW} and M.P., Valvani & Yalkowsky 1980)
82000 (quoted, Briggs 1981)
76514 (gen. col.-HPLC, Wasik et al. 1981)
93000 (quoted, Mabey et al. 1982)
83000 (20-25°C, quoted, Urano et al. 1982)
82000 (15°C, Verschueren 1983; Shigeoka et al. 1988)
85832, 79055 (quoted, calculated-K_{OW} and M.P., Yalkowsky et al. 1983)
150580 (20°C, shake flask-UV, Hashimoto et al. 1984)
128935 (20°C, calculated, Hashimoto et al. 1984)
82000 (quoted, extrapolated from Ericksen & Dobbert 1955, Leuenberger et al. 1985)
86600 (quoted, Riddick et al. 1986)
87000 (quoted, Howard 1989)
80200 (selected, Yaws et al. 1990)
316660 (quoted from Windholz 1983, Chessells et al. 1992)
94110 (calculated-K_{OW}, Müller & Klein 1992)
87830 (selected, Yalkowsky & Pinal 1993)

Vapor Pressure (Pa at 25°C):
55.54 (extrapolated-regression of tabulated data, Stull 1947)
70.75 (calculated-Antoine eqn., Dreisbach 1955)

45.71 (gas saturation and diaphragm manometer fitted to Antoine eqn., Biddiscombe & Martin 1958; quoted, Mabey et al. 1982)
45.70 (calculated-Antoine eqn., Andon et al. 1960)
70.7 (20°C, supercooled liq., Andon et al. 1960; quoted, Callahan et al 1979)
46.91 (extrapolated-Antoine eqn., Boublik et al. 1973)
83.95 (extrapolated subcooled liquid value, Antoine eqn., Weast 1973)
26.66 (20°C, Verschueren 1977, 1983; quoted, Mackay et al. 1982)
133.3 (40°C, Verschueren 1977,83)
71.93 (calculated-B.P., Mackay et al. 1982)
16.27 (extrapolated-Cox eqn., Chao et al. 1983)
54.74 (extrapolated-Antoine eqn., Boublik et al. 1984)
57.5 (Cheung 1984)
47.01 (extrapolated-Antoine eqn., Boublik et al. 1984)
47.00 (extrapolated-Antoine eqn., Dean 1985)
45.33 (Leuenberger et al. 1985)
47.0 (extrapolated-Antoine eqn., Dean 1985)
46.66 (Howard et al. 1986; quoted, Banerjee et al. 1990)
55.0 (quoted, Riddick et al. 1986)
45.7, 45.32 (interpolated-Antoine eqn., Stephenson & Malanowski 1987)
69.86 (quoted, Howard 1989)
46.7, 286.6 (quoted, calculated-solvatochromic p., Banerjee et al. 1990)
61.16 (extrapolated-Antoine eqn., Nesterova et al. 1990)
70.6 (quoted, Mackay 1991; Guinee & Heijungs 1993)

Henry's Law Constant (Pa·m^3/mol):
0.0402 (quoted exptl., Hine & Mookerjee 1975)
0.065 (calculated-group contribution, Hine & Mookerjee 1975)
1.082 (calculated-bond contribution, Hine & Mookerjee 1975)
0.046 (calculated-P/C, Mabey et al. 1982)
0.0274 (Petrasek et al. 1983; quoted, Barton 1987)
0.011 (8°C, Leuenberger et al. 1985)
0.0399 (27°C, estimated-P/C over diluted aq. soln., Abd-El-Bary et al. 1986)
0.0589 (25°C, estimated-P/C over diluted aq. soln., Abd-El-Bary et al. 1986)
0.0338 (quoted, Gaffney et al. 1987)
0.0402, 0.101 (quoted, calculated- χ , Nirmalakhandan & Speece 1988)
0.0402 (quoted, Howard 1989)
0.096 (quoted, Ince & Inel 1991)
0.0402, 0.0495 (quoted, calculated-molecular structure, Russell et al. 1992)
0.0539 (calculated-P/C, Shiu et al. 1994)

Octanol/Water Partition Coefficient, log K_{OW}:
1.46 (shake flask-AS, Fujita et al. 1964; quoted, Leo et al. 1969, 1971; Mirrlees et al. 1976; Rekker 1977; Roberts et al. 1977; Nagel & Urich 1980; Hersey et al. 1989)
1.46 $\pm$ 0.01 (shake flask-AS, Iwasa et al. 1965)
1.42 (shake flask-AS at pH 7.4, Rogers & Cammarata 1969)

1.60 (shake flask, Kiezyk & Mackay 1971)
1.47 (Leo et al. 1971; quoted, Berthod et al. 1988)
1.46 (Tute 1971; quoted, Fujisawa & Masuhara 1981)
1.51 (shake flask-AS at pH 7.45, Umeyama et al. 1971)
1.49 (shake flask, Korenman 1972)
1.48 (Roth & Seeman 1972; quoted, Rekker 1977)
1.46, 1.61 (LC-k', calculated-π const., Carlson et al. 1975)
1.46 (McCall 1975; quoted, Callahan et al. 1979)
1.54 (shake flask-AS, Davis et al. 1976)
1.45 (HPLC, Mirrlees et al. 1976)
1.48 $\pm$ 0.02 (shake flask at pH 7, Unger et al. 1978)
1.48 (Hansch & Leo 1979; quoted, Miyake & Terada 1982; Kaiser 1983; Kaiser et al. 1984; Iwase et al. 1985; Brooke et al. 1986; Miyake et al. 1987; Hodson & Williams 1988)
1.46, 1.48; 1.51 (quoted values; calculated-f const., Rekker 1977; quoted, Hanai & Hubert 1982)
1.48, 1.46, 1.49, 1.51, 0.62, 2.20 (quoted, Hansch & Leo 1979)
1.66 (calculated, McLeese et al. 1979)
1.46 (quoted, Valvani & Yalkowsky 1980; Paris et al. 1982)
1.46 (calculated-HPLC-k', Butte et al. 1981; quoted, Butte et al. 1987)
1.46 (calculated- χ , Kier & Hall 1981; quoted, Shigeoka et al. 1988)
1.55 (calculated-f const., Könemann 1981; Könemann & Musch 1981; quoted, Verhaar et al. 1992)
1.45 (gen. col.-HPLC, Wasik et al. 1981)
1.54 (RPHPLC-k', D'Amboise & Hanai 1982)
1.28, 1.54 (RP-LC, calculated-f const. as per Rekker 1977, Hanai & Hubert 1982)
1.54 (calculated-f const., Hermens & Leeuwangh 1982)
1.48, 1.54 (quoted, correlated-HPLC-k', Miyake & Terada 1982; Miyake et al. 1987)
1.62 (inter-lab. shake flask average, Eadsforth & Moser 1983)
1.16 (inter-lab. HPLC average, Eadsforth & Moser 1983)
1.46 (Petrasek et al. 1983; quoted, Barton 1987)
1.46 (quoted, Verschueren 1983)
1.46 (quoted from Hansch & Leo 1982, Banerjee et al. 1984)
1.49, 1.53 $\pm$ 0.09 (selected best lit. value, exptl.-ALPM, Garst 1984)
1.46, 1.55 $\pm$ 0.07 (selected best lit. value, exptl.-ALPM, Garst & Wilson 1984)
1.46, 1.08 (quoted, calculated-γ from UNIFAC, Campbell & Luthy 1985)
1.47 (quoted, Hardy et al. 1985)
1.48, 1.55 (quoted, calculated-TSA, Iwase et al. 1985)
1.76 (computed, Klopman et al. 1985; quoted, Pangrekar et al. 1994)
1.49 (quoted from Hansch & Leo 1979, Burkhard & Kuehl 1986)
1.46 (RPHPLC-RT standard, Chin et al. 1986)
1.00, 1.42 (HPLC-k', Eadsforth 1986)
1.47, 1.47 (quoted: shake flask, calculated, Eadsforth 1986)
1.49 (quoted, Schultz et al. 1987)
1.46 (quoted, Schultz & Cajina-Quezada 1987; Abernethy et al. 1988)
1.46 (shake flask-CPC, Berthod et al. 1988)
1.46 (RP-HPLC-capacity ratio, Minick et al. 1988)

1.46 (HPLC-RP, Shigeota et al. 1988; quoted, Saito et al. 1993)
1.49, 1.59 (quoted exptl., calculated-V_I and solvatochromic p., Kamlet et al. 1988)
1.49, 1.46, 1.475, 1.36 (quoted from Hansch & Leo 1979, CLOGP, calculated-CLOGP, calculated-MO, Bodor et al. 1989)
1.48, 1.52 ± 0.01 (calculated-pK_a, filter chamber-AS, Hersey et al. 1989)
1.50 (recommended, Sangster 1989, 1993)
1.46 (quoted, Howard 1989; Gluck & Martin 1990; El Tayar et al. 1991)
1.52, 1.58, 1.69 (CPC-RV, Gluck & Martin 1990)
1.47 (shake flask-AS, Kramer & Henze 1990)
1.46 (quoted, Ince & Inel 1991)
1.46 (quoted, Geyer et al. 1991)
1.37 ± 0.06 (liq.-liq. extraction-flow injection-AS, Kubáň 1991)
1.46 (quoted, Mackay 1991; Guinee & Heijungs 1993)
1.46 (quoted from Medchem software value, Chessells et al. 1992)
1.48 (calculated-K_{OW}, Müller & Klein 1992)
1.46 (calculated-CLOGP 3.51, Jaworska & Schultz 1993)
1.57 (shake flask-GC, Kishino & Kobayashi 1994)

Bioconcentration Factor, log BCF:
4.16, 4.99 (total ^{14}C in fathead minnows: observed, calculated, mean exposure level 0.0025 $\mu g \cdot mg^{-1}$, Call et al. 1980)
4.23, 5.09 (total ^{14}C in fathead minnows: observed, calculated, mean exposure level 0.0327 $\mu g \cdot mg^{-1}$, Call et al. 1980)
4.20 (total ^{14}C in fathead minnows, mean value, Call et al. 1980)
2.20 (goldfish, k_1/k_2, Nagel & Urich 1980)
0.97 (microorganisms-water, calculated-K_{OW}, Mabey et al. 1982)
1.30, 2.30, 3.34 (golden ide, algae, activated sludge, Freitag et al. 1985)
2.30 (*chlorella fusca*, Freitag et al. 1985; quoted, Howard 1989)
1.20 (algae, maximum apparent BCF, Hardy et al. 1985)
0.544 (algae, real BCF with biotransformation, Hardy et al. 1985)
3.14 (*daphnia magna*, estimated-^{14}C activity and on dry wt. basis, Dauble et al. 1986)
2.44 (*daphnia magna*, based on elimination phase, Dauble et al. 1986; quoted, Howard 1989)
1.28 (*daphnia magna*, Dauble et al. 1986; quoted, Geyer et al. 1991)
1.24 (zebrafish, Butte et al. 1987)
0.88 (calculated-K_{OW}, Howard 1989)

Sorption Partition Coefficient, log K_{OC}:
1.43 (soil, Kenaga & Goring 1980)
1.48 (organic carbon in soils, Briggs 1981)
1.15 (sediment-water, calculated-K_{OW}, Mabey et al. 1982)
1.57, 1.96 (silt loams, Scott et al. 1983, quoted, Howard 1989)
3.46 (untreated fine sediment, Isaacson & Frink 1984)
3.49 (untreated coarse sediment, Isaacson & Frink 1984)
1.35 (HPLC-k', Hodson & Williams 1988)
2.17 (soil, calculated-K_{OW}, Howard 1989)

2.4, 2.43 (soil: quoted, calculated- χ , Meylan et al. 1992)

Half-Lives in the Environment:

Air: half-life by reaction with OH radicals in air is 0.61 day (Howard 1989); 2.28 to 22.8 hours, based on measured reaction rate constant with OH radicals (Howard et al. 1991); degradation rate constant 0.0462 day^{-1} corresponding to a half-life of 360 hours in air (quoted from Howard 1989, Guinee & Heijungs 1993); atmospheric transformation lifetime was estimated to be < 1 day (Kelly et al. 1994).

Surface water: rate constant of $(1.3 \pm 0.2)x10^3$ $M^{-1} \cdot sec^{-1}$ for the reaction with ozone at pH 2.0-6.0 (Hoigné & Bader 1983); 46 hours (summer), 173 hours (winter) in distilled water and 39 hours (summer), 94 hours (winter) in estuarine water, based on phototransformation rate when exposed to full sunlight and microbes (Hwang et al. 1986); 43 hours (summer), 118 hours (winter) in poisoned esturaine water, based on phototransformation rate and 384 hours or 16 days (summer), 2640 hours or 110 days (winter) in poisoned esturaine water, based on photomineralization rate (Hwang et al. 1986); 384 hours or 16 days (summer), 4056 hours or 169 days (winter) in distilled water and 168 hours or 7 days (summer), 1752 hours or 73 days (winter) in estuarine water, based on photomineralization rate when exposed to full sunlight and microbes (Hwang et al. 1986); half-life of 2000 hours in water at pH 8 and (19 ± 2)°C for the reaction with singlet oxygen (Scully & Hoigné 1987); biodegradation half-lives: 11 days in river waters and 3 days in Superior harbor waters (Vaishnav & Babeu 1987); 5.3 to 56.5 hours, based on estimated aqueous aerobic biodegradation half-life (Howard et al. 1991); degradation rate constant 0.0217 day^{-1} corresponding to a half-life of 766 hours in water (quoted from Howard 1989, Guinee & Heijungs 1993).

Groundwater: biodegradation half-life of 20 days (Vashinav & Babeu 1987); 12 to 168 hours, based on estimated aqueous aerobic biodegradation half-life (Howard et al. 1991).

Soil: degradation in soil completed in 2-5 days even in subsurface soils and half-life of 2.70 and 3.51 hours of low concentration of phenol in 2 silt loam soils (Howard 1989); 24 to 240 hours, based on aerobic soil die-away data (Howard et al. 1991); half-life of 4.1 days in a slightly basic sandy loam soil containing 3.25% organic matter and 23 days in acidic clay soil with <1.0% organic matter, based on aerobic batch lab. microcosm experiments (Loehr & Matthews 1992); degradation rate constant 0.227 day^{-1} corresponding to a half-life of 73.3 hours in soil (quoted from Howard 1989, Guinee & Heijungs 1993).

Biota: elimination from goldfish within 4 hours (Nagel & Urich 1980); depuration half-lives of 336 hours (observed), 385 hours (calculated) for mean exposure level of 0.0025 $\mu g \cdot mL^{-1}$ and 438 hours (observed), 497 hours (calculated) for mean exposure level of 0.0375 $\mu g \cdot mL^{-1}$ (Call et al. 1980); and depuration half-lives was calculated to be 8 to 44 minutes in algae (Hardy et al. 1985); half-lives in fish <1 day for glodfish, 14-18 days for minnows (Niimi 1987) .

Environmental Fate Rate Constants or Half-lives:

Volatilization: estimated half-life for evaporation from water in 3.2 months (Howard 1989); half-life of 88 days, calculated for evaporation from a model river of 1 m deep with a current of 3 m/sec and with a wind velocity of 3 m/sec (Lyman et al. 1982; quoted, Howard 1989).

Photolysis: phototransformation rate of 0.015 h^{-1} with half-life of 46 hours in the summer (mean temp 24°C) and 0.0040 h^{-1} with half-life of 173 hours in the winter (mean temp 10°C) in distilled water; 0.018 h^{-1} with half-life of 39 hours in the summer and 0.0074 h^{-1} with half-life of 94 hours in the winter in estuarine water when exposed to full sunlight and microbes (Hwang et al. 1986); photomineralization rate of 0.04 h^{-1} with half-life of 16 days in the summer and 0.0041 h^{-1} with half-life of 169 days in the winter in distilled water; 0.095 h^{-1} with half-life of 7 days in the summer and 0.010 h^{-1} with half-life of 73 days in winter in estuarine water when exposed to full sunlight and microbes (Hwang et al. 1986); atmospheric half-life of 46 to 173 hours, based on reported half-life for photolysis under sunlight for phenol in distilled water in the summer and winter; aquatic photolysis half-life of 46 to 173 hours, based on reported half-life for photolysis under sunlight for phenol in distilled water in the summer and winter (Howard et al. 1991).

Oxidation: rate constant of $1x10^4$ M^{-1} sec^{-1} for the reaction with RO_2 radicals at 30°C in aquatic systems with half-life of 0.8 day (Howard 1972; Hendry et al. 1974; quoted, Mill 1982); rate constant of $<2x10^2$ M^{-1} sec^{-1} for the reaction with singlet oxygen at 25°C with half-life of >100 years (Foote 1976; Mill 1979; quoted, Mill 1982); $(2.0 \pm 0.4)x10^{-12}$ $cm^3 \cdot molecule^{-1} \cdot sec^{-1}$ for the reaction with NO_3 radicals in air at 300 ± 1 K with reference to reaction for NO_3 radicals with *cis*-2-butene (Graham & Johnston 1978; Japar & Niki 1975; quoted, Carter et al. 1981); $(2.59 \pm 0.52)x10^{-12}$ $cm^3 \cdot molecule^{-1} \cdot sec^{-1}$ for the reaction with NO_3 radicals in air at 300 ± 1 K with reference to reaction for NO_3 radicals with *cis*-2-butene (Carter et al. 1981; quoted, Atkinson 1991); dye-sensitized photooxidation first-order rate constant of $6.5x10^3$ sec^{-1} and second-order rate constants of $2.66x10^8$, $6.16x10^7$, $1.95x10^7$ $L \cdot mol^{-1} \cdot sec^{-1}$ at pH 10.3, 9, and 8 respectively of aqueous phenol (Okamoto et al. 1982); rate constant of $(1.3 \pm 0.2)x10^3$ $M^{-1} \cdot sec^{-1}$ for the reaction with ozone in water using 1 mM propyl alcohol as scavenger at pH 2.0-6.0 and 20-23°C (Hoigné & Bader 1983); rate constant of $(2.10 \pm 0.50)x10^{-12}$ $cm^3 \cdot molecule^{-1} \cdot sec^{-1}$ for the reaction with NO_3 radicals in air at 296 ± 2 K (Atkinson et al. 1984); $(3.64 \pm 0.14)x10^{-12}$ $cm^3 \cdot molecule^{-1} \cdot sec^{-1}$ for the reaction with NO_3 radicals in air at 294 K with reference to reaction for NO_3 radicals with 2-methyl-2-butene (Atkinson et al. 1984; quoted, Atkinson 1991); calculated rate constant of $45.5x10^{-12}$ $cm^3 \cdot molecule^{-1} \cdot sec^{-1}$ and observed rate constant of $28.0x10^{-12}$ $cm^3 \cdot molecule^{-1} \cdot sec^{-1}$ for the reaction with OH radicals in air at room temperature (Atkinson et al. 1985); calculated rate constant of $36x10^{-12}$ $cm^3 \cdot molecule^{-1} \cdot sec^{-1}$ and observed rate constant of $28.3x10^{-12}$ $cm^3 \cdot molecule^{-1} \cdot sec^{-1}$ for the reaction with OH radicals in air at room temperature (Atkinson 1985, 1987; quoted, Sabljic & Güsten 1990; Müller & Klein 1991); rate constant of (2 to 3)$x10^6$ $M^{-1} \cdot sec^{-1}$ at pH 8, 1.9 x 10^7 $M^{-1} \cdot sec^{-1}$ at pH 9, $4.6x10^7$ $M^{-1} \cdot sec^{-1}$ at pH 9.5, $9.0x10^7$ $M^{-1} \cdot sec^{-1}$ at pH 10 and $1.8x10^8$ $M^{-1} \cdot sec^{-1}$ at pH 11.5 for the reaction with singlet oxygen at (19 ± 2)°C in water (Scully & Hoigné 1987); rate constant of 3.63×10^{-12} $cm^3 \cdot molecule^{-1} \cdot sec^{-1}$ for the gas-phase reaction with NO_3 radicals in air at 298 K (Atkinson et al. 1988; quoted, Sabljic & Güsten 1990; Müller & Klein 1991); second-order rate constant of 26.3×10^{-12} $cm^3 \cdot molecule^{-1} \cdot sec^{-1}$ for the reaction with OH radicals in air at 296 $\pm$ 2 K with reference to reaction for OH radicals with 2-methyl-2-butene (Atkinson 1989; quoted, Atkinson 1990; Atkinson et al. 1992); half-life of 77 to 3840 hours in water, based on reported reaction rate constant for RO_2 radical with the phenols class, and 2.28 to 22.8 hours in air, based on measured reaction rate constant for OH radicals with phenol (Howard et al. 1991); rate constant of $(2.6 \pm 4) \times 10^6$ $M^{-1} \cdot sec^{-1}$ for the reaction with singlet oxygen in aqueous phosphate buffer at (27 $\pm$ 1)°C (Tratynyek & Hoigné 1991); atmospheric rate constants of 10.45×10^{-12} $cm^3 \cdot molecule^{-1} \cdot sec^{-1}$ for reaction with OH radicals and 11.44×10^{-12} $cm^3 \cdot molecule^{-1} \cdot sec^{-1}$ for reaction with NO_3 radicals (Müller & Klein 1991); rate constant of $(3.92 \pm 0.25) \times 10^{-12}$ $cm^3 \cdot molecule^{-1} \cdot sec^{-1}$ for the reaction with NO_3 radicals in air at 296 $\pm$ 2 K with reference to reaction for NO_3 radicals with 2-methyl-2-butene (Atkinson et al. 1992); 72.1 mg/L total organic carbon (TOC) with 94% degradation of TOC after 5 hours illumination with 250 watt tungsten lamp by photo-Fenton reaction (Ruppert et al. 1993); gas-phase hydroxyl radical rate constants of 26.3×10^{-12} $cm^3 \cdot molecule^{-1} \cdot sec^{-1}$ (exptl., quoted from Atkinson 1989) and 12.12×10^{-12} $cm^3 \cdot molecule^{-1} \cdot sec^{-1}$, calculated based on molecular orbital estimation method (Klamt 1993).

Hydrolysis: no hydrolyzable function group (Howard et al. 1991).

Biodegradation: 1 to 2 days for bacteria to utilize 95% of 300 ppm in parent substrate (Tabak et al. 1964); decomposition by soil microflora within one day (Alexander & Lustigman 1966; quoted, Verschueren 1983); complete disappearance in soil suspensions in two days (Woodcock 1971; quoted, Verschueren 1983); average rate of biodegradation 80.0 mg COD $g^{-1} \cdot h^{-1}$ based on measurements of COD decrease using activated sludge inoculum with 20 days of adaptation to the substrate (Pitter 1976; quoted, Scow 1982); biodegradation rate constants of 0.094 $\pm$ 0.003 h^{-1} at feed concentration of 180 mg/L at 20°C and 0.095 $\pm$ 0.007 h^{-1} at feed concentration of 360 mg/L at 20°C in a continuous stirred reactor (Beltrame et al. 1984); biodegradation rate constants of 0.035 day^{-1} with half-life of 20 days in ground water, and 0.065 day^{-1} with half-life of 11 days in Lester River water, 0.247 day^{-1} with half-life of 3 days in Superior harbor waters (Vaishnav & Babeu 1987); microbial degradation is the primary process for transformation with a rate of 0.03 h^{-1} and half-life of 28 hours for estuarine water in the summer (mean temp 24°C) and 0.011 h^{-1} with half-life of 62 hours in the winter (mean temp. 10°C); 0.4 h^{-1} with half-life of 2 days in the summer and 0.0051 h^{-1} with half-life of 146 days in the winter in darkness with microbes (Hwang et al. 1986); biodegradation rate constants of 0.041-0.028 h^{-1} in 10-100 mg/L sludge (Urano & Kato 1986a); biodegradation rate constants 0.025-0.028 h^{-1} (Urano & Kato 1986b); complete degradation within one day in water from 3 lakes, and degradation is somewhat slower in salt water with half-life of 9 days in estuarine river (Howard 1989); aerobic half-life of 6 to 84 hours, based on aerobic river and lake die-away study data and anaerobic half-life of 192 to 672 hours, based on aqueous anaerobic screening studies (Howard et al. 1991); average experimental rate constant of 0.0498 h^{-1} compared to using group contribution method predicted rate constants of 0.0545 $hour^{-1}$ (nonlinear) and 0.0503 h^{-1} (linear) (Tabak & Govind 1993).

Biotransformation: microbial transformation rate constant of $(7.1 \pm 1.3)x10^{-12}$ $L \cdot organism^{-1} \cdot h^{-1}$ (Paris et al. 1982); estimated bacterial transformation rate constant $3x10^{-6}$ $mL \cdot cell^{-1} \cdot h^{-1}$ in water (Mabey et al. 1982); microbial transformation rate constants in pond and river samples of $(2.0 \pm 1.5)x10^{-10}$ to $(4.8 \pm 3.1)x10^{-10}$ $L \cdot organism^{-1} \cdot h^{-1}$ at five different sites (Paris et al. 1983; quoted, Steen 1991); degradation rate constants of 1.08 x 10^{-16} $mol \cdot cell^{-1} \cdot h^{-1}$ from pure culture studies and $0.90x10^{-12}$ to $3.00x10^{-12}$ $mol \cdot cell^{-1} \cdot h^{-1}$ with microorganisms in Seneca River waters (Banerjee et al. 1984).

Bioconcentration, Uptake (k_1) and Elimination (k_2) Rate Constants:

k_1: 3.15 h^{-1} (goldfish, Nagel & Urich 1980)

k_2: 0.02 h^{-1} (goldfish, Nagel & Urich 1980)

Common Name: *o*-Cresol
Synonym: 2-hydroxytoluene, 2-methylphenol, *o*-cresylic acid, *o*-hydroxytoluene
Chemical Name: 2-methylphenol
CAS Registry No: 95-48-7
Molecular Formula: $C_6H_4(CH_3)OH$
Molecular Weight: 108.13
Melting Point (°C):
30.8 (Stull 1947)
31 (Ericksen & Dobbert 1955; Verschueren 1983)
30.99 (Andon et al. 1960)
30 (Valvani & Yalkowsky 1980)
30.85 (Mackay et al. 1982)
Boiling Point (°C):
190.8 (Stull 1947)
191 (Ericksen & Dobbert 1955; Biddiscombe & Martin 1958; Andon et al. 1960; Mackay et al. 1982; Verschueren 1983)
190.97 (Boublik et al. 1973)
191.27 (Boublik et al. 1984)
Density (g/cm^3 at 20°C):
1.5445 (supercooled liq., Ericksen & Dobbert 1955)
1.135 (25°C, Andon et al. 1960)
Acid Dissociation Constants, pK_a:
10.28 (Pearce & Simkins 1968)
10.1 (McLeese et al. 1979)
10.2 (Hoigné & Bader 1983; Weast 1982-83)
10.26 (Dean 1985)
10.287 (Riddick et al. 1986; Howard 1989)
Molar Volume (cm^3/mol):
125.6 (calculated-LeBas method)
127 (Andon et al. 1960)
98.80 (calculated-density, Jaworska & Schultz 1993)
Molecular Volume (A^3):
Total Surface Area, TSA (A^2):
248.6 (Iwase et al. 1985)
Heat of Fusion ΔH_{fus}, kcal/mol:
3.78 (Tsonopoulos & Prausnitz 1971)
Entropy of Fusion ΔS_{fus}, cal/mol·K (e.u.):
12.43 (Tsonopoulos & Prausnitz 1971)
Fugacity Ratio at 25°C (assuming ΔS_{fus} = 13.5 e.u.), F: 0.872

Water Solubility (g/m^3 or mg/L at 25°C):
24970 (shake flask-centrifuge, Booth & Everson 1948)
27161 (estimated, McGowan 1954)
25950 (shake flask-UV at pH 5.1, Blackman et al. 1955)
25000 (20°C, shake flask-optical, Ericksen & Dobbert 1955)
25000 (shake flask-AS, Roberts et al. 1977)

1300 (50°C, Haque et al. 1980)
24207, 19676 (quoted, calculated-K_{OW} and M.P., Valvani & Yalkowsky 1980; quoted, Müller & Klein 1992)
2725 (gen. col.-HPLC, Wasik et al. 1981)
25000 (20-25°C, Urano et al. 1982)
31000, 56000 (40°C, 100°C, Verschueren 1983)
26000 (Leuenberger et al. 1985)
30800 (40°C, Riddick et al. 1986; quoted, Howard 1989)
24660, 8110 (quoted, calculated- χ , Nirmalakhandan & Speece 1988)
24500 (selected, Yaws et al. 1990)
26800 (shake flask-HPLC/UV at pH 3.6, Varhanickova et al. 1995)

Vapor Pressure (Pa at 25°C):
58.16 (extrapolated-regression of tabulated data, Stull 1947)
38.6 (gas saturation and diaphragm manometer fitted to Antoine eqn., Biddiscombe & Martin 1958)
38.6 (calculated-Antoine eqn., Andon et al. 1960)
30.51 (extrapolated-Antoine eqn., Boublik et al. 1973, 1984)
32.0 (Verschueren 1977, 1983; quoted, Mackay et al. 1982)
133.3 (38.2°C, Haque et al. 1980)
95.23 (calculated-B.P., Mackay et al. 1982)
38.32 (extrapolated-Cox eqn., Chao et al. 1983)
22.50 (extrapolated-Antoine eqn., Boublik et al. 1984)
30.74 (extrapolated-Antoine eqn., Dean 1985)
38.7 (Leuenberger et al. 1985)
40.0 (Riddick et al. 1986; quoted, Howard 1989)
39.40 (extrapolated-Antoine eqn., Nesterova et al. 1990)
36.7 (selected P_L, Shiu et al. 1994)

Henry's Law Constant (Pa·m^3/mol):
0.124 (exptl., Hine & Mookerjee 1975)
0.070 (calculated-group contribution, Hine & Mookerjee 1975)
1.082 (calculated-bond contribution, Hine & Mookerjee 1975)
0.284 (8°C, Leuenberger et al. 1985)
0.162 (calculated, Leuenberger et al. 1985; quoted, Howard 1989)
0.122 (quoted, Gaffney et al. 1987)
0.124, 0.139 (quoted, calculated- χ , Nirmalakhandan & Speece 1988)
0.0852 (computed, Yaws et al. 1991)
0.1546 (calculated-P/C, Shiu et al. 1994)

Octanol/Water Partition Coefficient, log K_{OW}:
2.04 (shake flask-AS, Korenman & Pereshein 1970; quoted, Sangster 1989)
1.95 (quoted from Hansch & Dunn III unpublished result, Leo et al. 1971; quoted, Rekker 1977; Roberts et al. 1977; Iwase et al. 1985; Saito et al. 1993)

1.95 (LC-k', Carlson et al. 1975)
2.09 (calculated-π const., Carlson et al. 1975; quoted, Sangster 1989)
1.95, 2.03, 2.02 (quoted, calculated-f const., calculated-π const., Rekker 1977)
2.24 (McLeese et al. 1979)
1.95 (quoted, Valvani & Yalkowsky 1980)
2.17 (calculated-HPLC-k', Butte et al. 1981; quoted, Butte et al. 1987)
1.96 (gen. col.-HPLC, Wasik et al. 1981; quoted, Sangster 1989)
1.95 (quoted from Hansch & Leo 1979, Miyake & Terada 1982)
1.99 (RP-HPLC-k', Miyake & Terada 1982)
1.95 (Hansch & Leo 1985; quoted, Howard 1989)
1.95, 2.01 (quoted, calculated-TSA, Iwase et al. 1985)
1.98 (recommended, Sangster 1989)
1.97 (calculated, Müller & Klein 1992)
1.95 (EPA CLOGP Data Base, Hulzebos et al. 1993)
2.12 (calculated-CLOGP 3.51, Jaworska & Schultz 1993)
1.97, 1.98 (COMPUTOX data bank, Kaiser 1993)
2.12 (calculated-QSAR, Kollig 1993)

Bioconcentration Factor, log BCF:
1.26 (calculated-K_{OW}, Lyman et al. 1982; quoted, Howard 1989)
1.03 (zebrafish, Butte et al. 1987)

Sorption Partition Coefficient, log K_{OC}:
1.34 (Brookstone clay loam soil at pH 5.7, Boyd 1982; quoted, Howard 1989)
1.26 (calculated-S, Boyd 1982; quoted, Howard 1989)
1.76 (calculated-K_{OW}, Kollig 1993)

Half-Lives in the Environment:

Air: rate constant of 34×10^{-12} cm^3 molecule^{-1} sec^{-1} for the reaction with hydroxyl radicals in air with reference to the reaction of OH radicals with 2-methyl-2-butene at 300 $\pm$ 1 K (Atkinson et al. 1979; quoted, Carter et al. 1981); rate constant of $(12 \pm 2) \times 10^{-12}$ cm^3 molecule^{-1} sec^{-1} for the reaction with NO_3 radicals in air with reference to the reaction of NO_3 radicals with 2-methyl-2-butene at 300 $\pm$ 1 K (Japar & Niki 1975; Graham & Johnston 1978; quoted, Carter et al. 1981); photooxidation half-life of 1.6-16 hours in air, based on measured rate data for the vapor phase reaction with hydroxyl radicals in air (Atkinson 1985; selected, Howard et al. 1991); loss rate of 10.8×10^{-4} min^{-1} in outdoor Teflon chambers in the dark (Grosjean 1985); estimated photodegradation half-life of 9.6 hours in air, based on measured rate constant for the reaction with photochemically generated hydroxyl radicals (quoted, Howard 1989); rate constant of 42×10^{-12} cm^3 molecule^{-1} sec^{-1} for the reaction with OH radicals in air with reference to the reaction of OH radicals with 2-methyl-2-butene at 296 $\pm$ 2 K (Atkinson 1989; quoted, Atkinson et al. 1992); rate constant of $(13.7 \pm 0.9) \times 10^{-12}$ cm^3 molecule^{-1} sec^{-1} for the reaction with NO_3 radicals in air with reference to the reaction of NO_3 radicals with 2-methyl-2-butene at 296 $\pm$ 2 K (Atkinson et al. 1992); atmospheric transformation lifetime was estimated to be $<$ 1 day (Kelly 1994).

Surface water: photooxidation half-life of 66-3480 hours in water, based on reaction rate constants for OH and RO_2 radicals with the phenol class (Güesten et al. 1981; Mill & Mabey 1985; quoted, Howard et al. 1991); rate constant of $(1.2 \pm 0.3)x10^4\ M^{-1}\ sec^{-1}$ for the reaction with ozone at pH 1.5/2.0 (Hoigné & Bader 1983).

Ground water: estimated half-life for cresols, 0.01 year at Noordwijk (Zoeteman et al. 1981).

Sediment:

Soil: 24-168 hours, based on estimated unacclimated aqueous aerobic biodegradation half-life (Howard et al. 1991); 5.1 days in an acidic clay soil with <1.0% organic matter and 1.6 day in slightly basic sandy loam soil with 3.25% organic matter, based on aerobic batch lab. microcosm experiments (Loehr & Matthews 1992).

Biota:

Environmental Fate Rate Constants and Half-Lives:

Volatilization:

Photolysis:

Oxidation: rate constant of $34x10^{-12}\ cm^3 \cdot molecule^{-1} \cdot sec^{-1}$ for the reaction with OH radicals in air with reference to the reaction of OH radicals with 2-methyl-2-butene at 300 ± 1 K (Perry et al. 1977; Atkinson et al. 1979; quoted, Carter et al. 1981); rate constant of $(12 \pm 2)x10^{-12}\ cm^3 \cdot molecule^{-1} \cdot sec^{-1}$ for the reaction with NO_3 radicals in air with reference to the reaction of NO_3 radicals with 2-methyl-2-butene at 300 ± 1 K (Japar & Niki 1975; Graham & Johnston 1978; quoted, Carter et al. 1981); rate constant of $<2x10^2\ M^{-1}\ sec^{-1}$ for oxidation with singlet oxygen at 25°C in the aquatic system with half-life >100 years (Foote 1976; Mill 1979; quoted, Mill 1982) and $20x10^{-9}\ M^{-1}\ sec^{-1}$ for oxidation with OH radicals at 25°C in the atmosphere with half-life of 0.3 day (Hendry & Kenley 1979; quoted, Mill 1982); rate constant of $(13.9 \pm 2.4)x10^{-12}\ cm^3 \cdot molecule^{-1} \cdot sec^{-1}$ for the reaction with NO_3 radicals in air with reference to the reaction of NO_3 radicals with 2-methyl-2-butene at 300 ± 1 K (Carter et al. 1981; quoted, Atkinson 1991); photooxidation half-life of 66-3480 hours in water, based on reaction rate constants for OH and RO_2 radicals with the phenol class (Güesten et al. 1981; Mill & Mabey 1985; quoted, Howard et al. 1991); rate constant of $2.6x10^{-19}\ cm^3 \cdot molecule^{-1} \cdot sec^{-1}$ for the gas-phase reaction with O_3 in air at 296 ± 2 K (Atkinson et al. 1982, 1984; Atkinson 1985); rate constant of $2x10^{-8}\ M^{-1} \cdot sec^{-1}$ with half-life of 0.3 day for reaction with OH radical in the atmosphere (Mill 1982); rate constant $2.0x10^{13}\ cm_3 \cdot mol^{-1} \cdot sec^{-1}$ at 300 K (Lyman 1982); rate constant of $(1.2 \pm 0.3)x10^4\ M^{-1} \cdot sec^{-1}$ for the reaction with ozone in water at pH 1.5-2.0 (Hoigné & Bader 1983); rate constant of $4.1x10^{-11} \cdot cm^3$ $molecule^{-1} \cdot sec^{-1}$ for the gas-phase reaction with OH radicals at 298 K in air (Atkinson & Lloyd 1984; quoted, Carlier et al. 1986); rate constant of $1x10^{-11}$ $cm^3 \cdot molecule^{-1} \cdot sec^{-1}$ for the gas-phase reaction with NO_3 radicals at 300 K in air (Atkinson & Lloyd 1984; quoted, Carlier et al. 1986); second-order rate constant of $(1.20 \pm 0.34)x10^{-11}\ cm^3 \cdot molecule^{-1} \cdot sec^{-1}$ for the gas-phase reaction with NO_3 radicals at 296 ± 2 K in air (Atkinson et al. 1984); $(15.6 \pm 1.7)x10^{-12}$ $cm^3 \cdot molecule^{-1} \cdot sec^{-1}$ for the reaction with NO_3 radicals in air with reference to the reaction of NO_3 radicals with *m*-cresol at 298 ± 1 K (Atkinson et al. 1984; quoted, Atkinson 1991); photooxidation half-life of 1.6-16 hours in air, based on measured rate constant of 22×10^{-12} $cm^3 \cdot molecule^{-1} \cdot sec^{-1}$ for the vapor-phase reaction with 2.4×10^8 NO_3 radicals/cm^3 at room temperature in air for a loss rate of 450 day^{-1} (Atkinson et al. 1984, 1985; quoted, Atkinson 1985); 40×10^{-12} $cm^3 \cdot molecule^{-1} \cdot sec^{-1}$ for the vapor-phase reaction with 5×10^5 hydroxyl radicals/cm^3 at room temperature in air for a loss rate of 1.7 day^{-1} (Atkinson 1985; quoted, Howard et al. 1991); comparison of calculated reaction rate constant of 44.8×10^{-12} $cm^3 \cdot molecule^{-1} \cdot sec^{-1}$ with experimental rate constant of 37×10^{-12} $cm^3 \cdot molecule^{-1} \cdot sec^{-1}$ for the gas-phase reaction with OH radicals in air at room temperature (Atkinson 1985); comparison of calculated second-order rate constant of 56.2×10^{-12} $cm^3 \cdot molecule^{-1} \cdot sec^{-1}$ with observed value of 38.0×10^{-12} $cm^3 \cdot molecule^{-1} \cdot sec^{-1}$ for the gas-phase reaction with OH radicals in air at room temperature (Atkinson et al. 1985); calculated rate constant of 44×10^{-12} $cm^3 \cdot molecule^{-1} \cdot sec^{-1}$ and observed rate constant of 40×10^{-12} $cm^3 \cdot molecule^{-1} \cdot sec^{-1}$ for the gas-phase reaction with OH radicals in air at room temperature (Atkinson 1987; quoted, Sabljic & Güesten 1990; Müller & Klein 1991); rate constant of 42×10^{-12} $cm^3 \cdot molecule^{-1} \cdot sec^{-1}$ for the gas-phase reaction with NO_3 radicals in air at 298 K (Atkinson et al. 1988; quoted, Atkinson 1990; Sabljic & Güesten 1990; Müller & Klein 1991); rate constant of 20.9×10^{-12} $cm^3 \cdot molecule^{-1} \cdot sec^{-1}$ for the reaction with OH radicals in air with reference to the reaction of OH radicals with 2-methyl-2-butene at 296 ± 2 K (Atkinson 1989; quoted, Atkinson et al. 1992); rate constant of $(13.7 \pm 0.9) \times 10^{-12}$ $cm^3 \cdot molecule^{-1} \cdot sec^{-1}$ for the reaction with NO_3 radicals in air with reference to the reaction of NO_3 radicals with 2-methyl-2-butene at 296 ± 2 K (Atkinson et al. 1992); gas-phase hydroxyl radical rate constants of 42.0×10^{-12} $cm^3 \cdot molecule^{-1} \cdot sec^{-1}$ (exptl.), 30.36×10^{-12} $cm^3 \cdot molecule^{-1} \cdot sec^{-1}$, calculated based on molecular orbital estimation method (Klamt 1993).

Hydrolysis:

Biodegradation: 1 to 2 days for bacteria to utilize 95% of 300 ppm in the parent substrate (Tabak et al. 1964); half-life of 2 days at 20°C and 7 days at 4°C in river water (Ludzack & Ettinger 1960; quoted, Howard 1989); completely degraded by a soil microflora within one day (Alexander & Lustigman 1966; quoted, Verschueren 1983); average rate of biodegradation 54.0 mg COD $g^{-1} \cdot h^{-1}$ based on measurements of COD decrease using activated sludge inoculum with 20 days of adaptation to the substrate (Pitter 1976; quoted, Scow 1982); aqueous aerobic half-life of 24-168 hours, based on unacclimated aerobic screening test data (Takemoto et al. 1981; Urushigawa et al. 1983; quoted, selected, Howard et al. 1991); aqueous anaerobic half-life of 96-672 hours, based on estimated unacclimated aqueous aerobic biodegradation half-life (Howard et al. 1991); exptl. rate constant of 0.0679 h^{-1} compared to group contribution method predicted rate constants of 0.0728 h^{-1} (nonlinear) and 0.0567 $hour^{-1}$ (linear) (Tabak & Govind 1993).

Biotransformation: microbial transformation rate constant of $(2.7 \pm 1.3) \times 10^{-10}$ $L \cdot organism^{-1} \cdot h^{-1}$ (Paris et al. 1983; quoted, Steen 1991).

Bioconcentration Uptake (k_1) and Elimination (k_2) Rate Constants:

Common Name: *m*-Cresol
Synonym: *m*-cresylic acid, 1-hydroxy-3-methylbenzene, 3-hydroxytoluene, *m*-hydroxytoluene, 3-methylphenol, *m*-methylphenol
Chemical Name: *m*-cresol, 3-methylphenol
CAS Registry No: 108-39-4
Molecular Formula: $C_6H_4(CH_3)OH$
Molecular Weight: 108.13
Melting Point (°C):
10.9 (Stull 1947)
10-11 (Ericksen & Dobbert 1955)
12.22 (Andon et al. 1960; Tsonopoulos & Prausnitz 1971)
12 (Verschueren 1983)
Boiling Point (°C):
202.8 (Stull 1947)
202.5 (Ericksen & Dobbert 1955)
202.23 (Biddiscombe & Martin 1958; Andon et al. 1960)
202.12 (Boublik et al. 1973,84)
202.0 (Verschueren 1983)
201.89 (Boublik et al. 1984)
Density (g/cm^3):
1.5402 (Ericksen & Dobbert 1955)
1.0341 (Andon et al. 1960)
1.0380 (Verschueren 1983)
1.0302 (25°C, Riddick et al. 1986)
Dissociation Constants, pK_a:
10.09 (Pearce & Simkins 1968)
10.00 (Hoigné & Bader 1983; Dean 1985)
10.01 (Weast 1982-83)
10.09 (Riddick et al. 1986; Howard 1989)
Molar Volume (cm^3/mol):
127 (Andon et al. 1960)
105.6 (calculated-density, Rohrschneider 1973)
116.3 (Kamlet et al. 1986)
0.634 (instrinsic $V_I/100$, Kamlet et al. 1988)
141.0 (calculated-LeBas method, Abernethy et al. 1988)
64.0 (intrinsic volume V_I, Abernethy et al. 1988)
Molecular Volume, TMV ($Å^3$):
Total Surface Area, TSA ($Å^2$):
250.8 (Iwase et al. 1985)
Heat of Fusion ΔH_{fus}, kcal/mol:
2.56 (Tsonopoulos & Prausnitz 1971)
Entropy of Fusion ΔS_{fus}, cal/mol·K (e.u.):
8.97 (Tsonopoulos & Prausnitz 1971)
Fugacity Ratio at 25°C (assuming ΔS_{fus} = 13.5 e.u.), F: 1.0

Water Solubility (g/m^3 or mg/L at 25°C):

22664 (shake flask-centrifuge, Booth & Everson 1948)
21575 (estimated, McGowan 1954)
22000 (20°C, shake flask-optical, Ericksen & Dobbert 1955)
25000 (shake flask-AS, Roberts et al. 1977)
2700 (50°C, quoted, Haque et al. 1980)
21084, 21575 (quoted, calculated-K_{OW} and M.P., Valvani & Yalkowsky 1980)
2800 (gen. col.-HPLC, Wasik et al. 1981; Tewari et al. 1982)
22000 (20-25°C, Urano et al. 1982)
23500, 58000 (20°C, 100°C, Verschueren 1977,83)
23000 (Leuenberger et al. 1985; quoted, Howard 1989)
23656, 80158 (quoted, calculated-fragment const., Wakita et al. 1986)
21928, 8110 (quoted, calculated- χ , Nirmalakhandan & Speece 1988)
21800 (selected, Yaws et al. 1990)
21085 (quoted from Valvani & Yalkowsky 1980, Müller & Klein 1992)
19600 (shake flask-HPLC/UV at pH 4.45, Varhanickova et al. 1995)

Vapor Pressure (Pa at 25°C):

23.95 (extrapolated-regression of tabulated data, Stull 1947)
19.0 (gas saturation-diaphragm manometer fitted to Antoine eqn., Biddiscombe & Martin 1958)
19.0 (calculated-Antoine eqn., Andon et al. 1960)
22.19 (extrapolated-Antoine eqn., Boublik et al. 1973,84)
5.33 (20°C, quoted, Verschueren 1977,83)
16.0 (30°C, quoted, Verschueren 1977,83)
133.3 (52°C, Andon et al. 1960; Haque et al. 1980)
36.43 (calculated-Cox eqn., Chao et al. 1983)
11.89 (extrapolated-Antoine eqn., Boublik et al. 1984)
22.29 (extrapolated-Antoine eqn., Dean 1985)
18.7 (Leuenberger et al. 1985)
19.1 (Riddick et al. 1986; quoted, Howard 1989)
19.7 (extrapolated-Antoine eqn., Nesterova et al. 1990)

Henry's Law Constant ($Pa \cdot m^3/mol$):

0.0203 (8°C, Leuenberger et al. 1985)
0.0881 (calculated, Leuenberger et al. 1985; quoted, Howard 1989)
0.0875 (calculated-P/C, Meylan & Howard 1991)
0.0627 (estimated-bond contribution, Meylan & Howard 1991)
0.0718 (computed, Yaws et al. 1991)
0.1068 (calculated-P/C, Shiu et al. 1994)

Octanol/Water Partition Coefficient, log K_{OW}:

1.96 (shake flask-AS, Fujita et al. 1964; quoted, Leo et al. 1971; Rekker 1977; Roberts et al. 1977; Sangster 1989)

1.96, 2.01 (quoted, Leo et al. 1971; Verschueren 1983)
2.07 (calculated using data from Fujita et al. 1964, Umeyama et al. 1971)
1.96, 2.03 (quoted, calculated-f const., Rekker 1977)
1.95 (quoted, Valvani & Yalkowsky 1980)
1.94 (HPLC, Butte et al. 1981; quoted, Sangster 1989)
1.96 (gen. col.-HPLC, Wasik et al. 1981; Tewari et al. 1982; quoted, Sangster 1989)
1.96 (shake flask-UV, Saarikoski & Viluksela 1982)
1.96 (Hansch & Leo 1985; quoted, Howard 1989)
1.95, 2.06 (quoted, calculated-TSA, Iwase et al. 1985)
2.07 (computed, Klopman et al. 1985; quoted, Pangrekar et al. 1994)
1.95 (quoted, Kamlet et al. 1986)
1.97 (selected, Yoshida et al. 1986)
1.96, 2.11 (quoted exptl., calculated-V_I and solvatochromic p., Kamlet et al. 1988)
1.98 (recommended, Sangster 1989, 1993)
1.97 (calculated-CLOGP, Müller & Klein 1992)
1.94 (quoted, Saito et al. 1993)
1.96 (EPA CLOGP Data Base, Hulzebos et al. 1993)
1.96 (COMPUTOX databank, Kaiser 1993)
2.12 (calculated-QSAR, Kollig 1993)

Bioconcentration Factor, log BCF:
1.30 (golden ide, after 3 days, Freitag et al. 1985; quoted, Howard 1989)
3.69 (algae, after 1 day, Freitag et al. 1985; quoted, Howard 1989)
3.04 (activated sludge, after 5 days, Freitag et al. 1985)
1.03 (zebrafish, Butte et al. 1987)

Sorption Partition Coefficient, log K_{OC}:
1.54 (Brookstone clay loam soil, Boyd 1982; quoted, Howard 1989)
1.26 (calculated-S, Boyd 1982; quoted, Howard 1989)
1.76 (calculated-K_{OW}, Kollig 1993)

Half-Lives in the Environment:
Air: rate constant of $(7 \pm 1) \times 10^{-12}$ cm^3 $molecule^{-1}$ sec^{-1} for the reaction with NO_3 radicals in air with reference to the reaction of NO_3 radicals with 2-methyl-2-butene (Japar & Niki 1975; Graham & Johnston 1978; quoted, Carter et al. 1981); rate constant of 48×10^{-12} cm^3 $molecule^{-1}$ sec^{-1} for the reaction with OH radicals in air with reference to the reaction of OH radicals with 2-methyl-2-butene (Atkinson et al. 1979; quoted, Carter et al. 1981); half-life of 8.0 hours, based on rate constant for the gas phase reaction with hydroxyl radicals in air (Atkinson et al. 1979; quoted, Howard 1989); photooxidation half-life of 1.1-11.3 hours in air, based on measured rate data for the vapor phase reaction with hydroxyl radicals in air (Atkinson 1985; selected, Howard et al. 1991); rate constant of $(9.74 \pm 0.47) \times 10^{-12}$ cm^3 $molecule^{-1}$ sec^{-1} for the reaction with NO_3 radicals in air with reference to the reaction of NO_3 radicals with 2-methyl-2-

butene (Atkinson et al. 1992); rate constant of $64x10^{-12}$ cm^3 $molecule^{-1}$ sec^{-1} for the reaction with OH radicals in air with reference to the reaction of OH radicals with 2-methyl-2-butene (Atkinson 1989; quoted, Atkinson et al. 1992); atmospheric transformation lifetime was estimated to be < 1 day (Kelly et al. 1994).

Surface water: 48-696 hours, based on unacclimated marine water grab sample data (Pfaender & Batholomew 1982; selected, Howard et al. 1991); rate constant of $(1.3 \pm 0.3)x10^4$ M^{-1} sec^{-1} for the reaction with ozone at pH 1.5/2.0 (Hoigné & Bader 1983).

Ground water: estimated half-life for cresols, 0.01 year at Noordwijk (Zoeteman et al. 1981); 96-1176 hours, based on estimated unacclimated aqueous aerobic biodegradation half-life and aqueous anaerobic biodegradation half-life (Howard et al. 1991).

Sediment:

Soil: 48-696 hours, based on unacclimated marine water grab sample data (Pfaender & Batholomew 1982; selected, Howard et al. 1991); 11.3 days in an acidic clay soil with <1.0% organic matter and 0.6 day in a slightly basic sandy loam soil with 3.25% organic matter, based on aerobic batch lab. microcosm experiments (Loehr & Matthews 1992).

Biota:

Environmental Fate Rate Constants and Half-Lives:

Volatilization:

Photolysis: direct aqueous photolysis rate constant of 0.011 ± 0.001 $minute^{-1}$ with a half-life of 60.4 minutes (Stegeman et al. 1993).

Oxidation: rate constant of $(7 \pm 1)x10^{-12}$ cm^3 $molecule^{-1}$ sec^{-1} for the reaction with NO_3 radicals in air with reference to the reaction of NO_3 radicals with 2-methyl-2-butene (Japar & Niki 1975; Graham & Johnston 1978; quoted, Carter et al. 1981); rate constant of $48x10^{-12}$ cm^3 $molecule^{-1}$ sec^{-1} for the reaction with OH radicals in air with reference to the reaction of OH radicals with 2-methyl-2-butene (Atkinson et al. 1979; quoted, Carter et al. 1981); rate constant of $(8.10 \pm 1.16)x10^{-12}$ cm^3 $molecule^{-1}$ sec^{-1} for the reaction with NO_3 radicals at 300 ± 1 K in air with reference to the reaction of NO_3 radicals with 2-methyl-2-butene (Carter et al. 1981; quoted, Atkinson 1991); rate constant of $1.94x10^{-19}$ cm^3 $molecule^{-1}$ sec^{-1} for the gas-phase reaction with O_3 in air at 296 ± 2 K (Atkinson et al. 1982,84); photooxidation half-life of 66-3480 hours in water, based on reported reaction rate constants for OH and RO_2 radicals with the phenol class (Mill & Mabey 1985; Güesten et al. 1981; selected, Howard et al. 1991); rate constant of $(1.3 \pm 0.3)x10^4$ M^{-1} sec^{-1} for the reaction with ozone in water at pH 1.5/2.0 (Hoigné & Bader 1983); rate constant of $5.9x10^{-11}$ cm^3 $molecule^{-1}$ sec^{-1} for the gas-phase reaction with OH radicals in air at 298 K (Atkinson & Lloyd 1984; quoted, Carlier et al. 1986); rate constant of $1x10^{-11}$ cm^3 $molecule^{-1}$ sec^{-1} for the gas-phase reaction with NO_3 radicals in air at 300 K (Atkinson & Lloyd 1984; quoted, Carlier et al. 1986); second-order rate constant of $(9.20 \pm 2.4)x10^{-12}$ cm^3 $molecule^{-1}$ sec^{-1} for the gas-phase reaction with NO_3 radicals in air at 296 ± 2 K (Atkinson et al. 1984); $(15.9 \pm 1.7)x10^{-12}$ cm^3 $molecule^{-1}$ sec^{-1}

for the gas-phase reaction with NO_3 radicals in air at 298 ± 1 K with reference to the reaction of NO_3 radicals with phenol (Atkinson et al. 1984; quoted, Atkinson 1991); photooxidation half-life of 1.1-11.3 hours in air, based on measured rate data for the vapor phase reaction with hydroxyl radicals in air (Atkinson 1985; selected, Howard et al. 1991); comparison of calculated rate constant 93.9×10^{-12} cm^3 $molecule^{-1}$ sec^{-1} with observed rate constant of 54×10^{-12} cm^3 $molecule^{-1}$ sec^{-1} for the gas-phase reaction with OH radicals in air at room temperature (Atkinson 1985); comparison of calculated rate constant 123×10^{-12} cm^3 $molecule^{-1}$ sec^{-1} with observed rate constant of 56×10^{-12} cm^3 $molecule^{-1}$ sec^{-1} for the gas-phase reaction with OH radicals in air at room temperature (Atkinson et al. 1985); calculated rate constant of 94×10^{-12} cm^3 $molecule^{-1}$ sec^{-1} and observed rate constant of 57×10^{-12} cm^3 $molecule^{-1}$ sec^{-1} for the reaction with OH radicals in air at room temperature (Atkinson 1987; quoted, Sabljic & Güesten 1990; Müller & Klein 1991); rate constant of 15.8×10^{-12} cm^3 $molecule^{-1}$ sec^{-1} for the gas-phase reaction with NO_3 radicals in air at 298 K (Atkinson et al. 1988; quoted, Sabljic & Güesten 1990; Müller & Klein 1991); rate constant of $(9.74 \pm 0.74)\times10^{-12}$ $cm^3\cdot molecule^{-1}\cdot sec^{-1}$ for the reaction with NO_3 radicals in air with reference to the reaction of NO_3 radicals with 2-methyl-2-butene (Atkinson et al. 1992); rate constant of 64×10^{-12} $cm^3\cdot molecule^{-1}\cdot sec^{-1}$ for the reaction with OH radicals in air with reference to the reaction of OH radicals with 2-methyl-2-butene (Atkinson 1989; quoted, Atkinson 1991; Atkinson et al. 1992); exptl. gas-phase hydroxyl radical rate constants of 64.0×10^{-12} cm^3 $molecule^{-1}$ sec^{-1} compared to calculated value of 34.42×10^{-12} cm^3 $molecule^{-1}$ sec^{-1}, based on molecular orbital estimation method (Klamt 1993).

Hydrolysis:

Biodegradation: 1 to 2 days for bacteria to utilize 95% of 300 ppm in parent substrate (Tabak et al. 1964); completely degraded by a soil microflora within one day (Alexander & Lustigman 1966; quoted, Verschueren 1983); average rate of biodegradation 55.0 mg COD $g^{-1}\cdot h^{-1}$ based on measurements of COD decrease using activated sludge inoculum with 20 days of adaptation to the substrate (Pitter 1976; quoted, Scow 1982); aqueous aerobic half-life of 48-696 hours, based on unacclimated marine water grab sample data (Pfaender & Batholomew 1982; selected, Howard et al. 1991); aqueous anaerobic half-life of 360-1176 hours, based on anaerobic screening test data (Horowitz et al. 1982; Shelton & Tiedje 1981; selected, Howard et al. 1991).

Biotransformation:

Bioconcentration Uptake (k_1) and Elimination (k_2) Rate Constants:

Common Name: *p*-Cresol
Synonym: *p*-cresylic acid, 1-hydroxy-4-methylbenzene, 4-hydroxytoluene, 4-methylphenol, *p*-hydroxytoluene, *p*-methylphenol
Chemical Name: 4-methylphenol
CAS Registry No: 106-44-5
Molecular Formula: $CH_3C_6H_4OH$
Molecular Weight: 108.13
Melting Point (°C):
35.5 (Stull 1947)
35-36 (Ericksen & Dobbert 1955)
34.69 (Andon et al. 1960)
34.75 (Tsonopoulos & Prausnitz 1971)
35.00 (Valvani & Yalkowsky 1980)
34.80 (Verschueren 1983)
34.85 (Mackay et al. 1982)
Boiling Point (°C):
201.8 (Stull 1947)
202 (Ericksen & Dobbert 1955; Verschueren 1983; Banerjee et al. 1990)
201.94 (Biddiscombe & Martin 1958; Andon et al. 1960)
201.89 (Boublik et al. 1973,84)
201.85 (Mackay et al. 1982)
202.05 (Boublik et al. 1984)
Density (g/cm^3 at 20°C):
1.5398 (supercooled liq., Ericksen & Dobbert 1955)
1.154 (25°C, Andon et al. 1960; Riddick et al. 1986)
1.0347 (Verschueren 1983)
Dissociation Constants, pK_a:
10.26 (Pearce & Simkins 1968, Dean 1985, Riddick et al. 1986; Howard 1989)
10.28 (Serjeant & Dempsey 1979; Tratnyek & Hoigné 1991; Haderlein & Schwarzenbach 1993)
10.20 (Hoigné & Bader 1983; Scully & Hoigné 1987)
10.17 (Weast 1982-83)
10.27 (Miyake et al. 1987)
Molar Volume (cm^3/mol):
125.6 (calculated-LeBas method)
127 (Andon et al. 1960)
0.634 (intrinsic $V_I/100$, Kamlet et al. 1988)
99.0 (calculated-density, Jaworska & Schultz 1993)
Molecular Volume, TMV ($Å^3$):
Total Surface Area, TSA ($Å^2$):
251.2 (Iwase et al. 1985)
Heat of Fusion ΔH_{fus}, kcal/mol:
3.040 (Tsonopoulos & Prausnitz 1971)
Entropy of Fusion ΔS_{fus}, cal/mol K (e.u.):
9.873 (Tsonopoulos & Prausnitz 1971)
Fugacity Ratio at 25°C (assuming ΔS_{fus} = 13.5 e.u.), F: 0.800

Water Solubility (g/m^3 or mg/L at 25°C):

24234 (shake flask-centrifuge, Booth & Everson 1948)
17000 (20°C, shake flask-optical, Ericksen & Dobbert 1955)
23000 (quoted, Morrison & Boyd 1959)
1800 (quoted, Lange 1973; Smith et al. 1978)
21000 (shake flask-AS, Roberts et al. 1977)
17000 (35°C, quoted, Haque et al. 1980)
24000, 53000 (40°C, 100°C, Verschueren 1977,83)
16747, 17537 (quoted, calculated-K_{OW} and M.P., Valvani & Yalkowsky 1980)
19000 (20-25°C, Urano et al. 1982)
1800 (selected, Neely 1982; McKim et al. 1985)
1840 (recommended, Neely & Blau 1985)
18000 (Leuenberger et al. 1985)
22600 (40°C, Riddick et al. 1986; quoted, Howard 1989)
19543, 8110 (quoted, calculated- χ , Nirmalakhandan & Speece 1988)
19400 (selected, Yaws et al. 1990)
16760 (quoted from Valvani & Yalkowsky 1980, Müller & Klein 1992)
22000 (shake flask-HPLC/UV at pH 3.9, Varhanickova et al. 1995)

Vapor Pressure (Pa at 25°C):

21.89 (extrapolated-regression of tabulated data, Stull 1947)
15.91 (gas saturation-diaphragm manometer fitted to Antoine eqn., Biddiscombe & Martin 1958)
16.0 (calculated-Antoine eqn., Andon et al. 1960)
133.3 (53°C, Andon et al. 1960; Haque et al. 1980)
11.8 (extrapolated-Antoine eqn., Boublik et al. 1973,84)
14.7 (quoted, Verschueren 1977,83)
14.4 (quoted, Smith et al. 1978)
15.20, 51.67 (quoted, calculated-B.P., Mackay et al. 1982)
13.94 (extrapolated-Cox eqn., Chao et al. 1983)
17.46 (calculated-Antoine eqn., Boublik et al. 1984)
11.83 (extrapolated-Antoine eqn., Dean 1985)
11.75 (recommended, Neely & Blau 1985)
14.66 (Howard et al. 1986; quoted, Banerjee et al. 1990)
17.3 (Riddick et al. 1986; quoted, Howard 1989)
15.6 (calculated-Antoine eqn., Stephenson & Malanowski 1987)
14.7, 117.3 (quoted, calculated-solvatochromic p., Banerjee et al. 1990)
16.1 (extrapolated-Antoine eqn., Nesterova et al. 1990)

Henry's Law Constant ($Pa \cdot m^3/mol$):

0.0802 (exptl., Hine & Mookerjee 1975)
0.0699 (calculated-group contribution, Hine & Mookerjee 1975)
1.0820 (calculated-bond contribution, Hine & Mookerjee 1975)
0.866 (calculated-P/C, Smith & Bomberger 1980)
0.0223 (8°C, Leuenberger et al. 1985)

0.689 (calculated-P/C, Neely & Blau 1985)
0.0973 (calculated, Leuenberger et al. 1985; quoted, Howard 1989)
0.101 (quoted, Gaffney et al. 1987)
0.0802, 0.143 (quoted, calculated- χ , Nirmalakhandan & Speece 1988)
0.0397 (calculated-P/C, Yaws et al. 1991)
0.0653 (calculated-P/C, Shiu et al. 1994)

Octanol/Water Partition Coefficient, log K_{OW}:
1.94 (shake flask-AS, Fujita et al. 1964; quoted, Leo et al. 1969, 1971; Rekker 1977; Sangster 1989)
1.92, 1.94 (quoted, Leo et al. 1971; Verschueren 1983)
1.99 (shake flask-AS at pH 7.45, Umeyama et al. 1971; quoted, Sangster 1989)
1.94 (LC-k', Carlson et al. 1975)
1.97 (calculated-π const., Carlson et al. 1975; quoted, Sangster 1989)
1.94, 1.92, 1.95; 2.03 (quoted values; calculated-f const., Rekker 1977)
1.95 (quoted from Leo et al. 1971, Roberts et al. 1977)
2.00 (quoted, Hansch & Leo 1979)
1.96 (quoted, Valvani & Yalkowsky 1980; Miyake & Terada 1982)
1.97 (calculated-RPHPLC-k', Miyake & Terada 1982)
2.20 (selected, Neely 1982)
1.92, 1.98 $\pm$ 0.07 (selected best lit. value, exptl.-ALPM, Garst & Wilson 1984)
1.94 (quoted from Hansch & Leo 1979, Haky & Young 1984; Miyake et al. 1987)
1.62 (HPLC-k', Haky & Young 1984)
1.94, 1.73 (quoted, calculated-γ from UNIFAC, Campbell & Luthy 1985; quoted, Sangster 1989)
1.94 (Hansch & Leo 1985; quoted, Howard 1989; El Tayar et al. 1991)
2.07 (computed, Klopman et al. 1985; quoted, Pangrekar et al. 1994)
1.95 (recommended, Neely & Blau 1985)
1.95, 2.06 (quoted, calculated-TSA, Iwase et al. 1985)
2.0 (McKim et al. 1985)
1.91 (calculated-HPLC-k', Miyake et al. 1987)
1.94 (quoted, Schultz et al. 1987; Veith & Broderius 1987)
1.94 (RP-HPLC-capacity ratio, Minick et al. 1988)
1.94, 2.11 (quoted exptl., calculated-V_I and solvatochromic p., Kamlet et al. 1988)
1.97 (recommended, Sangster 1989, 1993)
1.97 (calculated-CLOGP, Müller & Klein 1992)
1.92 (quoted from Leo et al. 1971; Haderlein & Schwarzenbach 1993)
1.94 (calculated-CLOGP 3.51, Jaworska & Schultz 1993)
1.96 (quoted, Saito et al. 1993)
1.94, 1.97 (COMPUTOX databank, Kaiser 1993)
2.12 (calculated-QSAR, Kollig 1993)

Bioconcentration Factor, log BCF:
1.26 (estimated-K_{OW}, Lyman et al. 1982; quoted, Howard 1989)

Sorption Partition Coefficient, log K_{OC}:

2.81 (Coyote Creek sediment, Smith et al. 1978; quoted, Howard 1989)

1.69 (Brookstone clayloam soil, Boyd 1982; quoted, Howard 1989)

-0.046 (predicted-S, Boyd 1982; quoted, Howard 1989)

1.76 (calculated-K_{OW}, Kollig 1993)

Half-Lives in the Environment:

Air: rate constant of $(13 \pm 2)\times10^{-12}$ $cm^3 \cdot molecule^{-1} \cdot sec^{-1}$ for the reaction with NO_3 radicals in air at 300 ± 1 K (Japar & Niki 1975; Graham & Johnston 1978; quoted, Carter et al. 1981); rate constant of 38×10^{-12} $cm^3 \cdot molecule^{-1} \cdot sec^{-1}$ for the reaction with OH radicals in air at 300 ± 1 K (Atkinson et al. 1979; quoted, Carter et al. 1981); photooxidation half-life of 10 hours, based on measured rate data for the vapor phase reaction with hydroxyl radicals in air (Atkinson et al. 1979; quoted, Howard 1989); photooxidation half-life of 1.5-15 hours, based on measured rate data for the vapor phase reaction with hydroxyl radicals in air (Atkinson 1985; quoted, Howard et al. 1991); rate constant of 47×10^{-12} $cm^3 \cdot molecule^{-1} \cdot sec^{-1}$ for the reaction with OH radicals in air at 296 ± 2 K (Atkinson 1989; quoted, Atkinson et al. 1992); rate constant of $(10.7 \pm 1)\times10^{-12}$ $cm^3 \cdot molecule^{-1} \cdot sec^{-1}$ for the reaction with NO_3 radicals in air at 296 ± 2 K (Atkinson et al. 1992); atmospheric transformation lifetime was estimated to be < 1 day (Kelly et al. 1994).

Surface water: 0.55 hour in river, 12 hours in pond and eutrophic lake, and 2400 hours in oligotrophic lake for a point source continuously discharging 1.0 $\mu g\ ml^{-1}$ predicted by one compartment model for all processes including dilution (Smith et al. 1978; quoted, Howard 1989); rate constant of $(3.0 \pm 0.6)\times10^4$ $M^{-1} \cdot sec^{-1}$ for the reaction with ozone at pH 1.5/2.0 (Hoigné & Bader 1983); half-life of 1-16 hours, based on unacclimated marine and freshwater grab sample data (Van Veld & Spain 1983; Rogers et al. 1984; quoted, Howard et al. 1991); half-life of 500 hours for the reaction with singlet oxygen in water at pH 8 and (19 ± 2)°C (Scully & Hoigné 1987).

Ground water: estimated half-life for cresols, 0.01 year at Noordwijk (Zoeteman et al. 1981); 2-672 hours, based on estimated unacclimated aqueous aerobic biodegradation half-life and aqueous anaerobic biodegradation half-life (Howard et al. 1991).

Sediment:

Soil: 1-16 hours, based on unacclimated marine and freshwater grab sample data (Van Veld & Spain 1983; Rogers et al. 1984; quoted, Howard et al. 1991); 0.5 day in an acidic clay soil with <1.0% organic matter and 1.0 day in a slightly basic sandy loam soil with 3.25% organic matter, based on aerobic batch lab. microcosm experiments (Loehr & Matthews 1992).

Biota:

Environmental Fate Rate Constants and Half-Lives:

Volatilization: estimated half-life from a lake is 167 days as calculated from equations of Mackay & Wolkoff 1973 (Smith et al. 1978).

Photolysis: laboratory determined rate constant of 6.8×10^{-7} sec^{-1} under overcast weather of April at 25°C (Smith et al. 1978; quoted, Howard 1989); estimated half-life of 4800 hours in river, > 10000 hours in both eutrophic lake and pond and 2400 hours in oligotrophic lake, based on an average photolysis rate on a summer day at 40° latitude by the one compartment model (Smith et al. 1978; quoted, Howard 1989); photolytic half-life of 5800 hours in aquatics (Haque et al. 1980).

Oxidation: photooxidation half-life of 144-11325 hours in water, based on measured rate data for reactions with singlet oxygen and hydroxyl radicals in aqueous solution (Anbar & Neta 1967; Scully & Hoigne 1987; quoted, Howard et al. 1991); rate constant of $(13 \pm 2) \times 10^{-12}$ $cm^3 \cdot molecule^{-1} \cdot sec^{-1}$ for the reaction with NO_3 radicals in air at 300 ± 1 K (Japar & Niki 1975; Graham & Johnston 1978; quoted, Carter et al. 1981); rate constant of 38×10^{-12} $cm^3 \cdot molecule^{-1} \cdot sec^{-1}$ for the reaction with OH radicals in air at 300 ± 1 K (Atkinson et al. 1979; quoted, Carter et al. 1981); photooxidation half-life of 10 hours, based on measured rate data for the vapor phase reaction with hydroxyl radicals in air (Atkinson et al. 1979; quoted, Howard 1989); second-order rate constant of $(15 \pm 2.4) \times 10^{-12}$ $cm^3 \cdot molecule^{-1} \cdot sec^{-1}$ by relative technique with reference to 2-methyl-2-butene for the reaction with NO_3 radicals at 300 ± 1 K in air (Carter et al. 1981; quoted, Atkinson 1991); rate constant of 4.71×10^{-19} $cm^3 \cdot molecule^{-1} \cdot sec^{-1}$ for the gas-phase reaction with O_3 in air at 296 ± 2 K (Atkinson et al. 1982, 84); rate constant 4.50×10^{-11} $cm^3 \cdot molecule^{-1} \cdot sec^{-1}$ for the gas-phase reaction with OH radicals at 298 K in air (Atkinson & Lloyd 1984; quoted, Carlier et al. 1986); rate constant of 1×10^{-11} $cm^3 \cdot molecule^{-1} \cdot sec^{-1}$ for the gas-phase reaction with NO_3 radicals at 300 K in air (Atkinson & Lloyd 1984; quoted, Carlier et al. 1986); rate constant of $(1.27 \pm 0.36) \times 10^{-11}$ $cm^3 \cdot molecule^{-1} \cdot sec^{-1}$ for the gas-phase reaction with NO_3 radicals at 296 ± 2 K in air (Atkinson et al. 1984); rate constant of $(16.6 \pm 1.8) \times 10^{-12}$ $cm^3 \cdot molecule^{-1} \cdot sec^{-1}$ by relative technique with reference to *m*-cresol for the reaction with NO_3 radicals at 298 ± 1 K in air (Atkinson et al. 1984; quoted, Atkinson 1991); photooxidation half-life of 1.5-15 hours, based on measured rate data for the vapor phase reaction with hydroxyl radicals in air (Atkinson 1985; selected, Howard et al. 1991); rate constant of $(3.0 \pm 0.6) \times 10^4$ $M^{-1} \cdot sec^{-1}$ for the reaction with ozone in water at pH 1.5/2.0 (Hoigné & Bader 1983); photooxidation half-life of 1.5-15 hours, based on measured rate data for the vapor phase reaction with hydroxyl radicals in air (Atkinson 1985; selected, Howard et al. 1991); comparison of calculated reaction rate constant of 44.8×10^{-12} $cm^3 \cdot molecule^{-1} \cdot sec^{-1}$ with observed rate constant of 41×10^{-12} $cm^3 \cdot molecule^{-1} \cdot sec^{-1}$ for the gas-phase reaction with OH radicals in air at room temperature (Atkinson 1985); comparison of calculated rate constant of 56.2×10^{-12} $cm^3 \cdot molecule^{-1} \cdot sec^{-1}$ with observed rate constant of 42×10^{-12} $cm^3 \cdot molecule^{-1} \cdot sec^{-1}$ for the gas-phase reaction with OH radicals in air at room temperature (Atkinson et al. 1985; quoted, Sabljic & Güesten 1990; Müller & Klein 1991); calculated rate constant of 44×10^{-12} $cm^3 \cdot molecule^{-1} \cdot sec^{-1}$ and observed rate constant of 44×10^{-12} $cm^3 \cdot molecule^{-1} \cdot sec^{-1}$ for the gas-phase reaction with OH radicals in air at room temperature (Atkinson 1987; quoted, Sabljic & Güesten 1990); rate constant of $1.1x10^7$ M^{-1} sec^{-1} at pH 8.3, $2.4x10^7$ $M^{-1}\cdot sec^{-1}$ at pH 8.8, $1.6x10^8$ $M^{-1}\cdot sec^{-1}$ at pH 10 and $3.5x10^8$ $M^{-1}\cdot sec^{-1}$ at pH 11.5 for the reaction with singlet oxygen in water at (19 ± 2)°C (Scully & Hoigné 1987); rate constant of $21.9x10^{-12}$ $cm^3\cdot molecule^{-1}\cdot sec^{-1}$ for the gas-phase reaction with NO_3 radicals in air at 298 K (Atkinson et al. 1988; quoted, Sabljic & Güesten 1990; Müller & Klein 1991); aqueous rate constant of $(9.6 \pm 2.8)x10^6$ $M^{-1}\cdot sec^{-1}$ for the reaction with singlet oxygen in aqueous phosphate buffer at (27 ± 1)°C (Tratnyek & Hoigné 1991); rate constant of $47x10^{-12}$ $cm^3\cdot molecule^{-1}\cdot sec^{-1}$ for the reaction with OH radicals in air at 296 ± 2 K (Atkinson 1989; quoted, Atkinson 1991; Atkinson et al. 1992); $(10.7 \pm 1)x10^{-12}$ $cm^3\cdot molecule^{-1}\cdot sec^{-1}$ for the reaction with NO_3 radicals in air at 296 ± 2 K (Atkinson et al. 1992); gas-phase hydroxyl radical rate constants of $47.0x10^{-12}$ $cm^3\cdot molecule^{-1}\cdot sec^{-1}$ (quoted exptl.) and $30.83x10^{-12}$ $cm^3\cdot molecule^{-1}\cdot sec^{-1}$, calculated based on molecular orbital estimation method (Klamt 1993).

Hydrolysis: no hydrolyzable functional groups (Smith et al. 1978).

Biodegradation: 1 to 2 days for bacteria to utilize 95% of 300 ppm in parent substrate (Tabak et al. 1964); completely degraded by a soil microflora in one day (Alexander & Lustigman 1966; quoted, Verschueren 1983); average rate of biodegradation 55.0 mg COD $g^{-1}\cdot h^{-1}$ based on measurements of COD decrease using activated sludge inoculum with 20 days of adaptation to the substrate (Pitter 1976; quoted, Scow 1982); major environmental pathway with a half-life of 12 hours in eutrophic lake for a point source continuously discharging 1.0 $\mu g\cdot mL^{-1}$ predicted by one compartment model for all processes including dilution (Smith et al. 1978); laboratory determined rate constant of $5.2x10^{-7}$ $mL\cdot cell^{-1}\cdot h^{-1}$ at 25°C (Smith et al. 1978); calculated first-order rate constants (after a lag period) of 1.7 day^{-1} in river water, 0.8-4.7 day^{-1} in estuary water and 2.8-4.8 day^{-1} in marine water (Vashon & Schwab 1982; quoted, Battersby 1990); aqueous aerobic half-life of 1-16 hours, based on unacclimated marine and freshwater grab sample data (Van Veld & Spain 1983; Rogers et al. 1984; selected, Howard et al. 1991); aqueous anaerobic half-life of 240-672 hours, based on anaerobic screening test data (Boyd et al. 1983; Horowitz et al. 1982; selected, Howard et al. 1991); biodegradation rate constant of $1.72x10^{-17}$ $mol\cdot cell^{-1}\cdot h^{-1}$ in pure culture system (Banerjee et al. 1984).

Biotransformation: estimated half-life of 12 hours in river, eutrophic lake and pond and >10000 hours in oligotrophic lake, based on an average photolysis rate on a summer day at 40° latitude by the one compartment model (Smith et al. 1978).

Bioconcentration Uptake (k_1) and Elimination (k_2) Rate Constants:

Common Name: 2,4-Dimethylphenol
Synonym: 2,4-xylenol, *as-m*-xylenol, 1-hydroxy-2,4-dimethylbenzene
Chemical Name: 2,4-dimethylphenol
CAS Registry No: 105-67-9
Molecular Formula: $CH_3C_6H_3(CH_3)OH$
Molecular Weight: 122.17
Melting Point (°C):
22.8 (Ericksen & Dobbert 1955)
24.5 (Andon et al. 1960; Callahan et al. 1979)
26.0 (Valvani & Yalkowsky 1980; Verschueren 1983)
27-28 (Weast 1982-83; Mabey et al. 1982)
27.0 (Dean 1985)
Boiling Point (°C):
211 (Ericksen & Dobbert 1955; Andon et al. 1960; Callahan et al. 1979)
210 (Weast 1982-83)
210.5 (Verschueren 1983; Budavari 1989)
210-212 (Dean 1985)
Density (g/cm^3 at 20°C):
1.5398 (subcooled liquid, Ericksen & Dobbert 1955)
1.02017 (Andon et al. 1960)
0.9650 (Weast 1982-83)
1.0360 (Verschueren 1983)
1.0276 (14°C, Dean 1985)
Molar Volume (cm^3/mol):
134.5 (20°C, calculated-liquid density, Kamlet et al. 1986)
115.9 (calculated from density, Jaworska & Schultz 1993)
147.8 (calculated-LeBas method)
Molecular Volume ($Å^3$):
Total Surface Area, TSA ($Å^2$):
Dissociation Constant, pK_a:
10.6 (Herington & Kynaston 1957; quoted, Callahan et al. 1979)
10.58 (Dean 1985)
10.63 (Riddick et al. 1986; Howard 1989)
10.1 (Kollig 1993)
Heat of Fusion ΔH_{fus}, kcal/mol:
Entropy of Fusion ΔS_{fus}, cal/mol·K (e.u.):
Fugacity Ratio at 25°C (assuming ΔS_{fus} = 13.5 e.u.), F: 0.98

Water Solubility (g/m^3 or mg/L at 25°C):
8795 (shake flask-UV at pH 5.1, Blackman 1955)
6200 (shake flask-optical, extrapolated value, Ericksen & Dobbert 1957; quoted, Leuenberger et al. 1985; Howard 1989)
7868 (shake flask-LSC, Banerjee et al. 1980; quoted, Davies & Dobbs 1984; Banerjee 1985; Riddick et al. 1986)
7888 (shake flask-radioactive analysis, Veith et al. 1980)
1040 (calculated-K_{OW} as per Hansch et al. 1968, Veith et al. 1980)

4864, 10640 (quoted, calculated-K_{OW} and M.P., Valvani & Yalkowsky 1980)
7819 (gen. col.-HPLC, Wasik et al. 1981)
590 (calculated-K_{OW}, Mabey et al. 1982)
7943 (quoted, Isnard & Lambert 1988)
6866 (calculated as per Nirmalakhandan & Speece 1988, Müller & Klein 1992)
8200 (shake flask-HPLC/UV, Varhanickova et al. 1995)

Vapor Pressure (Pa at 25°C):
13.02 (calculated-Antoine eqn., Andon et al. 1960)
8.278 (supercooled liq., Andon et al. 1960; quoted, Callahan et al. 1979)
10.28 (extrapolated-Antoine eqn., Boublik et al. 1973, 1984)
8.266 (20°C, quoted, Mabey et al. 1982)
13.17 (calculated-Cox eqn., Chao et al. 1983)
20.22 (extrapolated-Antoine eqn., Boublik et al. 1984)
10.28 (extrapolated-Antoine eqn., Dean 1985)
13.06 (Leuenberger et al. 1985; quoted, Howard 1989)
12.96 (extrapolated-Antoine eqn., Nesterova et al. 1990)

Henry's Law Constant (Pa·m^3/mol):
1.722 (calculate-P/C, Mabey et al. 1982)
500, 532 (EPICS-GC, calculated-temp regression eqn., Ashworth et al. 1988)
0.0638 (8°C, Leuenberger et al. 1985; quoted, Howard 1989)
0.203 (calculated-P/C, Meylan & Howard 1991)
0.0692 (estimated-bond contribution, Meylan & Howard 1991)
0.1815 (calculated-P/C, Shiu et al. 1994)

Octanol/Water Partition Coefficient, log K_{OW}:
2.30 (20°C, shake flask-AS, CR, Korenman 1973; quoted, Hansch & Leo 1979; Mabey et al. 1982; Kamlet et al. 1986; Sangster 1989)
2.50 (calculated as per Tute 1971, Callahan et al. 1979; quoted, Westall et al. 1985)
2.42 (23 ± 1.5°C, shake flask-LSC, Banerjee et al. 1980; quoted, Arbuckle 1983; Davies & Dobbs 1984; Sangster 1989)
2.42 (measured, Veith et al. 1980)
1.99, 2.54 (estimated-RPHPLC-RT, quoted calculated value, Veith et al. 1980)
2.30 (quoted, Valvani & Yalkowsky 1980)
2.54 (35°C, shake flask-AS, Rogers & Wong 1980; quoted, Sangster 1989)
2.34 (gen. col.-HPLC, Wasik et al. 1981; quoted, Sangster 1989)
2.37 (calculated-γ from UNIFAC not considering mutual solubility of octanol & water, Arbuckle 1983)
2.95 (calculated-γ from UNIFAC by considering mutual solubility of octanol & water, Arbuckle 1983)
2.30 (quoted from Hansch & Leo 1979, Haky & Young 1984)
2.14, 2.49 (correlated-HPLC-k', Haky & Young 1984)
2.30 (Hansch & Leo 1985; quoted, Howard 1989)

2.39 (computed, Klopman et al. 1985; quoted, Pangrekar et al. 1994)
2.30, 1.83 (quoted, calculated-γ from UNIFAC, Banerjee & Howard 1988)
2.42 (quoted, Isnard & Lambert 1988)
2.35 (recommended, Sangster 1989)
2.47 (calculated-CLOGP, Müller & Klein 1992)
2.81 (quoted from Hansch & Leo 1979, Dallos et al. 1993)
2.48 (calculated-UNIFAC, Dallos et al. 1993)
2.30 (calculated-CLOGP 3.51, Jaworska & Schultz 1993)
2.46 (quoted, Saito et al. 1993)
2.30, 2.35 (COMPUTOX databank, Kaiser 1993)
2.80-3.20 (quoted lit. range, NCASI 1993)
2.77 (calculated-QSAR, Kollig 1993)

Bioconcentration Factor, log BCF:
1.18 (bluegill sunfish, Barrows et al. 1980; quoted, Howard 1989)
2.18 (bluegill sunfish, Veith et al. 1980; quoted, Davies & Dobbs 1984; Sabljic 1987a)
1.86 (microorganisms-water, calculated-K_{OW}, Mabey et al. 1982)
1.88 (calculated- χ , Sabljic 1987a)
2.18 (quoted, Isnard & Lambert 1988)

Sorption Partition Coefficient, log K_{OC}:
1.98 (sediment-water, calculated-K_{OW}, Mabey et al. 1982)
2.63 (soil, calculated-K_{OW}, Lyman et al. 1982; quoted, Howard 1989)
2.19 (activated carbon, Blum et al. 1994)
1.76 (calculated-K_{OW}, Kollig 1993)

Half-Lives in the Environment:
Air: half-life of 8.0 hours, based on reaction with photochemically produced hydroxyl radicals in air (GEMS 1986; quoted, Howard 1989); photooxidation half-life of 1.19-11.9 hours, based on estimated rate constant for the reaction with hydroxyl radicals in air (Howard et al. 1991).
Surface water: photooxidation half-life of 77-3840 hours in water, based on reported reaction rate constants for RO_2 radicals with the phenol class (Mill & Mabey 1985; quoted, Howard et al. 1991).
Groundwater: 48-336 hours, based on estimated aqueous aerobic biodegradation half-life (Howard et al. 1991).
Sediment:
Soil: 24-168 hours, based on estimated aqueous aerobic biodegradation half-life (Howard et al. 1991).
Biota:

Environmental Fate Rate Constants or Half-Lives:
Volatilization:
Photolysis:

Oxidation: aqueous oxidation rate constant of < 4×10^6 $M^{-1} \cdot h^{-1}$ for singlet oxygen, 1.1×10^8 $M^{-1} \cdot h^{-1}$ for peroxy radical at 25°C (Mabey et al. 1982); photooxidation half-life of 77-3840 hours in water, based on reported reaction rate constants for RO_2 radicals with the phenol class (Mill & Mabey 1985; selected, Howard et al. 1991); photooxidation half-life of 8.0 hours in air, based on reaction with photochemically produced hydroxyl radicals in air (GEMS 1986; selected, Howard 1989); photooxidation half-life of 1.19-11.9 hours, based on estimated rate constant for the reaction with hydroxyl radicals in air (Howard et al. 1991); gas-phase hydroxyl radical rate constants of 71.5×10^{-12} $cm^3 \cdot molecule^{-1} \cdot sec^{-1}$, calculated based on molecular orbital estimation method (Klamt 1993).

Hydrolysis:

Biodegradation: average rate of biodegradation 28.2 mg COD $g^{-1} \cdot h^{-1}$ based on measurements of COD decrease using activated sludge inoculum with 20 days of adaptation to the substrate (Pitter 1976; quoted, Scow 1982); first-order rate constant of 1.0 day^{-1} corresponding to a half-life of 0.7 day in adapted activated sludge under aerobic conditions (Mills et al. 1982); aqueous aerobic half-life of 24-168 hours, based on aqueous aerobic screening test data (Petrasek et al. 1983; Chambers et al. 1963; selected, Howard et al. 1991); aqueous anaerobic half-life of 96-672 hours, based on estimated aqueous aerobic biodegradation half-life (Howard et al. 1991; average exptl. rate constant of 0.0578 h^{-1} compared to group contribution method predicted rate constants of 0.0758 h^{-1} (nonlinear) and 0.0646 h^{-1} (linear) (Tabak & Govind 1993).

Biotransformation: rate constant for bacterial transformation of 1×10^7 mL $cell^{-1}$ h^{-1} in water (Mabey et al. 1982).

Bioconcentration, Uptake (k_1) and Elimination (k_2) Rate Constants:

Common Name: 2,6-Dimethylphenol
Synonym: 2,6-xylenol, *vic-m*-xylenol
Chemical Name: 2,6-dimethylphenol
CAS Registry No: 576-26-1
Molecular Formula: $CH_3C_6H_3(CH_3)OH$
Molecular Weight: 122.17
Melting Point (°C):
49.0 (Ericksen & Dobbert 1955; Weast 1982-83, Dean 1985)
45.62 (Andon et al. 1960)
42.0 (Valvani & Yalkowsky 1980)
Boiling Point (°C):
202 (Ericksen & Dobbert 1955)
201.03 (Andon et al. 1960; Boublik et al. 1973,84)
212 (Weast 1982-83)
211.528 (Boublik et al. 1984)
203 (Dean 1985)
Density (g/cm^3 at 20°C):
1.5371 (subcooled liq., Ericksen & Dobbert 1955)
1.1320 (25°C, Andon et al. 1960)
Molar Volume (cm^3/mol):
147.8 (calculated-LeBas method)
107.9 (calculated-density)
Molecular Volume (Å^3):
Total Surface Area, TSA (Å^2):
Dissociation Constant, pK_a:
10.6 (McLeese et al. 1979; Dean 1985)
10.63 (Riddick et al. 1986)
Heat of Fusion ΔH_{fus}, kcal/mol:
Entropy of Fusion ΔS_{fus}, cal/mol·K (e.u.):
Fugacity Ratio at 25°C (assuming ΔS_{fus} = 13.5 e.u.), F: 0.579

Water Solubility (g/m^3 or mg/L at 25°C):
6230 (shake flask-UV at pH 5.1, Blackman et al. 1955)
5333, 7030 (quoted, calculated-K_{OW} and M.P., Valvani & Yalkowsky 1980)
9651 (gen. col.-HPLC, Wasik et al. 1981)
5900 (Leuenberger et al. 1985)
2900 (8°C, Leuenberger et al. 1985)
5339 (quoted from Valvani & Yalkowsky 1980, Müller & Klein 1992)
6150 (shake flask-HPLC/UV at pH 6.3, Varhanickova et al. 1995)

Vapor Pressure (Pa at 25°C):
19.06 (Dreisbach 1955)
23.30 (ebulliometry/gas saturation, Andon et al. 1960)
24.31 (calculated-Antoine eqn., Andon et al. 1960)
34.41 (extrapolated-Antoine eqn., Boublik et al. 1973,84)

33.68 (extrapolated-Cox eqn., Chao et al. 1983)
10.37 (extrapolated-Antoine eqn., Boublik et al. 1984)
34.40 (extrapolated-Antoine eqn., Dean 1985)
24.0 (quoted, Leuenberger et al. 1985)
19.09 (Riddick et al. 1986)
24.0 (extrapolated-Antoine eqn., Stephenson & Malanowski 1987)
43.80 (extrapolated-Antoine eqn., Nesterova et al. 1990)

Henry's Law Constant (Pa·m^3/mol):
0.684 (modified gas-stripping, Hawthorne et al. 1985)
0.193 (8°C, Leuenberger et al. 1985)
0.377 (calculated-P/C, Shiu et al. 1994)

Octanol/Water Partition Coefficient, log K_{OW}:
2.36 (quoted unpublished result, Leo et al. 1971; Hansch & Leo 1979)
2.34 (LC-k', Carlson et al. 1975)
2.38 (calculated-π const., Carlson et al. 1975; quoted, Sangster 1989)
2.40 (35°C, shake flask-AS, Rogers & Wong 1980; quoted, Sangster 1989)
2.31 (gen. col.-HPLC, Wasik et al. 1981; quoted, Sangster 1989)
2.36 (quoted from Hansch & Leo 1979, Haky & Young 1984)
2.07 (HPLC-k', Haky & Young 1984; quoted, Sangster 1989)
2.67, 2.43 (correlated-HPLC-k', Haky & Young 1984)
2.51 (HPLC, Eadsforth 1986)
2.36 (quoted from unpublished CLOGP data, Sangster 1989)
2.36 (recommended, Sangster 1989)
2.47 (calculated-CLOGP, Müller & Klein 1992)

Bioconcentration Factor, log BCF:

Sorption Partition Coefficient, log K_{OC}:

Half-Lives in the Environment:
Air:
Surface water:
Groundwater:
Sediment:
Soil:
Biota:

Environmental Fate Rate Constants or Half-Lives:

Volatilization:

Photolysis:

Photooxidation: atmospheric reaction rate constants with hydroxyl radicals: 54.1×10^{-12} $cm^3 \cdot molecule^{-1} \cdot sec^{-1}$ (estimated from Atmospheric Oxidation Program), 65.9×10^{-12} $cm^3 \cdot molecule^{-1} \cdot sec^{-1}$ (exptl.) and 30.2×10^{-12} $cm^3 \cdot molecule^{-1} \cdot sec^{-1}$ (estimated from Fate of Atmospheric Pollutants Program) (Meylan & Howard 1993); gas-phase OH radical rate constants of 65.9×10^{-12} $cm^3 \cdot molecule^{-1} \cdot sec^{-1}$ (exptl.) and 49.48×10^{-12} $cm^3 \cdot molecule^{-1} \cdot sec^{-1}$, calculated based on molecular orbital estimation method (Klamt 1993).

Hydrolysis:

Biodegradation: 7 to 10 days for bacteria to utilize 95% of 200 ppm in parent substrate (Tabak et al. 1964); average rate of biodegradation 9.0 mg COD $g^{-1} \cdot h^{-1}$ based on measurements of COD decrease using activated sludge inoculum with 20 days of adaptation to the substrate (Pitter 1976; quoted, Scow 1982).

Biotransformation:

Bioconcentration, Uptake (k_1) and Elimination (k_2) Rate Constants:

Common Name: 3,4-Dimethylphenol
Synonym: 3,4-xylenol, *as-o*-xylenol
Chemical Name: 3,4-dimethylphenol
CAS Registry No: 95-65-8
Molecular Formula: $CH_3C_6H_3(CH_3)OH$
Molecular Weight: 122.17
Melting Point (°C):
67.0 (Ericksen & Dobbert 1955)
65.11 (Andon et al. 1960; Budavari 1989)
66-68 (Weast 1982-83)
65.0 (Verschueren 1983)
62.5 (Dean 1985)
Boiling Point (°C):
223 (Ericksen & Dobbert 1955)
226.947 (Andon et al. 1960; Boublik et al. 1973, 1984; Budavari 1989)
225 (Weast 1982-83; Verschueren 1983; Dean 1985)
225.391 (Boublik et al. 1984)
Density (g/cm^3 at 20°C):
1.5442 (subcooled liquid, Ericksen & Dobbert 1955)
1.1380 (25°C, Andon et al. 1960)
0.9830 (Weast 1982-83)
Molar Volume (cm^3/mol):
147.8 (calculated-LeBas method)
Molecular Volume ($Å^3$):
Total Surface Area, TSA ($Å^2$):
Dissociation Constant, pK_a:
10.40 (McLeese et al. 1979)
10.32 (Dean 1985)
10.36 (Riddick et al. 1986)
Heat of Fusion ΔH_{fus}, kcal/mol:
Entropy of Fusion ΔS_{fus}, cal/mol·K (e.u.):
Fugacity Ratio at 25°C (assuming ΔS_{fus} = 13.5 e.u.), F: 0.384

Water Solubility (g/m^3 or mg/L at 25°C):
5100 (shake flask-optical, extrapolated value, Ericksen & Dobbert 1957; quoted, Leuenberger et al. 1985)
5000 (shake flask-AS, Roberts et al. 1977)
4539, 4236 (quoted, calculated-K_{OW} and M.P., Valvani & Yalkowsky 1980)
4545 (quoted from Valvani & Yalkowsky 1980, Müller & Klein 1992)
7250 (shake flask-HPLC/UV at pH 6.25, Varhanickova et al. 1995)

Vapor Pressure (Pa at 25°C):
1.895 (calculated-Antoine eqn., Andon et al. 1960)
2.541 (extrapolated-Antoine eqn., Boublik et al. 1973,84)
3.621 (extrapolated-Cox eqn., Chao et al. 1983)

7.327 (extrapolated-Antoine eqn., Boublik et al. 1984)
2.54 (extrapolated-Antoine eqn., Dean 1985)
1.87 (Leuenberger et al. 1985)
5.16 (Riddick et al. 1986)
3.795 (extrapolated-Antoine eqn., Nesterova et al. 1990)

Henry's Law Constant (Pa m^3/mol):
0.00942 (Leuenberger et al. 1985)
0.0212 (calculated-P/C, Shiu et al. 1994)

Octanol/Water Partition Coefficient, log K_{OW}:
2.23 (20°C, shake flask-AS, CR, Korenman 1973; quoted, Hansch & Leo 1979; Sangster 1989)
2.35 (quoted from Leo et al. 1971, Roberts et al. 1977)
2.23 (recommended, Sangster 1989)
2.47 (calculated-CLOGP, Müller & Klein 1992)

Bioconcentration Factor, log BCF:

Sorption Partition Coefficient, log K_{OC}:

Half-Lives in the Environment:
Air:
Surface water:
Groundwater:
Sediment:
Soil:
Biota:

Environmental Fate Rate Constants or Half-Lives:
Volatilization:
Photolysis:
Hydrolysis:
Oxidation: gas-phase hydroxyl radical rate constants of 81.4×10^{-12} $cm^3 \cdot molecule^{-1} \cdot sec^{-1}$ (exptl.) and 55.1×10^{-12} $cm^3 \cdot molecule^{-1} \cdot sec^{-1}$, calculated based on molecular orbital estimation method (Klamt 1993).
Biodegradation: average rate of biodegradation 13.4 mg COD $g^{-1} \cdot h^{-1}$ based on measurements of COD decrease using activated sludge inoculum with 20 days of adaptation to the substrate (Pitter 1976; quoted, Scow 1982).
Biotransformation:
Bioconcentration, Uptake (k_1) and Elimination (k_2) Rate Constants:

Common Name: 2,3,5-Trimethylphenol
Synonym:
Chemical Name: 2,3,5-trimethylphenol
CAS Registry No: 697-82-5
Molecular Formula: $C_6H_2(CH_3)_3OH$
Molecular Weight: 136.2
Melting Point (°C): 93.78
Boiling Point (°C): 235.7
Density (g/cm^3):
Dissociation Constants, pK_a:

10.60 (Blackman et al. 1955)

Molar Volume (cm^3/mol):

170.0 (calculated-LeBas method)

Molecular Volume, TMV (Å^3):
Total Surface Area, TSA (Å^2):
Heat of Fusion, ΔH_{fus}, kcal/mol:
Entropy of Fusion, ΔS_{fus}, cal/mol·K (e.u.):
Fugacity Ratio at 25°C (assuming ΔS_{fus} = 13.5 e.u.), F: 0.213

Water Solubility (g/m^3 or mg/L at 25°C):

762 (shake flask-UV, Blackman et al. 1955)
855 (shake flask-HPLC/UV, Varhanickova et al. 1995)

Vapor Pressure (Pa at 25°C):

2.426 (extr. liquid, Handley et al. 1964)
4.45 (extrapolated-Cox eqn., Chao et al. 1983)
2.43 (extr. liquid, Antoine eqn., Boublik et al. 1984)
2.43 (extrapolated-Antoine eqn., Dean 1985)
2.44 (extr. liquid, Antoine eqn., Stephenson & Malanowski 1987)

Henry's Law Constant (Pa • m^3/mol):

0.404 (calculated-P/C, this work)

Octanol/Water Partition Coefficient, log K_{OW}:

3.06 (COMPUTOX databank, Kaiser 1993)

Bioconcentration Factor, log BCF: no data

Sorption Partition Coefficient, log K_{OC}: no data

Half-Lives in the Environmen/Environmental Fate Rate Constants and Half-Lives: no data

Common Name: 2,4,6-Trimethylphenol
Synonym:
Chemical Name: 2,4,6-Trimethylphenol
CAS Registry No: 527-60-6
Molecular Formula: $C_6H_2(CH_3)_3OH$
Molecular Weight: 136.2
Melting Point (°C): 72
Boiling Point (°C): 246
Density (g/cm^3):
Dissociation Constants, pK_a:
10.90 (Blackman et al. 1955)
10.88 (Dean 1985)
11.23 (Nendza & Seydel 1988)
Molar Volume (cm^3/mol):
170.0 (calculated-LeBas method)
Molecular Volume, TMV (Å^3):
Total Surface Area, TSA (Å^2):
Heat of Fusion, ΔH_{fus}, kcal/mol:
Entropy of Fusion, ΔS_{fus}, cal/mol·K (e.u.):
Fugacity Ratio at 25°C (assuming ΔS_{fus} = 13.5 e.u.), F: 0.343

Water Solubility (g/m^3 or mg/L at 25°C):
1007 (shake flask-UV, Blackman et al. 1955)
1420 (shake flask-HPLC/UV, Varhanickova et al. 1995)

Vapor Pressure (Pa at 25°C):
7.00 (extr. liquid, Antoine eqn., Boublik et al. 1984)
19.6 (extr. liquid, Antoine eqn., Stephenson & Malanowski 1987)

Henry's Law Constant (Pa·m^3/mol):
0.2512 (calculated-P/C, Shiu et al. 1994)

Octanol/Water Partition Coefficient, log K_{OW}:
2.73 (gen. col.-HPLC/UV, Wasik et al. 1981)
2.73 (quoted and recommended, Sangster 1989)
2.73 (COMPUTOX databank, Kaiser 1993)

Bioconcentration Factor, log BCF: no data

Sorption Partition Coefficient, log K_{OC}: no data

Half-Lives in the Environment/Environmental Fate Rate Constants and Half-Lives: no data

Common Name: 2-Ethylphenol
Synonym: *o*-ethylphenol
Chemical Name: 2-ethylphenol, *o*-ethylphenol
CAS Registry No: 90-00-6
Molecular Formula: $C_2H_5C_6H_4OH$
Molecular Weight: 122.17
Melting Point (°C):
-18 (Verschueren 1983)
-3.4 (Dean 1985)
Boiling Point (°C): 204.52
195-197 (Verschueren 1983)
207.7 (Boublik et al. 1984)
204.5 (Boublik et al. 1984; Dean 1985)
Density (g/cm^3 at 20°C):
1.01885 (Biddiscombe et al. 1963)
1.037 (Verschueren 1983; 25°C, Dean 1985)
Molar Volume (cm^3/mol): 147.8 (calculated-LeBas method, Shiu et al. 1994)
Molecular Volume ($Å^3$):
Total Surface Area, TSA ($Å^2$):
Dissociation Constant, pK_a: 10.02 (Dean 1985)
Heat of Fusion ΔH_{fus}, kcal/mol:
Entropy of Fusion ΔS_{fus}, cal/mol K (e.u.):
Fugacity Ratio at 25°C (assuming ΔS_{fus} = 13.5 e.u.), F: 1.0

Water Solubility (g/m^3 or mg/L at 25°C):
5333, 7361 (quoted, calculated-K_{OW} and M.P., Valvani & Yalkowsky 1980)
14042 (shake flask-HPLC/UV at pH 5.2, Varhanickova et al. 1995)

Vapor Pressure (Pa at 25°C):
20.4 (24.93°C, ebulliometry, Biddiscombe et al. 1963)
20.84 (calculated-Antoine eqn., Biddiscombe et al. 1963)
16.7, 29 (extrapolated-Antoine eqn., Boublik et al. 1984)
27.01 (extrapolated-Antoine eqn., Dean 1985)
19.55 (extrapolated-Antoine eqn., Nesterova et al. 1990)

Henry's Law Constant ($Pa \cdot m^3/mol$):
0.174 (calculated-P/C, this work)

Octanol/Water Partition Coefficient, log K_{OW}:
2.47 (shake flask, Hansch & Leo 1979; quoted, Valvani & Yalkowsky 1980)
2.64 (shake flask, Korenman 1980)
2.64 (HPLC, Butte et al. 1981)
2.47 (quoted from unpublished CLOGP Data Base and recommended, Sangster 1989)
2.46 (HPLC-RT, Ritter et al. 1994)

Bioconcentration Factor, log BCF/Sorption Partition Coefficient, log K_{OC}: no data
Half-Lives in the Environment/Environmental Fate Rate Constants or Half-Lives: no data

Common Name: 4-Ethylphenol
Synonym: *p*-ethylphenol
Chemical Name: 2-ethylphenol, *p*-ethylphenol
CAS Registry No: 123-07-9
Molecular Formula: $C_2H_5C_6H_4OH$
Molecular Weight: 122.17
Melting Point (°C):
47.0 (Valvani & Yalkowsky 1980; Dean 1985)
42-45 (Verschueren 1983)
Boiling Point (°C):
217.99, 219.99 (Boublik et al. 1984)
218-219 (Verschueren 1983; Dean 1985)
Density (g/cm^3 at 20°C):
1.054 (25°C, Biddiscombe et al. 1963)
1.011 (25°C, Dean 1985)
Molar Volume (cm^3/mol):
116.10 (calculated from density, Jaworska & Schultz 1993)
147.8 (calculated-LeBas method, Shiu et al. 1994)
Molecular Volume ($Å^3$):
Total Surface Area, TSA ($Å^2$):
Dissociation Constant, pK_a:
10.0 (Dean 1985)
Heat of Fusion ΔH_{fus}, kcal/mol:
Entropy of Fusion ΔS_{fus}, cal/mol K (e.u.):
Fugacity Ratio at 25°C (assuming ΔS_{fus} = 13.5 e.u.), F: 0.606

Water Solubility (g/m^3 or mg/L at 25°C):
5000 (shake flask-AS, Roberts et al. 1977)
5457, 5714 (quoted, calculated-K_{OW} and M.P., Valvani & Yalkowsky 1980; quoted, Müller & Klein 1992)
7980 (shake flask-HPLC/UV, Varhanickova et al. 1995)

Vapor Pressure (Pa at 25°C):
2.810 (exptl., Biddiscombe et al. 1963)
5.030 (calculated-Antoine eqn., Biddiscombe et al. 1963)
6.575 (extrapolated-Antoine eqn., Boublik et al. 1984)
7.530 (extrapolated-Antoine eqn., Dean 1985)
6.137 (extrapolated-Antoine eqn., Nesterova et al. 1990)

Henry's Law Constant (Pa m^3/mol):
0.132 (calculated-P/C, this work)

Octanol/Water Partition Coefficient, log K_{OW}:

2.75 (quoted average, Leo et al. 1971)
2.58 (shake flask-AS at pH 7.45, Umeyama et al. 1971; quoted, Sangster 1989)
2.26 (shake flask, Korenman 1972)
2.40 (quoted from Leo et al. 1971, Roberts et al. 1977)
2.26, 2.58 (quoted, Hansch & Leo 1979)
2.60 (shake flask-AS, Rogers & Wong 1980; quoted, Sangster 1989)
2.42 (quoted, Valvani & Yalkowsky 1980)
2.37 (HPLC, Butte et al. 1981; quoted, Sangster 1989)
2.26 (quoted from Hansch & Leo 1979, Miyake & Terada 1982)
2.12, 2.19 (RP-HPLC-k', Miyake & Terada 1982)
2.48 (calculated-π, Miyake & Terada 1982)
2.66, 2.81 (quoted, calculated, Verschueren 1983)
2.58 (selected best lit. value, Garst & Wilson 1984)
2.59 $\pm$ 0.07 (exptl.-ALPM, Garst & Wilson 1984)
2.50 (recommended, Sangster 1989)
2.50 (calculated-CLOGP, Müller & Klein 1992)
2.58 (calculated-CLOGP 3.51, Jaworska & Schultz 1993)
2.50 (COMPUTOX databank, Kaiser 1993)

Bioconcentration Factor, log BCF:

Sorption Partition Coefficient, log K_{OC}:

Half-Lives in the Environment:

Air:
Surface water:
Groundwater:
Sediment:
Soil:
Biota:

Environmental Fate Rate Constants or Half-Lives:

Volatilization:
Photolysis:
Hydrolysis:
Oxidation:
Biodegradation:
Biotransformation:
Bioconcentration, Uptake (k_1) and Elimination (k_2) Rate Constants:

Common Name: 4-*tert*-Butylphenol
Synonym: *p*-*tert*-butylphenol, 4-(α,α-dimethylehtylphenol)
Chemical Name: 4-*tert*-butylphenol, *p*-*tert*-butylphenol
CAS Registry No: 98-54-4
Molecular Formula: $(CH_3)_3CC_6H_4OH$
Molecular Weight: 150.22
Melting Point (°C):
    99.0 (Stull 1947; Verschueren 1983; Stephenson & Malanowski 1987)
Boiling Point (°C):
    238 (Stull 1947)
    236 (Verschueren 1983)
    240.0 (Stephenson & Malanowski 1987)
Density (g/cm^3 at 20°C):
Molar Volume (cm^3/mol):
    148.6 (calculated-density, Jaworska & Schultz 1993)
    192.2 (calculated-LeBas method, Shiu et al. 1994)
Molecular Volume (Å^3):
Total Surface Area, TSA (Å^2):
Dissociation Constnat, pK_a: 9.9 (McLeese et al. 1979)
Heat of Fusion ΔH_{fus}, kcal/mol:
Entropy of Fusion ΔS_{fus}, cal/mol·K (e.u.):
Fugacity Ratio at 25°C (assuming ΔS_{fus} = 13.5 e.u.), F:0.177

Water Solubility (g/m^3 or mg/L at 25°C):
    650 (20-25°C, Geyer et al. 1981)
    1000 (quoted, Thomas 1982)
    700 (Verschueren 1983)
    580 (Yalkowsky et al. 1987)
    1850 (shake flask-HPLC/UV at pH 6.05, Varhanickova et al. 1995)

Vapor Pressure (Pa at 25°C):
    6.402 (extrapolated-regression of tabulated data, Stull 1947)
    1.225 (extr. liquid, Antoine eqn., Handley et al. 1964)
    6.18 (quoted, Thomas 1982)
    5.072 (extrapolated-Cox eqn., Chao et al. 1983)
    1.24 (extr. liquid, Antoine eqn., Boublik et al. 1984)
    1.25 (extrapolated-Antoine eqn., Dean 1985)
    1.272 (extr. liquid, Antoine eqn., Stephenson & Malanowski 1987)
    0.492 (calculated-Antoine eqn., Stephenson & Malanowski 1987)

Henry's Law Constant (Pa·m^3/mol):
    0.113 (calculated-1/K_{AW}, C_W/C_A, reported as exptl., Hine & Mookerjee 1975; quoted, Nirmalakhandan & Speece 1988)
    0.139 (calculated-group contribution, Hine & Mookerjee 1975)
    3.750 (calculated-bond contribution, Hine & Mookerjee 1975)
    0.922 (calculated-P/C, Thomas 1982)
    0.113, 0.375 (quoted, calculated- χ , Nirmalakhandan & Speece 1988)

Octanol/Water Partition Coefficient, log K_{OW}:

3.31 (calculated-f const., McLeese et al. 1979)
3.31 (quoted, Verschueren 1983)
3.04 (shake flask, Geyer et al. 1984; quoted, Sangster 1989)
2.94, 3.14 (shake flask-OECD 1981 Guideline, Geyer et al. 1984)
3.14 (quoted, Freitag et al. 1985)
3.13 (selected, Yoshida et al. 1986)
3.41 (HPLC-RT, Butte et al. 1987)
3.04 (recommended, Sangster 1989)
3.31 (calculated-CLOGP 3.15, Jaworska & Schultz 1992)
3.31, 3.04 (quoted CLOGP, COMPUTOX, Kaiser 1993)

Bioconcentration Factor, log BCF:

1.53 (*Chlorella*, after exposure to 50 μg/L for 24 h, Geyer et al. 1981)
1.88 (calculated-S, Geyer et al. 1981)
1.53 (algae, Freitag et al. 1982)
2.08 (golden orfe, Freitag et al. 1982)
2.38 (activated sludge, Freitag et al. 1982,85)
1.53 (Alga *Chlorella fusca*, wet weight basis, Geyer et al. 1984)
2.23 (calculated-K_{OW}, Geyer et al. 1984)
1.48 (algae, Freitag et al. 1985)
2.07 (golden ide, Freitag et al. 1982)
1.86 (zebrafish, Butte et al. 1987)
2.38 (activated sludge, Freitag et al. 1987)
1.48 (algae, Freitag et al. 1987)
2.07 (fish, Freitag et al. 1987)

Sorption Partition Coefficient, log K_{OC}:

Half-Lives in the Environment:

Air:
Surface water:
Groundwater:
Sediment:
Soil:
Biota:

Environmental Fate Rate Constants or Half-Lives:

Volatilization: volatilization half-life of 117 hours from water body with depth of 1 meter (Thomas 1982).
Photolysis:
Oxidation:
Hydrolysis:
Biodegradation:
Biotransformation:
Bioconcentration, Uptake (k_1) and Elimination (k_2) Rate Constants:

Common Name: 4-Octylphenol
Synonym: *p*-octylphenol
Chemical Name: 4-octylphenol
CAS Registry No:
Molecular Formula: $C_8H_{17}C_6H_4OH$
Molecular Weight: 206.33
Melting Point (°C):
    41-42 (Weast 1982-83)
Boiling Point (°C):
    169.0 (at 10 mm Hg, Weast 1982-83)
Density (g/cm^3 at 20°C):
Molar Volume (cm^3/mol):
    281 (calculated-LeBas method, Shiu et al. 1994)
Molecular Volume (Å^3):
Total Surface Area, TSA (Å^2):
Dissociation Constant, pK_a:
Heat of Fusion ΔH_{fus}, kcal/mol:
Entropy of Fusion ΔS_{fus}, cal/mol K (e.u.):
Fugacity Ratio at 25°C (assuming ΔS_{fus} = 13.5 e.u.), F: 0.695

Water Solubility (g/m^3 or mg/L at 25°C):
    12.6 ± 0.50 (gen. col.-HPLC/fluo., Ahel & Giger 1993a)
    14.1 ± 0.60 (shake flask-HPLC/UV, Varhanickova et al. 1995)

Vapor Pressure (Pa at 25°C):
    0.071 (quoted, Shiu et al. 1994)

Henry's Law Constant (Pa m^3/mol):
    0.4916 (calculated-P/C, Shiu et al. 1994)

Octanol/Water Partition Coefficient, log K_{OW}:
    3.70 (McLeese et al. 1981; quoted, Ahel & Giger 1993b)
    4.12 ± 0.10 (shake flask-HPLC/fluo., Ahel & Giger 1993b)

Bioconcentration Factor, log BCF:

Sorption Partition Coefficient, log K_{OC}:

Half-Lives in the Environment:
Air:
Surface water:
Groundwater:
Sediment:
Soil:
Biota:

Environmental Fate Rate Constants or Half-Lives:
Volatilization:
Photolysis:
Hydrolysis:
Oxidation:
Biodegradation: average exptl. rate constant of 0.0894 h^{-1} compared to group contribution method predicted rate constants of 0.1124 h^{-1} (nonlinear) and 0.0982 h^{-1} (Tabak & Govind 1993).
Biotransformation:
Bioconcentration, Uptake (k_1) and Elimination (k_2) Rate Constants:

Common Name: 4-Nonylphenol
Synonym: *p*-nonylphenol
Chemical Name: 4-nonylphenol
CAS Registry No: 104-40-5
Molecular Formula: $C_9H_{19}C_6H_4OH$
Molecular Weight: 220.36
Melting Point (°C):
96.0
Boiling Point (°C):
293-297 (Budavari 1989)
Density (g/cm^3 at 20°C):
1.513 (Budavari 1989)
Molar Volume (cm^3/mol):
279.80 (calculated from density, Jaworska & Schultz 1993)
303.2 (calculated-LeBas method, Shiu et al. 1994)
Molecular Volume (Å^3):
Total Surface Area, TSA (Å^2):
Dissociation Constant, pK_a:
Heat of Fusion ΔH_{fus}, kcal/mol:
Entropy of Fusion ΔS_{fus}, cal/mol·K (e.u.):
Fugacity Ratio at 25°C (assuming ΔS_{fus} = 13.5 e.u.), F: 0.695

Water Solubility (g/m^3 or mg/L at 25°C):
5.0 (quoted, Geyer et al. 1982)
7.0 (Yalkowsky et al. 1987)
5.43 ± 0.17 (gen. col.-HPLC/fluo., Ahel & Giger 1993a)

Vapor Pressure (Pa at 25°C):
0.0701 (ebulliometry, extrapolated, Hon et al. 1976)
0.072 (extrapolated-Antoine eqn., Boublik et al. 1984)
0.0691 (extrapolated-Antoine eqn., Stephenson & Malanowski 1987)
0.174, 0.14, 0.109, 0.0908, 0.0802, 0.0556 (GC-RT, 7 isomers, Bidleman & Renberg 1985)

Henry's Law Constant (Pa m^3/mol):
1.5705 (calculated-P/C, Shiu et al. 1994)

Octanol/Water Partition Coefficient, log K_{OW}:
4.20 (McLeese et al. 1981; quoted, Ahel & Giger 1993b)
4.10 (quoted, Geyer et al. 1982)
5.90 (selected, Yoshida et al. 1986)
5.76 (HPLC-RT, Itokawa et al. 1989, quoted, Sangster 1993)
4.48 ± 0.12 (shake flask-HPLC/fluo., Ahel & Giger 1993b)

6.36 (calculated-CLOGP 3.51, Jaworska & Schultz 1993)
6.36 (COMPUTOX, Kaiser 1993)

Bioconcentration Factor, log BCF:
2.45 (salmon, McLeese et al. 1981)
1.0 (mussel *mytilus edulis*, Geyer et al. 1982)

Sorption Partition Coefficient, log K_{OC}:

Half-Lives in the Environment:
Air:
Surface water:
Groundwater:
Sediment:
Soil:
Biota: 4 days in salmon before excretion (McLeese et al. 1981).

Environmental Fate Rate Constants or Half-Lives:
Volatilization:
Photolysis:
Hydrolysis:
Oxidation:
Biodegradation: average exptl. rate constant of 0.0894 h^{-1} compared to group contribution method predicted rate constants of 0.1124 h^{-1} (nonlinear) and 0.0982 h^{-1} (linear) (Tabak & Govind 1993).
Biotransformation:
Bioconcentration, Uptake (k_1) and Elimination (k_2) Rate Constants:
k_1: 45.0 day^{-1} (McLeese et al. 1981)
k_2: 0.16 day^{-1} (McLeese et al. 1981)

Common Name: 1-Naphthol
Synonym: α-naphthol, 1-naphthalenol, 1-hydroxynaphthalene
Chemical Name: 1-naphthol
CAS Registry No: 90-15-3
Molecular Formula: $C_{10}H_7OH$
Molecular Weight: 144.17
Melting Point (°C):
96 (Stull 1947; Weast 1982-83; Verschueren 1983; Dean 1985)
94.9 (Tsonopoulos & Prausnitz 1971)
Boiling Point (°C):
282.5 (Stull 1947)
288 (sublime, Weast 1982-83; Dean 1985)
Density (g/cm^3):
1.0989 (99°C, Weast 1982-83)
1.0954 (99°C, Dean 1985)
Dissociation Constants, pK_a:
9.20 (McLeese et al. 1979)
9.30 (Dean 1985)
Molar Volume (cm^3/mol):
122.3 (calculated-density, Jaworska & Schultz 1993)
155 (calculated-LeBas method)
Molecular Volume, TMV ($Å^3$):
Total Surface Area, TSA ($Å^2$):
283.1 (Iwase et al. 1985)
161.6 (Sabljic 1987)
Heat of Fusion, ΔH_{fus}, kcal/mol: 5.61 (Tsonopoulos & Prausnitz 1971)
Entropy of Fusion, ΔS_{fus}, cal/mol·K: 15.2 (Tsonopoulos & Prausnitz 1971)
Fugacity Ratio at 25°C (assuming ΔS_{fus} = 13.5 kcal/mol·K), F: 0.199

Water Solubility (g/m^3 or mg/L at 25°C):
438 (quoted, Tsonopoulos & Prausnitz 1971)

Vapor Pressure (Pa at 25°C):
1.55 (extrapolated-regression of tabulated data, Stull 1947)
0.509 (extr. liquid, Antoine eqn., Boublik et al. 1984)
0.29 (extrapolated-Antoine eqn., Dean 1985)

Henry's Law Constant ($Pa \cdot m^3/mol$):
0.49 (calculated-P/C, this work)

Octanol/Water Partition Coefficient, log K_{OW}:
2.98 (shake flask-AS, Hansch & Anderson 1967)
2.98 (Tute 1971; quoted, Fujisawa & Masuhara 1981)
2.98, 2.84, 2.31 (quoted, Hansch & Leo 1979)
2.98 (quoted, McLeese et al. 1979)
2.98 (quoted, Verschueren 1983)
2.28 (HPLC-k', Haky & Young 1984)

2.98, 2.67 (quoted, calculated-TSA, Iwase et al. 1985)
2.71 (selected, Yoshida et al. 1986)
2.84 (recommended, Sangster 1989, 1993)
2.84 (CPC-RV, Gluck & Martin 1990)
3.13 (CPC, Menges et al. 1990)
2.81 (shake flask-HPLC, Menges et al. 1991)
3.02 (concurrent chromatography, Berthod et al. 1988)
2.84 (calculated-CLOGP 3.15, Jaworska & Schultz 1993)
2.98 (quoted, Satio et al. 1993)
2.84 (COMPUTOX, Kaiser 1993)
2.43 (HPLC-RT, Ritter et al. 1994)

Bioconcentration Factor, log BCF:

Sorption Partition Coefficient, log K_{OC}:
2.64 (soil and sediment, Hassett et al. 1981)
3.33 (soil, Means et al. 1982)
3.41 (soil, calculated- χ , Sabljic 1987)
2.10-2.77 (soil, calculated-K_{OW}, model of Karickhoff et al. 1979, Sabljic 1987)
2.63-2.99 (soil, calculated-K_{OW}, model of Kenaga & Goring 1980, Sabljic 1987)
1.84-2.19 (soil, calculated-K_{OW}, model of Briggs 1981, Sabljic 1987)
1.99-2.66 (soil, calculated-K_{OW}, model of Means et al. 1982, Sabljic 1987)
1.30-1.90 (soil, calculated-K_{OW}, model of Chiou et al. 1983, Sabljic 1987)
2.89 (soil, calculated- χ , Bahnick & Doucette 1988)
2.52-2.96, 2.53 (soil: quoted, calculated-K_{OW}, Xing et al. 1994)
1.92 - 2.64 (organic sorbent: celludlose, lignin, Xing et al. 1994)

Half-Lives in the Environmental Compartments:
Air:
Surface water:
Ground water:
Sediment:
Soil:
Biota:

Environmental Fate Rate Constants and Half-Lives:
Volatilization:
Photolysis:
Photooxidation:
Hydrolysis:
Biodegradation: average rate of biodegradation 38.4 mg COD $g^{-1} \cdot h^{-1}$ based on measurements of COD decrease using activated sludge inoculum with 20 days of adaptation to the substrate (Pitter 1976; quoted, Scow 1982).
Biotransformation:
Bioconcentration Uptake (k_1) and Elimination (k_2) Rate Constants:

Common Name: 2-Naphthol
Synonym: β-naphthol, 2-naphthalenol, 2-hydroxynaphthalene
Chemical Name: 2-naphthol
CAS Registry No: 135-19-3
Molecular Formula: $C_{10}H_7OH$
Molecular Weight: 144.17
Melting Point (°C):
- 122.5 (Stull 1947)
- 121.9 (Tsonopoulos & Prausnitz 1971)
- 123-4 (Weast 1982-83)

Boiling Point (°C):
- 288 (Stull 1947)
- 285 (Urano et al. 1982)
- 295 (Weast 1982-83; Verschueren 1983)

Density (g/cm^3 at 20°C):
- 1.28 (Weast 1982-83)

Dissociation Constants, pK_a:
- 9.57 (Dean 1985)

Molar Volume (cm^3/mol):
- 155 (calculated-LeBas method)

Molecular Volume, TMV ($Å^3$):
Total Surface Area, TSA ($Å^2$):
- 286.6 (Iwase et al. 1985)

Heat of Fusion, ΔH_{fus}, kcal/mol:
- 4.49 (Tsonopoulos & Prausnitz 1971)

Entropy of Fusion, ΔS_{fus}, cal/mol·K:
- 11.4 (Tsonopoulos & Prausnitz 1971)

Fugacity Ratio at 25°C (assuming ΔS_{fus} = 13.5 kcal/mol • K), F: 0.11

Water Solubility (g/m^3 or mg/L at 25°C):
- 700 (20-25°C, Seidell 1941; Lange 1971; quoted, Urano et al. 1982)
- 754 (quoted, Tsonopoulos & Prausnitz 1971)
- 1000 (shake flask-UV, Roberts et al. 1977)
- 740 (Verschueren 1983)

Vapor Pressure (Pa at 25°C):
- 1.43 (extrapolated-regression of tabulated data, Stull 1947)
- 0.386 (extr. liquid, Antoine eqn., Boublik et al. 1984)
- 0.16 (extrapolated-Antoine eqn., Dean 1985)

Henry's Law Constant (Pa m^3/mol):
- 0.280 (calculated-P/C, this work)

Octanol/Water Partition Coefficient, log K_{OW}:

2.84 (shake flask-AS, Hansch & Anderson 1967, quoted, Roberts et al. 1977)
2.70, 2.84, 2.89 (quoted, Hansch & Leo 1979)
2.26 (calculated-f const., McLeese et al. 1979)
2.84 (quoted, Verschueren 1983)
2.84, 2.74 (quoted, calculated-TSA, Iwase et al. 1985)
2.01, 2.46 (HPLC-k', Eadsforth 1986)
2.84, 3.46 (quoted: shake flask, calculated, Eadsforth 1986)
2.81 (selected, Yoshida et al. 1986)
2.84 (recommended, Sangster 1989)
2.84, 2.85 (quoted, CPC-RV, El Tayar et al. 1991)
2.84 (quoted, Saito et al. 1993)
2.70 (EPA CLOGP Data Base, Hulzebos et al. 1993)
2.70 (COMPUTOX, Kaiser 1993)

Bioconcentration Factor, log BCF:

Sorption Partition Coefficient, log K_{OC}:

3.33 (soil, Means et al. 1982)
3.41 (soil, calculated- χ , Sabljic 1987)
2.10-2.77 (soil, calculated-K_{OW} based on model of Karickhoff et al. 1979, Sabljic 1987)
2.63-2.99 (soil, calculated-K_{OW} based on model of Kenaga & Goring 1980, Sabljic 1987)
1.84-2.19 (soil, calculated-K_{OW} based on model of Briggs 1981, Sabljic 1987)
1.99-2.66 (soil, calculated-K_{OW} based on model of Means et al. 1982, Sabljic 1987)
1.30-1.90 (soil, calculated-K_{OW} based on model of Chiou et al. 1983, Sabljic 1987)

Half-Lives in the Environment:

Air:
Surface water:
Ground water:
Sediment:
Soil:
Biota:

Environmental Fate Rate Constants and Half-Lives:

Volatilization:
Photolysis:
Oxidation:
Hydrolysis:
Biodegradation: average rate of biodegradation 39.2 mg COD $g^{-1} \cdot h^{-1}$ based on measurements of COD decrease using activated sludge inoculum with 20 days of adaptation to the substrate (Pitter 1976; quoted, Scow 1982).
Biotransformation:
Bioconcentration Uptake (k_1) and Elimination (k_2) Rate Constants:

Common Name: 2-Phenylphenol
Synonym: *o*-phenylphenol, 2-hydroxybiphenyl
Chemical Name: 2-phenylphenol
CAS Registry No: 90-43-7
Molecular Formula: $C_6H_5C_6H_4OH$
Molecular Weight: 170.21
Melting Point (°C):
56.0 (Verschueren 1983)
57.0 (Dean 1985)
Boiling Point (°C):
278 (Boublik et al. 1984)
275 (Verschueren 1983)
282 (Dean 1985)
Density (g/cm³ at 20°C):
1.213 (Dean 1985)
1.20 (25°C, Verschueren 1983)
Molar Volume (cm³/mol):
147.50 (calculated-density, Jaworska & Schultz 1993)
192.0 (calculated-LeBas method)
Molecular Volume (Å³):
Total Surface Area, TSA (Å²):
Dissociation Constant, pK_a:
9.55 (Dean 1985)
Heat of Fusion ΔH_{fus}, kcal/mol:
Entropy of Fusion ΔS_{fus}, cal/mol·K (e.u.):
Fugacity Ratio at 25°C (assuming ΔS_{fus} = 13.5 e.u.), F: 0.494

Water Solubility (g/m³ or mg/L at 25°C):
700 (Verschueren 1983)

Vapor Pressure (Pa at 25°C):
2667, 13330 (163°C, 206°C, Verschueren 1983)
30.0 (Stephenson & Malanowski 1987)

Henry's Law Constant (Pa·m³/mol):

Octanol/Water Partition Coefficient, log K_{OW}:
3.09 (Hansch & Leo 1979)
3.36 (calculated-CLOGP 3.51, Jaworska & Schultz 1993)
3.36, 3.09 (quoted CLOGP, COMPUTOX, Kaiser 1993)
3.09 (quoted CLOGP and recommended, Sangster 1993)

Bioconcentration Factor, log BCF:

Sorption Partition Coefficient, log K_{OC}:

Half-Lives in the Environment:

Air: photooxidation half-life of 0.1-22 hours, based on estimated reaction rate constant with OH radicals (Atkinson 1987; quoted, Howard et al. 1991) and NO_3 radicals in nighttime air to react with the phenol and cresol classes (Howard et al. 1991).

Surface water: photooxidation half-life of 66-3840 hours in water, based on reported reaction rate constants for OH and RO_2 radicals to react with phenol class compounds (Mill & Mabey 1985; Güesten et al. 1981; quoted, Howard et al. 1991); 24-168 hours, based on estimated aqueous aerobic biodegradation half-life (Howard et al. 1991).

Groundwater: 48-336 hours, based on estimated aqueous aerobic biodegradation half-life (Howard et al. 1991).

Sediment:

Soil: 24-168 hours, based on estimated aqueous aerobic biodegradation half-life (Howard et al. 1991).

Biota:

Environmental Fate Rate Constants or Half-Lives:

Volatilization:

Photolysis:

Hydrolysis:

Oxidation: photooxidation half-life of 66-3840 hours in water, based on reported reaction rate constants for OH and RO_2 radicals to react with phenol class compounds (Mill & Mabey 1985; Guesten et al. 1981; quoted, Howard et al. 1991); photooxidation half-life of 0.1-22 hours, based on estimated reaction rate constant with OH radicals (Atkinson 1987; quoted, Howard et al. 1991) and NO_3 radicals in nighttime air to react with the phenol and cresol classes (Howard et al. 1991).

Biodegradation: 100% degradation under aerobic and anaerobic conditions after 3 weeks (504 h) at 22°C (Verschueren 1983); aqueous aerobic half-life of 24-168 hours, based on a river die-away study in which a 50% degradation was observed over a one week period (Gonsior et al. 1984; quoted, Howard et al. 1991); aqueous anaerobic half-life of 96-672 hours, based on estimated aqueous aerobic biodegradation half-life (Howard et al. 1991).

Biotransformation:

Bioconcentration, Uptake (k_1) and Elimination (k_2) Rate Constants:

Common Name: 4-Phenylphenol
Synonym: *p*-phenylphenol, 4-hydroxybiphenyl
Chemical Name: 4-phenylphenol
CAS Registry No: 92-69-3
Molecular Formula: $C_6H_5C_6H_4OH$
Molecular Weight: 170.21
Melting Point (°C):
164.0 (Dean 1985)
Boiling Point (°C):
307 (Boublik et al. 1984)
Density (g/cm^3 at 20°C):
Molar Volume (cm^3/mol):
147.50 (calculated from density, Jaworska & Schultz 1993)
192.0 (calculated-LeBas method)
Molecular Volume (Å^3):
Total Surface Area, TSA (Å^2):
Dissociation Constant, pK_a:
9.55 (Dean 1985)
Heat of Fusion ΔH_{fus}, kcal/mol:
Entropy of Fusion ΔS_{fus}, cal/mol K (e.u.):
Fugacity Ratio at 25°C (assuming ΔS_{fus} = 13.5 e.u.), F: 0.0422

Water Solubility (g/m^3 or mg/L at 25°C):
9.79 (CESARS 1988)

Vapor Pressure (Pa at 25°C):
0.0278 (liquid, extrapolated-Antoine eqn., Boublik et al. 1984)
0.020 (extrapolated-Antoine eqn., Dean 1985)
0.204 (extr. solid, Antoine eqn., Stephenson & Malanowski 1987)
8.71 (extr. liquid, Antoine eqn., Stephenson & Malanowski 1987)

Henry's Law Constant (Pa m^3/mol):

Octanol/Water Partition Coefficient, log K_{OW}:
3.20 (shake flask-AS, Norrington et al. 1975)
3.63 (HPLC-RV, Garst 1984)
3.20 (quoted, CESARS 1987)
2.88 (CPC, Terada et al. 1987)
3.20 (recommended, LOGKOW databank, Sangster 1993)
3.36 (quoted from CLOGP 3.51, Jaworska & Schultz 1993)
3.20 (COMPUTOX databank, Kaiser 1993)

Bioconcentration Factor, log BCF: no data
Sorption Partition Coefficient, log K_{OC}: no data
Half-Lives in the Environment/Environmental Fate Rate Constants or Half-Lives: no data

Common Name: 4-Chloro-*m*-cresol
Synonym: *p*-chloro-*m*-cresol, 2-chloro-5-hydroxytoluene, 4-chloro-3-methylphenol
Chemical Name: 4-chloro-*m*-cresol, 4-chloro-3-methylphenol
CAS Registry No: 59-50-7
Molecular Formula: $CH_3(Cl)C_6H_3OH$
Molecular Weight: 142.59
Melting Point (°C):
66 (Weast 1977; Callahan et al. 1979; Mabey et al. 1982)
68 (Dean 1985)
Boiling Point (°C):
235 (Weast 1977; Callahan et al. 1979; Dean 1985)
Density (g/cm^3):
Dissociation Constants, pK_a:
Molar Volume (cm^3/mol):
147.5 (calculated-LeBas method)
Molecular Volume, TMV ($Å^3$):
Total Surface Area, TSA ($Å^2$):
Heat of Fusion ΔH_{fus}, kcal/mol:
Entropy of Fusion ΔS_{fus}, cal/mol K (e.u.):
Fugacity Ratio at 25°C, (assuming ΔS_{fus} = 13.5 e.u.), F: 0.393

Water Solubility (g/m^3 or mg/L at 25°C):
3990 (shake flask-UV with buffer at pH 5.1, Blackman et al. 1955; quoted, IUPAC 1985)
3850 (20°C, Windholz 1976; quoted, Callahan et al. 1979)
3850 (20°C, Verschueren 1977; quoted, Mabey et al. 1982)
3846 (Merck Index 1989)

Vapor Pressure (Pa at 25°C):
6.67 (Mabey et al. 1982)

Henry's Law Constant (Pa m^3/mol):
0.253 (20°C, calculated-P/C, Mabey et al. 1982)

Octanol/Water Partition Coefficient, log K_{OW}:
2.18 (Hansch & Leo 1979)
2.95 (calculated as per Tute 1971, Callahan et al. 1979)
3.10 (Veith et al. 1979)
3.10 (quoted, Mabey et al. 1982)

Bioconcentration Factor, log BCF:
2.60 (microorganisms-water, calculated-K_{OW}, Mabey et al. 1982)

Sorption Partition Coefficient, log K_{OC}:

2.78 (sediment-water, calculated-K_{OW}, Mabey et al. 1982)

Half-Lives in the Environment:

Air:

Surface water:

Ground water:

Sediment:

Soil:

Biota:

Environmental Fate Rate Constants and Half-Lives:

Volatilization:

Photolysis:

Oxidation: rate constants of $<7 \times 10^5$ $M^{-1} \cdot hour^{-1}$ for singlet oxygen and 1×10^7 $M^{-1} \cdot hour^{-1}$ for peroxy radical (Mabey et al. 1982).

Hydrolysis:

Biodegradation:

Biotransformation: bacterial transformation rate constant of 3×10^{-9} $mL \cdot cell^{-1} \cdot hour^{-1}$ in water (Mabey et al. 1982).

Bioconcentration Uptake (k_1) and Elimination (k_2) Rate Constants:

Common Name: 2-Chlorophenol
Synonym: *o*-chlorophenol, 1-chloro-2-hydroxybenzene
Chemical Name: 2-chlorophenol
CAS Registry No: 95-57-8
Molecular Formula: $C_6H_4(OH)Cl$
Molecular Weight: 128.56
Melting Point (°C):
 7.0 (Stull 1947; Verschueren 1983; Renner 1990)
 8.4 (Drahonovsky & Vacek 1971; Callahan et al. 1979)
 9.0 (Weast 1978; Kaiser et al. 1984)
Boiling Point (°C):
 173 (Sidgwick & Turner 1922)
 174.5 (Stull 1947)
 175 (Krijgsheld & van der Gen 1986; Renner 1990)
Density (g/cm^3):
 1.257 (25°C, Krijgsheld & van der Gen 1986)
Dissociation Constants, pK_a:
 8.65 (Farquharson et al. 1958; Saarikoski & Viluksela 1982; Renner 1990)
 8.49 (Doedens 1967)
 8.48 (Pearce & Simkins 1968; Krijgsheld & van der Gen 1986)
 8.52 (Drahonovsky & Vacek 1971; Callahan et al. 1979; Könemann & Musch 1981; Ugland et al. 1981; Westall et al. 1985; Howard 1989; Renner 1990)
 8.29 (Sillén & Martell 1971; Kaiser et al. 1984; Shigeoka et al. 1988)
 8.55 (Serjeant & Dempsey 1979; Hammers et al. 1982; Dean 1985; Schultz & Cajina-Quezada 1987; Tratnyek & Hoigné 1991; Haderlein & Schwarzenbach 1993)
 8.30 (Hoigné & Bader 1983)
 8.48, 9.30 (quoted, HPLC, Miyake et al. 1987)
Molar Volume (cm^3/mol):
 95.0 (calculated-density, Jaworska & Schultz 1993)
 124.3 (calculated-LeBas method, Ma et al. 1993; Shiu et al. 1994)
Molecular Volume, TMV ($Å^3$):
Total Surface Area, TSA ($Å^2$):
Heat of Fusion ΔH_{fus}, kcal/mol:
 2.57 (Tsonopoulos & Prausnitz 1971)
Entropy of Fusion ΔS_{fus}, cal/mol·K (e.u.):
 9.11 (Tsonopoulos & Prausnitz 1971)
Fugacity Ratio, F (assuming ΔS_{fus} = 13.5 e.u.): 1.0

Water Solubility (g/m^3 or mg/L at 25°C):
 25000 (20°C, synthetic methods, Sidgwick & Turner 1922)
 29000 (20-25°C, Seidell 1941; Urano et al. 1982)
 24650 (20°C, shake flask-UV, Mully & Metcalf 1966; qquoted, Ma et al. 1993)
 22000 (shake flask-AS, Roberts et al. 1977)
 28500 (20°C, quoted, Verschueren 1977, 1983; Callahan et al. 1979; Mabey et al. 1982; LeBlanc 1984; Krijgsheld & van der Gen 1986; Shigeoka et al. 1988)

11480 (shake flask-LSC, Banerjee et al. 1980; quoted, Davies & Dobbs 1984; Ma et al. 1993)
11200 (shake flask-radioactive analysis, Veith et al. 1980)
1990 (calculated-K_{OW} as per Hansch et al. 1968, Veith et al. 1980)
20000 (recommanded, IUPAC 1985)
11480, 15456 (quoted, calculated-fragment const., Wakita et al. 1986)
28500 (selected, Suntio et al. 1988)
11480 (quoted, Isnard & Lambert 1988, 1989)
28000 (quoted, Howard 1989)
28155 (calculated-K_{OW}, Müller & Klein 1992)
23256 (shake flask-HPLC/UV at pH 4.8, Ma et al. 1993)

Vapor Pressure (Pa at 25°C):
316 (interpolated-regression of tabulated data, Stull 1947)
313 (extrapolated from Antoine eqn., Weast 1972-73)
267 (quoted, Chiou & Freed 1977)
5532 (80°C, quoted, Verschueren 1977,83)
293, 316 (quoted, Callahan et al. 1979)
236 (20°C, extrapolated from Antoine eqn., Mabey et al. 1982)
316 (extrapolated, Cheung 1984)
293, 180.7 (extrapolated, Antoine eqn., Dean 1985)
133 (12.1°C, quoted, Krijgsheld & van der Gen 1986)
131.8 (Antoine eqn., Stephenson & Malanowski 1987)
189 (quoted, Howard 1989)

Henry's Law Constant (Pa·m^3/mol):
1.065 (calculated-P/C, Mabey et al. 1982)
0.6884 (calculated-P/C, Shiu et al. 1994)

Octanol/Water Partition Coefficient, log K_{OW}:
2.15 (shake flask-AS, Fujita et al. 1964; quoted, Leo et al. 1971; Rekker 1977; Roberts et al. 1977; Suntio et al. 1988, Howard 1989)
2.15, 2.19 (Leo et al. 1971; quoted, Carson et al. 1975; Chiou & Freed 1977; Menges et al. 1990; Haderlein & Schwarzenbach 1993)
2.20 (calculated using data from Fujita et al. 1964, Umeyama et al. 1971)
2.15, 2.08 (LC-k', calculated-π const., Carson et al. 1975)
2.17 (quoted from Leo et al. 1971, Callahan et al. 1979; quoted, Westall et al. 1985)
2.27 (calculated-f const., Rekker 1977; Könemann 1981; Könemann & Musch 1981)
2.17 (Hansch & Leo 1979; quoted, Kaiser 1983; Kaiser et al. 1984; Chin et al. 1986)
2.16 (shake flask-LSC, Banerjee et al. 1980; quoted, Arbuckle 1983; Davies & Dobbs 1984)
2.16 (measured value from this study, Veith et al. 1980)
0.83 (estimated-RPHPLC-RT, Veith et al. 1980)
2.16 ± 0.03 (HPLC-k', Hammers et al. 1982)
2.12 (HPLC-k'correlated, Hammers et al. 1982)
2.15 (quoted, Saarikoski & Viluksela 1982; Schultz & Cajina-Quezada 1987)
2.15, 2.19 (quoted, Verschueren 1983)

2.15 (quoted from Hansch & Leo 1982, Banerjee et al. 1984; quoted, Banerjee 1987)
2.17 (quoted from Leo et al. 1971, Beltrame et al. 1984; quoted, Shigeoka et al. 1988; Saito et al. 1993)
2.23 (computed, Klopman et al. 1985; quoted, Pangrekar et al. 1994)
2.17, 2.32 (quoted, RPHPLC-RT, Chin et al. 1986)
1.56, 1.99 (HPLC-k', Eadsforth 1986)
2.17, 2.18 (quoted: shake flask, calculated, Eadsforth 1986)
2.15 (quoted, Krijgsheld & van der Gen 1986)
2.15, 2.13 (calculated-HPLC-k', Miyake et al. 1987)
2,15, 2.16 (quoted, shake flask-CPC, Berthod et al. 1988)
2.24 (batch equilibration-UV, Beltrame et al. 1988)
2.16 (quoted, Isnard & Lambert 1988)
2.25 (back flashing-CPC, Menges et al. 1990)
2.16 (quoted, Liu et al. 1991)
2.14, 2.05 (quoted, CPC-RV, El Tayar et al. 1991)
2.20 (calculated, Müller & Klein 1992)
2.24 (calculated-π const., Verhaar et al. 1992)
2.15 (calculated-CLOGP 3.51, Jaworska & Schultz 1993)
2.18, 2.17 (COMPUTOX databank, Kaiser 1993)
2.20 (calculated-QSAR, Kollig 1993)
2.29 (shake flask-GC, Kishino & Kobayashi 1994)

Bioconcentration Factor, log BCF:
2.33 (bluegill sunfish, Barrows et al. 1980; quoted, LeBlanc 1984; Howard 1989)
2.33 (bluegill sunfish, Veith et al. 1980; quoted, Davies & Dobbs 1984)
1.13 (calculated-K_{OW}, Veith et al. 1980)
1.61 (microorganisms-water, calculated-K_{OW}, Mabey et al. 1982)
2.33 (bluegill sunfish, Bysshe 1982)
0.81 (gold fish, Kobayashi et al. 1979; quoted, Verschueren 1983; Howard 1989)
0.81 (Isnard & Lambert 1988)
0.28-1.40 (estimated from K_{OW}, Howard 1989)

Sorption Partition Coefficient, log K_{OC}:
3.70, 3.60 (sediment: fine, coarse; Isaacson & Frink 1984; quoted, Howard 1989)
1.86 (sediment-water, calculated-K_{OW}, Mabey et al. 1982)
1.71 (clay loam soil, Boyd 1982; quoted, Howard 1989)
3.69 (untreated fine sediment, Isaacson & Frink 1984)
3.60 (untreated coarse sediment, Isaacson & Frink 1984)
3.98 (treated fine sediment, Isaacson & Frink 1984)
4.36 (treated coarse sediment, Isaacson & Frink 1984)
1.20-2.55 (estimated from K_{OW}, Howard 1989)
1.82 (calculated-K_{OW}, Kollig 1993)

Half-Lives in the Environment:
Air: half-life in the atmosphere was estimated to be 1.96 days (Howard 1989).
Surface water: half-life of 16.8 days (sludge) and 55.2 days (polluted river waters) (Mills et al. 1982); rate constant of $(1.1 \pm 0.3)\times10^3$ $M^{-1}\cdot sec^{-1}$ for the reaction

with ozone at pH 1.8-4.0 (Hoigné & Bader 1983); first-order photolysis disappearance rate constant of 3.01×10^{-2} minute^{-1} in the absence of DOM at 313 nm (Kawaguchi 1992).

Groundwater:

Sediment:

Soil: 7.2 days in an acidic clay soil with <1.0% organic matter and 1.7 day in a slightly basic sandy loam soil with 3.25% organic matter, based on aerobic batch Lab. microcosm experiments (Loehr & Matthews 1992).

Biota: <1 day in tissue of bluegill sunfish (Barrows et al. 1980).

Enrironmental Fate Rate Constants and Half-Lives:

Volatilization: 1.35, 1.60 hour from stirred and static water at 23.8°C (Chiou et al. 1980; Howard 1989); 73 days, based on estimation from a model river 1 m deep flowing at 1 m/sec with a wind velocity of 3 m/sec (Howard 1989).

Photolysis: vapor phase half-life of 47 hours (Howard 1989); first-order photolysis disappearance rate constant of 3.01×10^{-2} minute^{-1} in the absence of DOM at 313 nm (Kawaguchi 1992).

Oxidation: aqueous oxidation rate constant of < 7×10^5 M^{-1} h^{-1} for singlet oxygen and 1×10^7 M^{-1} h^{-1} for peroxy radical at 25 °C (Mabey et al. 1982); rate constant of $(1.1 \pm 0.3) \times 10^3$ $M^{-1} \cdot sec^{-1}$ for the reaction with ozone in water at pH 1.8-4.0 (Hoigné & Bader 1983); rate constant of $(9.2 \pm 9.4) \times 10^6$ $M^{-1} \cdot sec^{-1}$ for the reaction with singlet oxygen in aqueous phosphate buffer at (27 ± 1)°C (Tratnyek & Hoigné 1991).

Hydrolysis:

Biodegradation: average rate of biodegradation 25.0 mg COD $g^{-1} \cdot h^{-1}$ based on measurements of COD decrease using activated sludge inoculum with 20 days of adaptation to the substrate (Pitter 1976; quoted, Scow 1982); time necessary for complete degradation of 16 m/L in 14-25 days by wastewater (Haller 1978); aerobic degradation in a non-sterile clay loam soil: 91% loss at 0°C after 8 days at the termination of the experiment, 94% loss at 4°C after 8 days and 100% loss at 20°C after 10-15 days all under same exptl. conditions (Baker et al. 1980); rate constants of 1.0 day^{-1} with a half-life of 0.7 days in adopted activated sludge and 0.3 day^{-1} with a half-life of 2.3 days in soil suspension under aeeerobic conditions (Mills et al. 1982); completely degraded in soil suspensions in 14 days and by a soil microflora within 64 days (quoted, Verschueren 1983); approximately 48 hours in a column microcosm under aerobic conditions (Suflita & Miller 1985); degradation rate of 10 $\mu mol \cdot L^{-1} \cdot day^{-1}$ in freshwater and 8 $\mu mol \cdot L^{-1} \cdot day^{-1}$ in saline water with acclimated sulfidogenic sediment cultures (Häggblom & Young 1990).

Biotransformation: rate constant for bacterial transformation of 1×10^{-7} $mL \cdot cell^{-1} \cdot h^{-1}$ in water (Mabey et al. 1982); microbial transformation rate constant of $(7.1 \pm 1.6) \times 10^{-11}$ $L \cdot organism^{-1} \cdot h^{-1}$ (Paris et al. 1983; quoted, Steen 1991); degradation rate of 3.49×10^{-17} ($\pm 38\%$ SD) $mol \cdot cell^{-1} \cdot h^{-1}$ from pure culture studies (Banerjee et al. 1984).

Bioconcentration, Uptake (k_1) and Elimination (k_2) Rate Constants:

Common Name: 3-Chlorophenol
Synonym: *m*-chlorophenol, 1-chloro-3-hydroxybenzene
Chemical Name: 3-chlorophenol
CAS Registry No: 108-43-0
Molecular Formula: C_6H_4ClOH
Molecular Weight: 128.56
Melting Point (°C):
32.5 (Stull 1947)
36 (Weast 1978; Kaiser et al. 1984)
32.8 (Verschueren 1983)
33 (Krijgsheld & van der Gen 1986)
31.0 (Renner 1990)
Boiling Point (°C):
214 (Sidgwick & Turner 1922; Stull 1947; Verschueren 1983; Krijgsheld & van der Gen 1986)
216 (Renner 1990)
Density (g/cm^3, 20°C):
1.245 (45°C, Verschueren 1983)
1.245 (25°C, Krijgsheld & van der Gen 1986)
Dissociation Constants, pK_a:
9.12 (Farquharson et al. 1958; Renner 1990)
8.85 (Doedens 1967; Jones 1981; Bintein & Devillers 1994)
9.08 (Pearce & Simkins 1968)
8.78 (Sillén & Martell 1971; Kaiser et al. 1984; Shigeoka et al. 1988)
9.12 (Serjeant & Dempsey 1979; Howard 1989; Tratnyek & Hoigné 1991; Haderlein & Schwarzenbach 1993)
8.97 (Könemann 1981; Könemann & Musch 1981; Ugland et al. 1981; Renner 1990; Ma et al. 1993)
9.10 (Dean 1985; Schultz & Cajina-Quezada 1987)
9.02 (Krijgsheld & van der Gen 1986)
9.37 (quoted from Ugland et al. 1981, Lagas 1988)
Molar Volume (cm^3/mol):
0.626 (intrinsic $V_I/100$, Kamlet et al. 1988)
124.3 (calculated-LeBas method, Ma et al. 1993)
Molecular Volume, TMV ($Å^3$):
Total Surface Area, TSA ($Å^2$):
Heat of Fusion, ΔH_{fus}, kcal/mol:
2.8 (Tsonopoulos & Prausnitz 1971)
Entropy of Fusion, Δ_{fus}, cal/mol·K (e.u.):
9.13 (Tsonopoulos & Prausnitz 1971)
Fugacity Ratio at 25°C, F: 0.8334

Water Solubility (g/m^3 or mg/L at 25°C):
26000 (20°C, synthetic methods, Sidgwick & Turner 1922)
26000 (20-25°C, Seidell 1941; Urano et al. 1982; Shigeoka et al. 1988)
22420 (20°C, Mulley & Metcalf 1966)

26000 (20$_o$C, quoted, Verschueren 1983; Krijgsheld & van der Gen 1986)
22000 (recommended, IUPAC 1985)
26000 (selected, Suntio et al. 1988)
26000 (quoted, Howard 1989)
22190 (shake flask-HPLC/UV at pH 4.6, Ma et al. 1993)

Vapor Pressure (Pa at 25°C):
41.9 (extrapolated-regression of tabulated data, Stull 1947)
41.4 (extrapolated liquid value, Antoine eqn, Weast 1972-73)
29.3 (estimated, Chiou & Freed 1977)
667 (72°C, quoted, Verschueren 1983)
133 (44.2°C, quoted, Krijgsheld & van der Gen 1986)
1.03 (interpolated, solid value, Antoine eqn., Stephenson & Malanowski 1987)
33.4 (extrapolated, P_L, Antoine eqn., Stephenson & Malanowski 1987)
16.7 (quoted, Howard 1989)

Henry's Law Constant (Pa·m^3/mol):
0.0567 (calculated, Hine & Mookerjee 1975; quoted, Howard 1989)
0.2045 (calculated-P/C, Shiu et al. 1994)

Octanol/Water Partition Coefficient, log K_{OW}:
2.50 (shake flask-AS, Fujita et al. 1964; quoted, Leo et al. 1969,71; quoted, Chiou & Freed 1977; Rekker 1977; Haderlein & Schwarzenbach 1993)
2.55 (calculated using data from Fujita et al. 1964, Umeyama et al. 1971)
2.48, 2.17, 2.29 (quoted observed value, calculated-π const., calculated-f const., Rekker 1977)
2.50 (Hansch & Leo 1979; quoted, Kaiser 1983; Kaiser et al. 1984; Lagas 1988; Bodor et al. 1989; Howard 1989; Bintein & Devillers 1994)
1.98 (calculated, McLeese et al. 1979)
2.33 (calculated-HPLC-k', Butte et al. 1981; quoted, Butte et al. 1987)
2.27 (calculated-f const., Könemann 1981; Könemann & Musch 1981)
2.59 (calculated-π const., Verhaar et al. 1992)
2.50 ± 0.02 (HPLC-k', Hammers et al. 1982)
2.36 (HPLC-k'correlated, Hammers et al. 1982)
2.68, 2.55 (quoted, calculated-RP-HPLC-k', Miyake & Terada 1982)
2.47, 2.50 (quoted, Verschueren 1983)
2.50 (quoted from Hansch & Leo 1982, Banerjee et al. 1984; quoted, Banerjee 1987)
2.48 (quoted from Leo et al. 1971, Beltrame et al. 1984; quoted, Shigeoka et al. 1988)
2.23 (computed, Klopman et al. 1985; quoted, Pangrekar et al. 1994)
1.87, 2.29 (HPLC-k', Eadsforth 1986)
2.48, 2.18 (quoted: shake flask, calculated, Eadsforth 1986)
2.50 (quoted, Krijgsheld & van der Gen 1986; Lagas 1988)
2.50 (Schultz & Cajina-Quezada 1987)
2.55 (batch equilibration-UV, Beltrame et al. 1988; quoted, Beltrame et al. 1989)
2.50 (RP-HPLC-capacity ratio, Minick et al. 1988)

2.48 (RP-HPLC, Shigeoka et al. 1988; quoted, Saito et al. 1993)
2.49, 2.44 (quoted exptl., calculated-V_I and solvatochromic p., Kamlet et al. 1988)
2.42 (selected, Suntio et al. 1988)
2.50, 2.485, 1.78 (quoted from CLOGP, calculated-CLOGP, calculated-MO, Bodor et al. 1989)
2.50 (THOR 1989; quoted, Connell & Markwell 1990)
2.50 (quoted, Liu et al. 1991)
2.59 (calculated-π const., Verhaar et al. 1992)
2.48, 2.50 (COMPUTOX databank, Kaiser 1993)
2.64 (shake flask-GC, Kishino & Kobayashi 1994)

Bioconcentration Factor, log BCF:
1.30 (Golden ide, Freitag et al. 1985)
1.25 (zebrafish, Butte et al. 1987)
0.845, 1.23 (earthworms *e. fetida andrei*: Kooyenberg soil, Holten soil, van Gestel & Ma 1988)
2.01, 2.09 (earthworms *l. rubellus*: Kooyenberg soil, Holten soil, van Gestel & Ma 1988)
1.17-2.55 (estimated from K_{OW}, Howard 1989)
1.00, 1.32, 2.17, 2.18 (earthworm system, from literature, Connell & Markwell 1990)
1.10, 1.40, 10.1, 16.3 (earthworm system, derived data, Connell & Markwell 1990)

Sorption Partition Coefficient, log K_{OC}:
1.20-2.74 (estimated from K_{OW}, Howard 1989)

Half-Lives in the Environment:
Air: vapor phase half-life was estimated to be 1.96 days (Howard 1989).
Surface water:
Ground water:
Sediment: half-life was approximately 30 days in sediment from a farm stream at 20°C (Howard 1989).
Soil: disappearance half-life of 2.5 days from Kooyenburg soil, 5.4 days from Holten soil with earthworms *e. fetida andrei* and 2.6 days from Kooyenburg soil, 2.1 days from Holten soil with earthworms *l. rubblus* (van Gestel & Ma 1988); half-life of 15.1 days in an acidic clay soil with $<1.0\%$ organic matter and 21.8 days in a slightly basic sandy loam soil with 3.25% organic matter, based on aerobic batch lab. microcosm experiments (Loehr & Matthews 1992).
Biota:

Environmental Fate Rate Constants and Half-Lives:
Volatilization: estimated half-life of 73 days for evaporation from a model river of 1 m deep, flowing at 1 m/sec with a wind velocity of 3 m/sec (Lyman 1982; quoted, Howard 1989).
Photolysis: vapor phase half-life of 1.96 days (Howard 1989).
Oxidation: rate constant of $(5.4 \pm 1.0)x10^6\ M^{-1}\cdot sec^{-1}$ for the reaction with singlet oxygen in aqueous phosphate buffer at (27 ± 1)°C (Tratnyek & Hoigné 1991).
Hydrolysis: will not be an important degradation process (Howard 1989).

Biodegradation: 95% degradation in 3-6 days in mixed bacteria cultures (Tabak et al. 1964); completely degraded in soil suspensions within 72 days and by a soil microflora within 64 days (quoted, Verschueren 1983); time necessary for complete degradation of 16 mg per liter in 14-25 days by wastwater (Haller 1978); degradation rate of 15 $\mu mol \cdot L^{-1} \cdot day^{-1}$ in freshwater and 18 $\mu mol \cdot L^{-1} \cdot day^{-1}$ in saline water with acclimated sulfidogenic sediment cultures (Häggblom & Young 1990).

Biotransformation: degradation rate of 4.32×10^{-17} ($\pm$47% SD) $mol \cdot cell^{-1} \cdot h^{-1}$ from pure culture studies (Banerjee et al. 1984).

Bioconcentration Uptake (k_1) and Elimination (k_2) Rate Constants:

Common Name: 4-chlorophenol
Synonym: *p*-chlorophenol, 1-chloro-4-hydroxybenzene
Chemical Name: 4-chlorophenol
CAS Registry No: 106-48-9
Molecular Formula: ClC_6H_4OH
Molecular Weight: 128.56
Melting Point (°C):
42.0 (Stull 1947)
42.9 (Tsonopoulos & Prausnitz 1971)
43.70 (Weast 1982; Kaiser et al. 1984)
43.0 (Verschueren 1983; Krijgsheld & van der Gen 1986; Renner 1990)
Boiling Point (°C):
217 (Sidgwick & Turner 1922; Verschueren 1983; Krijgsheld & van der Gen 1986; Renner 1990)
220 (Stull 1947)
Density (g/cm^3 at 20°C):
1.306 (Verschueren 1983; Krijgsheld & van der Gen 1986)
Dissociation Constants, pK_a:
9.37 (Farquharson et al. 1958; Saarikoski & Viluksela 1982; Renner 1990)
9.18 (Doedens 1967)
9.42 (Pearce & Simkins 1968)
9.14 (Sillén & Martell 1971; Kaiser et al. 1984; Shigeoka et al. 1988)
9.40 (McLeese et al. 1979)
9.41 (Serjeant & Dempsey 1979; Howard 1989; Tratnyek & Hoigné 1991; Haderlein & Schwarzenbach 1993)
9.37 (Ugland et al. 1981; Renner 1990; Ma et al. 1993)
9.38 (Paris et al. 1982; Krijgsheld & van der Gen 1986)
9.20 (Hoigné & Bader 1983)
9.43 (Dean 1985; Schultz & Cajina-Quezada 1987)
9.50 (Nendza & Seydel 1988)
9.37 ± 0.01 (potentiometric partition, Hersey et al. 1989)
9.57 ± 0.01 (AS with pH profile, Hersey et al. 1989)
Molar Volume (cm^3/mol):
0.626 (intrinsic $V_I/100$, Kamlet et al. 1988)
95.50 (calculated-density, Jaworska & Schultz 1993)
124.3 (calculated-LeBas method, Ma et al. 1993; Shiu et al. 1994)
Molecular Volume, TMV ($Å^3$):
Total Surface Area, TSA ($Å^2$):
Heat of Fusion ΔH_{fus}, kcal/mol:
3.51 (Tsonopoulos & Prausnitz 1971)
Entropy of Fusion ΔS_{fus}, cal/mol·K:
11.1 (Tsonopoulos & Prausnitz 1971)
Fugacity Ratio at 25°C, (assuming ΔS_{fus} = 13.5 e.u.) F: 0.6637

Water Solubility (g/m^3 or mg/L at 25°C):

27000 (20°C, synthetic methods, Sidgwick & Turner 1922)
27000 (20-25°C, Seidell 1941; Urano et al. 1982)
27000 (shake flask-UV at pH 5.1, Blackman et al. 1955; quoted, Ma et al. 1993)
26250 (20°C, shake flask-UV, Mulley & Metcalf 1966; quoted, Ma et al. 1993)
24000 (shake flask-AS, Roberts et al. 1977)
28000 (Callahan et al. 1979; Neely 1984)
27000 (selected, Verschueren 1983; quoted, Shigeoka et al. 1988; Suntio et al. 1988)
27100 (20°C, quoted, Verschueren 1983; LeBlanc 1984; Krijgsheld & van der Gen 1986; Shigeoka et al. 1988)
27000 (recommended, IUPAC 1985)
26915 (quoted, Isnard & Lambert 1989)
27000 (selected, Howard 1989)
26390 (shake flask-HPLC/UV at pH 4.6, Ma et al. 1993)

Vapor Pressure (Pa at 25°C):

28.6 (extrapolated-regression of tabulated data, Stull 1947)
28 (extrapolated liquid value, Antoine eqn., Weast 1972-73)
17.3 (estimated, Chiou & Freed 1977)
20 (estimated, Cheung 1984)
13.3 (20°C, quoted, Verschueren 1977, 1983)
33.3 (30°C, quoted, Verschueren 1977, 1983)
133 (49.8°C, quoted, Krijgsheld & van der Gen 1986)
3.47 (extrapolated liquid value, Antoine eqn., Stephenson & Malanowski 1987)
11.6 (quoted, Howard 1989)

Henry's Law Constant ($Pa \cdot m^3/mol$):

0.0567 (calculated, Hine & Mookerjee 1975; Howard 1989)
0.0471 (calculated-P/C, Leuenberger et al. 1985)
0.0952 (calculated-P/C, Shiu et al. 1994)

Octanol/Water Partition Coefficient, log K_{OW}:

2.39 (shake flask-AS, Fujita et al. 1964; quoted, Leo et al. 1969, 1971; Mirrlees et al. 1976; Rekker 1977; Roberts et al. 1977)
2.39 (Leo et al. 1971; quoted, Carson et al. 1975; Chiou & Freed 1977)
2.53 (shake flask-AS at pH 7.45, Umeyama et al. 1971)
2.39, 2.37 (LC-k', calculated-π const., Carson et al. 1975)
2.40 (HPLC, Mirrlees et al. 1976)
2.39, 2.17, 2.29 (quoted exptl., calculated-π const., calculated-f const., Rekker 1977)
2.35 (Hansch & Leo 1979; quoted, Kaiser 1983; Kaiser et al. 1984)
2.51 (calculated, McLeese et al. 1979)
2.40 $\pm$ 0.03, 2.35 (HPLC-k', HPLC-k'correlated, Hammers et al. 1982)
2.65, 2.51 (quoted, calculated-RP-HPLC-k', Miyake & Terada 1982)

2.39 (quoted, Paris et al. 1982; Saarikoski & Viluksela 1982; Krijgsheld & van der Gen 1986)
2.39, 2.44 (quoted, Verschueren 1983)
2.39 (quoted from Hansch & Leo 1982, Banerjee et al. 1984; quoted, Banerjee 1987)
2.41 (quoted from Leo et al. 1971, Beltrame et al. 1984; quoted, Shigeoka et al. 1988)
2.39, 2.46 ± 0.06 (selected best lit. value, exptl.-ALPM, Garst & Wilson 1984)
2.23 (computed, Klopman et al. 1985; quoted, Pangrekar et al. 1994)
2.39 (quoted from Hansch & Leo 1979, Miyake et al. 1987; Minick et al. 1988)
2.39 (Schultz et al. 1987; Schultz & Cajina-Quezada 1987)
3.20, 2.88 (quoted, CPC-RV, Terada et al. 1987)
2.43 (batch equilibration-UV, Beltrame et al. 1988; quoted, Beltrame et al. 1989)
2.39 (RP-HPLC-capacity ratio, Minick et al. 1988)
2.41 (HPLC-RP, Shigeoka et al. 1988; quoted, Saito et al. 1993)
2.45, 2.47 (quoted exptl., calculated-V_I and solvatochromic p., Kamlet et al. 1988)
2.42 (selected, Suntio et al. 1988)
2.42 ± 0.03 (filter chamber-AS, Hersey et al. 1989)
2.46 ± 0.01 (potentiometric partition, Hersey et al. 1989)
2.34 ± 0.01, 2.45 ± 0.02 (AS with pH profile, Hersey et al. 1989)
2.39 (quoted, Howard 1989; El Tayar et al. 1991)
2.39 (quoted, Liu et al. 1991)
2.42 (quoted from Leo et al. 1971; Haderlein & Schwarzenbach 1993)
2.48 (calculated-CLOGP 3.51, Jaworska & Schultz 1993)
2.48, 2.35 (COMPUTOX databank, Kaiser 1993)
2.35-2.42 (quoted lit. range, NCASI 1993)
2.63 (shake flask-GC, Kishino & Kobayashi 1994)

Bioconcentration Factor, log BCF:
1.18 (goldfish, Kobayashi et al. 1979; quoted, Verschueren 1983; Howard 1989)
0.30-1.59 (estimated from K_{OW}, Howard 1989)
1.05-1.50 (estimated, NCASI 1993)
1.18 (quoted, goldfish, NCASI 1993)

Sorption Partition Coefficient, log K_{OC}:
1.85 (clay loam soil, Boyd 1982; Howard 1989)
1.20-2.68 (estimated from K_{OW}, Howard 1989)

Half-Lives in the Environment:
Air: half-life of 1.96 days in the atmosphere (Howard 1989)
Surface water: half-life of 475 hours for biodegradation in river waters with a calculated first-order rate constant of 0.035 day^{-1} (Lee & Ryan 1979; quoted, Battersby 1990); 216-624 hours and 72-1080 hours for 75 % degradation in mineral medium and seawater respectively (de Kreuk & Hanstveit 1981); rate constant of 600 ± 100 $M^{-1}\cdot sec^{-1}$ for the reaction with ozone at pH 1.5-6.0 (Hoigné & Bader 1983); 63 hours (summer at season temp., 25°C), 99 hours (winter at season temp., 14°C) in distilled water and 28 hours (summer), 63 hours (winter)

in estuarine surface water, based on photo-transformation rate (Hwang et al. 1986); 58 days (summer), 224 days (winter) in distilled water and 10 days (summer), 95 days (winter) in surface estuarine water, based on photo-mineralization rate (Hwang et al. 1986); half-life of 216 hours at 21°C in Skidway River water (Pritchard 1987); half-life of 20 days in water (Howard 1989); at a concentration of 1 mg/L, 4-CP was degraded completely within 15 days in river waters (NCASI 1993).

Ground water:

Sediment: 72 hours at 22°C in Skidway River water-sediment slurry (Pritchard 1987); half-life of 3 days in sediment and seawater (Howard 1989).

Soil: 2.5 days in an acidic clay soil with $<1.0\%$ organic matter and 1.0 day in a slightly basic sandy loam soil with 3.25% organic matter, based on aerobic batch lab. microcosm experiments (Loehr & Matthews 1992); 70% degradation with 1 to 2 days (NCASI 1993).

Biota:

Environmental Fate Rate Constants and Half-Lives:

Volatilization: half-life of 12.8 hours, 17.4 hours for evaporation from stirred and static water at a depth of 0.38 cm at 23.6°C (Chiou et al. 1980; Howard 1989); half-life of 73 days was estimated for evaporation from a river 1 m deep, flowing at 1 m/sec with a wind velocity of 3 m/s (Howard 1989).

Photolysis: photo-transformation rate constants: 0.011 h^{-1} with half-life of 63 hours for distilled water in summer (mean temp 25°C) and 0.007 h^{-1} with half-life of 99 hours in winter (mean temp 14°C); 0.03 h^{-1} with half-life of 28 hours for estuarine water in summer and 0.011 h^{-1} with half-life of 63 hours in winter exposed to full sunlight and microbes (Hwang et al. 1986); photomineralization rate constants: 0.012 h^{-1} with half-life of 58 days for distilled water in summer and 0.003 h^{-1} with half-life of 224 days in winter; 0.07 h^{-1} with half-life of 10 days in summer and 0.007 h^{-1} with half-life of 95 days in winter for surface estuarine water exposed to full sunlight and microbes (Hwang et al. 1986).

Oxidation: rate constant of 600 ± 100 $M^{-1}\cdot sec^{-1}$ for the reaction with ozone in water using PrOH as scavenger at pH 1.5-6.0 and 20-23°C (Hoigné & Bader 1983); rate constant of $(6.0 \pm 3.6) \times 10^6$ $M^{-1}\cdot sec^{-1}$ for the reaction with singlet oxygen in aqueous buffer at (27 ± 1)°C (Tratnyek & Hoigné 1991); 72.1 mg/L total organic carbon (TOC) degraded to 98% TOC after 5 hours illumination with 250 watt tungsten lamp by the photo-Fenton reaction (Ruppert et al. 1993).

Hydrolysis:

Biodegradation: 95% degradation in 3-6 days in a mixed bacteria cultures (Tabak et al. 1964); average rate of biodegradation 11.0 mg COD $g^{-1}\cdot h^{-1}$ based on measurements of COD decrease using activated sludge inoculum with 20 days of adaptation to the substrate (Pitter 1976; quoted, Scow 1982); time necessary for complete degradation of 16 mg per liter in 7-14 days by wastewater and in 14-25 days by soil (Haller 1978); half-life of 475 hours in river waters with a calculated first-order rate constant of 0.035 day^{-1} (Lee & Ryan 1979; quoted, Battersby 1990); aerobic degradation in a non-sterile clay-loam soil: 77% loss at 0°C at the termination of the experiment after 14 days, 84% loss at 4°C after 12 days and 100% loss at 20°C after 30 days all under same experimental conditions (Baker et al. 1980); 216-480 hours and 72-1080 hours for 75% degradation in mineral medium and seawater (de Kreuk & Hanstveit 1981); biodegradation first order rate constant 0.23 day^{-1} in aquatic systems (Scow 1982); completely degraded in soil suspensions in 9 days and by a soil microflora in 16 days (quoted, Verschueren 1983); microbial degradation is the primary transformation process; transformation rate constant of 0.06 h^{-1} with half-life of 11 hours for estuarine water in summer and rate constant of 0.006 h^{-1} with half-life of 116 hours in winter in the darkness (Hwang et al. 1986); mineralization rate constant of 0.293 h^{-1} with half-life of 2 days for estuarine water in summer and 0.003 h^{-1} with half-life of 231 days in winter in the darkness (Hwang et al. 1986); degradation rate constants of 0.035 day^{-1} with a half-life of 480 days in Skidway River water and 0.23 day^{-1} with a half-life of 72 hours in Skidway River water-sediment slurry (Pritchard 1987); degradation rate constant of 37 $\mu mol \cdot L^{-1} \cdot day^{-1}$ in freshwater and 22 $\mu mol \cdot L^{-1} \cdot day^{-1}$ in saline water with acclimated sulfindogenic sediment cultures (Häggblom & Young 1990); average transformation rate of 53 $\mu mol \cdot Liter^{-1} \cdot day^{-1}$ at 31°C for anaerobic degradation in freshwater sediments with an average adaptation time of 37 days (Zhang & Wiegel 1990); 70% degradation within 1-2 days in soil and completely degraded within 15 days in river water (NCASI 1993).

Biotransformation: microbial transformation rate constant of $(1.7 \pm 0.9) \times 10^{-12}$ $L \cdot organism^{-1} \cdot h^{-1}$ (Paris et al. 1982); microbial transformation rate constant of $(4.7 \pm 1.6 \times 10^{-11}$ $L \cdot organism^{-1} \cdot h^{-1}$ to $(9.0 \pm 1.7) \times 10^{-11}$ $L \cdot organism^{-1} \cdot h^{-1}$ in pond and river samples at five different sites (Paris et al. 1983); degradation rate of 5.44×10^{-17} ($\pm$32% SD) $mol \cdot cell^{-1} \cdot h^{-1}$ from pure culture studies and 0.3×10^{-12} $mol \cdot cell^{-1} \cdot h^{-1}$ with microorganisms in Seneca River waters (Banerjee et al. 1984).

Bioconcentration Uptake (k_1) and Elimination (k_2) Rate Constants:

Common Name: 2,4-dichlorophenol
Synonym: 2,4-DCP
Chemical Name: 2,4-dichlorophenol
CAS Registry No: 120-83-2
Molecular Formula: $C_6H_3Cl_2OH$
Molecular Weight: 163.0
Melting Point (°C):
45.0 (Stull 1947; Verschueren 1977, 1983; Weast 1978; Callahan et al. 1979; Kaiser et al. 1978; Bidleman & Renberg 1985; Krijgsheld & van der Gen 1986)
43.0 (Renner 1990)
Boiling Point (°C):
210 (Stull 1947; Verschueren 1977, 1983; Callahan et al. 1979; Krijgsheld & van der Gen 1986; Renner 1990)
209-210 (Dean 1985)
Density (g/cm^3):
1.383 (at 60/25°C, Verschueren 1983)
Dissociation Constants, pK_a:
7.80 (Blackman et al. 1955; Hoigné & Bader 1983; Scully & Hoigné 1987)
7.85 (Farquharson et al. 1958; Saarikoski & Viluksela 1982; Krijgsheld & van der Gen 1986; Renner 1990)
7.68 (Doedens 1967)
7.85 (Pearce & Simkins 1968; Callahan et al. 1979; Schellenberg et al. 1984; Dean 1985; Leuenberger et al. 1985; Westall et al. 1985; Lee et al. 1993)
7.89 (Sillén & Martell 1971; Serjeant & Dempsey 1979; Hammers et al. 1982; Kaiser et al. 1984; Tratnyek & Hoigné 1991)
7.90 (Könemann 1981; Könemann & Musch 1981; Ugland et al. 1981; Renner 1990; Ma et al. 1993)
7.70 (Xie 1983)
8.01, 8.04, 8.09 (measured values, Xie & Dyrssen 1984)
8.09 (Shigeoka et al. 1988)
7.91 (Nendza & Seydel 1988)
Molar Volume (cm^3/mol):
106.7 (calculated-density, Jaworska & Schultz 1993)
145.2 (calculated-LeBas method, Ma et al. 1993; Shiu et al. 1994)
Molecular Volume, TMV ($Å^3$):
Total Surface Area, TSA ($Å^2$):
148.4 (calculated- χ , Sabljic 1987b)
Heat of Fusion, Δ_{fus}, kcal/mol:
Entropy of Fusion, ΔS_{fus}, cal/mol·K (e.u.):
Fugacity Ratio at 25°C (assuming ΔS_{fus}= 13.5 e.u.), F: 0.6488

Water Solubility (g/m^3 or mg/L at 25°C):
4500 (gravimetric, Mosso 1887; quoted, Ma et al. 1993)
6194 (shake flask-UV at pH 5.1, Blackman et al. 1955; quoted, Ma et al. 1993)
5000 (shake flask-AS, Roberts et al. 1977)
4600 (20°C, quoted, Verschueren 1977, 1983; Callahan et al. 1979; Mabey et al. 1982; Shigeoka et al. 1988)

4500 (quoted, Verschueren 1983)
4500 (Leuenberger et al. 1985)
4500 (20°C, quoted, Krijgsheld & van der Gen 1986)
4500 (20°C, Yalkowsky et al. 1987)
4600 (selected, Suntio et al. 1988)
4470 (quoted, Isnard & Lambert 1988)
4483 (Nirmalakhandan & Speece 1989; quoted, Müller & Klein 1992)
5547 (shake flask-HPLC/UV at pH 5.1, Ma et al. 1993)

Vapor Pressure (Pa at 25°C):
20.6 (extrapolated-regression of tabulated data, Stull 1947)
11.9 (Weast 1973-74)
16.0 (quoted, Callahan et al. 1979)
18 (subcooled liquid value, Hamilton 1980; quoted, Bidleman & Renberg 1985)
7.87 (estimated, Mabey et al. 1982)
13 (estimated, Cheung 1984)
15.4 (capillary GC-RT, Bidleman & Renberg 1985)
2.40 (8°C, Leuenberger et al. 1985)
11.87 (extrapolated, Leuenberger et al. 1985)
133 (53°C, quoted, Krijgsheld & van der Gen 1986)

Henry's Law Constant (Pa·m^3/mol):
0.284 (calculated-P/C, Mabey et al. 1982)
0.11 (8°C, calculated-P/C, Leuenberger et al. 1985)
0.435 (calculated-P/C, Shiu et al. 1994)

Octanol/Water Partition Coefficient, log K_{OW}:
3.08 (Leo et al. 1971; quoted, Roberts et al. 1977; LeBlanc 1984)
3.06 (shake flask-AS, Stockdale & Selwyn 1971)
3.06, 3.08 (Hansch & Leo 1979; quoted, Kaiser 1983; Kaiser et al. 1987)
2.75 (selected, Callahan et al. 1979; quoted, Westall et al. 1985; Chin et al. 1986)
2.93 (calculated-f const., Könemann 1981; Könemann & Musch 1981; McKim et al. 1985)
2.90 (calculated-f const., Mabey et al. 1982)
3.08 (quoted, Saarikoski & Viluksela 1982; Saarikoski et al. 1986; Krijgsheld & van der Gen 1986)
3.08 (calculated as per Leo et al. 1971, Xie 1983)
2.92 (HPLC-k'correlated, Hammers et al. 1982)
3.08 (quoted from Hansch & Leo 1979, Miyake & Terada 1982)
3.14 (RP-HPLC-k', Miyake & Terada 1982)
3.23 (Verschueren 1983; quoted, Lee et al. 1993)
3.06 (quoted from Hansch & Leo 1982, Banerjee et al. 1984; Banerjee 1987)
3.41 (shake flask-UV, Beltrame et al. 1984)
3.23 (shake flask-HPLC/UV, Schellenberg et al. 1984)
3.20, 3.17, 3.17 (20°C, shake flask-GC-calculated-regression, Xie & Dyrssen 1984)
3.21, 3.23, 3.08, 2.99 (shake flask-GC, HPLC-k', calculated-π const., calculated-f const., Xie et al. 1984)

2.70 (computed, Klopman et al. 1985; quoted, Pangrekar et al. 1994)
3.23 (OECD 1981 guidelines, Leuenberger et al. 1985)
2.75, 2.87 (quoted, RPHPLC-RT, Chin et al. 1986)
2.92 (Veith & Broderius 1987)
3.08, 3.61 (quoted, CPC-RV, Terada et al. 1987)
3.21 (HPLC-RP, Shigeoka et al. 1988; quoted, Saito et al. 1993)
3.16 (batch equilibrium-UV, Beltrame et al. 1988)
3.20 (selected, Suntio et al. 1988)
3.08 (quoted, Isnard & Lambert 1988, 1989)
3.08 (quoted, Liu et al. 1991)
3.07 (calculated-CLOGP, Müller & Klein 1992)
3.17 (calculated-π const., Verhaar et al. 1992)
3.06 (EPA CLOGP Data Base, Hulzebos et al. 1993)
2.92 (calculated-CLOGP 3.51, Jaworska & Schultz 1993)
3.07, 3.08 (COMPUTOX databank, Kaiser 1993)
2.80-3.23 (quoted lit. range, NCASI 1993)
3.17 (recommended, Sangster 1993)
3.07 (calculated-QSAR, Kollig 1993)
3.20 (shake flask-GC, Kishino & Kobayashi 1994)

Bioconcentration Factor, log BCF:
1.53 (goldfish, Kobayashi 1979; quoted, Verschueren 1983)
1.0 (trout, Hattula et al. 1981)
2.27 (microorganism-water, calculated-K_{OW}, Mabey et al. 1982)
2.0 (golden ide, after 3 days, Freitag et al. 1985)
2.42 (algae, after 1 day, Freitag et al. 1985)
2.53 (activated sludge, after 5 days, Freitag et al. 1985)
1.0 (quoted, brown trout, Walden et al. 1986)
1.80 (correlated, Isnard & Lambert 1988)
1.41-1.65 (estimated, NCASI 1993)
1.53, 1.0, 2.41 (quoted lit. values: goldfish, trout, algae, NCASI 1993)

Sorption Partition Coefficient, log K_{OC}:
2.59 (sediment-water, calculated-K_{OW}, Mabey et al. 1982)
2.59-3.02 (soil, calculated-K_{OW}, model of Karickhoff et al. 1979, Sabljic 1987a,b)
2.89-3.12 (soil, calculated-K_{OW}, model of Kenaga & Goring 1980, Sabljic 1987a,b)
2.10-2.32 (soil, calculated-K_{OW}, model of Briggs 1981, Sabljic 1987a,b)
2.48-2.91 (soil, calculated-K_{OW}, model of Means et al. 1982, Sabljic 1987a,b)
1.74-2.13 (soil, calculated-K_{OW}, model of Chiou et al. 1983, Sabljic 1987a,b)
3.60, 3.50 (untreated fine and coarse sediment, Isaacson & Frink 1984)
3.71, 3.98 (treated fine and coarse sediment, Isaacson & Frink 1984)
2.75 (sediment, Schenllenberg et al. 1984; quoted, Sabljic 1987a,b)
2.76 (soil, calculated- χ , Sabljic 1987a,b)
2.49 (calculated-K_{OW}, Kollig 1993)

Half-Lives in the Environment:

Air: 21.2-212 hours, based on estimated rate constant for reaction with hydroxyl radicals in air (Howard et al. 1991).

Surface water: rate constant of $< 1.5 \times 10^3\ M^{-1} \cdot sec^{-1}$ for the reaction with ozone at pH 1.5-3.0 (Hoigné & Bader 1983); half-lives of 0.8 hour (summer), 3.0 hours (winter) for distilled water; 0.7 hour (summer), 2.0 hours (winter) for poisoned estuarine water; 0.6 hour (summer), 2.0 hours (winter) for estuarine water, all based on photo-transformation rate under full sunlight and microbes; half-lives of 8.0 days (summer), 14 days (winter) for distilled water; 6 days (summer), 14 days (winter) for poisoned estuarine water and 4 days (summer), 17 days (winter) for estuarine water, based on photo-mineralization rate under full sunlight and microbes (Hwang et al. 1986); half-life of 62 hours in water at pH 8 and (19 ± 2)°C for the reaction with singlet oxygen (Scully & Hoigné 1987); 2.5-2.6 hours for photochemical transformation in Xenotest 1200 (Svenson & Björndal 1988); half-lives range 0.8-3 hours based on measured rate of photolysis in distilled water under sunlight in summer and winter (Howard et al. 1991); complete biodegradation within 5 - 23 days in seawater (NCASI 1993).

Ground water: 133-1032 hours, based on unacclimated aqueous aerobic and anaerobic biodegradation half-lives (Howard et al. 1991).

Sediment: mean half-life of dechlorination: 116 days in July and 47 days in November (Hale et al. 1991).

Soil: 176-1680 hours, based on aerobic soil die-away test data (Baker et al. 1980; Haider et al. 1974; selected, Howard et al. 1991); 3.5 days in an acidic clay soil with <1.0% organic matter and 1.5 day in a slightly basic sandy loam soil with 3.25% organic matter, based on aerobic batch lab. microcosm experiments (Loehr & Mattews 1992).

Biota:

Environmental Fate Rate Constants and Half-Lives:

Volatilization:

Photolysis: the primary transformation process: photo-transformation rate constant of 0.82 h^{-1} with half-life of 0.8 hour for distilled water in summer (mean temp. 25°C) and 0.21 h^{-1} with half-life of 3 hours in winter (mean temp. 11°C); rate constant of 1.16 h^{-1} with half-life of 0.6 hour for estuarine water in summer and 0.44 h^{-1} with half-life of 2 hours in winter; 1.0 h^{-1} with half-life of 0.7 hour in summer and 0.38 h^{-1} with half-life of 2 hours in winter for poisoned estuarine water when exposed to full sunlight and microbes (Hwang et al. 1986); and photo-mineralization rate of 0.09 h^{-1} with half-life of 8 days in summer and 0.049 h^{-1} with half-life of 14 days in winter for distilled water; 0.20 h^{-1} with half-life of 4 days in summer and 0.04 h^{-1} with half-life of 17 days in winter for estuarine water; 0.12 h^{-1} with half-life of 6 days in summer and 0.05 h^{-1} with half-life of 14 days in winter for poisoned estuarine water when exposed to full sunlight and microbes (Hwang et al. 1986); phototransformation half-life of 2.6 hours in Xenotest 1200 (Svenson & Björndal 1988); atmospheric and aqueous photolysis half-lives of 0.8-3 hours based on measured rate of photolysis in distilled water under sunlight in summer and winter (Howard et al. 1991).

Oxidation: aqueous oxidation rate constant of < $7 \times 10^5\ M^{-1}h^{-1}$ for singlet oxygen and $1 \times 10^7\ M^{-1}h^{-1}$ for peroxy radical at 25 °C (Mabey et al. 1982); rate constant of $< 1.5 \times 10^3\ M^{-1} \cdot sec^{-1}$ for the reaction with ozone at pH 1.5-3.0 (Hoigné & Bader 1983); gas-phase photooxidation half-life of 21.2-212 hours, based on estimated rate constant for reaction with hydroxyl radicals in air (Howard et al. 1991); second-order rate constant of $7 \times 10^5\ M^{-1} \cdot sec^{-1}$ at pH 5.5, $2 \times 10^6\ M^{-1} \cdot sec^{-1}$ at pH 6.0, $10 \times 10^6\ M^{-1} \cdot sec^{-1}$ at pH 6.6, $15 \times 10^6\ M^{-1} \cdot sec^{-1}$ at pH 7.0, $76 \times 10^6\ M^{-1} \cdot sec^{-1}$ at pH 7.9, $120 \times 10^6\ M^{-1} \cdot sec^{-1}$ at pH 9.0 and pH 9.6 for the reaction with singlet oxygen in water at (19 ± 2)°C (Scully & Hoigné 1987); rate constant of (5.1 ± 4.7)x$10^6\ M^{-1} \cdot sec^{-1}$ for the reaction with singlet oxygen in aqueous phosphate buffer at (27 ± 1)°C (Tratnyek & Hoigné 1991); gas-phase rate constants of $1.06 \times 10^{-12}\ cm^3 \cdot molecule^{-1} \cdot sec^{-1}$ (exptl.) and $1.71 \times 10^{-12}\ cm^3 \cdot molecule^{-1} \cdot sec^{-1}$ for hydroxyl radical, calculated based on molecular orbital estimation method (Klamt 1993).

Hydrolysis: no hydrolyzable groups (Howard et al. 1991).

Biodegradation: 7 to 10 days for bacteria to utilize 95% of 200 ppm in parent substrate (Tabak et al. 1964); aqueous aerobic half-life of 66.7-199 hours, based on aerobic lake die-away test data (Aly & Faust 1964; selected, Howard et al. 1991); aqueous anaerobic half-life of 324-1032 hours, based on anaerobic lake die-away test data (Aly & Faust 1964; selected, Howard et al. 1991); completely degraded in soil suspensions in 9 days (Woodcock 1971; quoted, Verschueren 1983); average rate of biodegradation 10.5 mg COD $g^{-1} \cdot h^{-1}$ based on measurements of COD decrease using activated sludge inoculum with 20 days of adaptation to the substrate (Pitter 1976; quoted, Scow 1982); aerobic degradation in a non-sterile clay-loam soil: 79% loss at 0°C at the termination of the experiment after 14 days, 82% loss at 4°C after 12 days and 84% loss at 20°C after 40 days all under same experimental conditions (Baker et al. 1980); rate constants of 0.5 day^{-1} with a half-life of 1.4 days in adapted activated sludge and 0.1 day^{-1} with a half-life of 6 dyas in natural waters under aaerobic conditions (Mills et al. 1982); microbial degradation negligible in darkness (Hwang et al. 1986); rate constant of 0.223 h^{-1} for maximum removal by activated sludge microorganisms (Chudoba et al. 1989); biordegradation first-order rate of hydroxylation, 0.017 min^{-1} by *pseudomonas putida* Fl (Spain et al. 1989; quoted, Neilson et al. 1991); transformation rate of 245 $\mu mol \cdot L^{-1} \cdot day^{-1}$ at 31°C for anaerobic degradation in freshwater sediments with an average of 7 days adaptation time (Zhang & Wiegel 1990); complete biodegradation in water, seawater, sludge and lagoon within 16 to 23 days (NCASI 1993).

Biotransformation: rate constant for bacterial transformation of $1 \times 10^7\ mL \cdot cell^{-1} \cdot h^{-1}$ in water (Mabey et al. 1982); degradation rate of $3.76 \times 10^{19}\ mol \cdot cell^{-1} \cdot h^{-1}$ (±47% SD) from pure culture studies and 0.02×10^{-14} to $2 \times 10^{-14}\ mol \cdot cell^{-1} \cdot h^{-1}$ with microorganisms in Seneca River waters (Banerjee et al. 1984).

Bioconcentration Uptake (k_1) and Elimination (k_2) Rate Constants:

Common Name: 2,6-Dichlorophenol
Synonym: 2,6-DCP
Chemical Name: 2,6-dichlorophenol
CAS Registry No: 87-65-0
Molecular Formula: $C_6H_3Cl_2OH$
Molecular Weight: 163
Melting Point (°C):
69.0 (Weast 1978; Kaiser et al. 1984; Bidleman & Renberg 1985)
65-66 (Verschueren 1983)
66.0 (Renner 1990)
Boiling Point (°C):
220 (Stull 1947; Renner 1990)
218-220 (Verschueren 1983)
Density (g/cm^3):
Dissociation Constants, pK_a:
6.91 (Farquharson et al. 1958; Saarikoski & Viluksela 1982; Renner 1990)
6.80 (Doedens 1967)
6.79 (Pearce & Simkins 1968; Serjeant & Dempsey 1979; Hammers et al. 1982)
6.79 (Sillén & Martell 1971; Kaiser et al. 1984; Xie & Dyrssen 1984; Shigeoka et al. 1988)
7.80 (McLeese et al. 1979)
6.78 (Ugland et al. 1981; Dean 1985; Renner 1990; Ma et al. 1993)
6.54 (Nendza & Seydel 1988)
Molar Volume (cm^3/mol):
145.2 (calculated-LeBas method, Ma et al. 1993; Shiu et al. 1994)
Molecular Volume, TMV ($Å^3$):
Total Surface Area, TSA ($Å^2$):
Heat of Fusion ΔH_{fus}, kcal/mol:
Entropy of Fusion ΔS_{fus}, cal/mol·K:
Fugacity Ratio at 25°C (assuming ΔS_{fus} = 13.5 kcal/mol·K), F: 0.3842

Water Solubility (g/m^3 or mg/L at 25°C):
2625 (shake flask-HPLC/UV at pH 4.5, Ma et al. 1993)

Vapor Pressure (Pa at 25°C):
13.1 (extrapolated-regression of tabulated data, Stull 1947)
12.9 (extrapolated-Antoine eqn., Weast 1972-73)
12.2 (subcooled liq. value, Hamilton 1980; quoted, Bidleman & Renberg 1985)
12.7 (capillary GC-RT, Bidleman & Renberg 1985)
11.2 (extrapolated-Antoine eqn., Stephenson & Malanowski 1987)

Henry's Law Constant ($Pa \cdot m^3/mol$):
0.7376 (calculated-P/C, Shiu et al. 1994)

Octanol/Water Partition Coefficient, log K_{OW}:

2.86 (Hansch & Leo 1979; quoted, Kaiser et al. 1984)
2.57 (calculated, McLeese et al. 1979)
2.64 (HPLC-k'correlated, Hammers et al. 1982)
3.36 (shake flask-UV, Beltrame et al. 1984)
2.77 (calculated, Xie & Dyrssen 1984)
2.84 (quoted, Saarikoski & Viluksela 1982)
2.84, 2.92, 2.84, 2.99 (shake flask-GC, HPLC-k', calculated-π const., calculated-f const., Xie et al. 1984)
2.70 (computed, Klopman et al. 1985; quoted, Pangrekar et al. 1994)
2.84 (HPLC-RP, Shigeoka et al. 1988; quoted, Saito et al. 1993)
3.00 (estimated, Suntio et al. 1988)
2.80 (batch equilibration-UV, Beltrame et al. 1988)
2.73 (quoted, Liu et al. 1991)
2.89, 2.86 (COMPUTOX databank, Kaiser 1993)
2.80-3.00 (quoted lit. range, NCASI 1993)
2.92 (shake flask-GC, Kishino & Kobayashi 1994)

Bioconcentration Factor, log BCF:

1.44-1.56 (estimated, NCASI 1993)

Sorption Partition Coefficient, log K_{OC}:

Half-Lives in the Environment:

Air:

Surface water: 1920-∞ hours and 144-360 hours for 75% degradation in mineral medium and seawater batch experiments respectively (de Kreuk & Hanstveit 1981).

Ground water:

Sediment:

Soil: 16.2 days in an acidic clay soil with <1.0% organic matter and 2.4 days in a slightly basic sandy loam soil with 3.25% organic matter, based on aerobic batch lab. microcosm experiments (Loehr & Matthews 1992).

Biota:

Environmental Fate Rate Constants and Half-Lives:

Volatilization:

Photolysis:

Photooxidation:

Hydrolysis:

Biodegradation: 1920-∞ hours and 144-360 hours for 75 % degradation in mineral medium and seawater batch experiments respectively (de Kreuk & Hanstveit 1981).

Biotransformation:

Bioconcentration Uptake (k_1) and Elimination (k_2) Rate Constants:

Common Name: 3,4-Dichlorophenol
Synonym: 3,4-DCP
Chemical Name: 3,4-dichlorophenol
CAS Registry No: 95-77-2
Molecular Formula: $C_6H_3Cl_2OH$
Molecular Weight: 163
Melting Point (°C):
66-68 (Dean 1985)
68.0 (Weast 1978; Kaiser et al. 1984)
67.0 (Renner 1990)
Boiling Point (°C):
145-146 (Dean 1985)
253 (Renner 1990)
Density (g/cm^3):
Dissociation Constants, pK_a:
7.39 (Doedens 1967; Jones 1981; Bintein & Devillers 1994)
8.59 (Pearce & Simkins 1968; Serjeant & Dempsey 1979; Hammers et al. 1982)
8.59 (Sillén & Martell 1971; Kaiser et al. 1984)
8.62 (Ugland et al. 1981; Lagas 1988; Renner 1990; Ma et al. 1993)
8.68 (Xie & Dyrssen 1984; Shigeoka et al. 1988)
8.63 (Dean 1985)
Molar Volume (cm^3/mol):
145.2 (calculated-LeBas method, Ma et al. 1993; Shiu et al. 1994)
Molecular Volume, TMV ($Å^3$):
Total Surface Area, TSA ($Å^2$):
Heat of Fusion, ΔH_{fus}, kcal/mol:
Entropy of Fusion, ΔS_{fus}, cal/mol·K (e.u.):
Fugacity Ratio at 25°C, F: 0.3842

Water Solubility (g/m^3 or mg/L at 25°C):
9256 (shake flask-HPLC/UV at pH 5.1, Ma et al. 1993)

Vapor Pressure (Pa at 25°C):

Henry's Law Constant ($Pa \cdot m^3/mol$):

Octanol/Water Partition Coefficient, log K_{OW}:
3.37 (Hansch & Leo 1979; quoted, Kaiser et al. 1984)
3.05 (HPLC-k' correlated, Hammers et al. 1982)
3.33 (calculated-HPLC-RT, Banerjee et al. 1984; quoted, Banerjee 1987)
3.47 (shake flask-UV, Beltrame et al. 1984)
3.44, 3.41, 3.43, 2.99 (shake flask-GC, HPLV-k', calculated-π const., calculated-f const., Xie et al. 1984; quoted, Bintein & Devillers 1994)
3.38 (calculated, Xie & Dyrssen 1984)
2.70 (computed, Klopman et al. 1985; quoted, Pangrekar et al. 1994)
3.22 (quoted, Lagas 1988)

3.44 (HPLC-RP, Shigeoka et al. 1988; quoted, Saito et al. 1993)
3.68 (batch equilibration-UV, Beltrame et al. 1988)
3.20 (THOR 1989; quoted, Connell & Markwell 1990)
2.89, 3.37 (COMPUTOX databank, Kaiser 1993)

Bioconcentration Factor, log BCF:

1.43, 1.38 (earthworm *E. fetida andrei*: in Kooyenburg soil, Holten soil, van Gestel & Ma 1988)
1.30, 1.61 (earthworm *L. rubellus*: in Kooyenburg soil, Holten soil, van Gestel & Ma 1988)
1.79, 1.8, 1.94, 2.04 (earthworm system, collated from literature, Connell & Markwell 1990)
0.8, 1.3, 1.4, 1.80 (earthworm system, derived data, Connell & Markwell 1990)

Sorption Partition Coefficient, log K_{OC}:

Half-Lives in the Environment:

Air:
Surface water:
Ground water:
Sediment: mean half-life of dechlorination: 115 days in July and 86 days in November with relatively long half-lives of 66 days reported from site 1, 64 days from site 3 and 132 days from site 5 in January (Hale et al. 1991).
Soil: disappearance half-lives of 10.1 days from Kooyenburg soil, 11.2 days from Holten soil with earthworms *E. fetida andrei* and 24.7 days from Kooyenburg soil, 49.5 days from Holten soil with earthworms *L. rebellus* (van Gestel & Ma 1988); half-life of 18.3 days in an acidic clay soil with $<1.0\%$ organic matter and 3.2 days in a slightly basic sandy loam soil with 3.25% organic matter, based on aerobic batch lab. experiments (Loehr & Matthews 1992).
Biota:

Environmental Fate Rate Constants and Half-Lives:

Volatilization:
Photolysis:
oxidation:
Hydrolysis:
Biodegradation: biodegradation first-order rate of hydroxylation of 0.008 min^{-1} by *Pseudomonas putida* Fl (Spain et al. 1989; quoted, Neilson et al. 1991).
Biotransformation: degradation rate of 6.84×10^{-19} ($\pm 38\%$ SD) $mol \cdot cell^{-19} \cdot hour^{-1}$ from pure culture studies (Banerjee et al. 1984).
Bioconcentration Uptake (k_1) and Elimination (k_2) Rate Constants:

Common Name: 2,3,4-Trichlorophenol
Synonym:
Chemical Name: 2,3,4-trichlorophenol
CAS Registry No: 1595-06-0
Molecular Formula: $C_6H_2Cl_3OH$
Molecular Weight: 197.45
Melting Point (°C):
83.5 (Weast 1978; Kaiser et al. 1984)
77-79 (Dean 1985)
81.0 (Renner 1990)
Boiling Point (°C): sublimes (Dean 1985)
Density (g/cm^3):
Dissociation Constants, pK_a:
7.66 (Doedens 1967)
6.50 (Sillén & Martell 1971; McLeese et al. 1979; Kaiser et al. 1984; Shigeoka et al. 1988)
6.97 (Ugland et al. 1981; Dean 1985; Renner 1990; Ma et al. 1993)
7.18 (Schellenberg et al. 1984)
Molar Volume (cm^3/mol):
166.1 (calculated-LeBas method, Ma et al. 1993; Shiu et al. 1994)
Molecular Volume (A^3):
Total Surface Area, TSA (A^2):
Heat of Fusion, ΔH_{fus}, kcal/mol:
Entropy of Fusion, ΔS_{fus}, kcal/mol·K (e.u.):
Fugacity Ratio at 25°C, F: 0.287

Water Solubility (g/m^3 or mg/L at 25°C):
500 (estimated, Ma et al. 1990)
915 (shake flask-HPLC/UV at pH 5.1, Ma et al. 1993)

Vapor Pressure (Pa at 25°C):
1.0 (Ma et al. 1990)
3.48 (selected P_L, Shiu et al. 1994)

Henry's Law Constant (Pa·m^3/mol):
0.3959 (calculated-P/C, Shiu et al. 1994)

Octanol/Water Partition Coefficient, log K_{OW}:
4.07 (Hansch & Leo 1979; quoted, Kaiser et al. 1984)
3.49 (calculated, McLeese et al. 1979)
3.54 (HPLC-k'correlated, Hammers et al. 1982)
3.80 (shake flask-UV, Beltrame et al. 1984; quoted, Shigeoka et al. 1988)
3.16 (computed, Klopman et al. 1985; quoted, Pangrekar et al. 1994)

3.80 (HPLC-RP, Shigeoka et al. 1988; quoted, Saito et al. 1993)
3.82 (batch equilibration-UV, Beltrame et al. 1988)
3.70 (estimated, Ma et al. 1990)
4.07 (COMPUTOX databank, Kaiser 1993)

Bioconcentration Factor, log BCF:

Sorption Partition Coefficient, log K_{OC}:

Half-Lives in the Environment:

Air:

Surface water: 1.7 hour for photochemical transformation in Xenotest 1200 (Svenson & Björndal 1988).

Ground water:

Sediment:

Soil:

Biota:

Environmental Fate Rate Constants and Half-Lives:

Volatilization:

Photolysis: phototransformation half-life of 1.7 hour in Xenotest 1200 (Svenson & Björndal 1988).

Photooxidation:

Hydrolysis:

Biodegradation:

Biotransformation:

Bioconcentration Uptake (k_1) and Elimination (k_2) Rate Constants:

Common Name: 2,4,5-Trichlorophenol
Synonym: 245 TCP
Chemical Name: 2,4,5-trichlorophenol
CAS Registry No: 95-95-4
Molecular Formula: $C_6H_2Cl_3OH$
Molecular Weight: 197.45
Melting Point (°C):
62.0 (Stull 1947; Bidleman & Renberg 1985)
70.0 (Weast 1978; Kaiser et al. 1984)
61-63 (Verschueren 1983)
65.0 (Renner 1990)
Boiling Point (°C):
252 (Stull 1947; Verschueren 1983)
246 (Renner 1990)
Density (g/cm^3 at 20°C):
1.50 (75°C, Verschueren 1983)
Dissociation Constants, pK_a:
7.00 (Blackman et al. 1955, Sillén & Martell 1971; Kaiser et al. 1984)
7.07 (Farquharson et al. 1958; Saarikoski & Viluksela 1982; Renner 1990)
7.43 (Doedens 1967; Jones 1981; Bintein & Devillers 1994)
6.72 (Ugland et al. 1981; Dean 1985; Lagas 1988; Renner 1990; Ma et al. 1993)
6.90 (Hoigné & Bader 1983)
6.94 (Schellenberg et al. 1984)
6.83 (Nendza & Seydel 1988)
Molar Volume (cm^3/mol):
166.1 (calculated-LeBas method, Ma et al. 1993; Shiu et al. 1994)
Molecular Volume, TMV (Å^3):
Total Surface Area, TSA (Å^2):
165.5 (calculated- χ , Sabljic 1987b)
Heat of Fusion, ΔH_{fus}, kcal/mol:
Entropy of Fusion, ΔS_{fus}, cal/mol·K (e.u.):
Fugacity Ratio at 25°C (assuming ΔS_{fus} = 13.5 e.u.), F: 0.37

Water Solubility (g/m^3 or mg/L at 25°C):
948 (shake flask-UV at pH 5.1, Blackman et al. 1955)
1190 (quoted, Verschueren 1983)
982 (selected, Cheung 1984)
2000 (quoted, LeBlanc 1984)
990 (recommended at pH 5.1, IUPAC 1985)
1200 (quoted, Leunberger et al. 1985)
700 (8°C, Leuenberger et al. 1985)
960 (selected, Suntio et al. 1988)
1202 (quoted, Isnard & Lambert 1988)
982 (quoted, Howard 1991)
1200 (quoted, Müller & Klein 1992)
649 (shake flask-HPLC/UV at pH 4.9, Ma et al. 1993)

Vapor Pressure (Pa at 25°C):

6.69 (extrapolated-regression of tabulated data, Stull 1947)
3.05 (estimated, Chiou & Freed 1977)
2.48 (extrapolated, Cheung 1984)
6.60 (extrapolated liquid value, Antoine eqn, Weast 1972-73)
6.12 (subcooled liquid, Hamilton 1980; quoted, Bidleman & Renberg 1985)
2.66 (capillary GC-RT, Bidleman & Renberg 1985)
2.93 (selected, Leuenberger et al. 1985)
7.64 (extrapolated-Antoine eqn., Stephenson & Malanowski 1987)
2.93 (calculated, Howard 1991)

Henry's Law Constant (Pa·m^3/mol):

0.132 (8°C, Leuenberger et al. 1985)
0.59 (calculated, Howard 1991)
0.521 (calculated-P/C, Shiu et al. 1994)

Octanol/Water Partition Coefficient, log K_{OW}:

3.72 (quoted unpublished result, Leo et al. 1971; quoted, Chiou & Freed 1977; Veith et al. 1979; Bintein & Devillers 1994)
3.72 (Hansch & Leo 1979; quoted, Veith et al. 1979; Mackay 1982; Kaiser 1983; Kaiser et al. 1984; Veith & Kosian 1983; Isnard & Lambert 1988; Suntio et al. 1988; Howard 1991)
3.72, 2.39 (quoted, estimated-HPLC-RT, Veith et al. 1979)
3.72 (quoted unpublished value, Veith et al. 1979b)
3.63 (HPLC-k'correlated, Hammers et al. 1982)
3.80 (shake flask-GC, Saarikoski & Viluksela 1982)
3.72 (quoted from Hansch & Leo 1982, Banerjee et al. 1984; quoted, Banerjee 1987)
3.06, 3.72 (quoted, Verschueren 1983)
3.72 (quoted from Leo et al. 1971, Beltrame et al. 1984)
4.19 (shake flask-HPLC/UV, Schellenberg et al. 1984)
4.10, 3.96, 4.12, 3.73 (shake flask-GC, HPLC-k', calculated-π const., calculated-f const., Xie et al. 1984)
3.16 (computed, Klopman et al. 1985; quoted, Pangrekar et al. 1994)
3.72 (quoted, McKim et al. 1985; Isnard & Lambert 1988; Suntio et al. 1988)
3.53 (selected, Yoshida et al. 1986)
3.88 (quoted from Xie & Dyrssen 1984, Lagas 1988)
3.84 (batch equilibration-UV, Beltrame et al. 1988)
4.10 (HPLC-RP, Shigeoka et al. 1988; quoted, Saito et al. 1993)
3.72 (THOR 1989; quoted, Connell & Markwell 1990)
3.72 (quoted, Liu et al. 1991)
3.85 (calculated-CLOGP, Müller & Klein 1992)
3.85, 3.72 (COMPUTOX databank, Kaiser 1993)
3.72 (quoted, NCASI 1993)
3.85 (calculated-QSAR, Kollig 1993)
4.02 (shake flask-GC, Kishino Kobayasyi 1994)

Bioconcentration Factor, log BCF:

3.28 (fathead minnow-28 days exposure, quoted, Veith et al. 1979b)

3.28, 2.70 (total ^{14}C in fathead minnows, observed, calculated, mean exposure level 0.0048 $\mu g \cdot mL^{-1}$, Call et al. 1980)

3.26, 2.82 (total ^{14}C in fathead minnows, observed, calculated, mean exposure level 0.0493 $\mu g \cdot mL^{-1}$, Call et al. 1980)

3.27 (total ^{14}C in fathead minnows, mean value, Call et al. 1980)

3.28, 2.40 (quoted, calculated-K_{OW}, Mackay 1982)

3.28 (quoted from Mackay 1982, Sabljic 1987a)

2.88 (calculated- χ , Sabljic 1987a)

3.28 (quoted, Isnard & Lambert 1988)

1.79 (quoted, fish, van Gestel & Ma 1988)

1.81, 1.53 (earthworms *E. fetida andrei*: Kooyenburt soil, Holten soil, van Gestel & Ma 1988; quoted, Howard 1991)

2.04, 2.82 (earthworms *L. rubellus*: Kooyenburt soil, Holten soil, van Gestel & Ma 1988; quoted, Howard 1991)

2.04, 2.39, 2.61, 3.36 (earthworm system, collated from literature, Connell & Markwell 1990)

0.40, 1.50, 2.50, 8.40 (earthworm system, derived data, Connell & Markwell 1990)

3.26, 3.28; 3.61 (quoted, fathead minnows; fish, Howard 1991)

2.14 (estimated, NCASI 1993)

Sorption Partition Coefficient, log K_{OC}:

3.49-3.98 (soil, calculated-K_{OW}, model of Karickhoff et al. 1979, Sabljic 1987a,b)

3.38-3.64 (soil, calculated-K_{OW}, model of Kenaga & Goring 1980, Sabljic 1987a,b)

2.56-2.82 (soil, calculated-K_{OW}, model of Briggs 1981, Sabljic 1987a,b)

3.38-3.87 (soil, calculated-K_{OW}, model of Means et al. 1982, Sabljic 1987a,b)

2.55-2.99 (soil, calculated-K_{OW}, model of Chiou et al. 1983, Sabljic 1987a,b)

3.36 (sediment, Schellenberg et al. 1984; quoted, Sabljic 1987a,b)

2.99 (soil, calculated- χ , Sabljic 1987a,b)

3.25, 3.12, 3.38, 2.56 (quoted:lake sediment, river sediment, aquifer material, soil, Howard 1991)

3.34, 3.30 (soils, Howard 1991)

2.93 (calculated-K_{OW}, Kollig 1993)

Half-Lives in the Environment:

Air: 30.1-301 hours, based on an estimated rate constant for the vapor phase reaction with hydroxyl radicals in air (Howard et al. 1991); estimated half-life of 7.5 days for reactions with OH radicals (Howard 1991); atmospheric transformation lifetime was estimated to be > 5 days (Kelly et al. 1994).

Surface water: 840 hours for 75% degradation in mineral medium (de Kreuk & Hanstveit 1981); rate constant of $<3 \times 10^3$ $M^{-1} \cdot sec^{-1}$ for the reaction with ozone at pH 1.2-1.5 (Hoigné & Bader 1983); 0.5 hour (summer), 1.0 hour (winter) in distilled water and 0.5 hour (summer), 1.0 hour (winter) in surface estuarine water based on photo-transformation rate under full sunlight and microbes (Hwang et al.

1986); 7 days (summer), 14 days (winter) in distilled water and 3 days (summer), 8 days (winter) in estuarine water based on photo-mineralization rater under full sunlight and microbes (Hwang et al. 1986); 0.6 hour (summer), 1.0 hour (winter) in poisoned estuarine water based on photo-transformation rate and 6 days (summer), 14 days (winter) in poisoned estuarine water based on photomineralization rate (Hwang et al. 1986); half-life of 16560 hours at 21°C in Skidway River water (Pritchard 1987); photodegradation half-lives of 0.5 hour (summer), 1.0 hours (winter) in distilled water and 0.6 hours (summer), 1.0 hours (winter) in estuarine water under irradiation by natural sunlight (quoted from Hwang et al. 1987, Sanders et al. 1993); 0.5-336 hours, based on aqueous photolysis half-life (Howard et al. 1991); rapidly photolyze with a half-life of 0.6-1.0 hour at water surfaces, 690 days in water column (Howard 1991).

Ground water: 1104-43690 hours, based on estimated unacclimated aqueous aerobic biodegradation half-life and aqueous anaerobic biodegradation half-life (Howard et al. 1991).

Sediment: half-life of 23 days from calculated degradation rate constant of 0.0030 day^{-1} for radiolabelled 2,4,5-TCP in Skidway River water-sediment slurry (Lee & Ryan 1979; quoted, Pritchard 1987); biodegration half-life of 23 days in sediments (Howard 1991).

Soil: disappearance half-life of 3.4 days from Kooyenburg soil and 6.6 days from Holten soil with earthworms *E. fetida andrei* and 39.6 days from Kooyenburg soil, 13.9 days from Holten soil with earthworm *L. rubellus* (van Gestel & Ma 1988); half-life ranges 552-16560 hours, based on estimated unacclimated aqueous aerobic biodegradation half-life (Howard et al. 1991); under aerobic conditions biodegradation half-life of 15 days in a soil suspension (Howard 1991); half-life of 22.3 days in an acidic clay soil with $<1.0\%$ organic matter and 14.6 days in a slightly basic sandy loam soil with 3.25% organic matter, based on aerobic batch lab. microcosm experiments (Loehr & Matthews 1992).

Biota: depuration half-lives of 12 hours (observed), 9.2 hours (calculated) for mean exposure level of 0.0048 $\mu g \cdot mL^{-1}$ and 12 hours (observed), 6.6 hours (calculated) for mean exposure level of 0.0493 $\mu g \cdot mL^{-1}$ (fathead minnows, Call et al. 1980).

Environmental Fate Rate Constants and Half-Lives:

Volatilization: volatilization half-lives were estimated to be 9.0 days from a model river 1m deep, flowing 1 m/sec with a wind speed of 3 m/sec, and 207 days from a model pond (Howard 1991).

Photolysis: the primary transformation process; photo-transformation rate constant of 1.30 h^{-1} with half-life of 0.6 hour in distilled water in summer (mean temp. 25°C) and 0.61 h^{-1} with half-life of 1 hour in distilled water in winter (mean temp 18°C); 1.2 h^{-1} with half-life of 0.6 day for poisoned estuarine water in summer and 0.65 h^{-1} with half-life of 1.0 hour in winter and 1.4 h^{-1} with half-life of 0.5 hour for estuarine water in summer and 0.65 h^{-1} with half-life of 1.0 hour in winter under full sunlight and microbes (Hwang et al. 1986); photo-mineralization rate constant of 0.10 day^{-1} with half-life of 7 days in distilled water in summer and 0.05 h^{-1} with half-life of 14 days in winter and 0.12 h^{-1}

with half-life of 6 days in summer and 0.05 h^{-1} with half-life of 14 days in winter and 0.28 h^{-1} with half-life of 3 days in summer and 0.09 h^{-1} with half-life of 8 days in winter under full sunlight and microbes (Hwang et al. 1986); photodegradation rate constants of 1.30 h^{-1} corresponding to a half-life of 0.5 hour (summer), 0.61 h^{-1} corresponding to a half-life of 1.0 hour (winter) in distilled water and 1.20 h^{-1} corresponding to a half-life of 0.6 hour (summer), 0.65 h^{-1} corresponding to a half-life of 1.0 hour (winter) in estuarine water under irradiation by natural sunlight (quoted from Hwang et al. 1987, Sanders et al. 1993); 0.5-336 hours, based on photolysis rate constants for transformation and mineralization under summer and winter sunlight conditions (Howard et al. 1991); photolysis half-life in natural water (and distilled water) exposed to midday sunlight during summer and winter was 0.6 and 1.0 hour respectively (Howard 1991).

Oxidation: aqueous photooxidation half-life of 66-3480 hours in water, based on reported reaction constants for OH and RO_2 radicals with the phenol class (Mill & Mabey 1985; Güesten et al. 1981; selected, Howard et al. 1991); photooxidation rate constant of $<3x10^3$ $M^{-1}\cdot sec^{-1}$ for the reaction with ozone in water at pH 1.2-1.5 (Hoigné & Bader 1983); atmospheric photooxidation half-life of 30.1-301 hours, based on an estimated rate constant for the vapor phase reaction with hydroxyl radicals in air (Howard et al. 1991).

Hydrolysis: half-life of $>8x10^6$ years, based on hydrolysis rate constant is zero at pH 7.0 (Kollig et al. 1987; selected, Howard et al. 1991).

Biodegradation: decomposition in suspended soils: >72 days for complete disappearance (Woodcock 1971; quoted, Verschueren 1983); 840 hours for 75 % degradation in mineral medium (de Kreuk & Hanstveit 1981); aqueous aerobic half-life of 552-16560 hours, based on unacclimated aerobic river die-away test data (Lee & Ryan 1979; selected, Howard et al. 1991); aqueous anaerobic half-life of 3028-43690 hours, based on unacclimated anaerobic grab sample data for soil and ground water (Gibson & Suflita 1986; Baker & Mayfield 1980; selected, Howard et al. 1991); degradation with microbes in darkness negligible (Hwang et al. 1986); degradation rate constant of 0.00010 day^{-1} with a half-life of 16560 hours and 0.030 day^{-1} with a half-life of 552 hours for Skidway River water and water-sediment slurry respectively (Pritchard 1987); biodegradation half-life of 690 days in river water (Howard 1991).

Biotransformation: degradation rate of $1.43x10^{-20}$ ($\pm$77% SD) $mol\cdot cell^{-1}\cdot h^{-1}$ from pure culture studies and $5x10^{-15}$ $mol\cdot cell^{-1}\cdot h^{-1}$ with microorganisms in Seneca River waters (Banerjee et al. 1984).

Bioconcentration Uptake (k_1) and Elimination (k_2) Rate Constants:

Common Name: 2,4,6-Trichlorophenol
Synonym: 2,4,6 TCP
Chemical Name: 2,4,6-trichlorophenol
CAS Registry No: 88-06-2
Molecular Formula: $C_6H_2Cl_3OH$
Molecular Weight: 197.45
Melting Point (°C):

68.5 (Stull 1947; Bidleman & Renberg 1985)
69.5 (Weast 1978; Kaiser et al. 1984)
67 (Schmidt-Bleek et al. 1982)
68 (Mabey et al. 1982; Verschueren 1983; Renner 1990)

Boiling Point (°C):

246 (Stull 1947; Schmidt-Bleek et al. 1982; Renner 1990)
244.5 (Mabey et al. 1982; Verschueren 1983)

Density (g/cm^3):

1.675 (Schmidt-Bleek et al. 1982)
1.665 (Jaworska & Schultz 1993)

Dissociation Constants, pK_a:

6.10 (Blackman et al 1955; Sillén & Martell 1971; Kaiser et al. 1984; Hoigné & Bader 1983; Scully & Hoigné 1987)
6.22 (Farquharson et al. 1958; Saarikoski & Viluksela 1982)
5.99 (Callahan et al. 1979; Ugland et al. 1981; Mabey et al. 1982; Dean 1985; Westall et al. 1985; Renner 1990; Ma et al. 1993)
6.23 (Serjeant & Dempsey 1979; Tratnyek & Hoigné 1991)
6.00 (Xie 1983, Yoshida et al. 1987)
6.15 (Schellenberg et al. 1984; Leuenberger et al. 1985)

Molar Volume (cm^3/mol):

118.6 (calculated-density, Jaworska & Schultz 1993)
166.1 (calculated-LeBas method, Ma et al. 1993; Shiu et al. 1994)

Molecular Volume, TMV ($Å^3$):
Total Surface Area, TSA ($Å^2$):

165.5 (calculated- χ , Sabljic 1987b)

Heat of Fusion, ΔH_{fus}, kcal/mol:
Entropy of Fusion, ΔS_{fus}, cal/mol·K (e.u.):
Fugacity Ratio at 25°C (assuming ΔS_{fus} = 13.5 e.u.), F: 0.3630

Water Solubility (g/m^3 or mg/L at 25°C):

858 (gravimetric, Daccomo 1885)
434 (shake flask-UV at pH 5.1, Blackman et al. 1955; selected, Suntio et al. 1988)
900 (shake flask-AS, Roberts et al. 1977; quoted, Kilzer et al. 1979)
800 (Verschueren 1977, 1983; quoted, Kilzer et al. 1979; Callahan et al. 1979; Mabey et al. 1982; Neely 1984; Leuenberger et al. 1985; Shigeoka et al. 1988)
900 (quoted, Körte et al. 1978; Howard 1989)
420 (20-25°C, quoted, Geyer et al. 1982; Schmidt-Bleek et al. 1982)
1000 (quoted, LeBlanc 1984)
430 (recommended at pH 5.1, IUPAC 1985)
500 (8°C, Leuenberger et al. 1985)

312 (20°C, phenol form, shake flask-HPLC/UV, Yoshida et al. 1987)
312×(1 + $10^{pH-6.06}$) (20°C, measured at pH 4, 5, 6 and 7, shake flask-HPLC/UV, Yoshida et al. 1987)
300, 580, 270000 (river, oligotrophic lake, eutrophic lake, Yoshida et al. 1987)
900 (Aquasol database, Yalkowsky et al. 1987)
794 (quoted, Isnard & Lambert 1988)
804 (Nirmalakhandan & Speece 1989; quoted, Müller & Klein 1992)
708 (shake flask-HPLC/UV at pH 4.7, Ma et al. 1993)

Vapor Pressure (Pa at 25°C):
4.17 (extrapolated-regression of tabulated data, Stull 1947)
4.00 (20-25°C, extrapolated, Jordan 1954; quoted, Kilzer et al. 1979)
4.12 (extrapolated liquid, Antoine eqn, Weast 1972-73)
2.11 (estimated, Chiou & Freed 1977)
1.33 (quoted, Callahan et al. 1979)
3.83 (subcooled liquid, Hamilton 1980; quoted, Bidleman & Renberg 1985)
1.12 (gas satuaration, Politzki et al. 1982; selected, Howard 1989)
1.30 (20°C, Schmidt-Bleek et al. 1982)
1.60 (extrapolated, Mabey et al. 1982)
2.67 (Leuenberger et al. 1985)
3.26 (capillary GC-RT, P_L, Bidleman & Renberg 1985)
4.28 (extrapolated-Antoine eqn., Stephenson & Malanowski 1987)
0.827 (GC-RT, solid value, Yoshida et al. 1987)

Henry's Law Constant (Pa·m^3/mol):
0.405 (calculated-P/C, Mabey et al. 1982)
0.132 (8°C, calculated-P/C, Leuenberger et al. 1985)
0.523/(1 + $10^{pH+6.06}$) (20°C, calculated-P/C, Yoshida et al. 1987)
0.5687 (calculated-P/C, Shiu et al. 1994)

Octanol/Water Partition Coefficient, log K_{OW}:
3.06 (quoted, Leo et al. 1971)
3.69 (quoted unpublished result, Leo et al. 1971; quoted, Chiou & Freed 1977; Roberts et al. 1977; Kaiser 1983; Yoshida et al. 1986; Howard 1989)
3.69, 3.62, 4.02 (Hansch & Leo 1979; quoted, Kaiser et al. 1984)
3.38 (Callahan et al. 1979; quoted, Westall et al. 1985)
3.62 (quoted from Hansch & Leo 1979, Miyake & Terada 1982)
3.48 (RP-HPLC-k', Miyake & Terada 1982)
3.61 (calculated-f const., Mabey et al. 1982)
4.03 (shake flask-GC, Saarikoski & Viluksela 1982)
2.80 (quoted, Geyer et al. 1982; Schmidt-Bleek et al. 1982)
3.77 (calculated as per Leo et al. 1971, Xie 1983)
3.37 (quoted from Leo et al. 1971, Beltrame et al. 1984)
3.72 (shake flask-HPLC/UV, Schellenberg et al. 1984)
2.97, 2.80 (shake flask, OECD 1981 Guidelines, Geyer et al. 1984)
3.75, 3.69, 3.77, 3.73 (shake flask-GC, HPLC-k', calc-π const., calc-f const., Xie et al. 1984)

3.66 (calculated, Xie & Dyrssen 1984; quoted, Söderström et al. 1994)
2.80 (quoted, Freitag et al. 1985)
3.16 (computed, Klopman et al. 1985; quoted, Pangrekar et al. 1994)
3.72 (OECD 1981 guidelines, Leuenberger et al. 1985)
3.73 (quoted, Carlberg et al. 1986)
3.38, 3.29 (quoted, RPHPLC-RT, Chin et al. 1986)
3.75 (quoted of Xie et al. 1984, Xie et al. 1986)
3.62, 3.97 (quoted, CPC-RV, Terada et al. 1987)
2.97 (quoted, Isnard & Lambert 1988; Liu et al. 1991)
3.75 (HPLC-RP, Shigeoka et al. 1988; quoted, Saito et al. 1993)
3.68 (batch equilibration-UV, Beltrame et al. 1988)
3.75 (selected, Suntio et al. 1988)
3.62 (quoted, Hodson & Williams 1988)
3.57 (calculated-CLOGP, Müller & Klein 1992)
3.38-3.76 (quoted lit. range, NCASI 1993)
3.69 (EPA CLOGP Data Base, Hulzebos et al. 1993)
3.69 (calculated-CLOGP 3.51, Jaworska & Schultz 1993)
3.85, 3.72 (COMPUTOX databank, Kaiser 1993)
3.57 (calculated-QSAR, Kollig 1993)
2.67 (shake flask-GC, Kishino & Kobayashi 1994)

Bioconcentration Factor, log BCF:
2.00-2.32 (Landner et al. 1977; Laake 1982; quoted, Carlberg et al. 1986)
2.40 (fish, Körte et al. 1978)
1.60, 1.71, 2.49 (activated sludge, algae, golden orfe, Freitag et al. 1982)
1.71; 1.81 (algae: exptl; calculated, Geyer et al. 1981)
1.54-1.78, 1.60 (mussel *mytilus edulis*, Geyer et al. 1982)
2.92 (microorganisms-water, calculated-K_{OW}, Mabey et al. 1982)
3.24, 3.48, 3.1-4.09 (algae, snail, guppy, Virtanen & Hattula 1982)
1.71, 2.49, 1.60 (alage, fish, activated sludge, Klein et al. 1984)
1.71, 2.13 (algae: exptl., calculated-K_{OW}, Geyer et al. 1984)
1.78, 1.70, 2.49 (activated sludge, algae, golden ide, Freitag et al. 1985)
1.48-2.08 (rainbow trout, Oikari et al. 1985)
2.75; 2.84 (Atlantic salmon fry: humic water; lake water, Carlberg et al. 1986)
1.30 (Isnard & Lambert 1988)
1.94; 2.83 (flagfish: whole fish; fish lipid, Smith et al. 1990)
1.87-2.16 (estimated, NCASI 1993)
1.48-2.08, 3.01-4.09, 2.49, 3.48, 1.60, 1.70-3.24 (trout, guppy, fish, snail, mussel, algae, quoted from literature, NCASI 1993)
3.6, 5.4, 5.0 (perch bile to water, Söderström et al. 1994)

Sorption Partition Coefficient, log K_{OC}:
2.34 (sediment, Virtanen & Hattula 1982)
3.30 (sediment-water, calculated-K_{OW}, Mabey et al. 1982)
2.52 (soil, Seip et al. 1986)
3.41-3.84 (soil, calculated-K_{OW}, model of Karickhoff et al. 1979, Sabljic 1987a,b)
3.33-3.57 (soil, calculated-K_{OW}, model of Kenaga & Goring 1980, Sabljic 1987a,b)

2.52-2.75 (soil, calculated-K_{OW}, model of Briggs 1981, Sabljic 1987a,b)
3.30-3.73 (soil, calculated-K_{OW}, model of Means et al. 1982, Sabljic 1987a,b)
2.48-2.87 (soil, calculated-K_{OW}, model of Chiou et al. 1983, Sabljic 1987a,b)
3.02 (sediment, Schellenberg et al. 1984; quoted, Sabljic 1987a,b)
3.03 (sediment, Leuenberger et al. 1985)
2.99 (soil, calculated, Sabljic 1987a,b)
3.34, 2.79, 2.23 (average values, soil at pH 6, 7, and 7.7, Yoshida et al. 1987)
2.79, 3.34, 2.04 (river, oligotrophic lake, eutrophic lake, Yoshida et al. 1987)
2.50 (soil, calculated- χ , Bahnick & Doucette 1988)
2.92-3.12; 3.03 (quoted lit. range, mean value, Robinson & Novak 1994)
2.82-3.02 (log K_{OM}, organic matter, Robinson & Novak 1994)
2.81, 3.03 (log K_{HA}, humic acid, Robinson & Novak 1994)
2.25 (calculated-K_{OW}, Kollig 1993)

Half-Lives in the Environment:

Air: 123.4-1234 hours, based on an estimated rate constant for the vapor phase reaction with OH radicals (Atkinson 1987; quoted, Howard et al. 1991); atmospheric transfomation lifetime was estimated to be > 5 days (Kelly et al. 1994).

Surface water: 216-432 hours and 624-1560 hours for 75% degradation in mineral medium and seawater batch experiment respectively (de Kreuk & Hanstveit 1981); 1.2 hour, based on photochemical transformation in Xenotest 1200 (Svenson & Björndal 1988); half-life of 2-96 hours, based on aqueous photolysis half-life (Sugiura et al. 1984; quoted, Howard et al. 1991); half-life of 62 hours for the reaction with singlet oxygen at pH 8 and (19 $\pm$ 2)°C in water (Scully & Hoigné 1987); photolysis half-life of 24 hours (Paasivirta 1992).

Ground water: 336-43690 hours, based on estimated unacclimated aqueous aerobic biodegradation half-life and aqueous anaerobic biodegradation half-life (Howard et al. 1991).

Sediment:

Soil: 168-1680 hours, based on estimated unacclimated aqueous aerobic biodegradation half-life including soil grab sample data (Haider et al. 1974; quoted, Howard et al. 1991); 6.3 days in an acidic clay soil with <1.0% organic matter and 5.3 days in a slightly basic sandy loam soil with 3.25% organic matter, based on aerobic batch lab. microcosm experiments (Loehr & Matthews 1992); half-life of 6720 hours (Paasivirta 1992).

Biota: 0.15 day of clearance from whole flagfish; 0.13 day of clearance from flagfish lipid (Smith et al. 1990).

Environmental Fate Rate Constants and Half-Lives:

Volatilization: volatilization and sorption half-life estimated to be 192 h from natural pond (Schauerte et al. 1982); volatilization rate constant from water of $3.90\text{x}10^{-3}/(1 + 10^{pH-6.06})/L_w$ h^{-1} (L_w, depth of water phase), (Yoshida et al. 1987).

Photolysis: aqueous photolysis half-life of 2-96 hours, based on lab. photolysis rate constants for an environmental pond at noon under fall sunlight conditions at 40°N latitude (Sugiura et al. 1984; quoted, Howard et al. 1991); photodegradation rate constants $k_p = (0.156 + 9.16\text{x}10^{pH-5.86})/(1 + 10^{pH-5.86})$ h^{-1} in water, for summer, fine day, and $k_p = (0.042 + 1.40x10^{pH-6.09})/((1 + 10^{pH-6.09})$ h^{-1} in water, for winter, fine day; 0.72 day^{-1} for river, 2.2 day^{-1} for oligotrophic lake and 0.776 day^{-1} for eutrophic lake (Yoshida et al. 1987); phototransformation half-life of 1.2 hour in Xenotest 1200 (Svenson & Björndal 1988); photolysis half-life of 24 hours (Paasivirta 1992).

Oxidation: aqueous oxidation rate constant of < $7x10^4$ $M^{-1}h^{-1}$ for singlet oxygen and $1x10^6$ $M^{-1}h^{-1}$ for peroxy radical at 25 °C (Mabey et al. 1982); rate constant of $<1x10^4$ $M^{-1}\cdot sec^{-1}$ for the reaction with ozone in water at pH 1.3-1.5 (Hoigné & Bader 1983); photooxidation half-life of 123.4-1234 hours, based on an estimated rate constant for the vapor phase reaction with hydroxyl radicals in air (Atkinson 1987; quoted, Howard et al. 1991); photooxidation half-life of 20.3-2027 hours, based on measured rate data for reaction with singlet oxygen in aqueous solution (Scully & Hoigne 1987; quoted, Howard et al. 1991); rate constants: $2x10^6$ $M^{-1}\cdot sec^{-1}$ at pH 4.2, $6x10^6$ $M^{-1}\cdot sec^{-1}$ at pH 4.8, $1.5x10^7$ $M^{-1}\cdot sec^{-1}$ at pH 5.2, $2.6x10^7$ $M^{-1}\cdot sec^{-1}$ at pH 5.5, $5.5x10^7$ $M^{-1}\cdot sec^{-1}$ at pH 6.0, $9.50x10^7$ $M^{-1}\cdot sec^{-1}$ at pH 7 and $1.2x10^8$ $M^{-1}\cdot sec^{-1}$ at pH 9 for the reaction with singlet oxygen in aqueous solution at (19 ± 2)°C (Scully & Hoigné 1987); rate constant of (1.7 ± 0.7)$x10^7$ $M^{-1}\cdot sec^{-1}$for the reaction with singlet oxygen in aqueous phosphate buffer at (27 ± 1)°C (Tratnyek & Hoigné 1991).

Hydrolysis: half-life of $>8x10^6$ years, based on hydrolysis rate constant of zero at pH 7 (Kollig et al. 1987; quoted, Howard et al. 1991).

Biodegradation: 7 to 10 days for bacteria to utilize 95% of 300 ppm in parent substrate (Tabak et al. 1964); decompositions in soil suspensions: 5 days for complete disappearance (Woodcock 1971; quoted, Verschueren 1983); 216-432 hours and 624-1560 hours for 75% degradation in mineral medium and seawater batch experiment respectively (de Kreuk & Hanstveit 1981); aqueous aerobic half-life of 168-1680 hours, based on unacclimated aerobic river die-away test and soil grab sample data (Blades-Fillmore et al. 1982; Haider et al. 1974; quoted, Howard et al. 1991); aqueous anaerobic half-life of 4050-43690 hours, based on unacclimated anaerobic grab sample data for soil (Baker & Mayfield 1980; quoted, Howard et al. 1991); 92% aerobic biodegraded after 28 days by both Sturm of OECD and sealed vessel tests (Birch & Fletcher 1991); biodegradation rate constant of $3.5x10^{-11}$ $L/cell\cdot h^{-1}$ at 20°C and pH 7; biodegradation rate constants $8.4x10^{-5}$ day^{-1} (river), $8.2x10^{-6}$ day^{-1} (oligotrophic lake) and $8.4x10^{-4}$ day^{-1} (eutrophic lake) in water compartment, $8.4x10^{-4}$ day^{-1} (river), $8.2x10^{-5}$ day^{-1} (oligotrophic lake) and $8.4x10^{-3}$ day^{-1} (eutrophic lake) in soil compartment (Yoshida et al. 1987); up to 60% can be mineralized by acclimated bacteria *pseudomonas aeruginosa* in 48 hours but decreased with increasing humic acid concentration (Robinson & Novak 1994).

Biotransformation: rate constant for bacterial transformation of $3x10^9$ mL $cell^{-1}$ h^{-1} in water (Mabey et al. 1982).

Bioconcentration Uptake (k_1) and Elimination (k_2) Rate Constants:

k_1: 421 day^{-1}, 3573 day^{-1} (flagfish: whole fish, fish lipid, Smith et al. 1990)

k_2: 4.78 day^{-1}, 5.26 day^{-1} (flagfish: whole fish, fish lipid, Smith et al. 1990)

k_2: 4.78 day^{-1}, 2.54 day^{-1} (fish: bioconcentration based; toxicity based, Smith et al. 1990)

Common Name: 2,3,4,5-Tetrachlorophenol
Synonym:
Chemical Name: 2,3,4,5-tetrachlorophenol
CAS Registry No: 4901-51-3
Molecular Formula: C_6HCl_4OH
Molecular Weight: 231.89
Melting Point (°C):
117.0 (Weast 1978; Kaiser et al. 1984; Renner 1990)
95-98 (Verschueren 1983)
Boiling Point (°C):
Density (g/cm^3):
Dissociation Constants, pK_a:
6.96 (Doedens 1967)
5.30 (Sillén & Martell 1971; Kaiser et al. 1984)
5.64 (Könemann 1981; Könemann & Musch 1981; Ugland et al. 1981; Dean 1985; Renner 1990; Ma et al. 1993)
6.35 (Schellenberg et al. 1984)
6.61 (Xie & Dyrssen 1984)
6.12 (Nendza & Seydel 1988)
Molar Volume (cm^3/mol):
187 (calculated-LeBas method, Ma et al. 1993; Shiu et al. 1994)
Molecular Volume, TMV ($Å^3$):
Total Surface Area, TSA ($Å^2$):
182.6 (calculated- χ , Sabljic 1987b)
Heat of Fusion, ΔH_{fus}, kcal/mol:
Entropy of Fusin, ΔS_{fus}, cal/mol·K (e.u.):
Fugacity Ratio at 25°C, assuming ΔS_{fus} = 13.5 e.u., F: 0.1259

Water Solubility (g/m^3 or mg/L at 25°C):
100 (estimated, Ma et al. 1990)
166 (shake flask-HPLC/UV at pH 4.9, Ma et al. 1993)

Vapor Pressure (Pa at 25°C):
0.10 (estimated, Ma et al. 1990)

Henry's Law Constant (Pa m^3/mol):
0.140 (calculated-P/C, Shiu et al. 1994)

Octanol/Water Partition Coefficient, log K_{OW}:
4.95 (Hansch & Leo 1979; quoted, Kaiser et al. 1984)
4.42 (calculated-f const., Könemann 1981; Könemann & Musch 1981)
5.03 (correlated-HPLC-RT, Banerjee et al. 1984)
4.21 (shake flask-UV, Beltrame et al. 1984)
4.87 (Schellenberg et al. 1984)
4.82, 4.68, 5.16, 4.48 (shake flask-GC, HPLC-k', calculated-π const., calculated-f const., Xie et al. 1984)
4.71, 4.70, 4.71 (shake flask-GC-calculated-regression, Xie & Dyrssen 1984)

5.03 (selected, Suntio et al. 1988)
4.54 (batch equilibration-UV, Beltrame et al. 1988)
4.50 (THOR 1989; quoted, Connell & Markwell 1990)
4.30 (quoted, Liu et al. 1991)
4.95 (COMPUTOX databank, Kaiser 1993)

Bioconcentration Factor, log BCF:
1.61, 1.60 (earthworm *E. fetida andrei*: in Kooyenburg soil, Holten soil, van Gestel & Ma 1988)
2.55, 2.33 (earthworm *L. rubellus*: in Kooyenburg soil, Holten soil, van Gestel & Ma 1988)
2.61, 2.73, 3.49, 3.53 (earthworm system, collated from literature, Connell & Markwell 1990)
0.40, 0.50, 2.30, 4.10 (earthworm system, derived data, Connell & Markwell 1990)

Sorption Partition Coefficient, log K_{OC}:
4.21-4.66 (soil, calculated-K_{OW}, model of Karickhoff et al. 1979, Sabljic 1987a,b)
3.77-4.01 (soil, calculated-K_{OW}, model of Kenaga & Goring 1980, Sabljic 1987a,b)
2.94-3.17 (soil, calculated-K_{OW}, model of Briggs 1981, Sabljic 1987a,b)
4.11-4.55 (soil, calculated-K_{OW}, model of Means et al. 1982, Sabljic 1987a,b)
3.20-3.60 (soil, calculated-K_{OW}, model of Chiou et al. 1983, Sabljic 1987a,b)
4.12 (sediment, Schellenberg et al. 1984; quoted, Sabljic 1987a,b)
3.32 (soil, calculated- χ , Sabljic 1987a,b)

Half-Lives in the Environment:
Air:
Surface water: half-life of 0.88 hour, based on photochemical transformation in Xenotest 1200 (Svenson & Björndal 1988).
Ground water:
Sediment:
Soil: disappearance half-lives of 43.4 days from Kooyenburg soil, 29.1 days from Holten soil with earthworm *E. fetida andrei* and 26.8 days from Kooyenburg soil, 42.5 days from Holten soil with earthworm *L. rebellus* (van Gestel & Ma 1988).
Biota:

Environmental Fate Rate Constants and Half-Lives:
Volatilization:
Photolysis: phototransformation half-life of 0.88 hour in Xenotest 1200 (Svenson & Björndal 1988).
Oxidation:
Hydrolysis:
Biodegradation:
Biotransformation: degradation rate constant 7.14×10^{-22} ($\pm$41% SD) $mol\cdot cell^{-1}\cdot hour^{-1}$ from pure culture studies and 5×10^{-16} $mol\cdot cell^{-1}\cdot hour^{-1}$ with microorganisms in Seneca River waters (Banerjee et al. 1984).
Bioconcentration Uptake (k_1) and Elimination (k_2) Rate Constants:

Common Name: 2,3,4,6-Tetrachlorophenol
Synonym:
Chemical Name: 2,3,4,6-tetrachlorophenol
CAS Registry No: 58-90-3
Molecular Formula: C_6HCl_4OH
Molecular Weight: 231.89
Melting Point (°C):

69.5 (Stull 1947; Bidleman & Renberg 1985)
70 (Weast 1978; Kaiser et al. 1984)
69-70 (Verschueren 1983)
67 (Renner 1990)

Boiling Point (°C):

275 (Stull 1947)
164 (at 23 mm Hg, Verschueren 1983)

Density (g/cm^3):

1.60 (at 60°C, Verschueren 1983)

Dissociation Constants, pK_a:

5.40 (Blackman et al. 1955; Xie 1983; Schellenberg et al. 1984)
5.46 (Farquharson et al. 1958; Saarikoski & Viluksela 1982; Renner 1990)
5.38 (Doedens 1967; Jones 1981; Bintein & Devillers 1994)
5.30 (Sillén & Martell 1971; McLeese et al. 1979; Kaiser et al. 1984)
5.22 (Ugland et al. 1981; Dean 1985; Lagas 1988; Renner 1990; Ma et al. 1993)
5.62 (Xie & Dyrssen 1984; Xie et al. 1986; Shigeoka et al. 1988; Söderström et al. 1994)
5.39 (Leuenberger et al. 1985)

Molar Volume (cm^3/mol):

187 (calculated-LeBas method, Ma et al. 1993; Shiu et al. 1994)

Molecular Volume ($Å^3$):
Total Surface Area, TSA ($Å^2$):

182.6 (calculated- χ , Sabljic 1987b)

Heat of Fusion, ΔH_{fus}, kcal/mol:
Entropy of Fusion, ΔS_{fus}, cal/mol·K (e.u.):
Fugacity Ratio at 25°C, F: 0.3589

Water Solubility (g/m^3 or mg/L at 25°C):

183 (shake flask-UV at pH 5.1, Blackman et al. 1955)
140 (20°C, selected, Verschueren 1977; 1983; Shigeoka et al. 1988)
100 (20°C, Callahan et al. 1979)
180 (recommended at pH 5.1, IUPAC 1985)
100 (selected, Suntio et al. 1988)

Vapor Pressure (Pa at 25°C):

0.763 (extrapolated-regression of tabulated data, Stull 1947)
0.28 (estimated, Cheung 1984)
0.75 (extrapolated liquid, Antoine eqn., Weast 1972-73)

0.692 (subcooled liq. value, Hamilton 1980; quoted, Bidleman & Renberg 1985)
0.564 (capillary GC-RT, Bidleman & Renberg 1985)
0.810 (extrapolated-Antoine eqn., Stephenson & Malanowski 1987)

Henry's Law Constant (Pa m^3/mol):
0.3548 (calculated-P/C, Shiu et al. 1994)

Octanol/Water Partition Coefficient, log K_{OW}:
4.10 (Hansch & Leo 1979; quoted, Miyake & Terada 1982; Kaiser 1983; Kaiser et al. 1984; Terada et al. 1987)
4.20 (calculated, McLeese et al. 1979)
4.27 (RP-HPLC-k', Miyake & Terada 1982)
4.45 (shake flask-GC, Saarikoski & Viluksela 1982)
4.81 (calculated as per Leo et al. 1971, Xie 1983)
4.12 (shake flask-UV, Beltrame et al. 1984)
4.87 (shake flask-HPLC/UV, Schellenberg et al. 1984)
4.42, 4.34, 4.81, 4.48 (shake flask-GC, HPLC-k', calculated-π const., calculated-f const., Xie et al. 1984; quoted, Bintein & Devillers 1994)
4.31, 4.31, 4.30 (shake flask-GC-calculated-regression, Xie & Dyrssen 1984)
4.42 (OECD 1981 guidelines, Leuenberger et al. 1985)
4.42 (quoted from Xie et al. 1984, Xie et al. 1986)
4.10, 4.25 (quoted, CPC-RV, Terada et al. 1987)
4.42 (quoted from Xie & Dyrssen 1984, Lagas 1988)
4.42 (HPLC-RP, Shigeoka et al. 1988; quoted, Saito et al. 1993)
5.03 (selected, Suntio et al. 1988)
4.37 (batch equilibration-UV, Beltrame et al. 1988)
4.33, 4.10 (COMPUTOX databank, Kaiser 1993)
2.25-2.42 (quoted lit. range, NCASI 1993)
4.30 (quoted, Söderström et al. 1994)
4.24 (shake flask-GC, Kishino & Kobayashi 1994)

Bioconcentration Factor, log BCF:
2.65 (trout, Hattula et al. 1981)
2.55-2.69 (estimated, NCASI 1993)
3.3, 3.8, 4.4, 4.5 (perch bile to water, Söderström et al. 1994)

Sorption Partition Coefficient, log K_{OC}:
4.21 (soil, calculated-K_{OW}, model of Karickhoff et al. 1979, Sabljic 1987a,b)
3.77 (soil, calculated-K_{OW}, model of Kenaga & Goring 1980, Sabljic 1987a,b)
2.94 (soil, calculated-K_{OW}, model of Briggs 1981, Sabljic 1987a,b)
4.11 (soil, calculated-K_{OW}, model of Means et al. 1982, Sabljic 1987a,b)
3.20 (soil, calculated-K_{OW}, model of Chiou et al. 1983, Sabljic 1987a,b)
3.82 (sediment, Schellenberg et al. 1984; quoted, Sabljic 1987a,b)
3.35 (soil, Seip et al. 1986; quoted, Sabljic 1987a,b)

3.32 (soil, calculated- χ , Sabljic 1987a,b)
3.90 (calculated, Lagas 1988)
2.45, 2.70 (totally dissociated as phenolate-calculated, Lagas 1988)

Half-Lives in the Environment:

Air: half-lives of 364.4-3644 hours, based on estimated photooxidation half-life in air (Howard et al. 1991).

Surface water: 0.88 hour, based on photochemical transformation in Xenotest 1200 (Svenson & Björndal 1988); half-lives of 1-336 hours, based on estimated aqueous photolysis data for trichlorophenols and PCP (Howard et al. 1991).

Ground water: half-lives of 1344-8640 hours, based on estimated unacclimated aqueous aerobic biodegradation half-life (Howard et al. 1991).

Sediment:

Soil: half-lives of 672-4320 hours, based on estimated unacclimated aqueous aerobic biodegradation half-life (Howard et al. 1991).

Biota:

Environmental Fate Rate Constants and Half-Lives:

Volatilization:

Photolysis: aqueous photolysis half-life estimated between 1-336 hours, based on aqueous data for 24,5- 2,4,6-trichlorophenols and pentachlorophenol (Howard et al. 1991).

Photooxidation: atmospheric half-lives of 364.4-3644 hours, based on estimated rate constant for reaction with OH radicals and aqueous half-lives of 66.0-3480 hours, based on reaction rate constants with OH and $RO_2\cdot$ radicals with phenol class (Howard et al. 1991).

Hydrolysis: no hydrolyzable groups (Howard et al. 1991).

Biodegradation: decomposition in soil suspensions: >72 days for complete disappearance (Woodcock 1971; quoted, Verschueren 1983); aqueous aerobic half-lives of 672-4032 hours, based on acclimated aerobic screening test data and aqueous anerobic half-lives of 2688-16128 hours, based on unacclimated aerobic biodegradation half-lives (Howard et al. 1991).

Biotransformation:

Bioconcentration Uptake (k_1) and Elimination (k_2) Rate Constants:

Common Name: Pentachlorophenol
Synonym: chlorophen, PCP, penchlorol
Chemical Name: pentachlorophenol
CAS Registry No: 87-86-5
Molecular Formula: C_6Cl_5OH
Molecular Weight: 266.34
Melting Point (°C):
190 (Verschueren 1977, 1983; Callahan et al. 1979; Howard 1991; Lee et al. 1993)
191 (Firestone 1977; Weast 1978; Kaiser et al. 1984)
190.2 (Crosby 1981)
187 (Schmidt-Bleek et al. 1982)
Boiling Point (°C):
210 (dec., Firestone 1977)
310 (Verschueren 1977, 1983; Callahan et al. 1979; Renner 1990)
300.6 (Crosby 1981)
309-310 (dec., Howard 1991)
Density (g/cm^3):
1.987 (Firestone 1977)
1.978 (Schmidt-Bleek et al. 1982; Verschueren 1983)
Dissociation Constants, pK_a:
4.80 (Blackman et al. 1955; Sillén & Martell 1971; McLeese et al. 1979; Kaiser et al. 1984; Kollig 1993)
5.00 (Farquharson et al. 1958; Renner 1990)
4.92 (Doedens 1967; Jones 1981; Bintein & Devillers 1994)
4.74 (Drahonovsky & Vacek 1971; Callahan et al. 1979; Ugland et al. 1981; Könemann 1981; Könemann & Musch 1981; Dean 1985; Westall et al. 1985; Lagas 1988; Renner 1990; Howard 1991; Lee et al. 1991,93)
4.71 (spectrophotometric, Cessna & Grover 1978; Saarikoski & Viluksela 1982; Saarikoski et al. 1986; Tratnyek & Hoigné 1991)
5.30 (Gebefügi et al. 1979; Xie 1983)
4.70 (Crosby 1981; Hoigné & Bader 1983)
5.20 (Renberg 1981; Renner 1990; Larsson et al. 1993)
5.25 (Schellenberg et al. 1984)
4.90 (Xie & Dyrssen 1984; Xie et al. 1986; Shigeoka et sl. 1988; Söderström et al. 1994)
4.75 (Leuenberger et al. 1985)
4.60 (Nendza & Seydel 1988)
Molar Volume (cm^3/mol):
207.9 (calculated-LeBas method, Ma et al. 1993; Shiu et al. 1994)
Molecular Volume, TMV ($Å^3$):
Total Surface Area, TSA ($Å^2$):
199.7 (calculated- χ , Sabljic 1987b)
Heat of Fusion, ΔH_{fus}, kcal/mol:
Entropy of Fusion, ΔS_{fus}, cal/mol·K (e.u.):
Fugacity Ratio at 25°C, (assuming ΔS_{fus} = 13.5 e.u.) F: 0.0336

Water Solubility (g/m^3 or mg/L at 25°C):

15.4 (gravimetric, Carswell & Nason 1938; quoted, Ma et al. 1993)
18.0 (27°C, gravimetric, Carswell & Nason 1938)
9.59 (shake flask-UV, at pH 5.1, Blackman et al. 1955; quoted, Ma et al. 1993)
14 (Stephen & Stephen 1963; quoted, Lu & Metcalf 1975)
14.0 (20°C, shake flask-UV, at pH 3.0, Bevenue & Beckman 1967; quoted, Valsaraj et al. 1991)
10 (shake flask-gravimetric, at pH 5.0, Toyota & Kuwahara 1967; quoted, Ma et al. 1993)
14 (gravimetric at pH 5.0, Toyota & Kuwahara 1967; quoted, Ma et al. 1993)
20-25 (selected, Günther et al. 1968; quoted, Gebefügi et al. 1979; Crossland & Wolff 1985)
20 (30°C, Firestone 1977)
14 (20°C, Verschueren 1977, 1983; quoted, Callahan et al. 1979; Kilzer et al. 1979; Geyer et al. 1981; Crosby 1981; Kenaga & Goring 1980; Mabey et al 1982; LeBlanc 1984; McKim et al. 1985; Shigeoka et al. 1988; Howard 1991; Lee et al. 1993)
20 (20°C, Korte et al. 1978)
14, 20-25 (selected lit. values, Geyer et al. 1980, Geyer et al. 1984)
15 (23°C, Klöpffer et al. 1982)
20 (20°C, Schmidt-Bleek et al. 1982; quoted, Hattemer-Frey & Travis 1989)
15.4 (Cheung 1984)
5-10 (at pH 5-6 in contaminated water, Goerlitz et al. 1985)
14 (recommended at pH 4.5-5.5, IUPAC 1985; selected; Suntio et al. 1988)
14.1 (quoted, Isnard & Lambert 1988)
41 (predicted- χ , Nirmalakhandan & Speece 1988; quoted, Valsaraj et al. 1991)
8 $\pm$ 2 (shake flask-UV at pH 2.5, Valsaraj et al. 1991)
32 $\pm$ 3 (shake flask-UV at pH 5.0, Valsaraj et al. 1991)
19.0 (quoted, Müller & Klein 1992)
18.4 (shake flask-HPLC/UV, at pH 4.8, Ma et al. 1993)

Vapor Pressure (Pa at 25°C):

0.0227 (20°C, static method, Carswell & Nason 1938)
0.0147 (20°C, Bevenue & Beckman 1967; quoted, Callahan et al. 1979)
0.231 (subcooled liq. extrapolated-Antoine eqn., Weast 1976-77; quoted, Bidleman & Renberg 1985)
0.100 (Weast 1972-73)
0.0211 (quoted, Chiou & Freed 1977)
0.0213 (Firestone 1977)
0.0147-0.0227 (20°C, Goll 1954; Bevenue & Beckman 1967; Neumüller 1974; quoted, Gebefügi et al. 1979; Kilzer et al. 1979)
0.0147 (20°C, quoted, Klizer et al. 1979; Mabey et al. 1982; Verschueren 1983)
0.0956 (subcooled liq., Hamilton 1980; quoted, Bidleman & Renberg 1985)
0.0227 (20°C, quoted, Crosby 1981)
0.00415 (23°C, OECD, Klöpffer et al. 1982)
0.0093 (20°C, Schmidt-Bleek et al. 1982; quoted, Hattemer-Frey & Travis 1989)
0.1153 (extrapolated-Antoine eqn., Boublik et al. 1984)

0.50 (20°C, quoted, Crossland & Wolff 1985)
0.115 (capillary GC-RT, P_L, Bidleman & Renberg 1985)
0.127 (extrapolated-Antoine eqn., Stephenson & Malanowski 1987)
0.0147 (Howard 1991)

Henry's Law Constant (Pa·m³/mol):
0.284 (calculated-P/C, Mabey et al. 1982)
0.00248 (calculated-P/C, Hellmann 1987; quoted, Meylan & Howard 1991)
0.0127 (estimated-bond contribution, Hellmann 1987; quoted, Meylan & Howard 1991)
0.277 (calculated-P/C, Howard 1991)
0.079 (calculated-P/C, Shiu et al. 1994)

Octanol/Water Partition Coefficient, log K_{OW}:
5.01 (quoted unpublished result, Leo et al. 1971; quoted, Chiou & Freed 1977; Callahan et al. 1979; Beltrame et al. 1984; Crossland & Wolff 1985; Westall et al. 1985)
5.01 (Firestone 1977)
5.01, 5.12, 5.86, 3.81 (Hansch & Leo 1979)
5.01 (quoted, Veith et al. 1979; Mackay 1982; Chin et al. 1986)
5.01 (calculated, Veith et al. 1979b)
5.01 (quoted, McLeese et al. 1979)
4.16 (quoted, Rao & Davidson 1980)
5.10 (calculated-HPLC-k', Butte et al. 1981; quoted, Butte et al. 1987)
3.25 (quoted, Crosby 1981)
5.19 (calculated-f const., Könemann 1981; Könemann & Musch 1981)
4.00 (at pH 4, Renberg 1981; quoted, Larsson et al. 1993)
0.00 (at pH 8, Renberg 1981; quoted, Larsson et al. 1993)
5.12, 5.08 (quoted from Hansch & Leo 1979, correlated-RP-HPLC-k', Miyake & Terada 1982)
5.04 (calculated-f constant, Mabey et al. 1982)
5.15 (shake flask-GC, Saarikoski & Viluksela 1982; Saarikoski et al. 1986)
5.05 (quoted, Kaiser & Valdmanis 1982)
4.84 (shake flask-GC, apparent value at pH 1.2, Kaiser & Valdmanis 1982)
1.30 (shake flask-GC, apparent value at pH 10.5, Kaiser & Valdmanis 1982)
3.69 (Geyer et al. 1982, Schmidt-Bleek et al. 1982; quoted, Hattemer-Frey & Travis 1989)
3.29 (shake flask average, OECD/EEC Lab. comparison tests, Harnish et al. 1983)
5.01 (quoted from Hansch & Leo 1979, Kaiser 1983)
5.01 (Verschueren 1983; quoted, Lee et al. 1993)
5.85 (calculated as per Leo et al. 1971, Xie 1983)
5.86 (quoted from Hansch & Leo 1982, Banerjee et al. 1984)
5.01 (quoted from Hansch & Leo 1979, Davies & Dobbs 1984; Kaiser et al. 1984)
5.12 (selected best lit. value, Garst & Wilson 1984)
5.11 ± 0.07 (exptl.-ALPM, Garst & Wilson 1984)
3.69, 3.81 (shake flask, OECD 1981 Guidelines, Geyer et al. 1984)
5.24 (shake flask-HPLC/UV, Schellenberg et al. 1984)

5.04, 5.08, 5.85, 5.22 (shake flask-GC, HPLC-k', calculated-π const., calculated-f const., Xie et al. 1984; quoted, Bintein & Devillers 1994)
5.05 (calculated, Xie & Dyrssen 1984; quoted, Lagas 1988)
3.81 (quoted, Freitag et al. 1985)
5.12 (Hansch & Leo 1985; quoted, Howard 1991)
4.10 (computed, Klopman et al. 1985; quoted, Pangrekar et al. 1994)
5.24 (OECD 1981 guidelines, Leuenberger et al. 1985)
5.01, 4.71 (quoted, RPHPLC-RT, Chin et al. 1986)
5.04 (quoted from Xie et al. 1984, Xie et al. 1986; Söderström et al. 1994)
2.50 (at pH 4.7, Geyer et al. 1987)
5.10 (quoted, Butte et al. 1987)
4.47 (CPC-RV, Terada et al. 1987)
5.12 (quoted, Tereda et al. 1987; Hodson et al. 1988)
5.04 (HPLC-RP, Shigeoka et al. 1988; quoted, Saito et al. 1993)
5.05 (selected, Suntio et al. 1988)
5.00 (batch equilibration-UV, Beltrame et al. 1988)
5.01 (quoted, Isnard & Lambert 1988, 1989)
5.24 (THOR 1989; quoted, Connell & Markwell 1990)
5.01 (quoted, Liu et al. 1991)
5.06 (calculated-CLOGP, Müller & Klein 1992)
5.01, 5.3798 (quoted, calculated-original UNIFAC, Chen et al. 1993)
5.24 (EPA CLOGP Data Base, Hulzebos et al. 1993)
5.18 (recommended, LOGKOW databank, Sangster 1993)
5.06, 5.12 (COMPUTOX databank, Kaiser 1993)
5.06 (calculated-QSAR, Kollig 1993)
5.02 (shake flask-GC/ECD, Kishino & Kobayashi 1994)

Bioconcentration Factor, log BCF:
3.75 (fish, Statham et al. 1976; quoted, Hattemer-Frey & Travis 1989)
3.04 (fish, Körte et al. 1978)
2.89 (fathead minnow, Veith et al. 1979; quoted, Davies & Dobbs 1984)
2.89 (fathead minnow, calculated value, Veith et al. 1979b)
3.09, 2.64 (algae: exptl., calculated, Geyer et al. 1981)
2.00 (trout, Hattula et al. 1981)
3.04, 3.10, 3.02 (activated sludge, algae, golden orfe, Freitag et al. 1982)
2.54 (mussel *mytilus edulis*, Geyer et al. 1982)
2.89, 3.69 (quoted, calculated-K_{OW}, Mackay 1982)
4.20 (microorgansims-water, calculated-K_{OW}, Mabey et al. 1982)
1.60 (killifish, Trujillo et al. 1982)
1.86, 1.72, 1.60 (low-PCP flowing, high-PCP flowing, high-PCP static soft water; Brockway et al. 1984)
1.66, 1.62, 1.26 (low-PCP flowing, high-PCP flowing, high-PCP static hard water; Brockway et al. 1984)
3.10 (*alga chlorella fusca* in culture flasks, Geyer et al. 1984; quoted, Brockway et al. 1984)
3.10, 2.72 (algae: exptl, calculated-K_{OW}, Geyer et al. 1984)
3.10, 3.02, 3.04 (algae, fish, sludge, Klein et al. 1984)

3.00 (quoted, LeBlanc 1984)
3.04, 3.10, 2.42 (activated sludge, algae, golden ide, Freitag et al. 1985)
0.57 (human fat, Geyer et al. 1987)
2.99 (zebrafish, Butte et al. 1987)
0.46 (15°C, initial concn. 1.0 mg/L uptake by *allolobophora caliginosa* at 24 hours, Haque & Ebing 1988)
0.38 (15°C, initial concn. 10.0 mg/L uptake by *allolobophora caliginosa* at 24 hours, Haque & Ebing 1988)
0.80 (whole *allolobophora caliginosa*/soil, uptake from soil after 131 days exposure in outdoor lysimeters, Haque & Ebing 1988)
1.35 (whole *lumbricus terrestris*/soil, uptake from soil after 131 days exposure in outdoor lysimeters, Haque & Ebing 1988)
2.89 (quoted, Isnard & Lambert 1988)
2.80, 2.63 (earthworm *e. fetida andrei*: in Kooyenburg soil, Holten soil, van Gestel & Ma 1988)
2.80, 2.63 (earthworm *e. fetida andrei*: in Kooyenburg soil, Holten soil, van Gestel & Ma 1988)
-2.66 (daily intake/cow adipose tissue, Travis & Arms 1988; quoted, Hattemer-Frey & Travis 1989)
4.38, 4.50, 4.53, 4.90 (earthworm system, collated from literature, Connell & Markwell 1990)
4.00, 5.30, 3.40, 8.00 (earthworm system, derived data, Connell & Markwell 1990)
2.97 (*p. hoyi*, Landrum & Dupuis 1990)
2.11 (*m. relicta*, Landrum & Dupuis 1990)
2.16-2.53 (soft tissue of freshwater mussel, Mäkelä & Oikari 1990)
2.33; 3.21 (flagfish: whole fish; fish lipid, Smith et al. 1990)
2.78, 2.11, 1.72 (goldfish at pH 7, pH 8, pH 9, Stehly & Hayton 1990)
3.0, 3.4, 3.9, 4.0 (perch bile to water, Söderström et al. 1994)

Sorption Partition Coefficient, log K_{OC}:
2.95 (soil, calculated-K_{OW}, Kenaga & Goring 1980)
4.72 (sediment-water, calculated-K_{OW}, Mabey et al. 1982)
3.11-5.65 (soil, calculated-K_{OW}, model of Karickhoff et al. 1979, Sabljic 1987a,b)
3.17-4.54 (soil, calculated-K_{OW}, model of Kenaga & Goring 1980, Sabljic 1987a,b)
3.37-3.69 (soil, calculated-K_{OW}, model of Briggs 1981, Sabljic 1987a,b)
3.00-5.54 (soil, calculated-K_{OW}, model of Means et al. 1982, Sabljic 1987a,b)
2.21-4.49 (soil, calculated-K_{OW}, model of Chiou et al. 1983, Sabljic 1987a,b)
4.52 (sediment, Schellenberg et al. 1984)
3.73 (quoted average of Kenaga & Goring 1980 & Schellenberg et al. 1984 values, Sabljic 1987a,b)
3.46 (soil, calculated- χ , Sabljic 1987a,b)
2.95 (soil, calculated- χ , Bahnick & Doucette 1988)
4.04 (estimated, HPLC-k', Hodson & Williams 1988)
4.40 (calculated, Lagas 1988)
3.10, 3.26 (totally dissociated as phenolate-calculated, Lagas 1988)
5.27, 5.71 (Bluepoint soil at pH 7.8, pH 7.4, Bellin et al. 1990)
5.58, 5.52 (Glendale soil at pH 7.3, pH 4.3, Bellin et al. 1990)

3.49, 3.57 (Norfolk soil at pH 4.3, pH 4.4, Bellin et al. 1990)
4.32-4.65 (Norfolk + lime soil at pH 6.9, Bellin et al. 1990)
4.51, 4.54 (neutral form, silt-clay slurries, Jafvert & Weber 1991)
3.06 (calculated-K_{OW}, Kollig 1993)

Half-Lives in the Environment:

Air: 139.2-1392 hours, based on an estimated rate constant for the vapor phase reaction with hydroxyl radicals in air (Howard et al. 1991); photolysis half-life of 6.5 hours in noonday summer sunshine (Howard 1991); half-lives at latitude of 43.70 °N is 216 hours at noon in January to less than 24 hours in mid-summer for reaction with hydroxyl radicals (Bunce et al. 1991); atmospheric transformation lifetime was estimated to be > 5 days (Kelly et al. 1994).

Surface water: calculated photolysis half-life of 4.75 hours from a determined rate of 3.4×10^{-4} sec^{-1} for a depth of 300 cm at pH 7 with light intensity of 0.04 watts/cm^2 between 290 and 330 nm on a midsummer day at the latitude of Cleveland, Ohio (Hiatt et al. 1960; quoted, Callahan et al. 1979); photolysis half-life of 1.5 day was estimated from photolytic destruction by sunlight in an aqueous solution at Davis, California (Wong & Crosby 1978; quoted, Callahan et al. 1979); photolytic half-life of 10-15 days (Brockway et al. 1984); rate constant of $>> 3.0 \times 10^5$ M^{-1} sec^{-1} for the reaction with ozone at pH 2.0 (Hoigné & Bader 1983); estimated direct photolysis midday half-life of 20 minutes from experimentally determined rate constant of 2.1 h^{-1} (quoted unpublished result, Zepp et al. 1984); 1.5 to 3.0 days for direct photo-transformation from outdoor ponds (Crossland & Wolff 1985); half-lives of 1 hour (summer), 2 hours (winter) for distilled water; 2 hours (summer), 3 hours (winter) for estuarine water; 2 hours (summer), 3 hours (winter) for poisoned estuarine water, based on photo-transformation rate constants (Hwang et al. 1986); 6 days (summer), 14 days (winter) for distilled water; 3 days (summer), 7 days (winter) for estuarine water; 6 days (summer), 10 days (winter) for poisoned estuarine water, based on photo-mineralization rate constants (Hwang et al. 1986); 0.75 hour and 0.96 hour, based on photochemical transformation in Xenotest 1200 (Svenson & Björndal 1988); half-life of 1-110 hours, based on aqueous photolysis half-life (Howard et al. 1991); photodegradation half-lives ranging from hours to days, more rapid at the surface (Howard 1991); photodegradation half-lives of 1.0 hour (summer), 2.0 hours (winter) in distilled water and 2.0 hours (summer), 3.0 hours (winter) in estuarine water under irradiation by natural sunlight (quoted from Hwang et al. 1987, Sanders et al. 1993).

Ground water: 1104-36480 hours, based on estimated unacclimated aqueous aerobic sediment grab sample data (Delaune et at. 1983; selected, Howard et al. 1991) and unacclimated anaerobic grab sample data for ground water (Baker & Mayfield 1980; selected, Howard et al. 1991).

Sediment:

Soil: disappearance half-lives of 23.2 days from Kooyenburg soil, 47.9 days from Holten soil with earthworm *e. fetida andrei* and 27.4 days from Kooyenburg soil, 31.8 days from Holten soil with earthworm *l. rubellus* (van Gestel & Ma 1988); half-life 552-4272 hours, based on estimated unacclimated aqueous aerobic biodegradation half-life (Howard et al. 1991); 12.0 days in an acidic clay soil with < 1.0% organic matter and 6.7 days in a slightly basic sandy loam soil with 3.25% organic matter, based on aerobic batch lab. microcosm experiments (Loehr & Matthews 1992).

Biota: biological half-life in guppy *lebistes reticulatus* was estimated at about 30 days (Landner et al. 1977); elimination half-lives 23, 9.3, 6.9, and 6.2 hours for fat, liver muscle, and blood respectively (rainbow trout, Call et al. 1980); estimated half-life of 7.0 days in trout (Niimi & Cho 1983; quoted, Niimi & Palazzo 1985); clearance from flagfish: 0.68 day from whole fish and 0.68 day from fish lipid (Smith et al. 1990).

Environmental Fate Rate Constants and Half-Lives:

Volatilization/evaporation: half-life of 84 hours from the rate of loss experiment on watch glass for an exposure period of 192 hours (Dobbs & Grant 1980); stripping loss rate constant of 0.0076 day^{-1} (Moos et al. 1983); 0.028 day^{-1} for nondissociated PCP, assuming diffusion coefficient in air to be $7x10^{-6}$ m^2/sec and in water $7x10^{-10}$ m^2/sec with wind speed 0.1 m above the pond is 2 m/s and the average temperature is 15°C for water depth of 1 m (Crossland & Wolff 1985); calculated rate constant of $5x10^{-4}$ day^{-1} to $1x^{-7}$ day^{-1} for total PCP (Crossland & Wolff 1985).

Photolysis: calculated photolysis half-life of 4.75 hours from observed rate of $3.4x10^{-4}$ sec^{-1} for a depth of 300 cm at pH 7 with light intensity of 0.04 $watts/cm^2$ between 290 and 330 nm on a midsummer day at the latitude of Cleveland, Ohio (Hiatt et al. 1960; quoted, Callahan et al. 1979); photolysis half-life of 1.5 day was estimated from photolytic destruction by sunlight in an aqueous solution at Davis, California (Wong & Crosby 1978; quoted, Callahan et al. 1979); exposure of aqueous PCP solutions to either sunlight or laboratory ultraviolet light resulted in rapid degradation at pH 7.3 and slower degradation at pH 3.3 (Wong & Crosby 1981); photolytic half-life of 10-15 days (Brockway et al. 1984); 0.23 to 0.46 day^{-1} for direct photo-transformation, is the main loss process for PCP from ponds, with half-life of 1.5 to 3.0 days (Crossland & Wolff 1985); photo-transformation rate constants of 0.6 h^{-1} with half-life of 1 hour for distilled water in summer (mean temperature 25°C) and 0.37 h^{-1} with half-life of 2 hours in winter (mean temperature 11°C); 0.37 h^{-1} with half-life of 2 hours for both poisoned estuarine water and estuarine water in summer and 0.27 h^{-1} with half-life of 3 hours in winter during days when exposed to full sunlight and microbes (Hwang et al. 1986); photo-mineralization rate constants of 0.11 h^{-1} with half-life of 6 days for distilled water in summer (mean temperature 25°C) and 0.049 h^{-1} with half-life of 14 days in winter (mean temperature 11°C); 0.12 h^{-1} with half-life of 6 days for poisoned estuarine water in summer and 0.07 h^{-1} with half-life of 10 days in winter; 0.25 h^{-1} with half-life of 10 days for estuarine water in summer and 0.10 h^{-1} with half-life of 7 days for winter during days when exposed to full sunlight and microbes (Hwang et al. 1986); phototransformation half-life of 0.75 hour in Xenotest 1200 (Svenson & Björndal 1988); aqueous photolysis half-life of 1-110 hours (Hwang et al. 1986; Sugiura et al. 1984; selected, Howard et al. 1991); assuming a linear rate of photolysis during 96-hours period, they determined the half-life of PCP in their systems to be 7.43 days (Smith et al. 1987); photodegradation rate constants of 0.60 h^{-1} corresponding to a half-life of 1.0 hour (summer), 0.37 h^{-1} corresponding to a half-life of 2 hours (winter) in distilled water; and 0.37 h^{-1} corresponding to a half-life of 2 hours (summer), 0.27 h^{-1} corresponding to a half-life of 3.0 hours (winter) in estuarine water under irradiation by natural sunlight (quoted from Hwang et al. 1987, Sanders et al. 1993).

Oxidation: aqueous oxidation rate constant of < $7x10^3$ $M^{-1}h^{-1}$ for singlet oxygen and $1x10^5$ $M^{-1}h^{-1}$ for peroxy radical at 25°C (Mabey et al. 1982); rate constant of >>$3.0x10^5$ $M^{-1}\cdot sec^{-1}$ for the reaction with ozone in water at pH 2.0 (Hoigné & Bader 1983); photooxidation half-life of 66-3480 hours in water, based on reported reaction rate constants for reaction of OH and RO_2 radicals with phenol class in aqueous solution (Mill & Mabey 1985; Güesten et al. 1981; quoted, Howard et al. 1991); photooxidation half-life of 139.2-1392 hours, based on an estimated rate constant for the vapor phase reaction with hydroxyl radicals in air (Atkinson 1987; quoted, Howard et al. 1991); rate constant of $(0.2 \pm 5.5)x10^6$ $M^{-1}\cdot sec^{-1}$ for the reaction with singlet oxygen in aqueous phosphate buffer at (27 ± 1)°C (Tratnyek & Hoigné 1991); atmospheric half-lives vary from less than 24 h at noon in mid-summer to 216 h in January at latitude of 41.79°N for reaction with OH radicals (Bunce et al. 1991).

Hydrolysis: is not expected to occur (Crossland & Wolff 1985).

Biodegradation: 1800-2160 h and 480-∞ h to obtain 75% degradation in mineral medium and seawater respectively (de Kreuk & Hanstveit 1981); rate constants of 0.10 day^{-1} with a half-life of 7 days in unadapted Nutrient Broth and 1.0 day^{-1} with a half-life of 0.7 day in adapted Nutrient Broth under aerobic conditions (Mills et al. 1982); aqueous aerobic half-life of 552-4272 hours, based on unacclimated and acclimated aerobic sediment grab sample data (Delaune et al. 1983; Baker & Mayfield 1980; quoted, Howard et al. 1991); aqueous anaerobic half-life of 1008-36480 hours, based on unacclimated anaerobic grab sample data for soil and ground water (Ide et al. 1972; Baker & Mayfield 1980; quoted, Howard et al. 1991); aerobic degradation rate constant of 0.0017 liter $\mu g^{-1}\cdot day^{-1}$ (Moos et al. 1983); microbial degradation negligible in darkness (Hwang et al. 1986); degradation rate constant of 0.12 ± 0.01 h^{-1} in the absence of light (Minero et al. 1993).

Biotransformation: rate constant for bacterial transformation of $3x10^9$ $mL\cdot cell^{-1}\cdot h^{-1}$ in water (Mabey et al. 1982); degradation rate of $3x10^{14}$ $mol\cdot cell^{-1}\cdot h^{-1}$ with microorganisms in Seneca River waters (Banerjee et al. 1984).

Bioconcentration Uptake (k_1) and Elimination (k_2) Rate Constants:

k_1: 18.3 $hour^{-1}$, 19 $hour^{-1}$ (at 1 mM buffer concn.), 18.5 $hour^{-1}$ (at 10 mM buffer concn.) at pH 8 (guppy *p. reticulata* Peters, Saarikoski et al. 1986)

k_1: 222 day^{-1}, 1677 day^{-1} (flagfish: whole fish; fish lipid, Smith et al. 1990)

k_2: 1.03 day^{-1}, 1.03 day_{-1} (flagfish: whole fish; fish lipid, Smith et al. 1990)

k_2: 1.03 day^{-1}, 0.95 day_{-1} (flagfish: BCF based, toxicity based, Smith et al. 1990)

k_2: 0.00195 ± 0.00063 $hour^{-1}$ (*m. relicta*, Landrum & Dupuis 1990)

k_2: 0.00330 ± 0.00140 $hour^{-1}$ (*p. hoyi*, Landrum & Dupuis 1990)

Common Name: 2-Nitrophenol
Synonym: *o*-nitrophenol, 2-hydroxynitrobenzene
Chemical Name: 2-nitrophenol
CAS Registry No: 88-75-5
Molecular Formula: $C_6H_4(NO_2)OH$
Molecular Weight: 139.11
Melting Point (°C):

45.01 (Stull 1947; Morrison & Boyd 1959; Verschueren 1983; Stephenson & Malanowski 1987)
45.6 (Tsonopoulos & Prausnitz 1971)
45.3 (Weast 1977; Callahan et al. 1979; Mabey et al. 1982)
45.1 (Yalkowsky & Valvani 1980; Schmidt-Bleek et al. 1982)
45-46 (Schwarzenbach et al. 1988)

Boiling Point (°C):

214.5 (Stull 1947)
216.0 (Weast 1977; Callahan et al. 1979; Mabey et al. 1982; Stephenson & Malanowski 1987; Howard 1989)
217 (Morrison & Boyd 1959; Schmidt-Bleek et al. 1982)
214/217 (Verschueren 1983)

Density (g/cm^3 at 20°C):

1.495 (Schmidt-Bleek et al. 1982)
1.657 (Verschueren 1983)

Molar Volume (cm^3/mol):

100.60 (calculated from density, Jaworska & Schultz 1993)
131.9 (calculated-LeBas method)
67.6 (intrinsic V_I, Hickey & Passino-Reader 1991)

Molecular Volume ($Å^3$):
Total Surface Area, TSA ($Å^2$):
Dissociation Constant, pK_a:

7.21 (Pearce & Simkins 1968; quoted, Callahan et al. 1979)
7.23 (Serjeant & Dempsey 1979; quoted, Schwarzenbach et al. 1988; Howard 1989; Haderlein & Schwarzenbach 1993)
8.28 (Mabey et al. 1982)
7.222 (Dean 1985)

Heat of Fusion ΔH_{fus}, kcal/mol:

4.20 (Tsonopoulos & Prausnitz 1971)

Entropy of Fusion ΔS_{fus}, cal/mol·K (e.u.):

13.2 (Tsonopoulos & Prausnitz 1971)
11.70, 13.5 (observed, estimated, Yalkowsky & Valvani 1980)

Fugacity Ratio at 25°C (assuming ΔS_{fus} = 13.5 e.u.), F: 0.634

Water Solubility (g/m^3 or mg/L at 25°C):

2500 (Duff 1929; quoted, Tsonopoulos & Prausnitz 1971)
3300 (interpolated from Sidgwick et al. 1915, Palit 1947)
2000 (quoted, Morrison & Boyd 1959)

2100 (20°C, Verschueren 1977, 1983; quoted, Callahan et al. 1979; Mabey et al. 1982)
15250, 28400 (quoted, calculated-K_{OW}, Yalkowsky & Morozowich 1980)
1262 (20°C, Schmidt-Bleek et al. 1982)
1390, 1095 (20°C, shake flask-UV, calculated, Hashimoto et al. 1984)
1080 (20 ± 0.5°C, shake flask-UV with buffer at pH 1.5, Schwarzenbach et al. 1988; quoted, Müller & Klein 1992)
2500, 1080 (25°C, 20°C, quoted, Howard 1989)

Vapor Pressure (Pa at 25°C):
29.1 (extrapolated-regression of tabulated data, Stull 1947)
6.921 (20°C, Schmidt-Bleek et al. 1982)
20.13 (20°C, Mabey et al. 1982)
17.25 (extrapolated-Antoine eqn., Stephenson & Malanowski 1987)
10.61 (20°C, calculated-M.P., Schwarzenbach et al. 1988)
18.86 (subcooled liq. at 20°C, estimated-GC-RT, Schwarzenbach et al. 1988)
26.66 (quoted from Swarzenbach et al. 1988, Howard 1989)

Henry's Law Constant (Pa·m^3/mol):
0.766 (calculated-P/C, Mabey et al. 1982)
0.355 (Leuenberger et al. 1985; quoted, Howard 1989)
1.367 (20°C, calculated-P/C, Schwarzenbach et al. 1988)

Octanol/Water Partition Coefficient, log K_{OW}:
1.79 (shake flask-AS, Fujita et al. 1964; quoted, Leo et al. 1971; Rekker 1977; Callahan et al. 1979; Beltrame et al. 1984; Westall et al. 1985)
1.25 (calculated-f const., Rekker 1977)
1.18 (calculated-π const., Rekker 1977)
1.79 (shake flask, Korenman et al. 1977)
1.68 (shake flask at pH 7, Unger et al. 1978)
1.81 (QSAR, Scherrer & Howard 1979)
1.78 (Schmidt-Bleek et al. 1982)
1.75 (calculated-f const., Mabey et al. 1982)
1.47 (computed, Klopman et al. 1985; quoted, Pangrekar et al. 1994)
1.79 (quoted, Yoshida et al. 1986; Howard 1989)
1.89 (21.5 ± 0.5°C, shake flask-UV with buffer at pH 1.5, Schwarzenbach et al. 1988; quoted, Haderlein & Schwarzenbach 1993)
1.79 (quoted, Leahy et al. 1989)
1.76 (shake flask-AS, Kramer & Henze 1990)
1.72, 1.68 (quoted, CPC-RV at pH 7.4, El Tayar et al. 1991)
1.85 (calculated-CLOGP, Müller & Klein 1992)
1.85 (calculated-CLOGP 3.51, Jaworska & Schultz 1993)
11.77 (recommended, Sangster 1993)
1.85, 1.79 (COMPUTOX databank, Kaiser 1993)

Bioconcentration Factor, log BCF:

2.10 (bluegill sunfish, Veith et al. 1980; quoted, Bysshe 1982)
1.23 (microorganisms-water, calculated-K_{OW}, Mabey et al. 1982)
1.15 (calculated-K_{OW}, Hansch & Leo 1985; quoted, Howard 1989)

Sorption Partition Coefficient, log K_{OC}:

1.43 (sediment-water, calculated-K_{OW}, Mabey et al. 1982)
1.81 (soil, calculated-S, quoted, Howard 1989)

Half-Lives in the Environment:

Air: half-life of 14 hours for the gas phase reaction with hydroxyl radicals (GEMS 1986; quoted, Howard 1989); photooxidation half-life of 7-71 hours, based on measured rate data for the vapor phase reaction with hydroxyl radicals in air (Howard et al. 1991).

Surface water: 168-672 hours, based on aerobic river die-away test data (Ludzack et al. 1958; selected value, Howard et al. 1991); volatilization half-life of 12 days from a model river was estimated using Henry's law constant for a model river of 1 m deep with 1 m/sec current and a 3 m/sec wind and a half-life of 1-8 days in freshwater (Howard 1989); calculated half-life of 390 hours for a body of water with 1 m in depth (Schmidt-Bleek et al. 1982; quoted, Howard 1989); photooxidation half-life of 480 minutes in the presence of hydrogen peroxide, irradiated with a mercury-xenon lamp (Lipczynska-Kochany 1992).

Groundwater: 336-672 hours, based on estimated aqueous aerobic and anaerobic biodegradation half-life (Howard et al. 1991).

Sediment:

Soil: half-life of 10 days in flooded soil (Howard 1989); 168-672 hours, based on estimated aqueous aerobic biodegradation half-life and anaerobic soil grab sample data (Sudhakar-Barik & Sethunnathan 1978; selected, Howard et al. 1991).

Biota:

Environmental Fate Rate Constants or Half-Lives:

Volatilization: volatilization half-life of 296 hours from a model river was estimated using Henry's law constant for a model river of 1 m deep with 1 m/sec current and a 3 m/sec wind (Howard 1989)

Photolysis:

Oxidation: rate constant of $< 2x10^5$ $M^{-1}\cdot h^{-1}$ for singlet oxygen, and $2x10^6$ $M^{-1}\cdot h^{-1}$ for peroxy radical at 25°C (Mabey et al. 1982); rate constant of $0.922x10^{12}$ $cm^3\cdot molecule^{-1}\cdot sec^{-1}$ for the vapor-phase reaction with OH radicals at 296 K in the atmosphere (Becker et al. 1984; quoted, Carlier et al. 1986); observed second-order rate constant of $0.9x10^{-12}$ $cm^3\cdot molecule^{-1}\cdot sec^{-1}$ compared with calculated value of $3.4x10^{-12}$ $cm^3\cdot molecule^{-1}\cdot sec^{-1}$ for the reaction with hydroxyl radicals at room temperature in air (Atkinson 1985); observed rate constant of $0.9x10^{-12}$ $cm^3\cdot molecule^{-1}\cdot sec^{-1}$ compared with calculated value of $4.4x10^{-12}$ $cm^3\cdot$

molecule^{-1}·sec^{-1} for the reaction with hydroxyl radicals at room temperature in air (Atkinson 1987); photooxidation half-life of 7-71 hours, based on measured rate data for the vapor phase reaction with hydroxyl radicals in air (Howard et al. 1991); rate constant of $<0.02 \times 10^{-12}$ cm^3·molecule^{-1}·sec^{-1} for reaction with NO_2 at 296 K (Atkinson et al. 1992); aqueous photooxidation half-life of 480 minutes in the presence of hydrogen peroxide, irradiated with a mercury-xenon lamp (Lipczynska-Kochany 1992); phototransformation decay rate constant of 0.27 min^{-1} on 0.2 gram per liter of TiO_2 solution (Minero et al. 1993).

Hydrolysis: resistant to hydrolysis (Howard 1989).

Biodegradation: 95% degradation in 3-6 days in mixed bacteria cultures (Tabak et al. 1964); average rate of biodegradation 14.0 mg COD g^{-1}·h^{-1} based on measurements of COD decrease using activated sludge inoculum with 20 days of adaptation to the substrate (Pitter 1976; quoted, Scow 1982); time necessary for complete degradation of 16 mg/L in 3-5 days by wastewater and 7-14 days by soil (Haller 1978); aqueous aerobic half-life of 168-672 hours, based on unacclimated aerobic screening test data (Sasaki 1978; Gerike & Fischer 1979; selected, Howard et al. 1991); aqueous anaerobic half-life of 168-672 hours, based on anaerobic soil grab sample data (Sudhakar-Barik & Sethunnathan 1978; selected, Howard et al. 1991).

Biotransformation: estimated bacterial transformation rate constant of 2×10^{-9} mL·cell^{-1}·h^{-1} in water (Mabey et al. 1982).

Bioconcentration, Uptake (k_1) and Elimination (k_2) Rate Constants:

Common Name: 3-Nitrophenol
Synonym: *m*-nitrophenol
Chemical Name: 3-nitrophenol
CAS Registry No: 554-84-7
Molecular Formula: $C_6H_4(NO_2)OH$
Molecular Weight: 139.11
Melting Point (°C):
96.0 (Morrison & Boyd 1959; Verschueren 1983)
97.0 (Tsonopoulos & Prausnitz 1971; Howard 1989)
Boiling Point (°C):
194 (at 70 mm Hg, Verschueren 1983; Howard 1989)
Density (g/cm^3 at 20°C):
1.2797 (10°C, Verschueren 1983)
Molar Volume (cm^3/mol):
0.676 (intrinsic V_I, Hickey & Passino-Reader 1991)
131.9 (calculated-LeBas method)
Molecular Volume (Å^3):
Total Surface Area, TSA (Å^2):
Dissociation Constant, pK_a:
8.0 (Fieser & Fieser 1958)
8.36 (Serjeant & Dempsey 1979; Dean 1985; Howard 1989; Haderlein & Schwarzenbach 1993)
Heat of Fusion ΔH_{fus}, kcal/mol: 5.10 (Tsonopoulos & Prausnitz 1971)
Entropy of Fusion ΔS_{fus}, cal/mol·K (e.u.): 13.8 (Tsonopoulos & Prausnitz 1971)
Fugacity Ratio at 25°C (assuming ΔS_{fus} = 13.5 e.u.), F: 0.194

Water Solubility (g/m^3 or mg/L at 25°C):
30200 (interpolated from Sidgwick et al. 1915, Palit 1947)
14000 (quoted, Morrison & Boyd 1959)
13500 (Matsuguma 1963; Verschueren 1983; Howard 1989; Müller & Klein 1992)
13600 (selected, Tsonopoulos & Prausnitz 1971)
13000 (shake flask-AS, Roberts et al. 1977)
11546, 9460 (20°C, shake flask-UV, calculated, Hashimoto et al. 1984)

Vapor Pressure (Pa at 25°C):
99.98 (Weber et al. 1981; quoted, Howard 1989)

Henry's Law Constant (Pa·m^3/mol):
2.203×10^{-4} (quoted, Gaffney 1987)
1.033 (calculated-P/C, Howard 1989)

Octanol/Water Partition Coefficient, log K_{OW}:
2.00 (shake flask-AS, Fujita et al. 1964; quoted, Rekker 1977; Nagel & Urich 1980)
2.00 (Leo et al. 1971; quoted, Roberts et al. 1977; Beltrame et al. 1984; Haderlein & Schwarzenbach 1993)
2.00 (shake flask-AS at pH 7.45, Umeyama et al. 1971)

1.27, 1.18 (calculated-f const., calculated-π const., Rekker 1977)
2.00 (quoted, Yoshida et al. 1983; Howard 1989)
1.47 (computed, Klopman et al. 1985; quoted, Pangrekar et al. 1994)
2.05 (Beltrame et al. 1988, 1989)
1.52, 1.74 (quoted, CPC-RV at pH 7.4, Le Tayar et al. 1991)
1.85 (calculated-CLOGP, Müller & Klein 1992)
1.92, 1.51 (COMPUTOX databank, Kaiser 1993)

Bioconcentration Factor, log BCF:
1.40 (goldfish, ratio of rate constant k_1/k_2, Nagel & Urich 1980)
1.28 (estimated-K_{OW}, quoted, Howard 1989)

Sorption Partition Coefficient, log K_{OC}:
1.72 (Brookton clay loam, Boyd 1982; quoted, Howard 1989)
1.36 (soil, estimated-S, Howard 1989)

Half-Lives in the Environment:
Air: 14 hours for reaction with hydroxyl radicals in vapor phase (Howard 1989).
Surface water: half-life of 4.2 days in a model river (Howard 1989).
Groundwater:
Sediment:
Soil: 6.3 days in flooded soil under anaerobic conditions (Howard 1989).
Biota: elimination from goldfish within 4 hours (Nagel & Urich 1980).

Environmental Fate Rate Constants or Half-Lives:
Volatilization: volatilization half-life of 4.2 days from a model river was estimated using Henry's law constant for a model river of 1 m deep with 1 m/sec current and a 3 m/sec wind (Howard 1989).
Photolysis:
Oxidation: estimated photooxidation half-life of 14 hours, based on estimated rate data for the vapor phase reaction with hydroxyl radicals in air (Howard 1989); phototransformation decay rate constant of 0.18 min^{-1} on 0.2 gram per liter of TiO_2 solution (Minero et al. 1993).
Hydrolysis: resistant to hydrolysis (Howard 1989).
Biodegradation: 95% degradation in 3-6 days in mixed bacteria cultures (Tabak et al. 1964); decomposition by a soil microflora in 4 days (Alexander & Lustigman 1966; quoted, Verschueren 1983); time necessary for complete degration of 16 mg/L in 3-5 days by wastewater and 3-5 days by soil (Haller 1978); average rate of biodegradation 17.5 mg COD $g^{-1}\cdot h^{-1}$ based on measurements of COD decrease using activated sludge inoculum with 20 days of adaptation to the substrate (Pitter 1976; quoted, Scow 1982); half-life of 6.3 days or less in flooded soil under anaerobic conditions (Howard 1989).
Biotransformation:
Bioconcentration, Uptake (k_1) and Elimination (k_2) Rate Constants:
k_1: 1.54 h^{-1} (goldfish, Nagel & Urich 1980)
k_2: 0.06 h^{-1} (goldfish, Nagel & Urich 1980)

Common Name: 4-Nitrophenol
Synonym: *p*-nitrophenol, 4-hydroxynitrobenzene
Chemical Name: 4-nitrophenol
CAS Registry No: 100-02-7
Molecular Formula: $C_6H_4(NO_2)OH$
Molecular Weight: 139.11
Melting Point (°C):
113.9 (Tsonopoulos & Prausnitz 1971)
114.9 (Weast 1977; Callahan et al. 1979; Brecken-Folse et al. 1994; Howe et al. 1994)
114.0 (Fieser & Fieser 1958; Morrison & Boyd 1959; Schmidt-Bleek et al. 1982; Verschueren 1983; Stephenson & Malanowski 1987)
112-114 (Schwarzenbach et al. 1988)
113-114 (Howard 1989)
Boiling Point (°C):
279 (decomposes, Weast 1977; Verschueren 1983; Stephenson & Malanowski 1987; Howard 1989; Brecken-Folse et al. 1994; Howe et al. 1994)
311 (Schmidt-Bleek et al. 1982)
Density (g/cm^3 at 20°C):
1.479 (Schmidt-Bleek et al. 1982; Verschueren 1983)
Molar Volume (cm^3/mol):
67.7 (intrinsic V_I, Hickey & Passino-Reader 1991)
131.9 (calculated-LeBas method)
Molecular Volume ($Å^3$):
Total Surface Area, TSA ($Å^2$):
Dissociation Constant, pK_a:
7.08 (21.5°C, AS, Schwarzenbach et al. 1988)
7.17 (Fieser & Fieser 1958)
7.156 (Serjeant & Dempsey 1979; Howard 1989; Haderlein & Schwarzenbach 1993)
7.15 (Dean 1985; Miyake et al. 1987; Brecken-Folse et al. 1994; Howe et al. 1994)
Heat of Fusion ΔH_{fus}, kcal/mol:
3.80 (Tsonopoulos & Prausnitz 1971)
Entropy of Fusion ΔS_{fus}, cal/mol·K (e.u.):
9.82 (Tsonopoulos & Prausnitz 1971)
15.0, 13.5 (observed, estimated, Yalkowsky & Valvani 1980)
Fugacity Ratio at 25°C (assuming ΔS_{fus} = 13.5 e.u.), F: 0.135

Water Solubility (g/m^3 or mg/L at 25°C):
25000 (Duff 1929; quoted, Howard 1989)
32800 (interpolated from Sidgwick et al. 1915, Palit 1947)
17000 (quoted, Morrison & Boyd 1959)
15766 (selected, Tsonopoulos & Prausnitz 1971)
14000 (shake flask-AS, Roberts et al. 1977)
16000 (Verschueren 1977, 1983; quoted, Callahan et al. 1979; Mabey et al. 1982)
4012, 3576 (quoted, calculated-K_{OW}, Yalkowsky & Morozowich 1980)
12400 (Rott et al. 1982)
12400 (20°C, Schmidt-Bleek et al. 1982)

16000 (selected, Yoshida et al. 1983)
13490, 11140 (20°C, shake flask-UV, calculated, Hashimoto et al. 1984)
15850 (quoted, Isnard & Lambert 1988)
11570 (20 ± 0.5°C, shake flask-UV with buffer at pH 1.5, Schwarzenbach et al. 1988; quoted, Howard 1989; Müller & Klein 1992)
16000 (selected, Brecken-Folse et al. 1994; Howe et al. 1994)

Vapor Pressure (Pa at 25°C):
0.0054 (20°C, Schmidt-Bleek et al. 1982)
293.3 (146°C, quoted, Mabey et al. 1982)
0.0012 (selected, Yoshida et al. 1983)
0.0044 (extrapolated-Antoine eqn., Stephenson & Malanowski 1987)
0.111 (subcooled liq. at 20°C estimated by GC-RT, Schwarzenbach et al. 1988)
0.0131 (20°C, solid P_S, Schwarzenbach et al. 1988)
0.133 (quoted, Howard 1989)

Henry's Law Constant ($Pa \cdot m^3/mol$):
4.21×10^{-5} (quoted exptl., Hine & Mookerjee 1975)
5.55×10^{-4} (calculated-group contribution, Hine & Mookerjee 1975)
9.87×10^{-3} (calculated-bond contribution, Hine & Mookerjee 1975)
2.533 (calculated-P(at 146°C)/C(at 25°C), Mabey et al. 1982)
1.04×10^{-5} (calculated-P/C, Yoshida et al. 1983)
0.00335 (20°C, calculated-P/C, Schwarzenbach et al. 1988; quoted, Howard 1989)

Octanol/Water Partition Coefficient, log K_{OW}:
1.91 (shake flask-AS, Fujita et al. 1964; quoted, Leo et al. 1971; Carlson et al. 1975; Rekker 1977; Callahan et al. 1979; Beltrame et al. 1984; Westall et al. 1985)
1.73 (calculated-π const., Carlson et al. 1975)
1.18 (calculated-π const., Rekker 1977)
1.27 (calculated-f const., Rekker 1977)
1.96 (quoted from Leo et al. 1971, Roberts et al. 1977)
1.68 ± 0.01 (RP-HPLC-k', Unger et al. 1978)
1.91, 1.97 (quoted, calculated-f const., Chou & Jurs 1979)
2.91 (quoted unpublished data, Veith et al. 1979b)
1.91 (Hansch & Leo 1979; quoted, Miyake & Terada 1982; Brooke et al. 1986)
2.07 (calculated-RPHPLC-k', Miyake & Terada 1982)
1.97 (calculated-f const., Mabey et al. 1982)
1.85 (Schmidt-Bleek et al. 1982)
1.91 (quoted, Yoshida et al. 1983)
1.95, 1.27 (quoted, calculated-f const., Harnish et al. 1983)
1.86 (HPLC-ref. substances extrapolated, Harnish et al. 1983)
1.88 (shake flask average, OECD/EEC lab. comparison tests, Harnish et al. 1983)
1.47 (computed, Klopman et al. 1985; quoted, Pangrekar et al. 1994)
2.08 (selected, Yoshida et al. 1986)

1.91, 1.96 ± 0.08 (selected best lit. value, exptl.-ALPM, Garst 1984; Garst & Wilson 1984)
1.85, 1.92 (shake flask, OECD 1981 Guidelines, Geyer et al. 1984)
1.90, 0.70 (quoted, calculated-γ from UNIFAC, Campbell & Luthy 1985)
1.91, 1.90 (quoted, HPLC-RV, Brooke et al. 1986)
1.95 (quoted from Hansch & Leo 1979; Burkhard & Kuehl 1986)
1.97, 2.10 (quoted, calculated-HPLC-k', Miyake et al. 1987)
1.93 (batch equilbration-UV, Beltrame et al. 1988, 1989)
2.91 (quoted, Isnard & Lambert 1988)
2.91 (measured value quoted from Mackay 1982, Schüürmann & Klein 1988)
1.91 (measured value quoted from THOR 1986, Schüürmann & Klein 1988)
1.85 (calculated-CLOGP 3.42 of Leo 1986, Schüürmann & Klein 1988)
2.04 (21.5 ± 1.5°C, shake flask-UV with buffer at pH 1.5, Schwarzenbach et al. 1988; Haderlein & Schwarzenbach 1993)
1.91 (quoted, Howard 1989)
1.38, 1.77 (quoted, CPC-RV at pH 7.4, Tayar et al. 1991)
1.85 (calculated-CLOGP, Müller & Klein 1992)
1.91 (selected, Brecken-Folse et al. 1994)
1.85, 1.91 (COMPUTOX databank, Kaiser 1993)

Bioconcentration Factor, log BCF:
2.10 (fathead minnows, quoted, Veith et al. 1979b)
2.56 (mean whole body ^{14}C in fathead minnows, Call et al. 1980)
1.90, 2.34 (total ^{14}C in fathead minnows: observed, calculated, mean exposure level 0.0041 $\mu g \cdot mL^{-1}$, Call et al. 1980)
2.44, 2.65 (total ^{14}C in fathead minnows: observed, calculated, mean exposure level 0.0441 $\mu g \cdot mL^{-1}$, Call et al. 1980)
1.76 (golden orfe, Freitag et al. 1982; quoted, Howard 1989)
1.04 (green algae, Freitag et al. 1982; quoted, Howard 1989)
1.48 (activated sludge, Freitag et al. 1982)
1.43 (microorganisms-water, calculated-K_{OW}, Mabey et al. 1982)
0.301 (calculated-K_{OW}, Yoshida et al. 1983)
2.10 (Mackay 1982; quoted, Schüürmann & Klein 1988)
1.48 (alga *chlorella fusca*, wet wt. basis, Geyer et al. 1984)
1.45 (alga *chlorella fusca*, calculated-K_{OW}, Geyer et al. 1984)
1.60, 1.48, 1.48 (golden ide, algae, activated sludge, Freitag et al. 1985)
2.10 (quoted, Isnard & Lambert 1988)
1.90 (fathead minnow, quoted, Howard 1989)

Sorption Partition Coefficient, log K_{OC}:
1.65 (sediment-water, calculated-K_{OW}, Mabey et al. 1982)
2.41 (soil, calculated-K_{OW}, Yoshida et al. 1983)
1.74 (Brookston clay loam, Boyd 1982; quoted, Howard 1989)
1.32 (soil, calculated-S, Howard 1989)
2.37 (soil, quoted exptl., Meylan et al. 1992)
2.49 (soil, calculated- χ and fragment contribution, Meylan et al. 1992)

Half-Lives in the Environment:

Air: photooxidation half-life of 14.5-145 hours, based on estimated rate data for the vapor phase reaction with hydroxyl radicals in air (Howard et al. 1991); atmospheric transformation lifetime was estimated to be 1-5 to > 5 days (Kelly et al. 1994).

Surface water: photodegradation half-lives of 5.7 day at pH 5, 6.7 days at pH 7 and 13.7 days at pH 9 in water (Hustert et al. 1981); half-life 1-8 days in freshwater, 1-3 years in marine systems but decreased to 13-20 days with the presence of sediment, the mean half-life of 7.7 days estimated by a nonsteady-state equilibrium model, and photolysis half-lives of 2-14 days in clear surface waters (Howard 1989); 18.2-168 hours, based on estimated unacclimated aqueous aerobic biodegradation half-life (Howard et al. 1991); half-life of 440 minutes for photolysis in the presence of hydrogen peroxide, irradiated with a mercury-xenon lamp (Lipczynska-Kochany 1992).

Groundwater: 36.4-235 hours, based on pond die-away test data and anaerobic die-away data in two different flooded soils (Paris et al. 1983; Sudhakar-Barik & Sethunnathan 1978; quoted, Howard et al. 1991).

Sediment:

Soil: half-life of approximately 1 day in agricultural topsoil and 10 days in flooded soil, >40 days in subsoil under aerobic conditions and longer still under anaerobic conditions (Howard 1989); 17-29 hours, based on aerobic soil die-away test data (Loekke 1985; selected, Howard et al. 1991).

Biota: depuration half-lives of 72 hours (observed), 37 hours (calculated) for mean exposure level of 0.041 $\mu g \cdot mL^{-1}$ and 228 hours (observed), 206 hours (calculated) for mean exposure level of 0.041 $\mu g \cdot mL^{-1}$ (Call et al. 1980).

Environmental Fate Rate Constants or Half-Lives:

Volatilization:

Photolysis: atmospheric photolysis half-life of 3.1-329 hours, based on sunlight photolysis at pH 9 and pH 4 (Hustert et al. 1981; Lemaire et al. 1985; selected, Howard et al. 1991); aquatic atmospheric photolysis half-life of 3.1-329 hours, based on sunlight photolysis at pH 9 and pH 4 (Hustert et al. 1981; Lemaire et al. 1985; selected, Howard et al. 1991); rate constant of 3.96×10^{-3} hour^{-1} for the direct photolysis in water (Yoshida et al. 1983); photolysis half-lives range form hours to a week or more in atmosphere and 2-14 days in clear surface waters (Howard 1989).

Oxidation: oxidation rate constant < 2×10^{5} $M^{-1} \cdot h^{-1}$ for singlet oxygen, 2×10^{6} $M^{-1} \cdot h^{-1}$ for peroxy radical (Mabey et al. 1982); photooxidation half-life of 21 days to 5.6 years in water, based on rate constants for reaction in water with hydroxyl radicals (Dorfman & Adams 1973; Scully & Hoigné 1987; selected, Howard et al. 1991); rate constant of <50 $M^{-1} \cdot sec^{-1}$ for 0.01-14 mM to react with ozone in water using PrOH as scavenger at pH 1.7 and 20-23°C (Hoigné & Bader 1983); rate constant of 8.3×10^{-12} $cm^3 \cdot molecule^{-1} \cdot sec^{-1}$ for the gas-phase reaction with OH radicals at 296 K in the atmosphere (Becker et al. 1984; quoted, Carlier et al. 1986); photooxidation half-life of 14.5-145 hours in air, based on estimated rate data for the vapor phase reaction with hydroxyl radicals in air

(Howard et al. 1991); aqueous photooxidation half-life of 440 minutes in the presence of hydrogen peroxide, irradiated with a mercury-xenon lamp (Lipczynska-Kochany 1992); phototransformation decay rate constant of 0.15 min^{-1} on 0.20 g/L of TiO_2 solution (Minero et al. 1993).

Hydrolysis:

Biodegradation: 95% degradation in 3-6 days in mixture bacteria cultures (Tabak et al. 1964); decomposition by a soil microflora in 16 days (Alexander & Lustigman 1966; quoted, Verschueren 1983); average rate of biodegradation 17.5 mg COD $g^{-1}\cdot h^{-1}$ based on measurements of COD decrease using activated sludge inoculum with 20 days of adaptation to the substrate (Pitter 1976; quoted, Scow 1982); time necessary for complete degradation of 16 mg per liter in 3-5 days by wastewater and 7-14 days by soil (Haller 1978); aqueous aerobic half-life of 18.2-168 hours, based on pond die-away test data (Paris et al. 1983; Bourquin 1984; selected, Howard et al. 1991); aqueous anaerobic half-life of 163-235 hours, based on anaerobic die-away data in two different flooded soils (Sudhakar-Barik & Sethunnathan 1978; selected, Howard et al. 1991); was not biodegraded with activated sludge at a concn. of 94 mg/L but completely degraded with a decreased concn. of 19 mg/L in 28 days expt. (Kool 1984).

Biotransformation: estimated bacterial transformation rate constant of 1.00×10^{-7} $mL\cdot cell^{-1}\cdot h^{-1}$ in water (Mabey et al. 1982); rate constant of $(3.80 \pm 1.40)\times10^{-11}$ $L\cdot organism^{-1}\cdot h^{-1}$ (Paris et al. 1983; quoted, Steen 1991).

Bioconcentration, Uptake (k_1) and Elimination (k_2) Rate Constants:

Common Name: 2,4-Dinitrophenol
Synonym: 2,4-hydroxynitrobenzene, Aldifen, 2,4-DNP
Chemical Name: 2,4-dinitrophenol
CAS Registry No: 51-28-5
Molecular Formula: $C_6H_3(NO_2)_2OH$
Molecular Weight: 184.11
Melting Point (°C):

113 (Fieser & Fieser 1958; Morrison & Boyd 1959)
114 (Verschueren 1977; Callahan et al. 1979; Mabey et al. 1982; Brecken-Folse et al. 1994; Howe et al. 1994)
112-114 (Stephenson & Malanowski 1987)
111-113 (Schwarzenbach et al. 1988)

Boiling Point (°C):
Density (g/cm^3 at 20°C):
Molar Volume (cm^3/mol): 156.8 (LeBas method)
Molecular Volume ($Å^3$):
Total Surface Area, TSA ($Å^2$):
Dissociation Constant, pK_a:

4.0 (Fieser & Fieser 1958)
4.09 (Pearce & Simkins 1968; quoted, Callahan et al. 1979; Brecken-Folse et al. 1994; Howe et al. 1994)
3.94 (Schwarzenbach et al. 1988; Haderlein & Schwarzenbach 1993)
3.90 (Kollig 1993)

Heat of Fusion ΔH_{fus}, kcal/mol:
Entropy of Fusion ΔS_{fus}, cal/mol·K (e.u.):
Fugacity Ratio at 25°C (assuming ΔS_{fus} = 13.5 e.u.), F: 0.132

Water Solubility (g/m^3 or mg/L at 25°C):

6000 (quoted, Morrison & Boyd 1959)
5600 (18°C, Verschueren 1977; quoted, Callahan et al. 1979; Mabey et al. 1982)
335 (20 ± 0.5°C, shake flask-UV at pH 1.5, Schwarzenbach et al. 1988; quoted, Müller & Klein 1992)
5000 (selected, Brecken-Folse et al. 1994; Howe et al. 1994)
6000 (quoted, Howard 1989)

Vapor Pressure (Pa at 25°C):

1.987×10^{-3} (18°C, Mabey et al. 1982)
5.52×10^{-3} (calculated-Antoine eqn., Stephenson & Malanowski 1987)

Henry's Law Constant (Pa·m^3/mol):

6.54×10^{-5} (18°C, calculated-P/C, Mabey et al. 1982)
0.00335 (20°C, calculated-P/C, Swarzenbach et al. 1988)
8.88×10^{-5} (calculated-P/C, Howard 1989)

Octanol/Water Partition Coefficient, log K_{OW}:

1.53 (Leo et al. 1971; quoted, Callahan et al. 1979; Westall et al. 1985; Ryan et al. 1988)
1.50 (shake flask-AS, Stockdale & Selwyn 1971)
1.56 (shake flask, Korenman et al.1977)
1.52 (QSAR, Scherrer & Howard 1979)
1.54 (quoted, Mabey et al. 1982)
1.54 (selected best lit. value, Garst 1984)
1.79 ± 0.06 (exptl.-ALPM, Garst 1984)
1.18 (computed, Klopman et al. 1985; quoted, Pangrekar et al. 1994)
1.67 (Schwarzenbach et al. 1988; quoted, Haderlein & Schwarzenbach 1993)
1.52, 1.54, 1.92, 1.54 (quoted: Hansch & Leo 1979, CLOGP, calculated-CLOGP, calculated-MO, Bodor et al. 1989)
1.54 (quoted, Howard 1989)
1.91 (calculated-CLOGP, Müller & Klein 1992)
1.54 (recommended, Sangster 1993)
1.53 (selected, Brecken-Folse et al. 1994)
1.92 (calculated-QSAR, Kollig 1993)

Bioconcentration Factor, log BCF:

1.19 (microorganism-water, calculated-K_{OW}, Mabey et al. 1982)
<1.0 (calculated, Howard 1989)

Sorption Partition Coefficient, log K_{OC}:

1.22 (sediment-water, calculated-K_{OW}, Mabey et al. 1982)
1.56, 2.14 (calculated-S, K_{OW}, Howard 1989)
-0.09 (calculated-K_{OW}, Kollig 1993)
3.09 (activated carbon, Blum et al. 1994)

Half-Lives in the Environment:

Air: calculated vapor-phase half-life of 14 hours for reaction with photochemically generated OH radicals (Howard 1989); photooxidation half-life of 111-1114 hours, based on estimated rate data for the vapor phase reaction with hydroxyl radicals in air (Howard et al. 1991); atmospheric transformation lifetime was estimated to be > 5 days (Kelly et al. 19914).

Surface water: calculated half-life of 58 days for reaction alkylperoxy radicals in water (Howard 1989); 77-3840 hours, based on reported reaction rate constants for RO_2 radicals with the phenol class (Mill & Mabey 1985; selected, Howard et al. 1991); photooxidation half-life of 200 minutes in the presence of hydrogen peroxide, irradiated with a mercury-xenon lamp (Lipczynska-Kochany 1992).

Groundwater: 68-14624 hours, based on estimated aqueous aerobic biodegradation half-life and estimated aqueous anaerobic biodegradation half-life (quoted, Howard et al. 1991).

Sediment:

Soil: 1622-6312 hours, based on data from aerobic soil column studies (Kincannon & Lin 1985) and aerobic soil die-away test data (Sudhakar-Barik & Sethunnathan 1978; selected, Howard et al. 1991); a Class C compounds with a half-life of > 50 days in soil (Ryan et al. 1988).

Biota:

Environmental Fate Rate Constants or Half-Lives:

Volatilization:

Photolysis:

Oxidation: calculated reaction rate constant of 5×10^5 mol·L^{-1}·h^{-1} and ambient alkylperoxy radical concentration of 1×10^{-1} mol/L (Mabey et al. 1982); aqueous oxidation rate constants of 3×10^4 M^{-1}·h^{-1} for singlet oxygen, and 5×10^5 M^{-1}·h^{-1} for peroxy radical (Mabey et al. 1982); photooxidation half-life of 77-3840 hours in water, based on reported reaction rate constants for RO_2 with the phenol class (Mill & Mabey 1985; selected, Howard et al. 1991); photooxidation half-life of 111-1114 hours in air, based on the estimated reaction rate constant of 1.7×10^{-11} cm^3·molecule^{-1}·sec^{-1} with an ambient hydroxyl radical concentration of 8×10^5 molecules·cm^{-3} for the vapor phase reaction with hydroxyl radicals in air (Howard et al. 1991); aqueous photooxidation half-life of 200 minutes in the presence of hydrogen peroxide irradiated with a mercury-xenon lamp (Lipczynska-Kochany 1992).

Hydrolysis:

Biodegradation: 95% degradation in 7-10 days in mixed bacteria cultures (Tabak et al. 1964); average rate of biodegradation 6.0 mg COD g^{-1}·h^{-1} based on measurements of COD decrease using activated sludge inoculum with 20 days of adaptation to the substrate (Pitter 1976; quoted, Scow 1982); first-order rate constant of 0.2 day^{-1} corresponding to a half-life of 3.6 days in adapted activated sludge under aerobic conditions (Mills et al. 1982); aqueous aerobic half-life of 1622-6312 hours, based on data from aerobic soil column studies (Kincannon & Lin 1985; selected, Howard et al. 1991) and aerobic soil die-away test data (Sudhakar-Barik & Sethunnathan 1978; selected, Howard et al. 1991); aqueous anaerobic half-life of 68-170 hours, based on anaerobic flooded soil die-away tests (Sudhakar-Barik & Sethunnathan 1978; selected, Howard et al. 1991).

Biotransformation: bacterial transformation rate constant of 3×10^{-9} mL·cell^{-1}·h^{-1} in water (Mabey et al. 1982).

Bioconcentration, Uptake (k_1) and Elimination (k_2) Rate Constants:

Common Name: 2,4,6-Trinitrophenol
Synonym: picric acid
Chemical Name: 2,4,6-trinitrophenol
CAS Registry No: 88-89-1
Molecular Formula: $C_6H_2(NO_2)_3OH$
Molecular Weight: 229.11
Melting Point (°C):
122-123 (Weast 1982-83; Dean 1985)
121.8 (Verschueren 1983)
122.5 (Stephenson & Malanowski 1987)
Boiling Point (°C):
>300 (explodes, Weast 1982-83; Verschueren 1983; Dean 1985; Stephenson & Malanowski 1987)
Density (g/cm^3 at 20°C):
1.763 (Weast 1982-83; Verschueren 1983; Dean 1985)
Dissociation Constants, pK_a:
0.78 (Schwarzenbach et al. 1988)
Molar Volume (cm^3/mol):
188.9 (calculated-LeBas method)
Molecular Volume, TMV ($Å^3$):
Total Surface Area, TSA ($Å^2$):
Heat of Fusion, ΔH_{fus}, kcal/mol:
4.66 (Tsonopoulos & Prausnitz 1971)
Entropy of Fusion, ΔS_{fus}, cal/mol·K (e.u.):
11.8 (Tsonopoulos & Prausnitz 1971)
Fugacity Ratio at 25°C (assuming ΔS_{fus} = 13.5 e.u.), F: 0.110

Water Solubility (g/m^3 or mg/L at 25°C):
14000 (quoted, Morrison & Boyd 1959)
13750 (selected, Tsonopoulos & Prausnitz 1971)
11482, 4266 (quoted, calculated-K_{OW}, Yalkowsky & Morozowich 1980)
14000 (20°C, quoted, Verschueren 1983)
13000 (Dean 1985)

Vapor Pressure (Pa at 25°C):
4.68×10^{-5} (extrapolated-Antoine eqn., Stephenson & Malanowski 1987)

Henry's Law Constant ($Pa \cdot m^3/mol$):
7.66×10^{-7} (calculated-P/C, this work)

Octanol/Water Partition Coefficient, log K_{OW}:
2.03 (quoted, Verschueren 1983)
1.44 (shake flask, Korenman et al. 1977)

1.34 (QSAR, Scherrer & Howard 1979)
2.03, 1.46 (quoted lit. best value, HPLC-RV, Garst 1984)
0.00 (selected, Yoshida et al. 1986)
-0.97, 0.87 (quoted from CLOGP database: pH 2.7, pH 1, shake flask, Sangster 1993)
2.03, 1.33 (quoted CLOGP, recommended, Sangster 1993)
1.82 (COMPUTOX databank, Kaiser 1993)

Bioconcentration Factor, log BCF:

Sorption Partition Coefficient, log K_{OC}:

Half-Lives in the Environmental Compartments:

Air: 672-4320 hours, based on estimated photooxidation half-lives in air (Howard et al. 1991).

Surface water: 672-4320 hours, based on estimated aqueous aerobic biodegradation half-lives (Howard et al. 1991).

Ground water: 48-8640 hours, based on estimated both aqueous aerobic and aqueous anaerobic biodegradation half-lives (Howard et al. 1991).

Sediment:

Soil: 672-4320 hours, based on estimated photooxidation aqueous aerobic biodegradation half-lives (Howard et al. 1991).

Biota:

Environmental Fate Rate Constants and Half-Lives:

Volatilization:

Photolysis:

Oxidation: photooxidation half-life of 677-4320 hours, based on estimated reaction rates with OH and NO_3 radicals in air (Howard et al. 1991).

Hydrolysis: no hydrolyzable group (Howard et al. 1991).

Biodegradation: 95% degradation in 3-6 days in mixed bacteria cultures (Tabak et al. 1964); aqueous aerobic half-life of 672-8640 hours, based on aerobic biodegradation screening test (Howard et al. 1991); aqueous anaerobic half-life of 48-300 hours, based on aqueous anaerobic natural water die-away test data (Howard et al. 1991).

Biotransformation:

Bioconcentration Uptake (k_1) and Elimination (k_2) Rate Constants:

Common Name: 4,6-Dinitro-*o*-Cresol
Synonym: 2,4-dinitro-6-methylphenol, DNOC, 2-methyl-4,6-dinitrophenol, 6-mehtyl-2,4-dinitrophenol
Chemical Name: 4,6-dinitro-*o*-cresol, 2,4-dinitro-6-methylphenol
CAS Registry No: 534-52-1
Molecular Formula: $CH_3C_6H_2(NO_2)_2OH$
Molecular Weight: 198.13
Melting Point (°C):
85.8 (Verschueren 1977, 1983; Callahan et al. 1979)
87.5 (Dean 1985; Howard 1991)
86.5 (Stephenson & Malanowski 1987)
84-86 (Schwarzenbach et al. 1988)
Boiling Point (°C):
312 (Howard 1991)
Density (g/cm^3 at 20°C):
Dissociation Constants, pKa:
4.35 (Pearce & Simkins 1968; Callahan et al. 1979; Westall et al. 1985)
4.31 (AS, Schwarzenbach et al. 1988; Howard 1991)
4.46 (Jafvert 1990; quoted, Bintein & Devillers 1994)
Molar Volume (cm^3/mol):
182.6 (calculated-LeBas method)
Molecular Volume (A^3):
Total Surface Area, TSA (A^2):
Heat of Fusion, ΔH_{fus}, kcal/mol:
Entropy of Fusion, ΔS_{fus}, cal/mol·K (e.u.):
Fugacity Ratio at 25°C (assuming ΔS_{fus} = 13.5 e.u.), F: 0.250

Water Solubility (g/m^3 or mg/L at 25°C):
258, 250 (quoted, Günther et al. 1968)
290 (calculated-K_{OW}, Mabey et al. 1982)
198 (20°C, neutral species at pH 1.5 of buffer solution HCl/NaH_2PO_4, shake flask-UV, Schwarzenbach et al. 1988)
150 (20°C, quoted, Howard 1991)

Vapor Pressure (Pa at 25°C):
6.67 (20°C, estimated, Mabey et al. 1982)
0.243 (20°C, GC-RT, subcooled liquid value P_L, Schwarzenbach et al. 1988)
0.0432 (20°C, solid P_S, Schwarzenbach et al. 1988)
0.01107 (20°C, quoted, Howard 1991)

Henry's Law Constant (Pa·m^3/mol):
4.05 (calculated-P/C, Mabey et al. 1982)
43.22 (20°C, calculated-P/C, Schwarzenbach et al. 1988; quoted Howard 1991)

Octanol/Water Partition Coefficient, log K_{OW}:

2.70 (calculated-f const., Mabey et al. 1982)
2.85 (calculated as per Tute 1971; quoted, Callahan et al. 1979; Westall et al. 1985; Bintein & Devillers 1994)
2.12 (21.5 °C, neutral species, shake flask-UV, Schwarzenbach et al. 1988)
- 0.81 (21.5 °C, ionic species at pH 12, Schwarzenbach et al. 1988)
2.12 (quoted, Howard 1991)
2.12 (quoted, Sangster 1993)
2.56, 1.98 (COMPUTOX databank, Kaiser 1993)

Bioconcentration Factor, log BCF:

2.10 (microorganism-water, calculated-K_{OW}, Mabey et al. 1982)
1.38, 1.57 (calculated-S, K_{OW} regression eqns., Howard 1991)

Sorption Partition Coefficient, log K_{OC}:

2.38 (sediment-water, calculated-K_{OW}, Mabey et al. 1982)
3.57 $\pm$ 0.130 (natural sediment, Jafvert 1990)
2.44, 2.0-2.53 (calculated-S, K_{OW} regression eqns., Howard 1991)
2.41, 2.78 (soil, quoted, calc-χ and fragment contribution, Meylan et al. 1992)
3.28 (activated carbon, Blum et al. 1994)

Half-Lives in the Environment:

Air: photooxidation half-life of 310-3098 hours in air, based on estimated rate constant for the vapor phase reaction with hydroxyl radicals in air (Atkinson 1987; selected, Howard et al. 1991); half-life of 77 days for the vapor-phase reaction with OH radicals (concn. of 5×10^5 molecule·cm^{-3}) (Howard 1991); atmospheric transformation lifetime was estimated to be > 5 days (Kelly et al. 1994).

Surface water: photooxidation half-life of 77-3840 hours in water, based on reported reaction rate constants for RO_2 radicals with the phenol class (Mill & Mabey 1985; selected, Howard et al. 1991); half-life was estimated to be 58 days for photooxidation via peroxy radicals and 2600 years for reaction with singlet oxygen in water (Howard 1991).

Ground water: 68-1008 hours, based on estimated aqueous aerobic and anaerobic biodegradation half-lives (Howard et al. 1991).

Sediment:

Soil: 168-504 hours, based on data from a soil die-away study (Kincannon & Lin 1985; selected, Howard et al. 1991).

Biota:

Environmental Fate Rate Constants and Half-Lives:

Volatilization: nonvolatile from water, based on low HLC value (Howard 1991).

Photolysis:

Oxidation: aqueous oxidation rate constant of $3x10^4$ M^{-1} h^{-1} for singlet oxygen and $5x10^5$ M^{-1} h^{-1} for peroxy radical at 25°C (Mabey et al. 1982); photooxidation half-life of 77-3840 hours in water, based on reported reaction rate constants for RO_2 radicals with the phenol class (Mill & Mabey 1985; selected, Howard et al. 1991); photooxidation half-life of 310-3098 hours in air, based on estimated rate constant for the vapor phase reaction with hydroxyl radicals in air (Atkinson 1987; selected, Howard et al. 1991).

Hydrolysis:

Biodegradation: aqueous aerobic half-life of 168-504 hours, based on data from a soil die-away study (Kincannon & Lin 1985; selected, Howard et al. 1991); aqueous anaerobic half-life of 68-170 hours, based on flooded soil die-away tests for 2,4-dinitrophenol (Sudhakar-Barik & Sethunnathan 1978; selected, Howard et al. 1991).

Biotransformation: estimated bacterial transformation rate constant of $3x10^9$ $mL\cdot cell^{-1}\cdot h^{-1}$ in water (Mabey et al. 1982).

Bioconcentration Uptake (k_1) and Elimination (k_2) Rate Constants:

Common Name: Catechol
Synonym: 1,2-dihydroxybenzene, 1,2-benzenediol, pyrocatechol
Chemical Name: 1,2-dihydroxybenzene
CAS Registry No: 120-80-9
Molecular Formula: $C_6H_4(OH)_2$
Molecular Weight: 110.11
Melting Point (°C): 104
Boiling Point (°C): 245.5
Density (g/cm³):
Dissociation Constants, pK_a:
9.4 (Fieser & Fieser 1958)
9.50 (McLeese et al. 1979)
Molar Volume (cm³/mol):
110.8 (calculated-LeBas method)
Molecular Volume, TMV (Å³):
Total Surface Area, TSA (Å²):
237.0 (Iwase et al. 1985)
Heat of Fusion, ΔH_{fus}, kcal/mol:
5.44 (Tsonopoulos & Prausnitz 1971)
Entropy of Fusion, ΔS_{fus}, cal/mol·K (e.u.):
14.4 (Tsonopoulos & Prausnitz 1971)
14.39 (Yalkowsky & Valvani 1980)
Fugacity Ratio at 25°C (assuming ΔS_{fus} = 13.5 e.u.), F: 0.165

Water Solubility (g/m³ or mg/L at 25°C):
71166 (30°C, synthetic method, Walker et al. 1931)
45000 (quoted, Fieser & Fieser 1958; Morrison & Boyd 1959)
636190 (quoted, Tsonopoulos & Prausnitz 1971)
26414, 35630 (quoted, calculated-K_{OW}, Yalkowsky & Morozowich 1980)
42724, 38610 (calculated-ΔS_{fus} and M.P., estimated, Yalkowsky & Valvani 1980)
43000 (quoted, Dean 1985)
26400, 35600 (quoted, calculated-K_{OW}, Yalkowsky & Morozowich 1980)
446680 (quoted, Isnard & Lambert 1989)

Vapor Pressure (Pa at 25°C):
1.34 (extrapolated-Antoine eqn., Boublik et al. 1973)
5.44 (extrapolated-Antoine eqn., Boublik et al. 1984)
1.03 (extrapolated-Antoine eqn., Dean 1985)
1.06 (P_L, extrapolated-Antoine eqn., Stephenson & Malanowski 1987)

Henry's Law Constant (Pa·m³/mol):
0.022 (calculated-P/C, this work)

Octanol/Water Partition Coefficient, log K_{OW}:

0.86 (shake flask-AS, Fujuta et al. 1964)
0.95 (Leo et al. 1971; quoted, Veith et al. 1979)
0.85 (shake flask-UV, pH 5.6, Umeyama et al. 1971)
0.84 (shake flask, Korenman 1972)
0.88 (calculated-π const., Norrington et al. 1975)
0.86 (shake flask at pH 7, Unger et al. 1978)
0.88, 1.01, 0.85, 0.84 (quoted, Hansch & Leo 1979)
0.95 (Veith et al. 1979; quoted, Burkhard & Kuehl 1986)
0.88 (quoted, McLeese et al. 1979)
0.53 (HPLC-k', Haky & Young 1984)
0.88, 0.82 (quoted, calculated-TSA, Iwase et al. 1985)
1.28 (computed, Klopman et al. 1985; quoted, Pangrekar et al. 1994)
1.10 (HPLC-RT, Webster et al. 1985)
0.88 (EPA CLOGP Data Base, Hulzebos et al. 1993)
0.88 (COMPUTOX databank, Kaiser 1993)
0.88 (recommended, LOGKOW databank, Sangster 1993)

Bioconcentration Factor, log BCF:

Sorption Partition Coefficient, log K_{OC}:

Half-Lives in the Environmental Compartments:

Air: 24-168 hours, based on estimated photooxidation half-lives in air (Atkinson 1987; selected, Howard et al. 1991); atmospheric transformation lifetime was estimated to be 1 day (Kelly et al. 1994).

Surface water: 24-168 hours, based on estimated aqueous aerobic biodegradation half-lives (Howard et al. 1991).

Ground water: 48-336 hours, based on estimated aqueous aerobic biodegradation half-lives (Howard et al. 1991).

Sediment:

Soil: 24-168 hours, based on estimated aqueous aerobic biodegradation half-lives (Howard et al. 1991).

Biota:

Environmental Fate Rate Constants and Half-Lives:

Volatilization:

Photolysis:

Oxidation: $<2 \times 10^2$ M^{-1} sec^{-1} for the reaction with singlet oxygen at 25°C in aquatic systems with half-life >100 years (Foote 1976; Mill 1979; quoted, Mill 1982); photooxidation half-life of 2.6-26 hours in air, based on vapor-phase reaction rate constant with OH radicals in atmosphere (Howard et al. 1991); aqueous photooxidation half-life of 77-3840 hours, based on reaction rate constant for $RO_2\cdot$ in aqueous solution (Mill & Mabey 1985; selected, Howard et al. 1991).

Hydrolysis:

Biodegradation: 95% degradation in 1-2 days in mixed bacteria cultures (Tabak et al. 1964); average rate of biodegradation 55.5 mg COD $g^{-1} \cdot h^{-1}$ based on measurements of COD decrease using activated sludge inoculum with 20 days of adaptation to the substrate (Pitter 1976; quoted, Scow 1982); 2mM catechol rapidly degraded by strain Cat2 in batch culture in 9 days (Schnell et al. 1989); aqueous aerobic half-life of 24-168 hours, based on aerobic biological screening test data (Heukelekian & Rand 1955; Okey & Bogan 1965; Pitter 1976; Urushigawa et al. 1983; Gerike & Fischer 1979; selected, Howard et al. 1991); aqueous anaerobic half-life of 96-672 hours, based on estimated aqueous aerobic biodegradation half-lives (Howard et al. 1991).

Biotransformation:

Bioconcentration Uptake (k_1) and Elimination (k_2) Rate Constants:

Common Name: Resorcinol
Synonym: 1,3-benzenediol, *m*-dihydroxybenzene, *m*-hydroxyphenol, resorcin
Chemical Name: 1,3-dihydroxybenzene
CAS Registry No: 108-46-3
Molecular Formula: $C_6H_4(OH)_2$
Molecular Weight: 110.11
Melting Point (°C):
110.7 (Stull 1947)
110 (Fieser & Fieser 1958; Morrison & Boyd 1959; Stephenson & Malanowski 1987)
Boiling Point (°C):
276.5 (Stull 1947)
281 (Fieser & Fieser 1958; Morrison & Boyd 1959)
276-280 (Verschueren 1983)
277 (Stephenson & Malanowski 1987)
Density (g/cm^3):
Dissociation Constants, pK_a:
9.40 (Fieser & Fieser 1958; McLeese et al. 1979)
Molar Volume (cm^3/mol):
110.8 (calculated-LeBas method)
Molecular Volume, TMV (Å^3):
Total Surface Area, TSA (Å^2):
239.5 (Iwase et al. 1985)
Heat of Fusion, ΔH_{fus}, kcal/mol:
5.09 (Tsonopoulos & Prausnitz 1971)
Entropy of Fusion, ΔS_{fus}, cal/mol·K (e.u.):
13.3 (Tsonopoulos & Prausnitz 1971)
Fugacity Ratio at 25°C (assuming ΔS_{fus} = 13.5 e.u.), F: 0.144

Water Solubility (g/m^3 or mg/L at 25°C):
2188870 (33.6°C, synthetic method, Walker et al. 1931)
1230000 (quoted, Fieser & Fieser 1958; Morrison & Boyd 1959)
1309100 (quoted, Tsonopoulos & Prausnitz 1971)
840000, 2290000 (0, 30°C, Verschueren 1983)
110000 (Dean 1985)
831760 (quoted, Isnard & Lambert 1989)

Vapor Pressure (Pa at 25°C):
0.330 (extrapolated-regression of tabulated data, Stull 1947)
0.01177 (Knudsen's method, calculated-Antoine eqn., Hoyer & Peperle 1957)
0.028 (gas saturation 55-106°C, extrapolated-Antoine eqn., Bender et al. 1983)
0.0515 (extrapolated-Antoine eqn., Boublik et al. 1984)
0.030 (extrapolated-Antoine eqn., Dean 1985)
0.0118 (P_S, interpolated-Antoine eqn., temp. range 10-50°C, Stephenson & Malanowski 1987)

Henry's Law Constant (Pa·m^3/mol):
 $4.0x10^{-8}$ (calculated-P/C, this work)

Octanol/Water Partition Coefficient, log K_{OW}:
 0.80 (Leo et al. 1969)
 0.80, 0.77, 0.78 (quoted, Hansch & Leo 1979)
 0.70 (McLeese et al. 1979)
 0.80, 0.88 (quoted, shake flask-HPLC, Nahum & Horvath 1980)
 0.77, 0.80 (quoted, Verschueren 1983)
 0.80, 0.87 (quoted, calculated-TSA, Iwase et al. 1985)
 1.27 (computed, Klopman et al. 1985; quoted, Pangrekar et al. 1994)
 0.80 (quoted, Isnard & Lambert 1989)
 0.80 (COMPUTOX, Kaiser 1993)

Bioconcentration Factor, log BCF:

Sorption Partition Coefficient, log K_{OC}:

Half-Lives in the Environmental Compartments:
 Air:
 Surface water:
 Ground water:
 Sediment:
 Soil:
 Biota:

Environmental Fate Rate Constants and Half-Lives:
 Volatilization:
 Photolysis:
 Photooxidation:
 Hydrolysis:
 Biodegradation: 95% degradation in 1-2 days in mixed bacteria cultures (Tabak et al. 1964); average rate of biodegradation 57.5 mg COD $g^{-1}·h^{-1}$ based on measurements of COD decrease using activated sludge inoculum with 20 days of adaptation to the substrate (Pitter 1976; quoted, Scow 1982); 2 mM resorcinol solution degraded by strain Re10 within 4 days (Schnell et al. 1989).
 Biotransformation:
 Bioconcentration Uptake (k_1) and Elimination (k_2) Rate Constants:

Common Name: Hydroquinone
Synonym: 1,4-benzenediol, *p*-dihydroxybenzene, *p*-hydroxyphenol, quinol, hydroquinol
Chemical Name: 1,4-dihydroxybenzene
CAS Registry No: 123-31-9
Molecular Formula: $C_6H_4(OH)_2$
Molecular Weight: 110.11
Melting Point (°C): 172
Boiling Point (°C): 287
218.2 (Verschueren 1983)
Density (g/cm^3):
Dissociation Constants, pK_a:
9.90 (McLeese et al. 1979)
Molar Volume (cm^3/mol):
110.8 (calculated-LeBas method)
Molecular Volume, TMV (Å^3):
Total Surface Area, TSA (Å^2):
Heat of Fusion, ΔH_{fus}, kcal/mol:
6.48 (Tsonopoulos & Prausnitz 1971)
Entropy of Fusion, ΔS_{fus}, cal/mol·K:
14.5 (Tsonopoulos & Prausnitz 1971)
Fugacity Ratio at 25°C (assuming ΔS_{fus} = 13.5 e.u.), F: 0.0352

Water Solubility (g/m^3 or mg/L at 25°C):
73672 (synthetic method, Walker et al. 1931)
80135 (shake flask-interferometry, Korman & La Mer 1936)
80000 (quoted, Fieser & Fieser 1958; Morrison & Boyd 1959)
80750 (selected, Tsonopoulos & Prausnitz 1971)
70000 (20-25°C, Geyer et al. 1981)
70000 (quoted, Rott et al. 1982; Dean 1985)
70000 (quoted, Verschueren 1983)
86454 (Windholz 1983; quoted, Chessells et al. 1992)
70800 (quoted, Isnard & Lambert 1989)

Vapor Pressure (Pa at 25°C):
0.00276 (68-126°C, gas saturation, extr. Antoine eqn., Bender et al. 1983)
0.0394 (extr. liquid, Antoine eqn., Boublik et al. 1984)
0.00255 (P_S, interpolated, Antoine eqn., Stephenson & Malanowski 1987)

Henry's Law Constant (Pa·m^3/mol):
3.89×10^{-6} (quoted, Meylan & Howard 1991)
5.91×10^{-6} (estimated-bond contribution, Meylan & Howard 1991)
4.00×10^{-6} (calculated-P/C, this work)

Octanol/Water Partition Coefficient, log K_{OW}:

0.59 (Leo et al. 1971)
0.495 (shake flask-UV at pH 5.62, Umeyama et al. 1971)
0.59 (calculated-π constant, Norrington et al. 1975)
0.55, 0.59 (shake flask, OECD 1981 Guidelines, Geyer et al. 1984)
0.59, 0.50, 0.61 (quoted, Hansch & Leo 1979; quoted, HPLC, Nahum & Horvath 1980)
0.47 (calculated-π constant, McLeese et al. 1979)
0.56, 0.54 (quoted, shake flask-HPLC, Nahum & Horvath 1980)
0.50, 0.59 (quoted, Verschueren 1983)
1.28 (computed, Klopman et al. 1985; quoted, Pangrekar et al. 1994)
0.59 (quoted, Schultz et al. 1987)
0.59 (quoted, Isnard & Lambert 1989)
0.59 (quoted from Medchem software value, Chessells et al. 1992)
0.59 (COMPUTOX, Kaiser 1993)

Bioconcentration Factor, log BCF:

1.81 (green algae, *Chlorella*, exposure to 50 μg/L for 24 hours, Geyer et al. 1981)
0.95 (calculated-S, Geyer et al. 1981)
1.60, 2.72 (golden orfe, activated sludge, Freitag et al. 1982)
1.60, 1.60 (algae, fish, Freitag et al. 1984)
1.54 (algae, wet weight basis after 1 day, Geyer et al. 1984)
0.602 (calculated-K_{OW}, Geyer et al. 1984)
2.93 (activated sludge, Freitag et al. 1987)

Sorption Partition Coefficient, log K_{OC}:

Half-Lives in the Environmental Compartments:

Air: photooxidation half-life of 6-26.1 hours, based on estimated rate constant for the vapor-phase reaction with hydroxyl radicals in air (Howard et al. 1991); atmospheric transformation lifetime was estimated to be 1 to 5 days (Kelly et al. 1994).

Surface water: aqueous photooxidation half-life of 0.39-19.3 hours, based on measured rate data for the reaction with alkylperoxyl radicals in aqueous solution (Mill 1982; quoted, Howard et al. 1991).

Ground water: 48-336 hours, based on estimated unacclimated aqueous aerobic biodegradation half-life (Howard et al. 1991).

Sediment:

Soil: 24-168 hours, based on estimated unacclimated aqueous aerobic biodegradation half-life (Howard et al. 1991).

Biota:

Environmental Fate Rate Constants and Half-Lives:

Volatilization:

Photolysis:

Oxidation: $1x10^6$ M^{-1} sec^{-1} for the reaction with RO_2 radicals at 30°C in aquatic systems with half-life of 12 minutes (Howard 1972; Hendry et al. 1974; quoted, Mill 1982); $<2x10^2$ M^{-1} sec^{-1} for the reaction with singlet oxygen at 25°C in aquatic systems with half-life of >100 years (Foote 1976; Mill 1979; quoted, Mill 1982); aqueous photooxidation half-life of 0.39-19.3 hours in surface water, based on measured rate data for the reaction with alkylperoxyl radicals in aqueous solution (Mill 1982; selected, Howard et al. 1991); photooxidation half-life of 6-26.1 hours in air, based on estimated rate constant for the vapor-phase reaction with hydroxyl radicals in air (Atkinson 1987; selected, Howard et al. 1991); 72.1 mg/l of total organic carbon (TOC) degraded to 98% TOC after 5 hours of illumination with a 250 W tungsten lamp by photo-Fenton reaction (Ruppert et al. 1993).

Hydrolysis: no hydrolyzable group (Howard et al. 1991).

Biodegradation: 95% degradation in 1-2 days in mixed bacteria cultures (Tabak et al. 1964); average rate of biodegradation 54.2 mg COD $g^{-1} \cdot h^{-1}$ based on measurements of COD decrease using activated sludge inoculum with 20 days of adaptation to the substrate (Pitter 1976; quoted, Scow 1982); aqueous aerobic half-life of 24-168 hours, based on aqueous screening test data (Ludzack & Ettinger 1960; Belly & Goodhue 1976; Gerike & Fischer 1979; selected, Howard et al. 1991); aqueous anaerobic half-life of 96-672 hours, based on estimated unacclimated aqueous aerobic biodegradation half-life (Howard et al. 1991).

Biotransformation:

Bioconcentration Uptake (k_1) and Elimination (k_2) Rate Constants:

Common Name: Guaiacol
Synonym: 2-methoxyphenol, methylcatechol, *o*-hydroxyanisole, 1-hydroxy-2-methoxybenzene, pyrocatechol monomethylether
Chemical Name: *o*-methoxyphenol
CAS Registry No: 90-05-1
Molecular Formula: $C_6H_4(OCH_3)OH$
Molecular Weight: 124.13
Melting Point (°C): 27-28, 32
Boiling Point (°C): 204-206
Density (g/cm³): 1.129 (crystal); 1.112 (liq.) solidified at 28°C
Dissociation Constants, pK:
Molar Volume (cm³/mol):
118.9 (calculated-LeBas method)
Molecular Volume (A³):
Total Surface Area, TSA (A²):
Heat of Fusion ΔH_{fus}, kcal/mol:
Entropy of Fusion ΔS_{fus}, cal/mol·K (e.u.):
Fugacity Ratio at 25 °C, (assuming ΔS_{fus} = 13.5 e.u.) F: 0.85

Water Solubility (g/m³ or mg/L at 25°C):
15271, 98600 (quoted, calculated-K_{OW}, Yalkowsky & Morozowich 1980)
143000-16700 (Windholz 1983)
16000 (15°C, Verschueren 1983)
15000 (room temperature, Dean 1985)
18700, 1316 (15°C, 37 °C, Yalkowsky et al. 1987)
24100 (quoted, Sagebiel & Seiber 1993)
24800 (25°C, shake flask-HPLC/UV, Tam et al. 1994)

Vapor Pressure (Pa at 25°C):
133.3 (52.4°C, Weast 1982-83)
13.73 (Verschueren 1983)
8.88 (extrapolated-Antoine eqn., Boublik et al. 1984)
6.79 (extrapolated-Antoine eqn., Dean 1985)
24.5 (extrapolated-Antoine eqn., Stephenson & Malanowski 1987)
21.0 (quoted, Sagebiel & Seiber 1993)

Henry's Law Constant (Pa m³/mol):
0.30 (calculated-P/C, Sagebiel et al. 1992)
0.13 (gas stripping-GC, Sagebiel et al. 1992)
0.11 (head space-GC, Sagebiel et al. 1992)
0.132 (bubble chamber, Sagebiel & Seiber 1993)

Octanol/Water Partition Coefficient, log K_{OW}:

1.33 (shake flask-UV, Umeyama et al. 1971)
1.35 (shake flask, Korenman 1973)
1.33 (shake flask-UV, Norrington et al. 1975)
1.58 (Hansch & Leo 1979)
1.25 (HPLC-RT, Butte et al. 1981)
1.31,2.53 (quoted, Verschueren 1983)
1.32 (recommended, LOGKOW databank, Sangster 1993)

Bioconcentration Factor, log BCF:

Sorption Partition Coefficient, log K_{OC}:

Half-Lives in the Environment:

Air:
Surface water:
Ground water:
Sediment:
Soil:
Biota:

Environmental Fate Rate and Half-Lives:

Volatilization:
Photolysis:
Oxidation:
Hydrolysis:
Biodegradation:
Biotransformation:
Bioconcentration Uptake (k_1) and Elimination Rate Constants (k_2):

Common Name: 3-Methoxyphenol
Synonym: *m*-methoxyphenol, *m*-hydroxyanisole, resorsinol monomethylether
Chemical Name: 3-methoxyphenol
CAS Registry No: 150-19-6
Molecular Formula: $CH_3OC_6H_4OH$
Molecular Weight: 124.15
Melting Point (°C): -17
Boiling Point (°C): 244
Density (g/cm^3):
Dissociation Constants, pK_a:
Molar Volume (cm^3/mol):
118.9 (calculated-LeBas method)
106.8 (Jaworska & Schultz 1993)
Molecular Volume, TMV (Å^3):
Total Surface Area, TSA (Å^2):
Heat of Fusion, ΔH_{fus}, kcal/mol:
Entropy of Fusion, ΔS_{fus}, cal/mol·K (e.u.):
Fugacity Ratio at 25°C, (assuming ΔS_{fus} = 13.5 e.u.) F: 1.0

Water Solubility (g/m^3 or mg/L at 25°C):
67800 (shake flask-HPLC/UV, Varhanickova et al. 1995)

Vapor Pressure (Pa at 25°C):
0.262 (extrapolated-Antoine eqn., Stephenson & Malanowski 1987)

Henry's Law Constant (Pa • m^3/mol):
0.10 (estimated as per Sagebiel et al. 1992 data on 2-methoxyphenol)

Octanol/Water Partition Coefficient, log K_{OW}:
1.58 (Leo et al. 1969)
1.58 (quoted, Hansch & Leo 1979)
1.58 (quoted, Verschueren 1977, 1983)
1.47 (correlated from HPLC-k', Minick et al. 1988)
1.58 (quoted, El Tayar et al. 1991)
1.58 (quoted from CLOGP 3.51, Jaworska & Schultz 1993)
1.58 (COMPUTOX databank, Kaiser 1993)

Bioconcentration Factor, log BCF: no data

Sorption Partition Coefficient, log K_{OC}: no data

Half-Lives in the Environment/Environmental Fate Rate Constants and Half-Lives: no data

Common Name: 4-Methoxyphenol
Synonym: *p*-methoxyphenol, *p*-hydroxyanisole, hydroquinone monomethylether
Chemical Name: 4-methoxyphenol
CAS Registry No: 150-76-5
Molecular Formula: $CH_3OC_6H_4OH$
Molecular Weight: 124.15
Melting Point (°C): 53
Boiling Point (°C): 243
Density (g/cm^3):
Dissociation Constants, pK_a:
Molar Volume (cm^3/mol):
106.3 (calculated-density, Jaworska & Schultz 1993)
118.9 (calculated-LeBas method)
Molecular Volume, TMV ($Å^3$):
Total Surface Area, TSA ($Å^2$):
Heat of Fusion, ΔH_{fus}, kcal/mol:
Entropy of Fusion, ΔS_{fus}, cal/mol·K (e.u.):
Fugacity Ratio at 25°C, (assuming ΔS_{fus} =13.5 e.u.) F: 0.529

Water Solubility (g/m^3 or mg/L at 25°C):
40000 (quoted, Verschueren 1977, 1983)
40000 (quoted, Mailhot & Peters 1988)
19500 (shake flask-HPLC/UV, Varhanickova et al. 1995)

Vapor Pressure (Pa at 25°C):
0.033 (extrapolated-Antoine eqn., Boublik et al. 1984)
0.0394 (extrapolated-Antoine eqn., Stephenson & Malanowski 1987)

Henry's Law Constant ($Pa \cdot m^3/mol$):

Octanol/Water Partition Coefficient, log K_{OW}:
1.34 (shake flask-UV, Fujita et a. 1968; quoted, Leo et al. 1969)
1.37 (shake flask-UV at pH 7.45, Umeyama et al. 1971)
1.34, 1.33, 1.37 (quoted, Hansch & Leo 1979)
1.34 (quoted, Verschueren 1977, 1983)
1.47 (shake flask, Korenman et al. 1980)
1.57, 1.62 (quoted, correlated-HPLC-k', Miyake & Terada 1982)
1.34 (quoted, Schultz et al. 1987; Veith & Broderius 1987)
1.34 (quoted, Mailhot & Peters 1988)
1.24 (correlated from HPLC-k', Minick et al. 1988)
1.34 (calculated-CLOGP 3.15, Jaworska & Schultz 1993)
1.34 (recommended, LOGKOW databank, Sangster 1993)
1.34 (COMPUTOX databank, Kaiser 1993)

Bioconcentration Factor, log BCF: no data
Sorption Partition Coefficient, log K_{OC}: no data
Half-Lives in the Environment/Environmental Fate Rate Constants and Half-Lives: no data

Common Name: 4-Chloroguaiacol
Synonym: 4-chloromethoxyphenol
Chemical Name: 4-chloroguaiacol, 4-chloromethoxyphenol
CAS Registry No: 16766-30-6
Molecular Formula: $C_7H_7ClO_2$
Molecular Weight: 158.6
Melting Point (°C): liquid
Boiling Point (°C):
Density (g/cm³):
Dissociation Constants, pK_a:
Molar Volume, V_M (cm³/mol):

155.6 (calculated-LeBas method)

Molecular Volume, TMV (Å³):
Total Surface Area, TSA (Å²):
Heat of Fusion, ΔH_{fus}, (kcal/mol):
Entropy of Fusion, ΔS_{fus} (cal/mol·K or e.u.):
Fugacity Ratio at 25 °C, (assuming ΔS_{fus} = 13.5 e.u.) F: 1.0

Water Solubility (g/m³ or mg/L at 25°C):

5140 (shake flask-HPLC/UV, pH 2.8, Tam et al. 1994)
5604 (shake flask-GC/ECD, pH 2.8, Tam et al. 1994)
5370 (selected, Tam et al. 1994)

Vapor Pressure (Pa at 25°C):

Henry's Law Constant (Pa·m³/mol):

Octanol/Water Partition Coefficient, log K_{OW}:

2.15 (calculated-f const., Niimi et al. 1990)
2.11-2.52 (quoted, NCASI 1993)

Bioconcentration Factor, log BCF:

0.0 (trout, 1-21 d exposure, BCF < 1, Niimi et al. 1990)
0.86-1.19 (quoted, NCASI 1993)

Sorption Partition Coefficient, log K_{OC}: no data

Half-Lives in the Environment/Environmental Fate Rate Constants or Half-Lives: no data

Common Name: 4,5-Dichloroguaiacol
Synonym: 4,5-dichloro-2-methoxyphenol
Chemical Name: 4,5-dichloroguaiacol
CAS Registry No: 2460-49-3
Molecular Formula: $C_7H_6Cl_2O_2$
Molecular Weight: 193.04
Melting Point (°C): 69-70
Boiling Point (°C):
Density (g/cm^3):
Dissociation Constants, pK_a: 8.26, 8.52 (20°C, regressions, Xie & Dyrssen 1984)
Molar Volume, V_M (cm^3/mol): 176.5 (calculated-LeBas method)
Molecular Volume, TMV (Å^3):
Total Surface Area, TSA (Å^2):
Heat of Fusion, ΔH_{fus}, (kcal/mol):
Entropy of Fusion, ΔS_{fus} (cal/mol·K or e.u.):
Fugacity Ratio at 25 °C, (assuming ΔS_{fus} = 13.5 e.u.) F: 0.367

Water Solubility (g/m^3 or mg/L at 25°C):
584 (shake flask-HPLC/UV, pH 5.6, Tam et al. 1994)
565 (shake flask-GC/ECD, pH 5.6, Tam et al. 1994)
575 (selected, Tam et al. 1994)

Vapor Pressure (Pa at 25°C):
1.54 (subcooled liquid value, correlated-GC-RT, Bidleman & Renberg 1985)

Henry's Law Constant (Pa·m^3/mol):

Octanol/Water Partition Coefficient, log K_{OW}:
3.04 (calculated-f const., Rekker 1977)
3.28 (Hansch & Leo 1979)
3.26 (shake flask-UV, Saarikoski & Viluksela 1982)
3.41 (calculated-f const., Xie 1983)
3.18 (shake flask-GC, Xie et al. 1984)
3.28, 3.28, 3.04 (correlated-HPLC-k', calc-π const., f const., Xie et al. 1984)
3.20, 3.19, 3.19 (shake flask-GC, regressions, Xie & Dryssen 1984)
3.23 (selected, Suntio et al. 1988)
2.86 (calculated-f const., Niimi et al. 1990)
2.51-2.93 (quoted, NCASI 1993)
3.19, 3.26, 3.18; 3.26 (quoted values; recommended, Sangster 1993)

Bioconcentration Factor, log BCF or log K_B:
1.74-2.05, 2.03 (rainbow trout: 1-21 days exposure, mean value, Niimi et al. 1990)
1.18-1.51, 1.75 (quoted, estimated, NCASI 1993)

Sorption Partition Coefficient, log K_{OC}: no data
Half-Lives in the Environment/Environmental Fate Rate Constants or Half-Lives: no data

Common Name: 3,4,5-Trichloroguaiacol
Synonym: 3,4,5-trichloro-2-methoxyphenol
Chemical Name: 3,4,5-trichloroguaiacol
CAS Registry No: 57057-83-7
Molecular Formula: $C_7H_5Cl_3O_2$
Molecular Weight: 227.49
Melting Point (°C): 85-86
Boiling Point (°C):
Density (g/cm^3):
Dissociation Constants, pK_a:

7.55 (Könemann 1981)
7.43, 7.52, 7.56 (20°C, regressions, Xie & Dyrssen 1984)
7.90 (Leurenberger et al. 1985)
7.56 (Xie et al. 1986)

Molar Volume, V_M (cm^3/mol):

197.4 (calculated-LeBas method)

Molecular Volume, TMV (Å^3):
Total Surface Area, TSA (Å^2):
Heat of Fusion, ΔH_{fus}, (kcal/mol):
Entropy of Fusion, ΔS_{fus} (cal/mol·K or e.u.):
Fugacity Ratio at 25 °C, (assuming ΔS_{fus} = 13.5 e.u.) F: 0.255

Water Solubility (g/m^3 or mg/L at 25°C):

305 (shake flask-HPLC/UV, pH 5.9, Tam et al. 1994)
313 (shake flask-GC/ECD, pH 5.9, Tam et al. 1994)
310 (selected, Tam et al. 1994)

Vapor Pressure (Pa at 25°C):

0.64 (subcooled liquid value, correlated-GC-RT, Bidleman & Renberg 1985)

Henry's Law Constant (Pa·m^3/mol):

Octanol/Water Partition Coefficient, log K_{OW}:

3.77 (shake flask-UV, Saarikoski & Viluksela 1982)
4.11 (shake flask-GC, Xie et al. 1984; quoted, Xie et al. 1986)
4.18 (correlated-HPLC-k', Xie et al. 1984)
4.32 (calculated-π const., Xie et al. 1984)
3.78 (calculated-f const., Xie et al. 1984)
4.14, 4.14, 4.13 (shake flask-GC, regressions, Xie & Dyrssen 1984)
4.00 (Leuenberger et al. 1985)
4.15 (selected, Suntio et al. 1988)
3.57 (calculated-f const., Niimi et al. 1990)
4.15 (quoted, NCASI 1993)

3.77, 4.11, 4.13, 4,18; 3.77 (quoted values; recommended, Sangster 1993)
3.77 (COMPUTOX databank, Kaiser 1993)

Bioconcentration Factor, log BCF or log K_B:
1.60 (rainbow trout, Oikari et al. 1985, quoted, NCASI 1993)
2.06-2.51, 2.41 (rainbow trout: 1-21 days exposure, mean value, Niimi et al. 1990, quoted, NCASI 1993)
1.78-1.95 (field studies, bile of rainbow trout, Niimi et al. 1990)
2.47 (estimated, NCASI 1993)

Sorption coefficient, log K_{OC}:

Half-Lives in the Environment:
Air:
Water:
Soil:
Sediment:
Biota: 1-2 days in bleak and 2 days in trout (Niimi et al. 1990).

Environmental Fate Rate Constants or Half-Lives:
Volatilization:
Photolysis:
Oxidation:
Hydrolysis:
Biodegradation:
Biotransformation:
Bioconcentration, Uptake(k_1) and Elimination (k_2) Rate Constants Rates:

Common Name: 4,5,6-Trichloroguaiacol
Synonym: 2,3,4-trichloro-6-methoxyphenol
Chemical Name: 4,5,6-trichloroguaiacol
CAS Registry No: 2668-24-8
Molecular Formula: $C_7H_5Cl_3O_2$
Molecular Weight: 227.49
Melting Point (°C): 112-115
Boiling Point (°C):
Density (g/cm^3):
Dissociation Constants, pK_a:
7.07 (Leuenberger et al. 1985)
7.20 (20°C, regressions, Xie & Dyrssen 1984)
Molar Volume, V_M (cm^3/mol):
197.4 (calculated-LeBas method)
Molecular Volume, TMV (Å^3):
Total Surface Area, TSA (Å^2):
Heat of Fusion, ΔH_{fus}, (kcal/mol):
Entropy of Fusion, ΔS_{fus} (cal/mol K or e.u.):
Fugacity Ratio at 25 °C, (assuming ΔS_{fus} = 13.5 e.u.) F: 0.129

Water Solubility (g/m^3 or mg/L at 25°C):
57 (shake flask-HPLC/UV, pH 5.8, Tam et al. 1994)
50 (shake flask-GC/ECD, pH 5.8, Tam et al. 1994)
54 (selected, Tam et al. 1994)

Vapor Pressure (Pa at 25°C):
0.249 (subcooled liquid value, correlated-GC-RT, Bidleman & Renberg 1985)

Henry's Law Constant (Pa·m^3/mol):
8.82 (calculated-P/C, this work)

Octanol/Water Partition Coefficient, log K_{OW}:
3.74 (shake flask-GC, Xie et al. 1984)
3.92 (correlated-HPLC-k', Xie et al. 1984)
3.91 (calculated-π const., Xie et al. 1984)
3.78 (calc-f const., Xie et al. 1984)
3.73, 3.74. 3.72 (shake flask-GC, regressions, Xie & Dyrssen 1984)
3.83 (selected, Suntio et al. 1988, quoted, NCASI 1993)
3.57 (calculated-f const., Niimi et al. 1990)
3.72, 3.92 (quoted, Sangster 1993)
3.82 (COMPUTOX databank, Kaiser 1993)

Bioconcentration Factor, log BCF:

2.60 (bleaks, Renberg et al. 1980; quoted, Walden et al. 1986)

1.90-2.11 (rainbow trout, Oikari et al. 1985, quoted, NCASI 1993)

2.59 (bleaks, Walden et al. 1986)

1.88-2.05, 1.97 (rainbow trout: 1-21 days exposure, mean value, Niimi et al. 1990, quoted, NCASI 1993)

1.78-1.95 (field studies, bile of trout, Niimi et al. 1990)

4.81 (fourhorn sculpin *myoxocephalus quadricornis*, bile BCF in brackish water, during 6-20 days exposure under continous water-flow conditions, Wachtmeister et al. 1991)

2.22 (estimated, NCASI 1993)

Sorption Partition Coefficient, log K_{OC}:

3.11 (sediment, Remberger et al. 1986, quoted, Neilson et al. 1991)

2.80, 2.94 (soil, quoted, calculated- χ and fragment contribution, Meylan et al. 1992)

Half-Lives in the Environment:

Air:

Water:

Soil:

Sediment:

Biota: one to two days in bleak (Niimi et al. 1990).

Environmental Fate Rate Constants or Half-Lives:

Volatilization:

Photolysis:

Oxidation:

Hydrolysis:

Biodegradation:

Biotransformation:

Bioconcentration, Uptake(k_1) and Elimination (k_2) Rate Constants Rates:

Common Name: 3,4,5,6-Tetrachloroguaiacol
Synonym: 2,3,4,5-tetrachloro-6-methoxyphenol
Chemical Name: 3,4,5,6-tetrachloroguaiacol
CAS Registry No: 2539-17-5
Molecular Formula: $C_7H_4Cl_4O_2$
Molecular Weight: 261.93
Melting Point (°C): 121-122
Boiling Point (°C):
Density (g/cm³):
Dissociation Constants, pK_a:

6.26 (Leuenberger et al. 1985)
6.19, 6.12, 6.26 (20°C, regressions, Xie & Dyrssen 1984)

Molar Volume, V_M (cm³/mol):

218.3 (calculated-LeBas method)

Molecular Volume, TMV (Å³):
Total Surface Area, TSA (Å²):
Heat of Fusion ΔH_{fus}, (kcal/mol):
Entropy of Fusion ΔS_{fus} (cal/mol·K or e.u.):
Fugacity Ratio at 25 °C, (assuming ΔS_{fus} = 13.5 e.u.) F: 0.112

Water Solubility (g/m³ or mg/L at 25°C):

27 (shake flask-HPLC/UV, pH 4.2, Tam et al. 1994)
25 (shake flask-GC/ECD, pH 4.2, Tam et al. 1994)
26 (selected, Tam et al. 1994)

Vapor Pressure (Pa at 25°C):

0.138 (subcooled liquid value, correlated-GC-RT, Bidleman & Renberg 1985)

Henry's Law Constant (Pa·m³/mol):

1.013 (Leuenberger et al. 1985; quoted, Barton 1987)
1.390 (calculated-P/C, this work)

Octanol/Water Partition Coefficient, log K_{OW}:

4.29 (Saarikoski & Viluksela 1982)
4.45 (shake flask-GC, Xie et al. 1984; quoted, Xie et al. 1986; Barton 1987)
4.76 (correlated-HPLC-k', Xie et al. 1984)
5.01 (calculated-π const., Xie et al. 1984)
4.52 (calculated-f const., Xie et al. 1984)
4.41, 4.43, 4.42 (shake flask, regressions, Xie & Dyrssen 1984)
4.53 (Leuenberger et al. 1985)
4.61 (selected, Suntio et al. 1988, quoted, NCASI 1993)
4.28 (calculated-f const., Niimi et al. 1990)
4.59 (COMPUTOX databank, Kaiser 1993)

4.42, 4.76 (quoted, Sangster 1993)

Bioconcentration Factor, log BCF:

2.60 (bleaks, after 2 weeks exposure, Renberg et al. 1980; quoted, Walden et al. 1986)
1.60-2.18 (rainbow trout, Oikari et al. 1985, quoted, NCASI 1993)
1.78-1.95 (bile of rainbow trout, field studies, Niimi et al. 1990)
2.04-2.38, 2.26 (rainbow trout: 1-21 days exposure, mean value, Niimi et al. 1990, quoted, NCASI 1993)
3.2 (roach, Niimi et al. 1990, quoted, NCASI 1993)
2.84 (estimated, NCASI 1993)

Sorption Partition Coefficient, log K_{OC}:

2.85 (soil, Seip et al. 1986)
3.15 (sediment, Remberger et al. 1986, quoted, Neilson et al. 1991)
2.30 (soil, calculated- χ , Bahnick & Doucette 1988)
2.85, 3.17 (soil, quoted, calculated- χ and fragment contribution, Meylan et al. 1992)

Half-Lives in the Environment:

Air:
Water:
Soil:
Sediment:
Biota: less than 10 days in trout liver (Niimi et al. 1990); 1-2 days in bleak (Niimi et al. 1990).

Environmental Fate Rate Constants or Half-Lives:

Volatilization:
Photolysis:
Oxidation:
Hydrolysis:
Biodegradation:
Biotransformation:
Bioconcentration, Uptake(k_1) and Elimination (k_2) Rate Constants Rates:

4.2 Summary Tables and QSPR Plots

Table 4.1 Physical-chemical properties of phenolic compounds.

compound	CAS no.	formula	MW	mp, °C	bp, °C	fugacity ratio, F at 25 °C	molar volume, cm³/mol V_M from ρ	V_M LeBas	V_M (a)	TSA, Å² (b)	pK_a
Phenol and chlorophenols:											
phenol	108-95-2	C_6H_5OH	94.1	41	182	0.6946	90.4	103.4	83.0	224.2	9.89
2-chlorophenol	95-57-8	C_6H_4OHCl	128.56	9	175-176	1.0000	101.8	124.3	95.0		8.49
3-chlorophenol	108-43-0	C_6H_4OHCl	128.56	33	214	0.8334		124.3			8.85
4-chlorophenol	106-48-9	C_6H_4OHCl	128.56	43	220	0.6637	101.6	124.3	95.5		9.18
2,3-dichlorophenol	576-24-9	$C_6H_4Cl_2O$	163.00	59	206	0.4610		145.2			6.44
2,4-dichlorophenol	120-83-2	$C_6H_4Cl_2O$	163.00	44	209-210	0.6488		145.2	106.7		7.68
2,5-dichlorophenol	583-78-8	$C_6H_4Cl_2O$	163.00	58	211	0.4716		145.2			6.45
2,6-dichlorophenol	87-65-0	$C_6H_4Cl_2O$	163.00	67	218-220	0.3842		145.2			6.8
3,4-dichlorophenol	95-77-2	$C_6H_4Cl_2O$	163.00	67	145-146	0.3842		145.2			7.39
3,5-dichlorophenol	591-35-5	$C_6H_4Cl_2O$	163.00	68	233	0.3756		145.2			6.92
2,3,4-trichlorophenol	1595-06-0	$C_6H_3Cl_3O$	197.45	79-81		0.287		166.1			7.66
2,3,5-trichlorophenol	933-78-8	$C_6H_3Cl_3O$	197.45	62		0.4306		166.1			7.37
2,3,6-trichlorophenol	933-75-5	$C_6H_3Cl_3O$	197.45	58		0.4716		166.1			7.13
2,4,5-trichlorophenol	95-95-4	$C_6H_3Cl_3O$	197.45	68-70	244-250	0.37		166.1			7.43
2,4,6-trichlorophenol	88-06-2	$C_6H_3Cl_3O$	197.45	69.5	246	0.3630	132.5	166.1	118.6		7.42
3,4,5-trichlorophenol	609-19-8	$C_6H_3Cl_3O$	197.45	101		0.1771		166.1			7.74
2,3,4,5-tetrachlorophenol	4901-51-3	$C_6H_2Cl_4O$	231.89	116		0.1259		187			6.96
2,3,4,6-tetrachlorophenol	58-90-3	$C_6H_2Cl_4O$	231.89	70		0.3589		187			5.38
2,3,5,6-tetrachlorophenol	935-95-5	$C_6H_2Cl_4O$	231.89	115		0.1288		187			5.48
pentachlorophenol	87-86-5	C_6HCl_5O	266.34	174		0.0336		207.9			4.92

compound	CAS no.	formula	MW	mp, °C	bp, °C	fugacity ratio, F at 25 °C	molar volume, cm³/mol V_M from ρ	V_M LeBas	V_M (a)	TSA, Å² (b)	pK_a
Cresols and other alkylated phenols:											
o-cresol (2-methylphenol)	95-48-4	$C_6H_4(CH_3)OH$	108.13	31	191	0.872	104.4	125.6	98.8	248.6	10.26
m-cresol (3-methylphenol)	108-39-4	$C_6H_4(CH_3)OH$	108.13	12	203	1	104.6	125.6		250.8	10
p-cresol (4-methylphenol)	106-44-5	$C_6H_4(CH_3)OH$	108.13	34.8	202	0.800		125.6		251.2	10.26
2,6-di-*t*-butyl-4-methylphenol	128-37-0	$C_{15}H_{24}O$	220.36	71		0.351		303.2			
2,3-dimethylphenol	526-75-0	$C_8H_{10}O$	122.17	75	218	0.320		147.8			
2,4-dimethylphenol	105-67-9	$C_8H_{10}O$	122.17	26	211.5	0.977		147.8	115.9		10.6
2,5-dimethylphenol	105-67-9	$C_8H_{10}O$	122.17	75	212	0.320		147.8			10.3
2,6-dimethylphenol	576-26-1	$C_8H_{10}O$	122.17	49	203	0.579		147.8			10.6
3,4-dimethylphenol	95-65-8	$C_8H_{10}O$	122.17	67	225	0.384		147.8			10.4
3,5-dimethylphenol	108-68-9	$C_8H_{10}O$	122.17	68	219.5	0.376		147.8			10.2
2,3,5-trimethylphenol	697-82-5	$C_9H_{12}O$	136.19	93		0.213		170			10.6
2,3,6-trimethylphenol	2416-94-6	$C_9H_{12}O$	136.19	63		0.421		170			
2,4,5-trimethylphenol	496-78-6	$C_9H_{12}O$	136.19	72		0.343		170			
2,4,6-trimethylphenol	527-60-6	$C_9H_{12}O$	136.19	72		0.343		170			10.9
3,4,5-trimethylphenol	527-54-8	$C_9H_{12}O$	136.19	108		0.151		170			
3-methyl-5-ethylphenol		$C_9H_{12}O$	136.19					170			10.1
2-ethylphenol	90-00-6	$C_8H_{10}O$	122.17					147.8			
3-ethylphenol	620-17-7	$C_8H_{10}O$	122.17					147.8			
4-ethylphenol	123-07-9	$C_8H_{10}O$	122.17	42-45				147.8	116		
2-propylpheonol	644-35-0	$C_9H_{12}O$	136.19					170			
4-*n*-propylphenol	645-56-7	$C_9H_{12}O$	136.19					170			10.3
2-isopropylpheonol	88-69-7	$C_9H_{12}O$	136.19					170			
4-isopropylphenol	99-89-8	$C_9H_{12}O$	136.19	59-61				170			10.3
4-*n*-butylphenol	1638-22-8	$C_{10}H_{14}O$	150.22					192.2			10.3

compound	CAS no.	formula	MW	mp, °C	bp, °C	fugacity ratio, F at 25 °C	molar volume, cm³/mol V_M from ρ	V_M LeBas	V_M (a)	TSA, Å² (b)	pK_a
2-*sec*-butylphenol		$C_{10}H_{14}O$	150.22					192.2			
2-*tert*-butylphenol	88-18-6	$C_{10}H_{14}O$	150.22					192.2			
3-*tert*-butylphenol	585-34-2	$C_{10}H_{14}O$	150.22	41		0.695		192.2			10.1
4-*sec*-butylphenol	99-71-8	$C_{10}H_{14}O$	150.22	61	240-242	0.440		192.2			
4-*tert*-butylphenol	98-54-4	$C_{10}H_{14}O$	150.22	101	239.5	0.177		192.2	132.4		9.9
2-methyl-5-*tert*-butylphenol		$C_{11}H_{16}O$	164.25					214.4			
2,6-di-*sec*-butylphenol	128-39-2	$C_{14}H_{22}O$	206.22					281			
3,5-di-*tert*-butylphenol	1138-52-9	$C_{14}H_{22}O$	206.33					281			
3-pentylphenol		$C_{11}H_{16}O$	164.25					214.4			
4-hexylphenol	2446-69-7	$C_{12}H_{18}O$	178.28	32		0.853		236.6			
2-heptylphenol		$C_{13}H_{20}O$	176.30					258.8			
4-heptylphenol	1987-50-4	$C_{13}H_{20}O$	176.30					258.8			
4-octylphenol	1806-26-4	$C_{14}H_{22}O$	206.33	41		0.695		281			
4-*tert*-octylphenol	140-66-9	$C_{14}H_{22}O$	206.33	79-82				281			
4-nonylphenol	104-40-5	$C_{15}H_{24}O$	220.36	42		0.679		303.2	279.8		
1-naphthol	90-15-3	$C_{10}H_8O$	144.17	96	282.5	0.199		155	122.3	281.3	
2-naphthol	135-19-3	$C_{10}H_8O$	144.17	122		0.110		155		286.6	
2-phenylphenol	90-43-7	$C_{12}H_{10}O$	170.21	56		0.494		147.5	147.5		
3-phenylphenol	588-51-8	$C_{12}H_{10}O$	170.21	76-78		0.313		147.5			
4-phenylphenol	92-69-3	$C_{12}H_{10}O$	170.21	164		0.0422		147.5			
2-allylphenol	1745-81-9	$C_9H_{10}O$	134.18	-6	220	1		162.6	124.9		
Nitrophenols:											
2-nitrophenol	108-95-2	$C_6H_4OHNO_2$	139.11	45.3	216	0.630		131.9			7.23

compound	CAS no.	formula	MW	mp, °C	bp, °C	fugacity ratio, F at 25 °C	molar volume, cm³/mol V_M from ρ	V_M LeBas	V_M (a)	TSA, Å² (b)	pK_a
3-nitrophenol	88-75-5	$C_6H_4OHNO_2$	139.11	97		0.194		131.9			8.36
4-nitrophenol	554-84-7	$C_6H_4OHNO_2$	139.11	114	279	0.132		131.9			7.08
2,4-dinitrophenol	51-28-5	$C_6H_4N_2O_5$	184.1	114		0.132		160.4			4.09
2,6-dinitrophenol	329-71-5	$C_6H_4N_2O_5$	184.1	108		0.151		160.4			
2,4,6-trinitrophenol	88-89-1	$C_6H_3N_3N_7$	229.1	121.8	284	0.110		181.5			0.8
4,6-dinitro-*o*-cresol	534-52-1	$C_7H_6N_2O_5$	198.13	85.8		0.250		179			4.35
Other substituted phenols:											
catechol(1,2-dihydroxybenzene	120-80-9	$C_6H_4(OH)_2$	110.1	104	245	0.165		110.8		237.00	9.5
resorcinol (1,3-benzenediol)	108-46-3	$C_6H_4(OH)_2$	110.11	110	277	0.144		110.8		239.5	9.4
hydroquinone	123-31-9	$C_6H_4(OH)_2$	110.11	172	287	0.035		110.8			
4-methyoxphenol	150-76-5	$C_7H_8O_2$	124.14	53	243	0.529		125.6	106.3		
3-methyoxyphenol	150-19-6	$C_7H_8O_2$	124.14	-17	244	1		125.6	106.8		
guaiacol (2-methoxyphenol)	90-05-1	$C_7H_8O_2$	124.14	32	205	0.85		125.6			
4-chloroguaiacol	16766-30-6	$C_7H_7ClO_2$	158.6	liquid		1		153.9			
5-chloroguaiacol	3753-23-5	$C_7H_7ClO_2$	158.6	34-35		0.815		153.9			
4,5-dichloroguaiacol	2460-49-3	$C_7H_6Cl_2O_2$	193.04	69-79		0.37		174.8			8.52
4,6-dichloroguaiacol	2460-49-3	$C_7H_6Cl_2O_2$	193.04	63-64		0.421		174.8			
3,4,5-trichloroguaiacol	57057-83-7	$C_7H_5Cl_3O_2$	227.49	85-86		0.255		195.7			7.56
4,5,6-trichloroguaiacol	2668-24-8	$C_7H_5Cl_3O_2$	227.49	112-115		0.129		195.7			7.2
tetrachloroguaiacol	2539-17-5	$C_7H_4Cl_4O_2$	261.93	121-122		0.112		216.6			6.26

(a) Jaworska & Schultz 1993
(b) Iwase et al. 1985

Table 4.2 Summary and selected physical-chemical properties of phenolic compounds at 25 °C.

compound	selected properties						Henry's law const.
	P_S, Pa	P_L, Pa	S, g/m^3	C^S, mol/m^3	C_L, mol/m^3	log K_{OW}	$Pa{\cdot}m^3/mol$ calc-P/C
Phenol and chlorophenols:							
phenol	47	67.66	88360	939.0	1352	1.46	0.0539
2-chlorophenol	132	132	24650	191.74	191.74	2.17	0.6884
3-chlorophenol	35	41.99	22000	171.13	205.32	2.5	0.2045
4-chlorophenol	20	30.13	27000	210.02	316.44	2.4	0.0952
2,3-dichlorophenol			8215	50.40	109.32	3.15	
2,4-dichlorophenol	12	18.50	4500	27.61	42.55	3.2	0.4347
2,5-dichlorophenol						3.2	
2,6-dichlorophenol	12	12.70	2652	16.27	42.34	2.86	0.2999
3,4-dichlorophenol			9256	56.79	147.79	3.37	
3,5-dichlorophenol			7394	45.36	120.78	3.2	
2,3,4-trichlorophenol	1	3.48	500	2.53	8.82	3.8	0.3949
2,3,5-trichlorophenol	1	2.32	500	2.53	5.88	3.69	0.3949
2,3,6-trichlorophenol			450	2.28	4.83	3.8	
2,4,5-trichlorophenol	2.5	6.76	948	4.80	12.98	3.72	0.5207
2,4,6-trichlorophenol	1.25	3.44	434	2.20	6.06	3.69	0.5687
3,4,5-trichlorophenol						4.3	
2,3,4,5-tetrachlorophenol	0.1	0.79	166	0.72	5.69	4.8	0.1397
2,3,4,6-tetrachlorophenol	0.28	0.78	183	0.79	2.20	4.45	0.3548
2,3,5,6-tetrachlorophenol	0.1	0.78	100	0.43	3.35	4.9	0.2319
pentachlorophenol	0.00415	0.12	14	0.053	1.565	5.05	0.0790

compound	selected properties P_S, Pa	P_L, Pa	S, g/m^3	C^S, mol/m^3	C_L, mol/m^3	log K_{OW}	Henry's law const. Pa·m^3/mol calc-P/C
Cresols and other alkylated phenols:							
o-cresol (2-methylphenol)		41	26000	207.01	237.32	1.98	0.1559
m-cresol (3-methylphenol)		16	22000	175.16	175.16	1.98	0.0913
p-cresol (4-methylphenol)		13	20000	159.24	199.05	1.96	0.0653
2,6-di-*t*-butyl-4-methylphenol							
2,3-dimethylphenol		8.05	6000	47.77	149.17		0.0540
2,4-dimethylphenol		13.02	8795	70.02	71.64	2.35	0.1817
2,5-dimethylphenol		10.68	3176	25.29	78.96		0.1353
2,6-dimethylphenol		32.82	6230	49.60	85.68	2.36	0.3831
3,4-dimethylphenol		2.241	5100	40.61	105.68	2.23	0.0212
3,5-dimethylphenol		3.76	5500	43.79	116.59	2.35	0.0322
2,3,5-trimethylphenol		2.43	800	6.37	29.97		0.0811
2,3,6-trimethylphenol							
2,4,5-trimethylphenol		7.73				2.67	
2,4,6-trimethylphenol		19.58	1200	9.55	27.86	2.73	0.7027
3,4,5-trimethylphenol		2.144	1540	12.26	81.18		0.0264
3-methyl-5-ethylphenol		1.18	2314	18.42			
2-ethylphenol		20.4	14042			2.47	
3-ethylphenol		6.7				2.5	
4-ethylphenol		5	7980			2.5	
2-propylpheonol		11.2				2.93	
4-*n*-propylphenol		7.3	1728			3.2	
2-isopropylpheonol		12.45	4423			2.88	
4-isopropylphenol			3263				
4-*n*-butylphenol			617	4.91		3.65	

compound	selected properties					log K_{OW}	Henry's law const. Pa·m³/mol
	P_S, Pa	P_L, Pa	S, g/m³	C^S, mol/m³	C_L, mol/m³		calc-P/C
2-*sec*-butylphenol		5				2.8	
2-*tert*-butylphenol						2.7	
3-*tert*-butylphenol			2070	16.48	23.73	2.6	
4-*sec*-butylphenol		4	960	7.64	17.35	2.1	0.2305
4-*tert*-butylphenol		1.24	580	4.62	26.07	3.04	0.0476
2-methyl-5-*tert*-butylphenol		3.688	410	3.26			
2,6-di-*sec*-butylphenol						4.4	
3,5-di-*tert*-butyl-			14				
2,6-di-*tert*-butyl-4-methyl-		0.0338				4.17	
3-pentylphenol						3.13	
4-hexylphenol			391			3.6	
2-heptylphenol						4.45	
4-heptylphenol						4.15	
4-octylphenol		0.071	12.6	0.1003	0.1444	4.12	0.4916
4-*tert*-octylphenol						3.7	
2-nonylphenol						5.76	
3-nonylphenol						5.61	
4-nonylphenol		0.1	5.43	0.0432	0.0637	4.48	1.5705
1-naphthol		0.5	438	3.487	17.568	2.84	
2-naphthol		0.4	740	5.892	53.657	2.7	
2-phenylphenol		30	700	5.573	11.291	3.09	
3-phenylphenol							
4-phenylphenol		8.7	9.8	0.0780	1.849	3.2	
2-allylphenol						2.64	

compound	selected properties P_S, Pa	P_L, Pa	S, g/m³	C^S, mol/m³	C_L, mol/m³	log K_{OW}	Henry's law const. calc-P/C
Nitrophenols:							
2-nitrophenol		20	1080	8.599	13.652	1.8	
3-nitrophenol			11550	91.959	473.93	2	
4-nitrophenol			13500	107.48	815.83	1.91	
2,4-dinitrophenol			335			1.67	
2,6-dinitrophenol							
2,4,6-trinitrophenol			13750	109.475	992.466	1.33	
4,6-dinitro-*o*-cresol			222	1.768	7.058	2.12	
Other substituted compounds:							
catechol (1,2-dihydroxybenzene)		1.34	45000	408.72	2470.45	0.77	
resorcinol (1,3-benzenediol)	0.00118		110000	999.00	6922.46	0.8	
hydroquinone (1,4-benzenediol)	0.00255		70000	635.73	18079.16	0.59	
4-methyoxphenol			19500	155.81	294.81	1.34	
3-methyoxyphenol			67800	541.75	541.75	1.58	
guaiacol (2-methoxyphenol)	20.825	24.5	24800	198.16	233.13	1.33	
4-chloroguaiacol			5370	33.859	33.86	2.15	
5-chloroguaiacol			3960	24.968	30.64		
4,5-dichloroguaiacol	0.570	1.54	575	2.979	8.050	3.26	
4,6-dichloroguaiacol			708	3.668	8.712	2.86	
3,4,5-trichloroguaiacol	0.163	0.64	310	1.363	5.344	3.77	
3,4,6-trichloroguaiacol							
4,5,6-trichloroguaiacol	0.032	0.25	54	0.237	1.840	3.74	
tetrachloroguaiacol	0.016	0.14	26	0.099	0.886	4.45	

Table 4.3 Suggested half-life classes for phenolic compounds in various environmental compartments at 25 °C.

Compound	Air class	Water class	Soil class	Sediment class
phenol	2	3	4	5
2,4-dichloro-	3	3	5	6
2,4,5-trichloro-	4	4	6	7
2,4,6-trichloro-	4	4	6	7
2,3,4,6-tetrachloro-	5	5	6	7
pentachloro-	5	5	6	7
o-cresol	1	2	3	4
p-cresol	1	2	3	4
2,4-dimethylphenol	2	3	4	5
2-nitrophenol	3	3	5	6
4-nitrophenol	3	3	5	6
2,4-dinitrophenol	4	4	6	7
4,6-dinitro-*o*-cresol	4	4	6	7
catechol	2	3	4	5
guaiacol	2	3	4	5
tetrachloroguaiacol	4	4	6	7

where,

Class	Mean half-life (hours)	Range (hours)
1	5	< 10
2	17 (~ 1 day)	10-30
3	55 (~ 2 days)	30-100
4	170 (~ 1 week)	100-300
5	550 (~ 3 weeks)	300-1,000
6	1700 (~ 2 months)	1,000-3,000
7	5500 (~ 8 months)	3,000-10,000
8	17000 (~ 2 years)	10,000-30,000
9	~ 5 years	> 30,000

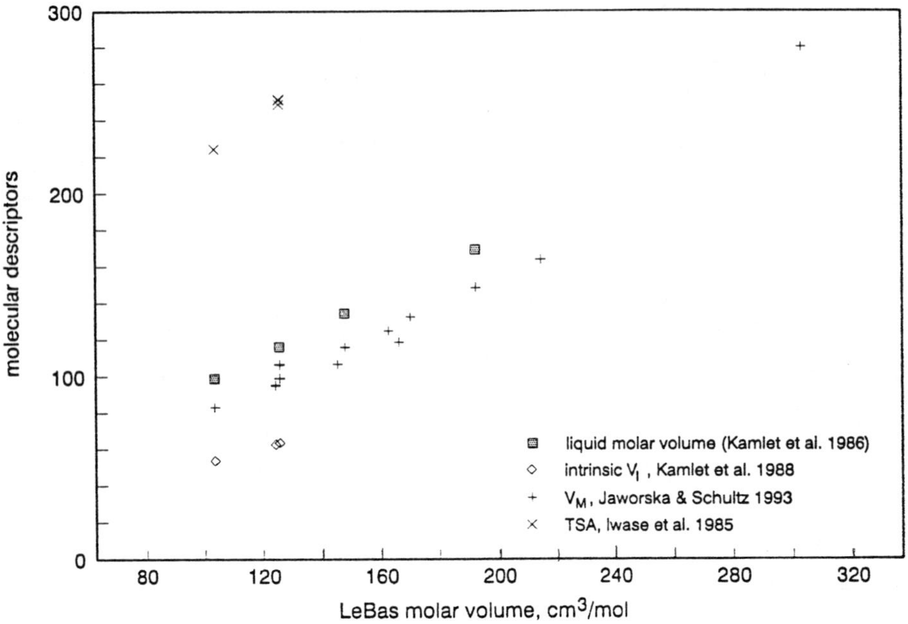

Figure 4.1 Plot of molecular descriptors versus LeBas molar volume

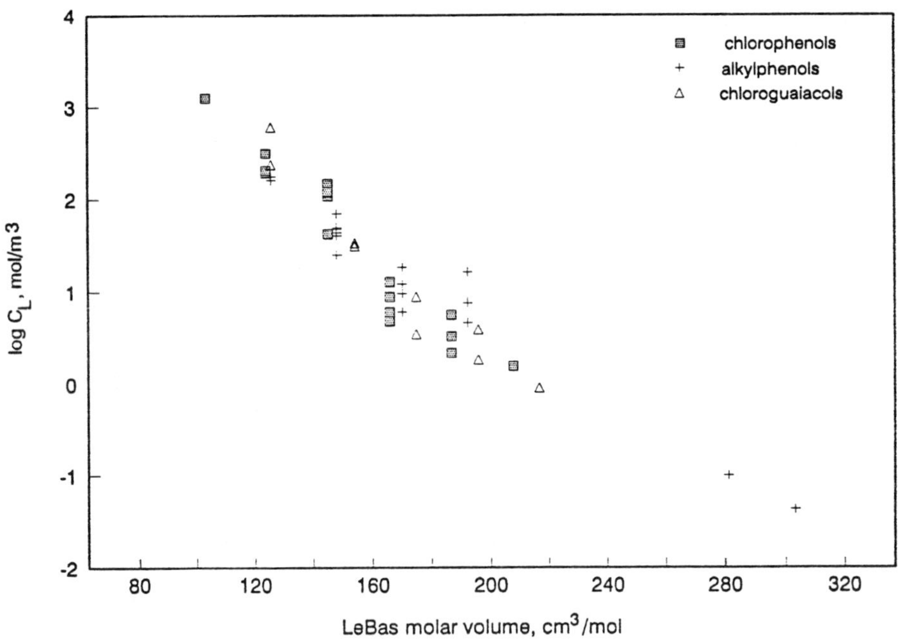

Figure 4.2 Plot of solubility (liquid or subcooled liquid) vs LeBas molar volume

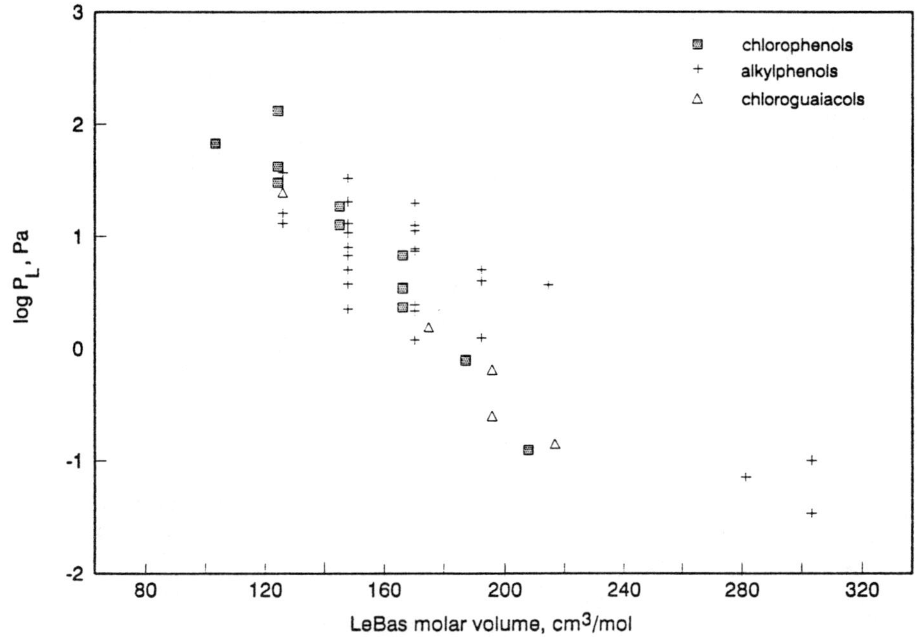

Figure 4.3 Plot of log vapour pressure (liquid or subcooled liquid) vs LeBas molar volume

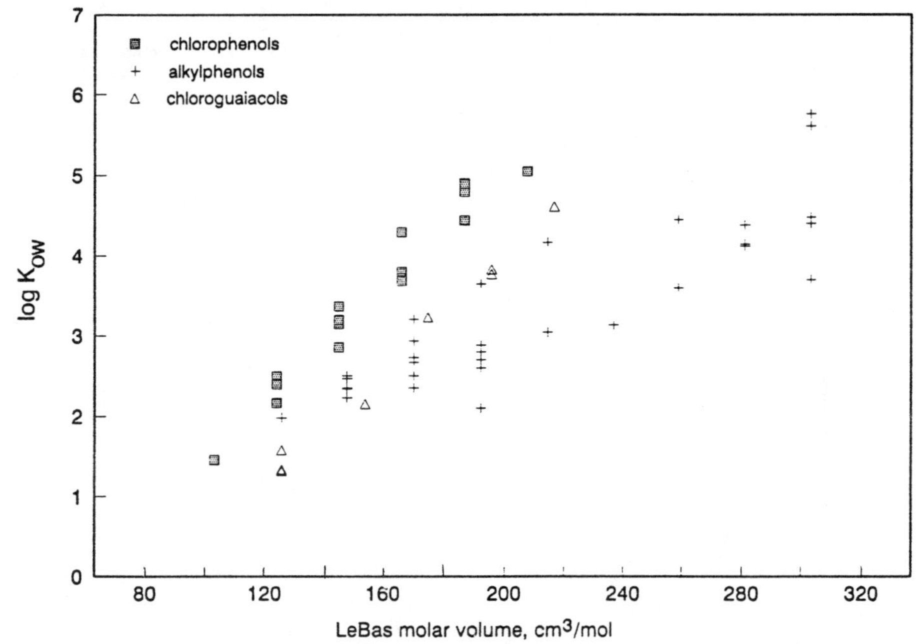

Figure 4.4 Plot of log K_{OW} versus LeBas molar volume

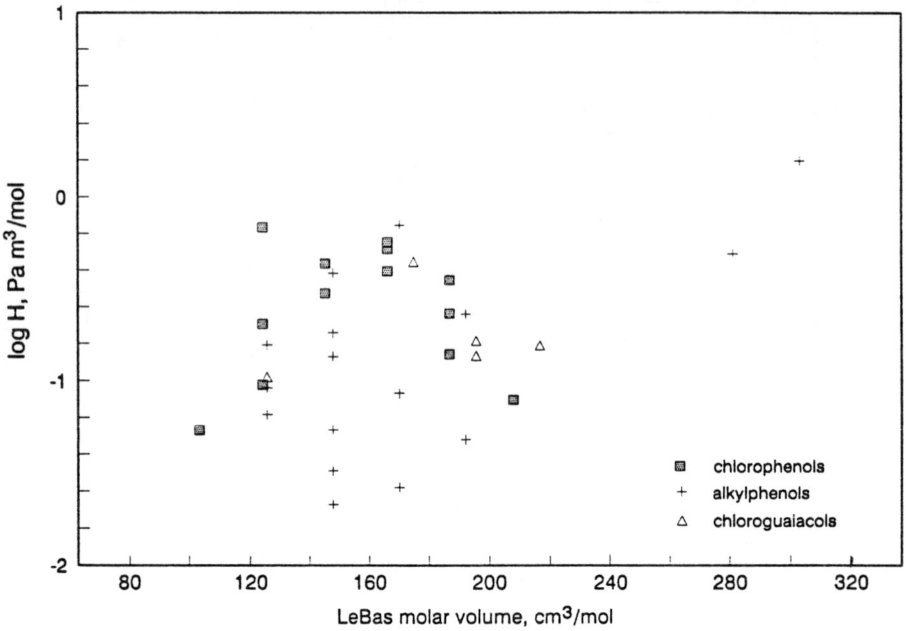

Figure 4.5 Plot of log Henry's law constant vs LeBas molar volume

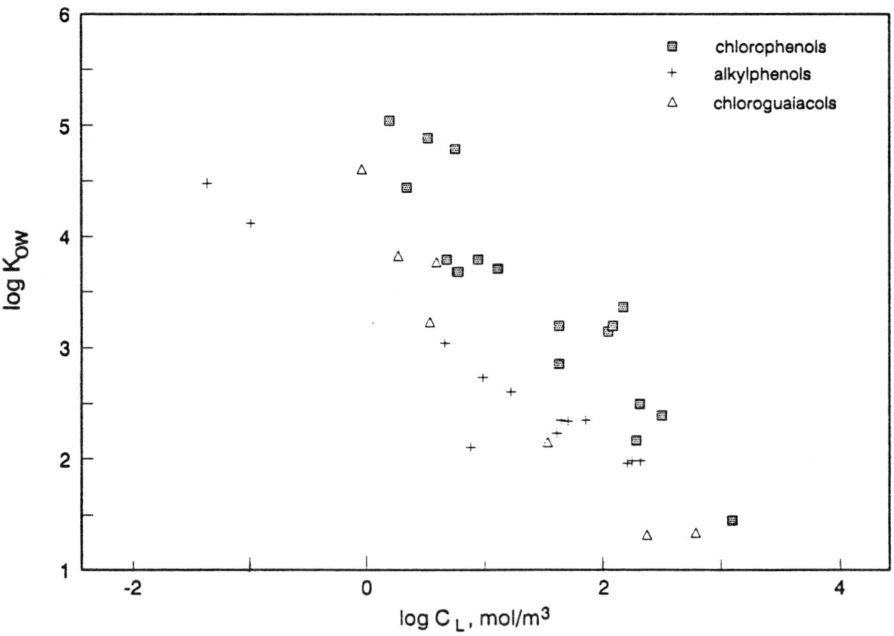

Figure 4.6 Plot of log K_{OW} versus log solubility (liquid or subcooled liquid)

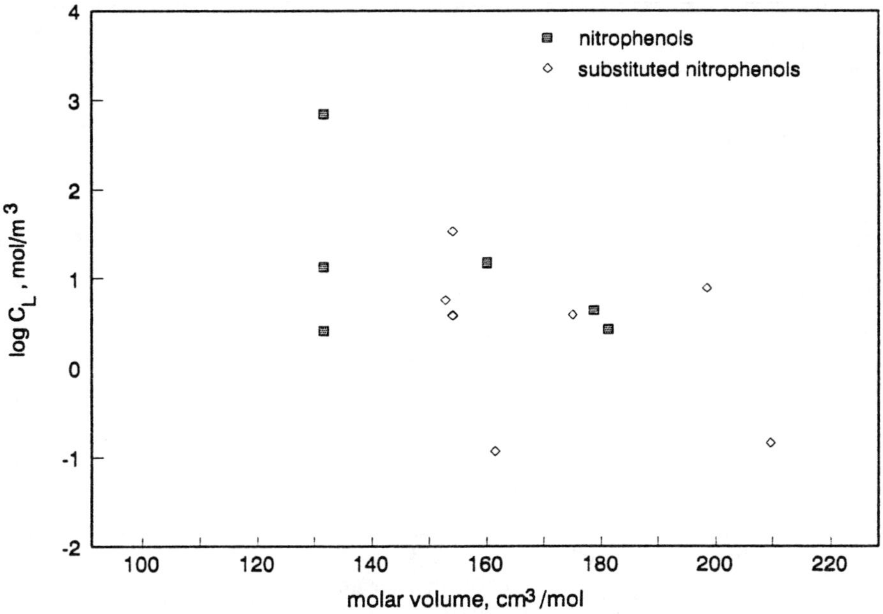

Figure 4.1a log solubility vs LeBas molar volume for nitrophenols

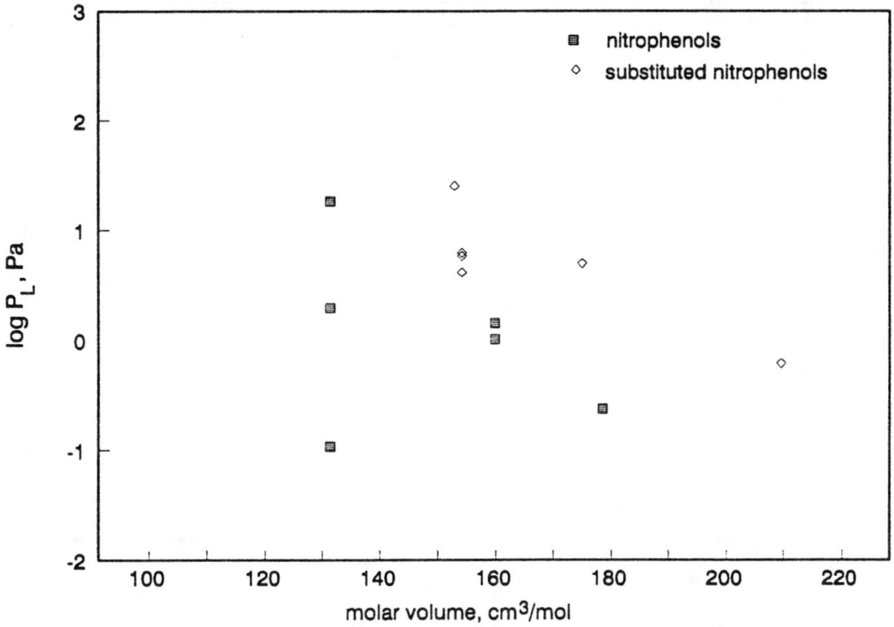

Figure 4.2a log vapor pressure vs LeBas molar volume for nitrophenols

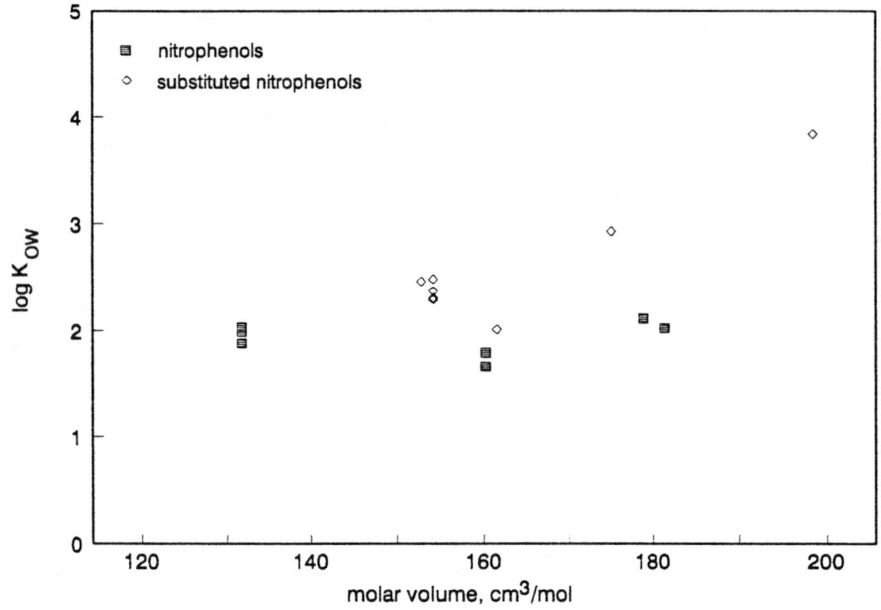

Figure 4.3a log K_{OW} versus LeBas molar volume for nitrophenols

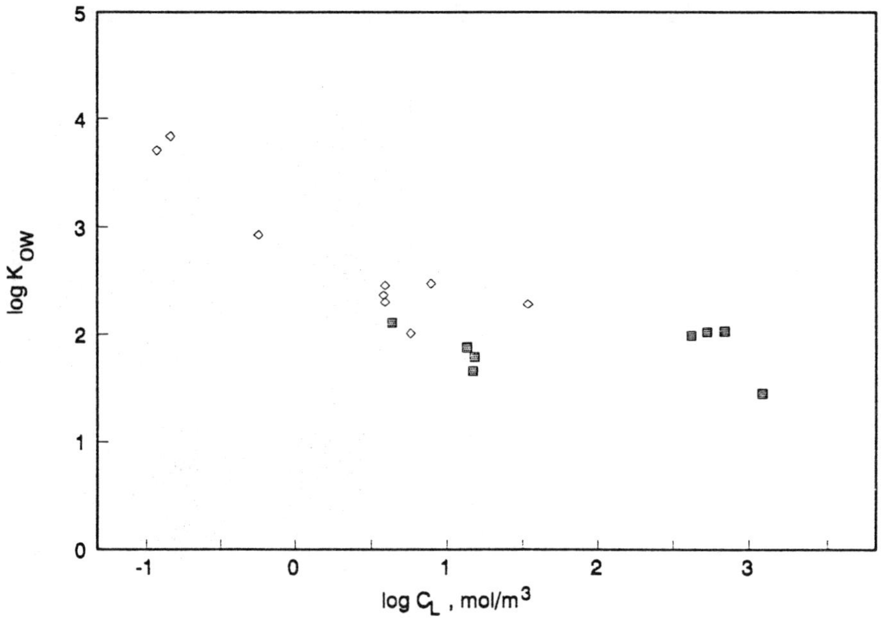

Figure 4.4a log K_{OW} versus log liquid solubility for nitrophenols

4.3 Illustrative Fugacity Calculations: Levels I, II, III

Chemical name: Phenol

Level I calculation: (six-compartment model)

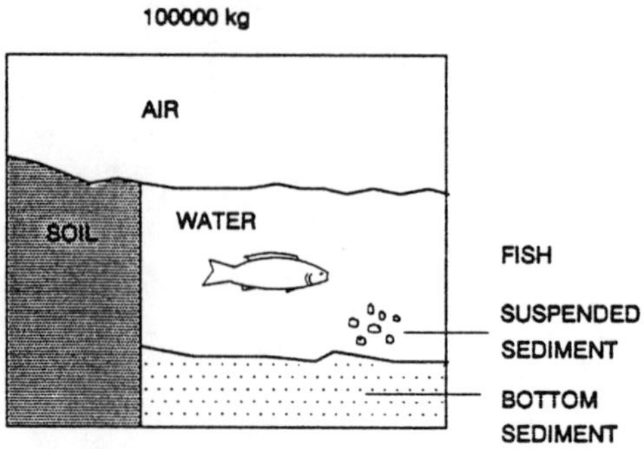

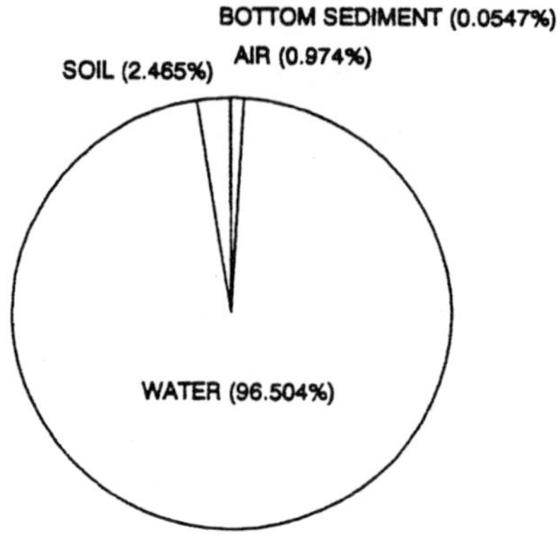

Distribution of mass

Level I calculation:

Physical-chemical properties:		Partition coeff.:		Z values in water:	
molecular wt., g/mol	94.10	HLC	5.01E-02	at data pH	5.1
melting point, C	41	Kaw	2.02E-05	neutral	19.978
solubility, g/m3	88360	Koc	11.8245	ionic	0.00
vapor pressure, Pa	4.70E+01	BCF	1.4420	total	19.98
log Kow	1.46	Ksw	0.5676	at environ. pH	5.1
fugacity ratio, F	0.6946	Ksd/w	1.1352	neutral	19.978
dissoc. const. pKa	10	Kssd/w	3.5474	ionic	0.00
		Kar/w	8.87E+04	total	19.98

Compartment	Z	Concentration			Amount	Amount
	mol/m3 Pa	mol/m3	mg/L (or g/m3)	ug/g	kg	%
Air	4.03E-04	1.04E-10	9.74E-09	8.22E-06	974.3238	9.74E-01
Water	2.00E+01	5.13E-06	4.83E-04	4.83E-04	96504	96.504
Soil	1.13E+01	2.91E-06	2.74E-04	1.14E-04	2465	2.4648
Biota (fish)	2.88E+01	7.39E-06	6.96E-04	6.96E-04	0.1392	1.39E-04
Suspended sediment	7.09E+01	1.82E-05	1.71E-03	1.14E-03	1.7117	1.71E-03
Bottom sediment	2.27E+01	5.82E-06	5.48E-04	2.28E-04	54.774	0.0548
	Total				100000	100

f = 2.567E-07 Pa

Chemical name: Phenol

Level II calculation: (six-compartment model)

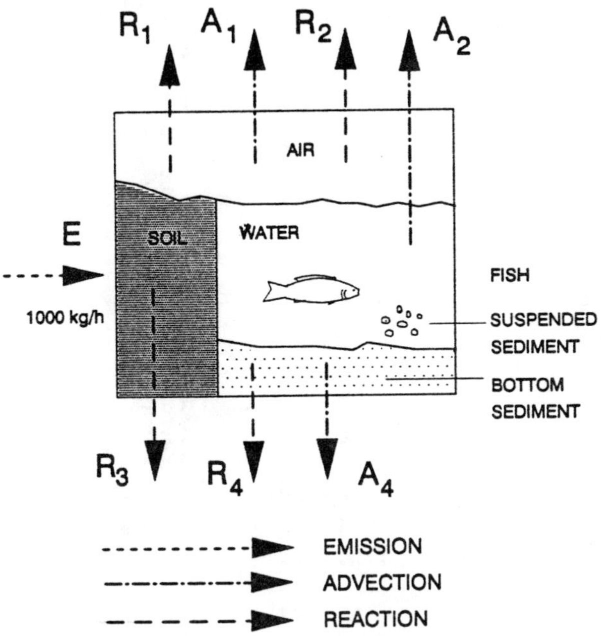

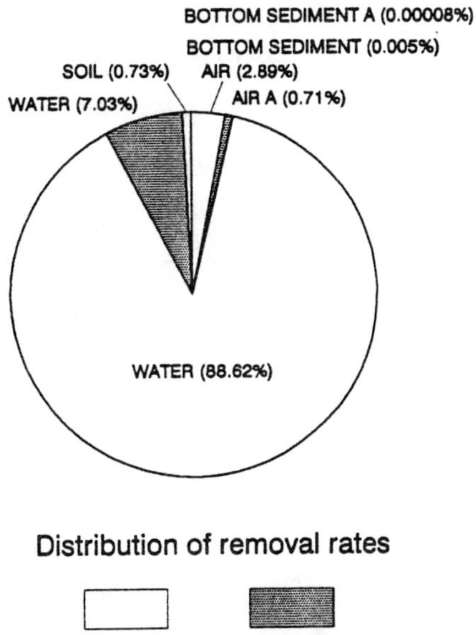

Distribution of removal rates

Level II Calculation: Compartment	Half-life h	D Values Reaction mol/Pa.h	D Values Advection mol/Pa.h	Conc'n mol/m3	Loss Reaction kg/h	Loss Advection kg/h	Removal %
Air	17	1.64E+09	4.03E+08	7.55E-11	28.9482	7.1013	3.6050
Water	55	5.03E+10	4.00E+09	3.74E-06	886.240	70.336	95.658
Soil	170	4.16E+08		2.12E-06	7.323		0.7323
Biota (fish)				5.39E-06			
Suspended sediment				1.33E-05			
Bottom sediment	550	2.86E+06	4.54E+04	4.24E-06	0.0503	7.98E-04	5.11E-03
	Total	5.24E+10	4.40E+09		922.56	77.439	100
	R + A		5.68E+10			1000	

f = 1.871E-07 Pa
Total amt = 72884 kg

Overall residence time = 72.88 h
Reaction time = 79.00 h
Advection time = 941.19 h

Fugacity Level III calculations: (four-compartment model)

Chemical name: Phenol

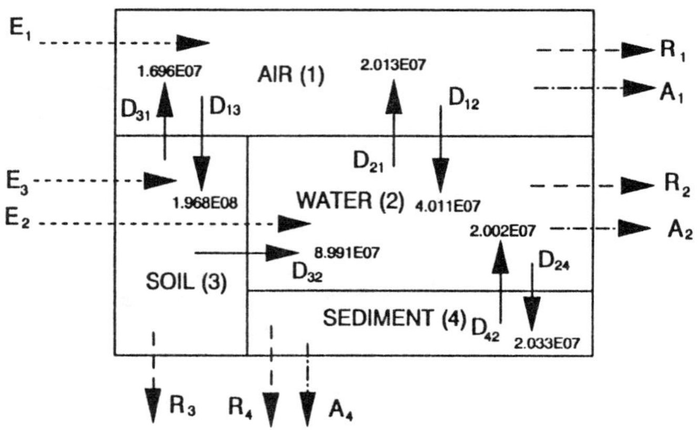

Phase Properties and Rates:

Compartment	Bulk Z mol/m3 Pa	Half-life h	D Values Reaction mol/Pa h	D Values Advection mol/Pa h
Air (1)	4.034E-04	17	1.64E+09	4.03E+08
Water (2)	1.998E+01	55	5.03E+10	4.00E+09
Soil (3)	1.166E+01	170	8.56E+08	
Sediment (4)	2.052E+01	550	1.29E+07	2.05E+05

	E(1)=1000	E(2)=1000	E(3)=1000	E(1,2,3)	
Overall residence time =	38.38	73.62	225.63	178.26	h
Reaction time =	46.73	79.47	227.91	184.60	h
Advection time =	214.75	1000.84	22520.61	5191.32	h

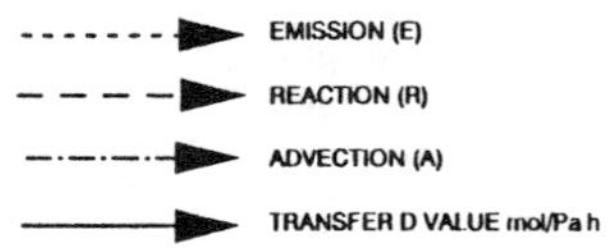

Phase Properties, Compositions, Transport and Transformation Rates:

Emission, kg/h			Fugacity, Pa				Concentration, g/m3				Amounts, kg				Total
E(1)	E(2)	E(3)	f(1)	f(2)	f(3)	f(4)	C(1)	C(2)	C(3)	C(4)	m(1)	m(2)	m(3)	m(4)	Amount, kg
1000	0	0	4.658E-06	5.011E-09	9.521E-07	3.073E-09	1.768E-07	9.421E-06	1.045E-03	5.933E-06	1.768E+04	1.884E+03	1.881E+04	2.967E+00	3.838E+04
0	1000	0	1.725E-09	1.955E-07	3.525E-10	1.199E-07	6.547E-11	3.675E-04	3.869E-07	2.314E-04	6.547E+00	7.349E+04	6.964E+00	1.157E+02	7.362E+04
0	0	1000	8.222E-08	1.834E-08	1.106E-05	1.125E-08	3.121E-09	3.449E-05	1.213E-02	2.172E-05	3.121E+02	6.897E+03	2.184E+05	1.086E+01	2.256E+05
50	250	700	2.909E-07	6.195E-08	7.787E-06	3.799E-08	1.104E-08	1.165E-04	8.546E-03	7.336E-05	1.104E+03	2.330E+04	1.538E+05	3.668E+01	1.783E+05

Emission, kg/h			Loss, Reaction, kg/h				Loss, Advection, kg/h			Intermedia Rate of Transport, kg/h T12	T21	T13	T31	T32	T24	T42
E(1)	E(2)	E(3)	R(1)	R(2)	R(3)	R(4)	A(1)	A(2)	A(4)	air-water	water-air	air-soil	soil-air	soil-water	water-sed	sed-water
1000	0	0	7.209E+02	2.374E+01	7.67E+01	3.738E-03	1.768E+02	1.884E+00	5.933E-05	1.758E+01	9.492E-03	8.625E+01	1.519E+00	8.056E+00	9.588E-03	5.790E-03
0	1000	0	2.669E-01	9.260E+02	2.84E-02	1.458E-01	6.547E-02	7.349E+01	2.314E-03	6.509E-03	3.702E-01	3.193E-02	5.626E-04	2.983E-03	3.740E-01	2.259E-01
0	0	1000	1.272E+01	8.691E+01	8.90E+02	1.368E-02	3.121E+00	6.897E+00	2.172E-04	3.103E-01	3.475E-02	1.522E+00	1.764E+01	9.354E+01	3.510E-02	2.120E-02
50	250	700	4.502E+01	2.935E+02	6.27E+02	4.622E-02	1.104E+01	2.330E+01	7.336E-04	1.098E+00	1.174E-01	5.386E+00	1.243E+01	6.588E+01	1.185E-01	7.159E-02

Level III Distribution

Chemical name: Phenol

Distribution of mass

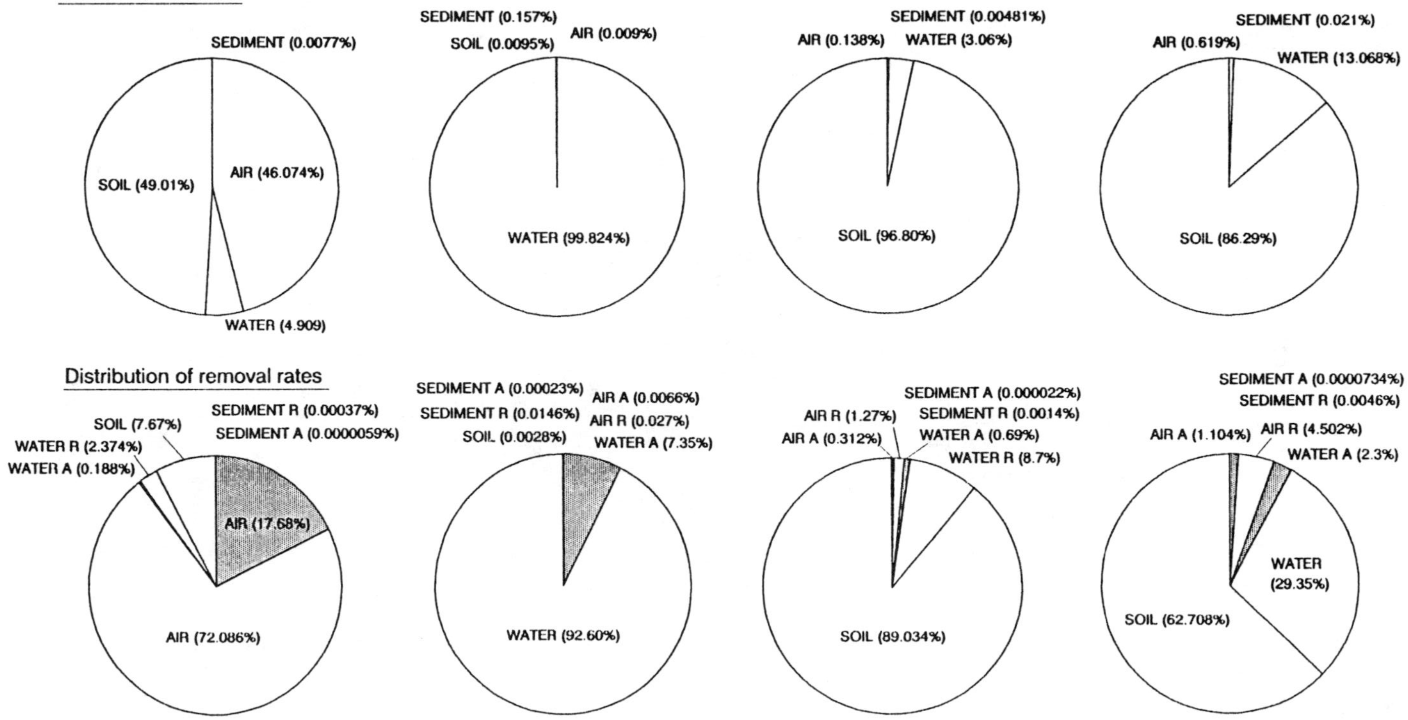

Emission rates:

E(1) = 1000 kg/h	E(1) = 0	E(1) = 0	E(1) = 50 kg/h
E(2) = 0	E(2) = 1000 kg/h	E(2) = 0	E(2) = 250 kg/h
E(3) = 0	E(3) = 0	E(3) = 1000 kg/h	E(3) = 700 kg/h

Reaction   Advection

Residence time:

t = 38.38 h	t = 73.62 h	t = 225.6 h	t = 178 h

Chemical name: 4-Chlorophenol

Level I calculation: (six-compartment model)

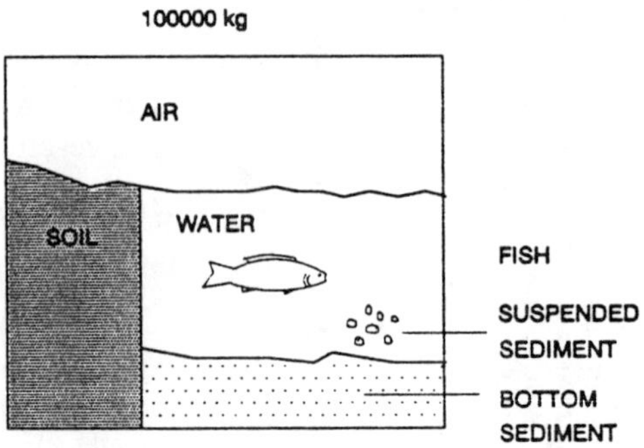

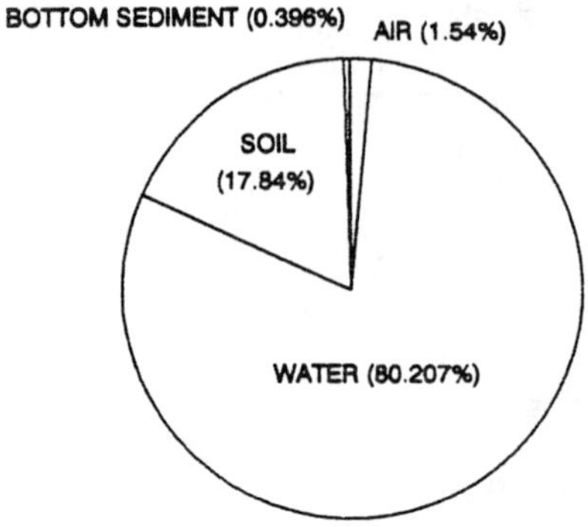

Distribution of mass

Level I calculation:

Physical-chemical properties:		Partition coeff.:		Z values in water:	
molecular wt., g/mol	128.56	HLC	0.09523	at data pH	5.1
melting point, C	43	Kaw	3.84E-05	neutral	10.501
solubility, g/m3	27000	Koc	102.9873	ionic	1.32E-04
vapor pressure, Pa	20.00	BCF	12.5594	total	10.501
log Kow	2.40	Ksw	4.9434	at environ. pH	5.1
fugacity ratio, F	0.6637	Ksd/w	9.8868	neutral	10.501
dissoc. const. pKa	10	Kssd/w	30.8962	ionic	1.32E-04
		Kar/w	1.99E+05	total	10.501

Compartment	Z	Concentration			Amount	Amount
	mol/m3 Pa	mol/m3	mg/L (or g/m3)	ug/g	kg	%
Air	4.03E-04	1.20E-10	1.54E-08	1.30E-05	1540.672	1.5407
Water	1.05E+01	3.12E-06	4.01E-04	4.01E-04	80207.14	80.2071
Soil	5.19E+01	1.54E-05	1.98E-03	8.26E-04	17842.29	17.8423
Biota (fish)	1.32E+02	3.92E-05	5.04E-03	5.04E-03	1.0074	1.01E-03
Suspended sediment	3.24E+02	9.64E-05	1.24E-02	8.26E-03	12.3905	1.24E-02
Bottom sediment	1.04E+02	3.08E-05	3.96E-03	1.65E-03	396.495	0.3965
	Total				100000	100

f = 2.971E-07 Pa

Chemical name: 4-Chlorophenol

Level II calculation: (six-compartment model)

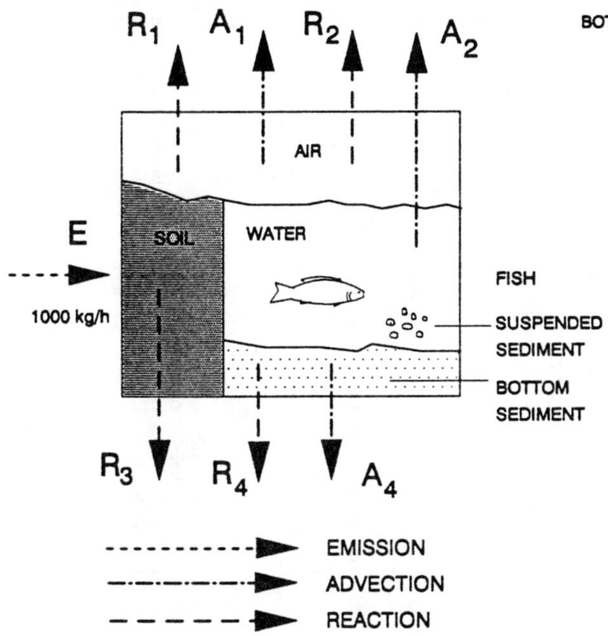

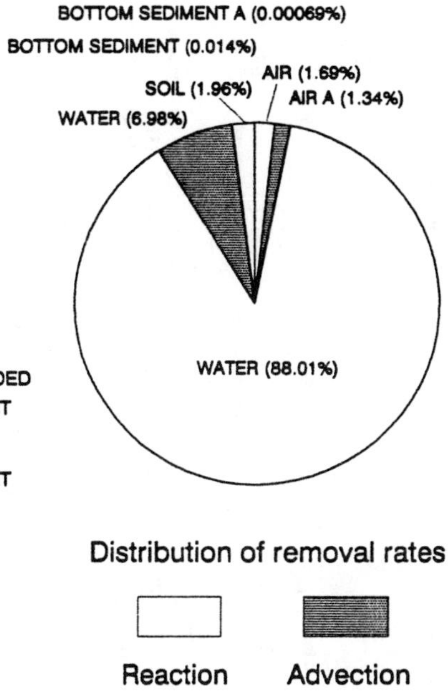

Level II Calculation:

Compartment	Half-life h	D Values Reaction mol/Pa.h	D Values Advection mol/Pa.h	Conc'n mol/m3	Loss Reaction kg/h	Loss Advection kg/h	Removal %
Air	55	5.08E+08	4.03E+08	1.04E-10	16.906	13.417	3.0323
Water	55	2.65E+10	2.10E+09	2.72E-06	880.102	69.849	94.995
Soil	550	5.89E+08		1.34E-05	19.578		1.9578
Biota (fish)				3.41E-05			
Suspended sediment				8.39E-05			
Bottom sediment	1700	4.23E+06	2.08E+05	2.69E-05	0.1408	6.91E-03	1.48E-02
	Total	2.76E+10	2.50E+09		916.73	83.273	100
	R + A		3.01E+10			1000	

f = 2.587E-07 Pa

Total amt = 87086 kg

Overall residence time = 87.09 h

Reaction time = 95.00 h

Advection time = 1045.79 h

Fugacity Level III calculations: (four-compartment model)

Chemical name: 4-Chlorophenol

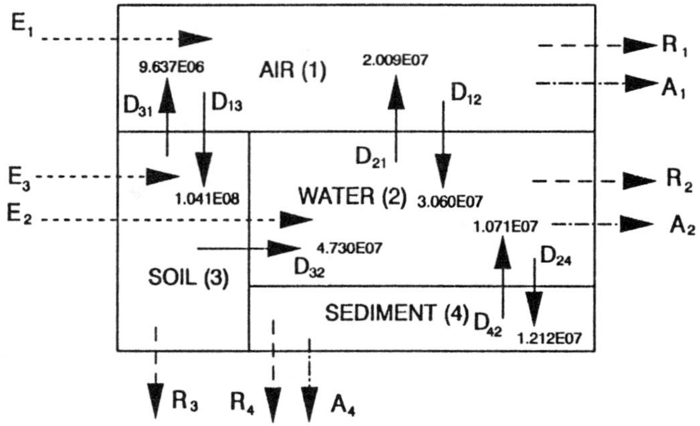

Phase Properties and Rates:

Compartment	Bulk Z mol/m3 Pa	Half-life h	D Values Reaction mol/Pa h	D Values Advection mol/Pa h
Air (1)	4.034E-04	55	5.08E+08	4.03E+08
Water (2)	1.050E+01	55	2.65E+10	2.10E+09
Soil (3)	2.911E+01	550	6.60E+08	
Sediment (4)	2.916E+01	1700	5.94E+06	2.92E+05

	E(1)=1000	E(2)=1000	E(3)=1000	E(1,2,3)
Overall residence time =	114.06	73.91	737.04	540.11 h
Reaction time =	186.58	79.80	744.55	565.52 h
Advection time =	293.48	1002.23	73059.47	12020.67 h

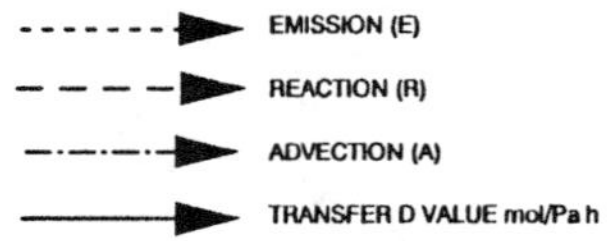

Phase Properties, Compositions, Transport and Transformation Rates:

Emission, kg/h			Fugacity, Pa				Concentration, g/m3				Amounts, kg				Total Amount, kg
E(1)	E(2)	E(3)	f(1)	f(2)	f(3)	f(4)	C(1)	C(2)	C(3)	C(4)	m(1)	m(2)	m(3)	m(4)	
1000	0	0	7.443E-06	9.753E-09	1.081E-06	6.978E-09	3.860E-07	1.317E-05	4.045E-03	2.616E-05	3.860E+04	2.634E+03	7.281E+04	1.308E+01	1.141E+05
0	1000	0	5.231E-09	2.721E-07	7.598E-10	1.946E-07	2.713E-10	3.673E-04	2.843E-06	7.298E-04	2.713E+01	7.347E+04	5.117E+01	3.649E+02	7.391E+04
0	0	1000	1.004E-07	1.808E-08	1.086E-05	1.293E-08	5.206E-09	2.441E-05	4.065E-02	4.849E-05	5.206E+02	4.882E+03	7.316E+05	2.425E+01	7.370E+05
50	250	700	4.437E-07	8.116E-08	7.658E-06	5.806E-08	2.301E-08	1.096E-04	2.865E-02	2.177E-04	2.301E+03	2.192E+04	5.158E+05	1.089E+02	5.401E+05

Emission, kg/h			Loss, Reaction, kg/h				Loss, Advection, kg/h			Intermedia Rate of Transport, kg/h T12	T21	T13	T31	T32	T24	T42
E(1)	E(2)	E(3)	R(1)	R(2)	R(3)	R(4)	A(1)	A(2)	A(4)	air-water	water-air	air-soil	soil-air	soil-water	water-sed	sed-water
1000	0	0	4.864E+02	3.319E+01	9.17E+01	5.333E-03	3.860E+02	2.634E+00	2.616E-04	2.928E+01	2.519E-02	9.966E+01	1.339E+00	6.574E+00	1.520E-02	9.606E-03
0	1000	0	3.418E-01	9.257E+02	6.45E-02	1.488E-01	2.713E-01	7.347E+01	7.298E-03	2.057E-02	7.028E-01	7.004E-02	9.413E-04	4.620E-03	4.240E-01	2.680E-01
0	0	1000	6.559E+00	6.151E+01	9.22E+02	9.884E-03	5.206E+00	4.882E+00	4.849E-04	3.948E-01	4.670E-02	1.344E+00	1.346E+01	6.605E+01	2.817E-02	1.781E-02
50	250	700	2.900E+01	2.761E+02	6.50E+02	4.437E-02	2.301E+01	2.192E+01	2.177E-03	1.745E+00	2.096E-01	5.941E+00	9.487E+00	4.657E+01	1.265E-01	7.994E-02

Level III Distribution

Chemical name: 4-Chlorophenol

Distribution of mass

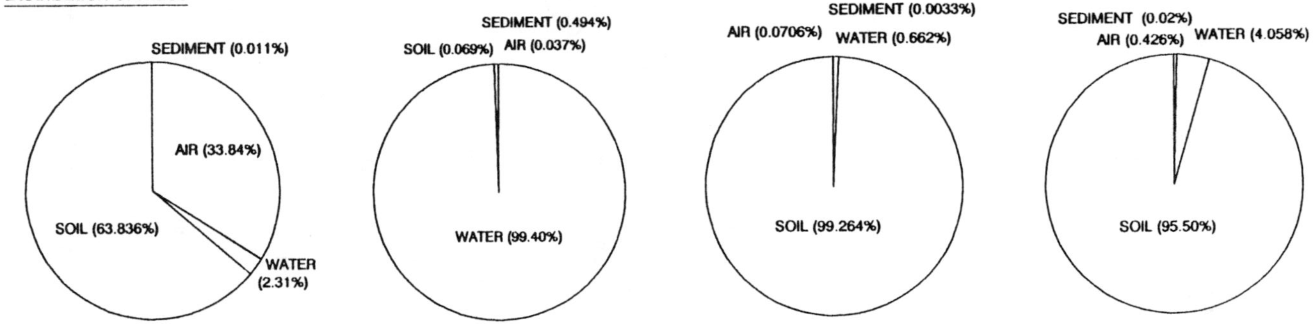

Distribution of removal rates

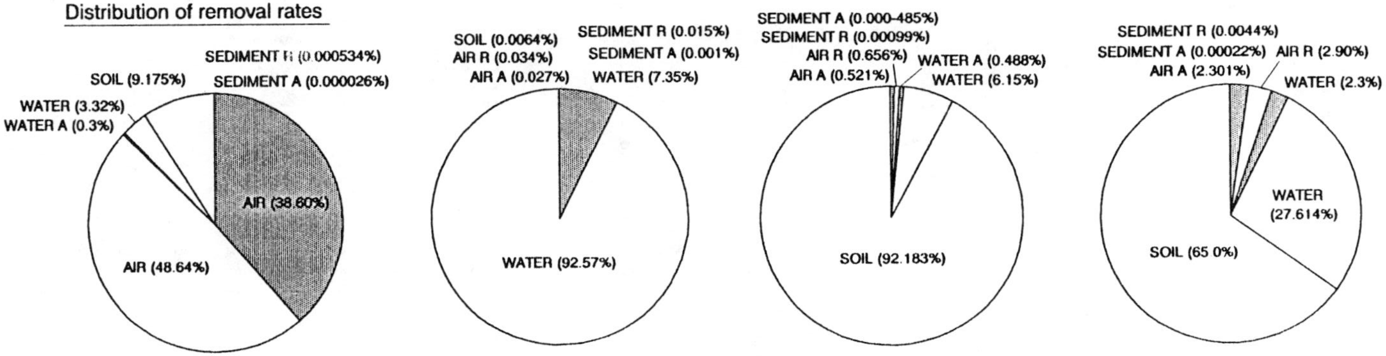

Emission rates:

E(1) = 1000 kg/h	E(1) = 0	E(1) = 0	E(1) = 50 kg/h
E(2) = 0	E(2) = 1000 kg/h	E(2) = 0	E(2) = 250 kg/h
E(3) = 0	E(3) = 0	E(3) = 1000 kg/h	E(3) = 700 kg/h

Residence time:

t = 114 h	t = 74 h	t = 737 h	t = 540 h

Chemical name: 2,4-Dichlorophenol

Level I calculation: (six-compartment model)

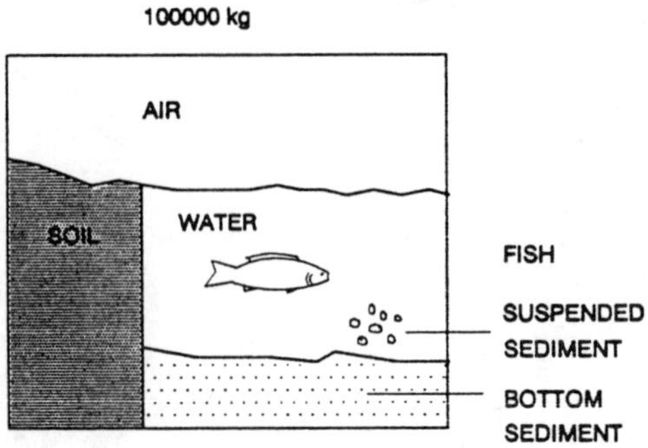

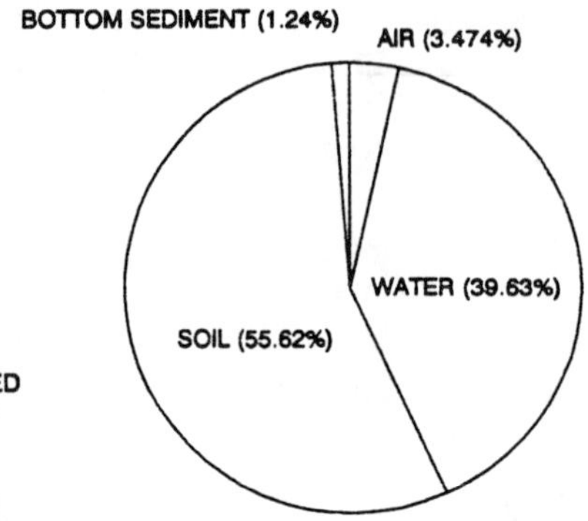

Distribution of mass

Level I calculation:

Physical-chemical properties:		Partition coeff.:		Z values in water:	
molecular wt., g/mol	163.00	HLC	0.43467	at data pH	5.1
melting point, C	44	Kaw	1.75E-04	neutral	2.297
solubility, g/m3	4500	Koc	649.8062	ionic	3.64E-03
vapor pressure, Pa	12.00	BCF	79.2447	total	2.301
log Kow	3.20	Ksw	31.1907	at environ. pH	5.1
fugacity ratio, F	0.6488	Ksd/w	62.3814	neutral	2.297
dissoc. const. pKa	7.9	Kssd/w	194.9419	ionic	3.64E-03
		Kar/w	3.24E+05	total	2.301

Compartment	Z	Concentration			Amount	Amount
	mol/m3 Pa	mol/m3	mg/L (or g/m3)	ug/g	kg	%
Air	4.03E-04	2.13E-10	3.47E-08	2.93E-05	3474.384	3.4744
Water	2.30E+00	1.22E-06	1.98E-04	1.98E-04	39627.47	39.6275
Soil	7.18E+01	3.79E-05	6.18E-03	2.58E-03	55620.38	55.6204
Biota (fish)	1.82E+02	9.63E-05	1.57E-02	1.57E-02	3.1403	3.14E-03
Suspended sediment	4.48E+02	2.37E-04	3.86E-02	2.58E-02	38.6253	3.86E-02
Bottom sediment	1.44E+02	7.58E-05	1.24E-02	5.15E-03	1236.008	1.2360
	Total				100000	100

f = 5.284E-07 Pa

Chemical name: 2,4-Dichlorophenol

Level II calculation: (six-compartment model)

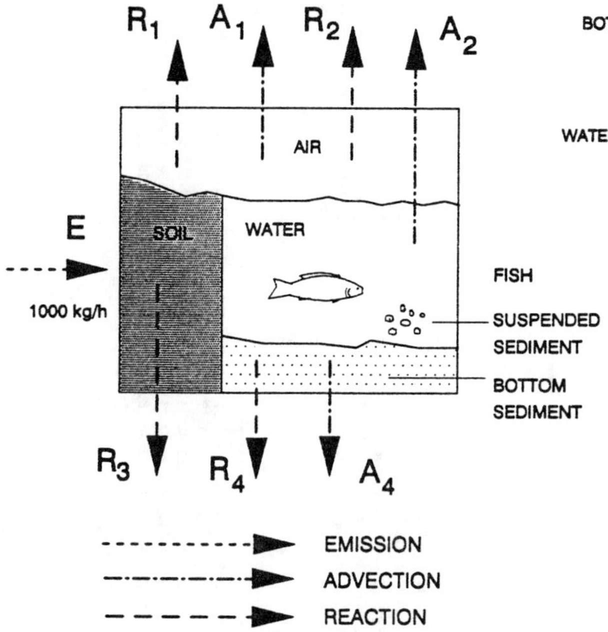

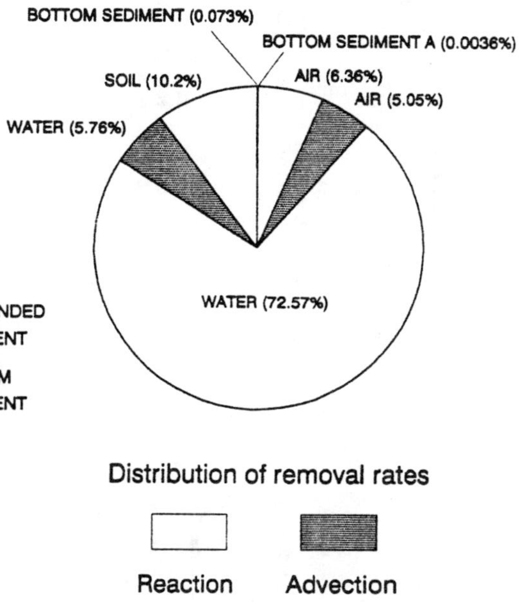

Distribution of removal rates

Level II Calculation:

Compartment	Half-life h	D Values Reaction mol/Pa.h	D Values Advection mol/Pa.h	Conc'n mol/m3	Loss Reaction kg/h	Loss Advection kg/h	Removal %
Air	55	5.08E+08	4.03E+08	3.10E-10	63.624	50.495	11.4119
Water	55	5.80E+09	4.60E+08	1.77E-06	725.667	57.593	78.326
Soil	550	8.14E+08		5.51E-05	101.853		10.1853
Biota (fish)				1.40E-04			
Suspended sediment				3.44E-04			
Bottom sediment	1700	5.85E+06	2.87E+05	1.10E-04	0.7323	3.59E-02	7.68E-02
	Total	7.12E+09	8.64E+08		891.88	108.124	100
	R + A		7.98E+09			1000	

f = 7.679E-07 Pa

Total amt = 145335 kg

Overall residence time = 145.34 h

Reaction time = 162.95 h

Advection time = 1344.16 h

Fugacity Level III calculations: (four-compartment model)

Chemical name: 2,4-Dichlorophenol

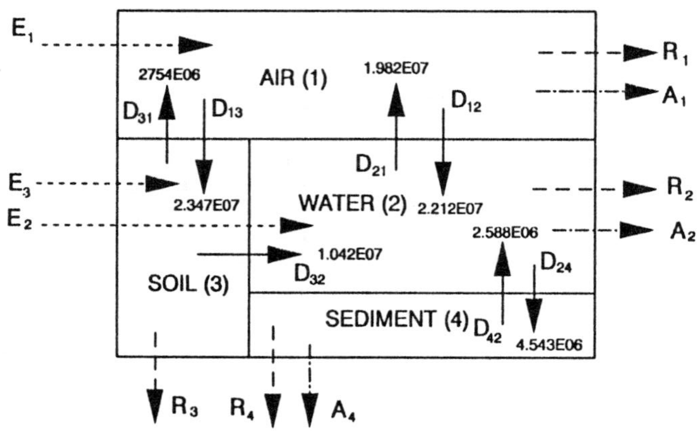

Phase Properties and Rates:

Compartment	Bulk Z mol/m3 Pa	Half-life h	D Values Reaction mol/Pa h	D Values Advection mol/Pa h
Air (1)	4.034E-04	55	5.08E+08	4.03E+08
Water (2)	2.303E+00	55	5.80E+09	4.61E+08
Soil (3)	3.657E+01	550	8.29E+08	
Sediment (4)	3.054E+01	1700	6.23E+06	3.05E+05

	E(1)=1000	E(2)=1000	E(3)=1000	E(1,2,3)
Overall residence time =	63.04	74.67	782.37	569.48 h
Reaction time =	109.30	80.69	784.18	594.09 h
Advection time =	148.98	1000.68	339286	13746.28 h

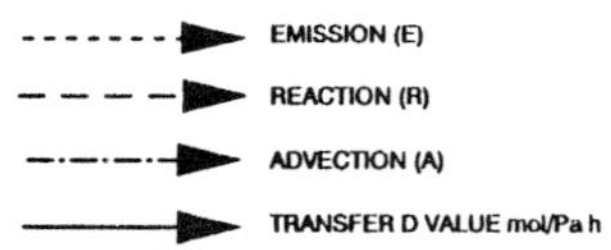

Phase Properties, Compositions, Transport and Transformation Rates:

Emission, kg/h			Fugacity, Pa				Concentration, g/m3				Amounts, kg				Total Amount, kg
E(1)	E(2)	E(3)	f(1)	f(2)	f(3)	f(4)	C(1)	C(2)	C(3)	C(4)	m(1)	m(2)	m(3)	m(4)	
1000	0	0	6.409E-06	2.285E-08	1.785E-07	1.138E-08	4.215E-07	8.578E-06	1.064E-03	5.668E-05	4.215E+04	1.716E+03	1.915E+04	2.834E+01	6.304E+04
0	1000	0	2.021E-08	9.758E-07	5.628E-10	4.862E-07	1.329E-09	3.663E-04	3.355E-06	2.421E-03	1.329E+02	7.327E+04	6.039E+01	1.210E+03	7.467E+04
0	0	1000	2.120E-08	1.214E-08	7.282E-06	6.048E-09	1.394E-09	4.557E-06	4.341E-02	3.011E-05	1.394E+02	9.115E+02	7.813E+05	1.506E+01	7.824E+05
50	250	700	3.404E-07	2.536E-07	5.106E-06	1.263E-07	2.238E-08	9.520E-05	3.044E-02	6.290E-04	2.238E+03	1.904E+04	5.479E+05	3.145E+02	5.695E+05

Emission, kg/h			Loss, Reaction, kg/h				Loss, Advection, kg/h			Intermedia Rate of Transport, kg/h T12	T21	T13	T31	T32	T24	T42
E(1)	E(2)	E(3)	R(1)	R(2)	R(3)	R(4)	A(1)	A(2)	A(4)	air-water	water-air	air-soil	soil-air	soil-water	water-sed	sed-water
1000	0	0	5.311E+02	2.162E+01	2.41E+01	1.155E-02	4.215E+02	1.716E+00	5.668E-04	2.311E+01	7.383E-02	2.452E+01	8.014E-02	3.031E-01	1.692E-02	4.802E-03
0	1000	0	1.674E+00	9.231E+02	7.61E-02	4.934E-01	1.329E+00	7.327E+01	2.421E-02	7.288E-02	3.153E+00	7.730E-02	2.527E-04	9.557E-04	7.226E-01	2.051E-01
0	0	1000	1.757E+00	1.148E+01	9.84E+02	6.138E-03	1.394E+00	9.115E-01	3.011E-04	7.646E-02	3.923E-02	8.110E-02	3.269E+00	1.236E+01	8.990E-03	2.551E-03
50	250	700	2.820E+01	2.399E+02	6.90E+02	1.282E-01	2.238E+01	1.904E+01	6.290E-03	1.227E+00	8.194E-01	1.302E+00	2.293E+00	8.671E+00	1.878E-01	5.329E-02

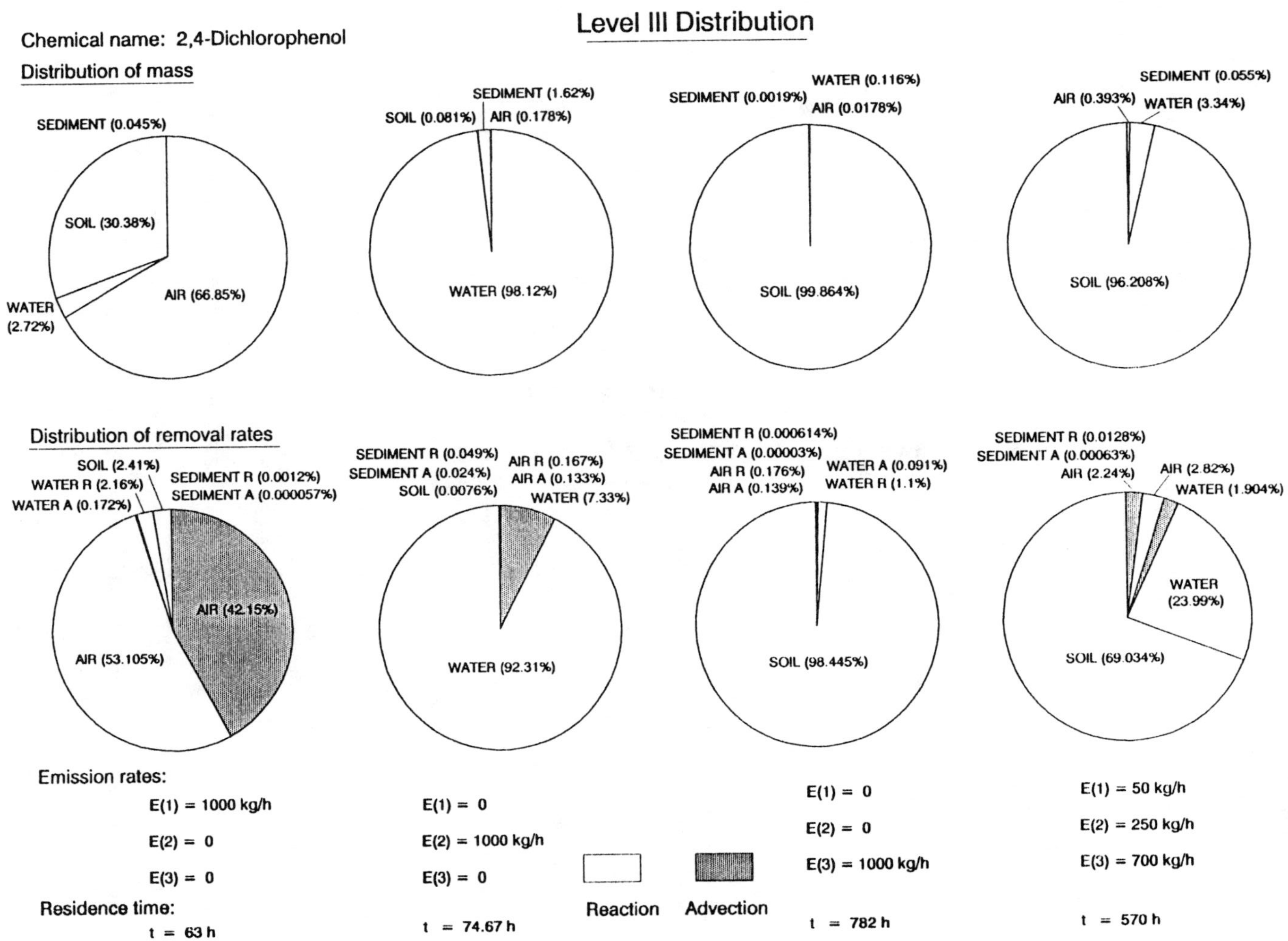
Level III Distribution
Chemical name: 2,4-Dichlorophenol
Distribution of mass
SEDIMENT (0.045%)
SOIL (30.38%)
WATER (2.72%)
AIR (66.85%)
SEDIMENT (1.62%)
SOIL (0.081%)
AIR (0.178%)
WATER (98.12%)
SEDIMENT (0.0019%)
WATER (0.116%)
AIR (0.0178%)
SOIL (99.864%)
SEDIMENT (0.055%)
AIR (0.393%)
WATER (3.34%)
SOIL (96.208%)
Distribution of removal rates
SOIL (2.41%)
WATER R (2.16%)
WATER A (0.172%)
SEDIMENT R (0.0012%)
SEDIMENT A (0.000057%)
AIR (42.15%)
AIR (53.105%)
SEDIMENT R (0.049%)
SEDIMENT A (0.024%)
SOIL (0.0076%)
AIR R (0.167%)
AIR A (0.133%)
WATER (7.33%)
WATER (92.31%)
SEDIMENT R (0.000614%)
SEDIMENT A (0.00003%)
AIR R (0.176%)
AIR A (0.139%)
WATER A (0.091%)
WATER R (1.1%)
SOIL (98.445%)
SEDIMENT R (0.0128%)
SEDIMENT A (0.00063%)
AIR (2.24%)
AIR (2.82%)
WATER (1.904%)
WATER (23.99%)
SOIL (69.034%)
Emission rates:
E(1) = 1000 kg/h
E(2) = 0
E(3) = 0
Residence time:
t = 63 h
E(1) = 0
E(2) = 1000 kg/h
E(3) = 0
t = 74.67 h
Reaction
Advection
E(1) = 0
E(2) = 0
E(3) = 1000 kg/h
t = 782 h
E(1) = 50 kg/h
E(2) = 250 kg/h
E(3) = 700 kg/h
t = 570 h

Chemical name: 2,4,6-Trichlorophenol

Level I calculation: (six-compartment model)

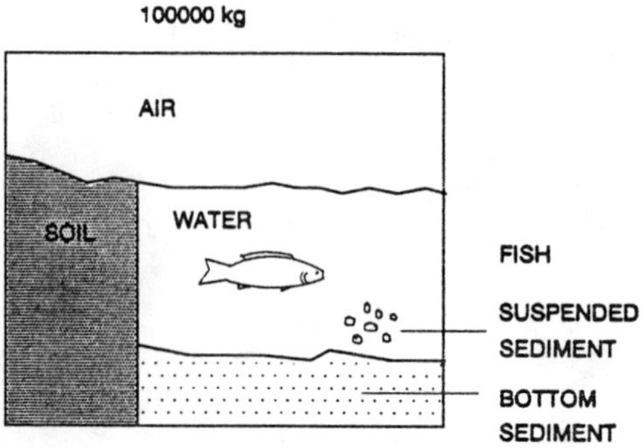

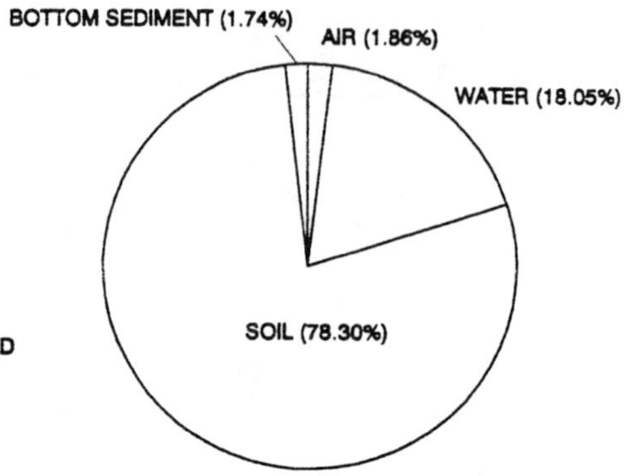

Distribution of mass

Level I calculation:

Physical-chemical properties:		Partition coeff.:		Z values in water:	
molecular wt., g/mol	197.45	HLC	0.50955	at data pH	5.1
melting point, C	69.5	Kaw	2.06E-04	neutral	1.784
solubility, g/m3	434	Koc	2008.0932	ionic	0.1784
vapor pressure, Pa	1.12	BCF	244.8894	total	1.963
log Kow	3.69	Ksw	96.3885	at environ. pH	5.1
fugacity ratio, F	0.3630	Ksd/w	192.7769	neutral	1.784
dissoc. const. pKa	6.1	Kssd/w	602.4279	ionic	0.1784
		Kar/w	1.94E+06	total	1.963

Compartment	Z	Concentration			Amount	Amount
	mol/m3 Pa	mol/m3	mg/L (or g/m3)	ug/g	kg	%
Air	4.03E-04	9.40E-11	1.86E-08	1.57E-05	1855.274	1.8553
Water	1.96E+00	4.57E-07	9.03E-05	9.03E-05	18050.84	18.0508
Soil	1.89E+02	4.41E-05	8.70E-03	3.62E-03	78295.20	78.2952
Biota (fish)	4.81E+02	1.12E-04	2.21E-02	2.21E-02	4.4205	4.42E-03
Suspended sediment	1.18E+03	2.75E-04	5.44E-02	3.62E-02	54.3717	5.44E-02
Bottom sediment	3.78E+02	8.81E-05	1.74E-02	7.25E-03	1739.893	1.7399
	Total				100000	100

f = 2.329E-07 Pa

Chemical name: 2,4,6-Trichlorophenol

Level II calculation: (six-compartment model)

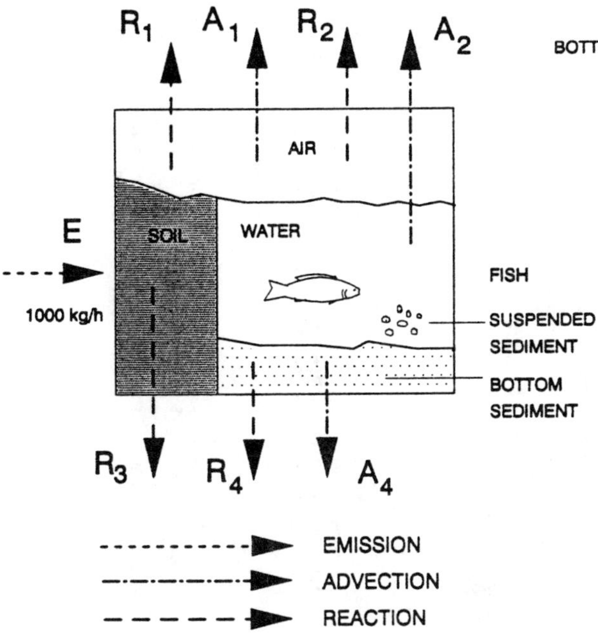

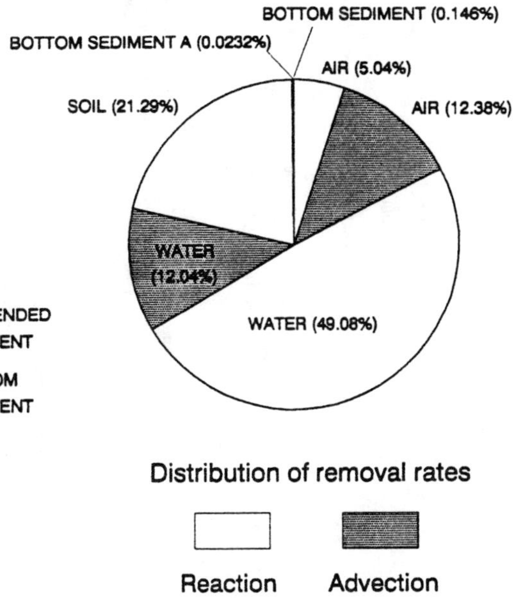

Level II Calculation:

Compartment	Half-life h	D Values Reaction mol/Pa.h	D Values Advection mol/Pa.h	Conc'n mol/m3	Loss Reaction kg/h	Loss Advection kg/h	Removal %
Air	170	1.64E+08	4.03E+08	6.27E-10	50.446	123.750	17.4196
Water	170	1.60E+09	3.93E+08	3.05E-06	490.816	120.402	61.122
Soil	1700	6.94E+08		2.94E-04	212.891		21.2891
Biota (fish)				7.47E-04			
Suspended sediment				1.84E-03			
Bottom sediment	5500	4.77E+06	7.57E+05	5.88E-04	1.4623	0.2321	0.1694
	Total	2.46E+09	7.96E+08		755.62	244.384	100
	R + A		3.25E+09			1000	

f = 1.554E-06 Pa

Total amt = 667017 kg

Overall residence time = 667.02 h

Reaction time = 882.75 h

Advection time = 2729.38 h

Fugacity Level III calculations: (four-compartment model)

Chemical name: 2,4,6-Trichlorophenol

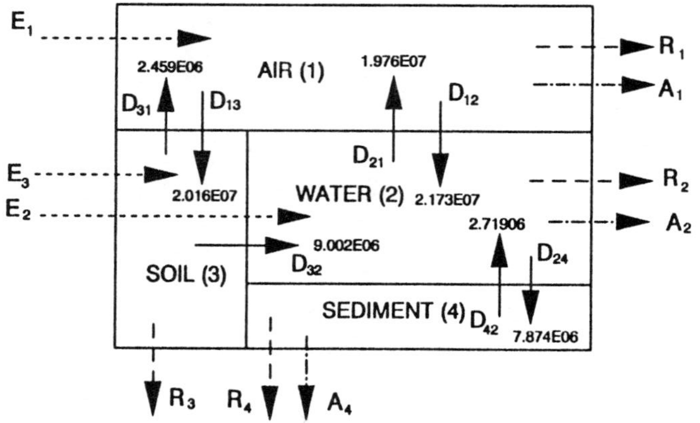

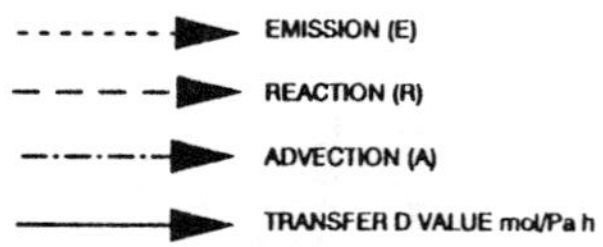

Phase Properties and Rates:

Compartment	Bulk Z mol/m3 Pa	Half-life h	D Values Reaction mol/Pa h	D Values Advection mol/Pa h
Air (1)	4.034E-04	170	1.64E+08	4.03E+08
Water (2)	1.969E+00	170	1.61E+09	3.94E+08
Soil (3)	9.517E+01	1700	6.98E+08	
Sediment (4)	7.724E+01	5500	4.87E+06	7.72E+05

	E(1)=1000	E(2)=1000	E(3)=1000	E(1,2,3)	
Overall residence time =	153.70	214.02	2416.74	1752.91	h
Reaction time =	464.26	268.01	2428.57	1920.40	h
Advection time =	229.77	1062.47	496069	20098.34	h

Phase Properties, Compositions, Transport and Transformation Rates:

Emission, kg/h			Fugacity, Pa				Concentration, g/m3				Amounts, kg				Total Amount, kg
E(1)	E(2)	E(3)	f(1)	f(2)	f(3)	f(4)	C(1)	C(2)	C(3)	C(4)	m(1)	m(2)	m(3)	m(4)	
1000	0	0	8.309E-06	9.026E-08	2.361E-07	8.504E-08	6.619E-07	3.509E-05	4.436E-03	1.297E-03	6.619E+04	7.018E+03	7.984E+04	6.485E+02	1.537E+05
0	1000	0	8.114E-08	2.503E-06	2.305E-09	2.358E-06	6.463E-09	9.731E-04	4.331E-05	3.596E-02	6.463E+02	1.946E+05	7.796E+02	1.798E+04	2.140E+05
0	0	1000	2.981E-08	3.206E-08	7.136E-06	3.020E-08	2.375E-09	1.246E-05	1.341E-01	4.606E-04	2.375E+02	2.492E+03	2.414E+06	2.303E+02	2.417E+06
50	250	700	4.566E-07	6.527E-07	5.008E-06	6.149E-07	3.637E-08	2.537E-04	9.410E-02	9.378E-03	3.637E+03	5.075E+04	1.694E+06	4.689E+03	1.753E+06

Emission, kg/h			Loss, Reaction, kg/h				Loss, Advection, kg/h			Intermedia Rate of Transport, kg/h T12	T21	T13	T31	T32	T24	T42
E(1)	E(2)	E(3)	R(1)	R(2)	R(3)	R(4)	A(1)	A(2)	A(4)	air-water	water-air	air-soil	soil-air	soil-water	water-sed	sed-water
1000	0	0	2.698E+02	2.861E+01	3.25E+01	8.170E-02	6.619E+02	7.018E+00	1.297E-02	3.566E+01	3.523E-01	3.308E+01	1.146E-01	4.195E-01	1.403E-01	4.566E-02
0	1000	0	2.635E+00	7.933E+02	3.18E-01	2.266E+00	6.463E+00	1.946E+02	3.596E-01	3.482E-01	9.768E+00	3.230E-01	1.119E-03	4.097E-03	3.891E+00	1.266E+00
0	0	1000	9.681E-01	1.016E+01	9.84E+02	2.902E-02	2.375E+00	2.492E+00	4.606E-03	1.279E-01	1.251E-01	1.187E-01	3.464E+00	1.268E+01	4.984E-02	1.622E-02
50	250	700	1.483E+01	2.069E+02	6.90E+02	5.908E-01	3.637E+01	5.075E+01	9.378E-02	1.959E+00	2.547E+00	1.818E+00	2.431E+00	8.900E+00	1.015E+00	3.302E-01

Level III Distribution

Chemical name: 2,4,6-Trichlorophenol

Distribution of mass

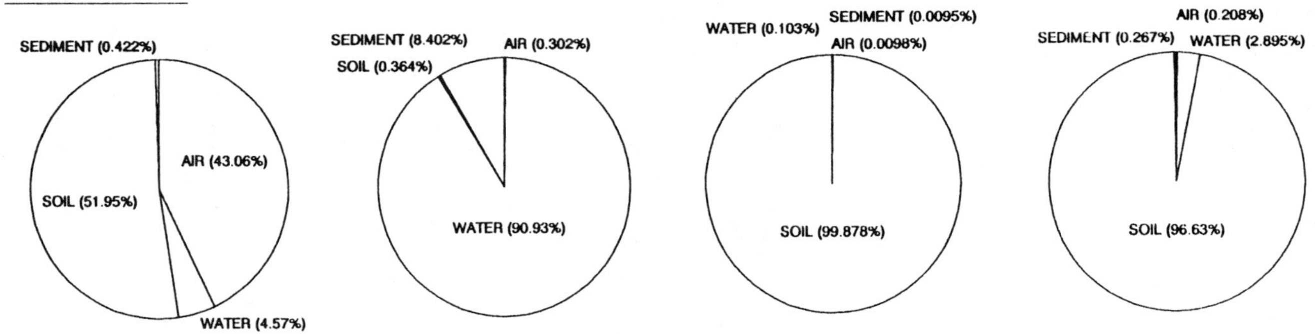

Distribution of removal rates

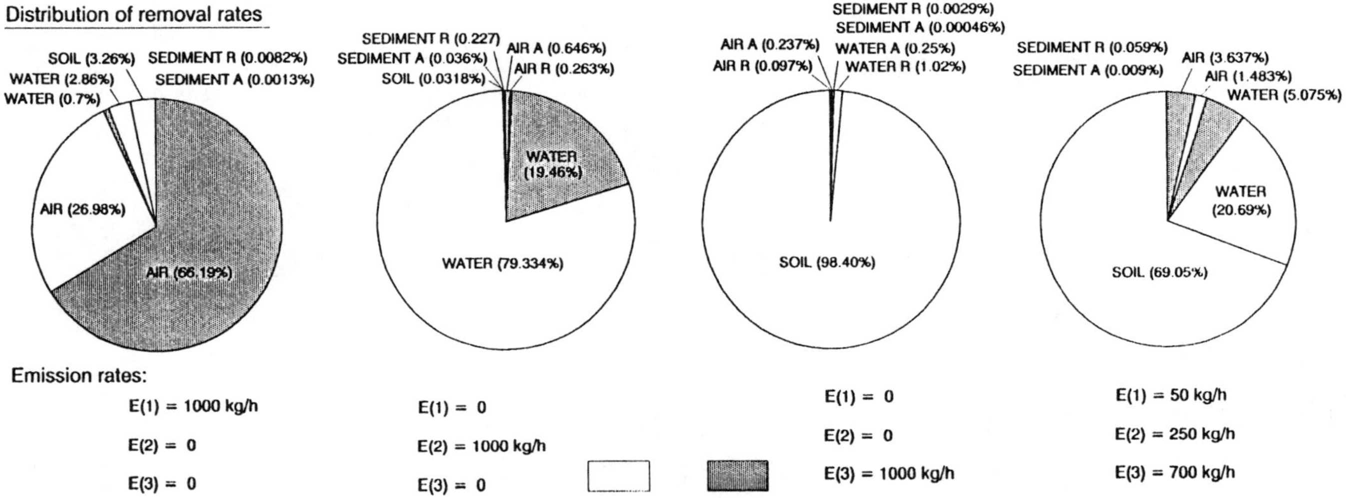

Emission rates:

E(1) = 1000 kg/h	E(1) = 0	E(1) = 0	E(1) = 50 kg/h
E(2) = 0	E(2) = 1000 kg/h	E(2) = 0	E(2) = 250 kg/h
E(3) = 0	E(3) = 0	E(3) = 1000 kg/h	E(3) = 700 kg/h

Residence time:

t = 154 h	t = 214 h	t = 2417 h	t = 1753 h

Chemical name: o-Cresol

Level I calculation: (six-compartment model)

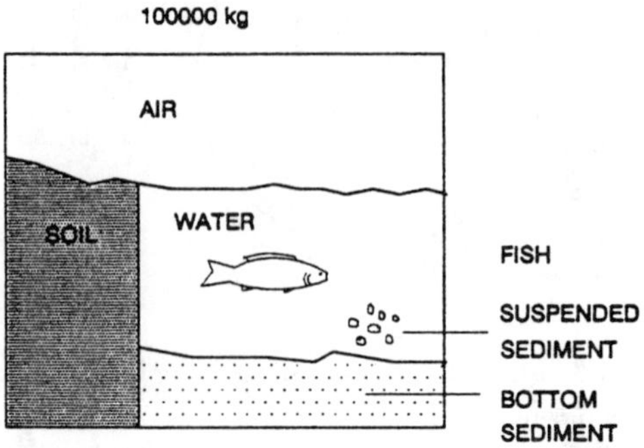

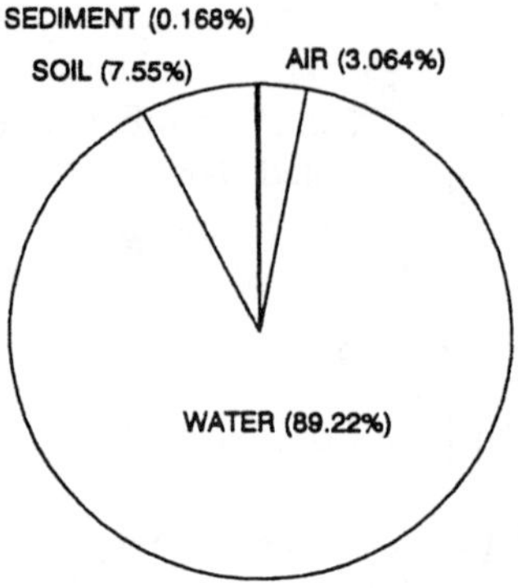

Distribution of mass

Level I calculation:

Physical-chemical properties:		Partition coeff.:		Z values in water:	
molecular wt., g/mol	108.13	HLC	0.17026	at data pH	5.1
melting point, C	31	Kaw	6.87E-05	neutral	5.873
solubility, g/m3	26000	Koc	39.1547	ionic	4.67E-05
vapor pressure, Pa	40.94	BCF	4.7750	total	5.873
log Kow	1.98	Ksw	1.8794	at environ. pH	5.1
fugacity ratio, F	0.8723	Ksd/w	3.7589	neutral	5.873
dissoc. const. pKa	10.2	Kssd/w	11.7464	ionic	4.67E-05
		Kar/w	1.28E+05	total	5.873

Compartment	Z	Concentration			Amount	Amount
	mol/m3 Pa	mol/m3	mg/L (or g/m3)	ug/g	kg	%
Air	4.03E-04	2.83E-10	3.06E-08	2.58E-05	3064.039	3.0640
Water	5.87E+00	4.13E-06	4.46E-04	4.46E-04	89217.15	89.2172
Soil	1.10E+01	7.75E-06	8.38E-04	3.49E-04	7545.46	7.5455
Biota (fish)	2.80E+01	1.97E-05	2.13E-03	2.13E-03	0.4260	4.26E-04
Suspended sediment	6.90E+01	4.85E-05	5.24E-03	3.49E-03	5.2399	5.24E-03
Bottom sediment	2.21E+01	1.55E-05	1.68E-03	6.99E-04	167.677	0.1677
	Total				100000	100

f = 7.024E-07 Pa

Chemical name: o-Cresol

Level II calculation: (six-compartment model)

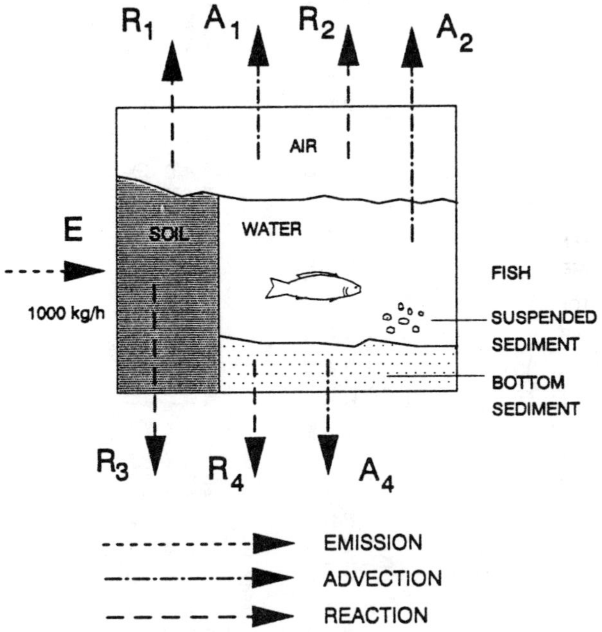

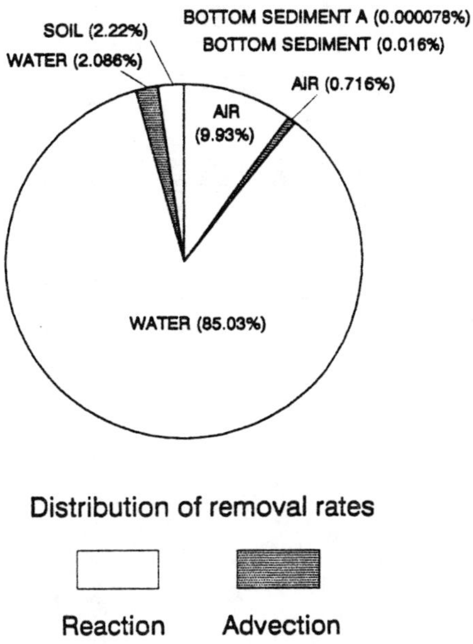

Distribution of removal rates

Level II Calculation:

Compartment	Half-life h	D Values Reaction mol/Pa.h	D Values Advection mol/Pa.h	Conc'n mol/m3	Loss Reaction kg/h	Loss Advection kg/h	Removal %
Air	5	5.59E+09	4.03E+08	6.63E-11	99.288	7.164	10.645
Water	17	4.79E+10	1.17E+09	9.65E-07	850.301	20.859	87.116
Soil	55	1.25E+09		1.81E-06	22.2278		2.2228
Biota (fish)				4.61E-06			
Suspended sediment				1.13E-05			
Bottom sediment	170	9.00E+06	4.42E+04	3.63E-06	0.1598	7.84E-04	1.61E-02
	Total	5.47E+10	1.58E+09		971.98	28.023	100
	R + A		5.63E+10			1000	

f = 1.642E-07 Pa

Total amt = 23380 kg

Overall residence time = 23.38 h

Reaction time = 24.05 h

Advection time = 834.30 h

Fugacity Level III calculations: (four-compartment model)

Chemical name: o-Cresol

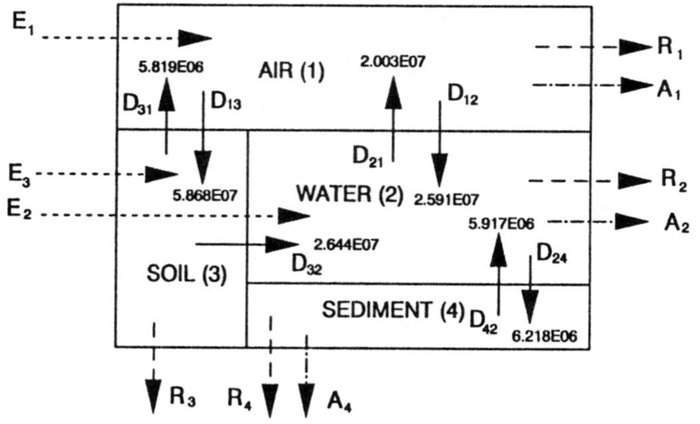

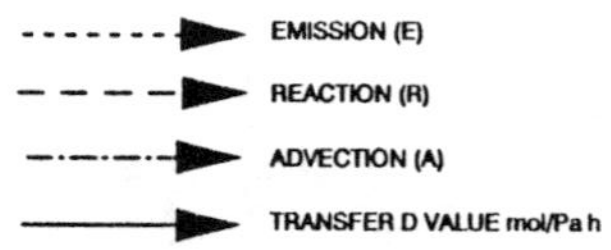

Phase Properties and Rates:

Compartment	Bulk Z mol/m3 Pa	Half-life h	D Values Reaction mol/Pa h	D Values Advection mol/Pa h
Air (1)	4.034E-04	5	5.59E+09	4.03E+08
Water (2)	5.874E+00	17	4.79E+10	1.17E+09
Soil (3)	7.281E+00	55	1.65E+09	
Sediment (4)	9.114E+00	170	1.86E+07	9.11E+04

	E(1)=1000	E(2)=1000	E(3)=1000	E(1,2,3)
Overall residence time =	7.49	23.96	78.25	61.14 h
Reaction time =	8.03	24.55	78.29	61.74 h
Advection time =	112.74	999.96	129123	6278.62 h

Phase Properties, Compositions, Transport and Transformation Rates:

Emission, kg/h			Fugacity, Pa				Concentration, g/m3				Amounts, kg				Total Amount, kg
E(1)	E(2)	E(3)	f(1)	f(2)	f(3)	f(4)	C(1)	C(2)	C(3)	C(4)	m(1)	m(2)	m(3)	m(4)	
1000	0	0	1.521E-06	8.315E-10	5.302E-08	2.103E-10	6.636E-08	5.281E-07	4.175E-05	2.072E-07	6.636E+03	1.056E+02	7.514E+02	1.036E-01	7.493E+03
0	1000	0	6.209E-10	1.884E-07	2.164E-11	4.765E-08	2.708E-11	1.197E-04	1.704E-08	4.696E-05	2.708E+00	2.393E+04	3.067E-01	2.348E+01	2.396E+04
0	0	1000	5.268E-09	2.962E-09	5.493E-06	7.491E-10	2.298E-10	1.881E-06	4.325E-03	7.382E-07	2.298E+01	3.762E+02	7.785E+04	3.691E-01	7.825E+04
50	250	700	7.991E-08	4.922E-08	3.848E-06	1.245E-08	3.486E-09	3.126E-05	3.029E-03	1.227E-05	3.486E+02	6.252E+03	5.453E+04	6.134E+00	6.114E+04

Emission, kg/h			Loss, Reaction, kg/h				Loss, Advection, kg/h			Intermedia Rate of Transport, kg/h T12	T21	T13	T31	T32	T24	T42
E(1)	E(2)	E(3)	R(1)	R(2)	R(3)	R(4)	A(1)	A(2)	A(4)	air-water	water-air	air-soil	soil-air	soil-water	water-sed	sed-water
1000	0	0	9.198E+02	4.305E+00	9.47E+00	4.224E-04	6.636E+01	1.056E-01	2.072E-06	4.262E+00	1.801E-03	9.653E+00	3.336E-02	1.516E-01	5.590E-04	1.346E-04
0	1000	0	3.754E-01	9.756E+02	3.86E-03	9.572E-02	2.708E-02	2.393E+01	4.696E-04	1.739E-03	4.081E-01	3.940E-03	1.362E-05	6.187E-05	1.267E-01	3.049E-02
0	0	1000	3.185E+00	1.533E+01	9.81E+02	1.505E-03	2.298E-01	3.762E-01	7.382E-06	1.476E-02	6.415E-03	3.343E-02	3.457E+00	1.570E+01	1.991E-03	4.793E-04
50	250	700	4.831E+01	2.548E+02	6.87E+02	2.500E-02	3.486E+00	6.252E+00	1.227E-04	2.238E-01	1.066E-01	5.070E-01	2.421E+00	1.100E+01	3.309E-02	7.965E-03

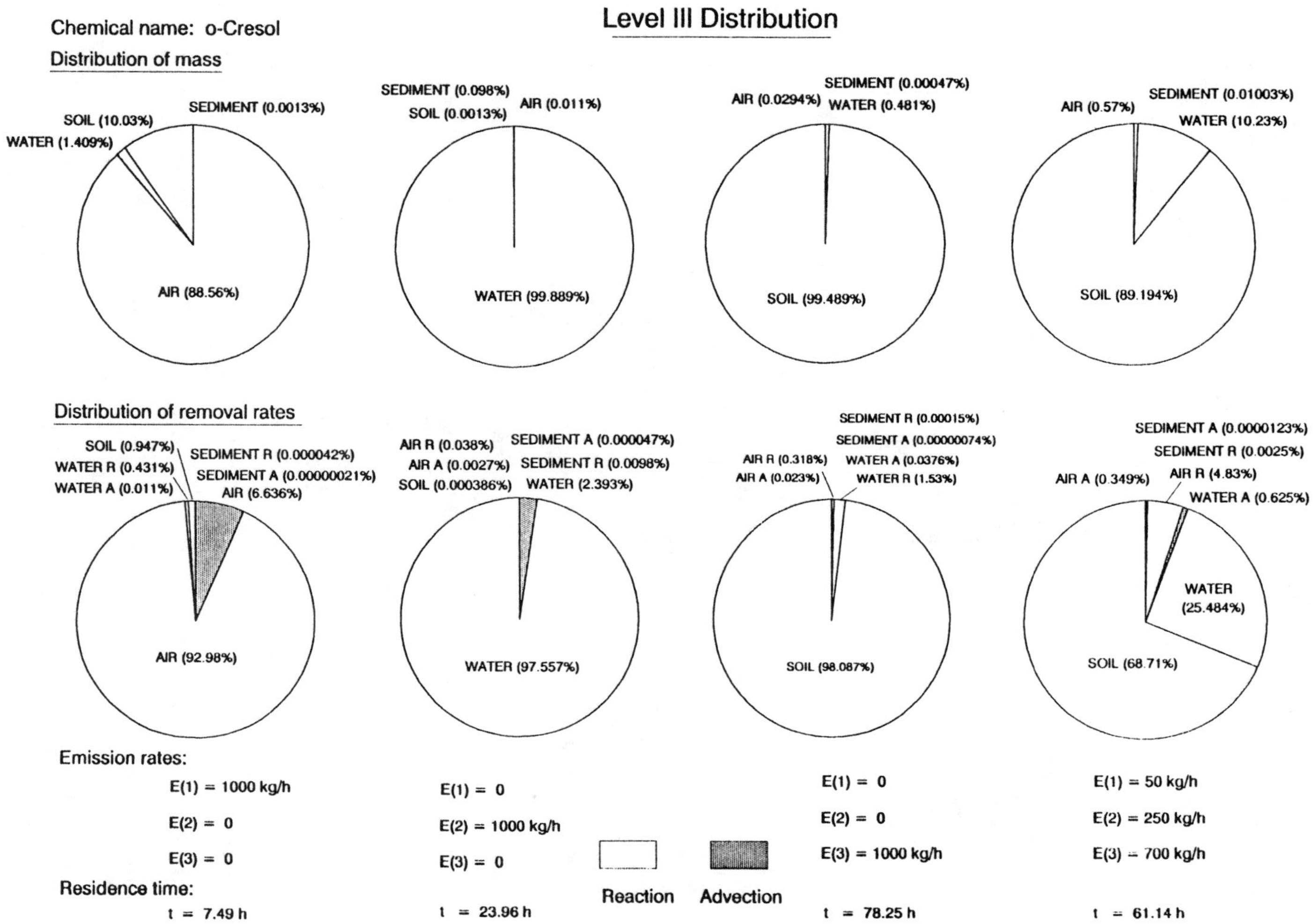
Level III Distribution
Chemical name: o-Cresol
Distribution of mass
SOIL (10.03%)
WATER (1.409%)
SEDIMENT (0.0013%)
AIR (88.56%)
SEDIMENT (0.098%)
SOIL (0.0013%)
AIR (0.011%)
WATER (99.889%)
AIR (0.0294%)
SEDIMENT (0.00047%)
WATER (0.481%)
SOIL (99.489%)
AIR (0.57%)
SEDIMENT (0.01003%)
WATER (10.23%)
SOIL (89.194%)
Distribution of removal rates
SOIL (0.947%)
WATER R (0.431%)
WATER A (0.011%)
SEDIMENT R (0.000042%)
SEDIMENT A (0.00000021%)
AIR (6.636%)
AIR (92.98%)
AIR R (0.038%)
AIR A (0.0027%)
SOIL (0.000386%)
SEDIMENT A (0.000047%)
SEDIMENT R (0.0098%)
WATER (2.393%)
WATER (97.557%)
SEDIMENT R (0.00015%)
SEDIMENT A (0.00000074%)
AIR R (0.318%)
AIR A (0.023%)
WATER A (0.0376%)
WATER R (1.53%)
SOIL (98.087%)
SEDIMENT A (0.0000123%)
SEDIMENT R (0.0025%)
AIR A (0.349%)
AIR R (4.83%)
WATER A (0.625%)
WATER (25.484%)
SOIL (68.71%)
Emission rates:
E(1) = 1000 kg/h
E(2) = 0
E(3) = 0
Residence time:
t = 7.49 h
E(1) = 0
E(2) = 1000 kg/h
E(3) = 0
t = 23.96 h
Reaction
Advection
E(1) = 0
E(2) = 0
E(3) = 1000 kg/h
t = 78.25 h
E(1) = 50 kg/h
E(2) = 250 kg/h
E(3) = 700 kg/h
t = 61.14 h

Chemical name: 2-Nitrophenol

Level I calculation: (six-compartment model)

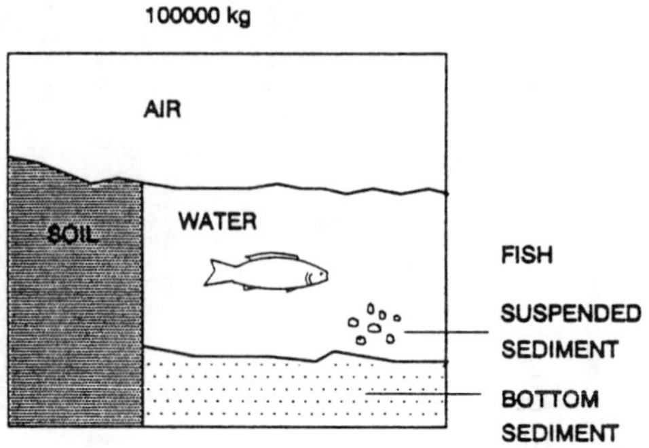

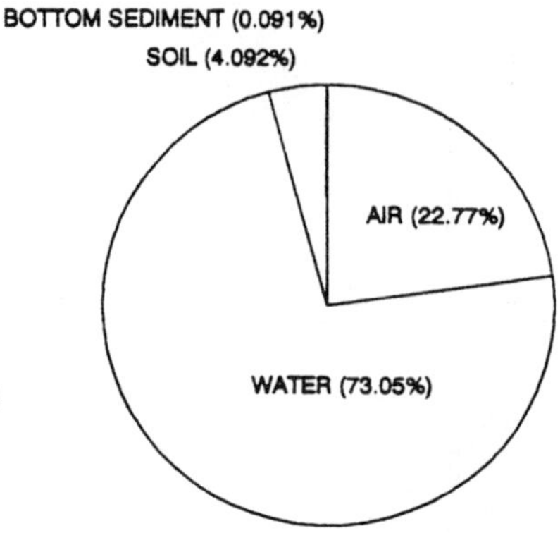

Distribution of mass

Level I calculation:

Physical-chemical properties:		Partition coeff.:		Z values in water:	
molecular wt., g/mol	139.11	HLC	1.5457	at data pH	6
melting point, C	45.3	Kaw	6.24E-04	neutral	0.611
solubility, g/m3	1080	Koc	25.8693	ionic	3.60E-02
vapor pressure, Pa	12.00	BCF	3.1548	total	0.647
log Kow	1.80	Ksw	1.2417	at environ. pH	6
fugacity ratio, F	0.6298	Ksd/w	2.4834	neutral	0.611
dissoc. const. pKa	7.23	Kssd/w	7.7608	ionic	3.60E-02
		Kar/w	3.15E+05	total	0.647

Compartment	Z	Concentration			Amount	Amount
	mol/m3 Pa	mol/m3	mg/L (or g/m3)	ug/g	kg	%
Air	4.03E-04	1.64E-09	2.28E-07	1.92E-04	22774.97	22.7750
Water	6.47E-01	2.63E-06	3.65E-04	3.65E-04	73049.43	73.0494
Soil	8.03E-01	3.26E-06	4.54E-04	1.89E-04	4081.83	4.0818
Biota (fish)	2.04E+00	8.28E-06	1.15E-03	1.15E-03	0.2305	2.30E-04
Suspended sediment	5.02E+00	2.04E-05	2.83E-03	1.89E-03	2.8346	2.83E-03
Bottom sediment	1.61E+00	6.52E-06	9.07E-04	3.78E-04	90.707	0.0907
	Total				100000	100

f = 4.058E-06 Pa

Chemical name: 2-Nitrophenol

Level II calculation: (six-compartment model)

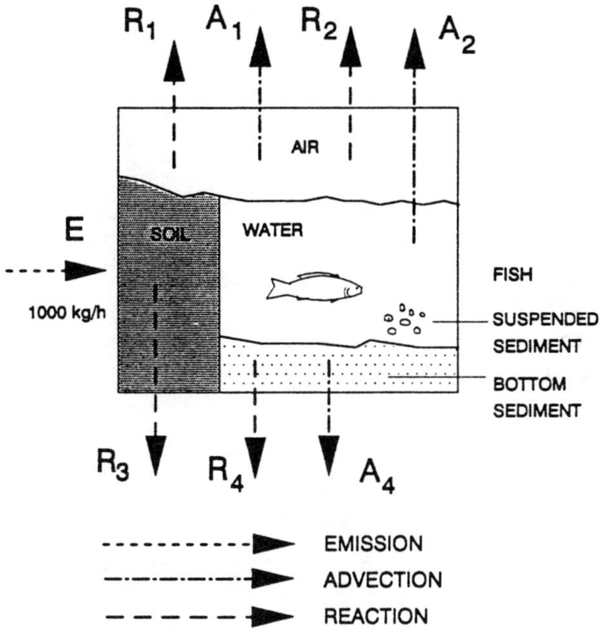

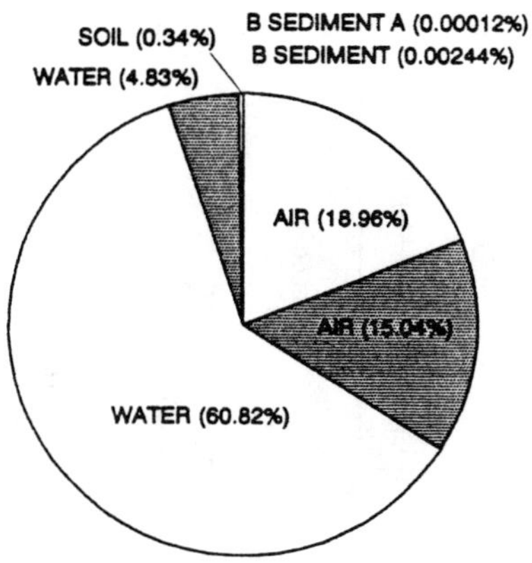

Distribution of removal rates

Level II Calculation:

Compartment	Half-life h	D Values Reaction mol/Pa.h	D Values Advection mol/Pa.h	Conc'n mol/m3	Loss Reaction kg/h	Loss Advection kg/h	Removal %
Air	55	5.08E+08	4.03E+08	1.08E-09	189.620	150.492	34.011
Water	55	1.63E+09	1.29E+08	1.73E-06	608.195	48.269	65.646
Soil	550	9.11E+06		2.15E-06	3.3984		0.3398
Biota (fish)				5.47E-06			
Suspended sediment				1.35E-05			
Bottom sediment	1700	6.55E+04	3.21E+03	4.31E-06	0.0244	1.20E-03	2.56E-03
	Total	2.15E+09	5.33E+08		801.24	198.763	100
	R + A		2.68E+09			1000	

f = 2.682E-06 Pa

Total amt = 66078 kg

Overall residence time = 66.08 h

Reaction time = 82.47 h

Advection time = 332.45 h

Fugacity Level III calculations: (four-compartment model)

Chemical name: 2-Nitrophenol

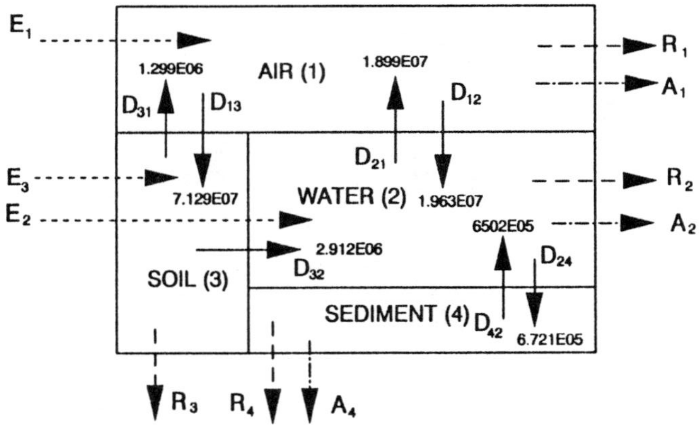

Phase Properties and Rates:

Compartment	Bulk Z mol/m3 Pa	Half-life h	D Values Reaction mol/Pa h	D Values Advection mol/Pa h
Air (1)	4.034E-04	55	5.08E+08	4.03E+08
Water (2)	6.470E-01	55	1.63E+09	1.29E+08
Soil (3)	5.959E-01	550	1.35E+07	
Sediment (4)	8.389E-01	1700	1.71E+05	8.39E+03

	E(1)=1000	E(2)=1000	E(3)=1000	E(1,2,3)	
Overall residence time =	49.24	73.46	620.77	455.37	h
Reaction time =	86.66	79.61	649.58	490.68	h
Advection time =	114.03	949.63	13995	6326.48	h

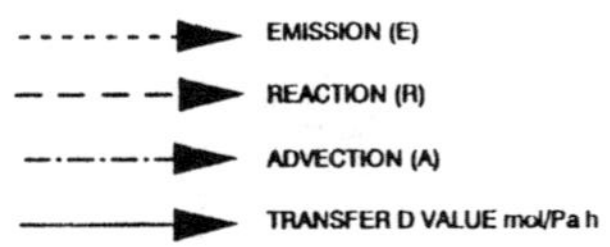

Phase Properties, Compositions, Transport and Transformation Rates:

Emission, kg/h			Fugacity, Pa				Concentration, g/m3				Amounts, kg				Total Amount, kg
E(1)	E(2)	E(3)	f(1)	f(2)	f(3)	f(4)	C(1)	C(2)	C(3)	C(4)	m(1)	m(2)	m(3)	m(4)	
1000	0	0	7.666E-06	8.965E-08	3.083E-06	7.263E-08	4.302E-07	8.069E-06	2.555E-04	8.477E-06	4.302E+04	1.614E+03	4.600E+03	4.238E+00	4.924E+04
0	1000	0	8.182E-08	4.042E-06	3.291E-08	3.275E-06	4.592E-09	3.638E-04	2.728E-06	3.821E-04	4.592E+02	7.276E+04	4.910E+01	1.911E+02	7.346E+04
0	0	1000	5.753E-07	6.706E-07	4.058E-04	5.433E-07	3.228E-08	6.036E-05	3.364E-02	6.340E-05	3.228E+03	1.207E+04	6.054E+05	3.170E+01	6.208E+05
50	250	700	8.064E-07	1.484E-06	2.842E-04	1.203E-06	4.526E-08	1.336E-04	2.356E-02	1.403E-04	4.526E+03	2.672E+04	4.240E+05	7.017E+01	4.554E+05

Emission, kg/h			Loss, Reaction, kg/h				Loss, Advection, kg/h			Intermedia Rate of Transport, kg/h T12	T21	T13	T31	T32	T24	T42
E(1)	E(2)	E(3)	R(1)	R(2)	R(3)	R(4)	A(1)	A(2)	A(4)	air-water	water-air	air-soil	soil-air	soil-water	water-sed	sed-water
1000	0	0	5.421E+02	2.033E+01	5.80E+00	1.728E-03	4.302E+02	1.614E+00	8.477E-05	2.094E+01	2.368E-01	7.602E+00	5.571E-01	1.249E+00	8.382E-03	6.570E-03
0	1000	0	5.785E+00	9.167E+02	6.19E-02	7.789E-02	4.592E+00	7.276E+01	3.821E-03	2.235E-01	1.068E+01	8.114E-02	5.947E-03	1.333E-02	3.779E-01	2.962E-01
0	0	1000	4.068E+01	1.521E+02	7.63E+02	1.292E-02	3.228E+01	1.207E+01	6.340E-04	1.571E+00	1.771E+00	5.705E-01	7.333E+01	1.644E+02	6.270E-02	4.914E-02
50	250	700	5.702E+01	3.367E+02	5.34E+02	2.860E-02	4.526E+01	2.672E+01	1.403E-03	2.203E+00	3.921E+00	7.997E-01	5.136E+01	1.151E+02	1.388E-01	1.088E-01

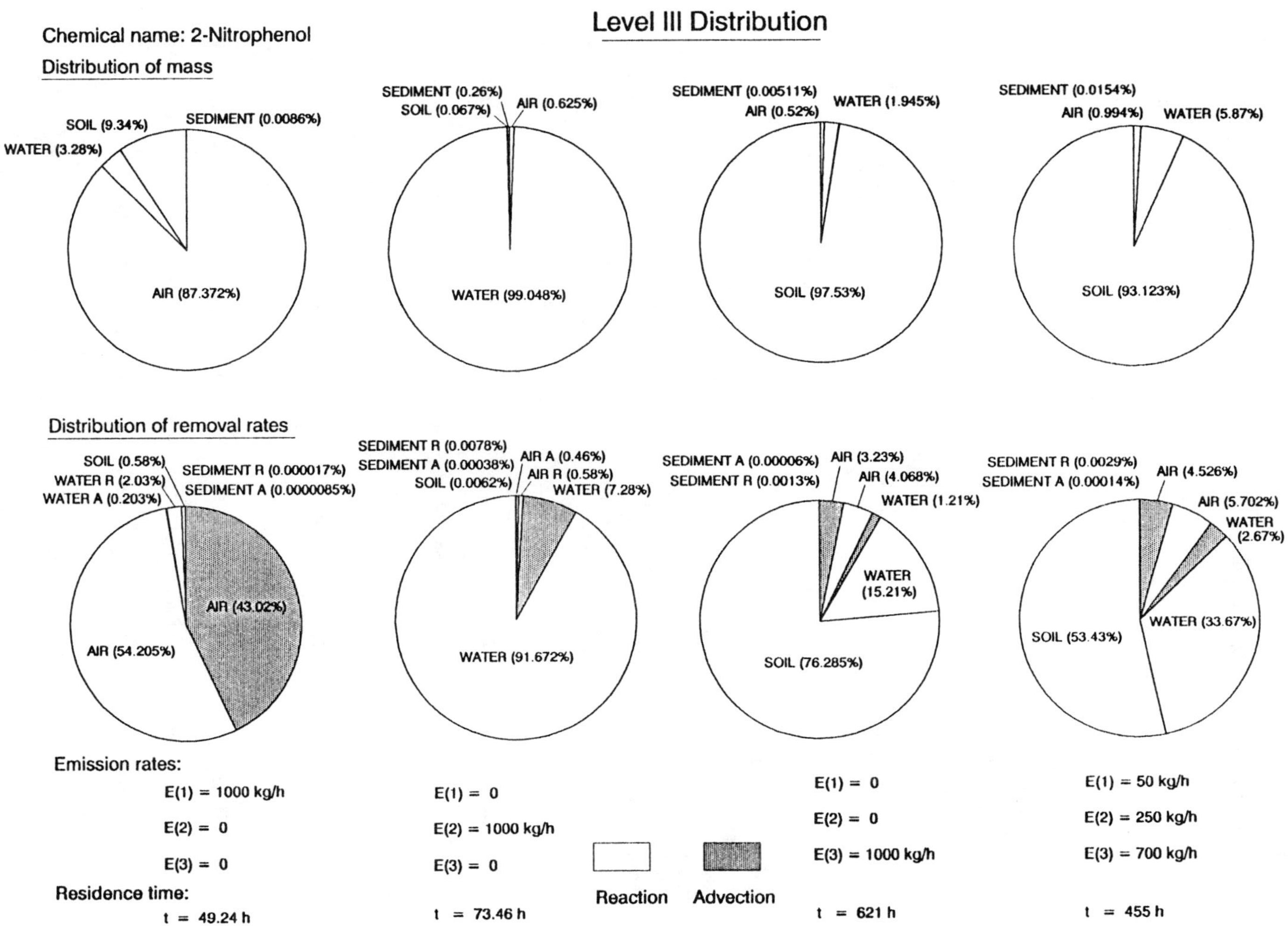

Level III Distribution
Chemical name: 2-Nitrophenol
Distribution of mass
SOIL (9.34%)
WATER (3.28%)
SEDIMENT (0.0086%)
AIR (87.372%)
SEDIMENT (0.26%)
SOIL (0.067%)
AIR (0.625%)
WATER (99.048%)
SEDIMENT (0.00511%)
AIR (0.52%)
WATER (1.945%)
SOIL (97.53%)
SEDIMENT (0.0154%)
AIR (0.994%)
WATER (5.87%)
SOIL (93.123%)
Distribution of removal rates
SOIL (0.58%)
WATER R (2.03%)
WATER A (0.203%)
SEDIMENT R (0.000017%)
SEDIMENT A (0.0000085%)
AIR (43.02%)
AIR (54.205%)
SEDIMENT R (0.0078%)
SEDIMENT A (0.00038%)
SOIL (0.0062%)
AIR A (0.46%)
AIR R (0.58%)
WATER (7.28%)
WATER (91.672%)
SEDIMENT A (0.00006%)
SEDIMENT R (0.0013%)
AIR (3.23%)
AIR (4.068%)
WATER (1.21%)
WATER (15.21%)
SOIL (76.285%)
SEDIMENT R (0.0029%)
SEDIMENT A (0.00014%)
AIR (4.526%)
AIR (5.702%)
WATER (2.67%)
WATER (33.67%)
SOIL (53.43%)
Emission rates:
E(1) = 1000 kg/h
E(2) = 0
E(3) = 0
Residence time:
t = 49.24 h
E(1) = 0
E(2) = 1000 kg/h
E(3) = 0
t = 73.46 h
Reaction
Advection
E(1) = 0
E(2) = 0
E(3) = 1000 kg/h
t = 621 h
E(1) = 50 kg/h
E(2) = 250 kg/h
E(3) = 700 kg/h
t = 455 h

Chemical name: 4-Nitrophenol

Level I calculation: (six-compartment model)

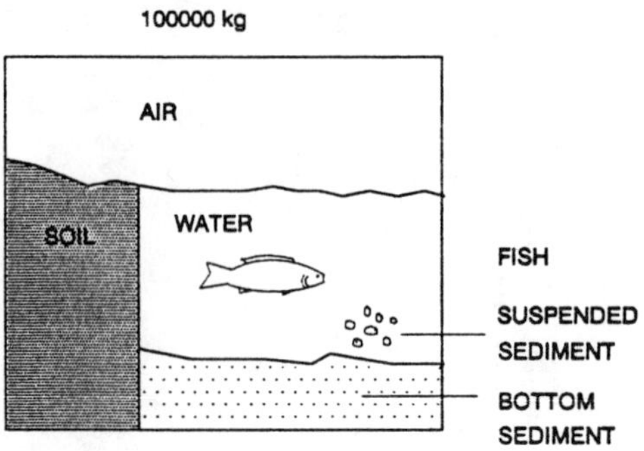

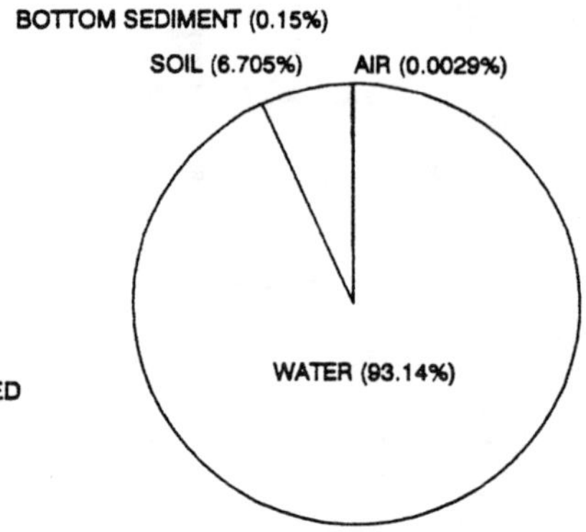

Distribution of mass

Level I calculation:

Physical-chemical properties:		Partition coeff.:		Z values in water:	
molecular wt., g/mol	139.11	HLC	1.55E-04	at data pH	6
melting point, C	114	Kaw	6.24E-08	neutral	5972.90
solubility, g/m3	13500	Koc	33.3261	ionic	496.80
vapor pressure, Pa	0.015	BCF	4.0642	total	6469.70
log Kow	1.91	Ksw	1.5997	at environ. pH	6
fugacity ratio, F	0.1317	Ksd/w	3.1993	neutral	5972.90
dissoc. const. pKa	7.08	Kssd/w	9.9978	ionic	496.80
		Kar/w	5.27E+07	total	6469.70

Compartment	Z	Concentration			Amount	Amount
	mol/m3 Pa	mol/m3	mg/L (or g/m3)	ug/g	kg	%
Air	4.03E-04	2.09E-13	2.90E-11	2.45E-08	2.9038	0.00290
Water	6.47E+03	3.35E-06	4.66E-04	4.66E-04	93138.56	93.1386
Soil	1.03E+04	5.36E-06	7.45E-04	3.10E-04	6704.51	6.7045
Biota (fish)	2.63E+04	1.36E-05	1.89E-03	1.89E-03	0.3785	3.79E-04
Suspended sediment	6.47E+04	3.35E-05	4.66E-03	3.10E-03	4.6559	4.66E-03
Bottom sediment	2.07E+04	1.07E-05	1.49E-03	6.21E-04	148.989	0.1490
	Total				100000	100

f = 5.174E-10 Pa

Chemical name: 4-Nitrophenol

Level II calculation: (six-compartment model)

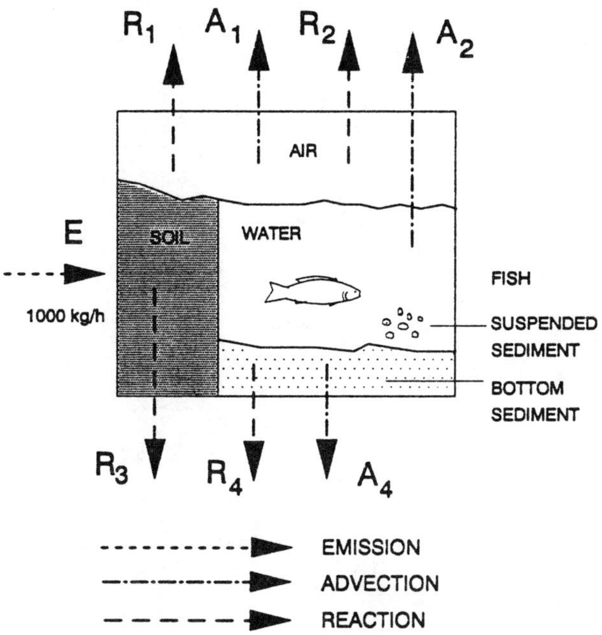

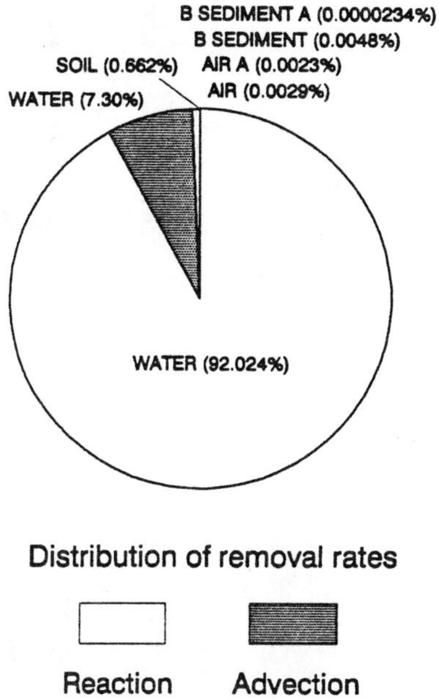

Distribution of removal rates

Level II Calculation:

Compartment	Half-life h	D Values Reaction mol/Pa.h	D Values Advection mol/Pa.h	Conc'n mol/m3	Loss Reaction kg/h	Loss Advection kg/h	Removal %
Air	55	5.08E+08	4.03E+08	1.64E-13	0.0287	0.0228	0.00515
Water	55	1.63E+13	1.29E+12	2.63E-06	920.239	73.035	99.327
Soil	550	1.17E+11		4.20E-06	6.6243		0.6624
Biota (fish)				1.07E-05			
Suspended sediment				2.62E-05			
Bottom sediment	1700	8.44E+08	4.14E+07	8.40E-06	0.0476	2.34E-03	5.00E-03
	Total	1.64E+13	1.29E+12		926.94	73.060	100
	R + A		1.77E+13			1000	

f = 4.057E-10 Pa
Total amt = 78415 kg

Overall residence time = 78.42 h
Reaction time = 84.60 h
Advection time = 1073.30 h

Fugacity Level III calculations: (four-compartment model)

Chemical name: 4-Nitrophenol

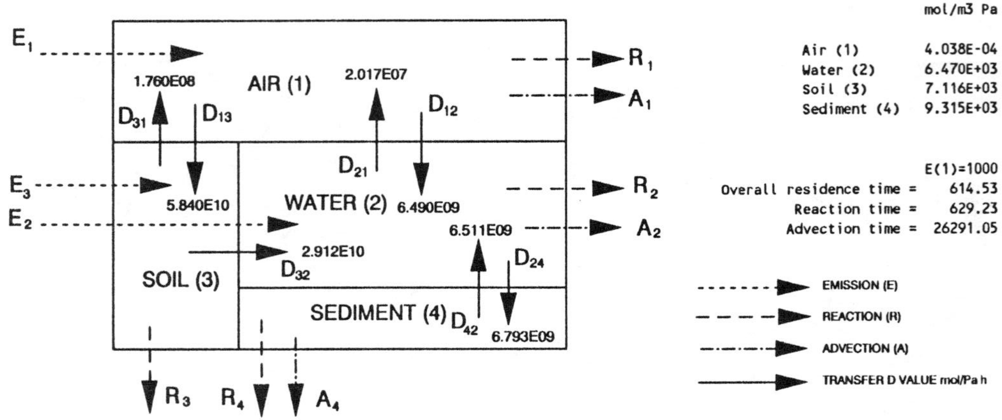

Phase Properties and Rates:

Compartment	Bulk Z mol/m3 Pa	Half-life h	D Values Reaction mol/Pa h	Advection mol/Pa h
Air (1)	4.038E-04	55	5.09E+08	4.04E+08
Water (2)	6.470E+03	55	1.63E+13	1.29E+12
Soil (3)	7.116E+03	550	1.61E+11	
Sediment (4)	9.315E+03	1700	1.90E+09	9.32E+07

	E(1)=1000	E(2)=1000	E(3)=1000	E(1,2,3)
Overall residence time =	614.53	73.73	683.53	527.63 h
Reaction time =	629.23	79.59	691.31	542.51 h
Advection time =	26291.05	1002.83	60750	19238.01 h

Phase Properties, Compositions, Transport and Transformation Rates:

Emission, kg/h			Fugacity, Pa				Concentration, g/m3				Amounts, kg				Total Amount, kg
E(1)	E(2)	E(3)	f(1)	f(2)	f(3)	f(4)	C(1)	C(2)	C(3)	C(4)	m(1)	m(2)	m(3)	m(4)	
1000	0	0	1.093E-07	9.572E-11	3.349E-08	7.647E-11	6.142E-09	8.616E-05	3.315E-02	9.910E-05	6.142E+02	1.723E+04	5.966E+05	4.955E+01	6.145E+05
0	1000	0	1.253E-13	4.084E-10	3.838E-14	3.263E-10	7.039E-15	3.676E-04	3.799E-08	4.229E-04	7.039E-04	7.352E+04	6.838E-01	2.114E+02	7.374E+04
0	0	1000	1.010E-10	6.247E-11	3.773E-08	4.991E-11	5.672E-12	5.623E-05	3.735E-02	6.467E-05	5.672E-01	1.125E+04	6.723E+05	3.234E+01	6.835E+05
50	250	700	5.537E-09	1.506E-10	2.809E-08	1.203E-10	3.111E-10	1.356E-04	2.780E-02	1.559E-04	3.111E+01	2.711E+04	5.004E+05	7.797E+01	5.276E+05

Emission, kg/h			Loss, Reaction, kg/h				Loss, Advection, kg/h			Intermedia Rate of Transport, kg/h T12	T21	T13	T31	T32	T24	T42
E(1)	E(2)	E(3)	R(1)	R(2)	R(3)	R(4)	A(1)	A(2)	A(4)	air-water	water-air	air-soil	soil-air	soil-water	water-sed	sed-water
1000	0	0	7.739E+00	2.171E+02	7.52E+02	2.020E-02	6.142E+00	1.723E+01	9.910E-04	9.870E+01	2.686E-04	8.882E+02	8.201E-01	1.357E+02	9.046E-02	6.927E-02
0	1000	0	8.869E-06	9.264E+02	8.62E-04	8.619E-02	7.039E-06	7.352E+01	4.229E-03	1.131E-04	1.146E-03	1.018E-03	9.399E-07	1.555E-04	3.860E-01	2.956E-01
0	0	1000	7.146E-03	1.417E+02	8.47E+02	1.318E-02	5.672E-03	1.125E+01	6.467E-04	9.115E-02	1.753E-04	8.202E-01	9.240E-01	1.529E+02	5.903E-02	4.520E-02
50	250	700	3.919E-01	3.416E+02	6.31E+02	3.178E-02	3.111E-01	2.711E+01	1.559E-03	4.999E+00	4.226E-04	4.499E+01	6.878E-01	1.138E+02	1.423E-01	1.090E-01

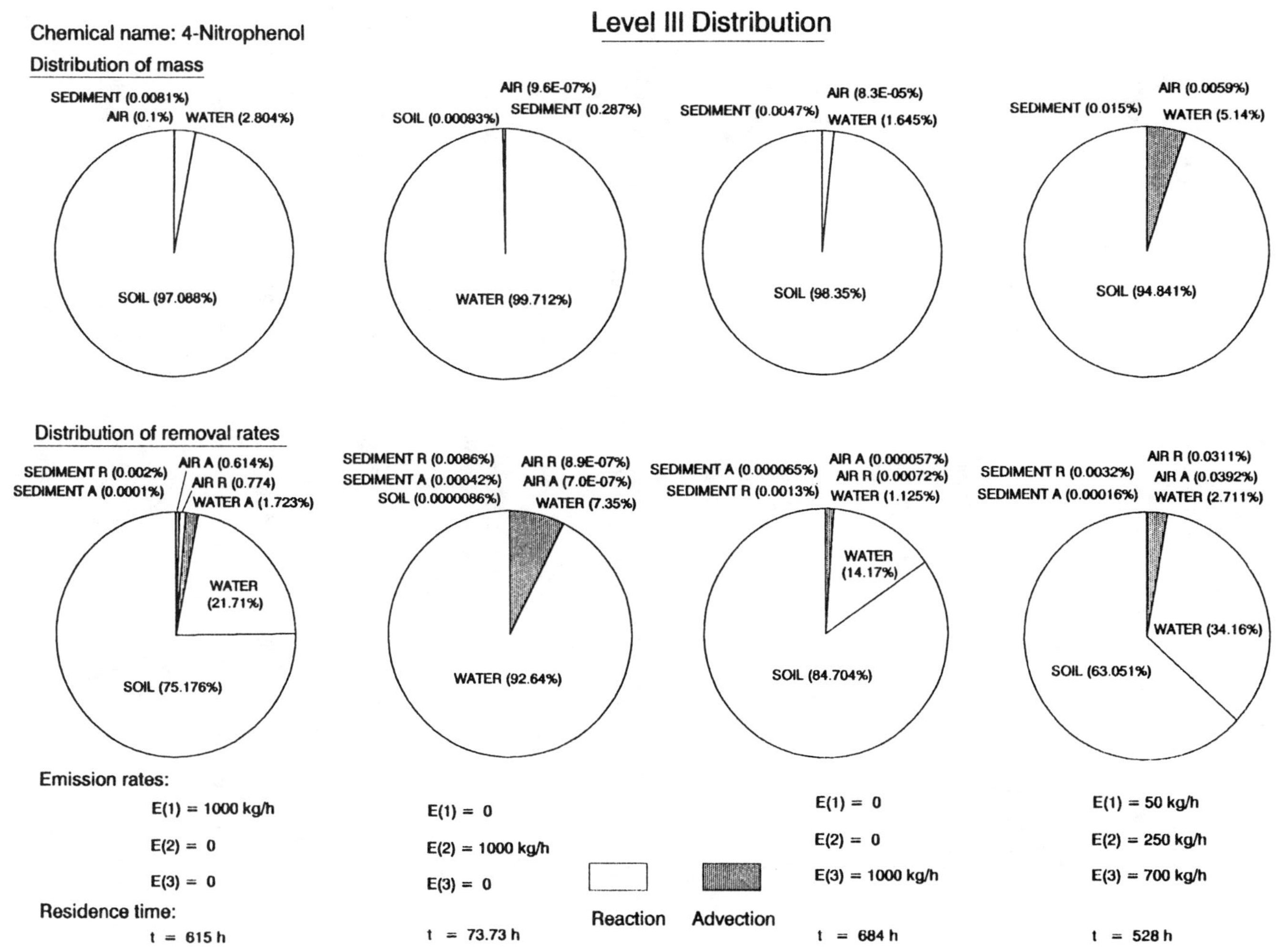

Level III Distribution
Chemical name: 4-Nitrophenol
Distribution of mass
SEDIMENT (0.0081%)
AIR (0.1%)
WATER (2.804%)
SOIL (97.088%)
AIR (9.6E-07%)
SOIL (0.00093%)
SEDIMENT (0.287%)
WATER (99.712%)
SEDIMENT (0.0047%)
AIR (8.3E-05%)
WATER (1.645%)
SOIL (98.35%)
SEDIMENT (0.015%)
AIR (0.0059%)
WATER (5.14%)
SOIL (94.841%)
Distribution of removal rates
SEDIMENT R (0.002%)
SEDIMENT A (0.0001%)
AIR A (0.614%)
AIR R (0.774)
WATER A (1.723%)
WATER (21.71%)
SOIL (75.176%)
SEDIMENT R (0.0086%)
SEDIMENT A (0.00042%)
SOIL (0.0000086%)
AIR R (8.9E-07%)
AIR A (7.0E-07%)
WATER (7.35%)
WATER (92.64%)
SEDIMENT A (0.000065%)
SEDIMENT R (0.0013%)
AIR A (0.000057%)
AIR R (0.00072%)
WATER (1.125%)
WATER (14.17%)
SOIL (84.704%)
SEDIMENT R (0.0032%)
SEDIMENT A (0.00016%)
AIR R (0.0311%)
AIR A (0.0392%)
WATER (2.711%)
WATER (34.16%)
SOIL (63.051%)
Emission rates:
E(1) = 1000 kg/h
E(2) = 0
E(3) = 0
E(1) = 0
E(2) = 1000 kg/h
E(3) = 0
E(1) = 0
E(2) = 0
E(3) = 1000 kg/h
E(1) = 50 kg/h
E(2) = 250 kg/h
E(3) = 700 kg/h
Reaction
Advection
Residence time:
t = 615 h
t = 73.73 h
t = 684 h
t = 528 h

Chemical name: 2,4-Dinitrophenol

Level I calculation: (six-compartment model)

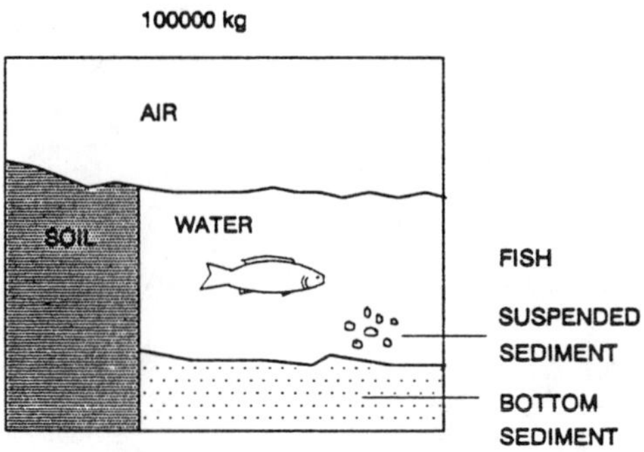

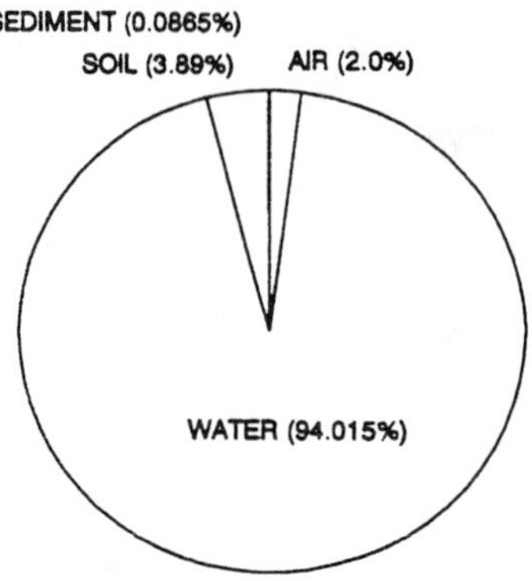

Distribution of mass

Level I calculation:

Physical-chemical properties:		Partition coeff.:		Z values in water:	
molecular wt., g/mol	184.11	HLC	0.10552	at data pH	6
melting point, C	114	Kaw	4.26E-05	neutral	0.1152
solubility, g/m3	335	Koc	19.1771	ionic	9.3617
vapor pressure, Pa	0.192	BCF	2.3387	total	9.4769
log Kow	1.67	Ksw	0.9205	at environ. pH	6
fugacity ratio, F	0.1317	Ksd/w	1.8410	neutral	0.1152
dissoc. const. pKa	4.09	Kssd/w	5.7531	ionic	9.3617
		Kar/w	4.12E+06	total	9.4769

Compartment	Z	Concentration			Amount	Amount
	mol/m3 Pa	mol/m3	mg/L (or g/m3)	ug/g	kg	%
Air	4.03E-04	1.09E-10	2.00E-08	1.69E-05	2001.045	2.00104
Water	9.48E+00	2.55E-06	4.70E-04	4.70E-04	94015.14	94.0151
Soil	8.72E+00	2.35E-06	4.33E-04	1.80E-04	3894.35	3.8944
Biota (fish)	2.22E+01	5.97E-06	1.10E-03	1.10E-03	0.2199	2.20E-04
Suspended sediment	5.45E+01	1.47E-05	2.70E-03	1.80E-03	2.7044	2.70E-03
Bottom sediment	1.74E+01	4.70E-06	8.65E-04	3.61E-04	86.541	0.0865
	Total				100000	100

f = 2.694E-07 Pa

Chemical name: 2,4-Dinitrophenol

Level II calculation: (six-compartment model)

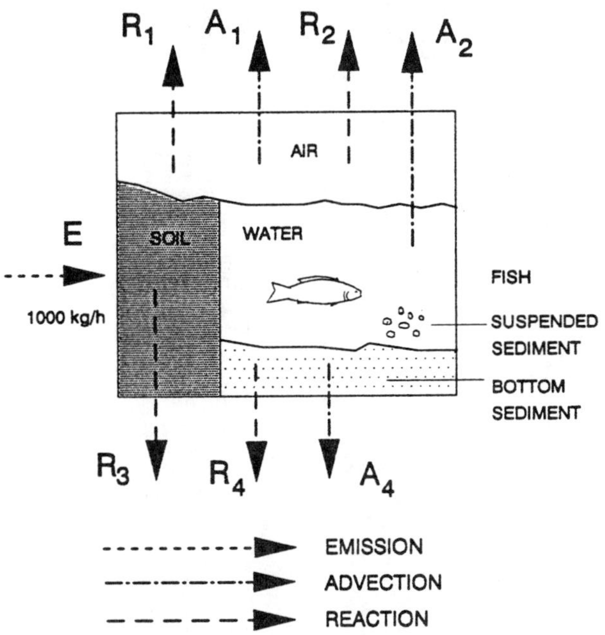

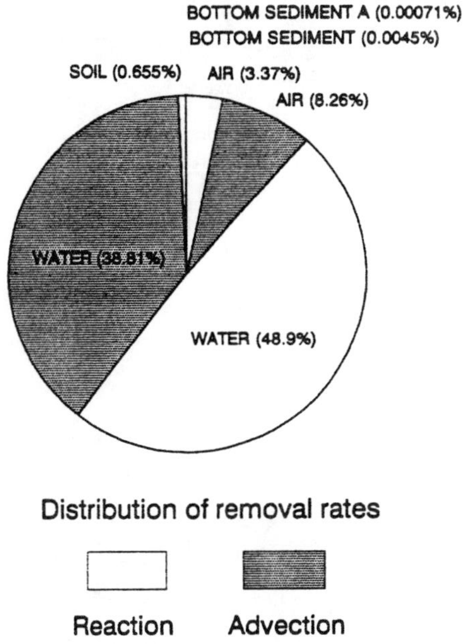

Distribution of removal rates

Level II Calculation:

Compartment	Half-life h	D Values Reaction mol/Pa.h	Advection mol/Pa.h	Conc'n mol/m3	Loss Reaction kg/h	Loss Advection kg/h	Removal %
Air	170	1.64E+08	4.03E+08	4.49E-10	33.674	82.605	11.628
Water	550	2.39E+09	1.90E+09	1.05E-05	489.011	388.104	87.712
Soil	1700	3.20E+07		9.70E-06	6.5535		0.6553
Biota (fish)				2.46E-05			
Suspended sediment				6.06E-05			
Bottom sediment	5500	2.20E+05	3.49E+04	1.94E-05	0.0450	7.15E-03	5.22E-03
	Total	2.58E+09	2.30E+09		529.28	470.717	100
	R + A		4.88E+09			1000	

f = 1.112E-06 Pa
Total amt = 412810 kg

Overall residence time = 412.81 h
Reaction time = 779.94 h
Advection time = 876.98 h

Fugacity Level III calculations: (four-compartment model)

Chemical name: 2,4-Dinitrophenol

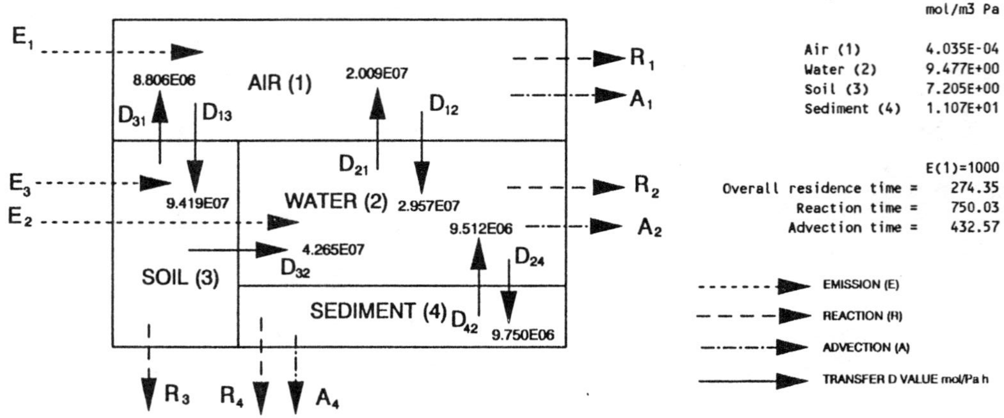

Phase Properties and Rates:

Compartment	Bulk Z mol/m3 Pa	Half-life h	D Values Reaction mol/Pa h	D Values Advection mol/Pa h
Air (1)	4.035E-04	170	1.64E+08	4.03E+08
Water (2)	9.477E+00	550	2.39E+09	1.90E+09
Soil (3)	7.205E+00	1700	5.29E+07	
Sediment (4)	1.107E+01	5500	6.97E+05	1.11E+05

	E(1)=1000	E(2)=1000	E(3)=1000	E(1,2,3)
Overall residence time =	274.35	442.83	1447.30	1137.53 h
Reaction time =	750.03	795.48	1891.36	1641.20 h
Advection time =	432.57	998.90	6164.37	3706.65 h

Phase Properties, Compositions, Transport and Transformation Rates:

Emission, kg/h			Fugacity, Pa				Concentration, g/m3				Amounts, kg				Total Amount, kg
E(1)	E(2)	E(3)	f(1)	f(2)	f(3)	f(4)	C(1)	C(2)	C(3)	C(4)	m(1)	m(2)	m(3)	m(4)	
1000	0	0	7.948E-06	1.257E-07	7.175E-06	1.188E-07	5.904E-07	2.193E-04	9.518E-03	2.421E-04	5.904E+04	4.387E+04	1.713E+05	1.210E+02	2.743E+05
0	1000	0	3.709E-08	1.262E-06	3.348E-08	1.193E-06	2.755E-09	2.203E-03	4.442E-05	2.431E-03	2.755E+02	4.406E+05	7.995E+02	1.216E+03	4.429E+05
0	0	1000	6.860E-07	5.267E-07	5.268E-05	4.976E-07	5.096E-08	9.191E-04	6.988E-02	1.014E-03	5.096E+03	1.838E+05	1.258E+06	5.071E+02	1.447E+06
50	250	700	8.869E-07	6.906E-07	3.724E-05	6.524E-07	6.588E-08	1.205E-03	4.940E-02	1.330E-03	6.588E+03	2.410E+05	8.893E+05	6.649E+02	1.138E+06

Emission, kg/h			Loss, Reaction, kg/h				Loss, Advection, kg/h			Intermedia Rate of Transport, kg/h T12	T21	T13	T31	T32	T24	T42
E(1)	E(2)	E(3)	R(1)	R(2)	R(3)	R(4)	A(1)	A(2)	A(4)	air-water	water-air	air-soil	soil-air	soil-water	water-sed	sed-water
1000	0	0	2.407E+02	5.527E+01	6.98E+01	1.525E-02	5.904E+02	4.387E+01	2.421E-03	4.327E+01	4.648E-01	1.378E+02	1.163E+01	5.635E+01	2.256E-01	2.080E-01
0	1000	0	1.123E+00	5.551E+02	3.26E-01	1.532E-01	2.755E+00	4.406E+02	2.431E-02	2.019E-01	4.669E+00	6.431E-01	5.429E-02	2.629E-01	2.266E+00	2.089E+00
0	0	1000	2.077E+01	2.316E+02	5.13E+02	6.390E-02	5.096E+01	1.838E+02	1.014E-02	3.735E+00	1.948E+00	1.190E+01	8.542E+01	4.137E+02	9.455E-01	8.714E-01
50	250	700	2.685E+01	3.037E+02	3.63E+02	8.378E-02	6.588E+01	2.410E+02	1.330E-02	4.829E+00	2.554E+00	1.538E+01	6.039E+01	2.925E+02	1.240E+00	1.143E+00

Level III Distribution

Chemical name: 2,4-Dinitrophenol

Distribution of mass

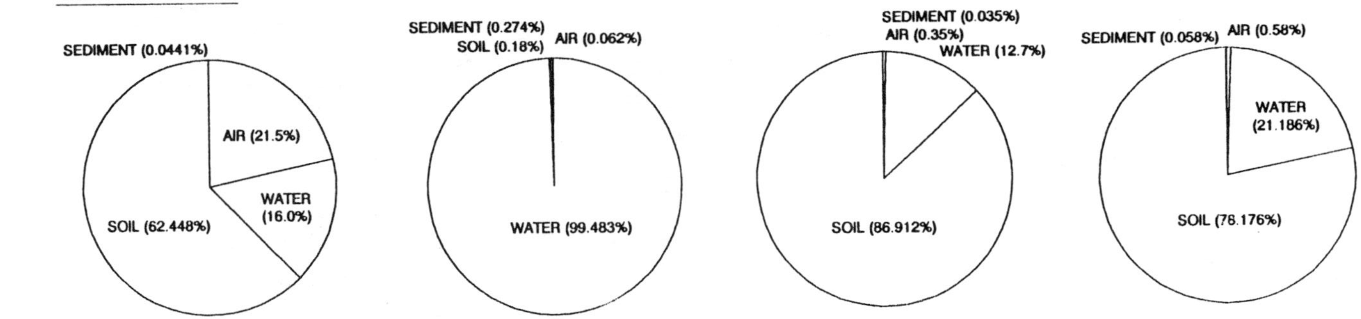

Distribution of removal rates

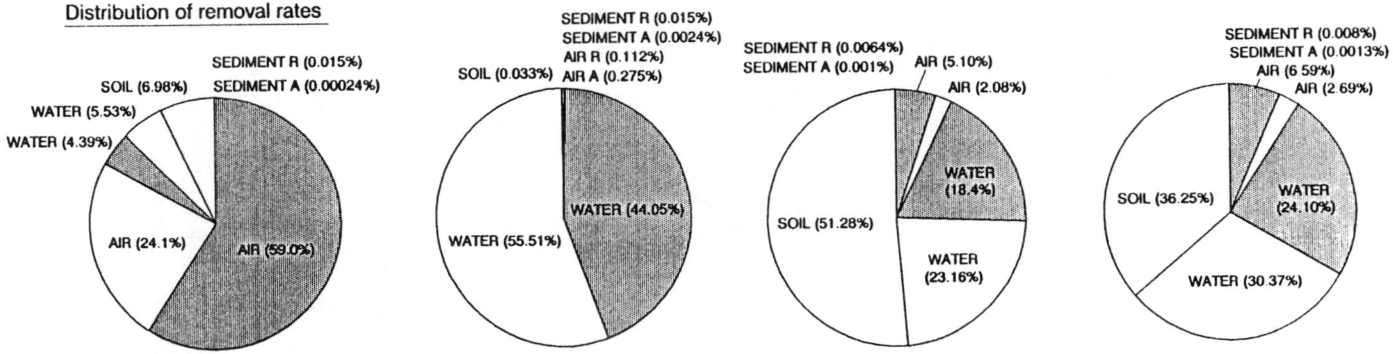

Emission rates:

E(1) = 1000 kg/h	E(1) = 0	E(1) = 0	E(1) = 50 kg/h
E(2) = 0	E(2) = 1000 kg/h	E(2) = 0	E(2) = 250 kg/h
E(3) = 0	E(3) = 0	E(3) = 1000 kg/h	E(3) = 700 kg/h

Residence time:

t = 274 h	t = 443 h	t = 1447 h	t = 1138 h

Reaction (white) — Advection (shaded)

4.4 Commentary on the Physical-Chemical Properties and Environmental Fate

Properties and Reactivity

The selected physical-chemical properties and reactivity classes of the phenolic compounds including the nitrophenols are given in Tables 4.1 to 4.3.

The chlorophenols and alkylphenols display the usual systematic changes in properties with substitution. On the other hand, the nitrophenols show a wide variation in properties of the individual isomers. The reason for this is the strong intra- and inter-molecular hydrogen bonding in aqueous solution which has been discussed in more detail by Morrison and Boyd (1959) and Schwarzenbach et al. (1988). Their partitioning behavior is described in the illustrative fugacity calculations.

QSPR Plots

A full discussion of the QSPR plots of the chloro- and alkylphenols has been given in Chapter 1 Section 1.5.3.

The highly structure-specific dependence of the properties of the nitrophenols suggests that simple QSPRs based on molar volume will be ineffective. It is recommended that for this series experimental data be used directly. There is also some doubt that whether the conventional expressions relating physical-chemical properties to environmental partition coefficients are valid.

The relatively low octanol-water partition coefficients, and Henry's law constants of the phenols suggest that this series of chemicals will tend to sorb relative little to solid phases and evaporation will be slight. The relatively high reactivity suggests that they will not be persistent.

Evaluative Calculations

The fugacity calculations illustrated in Figure 1.7 to Figure 1.12 and discussed earlier in Chapter 1 Section 1.5.4 indeed show that water is the primary medium of accumulation and of reactive loss. As with the alcohols they show little tendency for intermedia transport, although there can be appreciable deposition to soil and run-off from soil to water. Their overall persistence is generally in the range 70 to 700 hours with the longer times arising from accumulation in soils in which the nitrophenols are believed to have longer half-lives.

Overall, it appears that any environmental problems of the phenolic compounds will be restricted to local regions of high discharge rates, but these problems may be severe because of the high toxicity of this class of compounds.

4.5 REFERENCES

Abd-El-Bary, M.F., Hamoda, M.F., Tanisho, S., Wakao, N. (1986) Henry's constants for phenol over its diluted aqueous solution. *J. Chem. Eng. Data* 31, 229-230.

Abernethy, S.G., Mackay, D., McCarty, L.S. (1988) "Volume fraction" correlation for narcosis in aquatic organisms: The key role of partitioning. *Environ. Toxicol. Chem.* 7, 469-481.

Ahel, M., Giger, W. (1993a) Aqueous solubility of phenols and alkylphenol polyethoxylates. *Chemosphere* 26(8), 1461-1470.

Ahel, M., Giger, W. (1993b) Partitioning of alkylphenols and alkylphenol polyethoxylates between water and organic solvents. *Chemosphere* 26(8), 1471-1478.

Ahel, M., Scully, F.E., Jr., Hoignè, J., Giger, W. (1994) Photochemical degradation of nonylphenol and nonylphenol polyethoxylates in natural waters. *Chemosphere* 28(7), 1361-1368.

Ahlers, J., Benzing, M., Gles, A., Pauli, W., Rösick, E. (1988) Yeast as an unicellulor model system in ecotoxicology and xenobiochemistry. *Chemosphere* 17(8), 1603-1615.

Alexander, M., Lustigman, B.K. (1966) Effect of chemical structure on microbial degradation of substituted benzenes. *J. Agric Food Chem.* 14, 410-413.

Aly, O.M., Faust, S.D. (1964) Studies on the fate of 2,4-dichlorophenol and ester derivatives in natural surface waters. *J. Agric. Food Chem.* 12, 451-456.

Amidon, G.L., Williams, N.A. (1982) A solubility equation for non-electrolytes in water. *Int'l J. Pharm.* 11, 249-256.

Anbar, M., Neta, P. (1967) A compilation of specific bimolecular rate and hydroxyl radical with inorganic and organic compounds in aqueous solution. *Int'l. J. Appl. Radiation Isotopes* 18, 493-523.

Andon, R.J.L., Biddiscombe, D.P., Cox, J.D., Handley, R., Harrop, D., Herington, E.F.G., Martin, J.F. (1960) Thermodynamic properties of organic oxygen compounds. Part 1. Preparation and physical properties of pure phenol, cresols, and xylenols. *J. Chem. Soc.* 5246-5254.

Arbuckle, W.B. (1983) Estimating activity coefficients for use in calculating environmental parameters. *Environ. Sci. Technol.* 17(9), 537-542.

Arcand, Y., Hawari, J., Guiot, S.R. (1994) Solubility of pentachlorophenol in aqueous solutions: the pH effect. *Water Res.* in press.

Ashworth, R.A., Howe, G.B., Mullins, M.E., Rogers, T.N. (1988) Air-water partition coefficients of organics in dilute aqueous solutions. *J. Haz. Materials* 18, 25-36.

Atkinson, R. (1985) Kinetics and mechanisms of the gas-phase reactions of hydroxyl radical with organic compounds under atmospheric conditions. *Chem. Rev.* 85, 69-201.

Atkinson, R. (1987) Structure-activity relationship for the estimation of rate constants for the gas-phase reactions of OH radicals with organic compounds. *Int'l. J. Chem. Kinetics* 19, 799-828.

Atkinson, R. (1987) Estimation of OH radical rate constants and atmospheric lifetimes for polychlorobiphenyls, dibenzo-*p*-dioxins. *Environ. Sci. Technol.* 21, 305-307.

Atkinson, R. (1989) Kinetics and mechanisms of the gas-phase reactions of the hydroxyl radical with organic compounds. *J. Phys. Chem. Ref. Data* Monograph 1, 1-246.

Atkinson, R. (1990) Gas-phase tropospheric chemistry of organic compounds: A review. *Atmos. Environ.* 24A(1), 1-41.

Atkinson, R. (1991) Kinetics and mechanisms for the gas-phase reactions of the NO_3 radical with organic compounds. *J. Phys. Chem. Ref. Data* 20(2), 459-507.

Atkinson, R., Aschmann, S.M., Arey, J. (1992) Reactions of OH and NO_3 radicals with phenol, cresols, and 2-nitrophenol at 296 ± 2 K. *Environ. Sci. Technol.* 26(70), 1397-1403.

Atkinson, R., Aschmann, S.M., Fitz, D.R., Winer, A.M., Pitts, Jr., J.N. (1982) Rate constants for the gas-phase reactions of O_3 with a selected organics at 296 K. *Int'l. Chem. Kinetics* 14, 13-18.

Atkinson, R., Aschmann, S.M., Pitts, Jr., J.N. (1988) Rate constants for the gas-phase reactions of the NO_3 radicals with a series of organic compounds at 296 ± 2 K. *J. Phys. Chem.* 92, 3454-3457.

Atkinson, R., Aschmann, S.M., Pitts, Jr., J.N. (1985) Atmospheric gas phase loss processes for chlorobenzene, benzotrifluoride, and 4-chlorobenzotrifluoride, and generalization of predictive techniques for atmospheric lifetimes of aromatic compounds. *Arch. Environ. Contam. Toxicol.* 14, 417-425.

Atkinson, R., Aschmann, S.M., Winer, A.M. (1987) Kinetics of the reactions of NO_3 radicals with a series of aromatic compounds. *Environ. Sci. Technol.* 21, 1123-1126.

Atkinson, R., Carter, W.P.L. (1984) Kinetics and mechanisms of the gas-phase reactions of ozone with organic compounds under atmospheric conditions. *Chem. Rev.* 84, 437-470.

Atkinson, R., Carter, W.P.L., Plum C. N., Winer, A.M., Pitts, Jr., J.N. (1984) Kinetics of the gas-phase reactions of NO_3 radicals with a series of aromatics at 296 ± 2 K. *Int'l J. Chem. Kinetics* 16, 887-898.

Atkinson, R., Darnall, K.R., Lloyd, A.C., Winer, A.M., Pitts, J.N. Jr. (1979) Kinetics and mechanisms for the reaction of hydroxyl radicals with organic chemicals in the gas phase. *Adv. Photochem.* II, 375-488.

Atkinson, R., Lloyd, A.C. (1984) Evaluation of kinetic and mechanism data for modeling of photochemical smog. *J. Phys. Chem. Ref. Data* 13, 315-444.

Atkinson, R., Pitts, Jr., J.N. (1977) Kinetics and mechanisms of the gas-phase reactions of OH radicals with methoxybenzene and *o*-cresol over the temperature range 299-435 K. *J. Phys. Chem.* 81,1607-1611.

Atkinson, R., Plum C.N., Carter, W.P.L., Winer, A.M., Pitts, Jr., J.N. (1988) Rate constants for the gas-phase reactions of the nitrate radicals with a series of organics in air at 298 ± 1 K. *J. Phys. Chem.* 88, 1210-1215.

Bahnick, D.A., Doucette, W.J. (1988) Use of molecular connectivity indices to estimate soil sorption coefficients for organic chemicals. *Chemosphere* 17, 1703-1715.

Baker, M.D., Mayfield, C.I. (1980) Microbial and non-biological decomposition of chlorophenols and phenols in soil. *Water Air Soil Pollut.* 13, 411.

Baker, M.D., Mayfield, C.I., Inniss, W.E. (1980) Degradation of chlorophenols in soil, sediment and water at low temperature. *Water Res.* 14, 1765-1771.

Banerjee, S. (1987) Interrelationship between biodegradability, toxicity and structure of chlorophenols. In: *QSAR in Environmental Toxicology-II*, pp.17-23. Kaiser, K.L.E., Editor, D. Reidel Publishing Company, Dordrecht, the Netherlands.

Banerjee, S., Howard, P.H. (1988) Improved estimation of solubility and partitioning through correction of UNIFAC-derived activity coefficients. *Environ. Sci. Technol.* 22, 839-841.

Banerjee, S., Howard, P.H., Lande, S.S. (1990) General structure-vapor pressure relationships for organics. *Chemosphere* 21(10/11), 1173-1180.

Banerjee, S., Howard, P.H., Rosenberg, A.M., Dombrowski, A.E., Sikka, H., Tullis, D.L. (1984) Development of general kinetic model for biodegradation and its application to chlorophenols and related compounds. *Environ. Sci. Technol.* 18, 416-422.

Banerjee, S., Yalkowsky, S.H., Valvani, S.C. (1980) Water solubility and octanol/water partition coefficients of organics. Limitation of the solubility-partition coefficient correlation. *Environ. Sci. Technol.* 14, 1227-1229.

Barrows, M.E., Petrocelli, S.R., Macek, K.J. (1980) Bioconcentration and elimination of selected water pollutants by bluegill sunfish (*Lepomis macrochirus*). In: *Dynamic, Exposure, and Hazard Assessment of Toxic Chemicals*. Haque, R., Editor, pp.379-392, Ann Arbor Science Publisher Inc., Ann Arbor, Michigan.

Barton, D.A. (1987) Intermedia transport of organic compounds in biological wastewater treatment processes. *Environ. Progress* 6(4), 246-256.

Battersby, N.S. (1990) A review of biodegradation kinetics in the aquatic environment. *Chemosphere* 12, 1242-1284.

Becker, K.H., Biehl, H.M., Bruckmann, P., Fink, E.H., Führ, F., Klöpffer, W., Zellner, R., Zetsch, C. (1984) OH reaction constants and tropospheric lifetimes of selected environmental chemicals. *Kernforschungsanlage*. Jülich, GmbH. November ISSN 0343-7639.

Bellin, C.A., O'Connor, G.A., Jin, Y. (1990) Sorption and degradation of pentachlorophenol in sludge-amended soils. *J. Environ. Qual.* 19, 603-608.

Belly, R.T., Goodhue, C.T. (1976) A radiorespirometric technique for measuring the biodegradation of specific components in a complex effluent. In: *Proc. Int'l Biodegrad. Symposium* 3rd., pp. 1130-1137.

Beltrame, P., Beltrame, P.L., Cartini, P. (1984) Inhibiting action of chloro- and nitro-phenols on biodegradation of phenols: A structure-toxicity relationship. *Chemosphere* 13, 3-9.

Beltrame, P., Beltrame, P.L., Cartini, P. (1984) Influence of feed concentration on the kinetics of biodegradation of phenol in a continuous stirred reactor. *Water Res.* 18(4), 403-407.

Beltrame, P., Beltrame, P.L., Cartini, P. (1989) Lipophilic and electrophilic factors in the inhibiting action of substituted phenols on the biodegradation of phenol. *Chemosphere* 19(10/11), 1623-1627.

Beltrame, P., Beltrame, P.L., Cartini, P., Guardione, D., Lanzetta, C. (1988) Inhibiting action of chlorophenols on biodegradation of phenol and its correlation with structural properties of inhibitors. *Biotechn. & Bioeng.* 31, 821-828.

Beltrame, P., Beltrame, P.L., Cartini, P., Lanzetta, C. (1988) New results on the inhibiting action of chloro-and nitro-substituted phenols on the biodegradation of phenol. *Chemosphere* 17, 235-242.

Bender, R., Bieling, V., Maurer, G. (1983) The vapor pressures of solids: anthracene, hydroquinone, and resorcinol. *J. Chem. Thermodyn.* 15, 585-594.

Berthod, A., Han, Y.I., Armstrong, D.W. (1988) Centrifugal partition chromoatography. V. Octanol-water partition coefficients, direct and inderect determination. *J. Liq. Chromatogr.* 11, 1441-1456.

Bevenue, A., Beckman, H. (1967) Pentachlorophenol: A discussion of its properties and its occurence as a residue in human and animal tissues. *Residue Rev.* 19, 83-134.

Biddscombe, D.P., Martin, J.F. (1958) Vapor pressures of phenol and the cresols. *Trans. Faraday Soc.* 54, 1316-1322.

Biddscombe, D.P., Handley, R., Harrop, D., Head, J., Lewis, G.B., Martin, J.F., Sprake, H.S. (1963) Thermodynamic properties of organic oxygen compounds. Part XIII. Preparation and physical properties of ethyl phenols. *J. Chem. Soc.* 5764-5768.

Bidleman, T.F., Renberg, L. (1985) Determination of vapor pressures for chloroguaiacols, chloroveratroles, and nonylphenol by gas chromatography. *Chemosphere* 14, 1475-1481.

Bintein, S., Devillers, J. (1994) QSAR for organic chemical sorption in soils and sediments. *Chemosphere* 28(6), 1171-1188.

Birch, R.R., Fletcher, R.J. (1991) The application of dissolved inorganic carbon measurements to the study of aerobic biodegradability. *Chemosphere* 23(4), 507-524.

Blackman, G.E., Parke, M.H., Garton, G. (1955) The physiological activity of substituted phenols. I. Relationships between chemical structure and physiological activity. *Arch. Biochem. Biophys.* 54(1), 55-71.

Blades-Fillmore, L.A., Clement, W.H., Faust, S.D. (1982) The effect of sediment on the biodegradation of 2,4,6-trichlorophenol in Delaware River water. *J. Environ. Sci. Health* A17, 797-818.

Blum, D.J.W., Suffet, I.H., Duguet, J.P. (1994) Quantitative structure-activity relationship using molecular connectivity for the activated carbon adsorption of organic chemicals in water. *Water Res.* 28, 687-699.

Bodor, N., Gabanyi, Z., Wong, C.K. (1989) A new method for the estimation of partition coefficient. *J. Am. Chem. Soc.* 111, 3783-3786.

Booth, H.S., Everson, H.E. (1948) Hydrotropic solubilities. Solubilities in 40 per cent sodium xylenesulfonate. *Ind. & Eng. Chem.* 40(8), 1491-1493.

Boublik, T., Fried, V., Hala, E. (1973) *The Vapour Pressures of Pure Substances*. Elsevier, Amsterdam, The Netherlands.

Boublik, T., Fried, V., Hala, E. (1984) *The Vapour Pressures of Pure Substances*, 2nd ed., Elsevier, Amsterdam, The Netherlands.

Bourquin, A.W. (1984) *Biodegradation in the Estuarine-Marine Environments and the Genetically Altered Microbe*. EPA-600/D-84-051. NTIS PB84-151 315. U.S. EPA Environmental Research Lab., Gulf Breeze, Florida. 35pp.

Boyd, S.A. (1982) Adsorption of substituted phenols by soil. *Soil Sci.* 134, 337-343.

Boyd, S.A., Shelton, D.R., Berry, D., Tiedje, J.M. (1983) Anaerobic biodegradation of phenolic compounds in digested sludge. *Appl. Environ. Microbiol.* 46, 50-54.

Brecken-Folse, J.A., Mayer, F.L., Pedigo, L.E., Marking, L.L. (1994) Acute toxicity of 4-nitrophenol, 2,4-dinitrophenol, terbufos and trichlorfon to grass shrimp (*palaemonetes* spp.) and sheephead minnows (*cyprinodon variegatus*) as affected by salinity and temperature. *Environ. Toxicol. Chem.* 13, 67-77.

Briggs, G.G. (1981) Theoretical and experimental relationships between soil adsorption, octanol-water partition coefficients, water solubilities, bioconcentration factors and the parachor. *J. Agric. Food Chem.* 29, 1050-1059.

Brockway, D.L., Smith, P.D., Stancil, F.E. (1984) Fates and effects of pentachlorophenol in hard- and soft-water microcosms. *Chemosphere* 13(12), 1363-1377.

Brooke, D.N., Dobbs, A.J., Williams, N. (1986) Octanol/water partition coefficients (P): measurement, estimation, and interpretation, particularly for chemicals with $P > 10^5$. *Ecotoxicol. Environ. Safety* 11, 251-260.

Budavari, S., Editor (1989) *The Merck Index. An Encyclopedia of Chemicals, Drugs and Biologicals.* 11th Edition, Merck & Co. Inc., Rahway, New Jersey.

Bunce, N.J., Nakai, J.S., Yawching, M. (1991) A model for estimating the rate of chemical transformation of a VOC in the troposphere by two pathways: photolysis by sunlight and hydroxyl radical attack. *Chemosophere* 22, 305-315.

Burkhard, L.P., Kuehl, D.W. (1986) N-octanol/water partition coefficients by reverse phase liquid chromatography/mass spectrometry for eight tetrachlorinated planar molecules. *Chemosphere* 15(2), 163-167.

Butte, W., Blum, J.K. (1984) Calculation of bioconcentration factors from kinetic data by non-linear iterative least-square regression analysis using programable minicalculator. *Chemosphere* 12, 1459-1475.

Butte, W., Fooken, C., Klussman, R., Schulller, D. (1981) Evaluation of lipophilic properties for a series of phenols, using reversed-phase high-performance liquid chromatography and high-performance thin-layer chromatography. *J. Chromatogr.* 214, 59-67.

Butte, W., Willing, A., Zauke, G.-P. (1987) Bioaccumulation of phenols in zebrafish determined by a dynamic flow through test. In: *QSAR in Environmental Toxicology II*, pp.43-53. Kaiser, K.L.E., Editor, D. Reidel Publishing Company, Dordrecht, the Netherlands.

Bysshe, S.E. (1982) Bioconcentration factor in aquatic organisms. In: *Handbook of Chemical Property Estimation Methods*. Lyman, W.J., Reehl, W.F., Rosenblatt, D.H., Editors, Chapter 5, Ann Arbor Science, Ann Arbor, Michigan.

Call, D.J., Brooke, L.T., Lu, P.Y. (1980) Uptake, elimination and metabolism of three phenols by fathead minnows. *Arch. Environ. Contam. Toxicol.* 9, 699-714.

Callahan, M.A., Slimak, M.W., Gabel, N.W., May, I.P., Fowler, C.F., Freed, J.R., Jennings, P., Durfee, R.L., Whitmore, F.C., Maestri, B., Mabey, W.R., Holt, B.R., Gould, C. (1979) *Water Related Environmental Fate of 129 Priority Pollutants*. EPA-440-4-79-029a,b. Versar, Inc., Springfield, Virginia.

Campbell, J.R., Luthy, R.G. (1985) Prediction of aromatic solute partition coefficient using the UNIFAC group contribution model. *Environ. Sci. Technol.* 19, 980-985.

Carlberg, G.E., Martinsen, K., Kringstad, A., Gjessing, E., Grande, M., Källqvist, T., Skåre, J.U. (1986) Influence of aquatic humus on the bioavailability of chlorinated micropollutants in Atlantic salmon. *Arch. Environ. Comtam. Toxicol.* 15, 543-548.

Carlier, P., Hannachi, H., Mouvier, G. (1986) The chemistry of carbonyl compounds in the atmosphere-A review. *Atmos. Environ.* 20(11), 2079-2099.

Carlson, R.M., Carlson, R.E., Kopperman, H.L. (1975) Determination of partition coefficients by liquid chromatography. *J. Chromatogr.* 107, 219-223.

Carswell, T.G., Nason, H.K. (1938) Properties and uses of pentachlorophenol. *Ind. Eng. Chem.* 30, 622-626.

Carter, W.P., Winer, A.M., Pitts, J.N., Jr. (1981) Major atmospheric sink for phenol and the cresols. Reaction with nitrate radical. *Environ. Sci. Technol.* 15(7), 829-834.

CESARS (1987) Chemical Evaluation Search and Retrieval System. Ontario Ministry of the Environment and Michigan Department of Natural Resources.

CESARS (1988) Chemical Evaluation Search and Retrieval System. Ontario Ministry of the Environment and Michigan Department of Natural Resources.

Cessna, A.J., Grover, R. (1978) Spectrophotometric determination of dissociation constants of selected acidic herbicides. *J. Agric. Food Chem.* 26, 289.

Chambers, C.W., Tabak, H.H., Kabler, P.W. (1963) Degradation of aromatic compounds by phenol-adapted bacteria. *J. Water Pollut. Control Fed.* 35, 1517-1528.

Chao, J., Lin, C.T., Chung, T.H. (1983) Vapor pressure of coal chemicals. *J. Phys. Chem. Ref. Data* 12(4), 1034-1063.

Chen F., Holten-Andersen, J., Tyle, H. (1993) New developments of the UNIFAC model for environmental application. *Chemosphere* 26, 1325-1354.

Chessells, M., Hawker, D.W., Connell, D.W. (1992) influence of solubility in lipid on bioconcentration of hydrophobic compounds. *Ecotoxicol. Environ. Safety* 23, 260-273.

Cheung, B. (1984) M.A. Sc. Thesis, University of Toronto, Toronto, Ontario.

Chin, Y.-P., Weber Jr., W.J., Voice, T.C. (1986) Determination of partition coefficients and aqueous solubilities by reverse phase chromatography-II. *Water Res.* 20(11), 1443-1450.

Chiou, C.T., Freed, V.H. (1977) *Chemodynamic Studies on Bench Mark Industrial Chemicals.* NSF/RA-770286 prepared for Research Applied to National Needs, National Science Foundation, Washington, D.C.

Chiou, C.T., Freed, V.H., Peters, W., Kohnert, R.L. (1980) Evaporation of solutes from water. *Environ. Int'l* 3, 231-236.

Chiou, C.T., Porter, P.E., Schmedding, D.W. (1983) Partition equilibria of nonionic organic compounds between soil organic matter and water. *Environ. Sci. Technol.* 17, 227-231.

Choi, J., Aomine, S. (1974) Adsorption of pentachlorophenol by soil. *Soil Sci. Plant Nutr.* 20, 135-144.

Chou, J.T., Jurs, P.C. (1979) Chapter 5, Computation of partition coefficients from molecular structures by fragment addition method. In: *Physical and Chemical Properties of Drugs. Medical Research Series,* Vol 10. Yalkowsky, S.H., Sinkula, A.A., Valvani, S.C., Editors.

Chou, W.L., Speece, R.E., Siddiqi, R.H. (1979) Acclimation and degradation of petrochemical wastewater components by methane fermentation. *Biotechnol. Bioeng. Symp.* 8, 391-414.

Chudoba, J., Alboková, J., Lentge, B., Kümmel, R. (1989) Biodegradation of 2,4-dichlorophenol by activated sludge microorganisms. *Water Res.* 23(11), 1439-1442.

Chudoba, J., Prasil, M., Emmerova, H. (1968) Residue organic matter in activated sludge process effluents. III. Degradation of amino acids and phenols under continuous conditions. *Sb. Vys. Sk. Chem.-Technol. Praze, Technol. Vody.* 13, 45-63.

Connell, D.W., Markwell, R.D. (1990) Bioaccumulation in the soil to earthworm system. *Chemosphere* 20(1-2), 91-100.

CRC (1976-77) *Handbook of Chemistry and Physics.* 57th Edition, Weast, R.C., Editor, CRC Press, Boca Raton, Florida.

Crosby, T. (1981) Environmental chemistry of pentachlorophenol. *Pure Appl. Chem.* 53, 1051-1080.

Crossland, N.O., Wolff, C.J.M. (1985) Fate and biological effects of pentachlorophenol in outdoor ponds. *Environ. Toxicol. & Chem.* 4, 73-86.

Daccomo, C. (1885) *Ber. Deutsch. Chem. Ges.* 18, 1163-1164.

Dallos, A., Wienke, G., Ilchman, A., Gmehling, J. (1993) Vauausberechnung von octanol/wasser-vertilungskoeffizienten mit hilfe der UNIFAC-method. *Chem.-Ing.-Tech.* 65(2), 201-203.

D'Amboise, M., Hanai, T. (1982) Hydrophobicity and retention in reversed phased liquid chromatography. *J. Liq. Chromatogr.* 5(2), 229-244.

Dauble, D.D., Carlile, D.W., Hanf, Jr., R.W. (1986) Bioaccumulation of fossil fuel components during single-compound and complex-mixture exposures of *daphnia magna. Bull. Environ. Contam. Toxicol.* 37, 125-132.

Davies, R.P., Dobbs, A.J. (1984) The prediction of bioconcentration in fish. *Water Res.* 18(10), 1253-1262.

Davis, S.S., Elson, G., Tomlinson, E., Harrison, G., Dearden, J.C. (1976) The rapid determination of partition coefficient data using a continuous solvent extraction system. (AKUFVE). *Chem. Ind.* (London), 16, 677-683.

de Kreuk, J.F., Hanstveit, A.O. (1981) Determination of the biodegradability of the organic fraction of chemical wastes. *Chemosphere* 10, 561-575.

Dean, J. (1985) *Lange's Handbook of Chemistry.* 13th ed., McGraw-Hill, New York.

Delaune, R.D., Gambrell, R.P., Reddy, K.S. (1983) Fate of pentachlorophenol in estuarine sediment. *Environ. Pollut. Series B*6, 297-308.

Dobbs, A.J., Grant, C. (1980) Pesticide volatilisation rate: A new measurement of the vapour pressure of pentachlorophenol at room temperature. *Pestic. Sci.* 11, 29-32.

Doedens, J.D., Editor (1967) *Lange's Handbook of Chemistry.* McGraw-Hill, Inc., New York.

Dorfman, L.M., Adams, G.E. (1973) *Reactivity of the Hydroxyl radical in Aqueous Solution.* NSRD-NDB-46. NTIS COM-73-50623. National Bureau of Standards, Washington D.C. 51 pp.

Drahonovsky, J., Vacek, Z. (1971) Dissoziationskonstanten und austauscherchromatographie chlorierter phenole. *Coll. Czech. Chem. Commun.* 36(10), 3431-3440.

Dreisbach, P.R. (1955) *Physical Chemical Properties of Chemical Compounds.* No. 15 of the Adv. in Chemistry Series. American Chemical Society, Washington, DC.

Duff, J.C. (1929) solubilities of *o*- and *p*-nitrophenols in aqueous methanol solutions at 25° and 40°C. Formation of β-*p*-nitrophenol. *J. Chem. Soc.* 2789-2796.

Eadsforth, C.V. (1986) Application of reverse-phase HPLC for the determination of partition coefficients. *Pest. Sci.* 17(3), 311-325.

Eadsforth, C.V., Moser, P. (1983) Assessment of reversed phase chromatographic methods for determining partition coefficients. *Chemosphere* 12, 1459-1475.

El Tayar, N, Tsai, R.-S., Vallat, P., Altomare, C., Testa, B. (1991) Measurement of partition coefficients by various contrifugal partition chromatographic techniques. *J. Chromatogr.* 556, 181-194.

Ericksen, L.V., Dobbert, E. (1955) Das gegeseitige löslichkeitsverhalten von alkylphenolen und wasser. *Brennstoff-Chemie* 36(21/22), 338-345.

Farquharson, M.E., Gage, J.C., Northover, J. (1958) The biological action of chlorophenols. *Brit. J. Pharmacol.* 13, 20.

Firestone, D. (1977) Chemistry and analysis of pentachlorophenol and its contaminants. Division of Chemistry and Physics, Bureau of Foods. FDA By-Lines No.2, September 1977.

Fieser, L.F., Fieser, M. (1959) *Introduction to Organic Chemistry.* D.C. Heath & Co., Boston, Mass.

Foote, C.S. (1976) *Free Radicals in Biology.* Pryor, W.A., Editor, Academic Press, New York.

Freitag, D., Ballhorn, L., Geyer, H., Körte, F. (1985) Environmental hazard profile of organic chemicals. An experimental method for assessment of the behavior of chemicals in the ecosphere by simple laborarory tests with C-14 labelled chemicals. *Chemosphere* 14, 1589-1616.

Freitag, D., Geyer, H., Kraus, A., Viswanathan, R., Kotzias, D., Attar, A., Klein, W. Körte, F. (1982) Ecotoxicological profile analysis. VII. Screening chemicals for their environmental behavior by comparative evaluation. *Ecotoxicol. Environ. Safety* 14, 60-81.

Fujisawa, S., Masuhara, E. (1981) Determination of partition coefficients of acrylates, methacrylates, and vinyl monomers using high performance liquid chromatography (HPLC). *J. Biomed. Mat. Res.* 15, 787-793.

Fujita, T., Iwasa, J., Hansch, C. (1964) A new substituent constant derived from partition coefficients. *J. Am. Chem. Soc.* 86(23), 5175-5180.

Gaffney, J.S., Streit, G.E., Spall, W.D., Holl, J.H. (1987) Beyond acid rain. Do soluble oxidants and organic toxins interact with SO_2 and NO_2 to increase ecosystem effects? *Environ. Sci. Technol.* 21(6), 519-524.

Garst, J.E. (1984) Accurate, wide-range, automated, high-performance liquid chromatographic method for the estimation of octanol/water partition coefficients. II: Equilibrium in partition coefficient measurements, additivity of substituent constants, and correlation of biological data. *J. Pharm. Sci.* 73(11), 1623-1629.

Garst, J.E., Wilson, W.C. (1984) Accurate, wide-range, automated, high-performance liquid chromatographic method for the estimation of octanol/water partition coefficients. I: Effect of chromatographic conditions and procedure variables on accuracy and reproducibility of the method. *J. Pharm. Sci.* 73(11), 1616-1623.

Gebefügi, I., Parlar, H., Körte, F. (1979) Occurrence of pentachlorophenol in enclosed environments. *Ecotox. & Environ. Safety* 3, 269-300.

GEMS (1986) Graphical Exposure Modeling System. FAP. Fate of Atmospheric Pollution.

Gerike, P., Fischer, W.K. (1979) A correlation study of biodegradability determinations with various chemicals in various tests. *Ecotoxicol. Environ. Safety* 3, 159-173.

Geyer, H., Kraus, A.G., Klein, W. (1980) Relationship between water solubility and bioaccumulation potential of organic chemicals in rats. *Chemosphere* 9, 277-291.

Geyer, H., Viswanathan, R., Freitag, D., Körte, F. (1981) Relationship between water solubility of organic chemicals and their bioaccumulation by the alga *chlorella*. *Chemosphere* 10, 1307-1313.

Geyer, H., Sheehan, P., Kotzias, D., Freitag, D., Körte, F. (1982) Prediction of ecotoxicological behaviour of chemicals: relationship between physico-chemical properties and bioaccumulation of organic chemicals in the mussel *mytilus edulis*. *Chemosphere* 11, 1121-1134.

Geyer, H., Politzki, G., Freitag, D. (1984) Prediction of ecotoxicological behaviour of chemicals: relationship between n-octanol/water parition coefficient and bioaccumulation of organic chemicals by alga *chlorella*. *Chemosphere* 13, 269-284.

Geyer H.J., Scheunert, I., Körte, F. (1987) Correlation between the bioconcentration potential of organic environmental chemicals in humans and their n-octanol/water partition coefficients. *Chemosphere* 16, 239-252.

Geyer, H., Scheunert, I., Bruggemann, R., Steinberg, C., Körte, F., Dettrup, A. (1991) QSAR for organic chemical bioconcentration in *Daphnia*, Algae, and mussels. *Sci. Tot. Environ.* 109/110, 387-394.

Gibson, S.A., Suflita, J.M. (1986) Extrapolation of biodegradation results to ground water aquifers: Reductive dehalogenation of aromatic compounds. *Appl. Environ. Microbiol.* 52, 681-688.

Gluck, S.J., Martin, E.J. (1990) Extended octanol-water partition coefficient determination by dual-mode centrifugal partition chromatography. *J. Liq. Chromatogr.* 13(18), 3559-3570.

Goerlitz, D.F., Troutman, D.E., Godsy, E.M., Franks, B.J. (1985) Migration of wood-preserving chemicals in contaminated groundwater in sand aquifer at Pensacola, Florida. *Environ. Sci. Technol.* 19, 955-961.

Goll, O. (1954) Chlorophenol. In: *Ullmans Encyklopädie der Technischen Cheme.* Foerst, W., Editor, Urban & Schwarzenberg, Munich/Berlin. pp.494-499.

Gonsior, S.J., Bailey, R.E., Rhinehart, W.L., Spence, M.W. (1984) Biodegradation of o-phenylphenol in river water and avtivated sludge. *J. Agric. & Food Chem.* 32, 593-596.

Graham, R.A., Johnston, H.S. (1978) The photochemistry of the nitrate radical and the kinetics of the nitrogen pentoxide-ozone system. *J. Phys. Chem.* 82, 254.

Grosjean, D. (1985) Wall loss of gaseous pollutants in outdoor Teflon chambers. *Environ. Sci. Technol.* 19, 1059-1065.

Güesten, H., Filby, W.G., Schoop, S. (1981) Prediction of hydroxyl radical reaction rates with organic compounds in the gas-phase. *Atoms. Environ.* 15, 1763-1765.

Guinee, J., Heijungs, R. (1993) A proposal for the classification of toxic substances within the framework of life cycle assessment of products. *Chemosphere* 26(1)0, 1925-1944.

Günther, F.A., Westlake, W.E., Jaglan, P.S. (1968) Reported solubilities of 738 pesticide chemicals in water. *Residue Rev.* 20, 1-148.

Haderlein, S.B., Schwarzenbach, R.P. (1993) Adsorption of substituted nitrobenzenes and nitrophenols to mineral surfaces. *Environ. Sci. Technol.* 27, 316-326.

Häggblom, M.M., Young, L.Y. (1990) Chlorophenol degradation coupled to sulfate reduction. *Appl. & Environ. Microbiol.* 56(11), 3255-3260.

Haider, K., Jagnow, G., Kohnen, R., Lim, S.U. (1974) Degradation of chlorinated benzenes, phenols, and cyclohexane derivatives by benzene and phenol utilizing soil bacteria under aerobic conditions. *Arch. Microbiol.*96, 183-200.

Haky, J.E., Young, A.M. (1984) Evaluation of simple HPLC correlation method for the estimation of the octanol-water partition coefficients of organic compounds. *J. Chromatogr.* 7, 675-689.

Hale, D.D., Rogers, J.E., Wiegel, J. (1991) Environmental factors correlated to dichlorophenol dechlorination in anoxic freshwater sediments. *Environ. Toxicol. Chem.* 10, 1255-1265.

Haller, H. (1978) Degradation of mono-substituted benzoates and phenols by wastewater. *Journal WPCF* 2771-2777.

Hamilton, D.J. (1980) Gas chromatographic measurement of volatility of herbicide esters. *J. Chromatogr.* 195, 75-83.

Hammers, W.E., Meurs, G.J., DeLigny, C.L. (1982) Correlations between liquid chromatographic capacity factors ratio data on Lichrosorb RP-18 and partition coefficients in the octanol-water system. *J. Chromatogr.* 247, 1-13.

Hanai, T., Hubert, J. (1982) Hydrophobicity and chromatographic behaviour of aromatic acids found in urine. *J. Chromatogr.* 239, 527-536.

Handley, R., Harrop, D., Martin, J.F., Sprake, C.H.S. (1964) Thermodynamic properties of organic oxygen compounds. Part XIV. Purification and vapor pressures of some alkylphenols. *J. Chem. Soc.* 4404-4406.

Hansch, C., Anderson, S. (1967) The effect of intermolecular hydrophobic bonding on partition coefficients. *J. Org. Chem.* 32, 2583.

Hansch, C., Leo, A. (1979) *Substituent Constants for Correlation Analysis in Chemistry and Biology.* John Wiley & Sons, New York.

Hansch, C., Leo, A. (1982) *Medchem. Project*, Pomona College, Claremont, California.

Hansch, C., Leo, A. (1985) *Medchem. Project*, Pomona College, Claremont, California.

Hansch, C., Quinlan, J.E., Lawrence, G.L. (1968) The linear free-energy relationship between partition coefficients and aqueous solubility of organic liquids. *J. Org. Chem.* 33, 347-350.

Haque, A., Ebing, W. (1988) Uptake and accumulation of pentachlorophenol and sodium pentachlorophenate by earthworms from water and soil. *Sci. Total Environ.* 68, 113-125.

Haque, R., Falco, J., Cohen, S., Riordan, C. (1980) Role of transport and fate studies in exposure, assessment and screening toxic chemicals. In: *Dynamics, Exposure and Hazard Assessment of Toxic Chemicals.* Haque, R., Editor, pp.47-67. Ann Arbor Science Publishers Inc., Ann Arbor, Michigan.

Hardy, J.T., Dauble, D.D., Felice, L.J. (1985) Aquatic fate of synfuel residuals: Bioaccumulation of aniline and phenol by the freshwater phytoplankter *scenedesmus quadricauda*. *Environ. Toxicol. Chem.* 4, 29-35.

Harnish, M., Möckel, H.J., Schulze, G. (1983) Relationship between log P_{OW} shake-flask values and capacity factors derived from reversed-phase high-performance liquid chromatography of n-alkylbenzenes and some OECD reference substances. *J. Chromatogr.* 282, 315-332.

Hashimoto, Y., Tokura, K., Kishi, H., Strachan, W.M.J. (1984) Prediction seawater solubility of aromatic compounds. *Chemosphere* 13(8), 881-888.

Hassett, J.J., Banwart, W.L., Wood, S.G., Means, J.C. (1981) Sorption of α-naphthol: Implications concerning the limits of hydrophobic sorption. *Soil Sci. Soc. Am. J.* 45, 38-42.

Hattemer-Frey, H.A., Travis, C.C. (1989) Pentachlorophenol: Environmental partitioning and human exposure. *Arch. Environ. Contam. Toxicol.* 18, 482-489.

Hattula, M.L., Wasenius, V.-M., Reunanen, H., Arstila, A.U. (1981) Acute toxicity of some chlorinated phenols, catechols and cresols in trout. *Bull. Environ. Contam. Toxicol.* 26, 295-298.

Hawthorne, S.B., Sievers, R.E., Barkley, R.M. (1985) Organic emissions from shale oil wastewaters and their implications for air quality. *Environ. Sci. Technol.* 19, 992-997.

Hellmann, H. (1987) Model tests on volatilization of organic trace substances in surface waters. *Fresenius Z. Anal. Chem.* 328, 475-479.

Hendry, D.G., Kenley, R.A. (1979) *Atmospheric Reaction Products of Organic Compounds.* EPA Final Report. EPA 560/12-79-001.

Hendry, D.G., Mill, T., Piszkiewicz, L., Howard, J.A., Eigenmann, H.K. (1974) Critical review of hydrogen-atom transfer in the liquid phase. Chlorine atom, alkyltrichloromethyl, alkoxy, and alkyl peroxy radicals. *J. Phys. Chem. Ref. Data* 3, 937-978.

Herington, E.F.G., Kynaston, W. (1957) The ultraviolet absorption spectra and dissociation constants of certain phenols in aqueous solution. *Trans. Faraday Soc.* 53, 138-142.

Hermens, J., Leeuwangh, P. (1982) Joint toxicity of 8 and 24 chemicals to the guppy (*poecilia reticulata*). *Ecotoxicol. & Environ. Safety* 6, 302-310.

Hersey, A., Hill, A.P., Hyde, R.M., Livingstone, D.J. (1989) Principles of method selection in partition studies. *Quant. Struct.-Act. Relat.* 8, 288-296.

Heukelekian, H., Rand, M.C. (1955) Biochemical oxygen demand of pure organic componds. *J. Water Pollut. Control Assoc.* 29, 1040-1053.

Hiatt, C.W., Haskins, W.T., Olivier, L. (1960) The action of sunlight on sodium pentachlorophenate. *Am. J. Trop. Med. Hyg.* 9, 527-531.

Hickey, J.P., Passino-Reader, D.R. (1991) Linear solvation energy relationships: "Rule of Thumb" for estimation variable values. *Environ. Sci. Technol.* 25, 1753-1760.

Hine, J., Mookerjee, P.K. (1975) The intrinsic hydrophilic character of organic compounds. Correlations in terms of structural contributions. *J. Org. Chem.* 40, 292-298.

Hodson, J., Willams, N.A. (1988) The estimation of the adsorption coefficient (K_{OC}) for soils by high performance liquid chromatography. *Chemosphere* 17, 67-77.

Hoigné, J., Bader, H. (1983) Rate constants of reactions of ozone with organic and inorganic compounds in water - I. Non-dissociating organic compounds. *Water Res.* 17, 173-183.

Hoigné, J., Bader, H. (1983) Rate constants of reactions of ozone with organic and inorganic compounds in water - II. Dissociation organic compounds. *Water Res.* 17, 185-194.

Holmes, H.L., Lough, C.E. (1976) Effect of Intramolecular Hydrogen Bonding on Partition Coefficients. Suffield Technical Note No. DRES-TN-365. Defence Research Establishment Suffield-Information Canada. U.S. NTIS Report No. AD-A030683.

Hon, H.C., Singh, R.P., Kudchadker, A.P. (1976) Vapor pressure-boiling point measurement of five organic substances by twin ebulliometry. *J. Chem. Eng. Data* 21, 430-431.

Horowitz, A., Shelton, D.R., Cornell, C.P., Tiedje, J.M. (1982) Anaerobic degradation of aromatic compounds in sediment and digested sludge. *Dev. Ind. Microbiol.* 23, 435-444.

Horvath, A.L., Getzen, F.W., Editors (1985) *Solubility Data Series: Halogenated Benzenes, Toluenes and Phenols with Water.* Volume 20, Pergamon Press, Oxford, England.

Howard, J.A. (1972) Absolute rate constants for reactions of oxy radicals. *Adv. Free Radical Chem.* 4, 49-173.

Howard, P.H., Ed. (1989) *Handbook of Environmental Fate and Exposure Data. Vol. I. Large Production and Priority Pollutants.* Lewis Publishers, Chelsea, MI.

Howard, P.H., Ed. (1990) *Handbook of Fate and Exposure Data for Organic Chemicals. Vol. II, Solvents.* Lewis Publishers, Inc., Chelsea, Michigan.

Howard, P.H., Ed. (1991) *Handbook of Fate and Exposure Data for Organic Chemicals. Vol. III, Pesticides.* Lewis Publishers, Inc., Chelsea, Michigan.

Howard, P.H., Boethling, R.S., Jarvis, W.F., Meylan, W.M., Michalenko, E.M., Editors (1991) *Handbook of Environmental Degradation Rates.* Lewis Publishers Inc., Chelsea, Michigan.

Howard, P.H., Hueber, A.E., Mulesky, B.C., Crisman, J.S., Meylan, W., Crosbie, E., Gray, D.A., Sage, G.W., Howard, K.P., LaMacchia, A., Boethling, R.S., Troast, R. (1986) BILOG, BIODEG, and FATE/EXPOS: New files on microbial degradation and toxicity as well as environmental fate/exposure of chemicals. *Environ. Toxicol. Chem.* 5, 977-988.

Howe, G.E., Marking, L.L., Bills, T.D., Rach, J.J., Mayer, Jr., F.L. (1994) Effects of water temperature and pH on toxicity of terbufos, trichlorfon, 4-nitrophenol and 2,4-dinitrophenol to the amphipod *gammarus pseudolimnaeus* and rainbow trout *(oncorhynchus mykiss). Environ. Toxicol. Chem.* 13, 51-66.

Hoyer, von H., Peperle, W. (1957) Dampfdruckmessungen an organischen substanzen und ihre sublimations wärmen. *Zeit. Elektrochemie* 62(1), 61-66.

Hulzebos, E.M., Adema, D.M.M., Breemen, E.M. D.-van, Henzen, L., van Dis, W.A., Herbold, H.A., Hoekstra, J.A., Baerselman, R., van Gestel, C.A.M. (1993) Phytotoxicity studies with *lactuca sativa* in soil and nutrient solution. *Environ. Toxicol. Chem.* 12, 1079-1094.

Hustert, K., Mansour, M., Parlar, H., Körte, F. (1981) The EPA test: a method to determine the photochemical degradation of organic compounds in aquatic systems. *Chemosphere* 10, 995-998.

Hwang, H.-M., Hodson, R.E., Lee, R.F. (1986) Degradation of phenol and chlorophenols by sunlight and microbes in estuarine water. *Environ. Sci. Technol.* 20, 1002-1007.

Hwang, H.M., Hodson, R.E., Lee, R.F. (1987) Photolysis of phenol and chlorophenols in estuarine water. In: *Photochemistry of Environmental Aquatic Systems.* American Chemical Society, Washington, D.C.

Ide, A., Niki, Y., Sakamoto, F., Watanabe, I. (1972) Decomposition of pentachlorophenol in paddy soil. *Agric. Biol. Chem.* 36, 1937-1944.

Ilchmann, A., Wienke, G., Meyer, T., Gmehling, J. (1993) Bestimmung von verteilungskoeffizienten mit hilfe der flüssig/flüssig-gegenstrom-chromatographie. *Chem.-Ing.-Tech.* 65(1), 72-75.

Ince, N., Inel, Y. (1991) A semi-empirical approach to relate the volatilization rates of organic chemicals to their physical properties. *Water Res.* 25, 903-910.

Isaacson, P.J., Frink, C.R. (1984) Nonreversible sorption of phenolic compounds by sediment fractions: The role of sediment organic matter. *Environ. Sci. Technol.* 18, 43.

Isnard, P., Lambert, S. (1988) Estimating bioconcentration factors from octanol partition coefficient and aqueous solubility. *Chemosphere* 17, 21-34.

Isnard, P., Lambert, S. (1989) Aqueous solubility/*n*-octanol water partition coefficient correlations. *Chemosphere 18, 1837-1853.*

Itokawa, H., Totsuka, N., Hakahara, K., Meazuru, M., Takeya, K., Konda, M., Inamatsu, M., Morita, H. (1989) A quantitative structure-activity relationship for antitumor activity of long-chain phenols from *Ginkgo kiloba L. Chem. Pharm. Bull.* 36, 1619-1621.

IUPAC (1985) *Halogenated Benzenes, Toluenes and Phenols with Water, Solubility Data Series.* Vol. 20, Horvath, A.L., Getzen,. F.W., Eds., Pergamon Press, Oxford.

Iwasa, J. Fujita, T., Hansch, C. (1965) Substituent constants for aliphatic functions obtained from partition coefficients. *J. Med. Chem.* 8, 150-153.

Iwase, K, Komatsu, K., Jirono, S., Nakagawa, S., Moriguchi, I. (1985) Estimation of hydrophobicity based on the solvent-assessible surface area of molecules. *Chem. Pharm. Bull.* 33, 2114-2121.

Jafvert, C.T. (1990) Sorption of organic acid compounds to sediments: initial model development. *Environ. Toxicol. & Chem.* 9, 1259-1268.

Jafvert, C.T., Weber, E.J. (1991) *Sorption of Ionizable Organic Compounds to Sediments and Soils.* U.S. EPA Report EPA/600/3-91/017, U.S. Environ. Research Lab., Athens, GA.

Japar, S.M., Niki, H. (1975) Gas phase reactions of the nitrate radical with olefins. *J. Phys. Chem.* 79(16), 1629-1632.

Jaworska, J.S., Schultz, T.W. (1993) Quantitative relationships of structure-activity and volume fraction for selected nonpolar and polar narcotic chemicals. *SAR and QSAR in Environ. Res.* 1, 3-19.

Jones, D.C. (1929) The system *n*-butyl alcohol-water and *n*-butyl alcohol-acetone-water. *J. Chem. Soc.* (London) 799-816.

Jones, P.A. (1981) *Chlorophenols and Their Impurities in the Canadian Environment.* Environment Canada, Report SPE 3-EC-81-2F. p.322.

Jordan, T.E. (1954) *Vapor Pressure of Organic Compounds.* Interscience Publishers, Inc., New York.

Kahlbaum, G.W.A. (1898) Studien über dampfspunkraftmessungen. II. *Z. Phys. Chem.* 26, 577-658.

Kaiser, K.L.E. (1983) A non-linear function for the approximation of octanol/water partition coefficients of aromatic compounds with multiple chlorine substitution. *Chemosphere* 12, 1159-1167.

Kaiser, K.L.E. (1987) QSAR of acute toxicity of 1,4-di-substituted benzene derivatives and relationships with the acute toxicity of corresponding none-substituted benzene derivatives. In: *QSAR in Environmental Toxicology - II*, Kaiser, K.L.E., Ed., pp. 169-188, D. Reidel Publishing Co., Amsterdam, The Netherlands.

Kaiser, K.L.E. (1993) COMPUTOX database, National Water Reasearch Institue, Burlington, Ontario, Canada.

Kaiser, K.L.E., Dixon, D.G., Hodson, P.V. (1984) QSAR studies on chlorophenols, chlorobenzenes and para-substituted phenols. In: *QSAR in Experimental Toxicology.* Kaiser, K.L.E., Editor, pp.189-206, D. Reidel Publishing Co., Dordrecht, Netherlands.

Kaiser, K.L.E., Valdmanis, I. (1982) Apparent octanol/water partiton coefficients of pentachlorophenol as a function of pH. *Can. J. Chem.* 60, 2104-2106.

Kamlet, M.J., Doherty, R.M., Abboud, J-L. M., Abraham, M.H. (1986) Linear solvation energy relationship. 36. Molecular properties governing solubilities of organic nonelectrolytes in water. *J. Pharm. Sci.* 75(4), 338-349.

Kamlet, M.J., Doherty, R.M., Abraham, M.H., Marcus, Y., Taft, R.W. (1988) Linear solvation energy relationship. 46. An improved equation for correlation and prediction of octanol/water partition coefficients of organic nonelectrolytes (including strong hydrogen bond donor solutes). *J. Phys. Chem.* 92, 5244-5255.

Kamlet, M.J., Doherty, R.M., Taft, R.W., Abraham, M.H., Veith, G.D., Abraham, D. (1987) Solubility properties in polymers and biological media. 8. An analysis of the factors that influence toxicities of organic nonelectrolytes to the golden orfe fish (*luciscus idus melanotus*). *Environ. Sci. Technol.* 21, 149-155.

Karickhoff, S.W., Brown, D.S., Scott, T.A. (1979) Sorption of hydrophilic organic pollutants on natural sediments. *Water Res.* 13, 241-248.

Kawaguchi, H. (1992) Photooxidation of phenol in aqueous solution in the presence of hydrogen peroxide. *Chemosphere* 24, 1707-1712.

Kawaguchi, H. (1992) Determination of direct and indirect photolysis of 2-chlorophenol in humic acid solution and natural waters. *Chemosphere* 25(5), 635-641.

Kelly, T.J., Mukund, R., Spicer, C.W., Pollack, A.J. (1994) Concentrations and transformations of hazardous air pollutants. *Environ. Sci. Technol.* 28, 378a-387A.

Kenaga, E.E., Goring, C.A.I. (1980) In: *Aquatic Toxicology.* Eaton, J.G., Parrish, P.R., Hendricks, A.C., Editors, Am. Soc. for Testing and Materials, STP 707, pp.78-115.

Khan, A., Hassett, J.J., Banwart, W.L., Means, J.C., Wood, S.G. (1979) Sorption of acetophenone by sediments and soils. *Soil Sci.* 128, 297-302.

Kier, L.B., Hall, L.H. (1976) Molecular properties and molecular connectivity. In: *Molecular Connectivity in Chemistry and Drug Design.* Medicinal Chem. Vol.14, pp.123-167. Academic Press, New York.

Kier, L.B., Hall, L.H. (1981) Derivation and significance of valence molecular connectivity. *J. Pharmocol. Sci.* 70, 583-589.

Kiezyk, P.R., Mackay, D. (1971) Waste water treatment by solvent extraction. *Can. J. Chem. Eng.* 49(6), 747-752.

Kilzer, L., Scheunert, I., Geyer, H., Klein, W., Körte, F. (1979) Laboratory screening of the volatilization rates of organic chemicals from water and soil. *Chemosphere* 10, 751-761.

Kincannon, D.F., Lin, Y.S. (1985) Microbial degradation of hazardous wastes by land treatment. *Proc. Indust. Waste Conf.* 40, 607-619.

King, G.M. (1988) Dehalogenation in marine sediments containing natural sources of halophenols. *Appl. Environ. Microbiol.* 54, 3079-3085.

Kishino, T, Kobayashi, K. (1994) Relation between the chemical structures of chlorophenols and their dissociation constants and partition coefficients in several solvent-water systems. *Wat. Res.* 7, 1547-1552.

Klamt, A. (1993) Estimation of gas-phase hydroxyl radical rate constants of organic compounds from molecular orbital calculations. *Chemosphere* 26, 1273-1289.

Klein, W., Geyer, H., Freitag, D., Rohleder, H. (1984) Sensitivity of schemes for ecotoxicological hazard ranking of chemicals. *Chemosphere* 13, 203-211.

Klein, W., Kordel, W., Weiss, M., Poremski, H.J. (1988) Updating of the OECD test guideline 107 "Partition coefficient n-octanol/water." OECD laboratory intercomparison test on the HPLC method. *Chemosphere* 17, 361-386.

Klöpffer, W., Kaufmann, G., Rippen, G., Poremski, H.J. (1982) A laboratory method for testing the volatility from aqueous solution: First results and comparison with theory. *Ecotoxicol. Environ. Safety* 6, 545-559.

Klopman, G., Namboodiri, K., Schochet, M. (1985) Simple method of computing the partition coefficient. *J. Comput. Chem.* 6, 28-38.

Kobayasi, K.H. (1979) Metabolism of pentachlorophenol in fish. In: *Pesticide and Xenobiotic Metabolism in Aquatic Organisms*. American Chemical Society, Symposium Series 99.

Kobayasi, K.H., Akitake, H., Manabe, K. (1979) Relation between toxicity and accumulation of various chlorophenols in gold fish. *Bull. Japan Soc. Sci. Fish* 45, 173-175.

Kollig, H.P., Ed. (1993) *Environmental Fate Constants for Organic Chemicals Under Consideration for EPA's Hazardous Waste Identification Projects*. EPA Report EPA/600/R-93/132, Environmental Research Lab., U.S. EPA, Athens, GA.

Kollig, H.P., Ellington, J.J., Hamrick, K.J., Jafverts, C.T., Weber, E.J., Wolfe, N.L. (1987) *Hydrolysis Rate Constants, Partition Coefficients, and Water Solubilities for 129 Chemicals*. A summary of fate constants provided for the concentration-based Listing Program of 1987. USEPA Environ. Res. Lab., Off. Res. Development Prepublication. Athens, Georgia.

Könemann, W.H. (1981) Quantity structure-activity relationships in fish toxicity studies. Part 1: Relationship for 50 industrial pollutants. *Toxicology* 19, 209-221.

Könemann, H., Musch, A. (1981) Quantitative structure-activity relationships in fish toxicity studies. Part 2: The influence of pH on the QSAR of chlorophenols. *Toxicology* 19, 223-228.

Kool, H.J. (1984) Influence of microbial biomass on the biodegradability of organic compounds. *Chemosphere* 13, 751-761.

Korenman, Ya I. (1972) Phenol solvates in organic solvents. *Russ. J. Phys. Chem.* 46, 42-43.

Korenman, Ya I. (1973) Extraction of *p*-halophenols. *Zh. Prikl. Khim.* (Leningrad) 46(11), 2599-2600.

Korenman, Ya I., Nefedova, T.E., Byukova, R.I. (1977) Extraction of picramic acid. *Int. J. Pharmacol. Ther. Toxicol.* 51, 734-735.

Korenman, Ya I., Pereshein, V.V. (1970) Extractions of *o*-cresols from aqueous solutions. *J. Appl. Chem. USSR* (Engl. Transl.) 43, 1410-1412.

Korenman, Ya.I., Tishchenko, E.M., Sel'manschchuk, N.N. (1980) Isomerism and extraction of aklylphenol. *Russ. J. Phys. Chem.* 54, 745-746.

Korman, S., La Mer, V.K. (1936) Deuterium exchange equilibria in solution and the quinhydrone electrode. *J. Am. Chem. Soc.* 58, 1396-1403.

Körte, F., Freitag, D., Geyer, H., Klein, W., Kraus, A.G., Lahaniatis, E. (1978) Ecotoxicologic profile analysis: a concept for establishing ecotoxicologic priority lists for chemicals. *Chemosphere* No.1, 79-102.

Kraij, F., Sincic, D. (1980) Mutual solubilities of phenol, salicylaldehyde, phenol-salicylaldehyde mixture, and water with and without the presence of sodium chloride or sodium chloride plus sodium sulfate. *J. Chem. Eng. Data* 25, 335-338.

Kramer, C.-R., Henze, U. (1990) Partitioning properties of benzene derivatives. I. Temperature dependence of the partitioning of monosubstituted benzenes and nitrobenzenes in the octanol/water system. *Z. Phys. Chem.* (Leipzig) 271(3), 503-513.

Krijgsheld, K.R., van der Gen, A. (1986) Assessment of the impact of the emission of certain organochlorine compounds on the aquatic environment. Part I: Monochlorophenols and 2,4-dichlorophenol. *Chemosphere* 15(7), 825-860.

Kubáň, V. (1991) Determination of octa-1-ol-water partition coefficients by flow-injection extraction without phase separation. *Anal. Chim. Acta* 248, 493-499.

Laake, M., Editor (1982) *Ecotoxicological Methods for the Aquatic Environment* (in Norwegian). Nordforsk Miljøvårdsserien Publikation. pp.116-119.

Lagas, P. (1988) Sorption of chlorophenols in soil. *Chemosphere* 17(2), 205-216.

Lallai, A., Mura, G. (1989) pH variation during phenol biodegradation in mixed cultures of microorganisms. *Water Res.* 23, 1235-1238.

Landner, L., Lindström, K., Karlsson, M., Nordin, J., Sörensen, L. (1977) Bioaccumulation in fish of chlorinated phenols from kraft pulp mill bleachery effluents. *Bull. Environ. Contam. Toxicol.* 18, 663-673.

Landrum, P.F., Dupuis, W.S. (1990) Toxicity and toxicokinetics of pentachlorophenol and carbaryl to *pontoporeia hoyi* and *mysis relicta. Aquatic Toxicology & Risk Assessment, Thirteenth Volume ASTM STP* 1096. Landis, W.G., van der Schalie, W.H., Editors, American Society for Testing and Materials, Philadelphia.

Lange, N.A., Editor (1973) *Lange's Handbook of Chemistry.* McGraw-Hill, New York.

Larsson, P., Bremle, G., Okla, L. (1993) Uptake of pentachlorophenol in fish of acidified and nonacidified lakes. *Bull. Environ. Contam. Toxicol.* 50, 653-658.

Leahy, D.E., De Meere, A.L.J., Wait, A.R., Taylor, P.J., Tomenson, J.A. (1989) A general description of water-oil partitioning ratios using the rotating diffusion cell. *Int. J. Pharm.* 50, 117-132.

LeBlanc, G.A. (1984) Interspecies relationships in acute toxicity of chemicals to aquatic organisms. *Environ. Toxicol. Chem.* 3, 47-60.

Lee, L.S., Rao, P.S.C., Nkedl-Kizza, P., Delfino, J.J. (1990) Influence of solvent and sorbent characteristics on distribution of pentachlorophenol in octanol-water and soil-water systems. *Environ. Sci. Technol.* 24, 654-661.

Lee, L.S., Rao, P.S.C., Brusseau, M.L. (1991) Nonequilibrium sorption and transport of neutral and ionized chlorophenols. *Environ. Sci. Technol.* 25(4), 722-729.

Lee, R.F., Ryan, C. (1979) Microbial degradation of organochlorine compounds in estuarine water and sediments. In: *Microbial Degradation of Pollutants in Marine Environments.* Bourquin, A.W., Pritchard, P.H., Editors, EPA-600/9-79-012. U.S. Environmental Protection Agency, Washington, D.C.

Lemaire, J., Guth, J.A., Klais, O., Leahy, J., Merz, W., Philp, J., Wilmers, R., Wolff, C.J.M. (1985) Ring test of a method for assessing the phototransformation of chemicals in water. *Chemosphere* 14, 53-77.

Leo, A., Hansch, C., Church, C. (1969) Comparison of parameters currently used in the study of structure-activity relationships. *J. Med. Chem.* 12, 766-771.

Leo, A., Hansch, C., Elkins, D. (1971) Partition coefficients and their uses. *Chem. Rev.* 71, 525-616.

Leuenberger, C., Giger, W., Coney, R., Graydon, J.W., Molnar-Kubica, E. (1985a) Persistence chemicals in pulp mill effluents, occurrence and behaviour in an activated sludge treatment plant. *Water Res.* 19, 885-894.

Leuenberger, C., Ligocki, M.P., Pankow, J.F. (1985b) Trace organic compounds in rain. 4. Indentities, concentrations, and scavenging mechanisms, for phenols in urban air and rain. *Environ. Sci. Technol.* 19, 1053-1058.

Lichmann, A., Wienke, G. Meyer, T., Gmehling, J. (1993) Bestimmung von Berteilungskoeffizienten mit Hilfe der Flüssig-Gegenstrom-Chromatographie. *Chem.-Ing.-Tech.* 65, 72-75.

Lipczynska-Kochany, E. (1992) Degradation of nitrobenzene and nitrophenos by means of advanced oxidation processes in a homogenous phase: photolysis in the presence of hydrogen peroxide versus the Fenton reaction. *Chemosphere* 24, 1369-1380.

Liu, M.H., Kapila, S., Yanders, A.F., Clevenger, T.E., Elseewi, A.A. (1991) Role of entrainers in supercritical fluid extraction of chlorinated aromatics from soil. *Chemosphere* 23, 1085-1095.

Loehr, R.C., Matthews, J.E. (1992) Loss of organic chemicals in soil: Pure compound treatability studies. *J. Soil Contam.* 1(4), 339-360.

Loekke, H. (1985) Degradation of 4-nitrophenol in two Danish soils. *Environ. Pollut. Ser. A* 38, 171-181.

Lu, D., Thoman, K., Strachan, M.J., (1981) Biodegradation of pentachlorophenol in simulated aquatic environment. *Bull. Environ. Contam. Toxicol.* 26, 85-90.

Lu, P.-Y., Metcalf, R.L. (1975) Environmental fate and biodegradability of benzene derivatives as studied in a model aquatic ecosystem. *Environ. Health Perspec.* 10, 269-284.

Lu, P.-Y., Metacalf, R.L., Cole, L.K. (1978) The environmental fate ^{14}C-pentachlorophenol in laboratory model ecosystems. In: *Pentachlorophenol, Chemistry, Pharmacology and Environmental Toxicology.* Rao, K.R., Ed., pp. 53-63, Plenum Press, New York, N.Y.

Ludzack, F.J., Ettinger, M.B. (1960) Chemical structures resistant to aerobic biochemical stabilization. *J. Pollut. Control Fed.* 32, 1173-1200.

Ludzack, F.J., Schaffer, R.B., Bloomhuff, R.N., Ettinger, M.B. (1958) Biochemical oxidation of some commercially important organic cyanides. I. River oxidation. In: *Proc. 13th Indust. Waste Conf. Eng. Bull. Purdue Univ. Eng. Ext. Ser.* pp.297-312.

Lyman, W.J., Reehl, W.F., Rosenblatt, D.H. (1982) *Handbook on Chemical Property Estimation Methods. Environmental Behavior of Organic Compounds.* McGraw-Hill, New York.

Ma, K.C., Shiu, W.Y., Mackay, D. (1990) *A Critically Reviewed Compilation of Physical and Chemical Persistence Data for 110 Selected EMPPL Substances.* A report prepared for the Ministry of Environment, Water Resources Branch, Toronto, Ontario.

Ma, K.C., Shiu, W.Y., Mackay, D. (1993) Aqueous solubility of chlorophenols at 25°C. *J. Chem. Eng. Data*, 38, 364-366.

Mabey, W., Mill, T. (1978) Critical review of hydrolysis of organic compounds in water under environmental conditions. *J. Phys. Ref. Data* 7, 383-415.

Mabey, W., Smith, J.H., Podoll, R.T., Johnson, H.L., Mill, T., Chiou, T.W., Gate, J., Waight-Partridge, I., Jaber, H., Vandenberg, D. (1982) *Aquatic Fate Process for Organic Priority Pollutants.* EPA Report, No. 440/4-81-14.

Mackay, D. (1982) Correlation of bioconcentration factors. *Environ. Sci. Technol.* 16, 274-278.

Mackay, D. (1991) *Multimedia Environmental Models: The Fugacity Approach.* Lewis Publishers, Inc., Chelsea, MI.

Mackay, D., Bobra, A.M., Chan, D.W., Shiu, W.Y. (1982) Vapor pressure correlations for low-volatility environmental chemicals. *Environ. Sci. Technol.* 16(10), 645-649.

Mackay, D., Bobra, A.M., Shiu, W.Y., Yalkowsky, S.H. (1980) Relationships between aqueous solubility and octanol-water partition coefficient. *Chemosphere* 9, 701-711.

Mackay, D., Paterson, S., Shiu, W.Y. (1992a) Generic models for evaluating the regional fate of chemicals. *Chemosphere* 24, 695-717.

Mackay, D., Shiu, W.Y., Ma, K.C. (1992b) *Illustrated Handbook of Physical-Chemical Properties and Environmental Fate of Organic Chemicals, Vol. 1, Monoaromatic Hydrocarbons, Chlorobenzenes and PCBs*. Lewis Publishers, Inc., Chelsea, MI.

Mackay, D., Shiu, W.Y., Ma, K.C. (1992c) *Illustrated Handbook of Physical-Chemical Properties and Environmental Fate of Organic Chemicals, Vol. 2, Polynuclear Aromatic Hydrocarbons, Polychlorinated Dibenzo-p-dixoins and Dibenzofurans*. Lewis Publishers, Inc., Chelsea, MI.

Mackay, D., Shiu, W.Y., Ma, K.C. (1993) *Illustrated Handbook of Physical-Chemical Properties and Environmental Fate of Organic Chemicals, Vol. 3, Volatile Organic Chemicals*. Lewis Publishers, Inc., Chelsea, MI.

Mackay, D., Wolkoff, A.W. (1973) Rate of evaporation of low-solubility contaminants from water bodies to atmosphere. *Environ. Sci. Technol.* 7, 611-614.

Mailhot, H., Peters, R.H. (1988) Empirical relationships between the 1-octanol/water partition coefficient and nine physicochemical properties. *Environ. Sci. Technol.* 22, 1479-1488.

Mäkelä, P. Oikari, O.J. (1990) Uptake and body distribution of chlorinated phenolics in the freshwater mussel, *anodonta anatina* L. *Ecotox. Environ. Safety* 20, 354-362.

Matsuguma, H.J. (1963) In: *Kirk-Othmer Encyclopedia of Chemical Technology.* 2nd edition. pp. 888-894.

McCall, J.C. (1975) Liquid-liquid partition coefficients by high-pressure liquid chromatography. *J. Med. Chem.* 18(6), 549-552.

McGowan, J.C. (1954) The physical toxicity of chemicals. IV. Solubilities, partition coefficients and physical toxicities. *J. Appl. Chem.* 4, 41-47.

McKim, J., Schneider, P., Veith, G. (1985) Absorption dynamics of organic chemical transport across trout gills as related to octanol-water partition coefficient. *Toxicol. Appl. Pharmacolo.* 77, 1-10.

McKim, J.M., Schmieder, P.K., Erickson, R.J. (1986) Toxicokinetic modeling of ^{14}C-pentachlorophenol in the rainbow trout (*salmo gairdneri*). *Aqua. Toxicol.* 9, 59-80.

McLeese, D.W., Zitko, V., Peterson, M.R. (1979) Structure-lethality relationships for phenols, anilines and other aromatic compounds in shrimp and clams. *Chemosphere* 2, 53-57.

McLeese, D.W., Zitko, V., Sergent, D.B., Burridge, L., Metcalfe, C.D. (1981) Lethality and accumulation of alkylphenols in aquatic fauna. *Chemosphere* 10, 723-730.

Means, J.C., Woods, S.G., Hassett, J.J., Banwart, W.L. (1982) Sorption of amino- and carboxy-substituted polynuclear aromatic hydrocarbons by sediments and soils. *Environ. Sci. Technol.* 16, 93-98.

Menges, R.A., Bertrand, G.L., Armstrong, D.W. (1990) Direct measurement of octanol-water partition coefficient using centrifugal partition chromatograph with a back-flashing technique. *J. Liq. Chromatogr.* 13(15), 3061-3077.

Merck Index (1989) *The Merck Index. An Encyclopedia of Chemicals, Drugs and Biologicals.* Budavari, S., Editor, 11th Edition, Merck & Co., Inc., Rahway, New Jersey.

Meylan, W., Howard, P.H. (1991) Bond contribution method for estimating Henry's law constants. *Environ. Toxicol. & Chem.* 10, 1283-1293.

Meylan, W., Howard, P.H. (1993) Computer estimation of the atmospheric gas-phase reaction rate of organic compounds with hydroxyl radicals and ozone. *Chemosphere* 26, 2293-2299.

Meylan, W., Howard, P.H., Boethling, R.S. (1992) Molecular topology/fragment contribution method for predicting soil sorption coefficients. *Environ. Sci. Technol.* 26, 1560-1567.

Mill, T. (1979) *Structure Reactivity Correlations for Environmental Reactions.* EPA Final Report, EPA 560/11-79-012.

Mill, T. (1982) Hydrolysis and oxidation processes in the environment. *Environ. Toxicol. & Chem.* 1, 135-141.

Mill, T., Mabey, W. (1985) Photochemical transformations. *Environ. Exposure Chem.* 1, 175-216.

Mills, W.B., Dean, J.D., Porcella, D.B., Gherini, S.A., Hudson, R.J.M., Frick, W.E., Rupp, G.L., Bowie, G.L. (1982) *Water Quality Assessment: A Screening Procedure for Toxic and Conventional Pollutants*, Part 1, EPA Report No. EPA-600/6-82-004a, Environmental Research Lab., US EPA, Athens, GA.

Minero, C., Pelizzetti, E., Malato, S., Blanco, J. (1993) Large solar plant photocatalytic water decontamination: Degradation of pentachlorophenol. *Chemosphere* 26(12), 2103-2119.

Minick, D.J., Frenz, J.H., Patrick, M.A., Brent, D.A. (1988) A comprehensive method for determining hydrophobicity constants by reversed-phase high-performance liguid chromatography. *J. Med. Chem.* 31, 1923-1933.

Mirrlees, M.S., Moulton, J.J., Murphy, C.T., Taylor, P.J. (1976) Direct measurement of octanol-water partition coefficients by hgih pressure liquid chromatography. *J. Med. Chem.* 19, 615-619.

Miyake, K., Kitaura, F., Mizuno, N., Terada, H. (1987) Determination of partition coefficient and acid dissociation constant by high-performance liquid chromatography on porous polymer gel as a stationary phase. *Chem. Pharm. Bull.* 35(1), 377-388.

Miyake, K., Terada, H. (1982) Determination of partition coefficients of very hydrophobic compounds by high-performance liquid chromatography on glyceryl-coated controlled-pore glass. *J. Chromatogr.* 240, 9-20.

Moos, L.P., Kiesch, E.J., Wukasch, R.F., Grady, C.P.L., Jr. (1983) Pentachlorophenol biodegradation-I. Aerobic. *Water Res.* 17(11), 1575-1584.

Moos, L.P., Kiesch, E.J., Wukasch, R.F., Grady, C.P.L., Jr. (1984) Pentachlorophenol biodegradation-II. Anaerobic. *Water Res.* 17(11), 1575-1584.

Morrison, R.T., Boyd, R.N. (1959) *Organic Chemistry.* Allyn and Bacon, Inc., Boston.

Morrison, R.T., Boyd, R.N. (1973) *Organic Chemistry.* 3rd edition, Allyn and Bacon, Inc., Boston.

Mosso, U. (1887) *Jahresber. Fortacgr. Chem.*, 1800.

Müller, M., Klein, W. (1991) Estimating atmospheric degradation processes by SARs. *Sci. Total Environ.* 109/110, 261-273.

Müller, M., Klein, W. (1992) Comparative evaluation of methods predicting water solubility for organic compounds. *Chemosphere* 25, 769-782.

Mulley, B.A., Metcalf, A.D. (1966) Solubilization of phenols by non-ionic surface-active agents. *Sci. Pharm.* 8, 481-488.

Nagel, R., Urich, K. (1980) Kinetic studies on the elimination of different substituted phenols by goldfish (*carassius auratus*). *Bull. Environ. Contam. Toxicol.* 24, 374-378.

Nahum, A., Horvath, C. (1980) Evaluation of octanol-water partition coefficients by using high-performance liquid chromatography. *J. Chromatogr.* 192, 315-322.

NCASI (1993) A compilation of data on chemical and biological properties of 28 chlorinated phenolic compounds. Special Report No. 92-12, National Council of the Paper Industry for Air Stream Improvement, Inc., New York, N.Y.

Neely, W.B. (1982) Organizing data for environmental studies. *Environ. Toxicol. Chem.* 1, 256-266.

Neely, W.B. (1984) An analysis of aquatic toxicity data: water solubility and acute LC50 fish data. *Chemosphere* 13, 813-819.

Neely, W.B., Blau, G.E., Editors (1985) *Environmental Exposure from Chemicals.* Vol. I, CRC Press, Boca Raton, Florida.

Neilson, A.H., Allard, A.-S., Hynning, P.-Å., Remberger, M. (1983) Bacterial methylation of chlorinated phenols and guaiacols: Formation of veratroles from guaiacols and high molecular weight chlorinated lignin. *Appl. Environ. Microbiol.* 45, 774-783.

Neilson, A.H., Allard, A.-S., Hynning, P.-Å., Remberger, M. (1991) Distribution, fate and persistence of organochlorine compounds formed during production of bleached pulp. *Toxicol. Environ. Chem.* 30, 3-41.

Nendza, M., Seydel, J.K. (1988) Quantitative structure-toxicity relationship for ecotoxicologically relevant biotestsystems and chemicals. *Chemosphere* 17, 1585-1602.

Nesterova, T.N., Nazimudinov, A.G., Tsvetkov, V.S., Rozhnov, A.M., Roshchupkina, I.Y. (1990) Vapour pressures and enthalpies of vaporization of alkylphenols. *J. Chem. Thermodyn.* 22, 365-377.

Neumüller, O.A. (1974) *Römpp's Chemie-Lexikon.* p.2538, Frank'sche Verlagsbuchhandlung, Stuttgart.

Niimi, A.J. (1987) Biological half-lives of chemicals in fishes. *Rev. Environ. Contam. Toxicol.* 99, 1-46.

Niimi, A.J., Cho, C.Y. (1983) Laboratory and field analysis of pentachlorophenol (PCP) accumulation by salmonids. *Water Res.* 17(12), 1791-1795.

Niimi, A.J., Lee, H.B., Kissoon, G.P. (1990) Kinetics of chloroguaiacols and other chlorinated phenolic derivatives in rainbow trout (*salmo gairdneri*). *Environ. Toxicol. Chem.* 9, 649-653.

Niimi, A.J., McFadden, C.A. (1982) Uptake of sodium pentachlorophenate (NaPCP) from water by rainbow trout (*salmo gairdneri*) exposed to concentrations in the ng/L range. *Bull. Environ. Contam. Toxicol.* 28, 11-19.

Niimi, A.J., Palazzo, V. (1985) Temperature effect on the elimination of pentachlorophenol, hexachlorobenzene and mirex by rainbow trout (*salmo gairdneri*). *Water Res.* 19(2), 205-207.

Nirmalakhandan, N.N., Speece, R.E. (1988) QSAR model for predicting Henry's law constant. *Environ. Sci. Technol.* 22, 1349-1357.

Nitka, D.M., Paramentier, M., Easty, D.B. (1982) Sorption of pentachlorophenol by unbleached wood pulp fibers. *Bull. Environ. Contam. Toxicol.* 28, 605-610.

Norrington, F.E., Hyde, R.M., Williams, S.G., Wootton, R. (1975) Physicochemical-activity relations in practice. 1. A rational and self-consistent data bank. *J. Med. Chem.* 18, 604-607.

NRC (1982) *Chlorinated Phenols: Criteria for Environmental Quality.* National Research Council of Canada, NRC Associate Committee on Scientific Criteria for Environmental Quality. NRCC No. 18578, NRCC Publications. Ottawa, Canada.

OECD (1979) OECD Chemical Group, Chemicals Testing Programme, Laboratory Intercomparison Testing, Part 1, Results on one laboratory, obtained with the gas saturation method.

Oikari, A., Anäs, E. (1985) Chlorinated phenols and their conjugates in the bile of trout (*salmo gairdneri*) exposed to contaminated water. *Bull. Environ. Contam. Toxicol.* 35, 802-809.

Okamoto, K., Hondo, F., Itaya, A., Kusabayashi, S. (1982) Kinetics of dye-sensitized photodegradation of aqueous phenol. *J. Chem. Eng. Jap.* 15, 368-275.

Okey, R.W., Bogan, R.H. (1965) Apparent involvement of electronic mechanisms in limiting microbial metabolism of pesticides. *J. Water Pollut. Control Fed.* 37, 692-712.

Paasivirta, J. (1992) Predicted and observed fate of selected persistent chemicals in the environment. *Organohalogen Compounds* 1, 367-375.

Palit, S.R. (1947) Electronic interpretations of organic chemistry. II. Interpretations of the solubility of organic chemistry. *J. Phys. Chem.* 51, 837-857.

Pangrekar, J., Klopman, G., Rosenkranz, H.S. (1994) Expert system comparison of structural determinants of chemical toxicity to environmental bacteria. *Environ. Toxicol. & Chem.* 13(6), 979-1001.

Paris, D.F., Wolfe, N.L., Steen, W.C. (1982) Structure-activity relationships in microbial transformation of phenols. *App. Environ. Microbiol.* 44(1), 153-158.

Paris, D.F., Wolfe, N.L., Steen, W.C., Baughman, G.L. (1983) Effect of phenol molecular structure on bacterial transformation rate constants in pond and river samples. *App. Environ. Microbiol.* 45(3), 1153-1155.

Pearce, P.J., Simkins, R.J.J. (1968) Acid strengths of some substituted picric acids. *Can. J. Chem.* 46(2), 241-248.

Perrin, D.D., Dempsey, B., Serjeant, E.P. (1981) *pK_a Prediction for Organic Acids and Bases.* Chapman and Hall, New York, N.Y.

Perry, R.A., Atkinson, R., Pitts, J.N. (1977) Kinetics and mechanisms of the gas phase reaction of OH radicals with methoxybenzene and *o*-cresol over temperature range 299-435 K. *J. Phys. Chem.* 81, 1607-1611.

Petrasek, A.C., Kugelman, I.J., Austern, B.M., Pressley, T.A., Winslow, L.A., Wise, R.H. (1983) Fate of toxic organic compounds in wastewater treatment plants. *J. Water Pollut. Control Fed.* 55, 286-296.

Pfaender, F.K., Bartholomew, G.W. (1982) Measurements of aquatic biodegradation rates by determining heterotropic uptake of radiolabelled pollutants. *Appl. Environ. Microbiol.* 44, 159-164.

Pitter, P. (1976) Determination of biological degradability of organic substances. *Water Res.* 10, 231-235.

Platford, R.F. (1983) The octanol-water partitioning of some hydrophobic and hydrophilic compounds. *Chemosphere* 12, 1107-1111.

Politzki, G.R., Bieniek, D., Lahaniatis, E.S., Scheunert, I., Klein, W., Körte, F. (1982) Determination of vapour pressures of nine organic chemicals adsorbed on silica gel. *Chemosphere* 11, 1217-1229.

Price, K.S., Waggy, G.T., Conway, R.A. (1974) Brine shrimp bioassay and seawater BOD of petrochemicals. *J. Water Pollut. Control Fed.* 46, 63-77.

Pritchard, P.H. (1987) Assessing the Biodegradation of Sediment Associated Chemicals. In: *Fate and Effects of Sediment-Bound Chemicals in Aquatic Systems*. Dickson, K.L., Maki, A.W., Brungs, W.A., Eds., pp. 109-135, Pergamon Press, New York.

Rao, P.S.C., Davidson, J.M. (1980) Estimation of pesticide retention and transformation parameters required in nonpoint source pollution models. In: *Environmental Impact of Nonpoint Source Pollution*. Ann Arbor Scince Publishers, Inc., Ann Arbor, Michigan.

Reid, R.C., Prausnitz, J.M., Poling, B.E. (1977) *The Properties of Gases and Liquids*. 3rd Edition, McGraw-Hill Book Co., New York, N.Y.

Reid, R.C., Prausnitz, J.M., Poling, B.E. (1987) *The Properties of Gases and Liquids*, 4th Edition, McGraw-Hill Book Co., New York, N.Y.

Rekker, R.F. (1977) *The Hydrophobic Fragmental Constant. Its Derivation and Application. A Means of Characterizing Membrane Systems*. Elsevier Sci. Publishing Co., Amsterdam, the Netherlands.

Remberger, M., Allard, A.-S., Neilson, A.H. (1986) Biotransformations of chloroguaiacols, chlorocatechols, and chloroveratroles in sediments. *Appl. Environ. Microbiol.* 51(3), 552-558.

Renberg, L. (1981) Gas chromatographic determination of chlorophenols in environmental samples. *National Swedish Envirnment Protection Board Report* 1410, 135pp.

Renberg, L., Svanberg, O., Bengtson, B.E., Sundström, G. (1980) Chlorinated guaiacols and catechols bioaccumulation potential in bleaks (*alburnus alburnus Pisces*) and reproductive and toxic effects on the harpaticoid Nitocra spinipes (*Crustacea*). *Chemosphere* 9, 143-150.

Renner, G. (1990) Gas chromatographic studies of chlorinated phenols, chlorinated anisoles, and chlorinated phenylacetates. *Toxicol. Environ. Chem.* 27, 217-224.

Riddick, J., Bunger, W.B., Sakano, T.K. (1986) *Organic Solvents: Physical Properties and Method of Purification*. 4th Edition, John Wiley & Sons, New York, New York.

Ritter, S., Hauthal, W.H., Maurer, G. (1994) Partition coefficients of some environmentally important organic compounds between 1-octanol and water from reversed-phase high-performance liquid chromatography. *J. Chem. Eng. Data* 39, 414-417.

Roberts, M.S., Anderson, R.A., Swarbrick, J. (1977) Permeability of human epidermis to phenolic compounds. *J. Pharm. Parmac.* 29, 677-683.

Robinson, K., Novak, J.T. (1994) Fate of 2,4,6-trichloro-(^{14}C)-phenol bound to dissolved humic acid. *Water Res.* 445-452.

Rogers, J.A., Wong, A. (1980) The temperature dependence and thermodynamics of partitioning of phenols in the *n*-octanol-water system. *Int'l J. Pharmaceut.* 6, 339.

Rogers, J.E., Li, S.W., Felice, L.J. (1984) *Microbial Transformation Kinetics of Xenobiotics in Aquatic Environment*. EPA-600/3-84-043. (NTIS PB84-162866). Richland, WA: Battelle Pacific Northwest Labs. 105pp.

Rogers, K.S., Cammarata, A. (1969) Molecular orbital description of the partitioning aromatic compounds between polar and nonpolar phases. *Biochim. Biophys. Acta* 193(1), 22-29.

Rohrschneider, L. (1973) Solvent characterization of gas-liquid partition coefficients of selected solutes. *Anal. Chem.* 45(7), 1241-1247.

Roth, S., Seeman, P. (1972) Membrane concentrations of neutral and positive anesthetics (alcohols, chlorpromazine, morphine) fit the Meyer-Overton rule of anethesia. Negative narcotics do not. *Biochem. Biophys. Acta* 255, 207-219.

Rott, B., Viswanathan, R., Freitag, D., Korte, F. (1982) Vergleichende untersuchung der anwendbrakeit von unweltchemikalien. *Chemosphere* 11, 531-538.

Ruppert, G., Bauer, R., Heisler, G., Vovalic, S. (1993) Mineralization of cyclic organic water contaminants by the photo-Fenton reaction-Influence of structure and substituents. *Chemosphere* 27(8), 1339-1347.

Russell, C.J., Dixon, S.L., Jurs, P.C. (1992) Computer-assisted study of the relationship between molecular structure and Henry's law constant. *Anal. Chem.* 64, 1350-1355.

Ryan, J.A., Bell, R.M., Davidson, J.M., O'Connor, G.A. (1988) Plant uptake of non-ionic organic chemicals from soils. *Chemosphere* 17(12), 2299-2323.

Saarikoski, J., Viluksela, M. (1981) Influence of pH on the toxicity of substituted phenols to fish. *Arch. Environ. Contam. Toxicol.* 10, 747-753.

Saarikoski, J., Viluksela, M. (1982) Relation between physicochemical properties of phenols and their toxicity and accumulation in fish. *Ecotoxicol. Environ. Safty* 6, 501-512.

Saarikoski, J., Lindström, R., Tyynelä, M., Viliksela, M. (1986) Factors affecting the absorption of phenolics and carboxylic acids in the guppy (*Poecilia reticulata*). *Ecotoxicol. Environ. Saf.* 11, 158-173.

Sabljic, A. (1987a) Nonempirical modeling of environmental distribution and toxicity of major organic pollutants. In: *QSAR in Environmental Toxicology-II*. pp.309-332. Kaiser, K.L.E., Editor, D. Reidel Publishing Company, Dordrecht, The Netherlands.

Sabljic, A. (1987b) On the prediction of soil sorption coefficients of organic pollutants from molecular structure: Application of molecular topology model. *Environ. Sci. Technol.* 21(4), 358-366.

Sabljic, A., Güesten, H. (1990) Predicting the night-time NO_3 radical reactivity in the troposphere. *Atmos. Environ.* 24A(1), 73-78.

Sagebiel, J.C., Seiber, J.N. (1993) Studies on the occurence and distribution of wood smoke marker compounds in foggy atmospheres. *Environ. Toxicol. Chem.* 12, 813-822.

Sagebiel, J.C., Seiber, J.N., Woodrow, J.E. (1992) Comparison of headspace and gas-stripping methods for determing the Henry's law constant (H) for organic compounds of low to intermediate H. *Chemosphere* 25(12), 1763-1768.

Saito, H., Koyasu, J., Yoshida, K., Shigeoka, T., Koike, S. (1993) Cytotoxicity of 109 chemicals to goldfish GFS cells and relationships with 1-octanol/water partition coefficients. *Chemosphere* 26(5), 1015-1028.

Sanders, G., Jones, K.C., Hamliton-Taylor, J. (1993) A simple method to assess the susceptibility of polynuclear aromatic hydrocarbons to photolytic decomposition. *Atmos. Environ.* 27A, 139-144.

Sangster, J. (1989) Octanol-water partition coefficients of simple organic compounds. *J. Phys. Chem. Ref. Data* 18, 1111-1230.

Sangster, J. (1993) LOGKOW databank, Sangster Research Laboratory, Montreal, Quebec.

Sasaki, S. (1978) The scientific aspects of the chemical substance control law in Japan. In: *Aquatic Pollutants: Transformation and Biological Effects*. Hutzinger, O. et al., Editors, Pergamon Press, Oxford, U.K. pp.283-298.

Schauerte, W., Lay, J.P., Klein, W., Körte, F. (1982) Influence of 2,4,6-trichlorophenol and pentachlorophenol on the biota of aquatic systems. *Chemosphere* 11, 71-79.

Schellenberg, K., Leuenberger, C., Schwarzenbach, R.P. (1984) Sorption of chlorinated phenols by natural sediments and aquifer materials. *Environ. Sci. Technol.* 18, 652-657.

Scherrer, R.A., Howard, S.M. (1979) The analysis of electronic factors in quantitative structure-activity relationships using distribution coefficients. *Adv. Chem. Ser.* 112, 507-526.

Schmidt-Bleek, F., Haberland, W., Klein, A.W., Caroli, S. (1982) Steps toward environmental hazard assessment of new chemicals (including a hazard ranking scheme, based upon directive 78/831/EEC). *Chemosphere* 11(4), 383-415.

Schnell, S., Bak, F., Pfennig, N. (1989) Anaerobic degradation of aniline and dihydroxybenzenes by newly isolated sulfate-reducing bacteria and description of *desulfobacterium anilini. Arch. Microbiol.* 152, 556-563.

Schultz, T.W., Cajina-Quezada, M. (1987) Structure-activity relationships for mono alkylated or halogenated phenols. *Toxicol. Lett.* 37(2), 121-130.

Schultz, T.W., Riggin, G.W., Wesley, S.K. (1987) Structure-activity relationships for para-substituted phenols. In: *QSAR in Environmental Toxicology - II.* Kaiser, K.L.E., Ed., pp.333-345, D. Reidel Publishing Co., Dordrecht, The Netherlands.

Schüürmann, G., Klein, W. (1988) Advances in bioconcentration prediction. *Chemosphere* 17(8), 1551-1574.

Schwarzenbach, R.P., Westall, J. (1985) Sorption of hydrophobic trace organic compounds in groundwater systems. *Wat. Sci. Technol.* 17, 35-55.

Schwarzenbach, R.P., Stierli, R., Folsom, B.R., Zeyer, J. (1988) Compound properties relevant for assessing the environmental partitioning of nitrophenols. *Environ. Sci. Technol.* 22(1), 83-92.

Scott, H.D., Wolf, D.C., Lavy, T.L. (1983) Adsorption and degradation of phenol at low concentrations in soil. *J. Environ. Qual.* 12, 91-95.

Scow, K.M. (1982) Chapter 9, Rate biodegradation. In: *Handbook of Chemical Property Estimation Methods. Environmental Behaviour of Organic Compounds.* Lyman, W.J., Reehl, W.F., Rosenblatt, D.H., Editors, McGraw-Hill Book Co., New York, New York.

Scully, F.E., Jr., Hoigné, J. (1987) Rate constants for reactions of singlet oxygen with phenols and other compounds in water. *Chemosphere* 16, 681-694.

Seidell, A. (1941) *Solubilities of Organic Compounds.* Vol. 2, Van Nostrand, New York.

Seip, H.M., Alstad, J., Carlberg, G.E., Martinsen, K., Skaane, R. (1986) Measurement of mobility of organic compounds in soils. *Sci. Total Environ.* 50, 87-101.

Serjeant, E.P., Dempsey, B. (1979) *Ionization Constants of Organic Acids in Aqueous Solution.* Pergamon, Oxford, U.K.

Shelton, D.R., Tiedje, J.M. (1981) *Development Tests for Determining Anaerobic Degradation Potential.* EPA-560/5-81-013. NTIS PB84-166495. Mich. State Univ., Dept. Crop Soil Science, East Lansing, Michigan. 92pp.

Shigeoka, T., Sato, Y., Takeda, Y., Yoshida, K., Yamauchi, F. (1988) Acute toxicity of chlorophenols to green algae, *selenastrum capricornutum* and *chlorella vulgaris*, and quantitative structure-activity relationships. *Environ. Toxico. & Chem.* 7, 847-854.

Shigeoka, T., Yamagata, T., Minoda, T., Yamauchi, F. (1988) *Jpn. J. Toxicol. Environ. Health* 34, 343-349.

Shiu, W.Y., Ma, K.C., Varhanickova, D., Mackay, D. (1994) Chlorophenols and alkylphenols: A review and correlation of environmentally relevant properties and fate in an evaluative environment. *Chemosphere* 29(6), 1155-1224.

Sidgwick, N.V., Spurrell, W.J. Davies, T.E. (1915) CXXXII. The solubility of nitrophenols and other isomeric disubstitution products of benzene. *J. Chem. Soc.* 1202.

Sidgwick, N.V., Turner, S.L. (1922) CCLXXI.-The solubility of chlorophenols. *J. Chem. Soc., (London)* 2256-2263.

Sillén, L.G., Martell, A.E. (1971) *Stability Constants of Metal-Ion Complexes.* Supplement No.1, Spec. Publ. No.25, The Chemical Society, London.

Smejtek, P., Wang, S. (1993) Distribution of hydrophobic ionizable xenobiotics between water and liqid membranes: pentachlorophenol and pentachlorophenate. A comparison with octanol-water partition. *Arch. Environ. Contam. Toxicol.* 25, 394-404.

Smith, A.D., Bharath, A., Mallard, C., Orr, D., McCarty, L.S., Ozbum, G.W. (1990) Bioconcentration kinetics of some chlorinated benzenes and chlorinated phenols in American flagfish, *Jordanella floridae* (Goode and Bean). *Chemosphere* 20, 379-386.

Smith, J.D., Bomberger, D.C. (1980) Prediction of volatilization rate of chemicals in water. In: *Hydrocarbons and Halogenated Hydrocarbons in the Environment*, Afghan, B.K., Mackay, D., Eds., pp. 445-451, Plenum Press, New York, N.Y.

Smith, J.H., Mabey, W.R., Bohonos, N., Holt, B.R., Lee, S.S., Chou, T.W., Venberger, D.C., Mill, T. (1978) *Environmental Pathways of Selected Chemicals in Freshwater System. Part II. Laboratory Studies*. EPA-600/7-78-074. U.S. EPA, Athens, Georgia.

Smith, P.D., Brockway, D.L., Stancil, F.E., Jr. (1987) Effects of hardness, alkalinity and pH on the toxicity of pentachlorophenol to *selenastrum capricornutum* (printz). *Environ. Toxicol. Chem.* 6, 891-900.

Söderström, M., Wachtmeister, C.A., Förlin, L. (1994) Analysis of chlorophenolics from bleach Kraft Mill effluents (BKME) in bile of perch (*perca fluviatilis*) from the Baltic Sea and development of an analytical procedure also measuring chlorocatechols. *Chemosphere* 28(9), 1701-1719.

Southworth, G.R. (1982) Evaporation of phenols from synthetic liquid fuel spills. *Environ. Intl.* 7, 203-206.

Spain, J.C., Zylstra, G.J., Blake, C.K., Gibson, D.T. (1989) Monohydroxylation of phenol and 2,5-dichlorophenol by toluene dioxygenase in *pseudomonas putida* F1. *Appl. Environ. Microbiol.* 55, 2648-2652.

Statham, C.N., Melancon, M.J., Jr., Leck, J.L. (1976) Bioconcentration of xenobiotics in trout bile: A proposed monitoring aid for some water borne chemicals. *Science* 193, 680-681.

Steen, W.C. (1991) *Microbial Transformation Rate Constnnts of Structurally Diverse Man-made Chemicals*. EPA 600/3-91/016, PB91-181958. Environmental Research Lab., Office of Research and Development, U.S. Environmental Protection Agency, Athens, Georgia.

Stegeman, M.H.L., Peijnenberg, W.J.G.M., Verboom, H. (1993) A quality structure-activity relationship for the direct photohydrolysis of meta-substituted halobenzene derivatives in water. *Chemosphere* 26(5), 837-849.

Stehly, G.R., Hayton, W.L. (1990) Effect of pH on the accumulation kinetics of pentachlorophenol in goldfish. *Arch. Environ. Contam. Toxicol.* 19, 464-470.

Stephen, H., Stephen, Y. (1963) *Solubilities of Inorganic and Organic Compounds*. Vol. 1 and 2. Pergamon Press, Oxford.

Stephenson, R.M., Malanowski, A. (1987) *Handbook of the Thermodynamics of Organic Compounds*, Elsevier, New York.

Stockdale, M., Selwyn, M.J. (1971) Effects of ring substituents on the activity of phenols as inhibitors and uncouplers of mitochondrial respiration. *Eur. J. Biochem.* 21, 565-574.

Stolwijk, T.B., Vos, L.C., Sudholter, E.J.R., Reinhoudt, D.N. (1989) Partition coefficients of crown ethers. *Recl. Trav. Chim. Pays-Bas.* 108(3), 103-108.

Stull, D.R. (1947) Vapor pressure of pure substances: Organic compounds. *Ind. Eng. Chem.* 39(4), 517-560.

Sudhakar-Barik, Sethunnathan, N. (1978) Metabolism of nitrophenols in flooded soils. *J. Environ. Quality* 7, 349-352.

Suflita, J.M., Miller, G.D. (1985) Microbial metabolism of chlorophenolic comounds in groundwater aquifers. *Environ. Toxicol. Chem.* 4, 751-758.

Sugiura, K., Aoki, M., Kaneko, S., Daisaku, I., Komatsu, Y., Shibuya, H., Suzuki, H., Goto, M. (1984) Fate of 2,4,6-trichlorophenol, pentachlorophenol, p-chlorobiphenyl, and hexachlorobenzene in an outdoor experimental pond: Comparison between observations and predictions based on laboratory data. *Arch. Environ. Contam. Toxicol.* 13, 745-758.

Suntio, L.R., Shiu, W.Y., Mackay, D. (1988) A review of the nature and properties of chemicals present in pulp mill effluents. *Chemosphere* 17, 1249-1290.

Svenson, S., Björndal, H. (1988) A convenient test method for photochemical transformation of pollutants in the aquatic environment. *Chemosphere* 17, 2397-2405.

Tabak, H.H., Chambers, C.W., Kabler, P.W. (1964) Microbial metabolism of aromatic compounds. I. Decomposition of phenolic compounds and aromatic hydrocarbons by phenol-adapted bacteria. *J. Bacteriology* 87(4), 910-919.

Tabak, H.H., Govind, R. (1993) Prediction of biodegradation kinetics using a nonlinear group contribution method. *Environ. Toxicol. Chem.* 12, 251-260.

Takemoto, S., Kuge, Y., Nakamoto, M. (1981) The measurement of BOD in seawater. *Suishitsu Odaku Kenkyu.* 4, 80-90.

Tam, D., Varhanickova, D., Shiu, W.Y., Mackay, D. (1994) Aqueous solubility of chloroguaiacols. *J. Chem. Eng. Data* 39, 83-86.

Terada, H., Kosuge, Y., Murayama, W., Nakaya, N., Nunogaki, Y., Nunogaki, K.-I. (1987) Correlation of hydrophobic parameters of organic compounds determined by centrifugal partition chromatography with partition coefficients between octanol and water. *J. Chromatogr.* 400, 343-351.

Tewari, Y.B., Miller, M.M., Wasik, S.P., Martire, D.E. (1982) Aqueous solubility and octanol/water partition coefficient of organic compounds at 25.0°C. *J. Chem. Eng. Data* 27, 451-454.

Thomas, R.G. (1982) Chapter 15, Volatilization from water. In: *Handbook of Chemical Property Estimation Methods. Enveromental Behaviour of Organic Compounds.* Lyman, W.J., Reehl, W.F., Rosenblatt, D.H., Editors, McGraw-Hill Book Co., New York, New York.

THOR (1989) MedChem Release 3.54, Daylight Chemical Information System Inc., Claremont, California.

Toyota, H., Kuwahara, M. (1967) The study on production of PCP chemical fertilizer and its effect as herbicide and fertilizer, the solubility in water of PCP in PCP chemical fertilizer. *Nippon Dojohiryogaku Zasshi* 38, 93-97.

Tratnyek, P.G., Hoigné, J. (1991) Oxidation of substituted phenols in the environment: A QSAR analysis of rate constants for reaction with singlet oxygen. *Environ. Sci. Technol.* 25(9) 626-631.

Travis, C.C., Arms, A.D. (1988) Bioconcentration of organics in beef, milk, and vegetation. *Environ. Sci. Technol.* 22(3), 271-274.

Trevors, J.T. (1985) Effect of temperature on the degradation of pentachlorophenol *pseudomonas* species. *Chemosphere* 11, 471-475.

Trujillo, D.A., Ray, L.E., Murray, H.E., Giam, C.S. (1982) Bioaccumulation of pentachlorophenol by killifish (*fundulus similus*). *Chemosphere* 11, 25-31.

Tsonopoulos, C., Prausnitz, J.M. (1971) Activity coefficients of aromatic solutes in dilute aqueous solutions. *I & EC Fundam.* 10, 593-600. Supplementary materials.

Tute, M.S. (1971) Principle and practice of Hansch analysis: A guide structure-activity correlation for the medical chemist. *Adv. Drug Res.* 6, 1-77.

Ugland, K., Lundanes, E., Greibrokk, T., Bjorseth, A. (1981) Determination of chlorinated phenols by high-performance liquid chromatography. *J. Chromatogr.* 213, 83-90.

Umeyama, H., Nagai, T., Nogami, H. (1971) Mechanism of adsorption of phenols by carbon black from aqueous solution. *Chem. Pharm. Bull.* 19(8), 1714-1721.

Unger, S.H., Cook, J.R., Hollenberg, J.S. (1978) Simple procedure for determining octanol-aqueous partition, distribution, and ionization coefficients by reversed-phase high-pressure liquid chromatography. *J. Pharm. Sci.* 67, 1364-1367.

Urano, K., Kato, Z. (1986) A method to classify biodegradability of organic compounds. *J. Hazard. Material* 13, 135-145.

Urano, K., Kato, Z. (1986) Evaluation of biodegradation ranks priority organic compounds. *J. Hazard. Material* 13, 147-159.

Urano, K., Maeda, H., Ogura, K., Wada, H. (1982) Direct analytical method for aromatic compounds in water by steam carrier gas chromatography. *Water Res.* 16, 323-327.

Urushigawa, Y., Yonezawa, Y. (1979) Chemical-biological interactions in biological purification systems. V. Relation between biodegradation rate constants of aliphatic alcohols by activated sludge and their partition coefficients in a 1-octanol-water system. *Chemosphere* No.5, 139-142.

Urushigawa, Y., Yonezawa, Y., Masunaga, S., Matsue, Y., Masunaga, H., Yanagishi, T. (1984) Biodegradation of organic compounds by activated sludge. 13. Spectrophotochemical characteristics of mono-substituted phenols in biodegradation by activated sludge adapted to aniline, phenol, and *m*-cresol. *Kogai Shigan Kenkyusho Iho.* 14, 45-64.

Urushigawa, Y., Yonezawa, Y., Masunaga, S., Tashiro, T., Hirai, M., Tanaka, M. (1983) Biodegradability of mono-substituted phenols. I. Biodegradability by non-acclimated activated sludge microorganisms. *Kogai Shigen Kenkyuso Iho.* 12, 37-46.

Vaishnav, D.D., Babeu, L. (1987) Comparison of occurrence and rates of chemical biodegradation in natural waters. *Bull. Environ. Contam. Toxicol.* 39, 237-244.

Valsaraj, K.T., Thibodeaux, L.J., Lu, X.-Y. (1991) Studies in batch and continuous solvent sublation. III. Solubility of pentachlorophenol in alcohol-water mixtures and its effects on solvent sublation. *Sep. Sci. Technol.* 26(4), 529-538.

Valvani, S.C., Yalkowsky, S.H. (1980) Solubility and partitioning in drug design. In: *Physical Chemical Properties of Drug.* Medical Research Series, Vol. 10, Yalkowsky, S.H., Sinkinla, A.A., Valvani, S.C., Editors, Marcel Dekker Inc., New York, New York. pp.201-229.

Van Gestel, C.A.M., Ma, W.-C. (1988) Toxicity and bioaccumulation of chlorophenols in earthworms, in relation to bioavailability in soil. *Ecotoxicol. Environ. Saf.* 16, 289-297.

Van Veld, P.A., Spain, J.C. (1983) Degradation of selected xenobiotic compounds in three types of aquatic test systems. *Chemosphere* 12, 1291-1305.

Varhanickova, D., Shiu, W.Y., Mackay, D. (1995) Aqueous solubilities of alkylphenols and methoxyphenols at 25°C. *J. Chem. Eng. Data* in press.

Vashon, R.D., Schwab, B.S. (1982) Mineralization of linear alcohol ethoxylates and linear alcohol ethoxy sulfates at trace concentrations in estuarine water. *Environ. Sci. Technol.* 16, 433-436.

Veith, G.D., Austin, N.M., Morris, R.T. (1979a) A rapid method for estimating log P for organic chemicals. *Water Res.* 13, 43-47.

Veith, G.D., Broderius, S.J. (1987) Structure toxicity relationships for industrial chemicals causing type (II) narcosis syndrome. In: *QSAR in Environmental Toxicology-II.* pp.385-391. Kaiser, K.L.E., Editor, D. Reidel Publishing Company, Dordrecht, The Netherlands.

Veith, G.D., DeFoe, D.L., Bergstedt, B.V. (1979b) Measuring and estimating the bioconcentration factors of chemicals in fish. *J. Fish. Res. Board. Can.* 26, 1040-1048.

Veith, G.D., Kosian, P. (1983) Estimating bioconcentration potential from octanol/water partition coefficients. In: *Physical Behaviour of PCBs in the Great Lakes.* Mackay, D., Patterson, S., Eisenreich, S.J., Simmons, M.S., Editors, Ann Arbor Science Publishers, Ann Arbor, Michigan.

Veith, G.D., Macek, K.J., Petrocelli, S.R., Carroll, J. (1980) An evaluation of using partition coefficients and solubility to estimate bioconcentration factors for organic chemicals in fish. In: *Aquatic Toxicology.* pp.116-129. ASTM STP 707.

Verhaar, H.J.M., van Leeuwen, C.J., Hermens, L.M. (1992) Classifying environmental pollutants. 1: Structure-activity relationships for prediction of aquatic toxicity. *Chemosphere* 25, 471-491.

Verschueren, K. (1977) *Handbook of Environmental Data on Organic Chemicals.* Van Nostrand Reinhold Co., New York.

Verschueren, K. (1983) *Handbook of Environmental Data on Organic Chemiocals,* 2nd ed., Van Nostrand Reinhold Co., New York.

Virtanen, M.T., Hattula, M.L. (1982) The fate of 2,4,6-trichlorophenol in aquatic continuous-flow system. *Chemosphere* 11, 641-649.

Wachtmeister, C.A., Förlin, L., Arnoldsson, K.C., Larsson, J. (1991) Fish bile as a tool for monitoring aquatic pollutants: studies with radioactivity labelled 4,5,6-trichloroguaiacol. *Chemosphere* 22, 39-46.

Wagner, R. (1976) Investigation of the biodegradation of organic substances using the dilution bottle method. II. The degradation of test substances. *Vom. Wasser* 47, 241-265.

Wakita, K., Yoshimoto, M., Miyamoto, S. (1986) A method for calculation of the aqueous solubility of organic compounds by using new fragment solubility constants. *Chem. Pharm. Bull.* 34, 4663-4681.

Walden, C.C., McLeay, D.J., McKague, A.B. (1986) Cellulose production processes. In: *Handbook of Environmental Chemistry, Volume 3/Part D: Anthropogenic Compounds.* Hutzinger, O., Ed., pp. 1-34, Springer-Verlag, Berlin.

Walker, W.H., Collett, A.R., Lazzell, C.L. (1931) The solubility relations of the isomeric dihydroxybenzenes. *J. Phys. Chem.* 35, 3259.

Wang, L, Shao, Y., Hong, G. (1992) Predicting aqueous solubility and octanol/water partition coefficients of organic chemicals from molar volume. *Environ. Chem.* 11, 55-70.

Wang, L., Wang, X., Xu, O., Tian, L. (1986) Determination of the *n*-octanol/water partition coefficients of polycyclic aromatic hydrocarbons and estimation of aqueous solubilities. *Hungjing Kexue Xuebao* 6, 491-497.

Warner, H.P., Cohen, J.M., Ireland, J.C. (1987) *Determination Henry's Law Constants of Selected Priority Pollutants.* EPA/600/D-87/229, U.S. EPA, Cincinnati, Ohio.

Wasik, S.P., Tewari, Y.B., Miller, M.M., Martire, D.E. (1981) *Octanol/Water Partition Coefficients and Aqueous Solubilities of Organic Compounds.* NBSIR No.81-2406. U.S. Dept. of Commerce, Washington.

Weast, R.C., Editor (1972-73) *Handbook of Chemistry and Physics.* 53rd Edition, CRC Press, Cleveland, Ohio.

Weast, R.C., Editor (1973-74) *Handbook of Chemistry and Physics.* 54th Edition, CRC Press, Cleveland Ohio.

Weast, R.C., Editor (1976-77) *Handbook of Chemistry and Physics.* 57th Edition, CRC Press, Cleveland, Ohio.

Weast, R.C., Editor (1978) *Handbook of Chemistry and Physics.* 58th Edition, CRC Press, Inc., West Palm Beach, Florida.

Weast, R.C., Editor (1983-84) *Handbook of Chemistry and Physics.* 63th ed., CRC Press, Boca Raton, Florida.

Weber, R.C., et al. (1981) *Vapor Pressure Distribution of Selected Organic Chemicals.* pp. 39. U.S. EPA 600/2-81-021.

Westall, J.C., Leuenberger, C., Schwarzenbach, R.P. (1985) Influence of pH and ionic strength on the aqueous-nonaqueous distribution of chlorinated phenols. *Environ. Sci. Technol.* 19(2), 193-198.

Windholz, M., Editor (1983) *The Merck Index.* 10th Edition, Merck Co., Rahway, New Jersey.

Wong, A.S., Crosby, D.B. (1978) Photolysis of pentachlorophenol in water. In: *Pentachlorophenol: Chemistry, Pharmocology, and Environmental Toxicology.* Rao, K.R., Editor, pp. 19-25, Plenum Press, New York.

Wong, A.S., Crosby, D.B. (1981) Photodecomposition of pentachlorophenol in water. *J. Agric. Food Chem.* 29, 125-130.

Woodcock, D. (1971) Metabolism of fungicides and nematocides in soils. In: *Soil Biochemistry,* Vol. 2, Marcel Dekker Inc., New York.

Xie, T.M. (1983) Determination of trace amounts of chlorophenols and chloroguaiacols in sediment. *Chemosphere* 12(9/10), 1183-1191.

Xie, T.M., Abrahamsson, K., Fogelqvist, E., Josefsson, B. (1986) Distribution of chlorophenolics in a marine environment. *Environ. Sci. Technol.* 20, 457-463.

Xie, T.M., Dyrssen, D. (1984) Simultaneous determination of partition coefficients and acidity constants of chlorinated phenols and guaiacols by gas chromatography. *Anal. Chem. Acta* 160, 21-30.

Xie, T.M., Hulthe, B., Folestad, S. (1984) Determination partition coefficients of chlorinated phenols, guaiacols and catechols by shake-flask GC and HPLC. *Chemosphere* 13, 445-459.

Xing, B., McGill, W.B., Dudas, M.J. (1994) Sorption of α-naphthol onto organic sorbents varying in polarity and aromaticity. *Chemosphere* 28, 145-153.

Yahuhara, A., Otsuki, Fuwa, K. (1977) Photodecomposition of odorous chlorophenols. *Chemosphere* 6, 659-664.

Yalkowsky, S.H. (1979) Estimation of entropies of fusion of organic compounds. *Ind. Eng. Chem. Fundam.* 18, 108-111.

Yalkowsky, S.H., Morozowich, W. (1980) Chapter 3, A physical chemical basis for the design of orally active prodrugs. In: *Drug Design.* Vol. IX. Academic Press, Inc., New York.

Yalkowsky, S.H., Pinal, R. (1993) Estimation of the aqueous solubility of complex organic compounds. *Chemosphere* 26, 1239-1261.

Yalkowsky, S.H., Valvani, S.C. (1980) Solubility and partitioning I: Solubility of nonelectrolytes in water. *J. Pharm. Sci.* 69(8), 912-922.

Yalkowsky, S.H., Valvani, S.C., Mackay, D. (1983) Estimation of the aqueous solubility of some aromatic compounds. *Residue Rev.* 85, 43-55.

Yalkowsky, S.H., Valvani, S.C., Kuu, W.-Y., Dannenfelser, R.M., Editors (1987) Arizona *Database of Aqueous Solubility for Organic Compounds.* College of Pharmacy, University of Arizona, Tucson, Arizona.

Yaws, C.L., Yang., H.-C., Hopper, J.R., Hansen, K.C. (1990) Organic chemicals: water solubility data. *Chem. Eng.* July, 115-118.

Yaws, C., Yang, H.C., Pan, X. (1991) Henry's law constants for 362 organic compounds in water. *Chem. Eng.* 98(11), 179-185.

Yoshida, K., Mizuno, T., Ose, Y., Sato, T. (1986) The estimation for toxicity of chemicals on fish by physico-chemical properties. *Chemosphere* 15(2), 195-203.

Yoshida, K., Shigeoka, T., Yamauchi, F. (1983) Non-steady-state equilibrium model for the preliminary prediction of the fate of chemicals in the environment. *Ecotox. Environ. Saf.* 7, 179-190.

Yoshida, K., Shigeoka, T., Yamauchi, F. (1987) Evaluation of aquatic environmental fate of 2,4,6-trichlorophenol with a mathematical model. *Chemosphere* 16, 2531-2544.

Zepp, R.G., Schlotzhauer, P.F., Simmons, M.S., Miller, G.C., Baughman, G.L., Wolfe, N.L. (1984) Dynamics of pollutant photoreactions in the hydrosphere. *Fresenius Z. Anal. Chem.* 319, 119-125.

Zhang, X., Wiegel, J. (1990) Sequential anaerobic degradation of 2,4-dichlorophenol in freshwater sediments. *Appl. Environ. Microbiol.* 56(4), 1119-1127.

Zoeteman, B.C.J., De Greef, E., Brinkmann, F.J.J. (1981) Persistency of organic contaminats in grounwater, lessons from soil pollution incidents in the Neterlands. *Sci. Total. Environ.* 21, 187-192.

Chapter 5 Carboxylic Acids

5.1 List of Chemicals and Data Compilations

Common Name: Formic acid
Synonym: methanoic acid
Chemical Name: formic acid
CAS Registry No: 64-18-6
Molecular Formula: HCOOH
Molecular Weight: 46.026
Melting Point (°C):
8.40 (Weast 1982-83; Verschueren 1983; Stephenson & Malanowski 1987)
8.5 (Dean 1985)
8.27 (Riddick et al. 1986)
Boiling Point (°C):
100.7 (Weast 1982-83; Boublik et al. 1984; Stephenson & Malanowski 1987)
101.0 (Verschueren 1983)
100.8 (Dean 1985)
100.6 (Riddick et al. 1986)
100.0 (Banerjee et al. 1990)
Density (g/cm^3 at 20°C):
1.220 (Weast 1982-83; Verschueren 1983; Dean 1985)
1.21405 (25°C, Riddick et al. 1986)
Molar Volume (cm^3/mol): 46.2 (calculated-LeBas method)
Molecular Volume (Å^3):
Total Surface Area, TSA (Å^2):
145.1 (Iwase et al. 1985)
Dissociation Constants, pK:
3.751 (pK_a, Dean 1985)
3.752 (pK_a, Riddick et al. 1986)
5.770 (pK_s, Riddick et al. 1986)
3.740 (pK_a, Sangster 1989)
3.80 (Kollig 1993)
Heat of Fusion ΔH_{fus}, kcal/mol:
3.031 (Riddick et al. 1986)
Entropy of Fusion ΔS_{fus}, cal/mol·K (e.u.):
Fugacity Ratio at 25°C (assuming ΔS_{fus} = 13.5 e.u.), F: 1.0

Water Solubility (g/m^3 or mg/L at 25°C):
miscible (Dean 1985)
infinite (Riddick et al. 1986)

Vapor Pressure (Pa at 25°C):
4666 (20°C, quoted, Verschueren 1983)
7198 (30°C, quoted, Verschueren 1983)
4344 (24.6°C, quoted exptl., Boublik et al. 1984)
5773 (25.5°C, quoted exptl., Boublik et al. 1984)
5068 (calculated average-Antoine eqn., Boublik et al. 1984)
5744 (extrapolated-Antoine eqn., Boublik et al. 1984)
5744 (extrapolated-Antoine eqn., Dean 1985)

5750 (Riddick et al. 1986)
5711 (calculated-Antoine eqn., Stephenson & Malanowski 1987)
5679, 67850 (measured, calculated-solvatochromic p., Banerjee et al. 1990)

Henry's Law Constant (Pa·m³/mol):
0.017 (quoted, pH 4, Gaffney et al. 1987)
0.112 (computed, Yaws et al. 1991)
0.0183 (equilibrium partial pressure, pH 1.6-1.9, Khan & Brimblecombe 1992)

Octanol/Water Partition Coefficient, log K_{OW}:
-0.54 (shake flask-TN, Collander 1951; quoted, Leo et al. 1971; Rekker & de Kort 1979; Iwase et al. 1985)
-0.54 (shake flask-T, Whitehead & Geankoplis 1955)
-0.54 (Leo et al. 1971; quoted, Hansch et al. 1972)
-0.54, -0.46 (quoted, calculated-f const., Rekker & de Kort 1979)
-1.55, -0.22 (calculated, Verschueren 1983)
-0.56 (calculated-TSA, Iwase et al. 1985)
-0.54 (quoted from Hansch & Leo 1979, Bodor et al. 1989)
-0.54 (quoted from CLOGP, Bodor et al. 1989)
-0.67 (calculated-MO, Bodor et al. 1989)
-0.54 (quoted and recommended, Sangster 1989)
-0.70 (quoted, Kollig 1993)

Bioconcentration Factor, log BCF:

Sorption Partition Coefficient, log K_{OC}:
-0.27 (calculated-K_{OW}, Kollig 1993)

Half-Lives in the Environment:
Air:
Surface water:
Groundwater:
Sediment:
Soil:
Biota:

Environmental Fate Rate Constants or Half-Lives:
Volatilization:
Photolysis:
Hydrolysis:
Oxidation: measured rate constant of 3.7×10^{-13} cm³ molecule⁻¹ sec⁻¹ for the gas-phase reaction with hydroxyl radicals in air and 2.2×10^{-13} cm³ molecule⁻¹ sec⁻¹ for the solution-phase reaction with hydroxyl radicals in aqueous solution (Dagaut et al. 1988; quoted, Wallington et al. 1988).
Biodegradation:
Biotransformation:
Bioconcentration, Uptake (k_1) and Elimination (k_2) Rate Constants:

Common Name: Acetic acid
Synonym: ethanoic acid, methanecarboxylic acid, glacial acetic, vinegar acid
Chemical Name: acetic acid
CAS Registry No: 64-19-7
Molecular Formula: CH_3COOH
Molecular Weight: 60.052
Melting Point (°C):
16.7 (Stull 1947; Verschueren 1983; Riddick et al. 1986)
16.6 (Weast 1982-83; Dean 1985; Howard 1990)
16.6 (Stephenson & Malanowski 1987)
Boiling Point (°C):
118.1 (Stull 1947; Verschueren 1983)
117.9 (Hoy 1970; Weast 1982-83; Boublik et al. 1984; Dean 1985)
117.9 (Riddick et al. 1986; Stephenson & Malanowski 1987)
117.0 (Banerjee et al. 1990)
118.5 (Howard 1990)
Density (g/cm^3 at 20°C):
1.04920 (Weast 1982-83; Verschueren 1983; Dean 1985)
1.04955 (Riddick et al. 1986)
Molar Volume (cm^3/mol):
64.1 (exptl., Lee et al. 1972; quoted, Reid et al. 1977)
57.1 (calculated-density, Rohrschneider 1973)
68.4 (calculated-LeBas method, Reid et al. 1977)
57.5 (Kamlet et al. 1987)
Molecular Volume ($Å^3$):
Total Surface Area, TSA ($Å^2$):
173.6 (Iwase et al. 1985)
Dissociation Constant, pK:
4.733 (pK_a, Korman & La Mer 1936)
4.760 (pK_a, Fieser & Fieser 1958)
4.75 (pK_a, Weast 1982-83)
4.756 (pK_a, Dean 1985; Riddick et al. 1986)
14.45 (pK_s, Riddick et al. 1986)
4.760 (pK_a, Sangster 1989)
4.750 (pK_a, Howard 1990)
Heat of Fusion ΔH_{fus}, kcal/mol:
2.801 (Riddick et al. 1986)
Entropy of Fusion ΔS_{fus}, cal/mol·K (e.u.):
Fugacity Ratio at 25°C (assuming ΔS_{fus} = 13.5 e.u.), F: 1.0

Water Solubility (g/m^3 or mg/L at 25°C):
miscible (Dean 1985)
infinite (Riddick et al. 1986)
6029000 (Yalkowsky et al. 1987; quoted, Howard 1990)
miscible (Yaws et al. 1990)

Vapor Pressure (Pa at 25°C):

2040 (interpolated-regression of tabulated data, Stull 1947)
2666 (29.9°C, Stull 1947)
2030 (Hoy 1970)
2080 (ebulliometry, fitted to Antoine eqn., Ambrose et al. 1977)
1520, 2666 (20°C, 30°C, Verschueren 1983)
2666 (29.8°C, quoted exptl., Boublik et al. 1984)
2088 (extrapolated average-Antoine eqn., Boublik et al. 1984)
2050 (calculated-Antoine eqn., Dean 1985)
1520 (Howard et al. 1986; quoted, Banerjee et al. 1990)
2070 (Riddick et al. 1986)
2114 (calculated-Antoine eqn., Stephenson & Malanowski 1987)
1520, 32660 (measured, calculated-solvatochromic p., Banerjee et al. 1990)

Henry's Law Constant (Pa·m^3/mol):

0.0303 (partial pressure, Butler & Ramchandani 1935)
0.0305 (quoted exptl., Hine & Mookerjee 1975)
0.030 (calculated-group contribution, Hine & Mookerjee 1975)
0.028 (calculated-bond contribution, Hine & Mookerjee 1975)
0.0101 (effective Henry's law constant, pH 4, Gaffney et al. 1987)
0.0305, 0.0254 (quoted, calculated- χ , Nirmalakhandan & Speece 1988)
0.121 (computed, Yaws et al. 1991)
0.0101, 1.01×10^{-4} (at pH 4, pH 7, Howard 1990)
0.0183 (equilibrium partial pressure, pH 1.6-1.9, Khan & Brimblecombe 1992)
0.0184, 0.0285 (quoted: exptl., calculated value by bond contribution, Brimblecombe et al. 1992)
0.0305, 0.0431 (quoted, calculated-molecular structure, Russell et al. 1992)

Octanol/Water Partition Coefficient, log K_{OW}:

-0.30 (shake flask-TN, Leonard et al. 1948; quoted, Sangster 1989)
-0.31 (shake flask-TN, Collander 1951; quoted, Leo et al. 1971; Hansch et al. 1972; Rekker 1977; Sangster 1989)
-0.17 (quoted unpublished result, Rekker 1977)
-0.29 (calculated-f const., Rekker 1977)
-0.17 (calculated-π const., Rekker 1977)
-0.29 (shake flask-RC, Wolfenden 1978; quoted, Sangster 1989)
-0.24 (Hansch & Leo 1979; quoted, Bodor et al. 1989)
-0.31, -0.17 (quoted, Verschueren 1983)
-0.17 (Hansch & Leo 1985; quoted, Howard 1990)
-0.17, -0.02 (quoted, calculated-TSA, Iwase et al. 1985)
-0.17 (quoted from CLOGP, Bodor et al. 1989)
-0.234 (calculated-CLOGP, Bodor et al. 1989)
-0.37 (calculated-M.O. indices, Bodor et al. 1989)
-0.17 (quoted from Pomona College unpublished data, Sangster 1989)
-0.17 (recommended, Sangster 1989)

Bioconcentration Factor, log BCF:

< 0.0 (estimated-K_{OW}, Lyman et al. 1982; quoted, Howard 1990)

Sorption Partition Coefficient, log K_{OC}:

no detectable sorption (Podzol soil, Alfisol soil, sediment, von Oepen et al. 1991)

0.00 (soil, quoted exptl., Meylan et al. 1992)

-0.21 (soil, calculated- χ and fragment contribution, Meylan et al. 1992)

Half-Lives in the Environment:

Air: half-life of 26.7 days, based on measured rate constant of 0.6×10^{-12} cm^3 $molecule^{-1}$ sec^{-1} for the vapor-phase reaction with 5×10^5 hydroxyl radicals/cm^3 in air at 25°C (Atkinson 1985; quoted, Howard 1990).

Surface water: half-life of 26-46 years, based on OH radical concn. in sunlit natural water of 1×10^{-17} mol/L (Howard 1990).

Groundwater:

Sediment: >90% degradation in 14 days using Rhine River sediment as inocula (Kool 1984; quoted, Howard 1990).

Soil:

Biota:

Environmental Fate Rate Constants or Half-Lives:

Volatilization:

Photolysis:

Oxidation: photooxidation half-life of 26.7 days in air, based on measured rate constant of $(0.6 \pm 0.2) \times 10^{-12}$ $cm^3 \cdot molecule^{-1} \cdot sec^{-1}$ by the flash photolysis-resonance fluorescence technique (Zetzsch & Stuhl 1982; quoted, Atkinson 1985) for the vapor-phase reaction with 5×10^5 hydroxyl radicals/cm^3 in air at 25°C (Atkinson 1985; quoted, Howard 1990); rate constant of 2.40×10^{-13} $cm^3 \cdot molecule^{-1} \cdot sec^{-1}$ was observed for the vapor-phase reaction with hydroxyl radicals in air and 2.70×10^{-14} $cm^3 \cdot molecule^{-1} \cdot sec^{-1}$ was observed for the solution-phase reaction with hydroxyl radicals in aqueous solution (Dagaut et al. 1988; quoted, Wallington et al. 1988); rate constant of 0.74×10^{-12} $cm^3 \cdot molecule^{-1} \cdot sec^{-1}$ was observed for the vapor-phase reaction with hydroxyl radicals in air at 298 K (Atkinson 1990).

Hydrolysis:

Biodegradation: >90% degradation in 3 days using an activated sludge inoculum; in 24 hours in batch aeration in sewage and 14 days using sediment from the Rhine river as inocula (Howard 1990).

Biotransformation:

Bioconcentration, Uptake (k_1) and Elimination (k_2) Rate Constants:

Common Name: Propionic acid
Synonym: methylacetic acid, propanoic acid
Chemical Name: propanoic acid, propionic acid
CAS Registry No: 79-09-4
Molecular Formula: CH_3CH_2COOH
Molecular Weight: 74.079
Melting Point (°C):
-22.0 (Stull 1947; Verschueren 1983)
-20.8 (Weast 1982-83)
-20.7 (Riddick et al. 1986)
Boiling Point (°C):
141.25 (Butler & Ramchandani 1935)
141.1 (Stull 1947)
141.0 (Weast 1982-83; Verschueren 1983; Dean 1985; Banerjee et al. 1990)
141.2 (Boublik et al. 1984; Riddick et al. 1986)
Density (g/cm^3 at 20°C):
0.9930 (Weast 1982-83)
0.9920 (Verschueren 1983)
0.9934 (Dean 1985)
0.99349 (Riddick et al. 1986)
Molar Volume (cm^3/mol):
74.60 (calculated-density)
81.4 (calculated-LeBas method)
Molecular Volume ($Å^3$):
Total Surface Area, TSA ($Å^2$):
197.0 (Iwase et al. 1985)
Dissociation Constant, pK_a:
-6.80 (pK value, Perrin 1972)
4.874 (Dean 1985; Riddick et al. 1986)
4.87 (Sangster 1989)
Heat of Fusion ΔH_{fus}, kcal/mol:
2.548 (Riddick et al. 1986)
Entropy of Fusion ΔS_{fus}, cal/mol·K (e.u.):
Fugacity Ratio at 25°C (assuming ΔS_{fus} = 13.5 e.u.), F: 1.0

Water Solubility (g/m^3 or mg/L at 25°C):
miscible (Dean 1985)
infinite (Riddick et al. 1986)

Vapor Pressure (Pa at 25°C):
550 (interpolated-regression of tabulated data, Stull 1947)
666.5 (28.0°C, Stull 1947)
435 (ebulliometry, fitted to Antoine eqn., Ambrose et al. 1981)
387 (20°C, quoted, Verschueren 1983)

471 (extrapolated average-Antoine eqn., Boublik et al. 1984)
257 (extrapolated-Antoine eqn., Dean 1985)
453 (Howard et al. 1986; quoted, Banerjee et al. 1990)
451 (Riddick et al. 1986)
453, 8758 (measured, calculated-solvatochromic p., Banerjee et al. 1990)

Henry's Law Constant ($Pa \cdot m^3/mol$):
0.0445 (partial pressure, Butler & Ramchandani 1935)
0.045 (quoted exptl., Hine & Mookerjee 1975; quoted, Nirmalakhandan & Speece 1988)
0.042 (calculated-group contribution, Hine & Mookerjee 1975)
0.043 (calculated-bond contribution, Hine & Mookerjee 1975)
0.0451, 0.0298 (quoted, calculated- χ , Nirmalakhandan & Speece 1988)
0.018 (equilibrium partial pressure, Khan & Brimblecombe 1992)
0.018, 0.0431 (quoted: exptl., calc-bond contribution, Brimblecombe et al. 1992)

Octanol/Water Partition Coefficient, log K_{OW}:
0.25 (shake flask-TN, Collander 1951; quoted, Leo et al. 1971; Rekker 1977; Sangster 1989)
0.33 (quoted unpublished result, Rekker 1977)
0.33 (calculated-π const., Rekker 1977)
0.23, 0.24 (calculated-f const., Rekker 1977)
0.27 (shake flask-T, Umland 1983; quoted, Sangster 1989)
0.25, 0.33 (quoted, Verschueren 1983)
0.33, 0.42 (quoted, calculated-TSA, Iwase et al. 1985)
0.29 (quoted from Hansch & Leo 1979, Bodor et al. 1989)
0.33 (quoted from CLOGP, Bodor et al. 1989)
0.295 (calculated-CLOGP, Bodor et al. 1989)
0.19 (calculated-M.O., Bodor et al. 1989)
0.33 (quoted from Pomona College unpublished data, Sangster 1989)
0.33 (recommended, Sangster 1989)

Bioconcentration Factor, log BCF:

Sorption Partition Coefficient, log K_{OC}:

Half-Lives in the Environment:
Air:
Surface water:
Groundwater:
Sediment:

Soil:
Biota:

Environmental Fate Rate Constants or Half-Lives:

Volatilization:

Photolysis:

Oxidation: rate constant of $(1.6 \pm 0.5)\times10^{-12}$ cm^3 $molecule^{-1}$ sec^{-1} for the gas-phase reaction with OH radicals at 298 K in air was determined by the flash photolysis-resonance fluorescence technique (Zetzsch & Stuhl 1982; quoted, Atkinson 1985); rate constant of 1.22×10^{-12} cm^3 $molecule^{-1}$ sec^{-1} for the gas-phase reaction with OH radicals in air and 1.00×10^{-12} cm^3 $molecule^{-1}$ sec^{-1} for the solution-phase reaction with OH radicals in aqueous solution (Dagaut et al. 1988; quoted, Wallington et al. 1988).

Hydrolysis:

Biodegradation:

Biotransformation:

Bioconcentration, Uptake (k_1) and Elimination (k_2) Rate Constants:

Common Name: Butyric acid
Synonym: butanoic acid, *n*-butyric acid, ethylacetic acid
Chemical Name: *n*-butyric acid, butyric acid
CAS Registry No: 107-92-6
Molecular Formula: $CH_3CH_2CH_2COOH$
Molecular Weight: 88.106
Melting Point (°C):
-4.7 (Stull 1947)
-5.5/-8.0 (Verschueren 1983)
-5.3 (Dean 1985)
-5.2 (Riddick et al. 1986)
Boiling Point (°C):
163.5 (Stull 1947; Boublik et al. 1984)
163.7 (Verschueren 1983; Riddick et al. 1986)
163.3 (Dean 1985)
Density (g/cm³ at 20°C):
0.9590 (Verschueren 1983)
0.9582 (Dean 1985; Riddick et al. 1986)
Molar Volume (cm³/mol):
92.5 (Kamlet et al. 1987)
91.87 (calculated-density)
103.6 (calculated-LeBas method)
Molecular Volume (Å³):
Total Surface Area, TSA (Å²): 224.2 (Iwase et al. 1985)
Dissociation Constant, pK_a:
4.817 (Dean 1985)
4.822 (Riddick et al. 1986)
4.82 (Sangster 1993)
Heat of Fusion ΔH_{fus}, kcal/mol:
2.77 (Riddick et al. 1986)
Entropy of Fusion ΔS_{fus}, cal/mol·K (e.u.):
Fugacity Ratio at 25°C (assuming ΔS_{fus} = 13.5 e.u.), F: 1.0

Water Solubility (g/m³ or mg/L at 25°C):
12445, 18837 (quoted, calculated-K_{OW}, Yalkowsky & Morozowich 1980)
miscible (Dean 1985)
infinite (Riddick et al. 1986)

Vapor Pressure (Pa at 25°C):
133.8 (extrapolated-regression of tabulated data, Stull 1947)
133.3 (25.5°C, Stull 1947)
83.95 (ebulliometry, fitted to Antoine eqn., Ambrose et al. 1981)
57.32 (20°C, quoted, Verschueren 1983)
186.6 (30°C, quoted, Verschueren 1983)
98.25 (extrapolated average-Antoine eqn., Boublik et al. 1984)
104.3 (extrapolated-Antoine eqn., Dean 1985)
102 (Riddick et al. 1986)

Henry's Law Constant ($Pa \cdot m^3/mol$):

0.0542 (partial pressure, Butler & Ramchandani 1935)
0.0542 (quoted exptl., Hine & Mookerjee 1975)
0.0590 (calculated-group contribution, Hine & Mookerjee 1975)
0.0650 (calculated-bond contribution, Hine & Mookerjee 1975)
0.0542, 0.0375 (quoted, calculated- χ , Nirmalakhandan & Speece 1988)
0.0222 (equilibrium partial pressure, Khan & Brimblecombe 1992)
0.0222, 0.0654 (quoted: exptl., calculated-bond contribution, Brimblecombe et al. 1992)
0.0542, 0.358 (quoted, calculated-molecular structure, Russell et al. 1992)

Octanol/Water Partition Coefficient, log K_{OW}:

0.79 (shake flask-TN, Collander 1951; quoted, Leo et al. 1971; Iwase et al. 1985)
0.79 (quoted, Verschueren 1983)
0.94 (calculated-TSA, Iwase et al. 1985)
0.79, 0.824, 0.70 (quoted fron Hansch & Leo 1979, calculated-CLOGP, calculated-M.O., Bodor et al. 1989)
0.79 (quoted and recommended, Sangster 1989, 1993)

Bioconcentration Factor, log BCF:

Sorption Partition Coefficient, log K_{OC}:

Half-Lives in the Environment:

Air:
Surface water:
Groundwater:
Sediment:
Soil:
Biota:

Environmental Fate Rate Constants or Half-Lives:

Volatilization:
Photolysis:
Hydrolysis:
Oxidation: rate constant of $(2.4 \pm 0.7)\times10^{-12}$ cm^3 $molecule^{-1}$ sec^{-1} for the gas-phase reaction with OH radicals in air at 298 K was determined using flash photolysis-resonance fluorescence technique (Zetzsch & Stuhl 1981; quoted, Atkinson 1985); rate constant of 1.80×10^{-12} cm^3 $molecule^{-1}$ sec^{-1} for the gas-phase reaction with OH radicals in air and 3.70×10^{-12} cm^3 $molecule^{-1}$ sec^{-1} for the solution-phase reaction with OH radicals in aqueous solution (Zetzsch & Stuhl 1982; quoted, Wallington et al. 1988).
Biodegradation:
Biotransformation:
Bioconcentration, Uptake (k_1) and Elimination (k_2) Rate Constants:

Common Name: Isobutyric acid
Synonym: isobutanoic acid, *i*-butyric acid, dimethylacetic acid, 2-methylpropanoic acid, 2-methylpropionic acid, isopropylformic acid
Chemical Name: *i*-butyric acid, isobutyric acid
CAS Registry No: 79-31-2
Molecular Formula: $(CH_3)_2CHCOOH$
Molecular Weight: 88.106
Melting Point (°C):
-47.0 (Stull 1947; Verschueren 1983)
-46.0 (Dean 1985; Riddick et al. 1986)
Boiling Point (°C):
154.5 (Stull 1947)
154.4 (Verschueren 1983)
152.0 (Boublik et al. 1984)
154.0 (Dean 1985)
154.7 (Riddick et al. 1986)
Density (g/cm^3 at 20°C):
0.9490 (Verschueren 1983)
0.96815 (Riddick et al. 1986)
Molar Volume (cm^3/mol):
109 (exptl. at normal B.P., Lee et al. 1972; quoted, Reid et al. 1977)
112.8 (calculated-LeBas method, Reid et al. 1977)
Molecular Volume ($Å^3$):
Total Surface Area, TSA ($Å^2$):
Dissociation Constant, pK_a:
4.860 (Riddick et al. 1986)
Heat of Fusion ΔH_{fus}, kcal/mol:
1.200 (Riddick et al. 1986)
Entropy of Fusion ΔS_{fus}, cal/mol·K (e.u.):
Fugacity Ratio at 25°C (assuming ΔS_{fus} = 13.5 e.u.), F: 1.0

Water Solubility (g/m^3 or mg/L at 25°C):
228000 (20°C, synthetic method, Jones 1929)
200000 (20°C, quoted, Verschueren 1983)
170000 (Dean 1985)
228000 (20°C, Riddick et al. 1986)

Vapor Pressure (Pa at 25°C):
273.8 (interpolated-regression of tabulated data, Stull 1947)
133.3 (14.7°C, Stull 1947; quoted, Verschueren 1983)
5.808 (extrapolated-Antoine eqn., Boublik et al. 1984)
8.83 (extrapolated-Antoine eqn., Dean 1985)
185 (Riddick et al. 1986)

Henry's Law Constant (Pa·m^3/mol):

Octanol/Water Partition Coefficient, log K_{OW}:

0.79 (shake flask-TN, Collander 1951; quoted Hansch & Leo 1979)

0.50, 1.13 (calculated, Verschueren 1983)

0.79, 0.824, 0.70 (quoted Hansch & Leo 1979; calculated-CLOGP, calculated-M.O. indices, Bodor et al. 1989)

0.79 (quoted and recommended, Sangster 1989)

Bioconcentration Factor, log BCF:

Sorption Partition Coefficient, log K_{OC}:

Half-Lives in the Environment:

Air:

Surface water:

Groundwater:

Sediment:

Soil:

Biota:

Environmental Fate Rate Constants or Half-Lives:

Volatilization:

Photolysis:

Hydrolysis:

Oxidation:

Biodegradation:

Biotransformation:

Bioconcentration, Uptake (k_1) and Elimination (k_2) Rate Constants:

Common Name: *n*-Valeric acid
Synonym: pentanoic acid, valeric acid
Chemical Name: *n*-valeric acid, valeric acid
CAS Registry No: 109-52-4
Molecular Formula: $CH_3CH_2CH_2CH_2COOH$
Molecular Weight: 102.13
Melting Point (°C):
-34.5 (Stull 1947; Stephenson & Malanowski 1987)
-33.8 (Weast 1982-83)
-34/-58 (Verschueren 1983)
-33.7 (Dean 1985; Riddick et al. 1986)
Boiling Point (°C):
184.4 (Stull 1947)
186.0 (Weast 1982-83)
187.0 (Verschueren 1983; Stephenson & Malanowski 1987)
185.5 (Dean 1985; Riddick et al. 1986)
Density (g/cm^3 at 20°C):
0.9391 (Weast 1982-83)
0.9420 (Verschueren 1983)
0.9390 (Dean 1985; Riddick et al. 1986)
Molar Volume (cm^3/mol):
125.8 (calculated-LeBas method)
108.75 (calculated-density)
Molecular Volume (Å^3):
Total Surface Area, TSA (Å^2):
303.4 (Amidon et al. 1975)
Dissociation Constant, pK_a:
4.82 (18°C, Weast 1982-83)
4.860 (Riddick et al. 1986)
4.830 (Sangster 1989, 1993)
Heat of Fusion ΔH_{fus}, kcal/mol:
3.385 (Riddick et al. 1986)
Entropy of Fusion ΔS_{fus}, cal/mol·K (e.u.):
Fugacity Ratio at 25°C (assuming ΔS_{fus} = 13.5 e.u.), F: 1.0

Water Solubility (g/m^3 or mg/L at 25°C):
26000 (20°C, quoted, Amidon et al. 1975)
24000 (Verschueren 1983; Dean 1985)
24000 (20°C, Reddick et al. 1986)

Vapor Pressure (Pa at 25°C):
39.6 (extrapolated-regression of tabulated data, Stull 1947)
20.0 (20°C, quoted, Verschueren 1983)
5.35 (extrapolated average-Antoine eqn., Boublik et al. 1984)
19.0 (Riddick et al. 1986)
16.85 (extrapolated-Antoine eqn., Stephenson & Malanowski 1987)

Henry's Law Constant ($Pa \cdot m^3/mol$):

0.0478 (equilibrium partial pressure, pH 1.6-1.9, Khan & Brimblecombe 1992)

0.0478, 0.0618, 0.0989 (quoted exptl., calculated-P/C, calculated-bond contribution, Brimblecombe et al. 1992)

Octanol/Water Partition Coefficient, log K_{OW}:

1.42 (Scherrer & Howard 1979; quoted, Sangster 1993)

0.99, 1.69 (calculated, Verschueren 1983)

1.39 (recommended, Sangster 1989,93)

Bioconcentration Factor, log BCF:

Sorption Partition Coefficient, log K_{OC}:

Half-Lives in the Environment:

Air:

Surface water:

Groundwater:

Sediment:

Soil:

Biota:

Environmental Fate Rate Constants or Half-Lives:

Volatilization:

Photolysis:

Oxidation:

Hydrolysis:

Biodegradation:

Biotransformation:

Bioconcentration, Uptake (k_1) and Elimination (k_2) Rate Constants:

Common Name: Hexanoic acid
Synonym: butylacetic acid, caproic acid, *n*-hexoic acid
Chemical Name: butylacetic acid, hexanoic acid, *n*-hexanoic acid
CAS Registry No: 142-62-1
Molecular Formula: $CH_3(CH_2)_3CH_2COOH$
Molecular Weight: 116.16
Melting Point (°C):
  -6/-2 (Verschueren 1983)
  -3.44 (Riddick et al. 1986)
Boiling Point (°C):
  204/208 (Verschueren 1983)
  205.02 (Riddick et al. 1986)
Density (g/cm^3 at 20°C):
Molar Volume (cm^3/mol):
  125.9 (Hoy 1970; Amidon & Williams 1982)
  157.2 (calculated-LeBas method)
Molecular Volume ($Å^3$):
Total Surface Area, TSA ($Å^2$):
Dissociation Constant, pK_a:
  4.85 (Dean 1985; Bintein & Devillers 1994)
  4.879 (Riddick et al. 1986)
  4.87 (Sangster 1989, 1993)
Heat of Fusion ΔH_{fus}, kcal/mol:
  3.68 (Riddick et al. 1986)
Entropy of Fusion ΔS_{fus}, cal/mol·K (e.u.):
Fugacity Ratio at 25°C (assuming ΔS_{fus} = 13.5 e.u.), F: 1.0

Water Solubility (g/m^3 or mg/L at 25°C):
  11887, 14624 (quoted, calculated-K_{OW}, Yalkowsky & Morozowich 1980)
  9887 (Valvani et al. 1981; quoted, Banerjee 1985)
  9887, 6238 (quoted, calculated-K_{OW} and M.P., Amidon & Williams 1982)
  11000 (quoted, Verschueren 1983)
  10816 (Windholz 1983; quoted, Chessells et al. 1992)
  6391 (calculated- γ from UNIFAC, Banerjee 1985)
  9580 (20°C, Riddick et al. 1986)

Vapor Pressure (Pa at 25°C):
  4.40 (ebulliometry, fitted to Antoine eqn., Ambrose et al. 1981)
  26.7 (20°C, quoted, Verschueren 1983)
  5.0 (Riddick et al. 1986)

Henry's Law Constant (Pa m^3/mol):
  0.0768 (equilibrium partial pressure, pH 1.6-1.9, Khan & Brimblecombe 1992)
  0.0768, 0.0873, 0.149 (quoted: exptl., calculated-P/C, calculated-bond contribution,

Brimblecombe et al. 1992)

Octanol/Water Partition Coefficient, log K_{OW}:

1.87 (HPLC, D'Amboise & Hanai 1982; quoted, Sangster 1989,93)
1.90 (quoted from Valvani et al. 1981, Amidon & Williams 1982)
1.92 (shake flask-TN, Umland 1983; quoted, Sangster 1989,93)
1.88, 1.92 (quoted, Verschueren 1983)
2.09 (shake flask-fluorescence, Nishimura et al. 1985; quoted, Sangster 1989,93)
1.92, 1.32 (quoted, calculated- γ from UNIFAC, Banerjee & Howard 1988; quoted, Sangster 1993)
2.07, 1.88, 2.03±0.01 (quoted, calculated, potentiometric titration, Hersey et al. 1989; quoted, Sangster 1993)
1.88, 1.92 (quoted from Pomona College unpublished data, Sangster 1989)
1.92 (recommended, Sangster 1989, 1993)
1.92 (quoted, Chessells et al. 1992; El Tayar et al. 1992; Bintein & Devillers 1994)
1.92, 1.564 (quoted, calculated-original UNIFAC, Chen et al. 1993)

Bioconcentration Factor, log BCF:

Sorption Partition Coefficient, log K_{OC}:

1.38 (Alfisol, von Oepen et al. 1991)
1.42 (Podzol, von Oepen et al. 1991)
1.57 (sediment, von Oepen et al. 1991)
1.46 (soil, quoted exptl., Meylan et al. 1992)
0.88 (soil, calculated- χ and fragment contribution, Meylan et al. 1992)

Half-Lives in the Environment:

Air:
Surface water:
Groundwater:
Sediment:
Soil:
Biota:

Environmental Fate Rate Constants or Half-Lives:

Volatilization:
Photolysis:
Oxidation:
Hydrolysis:
Biodegradation:
Biotransformation:
Bioconcentration, Uptake (k_1) and Elimination (k_2) Rate Constants:

Common Name: Stearic acid
Synonym: octadecanoic acid
Chemical Name: stearic acid, ocatadecanoic acid, *n*-octadecylic acid
CAS Registry No: 57-11-4
Molecular Formula: $CH_3(CH_2)_{16}COOH$
Molecular Weight: 284.48
Melting Point (°C):
68.8 (Yalkowsky & Valvani 1980)
71.2 (Weast 1982-83)
69.4 (Verschueren 1983)
69.7 (Stephenson & Malanowski 1987)
Boiling Point (°C):
360 (dec., Weast 1982-83)
383 (Verschueren 1983)
376 (Stephenson & Malanowski 1987)
Density (g/cm^3 at 20°C):
0.9408 (Weast 1982-83)
Molar Volume (cm^3/mol):
423.6 (calculated-LeBas method)
Molecular Volume (Å^3):
Total Surface Area, TSA (Å^2):
Dissociation Constant, pK_a:
4.50 (estimated, Sangster 1989)
Heat of Fusion ΔH_{fus}, kcal/mol:
Entropy of Fusion ΔS_{fus}, cal/mol·K (e.u.):
39.57 (observed, Yalkowsky & Valvani 1980)
43.50 (estimated, Yalkowsky & Valvani 1980)
Fugacity Ratio at 25°C (assuming ΔS_{fus} = 13.5 e.u.), F: 0.3671

Water Solubility (g/m^3 or mg/L at 25°C):
340 (quoted, Verschueren 1983)

Vapor Pressure (Pa at 25°C):
1.69×10^{-12} (extrapolated-Antoine eqn., Stephenson & Malanowski 1987)

Henry's Law Constant (Pa m^3/mol):

Octanol/Water Partition Coefficient, log K_{OW}:
8.23 (HPLC, D'Amboise & Hanai 1982)
8.23 (quoted and recommended, Sangster 1989, 1993)

Bioconcentration Factor, log BCF: no data

Sorption Partition Coefficient, log K_{OC}: no data

Half-Lives in the Environment/Environmental Fate Rate Constants or Half-Lives: no data

Common Name: Oleic acid
Synonym: *cis*-9-octadecenoic acid, (Z)-9-octadecenoic acid
Chemical Name: oleic acid
CAS Registry No: 112-80-1
Molecular Formula: $CH_3(CH_2)_7CH=CH(CH_2)_6CH_2COOH$
Molecular Weight: 282.47
Melting Point (°C):
 13.38 (Riddick et al. 1986)
Boiling Point (°C):
 360.0 (dec., Riddick et al. 1986)
Density (g/cm^3 at 20°C):
 0.8870 (25°C, Riddick et al. 1986)
Molar Volume (cm^3/mol):
 416.2 (calculated-LeBas method)
 318.5 (25°C, calculated-density)
Molecular Volume (Å^3):
Total Surface Area, TSA (Å^2):
Dissociation Constant, pK_a:
 5.02 (Riddick et al. 1986)
 4.50 (Sangster 1989)
Heat of Fusion ΔH_{fus}, kcal/mol:
Entropy of Fusion ΔS_{fus}, cal/mol·K (e.u.):
Fugacity Ratio at 25°C (assuming ΔS_{fus} = 13.5 e.u.), F: 1.0

Water Solubility (g/m^3 or mg/L at 25°C):
 insoluble (McBain & Richards 1946)
 insoluble (Riddick et al. 1986)

Vapor Pressure (Pa at 25°C):
 0.00144 (extrapolated-Antoine eqn., Weast 1972-73)
 0.00113 (extrapolated-Antoine eqn., Stephenson & Malanowski 1987)

Henry's Law Constant (Pa·m^3/mol):

Octanol/Water Partition Coefficient, log K_{OW}:
 7.64 (RPHPLC-k', D'Amboise & Hanai 1982; quoted, Sangster 1989)
 7.64 (recommended, Sangster 1989)

Bioconcentration Factor, log BCF: no data

Sorption Partition Coefficient, log K_{OC}: no data

Half-Lives in the Environment/Environmental Fate Rate Constants or Half-Lives: no data

Common Name: Acrylic acid
Synonym: acroleic acid, ethylenecarboxylic acid, 2-propenoic acid, propenoic acid
Chemical Name: acrylic acid, 2-propenoic acid, propenoic acid
CAS Registry No: 79-10-7
Molecular Formula: $CH_2{=}CHCOOH$
Molecular Weight: 72.06
Melting Point (°C):
14.0 (Stull 1947)
13.0 (Weast 1982-83; Dean 1985)
12-14 (Verschueren 1983)
13.5 (Riddick et al. 1986; Stephenson & Malanowski 1987)
Boiling Point (°C):
141.0 (Stull 1947; Verschueren 1983; Stephenson & Malanowski 1987)
141.6 (Weast 1982-83)
141-142 (Dean 1985)
141.2 (Riddick et al. 1986)
139.0 (Banerjee et al. 1990)
Density (g/cm^3 at 20°C):
1.0511 (Weast 1982-83; Dean 1985; Riddick et al. 1986)
Molar Volume (cm^3/mol):
83.2 (calculated-LeBas method)
Molecular Volume ($Å^3$):
Total Surface Area, TSA ($Å^2$):
Dissociation Constant, pK:
-4.41 (Perrin 1972)
4.25 (pK_a, Weast 1982-83)
4.26 (pK_a, Dean 1985)
4.255 (pK_a, Riddick et al. 1986)
Heat of Fusion ΔH_{fus}, kcal/mol:
2.66 (Riddick et al. 1986)
Entropy of Fusion ΔS_{fus}, cal/mol·K (e.u.):
Fugacity Ratio at 25°C (assuming ΔS_{fus} = 13.5 e.u.), F: 1.0

Water Solubility (g/m^3 or mg/L at 25°C):
miscible (Dean 1985)
infinite (Riddick et al. 1986)
miscible (Yaws et al. 1990)

Vapor Pressure (Pa at 25°C):
570.8 (interpolated-regression of tabulated data, Stull 1947)
666.5 (27.3°C, Stull 1947)
506.5 (Hoy 1970)
426.6 (20°C, quoted, Verschueren 1983)
533 (Howard et al. 1986; quoted, Banerjee et al. 1990)
1030 (20°C, Riddick et al. 1986)
581.7 (Extrapolated-Antoine eqn., Stephenson & Malanowski 1987)

533, 12530 (measured, calculated-solvatochromic p., Banerjee et al. 1990)

Henry's Law Constant (Pa·m^3/mol):
0.042 (computed, Yaws et al. 1991)

Octanol/Water Partition Coefficient, log K_{OW}:
0.43 (Dermer et al. 1941; quoted, Leo et al. 1971)
0.31, 0.43 (calculated, Verschueren 1983)

Bioconcentration Factor, log BCF:

Sorption Partition Coefficient, log K_{OC}:

Half-Lives in the Environment:
Air: photooxidation half-life of 2.5-23.8 hours, based on an estimated rate constant for the vapor-phase reaction with hydroxyl radicals and ozone in air (Atkinson & Carter 1984; Atkinson 1987; selected, Howard et al. 1991); atmospheric transformation lifetime was estimated to be < 1 day (Kelly et al. 1994).
Surface water: 24-168 hours, based on estimated unacclimated aqueous aerobic biodegradation half-life (Howard et al. 1991).
Groundwater: 48-4320 hours, based on estimated unacclimated aqueous aerobic and anaerobic biodegradation half-lives (Howard et al. 1991).
Sediment:
Soil: 24-168 hours, based on estimated unacclimated aqueous aerobic biodegradation half-life (Howard et al. 1991).
Biota:

Environmental Fate Rate Constants or Half-Lives:
Volatilization:
Photolysis:
Hydrolysis: no hydrolyzable groups (Howard et al. 1991).
Oxidation: photooxidation half-life of 2.5-23.8 hours in air, based on an estimated rate constant for the vapor-phase reaction with hydroxyl radicals and ozone in air (Atkinson & Carter 1984; Atkinson 1987; selected, Howard et al. 1991).
Biodegradation: aqueous aerobic half-life of 24-168 hours, based on unacclimated aqueous screening test data (Dore et al. 1975; Sasaki 1978; selected, Howard et al. 1991); aqueous anaerobic half-life of 672-4320 hours, based on unacclimated anaerobic reactor test data (Chou et al. 1979; selected, Howard et al. 1991).
Biotransformation:
Bioconcentration, Uptake (k_1) and Elimination (k_2) Rate Constants:

Common Name: Chloroacetic acid
Synonym:
Chemical Name: chloroacetic acid (α or β)
CAS Registry No: 79-11-8
Molecular Formula: $ClCH_2COOH$
Molecular Weight: 94.50
Melting Point (°C):
61.2 (α, Stull 1947; Yalkowsky & Valvani 1980)
56.0 (β, Yalkowsky & Valvani 1980)
56-63 (Weast 1982-83; Dean 1985)
Boiling Point (°C):
189.5 (Stull 1947)
188 (Weast 1982-83)
188.5 (Boublik et al. 1984)
189 (Dean 1985)
Density (g/cm^3 at 20°C):
1.4043 (40°C, Weast 1982-83)
Molar Volume (cm^3/mol): 89.3 (LeBas method)
Molecular Volume (Å^3):
Total Surface Area, TSA (Å^2):
Dissociation Constant, pK_a:
2.85 (Weast 1982-83)
Heat of Fusion ΔH_{fus}, kcal/mol:
Entropy of Fusion ΔS_{fus}, cal/mol·K (e.u.):
13.5 (estimated, Yalkowsky & Valvani 1980)
8.78 (observed for α, Yalkowsky & Valvani 1980)
10.09 (observed for β, Yalkowsky & Valvani 1980)
Fugacity Ratio at 25°C (assuming ΔS_{fus} = 13.5 e.u.), F: 0.461

Water Solubility (g/m^3 or mg/L at 25°C):
89170 (α, calculated-ΔS_{fus} and M.P., Yalkowsky & Valvani 1980)
120835 (α, calculated-M.P., Yalkowsky & Valvani 1980)
88960 (β, calculated-ΔS_{fus} and M.P., Yalkowsky & Valvani 1980)
107200 (β, calculated-M.P., Yalkowsky & Valvani 1980)

Vapor Pressure (Pa at 25°C):
39.70 (extrapolated-regression of tabulated data, Stull 1947)
15.71 (extrapolated average-Antoine eqn., Boublik et al. 1984)
18.52 (extrapolated-Antoine eqn., Dean 1985)

Henry's Law Constant (Pa·m^3/mol):

Octanol/Water Partition Coefficient, log K_{OW}:
0.22 (shake flask, Hansch & Leo 1987; quoted, Sangster 1993)
0.22 (recommended, Sangster 1993)

Bioconcentration Factor, log BCF:

Sorption Partition Coefficient, log K_{OC}:

Half-Lives in the Environment:

Air: photooxidation half-life of 230-2300 hours, based on estimated rate constant for the reaction with hydroxyl radicals in air (Atkinson 1987; quoted, Howard et al. 1991); atmospheric transformation lifetime was estimated to be 5 days (Kelly et al. 1994).

Surface water: aqueous photolysis half-life of 1900-19000 hours, based on experimental photolysis data utilizing an artificial light source (Draper & Crosby 1983; quoted, Howard et al. 1991); 24-168 hours, based on estimated aqueous aerobic biodegradation half-life (Howard et al. 1991).

Groundwater: 48-336 hours, based on estimated aqueous aerobic biodegradation half-life (Howard et al. 1991).

Sediment:

Soil: 24-168 hours, based on estimated aqueous aerobic biodegradation half-life (Howard et al. 1991).

Biota:

Environmental Fate Rate Constants or Half-Lives:

Volatilization:

Photolysis: aqueous photolysis half-life of 1900-19000 hours, based on experimental photolysis data utilizing an artificial light source (Draper & Crosby 1983; quoted, Howard et al. 1991); atmospheric photolysis half-life of 1900-19000 hours, based on estimated aqueous photolysis half-life (Howard et al. 1991); photocatalyzed mineralization by the presence of TiO_2 with the rate of 5.5 ppm/minute per gram of catalyst (Ollis 1985).

Hydrolysis: first-order hydrolysis half-life of 23000 hours, based on losses in dark control tests during photolysis experiments (Draper & Crosby 1983; quoted, Howard et al. 1991).

Oxidation: photooxidation half-life of 230-2300 hours, based on estimated rate constant for the reaction with hydroxyl radicals in air (Atkinson 1987; quoted, Howard et al. 1991).

Biodegradation: aqueous aerobic half-life of 24-168 hours, based on river die-away tests using radio-labelled material (Boethling & Alexander 1979; quoted, Howard et al. 1991); aqueous anaerobic half-life of 96-672 hours, based on estimated aqueous aerobic biodegradation half-life (Howard et al. 1991).

Biotransformation:

Bioconcentration, Uptake (k_1) and Elimination (k_2) Rate Constants:

Common Name: Dichloroacetic acid
Synonym: dichloroethanoic acid
Chemical Name: dichloroacetic acid
CAS Registry No: 79-43-6
Molecular Formula: $Cl_2CHCOOH$
Molecular Weight: 128.94
Melting Point (°C):
9.7 (Stull 1947; Dean 1985)
13.5 (Weast 1982-83)
5-6 (Verschueren 1983)
Boiling Point (°C):
194.4 (Stull 1947)
194 (Weast 1982-83; Verschueren 1983)
193-194 (Dean 1985)
Density (g/cm^3 at 20°C):
1.5634 (Weast 1982-83)
1.5630 (Verschueren 1983; Dean 1985)
Molar Volume (cm^3/mol):
110.2 (calculated-LeBas method)
Molecular Volume ($Å^3$):
Total Surface Area, TSA ($Å^2$):
Dissociation Constant, pK_a:
1.26 (Dean 1985)
Heat of Fusion ΔH_{fus}, kcal/mol:
Entropy of Fusion ΔS_{fus}, cal/mol·K (e.u.):
Fugacity Ratio at 25°C (assuming ΔS_{fus} = 13.5 e.u.), F: 1.0

Water Solubility (g/m^3 or mg/L at 25°C):
86300 (Verschueren 1983)
miscible (Dean 1985)

Vapor Pressure (Pa at 25°C):
37.94 (extrapolated-regression of tabulated data, Stull 1947)
133.3 (44°C, Stull 1947; quoted, Verschueren 1983)

Henry's Law Constant ($Pa{\cdot}m^3/mol$):

Octanol/Water Partition Coefficient, log K_{OW}:
-0.34, 1.39 (quoted, Verschueren 1983)
0.92 (shake flask, Hansch & Leo 1987)
0.92 (quoted and recommended, Sangster 1993)

Bioconcentration Factor, log BCF:

Sorption Partition Coefficient, log K_{OC}:

Half-Lives in the Environment:

Air:
Surface water:
Groundwater:
Sediment:
Soil:
Biota:

Environmental Fate Rate Constants or Half-Lives:

Volatilization:
Photolysis: photocatalyzed mineralization by the presence of TiO_2 with the rate of 8.5 ppm/minute per gram catalyst (Ollis 1985).
Hydrolysis:
Oxidation:
Biodegradation:
Biotransformation:
Bioconcentration, Uptake (k_1) and Elimination (k_2) Rate Constants:

Common Name: Trichloroacetic acid
Synonym: TCA
Chemical Name: trichloroacetic acid (α or β)
CAS Registry No: 76-03-9
Molecular Formula: Cl_3CCOOH
Molecular Weight: 163.39
Melting Point (°C):
  57.0 (Stull 1947; Verschueren 1983)
  57.5 (Yalkowsky & Valvani 1980)
  50-58 (Weast 1982-83)
  57-58 (Dean 1985)
  51.0 (Banerjee et al. 1990)
Boiling Point (°C):
  195.6 (Stull 1947)
  197.5 (Weast 1982-83; Verschueren 1983)
  197.6 (Boublik et al. 1984)
  196-197 (Dean 1985)
  126.0 (Banerjee et al. 1990)
Density (g/cm^3 at 20°C):
  1.620 (25°C, Weast 1982-83)
Molar Volume (cm^3/mol):
  131.1 (calculated-LeBas method)
Molecular Volume ($Å^3$):
Total Surface Area, TSA ($Å^2$):
Dissociation Constant, pK_a:
  0.70 (Weast 1982-83)
  0.52 (Dean 1985)
Heat of Fusion ΔH_{fus}, kcal/mol:
Entropy of Fusion ΔS_{fus}, cal/mol K (e.u.):
  4.25 (observed, Yalkowsky & Valvani 1980)
  13.5 (estimated, Yalkowsky & Valvani 1980)
Fugacity Ratio at 25°C (assuming ΔS_{fus} = 13.5 e.u.), F: 0.4825

Water Solubility (g/m^3 or mg/L at 25°C):
  114550 (calculated-ΔS_{fus} and M.P., Yalkowsky & Valvani 1980)
  191860 (calculated-M.P., Yalkowsky & Valvani 1980)
  13000 (quoted, Verschueren 1983)
  1200000 (Dean 1985)

Vapor Pressure (Pa at 25°C):
  21.88 (extrapolated-regression of tabulated data, Stull 1947)
  10.63 (extrapolated-Antoine eqn., Boublik et al. 1984)
  66.14 (extrapolated-Antoine eqn., Dean 1985)
  133.3 (Howard et al. 1986; quoted, Banerjee et al. 1990)

133.3, 491 (measured, calculated-solvatochromic p., Banerjee et al. 1990)

Henry's Law Constant ($Pa \cdot m^3/mol$):

Octanol/Water Partition Coefficient, log K_{OW}:

0.10, 1.96 (calculated, Verschueren 1983)

1.33 (shake flask, Hansch & Leo 1987)

1.33 (quoted and recommended, Sangster 1993)

Bioconcentration Factor, log BCF:

Sorption Partition Coefficient, log K_{OC}:

Half-Lives in the Environment:

Air:

Surface water:

Groundwater:

Sediment:

Soil:

Biota:

Environmental Fate Rate Constants or Half-Lives:

Volatilization:

Photolysis: no photocatalyzed mineralization by the presence of TiO_2 as compared to both dichloroacetic acid and trichloroacetic acid (Ollis 1985).

Hydrolysis:

Oxidation:

Biodegradation:

Biotransformation:

Bioconcentration, Uptake (k_1) and Elimination (k_2) Rate Constants:

Common Name: Benzoic acid
Synonym:
Chemical Name: benzoic acid
CAS Registry No: 65-85-0
Molecular Formula: C_6H_5COOH
Molecular Weight: 122.13
Melting Point (°C):

121.7 (Stull 1947; Verschueren 1983)
122.4 (Tsonopoulos & Prausnitz 1971; Yalkowsky & Valvani 1980; Dean 1985; Howard 1989)
122.1 (Weast 1982-83)
122.0 (Yalkowsky et al. 1983; Lee et al. 1993)

Boiling Point (°C):

249.2 (Stull 1947; Howard 1989)
249 (Weast 1982-83; Verschueren 1983)

Density (g/cm^3 at 20°C):

1.2659 (15°C, Weast 1982-83)
1.27 (Verschueren 1983)
1.080 (Dean 1985)

Molar Volume (cm^3/mol):

112.4 (calculated-LeBas method)
0.650 (V_I/100, Kamlet et al. 1988)

Molecular Volume (Å^3):
Total Surface Area, TSA (Å^2):

255.6 (Iwase et al. 1985)

Dissociation Constant, pK_a:

4.20 (McDaniell & Brown 1958; quoted, Pearce & Simkins 1968)
4.20, 4.33 ± 0.02 (quoted, HPLC, Unger et al. 1978)
4.18, 4.38 ± 0.03 (quoted, HPLC, Unger et al. 1978)
4.19 (Weast 1982-83, Klein et al. 1988)
4.17, 4.08 (quoted, shake flask-TN, Clarke 1984)
4.20, 4.05±0.01 (quoted, equilibrium titration, Clarke & Cahoon 1987)
4.204 (Dean 1985; Lee et al. 1993)
4.18, 4.14 (quoted, HPLC, Miyake et al. 1987)
4.205 (Howard 1989)
4.200 (Sangster 1989, Hersey et al. 1989))

Heat of Fusion ΔH_{fus}, kcal/mol:

4.30 (Tsonopoulos & Prausnitz 1971)

Entropy of Fusion ΔS_{fus}, cal/mol·K (e.u.):

10.90 (Tsonopoulos & Prausnitz 1971)
10.47 (Yalkowsky & Valvani 1980)

Fugacity Ratio at 25°C (assuming ΔS_{fus} = 13.5 e.u.), F: 0.109

Water Solubility (g/m^3 or mg/L at 25°C):

2700 (18°C, Hodgman 1952; quoted, Chiou et al. 1977; Freed et al. 1977)
3404 (selected, Tsonopoulos & Prausnitz 1971)

4200 (shake flask-liquid scintillation counting, Lu & Metcalf 1975)
3442 (quoted, Yalkowsky & Morozowich 1980)
3140 (calculated-K_{OW}, Yalkowsky & Morozowich 1980)
3442, 520 (quoted, calculated, Amidon & Williams 1981)
3604 (shake flask-UV, Yalkowsky et al. 1983)
1114 (calculated-K_{OW} and M.P., Yalkowsky et al. 1983)
2700 (18°C, Mackay et al. 1980; quoted, Howard 1989)
2700 (20-25°C, Urano et al. 1982)
2900 (Verschueren 1983; quoted, Lee et al. 1993)
2700 (18°C, quoted, Verschueren 1983)
3400 (Windholz 1983; quoted, Chessells et al. 1992)
2900 (quoted, Dean 1985)
3162 (quoted, Isnard & Lambert 1989)
3390 (selected, Yaws et al. 19900
2700 (quoted, Müller & Klein 1992)

Vapor Pressure (Pa at 25°C):
0.840 (extrapolated-regression of tabulated data, Stull 1947)
0.153 (quoted lit. average, Malaspina et al. 1973)
0.160 (effusion/DTA, fitted to Antoine eqn., Malaspina et al. 1973)
0.100 (selected, Cox 1974)
0.110 (effusion method, fitted to Antoine eqn., Colomina et al. 1982)
0.0907 (20°C, evaporation method, Gückel et al. 1982)
0.105 (manometer, extrapolated, de Kruif & Block 1982)
0.096 (quoted from literature, de Kruif & Block 1982)
0.9505 (extrapolated-Antoine eqn., Dean 1985)
0.0998 (extrapolated, P_S, Antoine eqn., Stephenson & Malanowski 1987)
0.772 (extrapolated, P_L, Antoine eqn., Stephenson & Malanowski 1987)
0.600 (20°C, OECD 1981; quoted, Howard 1989)

Henry's Law Constant (Pa·m^3/mol):
0.00709 (quoted, Howard 1989)
0.00575 (calculated-P/C, Meylan & Howard 1991)
0.0110 (calculated-bond contribution, Meylan & Howard 1991)
4.15×10^{-3} (computed, Yaws et al. 1991)

Octanol/Water Partition Coefficient, log K_{OW}:
1.87 (shake flask-AS, Fujita et al. 1964; quoted, Leo et al. 1971; Chiou et al. 1977; Freed et al. 1977; Chiou et al. 1977; Unger et al. 1978; Verschueren 1983; Yalkowsky et al. 1983; Iwase et al. 1985; Isnard & Lambert 1989; Sangster 1989)
1.68 (shake flask-AS, Halmekoski & Hannikainen 1964; quoted, Sangster 1989)
1.85 ± 0.01 (shake flask-AS, Iwasa et al. 1965)
2.03 (shake flask-LSC, Lu & Metcalf 1975)

1.78 ± 0.01 (HPLC-k', Unger et al. 1978)
1.77 ± 0.01 (HPLC-k', Unger et al. 1978)
2.0 (quoted, Amidon & Williams 1981)
1.79 (RPHPLC-k', D'Amboise & Hanai 1982)
1.86, 1.87 (Rekker 1977; quoted, Hanai & Hubert 1982)
1.79 (calculated-f const. as per Rekker 1977, Hanai & Hubert 1982)
1.94 (RP-HPLC, Hanai & Hubert 1982)
1.87 (quoted from Hansch & Leo 1979, Miyake & Terada 1982; Haky & Young 1984; Miyake et al. 1987)
1.95 (calculated-HPLC-k', Miyake & Terada 1982)
1.87 (Verschueren 1983; quoted, Lee et al. 1993)
1.47 (estimated from solubility ratio, Yalkowsky et al. 1983; quoted, Sangster 1989)
1.87, 1.87 (quoted, microelectrometric titration, Clarke 1984; quoted, Sangster 1989)
2.03 (selected best lit. value, Garst & Wilson 1984)
2.18 ± 0.03 (exptl.-ALPM, Garst & Wilson 1984)
1.44, 1.87 (correlated-HPLC-k', Haky & Young 1984)
1.87 (Hansch & Leo 1985; quoted, Howard 1989; Chessells et al. 1992)
1.84 (calculated-TSA, Iwase et al. 1985)
1.88 (shake flask-AS at pH 0.5, Nishimura et al. 1985; quoted, Sangster 1989)
1.95 (quoted, Yoshida et al. 1986)
1.97 (microelectrometric titration, Clarke & Cahoon 1987; quoted, Sangster 1989)
1.94 (shake flask-RC at pH 0.5, Laznicek et al. 1987; quoted, Sangster 1989)
1.93 (calculated-HPLC-k', Miyake et al. 1987)
1.87, 1.88 (quoted, CPC, Berthod et al. 1988)
1.87 (quoted, Isnard & Lambert 1989)
1.95, 1.93 (quoted, calculated-V_I, solvatochromic parameters, Kamlet et al. 1988)
1.90 (selected, Klein et al. 1988)
1.95 (quoted from Hansch & Leo 1979, Bodor et al. 1989)
1.87, 1.885, 1.57 (quoted, calculated-CLOGP, M.O. indices, Bodor et al. 1989)
1.87, 1.89, 1.85±0.04 (quoted, calculated, "Filter Chamber"-AS, Hersey et al. 1989)
1.87 (recommended, Sangster 1989)
1.87, 1.88 (quoted, back flashing-CPC, Menges et al. 1990)

Bioconcentration Factor, log BCF:
1.32, 2.00, 2.14 (fish, algae, mosquito, Lu & Metcalf 1975; quoted, Howard 1989)
3.26, 3.45 (daphania, snail, Lu & Metcalf 1975; quoted, Howard 1989)
0.48 (alga *chlorella fusca*, wet wt. basis, Geyer et al. 1984)
1.43 (alga *chlorella fusca*, calculated-K_{OW}, Geyer et al. 1984)
<1.0 (golden ide, Freitag et al. 1985; quoted, Howard 1989)
<1.0 (algae, Freitag et al. 1985)
3.11 (activated sludge, Freitag et al. 1985)
1.15 (trout muscle, calculated-K_{OW}, Branson 1978; quoted, Howard 1989)
1.32 (quoted, Isnard & Lambert 1989)

Sorption Partition Coefficient, log K_{OC}:

1.26 (Alfisol soil, von Oepen et al. 1991)
1.86 (Podzol soil, von Oepen et al. 1991)
0.602 (sediment, von Oepen et al. 1991)
1.50 (soil, quoted exptl., Meylan et al. 1992)
1.16 (soil, calculated- χ and fragment contribution, Meylan et al. 1992)

Half-Lives in the Environment:

Air: free vapor reacts with photochemically produced hydroxyl radicals with an estimated half-life of 2.0 days (Howard 1989).
Surface water: if released into water, should readily biodegrade with an estimated half-life of 0.2-3.6 days (Howard 1989).
Groundwater:
Sediment:
Soil:
Biota:

Environmental Fate Rate Constants or Half-Lives:

Volatilization: rate of evaporation at 20°C of 1.89×10^{-8} mol·cm^{-2}·h^{-1} determined by evaporation method (Gückel et al. 1982).
Photolysis:
Oxidation: photooxidation half-life of 2.0 days, based on estimated vapor-phase reaction with hydroxyl radicals in air (Howard 1989).
Hydrolysis:
Biodegradation: completely degraded by a soil microflora after 24 hours (Alexander & Lustigman 1966; quoted, Verschueren 1983); completely degraded for 16 mg per liter concn. within one day by soil and by wastewater (Haller 1978); if released into water, should readily biodegrade with an estimated half-life of 0.2-3.6 days (Howard 1989); average experimental rate constant of 0.11533 hour^{-1} compared to group method predicted rate constants of 0.0993 h^{-1} (nonlinear) and 0.0263 h^{-1} (linear) (Tabak & Govind 1993).
Biotransformation:
Bioconcentration, Uptake (k_1) and Elimination (k_2) Rate Constants:

Common Name: *o*-Toluic acid
Synonym: 2-methyl benzoic acid
Chemical Name: *o*-toluic acid, 2-methyl benzoic acid
CAS Registry No: 118-90-1
Molecular Formula: $CH_3C_6H_4COOH$
Molecular Weight: 136.16
Melting Point (°C):
103.9 (Tsonopoulos & Prausnitz 1971)
103.7 (Yalkowsky & Valvani 1980; Verschueren 1983)
107-108 (Weast 1982-83; Dean 1985)
Boiling Point (°C):
258-259 (at 751 mm Hg, Weast 1982-83)
259.2 (Verschueren 1983)
258-259 (Dean 1985)
Density (g/cm^3 at 20°C):
1.062 (115°C, Weast 1982-83; Verschueren 1983)
Molar Volume (cm^3/mol):
157.0 (calculated-LeBas method)
Molecular Volume (Å^3):
Total Surface Area, TSA (Å^2):
Dissociation Constant, pK_a:
3.91 (McDaniel & Brown 1958; quoted, Pearce & Simkins 1968)
3.91 (Weast 1982-83)
3.90 (Dean 1985)
Heat of Fusion ΔH_{fus}, kcal/mol:
4.82 (Tsonopoulos & Prausnitz 1971)
Entropy of Fusion ΔS_{fus}, cal/mol·K (e.u.):
12.8 (Tsonopoulos & Prausnitz 1971)
12.79 (observed, Yalkowsky & Valvani 1980)
13.5 (estimated, Yalkowsky & Valvani 1980)
Fugacity Ratio at 25°C (assuming ΔS_{fus} = 13.5 e.u.), F: 0.1662

Water Solubility (g/m^3 or mg/L at 25°C):
1185 (Führer 1924; quoted, Tsonopoulos & Prausnitz 1971)
414 (calculated-ΔS_{fus} and M.P., Yalkowsky & Valvani 1980)
463 (calculated-M.P., Yalkowsky & Valvani 1980)
1074 (shake flask-UV, Sugunan & Thomas 1993)

Vapor Pressure (Pa at 25°C):

Henry's Law Constant (Pa·m^3/mol):

Octanol/Water Partition Coefficient, log K_{OW}:

2.54, 2.83 (Smith & White 1929; quoted, Leo et al. 1971; Verschueren 1983)
2.18 (shake flask-AS, Tomida et al. 1978; quoted, Sangster 1993)
2.81 (Scherrer & Howard 1979; quoted, Sangster 1993)
2.46 (shake flask at pH 1, Hansch & Leo 1987; quoted, Sangster 1993)
2.27 (shake flask-AS at pH 2, Da et al. 1992; quoted, Sangster 1993)
2.46 (recommended, Sangster 1993)

Bioconcentration Factor, log BCF:

Sorption Partition Coefficient, log K_{OC}:

Half-Lives in the Environment:

Air:
Surface water:
Groundwater:
Sediment:
Soil:
Biota:

Environmental Fate Rate Constants or Half-Lives:

Volatilization:
Photolysis:
Oxidation:
Hydrolysis:
Biodegradation: decomposition by a soil microflora in 16 days (Alexander & Lustigman 1966; quoted, Verschueren 1983).
Biotransformation:
Bioconcentration, Uptake (k_1) and Elimination (k_2) Rate Constants:

Common Name: *m*-Toluic acid
Synonym: 3-methyl benzoic acid
Chemical Name: *m*-toluic acid, 3-methyl benzoic acid
CAS Registry No: 99-04-7
Molecular Formula: $CH_3C_6H_4COOH$
Molecular Weight: 136.16
Melting Point (°C):
110.9 (Tsonopoulos & Prausnitz 1971)
108.8 (Yalkowsky & Valvani 1980; Verschueren 1983)
111-113 (Weast 1982-83; Dean 1985)
Boiling Point (°C):
263 (Urano et al. 1982; Verschueren 1983; Dean 1985)
263 (sublime, Weast 1982-83)
Density (g/cm^3 at 20°C):
1.054 (112°C, Weast 1982-83; Verschueren 1983)
Molar Volume (cm^3/mol):
157.0 (calculated-LeBas method)
Molecular Volume (Å^3):
Total Surface Area, TSA (Å^2):
Dissociation Constant, pK_a:
4.27 (Weast 1982-83)
4.269 (Dean 1985)
4.22 (Sangster 1993)
Heat of Fusion ΔH_{fus}, kcal/mol:
3.76 (Tsonopoulos & Prausnitz 1971)
Entropy of Fusion ΔS_{fus}, cal/mol·K (e.u.):
9.79 (Tsonopoulos & Prausnitz 1971)
9.84 (observed, Yalkowsky & Valvani 1980)
13.5 (estimated, Yalkowsky & Valvani 1980)
Fugacity Ratio at 25°C (assuming ΔS_{fus} = 13.5 e.u.), F: 0.1425

Water Solubility (g/m^3 or mg/L at 25°C):
982 (Föhner 1924; quoted, Tsonopoulos & Prausnitz 1971)
305 (calculated-ΔS_{fus} and M.P., Yalkowsky & Valvani 1980)
521 (calculated-M.P., Yalkowsky & Valvani 1980)
1700 (20-25°C, shake flask-GC, Urano et al. 1982)
850 (15°C, quoted, Verschueren 1983)
900 (Dean 1985)
1246 (shake flask-UV, Sugunan & Thomas 1993)

Vapor Pressure (Pa at 25°C):

Henry's Law Constant (Pa·m^3/mol):

Octanol/Water Partition Coefficient, log K_{OW}:

- 2.37 (shake flask-AS, Fujita et al. 1964; quoted, Leo et al. 1971; Hansch & Leo 1979; Scherrer & Howard 1979; Sangster 1993)
- 2.37 (quoted from Hansch & Leo 1979, Miyake & Terada 1982; Terada et al. 1987)
- 2.43 (calculated-HPLC-k', Miyake & Terada 1982)
- 2.37 (quoted, Verschueren 1983)
- 2.41 (shake flask-AS, Miyake et al. 1987; quoted, Sangster 1993)
- 2.39 (CPC-RV, Terada et al. 1987; quoted, Sangster 1993)
- 2.44 (shake flask at pH 1, Wang et al. 1989; quoted, Sangster 1993)
- 2.47 (correlated-HPLC-RT, Wang et al. 1989; quoted, Sangster 1993)
- 2.38 (shake flask-AS at pH 2, Da et al. 1992; quoted, Sangster 1993)
- 2.37 (recommended, Sangster 1993)

Bioconcentration Factor, log BCF:

Sorption Partition Coefficient, log K_{OC}:

Half-Lives in the Environment:

Air:
Surface water:
Groundwater:
Sediment:
Soil:
Biota:

Environmental Fate Rate Constants or Half-Lives:

Volatilization:
Photolysis:
Oxidation:
Hydrolysis:
Biodegradation: decomposition by a soil microflora in two days (Alexander & Lustigman 1966; quoted, Verschueren 1983).
Biotransformation:
Bioconcentration, Uptake (k_1) and Elimination (k_2) Rate Constants:

Common Name: *p*-Toluic acid
Synonym: 4-methyl benzoic acid
Chemical Name: *p*-toluic acid, 4-methyl benzoic acid
CAS Registry No: 99-94-5
Molecular Formula: $CH_3C_6H_4COOH$
Molecular Weight: 136.16
Melting Point (°C):
179.9 (Tsonopoulos & Prausnitz 1971)
179.6 (Yalkowsky & Valvani 1980; Verschueren 1983)
182 (Weast 1982-83)
180-182 (Dean 1985)
Boiling Point (°C):
275 (sublime, Weast 1982-83; Verschueren 1983)
274-275 (Dean 1985)
Density (g/cm³ at 20°C):
Molar Volume (cm³/mol):
157.0 (calculated-LeBas method)
Molecular Volume (Å³):
Total Surface Area, TSA (Å²):
279.7 (Iwase et al. 1985)
Dissociation Constant, pK_a:
4.37, 4.30 ± 0.09 (quoted, HPLC, Unger et al. 1978)
4.37, 4.41 ± 0.01 (quoted, HPLC, Unger et al. 1978)
4.36 (Weast 1982-83)
4.37, 4.26 (quoted, shake flask-TN, Clarke 1984)
4.362 (Dean 1985)
4.39 (Sangster 1993)
Heat of Fusion ΔH_{fus}, kcal/mol:
5.43 (Tsonopoulos & Prausnitz 1971)
Entropy of Fusion ΔS_{fus}, cal/mol K (e.u.):
12.0 (Tsonopoulos & Prausnitz 1971)
12.0 (observed, Yalkowsky & Valvani 1980)
13.5 (estimated, Yalkowsky & Valvani 1980)
Fugacity Ratio at 25°C (assuming ΔS_{fus} = 13.5 e.u.), F: 0.0293

Water Solubility (g/m³ or mg/L at 25°C):
383 (selected, Tsonopoulos & Prausnitz 1971)
1737 (calculated-ΔS_{fus} and M.P., Yalkowsky & Valvani 1980)
2660 (calculated-M.P., Yalkowsky & Valvani 1980)
331 (shake flask-UV, Sugunan & Thomas 1993)

Vapor Pressure (Pa at 25°C):

Henry's Law Constant (Pa·m³/mol):

Octanol/Water Partition Coefficient, log K_{OW}:

2.27 (shake flask-AS, Fujita et al. 1964; quoted, Leo et al. 1971; Unger et al. 1978; Scherrer & Howard 1979; Verschueren 1983; Iwase et al. 1985; Sangster 1993)
2.22 ± 0.02 (HPLC, Unger et al. 1978)
2.26 ± 0.01 (HPLC, Unger et al. 1978)
2.36 (shake flask-AS at pH 2, Ezumi & Kubota 1980; quoted, Sangster 1993)
2.43 (calculated-HPLC-k', Miyake & Terada 1982)
2.27, 2.34 (quoted, electrometric titration, Clarke 1984; quoted, Sangster 1993)
2.67 (correlated-HPLC-RT, Garst 1984; quoted, Sangster 1993)
2.29 (calculated-TSA, Iwase et al. 1985)
2.66 (CPC, Terada et al. 1987; quoted, Sangster 1993)
2.38 (shake flask at pH 1, Wang et al. 1989; quoted, Sangster 1993)
2.41 (correlated-HPLC-RT, Wang et al. 1989; quoted, Sangster 1993)
2.35 (correlated-HPLC-RT, Jenke et al. 1990; quoted, Sangster 1993)
2.26 (shake flask-AS at pH 2, Da et al. 1992; quoted, Sangster 1993)
2.34 (recommended, Sangster 1993)

Bioconcentration Factor, log BCF:

Sorption Partition Coefficient, log K_{OC}:

2.15, 1.15, 1.30 (Podzol soil, Alfisol soil, sediment, von Oepen et al. 1991)
1.77 (soil, quoted exptl., Meylan et al. 1992)
1.37 (soil, calculated- χ and fragment contribution, Meylan et al. 1992)

Half-Lives in the Environment:

Air:
Surface water:
Groundwater:
Sediment:
Soil:
Biota:

Environmental Fate Rate Constants or Half-Lives:

Volatilization:
Photolysis:
Oxidation:
Hydrolysis:
Biodegradation: decomposition by a soil microflora in 8 days (Alexander & Lustigman 1966; quoted, Verschueren 1983).
Biotransformation:
Bioconcentration, Uptake (k_1) and Elimination (k_2) Rate Constants:

Common Name: Phenylacetic acid
Synonym: phenylethanoic acid, α-toluic acid
Chemical Name: phenylacetic acid
CAS Registry No: 103-82-2
Molecular Formula: $C_6H_5CH_2COOH$
Molecular Weight: 136.15
Melting Point (°C):
76.5 (Stull 1947)
77.0 (Tsonopoulos & Prausnitz 1971; Weast 1982-83; Dean 1985)
76.7 (Yalkowsky & Valvani 1980)
77-78.5 (Verschueren 1983; Stephenson & Malanowski 1987)
Boiling Point (°C):
265.5 (Stull 1947; Weast 1982-83; Dean 1985; Stephenson & Malanowski 1987)
265.0 (Verschueren 1983)
Density (g/cm^3 at 20°C):
1.081 (Verschueren 1983)
Molar Volume (cm^3/mol):
157.0 (calculated-LeBas method)
Molecular Volume (Å^3):
Total Surface Area, TSA (Å^2):
Dissociation Constant, pK_a:
4.28 (18°C, Weast 1982-83)
4.31 (Sangster 1989)
Heat of Fusion ΔH_{fus}, kcal/mol:
4.09 (Tsonopoulos & Prausnitz 1971)
Entropy of Fusion ΔS_{fus}, cal/mol·K (e.u.):
11.7 (Tsonopoulos & Prausnitz 1971)
9.90 (observed, Yalkowsky & Valvani 1980)
13.5 (estimated, Yalkowsky & Valvani 1980)
Fugacity Ratio at 25°C (assuming ΔS_{fus} = 13.5 e.u.), F: 0.306

Water Solubility (g/m^3 or mg/L at 25°C):
16600 (20°C, Hodgman 1952; quoted, Freed et al. 1977)
17790 (selected, Tsonopoulos & Prausnitz 1971)
179796 (calculated-ΔS_{fus} and M.P., Yalkowsky & Valvani 1980)
248760 (calculated-M.P., Yalkowsky & Valvani 1980)
17300, 11588 (quoted exptl., calculated, Amidon & Williams 1981)
16600 (20°C, quoted, Verschueren 1983)

Vapor Pressure (Pa at 25°C):
0.835 (extrapolated-regression of tabulated data, Stull 1947)
0.827 (extrapolated-Antoine eqn., Stephenson & Malanowski 1987)

Henry's Law Constant (Pa·m^3/mol):
0.0056 (calculated-P/C, this work)

Octanol/Water Partition Coefficient, log K_{OW}:

1.41 (shake flask-AS, Fujita et al. 1964; quoted, Leo et al. 1971; Freed et al. 1977; Sangster 1989)
1.41 ± 0.01 (shake flask-AS, Iwasa et al. 1965)
1.51 (shake flask-AS, Yaguzhinskii et al. 1973; quoted, Sangster 1989)
1.46 (quoted, Amidon & Williams 1981)
1.41, 1.42, 1.45, 1.96 (quoted from Rekker 1977, Hanai & Hubert 1982)
1.94 (RP-LC, Hanai & Hubert 1982)
1.75 (calculated-f const. as per Rekker 1977, Hanai & Hubert 1982)
1.45 (shake flask-AS at pH 3.5, Kuchar et al. 1982; quoted, Sangster 1989)
1.41 (quoted, Verschueren 1983)
1.51 (selected best lit. value, Garst 1984)
1.95 ± 0.04 (exptl.-ALPM, Garst 1984)
1.46, 1.41, 1.414, 2.05 (quoted, quoted CLOGP, calculated-CLOGP, calculated-M.O., Bodor et al. 1989)
1.41 (recommended, Sangster 1989)

Bioconcentration Factor, log BCF:

Sorption Partition Coefficient, log K_{OC}:

1.49, 1.42, 1.45 (Podzol soil, Alfisol soil, sediment, von Oepen et al. 1991)
1.45 (soil, quoted exptl., Meylan et al. 1992)
1.42 (soil, calculated- χ and fragment contribution, Meylan et al. 1992)

Half-Lives in the Environment:

Air:
Surface water:
Groundwater:
Sediment:
Soil:
Biota:

Environmental Fate Rate Constants or Half-Lives:

Volatilization:
Photolysis:
Oxidation:
Hydrolysis:
Biodegradation:
Biotransformation:
Bioconcentration, Uptake (k_1) and Elimination (k_2) Rate Constants:

Common Name: Phthalic acid
Synonym: 1,2-benzene dicarboxylic acid, *o*-phthalic acid
Chemical Name: 1,2-benzene dicarboxylic acid, *o*-phthalic acid
CAS Registry No: 88-99-3
Molecular Formula: C_6H_4-1,2-$(COOH)_2$
Molecular Weight: 166.13
Melting Point (°C):

210-211 (dec., Weast 1982-83)
206-208 (dec., Verschueren 1983; Dean 1985)

Boiling Point (°C):

>191 (dec., Verschueren 1983)

Density (g/cm³ at 20°C):

1.593 (Weast 1982-83; Verschueren 1983; Dean 1985)

Molar Volume (cm³/mol):

199.2 (calculated-LeBas method)

Molecular Volume (Å³):
Total Surface Area, TSA (Å²):
Dissociation Constant, pK_a:

2.890 (pK_1, Weast 1982-83)
5.510 (pK_2, Weast 1982-83)
2.950 (pK_1, Dean 1985)
5.408 (pK_2, Dean 1985)

Heat of Fusion ΔH_{fus}, kcal/mol:
Entropy of Fusion ΔS_{fus}, cal/mol·K (e.u.):
Fugacity Ratio at 25°C (assuming ΔS_{fus} = 13.5 e.u.), F: 0.0148

Water Solubility (g/m³ or mg/L at 25°C):

6768, 576 (quoted, calculated-K_{OW}, Yalkowsky & Morozowich 1980)
5400 (14°C, Verschueren 1983)
6300 (Dean 1985)
7024 (Apelblat & Manzurola 1989)

Vapor Pressure (Pa at 25°C):

Henry's Law Constant (Pa m³/mol):

Octanol/Water Partition Coefficient, log K_{OW}:

0.79 (shake flask-AS, Tomida et al. 1978; quoted, Sangster 1993)
0.15 (electrometric titration, Freese et al. 1979; quoted, Sangster 1993)
0.10, 0.41 (calculated, Verschueren 1983)
0.73 (shake flask at pH 1, Hansch & Leo 1987; quoted, Sangster 1993)
0.71 (30°C, shake flask-AS at pH 1, Patrunkey & Pflegel 1992; quoted, Sangster 1993)

Bioconcentration Factor, log BCF:

Sorption Partition Coefficient, log K_{OC}:

0.301, 1.49, 0.301 (sediment, Podzol soil, Alfidol soil, von Oepen et al. 1991)
1.07 (soil, quoted exptl., Meylan et al. 1992)
1.87 (soil, calculated- χ and fragment contribution, Meylan et al. 1992)

Half-Lives in the Environment:

Air:
Surface water:
Groundwater:
Sediment:
Soil:
Biota:

Environmental Fate Rate Constants or Half-Lives:

Volatilization:
Photolysis:
Hydrolysis:
Oxidation:
Biodegradation: decomposition by a soil microflora in two days (Alexander & Lustigman 1966; quoted, Verschueren 1983); average rate of biodegradation 78.4 mg COD $g^{-1} \cdot h^{-1}$ based on measurements of COD decrease using activated sludge inoculum with 20 days of adaptation to the substrate (Pitter 1976; quoted, Scow 1982).
Biotransformation:
Bioconcentration, Uptake (k_1) and Elimination (k_2) Rate Constants:

Common Name: 2-Chlorobenzoic acid
Synonym: *o*-chlorobenzoic acid
Chemical Name: 2-chlorobenzoic acid, *o*-chlorobenzoic acid
CAS Registry No: 118-91-2
Molecular Formula: ClC_6H_4COOH
Molecular Weight: 156.57
Melting Point (°C):
140.9 (Tsonopoulos & Prausnitz 1971)
140.2 (Yalkowsky & Valvani 1980)
142 (Weast 1982-83; Verschueren 1983; Dean 1985)
Boiling Point (°C): sublimes (Verschueren 1983)
Density (g/cm³ at 20°C):
1.5440 (Weast 1982-83, Verschueren 1983)
1.5440 (25°C, Dean 1985)
Molar Volume (cm³/mol):
155.7 (calculated-LeBas method)
101.4 (calculated-density)
Molecular Volume (Å³):
Total Surface Area, TSA (Å²):
Dissociation Constant, pK_a:
2.92 (McDaniell & Brown 1958; quoted, Pearce & Simkins 1968)
2.92 (Weast 1982-83)
2.877 (Dean 1985)
3.85 (Sangster 1993)
Heat of Fusion ΔH_{fus}, kcal/mol:
6.15 (Tsonopoulos & Prausnitz 1971)
Entropy of Fusion ΔS_{fus}, cal/mol·K (e.u.):
14.9 (Tsonopoulos & Prausnitz 1971)
14.88 (observed, Yalkowsky & Valvani 1980)
13.50 (estimated, Yalkowsky & Valvani 1980)
Fugacity Ratio at 25°C (assuming ΔS_{fus} = 13.5 e.u.), F: 0.696

Water Solubility (g/m³ or mg/L at 25°C):
2130 (Osol & Kilpatrick 1933; quoted, Palit 1947)
2114 (selected, Tsonopoulos & Prausnitz 1971)
1572 (calculated-ΔS_{fus} and M.P., Yalkowsky & Valvani 1980)
1234 (calculated-M.P., Yalkowsky & Valvani 1980)
2100 (quoted, Verschueren 1983)
1100 (Dean 1985)

Vapor Pressure (Pa at 25°C):

Henry's Law Constant (Pa·m³/mol):

Octanol/Water Partition Coefficient, log K_{OW}:

1.98 (quoted unpublished result, Leo et al. 1971; quoted, Rekker 1977)
1.98 (quoted unpublished result, quoted, Rekker 1977)
2.63 (calculated-f const., Rekker 1977)
2.56 (calculated-π const., Rekker 1977)
2.66 (shake flask-AS, Tomida et al. 1978; quoted, Sangster 1993)
2.63 (calculated-chlorine substituents, Kaiser 1983)
1.98 (quoted, Verschueren 1983)
1.98 (shake flask, Hansch & Leo 1987; quoted, Sangster 1993)
1.99 (CPC, Berthod et al. 1988; quoted, Sangster 1993)
2.02 (shake flask-AS at pH 2, Da et al. 1992; quoted, Sangster 1993)
1.99 (recommended, Sangster 1993)

Bioconcentration Factor, log BCF:

Sorption Partition Coefficient, log K_{OC}:

Half-Lives in the Environment:

Air:
Surface water:
Groundwater:
Sediment:
Soil:
Biota:

Environmental Fate Rate Constants or Half-Lives:

Volatilization:
Photolysis:
Oxidation:
Hydrolysis:
Biodegradation: decomposition by s soil microflora: >64 days, biodegradation by waste water at pH 7.3 and 30°C: >25 days, and degradation by soil suspension at pH 7.3 and 30°C: 7-14 days (Alexander & Lustigman 1966; quoted, Verschueren 1983); complete degradation of 16 mg per liter by soil in 7-14 days (Haller 1978).
Biotransformation:
Bioconcentration, Uptake (k_1) and Elimination (k_2) Rate Constants:

Common Name: 3-Chlorobenzoic acid
Synonym: *m*-chlorobenzoic acid
Chemical Name: 3-chlorobenzoic acid, *m*-chlorobenzoic acid
CAS Registry No: 535-80-8
Molecular Formula: ClC_6H_4COOH
Molecular Weight: 156.57
Melting Point (°C):
158 (Weast 1982-83; Verschueren 1983)
157-158 (Dean 1985)
Boiling Point (°C):
sublime (Weast 1982-83; Verschueren 1983)
Density (g/cm³ at 20°C):
1.4960 (25°C, Weast 1982-83; Verschueren 1983; Dean 1985)
Molar Volume (cm³/mol):
155.7 (calculated-LeBas method)
104.7 (calculated-density)
Molecular Volume (Å³):
Total Surface Area, TSA (Å²):
Dissociation Constant, pK_a:
3.82 (Weast 1982-83)
3.83 (Dean 1985)
Heat of Fusion ΔH_{fus}, kcal/mol:
5.70 (Tsonopoulos & Prausnitz 1971)
Entropy of Fusion ΔS_{fus}, cal/mol·K (e.u.):
13.3 (Tsonopoulos & Prausnitz 1971)
Fugacity Ratio at 25°C (assuming ΔS_{fus} = 13.5 e.u.), F: 0.0484

Water Solubility (g/m³ or mg/L at 25°C):
385 (Osol & Kilpatrick 1933; quoted, Palit 1947)
398 (selected, Tsonopoulos & Prausnitz 1971)
400 (0°C, quoted, Verschueren 1983)
400 (quoted, Dean 1985)

Vapor Pressure (Pa at 25°C):

Henry's Law Constant (Pa·m³/mol):

Octanol/Water Partition Coefficient, log K_{OW}:
2.68 (shake flask-AS, Fujita et al. 1964; quoted, Leo et al. 1971; Rekker 1977; Hansch & Leo 1979; Veith et al. 1979; Sangster 1993)
2.65 (calculated-f const., Rekker 1977)
2.52 (calculated-π const., Rekker 1977)
0.89 (estimated-HPLC-RT, Veith et al. 1979; quoted, Sangster 1993)

2.68	(quoted from Hansch & Leo 1979, Miyake & Terada 1982; Terada et al. 1987)
2.57	(calculated-HPLC-k', Miyake & Terada 1982)
2.63	(calculated-chlorine substituents, Kaiser 1983)
2.68	(quoted, Veuschueren 1983)
2.62	(CPC-RV, Terada et al. 1987; quoted, Sangster 1993)
2.51	(shake flask at pH 1, Wang et al. 1989; quoted, Sangster 1993)
2.58	(correlated-HPLC-RT, Wang et al. 1989; quoted, Sangster 1993)
2.72	(shake flask-AS at pH 2, Da et al. 1992; quoted, Sangster 1993)
2.60	(recommended, Sangster 1993)

Bioconcentration Factor, log BCF:

Sorption Partition Coefficient, log K_{OC}:

Half-Lives in the Environment:

Air:

Surface water:

Groundwater:

Sediment:

Soil:

Biota:

Environmental Fate Rate Constants or Half-Lives:

Volatilization:

Photolysis: rate constant of direct aqueous photolysis of 0.013 ± 0.001 minute^{-1} with a half-life of 55 minutes (Stegeman et al. 1993).

Oxidation:

Hydrolysis:

Biodegradation: decomposed by soil microflora, 32 days (Alexander & Lustigman 1966; quoted, Verschueren 1983); complete degradation of 16 mg per liter by soil and by wastewater in 7-14 days (Haller 1978).

Biotransformation:

Bioconcentration, Uptake (k_1) and Elimination (k_2) Rate Constants:

Common Name: 4-Chlorobenzoic acid
Synonym: *p*-chlorobenzoic acid
Chemical Name: 4-chlorobenzoic acid, *p*-chlorobenzoic acid
CAS Registry No: 74-11-3
Molecular Formula: ClC_6H_4COOH
Molecular Weight: 156.57
Melting Point (°C):
239.7 (Yalkowsky & Valvani 1980)
243.0 (Weast 1982-83; Verschueren 1983)
241-243 (Dean 1985)
Boiling Point (°C):
sublimes (Verschueren 1983)
Density (g/cm³ at 20°C):
1.541 (24°C, Verschueren 1983)
Molar Volume (cm³/mol):
155.7 (calculated-LeBas method)
Molecular Volume (Å³):
Total Surface Area, TSA (Å²):
Dissociation Constant, pK_a:
3.98 (McDaniel & Brown 1958; quoted, Pearce & Simkins 1968)
3.98 (Weast 1982-83)
3.98, 3.85 (quoted, shake flask-TN, Clarke 1984)
3.986 (Dean 1985)
3.85 (Sangster 1993)
Heat of Fusion ΔH_{fus}, kcal/mol:
7.71 (Tsonopoulos & Prausnitz 1971)
Entropy of Fusion ΔS_{fus}, cal/mol·K (e.u.):
15.0 (Tsonopoulos & Prausnitz 1971)
15.03 (observed, Yalkowsky & Valvani 1980)
13.50 (estimated, Yalkowsky & Valvani 1980)
Fugacity Ratio at 25°C (assuming ΔS_{fus} = 13.5 e.u.), F: 0.0070

Water Solubility (g/m³ or mg/L at 25°C):
68.0 (Osol & Kilpatrick 1933; quoted, Palit 1947)
72.6 (selected, Tsonopoulos & Prausnitz 1971)
77.0 (quoted, Verschueren 1983)
200 (quoted, Dean 1985)

Vapor Pressure (Pa at 25°C):

Henry's Law Constant (Pa·m³/mol):

Octanol/Water Partition Coefficient, log K_{OW}:
2.65 (shake flask-AS, Fujita et al. 1964; quoted, Leo et al. 1971; Rekker 1977; Hansch & Leo 1979; Sangster 1993)
2.65 (calculated-f const., Rekker 1977)

2.52 (calculated-π const., Rekker 1977)
2.65 (quoted from Hansch & Leo 1979, Miyake & Terada 1982; Terada et al. 1987)
2.53 (calculated-HPLC-k', Miyake & Terada 1982)
2.63 (calculated-chlorine substituents, Kaiser 1983)
2.65 (quoted, Verschueren 1983)
2.65, 2.66 (quoted, electrometic titration, Clarke 1984; quoted, Sangster 1993)
2.65, 2.72 (quoted, shake flask-OECD 1981 Guidelines, Geyer et al. 1984)
2.65 (CPC-RV, Terada et al. 1987)
2.60 (shake flask at pH 1, Wang et al. 1989; quoted, Sangster 1993)
2.67 (correlated-HPLC-RT, Wang et al. 1989; quoted, Sangster 1993)
2.65 (correlated-HPLC-RT, Hayward et al. 1990; quoted, Sangster 1993)
2.71 (CC, average value, Vallat et al. 1990; quoted, Sangster 1993)
2.65, 2.66 (quoted, CPC-RV, El Tayar et al. 1991)
2.62 (shake flask-AS at pH 2, Da et al. 1992; quoted, Sangster 1993)
2.65 (recommended, Sangster 1993)

Bioconcentration Factor, log BCF:
<1.0 (golden orfe, Freitag et al. 1982)
1.80 (algae, Freitag et al. 1982)
2.23 (activated sludge, Freitag et al. 1982)
1.80 (alga *chlorella fusca*, wet wt. basis, Geyer et al. 1984)
1.99 (alga *chlorella fusca*, calculated-K_{OW}, Geyer et al. 1984)
<1.0 (golden ide, Freitag et al. 1985)
1.78 (algae, Freitag et al. 1985)
2.23 (activated sludge, Freitag et al. 1985)

Sorption Partition Coefficient, log K_{OC}:

Half-Lives in the Environment:
Air:
Surface water:
Groundwater:
Sediment:
Soil:
Biota:

Environmental Fate Rate Constants or Half-Lives:
Volatilization:
Photolysis:
Oxidation:
Hydrolysis:
Biodegradation: decomposition by a soil microflora, 64 days; degradation by waste water or soil suspension at pH 7.3 and 30°C, >25 days (quoted, Verschueren 1983).
Biotransformation:
Bioconcentration, Uptake (k_1) and Elimination (k_2) Rate Constants:

Common Name: Salicylic acid
Synonym: 2-hydroxybenzoic acid, *o*-hydroxybenzoic acid
Chemical Name: salicylic acid, 2-hydroxybenzoic acid
CAS Registry No: 69-72-7
Molecular Formula: HOC_6H_4COOH
Molecular Weight: 138.12
Melting Point (°C):
159.0 (Weast 1982-83)
158.0 (Verschueren 1983; Yalkowsky et al. 1983)
158.6 (Stephenson & Malanowski 1987)
Boiling Point (°C):
211 (sublime at 20 mm Hg, Weast 1982-83)
256 (Verschueren 1983)
Density (g/cm³ at 20°C):
1.443 (Weast 1982-83; Verschueren 1983)
Molar Volume (cm³/mol):
142.2 (calculated-LeBas method)
96.0 (calculated-density)
Molecular Volume (Å³):
Total Surface Area, TSA (Å²):
Dissociation Constant, pK_a:
3.009 (Korman & La Mer 1936)
3.00 (quoted, Unger et al. 1978)
3.52 ± 0.03 (HPLC, Unger et al. 1978)
3.29 ± 0.03 (HPLC, Unger et al. 1978)
2.96 (equilibrium titration, Clarke & Cahoon 1987)
2.97 (Sangster 1993)
Heat of Fusion ΔH_{fus}, kcal/mol:
Entropy of Fusion ΔS_{fus}, cal/mol·K (e.u.):
Fugacity Ratio at 25°C (assuming ΔS_{fus} = 13.5 e.u.), F: 0.0473

Water Solubility (g/m³ or mg/L at 25°C):
1800 (20°C, Hodgman 1952; quoted, Freed et al. 1977)
2189, 617 (quoted, calculated-K_{OW}, Yalkowsky & Morozowich 1980)
1800 (20°C, quoted, Verschueren 1983)
1550 (shake flask-UV, Yalkowsky et al. 1983)
240 (calculated-K_{OW} and M.P., Yalkowsky et al. 1983)
2558 (shake flask-TN, Apelblat & Manzurola 1989)

Vapor Pressure (Pa at 25°C):
0.0208 (extrapolated-Antoine eqn., Stephenson & Malanowski 1987)

Henry's Law Constant (Pa·m³/mol):
0.00144 (calculated-P/C, this work)

Octanol/Water Partition Coefficient, log K_{OW}:

2.26 (shake flask-GC, Hansch & Anderson 1967; quoted, Leo et al. 1971; Freed et al. 1977; Sangster 1993)
2.21 (quoted unpublished result, shake flask, Leo et al. 1971)
2.38 (shake flask-AS, Tomida et al. 1978; quoted, Sangster 1993)
2.23 ± 0.03 (quoted lit average, Unger et al. 1978)
2.00 ± 0.01, 2.18 ± 0.01 (HPLC, Unger et al. 1978)
2.24, 1.97 (quoted average, calculated-f const., Chou & Jurs 1979)
1.75 (correlated-S, Mackay et al. 1980; quoted, Sangster 1993)
2.21, 2.26 (quoted, Verschueren 1983)
2.23 (quoted from Hansch & Leo 1979, Yalkowsky et al. 1983; Haky & Young 1984)
2.21 (selected best lit. value, Garst & Wilson 1984)
2.25 ± 0.03 (exptl.-ALPM, Garst & Wilson 1984)
1.13, 1.92 (correlated-HPLC-k', Haky & Young 1984)
2.21 (shake flask-HPLC at pH 2, Bundgaard et al. 1986; Bundgaard & Nielsen 1988; quoted, Sangster 1993)
2.34 (electrometric titration, Clarke & Cahoon 1987; quoted, Sangster 1993)
2.24 (shake flask-RC at pH 0.5, Laznicek et al. 1987; quoted, Sangster 1993)
2.64 (CC, Ilchmann et al. 1993; quoted, Sangster 1993)
2.26 (recommended, Sangster 1993)

Bioconcentration Factor, log BCF:

Sorption Partition Coefficient, log K_{OC}:

Half-Lives in the Environment:

Air:
Surface water:
Groundwater:
Sediment:
Soil:
Biota:

Environmental Fate Rate Constants or Half-Lives:

Volatilization:
Photolysis:
Oxidation: rate constant of <600 M^{-1} sec^{-1} for 0.1-1 mM to react with ozone in water using 4 mM *t*-BuOH as scavenger at pH 1.3-3.0 and 20-23°C (Hoigné & Bader 1983).
Hydrolysis:
Biodegradation: decomposition by a soil microflora in two days (Alexander & Lustigman 1966; quoted, Verschueren 1983); average rate of biodegradation 95.0 mg COD $g^{-1} \cdot h^{-1}$ based on measurements of COD decrease using activated sludge inoculum with 20 days of adaptation to the substrate (Pitter 1976; quoted, Scow 1982).
Biotransformation:
Bioconcentration, Uptake (k_1) and Elimination (k_2) Rate Constants:

Common Name: 2,4-Dichlorophenoxyacetic acid
Synonym: 2,4-D
Chemical Name: 2,4-dichlorophenoxyacetic acid
CAS Registry No: 94-75-7
Molecular Formula: $Cl_2C_6H_3OCH_2COOH$
Molecular Weight: 221.04
Melting Point (°C):
    136-140 (Verschueren 1983)
    138 (Dean 1985)
    140.5 (Agrochemicals Handbook 1987; Howard 1991)
Boiling Point (°C):
    160 (at 0.4 mmHg, Dean 1985; Howard 1991)
    215 (Neely & Blau 1985)
Density (g/cm^3 at 30°C): 1.565 (Neely & Blau 1985)
Molar Volume (cm^3/mol):
    206.2 (calculated-LeBas method)
Molecular Volume (Å^3):
Total Surface Area, TSA (Å^2):
Dissociation Constant, pK_a:
    2.77 (potentiometric, Nelson & Faust 1969)
    2.87 (spectrophotometric, Cessna & Grover 1978)
    2.80 (Reinert & Rogers 1984)
    2.87 (McCall et al. 1980)
    2.64-3.31 (Howard 1991)
    2.80 (selected, Wauchope et al. 1992)
    2.97 (Sangster 1993)
    3.1 (Kollig 1993)
Heat of Fusion ΔH_{fus}, kcal/mol:
Entropy of Fusion ΔS_{fus}, cal/mol·K (e.u.):
Fugacity Ratio at 25°C (assuming ΔS_{fus} = 13.5 e.u.), F: 0.036 (Suntio et al. 1988)

Water Solubility (g/m^3 or mg/L at 25°C):
    522 (shake flask-UV, Leopold et al. 1960; quoted, Shiu et al. 1990)
    725 (Bailey & White et al. 1965; quoted, Shiu et al. 1990)
    725, 400, 900, 550 (quoted, Günther et al. 1968)
    890 (Hamaker 1975)
    900 (Kenaga 1980a,b; Kenaga & Goring 1980)
    600 (20°C, Khan 1980)
    470 (20-25°C, pH 5.6, Geyer et al. 1981)
    620 (20°C, Kidd & Hartley 1983, 1987)
    890 (Verschueren 1983; quoted, Shiu et al. 1990)
    620 (Worthing 1983; quoted, Shiu et al. 1990)
    891 (quoted, Garten & Trabalka 1983)
    690 (recommended, Neely & Blau 1985)
    609 (Gerstl & Helling 1987)
    682 (Yalkowsky et al. 1987; quoted, Howard 1991)

400 (20°C, selected, Suntio et al. 1988)
890 (quoted, Isnard & Lambert 1988)
703 (Gustafson 1989)
900 (selected, Isensee 1991; Mackay & Stiver 1991)
900, 600, 890, 703, 1072 (quoted, Wauchope et al. 1992)
890 (20-25°C, selected, Wauchope et al. 1992)

Vapor Pressure (Pa at 25°C):
$8x10^{-5}$ (Hamaker 1975)
0.180-1.69 (transpiration method, Spencer 1976; quoted, Suntio et al. 1988)
53 (160°C, Kidd & Hartley 1983, 1987)
$8.0x10^{-5}$ (recommended, Neely & Blau 1985; Lyman 1985)
1.0 (20°C, selected, Suntio et al. 1988)
0.20, 0.0032 (quoted, estimated from Henry's law constant, Howard 1991)
$5.6x10^{-5}$ (selected, Mackay & Stiver 1991)
$1.33x10^{-5}$, $8.0x10^{-5}$, $1,07x10^{-3}$ (quoted, Wauchope et al. 1992)
0.00107 (20-25°C, seleted, Wauchope et al. 1992)

Henry's Law Constant (Pa·m^3/mol):
0.55 (calculated-P/C, Suntio et al. 1988)
$1.39x10^{-5}$, $1.03x10^{-3}$ (quoted, calculated-bond contribution, Howard 1991)

Octanol/Water Partition Coefficient, log K_{OW}:
2.81 (shake flask-AS, Fujita et al. 1964; quoted, Garten & Trabalka 1983; Sangster 1993)
2.59 (electrometric titration, Freese et al. 1979; quoted, Sangster 1993)
1.57 (Kenaga & Goring 1980, Kenaga 1980b)
2.49 (correlated-S, Mackay et al. 1980; quoted, Sangster 1993)
2.81 (20°C, quoted, Verschueren 1983)
1.57, 4.88 (shake flask-OECD 1981 Guidelines, Geyer et al. 1984)
2.81 (Hansch & Leo 1985; quoted, Howard 1991)
2.81 (recommended, Neely & Blau 1985)
-1.36 (Gerstl & Helling 1987)
2.65 (shake flask, Hansch & Leo 1987; quoted, Sangster 1993)
3.00 (quoted, Isnard & Lambert 1988)
2.65 (CC, Ilchmann et al. 1993; quoted, Sangster 1993)
2.81 (recommended, Sangster 1993)
2.68 (calculated-QSAR, Kollig 1993)

Bioconcentration Factor, log BCF:
1.11, -0.097 (calculated-S, K_{OW}, Kenaga 1980a)
-2.46, 1.30 (beef fat, fish, Kenaga 1980b)
0.778, 1.94 (alga *chlorella*: exptl. 24 h exposure, calculated-S, Geyer et al. 1981)

0.778 (algae, Freitag et al. 1982)
<1.00 (golden orfe, Freitag et al. 1982)
1.230 (activated sludge, Freitag et al. 1982)
0.0 (fish, microcosm conditions, Garten & Trabalka 1983)
0.778 (algae, Geyer et al. 1984)
1.23 (algae, calculated-K_{OW}, Geyer et al. 1984)
0.85 (quoted, Isnard & Lambert 1988)
1.11 (calculated, Isensee 1991)
-5.0 (bluegill sunfish and channel catfish, Howard 1991)
-2.70 (frog tadpoles, Howard 1991)
-3.0, -2.52 (pH 7.8, seaweeds, Howard 1991)
0.778, 0.85 (quoted: alga, fish, Howard 1991)

Sorption Partition Coefficient, log K_{OC}:
1.30, 2.0 (quoted, calculated, Kenaga 1980a)
1.30, 2.11 (quoted, Kenaga & Goring 1980)
1.30 (quoted, Kenaga 1980b)
1.68, 1.86, 1,68; 1.76 (commerce soil, Tracy soil, Catlin soil; average value of 3 soils, McCall et al. 1981)
1.29 (soil, Neely & Blau 1985)
1.61 (soil, quoted, Sabljic 1987)
1.75, 2.00 (quoted, calculated-χ , Gerstl & Helling 1987)
1.29-2.13 (soil, quoted values, Howard 1991)
1.30 (selected, Mackay & Stiver 1991)
1.00 (sediment, von Oepen et al. 1991)
1.23 (Alfisol soil, von Oepen et al. 1991)
2.29 (Podzol soil, von Oepen et al. 1991)
1,30, 1.78, 1.51, 1,26,1.72, 1.75, 1.76 (soil, quoted, Wauchope et al. 1992)
1.30 (soil, selected, Wauchope et al. 1992)
0.68 (calculated-K_{OW}, Kollig 1993)

Half-Lives in the Environment:
Air: 1.8-18 hours, based on estimated rate constant for the vapor-phase reaction with hydroxyl radicals in air (Howard et al. 1991); photooxidation half-life of 23.9 hours for reactions with hydroxyl radicals in air (Howard 1991).
Surface water: 48-96 hours, based on reported photolysis half-lives for aqueous solution irradiated at UV wavelength of 356 nm (Baur & Bovey 1974; selected, Howard et al. 1991); degradation half-life of 14 days in sensitized, filtered and sterilized river water, based on sunlight photolysis test of 1 $\mu g\ mL^{-1}$ in distilled water (Zepp et al. 1975; quoted, Cessna & Muir 1991); typical biodegradation half-lives of 10 to <50 days with longer expected in oligotrophic waters, photolysis half-life of 29-43 days for water solutions irradiated at sunlight (Howard 1991).
Groundwater: 480-4320 hours, based on estimated unacclimated aqueous aerobic and anaerobic biodegradation half-lives (Howard et al. 1991).

Sediment: half-life of < 1 day for degradation in sediments and lake muds (Howard 1991).

Soil: 240-1200 hours, based on estimated unacclimated aqueous aerobic biodegradation half-life (Howard et al. 1991); biodegradation half-lives ranging from < 1 day to several weeks, half-life of 3.9 and 11.5 days in 2 moist soils and 9.4 to 254 days in the same soils under dry conditions (Howard 1991); field half-lives range from 2-16 days, with a selected value of 10 days (Wauchope et al. 1992).

Biota: depuration half-life of 13.8 hours in daphnids, 1.32 days in catfish (Ellgehausen et al. 1980).

Environmental Fate Rate Constants or Half-Lives:

Volatilization: volatilization from water is negligible, calculated volatilization half-lives from soil of 660 days (from 1 cm)and 7.1 years (from 10 cm) (Howard 1991).

Photolysis: aqueous photolysis half-lives of 2-4 days when irradiated at 356 nm, with a of 50 minutes in water when irradiated at 254 nm with a half-life of 29-43 days when exposed to September sunlight (Howard 1991).

Oxidation: photooxidation half-life of 1.8-18 hours, based on estimated rate constant for the vapor-phase reaction with hydroxyl radicals in air (Howard et al. 1991).

Hydrolysis: no hydrolyzable groups and rate constant at neutral pH is zero (Kollig et al. 1987; selected, Howard et al. 1991); generally resistant to hydrolysis, may become important at pH > 8 (Howard 1991).

Biodegradation: degradation kinetics not first-order, time for 50% decomposition in six soils: Commerce 5 days, Catlin 1.5 days, Keith 3.9 days, Cecil 3.0 days, Walla-Walla 2.5 days and Fargo 8.5 days, with an average time of 4 days (McCall et al. 1981); easily degraded under aerobic conditions with a half-life of 1.8 and 3.1 days for cometabolism and metabolism repectively, under anaerobic conditions the degradation rate decreases and the half-lives are 69 and 135 days (Liu et al. 1981; quoted, Muir 1991); second-order rate constants of (3.6-28.8)x10^{-6} mL·cell^{-1}·day^{-1} in natural water (Paris et al. 1981; quoted, Klečka 1985); first-order rate constant of <0.14-0.07 day^{-1} in river water at 25°C (Nesbitt & Watson 1980; quoted, Klečka 1985); rate constant of 0.058±0.006 day^{-1} in lake water at 29°C (Subba-Rao et al. 1982; quoted, Klečka 1985; quoted, Muir 1991); rate constant of 0.08-0.46 day^{-1} in soil at 25°C (McCall et al. 1981; quoted, Klečka 1985); aqueous aerobic half-life of 240-1200 hours, based on unacclimated aerobic river die-away test data (Nesbitt & Watson 1980a,b; quoted, Howard et al. 1991; Muir 1991); aqueous anaerobic half-life of 672-4320 hours, based on unacclimated aqueous screening test data (Liu et al. 1981; selected, Howard et al. 1991); first-order biodegradation rate constant of 0.035 day^{-1} (die-away test), 0.029 day^{-1} (CO_2 evolution test) in soil and 6.9x10^{-1} mL·(g bacteria)$^{-1}$·day^{-1} by activated sludge cultures (Scow 1982); biodegradation half-lives in river water of 18 to over 50 days (clear water) and 10 to 25 days (muddy water) with lag times of 6 to 12 days; degradation with a mixture of microorganisms from activated sludge, soil, and sediments lead to half-lives or 1.8-3.1 days under aerobic conditions and 69-135 days under anaerobic conditions (Howard 1991).

Biotransformation:

Bioconcentration, Uptake (k_1) and Elimination (k_2) Rate Constants:

5.2 Summary Tables and QSPR Plots

Table 5.1*a* Physical-chemical properties of carboxylic acids at 25 °C.

Compound	CAS	formula	MW	mp, °C	bp, °C	fugacity ratio F at 25 °C	pK_a
Aliphatic:							
formic acid	64-18-6	$HCOOH$	46.026	8.27	100.56	1	3.75
acetic acid	64-19-7	CH_3COOH	60.052	16.66	117.85	1	4.75
propionic acid	79-09-4	C_2H_5COOH	74.079	-20.7	141.163	1	4.87
butanoic acid	107-92-6	C_3H_7COOH	88.106		163.72		4.82
isobutyric acid	79-31-2	$(CH_3)_2CHCOOH$	88.11	-47	154	1	4.85
pentanoic acid	109-52-4	C_4H_9COOH	102.13	-33.67	185.5	1	4.83
3-methylbutanoic	503-74-2	$C_5H_{10}O_2$	102.13	-29.3	176.5	1	4.777
trimethylacetic acid	75-98-9	$(CH_3)_3CCOOH$	124.12	35.5	163.8	0.787	4.81
hexanoic acid	142-62-1	$C_5H_{11}COOH$	116.1	-3.44	205	1	4.87
octanoic acid	124-07-2	$C_7H_{15}COOH$	144.21	16.55	239.9	1	4.85
stearic acid	57-11-4	$C_{17}H_{35}COOH$	284.47	71.2	360	0.349	4.5
acrylic acid	79-10-7	$CH_2{=}CHCOOH$	72.06	13.5	141.2	1	4.255
2-butenoic (*cis*-)	3724-65-0	$CH_3CH{=}CHCOOH$	86.09	15.5	169	1	4.5
2-butenoic (*trans*-)	107-93-7	$CH_3CH{=}CHCOOH$	86.09	71.5	185	0.347	
vinylacetic acid	625-38-7	$CH_2{=}CHCH_2COOH$	86.09	-39	163	1	4.35
2-methylpropenoic acid	79-41-4	$C_4H_6O_2$	86.09	15.5	160.5	1	4.66
chloroacetic acid	79-11-8	$ClCH_2COOH$	94.5	50		0.566	2.85
dichloroacetic acid	79-43-6	$Cl_2CHCOOH$	128.95	9.7		1	1.48
trichloroacetic acid	76-03-9	Cl_3CCOOH	163.4	57		0.483	0.7

Compound	CAS	formula	MW	mp, °C	bp, °C	fugacity ratio F at 25 °C	pK_a
Aromatic:							
benzoic acid	65-85-0	C_6H_5COOH	122.13	122.4		0.109	4.19
2-methylbenzoic acid	118-901	$CH_3C_6H_4COOH$	136.15	103.75	259.2	0.167	3.92
3-methyl benzoic acid	99-04-7	$CH_3C_6H_4COOH$	136.15	108.8	263	0.148	4.27
4-methylbenzoic acid	99-94-5	$CH_3C_6H_4COOH$	136.15	179.6	275	0.0296	4.39
phenylacetic acid	103-82-2	$C_6H_5CH_2COOH$	136.15	77	265	0.306	4.31
phthalic acid	88-993	C_6H_4-1,2-$(COOH)_2$	166.13	210		0.0148	
2-chlorobenzoic acid	118-91-2	ClC_6H_4COOH	156.57	140.2	sublime	0.0725	
3-chlorobenzoic acid	535-80-8	ClC_6H_4COOH	156.57	158	sublime	0.0484	3.82
4-chlorobenzoic acid	74-11-3	ClC_6H_4COOH	156.57	239.7		0.00752	
salicylic acid	69-72-7	HOC_6H_4COOH	138.12	159	211	0.0473	
phenoxyacetic acid	122-59-8	$C_6H_5OCH_2COOH$	152.15	98-100			
2,4-D	94-75-7	$Cl_2C_6H_3OCH_2COOH$	221.04	138	215	0.0763	
2,4,5-T	93-76-5	$Cl_3C_6H_2OCH_2COOH$	255.49	158		0.0484	

Table 5.1*b* Physical-chemical properties of carboxylic acids at 25 °C

Compound	molar volume, cm^3/mol				surface area, $Å^2$		pK_a
	V_M from ρ	V_M (a)	V_I (a)	V_M LeBas	TSA (b)	TSA(c)	
Aliphatic:							
formic acid	37.91		22.4	46.2		145.1	3.75
acetic acid	57.53	57.5	32.3	68.4		173.6	4.75
propionic acid	74.97		42.1	90.6		197	4.87
butanoic acid	92.43	92.5	51.9	112.8		224.2	4.82
isobutyric acid	91.00			112.8			4.85
pentanoic acid	109.29		61.7	135	303.4	303.4	4.83
3-methylbutanoic				135			4.78
trimethylacetic acid				135			4.81
hexanoic acid	125.79		71.5	157.2			4.87
octanoic acid	159.07			201.6			4.85
stearic acid				349.6			4.5
acrylic acid	68.56			83.2			4.26
2-butenoic (*cis*-)	83.85			105.4			4.5
2-butenoic (*trans*-)	78.14						
vinylacetic acid	84.99						4.35
2-methylpropenoic acid							4.66
chloroacetic acid				89.3			2.85
dichloroacetic acid	82.50			110.2			1.48
trichloroacetic acid	100.86			131.1			0.7

Compound	molar volume, cm^3/mol				surface area, $Å^2$		pK_a
	V_M from ρ	V_M (a)	V_I (a)	V_M LeBas	TSA (b)	TSA (c)	
Aromatic:							
benzoic acid	96.17		65	112.4		255.6	4.19
2-methylbenzoic acid				157			3.92
3-methyl benzoic acid			74.8	157			4.27
4-methylbenzoic acid			74.8	157		279.7	4.39
phenylacetic acid	125.95			157			4.31
phthalic acid	104.29			173.6			
terephthalic acid							
2-chlorobenzoic acid	101.41		74	155.7			
3-chlorobenzoic acid	104.66		74	155.7			3.82
4-chlorobenzoic acid	101.60			155.7			
salicylic acid	95.72			146.2			
phenoxyacetic acid							
2,4-D				206.2			
2,4,5-T				227.1			

(a) Kamlet et al. 1987, 1988; (b) Amidon et al. 1975; (c) Iwase et al. 1985.

Table 5.2 Summary and selected physical-chemical properties of carboxylic acids at 25 °C

Compound	selected properties:						Henry's law constant, Pa·m³/mol		
	S, g/m³	C^S mol/m³	C_L mol/m³	P^S Pa	P_L Pa	log K_{OW}	calc-P/C	exptl. (a)	calculated (b)
Aliphatic:									
formic acid	miscible			5750	5750	-0.54		0.01832	
acetic acid	miscible			2079	2079	-0.31		0.01825	0.0285
propionic acid	miscible			435	435	0.26		0.018	0.0431
butanoic acid	miscible			84	84	0.79		0.0223	0.0654
isobutyric acid	22800	258.78	258.78	185	185	0.79		0.0897	
pentanoic acid	24000	234.99	234.99	19	19	1.39	0.081	0.0478	0.099
3-methylbutanoic acid	4100	40.14	40.14	25	25		0.623	0.0844	
trimethylacetic acid									
hexanoic acid	958	8.252	8.252	5	5	1.92	0.606	0.0768	0.149
octanoic acid	79.8	0.553	0.553	3.72	3.72		6.723		0.338
stearic acid	340	1.195	3.425		1.69×10^{-6}	8.23	4.05×10^{-6}		
acrylic acid	miscible			550	550	0.43			
2-butenoic *cis*-						0.72			
2-butenoic *trans*-									
vinylacetic acid									
2-methylpropenoic acid	8900	103.38	103.38	100	100	0.93			
chlorodehydroabietic									
chloroacetic acid					18.7	0.22			
dichloroacetic acid	miscible			35		0.92			
trichloroacetic acid					11	1.33			

Compound	selected properties:						Henry's law constant, $Pa \cdot m^3/mol$		
	S, g/m^3	C^S mol/m^3	C_L mol/m^3	P^S Pa	P_L Pa	log K_{OW}	calc-P/C	exptl. (a)	calculated (b)
Aromatic:									
benzoic acid	3400	27.839	255.405	0.11	1.468	1.89	0.484		
2-methylbenzoic acid	1074	7.888	47.236			2.46			
3-methyl benzoic acid	1246	9.152	61.836	1.39	9.392	2.37	6.934		
4-methylbenzoic acid	331	2.431	82.133			2.34			
phenylacetic acid	16600	121.92	398.45	0.83	2.703	1.41	0.072		
phthalic acid	7000	42.14	2847.01			0.73			
2-chlorobenzoic acid	2100	13.41	185.00			1.99			
3-chlorobenzoic acid	400	2.555	52.784			2.60			
4-chlorobenzoic acid	70	0.447	59.453			2.65			
salicylic acid	2300	16.65	352.05	0.0208	0.440	2.20	0.558		
phenoxyacetic acid	12000	78.87	3617.87			2.81			
2,4-D	890	4.026	52.771	8.0×10^{-5}	0.001	3.13	3.41×10^{-3}		
2,4,5-T	278	1.088	22.482						

(a) Brimblecombe et al. 1992; (b) Hine & Mookerjee 1975

Table 5.3 Suggested half-life classes for carboxylic acids at 25°C in various environmental compartments.

Compounds	Air class	Water class	Soil class	Sediment class
Aliphatic:				
formic acid	4	3	4	5
acetic acid	3	3	4	5
butanoic acid	3	3	4	5
hexanoic acid	3	3	4	5
stearic acid	3	3	4	5
acrylic acid	2	3	4	5
vinylacetic acid	2	3	4	5
chloroacetic acid	5	4	5	6
Aromatic:				
benzoic acid	3	3	4	5
2-methylbenzoic acid	3	3	4	5
phenylacetic acid	3	3	4	5
phthalic acid	3	3	4	5
2-chlorobenzoic acid	4	5	6	7
salicylic acid	3	3	4	5
2,4-D	2	3	5	6
2,4,5-T	3	4	6	7

where,

Class	Mean half-life (hours)	Range (hours)
1	5	< 10
2	17 (~ 1 day)	10-30
3	55 (~ 2 days)	30-100
4	170 (~ 1 week)	100-300
5	550 (~ 3 weeks)	300-1,000
6	1700 (~ 2 months)	1,000-3,000
7	5500 (~ 8 months)	3,000-10,000
8	17000 (~ 2 years)	10,000-30,000
9	~ 5 years	> 30,000

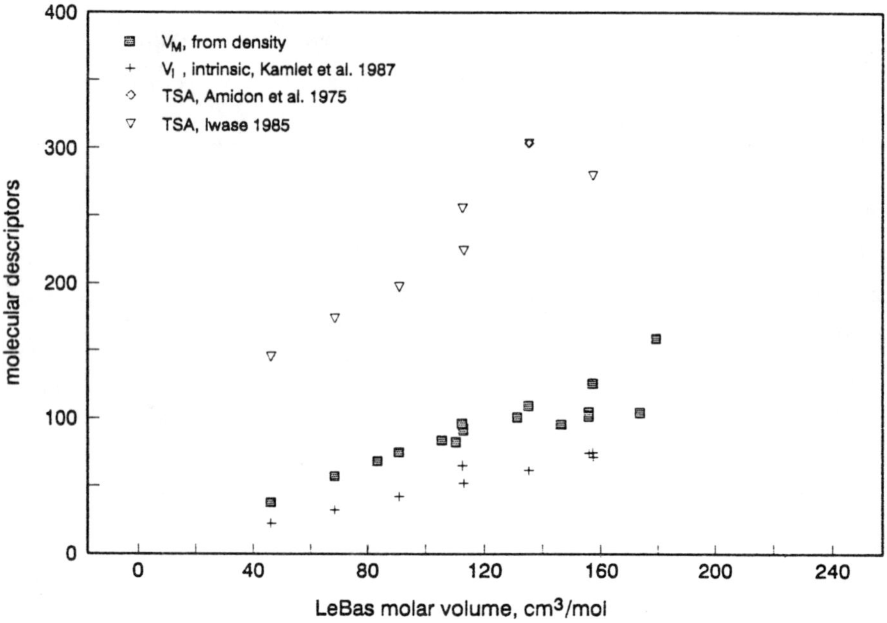

Figure 5.1 Plot of molecular descriptors vs LeBas molar volume

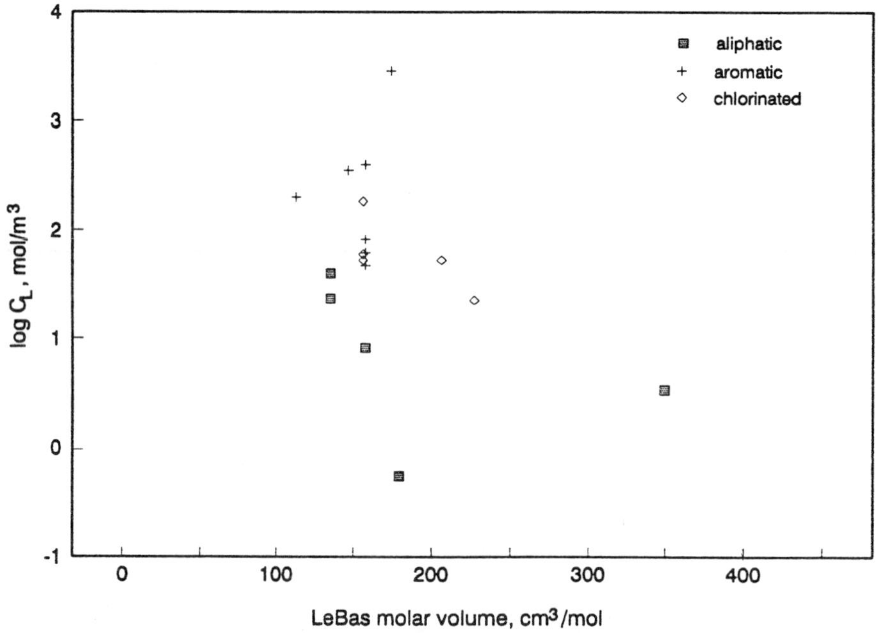

Figure 5.2 Plot of log liquid solutility vs LeBas molar volume

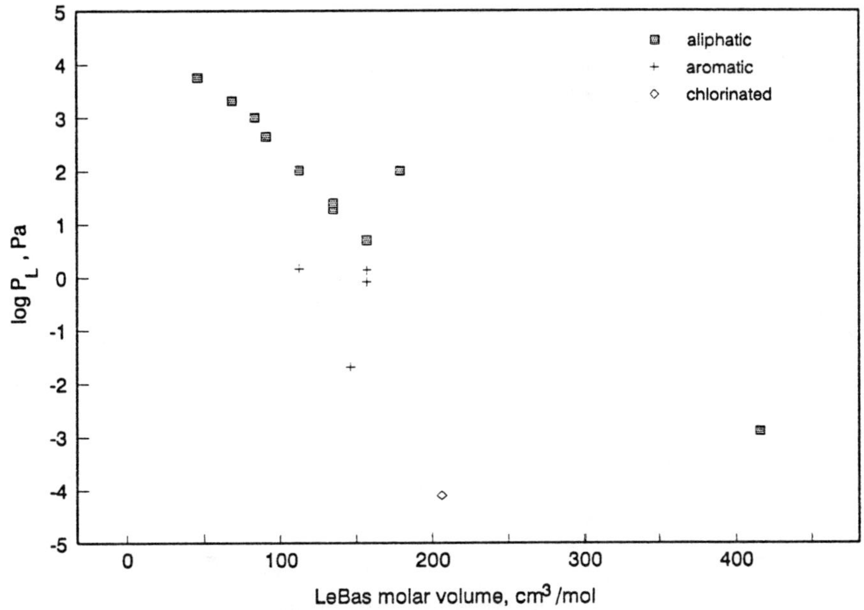

Figure 5.3 Plot of log vapor pressure versus LeBas molar volume

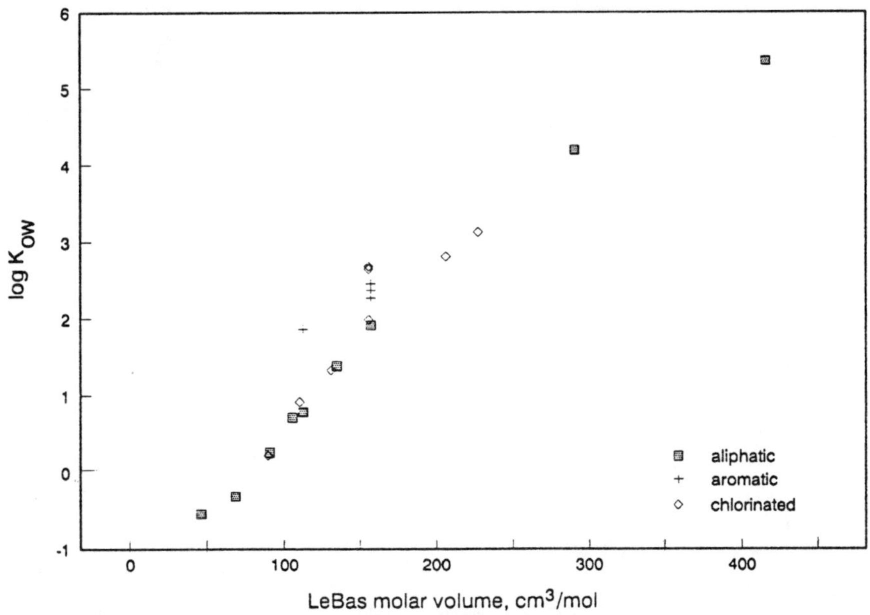

Figure 5.4 Plot of log K_{OW} versus LeBas molar volume

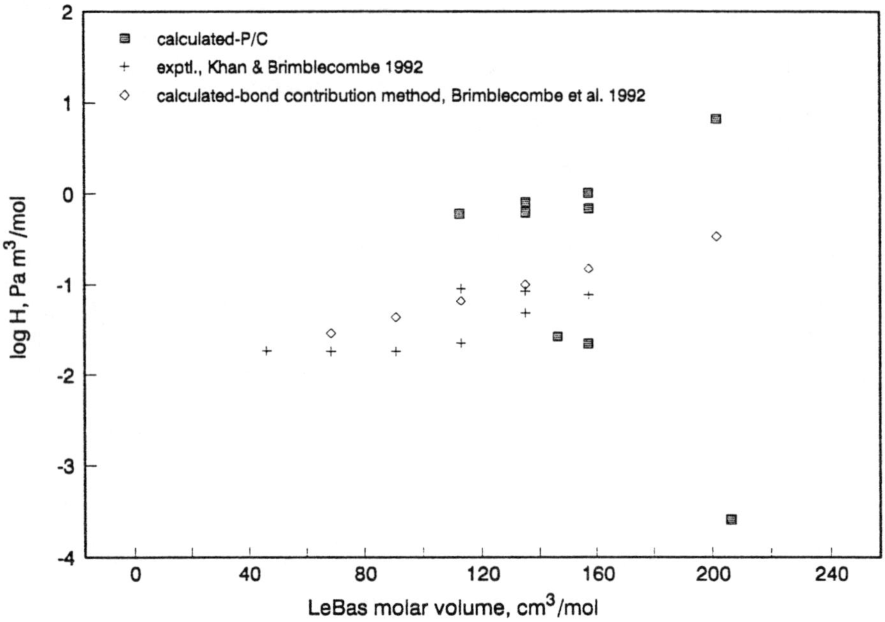

Figure 5.5 Plot of log H versus LeBas molar volume

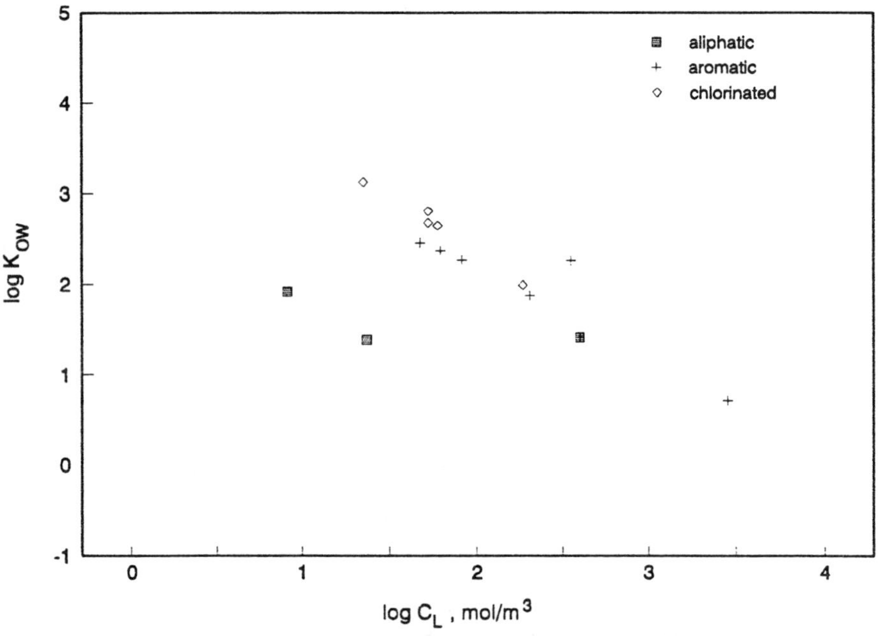

Figure 5.6 Plot of log K_{OW} versus log liquid solubility

5.3 Illustrative Fugacity Calculations: Levels I, II, III

Chemical name: Acetic acid

Level I calculation: (six-compartment model)

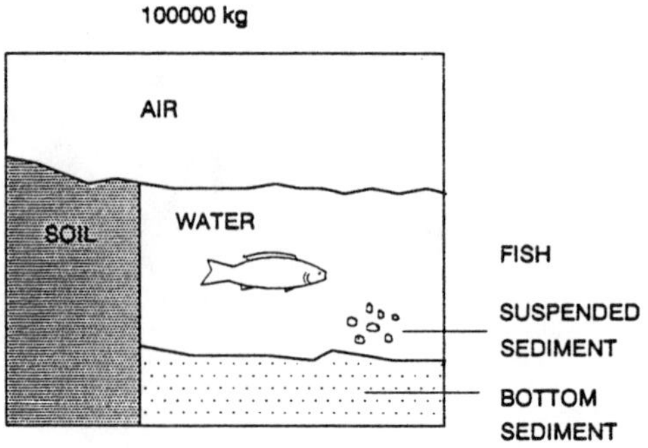

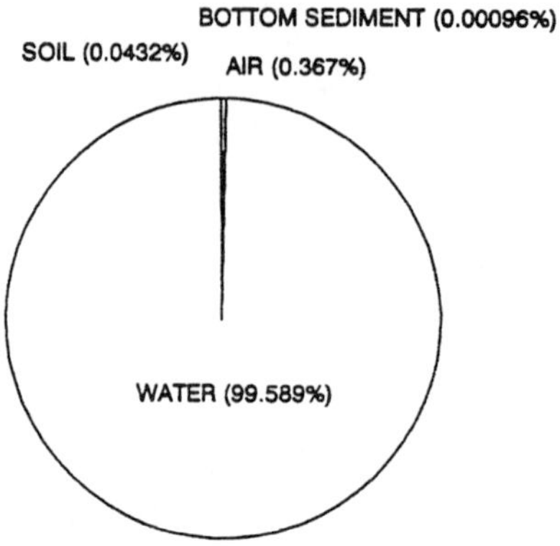

Distribution of mass

Level I calculation:

Physical-chemical properties:		Partition coeff.:		Z values in water:	
molecular wt., g/mol	60.05	HLC	0.01825	at data pH	6
melting point, C	16.6	Kaw	7.36E-06	neutral	0.306
solubility, g/m3	6841000	Koc	0.2008	ionic	54.490
vapor pressure, Pa	2079.000	BCF	0.0245	total	54.796
log Kow	-0.31	Ksw	0.0096	at environ. pH	6
fugacity ratio, F	1.0000	Ksd/w	0.0193	neutral	0.306
dissoc. const. pKa	3.75	Kssd/w	0.0602	ionic	54.490
		Kar/w	2.89E+03	total	54.796

Compartment	Z	Concentration			Amount	Amount
	mol/m3 Pa	mol/m3	mg/L (or g/m3)	ug/g	kg	%
Air	4.03E-04	6.10E-11	3.67E-09	3.09E-06	366.594	0.3666
Water	5.48E+01	8.29E-06	4.98E-04	4.98E-04	99589	99.5892
Soil	5.28E-01	7.99E-08	4.80E-06	2.00E-06	43.197	0.0432
Biota (fish)	1.34E+00	2.03E-07	1.22E-05	1.22E-05	0.00244	2.44E-06
Suspended sediment	3.30E+00	5.00E-07	3.00E-05	2.00E-05	0.0300	3.00E-05
Bottom sediment	1.06E+00	1.60E-07	9.60E-06	4.00E-06	0.960	9.60E-04
	Total				100000	100

f = 1.513E-07 Pa

* pseudo-solubility of a miscible compound

Chemical name: Acetic acid

Level II calculation: (six-compartment model)

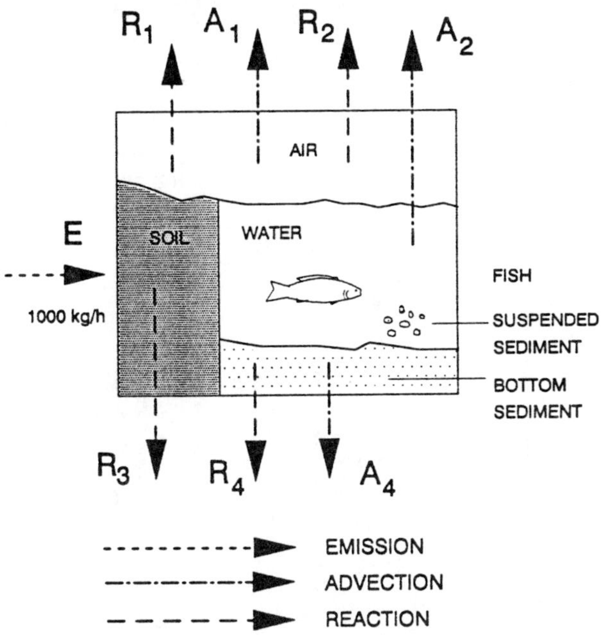

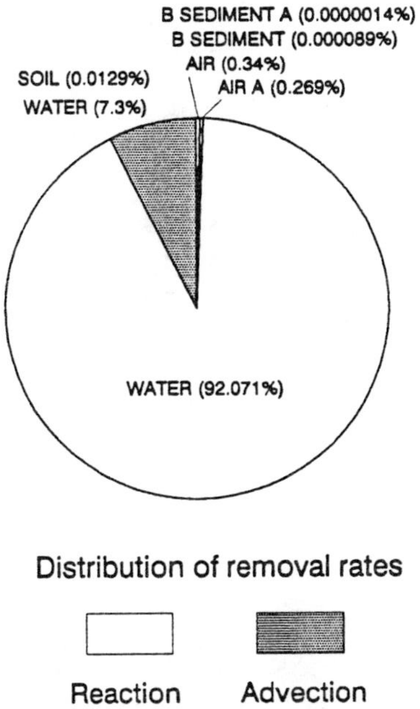

Level II Calculation:

Compartment	Half-life h	D Values Reaction mol/Pa.h	D Values Advection mol/Pa.h	Conc'n mol/m3	Loss Reaction kg/h	Loss Advection kg/h	Removal %
Air	55	5.08E+08	4.03E+08	4.48E-11	3.389	2.690	0.608
Water	55	1.38E+11	1.10E+10	6.08E-06	920.718	73.073	99.379
Soil	170	1.94E+07		5.86E-08	0.1292		0.0129
Biota (fish)				1.49E-07			
Suspended sediment				3.67E-07			
Bottom sediment	550	1.33E+05	2.11E+03	1.17E-07	8.87E-04	1.41E-05	9.02E-05
	Total	1.39E+11	1.14E+10		924.24	75.763	100
	R + A		1.50E+11			1000	

f = 1.110E-07 Pa

Total amt = 73374 kg

Overall residence time = 73.37 h

Reaction time = 79.39 h

Advection time = 968.47 h

Fugacity Level III calculations: (four-compartment model)

Chemical name: Acetic acid

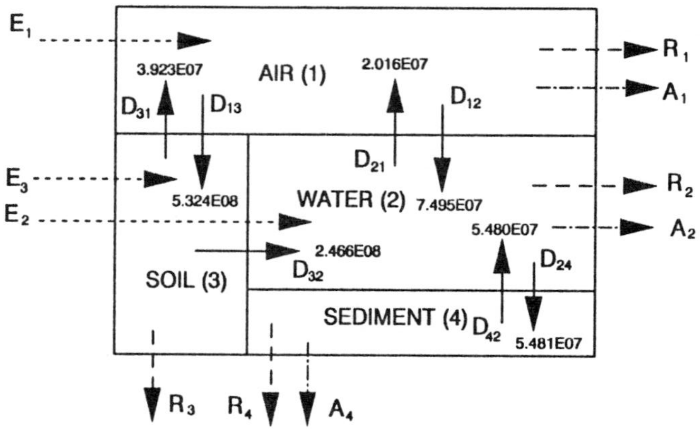

Phase Properties and Rates:

Compartment	Bulk Z mol/m3 Pa	Half-life h	D Values Reaction mol/Pa h	D Values Advection mol/Pa h
Air (1)	4.034E-04	55	5.08E+08	4.03E+08
Water (2)	5.480E+01	55	1.38E+11	1.10E+10
Soil (3)	1.670E+01	170	1.23E+09	
Sediment (4)	4.405E+01	550	2.78E+07	4.40E+05

	E(1)=1000	E(2)=1000	E(3)=1000	E(1,2,3)	
Overall residence time =	105.07	73.62	213.66	173.22	h
Reaction time =	145.11	79.47	217.83	181.50	h
Advection time =	380.82	1000.99	11151	3799.08	h

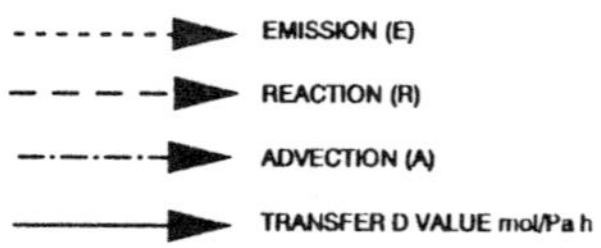

Phase Properties, Compositions, Transport and Transformation Rates:

Emission, kg/h			Fugacity, Pa				Concentration, g/m3				Amounts, kg				Total Amount, kg
E(1)	E(2)	E(3)	f(1)	f(2)	f(3)	f(4)	C(1)	C(2)	C(3)	C(4)	m(1)	m(2)	m(3)	m(4)	
1000	0	0	1.106E-05	1.201E-08	3.897E-06	7.931E-09	2.680E-07	3.951E-05	3.909E-03	2.098E-05	2.680E+04	7.902E+03	7.036E+04	1.049E+01	1.051E+05
0	1000	0	1.496E-09	1.117E-07	5.269E-10	7.378E-08	3.623E-11	3.676E-04	5.285E-07	1.951E-04	3.623E+00	7.351E+04	9.512E+00	9.757E+01	7.362E+04
0	0	1000	2.874E-07	1.854E-08	1.112E-05	1.224E-08	6.962E-09	6.099E-05	1.115E-02	3.238E-05	6.962E+02	1.220E+04	2.007E+05	1.619E+01	2.137E+05
50	250	700	7.547E-07	4.150E-08	7.978E-06	2.741E-08	1.828E-08	1.366E-04	8.002E-03	7.250E-05	1.828E+03	2.731E+04	1.440E+05	3.625E+01	1.732E+05

Emission, kg/h			Loss, Reaction, kg/h				Loss, Advection, kg/h			Intermedia Rate of Transport, kg/h T12	T21	T13	T31	T32	T24	T42
E(1)	E(2)	E(3)	R(1)	R(2)	R(3)	R(4)	A(1)	A(2)	A(4)	air-water	water-air	air-soil	soil-air	soil-water	water-sed	sed-water
1000	0	0	3.377E+02	9.957E+01	2.87E+02	1.322E-02	2.680E+02	7.902E+00	2.098E-04	4.979E+01	1.453E-02	3.537E+02	9.180E+00	5.770E+01	3.952E-02	2.610E-02
0	1000	0	4.566E-02	9.263E+02	3.88E-02	1.229E-01	3.623E-02	7.351E+01	1.951E-03	6.732E-03	1.352E-01	4.782E-02	1.241E-03	7.802E-03	3.677E-01	2.428E-01
0	0	1000	8.772E+00	1.537E+02	8.18E+02	2.040E-02	6.962E+00	1.220E+01	3.238E-04	1.294E+00	2.243E-02	9.188E+00	2.619E+01	1.646E+02	6.101E-02	4.029E-02
50	250	700	2.304E+01	3.441E+02	5.87E+02	4.568E-02	1.828E+01	2.731E+01	7.250E-04	3.397E+00	5.023E-02	2.413E+01	1.879E+01	1.181E+02	1.366E-01	9.020E-02

Level III Distribution

Chemical name: Acetic acid

Distribution of mass

Distribution of removal rates

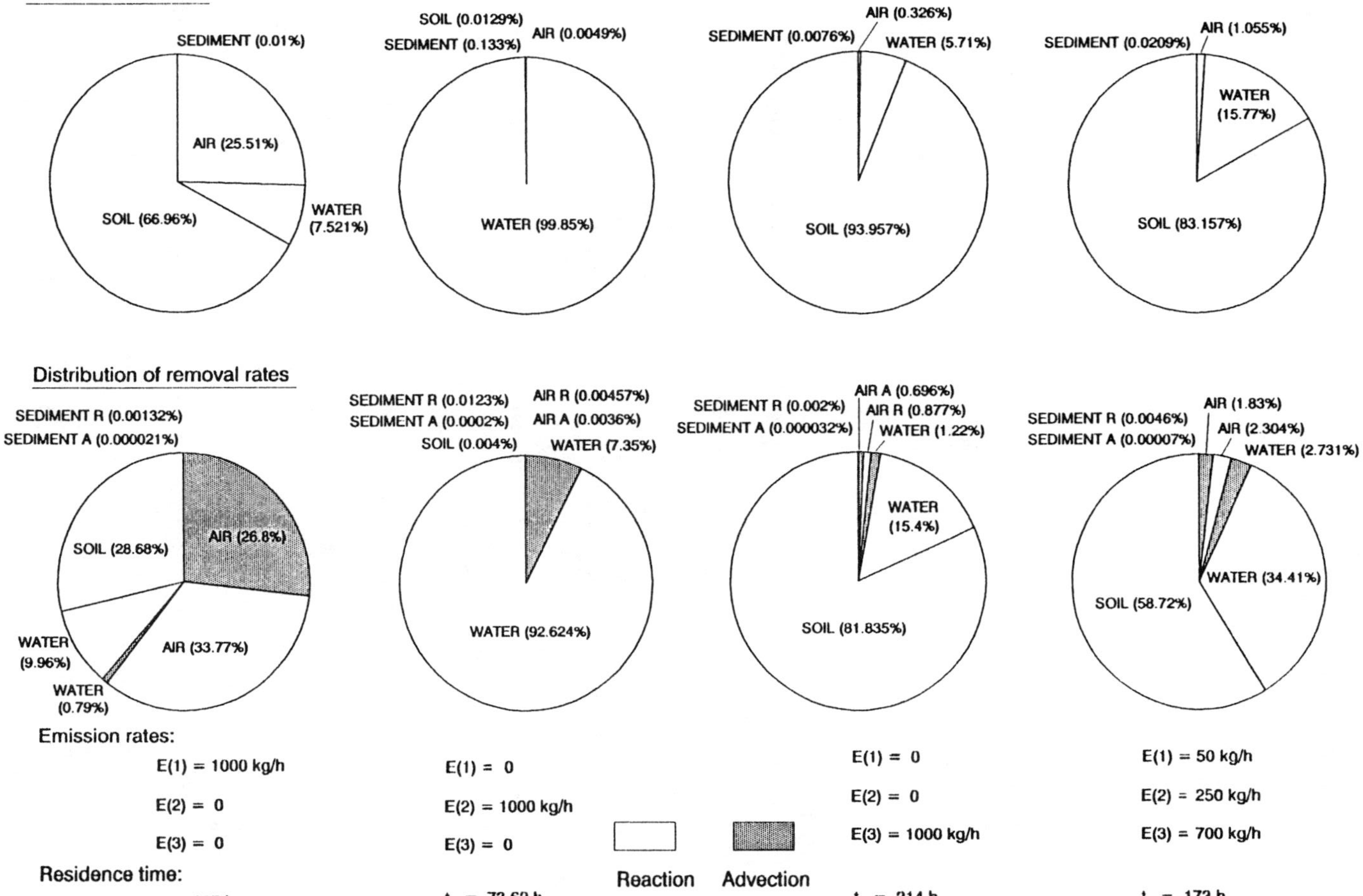

Emission rates:

E(1) = 1000 kg/h	E(1) = 0	E(1) = 0	E(1) = 50 kg/h
E(2) = 0	E(2) = 1000 kg/h	E(2) = 0	E(2) = 250 kg/h
E(3) = 0	E(3) = 0	E(3) = 1000 kg/h	E(3) = 700 kg/h

Residence time:

t = 105 h	t = 73.62 h	t = 214 h	t = 173 h

Chemical name: Isobutyric acid

Level I calculation: (six-compartment model)

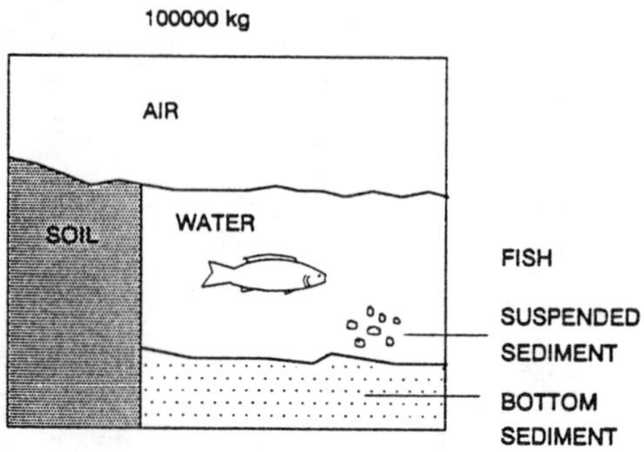

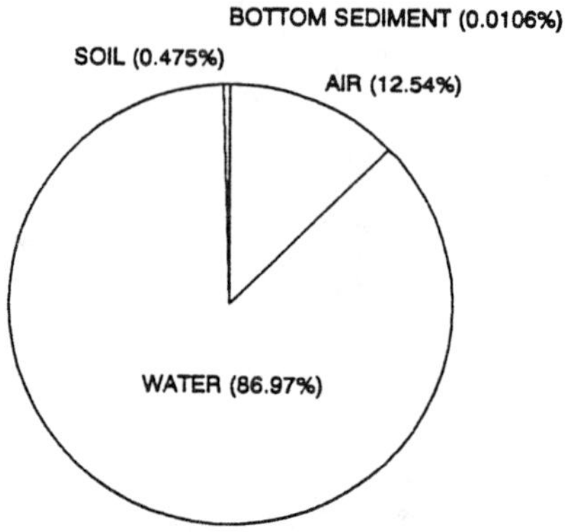

Distribution of mass

Level I calculation:

Physical-chemical properties:		Partition coeff.:		Z values in water:	
molecular wt., g/mol	88.11	HLC	0.71493	at data pH	6
melting point, C	-47	Kaw	2.88E-04	neutral	0.0867
solubility, g/m3	22800	Koc	2.5280	ionic	1.3121
vapor pressure, Pa	185.000	BCF	0.3083	total	1.3987
log Kow	0.79	Ksw	0.1213	at environ. pH	6
fugacity ratio, F	1.0000	Ksd/w	0.2427	neutral	0.0867
dissoc. const. pKa	4.82	Kssd/w	0.7584	ionic	1.3121
		Kar/w	3.24E+04	total	1.3987

Compartment	Z	Concentration			Amount	Amount
	mol/m3 Pa	mol/m3	mg/L (or g/m3)	ug/g	kg	%
Air	4.03E-04	1.42E-09	1.25E-07	1.06E-04	12542.02	12.5420
Water	1.40E+00	4.94E-06	4.35E-04	4.35E-04	86972	86.9722
Soil	1.70E-01	5.99E-07	5.28E-05	2.20E-05	474.917	0.4749
Biota (fish)	4.31E-01	1.52E-06	1.34E-04	1.34E-04	0.02681	2.68E-05
Suspended sediment	1.06E+00	3.74E-06	3.30E-04	2.20E-04	0.3298	3.30E-04
Bottom sediment	3.39E-01	1.20E-06	1.06E-04	4.40E-05	10.554	1.06E-02
	Total				100000	100

f = 3.528E-06 Pa

Chemical name: Isobutyric acid

Level II calculation: (six-compartment model)

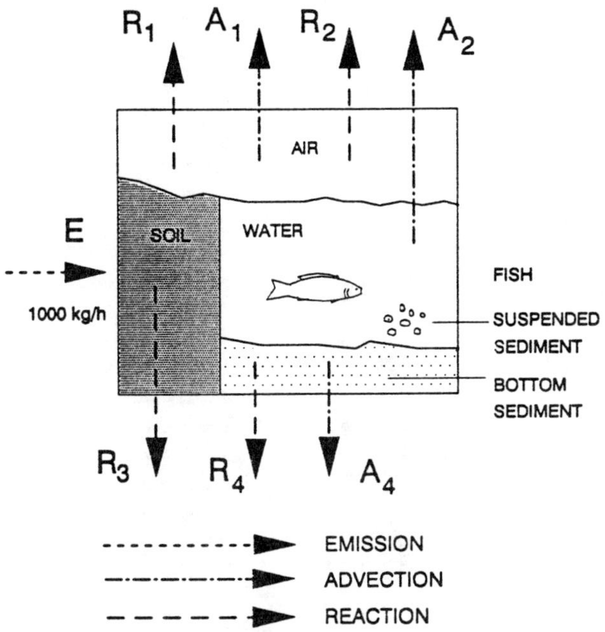

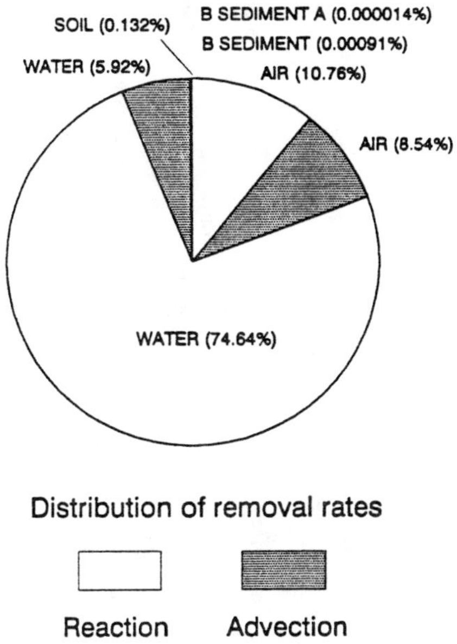

Distribution of removal rates

Level II Calculation:

Compartment	Half-life h	D Values Reaction mol/Pa.h	D Values Advection mol/Pa.h	Conc'n mol/m3	Loss Reaction kg/h	Loss Advection kg/h	Removal %
Air	55	5.08E+08	4.03E+08	9.70E-10	107.633	85.423	19.306
Water	55	3.52E+09	2.80E+08	3.36E-06	746.379	59.236	80.562
Soil	170	6.23E+06		4.08E-07	1.3186		0.1319
Biota (fish)				1.04E-06			
Suspended sediment				2.55E-06			
Bottom sediment	550	4.28E+04	6.79E+02	8.16E-07	9.06E-03	1.44E-04	9.20E-04
	Total	4.04E+09	6.83E+08		855.34	144.660	100
	R + A		4.72E+09			1000	

f = 2.403E-06 Pa
Total amt = 68110 kg

Overall residence time = 68.11 h
Reaction time = 79.63 h
Advection time = 470.83 h

Fugacity Level III calculations: (four-compartment model)

Chemical name: Isobutyric acid

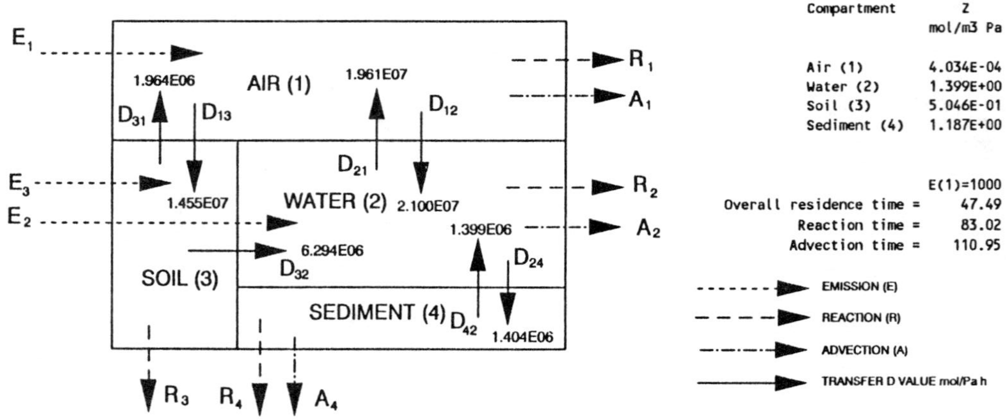

Phase Properties and Rates:

Compartment	Bulk Z mol/m3 Pa	Half-life h	D Values Reaction mol/Pa h	D Values Advection mol/Pa h
Air (1)	4.034E-04	55	5.08E+08	4.03E+08
Water (2)	1.399E+00	55	3.52E+09	2.80E+08
Soil (3)	5.046E-01	170	3.70E+07	
Sediment (4)	1.187E+00	550	7.48E+05	1.19E+04

	E(1)=1000	E(2)=1000	E(3)=1000	E(1,2,3)	
Overall residence time =	47.49	73.49	212.85	169.74	h
Reaction time =	83.02	79.47	219.21	180.68	h
Advection time =	110.95	975.42	7332	2803.01	h

Phase Properties, Compositions, Transport and Transformation Rates:

Emission, kg/h			Fugacity, Pa				Concentration, g/m3				Amounts, kg				Total Amount, kg
E(1)	E(2)	E(3)	f(1)	f(2)	f(3)	f(4)	C(1)	C(2)	C(3)	C(4)	m(1)	m(2)	m(3)	m(4)	
1000	0	0	1.199E-05	7.219E-08	3.854E-06	4.695E-08	4.262E-07	8.897E-06	1.713E-04	4.910E-06	4.262E+04	1.779E+03	3.084E+03	2.455E+00	4.749E+04
0	1000	0	6.146E-08	2.968E-06	1.975E-08	1.930E-06	2.185E-09	3.658E-04	8.782E-07	2.018E-04	2.185E+02	7.315E+04	1.581E+01	1.009E+02	7.349E+04
0	0	1000	5.285E-07	4.157E-07	2.508E-04	2.703E-07	1.879E-08	5.123E-05	1.115E-02	2.827E-05	1.879E+03	1.025E+04	2.007E+05	1.413E+01	2.128E+05
50	250	700	9.848E-07	1.037E-06	1.758E-04	6.741E-07	3.501E-08	1.277E-04	7.814E-03	7.049E-05	3.501E+03	2.555E+04	1.407E+05	3.525E+01	1.697E+05

Emission, kg/h			Loss, Reaction, kg/h				Loss, Advection, kg/h			Intermedia Rate of Transport, kg/h T12	T21	T13	T31	T32	T24	T42
E(1)	E(2)	E(3)	R(1)	R(2)	R(3)	R(4)	A(1)	A(2)	A(4)	air-water	water-air	air-soil	soil-air	soil-water	water-sed	sed-water
1000	0	0	5.370E+02	2.242E+01	1.26E+01	3.093E-03	4.262E+02	1.779E+00	4.910E-05	2.219E+01	1.247E-01	1.537E+01	6.667E-01	2.137E+00	8.931E-03	5.789E-03
0	1000	0	2.753E+00	9.217E+02	6.44E-02	1.272E-01	2.185E+00	7.315E+01	2.018E-03	1.138E-01	5.127E+00	7.881E-02	3.418E-03	1.096E-02	3.672E-01	2.380E-01
0	0	1000	2.367E+01	1.291E+02	8.18E+02	1.781E-02	1.879E+01	1.025E+01	2.827E-04	9.781E-01	7.181E-01	6.777E-01	4.339E+01	1.391E+02	5.142E-02	3.333E-02
50	250	700	4.411E+01	3.219E+02	5.73E+02	4.441E-02	3.501E+01	2.555E+01	7.049E-04	1.823E+00	1.791E+00	1.263E+00	3.041E+01	9.748E+01	1.282E-01	8.311E-02

Level III Distribution

Chemical name: Isobutyric acid

Distribution of mass

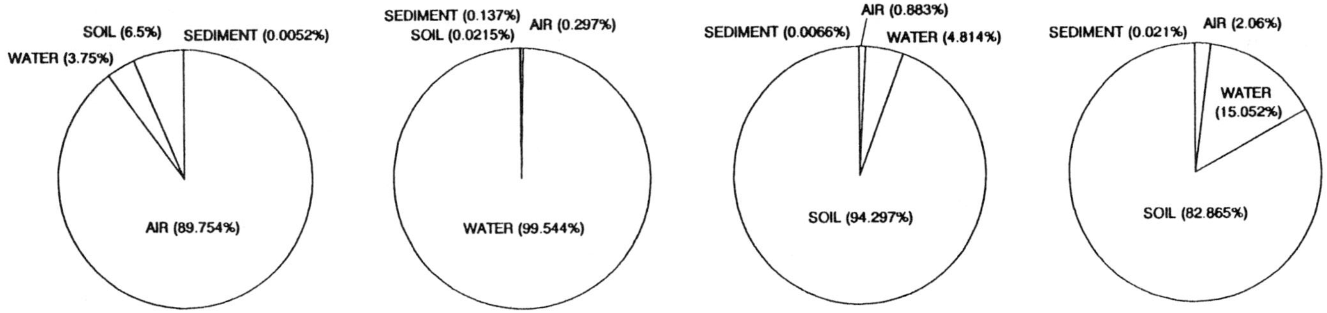

Distribution of removal rates

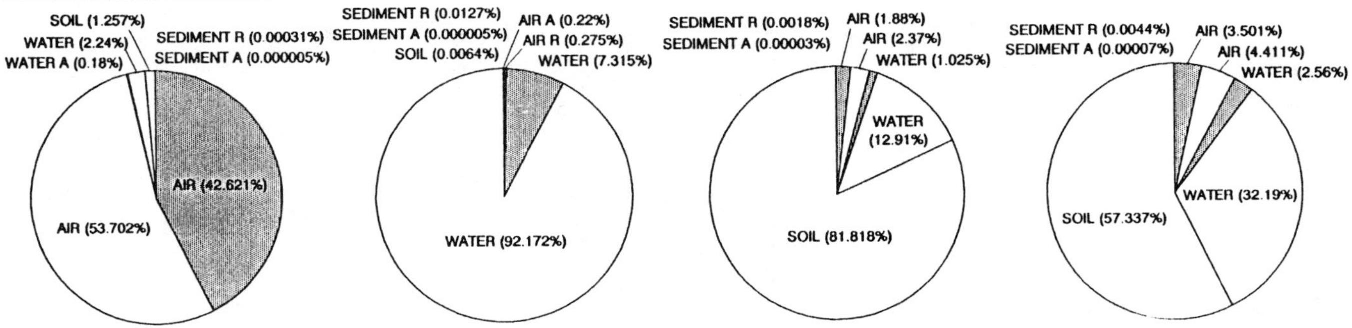

Reaction | Advection

Emission rates:

E(1) = 1000 kg/h	E(1) = 0	E(1) = 0	E(1) = 50 kg/h
E(2) = 0	E(2) = 1000 kg/h	E(2) = 0	E(2) = 250 kg/h
E(3) = 0	E(3) = 0	E(3) = 1000 kg/h	E(3) = 700 kg/h

Residence time:

t = 47.49 h	t = 73.49 h	t = 213 h	t = 170 h

Chemical name: Hexanoic acid

Level I calculation: (six-compartment model)

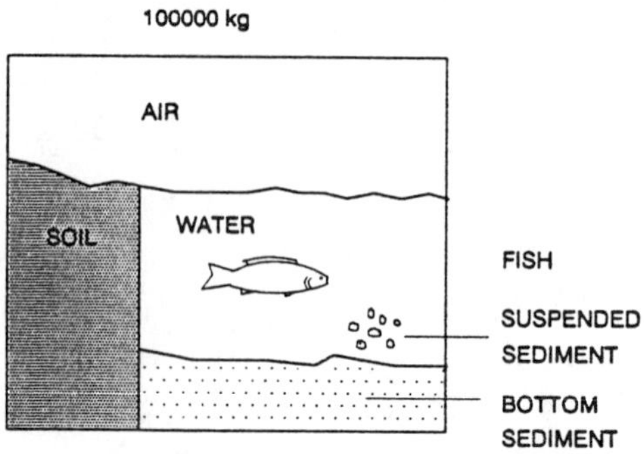

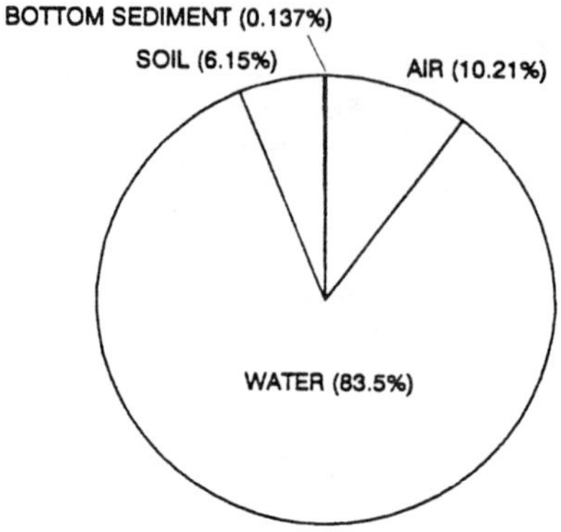

Distribution of mass

Level I calculation:

Physical-chemical properties:		Partition coeff.:		Z values in water:	
molecular wt., g/mol	116.10	HLC	0.60595	at data pH	6
melting point, C	-3.4	Kaw	2.44E-04	neutral	0.1139
solubility, g/m3	958	Koc	34.1023	ionic	1.5364
vapor pressure, Pa	5.000	BCF	4.1588	total	1.6503
log Kow	1.92	Ksw	1.6369	at environ. pH	6
fugacity ratio, F	1.0000	Ksd/w	3.2738	neutral	0.1139
dissoc. const. pKa	4.87	Kssd/w	10.2307	ionic	1.5364
		Kar/w	1.20E+06	total	1.6503

Compartment	Z	Concentration			Amount	Amount
	mol/m3 Pa	mol/m3	mg/L (or g/m3)	ug/g	kg	%
Air	4.03E-04	8.79E-10	1.02E-07	8.61E-05	10206.05	10.206
Water	1.65E+00	3.60E-06	4.18E-04	4.18E-04	83502	83.502
Soil	2.70E+00	5.89E-06	6.83E-04	2.85E-04	6150.83	6.1508
Biota (fish)	6.86E+00	1.50E-05	1.74E-03	1.74E-03	0.3473	3.47E-04
Suspended sediment	1.69E+01	3.68E-05	4.27E-03	2.85E-03	4.2714	4.27E-03
Bottom sediment	5.40E+00	1.18E-05	1.37E-03	5.70E-04	136.685	0.1367
	Total				100000	100

f = 2.179E-06 Pa

Chemical name: Hexanoic acid

Level II calculation: (six-compartment model)

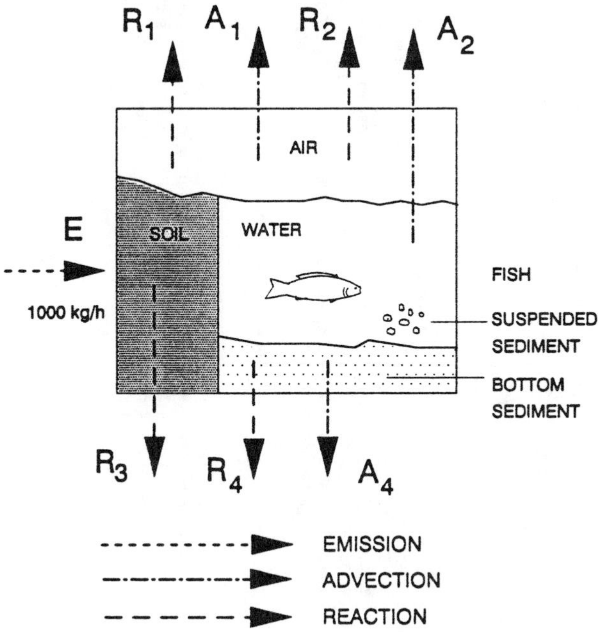

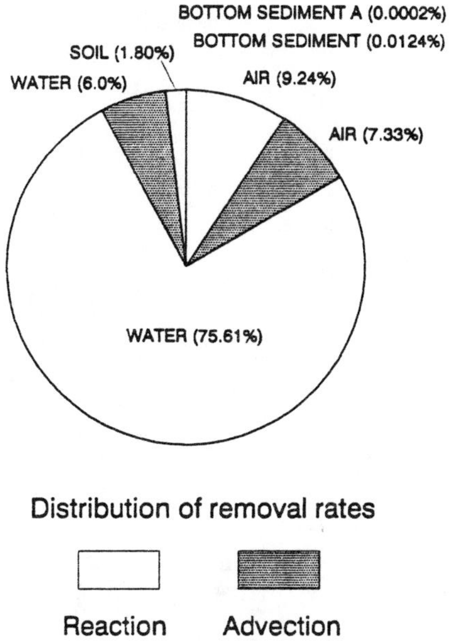

Distribution of removal rates

Level II Calculation:

Compartment	Half-life h	D Values Reaction mol/Pa.h	D Values Advection mol/Pa.h	Conc'n mol/m3	Loss Reaction kg/h	Loss Advection kg/h	Removal %
Air	55	5.08E+08	4.03E+08	6.32E-10	92.414	73.344	16.576
Water	55	4.16E+09	3.30E+08	2.58E-06	756.091	60.007	81.610
Soil	170	9.91E+07		4.23E-06	18.0188		1.8019
Biota (fish)				1.07E-05			
Suspended sediment				2.64E-05			
Bottom sediment	550	6.81E+05	1.08E+04	8.46E-06	0.1238	1.96E-03	1.26E-02
	Total	4.77E+09	7.33E+08		866.65	133.353	100
	R + A		5.50E+09			1000	

f = 1.566E-06 Pa

Total amt = 71863 kg

Overall residence time = 71.86 h

Reaction time = 82.92 h

Advection time = 538.89 h

Fugacity Level III calculations: (four-compartment model)

Chemical name: Hexanoic acid

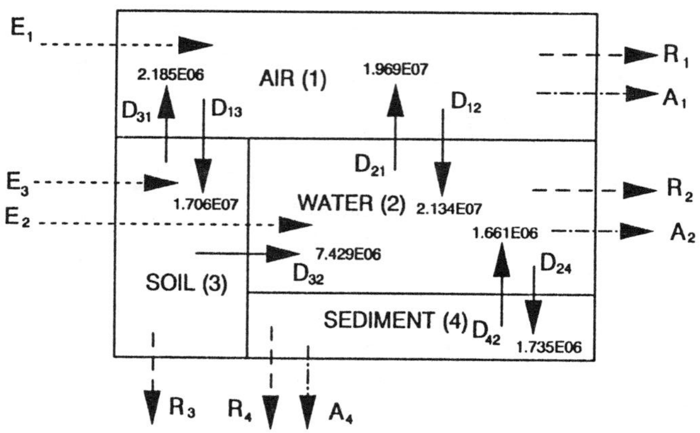

Phase Properties and Rates:

Compartment	Bulk Z mol/m3 Pa	Half-life h	D Values Reaction mol/Pa h	Advection mol/Pa h
Air (1)	4.034E-04	55	5.08E+08	4.03E+08
Water (2)	1.650E+00	55	4.16E+09	3.30E+08
Soil (3)	1.846E+00	170	1.35E+08	
Sediment (4)	2.401E+00	550	1.51E+06	2.40E+04

	E(1)=1000	E(2)=1000	E(3)=1000	E(1,2,3)
Overall residence time =	48.31	73.55	233.55	184.29 h
Reaction time =	84.22	79.52	235.97	193.43 h
Advection time =	113.27	979.89	22747	3898.15 h

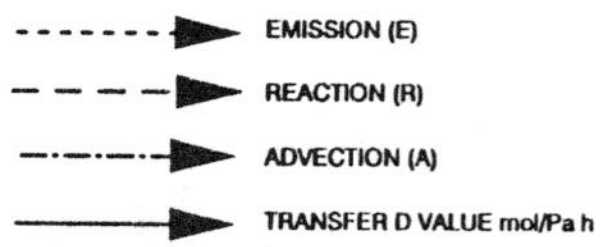

Phase Properties, Compositions, Transport and Transformation Rates:

Emission, kg/h			Fugacity, Pa				Concentration, g/m3				Amounts, kg				Total Amount, kg
E(1)	E(2)	E(3)	f(1)	f(2)	f(3)	f(4)	C(1)	C(2)	C(3)	C(4)	m(1)	m(2)	m(3)	m(4)	
1000	0	0	9.069E-06	4.468E-08	1.067E-06	2.424E-08	4.248E-07	8.560E-06	2.286E-04	6.756E-06	4.248E+04	1.712E+03	4.115E+03	3.378E+00	4.831E+04
0	1000	0	3.959E-08	1.910E-06	4.658E-09	1.036E-06	1.855E-09	3.660E-04	9.982E-07	2.888E-04	1.855E+02	7.320E+04	1.797E+01	1.444E+02	7.355E+04
0	0	1000	1.386E-07	9.850E-08	5.939E-05	5.344E-08	6.492E-09	1.887E-05	1.273E-02	1.489E-05	6.492E+02	3.775E+03	2.291E+05	7.447E+00	2.335E+05
50	250	700	5.604E-07	5.487E-07	4.163E-05	2.977E-07	2.625E-08	1.051E-04	8.922E-03	8.298E-05	2.625E+03	2.103E+04	1.606E+05	4.149E+01	1.843E+05

Emission, kg/h			Loss, Reaction, kg/h				Loss, Advection, kg/h			Intermedia Rate of Transport, kg/h						
										T12	T21	T13	T31	T32	T24	T42
E(1)	E(2)	E(3)	R(1)	R(2)	R(3)	R(4)	A(1)	A(2)	A(4)	air-water	water-air	air-soil	soil-air	soil-water	water-sed	sed-water
1000	0	0	5.352E+02	2.157E+01	1.68E+01	4.256E-03	4.248E+02	1.712E+00	6.756E-05	2.247E+01	1.021E-01	1.797E+01	2.706E-01	9.201E-01	8.998E-03	4.674E-03
0	1000	0	2.337E+00	9.224E+02	7.32E-02	1.820E-01	1.855E+00	7.320E+01	2.888E-03	9.811E-02	4.367E+00	7.844E-02	1.181E-03	4.017E-03	3.847E-01	1.999E-01
0	0	1000	8.180E+00	4.756E+01	9.34E+02	9.383E-03	6.492E+00	3.775E+00	1.489E-04	3.435E-01	2.252E-01	2.746E-01	1.507E+01	5.123E+01	1.984E-02	1.031E-02
50	250	700	3.307E+01	2.650E+02	6.55E+02	5.227E-02	2.625E+01	2.103E+01	8.298E-04	1.389E+00	1.254E+00	1.110E+00	1.056E+01	3.591E+01	1.105E-01	5.741E-02

Level III Distribution

Chemical name: Hexanoic acid

Distribution of mass

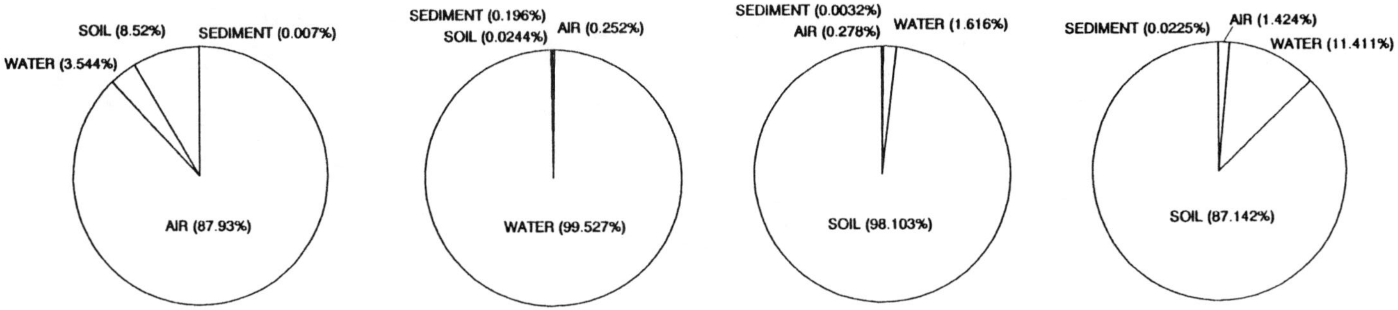

Distribution of removal rates

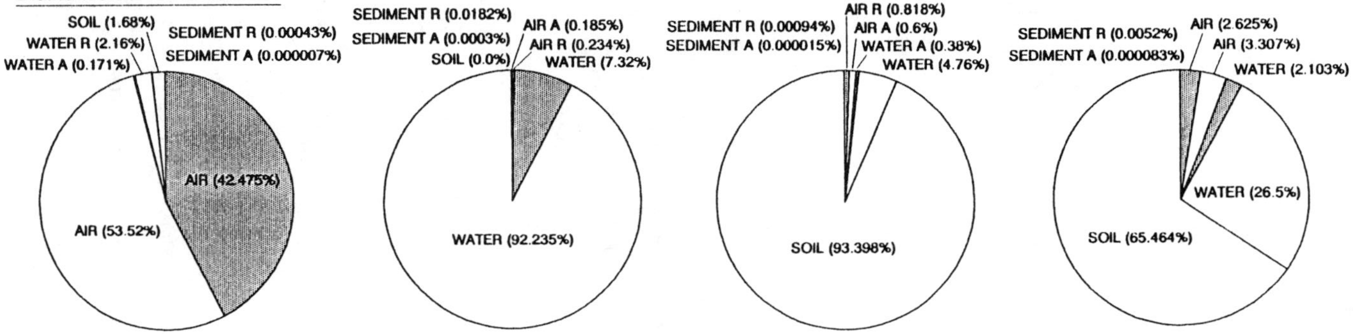

Reaction   Advection

Emission rates:

E(1) = 1000 kg/h	E(1) = 0	E(1) = 0	E(1) = 50 kg/h
E(2) = 0	E(2) = 1000 kg/h	E(2) = 0	E(2) = 250 kg/h
E(3) = 0	E(3) = 0	E(3) = 1000 kg/h	E(3) = 700 kg/h

Residence time:

t = 48.31 h	t = 73.55 h	t = 234 h	t = 184 h

Chemical name: Benzoic acid

Level I calculation: (six-compartment model)

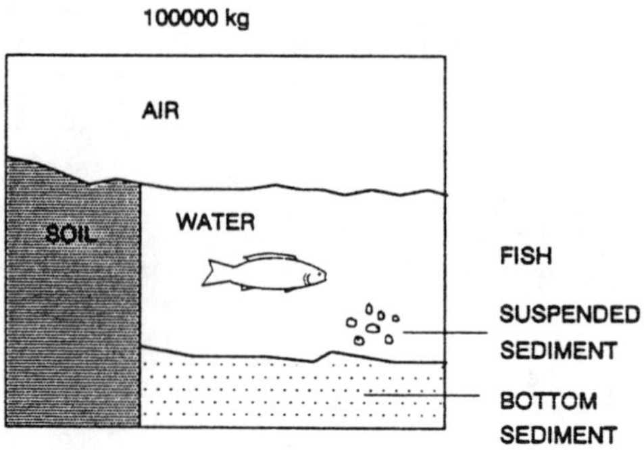

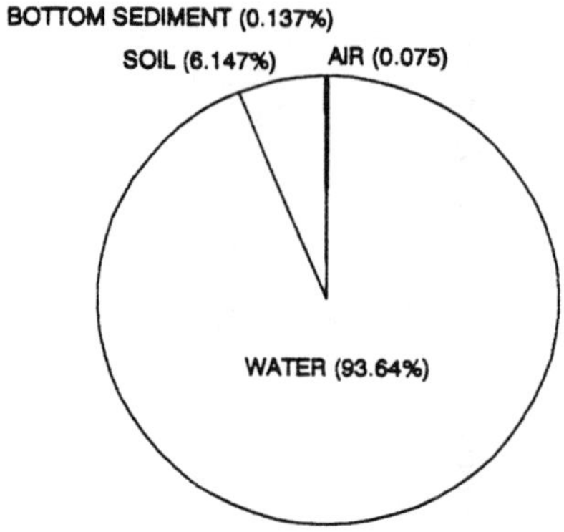

Distribution of mass

Level I calculation:

Physical-chemical properties:		Partition coeff.:		Z values in water:	
molecular wt., g/mol	122.13	HLC	0.00395	at data pH	6
melting point, C	122.4	Kaw	1.59E-06	neutral	3.860
solubility, g/m3	3400	Koc	30.3937	ionic	249.224
vapor pressure, Pa	0.110	BCF	3.7066	total	253.084
log Kow	1.87	Ksw	1.4589	at environ. pH	6
fugacity ratio, F	0.1088	Ksd/w	2.9178	neutral	3.860
dissoc. const. pKa	4.19	Kssd/w	9.1181	ionic	249.224
		Kar/w	5.94E+06	total	253.084

Compartment	Z	Concentration			Amount	Amount
	mol/m3 Pa	mol/m3	mg/L (or g/m3)	ug/g	kg	%
Air	4.03E-04	6.11E-12	7.46E-10	6.30E-07	74.6291	0.0746
Water	2.53E+02	3.83E-06	4.68E-04	4.68E-04	93637	93.637
Soil	3.69E+02	5.59E-06	6.83E-04	2.85E-04	6147.30	6.1473
Biota (fish)	9.38E+02	1.42E-05	1.74E-03	1.74E-03	0.3471	3.47E-04
Suspended sediment	2.31E+03	3.50E-05	4.27E-03	2.85E-03	4.2690	4.27E-03
Bottom sediment	7.38E+02	1.12E-05	1.37E-03	5.69E-04	136.607	0.1366
	Total				100000	100

f = 1.515E-08 Pa

Chemical name: Benzoic acid

Level II calculation: (six-compartment model)

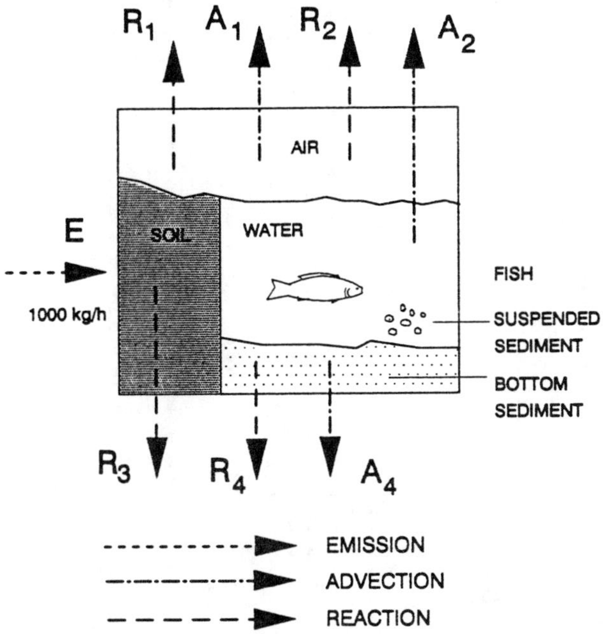

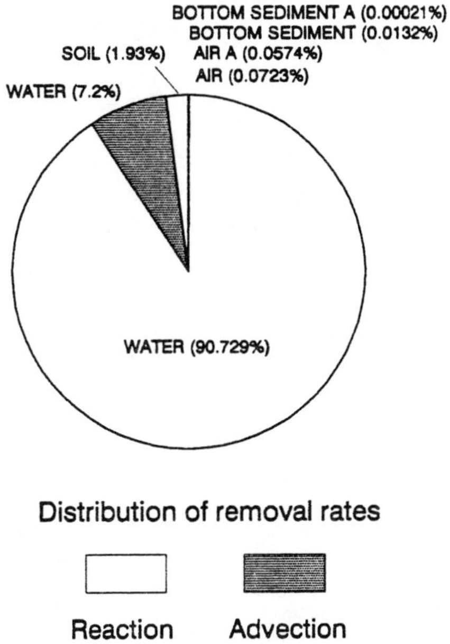

Level II Calculation:

Compartment	Half-life h	D Values Reaction mol/Pa.h	D Values Advection mol/Pa.h	Conc'n mol/m3	Loss Reaction kg/h	Loss Advection kg/h	Removal %
Air	55	5.08E+08	4.03E+08	4.70E-12	0.723	0.574	0.130
Water	55	6.38E+11	5.06E+10	2.95E-06	907.291	72.007	97.930
Soil	170	1.35E+10		4.30E-06	19.2707		1.9271
Biota (fish)				1.09E-05			
Suspended sediment				2.69E-05			
Bottom sediment	550	9.30E+07	1.48E+06	8.60E-06	0.1324	2.10E-03	1.34E-02
	Total	6.52E+11	5.10E+10		927.42	72.583	100
	R + A		7.03E+11			1000	

f = 1.165E-08 Pa

Total amt = 76900 kg

Overall residence time = 76.90 h

Reaction time = 82.92 h

Advection time = 1059.48 h

Fugacity Level III calculations: (four-compartment model)

Chemical name: Benzoic acid

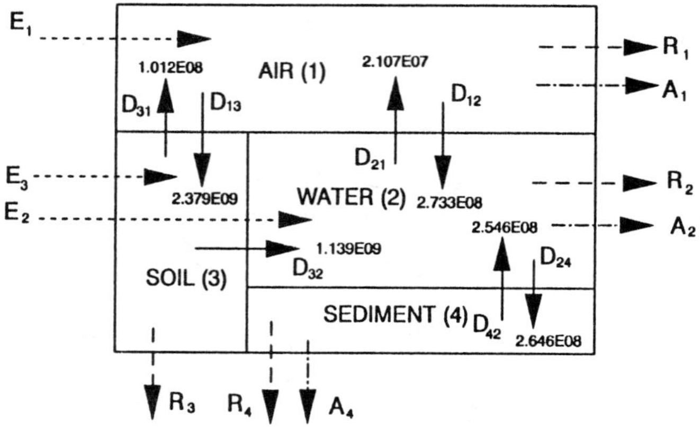

Phase Properties and Rates:

Compartment	Bulk Z mol/m3 Pa	Half-life h	D Values Reaction mol/Pa h	D Values Advection mol/Pa h
Air (1)	4.035E-04	55	5.08E+08	4.03E+08
Water (2)	2.531E+02	55	6.38E+11	5.06E+10
Soil (3)	2.605E+02	170	1.91E+10	
Sediment (4)	3.502E+02	550	2.21E+08	3.50E+06

	E(1)=1000	E(2)=1000	E(3)=1000	E(1,2,3)
Overall residence time =	174.06	73.66	235.35	191.86 h
Reaction time =	198.25	79.50	236.47	197.35 h
Advection time =	1426.87	1001.89	49858	6905.58 h

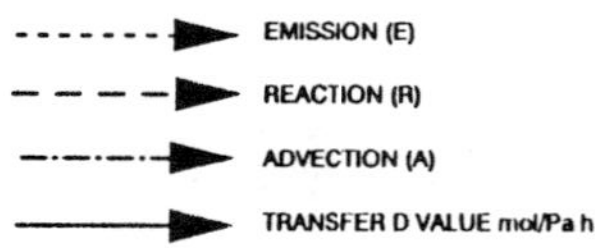

Phase Properties, Compositions, Transport and Transformation Rates:

Emission, kg/h E(1)	E(2)	E(3)	Fugacity, Pa f(1)	f(2)	f(3)	f(4)	Concentration, g/m3 C(1)	C(2)	C(3)	C(4)	Amounts, kg m(1)	m(2)	m(3)	m(4)	Total Amount, kg
1000	0	0	2.305E-06	1.360E-09	2.694E-07	7.521E-10	1.136E-07	4.205E-05	8.571E-03	3.216E-05	1.136E+04	8.410E+03	1.543E+05	1.608E+01	1.741E+05
0	1000	0	6.751E-11	1.189E-08	7.890E-12	6.574E-09	3.327E-12	3.676E-04	2.510E-07	2.811E-04	3.327E-01	7.351E+04	4.519E+00	1.406E+02	7.366E+04
0	0	1000	1.146E-08	6.722E-10	4.035E-07	3.716E-10	5.646E-10	2.078E-05	1.284E-02	1.589E-05	5.646E+01	4.156E+03	2.311E+05	7.946E+00	2.354E+05
50	250	700	1.233E-07	3.511E-09	2.960E-07	1.941E-09	6.075E-09	1.085E-04	9.417E-03	8.302E-05	6.075E+02	2.171E+04	1.695E+05	4.151E+01	1.919E+05

Emission, kg/h E(1)	E(2)	E(3)	Loss, Reaction, kg/h R(1)	R(2)	R(3)	R(4)	Loss, Advection, kg/h A(1)	A(2)	A(4)	Intermedia Rate of Transport, kg/h T12 air-water	T21 water-air	T13 air-soil	T31 soil-air	T32 soil-water	T24 water-sed	T42 sed-water
1000	0	0	1.431E+02	1.060E+02	6.29E+02	2.026E-02	1.136E+02	8.410E+00	3.216E-04	7.693E+01	3.351E-03	6.697E+02	3.328E+00	3.748E+01	4.397E-02	2.338E-02
0	1000	0	4.192E-03	9.263E+02	1.84E-02	1.771E-01	3.327E-03	7.351E+01	2.811E-03	2.253E-03	2.929E-02	1.962E-02	9.748E-05	1.098E-03	3.843E-01	2.044E-01
0	0	1000	7.114E-01	5.236E+01	9.42E+02	1.001E-02	5.646E-01	4.156E+00	1.589E-04	3.824E-01	1.656E-03	3.329E+00	4.986E+00	5.615E+01	2.172E-02	1.155E-02
50	250	700	7.654E+00	2.735E+02	6.91E+02	5.230E-02	6.075E+00	2.171E+01	8.302E-04	4.115E+00	8.649E-03	3.582E+01	3.657E+00	4.118E+01	1.135E-01	6.035E-02

Level III Distribution

Chemical name: Benzoic acid

Distribution of mass

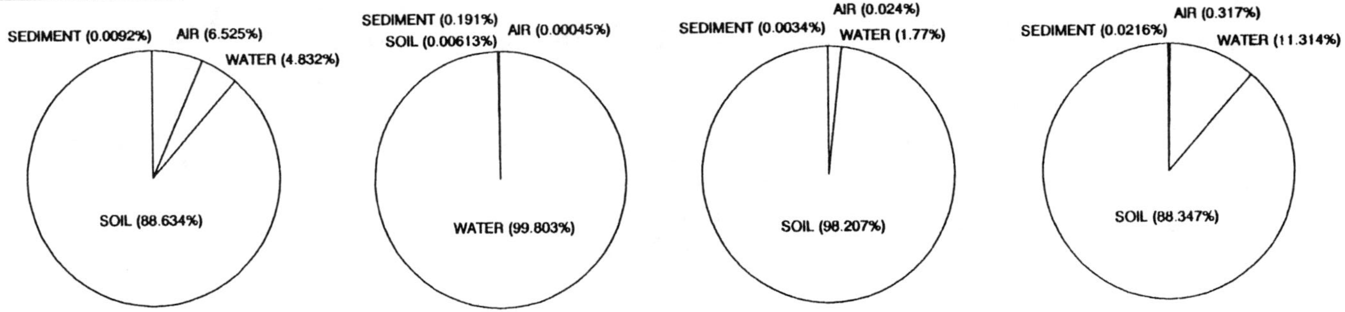

Distribution of removal rates

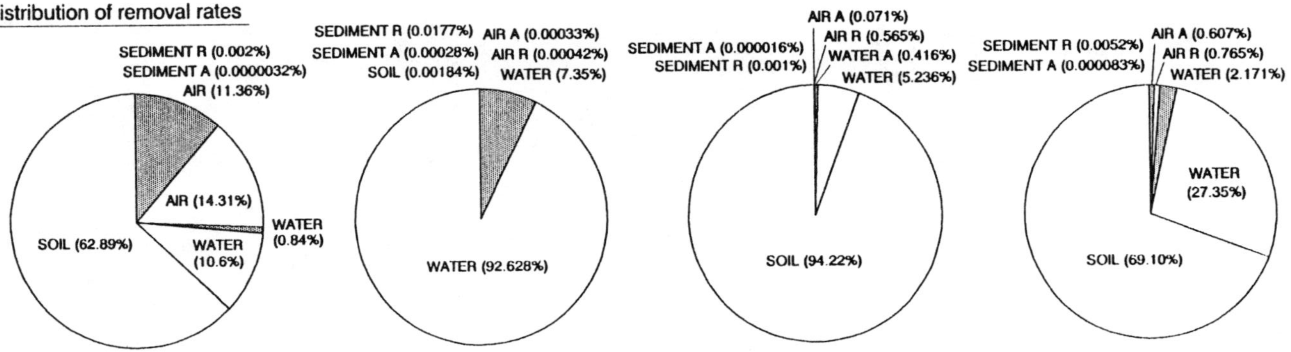

Emission rates:

E(1) = 1000 kg/h	E(1) = 0	E(1) = 0	E(1) = 50 kg/h
E(2) = 0	E(2) = 1000 kg/h	E(2) = 0	E(2) = 250 kg/h
E(3) = 0	E(3) = 0	E(3) = 1000 kg/h	E(3) = 700 kg/h

Residence time:

t = 174 h	t = 73.66 h	t = 235 h	t = 192 h

Reaction   Advection

Chemical name: 2,4-D

Level I calculation: (six-compartment model)

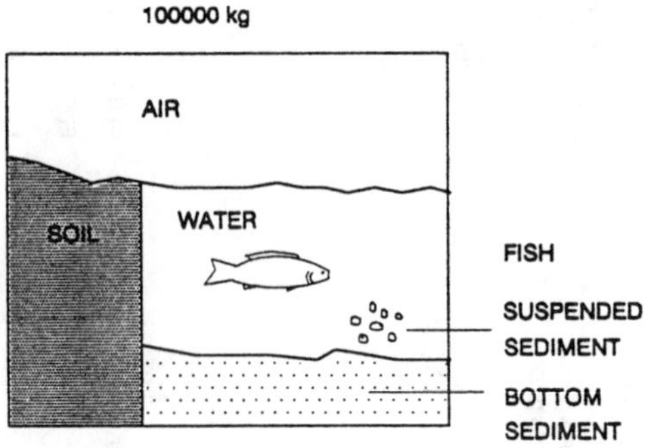

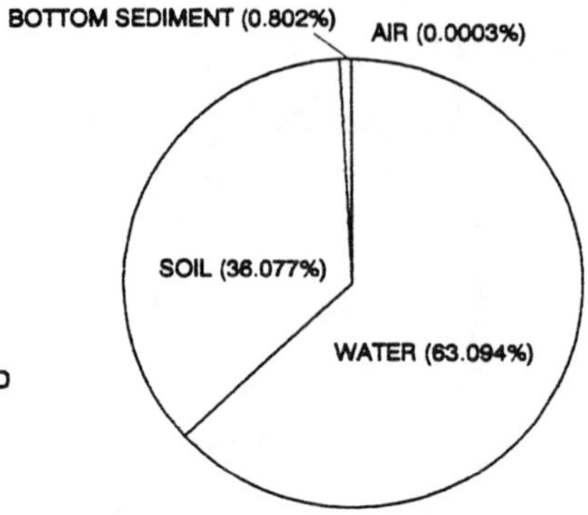

Distribution of mass

Level I calculation:

Physical-chemical properties:		Partition coeff.:		Z values in water:	
molecular wt., g/mol	255.49	HLC	2.30E-05	at data pH	6
melting point, C	158	Kaw	9.26E-09	neutral	27.457
solubility, g/m3	890	Koc	264.7182	ionic	43516.32
vapor pressure, Pa	8.00E-05	BCF	32.2827	total	43543.78
log Kow	2.81	Ksw	12.7065	at environ. pH	6
fugacity ratio, F	0.0484	Ksd/w	25.4130	neutral	27.457
dissoc. const. pKa	2.8	Kssd/w	79.4155	ionic	43516.32
		Kar/w	3.63E+09	total	43543.78

Compartment	Z	Concentration			Amount	Amount
	mol/m3 Pa	mol/m3	mg/L (or g/m3)	ug/g	kg	%
Air	4.03E-04	1.14E-14	2.92E-12	2.47E-09	0.2923	2.92E-04
Water	4.35E+04	1.23E-06	3.15E-04	3.15E-04	63094	63.094
Soil	5.53E+05	1.57E-05	4.01E-03	1.67E-03	36077	36.0767
Biota (fish)	1.41E+06	3.99E-05	1.02E-02	1.02E-02	2.0369	2.04E-03
Suspended sediment	3.46E+06	9.81E-05	2.51E-02	1.67E-02	25.0533	2.51E-02
Bottom sediment	1.11E+06	3.14E-05	8.02E-03	3.34E-03	801.705	0.8017
	Total				100000	100

f = 2.836E-11 Pa

Chemical name: 2,4-D

Level II calculation: (six-compartment model)

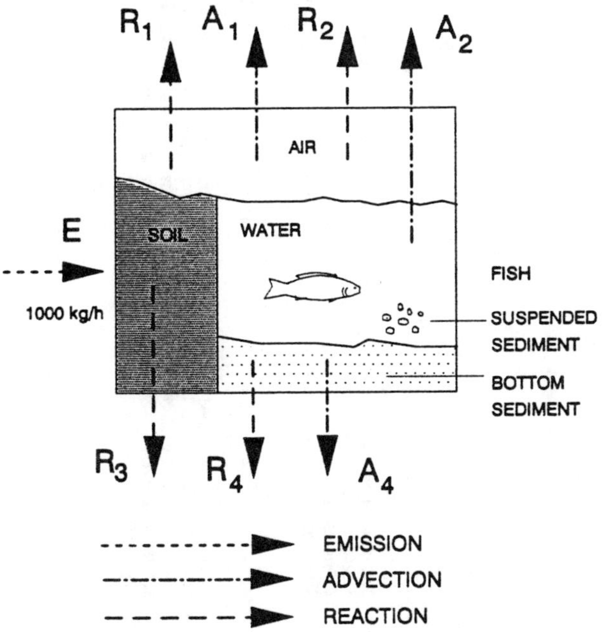

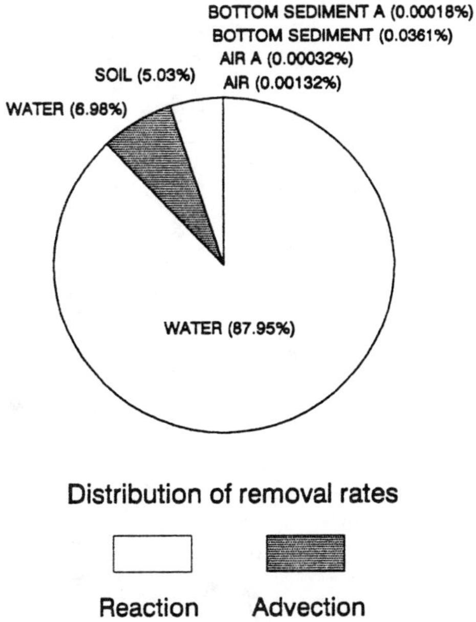

Level II Calculation:

Compartment	Half-life h	D Values Reaction mol/Pa.h	D Values Advection mol/Pa.h	Conc'n mol/m3	Loss Reaction kg/h	Loss Advection kg/h	Removal %
Air	17	1.64E+09	4.03E+08	1.27E-14	0.0132	0.0032	0.0016
Water	55	1.10E+14	8.71E+12	1.37E-06	879.512	69.803	94.931
Soil	550	6.27E+12		1.74E-05	50.290		5.0290
Biota (fish)				4.41E-05			
Suspended sediment				1.08E-04			
Bottom sediment	1700	4.51E+10	2.21E+09	3.47E-05	0.3616	1.77E-02	3.79E-02
	Total	1.16E+14	8.71E+12		930.18	69.824	100
	R + A		1.25E+14			1000	

f = 3.137E-11 Pa
Total amt = 110632 kg

Overall residence time = 110.63 h
Reaction time = 118.94 h
Advection time = 1584.46 h

Fugacity Level III calculations: (four-compartment model)

Chemical name: 2,4-D

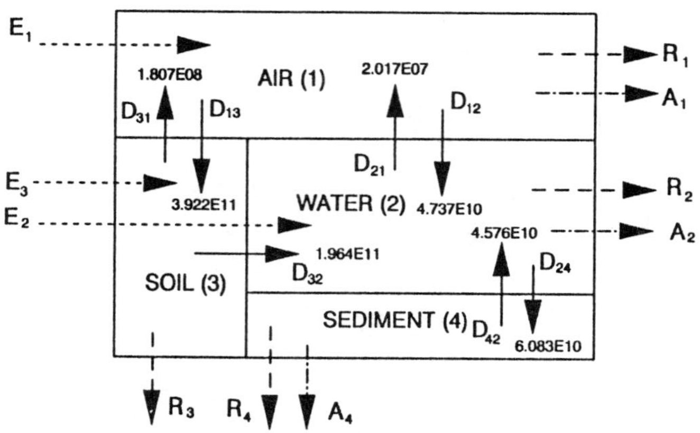

Phase Properties and Rates:

Compartment	Bulk Z mol/m3 Pa	Half-life h	D Values Reaction mol/Pa h	D Values Advection mol/Pa h
Air (1)	4.327E-04	17	1.76E+09	4.33E+08
Water (2)	4.356E+04	55	1.10E+14	8.71E+12
Soil (3)	2.897E+05	550	6.57E+12	
Sediment (4)	2.562E+05	1700	5.22E+10	2.56E+09

	E(1)=1000	E(2)=1000	E(3)=1000	E(1,2,3)	
Overall residence time =	699.48	74.16	772.76	594.45	h
Reaction time =	706.69	80.05	774.42	606.82	h
Advection time =	68476.48	1008.72	362027	29160.48	h

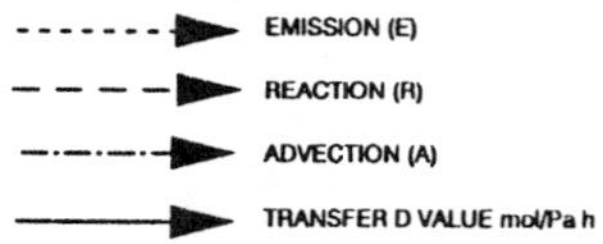

Phase Properties, Compositions, Transport and Transformation Rates:

Emission, kg/h E(1)	E(2)	E(3)	Fugacity, Pa f(1)	f(2)	f(3)	f(4)	Concentration, g/m3 C(1)	C(2)	C(3)	C(4)	Amounts, kg m(1)	m(2)	m(3)	m(4)	Total Amount, kg
1000	0	0	8.938E-09	4.144E-12	5.179E-10	2.508E-12	9.881E-10	4.613E-05	3.834E-02	1.641E-04	9.881E+01	9.225E+03	6.901E+05	8.206E+01	6.995E+05
0	1000	0	1.521E-15	3.302E-11	8.815E-17	1.998E-11	1.682E-16	3.675E-04	6.524E-09	1.308E-03	1.682E-05	7.351E+04	1.174E-01	6.539E+02	7.416E+04
0	0	1000	2.387E-13	9.588E-13	5.784E-10	5.802E-13	2.639E-14	1.067E-05	4.281E-02	3.797E-05	2.639E-03	2.134E+03	7.706E+05	1.898E+01	7.728E+05
50	250	700	4.471E-10	9.134E-12	4.308E-10	5.528E-12	4.942E-11	1.017E-04	3.188E-02	3.617E-04	4.942E+00	2.033E+04	5.739E+05	1.809E+02	5.944E+05

Emission, kg/h E(1)	E(2)	E(3)	Loss, Reaction, kg/h R(1)	R(2)	R(3)	R(4)	Loss, Advection, kg/h A(1)	A(2)	A(4)	Intermedia Rate of Transport, kg/h T12 air-water	T21 water-air	T13 air-soil	T31 soil-air	T32 soil-water	T24 water-sed	T42 sed-water
1000	0	0	4.028E+00	1.162E+02	8.69E+02	3.345E-02	9.881E-01	9.225E+00	1.641E-03	9.950E+01	2.136E-05	8.955E+02	2.391E-02	2.600E+01	6.441E-02	2.932E-02
0	1000	0	6.855E-07	9.262E+02	1.48E-04	2.666E-01	1.682E-07	7.351E+01	1.308E-02	1.693E-05	1.702E-04	1.524E-04	4.070E-09	4.424E-06	5.133E-01	2.336E-01
0	0	1000	1.076E-04	2.689E+01	9.71E+02	7.739E-03	2.639E-05	2.134E+00	3.797E-04	2.657E-03	4.941E-06	2.392E-02	2.670E-02	2.903E+01	1.490E-02	6.783E-03
50	250	700	2.015E-01	2.562E+02	7.23E+02	7.373E-02	4.942E-02	2.033E+01	3.617E-03	4.977E+00	4.707E-05	4.479E+01	1.989E-02	2.162E+01	1.420E-01	6.462E-02

Level III Distribution

Chemical name: 2,4-D

Distribution of mass

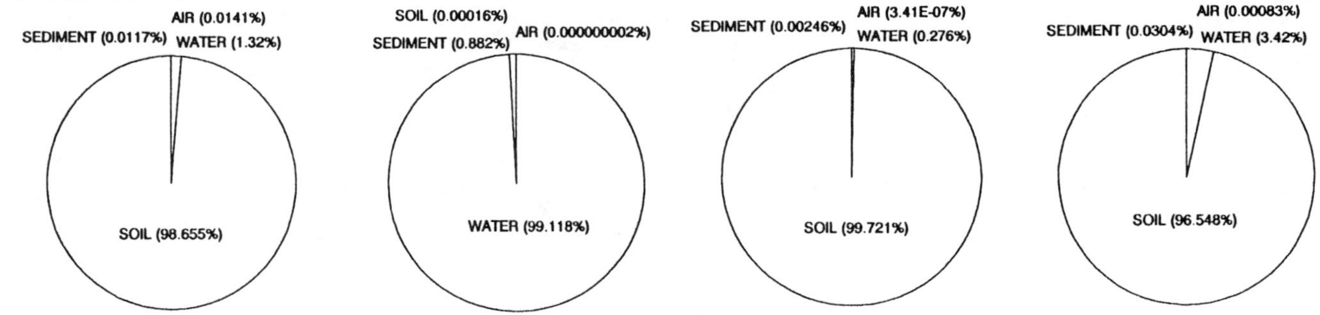

Distribution of removal rates

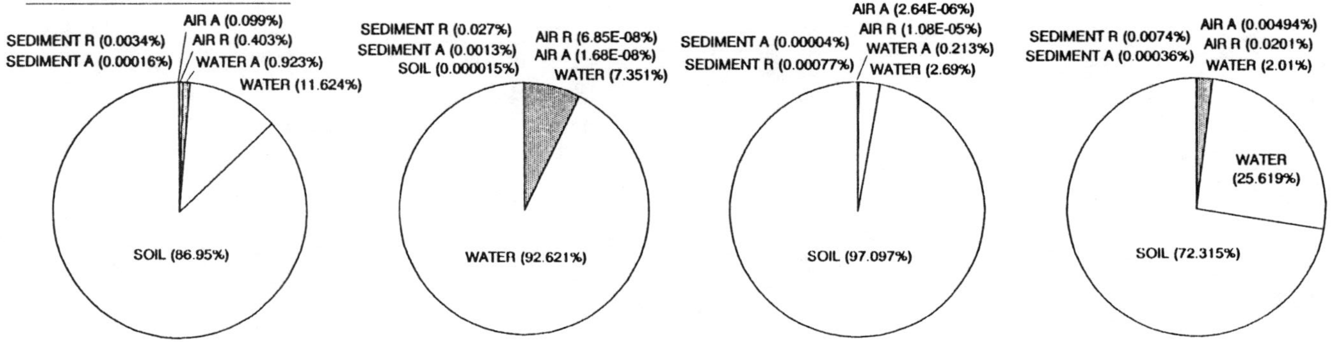

Reaction — Advection

Emission rates:				
E(1)	= 1000 kg/h	= 0	= 0	= 50 kg/h
E(2)	= 0	= 1000 kg/h	= 0	= 250 kg/h
E(3)	= 0	= 0	= 1000 kg/h	= 700 kg/h
Residence time: t	= 700 h	= 74.16 h	= 773 h	= 595 h

5.4 Commentary on the Physical-Chemical Properties and Environmental Fate

Properties and Reactivity

The selected physical-chemical properties of the carboxylic acids are given in Tables 5.1 and 5.2 including pK_a values which are generally below environmental pH, thus dissociation is extensive. Solubility and K_{OW} are thus sensitive to pH. Table 5.3 gives the estimated half-lives. These half-lives shoud be regarded as only tentative, average values or ranges. From reviews of the literature it is suggested that air, water and soil half-lives are typically Class 3 or 4, i.e., 30 to 300 hours with the sediment half-lives being somewhat longer. Chlorinated aromatic acids display slower reactions presumably because of the stability imported by the chlorine. This class of organic chemicals is more oxidized than the alcohols and carbonyl compounds and is thus less susceptible to further oxidation, but biodegradation is likely to be quite rapid in most cases, especially when the substance is soluble.

QSPR Plots

Figures 5.1 to 5.6 give the QSPR plots. The various molecular descriptors in Fig. 5.1 are clearly related to molar volume and to each other. The solubility data in Fig. 5.2 are very scattered largely because of the variety of acids included, but also because of variable dissociation. The vapor pressure data in Fig. 5.3 are better behaved especially for the aliphatic acids which are similar in structure. The log K_{OW} data in Figure 5.4 are surprisingly well-behaved, suggest an increase of 0.5 log units per methylene or a factor of 3 but there is more scatter of aromatic and longer chain acids.

The Henry's law constant data in Figure 5.5 show considerable scatter. The values are mostly below 1.0 $Pa \cdot m^3/mol$, thus appreciable evaporation is not expected and accurate values of H may not be needed. It is likely the H calculated from vapor pressure and solubility can be in substantial error for the carboxylic acids thus the experimental data should be assigned greater reliabililty.

Figure 5.6 shows the usual near inverse relationship between octanol-water partition coefficient and solubility.

These plots can be used with appropriate caution to help in estimating the properties of other acids, but the strong interaction with water renders the use of molar volume as a simple single descriptor suspect. More complex QSPR analyses are obviously needed for the acids to obtain reliable results.

Evaluative Calculations

The evaluative fugacity calculations show that most aliphatic acids partition primarily into water with little tendency to evaporate, because of the low Henry's law constant and little tendency to sorb to organic matter (because of the low K_{OW}).

The Level II calculations suggest that reaction in water will be the dominant mechanism of loss.

The Level III calculations show that these substances, when discharged to water or soil, are relatively immobile, tending to remain and react in these media. They are not very persistent, having half-lives of 5 to 10 days as controlled by degrading reactions. They are thus likely to be fairly localized, short lived contaminants. Their behavior will of course be affected by pH and the presence of other ionic species.

The aromatic carboxylic acids are generally similar to the aliphatic acids but the chlorinated acids, e.g., 2,4-D are more persistent. This class of chemical is often used as herbicides thus discharge is primarily to soil or water. There is negligible tendency to evaporate, but soil to water run-off can be appreciable. The key parameters are the reaction half-lives in soil and water. The relatively low K_{OW}s suggest that bioconcentration will be minimal.

This class of chemical is thus likely to be quite localized in its effects, being relatively non-persistent and immobile.

5.5 REFERENCES:

Agrochemicals Handbook (1987) Hartley, D., Kidd, H., Eds., second edition, Royal Society of Chemistry, Nottingham, England.

Alexander, M., Lustigman, B.K. (1966) Effect of chemical structure on microbial degradation of substituted benzenes. *J. Agric. Food Chem.* 14, 410-413.

Ambrose, D., Ellender, J.H., Sprake, C.H.S., Townsend, R. (1977) Thermodynamic properties of organic oxygen compounds. XLV. The vapour pressures of acetic acid. *J. Chem. Thermodyn.* 9, 735-741.

Ambrose, D., Ellender, J.H., Gundry, H.A., Lee, D.A., Townsend, R. (1981) Thermodynamic properties of organic oxygen compounds. LI. The vapour pressures of some esters and fatty acids. *J. Chem. Thermodyn.* 13, 795-802.

Amidon, G.L., Williams, N.A. (1982) A solubility equation for non-electrolytes in water. *Int'l J. Pharm.* 11, 249-256.

Amidon, G.L., Yalkowsky, S.H., Anik, S.T., Valvani, S.C. (1975) Solubility of nonelectrolytes in polar solvents. V. Estimation of the solubility of aliphatic monofunctional compounds in water using a molecular surface area approach. *J. Phys. Chem.* 7992), 2239-2246.

Apelblat, A., Manzurola, E. (1987) Solubility of oxalic acid, malonic, succinic, adipic, maleic, malic, citric and tartaric acids in water from 278.15 to 338.15 K. *J. Chem. Thermodyn.* 19, 317-320.

Apelblat, A., Manzurola, E. (1989) Solubility of ascorbic acid, 2-furancarboxylic, glutaric, pimelic, salicylic, and *o*-phthalic aicds in water from 279.15 to 342.15 K, and apparent molar volumes of ascorbic, glutaric, and pimelic acids in water at 298.15 K. *J. Chem. Thermodyn.* 21, 1005-1008.

Atkinson, R. (1985) Kinetics and mechanisms of the gas-phase reactions of hydroxyl radicals with organic compounds under atmospheric conditions. *Chem. Rev.* 85, 69-201.

Atkinson, R. (1987) Structure-activity relationship for the estimation of rate constants for the gas-phase reactions of OH radicals with organic compounds. *Int'l J. Chem. Kinetics* 19, 799-828.

Atkinson, R. (1990) Gas-phase tropospheric chemistry of organic compounds: A review. *Atmos. Environ.* 24A(1), 1-41.

Atkinson, R., Carter, W.P.L. (1984) Kinetics and mechanisms of the gas-phase reactions of ozone with organic compounds under atmospheric conditions. *Chem. Rev.* 84, 437-470.

Bailey, G.W., White, J.L. (1965) Herbicides-A compilation of their physical, chemical, and biological properties. *Residue Rev.* 10, 97.

Banerjee, S. (1985) Calculation of water solubility of organic compounds with UNIFAC-derived parameters. *Environ. Sci. Technol.* 19, 369-370.

Banerjee, S., Howard, P.H. (1988) Improved estimation of solubility and partitioning through correction of UNIFAC-derived activity coefficients. *Environ. Sci. Technol.* 22, 839-841.

Banerjee, S., Howard, P.H., Lande, S.S. (1990) General structure-vapor pressure relationships for organics. *Chemosphere* 21(10-11), 1173-1180.

Baur, J.R., Bovey, R.W. (1974) Ultraviolet and volatility loss of herbicides. *Arch. Environ. Contam. Toxicol.* 2, 275-288.

Berthod, A., Han, Y.I., Armstrong, D.W. (1988) Centrifugal partition chromatography. V. octanol-water partition coefficients, direct and indirect determination. *J. Liq. Chromatogr.* 11(7), 1441-1456.

Berthod, A., Menges, R.A., Armstrong, D.W. (1992) Direct octanol-water partition coefficient determination using counter-current chromatography. *J. Liq. Chromatogr.* 15(15-16), 2769-2785.

Bintein, S., Devillers, J. (1994) QSAR for organic chemical sorption in soils and sediments. *Chemosphere* 28(6), 1171-1188.

Bodor, N., Gabanyi, Z., Wong, C.K. (1989) A new method for the estimation of partition coefficient. *J. Am. Chem. Soc.* 111, 3783-3786.

Boethling, R.S., Alexander, M. (1979) Effects of concentration of organic chemicals on their biodegradation by natural microbial communities. *Appl. Environ. Microbiol.* 37, 1211-1216.

Boethling, R.S., Gregg, B., Frederick, R., Gabel, N.W., Campbell, S.E., Sabljic, A. (1989) Expert systems survey on biodegradation of xenobiotic chemicals. *Ecotox. Environ. Saf.* 18, 252-267.

Boublik, T., Fried, V., Hala, E. (1973) *The Vapour Pressures of Pure Substances*. Elsevier, Amsterdam, The Netherlands.

Boublik, T., Fried, V., Hala, E. (1984) *The Vapour Pressures of Pure Substances*. 2nd Edition, Elsevier, Amsterdam, The Netherlands.

Branson, D.R. (1978) In: *Predicting the Fate of Chemicals in the Aquatic Environment from Laboratory Data*. ASTM STP 657. pp.55-70. American Society for Testing and Materials, Philadephia, Pa.

Brimblecombe, P., Clegg, S.L., Khan, I. (1992) Thermodynamic properties of carboxylic acids relevant to their solubility in aqueous solutions. *J. Aerosol Sci.* 23, suppl. 1, S901-S904.

Brooke, D., Nielsen, I., De Bruijn, J., Hermens, J. (1990) An interlaboratory evaluation of the stir-flask method for the determination of octanol-water partition coefficients (LOG POW). *Chemosphere* 21(1/2), 119-133.

Bundgaard, H., Falch, E., Larsen, C., Mosher, G.L., Mikkelson, T.J. (1986) Pilocarpine prodrugs. II. Synthesis, stability, bioconversion and physicochemical properties of sequentially labile pilocarpine acid diesters. *J. Pharm. Sci.* 75(8), 775-783.

Bundgaard, H., Nielsen, N.M. (1988) Glycolamide esters as a novel biolabile prodrug type for non-steroidal anti-inflammatory carboxylic acid drugs. *Int'l J. Pharm.* 43(1-2), 101-110.

Butler, J.A.V., Ramchandani, C.N. (1935) The solubility of non-electrolytes. Part II. The influence of the polar group on the free energy of hydration of aliphatic compounds. *J. Chem. Soc.* 952-955.

Byast, T.H., Hance, R.J. (1975) Degradation of 2,4,5-T by South Vietnamese soils incubated in the laboratory. *Bull. Environ. Contam. Toxicol.* 14, 71-76.

Callis-Van Ginkel, C.H.D., Calis, G.H.M., Timmermans, C.W.M., de Kruif, C.G., Oonk, H.A.J. (1978) Enthalpies of sublimation and dimerization in the vapour phase of formic, acetic, propanoic, and butanoic acids. *J. Chem. Thermodyn.* 10, 1083-1088.

Casserly, D.M., Davis, E.M., Downs, T.D., Guthrie, R.K. (1983) Sorption of organics by *selenastrum capricornutum*. *Water Res.* 17(11), 1591-1594.

Cessna, A.J., Grover, R. (1978) Spectroscopic determination of dissociation constants of selected acidic herbicides. *J. Agric. Food Chem.* 26(1), 289-292.

Cessna, A.J., Muir, D.C.G. (1991) Chapter 6, Photochemical transformations. In: *Environmental Chemistry of Herbicides*. Vol. II., Grover, R., Cessna, A.J., Editors, CRC Press, Boca Raton, Florida.

Chen, F., Holten-Andersen, J., Tyle, H. (1993) New development of the UNIFAC model for environmental application. *Chemosphere* 26(7), 1325-1354.

Chessells, M., Hawker, D.W., Connell, D.W. (1992) Influence of solubility in lipid on bioconcentration of hydrophobic compounds. *Ecotox. Environ. Saf.* 23, 260-273.

Chiou, C.T., Freed, V.H., Schmedding, D.W., Kohnert, R.L. (1977) Partition coefficient and bioaccumulation of selected organic chemicals. *Environ. Sci. Technol.* 11(5), 475-478.

Chou, J.T., Jurs, P.C. (1979) Chapter 5. Computation of partition coefficients from molecular structures by a fragment addition method. In: *Physical Chemical Properties of Drugs.* Medical Research Series, Vol. 10, Yalkowsky, S.H., Sinkula, A.A., Valvani, S.C., Editors.

Chou, W.L., Speece, R.E., Siddiqi, R.H. (1979) Acclimation and degradation of petrochemical wastewater components by methane fermentation. *Biotechnol. Bioeng. Symp.* 8, 391-414.

Clarke, F.H. (1984) Ionization constants by curve-fitting: application to the determination of partition coefficients. *J. Pharmaceut. Sci.* 73(2), 226-230.

Clarke, F.H., Cahoon, N.M. (1987) Ionization constants by curve-fitting: Determination of partition and distribution coefficients of acids and bases and their ions. *J. Pharmaceut. Sci.* 76, 611-620.

Collander, R. (1949) The distribution of organic compounds between ethyl ether and water. *Acta. Chem. Scand.* 3, 717.

Collander, R. (1950) The distribution of organic compounds between 1-butanol and water. *Acta. Chem. Scand.* 4, 1085.

Collander, R. (1951) Partition of organic compounds between higher alcohols and water. *Acta. Chem. Scand.* 5, 774-780.

Colomina, M., Jiménez, P., Turriôn, C. (1982) Vapour pressures and enthalpies of sublimation of naphthalene and benzoic acid. *J. Chem. Thermodyn.* 14, 779-784.

Cox, J.D. (1974) Recommended reference materials. *Pure & Appl. Chem.* 40, 424-425.

Da, Y.Z., Ito, K., Fujiwara, H. (1992) Energy aspects of oil/water partition leading to the novel hydrophobic parameters for the analysis of quantitative structure-activity relationships. *J. Med. Chem.* 35(18), 3382-3387.

Dagaut, P., Wallington, T.J., Liu, R., Kurylo, M.J. (1988) The gas-phase reactions of hydroxyl radicals with a series of carboxylic acids over the temperature range 240-440 K. *Int'l J. Chem. Kinetics* 20, 331-338.

D'Amboise, M., Hanai, T. (1982) Hydrophobicity and retention in reversed phase liquid chromatography. *J. Liq. Chromatogr.* 5(2), 229-244.

Davies, R.P., Dobbs, A.J. (1984) The prediction of bioconcentration in fish. *Water Res.* 18(10), 1253-1262.

Dean, J.D., Editor (1985) *Lange's Handbook of Chemistry.* 13th Edition, McGraw-Hill, Inc., New York.

De Kock, A.C., Lord, D.A. (1987) A simple procedure for determining octanol-water partition coefficients using reversed phase high performance liquid chromatography (RPHPLC). *Chemosphere* 16(1), 133-142.

de Kruif, C.G., Block, J.G. (1982) The vapour pressure of benzoic acid. *J. Chem. Thermodyn.* 14, 201-206.

Dermer, O.C., Dermer, V.H. (1943) Identification of organic acids by partition between ethyl ether and water. *J. Am. Chem. Soc.* 65, 1653-1654.

Dermer, O.C., Markham, W.G., Trimble, H.M. (1941) Partition ratios of some organic acids between water and ethyl ether. *J. Am. Chem. Soc.* 63, 3524-3525.

Dobbs, A.J., Cull, M.R. (1982) Volatilisation of chemicals-relative loss rates and the estimation of vapour pressures. *Environ. Pollut.* 3, 289-298.

Dojlido, J.R. (1979) *Investigations of Biodegradability and Toxicity of Organic Compounds.* Final Report 1975-79. U.S. EPA-600/2-79-163.

Dore, M., Brunet, N., Legube, B. (1975) Participation of various organic compounds in the evaluation of global pollution criteria. *Trib. Cebedeau.* 28, 3-11.

Dorfman, L.M., Adams, G.E. (1973) *Reactivity of the Hydroxyl Radical in Aqueous Solution.* NSRD-NDB-46. NTIS COM-73-50623. National Bureau of Standards, Wasington DC.

Dow Chemical (1979) Personal communication from M. Thomas of Dow Chemical Company to N.W. Gabel of Versar, Inc.

Draper, W.M., Crosby, D.G. (1983) Photochemical generation of superoxide radical anion in water. *J. Agric. Food Chem.* 31, 734-737.

Dreisbach, R.R. (1952) *Pressure-Volume-Temperature Relationships of Organic Compounds.* Handbook Publishers Inc., Sandusky, Ohio.

Ellgehausen, H., Guth, J.A., Esser, H.O. (1980) Factors determining the bioaccumulation potential of pesticides in the individual compartments of aquatic food chains. *Ecotox. Environ. Saf.* 4, 134-157.

Ellington, J.J. (1989) *Hydrolysis Rate Constants for Enhancing Property-Reactivity Relationships.* EPA/600/3-89/063. NTIS PB89-220479. U.S. Environmental Protection Agency, Environmental Research Laboratory, Athens, Georgia.

Ellington, J.J., Stancek, F.E., Payne, W.D., Trusty, C.D. (1987) *Measurement of Hydrolysis Rate Constants for Evaluation of Hazardous Waste Land Disposal. Vol. I. Data on 54 Chemicals.* U.S. EPA-600/3-86-043.

Ellington, J.J., Stancil, F.E., Jr., Payne, W.D., Trusty, C.D. (1987) *Measurements of Hydrolysis Rate Constants for Evaluation of Hazardous Waste Land Disposal: Data on 54 Chemicals.* Volume II. EPA/600/53-87/019. U.S. Environmental Protection Agency, Athens, Georgia.

El Tayar, N.E., Testa, B., Carrupt, P.A. (1992) Polar intermolecular interactions encoded in partition coefficients: An indirect estimation of hydrogen-bond parameters of polyfunctional solutes. *J. Phys. Chem.* 96, 1455-1459.

El Tayar, N.E., Tsai, R.-S., Vallat, P., Altomare, C., Testa, B. (1991) Measurements of partition coefficients by various centrifugal partition chromatographic techniques. *J. Chromatogr.* 556, 181-194.

Elzerman, A.W., Coates, J.T. (1987) Hydrophobic organic compounds on sediments: Equilibria and kinetics of sorption. In: *Sources and Fates of Aquatic Pollutants.* Hites, R.A., Eisenreich, S.J., Editors, Advances in Chemistry Series 216, American Chemical Society, Washington, DC.

Ezumi, K., Kubota, T. (1980) Simultaneous determination of acid dissociation constants and true partition coefficients by analyses of the apparent partition coefficient. *Chem. Pharm. Bull.* 28(1), 85-91.

Fieser, L.F., Fieser, M. (1959) *Textbook of Organic Chemistry.* D.C. Heath and Company, Boston.

Freed, V.H., Chiou, C.T., Haque, R. (1977) Chemodynamics: Transport and behavior of chemicals in the environment-A problem in environmental health. *Environ. Health Perspectives* 220, 55-70.

Freese, E., Levin, B.C., Pearce, R., Sreevalson, T., Kaufman, J.J., Koski, W.S., Semo, N.M. (1979) Correlation between the growth inhibitory effects, partition coefficients and teratogenic effects of lipophilic acids. *Teratology* 20(3), 413-440.

Freitag, D., Ballhorn, L., Geyer, H., Körte, F. (1985) Environmental hazard profile of organic chemicals. An experimental method for the assessment of the behaviour of chemicals in the ecosphere by simple laboratory tests with C-14 labelled chemicals. *Chemosphere* 14, 1589-1616.

Freitag, D., Geyer, H., Kraus, A., Viswanathan, R., Kotzias, D., Attar, A., Klein, W., Körte, F. (1982) Ecotoxicological profile analysis. VII. Screening chemicals for their environmental behavior by comparative evaluation. *Ecotoxicol. & Environ. Safety* 6, 60-81.

Führer, H. (1924) Water solubilities in homologous series. *Ber. Deutsch Chem. Gse.* 57, 510-515.

Fujisawa, S., Masuhara, E. (1980) Binding of methylmethacrylate to bovine serum albumin. *J. Den. Res.* 59, 2056-2061.

Fujita, T., Iwasa, J., Hansch, C. (1964) A new substituent constant derived from partition coefficients. *J. Am. Chem. Soc.* 86(23), 5175-5180.

Gaffney, J.S., Streit, G.E., Spall, W.D., Hall, J.H. (1987) Beyond acid rain. Do soluble oxidants and organic toxins interact with SO_2 and NO_2 to increase ecosystem effects? *Environ. Sci. Technol.* 21, 519-524.

Garst, J.E. (1984) Accurate, wide range, automated, high-performance liquid chromatographic method for the estimation of octanol/water partition coefficients II: Equilibrium in partition coefficient measurements, additivity of substituent constants, and correlation of biological data. *J. Pharm. Sci.* 73(11), 1623-1629.

Garst, J.E., Wilson, W.C. (1984) Accurate, wide range, automated, high-performance liquid chromatographic method for the estimation of octanol/water partition coefficients I: Effect of chromatographic conditions and procedure variables on accuracy and reproducibility of the method. *J. Pharm. Sci.* 73(11), 1616-1623.

Garten, Jr., C.T., Trabalka, J.R. (1983) Evaluation of models for predicting terrestrial food chain behavior of xenobiotics. *Environ. Sci. Technol.* 17, 590-595.

GEMS (1984) Graphical Exposure Modeling System. FAP, Fate Atmospheric Pollution. Data Base, Office of Toxic Substances, U.S. Environmental Protection Agency.

GEMS (1986) Graphical Exposure Modeling System. FAP, Fate Atmospheric Pollution. Data Base, Office of Toxic Substances, U.S. Environmental Protection Agency.

Gerstl, Z., Helling, C.S. (1987) Evaluation of molecular connectivity as a predictive method for the adsorption of pesticides by soils. *J. Environ. Sci. & Health* B22(1), 55-69.

Geyer, H., Kraus, A.G., Klein, W., Richter, E., Körte, F. (1980) Relationship between water solubility and bioaccumulation potential of organic chemicals in rats. *Chemosphere* 9, 277-291.

Geyer, H., Politzki, G., Freitag, D. (1984) Prediction of ecotoxicological behavior of chemicals: Relationship between *n*-octanol/water partition coefficient and bioaccumulation of organic chemicals by alga *chlorella*. *Chemosphere* 113(2), 269-284.

Geyer, H., Sheehan, P., Kotzias, D., Freitag, D., Körte, F. (1982) Prediction of ecotoxicological behaviour of chemicals: relationship between phisico-chemical properties and bioaccumulation of organic chemicals in the mussel *mytilus edulis*. *Chemosphere* 11(11), 1121-1134.

Geyer, H., Visvanathan, R., Freitag, D., Körte, F. (1981) Relationship between water solubility of organic chemicals and their bioaccumulation by the *alga chlorella*. *Chemosphere* 10, 1307-1313.

Gückel, W., Kästel, R., Lewerenz, J., Synnatschke, G. (1982) A method for determining the volatility of active ingredients used in plant protection. Part III: The temperature relationship between vapour pressure and evaporation rate. *Pestic. Sci.* 13, 161-168.

Günther, F.A., Westlake, W.E., Jaglan, P.S. (1968) Reported solubilities of 738 pesticide chemicals in water. *Residue rev.* 20, 1-148.

Gustafson, D.I. (1989) Groundwater ubiquity score: A simple method for assessing pesticide leachability. *Environ. Toxicol. Chem.* 8, 339-357.

Haag, W.R., Yao, D.C.C. (1992) Rate constants for reaction of hydroxyl radicals with several drinking water contaminants. *Environ. Sci. Technol.* 26(5), 1005-1013.

Haky, J.E., Young, A.M. (1984) Evaluation of a simple HPLC correlation method for the estimation of the octanol-water partition coefficients of organic compounds. *J. Liq. Chromatogr.* 7, 675-689.

Haller, H.D. (1978) Degradation of mono-substituted benzoates and phenols by wastewater. *J. WPCF* 2771-2777.

Halmekoski, J., Hannikainen, H. (1964) R_M values of hydroxy-substituted benzoic acids and partition coefficients of the acids between water and aliphatic alcohols. *Suom. Kemistil.* B37(12), 221-226.

Hamaker, J.W. (1975) In: *Environmental Dynamics of Pesticides.* Haque, R., Freed, V.H., Editors, Plenum Press, New York. pp.115-153.

Hanai, T., Hubert, J. (1982) Hydrophobicity and chromatographic behaviour of aromatic acids found in urine. *J. Chromatogr.* 239, 527-536.

Hansch, C., Anderson, S.M. (1967) The effect of intramolecular hydrophobic bonding on partition coefficients. *J. Org. Chem.* 32, 2583.

Hansch, C., Leo, A. (1979) *Substituents Constants for Correlation Analysis in Chemistry and Biology.* Wiley, New York.

Hansch, C., Leo, A.J. (1982) *Medchem Project* Issue No.20, Pomona College, Claremont, California.

Hansch, C., Leo, A. (1985) *Medchem Project* Issue No. 26. Pomona College, Claremont, California.

Hansch, C., Leo, A. (1987) *Medchem Project* Issue No. 28. Pomona College, Claremont, California.

Hansch, C., Leo, A., Nikaitani, D. (1972) On the additive-constitutive character of partition coefficients. *J. Org. Chem.* 37, 3090-3092.

Hansch, C., Quinlan, J.E., Lawrence, G.L. (1968) The linear free-energy relationship between partition coefficients and aqueous solubilities of organic liquids. *J. Org. Chem.* 33, 347-350.

Hartley, D., Kidd, H. (1987) *The Agrochemicals Handbook*, 2nd ed., The Royal Society of Chemistry, The University, Nottingham, England.

Hayward, D.S., Kenley, R.A., Jenke, D.R. (1990) Interactions between polymer containers and parental solutions: the correlation of equilibrium constants for polymer-water partitioning with octanol-water partition coefficients. *Int'l J. Pharm.* 59(3), 245-253.

Hersey, A., Hill, A.P., Hyde, R.M., Livingstone, B.J. (1989) Principles of method selection in partition studies. *Quant. Struct.-Act. Relat.* 8(4), 288-296.

Heukelekian, H., Rand, M.C. (1955) Biochemical oxygen demand of pure organic compounds. *J. Water Pollut. Control Assoc.* 29, 1040-1053.

Hine, J., Mookerjee, P.K. (1975) The intrinsic hydrophilic character of organic compounds. Correlations in terms of structural contributions. *J. Org. Chem.* 40, 292-298.

Hodgman, C.R., Editor (1952) *Handbook of Chemistry and Physics.* 34th Ed., Chemical Rubber Publishing Co., Cleveland, Ohio.

Hoigné, J., Bader, H. (1983) Rate constants of reactions of ozone with organic and inorganic compounds in water-I. Non-dissociating organic compounds. *Water Res.* 17, 173-183.

Horowitz, A., Shelton, D.R., Cornell, C.P., Tiedje, J.M. (1982) Anaerobic degradation of aromatic compounds in sediment and digested sludge. *Dev. Ind. Microbiol.* 23, 435-444.

Howard, P.H., Ed. (1989) *Handbook of Fate and Exposure Data for Organic Chemicals. Vol.I-Large Production and Priority Pollutants.* Lewis Publishers, Chelsea, Michigan.

Howard, P.H., Ed. (1990) *Handbook of Fate and Exposure Data for Organic Chemicals. Vol. II. Solvents.* Lewis Publishers, Chelsea, Michigan.

Howard, P.H., Ed. (1991) *Handbook of Fate and Exposure Data for Organic Chemicals. Vol. III. Pesticides.* Lewis Publishers, Chelsea, Michigan.

Howard, P.H., Hueber, A.E., Mulesky, B.C., Crisman, J.S., Meylan, W., Crosbie, E., Gray, D.A., Sage, G.W., Howard, K.P., LaMacchia, A., Boethling, R.S., Troast, R. (1986) BIOLOG, BIODEG, and FATE/EXPOS: New files on microbial degradation and toxicity as well as environmental fate/exposure of chemicals. *Environ. Toxicol. Chem.* 5, 977-988.

Howard, P.H., Boethling, R.S., Jarvis, W.F., Meylan, W.M., Michalenko, E.M. (1991) *Handbook of Environmental Degradation Rates.* Lewis Publishers, Chelsea, Michigan.

Hoy, K.L. (1970) New values of the solubility parameters from vapor pressure data. *J. Paint Technol.* 42(541), 76-118.

Hulzebos, E.M., Adema, D.M.M., Breemen, E.M. D.-van, Henzen, L., van Dis, W.A., Herbold, H.A., Hoekstra, J.A., Baerselman, R., van Gestel, C.A.M. (1993) Phytotoxicity studies with *lactuca sativa* in soil and nutrient solution. *Environ. Toxicol. & Chem.* 12, 1079-1094.

Ilchmann, A., Wienke, G., Meyer, T., Gmehling, J. (1993) Countercurrent liquid/liquid chromatography-A reliable method for the determination of partition coefficients. *Chem.-Ing.-Tech.*, 65(1), 72-75.

Isensee, A.R. (1991) Chapter 5, Bioaccumulation and food chain accumulation. In: *Environmental Chemistry of Herbicides.* Vol. II., Grover, R., Cessna, A.J., Eds., pp.187-197, CRC Press, Boca Raton, Florida.

Isnard, P., Lambert, S. (1988) Estimating bioconcentration factors from octanol-water partition coefficient and aqueous solubility. *Chemosphere* 17(1), 21-34.

Isnard, P., Lambert, S. (1989) Aqueous solubility/*n*-octanol water partition coefficient correlations. *Chemosphere* 18, 1837-1853.

Iwasa, J., Fujita, T., Hansch, C. (1965) Substituent constants for aliphatic functions obtained from partition coefficients. *J. Med. Chem.* 8, 150-153.

Iwase, K., Komatsu, K., Hirono, S., Nakagawa, S., Moriguchi, I. (1985) Estimation of hydropbobicity based on the solvent-accessible surface area of molecules. *Chem. Pharm. Bull.* 33(5), 2114-2121.

Jenke, D.R., Hayward, D.S., Kenley, R.A. (1990) Liquid chromatographic measurement of solute/solvent partition coefficients: application to solute/container interactions. *J. Chromatogr. Sci.* 28(12), 609-612.

Jones, D.C. (1929) The systems *n*-butyl alcohol-water and *n*-butyl alcohol-acetone-water. *J. Chem. Soc.* (London) 799-816.

Jordan, T.E. (1954) *Vapor Pressure of Organic Compounds.* Interscience, New York.

Kahlbaum, G.W.A. (1898) Studien über dumpfspannkraftmessungen. II. *Z. Phys. Chem.* 26, 577-658.

Kaiser, K.L.E. (1983) A non-linear function for the approximation of octanol/water partition coefficients of aromatic compounds with multiple chlorine substitution. *Chemosphere* 12(9/10), 1159-1167.

Kamlet, M.J., Doherty, R.M., Abraham, M.H., Marcus, Y., Taft, R.W. (1988) Linear solvation energy relationships. 46. An improved equation for correlation and prediction of octanol/water partition coefficients of organic non-electrolytes (including strong hydrogen bond donor solutes). *J. Phys. Chem.* 92, 5244-5255.

Kamlet, M.J., Doherty, R.M., Taft, R.W., Abraham, M.H., Veith, G.D., Abraham, D.J. (1987) Solubility properties in polymers and biological media. 8. An analysis of the factors that influence toxicities of organic nonelectrolytes to the golden orfe fish (*leuciscus idus melanotus). Environ. Sci. Technol.* 21(2), 149-155.

Karickhoff, S.W. (1985) Chapter 3, Pollutant sorption in environmental systems. In: *Environmental Exposure From Chemicals.* Volume I. Neely, W.B., Blau, G.E., Editors, CRC Press, Inc., Boca Raton, Florida. pp.49-64.

Kelly, T.J., Mukund, R., Spicer, C.W., Pollack, A.J. (1994) Concentrations and transformations of hazardous air pollutants. *Environ. Sci. Technol.* 28, 278A-387A.

Kenaga, E.E. (1975) Partitioning and uptake of pesticides in biological systems. In: *Environmental Dynamics of Pesticides.* Haque, R., Freed, V.H. Editors, Plenum Press, New York. pp.217-273.

Kenaga, E.E. (1980) Predicted bioconcentration factors and soil sorption coefficients of pesticides and other chemicals. *Ecotoxicol. Environ. Safety* 4, 26-38.

Kenaga, E.E., Goring, C.A.I. (1980) In: *Aquatic Toxicology.* Eaton, J.G., Parrish, P.R., Hendricks, A.C., Eds., American Society for Testing and Materials, STP 707, pp.78-115.

Khan, I., Brimblecombe, P. (1992) Henry's law constants of low molecular weight (<130) organic acids. *J. Aerosol Sci.* 23, Suppl. 1, S897-S900.

Khan, U.K. (1980) *Pesticides in the Soil Environment, Fudamental Aspects of Pollution Control and Environmental Series 5.* Elsevier, Amsterdam, The Netherlands.

Kidd, H., Hartley, D. (1983) *Pesticide Index.* Royal Society of Chemistry, Nottingham, England.

Kidd, H., Hartley, D. (1987) *Pesticide Index.* Royal Society of Chemistry, Nottingham, England.

Kincannon, D.F., Lin, Y.S. (1985) Microbial degradation of hazardous wastes by land treatment. In: *Proc. Indust. Waste Conf.* 40, 607-619.

Klein, W., Kordel, W., Weiss, M., Poremski, H.J. (1988) Updating of the OECD test guideline 107 "partition coefficient n-octanol/water": OECD laboratory intercomparison test on HPLC method. *Chemosphere* 17, 361-386.

Klecka, G.M. (1985) Chapter 6, Biodegradation. In: *Environmental Exposure From Chemicals.* Volume I. Neely, W.B., Blau, G.E., Editors, CRC Press, Inc., Boca Raton, Florida. pp.110-155.

Kleopfer, R.D., Fairless, B.J. (1972) Characterization of organic compounds in a municipal water supply. *Environ. Sci. Technol.* 6, 1036-1037.

Klöpffer, W., Kaufmann, G., Rippen, G., Poremski, H-J. (1982) A laboratory method for testing the volatility from aqueous solution: First results and comparison with theory. *Ecotoxicol. & Environ. Safety* 6, 545-559.

Kollig, H.P., Ed. (1993) *Environmental Fate Constants for Organic Chemicals Under Consideration for EPA's Waste Identification Projects*. EPA Report EPA/600/R-93/132, Environmental Research Lab., USEPA, Athens, GA.

Kollig, H.P., Parrish, R.S., Holm, H.W. (1987) An estimate of the variability in biotransformation kinetics of xenobiotics in natural waters by aufwuchs communities. *Chemosphere* 16, 49-60.

Kool, H.J. (1984) Influence of microbial biomass on the biodegradability of organic compounds. *Chemosphere* 13, 751-761.

Korenman, I.M., Lunicheva, E.V., Pomerantseva, E.G., Khairulina, F.D. (1974) Distribution of methacrylic acid between organic solvents and water. *Zh. Prikl. Khim.* (Lenningrad), 47(10), 2365-2368.

Korenman, Ya.I., Danilov, U.N., Zhilinskaya, K.I. (1990) Regularities and prediction of quantitative characteristics of isomer extraction. *Ukr. Khim. Zh.* (Russ. Ed.) 56(3), 291-295.

Korman, S., La Mer, V.K. (1936) Deuterium equilibria in solution and the quinhydrone electrode. *J. Am. Chem. Soc.* 58, 1396-1403.

Kuchar, M., Rejholec, V., Brunova, B., Grimova, J., Matousova, O., Nemecek, O., Cepelakova, H. (1982) Benzyloxyaryl aliphatic acids: Synthesis and quantitative relations between structure and anti-inflamatory activity. *Coll. Czech. Chem. Commun.* 47, 2514.

Laznicek, M., Kvetina, J., Mazak, J., Krch, V. (1987) Plasma protein binding-lipophilicity relationships: Interspecies comparison of some organic acids. *J. Pharm. Pharmacol.* 39(2), 79-83.

Leahy, D.E. (1986) Intrinsic molecular volume as a measure of cavity term in linear solvation energy relationships: Octanol-water partition coefficients and aqueous solubilities. *J. Pharm. Sci.* 75, 629-636.

Lee, B.I., Erbar, J.H., Edmister, W.C. (1972) Thermodynamic properties at low temperatures. *Chem. Eng. Prog.* 68(9), 83-84.

Lee, L.S., Bellin, C.A., Pinal, R., Rao, P.S.C. (1993) Cosolvent effects on sorption of organic acids by soils from mixed solvents. *Environ. Sci. Technol.* 27, 165-171.

Leo, A. (1986) CLOGP-3.42 *Medchem Software*. Medicinal Chemistry Project, Pomona College, Claremont, California.

Leo, A., Hansch, C., Elkins, D. (1971) Partition coefficients and their uses. *Chem. Rev.* 71, 525-612.

Leonard, R.H., Peterson, W.H., Johnson, M.J. (1948) Lactic acid from fermentation of sulfite waste liquor. *Ind. Eng. Chem.* 40, 57-67.

Leopold, A.C., van Schaik, P., Neal, M. (1960) Molecular structure and herbicide adsorption. *Weeds* 8, 48.

Liu, D., Strachan, W.M.J., Thomson, K., Kwasniewska, K. (1981) Determination of the biodegradability of organic compounds. *Environ. Sci. Technol.* 15, 788-793.

Lu, P.Y., Metcalf, R.L. (1975) Environmental fate and biodegradability of benzene derivatives as studied in a model aquatic ecosystem. *Environ. Health Pespect.* 10, 269-284.

Ludzack, F.J., Ettinger, M.B. (1963) Biodegradability of organic chemicals isolated from rivers. *Purdue Univ., Eng. Bull. Ext. Ser.* 115, 278-282.

Lyman, W.J. (1985) Chapter 2, Estimation of physical properties. In: *Environmental Exposure From Chemicals*. Volume I. Neely, W.B., Blau, G.E., Editors, CRC Press, Inc., Boca Raton, Florida. pp.13-48.

Lyman, W.J., Reehl, W.F., Rosenblatt, D.H., Eds. (1982) *Handbook on Chemical Property Estimation Methods. Environmental Behavior of Organic Compounds.* McGraw-Hill, New York.

Ma, K.C., Shiu, W.Y., Mackay, D. (1990) *A Critically Reviewed Compilation of Physical, Chemical and Persistence Data for 110 Selected EMPPL Substances.* A report prepared for the Ontario Ministry of Environment, Water Resources Branch, Toronto, Ontario.

Mabey, W.R., Mill, T. (1978) Critical review of hydrolysis of organic compounds in water under environmental conditions. *J. Phys. Chem. Ref. Data* 7, 383-415.

Mackay, D. (1982) Correlation of bioconcentration factors. *Environ. Sci. Technol.* 16, 274-278.

Mackay, D. (1985) Chapter 5, Air/water exchange coefficients. In: *Environmental Exposure From Chemicals.* Volume I. Neely, W.B., Blau, G.E., Eds., pp.91-108. CRC Press, Inc., Boca Raton, Florida.

Mackay, D., Bobra, A., Shiu, W.Y., Yalkowsky, S.H. (1980) Relationships between aqueous solubility and octanol-water partition coefficient. *Chemosphere* 9, 701-711.

Mackay, D., Stiver, W. (1991) Chapter 8, Predictability and environmental chemistry. In: *Environmental Chemistry of Herbicides.* Vol. II., Grover, R., Cessna, A.J., Eds., pp. 281-297. CRC Press, Inc., Boca Raton, Florida.

Mackay, D., Wolkoff, A.W. (1973) Rate of evaporation of low-solubility contaminants from water bodies to atmosphere. *Environ. Sci. Technol.* 7(7), 611-614.

Malaspina, L., Gigli, R., Bardi, G. (1973) Microcalorimetric determination of the enthalpy of sublimation of benzoic acid and anthracene. *J. Chem. Phys.* 59(1), 387-394.

McBain, J.W., Richards, P.H. (1946) Solubilization of insoluble organic liquids by detergents. *Ind. & Eng. Chem.* 38(6), 642-644.

McCall, P.J., Swann, R.L., Laskowski, D.A., Unger, S.M., Vrona, S.A., Dishburger, H.J. (1980) Estimation of chemical mobility in soil from liquid chromatograpic retention times. *Bull. Environ. Contam. Toxicol.* 24, 190-195.

McCall, P.J., Swann, R.L., Laskowski, D.A., Vrona, S.A., Unger, S.M., Dishburger, H.J. (1981) Prediction of chemical mobility in soil from sorption coefficients. In: *Aquatic Toxicologh and Hazard Assessment. ASTM STP 737*, Branson, D.R., Dickson, K.L., Eds., pp. 49-58, American Society for Testing and Materials, Phildelphia.

McCall, P.J., Vrona, S.A., Kelley, S.S. (1981) Fate of uniformly carbon-14 ring labeled 2,4,5-trichlorophenoxyacetic acid and 2,4-dichlorophenoxyacetic acid. *J. Agric. Food Chem.* 29, 100-107.

McDaniel, D.H., Brown, H.C. (1958) Extended table of Hammett substituent constants based on the ionization of substituted benzoic acids. *J. Org. Chem.* 23, 420.

McDuffie, B. (1981) Estimation of octanol/water partition coefficients for organic pollutants using reverse-phase HPLC. *Chemosphere* 10, 73-83.

Menges, R.A., Bertrand, G.L., Armstrong, D.W. (1990) Direct measurement of octanol-water partition coefficient using centrifugal partition chromatograph with back-flashing technique. *J. Liq. Chromatogr.* 13(15), 3061-3077.

Meylan, W., Howard, P.H. (1991) Bond contribution method for estimating Henry's law constants. *Environ. Toxicol. & Chem.* 10, 1283-1293.

Meylan, W., Howard, P.H., Boethling, R.S. (1992) Molecular topology/fragment contribution method for predicting soil sorption coefficient. *Environ. Sci. Technol.* 26(8), 1560-1567.

Miyake, K., Kitaura, F., Mizuno, N., Terada, H. (1987) Determining of partition coefficient and acid dissociation constant by high-performance liquid chromatography on polymer gel as stationary phase. *Chem. Pharm. Bull.* 35(1), 377-388.

Miyake, K., Terada, H. (1982) Determination of partition coefficients of very hydrophobic compounds by high-performance liquid chromatography on glyceryl-coated controlled-glass. *J. Chromatogr.* 240, 9-20.

Moriguchi, I. (1975) Quantitative structure activity studies. Parameters relating to hydrophobicity. *Chem. Pharm. Bull.* 23(2), 247-257.

Muir, D.C.G. (1991) Chapter 1, Dissipation and transformation in water and sediment. In: *Environmental Chemistry of Herbicides.* Vol. II., Grover, R., Cessna, A.J., Eds., pp. 1-87. CRC Press, Inc., Boca Raton, Florida.

Müller, M., Klein, W. (1992) Comparitive evaluation of methods predicting water solubilty for organic compounds. *Chemosphere* 25(6), 769-782.

Neely, W.B. (1979) A preliminary assessment of the environmental exposure to be expected from the addition of a chemical to a simulated ecosystem. *J. Environ. Stud.* 13, 101.

Neely, W.B. (1984) An analysis of aquatic toxicity data: Water solubility and acute LC 50 fish data. *Chemosphere* 13(7), 813-819.

Neely, W.B. (1985) Chapter 7, Hydrolysis. In: *Environmental Exposure From Chemicals.* Volume I. Neely, W.B., Blau, G.E., Eds., pp.157-174. CRC Press, Inc., Boca Raton, Florida.

Neely, W.B., Blau, G.E. (1985) Chapter 1, Introduction to environmental exposure from chemicals. In: *Environmental Exposure From Chemicals.* Volume I. Neely, W.B., Blau, G.E., Eds., pp. 1-11, CRC Press, Inc., Boca Raton, Florida.

Nelson, N.H., Faust, S.D. (1969) Acidic dissociation constants of selected herbicides. *Environ. Sci. Technol.* 3, 1186-1188.

Nesbitt, H.J., Watson, J.R. (1980a) Degradation of the herbicide 2,4-D in river water- I. Description of study area and survey of rate determining factors. *Water Res.* 14, 1683-1688.

Nesbitt, H.J., Watson, J.R. (1980b) Degradation of the herbicide 2,4-D in river water- II. The role of suspended sediment. Nutrients and water temperature. *Water Res.* 14, 1689-1694.

Neuhauser, E.F., Lorhr, R.C., Malecki, M.R., Milligan, D.L., Durkin, P.R. (1985) The toxicity of selected organic chemicals to the earth worm *eisenia fetida*. *J. Environ. Qual.* 14(3), 383-388.

Nirmalakhandan, N.N., Speece, R.E. (1988) QSAR model for predicting Henry's constant. *Environ. Sci. Technol.* 22, 1349-1357.

Nishimura, K., Nozaki, Y., Yoshimi, A., Nakamura, S., Kitagawa, M., Kakeya, N., Kitao, K. (1985) Studies on the promoting effect of carboxylic acid derivatives on the rectal absorption of beta-lactam antibiotics in rats. *Chem. Pharm. Bull.* 33(1), 282-291.

Nkedi-Kizza. P., Rao, P.S.C., Johnson, J.W. (1983) Adsorption of diuron and 2,4,5-T on soil particle-size separates. *J. Environ. Qual.* 12, 195-197.

OECD (1979) OECD, Environment Committee, Chemicals Group, OECD Chemicals Testing Programme, Expert Group, Physical Chemistry, Final Report Vol.I, Part 1 and 2, Summary of the OECD Laboratory Intercomparison, Testing Programme Part I - on the physico-chemical properties p.33, Umweltbundesamt, Bundesrepublik Deutschland, Berlin, Dec. 1979.

OECD (1981) OECD Guidelines for Testing of Chemicals. Section 107, p.3. "Partition coefficient (*n*-octanol/water)-flask shaking method," OECD, Paris.

Oepen, B.V., Kördel, W., Klein, W. (1991) Sorption of nonpolar and polar compounds to soils: Processes, measurements and experience with the applicability of the modified OECD-Guideline 106. *Chemosphere* 22(3/4), 285-304.

Ollis, D.F. (1985) Contaminant degradation in water. *Environ. Sci. Technol.* 19(6), 480-484.

Osol, A., Kilpatrick, M. (1933) The "salting-out" and "salting-in" of weak acids. I. The activity coefficients of the molecules of ortho, meta, and para chlorobenzoic acids in aqueous salt solutions. *J. Am. Chem. Soc.* 55, 4430-4440.

Palit, S.R. (1947) Electronic interpretations of organic chemistry. II. Interpretation of the solubility of organic compounds. *J. Phys. Chem.* 51, 837-857.

Paris, D.F., Steen, W.C., Baughman, G.L., Barnett, J.T., Jr. (1981) Second-order model to predict microbial degradation of organic compounds in natural waters. *Appl. Environ. Microbiol.* 41, 603-609.

Patrunkey, M., Pflegel, P. (1992) Influence partition coefficients of dibasic organic acids on permeation studies with polyacrylate membranes. *Pharmazie* 47(4), 306-307.

Pavlou, S.P., Weston, D.P. (1983) Initial evaluation of alternatives for development of sediment related criteria for toxic contaminants in marine waters (Puget Sound). Phase I: Development of conceptual framework. Prepared by JRB Associates for the U.S. Environmental Protection Agency under EPA Contract No. 68-01-6388.

Pavlou, S.P., Weston, D.P. (1984) Initial evaluation of alternatives for development of sediment related critiria for toxic contaminants in marine waters (Puget Sound). Phase II: Development and testing of the sediment/water equilibrium-partitioning approach. Prepared by JRB Associates for the U.S. Environmental Protection Agency under EPA Contract No. 68-01-6388.

Pearce, P.J., Simkins, R.J.J. (1968) Acid strengths of some substituted picric acids. *Can. J. Chem.* 46, 241-248.

Peddicord, R.K. (1987) Chapter 20, Overview of the influence of dredging and dredged material disposal on the fate and effects of sediment-associated chemicals. In: *Fate and Effects of Sediment-Bound Chemicals in Aquatic Systems.* Dickson, K.L., Maki, A.W., Brungs, W.A., Eds., pp. 317-412, Pergamon Press, New York.

Perrin, D.D. (1972) *Dissociation Constants of Organic Bases in Aqueous Solutions.* IUPAC Chemical Data Series; Supplement, Butterworth, London.

Perry, R.H., Chilton, C.H., Editors (1973) *Chemical Engineer's Handbook.* 5th Edition, McGraw-Hill Book Company, New York.

Petrasek, A.C., Kugeman, I.J., Austern, B.M., Pressley, T.A., Winslow, L.A., Wise, R.H. (1983) Fate of toxic organic compounds in wastewater treatment plants. *J. Water Pollut. Control Fed.* 55, 286-296.

Pitter, P. (1976) Determination of biological degradability of organic substances. *Water Res.* 10, 231-235.

Quayle, O.R. (1953) The parachors of organic compounds. *Chem. Rev.* 53, 439-589.

Radding, S.B., Liu, D.H., Johnson, H.L., Mill, T. (1977) *Review of the Environmental Fate of Selected Chemicals.* EPA Report No.560/5-77-003. U.S. Environmental Protection Agency, Office of Toxic Substances, Washington, D.C.

Reid, R.C., Prausnitz, J.M., Sherwood, T.K. (1977) *The Properties of Gases and Liquids.* 3rd Edition, McGraw-Hill Book Company, New York.

Reinert, K.H., Rogers, J.H. (1984) Influence of sediment types on the sorption of Endothall. *Bull. Environ. Contam. Toxicol.* 32, 557-564.

Rekker, R.F. (1977) *The Hydrophobic Fragmental Constants. Its Derivation and Application, a Means of Characterizing Membrane Systems*. Elsevier Scientific Publishing Company, Oxford, England.

Rekker, R.F., de Kort, H.M. (1979) The hydrophobic fragmental constants: An extension to a 1000 data point set. *Eur. J. Med. Chem.* 14, 479-488.

Renberg, L.O., Sundström, S.G., Rosén-Olofsson, A-C. (1985) The determination of partition coefficients of organic compounds in technical products and waste waters for the estimation of their bioaccumulation potential using reversed phase thin layer chromatography. *Toxicol. Environ. Chem.* 10, 333-349.

Riddick, J., Bunger, W.B., Sakano, T.K. (1986) *Organic Solvents: Physical Properties and Method of Purification.* 4th Edition, John Wiley & Sons, New York, N.Y.

Rohrschneider, L. (1973) Solvent characterization of gas-liquid partition coefficients of selected solutes. *Anal. Chem.* 45(7), 1241-1247.

Rosenberg, A., Alexander, M. (1980) 2,4,5-Trichlorophenoxyacetic acid (2,4,5-T) decomposition in tropical soil and its cometabolism by bacteria in vitro. *J. Agric. Food Chem.* 28, 705-709.

Russell, C.J., Dixon, S.L., Jurs, P.C. (1992) Computer-assisted study of the relationship between molecular structure and Henry's law constant. *Anal. Chem.* 64, 1350-1355.

Ryan, J.A., Bell, R.M., Davidson, J.M., O'Connor, G.A. (1988) Plant uptake of non-ionic organic chemicals from soils. *Chemosphere* 17(12), 2299-2323.

Sabljic, A. (1987) Nonempirical modeling of environmental distribution and toxicity of major organic pollutants. In: *QSAR in Environmental Toxicology-II*. Kaiser, K.L.E., Ed., pp. 309-332, D. Reidel Publishing Co., Dordrecht, The Netherlands.

Sabljic, A., Güsten, H., Schönherr, J., Riederer, M. (1990) Modeling plant uptake of airborne organic chemicals. 1. Plant cuticle/water partitioning and molecular connectivity. *Environ. Sci. Technol.* 24, 1321-1326.

Saito, S., Tanoue, A., Matsuo, M. (1992) Applicability of the i/o-characters to a quantitative description of bioconcentration of organic chemicals in fish. *Chemosphere* 24(1), 81-87.

Sangster, J. (1989) Octanol-water partition coefficients of simple organic compounds. *J. Phys. Chem. Ref. Data* 18(3), 1111-1229.

Sangster, J. (1993) LOGKOW Databank. Sangster Research Labs. Montreal, Canada.

Sasaki, S. (1978) The scientific aspects of the chemical substance control law in Japan. In: *Aquatic Pollutants: Transformation and Biological Effects*. Hutzinger, O., et al. Editors, Pergamon Press, Oxford, U.K. pp. 283-298.

Scherrer, R.A., Howard, S.M. (1979) The analysis of electronic factors in quantitative structure-activity relationships using distribution coefficients. *Adv. Chem. Ser.* 112, 507-526.

Schüürmann, G., Klein, W. (1988) Advances in bioconcentration prediction. *Chemosphere* 17(8), 1551-1574.

Schwarz, F.P., Miller, J. (1980) Determination of the aqueous solubilities of organic liquids at 10.0, 20.0, 30.0 °C by elution chromatography. *Anal. Chem.* 52, 2162-2164.

Scow, K.M. (1982) Rate biodegradation. In: *Handbook on Chemical Property Estimation Methods. Environmental Behavior of Organic Compounds*. Lyman, W.J., Reehl, W.F., Rosenblatt, D.H., Editors, McGraw-Hill, New York.

Shimada, T., Nagasaki, Y., Yoshida, T. (1983) The acute toxicity and the accumulation of phenol and diethyl phthalate in *mugil cephalus*. *Kenkyu Hokuku-Kanagawa-Ken Kogai Senta* 5, 45-48.

Shiu, W.Y., Ma, K.C., Mackay, D. (1990) Solubilities of pesticides in water. Part 1, Environmental physical chemistry and Part 2, Data compilation. *Rev. Environ. Contam. Toxicol.* 116, 1-187.

Skurlatov, Y.I., Zepp, R.G., Baughaman, G.L. (1983) Photolysis rates of (2,4,5-trichlorophenoxy)acetic acid and 4-amino-3,5,6-trichloropicolinic acid in natural waters. *J. Agric. Food Chem.* 31, 1065-1071.

Smith, H.W., White, T.A. (1929) The distribution ratios of some organic acids between water and organic liquids. *J. Phys. Chem.* 33, 1953-1974.

Spencer, W.F. (1976) Vapor pressure and vapor loss. In: *A Literature Survey of Benchmark Pesticides.* The George Washington University Medical Center, Department of Medical and Public Affairs, Science Communication Division, Washington DC. pp.72-165.

Steen, W.C. (1991) *Microbial Transformation Rate Constants of Structurally Diverse Man Made Chemicals.* NTIS PB91-181958. U.S. Environmental Protection Agency, Environmental Research Laboratory, Athens, Georgia.

Steen, W.C., Paris, D.F., Baughman, G.L. (1979) 177th National Meeting of American Chemical Society, Honolulu, Hawaii.

Steen, W.C., Paris, D.F., Baughman, G.L. (1979) Effects of sediment sorption on microbial degradation of toxic substances. pp.168-173. In: *Proceedings of the Symposium on Processes Involving Contaminants and Sediments.* Baker, R., Editor, American Chemical Society, Washington, DC.

Stegeman, M.H.L., Peijnenburg, W.J.G.M., Verboom, H. (1993) A quantitative structure-activity relationship for the direct photolysis of meta-substituted halobenzene derivatives in water. *Chemosphere* 26(5), 837-849.

Stephenson, R. M., Malanowski S. (1987) *Handbook of the Thermodynamics of Organic Compounds.* Elsevier Science Publishing Co., Inc., New York.

Stull, D.R. (1947) Vapor pressure of pure substances: Organic compounds. *Ind. Eng. Chem.* 39(4), 517-560.

Subba-Rao, R.V., Rubin, H.E., Alexander, M. (1982) Kinetics and extent of mineralization of organic chemicals at trace levels in freshwater and sewage. *Appl. Environ. Microbiol.* 43(5), 1139-1150.

Sugunan, S., Thomas, B. (1993) Salting coefficients of 2,-, 3-, and 4-methylbenzoic acids. *J. Chem. Eng. Data* 38, 520-521.

Suntio, L.R., Shiu, W.Y., Mackay, D., Seiber, J.N., Goltfelty, D. (1988) Critical review of Henry's law constants for pesticides. *Rev. Environ. Contam. Toxicol.* 103, 1-59.

Tabak, H.H., Govind, R. (1993) Prediction of biodegradation kinetics using a nonlinear group contribution method. *Environ. Toxicol. Chem.* 12, 251-260.

Tabak, H.H., Quave, S.A., Mashni, C.I., Barth, E.F. (1981) Biodegradability studies with organic priority pollutant compounds. *J. Water Pollut. Control Fed.* 53, 1503-1518.

Terada, H., Kosuge, Y., Murayama, W., Nakaya, N., Nunogaki, Y., Nunogaki, K.-I. (1987) Correlation of hydrophobic parameters of organic compounds determined by centrifugal partition chromatography with partition coefficients between octanol and water. *J. Chromatogr.* 400, 343-351.

THOR (1986) THOR Database, Medicinal Chemistry Project, Pomona College, Claremont, CA.

Tomida, H., Yotsuyanagi, T., Ikeda, K. (1978) Solubilization of benzoic acid derivatives by polyoxyethylene lauryl ether. *Chem. Pharm. Bull.* 26(9), 2824-2831.

Tsonopoulos, C., Prausnitz, J.M. (1971) Activity coefficients of aromatic solutes in dilute aqueous solutions. *Ind. & Eng. Chem. Fundamentals* 10, 593-600.

Umland, J.B. (1983) The determination of 1-octanol/water partition ratios: An organic chemistry laboratory experiment. *J. Chem. Ed.* 60(12), 1081.

Unger, S.H., Cook, J.R., Hollenberg, J.S. (1978) Simple procedure for determining octanol-aqueous partition, distribution, and ionization coefficients by reversed-phase high pressure liquid chromatography. *J. Pharm. Sci.* 67(10), 1364-1367.

Urano, K., Maeda, H., Ogura, K., Wada, H. (1982) Direct analytical method for aromatic compounds in water by steam carrier gas chromatography. *Water Res.* 16, 323-327.

Vallat, P., El Tayar, N., Testa, B. , Slacanin, I., Marston, A., Hostettmann, K. (1990) Centrifugal counter-current chromatography. A promising means of measuring partition coefficients. *J. Chromatogr.* 504(2), 411-419.

Valvani, S.C., Yalkowsky, S.H., Roseman, T.J. (1981) Solubility and partitioning. IV. Aqueous solubility and octanol-water partition coefficients of liquid nonelectrolytes. *J. Pharm. Sci.* 70, 502-507.

Veith, G.D., Austin, N.M., Morris, R.T. (1979) A rapid method for estimating LOG P for organic chemicals. *Water Res.* 13, 43-47.

Veith, G.D., Defoe, D.L., Bergstedt, B.V. (1979) Measuring and estimating the bioconcentration factor of chemicals in fish. *J. Fish Res. Board Can.* 36, 1040-1048.

Veith, G.D., Macek, K.J., Petrocelli, S.R., Carroll, J. (1980) An evaluation of using partition coefficients and water solubility to estimate bioconcentration factors for organic chemicals in fish. In: *Aquatic Toxicology, ASTM STP 707.* Eaton, J.G., Parrish, P.R., Hendricks, A.C., Eds., pp. 116-129, American Society for Testing and Materials, philadelphia, Pa.

Veith, G.D., Morris, R.T. (1978) *A Rapid Method for Estimating LOG P for Organic Chemicals.* EPA-600/3-78-049. U.S. Environmental Protection Agency, Ecological Research Series.

Verschueren, K. (1977) *Handbook of Environmental Data on Organic Chemicals.* Van Nostrand Reinhold, New York.

Verschueren, K. (1983) *Handbook of Environmental Data on Organic Chemicals.* 2nd Edition, Van Nostrand Reinhold, New York.

von Oepen, B.V., Kördel, W., Klein, W. (1991) Sorption of nonpolar and polar compounds to soils: Processes, measurements and experience with the applicability of the modified OECD-Guideline 106. *Chemosphere* 22(3/4), 285-304.

Wallington, T.J., Dagaut, P., Kurylo, M.J. (1988) Correlation between gas-phase and solution-phase reactivities of hydroxyl radicals toward saturated organic compounds. *J. Phys. Chem.* 92, 5024-5028.

Wang, L., Xu, L., Xu, O., Tian, L., Zhang, Z. (1989) Determination of partition coefficients of organic acids and bases and the correlation of partition coefficients in different systems. *Huanjing Kexue Xuebao* 9(4), 418-424.

Wauchope, R.D., Buttler, T.M., Hornsby, A.G., Augustijn-Beckers, P.W.M., Burt, J.P. (1992) The SCS/ARS/CES pesticide properties database for environmental decision-making. *Rev. Environ. Contam. Toxicol.* 123, 1-155.

Weast, R.C. (1977) *CRC Handbook of Chemistry and Physics.* 58th Edition. CRC Press, Inc., Cleveland, Ohio.

Weast, R.C. (1982-83) *CRC Handbook of Chemistry and Physics.* 63rd Edition, CRC Press, Boca Raton, Florida.

Whitehead, K.E., Geankoplis, C.J. (1955) Separation of formic and sulfuric acids by extraction. *Ind. Eng. Chem.* 47, 2114-2122.

Windholz, M., Editor (1983) *The Merck Index.* 10th Ed., Merck Co., Rahway, New Jersey.

Wofford, H.W., Wilsey, C.D., Neff, G.S., Giam, C.S., Neff, J.M. (1981) Bioaccumulation and metabolism of phthalate esters by oysters, brown shrimp and sheepshead minnows. *Ecotox. Environ. Saf.* 5, 202-210.

Wolfe, N.L., Paris, D.F., Steen, W.C., Baughman, G.L. (1980c) Correlation of microbial degradation rates with chemical structure. *Environ. Sci. Technol.* 14, 1143-1144.

Wolfenden, R.V. (1978) Interaction of the peptide bond with solvent water: a vapor phase analysis. *Biochem.* 17, 201-204.

Wolosin, J.M., Ginsberg, H. (1975) The permeation of organic acids through lecithin bilayers. Resemblance to diffusion in polymers. *Biochim. Biophys. Acta* 389, 20-33.

Worthing, C.R., Editor (1983) *The Pesticide Manual.* (A World Compendium) 7th Edition, The British Crop Protection Council, Croydon, U.K.

Yalkowsky, S.H., Morozowich, W. (1980) Chapter 3, A physical chemical basis for the design of orally active prodrugs. In: *Drug Design*. Vol IX. Academic Press, Inc., New York.

Yalkowsky, S.H., Valvani, S.C. (1980) Solubility and partitioning I: Solubility of nonelectrolytes in water. *J. Pharm. Sci.* 69(8), 912-922.

Yalkowsky, S.H., Valvani, S.C., Kun, W.-Y., Dannenfelser, R.M., Eds. (1987) *Arizona Database of Aqueous Solubility for Organic Compounds.* College of Pharmacy, University of Arizona, Tucson, Arizona.

Yalkowsky, S.H., Valvani, S.C., Roseman, T.J. (1983) Solubility and partitioning. VI. Octanol solubility and octanol-water partition coefficients. *J. Pharm. Sci.* 72(8), 866-870.

Yaguzhinskii, L.S., Smirnova, E.G., Ratnikova, L.A., Kolesova, G.M., Krasinskaya, I.P. (1973) Hydrophobic sites of the mitochondrial electron transfer system. *J. Bioenerg.* 5, 163-174.

Yaws, C.L., Yang, H.C., Hopper, J.R., Hansen, K.C. (1990) Organic chemicals: Water solubility data. *Chem. Eng.* 97, 115-118.

Yaws, C.L., Yang, H.C., Pan, X. (1991) Henry's law constants for 362 organic compounds in water. *Chem. Eng.* 98(11), 179-185.

Yoshida, Y., Mizuno, T., Ose, Y., Sato, T. (1986) The estimation for toxicity of chemicals on fish by physico-chemical properties. *Chemosphere* 15(2), 195-203.

Zaroogian, G.E., Heltshe, J.F., Johnson, M. (1985) Estimation of bioconcentration in marine species using structure-activity models. *Environ. Toxicol. Chem.* 4, 3-12.

Zeep, R.G., Wolfe, N.L., Gordon, J.A., Baughman, G.L. (1975) Dynamics of 2,4-D esters in surface waters. Hydrolysis, photolysis and vaporization. *Environ. Sci. Technol.* 9, 1144.

Zetzsch, C., Stuhl, F. (1982) *Proceedings of 2nd European Symposium on Physico-Chemical Behavior of Atmospheric Pollutants*, pp. 129-137, Sept. 29 to Oct. 1 of 1981, Varese, Italy. Reidel Publishing Company, Dordrecht, The Netherlands.

Chapter 6 Esters

6.1 List of Chemicals and Data Compilations

Common Name: Methyl formate
Synonym: formic acid methyl ester, methyl methanoate
Chemical Name: methyl formate
CAS Registry No: 107-31-3
Molecular Formula: $HCOOCH_3$
Molecular Weight: 60.052
Melting Point (°C):
-99 (Weast 1982-83; Dean 1985; Riddick et al. 1986; Howard 1993)
-99.8 (Verschueren 1983)
Boiling Point (°C):
31.5 (Weast 1982-83; Dean 1985; Howard 1993)
32.0 (Verschueren 1983)
32.08 (Boublik et al. 1984)
31.75 (Riddick et al. 1986)
Density (g/cm^3 at 20°C):
0.9742 (Weast 1982-83)
0.9750 (Verschueren 1983)
Molar Volume (cm^3/mol):
62.8 (exptl. at normal B.P., Lee et al. 1972; quoted, Reid et al. 1977)
62.6 (calculated-LeBas method, Reid et al. 1977)
32.5 (intrinsic volume V_I, Leahy 1986; Kamlet et al. 1987)
61.6 (Leahy 1986; Kamlet et al. 1987)
62.1 (calculated-density)
Molecular Volume (Å^3):
Total Surface Area, TSA (Å^2):
Dissociation Constant pK_a:
Heat of Fusion, ΔH_{fus}, kcal/mol: 1.78 (Riddick et al. 1986)
Entropy of Fusion, ΔS_{fus}, cal/mol·K (e.u.):
Fugacity Ratio at 25°C (assuming ΔS_{fus} = 13.5 e.u.), F: 1.0

Water Solubility (g/m^3 or mg/L at 25°C):
230000 (USEPA 1975; quoted, Howard 1993)
304000 (20°C, quoted, Verschueren 1983)
230000 (Dean 1985; Riddick et al. 1986)
165400, 353610 (quoted, calculated-V_I and solvatochromic p., Leahy 1986)
165400, 345560 (quoted, calculated-V_I and solvatochromic p., Kamlet et al. 1987)
165200, 638260 (quoted, predicted- χ , Nirmalakhandan & Speece 1988)
185580 (calculated-C_A/K_{AW}, Hoff et al. 1993)

Vapor Pressure (Pa at 25°C):
63984 (20°C, quoted, Verschueren 1983)
93310 (30°C, quoted, Verschueren 1983)
83260 (quoted exptl., Boublik et al. 1984)
80840 (calculated-Antoine eqn., Boublik et al. 1984)
83440 (calculated-Antoine eqn., Dean 1985)
78060 (Riddick et al. 1986)
78074 (Daubert & Danner 1989; quoted, Howard 1993)

Henry's Law Constant (Pa·m^3/mol):

22.61 (quoted exptl., Hine & Mookerjee 1975; Howard 1993)
20.149 (calculated-group contribution, Hine & Mookerjee 1975)
17.96 (calculated-bond contribution, Hine & Mookerjee 1975)
22.61, 15.64 (quoted, calculated- χ , Nirmalakhandan & Speece 1988)
26.02 (quoted, Hoff et al. 1993)
20.365 (calculated-P/C, this work)

Octanol/Water Partition Coefficient, log K_{OW}:

-0.01 (calculated-V_I and solvatochromic p., Leahy 1986)
-0.264 (estimated, USEPA 1988; quoted, Howard 1993)

Bioconcentration Factor, log BCF:

-0.222 (estimated-S, Lyman et al. 1990; quoted, Howard 1993)

Sorption Partition Coefficient, log K_{OC}:

0.699 (soil, estimated-S, Lyman et al. 1990; quoted, Howard 1993)

Half-Lives in the Environment:

Air: estimated half-life of 74 days, based on experimental reaction rate constant for the vapor-phase reaction with photochemically produced hydroxyl radicals in the atmosphere at 23°C (Atkinson 1989; quoted, Howard 1993).
Surface water:
Groundwater:
Sediment:
Soil:
Biota:

Environmental Fate Rate Constants or Half-Lives:

Volatilization: based on Henry's law constant, half-life from a model river of one meter deep flowing at 1 m/sec with a wind velocity of 3 m/sec was estimated to be 5.3 hours (Lyman et al. 1990; quoted, Howard 1993).
Photolysis:
Oxidation: measured rate constant of $(2.27\pm0.34)x10^{12}$ cm^3 $molecule^{-1}$ sec^{-1} for the gas-phase reactions of OH radicals at 296 K (Wallington et al. 1988); exptl. photooxidation rate constant of $2.27x10^{-12}$ cm^3 $molecule^{-1}$ sec^{-1} for the vapor-phase reaction with photochemically produced $5x10^5$ hydroxyl radicals/cm^3 in the atmosphere at 23°C (Atkinson 1989; quoted, Howard 1993).
Hydrolysis: aqueous base-catalyzed hydrolysis rate constant of 36.6 M^{-1} sec^{-1} at 25°C corresponds to half-lives of 21.9 days, 2.19 days, 9.1 hours and 0.91 hour at respective pH of 6, 7, 8, 9 (Mabey & Mill 1978; quoted, Howard 1993).
Biodegradation:
Biotransformation:
Bioconcentration, Uptake (k_1) and Elimination (k_2) Rate Constants:

Common Name: Ethyl formate
Synonym: ethyl methanoate, formic acid ethyl ester
Chemical Name: ethyl formate
CAS Registry No: 109-94-4
Molecular Formula: $HCOOCH_2CH_3$
Molecular Weight: 74.079
Melting Point (°C):
-79.0 (Stull 1947; Verschueren 1983)
-80.5 (Weast 1982-83)
-79.4 (Dean 1985)
Boiling Point (°C):
54.3 (Stull 1947; Riddick et al. 1986)
54.5 (Weast 1982-83)
54.2 (Dean 1985)
Density (g/cm^3 at 20°C):
0.9168 (Weast 1982-83)
0.9240 (25°C, Verschueren 1983)
0.9220, 0.9153 (20°C, 25°C, Riddick et al. 1986)
Molar Volume (cm^3/mol):
80.3 (Leahy 1986; Kamlet et al. 1987)
42.4 (intrinsic volume V_I, Leahy 1986; Kamlet et al. 1987)
86.4 (calculated-LeBas method)
80.9 (calculated-density)
Molecular Volume ($Å^3$):
Total Surface Area, TSA ($Å^2$): 242.1 (Amidon et al. 1975)
Dissociation Constant pK_a:
Heat of Fusion, ΔH_{fus}, kcal/mol: 2.20 (Riddick et al. 1986)
Entropy of Fusion, ΔS_{fus}, cal/mol·K (e.u.):
Fugacity Ratio at 25°C (assuming ΔS_{fus} = 13.5 e.u.), F: 1.0

Water Solubility (g/m^3 or mg/L at 25°C):
88250 (Seidell 1941; quoted, Hansch et al. 1968)
93260 (estimated, McGowan 1954)
52440 (quoted, Deno & Berkheimer 1960)
275860 (calculated-K_{OW}, Hansch et al. 1968)
89063 (Hansch et al. 1968; quoted, Hine & Mookerjee 1975; Müller & Klein 1992)
18810 (20°C, quoted, Amidon et al. 1975)
118000 (quoted, Verschueren 1983; Riddick et al. 1986)
117000 (quoted, Dean 1985)
72390, 102260 (quoted, calculated-V_I and solvatochromic p., Leahy 1986)
72390, 97660 (quoted, calculated-V_I and solvatochromic p., Kamlet et al. 1987)
72280, 203240 (quoted, predicted- χ , Nirmalakhandan & Speece 1988)
3425 (calculated-C_A/K_{AW}, Hoff et al. 1993)

Vapor Pressure (Pa at 25°C):
33054 (interpolated-regression of tabulated data, Stull 1947)
26660 (20.0°C, Stull 1947)

34050 (Hine & Mookerjee 1975)
25590, 39990 (20°C, 30°C, quoted, Verschueren 1983)
39990 (30°C, quoted, Verschueren 1983)
32610 (quoted, Abraham 1984)
30844, 32610 (23.7°C, quoted exptl., interpolated-Antoine eqn., Boublik et al. 1984)
32540 (interpolated-Antoine eqn., Dean 1985)
32370 (quoted, Riddick et al. 1986)

Henry's Law Constant ($Pa \cdot m^3/mol$):
28.46 (exptl., Hine & Mookerjee 1975)
27.18 (calculated-group contribution, Hine & Mookerjee 1975)
25.96 (calculated-bond contribution, Hine & Mookerjee 1975)
28.46, 17.55 (quoted, calculated- χ , Nirmalakhandan & Speece 1988)
702 (quoted, Hoff et al. 1993)

Octanol/Water Partition Coefficient, log K_{OW}:
-0.38 (calculated, Iwasa et al. 1965)
0.23 (calculated-S, Hansch et al. 1968)
0.33 (Leo et al. 1971)
0.55 (calculated-V_I and solvatochromic p., Leahy 1986)
0.26 (calculated-CLOGP, Müller & Klein 1992)

Bioconcentration Factor, log BCF:

Sorption Partition Coefficient, log K_{OC}:

Half-Lives in the Environment:
Air:
Surface water:
Groundwater:
Sediment:
Soil:
Biota:

Environmental Fate Rate Constants or Half-Lives:
Volatilization:
Photolysis:
Oxidation: second-order rate constant of $(10.2 \pm 1.4) \times 10^{12}$ $cm^3 \cdot molecule^{-1} \cdot sec^{-1}$ for the gas phase reactions with hydroxyl radicals at 296 K (Wallington et al. 1988a,b); 6.5×10^{-13} $cm^3 \cdot molecule^{-1} \cdot sec^{-1}$ for the solution-phase reactions with hydroxyl radicals in aqueous solution (Wallington et al. 1988b).
Hydrolysis:
Biodegradation:
Biotransformation:
Bioconcentration, Uptake (k_1) and Elimination (k_2) Rate Constants:

Common Name: Propyl formate
Synonym: formic acid propyl ester, methanoic acid *n*-propyl ester, propyl methanoate
Chemical Name: *n*-propyl formate, propyl formate
CAS Registry No: 110-74-7
Molecular Formula: $HCOOCH_2CH_2CH_3$
Molecular Weight: 88.10
Melting Point (°C):
-92.9 (Stull 1947; Weast 1982-83; Dean 1985; Riddick et al. 1986)
Boiling Point (°C):
81.3 (Stull 1947; Weast 1982-83)
80.54 (Boublik et al. 1984)
80.9 (Dean 1985)
80.82 (Riddick et al. 1986)
Density (g/cm^3 at 20°C):
0.9058 (Weast 1982-83)
0.9006 (Dean 1985)
0.9055 (Riddick et al. 1986)
0.8996 (25°C, Riddick et al. 1986)
Molar Volume (cm^3/mol):
110.8 (calculated-LeBas method)
98.0 (calculated-density)
Molecular Volume (Å^3):
Total Surface Area, TSA (Å^2):
Dissociation Constant pK_a:
Heat of Fusion, ΔH_{fus}, kcal/mol:
Entropy of Fusion, ΔS_{fus}, cal/mol·K (e.u.):
Fugacity Ratio at 25°C (assuming ΔS_{fus} = 13.5 e.u.), F: 1.0

Water Solubility (g/m^3 or mg/L at 25°C):
28445 (Seidell 1941; quoted, Hansch et al. 1968)
27230 (quoted, Deno & Berkheimer 1960)
81100 (calculated-K_{OW}, Hansch et al. 1968)
27300 (Hansch et al. 1968; quoted, Hine & Mookerjee 1975; Müller & Klein 1992)
26606 (quoted average, Valvani et al. 1981)
41210 (predicted-K_{OW}, Valvani et al. 1981)
20500 (Dean 1985; Riddick et al. 1986)
28580, 69343 (quoted, predicted- χ , Nirmalakhandan & Speece 1988)

Vapor Pressure (Pa at 25°C):
10620 (interpolated-regression of tabulated data, Stull 1947)
13330 (29.5°C, Stull 1947)
11440 (Hine & Mookerjee 1975)
11760 (26.2°C, quoted exptl., Boublik et al. 1984)
11010 (calculated-Antoine eqn., Boublik et al. 1984)
10710 (extrapolated-Antoine eqn., Dean 1985)
11030 (Riddick et al. 1986)

Henry's Law Constant ($Pa \cdot m^3/mol$):

37.52 (quoted exptl., Hine & Mookerjee 1975)
38.392 (calculated-group contribution, Hine & Mookerjee 1975)
38.39 (calculated-bond contribution, Hine & Mookerjee 1975)
37.52, 22.61 (quoted, calculated- χ , Nirmalakhandan & Speece 1988)

Octanol/Water Partition Coefficient, log K_{OW}:

0.73 (calculated-S, Hansch et al. 1968)
0.83 (quoted unpublished result, Leo et al. 1971)
0.83 (exptl., Valvani et al. 1981)
0.83 (quoted from Hansch & Leo 1979, Bodor et al. 1989)
0.83 (quoted from CLOGP, Bodor et al. 1989)
0.794 (calculated-CLOGP, Bodor et al. 1989)
0.61 (calculated-M.O., Bodor et al. 1989)
0.83 (quoted from Pomona College unpublished data, Sangster 1989)
0.83 (recommended, Sangster 1989)
0.79 (calculated-CLOGP, Müller & Klein 1992)

Bioconcentration Factor, log BCF:

Sorption Partition Coefficient, log K_{OC}:

Half-Lives in the Environment:

Air:
Surface water:
Groundwater:
Sediment:
Soil:
Biota:

Environmental Fate Rate Constants or Half-Lives:

Volatilization:
Photolysis:
Oxidation: second-order rate constant of $(23.8 \pm 2.7) \times 10^{-12}$ $cm^3 \cdot molecule^{-1} \cdot sec^{-1}$ for the gas phase reactions with hydroxyl radicals at 296 K (Wallington et al. 1988).
Hydrolysis:
Biodegradation:
Biotransformation:
Bioconcentration, Uptake (k_1) and Elimination (k_2) Rate Constants:

Common Name: Methyl acetate
Synonym: acetic acid methyl ester, ethanoic acid methyl ester, methyl ethanoate
Chemical Name: methyl acetate
CAS Registry No:79-20-9
Molecular Formula: CH_3COOCH_3
Molecular Weight: 74.08
Melting Point (°C):
-98.7 (Stull 1947; Verschueren 1983)
-98.1 (Weast 1982-83; Dean 1985; Riddick et al. 1986; Howard 1993)
Boiling Point (°C):
57.8 (Stull 1947; Verschueren 1983)
57.0 (Weast 1982-83; Howard 1993)
56.9 (Boublik et al. 1984; Riddick et al. 1986)
56.3 (Dean 1985)
Density (g/cm^3 at 20°C):
0.9330 (Weast 1982-83)
0.9342 (Dean 1985; Riddick et al. 1986)
0.9279 (25°C, Riddick et al. 1986)
Molar Volume (cm^3/mol):
79.88 (Hoy 1970; Amidon & Williams 1982)
79.8 (Taft et al. 1985)
79.4 (Kamlet et al. 1986,87; Leahy 1986)
42.4 (intrinsic volume V_I, Leahy 1986; Kamlet et al. 1987)
84.8 (calculated-LeBas method)
79.8 (calculated-density)
Molecular Volume (Å^3):
Total Surface Area, TSA (Å^2):
244.9 (Amidon et al. 1975)
204.3 (Iwase et al. 1985)
Dissociation Constant:
22.50 (pK_s, Riddick et al. 1986)
Heat of Fusion ΔH_{fus}, kcal/mol:
Entropy of Fusion ΔS_{fus}, cal/mol·K (e.u.):
Fugacity Ratio at 25°C (assuming ΔS_{fus} = 13.5 e.u.), F: 1.0

Water Solubility (g/m^3 or mg/L at 25°C):
243730 (20°C, shake flask-turbidity, Fühner 1924)
243610 (Fühner 1924; quoted, Hansch et al. 1968)
294920 (estimated, McGowan 1954)
243500 (20°C, Stephen & Stephen 1963; quoted, Howard 1993)
275870 (calculated-K_{OW}, Hansch et al. 1968)
10870 (20°C, quoted, Amidon et al. 1975)
208790 (quoted average, Valvani et al. 1981; quoted, Banerjee 1985)
158380 (predicted-K_{OW}, Valvani et al. 1981; quoted, Banerjee 1985)
281640, 301790 (quoted, calculated-K_{OW} and M.P., Amidon & Williams 1982)
240000/319000 (20°C, quoted, Verschueren 1983)

240000 (Dean 1985)
245000 (20°C, Riddick et al. 1986)
213650, 162070 (quoted, calculated-V_I and solvatochromic p., Kamlet et al. 1986)
213650, 194850 (quoted, calculated-V_I and solvatochromic p., Leahy 1986)
213650, 194850 (quoted, calculated-V_I and solvatochromic p., Kamlet et al. 1987)
213300, 249460 (quoted, predicted- χ , Nirmalakhandan & Speece 1988)
243720 (quoted, Müller & Klein 1992)

Vapor Pressure (Pa at 25°C):
27980 (interpolated-regression of tabulated data, Stull 1947)
26660 (24.0°C, Stull 1947)
28830 (ebulliometry-Antoine eqn., Ambrose et al. 1981; quoted, Howard 1993)
31330 (quoted, Verschueren 1983)
28830 (quoted, Abraham 1984)
25670 (22.4°C, quoted exptl., Boublik et al. 1984)
28830 (calculated-Antoine eqn., Boublik et al. 1984)
28190 (calculated-Antoine eqn., Dean 1985)
28880 (quoted, Kamlet et al. 1986)
28830 (Riddick et al. 1986)

Henry's Law Constant (Pa·m³/mol):
9.19 (volatility-partial pressure, Butler & Ramchandani 1935)
11.65 (volatility-partial pressure-GC, Buttery et al. 1969)
8.795 (quoted exptl., Hine & Mookerjee 1975)
11.59 (calculated-group contribution, Hine & Mookerjee 1975)
9.210 (calculated-group contribution, Hine & Mookerjee 1975)
13.06 (vapor-liquid equilibrium, vapor phase-GC, Kieckbusch & King 1979)
9.210, 25.96 (quoted, calculated- χ , Nirmalakhandan & Speece 1988)

Octanol/Water Partition Coefficient, log K_{OW}:
0.18 (shake flask-CR, Collander 1951; quoted, Sangster 1989)
0.20 (selected, McGowan 1954)
-0.38 (calculated, Iwasa et al. 1965)
0.23 (calculated-S, Hansch et al. 1968; quoted, Leo et al. 1969)
0.18 (exptl., Valvani et al. 1981; quoted, Amidon & Williams 1982)
0.18 (quoted, Verschueren 1983)
0.18 (Hansch & Leo 1985; quoted, Howard 1993)
0.18, 0.17 (quoted, calculated-TSA, Iwase et al. 1985)
0.18, 0.36 (quoted, calculated-V_I and solvatochromic p., Taft et al. 1985)
0.18, 0.35 (quoted, calculated-V_I and solvatochromic p., Leahy 1986)
0.18, 0.70 (quoted, calculated-γ from UNIFAC, Banerjee & Howard 1988)
0.18 (quoted from Hansch & Leo 1979, Bodor et al. 1989)
0.18, 0.142 (quoted-CLOGP, calculated-CLOGP, Bodor et al. 1989)
0.05 (calculated-M.O. indices, Bodor et al. 1989)

0.18 (recommended, Sangster 1989)
0.14 (calculated-CLOGP, Müller & Klein 1992)
0.18, 0.4341 (quoted, calculated-original UNIFAC, Chen et al. 1993)

Bioconcentration Factor, log BCF:
-0.224 (calculated-S, Lyamn et al. 1982; quoted, Howard 1993)
-0.092 (calculated-K_{OW}, Lyamn et al. 1982; quoted, Howard 1993)

Sorption Partition Coefficient, log K_{OC}:
0.681 (soil, calculated-S, Lyman et al. 1982; quoted, Howard 1993)
1.474 (soil, calculated-K_{OW}, Lyman et al. 1982; quoted, Howard 1993)

Half-Lives in the Environment:
Air: photooxidation half-life of 2.4-24 hours for the gas-phase reaction with hydroxyl radicals in air, based on the rate of disappearance of hydrocarbon due to reaction with hydroxyl radicals (Darnall et al. 1976); atmospheric half-lives of 47-94 days, based on exptl. rate constants for the vapor-phase reaction with photochemically produced hydroxyl radicals in air (Atkinson et al. 1979; Güesten et al. 1984; Atkinson 1987; Wallington et al. 1988; quoted, Howard 1993).
Surface water:
Groundwater:
Sediment:
Soil:
Biota:

Environmental Fate Rate Constants or Half-Lives:
Volatilization: based on calculated Henry's law constant, half-life from a model river of 1 meter deep flowing at 1 m/sec with wind velocity of 3 m/sec was estimated to be 9.1 hours (Lyman et al. 1982; quoted, Howard 1993).
Photolysis:
Oxidation: photooxidation half-life of 2.4-24 hours for the gas-phase reaction with hydroxyl radicals in air, based on the rate of disappearance of hydrocarbon due to reaction with hydroxyl radicals (Darnall et al. 1976); using relative rate technique for *n*-butane, rate constant of $(0.17 \pm 0.05)\times10^{-12}$ $cm^3\cdot molecule^{-1}\cdot sec^{-1}$ was determined for the gas-phase reaction with OH radicals at 292 K in air (Campbell & Parkinson 1978; quoted, Atkinson 1985); exptl. rate constants 1.3 to 3.41×10^{-13} $cm^3\cdot molecule^{-1}\cdot sec^{-1}$ for the vapor-phase reaction with photochemically produced hydroxyl radicals in air (Atkinson et al. 1979; Güesten et al. 1984; Atkinson 1987; Wallington et al. 1988a,b; quoted, Howard 1993); rate constant of 2.00×10^{13} $cm^3\cdot molecule^{-1}\cdot sec^{-1}$ for the solution-phase reaction with hydroxyl radicals in aqueous solution (Wallington et al. 1988b).
Hydrolysis: alkaline hydrolysis rate constant of 0.182 M^{-1} sec^{-1} at 25°C in water (Salmi & Leimu 1947; quoted, Drossman et al. 1988).
Biodegradation:
Biotransformation:
Bioconcentration, Uptake (k_1) and Elimination (k_2) Rate Constants:

Common Name: Vinyl acetate
Synonym: ethanoic acid ester, ethenyl acetate, ethenyl ethanoate
Chemical Name: ethenyl acetate, vinyl acetate
CAS Registry No: 108-05-4
Molecular Formula: $CH_3COOCH=CH_2$
Molecular Weight: 86.09
Melting Point (°C):
-93.2 (Weast 1982-83; Howard 1989)
-92.8 (Dean 1985; Riddick et al. 1986)
Boiling Point (°C):
72.5 (Stull 1947; Dean 1985)
72.3 (Weast 1982-83)
Density (g/cm^3 at 20°C):
0.9317 (Weast 1982-83)
0.9318 (Dean 1985)
0.9312 (Riddick et al. 1986)
Molar Volume (cm^3/mol):
96.2 (calculated-LeBas method)
92.4 (calculated-density)
Molecular Volume ($Å^3$):
Total Surface Area, TSA ($Å^2$):
Dissociation Constant, pK_a:
Heat of Fusion ΔH_{fus}, kcal/mol:
Entropy of Fusion ΔS_{fus}, cal/mol·K (e.u.):
Fugacity Ratio at 25°C (assuming ΔS_{fus} = 13.5 e.u.), F: 1.0

Water Solubility (g/m^3 or mg/L at 25°C):
19722 (Deno & Berkheimer 1960)
25000 (Verschueren 1983)
20000 (Dean 1985; Riddick et al. 1986)
20000 (20°C, quoted, Howard 1989)

Vapor Pressure (Pa at 25°C):
14454 (interpolated-regression of tabulated data, Stull 1947)
13330 (23.3°C, Stull 1947)
15330 (Verschueren 1983)
14100 (Riddick et al. 1986)
11320 (20°C, quoted, Howard 1989)

Henry's Law Constant ($Pa \cdot m^3/mol$):
48.74 (calculated-P/C, Howard 1989)

Octanol/Water Partition Coefficient, log K_{OW}:
0.60 (shake flask, Fujisawa & Masuhara 1980; quoted, Sangster 1989)
0.21 (calculated-HPLC-RT, Fujisawa & Masuhara 1981)
0.73 (quoted and recommended, Sangster 1989)
0.73 (quoted, Howard 1989)

Bioconcentration Factor, log BCF:

0.32-0.37 (estimated-K_{OW}, Howard 1989)

Sorption Partition Coefficient, log K_{OC}:

1.28-1.77 (estimated-$_{OW}$, Howard 1989)

Half-Lives in the Environment:

Air: estimated half-life of 12 hours with photochemically produced hydroxyl radicals in an average atmosphere (Howard 1989).

Surface water: hydrolysis degradation half-life of 7.3 days at 25°C and pH 7; half-lives of 13 and 8 days for reactions with hydroxyl radicals and singlet oxygen respectively in natural water (Howard 1989).

Groundwater:

Sediment:

Soil:

Biota:

Environmental Fate Rate Constants or Half-Lives:

Volatilization: estimated half-lives of 4.4 hours from a model river of 1 m deep flowing at 1 m/sec with a wind velocity of 3 m/sec, and 2.2 days from an environmental pond (Howard 1989).

Photolysis:

Oxidation: half-lives of 13 and 8 days for reactions with OH radicals and singlet oxygen in sunlit natural water (Howard 1989).

Hydrolysis: dagradation half-life of 7.3 days at 25 °C and pH 7, the hydrolysis rate will increase as the pH increases (Howard 1989).

Biodegradation:

Biotransformation:

Bioconcentration, Uptake (k_1) and Elimination (k_2) Rate Constants:

Common Name: Ethyl acetate
Synonym: acetic acid ethyl ester, acetic ether, ethanoic acid ester, ethyl ethanoate
Chemical Name: *n*-ethyl acetate, ethyl acetate
CAS Registry No: 141-78-6
Molecular Formula: $CH_3COOCH_2CH_3$
Molecular Weight: 88.11
Melting Point (°C):
-82.4 (Stull 1947)
-83.6 (Weast 1982-83; Verschueren 1983; Riddick et al. 1986)
-82.0 (Verschueren 1983)
-83.0 (Howard 1990)
Boiling Point (°C):
77.1 (Stull 1947; Boublik et al. 1984; Riddick et al. 1986)
77.06 (Weast 1982-83)
77.0 (Verschueren 1983; Banerjee et al. 1990; Howard 1990)
Density (g/cm^3 at 20°C):
0.9003 (Weast 1982-83)
0.9010 (Verschueren 1983)
0.9006 (Riddick et al. 1986)
0.89455 (25°C, Riddick et al. 1986)
Molar Volume (cm^3/mol):
99.53 (Hoy 1970; Amidon & Williams 1982)
106.0 (exptl. at normal B.P., Lee et al. 1972; quoted, Reid et al. 1977)
98.1 (calculated-density, Rohrschneider 1973)
108.6 (calculated-LeBas method, Reid et al. 1977)
97.8 (Taft et al. 1985; Leahy 1986; Kamlet et al. 1986,87)
52.1 (intrinsic volume V_I, Leahy 1986; Kamlet et al. 1987)
Molecular Volume (Å^3):
Total Surface Area, TSA (Å^2):
233.8 (Iwase et al. 1985)
Dissociation Constnat:
22.83 (pK_s, Riddick et al. 1986)
Heat of Fusion ΔH_{fus}, kcal/mol:
2.51 (Riddick et al. 1986)
Entropy of Fusion ΔS_{fus}, cal/mol·K (e.u.):
Fugacity Ratio at 25°C (assuming ΔS_{fus} = 13.5 e.u.), F: 1.0

Water Solubility (g/m^3 or mg/L at 25°C):
64408 (20°C, shake flask-turbidity, Fühner 1924)
85300 (20°C, synthetic method, Jones 1929)
78720 (shake flask-centrifuge, Booth & Everson 1948)
80100 (turbidimetric method, Altshuller & Everson 1953)
80400 (average, Altshuller & Everson 1953)
79780 (shake flask-interferometry, Donahue & Bartell 1952)
69990 (estimated, McGowan 1954)
82220 (quoted, Deno & Berkheimer 1960)

80350 (shake flask-AS, Hansch et al. 1968; quoted, Wasik et al. 1981, 1982; Tewari et al. 1982; Müller & Klein 1992)
81095 (calculated-K_{OW}, Hansch et al. 1968)
63960 (gen. col.-GC, Wasik et al. 1981, 1982; Tewari et al. 1982; quoted, Howard 1990)
80360, 76740 (quoted, calculated-K_{OW} & M.P., Amidon & Williams 1982)
82500 (20°C, quoted average, Verschueren 1983)
74000 (35°C, quoted, Verschueren 1983)
65320 (Banerjee 1984; quoted, Banerjee 1985)
80000 (shake flask-HPLC, Banerjee 1984)
76740, 56890 (quoted average, predicted-K_{OW}, Valvani et al. 1981)
97000 (Dean 1985)
95000 (radiometric, Lo et al. 1986)
80800 (Riddick et al. 1986)
77900 (20.2°C, shake flask-GC/TC, Stephenson & Stuart 1986)
68100 (29.7°C, shake flask-GC/TC, Stephenson & Stuart 1986)
78530, 65320 (quoted, calculated-V_I & solvatochromic p., Kamlet et al. 1986)
78530, 78530 (quoted, calculated-V_I & solvatochromic p., Leahy 1986)
78530, 80360 (quoted, calculated-V_I & solvatochromic p., Kamlet et al. 1987)
78340, 82600 (quoted, predicted-χ, Nirmalakhandan & Speece 1988)
73700 (selected, Yaws et al. 1990)
98180 (calculated-C_A/K_{AW}, Hoff et al. 1993)

Vapor Pressure (Pa at 25°C):
11860 (interpolated-regression of tabulated data, Stull 1947)
13330 (27.0°C, Stull 1947)
12600 (ebulliometry, fitted to Antoine eqn., Ambrose et al. 1981)
9704 (20°C, quoted, Verschueren 1983)
13040 (25.7°C, quoted exptl., Boublik et al. 1984)
11220 (calculated-Antoine eqn., Boublik et al. 1984)
12605 (quoted, Abraham 1984)
12620 (calculated-Antoine eqn., Dean 1985)
3240 (Howard et al. 1986; quoted, Banerjee et al. 1990)
12610 (quoted, Kamlet et al. 1986)
12600 (quoted, Riddick et al. 1986)
3240, 5240 (measured, calculated-solvatochromic p., Banerjee et al. 1990)

Henry's Law Constant (Pa·m^3/mol):
13.42 (partial pressure, Butler & Ramchandani 1935)
17.78 (28°C, concn. ratio-GC, Nelson & Hoff 1968)
13.62 (quoted exptl., Hine & Mookerjee 1975)
15.64 (calculated-group contribution, Hine & Mookerjee 1975)
13.94 (calculated-group contribution, Hine & Mookerjee 1975)
17.20 (vapor-liquid equilibrium, vapor phase-GC, Kieckbusch & King 1979)
13.62, 24.22 (quoted, calculated-χ, Nirmalakhandan & Speece 1988)
11.33 (quoted, Hoff et al. 1993)
13.62, 6.23 (quoted, calculated-molecular structure, Russell et al. 1992)

Octanol/Water Partition Coefficient, log K_{OW}:

0.66 (shake flask-CR, Collander 1951; quoted, Rekker 1977; Hansch & Leo 1979; Sangster 1989)
0.70 (selected, McGowan 1954)
0.14 (calculated, Iwasa et al. 1965)
0.73 (shake flask-GC, Hansch & Anderson 1967; quoted, Leo et al. 1969,71; Rekker 1977; Hansch & Leo 1979; Iwase et al. 1985; Sangster 1989)
0.73 (shake flask-UV, Hansch et al. 1968; quoted, Hansch & Leo 1979; Wasik et al. 1981,82; Tewari et al. 1982; quoted, Sangster 1989; Howard 1990)
0.73 (calculated-π, Rekker 1977)
0.65, 0.66 (calculated-f const., Rekker 1977)
0.70 (exptl., Valvani et al. 1981; quoted, Amidon & Williams 1982)
0.53 (calculated-γ, Wasik et al. 1981; quoted, Sangster 1989)
0.68 (gen. col.-GC, Wasik et al. 1981, 1982; Tewari et al. 1982; quoted, Sangster 1989)
0.66, 0.73 (quoted, Verschueren 1983)
0.73 (quoted, Geyer et al. 1984)
0.73 (calculated-TSA, Iwase et al. 1985)
0.73, 0.81 (quoted, calculated-V_I and solvatochromic p., Taft et al. 1985)
0.73, 0.80 (quoted, calculated-V_I and solvatochromic p., Leahy 1986)
0.73, 1.02 (quoted, calculated-γ from UNIFAC, Banerjee & Howard 1988)
0.70, 0.73, 0.671, 0.59 (quoted from Hansch & Leo 1979, quoted from CLOGP, calculated-CLOGP, calculated-M.O. indices, Bodor et al. 1989)
0.73 (recommended, Sangster 1989)
0.67 (calculated-CLOGP, Müller & Klein 1992)
0.671 (calculated-QSAR, Kollig 1993)

Bioconcentration Factor, log BCF:

4.13 (alga *chlorella fusca*, wet wt. basis, Geyer et al. 1984)
0.70 (alga *chlorella fusca*, calculated-K_{OW}, Geyer et al. 1984)
1.48 (golden ide, Freitag et al. 1985)
4.13 (algae, Freitag et al. 1985)
3.52 (activated sludge, Freitag et al. 1985)

Sorption Partition Coefficient, log K_{OC}:

0.361 (calculated-K_{OW}, Kollig 1993)

Half-Lives in the Environment:

Air: photooxidation half-life of 2.75 -110 years, based on measured rate constant for the reaction with hydroxyl radicals in water (Dorfman & Adams 1973; selected, Howard et al. 1991); half-life of 2.4-24 hours for the gas-phase reaction with hydroxyl radicals in air, based on the rate of disappearance of hydrocarbon due to reaction with hydroxyl radicals (Darnall et al. 1976); photooxidation half-life of 35.3-353 hours, based on measured rate constant for the reaction with hydroxyl radicals in air (Atkinson 1985; selected, Howard et al. 1991).

Surface water: 24-168 hours, based on estimated unacclimated aqueous aerobic biodegradation half-life (Howard et al. 1991).

Groundwater: 24-168 hours, based on estimated unacclimated aqueous aerobic biodegradation half-life (Howard et al. 1991).

Sediment:

Soil: 24-168 hours, based on estimated unacclimated aqueous aerobic biodegradation half-life (Howard et al. 1991).

Biota:

Environmental Fate Rate Constants or Half-Lives:

Volatilization: using Henry's law constant, a half-life of 10.1 hours was estimated for a model river of 1 m deep flowing at 1 m/sec with a wind velocity of 3 m/sec (Lyman et al. 1982; quoted, Howard 1990).

Photolysis:

Oxidation: photooxidation half-life of 2.75-110 years in water, based on measured rate constant for the reaction with hydroxyl radicals in water (Dorfman & Adams 1973; selected, Howard et al. 1991); photooxidation half-life of 2.4-24 hours for the gas-phase reaction with hydroxyl radicals in air, based on the rate of disappearance of hydrocarbon due to reaction with hydroxyl radicals (Darnall et al. 1976); using relative rate technique for *n*-butane, rate constant of $(1.82 \pm 0.36)\times10^{-12}$ cm^3 $molecule^{-1}$ sec^{-1} was determined for the gas-phase reaction with OH radicals at 292 K in air (Campbell & Parkinson 1978; quoted, Atkinson 1985); photooxidation half-life of 35.3-353 hours in air, based on measured rate constant for the reaction with hydroxyl radicals in air (Atkinson 1985; selected, Howard et al. 1991); rate constant of $(15.1 \pm 1.4)\times10^{-13}$ cm^3 $molecule^{-1}$ sec^{-1} for the gas phase reactions of OH radicals at 296K (Wallington et al. 1988a,b); rate constant of 6.60×10^{-13} cm^3 $molecule^{-1}$ sec^{-1} for the solution-phase reaction with OH radicals in aqueous solution (Wallington et al. 1988b).

Hydrolysis: alkaline hydrolysis rate constant of 0.110 M^{-1} sec^{-1} at 25°C in water (Salmi & Leimu 1947; quoted, Drossman et al. 1988); half-life of 2.02 years, based on measured rate constants for acid and base catalyzed and neutral hydrolysis at pH 7 and 20°C (Mabey & Mill 1978; quoted, Howard 1990; Howard et al. 1991; Kollig 1993).

Biodegradation: aqueous aerobic half-life of 24-168 hours, based on unacclimated aqueous aerobic biodegradation screening test data (Heukelekian & Rand 1955; Price et al. 1974; selected, Howard et al. 1991); aqueous anaerobic half-life of 96-672 hours, based on unacclimated aqueous aerobic biodegradation half-life (Howard et al. 1991).

Biotransformation:

Bioconcentration, Uptake (k_1) and Elimination (k_2) Rate Constants:

Common Name: Propyl acetate
Synonym: acetic acid propyl ester, propyl ethanoate, *n*-propyl acetate
Chemical Name: propyl acetate, *n*-propyl acetate
CAS Registry No: 109-60-4
Molecular Formula: $CH_3COOCH_2CH_2CH_3$
Molecular Weight: 102.133
Melting Point (°C):

-92.5 (Stull 1947)
-95.0 (Weast 1982-83; Verschueren 1983; Riddick et al. 1986; Howard 1993)
-92.0 (Verschueren 1983; Dean 1985)

Boiling Point (°C):

101.8 (Stull 1947; Verschueren 1983)
101.6 (Weast 1982-83; Dean 1985; Howard 1993)
101.5 (Boublik et al. 1984; Riddick et al. 1986)

Density (g/cm³ at 20°C):

0.8878 (Weast 1982-83)
0.8900 (Verschueren 1983)
0.8360 (Dean 1985)
0.88303 (25°C, Riddick et al. 1986)

Molar Volume (cm³/mol):

62.2 (intrinsic volume V_I, Leahy 1986; Kamlet et al. 1987)
115.0 (Leahy 1986; Kamlet et al. 1986,87)
133.0 (calculated-LeBas method)
115.7 (calculated-density)

Molecular Volume (Å³):
Total Surface Area, TSA (Å²):
Dissociation Constant, pK:

-7.18 (Perrin 1972)

Heat of Fusion ΔH_{fus}, kcal/mol:
Entropy of Fusion ΔS_{fus}, cal/mol·K (e.u.):
Fugacity Ratio at 25°C (assuming ΔS_{fus} = 13.5 e.u.), F: 1.0

Water Solubility (g/m³ or mg/L at 25°C):

18895 (20°C, shake flask-turbidity, Fühner 1924)
18890 (Fühner 1924; quoted, Hansch et al. 1968)
20380 (estimated, McGowan 1954)
18160 (quoted, Deno & Berkheimer 1960)
18900 (Stephen & Stephen 1963; quoted, Howard 1993)
23240 (calculated-K_{OW}, Hansch et al. 1968)
18895 (shake flask-AS, Hansch et al. 1968; quoted, Wasik et al. 1981,82; Tewari et al. 1982; Müller & Klein 1992)
22340 (quoted, Hine & Mookerjee 1975)
20430 (gen. col.-GC, Wasik et al. 1981,82; Tewari et al. 1982)
22450 (20°C, quoted average, Verschueren 1983)
23000 (Dean 1985)
23000 (20°C, Riddick et al. 1986)

22600 (20°C, shake flask-GC/TC, Stephenson & Stuart 1986)
19800 (30°C, shake flask-GC/TC, Stephenson & Stuart 1986)
18585, 19460 (quoted, calculated-V_I and solvatochromic p., Kamlet et al. 1986)
18585, 22340 (quoted, calculated-V_I and solvatochromic p., Leahy 1986)
18585, 22865 (quoted, calculated-V_I and solvatochromic p., Kamlet et al. 1987)
18580, 26300 (quoted, predicted- χ , Nirmalakhandan & Speece 1988)

Vapor Pressure (Pa at 25°C):
4152 (interpolated-regression of tabulated data, Stull 1947)
5332 (28.8°C, Stull 1947)
4400 (Hine & Mookerjee 1975)
4478 (ebulliometry, fitted to Antoine eqn., Ambrose et al. 1981; quoted, Howard 1993)
4666 (quoted, Verschueren 1983)
4596 (25.4°C, quoted exptl., Boublik et al. 1984)
4494 (calculated-Antoine eqn., Boublik et al. 1984)
4497 (quoted, Abraham 1984)
4486 (extrapolated-Antoine eqn., Dean 1985)
4497 (quoted, Riddick et al. 1986)

Henry's Law Constant (Pa·m³/mol):
20.15 (exptl., Hine & Mookerjee 1975)
22.092 (calculated-group contribution, Hine & Mookerjee 1975)
20.62 (calculated-group contribution, Hine & Mookerjee 1975)
22.09 (vapor-liquid equilibrium, vapor phase-GC, Kieckbusch & King 1979)
20.15, 30.50 (quoted, calculated- χ , Nirmalakhandan & Speece 1988)
20.15, 10.10 (quoted, calculated-molecular structure, Russell et al. 1992)

Octanol/Water Partition Coefficient, log K_{OW}:
0.66 (calculated, Iwasa et al. 1965)
1.23 (calculated-S, Hansch et al. 1968; quoted, Leo et al. 1969; Wasik et al. 1981,82; Tewari et al. 1982)
1.11 (calculated-γ, Wasik et al. 1981; quoted, Sangster 1989)
1.24 (gen. col.-GC, Wasik et al. 1981,82; Tewari et al. 1982; quoted, Sangster 1989)
1.39, 1.60 (calculated, Verschueren 1983)
1.23 (Hansch & Leo 1985; quoted, Howard 1993)
1.24, 1.36 (quoted, calculated-V_I and solvatochromic p., Leahy 1986)
1.24 (recommended, Sangster 1989)
1.20 (calculated-CLOGP, Müller & Klein 1992)

Bioconcentration Factor, log BCF:
0.380 (calculated-S, Lyman et al. 1982; quoted, Howard 1993)
0.708 (calculated-K_{OW}, Lyman et al. 1982; quoted, Howard 1993)

Sorption Partition Coefficient, log K_{OC}:

1.288 (soil, calculated-S, Lyman et al. 1982; quoted, Howard 1993)

2.045 (soil, calculated-K_{OW}, Lyman et al. 1982; quoted, Howard 1993)

Half-Lives in the Environment:

Air: photooxidation half-life of 2.4-24 hours for the gas-phase reaction with hydroxyl radicals in air, based on the rate of disappearance of hydrocarbon due to reaction with hydroxyl radicals (Darnall et al. 1976); half-lives of 3.39-6.69 days, based on exptl. rate constant for the vapor-phase reaction with the photochemically produced hydroxyl radicals in the atmosphere (Conkle et al. 1975; Atkinson et al. 1979; Ambrose et al. 1981; Atkinson 1985,87; Drossman et al. 1987; quoted, Howard 1993).

Surface water:

Groundwater:

Sediment:

Soil:

Biota:

Environmental Fate Rate Constants or Half-Lives:

Volatilization: based on Henry's law constant, half-life from a model river of one meter deep flowing at 1 m/sec with a wind velocity of 3 m/sec was estimated to be 6.5 hours (Lyman et al. 1982; quoted, Howard 1993).

Photolysis:

Oxidation: exptl. photooxidation rate constant of 2.4 to 4.1×10^{-12} cm^3 $molecule^{-1}$ sec^{-1} for the vapor-phase reaction with the photochemically produced hydroxyl radicals in the atmosphere (Conkle et al. 1975; Atkinson et al. 1979; Ambrose et al. 1981; Atkinson 1985,87; Drossman et al. 1987; quoted, Howard 1993); photooxidation half-life of 2.4-24 hours for the gas-phase reaction with hydroxyl radicals in air, based on the rate of disappearance of hydrocarbon due to reaction with hydroxyl radicals (Darnall et al. 1976); using relative rate technique for 2-methylpropene, rate constant of 4.2×10^{-12} cm^3 $molecule^{-1}$ sec^{-1} was determined for the gas-phase reaction with OH radicals at 305 ± 2 K in air (Winer et al. 1977; quoted, Atkinson 1985); rate constant of $(1.7 \pm 0.2) \times 10^{-12}$ cm^3 $molecule^{-1}$ sec^{-1} for the gas-phase reaction with OH radicals at 296 in air (Zetzsch 1982; quoted, Atkinson 1985); measured rate constant of $(34.5 \pm 3.4) \times 10^{-13}$ cm^3 $molecule^{-1}$ sec^{-1} for the gas phase reaction with OH radicals at 296 K in air (Wallington et al. 1988a,b) and 2.30×10^{-13} cm^3 $molecule^{-1}$ sec^{-1} for the solution-phase reaction with OH radicals in aqueous solution (Wallington et al. 1988b).

Hydrolysis: alkaline hydrolysis rate constant of 0.087 M^{-1} sec^{-1} at 25°C in water (Salmi & Leimu 1947; quoted, Drossman et al. 1988).

Biodegradation:

Biotransformation:

Bioconcentration, Uptake (k_1) and Elimination (k_2) Rate Constants:

Common Name: Butyl acetate
Synonym: acetic acid butyl ester, butyl ethanoate, *n*-butyl acetate, ethanoic acid butyl ester
Chemical Name: butyl acetate, *n*-butyl acetate
CAS Registry No: 123-86-4
Molecular Formula: $CH_3COOCH_2CH_2CH_2CH_3$
Molecular Weight: 116.16
Melting Point (°C):
-77.9 (Weast 1982-83)
-76.8 (Verschueren 1983)
-73.5 (Dean 1985; Riddick et al. 1986)
-77.0 (Howard 1990)
Boiling Point (°C):
126.5 (Weast 1982-83)
124/127 (Verschueren 1983)
126.1 (Boublik et al. 1984; Dean 1985; Riddick et al. 1986)
125.0 (Banerjee et al. 1990)
125-126 (Howard 1990)
Density (g/cm^3 at 20°C):
0.8825 (Weast 1982-83)
0.882 (Verschueren 1983)
0.8813 (Dean 1985)
0.87636 (Riddick et al. 1986)
Molar Volume (cm^3/mol):
132.6 (Leahy 1986; Kamlet et al. 1987)
71.6 (intrinsic volume V_I, Leahy 1986; Kamlet et al. 1987)
155.2 (calculated-LeBas method)
132.0 (calculated-density)
Molecular Volume ($Å^3$):
Total Surface Area, TSA ($Å^2$):
344.6 (Amidon et al. 1975)
Dissociation Constant:
23.28 (pK_s, Riddick et al. 1986)
Heat of Fusion ΔH_{fus}, kcal/mol:
Entropy of Fusion ΔS_{fus}, cal/mol·K (e.u.):
Fugacity Ratio at 25°C (assuming ΔS_{fus} = 13.5 e.u.), F: 1.0

Water Solubility (g/m^3 or mg/L at 25°C):
23550 (Seidell 1941; quoted, Hansch et al. 1968)
23180 (estimated, McGowan 1954)
4955 (quoted, Deno & Berkheimer 1960)
6532 (calculated-K_{OW}, Hansch et al. 1968)
23580 (shake flask-AS, Hansch et al. 1968; quoted, Wasik et al. 1981,82; Tewari et al. 1982)
7040 (20°C, quoted, Amidon et al. 1975)
6702 (gen. col.-GC, Wasik et al. 1981,82; Tewari et al. 1982; quoted, Howard 1990; Müller & Klein 1992)

3936 (quoted, Hine & Mookerjee 1975)
5000 (quoted, Verschueren 1983)
14000 (20°C, quoted, Verschueren 1983)
4300 (Dean 1985)
6800 (20°C, Riddick et al. 1986)
6400 (19.7°C, shake flask-GC/TC, Stephenson & Stuart 1986)
5200 (30.3°C, shake flask-GC/TC, Stephenson & Stuart 1986)
5070, 6238 (quoted, calculated-V_I and solvatochromic p., Leahy 1986)
5070, 6840 (quoted, calculated-V_I and solvatochromic p., Kamlet et al. 1987)
5058, 8974 (quoted, predicted- χ , Nirmalakhandan & Speece 1988)

Vapor Pressure (Pa at 25°C):
1505 (Hoy 1970)
1440 (Hine & Mookerjee 1975)
1333 (20.0°C, quoted, Verschueren 1983)
1560 (24.5°C, quoted exptl., Boublik et al. 1984)
1530 (extrapolated-Antoine eqn., Boublik et al. 1984)
1138 (calculated-Antoine eqn., Boublik et al. 1984)
1470 (quoted, Abraham 1984)
1529 (extrapolated-Antoine eqn., Dean 1985)
2000 (Howard et al. 1986; quoted, Banerjee et al. 1990)
1664 (Riddick et al. 1986; quoted, Howard 1990)
2000, 635 (measured, calculated-solvatochromic p., Banerjee et al. 1990)

Henry's Law Constant (Pa·m^3/mol):
33.44 (quoted exptl., Hine & Mookerjee 1975; quoted, Howard 1990)
30.496 (calculated-group contribution, Hine & Mookerjee 1975)
31.21 (calculated-group contribution, Hine & Mookerjee 1975)
28.51 (vapor-liquid equilibrium, vapor phase-GC, Kieckbusch & King 1979)
33.44, 38.39 (quoted, predicted- χ , Nirmalakhandan & Speece 1988)

Octanol/Water Partition Coefficient, log K_{OW}:
1.18 (calculated, Iwasa et al. 1965)
1.73 (calculated-S, Hansch et al. 1968; quoted, Leo et al. 1969; Wasik et al. 1981,82; Tewari et al. 1982)
1.69 (calculated-γ, Wasik et al. 1981; quoted, Sangster 1989)
1.82 (gen. col.-GC, Wasik et al. 1981, 1982; Tewari et al. 1982; quoted, Sangster 1989; Howard 1990)
1.82, 1.92 (quoted, calculated-V_I and solvatochromic p., Leahy 1986)
1.73 (quoted, Isnard & Lambert 1988)
1.82 (recommended, Sangster 1989)
1.73 (calculated-CLOGP, Müller & Klein 1992)
1.73 (quoted from U.S. EPA CLOGP Data Base, Hulzebos et al. 1993)

Bioconcentration Factor, log BCF:

0.602 (calculated-S, Lyman et al. 1982; quoted, Howard 1990)
1.146 (calculated-K_{OW}, Lyman et al. 1982; quoted, Howard 1990)

Sorption Partition Coefficient, log K_{OC}:

1.531 (soil, calculated-S, Lyman et al. 1982; quoted, Howard 1990)
2.367 (soil, calculated-K_{OW}, Lyman et al. 1982; quoted, Howard 1990)

Half-Lives in the Environment:

Air: half-life of 2.4-24 hours for the gas-phase reaction with hydroxyl radicals in air, based on the rate of disappearance of hydrocarbon due to reaction with hydroxyl radicals (Darnall et al. 1976); half-life of 6.0 days was estimated for the vapor-phase reaction with an average atmospheric hydroxyl radicals of 8×10^5 per cm^3 (GEMS 1986; quoted, Howard 1990).

Surface water:

Groundwater:

Sediment:

Soil:

Biota:

Environmental Fate Rate Constants or Half-Lives:

Volatilization: using Henry's law constant, a half-life of 6.1 hours was estimated for a model river 1 m deep flowing at 1 m/sec with a wind velocity of 3 m/sec and an estimated half-life of 7.4 days for a 10 m deep similar model river (Lyman et al. 1982; quoted, Howard 1990).

Photolysis:

Oxidation: photooxidation half-life of 2.4-24 hours for the gas-phase reaction with hydroxyl radicals in air, based on the rate of disappearance of hydrocarbon due to reaction with hydroxyl radicals (Darnall et al. 1976); photooxidation half-life of 6.0 days in air was estimated for the vapor-phase reaction with an average atmospheric hydroxyl radicals of $8 \times 10^5/cm^3$ (GEMS 1986; quoted, Howard 1990); experimental rate constant of $(41.5 \pm 3.0) \times 10^{13}$ $cm^3 \cdot molecule^{-1} \cdot sec^{-1}$ for the gas-phase reaction with OH radicals at 296 K (Wallington et al. 1988).

Hydrolysis: alkaline hydrolysis rate constant of 0.087 M^{-1} sec^{-1} at 25°C in water (Salmi & Leimu 1947; quoted, Drossman et al. 1988); based on observed acid and base-catalyzed rate constants at 20°C, half-lives of 3.1 years, 114 days, 11.4 days at pH 7.0, 8.0, 9.0 were estimated respectively (Mabey & Mill 1978; quoted, Howard 1990).

Biodegradation:

Biotransformation:

Bioconcentration, Uptake (k_1) and Elimination (k_2) Rate Constants:

Common Name: Pentyl acetate
Synonym: acetic acid pentyl ester, amyl acetate, amylacetic ester, pentyl ethanoate, *n*-pentyl acetate, ethanoic acid pentyl ester
Chemical Name: pentyl acetate, *n*-pentyl acetate
CAS Registry No: 628-63-7
Molecular Formula: $CH_3COOCH_2CH_2CH_2CH_2CH_3$
Molecular Weight: 130.19
Melting Point (°C):
-70.8 (Weast 1982-83; Riddick et al. 1986)
Boiling Point (°C):
149.25 (Weast 1982-83)
149.2 (Dean 1985; Riddick et al. 1986)
Density (g/cm^3 at 20°C):
0.8756 (Weast 1982-83)
0.879 (Verschueren 1983)
0.8753 (Dean 1985)
0.8766 (Riddick et al. 1986)
Molar Volume (cm^3/mol):
81.3 (intrinsic volume V_I, Leahy 1986; Kamlet et al. 1987)
148.7 (Leahy 1986; Kamlet et al. 1987)
177.4 (calculated-LeBas method)
149.3 (calculated-density)
Molecular Volume ($Å^3$):
Total Surface Area, TSA ($Å^2$):
Dissociation Constant:
24.25 (pK_s, Riddick et al. 1986)
Heat of Fusion ΔH_{fus}, kcal/mol:
Entropy of Fusion ΔS_{fus}, cal/mol·K (e.u.):
Fugacity Ratio at 25°C (assuming ΔS_{fus} = 13.5 e.u.), F: 1.0

Water Solubility (g/m^3 or mg/L at 25°C):
1730 (shake flask-turbidimeter, McBain & Richards 1946)
1800 (quoted, Hine & Mookerjee 1975)
1800 (20°C, quoted, Verschueren 1983)
1700 (Dean 1985)
1700 (20°C, Riddick et al. 1986)
2200 (19.7°C, shake flask-GC/TC, Stephenson & Stuart 1986)
1600 (30.3°C, shake flask-GC/TC, Stephenson & Stuart 1986)
2060, 1926 (quoted, calculated-V_I and solvatochromic p., Leahy 1986)
2060, 1970 (quoted, calculated-V_I and solvatochromic p., Kamlet et al. 1987)
2060, 2980 (quoted, predicted- χ , Nirmalakhandan & Speece 1988)

Vapor Pressure (Pa at 25°C):
547 (quoted, Hine & Mookerjee 1975)
546 (quoted, Abraham 1984)
1290 (Riddick et al. 1986)

Henry's Law Constant (Pa·m^3/mol):

39.29 (quoted exptl., Hine & Mookerjee 1975)
42.096 (calculated-group contribution, Hine & Mookerjee 1975)
46.16 (calculated-group contribution, Hine & Mookerjee 1975)
35.94 (vapor-liquid equilbrium, vapor phase-GC, Kieckbusch & King 1979)
39.29, 48.33 (quoted, calculated- χ , Nirmalakhandan & Speece 1988)
29.29, 38.39 (quoted, calculated-molecular structure, Russell et al. 1992)

Octanol/Water Partition Coefficient, log K_{OW}:

1.70 (calculated, Iwasa et al. 1965)
2.23 (Leo et al. 1969)
2.42 (calculated-V_I and solvatochromic p., Leahy 1986)

Bioconcentration Factor, log BCF:

Sorption Partition Coefficient, log K_{OC}:

Half-Lives in the Environment:

Air: photooxidation half-life of 2.4-24 hours for the gas-phase reaction with hydroxyl radicals in air, based on the rate of disappearance of hydrocarbon due to reaction with hydroxyl radicals (Darnall et al. 1976).
Surface water: biodegradation half-life of 13 days in Lester river and 10 days in Superior habor waters (Vaishnav & Babeu 1987).
Groundwater: biodegradation half-life of 50 days (Vaishnav & Babeu 1987).
Sediment:
Soil:
Biota:

Environmental Fate Rate Constants or Half-Lives:

Volatilization:
Photolysis:
Oxidation: photooxidation half-life of 2.4-24 hours for the gas-phase reaction with hydroxyl radicals in air, based on the rate of disappearance of hydrocarbon due to reaction with hydroxyl radicals (Darnall et al. 1976).
Hydrolysis:
Biodegradation: biodegradation rates of 0.054 day^{-1} with half-life of 13 days in Lester river, 0.069 day^{-1} 10 days in Superior habor waters and 0.014 day^{-1} with half-life of 50 days in groundwater (Vaishnav & Babeu 1987).
Biotransformation:
Bioconcentration, Uptake (k_1) and Elimination (k_2) Rate Constants:

Common Name: Methyl acrylate
Synonym: acrylic acid methyl ester, methyl 2-propenoate
Chemical Name: methyl acrylate
CAS Registry No: 96-33-3
Molecular Formula: $CH_2=CHCOOCH_3$
Molecular Weight: 86.09
Melting Point (°C):
  <-75 (Weast 1982-83; Verschueren 1983; Riddick et al. 1986)
  -76.5 (Dean 1985)
Boiling Point (°C):
  80.2 (Stull 1947; Dean 1985; Riddick et al. 1986)
  80.5 (Weast 1982-83)
  80.0 (Verschueren 1983)
Density (g/cm^3 at 20°C):
  0.9535 (Weast 1982-83; Riddick et al. 1986)
  0.9580 (Verschueren 1983)
  0.9561 (Dean 1985)
Molar Volume (cm^3/mol):
  90.2 (Kamlet et al. 1987)
  99.6 (calculated-LeBas method)
  93.1 (calculated-density)
Molecular Volume ($Å^3$):
Total Surface Area, TSA ($Å^2$):
Dissociation Constant, pK_a:
Heat of Fusion ΔH_{fus}, kcal/mol:
Entropy of Fusion ΔS_{fus}, cal/mol·K (e.u.):
Fugacity Ratio at 25°C (assuming ΔS_{fus} = 13.5 e.u.), F: 1.0

Water Solubility (g/m^3 or mg/L at 25°C):
  60000 (Dean 1985)
  49400 (Riddick et al. 1986)

Vapor Pressure (Pa at 25°C):
  11250 (interpolated-regression of tabulated data, Stull 1947)
  13330 (28.0°C, Stull 1947)
  9330 (20°C, quoted, Verschueren 1983)
  14660 (30°C, quoted, Verschueren 1983)
  11000 (Riddick et al. 1986)

Henry's Law Constant (Pa·m^3/mol):

Octanol/Water Partition Coefficient, log K_{OW}:
  0.21 (Tute 1971; Laurence et al. 1972; quoted, Fujisawa & Masuhara 1981)

0.36 (calculated-HPLC-RT, Fujisawa & Masuhara 1981; quoted, Sangster 1989)
0.80 (shake flask-GC, Tanii & Hashimoto 1982; quoted, Sangster 1989)
0.80 (recommended, Sangster 1989)

Bioconcentration Factor, log BCF:

Sorption Partition Coefficient, log K_{OC}:

Half-Lives in the Environment:

Air: photooxidation half-life of 2.7-27 hours, based on estimated rate constants for the vapor phase reaction with hydroxyl radicals (Atkinson 1987; quoted, Howard et al. 1991); atmospheric transformation lifetime was estimated to be <1 to 1 - 5 days (Kelly et al. 1994).

Surface water: 24-168 hours, based on estimated aqueous aerobic biodegradation half-life (Howard et al. 1991).

Groundwater: 48-336 hours, based on measured hydrolysis data and estimated aqueous aerobic biodegradation half-life (Howard et al. 1991).

Sediment:

Soil: 24-168 hours, based on estimated aqueous aerobic biodegradation half-life (Howard et al. 1991).

Biota:

Environmental Fate Rate Constants or Half-Lives:

Volatilization:

Photolysis:

Hydrolysis: half-life of 2.8 years at pH 7 and 25°C, based on measured base rate constant of 0.0779 M^{-1} h^{-1} at pH 9 and 25°C (Roy 1972; quoted, Howard et al. 1991).

Oxidation: photooxidation half-life of 2.7-27 hours in air, based on estimated rate constants for the vapor phase reaction with hydroxyl radicals (Atkinson 1987; quoted, Howard et al. 1991); experimental rate constant of $(30.4 \pm 3.3)x10^{-13}$ cm^3 $molecule^{-1}$ sec^{-1} for the gas-phase reaction with hydroxyl radicals at 296 K in air (Wallington et al. 1988a,b) and $28.0x10^{-13}$ cm^3 $molecule^{-1}$ sec^{-1} for the solution-phase reaction with hydroxyl radicals in aqueous solution at room temperature (Wallington et al. 1988b).

Biodegradation: aqueous aerobic half-life of 24-168 hours, based on biological screening test data (Sasaki 1978; quoted, Howard et al. 1991); aqueous anaerobic half-life of 96-672 hours, based on aqueous aerobic biodegradation half-life (Howard et al. 1991).

Biotransformation:

Bioconcentration, Uptake (k_1) and Elimination (k_2) Rate Constants:

Common Name: Ethyl acrylate
Synonym: acrylic acid ethyl ester, ethyl 2-propenoate, propenoic acid ethyl ester
Chemical Name: ethyl acrylate, *n*-ethyl acrylate
CAS Registry No: 140-88-5
Molecular Formula: $CH_2{=}CHCOOCH_2CH_3$
Molecular Weight: 100.12
Melting Point (°C):
-71.2 (Stull 1947; Weast 1982-83; Dean 1985; Riddick et al. 1986)
Boiling Point (°C):
99.5 (Stull 1947; Dean 1985; Riddick et al. 1986)
99.8 (Weast 1982-83)
100 (Verschueren 1983)
Density (g/cm^3 at 20°C):
0.9234 (Weast 1982-83; Riddick et al. 1986)
0.9240 (Verschueren 1983)
0.9405 (Dean 1985)
Molar Volume (cm^3/mol): 121.8 (calculated-LeBas method)
Molecular Volume ($Å^3$):
Total Surface Area, TSA ($Å^2$):
Dissociation Constnat, pK_a:
Heat of Fusion ΔH_{fus}, kcal/mol:
Entropy of Fusion ΔS_{fus}, cal/mol·K (e.u.):
Fugacity Ratio at 25°C (assuming ΔS_{fus} = 13.5 e.u.), F: 1.0

Water Solubility (g/m^3 or mg/L at 25°C):
20000 (quoted, Verschueren 1983)
15000 (Dean 1985; Riddick et al. 1986)

Vapor Pressure (Pa at 25°C):
4860 (interpolated-regression of tabulated data, Stull 1947)
5332 (26.0°C, Stull 1947)
3866 (20°C, quoted, Verschueren 1983)
6532 (30°C, quoted, Verschueren 1983)
5100 (Riddick et al. 1986)

Henry's Law Constant ($Pa \cdot m^3/mol$):

Octanol/Water Partition Coefficient, log K_{OW}:
0.73 (Tute 1971; Laurence et al. 1972; quoted, Fujisawa & Masuhara 1981)
0.88 (calculated-f const. per Rekker 1977, Hermens & Leeuwangh 1982)
0.66 (calculated-HPLC-RT, Fujisawa & Masuhara 1981)
1.33 (shake flask-GC, Tanii & Hashimoto 1982; quoted, Sangster 1989)
1.32 (quoted from Pomona College unpublished data, Sangster 1989)
1.32 (recommended, Sangster 1989)

Bioconcentration Factor, log BCF:

Sorption Partition Coefficient, log K_{OC}:

Half-Lives in the Environment:

Air: photooxidation half-life of 2.37-22.7 hours, based on estimated rate constants for the vapor phase reaction with hydroxyl radicals (Atkinson 1987; quoted, Howard et al. 1991).

Surface water: 24-168 hours, based on estimated unacclimated aqueous aerobic biodegradation half-life (Howard et al. 1991).

Groundwater: 48-336 hours, based on estimated unacclimated aqueous aerobic biodegradation half-life (Howard et al. 1991).

Sediment:

Soil: 24-168 hours, based on estimated unacclimated aqueous aerobic biodegradation half-life (Howard et al. 1991).

Biota:

Environmental Fate Rate Constants or Half-Lives:

Volatilization:

Photolysis:

Hydrolysis: half-life of 2.8 years at pH 7 and 25°C, based on measured base rate constant of 0.0779 M^{-1} h^{-1} at pH 9 and 25°C (Mabey & Mill 1978; quoted, Howard et al. 1991).

Oxidation: photooxidation half-life of 2.37-22.7 hours in air, based on estimated rate constants for the vapor phase reaction with hydroxyl radicals (Atkinson 1987; quoted, Howard et al. 1991).

Biodegradation: aqueous aerobic half-life of 24-168 hours, based on aqueous aerobic screening test data (Price et al. 1974; Sasaki 1978; quoted, Howard et al. 1991); aqueous anaerobic half-life of 96-672 hours, based on estimated unacclimated aqueous aerobic biodegradation half-life (Howard et al. 1991).

Biotransformation:

Bioconcentration, Uptake (k_1) and Elimination (k_2) Rate Constants:

Common Name: Methyl methacrylate
Synonym: methyl 2-methyl-2-propenoate, methyl ester methacrylic acid, methacrylic acid methyl ester, MMA
Chemical Name: 2-methyl-2-propenoic acid methyl ester
CAS Registry No: 80-62-6
Molecular Formula: $H_2C{=}C(CH_3)COOCH_3$
Molecular Weight: 100.13
Melting Point (°C):
-48.0 (Perry & Chilton 1973; Weast 1982-83; Weast & Grasselli 1989)
-50.0 (Verschueren 1983)
-48.2 (Riddick et al. 1986)
Boiling Point (°C):
101.0 (Stull 1947; Verschueren 1983)
100.3 (Perry & Chilton 1973)
100-101 (Weast 1982-83; Weast & Grasselli 1989)
100.6 (Boublik et al. 1984)
100.3 (Dean 1985; Riddick et al. 1986)
Flash Point (°C): 10
Density (g/cm^3 at 20 °C):
0.9433 (Dean 1985)
0.9440 (Weast 1982-83; Weast & Grasselli 1989)
0.9360 (Verschueren 1983)
0.9433 (Riddick et al. 1986)
Molar Volume (cm^3/mol):
106.6 (25°C, Stephenson & Malanowski 1987)
106 (Kamlet et al. 1987)
121.8 (calculated-LeBas method)
106 (calculated-density)
Molecular Volume (Å^3):
Total Surface Area, TSA (Å^2):
Dissociation Constnat, pK_a:
Heat of Fusion ΔH_{fus}, kcal/mol:
Entropy of Fusion ΔS_{fus}, cal/mol·K (e.u.):
Fugacity Ratio at 25°C (assuming ΔS_{fus} = 13.5 e.u.), F: 1.0

Water Solubility (g/m^3 or mg/L at 25°C):
16000 (Dean 1985)
15000 (USEPA 1985; ENVIROFATE; ISHOW)
15900 (quoted, Yalkowsky et al. 1987)
15600 (20°C, Riddick et al. 1986, quoted, Howard 1989; Müller & Klein 1992)

Vapour Pressure (Pa at 25°C):
4888 (interpolated-regression of tabulated data, Stull 1947)
5332 (25.5 °C, Stull 1947)
3906, 4932 (20°C, ENVIROFATE)

3732 (20°C, quoted, Verschueren 1977, 1983)
5333 (26°C, quoted, Verschueren 1977, 1983)
5333 (25.5 °C, Weast 1982-83)
4440 (extrapolated average-Antoine eqn., Boublik et al. 1984)
4846 (extrapolated-Antoine equation, Dean 1985)
3999 (20°C, ACGIH 1986)
5100 (Riddick et al. 1986)
5081, 5020 (interpolated-Antoine eqns., Stephenson & Malanowski 1987)
4640 (quoted as 38.4 torr, Howard 1989)

Henry's Law Constant (Pa·m^3/mol):
24.31 (calculated-P/C, USEPA 1985)
32.823 (calculated-P/C, Howard 1989)
32.090 (calculated-P/C, this work)

Octanol/Water Partition Coefficient, log K_{OW}:
0.73 (Tute 1971; Laurence et al. 1972; quoted, Fujisawa & Masuhara 1981)
0.70 (shake flask, Fujisawa & Masuhara 1981)
0.67 (HPLC, Fujisawa & Masuhara 1981)
1.36 (Hansch & Leo 1982; LOGP; quoted, Howard 1989)
1.38 (shake flask-GC, Tanii & Hashimoto 1982)
0.79 (calculated-f const. as per Lyman et al. 1982, USEPA 1985)
1.38 (recommended, Sangster 1989)
1.03 (Hazardous Substances Databank)
0.95 (ISHOW)
1.06 (calculated-CLOGP, Müller & Klein 1992)
1.06 (calculated-QSAR, Kollig 1993)

Bioconcentration Factor, log BCF:
-0.03 (correlated as per Veith et al. 1979 for aquatic organisms, USEPA 1985)
0.55 (calculated-K_{OW}, Howard 1989)

Sorption Partition Coefficient, log K_{OC}:
1.80 (calculated-K_{OW} as per Lyman et al. 1982, USEPA 1985)
1.94 (calculated-K_{OW}, Howard 1989)
0.74 (calculated-K_{OW}, Kollig 1993)

Half-Lives in the Environment:
Air: photooxidation half-life of 1.1-9.7 hours, based on estimated rate constants for the vapor phase reaction with hydroxyl radicals (Atkinson 1987; quoted, Howard et al. 1991); atmospheric transformation lifetime was estimated to be < 1 to 1 - 5 days (Kelly et al. 1994).

Surface water: 168-672 hours, based on estimated aqueous unacclimated aerobic biodegradation half-life (Howard et al. 1991).

Groundwater: 336-1344 hours, based on estimated aqueous unacclimated aerobic biodegradation half-life (Howard et al. 1991).

Sediment:

Soil: 168-672 hours, based on estimated aqueous unacclimated aerobic biodegradation half-life (Howard et al. 1991).

Biota:

Environmental Fate Rate Constants or Half-Lives:

Volatilization: half-life of 6.3 hours from model river water (Lyman et al. 1982; quoted, Howard 1989).

Photolysis: photodegradation half-life of 2.7 hours in air of urban areas and > 3 hours in rural areas (Joshi et al. 1982; quoted, Howard 1989) (CHEMFATE; Hazardous Substance Databank).

Hydrolysis: alkaline hydrolysis rate constant at 25°C is determined as 171 M^{-1} $hour^{-1}$ (Sharma & Sharma 1970; quoted, Ellington et al. 1987; Ellington 1989); alkaline catalyzed rate constant is 200 M^{-1} h^{-1} at 25°C (experimental result) with estimated half-life of 3.9 years at pH 7 and 14 days at pH 9 (Ellington et al. 1987; quoted, Ellington 1989; Howard 1989; Howard et al. 1991; Kollig 1993).

Oxidation: photooxidation half-life of 1.1-9.7 hours in air, based on estimated rate constants for the vapor phase reaction with hydroxyl radicals (Atkinson 1987; quoted, Howard et al. 1991).

Biodegradation: completely degraded by activated sludge in approximately 20 hours (Slave et al. 1974; quoted, EPA 1985; Howard 1989; Hazardous Substance Databank); expert systems survey found that both aerobic ultimate degradation in receiving waters and anaerobic ultimate degradation were within a month and aerobic primary degradation in receiving waters was within few days (Boethling et al. 1989); aqueous aerobic half-life of 168-672 hours, based on unacclimated screening test data (Pahren & Bloodgood 1961; Sasaki 1978; quoted, Howard et al. 1991); aqueous anaerobic half-life of 672-2688 hours, based on unacclimated aerobic biodegradation half-life (Howard et al. 1991).

Biotransformation:

Bioconcentration, Uptake (k_1) and Elimination (k_2) Rate Constants:

Common Name: Methyl benzoate
Synonym: benzoic acid methyl ester
Chemical Name:
CAS Registry No: 93-58-3
Molecular Formula: $C_6H_5COOCH_3$
Molecular Weight: 136.13
Melting Point (°C): -12.1
Boiling Point (°C): 199.5
Density (g/cm^3 at 20°C):
Molar Volume (cm^3/mol): 151.2 (calculated-LeBas method)
Molecular Volume (Å^3):
Total Surface Area, TSA (Å^2):
Dissociation Constant, pK_a:
Heat of Fusion ΔH_{fus}, kcal/mol:
Entropy of Fusion ΔS_{fus}, cal/mol·K (e.u.):
Fugacity Ratio at 25°C (assuming ΔS_{fus} = 13.5 e.u.), F: 1.0

Water Solubility (g/m^3 or mg/L at 25°C):
- 1524 (titration, Booth & Everson 1948)
- 4018 (quoted, Hine & Mookerjee 1975)
- 2100 (quoted, Riddick et al. 1986)
- 2042, 1600 (quoted, calculated- χ , Nirmalakhandan & Speece 1988)

Vapor Pressure (Pa at 25°C):
- 53.0 (quoted, Hine & Mookerjee 1975, Riddick et al. 1986)
- 77.58 (extrapolated-Antoine eqn., Dean 1985)

Henry's Law Constant (Pa·m^3/mol):
- 1.80 (calculated-C_W/C_A, Hine & Mookerjee 1975)
- 2.48 (calculated-bond contribution, Hine & Mookerjee 1975)

Octanol/Water Partition Coefficient, log K_{OW}:
- 2.15 (HPLC-RT, D'Amboise & Hanai 1982)
- 2.22 (HPLC-k', Haky & Young 1984)
- 2.14 (shake flask-HPLC, Nielson & Bundgaard 1988)
- 2.12, 2.10 (quoted most, recommended, Klein et al. 1988)
- 2.20 (recommended, Sangster 1989)

Bioconcentration Factor, log BCF:

Sorption Partition Coefficient, log K_{OC}:
- 2.30 (Podzol soil, von Oepen et al. 1991)
- 2.01 (Alfisol soil, von Oepen et al. 1991)
- 1.98 (sediment, von Oepen et al. 1991)
- 2.10, 1.89 (quoted exptl., calculated- χ , Meylan et al. 1992)

Half-Lives in the Environment/Environmental Fate Rate Constants or Half-Lives: no data

Common Name: Ethyl benzoate
Synonym: benzoic acid ethyl ester, ethyl benzenecarboxylate
Chemical Name: benzoic acid ethyl ester, ethyl benzenecarboxylate, ehtyl benzoate
CAS Registry No: 93-89-0
Molecular Formula: $C_6H_5COOC_2H_5$
Molecular Weight: 150.18
Melting Point (°C):
-34.6 (Stull 1947)
-34.7 (Riddick et al. 1986)
Boiling Point (°C):
213.4 (Stull 1947)
212.4 (Riddick et al. 1986)
Density (g/cm^3 at 20°C):
1.05112 (15°C, Riddick et al. 1986)
1.03718 (30°C, Riddick et al. 1986)
Molar Volume (cm^3/mol):
144.2 (Hoy 1970; Amidon & Williams 1982)
83.4 (intrinsic volume V_I, Leahy 1986)
153.0 (Leahy 1986)
83.6 (intrinsic volume V_I, Kamlet et al. 1987)
153.6 (Kamlet et al. 1987)
175 (calculated-LeBas method)
Molecular Volume ($Å^3$):
Total Surface Area, TSA ($Å^2$):
Dissociation Constant, pK_a:
Heat of Fusion ΔH_{fus}, kcal/mol:
Entropy of Fusion ΔS_{fus}, cal/mol·K (e.u.):
Fugacity Ratio at 25°C (assuming ΔS_{fus} = 13.5 e.u.), F: 1.0

Water Solubility (g/m^3 or mg/L at 25°C):
720 (Seidell 1941; quoted, Andrews & Keefer 1950)
598 (quoted, Deno & Berkheimer 1960)
788 (quoted from Seidell 1941, Hansch et al. 1968)
702 (calculated-K_{OW}, Hansch et al. 1968)
1015, 509 (quoted, calculated-K_{OW}, Yalkowsky & Morozowich 1980)
905, 2026 (quoted, calculated-K_{OW} and M.P., Amidon & Williams 1982),
905, 807 (quoted, calculated-V_I and solvatochromic p., Leahy 1986)
905, 1088 (quoted, calculated-V_I and solvatochromic p., Kamlet et al. 1987)
905, 464 (quoted, calculated-V_I and M.P., Kamlet et al. 1987)
500 (20°C, Riddick et al. 1986)
788, 926 (quoted, calculated-fragment const., Wakita et al. 1986)
902, 398 (quoted, calculated- χ , Nirmalakhandan & Speece 1988)

Vapor Pressure (Pa at 25°C):
42.1 (extrapolated-regression of tabulated data, Stull 1947)

24.0 (Riddick et al. 1986)
27.5 (interpolated-Antoine eqn., Stephenson & Malanowski 1987)

Henry's Law Constant ($Pa \cdot m^3/mol$):

Octanol/Water Partition Coefficient, log K_{OW}:
2.62 (calculated-S, Hansch et al. 1968)
2.20 (shake flask-AS, Yaguzhinskii et al. 1973; quoted Sangster 1989)
2.64 (quoted from Valvani et al. 1981, Amidon & Williams 1982)
2.64, 2.83 (quoted, calculated-V_I and solvatochromic p., Leahy 1986)
2.64 (shake flask-HPLC, Nielsen & Bundgaard 1988; quoted Sangster 1989)
2.64, 2.60 (quoted most, recommended, Klein et al. 1988)
2.64 (quoted unpublished data from Pomona College, Sangster 1989)
2.64 (recommended, Sangster 1989)
2.64 (quoted, Gluck & Martin 1990)
2.66, 2.89, 2.90 (CPC-RV, Gluck & Martin 1990)

Bioconcentration Factor, log BCF:

Sorption Partition Coefficient, log K_{OC}:
2.25 (sediment, von Oepen et al. 1991)
2.17 (Alfisol soil, von Oepen et al. 1991)
2.43 (Podzol soil, von Oepen et al. 1991)
2.30, 2.16 (quoted exptl., calculated- χ and fragment contribution, Meylan et al. 1992)

Half-Lives in the Environment:
Air:
Surface water:
Groundwater:
Sediment:
Soil:
Biota:

Environmental Fate Rate Constants or Half-Lives:
Volatilization:
Photolysis:
Hydrolysis:
Oxidation:
Biodegradation:
Biotransformation:
Bioconcentration, Uptake (k_1) and Elimination (k_2) Rate Constants:

Common Name: *n*-Propyl benzoate
Synonym: benzoic acid *n*-propyl ester, propyl benzenecarboxylate
Chemical Name: benzoic acid *n*-propyl ester, propyl benzenecarboxylate, propyl benzoate
CAS Registry No: 2315-68-6
Molecular Formula: $C_6H_5COOCH_2CH_2CH_3$
Molecular Weight: 164.2
Melting Point (°C):
-51.6 (Riddick et al. 1986)
Boiling Point (°C):
231.2 (Riddick et al. 1986)
Density (g/cm^3 at 20°C):
1.0232 (Riddick et al. 1986)
Molar Volume (cm^3/mol):
93.6 (intrinsic volume V_I, Kamlet et al. 1987)
199.4 (calculated-LeBas method)
160.5 (caalculated-density)
Molecular Volume ($Å^3$):
Total Surface Area, TSA ($Å^2$):
Dissociation Constant, pK_a:
Heat of Fusion ΔH_{fus}, kcal/mol:
Entropy of Fusion ΔS_{fus}, cal/mol·K (e.u.):
Fugacity Ratio at 25°C (assuming ΔS_{fus} = 13.5 e.u.), F: 1.0

Water Solubility (g/m^3 or mg/L at 25°C):
350 (quoted exptl., Kamlet et al. 1987)
292 (calculated-V_I and solvatochromic p., Kamlet et al. 1987)
140 (calculated-V_I and M.P., Kamlet et al. 1987)
350, 966 (quoted, calculated- χ , Nirmalakhandan & Speece 1988)

Vapor Pressure (Pa at 25°C):
100 (Riddick et al. 1986)

Henry's Law Constant ($Pa{\cdot}m^3/mol$):

Octanol/Water Partition Coefficient, log K_{OW}:
1.80 (shake flask-HPLC, Nielsen & Bundgaard 1988; quoted, Sangster 1989)
3.18 (recommended, Sangster 1989)

Bioconcentration Factor, log BCF: no data

Sorption Partition Coefficient, log K_{OC}: no data

Half-Lives in the Environment/Environmental Fate Rate Constants or Half-Lives: no data

Common Name: Benzyl benzoate
Synonym: benzoic acid benzyl ester, phenylmethyl benzoate, benzyl benzenecarboxylate, benzyl phenylformate, benzenoate
Chemical Name: benzyl benzoate, phenylmethyl benzoate, benzyl benzenecarboxylate
CAS Registry No: 120-51-4
Molecular Formula: $C_6H_5COOCH_2C_6H_5$
Molecular Weight: 212.25
Melting Point (°C):
19.4 (Riddick et al. 1986)
Boiling Point (°C):
323.24 (Riddick et al. 1986)
Density (g/cm^3 at 20°C):
1.1200 (19.5°C, Riddick et al. 1986)
1.1121 (25°C, Riddick et al. 1986)
Molar Volume (cm^3/mol):
200.8 (Taft et al. 1985; Leahy 1986)
113.9 (intrinsic volume V_I, Leahy 1986)
228.8 (calculated-LeBas method)
191.2 (calculated-density)
189.8 (20°C, Stephenson & Malanowski 1987)
Molecular Volume (Å^3):
Total Surface Area, TSA (Å^2):
Dissociation Constant, pK_a:
Heat of Fusion ΔH_{fus}, kcal/mol:
Entropy of Fusion ΔS_{fus}, cal/mol·K (e.u.):
Fugacity Ratio at 25°C (assuming ΔS_{fus} = 13.5 e.u.), F: 1.0

Water Solubility (g/m^3 or mg/L at 25°C):
15.73 (Seidell 1941; quoted, Deno & Berkheimer 1960)
61.21 (calculated-V_I and solvatochromic p., Leahy 1986)

Vapor Pressure (Pa at 25°C):
461 (Riddick et al. 1986)
0.043 (extrapolated-Antoine eqn., Stephenson & Malanowski 1987)

Henry's Law Constant (Pa·m^3/mol):

Octanol/Water Partition Coefficient, log K_{OW}:
3.97, 3.86 (quoted, calculated-V_I and solvatochromic p., Taft et al. 1985)
3.97, 4.00 (quoted, calculated-V_I and solvatochromic p., Leahy 1986)
3.97 (quoted and recommended value, Sangster 1989)

Bioconcentration Factor, log BCF:

Sorption Partition Coefficient, log K_{OC}:

Half-Lives in the Environment/Environmental Fate Rate Constants or Half-Lives: no data.

Common Name: Dimethyl phthalate
Synonym: 1,2-benzenedicarboxylic acid dimethyl ester, dimethyl-1,2-benzenedicarboxylate, DMP, methyl phthalate, *o*-dimethylphthalate, phathalic acid dimethyl ester
Chemical Name: dimethyl phthalate, methyl phathalate
CAS Registry No: 131-11-3
Molecular Formula: o-$C_6H_4(COOCH_3)_2$
Molecular Weight: 194.18
Melting Point (°C):

0.0 (Fishbein & Albro 1972; Verschueren 1977, 1983; Callahan et al. 1979; Mabey et al. 1982)
5.5 (Howard 1989)

Boiling Point (°C):

283.7 (Stull 1947; Howard 1989)
282 (Fishbein & Albro 1972; Peakall 1975; Verschueren 1977, 1983; Callahan et al. 1979; Mabey et al. 1982; Banerjee et al. 1990)

Density (g/cm^3 at 20°C):

1.189 (25°C, Fishbein & Albro 1972)
1.190 (25°C, Verschueren 1983)

Molar Volume (cm^3/mol):

173.2 (Taft et al. 1985; Leahy 1986)
95.3 (intrinsic volume V_I, Leahy 1986; Kamlet et al. 1987)
173.6 (Kamlet et al. 1987)
206.4 (calculated-LeBas method)
162.8 (calculated-density, Stephenson & Malanowski 1987)
96.5 (intrinsic volume V_I, Leggett & Parker 1994)

Molecular Volume (A^3):
Total Surface Area, TSA (A^2):
Heat of Fusion ΔH_{fus}, (kcal/mol):
Entropy of Fusion ΔS_{fus} (cal/mol·K or e.u.):
Fugacity Ratio at 25°C, (assuming ΔS_{fus} = 13.5 e.u.) F: 1.0

Water Solubility (g/m^3 or mg/L at 25°C):

3965 (quoted, Deno & Berkheimer 1960)
5000 (20°C, Fishbein & Albro 1972; quoted, Callahan et al. 1979; Wolfe et al. 1980b; Mabey et al. 1982; Leyder & Boulanger 1983)
4000 (32°C, quoted from Monsanto Chemical Co. data sheets, Peakall 1975)
4320 (Wolfe et al. 1979; quoted, Callahan et al. 1979; Zaroogian et al. 1985)
4320 ± 37 (shake flask-GC, Wolfe et al. 1980b; quoted, Wolfe et al. 1980a)
5000 (quoted, Cowen & Baynes 1980)
4248 (shake flask-LSC, Veith et al. 1980; quoted, Davies & Dobbs 1984)
15073 (calculated-π, as per Hansch et al. 1968, Veith et al. 1980)
1507, 1542 (quoted, calculated-K_{OW}, Yalkowsky & Morozowich 1980)
4290 (20°C, shake flask-UV, Leyder & Boulanger 1983)
4520, 45000 (quoted lit. values, Leyder & Boulanger 1983)
5000, 1744 (20°C, quoted, Verschueren 1983)
4000 ± 60 (shake flask-HPLC/UV, Howard et al. 1985; quoted, Howard 1989)

4000-5000 (quoted, Russell et al. 1985)
5000000 (quoted from Callahan et al. 1979, misquoted in unit? Neuhauser et al. 1985)
39646 (calculated-V_I and solvatochromic p., Leahy 1986)
3965, 23890 (quoted, calculated-V_I and solvatochromic p., Kamlet et al. 1987)
3965, 3616 (quoted, calculated-V_I and M.P., Kamlet et al. 1987)
3980 (quoted, Isnard & Lambert 1988, 1989)
2810 (21°C, shake flask-HPLC/UV, Nielsen & Bundgaard 1989)
4320 (selected, Aichinger et al. 1992)

Vapor Pressure (Pa at 25°C):
0.863 (extrapolated-regression of tabulated data, Stull 1947)
0.245 (effusion method, extrapolated-Antoine eqn., Small et al. 1948; quoted, Hinckley et al. 1990)
<1.33 (20°C, Patty 1963; Fishbein & Albro 1972; quoted, Callahan et al. 1979; Neuhauser et al. 1985)
0.559 (20°C, calculated, Weast 1972-73; quoted, Mabey et al. 1982)
0.719 (quoted, Cowen & Baynes 1980)
0.220 (gas saturation-HPLC/UV, Howard et al. 1985; quoted, Howard 1989)
0.220 (Howard et al. 1986; quoted, Banerjee et al. 1990)
0.245 (extrapolated, Antoine eqn., Stephenson & Malanowski 1987)
0.220, 1.346 (quoted, calculated-solvatochromic p., Banerjee et al. 1990)
0.227 (quoted from Howard et al. 1985, Hinckley et al. 1990)
1.19 (GC-RT, Hinckley et al. 1990)

Henry's Law Constant (Pa m^3/mol):
0.111 (calculated-P/C, Fishbein & Albro 1972)
0.111 (calculated-P/C, Wolfe et al. 1980a)
0.0111 (calculated as per Lyman et al. 1982; quoted, Howard 1989)
0.218 (20°C, calculated-P/C, Mabey et al. 1982)

Octanol/Water Partition Coefficient, log K_{OW}:
2.12 (calculated as per Leo et al. 1971, Callahan et al. 1979; quoted, Neuhauser et al. 1985; Ryan et al. 1988)
2.12 (calculated, Wolfe et al. 1979; quoted, Callahan et al. 1979)
2.11, 3.40 (quoted, estimated-HPLC-RP, Veith et al. 1979)
1.61 (shake flask-LSC, Veith et al. 1980; Veith & Kosian 1983; quoted, Davies & Dobbs 1984; Zaroogian et al. 1985; Peddicord 1987; Saito et al. 1992)
1.56 (estimated-HPLC-RT, Veith et al. 1980)
1.61; 1.62, 1.82 (quoted; correlated-HPLC-k', McDuffie 1981)
1.56 (calculated, Mabey et al. 1982)
1.53 (20°C, shake flask-UV, Leyder & Boulanger 1983)
2.34 (calculated lit. value, Leyder & Boulanger 1983)
1.47, 1.90 (shake flask-HPLC/UV, HPLC-RT, Howard et al. 1985)
1.53, 1.56 (quoted, Howard et al. 1985)

1.56 (calculated, Hansch & Leo 1985; quoted, Howard 1989)
1.66 (shake flask, average from interlaboratory study, Renberg et al. 1985)
1.74 (HPLC, average from interlaboratory study, Renberg et al. 1985)
1.61 (HPTLC, average from interlaboratory study, Renberg et al. 1985)
1.5-2.1 (quoted, Russell et al. 1985)
2.12 (quoted from USEPA 1979, Neuhauser 1985)
1.62 (HPLC-RT, Eadsforth 1986)
1.86, 2.50 (quoted: shake flask, calculated, Eadsforth 1986)
1.53, 1.65 (quoted, calculated-solvatochromic p., Taft et al. 1985)
1.53, 1.63 (quoted, calculated-solvatochromic p., Leahy 1986)
1.56 (shake flask, log P database, Hansch & Leo 1987)
2.00 (quoted, Isnard & Lambert 1988, 1989)
1.46 (shake flask-HPLC/UV, Nielsen & Bundgaard 1989)
2.12 (quoted, Govind et al. 1991)
1.52 (calculated, Matthiessen et al. 1993)
1.56 (recommended, Sangster 1993)
1.52 (calculated-QSAR, Kollig 1993)

Bioconcentration Factor, log BCF:
0.67 (calculated-K_{OW}, Veith et al. 1979, 1980)
1.76 (bluegill sunfish, Barrows et al. 1980; quoted, Howard 1989)
1.76 (bluegill sunfish, Veith et al. 1980; Veith & Kosian 1983; quoted, Davies & Dobbs 1984; Sabljic 1987; Saito et al. 1992)
1.415 (bacteria, calculated-K_{OW}, Wolfe et al. 1980a)
0.49-0.8, 0.67-0.78 (shrimp, fish, Wofford et al. 1981)
1.204 (microorganisms-water, calculated-K_{OW}, Mabey et al. 1982)
0.77 (sheepshead minnows, quoted from Wofford et al. 1981, Zaroogian et al. 1985)
1.76, 2.22 (quoted, calculated- χ , Sabljic 1987)
1.76 (quoted, Isnard & Lambert 1988)
0.67, 0.73 (brown shrimp, sheepshead minnows, quoted, Howard 1989)
1.76 (fish, highest fish BCF, Matthiessen et al. 1993)

Sorption Partition Coefficient, log K_{OC}:
1.64 (soil, estimated, Kenaga 1980; quoted, Howard 1989)
2.20 (soil, estimated-K_{OW}, Wolfe et al. 1980a; quoted, Howard 1989)
1.24 (sediment-water, calculated-K_{OW}, Mabey et al. 1982)
1.52 (sediment, calculated-K_{OW}, Pavlou & Weston 1983, 1984; quoted, Peddicord 1987)
2.69 (activated carbon, calculated-χ , Blum et al. 1994)
1.20 (calculated-K_{OW}, Kollig 1993)

Half-Lives in the Environment:
Air: atmospheric half-life of 23.8 hours for reaction with hydroxyl radicals (Howard 1989); 112-1118 hours, based on estimated rate data for the reaction with hydroxyl radicals in air (Atkinson 1987; quoted, Howard et al. 1991); atmospheric transformation lifetime was estimated to be 1 to 5 days (Kelly et al. 1994).

Surface water: estimated half-life of <0.3 day in river waters (Zoeteman et al. 1980); mineralization half-life about 7 days in municipal digested sludge (Horowitz et al. 1982); biodegradation half-life of 1.90 days in an acclimated shake flask CO_2 evolution test (Sugatt et al. 1984); half-life less than 192-264 hours by biodegradation in fresh river water, 12 hours on the Rhine River, an estimated half-life of 13-27 hours in a modelling study of simulated ecosystem, abiotic degradation half-life 12.7 hours in pure water and 2.8 hours in the presence of nitrogen dioxide when irradiated with a UV lamp (Howard 1989); overall degradation half-life of 24-168 hours, based on unacclimated river die-away test data (Howard et al. 1991).

Ground water: 48-336 hours, based on estimated unacclimated aqueous aerobic biodegradation half-life (Howard et al. 1991).

Sediment:

Soil: biodegradation half-life < 5 days (aerobic) and about 20 days (anaerobic) in a garden soil (Shanker et al. 1985); 10-50 days, via volatilization subject to plant uptake from soil (Ryan et al. 1988); 24-168 hours, based on unacclimated aerobic river die-away test data and acclimated aerobic soil grab sample data (Howard et al. 1991).

Biota: half-life >24 and <48 hours (depuration half-life in tissues of bluefill sunfish in 21 day-exposure experiment (Barrows et al. 1980; quoted, Howard 1989).

Environmental Fate Rate Constant and Half-Lives:

Volatilization: estimated half-life of 46 days from a river of 1 meter deep with 1.0 m/sec current and a 3.0 m/sec wind using calculated Henry's law constant and considering the volatilization rate being controlled by the diffusion through the air layer (Lyman et al. 1982; quoted, Howard 1989).

Photolysis: abiotic degradation half-lives were reported as 3500 hours for direct sunlight in surface waters and 12.7 hours in pure water but reduced to 2.8 hours in the presence of NO_2 when irradiated with a UV lamp through a Pyrex filter (Howard 1989).

Oxidation: <<360 $M^{-1}\cdot h^{-1}$ for singlet oxygen and 0.05 $M^{-1}\cdot h^{-1}$ for peroxy radical (Mabey et al. 1982); photooxidation half-life was estimated to be 23.8 hours (Howard 1989); atmospheric half-life 112-1118 hours, based on estimated rate data for the reaction with hydroxyl radicals in air (Howard et al. 1991); calculated rate constant of 4×10^9 $M^{-1}\cdot sec^{-1}$ for the reaction with OH radicals in aqueous solution (Haag & Yao 1992).

Hydrolysis: second-order alkaline hydrolysis rate constant of 6.9×10^{-2} $M^{-1}\cdot sec^{-1}$ at pH 8 and 30°C in water, and a calculated half-life of 4 months (Wolfe et al. 1980b; quoted, Kollig 1993); the hydrolysis half-life was estimated to be 3.2 years under natural conditions at 30°C, 11.6 days (30°C) and 25 days (18°C) at pH 9 (Howard 1989); first-order hydrolysis half-life of 1163 days based on a second-order rate constant at pH 7 and 30°C and a base rate constant of 111.6 $M^{-1}\cdot h^{-1}$ corresponding to a half-life of 1.2 days at pH 9 and 18°C (Howard et al. 1991).

Biodegradation: calculated rate constant of $(9.49 \pm 0.41)\times10^2$ min^{-1} from retention times in reverse phase chromatography (Urushigawa & Yonezawa 1979; quoted, O'grady et al. 1985); significant degradation with rapid adaption within 7 days in an aerobic environment with a rate of >0.5 day^{-1} (Tabak et al. 1981; quoted,

Mills et al. 1982); biodegradation rate constant of 1.2×10^{-4} $mL\cdot cell^{-1}\cdot day^{-1}$ in river die-away test (Scow 1982); 58-88% mineralization in 7 days in municipal digested sludge (Horowitz et al. 1982); rate constant of 0.364 day^{-1} which corresponds to a half-life of 1.90 days in shake flask biodegration experiments (Sugatt et al. 1984); greater than 90% of DMP was degraded within 40 days in digested sludge (Shelton et al. 1984); biodegraded in excess of 90% in activated sludge systems in less than 24 hours (O'grady et al. 1985); soil-water biodegradation studies showed that 85% loss in Broome County soil after 120 hours and 75% loss in leachate sprayed soil after 48 hours (Russell et al. 1985); anaerobic digestion of sludge with a first-order rate constant of 8.9×10^{-3} h^{-1} and a half-life of 78 hours (Ziogou et al. 1989); aqueous aerobic half-life of 24-168 hours, based on unacclimated aerobic river die-away test data and acclimated aerobic soil grab sample data; aqueous anaerobic half-life of 96-672 hours, based on estimated unacclimated aqueous aerobic biodegradation half-life (Howard et al. 1991).

Biotransformation: estimated rate constant to be 5.2×10^{-6} $mL\cdot cell^{-1}\cdot h^{-1}$ for bacterial transformation in water (Wolfe et al. 1980a; quoted, Mabey et al. 1982).

Bioconcentration, Uptake (k_1) and Elimination (k_2) Rate Constants or Half-Lives: depuration half-life of 24-48 hours from bluegill sunfish (Barrows et al. 1980).

Common Name: Diethyl phthalate
Synonym: 1,2-benzenedicarboxylic acid ethyl ester, DEP, ethyl phthalate
Chemical Name: diethyl phthalate, ethyl phthalate
CAS Registry No: 84-66-2
Molecular Formula: $C_6H_4(COOC_2H_5)_2$
Molecular Weight: 222.26
Melting Point (°C):
- -40.5 (Patty 1963; Fishbein & Albro 1972; Callahan et al. 1979; Mabey et al. 1982; Verschueren 1983; Howard 1989)

Boiling Point (°C):
- 294 (Stull 1947)
- 296.1 (Fishbein & Albro 1972; Peakall 1975)
- 298 (Verschueren 1977, 1983; Callahan et al. 1979; Mabey et al. 1982)
- 295 (Howard 1989)
- 299 (Banerjee et al. 1990)

Density (g/cm^3 at 20°C):
- 1.123 (25°C, Fishbein & Albro 1972)
- 1.120 (25°C, Verschueren 1983)

Molar Volume (cm^3/mol):
- 208.9 (Kamlet et al. 1987)
- 115.3 (intrinsic volume V_I, Kamlet et al. 1987)
- 198.9 (calculated-density, Stephenson & Malanowski 1987)
- 254.0 (calculated-LeBas method)
- 116.1 (intrinsic V_I, Leggett & Parker 1994)

Molecular Volume (A^3):
Total Surface Area, TSA (A^2):
Heat of Fusion, ΔH_{fus}, (kcal/mol):
Entropy of Fusion, ΔS_{fus} (cal/mol·K or e.u.):
Fugacity Ratio at 25°C, F: 1.0

Water Solubility (g/m^3 or mg/L at 25°C):
- 598 (quoted, Deno & Berkheimer 1960)
- 1000 (20°C, Fishbein & Albro 1972; quoted, Wolfe et al. 1980)
- 1000 (32°C, quoted from Monsanto Chemical Co. data sheets, Peakall 1975; quoted, Callahan et al. 1979)
- 896 (Wolfe et al. 1979, 1980a; quoted, Callahan et al. 1979; Mabey et al. 1982; Neely 1984)
- 896 ± 20 (shake flask-GC, Wolfe et al. 1980b)
- 1200, 1240, 1160 (20°C: quoted average, elution chromatography, UV, Schwarz & Miller 1980)
- 1370, 1400, 1340 (30°C: quoted average, elution chromatography, UV, Schwarz & Miller 1980)
- 7028 (shake flask-LSC, Veith et al. 1980; quoted, Davies & Dobbs 1984)
- 27980 (calculated-π const., as per Hansch et al. 1968, Veith et al. 1980)
- 307, 1610 (quoted, calculated-K_{OW}, Yalkowsky & Morozowich 1980)
- 928 (20°C, shake flask-UV, Leyder & Boulanger 1983)

1020, 1200 (quoted lit. values, Leyder & Boulanger 1983)
210 (quoted, Verschueren 1983)
1080 ± 40 (shake flask-HPLC/UV, Howard et al. 1985; quoted, Howard 1989)
896, 928 (quoted, Howard et al. 1985)
1000000 (quoted from Callahan et al. 1979, misquoted in unit? Neuhauser et al. 1985)
900-1000 (quoted, Russell et al. 1985)
680 (measured, Russell & McDuffie 1986)
598, 2611 (quoted, calculated-solvatochromic p., Kamlet et al. 1987)
598, 443 (quoted, calculated-V_I and M.P., Kamlet et al. 1987)
1000 (quoted, Isnard & Lambert 1988, 1989)
896 (selected, Aichinger et al. 1992)

Vapor Pressure (Pa at 25°C):
0.467 (extrapolated-regression of tabulated data, Stull 1947)
0.086 (effusion method, extrapolated-Antoine eqn., Small et al. 1948; quoted, Hinckley et al. 1990)
0.052 (20°C, torsion-effusion method, Balson 1958; quoted, Grayson & Fosbraey 1982)
0.467 (calculated, Weast 1972-73; quoted, Mabey et al. 1982)
0.046 (20°C, gas-saturation method, Grayson & Fosbraey 1982)
0.0819 (gas-saturation method, Grayson & Fosbraey 1982)
0.220 (gas saturation-HPLC/UV, Howard et al. 1985; quoted, Howard 1989)
0.220 (Howard et al. 1986; quoted, Banerjee et al. 1990)
0.0064 (extrapolated, Antoine eqn., Stephenson & Malanowski 1987)
0.220, 0.392 (quoted, calculated-solvatochromic p., Banerjee et al. 1990)
0.0810 (quoted from Grayson & Fosbraey 1982, Hinckley et al. 1990)
0.230 (quoted from Howard et al. 1985, Hinckley et al. 1990)
0.280 (GC-RT, Hinckley et al. 1990)

Henry's Law Constant (Pa·m^3/mol):
0.0020 (calculated-P/C, Wolfe et al. 1980a)
0.0486 (Lyman et al. 1982; quoted, Howard 1989)
0.122 (calculated-P/C, Mabey et al. 1982)
4.710 (quoted from WERL Treatability Data, Ryan et al. 1988)

Octanol/Water Partition Coefficient, log K_{OW}:
3.22 (calculated as per Leo et al. 1971, Callahan et al. 1979; quoted, Neuhauser et al. 1985; Ryan et al. 1988)
3.15 (estimated-RPHPLC, Veith et al. 1979)
3.22 (Wolfe et al. 1979; quoted, Callahan et al. 1979)
1.40 (shake flask-LSC, Veith et al. 1980)
2.67 (estimated-RPHPLC, Veith et al. 1980)
2.21; 2.29, 2.35, 2.97 (correlated HPLC-RT; calculated-π const., f-const., aqueous solubility correlation, McDuffie 1981)

2.70 (quoted from Veith et al. 1980, Veith & Kosian 1983; Davies & Dobbs 1984; Saito et al. 1992)
2.47 (calculated, Mabey et al. 1982)
2.35 (20°C, shake flask-UV, Leyder & Boulanger 1983)
3.42 (calculated lit. value, Leyder & Boulanger 1983)
2.24, 2.29 (shake flask-HPLC/UV, HPLC-RT, Howard et al. 1985)
2.35, 2.47 (quoted, Howard et al. 1985)
2.4-3.2 (quoted, Russell et al. 1985)
2.47 (Hansch & Leo 1985; quoted, Howard 1989; Hulzebos et al. 1993)
3.20 (quoted from USEPA 1979, Neuhauser et al. 1985)
2.88, 2.82 (quoted, calculated-f const., De Kock & Lord 1987)
3.00 (predicted-RPHPLC, De Koch & Lord 1987)
1.40 (quoted, Peddicord 1987)
2.47 (shake flask, log P database, Hansch & Leo 1987)
3.00 (quoted, Isnard & Lambert 1988, 1989)
3.22 (selected, Govind et al. 1991)
2.57 (calculated, Matthiessen et al. 1993)
2.47 (recommended, Sangster 1993)
2.57 (calculated-QSAR, Kollig 1993)

Bioconcentration Factor, log BCF:
0.49-1.56 (calculated-K_{OW}, Veith et al. 1979, 1980)
2.07 (bluegill sunfish, Veith et al. 1980; Barrows et al. 1980; Veith & Kosian 1983; quoted, Davies & Dobbs 1984; Sabljic 1987; Howard 1989; Saito et al. 1992)
1.86 (bacteria, calculated-K_{OW}, Wolfe et al. 1980a)
1.08 (calculated-S, Lyman et al. 1982; quoted, Howard 1989)
1.64 (calculated-K_{OW}, Lyman et al. 1982; quoted, Howard 1989)
2.04 (microorganisms-water, calculated-K_{OW}, Mabey et al. 1982)
1.19 (mullet, Shimada et al. 1983; quoted, Howard 1989)
2.10 (Davies & Dobbs 1984; quoted, Sabljic 1987)
2.10, 2.80 (quoted, calculated- χ , Sabljic 1987)
2.07 (quoted, Isnard & Lambert 1988)
2.07 (fish, highest fish BCF, Matthiessen 1993)

Sorption Partition Coefficient, log K_{OC}:
2.65 (calculated-K_{OW}, Wolfe et al. 1980a)
2.72 (soil, calculated-S, Lyman et al. 1982; quoted, Howard 1989)
1.97 (soil, calculated-K_{OW}, Lyman et al. 1982; quoted, Howard 1989)
2.15 (sediment-water, calculated-K_{OW}, Mabey et al. 1982)
1.34 (sediment, calculated-K_{OW}, Pavlou & Weston 1983, 1984; quoted, Peddicord 1987)
1.84 (Broome County of New York soil, shake flask-GC, Russell & McDuffie 1986)
3.00 (sediment, von Oepen et al. 1991)
2.85 (Alfisol, von Oepen et al. 1991)
3.24 (Podzol, von Oepen et al. 1991)
2.03 (activated carbon, calculated-χ , Blum et al. 1994)
1.99 (calculated-K_{OW}, Kollig 1993)

Half-Lives in the Environment:

Air: an estimated half-life of 22.2 hours for the vapor reaction with photochemically generated $8x10^5$ molecules·cm^{-3} hydroxyl radicals in air at 25°C with an estimated rate constant of $1.08x10^{-11}$ cm^3·molecule^{-1}·sec^{-1} (GEMS 1986; quoted, Howard 1989); 21-212 hours, based on an estimated rate constant for the vapor phase reaction with hydroxyl radicals in air (quoted, Howard et al. 1991).

Surface water: biodegradation half-life of 2.21 days in an acclimated shake flask CO_2 evolution test (Sugatt et al. 1984); half-life of 2 days to >2 weeks for aerobic biodegradation in water; 3 days when incubated in dirty river water (Howard 1989); overall degradation half-life of 72-1344 hours, based on aerobic aqueous grab sample data for fresh and marine water (Howard et al. 1991).

Ground water: 144-2688 hours, based on estimated unacclimated aqueous aerobic biodegradation half-life (Howard et al. 1991).

Sediment: less than 14 days in freshwater sediment (Johnson et al. 1984).

Soil: degradation half-life of 10.5 days by microorganisms isolated from soil or waste water at 30°C (Kurane et al. 1977); 10-50 days via volatilization subject to plant uptake from soil (Ryan et al. 1988); 72-1344 hours, based on estimated unacclimated aqueous aerobic biodegradation half-life (Howard et al. 1991).

Biota: half-life >24 <48 hours in tissues of bluegill sunfish (Barrows et al. 1980).

Environmental Fate Rate Constant and Half-Lives:

Volatilization:

Photolysis:

Oxidation: oxidation rate constant of 18 M·sec^{-1} in aquatic environment (Wolfe et al. 1980a); rate constant <<360 M^{-1}·h^{-1} for singlet oxygen and 1.4 M^{-1}·h^{-1} for peroxy radical (Mabey et al. 1982); the estimated atmospheric reaction rate constant of $1.08x10^{-11}$ cm^3·molecule^{-1}·sec^{-1} with hydroxyl radicals corresponding to a half-life of 22.2 hours and the alkyl peroxy reaction half-life was calculated to be 6.5 years (Howard 1989); photooxidation half-life of 21-212 hours in air, based on an estimated rate constant for the vapor phase reaction with hydroxyl radicals in air (Howard et al. 1991); photooxidation half-life of 2.4-12.2 years in water, based on estimated rate data for the reaction with alkyloxyl radicals in aqueous solution (Howard et al. 1991); calculated oxidation rate constant of $4x10^9$ M^{-1}·sec^{-1} for the reaction with OH radicals in aqueous solution (Haag & Yao 1992).

Hydrolysis: phthalates are susceptible to alkaline hydrolysis, with theoretical half-lives at pH 8 and 30°C ranging from 4 months to 100 years, hydrolysis rate constant of $7.9x10^1$ M^{-1}·h^{-1} (Wolfe et al. 1980a); second-order alkaline hydrolysis rate constant of $2.5x10^{-2}$ M^{-1}·sec^{-1} at 30°C (Wolfe et al. 1980b; quoted, Kollig 1993); alkaline hydrolysis rate constant of $2.5x10^{-2}$ mol-sec^{-1} at 30°C and $1.2x10^{-2}$ mol-sec^{-1} at pH 10-12 (Wolfe et al. 1979; quoted, Callahan et al. 1979; quoted, Howard 1989) both at 30°C and pH 7, half-lives of 8.8 and 18 years at pH 7 were estimated (calculated as per Radding et al. 1977; quoted, Callahan et al. 1979; Howard 1989); first-order hydrolysis half-life of 8.8 years at pH 7 and 32 days at pH 10-12 based on base rate constant at 30°C (Howard et al. 1991).

Biodegradation: microbial degradation half-life of 10.5 days by microorganisms isolated from soil and waste water at 30°C (Kurane et al. 1977, quoted, Russell et al. 1985); estimated rate constant of $(6.59 \pm 0.43)x10^2\ min^{-1}$ from retention times of reverse phase chromatography (Urushigawa & Yonezawa 1979; quoted, O'grady et al. 1985); microbial degradation rate constant was $2x10^{-4}\ h^{-1}$ in an aquatic environment (Wolfe et al. 190a); 0 to 32% mineralization in >8 weeks in municipal digested sludge (Horowitz et al. 1982); complete degradation in 7.0 days for 5 and 10 ppm DEP in domestic waste water under aerobic conditions at 25°C (Tabak et al. 1981; quoted, Howard 1989); rate constant of 0.315 day^{-1} corresponding to a half-life of 2.21 days with a mixed microbial population and underwent >99% degradation in 28 days in shake flask biodegradation experiment (Sugatt et al. 1984); greater than 90% of DEP was degraded within 40 days in digested sludge (Shelton et al. 1984); 85% biodegradated after 14 days incubation in aerobic freshwater sediments at 22°C (Johnson et al. 1984); biodegraded in excess of 90% in less than 24 hours in activated sludge systems (O'grady et al. 1985); soil-water biodegradation studies showed the 86% loss after 120 hours in Broome County soils and 67% loss in leachate sprayed soil after 48 hours (Russell et al. 1985); anaerobic digestion of sludge with first-order rate constant of $6.0x10^{-3}\ h^{-1}$ and half-life of 115 hours (Ziogou et al. 1989); aqueous aerobic half-life of 72-1344 hours, based on aerobic aqueous grab sample data for fresh and marine water (Howard et al. 1991); aqueous anaerobic half-life of 672-5376 hours, based on estimated unacclimated aqueous aerobic biodegradation half-life and anaerobic screening test data (Howard et al. 1991).

Biotransformation: experimentally determined microbial rate constant of $3.2x10^{-9}$ $mL{\cdot}organism^{-1}{\cdot}\ h^{-1}$ (Wolfe et al. 1980a,c); estimated rate constant to be $1x10^{-7}$ $mL{\cdot}cell^{-1}{\cdot}h^{-1}$ for bacterial transformation in water (Mabey et al. 1982); mean microbial rate constant of $(7.2 \pm 15.4)x10^{-7}\ mL{\cdot}cell^{-1}{\cdot}h^{-1}$ for 54 batch aufwuchs cultures (Lewis & Holm 1981; Lewis et al. 1984).

Bioconcentration, Uptake (k_1) and Elimination (k_2) Rate Constants or Half-Lives: depuration half-life of <24 and >48 hours in tissues of bluegill sunfish (Barrows et al. 1980).

Common Name: Dibutyl phthalate
Synonym: *n*-butyl phathalate, di-*n*-butyl phthalate, DBP, dibutyl *o*-phthalate, *o*-benzenedicarboxylic acid dibutyl ester, benzene-*o*-dicarboxylic acid dibutyl ester
Chemical Name: *n*-butyl phathalate, di-*n*-butyl phthalate
CAS Registry No: 84-74-2
Molecular Formula: $o\text{-}C_6H_4(COOC_4H_9)_2$
Molecular Weight: 278.34
Melting Point (°C):
-35.0 (Fishbein & Albro 1972; Verschueren 1977, 1983; Callahan et al. 1979; Mabey et al. 1982; Dean 1985; Howard 1989)
Boiling Point (°C):
340 (Stull 1947; Patty 1963; Fishbein & Albro 1972; Peakall 1975; Callahan et al. 1979; Mabey et al. 1982; Verschueren 1983; Dean 1985; Howard 1989; Banerjee et al. 1990)
Density (g/cm^3 at 20°C):
1.0465 (21°C, Fishbein & Albro 1972)
1.0465 (Verschueren 1983; Dean 1985)
Molar Volume (cm^3/mol):
275.8 (Taft et al. 1985; Leahy 1986)
153.5 (intrinsic volume V_I, Leahy 1986)
347.2 (calculated-LeBas method)
265.1 (calculated-density)
Molecular Volume (A^3):
Total Surface Area, TSA (A^2):
Heat of Fusion, ΔH_{fus}, (kcal/mol):
Entropy of Fusion, ΔS_{fus} (cal/mol·K or e.u.):
Fugacity Ratio at 25°C, (assuming ΔS_{fus} = 13.5 e.u.) F: 1.0

Water Solubility (g/m^3 or mg/L at 25°C):
11.08 (quoted, Deno & Berkheimer 1960)
4500 (Patty 1967; quoted, Peakall 1975)
4500 (Fishbein & Albro 1972; quoted, Wolfe et al. 1980b)
10.0 (30°C, quoted from Monsanto Chemical Co. data sheets, Peakall 1975)
13.0 (Wolfe et al. 1979; quoted, Callahan et al. 1979; Mabey et al. 1982; Zaroogian et al. 1985)
10.8, 11.1, 10.5 (20°C: quoted average, elution chromatography, UV, Schwarz & Miller 1980)
11.4, 11.5, 11.2 (30°C: quoted average, elution chromatography, UV, Schwarz & Miller 1980)
13 ± 1.6, 4.45 (shake flask-GC, quoted, Wolfe et al. 1980b)
13.0 (quoted, Wolfe et al. 1980a; DeFoe et al. 1990)
3.25 (solubility in 35 liter instant ocean, Giam et al. 1980)
10.1 (20°C, shake flask-UV, Leyder & Boulanger 1983; quoted, DeFoe et al. 1990)
13, 25 (quoted, lit. values, Leyder & Boulanger 1983)
400, 4500 (quoted, Verschueren 1983)
398 (quoted, Garten & Trabalka 1983)

28 (26°C, quoted, Verschueren 1983)
11.2±0.3 (shake flask-HPLC/UV, Howard et al. 1985)
100 (quoted, Dean 1985)
10-13 (quoted, Russell et al. 1985)
13000 (quoted Callahan et al. 1979; Neuhauser et al. 1985)
11.2 (Howard et al. 1985; quoted, Howard 1989; DeFoe et al. 1990)
19.7 (calculated-solvatochromic p., Leahy 1986)
9.40 (best estimate by turbidity inflection, DeFoe et al. 1990)
8.70 (centrifugation, DeFoe et al. 1990)
13.0 (selected, Aichinger et al. 1992)

Vapor Pressure (Pa at 25°C):
0.0217 (extrapolated-regression of tabulated data, Stull 1947)
0.00253 (from boiling point measurement, Burrows 1948)
0.00374 (effusion method, average extrapolated-Antoine eqn., Small et al. 1948)
0.00270 (Small et al. 1948; quoted, Hinckley et al. 1990)
0.00590 (Perry & Weber 1949; quoted, Hinckley et al. 1990)
0.00187 (dew-point method at temp. range of 72-185°C, extrapolated-Antoine eqn., Werner 1952; quoted, Giam et al. 1980; Howard 1989)
0.00157; 0.00271, 0.00592 (quoted lit. value; calculated values, Hamilton 1980)
0.00277 (20°C, evaporation rate method, Gückel et al. 1982)
0.00540 (calculated-Antoine eqn., Gückel et al. 1982)
0.00133 (Jaber 1982; quoted, Mabey et al. 1982)
0.0213; 0.40, 0.0653, 1.0 (quoted, structure-based estimation methods, Tucker et al. 1983)
1.55×10^{-4} (extrapolated-Antoine eqn., Boublik et al. 1984)
0.00230 (quoted from OECD 1981 Guidelines, Dobbs et al. 1984)
0.0011, 0.00115 (saturated column-HPLC/UV, quoted, Howard et al. 1985)
0.00355 (extrapolated, Antoine eqn., Stephenson & Malanowski 1987)
0.0097, 0.0244 (quoted, calculated-solvatochromic p., Banerjee et al. 1990)
7.6×10^{-5} (quoted from Werner 1952, Hinckley et al. 1990)
0.0097 (quoted from Howard et al. 1985, Hinckley et al. 1990)
0.0041 (quoted from OECD interlab. studies, Hinckley et al. 1990)
0.0056 (GC-RT, Hinckley et al. 1990)

Henry's Law Constant (Pa·m^3/mol):
0.132 (calculated-P/C, Wolfe et al. 1980a)
0.0466 (Lyman et al. 1982; quoted, Howard 1989)
0.0284 (calculated-P/C, Mabey et al. 1982)
0.1835 (Atlas et al. 1983)
0.456, 0.446 (calculated-P/C, calculated-group contribution, Tucker et al. 1983)
0.0297 (quoted from WERl Treatability Data, Ryan et al. 1988)

Octanol/Water Partition Coefficient, log K_{OW}:

5.20 (calculated as per Leo et al. 1971, Callahan et al. 1979; quoted, Neuhauser et al. 1985; Peddicord 1987; Ryan et al. 1988)
5.15 (estimated, Veith & Morris 1978; quoted, Zaroogian et al. 1985)
5.15 (estimated-RPHPLC, Veith et al. 1979)
5.20 (Wolfe et al. 1979; quoted, Callahan et al. 1979)
4.13; 4.29, 4.47, 4.12 (correlated HPLC-RT; calculated-π const., f-const., solubility correlation, McDuffie 1981)
5.56 (calculated, Mabey et al. 1982)
5.43 (calculated as per Rekker 1977, Harnish et al. 1983)
4.08 (shake flask average, OECD/EEC lab. comparison tests, Harnish et al. 1983)
4.39, 4.56 (HPLC-extrapolated, Harnish et al. 1983)
4.57 (20°C, shake flask-UV, Leyder & Boulanger 1983)
5.38 (quoted lit. calculated value, Leyder & Boulanger 1983)
3.31 (Garten & Trabalka 1983)
4.79, 3.74 (shake flask-HPLC/UV, HPLC-RT, Howard et al. 1985)
4.11 (quoted OECD value, Howard et al. 1985)
4.6-5.2 (quoted, Russell et al. 1985)
4.72 (calculated, Hansch & Leo 1985; quoted, Howard 1989)
4.57 (HPLC-k', Eadsforth 1986)
5.15, 5.74 (quoted: shake flask, calculated, Eadsforth 1986)
4.30 (HPLC-RT, Haky & Leja 1986)
4.57, 4.48 (quoted, calculated-solvatochromic p., Taft et al. 1985)
4.57, 4.73 (quoted, calculated-solvatochromic p., Leahy 1986)
4.72 (shake flask, log P database, Hansch & Leo 1987)
4.72 (measured value, DeFoe et al. 1990)
5.56 (selected, Govind et al. 1991)
4.69 (calculated, Matthiessen et al. 1993)
4.72 (recommended, Sangster 1993)
4.69 (calculated-QSAR, Kollig 1993)

Bioconcentration Factor, log BCF:

3.15 (fish, Mayer & Sanders 1973; quoted, Howard 1989)
3.82, 3.70, 3.83, 3.28, 3.70, 3.43 (midge larvae, waterflea, scud, mayfly, grass shrimp, damselfly, Sanders et al. 1973)
3.83 (fish, Sanders et al. 1973; quoted, Howard 1989)
1.32 (calculated, Kenaga 1980; quoted, Howard 1989)
3.90 (bacteria, calculated-K_{OW}, Wolfe et al. 1980a)
1.32-1.62, 0.46-1.49, 1.07 (oyster, shrimp, fish, Wofford et al. 1981)
1.50, 1.22, 1.07 (American oyster, brown shrimp, sheephead minnow, Wofford et al. 1981; quoted, Howard 1989)
4.67 (microorganisms-water, calculated-K_{OW}, Mabey et al. 1982)
4.36 (*selenastrum capricornutum*, Casserly et al. 1983)
1.32 (oyster, quoted from Wofford et al. 1981, Zaroogian et al. 1985)
3.68, 3.68 (oyster, estimated values, Zaroogian et al. 1985)
1.08 (fish, highest BCF, Matthiessen et al. 1993)

Sorption Partition Coefficient, log K_{OC}:

2.20 (soil, estimated-S, Kenega 1980; quoted, Howard 1989)

3.81 (soil, estimated-S, Wolfe et al. 1980a; quoted, Howard 1989)

2.17 (marine sediment/seawater with 1% organic carbon, Bouwer et al. 1981; quoted, Howard 1989)

1.60, 1.30, 0.602 (between montmorillonite, kaolinite, calcium montmorillonite and seawater at concn. of 3-4 ppm, Sullivan et al. 1981; quoted, Howard 1989)

0.301, 0.602, 1.556 (between montmorillonite, kaolinite, calcium montmorillonite and seawater at concn. of 20 ppb, Sullivan et al. 1981; quoted, Howard 1989)

5.23 (sediment-water, calculated-K_{OW}, Mabey et al. 1982).

4.54 (sediment, calculated-K_{OW}, Pavlou & Weston 1983,84; quoted, Peddicord 1987)

3.14 (Broome County soil in New York, shake flask-GC, Russell & McDuffie 1986)

4.37 (calculated-K_{OW}, Kollig 1993)

Half-Lives in the Environment:

Air: estimated half-life of 18 hours for the vapor phase reaction with hydroxyl radicals in air (Howard 1989); 7.4-74 hours, based on estimated rate data for the vapor phase reaction with hydroxyl radicals in air (Howard et al. 1991); atmospheric transformation lifetime was estimated to be < 1 day (Kelly et al. 1994).

Surface water: biodegradation half-life of 15.4 days in an acclimated shake flask CO_2 evolution test (Sugatt et al. 1984); half-life of 2-12 days in water alone (Howard 1989); 24-336 hours, based on unacclimated aerobic river die-away test and fresh water/sediment grab sample data (Howard et al. 1991).

Ground water: 48-552 hours, based on estimated unacclimated aqueous aerobic biodegradation half-life (Howard et al. 1991).

Sediment: 97% degradation in 5 days in an aerobic pond water-sediment mixture; 7-30 days under anaerobic conditions, 50% biodegraded in 1-5 days in sediment-water systems of estuarine and freshwater sites (Howard 1989).

Soil: degradation half-life of 3 days by microorganisms isolated from soil or waste water at 30°C (Kurane et al. 1977); degradation half-lives ranging from 11 to 53 days as affected by soil type, pH, temperature, aeration status and sterilization (Inman et al. 1984); biodegradation half-life < 5 days (aerobic) and about 20 days (anaerobic) in a garden soil (Shanker et al. 1985); 10-50 days via volatilization subject to plant uptake from soil (Ryan et al. 1988); overall half-life 48-552 hours, based on unacclimated aerobic soil grab sample data (Howard et al. 1991).

Biota: elimination half-life of 3 days for waterfleas *Daphnia magna* (Mayer & Sanders 1973).

Environmental Fate Rate Constant and Half-Lives:

Volatilization: half-life from a stirred seawater solution 1.0 m deep is 28 days (Atlas et al. 1982; quoted, Howard 1989); evaporation rate of 3.42×10^{-10} $mol \cdot cm^{-2} \cdot h^{-1}$ at 20°C (Gückel et al. 1982); estimated half-life of 47 days in a river of 1.0 m deep with 1 m/sec current and a 3 m/sec wind using Henry's law constant while the rate of volatilization being controlled by the diffusion through air (Lyman et al. 1982; quoted, Howard 1989).

Photolysis: direct photolysis (near surface) rate constant is estimate to be $2x10^{-4}$ h^{-1} in natural water (Wolfe et al. 1980a).

Oxidation: the free radical oxidation rate constant is estimated to be 18 $M^{-1}·sec^{-1}$ (Wolfe et al. 1980a); photooxidation half-life of 2.4-12.2 years in water, based on estimated rate data for the reaction with alkyloxyl radicals in aqueous solution (Wolfe et al. 1980a; quoted, Howard et al. 1991); << 360 $M^{-1}·h^{-1}$ for singlet oxygen and 1.4 $M^{-1}·h^{-1}$ for peroxy radical (Mabey et al. 1982); photooxidation half-life of 18.4 hours with reaction with OH radicals (Howard 1989); photooxidation half-life of 7.4-74 hours in air, based on estimated rate data for the vapor phase reaction with hydroxyl radicals in air (Howard et al. 1991).

Hydrolysis: second-order alkaline hydrolysis rate constant of $2.2x10^{-2}$ $M^{-1}·sec^{-1}$ for pH 10-12 at 30°C (Wolfe et al. 1979; quoted, Callahan et al. 1979) with calculated half-life of 10 years (calculated as per Radding et al. 1977, Callahan et al. 1979); hydrolysis rate constant of $3.8x10^{1}$ $M^{-1}·h^{-1}$ in an aquatic environment (Wolfe et al. 1980a); overall rate constant of $1.0x10^{-2}$ $mol^{-1}·sec^{-1}$ at pH 7 and 30°C with first-order hydrolysis half-life of 10 years (Wolfe et al. 1980b; quoted, Kollig 1993); estimated half-life of 76 days at pH 9 and 10 years at neutral pH (Howard 1989); first-order hydrolysis half-life of 10 years, based on the overall hydrolysis rate constant (Howard et al. 1991).

Biodegradation: degradation half-life of 3 days by microorganisms isolated from soil or waste water at 30°C (Kurane et al. 1977); for the degradation freshwater hydrosoil, under aerobic conditions, DBP was rapidly degraded, 3% left after 5 days, but under anaerobic conditions required 30 days (Johnson et al. 1979, quoted, Russell et al 1985); a 90% loss in 3 days from river water (Brinkman et al. 1979, quoted, Russell et al. 1985); estimated rate constant of (13.26 ± 0.73)x10^{2} min^{-1} from retention times of reverse phase chromatography (Urushigawa & Yonezawa 1979; quoted, O'grady et al. 1985); microbial degradation rate constant of $2.9x10^{-8}$ $mL·organism^{-1}·h^{-1}$ (Wolfe et al. 1980a); significant degradation with rapid adaptation within 7 days in an aerobic environment with a rate of >0.5 day^{-1} (Tabak et al. 1981; quoted, Mills et al. 1982); biodegradation of $7.4x10^{-7}$ $mL·cell^{-1}·day^{-1}$ (river die-away test, Scow 1982); 32-85% degraded after 2-3 weeks in municipal digested sludge (Horowitz et al. 1982); approximately 90% added to soils decomposed within 80 days under both aerobic and anaerobic conditions (Inman et al. 1984); biodegradation first-order rate constants range 0.1-7.6$x10^{-4}$ h^{-1}, 4.9-36.5$x10^{-4}$ h^{-1} for water and water/sediment systems (under sterile conditions); and 29.3-409$x10^{-4}$ h^{-1}, 38.2-456$x10^{-4}$ h^{-1} for water and water/sediment systems (under active conditions) samples from six estuarine and freshwater sites, actual half-lives: 1.0-4.8 days for active sediment treatment, 1.7-13.0 days for active water treatment and 8-23 days for sterile sediment (Walker et al. 1984); rate constant of 0.050 day^{-1} corresponding to a half-life of 15.4 days in a shake flask biodegradation experiment (Sugatt et al. 1984); greater than 90% of DBP was degraded within 40 days in digested sludge (Shelton et al. 1984); biodegraded in excess of 90% in less than 24 hours in activated sludge systems (O'grady et al. 1985); soil-water biodegradation studies showed that 81% loss in Broome County soils after 24 hours and 70% loss in leachate sprayed soil after 24 hours (Russell et al. 1985); anaerobic digestion of sludge first-order rate constant of $10.6x10^{-3}$ h^{-1} and half-life of 65 hours (Ziogou et al. 1989); biodegradation half-life <5 days under aerobic conditions and 20 days under anaerobic conditions in a garden soil (Shanker et al. 1985); aqueous aerobic half-life of 24-552 hours, based on unacclimated aerobic river die-away test and soil grab sample data (Howard et al. 1991); aqueous anaerobic half-life of 48-552 hours, based on unacclimated anaerobic grab sample data for soil and sediment (Johnson & Lulves 1975; Verschueren 1983; Howard et al. 1991).

Biotransformation: $(1.9\text{-}4.4)\times10^{-8}$ $mL\cdot cell^{-1}\cdot h^{-1}$ for bacterial transformation in water (Steen et al. 1979; quoted, Mabey et al. 1982; Steen 1991); 2.9×10^{-8} $ml\cdot cell^{-1}\cdot h^{-1}$ for bacterial transformation in water (Wolfe et al. 1980a).

Bioconcentration, Uptake (k_1) and Elimination (k_2) Rate Constants or half-Lives: half-life of 3 days in *Daphnia magna* (Mayer & Sanders 1973).

Common Name: *n*-Dioctyl phthalate
Synonym: di-*n*-octyl phthalate, DOP, *o*-benzenedicarboxylic acid dioctyl ester, *n*-dioctyl phthalate, octyl phthalate, dioctyl-*o*-benzenedicarboxylate
Chemical Name: di-*n*-octyl phthalate
CAS Registry No: 117-84-0
Molecular Formula: $C_6H_4(COOC_8H_{17})_2$
Molecular Weight: 390.56
Melting Point (°C):
-25.0 (Patty 1963; Fishbein & Albro 1972; Callahan et al. 1979; Mabey et al. 1982)
Boiling Point (°C):
220.0 (Patty 1963; Fishbein & Albro 1972; Callahan et al. 1979; Mabey et al. 1982)
Density (g/cm^3 at 20°C):
0.978 (Fishbein & Albro 1972)
Molar Volume (cm^3/mol):
524.8 (calculated-LeBas method)
399.3 (calculated-density)
Molecular Volume (A^3):
Total Surface Area, TSA (A^2):
Heat of Fusion, ΔH_{fus}, (kcal/mol):
Entropy of Fusion, ΔS_{fus} (cal/mol·K or e.u.):
Fugacity Ratio at 25°C, F: 1.0

Water Solubility (g/m^3 or mg/L at 25°C):
50.0 (20°C, Fishbein & Albro 1972, quoted, Wolfe et al. 1980)
3.0 (shake flask-GC, Wolfe et al. 1979; quoted, Callahan et al. 1979; Mabey et al. 1982)
3.0 ± 0.5 (shake flask-GC, Wolfe et al. 1980b)
3.0 (shake flask-GC, Wolfe et al. 1980a; quoted, DeFoe et al. 1990)
0.285 (24°C, tech. grade, Verschueren 1983; quoted, DeFoe et al. 1990)
0.020 (best estimate of synthesized phthalate by turbidity inflection, DeFoe et al. 1990)
3.0 (selected, Aichinger et al. 1992)

Vapor Pressure (Pa at 25°C):
0.0253 (dew-point method, temp. range of 110-210°C, extrapolated from Antoine eqn., Werner 1952)
0.0187 (estimated by analogy to Henry's law constant, Mabey et al. 1982)
2.92×10^{-5} (extrapolated, Antoine eqn., Stephenson & Malanowski 1987)

Henry's Law Constant ($Pa \cdot m^3/mol$):
0.557 (calculated-P/C, Wolfe et al. 1980a)
1.722 (calculated-P/C, Mabey et al. 1982)
0.0297 (quoted from WERL Treatability Data, Ryan et al. 1988)

Octanol/Water Partition Coefficient, log K_{OW}:

9.2 (calculated as per Leo et al. 1971, Callahan et al. 1979; quoted, Neuhauser et al. 1985; Peddicord 1987; Ryan et al. 1988)

9.2 (calculated, Wolfe et al. 1979; quoted, Callahan et al. 1979)

8.06; 8.29, 8.71, 4.60 (correlated HPLC-RT; calculated-π const., f-const., aqueous solubility correlation, McDuffie 1981)

9.87 (calculated, Mabey et al. 1982)

8.92 (calculated-CLOG*P* for synthesized phthalate, DeFoe et al. 1990)

5.22 (shake flask, log P database, Hansch & Leo 1987; quoted, Sangster 1993)

7.90 (calculated-QSAR, Kollig 1993)

Bioconcentration Factor, log BCF:

2.82, 3.97, 0.064, 3.72, 2.64 (in lab. model ecosystem after 3 days: alga, daphnia, mosquito fish, mosquito larvae, snails, Sanborn et al. 1975)

4.45, 3.41, 3.97, 3.97, 4.13 (in lab. model ecosystem after 33 days: alga, daphnia, mosquito fish, mosquito larvae, snails, Sanborn et al. 1975)

3.46 (bacteria, calculated-K_{OW}, Wolfe et al. 1980a)

8.59 (microorganisms-water, calculated-K_{OW}, Mabey et al. 1982)

Sorption Partition Coefficient, log K_{OC}:

4.28 (calculated, Wolfe et al. 1980a)

9.56 (sediment-water, calculated-K_{OW}, Mabey et al. 1982)

7.91 (sediment, calculated-K_{OW}, Pavlou & Weston 1983, 1984; quoted, Peddicord 1987)

7.60 (calculated-K_{OW}, Kollig 1993)

Half-Lives in the Environment:

Air: 4.5-44.8 hours, based on estimated rate constant for the vapor phase reaction with hydroxyl radicals in air (Atkinson 1987; quoted, Howard et al. 1991).

Surface water: degradation half-life 5 days in an aquatic model ecosystem (Sanborn et al. 1975); overall degradation half-life 168-672 hours, based on estimated aqueous aerobic biodegradation half-life (Howard et al. 1991).

Ground water: 336-8760 hours, based on estimated aqueous aerobic and anaerobic biodegradation half-lives (Howard et al. 1991).

Sediment:

Soil: degradation half-life of 8 days by microorganisms isolated from soil or waste water at 30°C (Kurane et al. 1977); 10-50 days via volatilization subject to plant uptake from the soil (Ryan et al. 1988); degradation half-life 168-672 hours, based on estimated aqueous aerobic biodegradation half-life (Howard et al. 1991).

Biota:

Environmental Fate Rate Constant and Half-Lives:

Volatilization:

Photolysis:

Oxidation: the free radicals oxidation rate constant of 18 $M \cdot sec^{-1}$ in an aquatic environment (Wolfe et al. 1980a); rate constant <<360 $M^{-1} \cdot h^{-1}$ for singlet oxygen and 1.4 $M^{-1} \cdot h^{-1}$ for peroxy radical (Mabey et al. 1982); photooxidation half-life of 4.5-44.8 hours in air, based on estimated rate constant for the vapor phase reaction with hydroxyl radicals in air (Howard et al. 1991).

Hydrolysis: phthalates are susceptible to alkaline hydrolysis, with theoretical half-lifes at pH 8 and 30°C ranging from 4 months to 100 years (wolfe et al. 1980a; quoted, Kollig 1993); based rate constant 7.4 $M^{-1} \cdot h^{-1}$ at pH 7 and 25°C (Ellington et al. 1987); the first-order hydrolysis half-life was estimated to be 107 years at pH 7 and 1 year at pH 9 and 25°C (Howard et al. 1991).

Biodegradation: the pseudo first-order degradation rate constant of 0.14 day^{-1} corresponding to a half-life of 5 days in a model ecosystem (Sanborn et al. 1975); degradation half-life of 8 days by microorganisms isolated from soil or waste water at 30°C (Kurane et al. 1977); estimated rate constant of $(1.57 \pm 0.17) \times 10^{2}$ min^{-1} from retention times of reverse phase chromatography (Urushigawa & Yonezawa 1979); microbial degradation rate constant 2×10^{-4} h^{-1} in an aquatic environment (Wolfe et al. 1980a); significant degradation with gradual adaptation from 7 to 21 days in an aerobic environment with a rate of < 0.5 day^{-1} (Tabak et al. 1981; quoted, Mills et al. 1982); biodegradation rate constant of 7.4×10^{-9} $ml \cdot cell^{-1} \cdot day^{-1}$ at 30°C in water (Scow 1982); aqueous aerobic half-life of 168-672 hours, based on unacclimated and acclimated aqueous screening test data; aqueous anaerobic half-life of 4320-8760 hours, based on acclimated anaerobic screening test data (Howard et al. 1991);.

Biotransformation: 3.1x10-10 $ml \cdot cell^{-1} \cdot h^{-1}$ for bacterial transformation in water (Wolfe et al. 1980a; quoted, Mabey et al. 1982); microbial transformation of $(3.7 \pm 0.6) \times 10^{-13}$ $L \cdot organism^{-1} \cdot h^{-1}$ in natural aquatic systems (Steen et al. 1979; quoted, Steen 1991).

Bioconcentration, Uptake (k_1) and Elimination (k_2) Rate Constants or Half-Lives:

Common Name: Bis(2-ethylhexyl) phthalate
Synonym: bis(2-ethylhexyl) phathalic acid ester, di(2-ethylhexyl)orthophthalate, di(2-ethylhexyl) phthalate, di-*sec*-octyl phthalate, 2-ethylhexyl phthalate, 1,2-benzenedicarboxylic acid bis(2-ethylhexyl) ester, DEHP
Chemical Name: bis(2-ethylhexyl) phthalate, di-*sec*-octyl phthalate, 2-ethylhexyl phthalate
CAS Registry No: 117-81-7
Molecular Formula: o-$C_6H_4(COOCH_2CH(C_2H_5)C_4H_9)_2$
Molecular Weight: 390.54
Melting Point (°C):
- -50.0 (Patty 1963; Fishbein & Albro 1972; Callahan et al. 1979; Mabey et al. 1982; Dean 1985; Howard 1989)
- -55 (Verschueren 1983)

Boiling Point (°C):
- 386.9 (at 5mm Hg, Fishbein & Albro 1972; Peakall 1975; Mabey et al. 1982)
- 385 (Verschueren 1983)
- 384 (Dean 1985; Riddick et al. 1986)

Density (g/cm^3 at 20°C):
- 0.9850 (Fishbein & Albro 1972)
- 0.983 (Neely & Blau 1985)
- 0.9843 (Riddick et al. 1986)

Molar Volume (cm^3/mol):
- 396.1 (calculated-density, Stephenson & Malanowski 1987)
- 524.8 (calculated-LeBas method)

Molecular Volume (A^3):
Total Surface Area, TSA (A^2):
Heat of Fusion, ΔH_{fus}, (kcal/mol):
Entropy of Fusion, ΔS_{fus} (cal/mol·K or e.u.):
Fugacity Ratio at 25°C, F: 1.0

Water Solubility (g/m^3 or mg/L at 25°C):
- 0.285 (shake flask-nephelometry on technical grade DEHP, Hollifield 1979; quoted, DeFoe et al. 1990)
- 100 (20°C, Fishbein & Albro 1972)
- 50.0 (quoted from Monsanto Chemical Co. data sheets, Peakall 1975)
- 0.60 (Branson 1978)
- 100 (quoted from Metcalf & Lu 1973, Hollifield 1979; Graten & Trabalka 1983)
- 2.49 (Neely 1979; quoted, Lyman 1985; Elzerman & Coates 1987)
- 0.40 (shake flask-GC, Wolfe et al. 1979, 1980a; quoted, Callahan et al. 1979; Mabey et al. 1982; Geyer et al. 1982; Zaroogian et al. 1985)
- 0.40 $\pm$ 0.05 (shake flask-GC, Wolfe et al. 1980b)
- 0.40 (shake flask-GC, Wolfe et al. 1980a; quoted, DeFoe et al. 1990)
- 1.16 (solubility in 35 liter instant ocean solution, Giam et al. 1980)
- 0.60 (Kenaga & Goring 1980; quoted, Geyer et al. 1980)
- 0.40 (Wolfe et al. 1980; quoted, Geyer et al. 1982)
- 0.047 (quoted from OECD 1979/80, Klöpffer et al. 1982)
- 0.041 (20°C, shake flask-UV, Leyder & Boulanger 1983)
- 0.285 (24°C, Verschueren 1983)

0.34 ± 0.04 (shake flask-HPLC/UV, Howard et al. 1985; quoted, Howard 1989; DeFoe et al. 1990)
0.40, 0.041 (quoted, Howard et al. 1985)
2.49 (recommended, Neely & Blau 1985)
2.49, 3.36 (quoted, estimated-K_{OW} and M.P., Lyman 1985)
1.31 (quoted, Karickhoff 1985)
400 (quoted from Callahan et al. 1979, misquoted in unit μg/L? Neuhauser et al. 1985)
0.04-1.3 (quoted, Russell et al. 1985)
0.40 (Russell & McDuffie 1986)
< 100 (quoted, Riddick et al. 1986)
0.603 (quoted, Isnard & Lambert 1989)
0.27 (centrifugation, DeFoe et al. 1990)
0.36 (best estimate by turbidity inflection, DeFoe et al. 1990)
$9.8x10^{-6}$ to 0.633 (quoted lit. values, Sabljic et al. 1990)
0.40 (selected, Aichinger et al. 1992)

Vapor Pressure (Pa at 25°C):
$2.54x10^{-5}$ (effusion method, average extrapolated-Antoine eqn., Small et al. 1948)
$1.90x10^{-5}$ (Small et al. 1948; quoted, Hinckley et al. 1990)
$8.26x10^{-6}$ (measured by dew-point and tensimeter methods, temp range 120-225°C, extrapolated from Antoine eqn., Werner 1952; quoted, Giam et al. 1980; Hinckley et al. 1990)
$1.90x10^{-5}$ (Jordan 1954; quoted, Klöpffer et al. 1982)
$2.70x10^{-5}$ (20°C, calculated, Weast 1972-73; quoted, Mabey et al. 1982)
$2.70x19^{-5}$ (20°C, Hirzy et al. 1978; quoted, Callahan et al. 1979; Karickhoff 1985; Lyman 1985; Neely & Blau 1985)
$3.17x10^{-5}$ (quoted, Branson 1978)
$6\text{-}13x10^{-6}$ (20°C, quoted lit. values of OECD 1979, Dobbs & Cull 1982)
$6.0x10^{-6}$, $1.3x10^{-5}$ (20°C, vapor pressure balance, gas saturation method, OECD 1981)
$5.50x10^{-6}$ (20°C, estimated-evaporation rate, Dobbs & Cull 1982)
$2.67x10^{-5}$; 0.00293, $1.87x10^{-6}$, $7.47x10^{-4}$ (quoted; estimation-structure based methods, Tucker et al. 1983)
$9.50x10^{-5}$ (20°C, quoted from OECD 1981 Guidelines, Dobbs et al. 1984)
$8.6(\pm 6.6)x10^{-4}$ (gas saturation-HPLC/UV, Howard et al. 1985; quoted, Howard 1989; Hinckley et al. 1990)
$2.63x10^{-5}$ (recommended, Neely & Blau 1985, quoted, Lyman 1985; Mackay 1985)
70, 700 (quoted values, Riddick et al. 1986)
$5.08x10^{-5}$ (extrapolation, Antoine eqn., Stephenson & Malanowski 1987)
$(1.20\text{-}2.40)x10^{-5}$ (quoted from OECD interlab. studies, Hinckley et al. 1990)
$1.90x10^{-5}$ (GC-RT, Hinckley et al. 1990)

Henry's Law Constant (Pa·m³/mol):
0.0446 (calculated-P/C, Wolfe et al. 1980a)
1.114 (calculated, Lyman et al. 1982)
0.0304 (20-25°C, calculated-P/C, Mabey et al. 1982)
0.0263, 3.14 (calculated-P/C, calculated-group contribution, Tucker et al. 1983)

0.00415 (estimated-P/C, Lyman 1985)
0.101 (estimated as per Hine & Mookerjee 1975, Lyman 1985)
0.00413 (calculated-P/C, Mackay 1985, Neely & Blau 1985)
0.0297 (quoted from WERL Treatability Data, Ryan et al. 1988)
1.494 (calculated-P/C, Meylan & Howard 1991)
1.197 (estimated-bond contribution, Meylan & Howard 1991)

Octanol/Water Partition Coefficient, log K_{OW}:
3.58 (Lu & Metcalf 1975; quoted, Branson 1978)
4.20 (Mayer 1976; quoted, Veith et al. 1979; Mackay 1982; Graten & Trabalka 1983; Zaroogian et al. 1985; Isnard & Lambert 1988, 1989; Schüürmann & Klein 1988)
4.20 (estimated from solubility, Branson 1978)
5.30 (Hirzy et al. 1978; quoted, Callahan et al. 1979; Karickhoff 1985; Lyman 1985; Neely & Blau 1985)
8.73 (calculated, Wolfe et al. 1979; quoted, Callahan et al. 1979; Neuhauser et al. 1985; Ryan et al. 1988)
5.11 (quoted from OECD shake flask value, Geyer et al. 1982, 1984; Davies & Dobbs 1984; Howard et al. 1985; Howard 1989)
3.98 (Kenaga & Goring 1980; selected, Govind et al. 1991)
9.61 (calculated, Mabey et al. 1982)
9.70 (calculated as per Rekker 1977, Harnish et al. 1983)
8.15 (HPLC-extrapolated from ref. substances, Harnish et al. 1983)
7.86 (HPLC-extrapolated from alkylbenzenes, Harnish et al. 1983; quoted, Klein et al. 1988; Brooke et al. 1990)
5.03 (shake flask average, OECD/EEC lab. comparison tests, Harnish et al. 1983)
9.64 (quoted lit. calculated value, Leyder & Boulanger 1983)
7.94 (HPLC-RT, Howard et al. 1985)
5.11, 4.66-5.45 (shake flask method: mean, range, OECD 1981)
5.3-8.7 (quoted, Russell et al. 1985)
5.30 (recommended, Neely & Blau 1985; quoted, Karickhoff 1985; Elzerman & Coates 1987)
4.31, 5.27, 4.90, 4.98 (estimated-K_{OW}, Karickhoff 1985)
4.24, 5.06 (estimated-S, Karickhoff 1985)
5.10 (best estimated value, Karickhoff 1985)
5.03 (THOR 1986; quoted, Schüürmann & Klein 1988)
8.66 (calculated-CLOGP 3.42 of Leo 1986, Schüürmann & Klein 1988)
7.80, 8.90 (HPLC-RT, TLC-RT, Klein et al. 1988)
7.453 ± 0.061 (slow-stirring-GC, De Bruijn et al. 1989; quoted, Kollig 1993)
4.20 (shake flask, quoted from Klein et al. 1988, Brooke et al. 1990)
4.20, 5.22 (quoted, shake flask method, Brooke et al. 1990)
7.86, 9.68 (HPLC method, calculated, Brooke et al. 1990)
7.137 ± 0.153 (stir-flask method by BRE, Brooke et al. 1990)
7.453 ± 0.061 (stir-flask method by RITOX, Brooke et al. 1990)
8.66 (calculated-CLOGP, DeFoe et al. 1990)
4.20-7.68 (quoted lit. values, Sabljic et al. 1990)
8.66 (calculated, Matthiessen et al. 1993)

7.88 (recommended, Sangster 1993)

Bioconcentration Factor, log BCF:

2.11 (fish, Metcalf et al. 1973)

5.03 (mosquito larvae, Metcalf et al. 1973)

4.13, 3.49, 3.72, 3.36, 2.36 (scud, midge larvae, waterflea, mayfly, sowbug, Sanders et al. 1973)

3.56, 2.54, 2.62, 2.76, 2.66 (scud, midge, waterflea, mayfly, fathead minow, 14 day exposure, Mayer et al. 1973)

3.14, (fathead minnows, 28 day exposure, Mayer et al. 1973)

2.95-2.19 (fathead minnows, 56 day exposure, Mayer 1976)

2.76 (fathead minnows, 56 day exposure, Mayer 1976; quoted, Davies & Dobbs 1984)

2.93 (fathead minnows, Veith et al. 1979; quoted, Zaroogian et al. 1985; Schüürmann & Klein 1988; Howard 1989)

2.90, 2.15 (fathead minnows, quoted, Branson 1978)

2.06 (bluegill sunfish, Barrows et al. 1980; quoted, Howard 1989)

-1.92 (adipose tissue of female Albino rats, Geyer et al. 1980)

3.95 (bacteria, calculated-K_{OW}, Wolfe et al. 1980a)

0.84-1.05, 1.01-1.22, 1.03-1.13 (oyster, shrimp, fish, Wofford et al. 1981)

1.04 (oyster, Wofford et al. 1981; quoted, Zaroogian et al. 1985)

2.93, 2.88 (fish: quoted, calculated-K_{OW}, Mackay 1982)

2.32 (*Daphnia maga*, Brown & Thompson 1982a)

2.72. 2.62, 3.38 (quoted, Brown & Thompson 1982a)

3.37, 3.42, 3.40 (mussel *mytilus edulis*, Brown & Thompson 1982b; quoted, Geyer et al. 1982)

8.36 (microorganisms-water, calculated-K_{OW}, Mabey et al. 1982)

2.49, 2.11 (fish: flowing water, microcosm conditions, Graten & Trabalka 1983)

3.73, 3.57 (alga *chlorella*, calculated-K_{OW}, Geyer et al. 1984)

2.80 (sheephead minnows, predicted-pharmacokinetic model, Karara & Hayton 1984)

2.66, 3.73, 3.48 (golden orfe, algae, activated sludge, Freitag et al. 1982)

1.60, 3.73, 3.48 (golden ide, algae, activated sludge, Freitag et al. 1985)

2.96, 2.87 (oyster, estimated values, Zaroogian et al. 1985)

2.96, 2.87 (sheephead minnows, estimated values, Zaroogian et al. 1985)

2.76 (quoted, Isnard & Lambert 1989)

0.204-1.71 (rainbow trout, BCF to decline as body wt. increased, Tarr et al. 1990)

2.94 (fish, highest fish BCF, Matthiessen et al. 1993)

Sorption Partition Coefficient, log K_{OC}:

4.0-5.0 (soil, calculated values, Kenaga 1980; Wolfe et al. 1980a; quoted, Howard 1989)

4.756 (sediment-water, calculated-K_{OW} and S, Wolfe et al.1980a)

9.301 (sediment-water, calculated-K_{OW}, Mabey et al. 1982)

4.31, 5.27, 4.90, 4.98 (estimated-K_{OW}, Karickhoff 1985)

4.24, 5.06 (estimated-S, Karickhoff 1985)

5.10 (best estimate, expected at low sediment concn. of $< 10^{-4}$/mL, Karickhoff 1985)

5.0 (soil/sediment, Neely & Blau 1985; quoted, Elzerman & Coates 1987)

4.94 (Broome County soil in New York, shake flask-GC, Russell & McDuffie 1986; quoted, Howard 1989)

4.94, 5.22 (soil, quoted, calculated-χ and fragment contribution, Meylan et al. 1992)
2.60 (activated carbon, calculated-χ , Blum et al. 1994)
7.13 (calculated-K_{OW}, Kollig 1993)

Half-Lives in the Environment:

Air: half-life of 2.9-29 hours, based on estimated rate constant for the reaction with hydroxyl radicals in air (Howard et al. 1991); atmospheric transformation lifetime was estimated to be < 1 day (Kelly et al. 1994).

Surface water: biodegradation half-life of about 4 to 5 weeks in river water by naturally occurring, mixed microbial populations (Saeger & Tucker 1976); half-life of 5 days in a model ecosystem (Verschueren 1983); half-life of 5.25 days in an acclimated shake flask CO_2 evolution biodegradation experiment (Sugatt et al. 1984); half-life of 120-550 hours, based on unacclimated aqueous aerobic biodegradation half-life (Howard et al. 1991).

Ground water: 240-9336 hours, based on unacclimated aqueous aerobic and anaerobic biodegradation half-life (Howard et al. 1991).

Sediment:

Soil: degradation half-life of 1.3 days by microorganisms isolated from soil or waste water at 30°C (Kurane et al. 1977); biodegradation half-lives <20 days under anerobic conditions in a garden soil (Shanker et al. 1985); 10-50 days via volatilization subject to plant uptake from the soil (Ryan et al. 1988); overall half-life 120-550 hours, based on unacclimated aqueous aerobic biodegradation half-life (Howard et al. 1991).

Biota: elimination and degradation half-liives were < 4 days after 7 days for waterfleas *Daphnia magna* and 7 days for fathead minnows (Mayer & Sanders 1973); fish elimination half-life 12.2 days (fathead minnows *Pimephales promelas*, Mayer 1976); depuration half-life of 3 days (in tissues of bluegill sunfish continuously for 42-day exposure, Barrows et al. 1980); depuration half-life of 38 days from sheephead minnow (Karara & Hayton 1984).

Environmental Fate Rate Constant and Half-Lives:

Volatilization: calculated evaporation rate of 0.00052 cm/h corresponding to a half-life of 132600 hours for a pond of 1 m deep (Branson 1978).

Photolysis: rate constant for direct photolysis $2x10^{-4}$ h^{-1} in natural water (Wolfe et al. 1980a); half-life in water is estimated to be 143 days (Wolfe et al. 1980a; quoted, Howard 1989); atmospheric half-life of 3500-4800 hours and aqueous half-life of 3500-4800 hours, based on measured rate of aqueous photolysis for DMP (Howard 1991).

Oxidation: the free radical oxidation rate constant of 18 $M \cdot sec^{-1}$ (Wolfe et al. 1980b); photooxidation half-life of 44-584 days in water, based on estimated rate constant for the reaction with hydroxyl radicals in water (Wolfe et al. 1980; quoted, Howard et al. 1991); rate constant << 360 $M^{-1} \cdot h^{-1}$ for singlet oxygen and 7.2 $M^{-1} \cdot h^{-1}$ for peroxy radical (Mabey et al. 1982); photooxidation half-life of 2.9-29 hours in air, based on estimated rate constant for the reaction with hydroxyl radicals in air (Atkinson 1987; quoted, Howard et al. 1991).

Hydrolysis: second order alkaline hydrolysis rate constant of $1.1x10^{-4}$ $M^{-1} \cdot sec^{-1}$ for pH 10-12 at 30°C in water (Wolfe et al. 1979, 1980b; quoted, Callahan et al. 1979);

the half-life for hydrolytic degradation alone in the eutrophic lake ecosystem at steady-state increase from 100 to 2000 years (Wolfe et al. 1080a); a calculated half-life of 2000 years at pH 7 in water (calculated per Radding et al. 1977; quoted, Callahan et al. 1979; Howard 1989; Howard et al. 1991); exptl. rate constant of 9.5×10^{-7} day^{-1} at pH 7 and 25°C while estimated rate constant of 1.46×10^{-6} day^{-1} at pH 7 and 30°C (Neely 1985); first-order hydrolysis half-life of 2000 years at pH 7, based on measured base catalyzed hydrolysis rate constant (Howard et al. 1991).

Biodegradation: approximately calculated rate constant of 0.091 day^{-1} in pond water plus sediment incubated under aerobic conditions based on time for 50% degradation (Johnson & Lulves 1975; quoted, Klečka 1985); estimated biodegradation half-life of 4-5 weeks by naturally occurring, mixed microbial populations in river water (Saeger & Tucker 1976); degradation half-life or 1.3 days by microorganisms isolated from soil or waste water at 30°C (Kurane et al. 1977); rate constant of 0.023 day^{-1} in river water incubated under aerobic conditions (Klečka 1985); rate constant (mineralization rate divided by initial substrate concn.) of 0.028 $\pm$ 0.013 day^{-1} in lake water incubated under aerobic conditions (Subba-Rao et al. 1982; quoted, Klečka 1985); half-lives of 2.0-3.0 weeks, based on river die-away test data (Hattori et al. 1975; Saeger & Tucker 1976; Wolfe et al. 1980; estimated, Howard 1989); half-life of 0.8 day in activated sludge (Saeger & Tucker 1976; estimated, Howard 1989); microbial degradation rate constant of 4.2×10^{-12} mL·organism^{-1}·sec^{-1} (Wolfe et al. 1980a); significant degradation with gradual adaptation from 7 to 21 days in an aerobic environment with a rate of $<$ 0.5 day^{-1} (Tabak et al. 1981; quoted, Mills et al. 1982); aqueous aerobic half-life of 120-550 hours, based on grab sample die-away test data (Schouten et al. 1979 and Johnson & Lulves 1975; estimated, Howard et al. 1991); aqueous anaerobic half-life of 980-9336 hours, based on anaerobic die-away test data (Howard et al. 1991) and anaerobic aqueous screening studies (Horowitz et al. 1982 and Shelton et al. 1984; estimated, Howard et al. 1991); mean rate constant of 0.136 day^{-1} corresponding to 5.25 days in shake flask biodegradation experiment (Sugatt et al. 1984); more than 90% of initial 3.3 mg/liter will be degraded in activated sludge systems in 2-5 days (O'grady et al. 1985); greater than 50% loss in microbial degradation in less than 20 days under aerobic conditions, very low under anaerobic conditions in garden soil (Shanker et al. 1985); readily metabolized in uninoculated Erie slit loam with the absence of nonaqueous-phase liquids (NAPLs) for about 20 days (Efroymson & Alexander 1994).

Biotransformation: 4.2×10^{-12} ml·cell^{-1}·h^{-1} for bacterial transformation in water (Wolfe et al. 1980a; quoted, Mabey et al. 1982); microbial transformation of $(4.2 \pm 0.7) \times 10^{-15}$ L·organism^{-1}·h^{-1} in natural aquatic systems (Steen et al. 1979; quoted, Steen 1991).

Bioconcentration, Uptake (k_1) and Elimination (k_2) Rate Constants or Half-Lives: half-life of 7 days for elimination of DEHP from fathead minnows (Mayer et al. 1973); fish elimination half-life from 6.2 to 18.3 days with a mean value of 12.2 days after 56 day exposure period (fathead minnows, Mayer 1976); uptake rate constant by fish 76 mL/g·h (Branson 1978); depuration half-life of 72 hours (tissues of bluegill sunfish, Barrows et al. 1980).

Common Name: Butyl benzyl phthalate
Synonym: BBP, benzyl butyl phthalate
Chemical Name: butyl benzyl phthalate, benzyl butyl phthalate
CAS Registry No: 85-68-7
Molecular Formula: $C_6H_5COOCH_2C_6H_4COOC_4H_9$
Molecular Weight: 312.36
Melting Point (°C):
-35.0 (Callahan et al. 1979; Mabey et al. 1982; Howard 1989)
<-35 (Verschueren 1983)
Boiling Point (°C):
377.0 (Callahan et al. 1979; Mabey et al. 1982)
370.0 (Verschueren 1983; Howard 1989; Banerjee et al. 1990)
Density (g/cm^3 at 20°C):
Molar Volume (cm^3/mol):
384.2 (LeBas method)
Molecular Volume (A^3):
Total Surface Area, TSA (A^2):
Heat of Fusion, ΔH_{fus}, (kcal/mol):
Entropy of Fusion, ΔS_{fus} (cal/mol·K or e.u.):
Fugacity Ratio at 25°C, F: 1.0

Water Solubility (g/m^3 or mg/L at 25°C):
2.90 (Hirzy et al. 1978; quoted, Callahan et al. 1979; Mabey et al. 1982)
0.71 (shake flask-nephelometry, practical grade, Hollifield 1979)
2.9 ± 1.2 (shake flask-GC, Gledhill et al. 1980; quoted, Davies & Dobbs 1984)
40.2 (shake flask-LSC, Veith et al. 1980)
82.2 (calculated as per Hansch et al. 1968, Veith et al. 1980)
2.82 (20°C, shake flask-UV, Leyder & Boulanger 1983)
2.90 ± 1.2 (Verschueren 1983)
2.69 ± 0.15 (shake flask-HPLC/UV, Howard et al. 1985; quoted, Howard 1989)
2.90, 2.82 (quoted, Howard et al. 1985)
2.88 (quoted, Isnard & Lambert 1988)
2.69 (Marchini et al. 1988; quoted, Müller & Klein 1992)

Vapor Pressure (Pa at 25°C):
0.00115 (20°C, Gledhill et al. 1980)
0.00799 (calculated using Truton's rule, Mabey et al. 1982)
0.00115 (Petrasek et al. 1983; quoted, Howard 1989)
0.00115 (20°C, Verschueren 1983)
0.0011, 0.00133 (quoted, calculated-solvatochromic parameters and UNIFAC, Banerjee et al. 1990)

Henry's Law Constant (Pa·m^3/mol):
0.132 (calculated-P/C, Wolfe et al. 1980a)

0.132 (calculated, Lyman et al. 1982; quoted, Howard 1989)
0.841 (calculated-P/C, Mabey et al. 1982)
0.104 (quoted from WERL Treatability Data, Ryan et al. 1988)

Octanol/Water Partition Coefficient, log K_{OW}:

4.80 (Hirzy et al. 1978; quoted, Callahan et al. 1979)
5.80 (calculated as per Leo et al. 1971, Callahan et al. 1979; quoted, Ryan et al. 1988)
5.80 (calculated, Wolfe et al. 1979; quoted, Callahan et al. 1979)
4.77 (shake flask-GC analysis of both phases, Gledhill et al. 1980; quoted, Howard 1989)
4.05 (bluegill sunfish, shake flask-LSC, Veith et al. 1980; Veith & Kosian 1982; quoted, Davies & Dobbs 1984; Peddicord 1987; Saito et al. 1992)
4.75 (estimated-RPHPLC, Veith et al. 1980)
3.97; 4.44, 4.08 (correlated HPLC-k'; calculated-π const., f-const., McDuffie 1981)
4.11, 3.23-4.45 (shake flask method: mean, range, OECD 1981)
5.56 (calculated, Mabey et al. 1982)
4.91 (20°C, shake flask-UV, Leyder & Boulanger 1983)
5.63 (quoted lit. calculated value, Leyder & Boulanger 1983)
4.78 (quoted, Verschueren 1983)
3.57 (HPLC-RT, Howard et al. 1985)
4.91 (calculated, Hansch & Leo 1985; quoted, Howard 1989)
4.05 (quoted, Isnard & Lambert 1988; quoted, Banerjee & Baughman 1991)
5.80 (selected, Govind et al. 1991)
4.87 (calculated-CLOGP, Müller & Klein 1992)
5.06 (quoted, Kollig 1993)

Bioconcentration Factor, log BCF:

2.74-3.34 (calculated-K_{OW}, Veith et al. 1979, 1980)
2.82 (bluegill sunfish, Barrows et al. 1980; quoted, Howard 1989)
2.71 (calculated-K_{OW}, Gledhill et al. 1980)
2.82 (bluegill sunfish, quoted, Gledhill et al. 1980; quoted, Verschueren 1983)
2.19-2.95 (fathead minnows, quoted, Gledhill et al. 1980)
2.89 (bluegill sunfish, Veith et al. 1980; Veith & Kosian 1982; quoted, Davies & Dobbs 1984; Saito et al. 1992)
3.90 (calculated-K_{OW}, Wolfe et al. 1980a)
4.67 (microorganisms-water, calculated-K_{OW}, Mabey et al. 1982)
3.57 (sediment, calculated-K_{OW}, Pavlou & Weston 1983, 1984; quoted, Peddicord 1987)
2.89 (quoted, Isnard & Lambert 1988; quoted, Banerjee & Baughman 1991)
2.72 (calculated-K_{OW} & $S_{OCTANOL}$, Banerjee & Baughman 1991)

Sorption Partition Coefficient, log K_{OC}:

1.83-2.54 (soil, batch equilibration-GC, Gledhill et al. 1980; quoted, Howard 1989)
3.81 (calculated-K_{OW}, Wolfe et al. 1980a)

5.23 (sediment-water, calculated-K_{OW}, Mabey et al. 1982)
4.23 (Broome County soil in New York, Russell & McDuffie 1986; quoted, Howard 1989; Meylan et al. 1992; Kollig 1993)
4.23, 3.97 (soil, quoted, calculated- χ and fragment contribution, Meylan et al. 1992)
2.60 (activated carbon, calculated- χ , Blum et al. 1994)

Half-Lives in the Environment:

Air: half-life of 1.5 day in the atmosphere (GEMS 1984; quoted, Howard 1989); 6-60 hours, based on estimated rate data for the reaction with hydroxyl radicals in air (Atkinson 1987; quoted, Howard et al. 1991).

Surface water: biodegradation half-life was estimated to be about 2 hours by naturally occurring, mixed microbial populations in river water (Saeger & Tucker 1976); the hydrolytic half-lives ranging from 80 to 4000 days at pH 8 and 30°C (Wolfe et al. 1980a; quoted, Gledhill et al. 1980); half-life of 2 days for river water, < 4 days for lake water-sediment microcosm, and < 100 days for chemical degradation (hydrolysis) (Gledhill et al. 1980); biodegradation half-life of 19.4 days in an acclimated shake flask CO_2 evolution test (Sugatt et al. 1984); overall half-life of 24-168 hours, based on unacclimated river die-away test (Howard et al. 1991).

Ground water: 48-4320 hours, based on estimated unacclimated aqueous aerobic and anaerobic biodegradation half-lives (Howard et al. 1991).

Sediment:

Soil: degradation half-life of 3 days by microorganisms isolated from soil or waste water at 30°C (Kurane et al. 1977); 10-50 days via volatilization subject to plant uptake from the soil (Ryan et al. 1988); 24-168 hours, based on estimated unacclimated aqueous aerobic biodegradation half-life (Howard et al. 1991).

Biota: depuration half-life <24 and >48 hour (tissues of bluegill sunfish in 21-day exposure, Barrows et al. 1980) and half-life of < 2 days from bluegill sunfish (Gledhill et al. 1980; quoted, Verschueren 1983).

Environmental Fate Rate Constant and Half-Lives:

Volatilization:

Photolysis: direct photolysis (near surface) rate constant of $2x10^{-4}$ h^{-1} in natural water (Wolfe et al 1980a); photodegradation half-life is more than 100 days (Gledhill et al. 1980; quoted, Verschueren 1983; Howard 1989).

Oxidation: the free radical oxidation rate constant of 18 $M \cdot sec^{-1}$ for reaction with peroxy radicals (Wolfe et al. 1980a); rate constant $<<360$ $M^{-1} \cdot h^{-1}$ for singlet oxygen and 280 $M^{-1} \cdot h^{-1}$ for peroxy radical (Mabey et al. 1982); photooxidation half-life of 6-60 hours, based on estimated rate data for the reaction with hydroxyl radicals in air (Atkinson 1987; quoted, Howard et al. 1991).

Hydrolysis: the alkaline hydrolysis rate constant was estimated to be 38 $M^{-1} \cdot h^{-1}$; hydrolytic half-lives ranging from 80 to 4000 days at pH 8 and 30°C (Wolfe et al. 1980a; quoted, Gledhill et al. 1980); chemical degradation (hydrolysis) half-life is more than 100 days (Gledhill et al. 1980; quoted, Verschueren 1983; Howard 1989).

Biodegradation: degradation half-life was estimated to be approximately 2 days in river water and 2 hours in activated sludge (Saeger & Tucker 1976); degradation half-life of 3 days by microorganisms isolated from soil or waste water at 30°C (Kurane et al. 1977); microbial degradation rate constant of $2.9x10^{-8}$ mL·organism^{-1}·sec^{-1} (Wolfe et al. 1981a); 0-24% mineralization in >8 weeks in municipal digested sludge (Horowitz et al. 1982); primary degradation accounted for >95% loss in 7 days with an initial concentration of 1.0 mg/L in a lake water microcosm with half-life of less than 4 days and 100% primary degradation was observed after 9 days while the half-life was 2.0 days (Gledhill et al. 1980; quoted, Verschueren 1983; Howard 1989); significant degradation with rapid adaptation within 7 days in an aerobic environment with a rate of >0.5 day^{-1} (Tabak et al. 1981; quoted, Mills et al. 1982); degradation rate constant of 0.043 day^{-1} corresponding to a half-life of 19.4 days in a shake flask biodegradation experiment (Sugatt et al. 1984); greater than 90% degraded within 40 days in digested sludge (Shelton et al. 1984); 99% degraded in activated sludge systems in 48 hours (O'grady et al. 1985); anaerobic digestion of sludge with first-order rate constant of $6.5x10^{-3}$ h^{-1} and half-life of 107 hours (Ziogou et al. 1989); aqueous aerobic half-life of 24-168 hours, based on unacclimated river die-away test and aqueous anaerobic half-life of 672-4320 hours, based on unacclimated anaerobic screening test data (Howard et al. 1991).

Biotransformation: estimated rate constant of $3x10^{-9}$ mL·cell^{-1}·h^{-1} for bacterial transformation in water (Mabey et al. 1982); microbial transformation of (3.1 ± 0.8)$x10^{-11}$ L·organism^{-1}·h^{-1} in natural aquatic systems (Steen et al. 1979; quoted, Steen 1991).

Bioconcentration, Uptake (k_1) and Elimination (k_2) Rate Constants or Half-Lives: depuration half-life <24 and >48 hours from tissues of bluegill sunfish (Barrows et al. 1980).

6.2 Summary Tables and QSPR Plots

Table 6.1 Physical-chemical properties of esters at 25 °C.

Compound	CAS no.	formula	MW	mp, °C	bp, °C	fugacity ratio, F at 25 °C	molar volume, V_M, cm³/mol V_M *	V_M (a)	V_I (a)	LeBas	surface area TSA, Å² TSA (b)	TSA (c)
Alphatic esters:												
methyl formate	107-31-3	HCO_2CH_3	60.052	-99	31.75	1	62.140	61.6	32.5	62.6		
ethyl formate	109-94-4	$HCO_2C_2H_5$	74.079	-79.6	54.31	1	80.934	80.4	42.4	86.4	242.1	
propyl formate	110-74-4	$C_4H_8O_2$	88.106	-92.9	80.82	1	97.939			110.8		
butyl formate	592-84-1	$C_5H_{10}O_2$	102.13	-91.9	106.1	1	115.154			133		
isobutyl formate	542-55-2	$C_5H_{10}O_2$	102.13	-95.8	98.07	1	114.900			133		
methyl acetate	79-20-9	$(CH_3)_2CO_2$	74.08	-98.05	56.87	1	79.836	79.4	42.4	84.8	244.9	204.3
vinyl acetate	108-05-4	$C_4H_6O_2$	86.09	-92.8	72.5	1	92.4			96.2		
allyl acetate	591-87-7	$C_5H_8O_2$	100.18		103.5	1	0.1079					
ethyl acetate	141-78-6	$C_4H_8O_2$	88.104	-41.5	77.11	1	98.490	97.8	52.1	108.6		233.8
propyl acetate	109-60-4	$C_5H_{10}O_2$	102.13	-95	101.54	1	115.659	115	62.2	133		
butyl acetate	123-86-4	$C_6H_{12}O_2$	116.16	-73.5	126.06	1	132.542	131.6	71.6	155.2	344.6	
isobutyl acetate	110-19-0	$C_6H_{12}O_2$	116.16	-98.85	116.6	1	133.871			155.2		
pentyl acetate	628-63-7	$C_7H_{14}O_2$	130.186	-70.8	149.2	1	149.313	148.7	81.3	177.4		
isopentyl acetate	123-92-2	$C_7H_{14}O_2$	130.186	-78.5	142.1	1	150.261			177.4		
hexyl acetate	142-92-7	$C_8H_{16}O_2$	144.213	-80.9	170.5	1	166.125			199.6		
cyclohexyl acetate	622-45-7	$C_8H_{14}O_2$	142.2				146.6					
2-ethylhexyl acetate	103-09-3	$C_{10}H_{20}O_2$	172.267		198.6	1	197.599					
benzyl acetate	140-11-4	$C_9H_{10}O_2$	150.177		215.5	1	142.822					
ethyl propionate	105-37-3	$C_5H_{10}O_2$	102.133		99.1	1	115.535	114.6	62.2	130.8	321.5	
ethyl butyrate	105-54-4	$C_7H_{12}O_2$	116.16	-98	121.55	1	132.915	132.2	71.6	133.4		
methyl acrylate	96-33-3	$C_4H_6O_2$	86.09	-75	80.2	1	93.1		90.2	99.6		
ethyl acrylate	140-88-5	$C_5H_8O_2$	100.12	-71.2	99.5	1	109.2			123.4		
methyl methacrylate	80-62-6	$C_5H_8O_2$	100.12	-42.8	100.3	1	106.138			122.2		
ethyl methacrylate	97-63-2	$C_6H_{10}O_2$	114.15		118-119		125.6			145.6		

Compounds	CAS no.	formula	MW	mp, °C	bp, °C	fugacity ratio, F at 25 °C	molar volume, V_M, cm^3/mol V_M from ρ	V_M (a)	V_I (a)	LeBas	surface TSA, Å^2 TSA (b)	TSA (c)
Aromatic esters:												
methyl benzoate	93-58-3	$C_8H_8O_2$	136.15	-12.1	199.5	1		135.6	73.6	151.2		
ethyl benzoate	93-89-0	$C_2H_{10}O_2$	150.177		212.4	1		153.6	83.4	175		
propyl benzoate	2315-68-6	$C_{10}H_{12}O_2$	164.204		231.2	1	160.5			199.4		
phenyl benzoate	93-99-2	$C_{13}H_{10}O_2$	198.24	46	314	0.62				206.8		
benzyl benzoate	120-51-4	$C_{14}H_{12}O_2$	212.25	19.4	323.24	1	189.8		113.9	243.6		
Phthalate esters:												
dimethyl phthalate (DMP)	131-11-3	$C_{10}H_{10}O_4$	194.2	5	283.7	1	162.8	173.6	95.3	206.4		
diethyl phthalate (DEP)	84-66-2	$C_{12}H_{14}O_4$	222.26	-40.5	295	1	198.9	208.9	115.3	254		
dibutyl phthalate (DBP)	84-74-2	$C_{16}H_{22}O_4$	278.35	-35	340	1	265.1	275.8	153.5	347.2		
dioctyl phthalate (DOP)	117-84-0	$C_{24}H_{38}O_4$	390.56	-25		1	399.3			524.8		
butyl benzyl phthalate	85-68-7	$C_{19}H_{20}O_4$	312.39	-35	370	1				384.2		
di(2-ethylhexyl)-phthalate	117-81-7	$C_{24}H_{38}O_4$	390.54	-50	384	1	396.1			524.8		

* molar volume calculated from density
(a) liquid molar volume V_M and intrinsic V_I, Kamlet et al. 1987, 1988
(b) Amidon et al. 1975
(c) Iwase et al. 1985

Table 6.2 Summary and selected physical-chemical properties of esters and phthalate esters at 25 °C

Compounds	selected properties:						Henry's law constant, Pa·m³/mol		
	P^S Pa	P_L Pa	S, g/m³	C^S mol/m³	C_L mol/m³	log K_{OW}	calcd. P/C	exptl. (a)	exptl. (b)
Aliphatic esters:									
methyl formate	78060	78000	230000	3830.01	3830.01	-0.26	20.365		
ethyl formate	32370	32600	118000	1592.89	1592.89	0.23	20.466		
propyl formate	11030	11030	20500	232.67	232.67	0.73			
butyl formate	3840	3840							
isobutyl formate	5343	5400	10000	97.91	97.91		55.150		
methyl acetate	28800	28800	245000	3307.24	3307.24	0.23	8.708	0.65	13.06
vinyl acetate	14100	14100	20000	232.32	232.32	0.73	60.693		
allyl acetate	6130	6131							
ethyl acetate	12600	12600	80800	917.10	917.10		13.739		17.2
propyl acetate	4500	4500	21000	205.62	205.62	1.24	21.885		22.09
butyl acetate	1600	1600	6000	51.65	51.65	1.82	30.976		29.506
isobutyl acetate	2860	2860	6300	54.24	54.24	1.6	52.733		
pentyl acetate	550	550	1700	13.06	13.06	2.42	42.119		35.94
isopentyl acetate	600	600	2000	15.36	15.36	2.13	39.056		
hexyl acetate	670	670	500	3.47	3.47		193.245		
2-ethylhexyl acetate	53	53	98.4	0.57	0.57	3.72	92.786		
phenyl acetate						1.49			
benzyl acetate	189	189	sl. sol.	sl. sol.		1.96			
methyl propionate	11600	11600	62370	707.91	707.91		16.386	17.6	
ethyl propionate	4966	4960	19200	187.99	187.99	1.21	26.384		
methyl butyrate								20.82	
ethyl butyrate	2300	2300	6500	55.96	55.96	1.73	41.103		
methyl pentanoate								32.22	

Compounds	selected properties:						Henry's law constant, Pa·m³/mol		
	P^S Pa	P_L Pa	S, g/m³	C^S mol/m³	C_L mol/m³	log K_{OW}	calcd. P/C	exptl. (a)	exptl. (b)
methyl hexanoate								37.18	
methyl octanoate								79.32	
methyl acrylate	11000	11000	49400	573.82	573.82	0.8	19.170		
ethyl acrylate	5100	5100	15000	149.82	149.82	1.33	34.041		
methyl methacrylate	5100	5000	15600	155.81	155.81	1.38	32.090		
ethyl methacrylate						1.94			
Aromatic:									
methyl benzoate	52.58	52.28	2100	15.42	15.424	2.2	3.389		
ethyl benzoate	24	24	350	2.33	2.331	2.64	10.298		
propyl benzoate	100	17.6				3.18			
phenyl benzoate		0.416				3.59			
benzyl benzoate	0.043	0.043	16	0.075	0.075	3.97	0.570		
Phthalate esters:									
dimethyl phthalate (DMP)	0.22	0.22	4000	20.597	20.597	2.12	0.011		
diethyl phthalate (DEP)	0.22	0.05	1080	4.859	4.859	2.47	0.010		
dibutyl phthalate (DBP)	0.00187	0.002	11.2	0.040	0.040	4.72	0.050		
dioctyl phthalate (DOP)	0.187	2.92×10^{-5}	0.02	5.12×10^{-5}	5.12×10^{-5}	9.87	0.570		
butyl benzyl phthalate	0.00115	0.0011	2.69	0.0086	0.0086	4.91	0.128		
di(2-ethylhexyl)-phthalate	0.00086	0.00002	0.285	0.00073	0.00073	5.11	0.027		

(a) Buttery et al. 1969
(b) Kieckbusch & King 1979

Table 6.3 Suggested half-life classes for esters and phthalate esters at 25°C in various environmental compartments.

Compound	Air class	Water class	Soil class	Sediment class
Aliphatic esters:				
methyl formate	3	3	4	5
ethyl formate	3	3	4	5
butyl formate	3	3	4	5
methyl acetate	3	3	4	5
vinyl acetate	3	3	4	5
allyl acetate	3	3	4	5
ethyl acetate	3	3	4	5
propyl acetate	3	3	4	5
butyl acetate	3	3	4	5
pentyl acetate	3	3	4	5
methyl methacrylate	2	3	4	5
Phthalate esters:				
dimethyl phthalate (DMP)	4	4	5	6
diethyl phthalate (DEP)	4	4	5	6
dibutyl phthalate (DBP)	3	4	5	6
butyl benzyl phthalate	3	4	5	6
di(2-ethylhexyl)-phthalate	3	4	5	6

where,

Class	Mean half-life (hours)	Range (hours)
1	5	< 10
2	17 (~ 1 day)	10-30
3	55 (~ 2 days)	30-100
4	170 (~ 1 week)	100-300
5	550 (~ 3 weeks)	300-1,000
6	1700 (~ 2 months)	1,000-3,000
7	5500 (~ 8 months)	3,000-10,000
8	17000 (~ 2 years)	10,000-30,000
9	~ 5 years	> 30,000

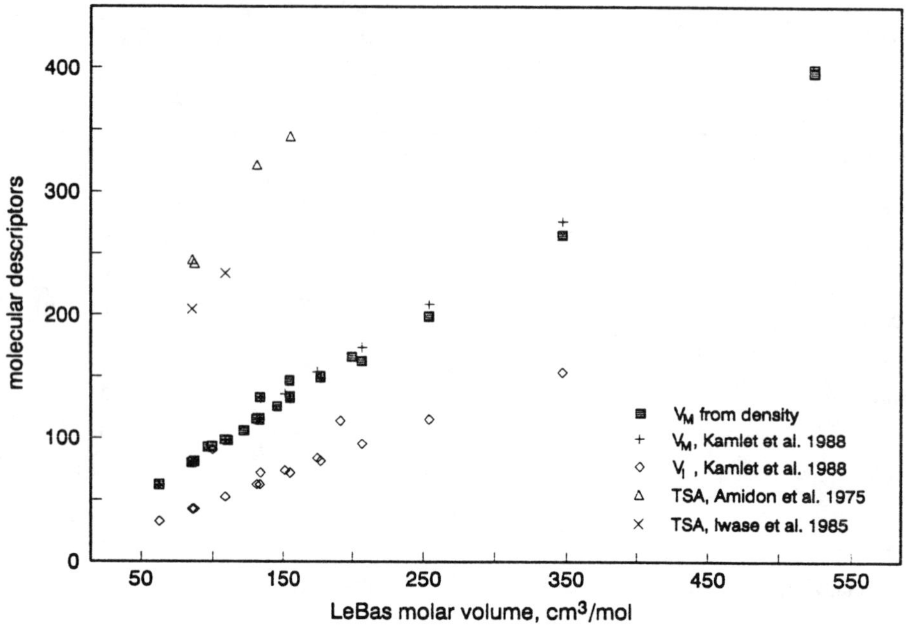

Figure 6.1 Plot of molecular descriptors versus LeBas molar volume

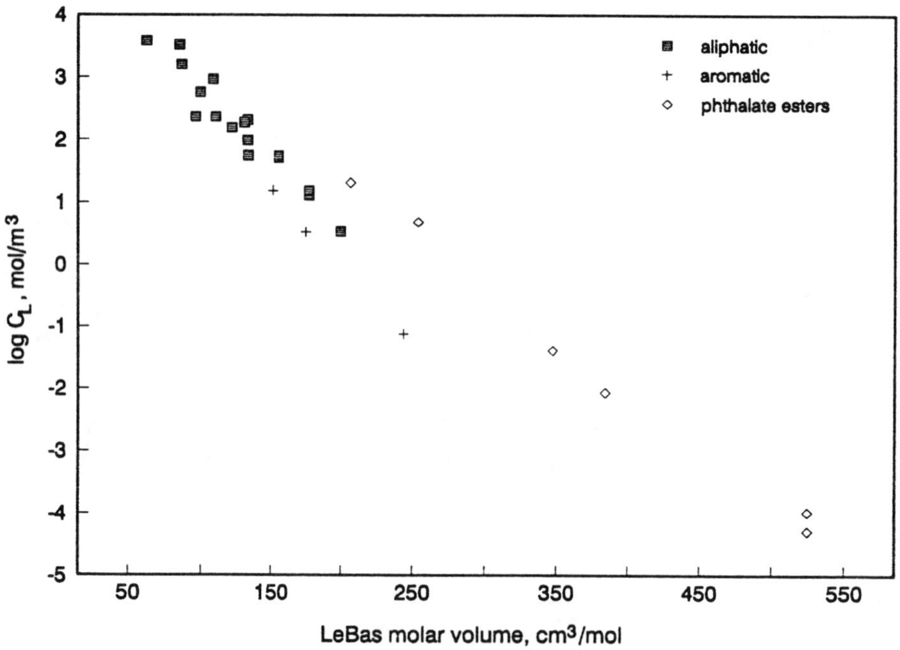

Figure 6.2 Plot of log liquid solubility versus LeBas molar volume

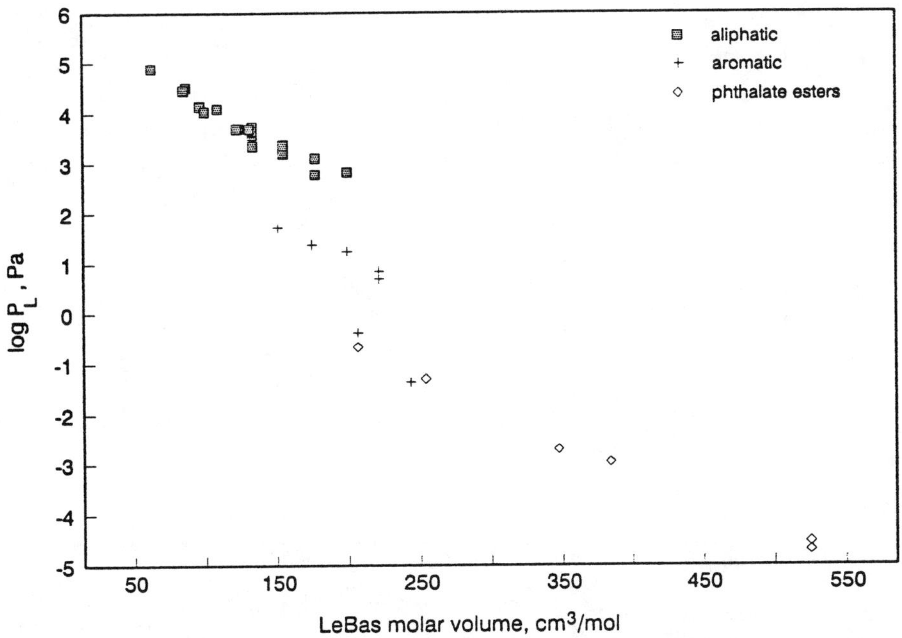

Figure 6.3 Plot of log liquid vapor pressure vs LeBas molar volume

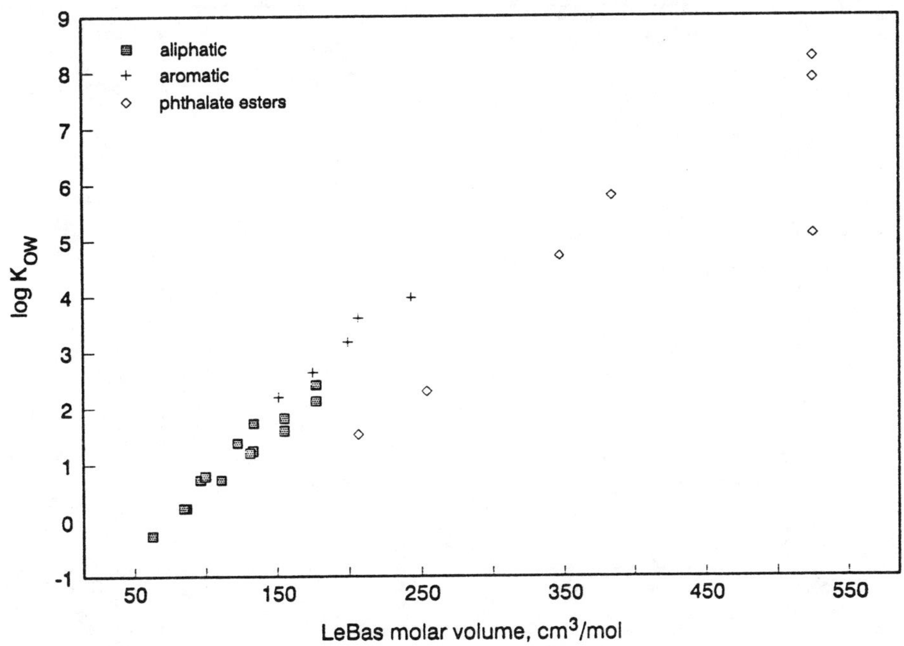

Figure 6.4 Plot of log K_{OW} versus LeBas molar volume

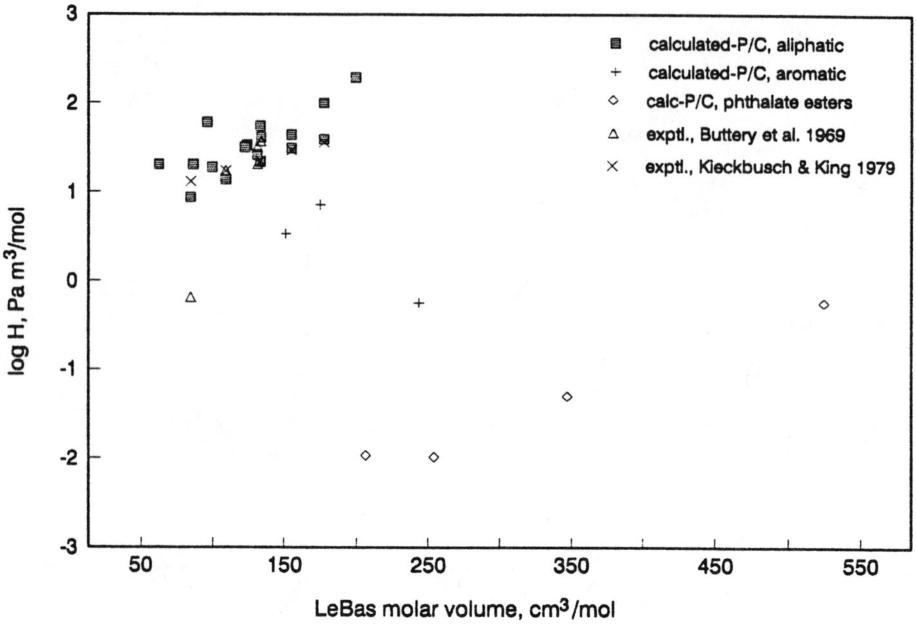

Figure 6.5 Plot of log H vs LeBas molar volume

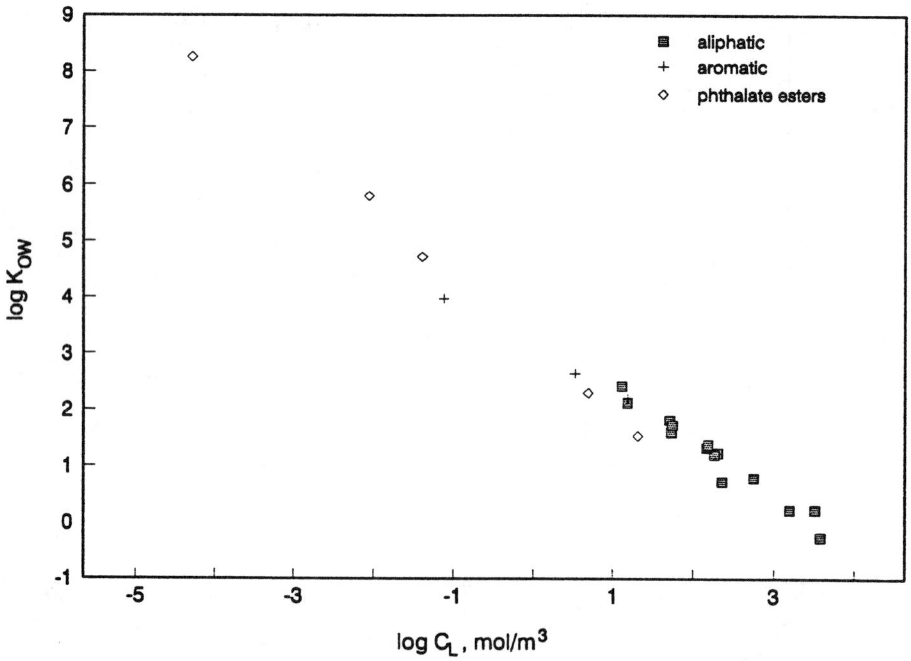

Figure 6.6 Plot of log K_{OW} versus log liquid solubility

6.3 Illustrative Fugacity Calculations: Levels I, II, III

Chemical name: Ethyl acetate

Level I calculation: (six-compartment model)

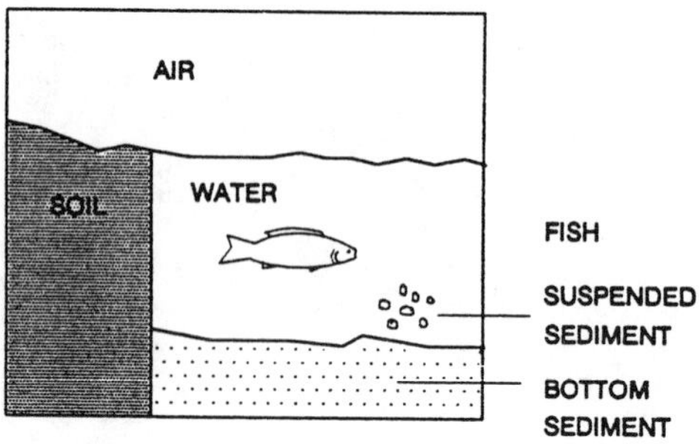

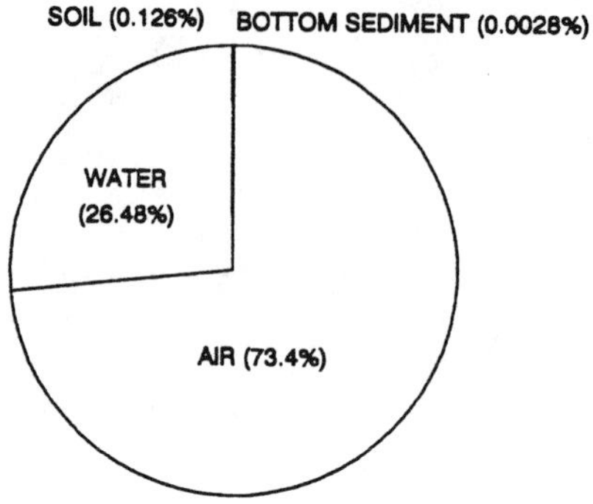

Distribution of mass

Level I calculation:

Physical-chemical properties:	
molecular wt., g/mol	88.10
melting point, C	-41.5
solubility, g/m3	80800
vapor pressure, Pa	12600
log Kow	0.73
fugacity ratio, F	1.0000
dissoc. const. pKa	

Partition coeff.:	
HLC	13.73837
Kaw	5.54E-03
Koc	2.2018
BCF	0.2685
Ksw	1.06E-01
Ksd/w	0.2114
Kssd/w	0.6605
Kar/w	476.190

Compartment	Z	Concentration			Amount	Amount
	mol/m3 Pa	mol/m3	mg/L (or g/m3)	ug/g	kg	%
Air	4.03E-04	8.33E-09	7.34E-07	6.19E-04	73388.22	73.3882
Water	7.28E-02	1.50E-06	1.32E-04	1.32E-04	26483	26.483
Soil	7.69E-03	1.59E-07	1.40E-05	5.83E-06	125.952	0.1260
Biota (fish)	1.95E-02	4.04E-07	3.56E-05	3.56E-05	0.00711	7.11E-06
Suspended sediment	4.81E-02	9.93E-07	8.75E-05	5.83E-05	0.0875	8.75E-05
Bottom sediment	1.54E-02	3.18E-07	2.80E-05	1.17E-05	2.7989	2.80E-03
	Total				100000	100

f = 2.065E-05 Pa

Chemical name: Ethyl acetate

Level II calculation: (six-compartment model)

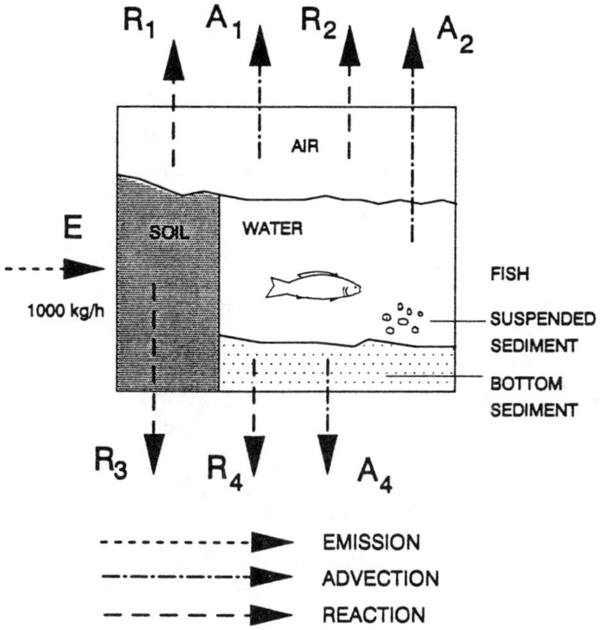

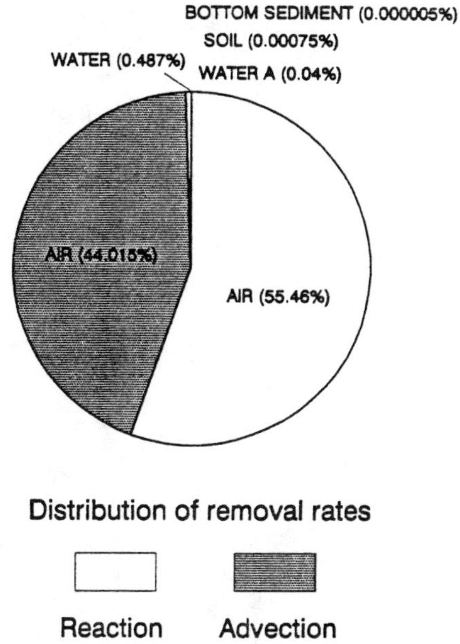

Level II Calculation:

Compartment	Half-life h	D Values Reaction mol/Pa.h	D Values Advection mol/Pa.h	Conc'n mol/m3	Loss Reaction kg/h	Loss Advection kg/h	Removal %
Air	55	5.08E+08	4.03E+08	4.13E-09	457.936	363.441	82.138
Water	55	1.83E+08	1.46E+07	7.44E-07	165.251	13.115	17.837
Soil	170	2.82E+05		7.87E-08	0.2543		0.0254
Biota (fish)				2.00E-07			
Suspended sediment				4.92E-07			
Bottom sediment	550	1.94E+03	3.08E+01	1.57E-07	1.75E-03	2.77E-05	1.77E-04
	Total	6.92E+08	4.18E+08		623.44	376.557	100
	R + A		1.11E+09			1000	

f = 1.023E-05 Pa
Total amt = 49523 kg

Overall residence time = 49.52 h
Reaction time = 79.43 h
Advection time = 131.52 h

Fugacity Level III calculations: (four-compartment model)

Chemical name: Ethyl acetate

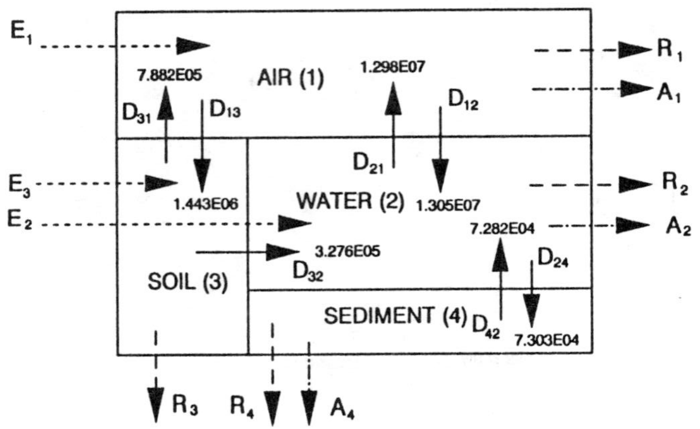

Phase Properties and Rates:

Compartment	Bulk Z mol/m3 Pa	Half-life h	D Values Reaction mol/Pa h	D Values Advection mol/Pa h
Air (1)	4.034E-04	55	5.08E+08	4.03E+08
Water (2)	7.279E-02	55	1.83E+08	1.46E+07
Soil (3)	2.576E-02	170	1.89E+06	
Sediment (4)	6.131E-02	550	3.86E+04	6.13E+02

	E(1)=1000	E(2)=1000	E(3)=1000	E(1,2,3)
Overall residence time =	44.84	71.85	173.85	141.90 h
Reaction time =	79.66	79.47	198.69	163.73 h
Advection time =	102.58	749.33	1390	1063.99 h

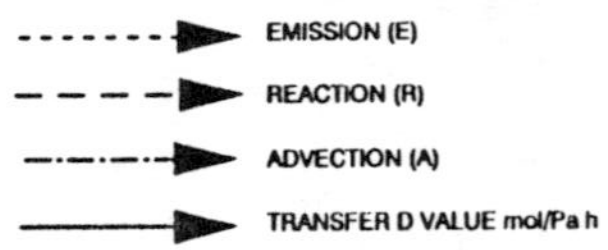

Phase Properties, Compositions, Transport and Transformation Rates:

Emission, kg/h			Fugacity, Pa				Concentration, g/m3				Amounts, kg				Total Amount, kg
E(1)	E(2)	E(3)	f(1)	f(2)	f(3)	f(4)	C(1)	C(2)	C(3)	C(4)	m(1)	m(2)	m(3)	m(4)	
1000	0	0	1.227E-05	7.682E-07	5.891E-06	5.006E-07	4.361E-07	4.926E-06	1.337E-05	2.704E-06	4.361E+04	9.852E+02	2.407E+02	1.352E+00	4.484E+04
0	1000	0	7.548E-07	5.385E-05	3.624E-07	3.509E-05	2.683E-08	3.453E-04	8.225E-07	1.895E-04	2.683E+03	6.906E+04	1.481E+01	9.477E+01	7.185E+04
0	0	1000	3.300E-06	6.068E-06	3.777E-03	3.955E-06	1.173E-07	3.891E-05	8.574E-03	2.136E-05	1.173E+04	7.783E+03	1.543E+05	1.068E+01	1.738E+05
50	250	700	3.112E-06	1.775E-05	2.645E-03	1.157E-05	1.106E-07	1.138E-04	6.002E-03	6.247E-05	1.106E+04	2.276E+04	1.080E+05	3.124E+01	1.419E+05

Emission, kg/h			Loss, Reaction, kg/h				Loss, Advection, kg/h			Intermedia Rate of Transport, kg/h T12	T21	T13	T31	T32	T24	T42
E(1)	E(2)	E(3)	R(1)	R(2)	R(3)	R(4)	A(1)	A(2)	A(4)	air-water	water-air	air-soil	soil-air	soil-water	water-sed	sed-water
1000	0	0	5.495E+02	1.241E+01	9.81E-01	1.703E-03	4.361E+02	9.852E-01	2.704E-05	1.411E+01	8.783E-01	1.560E+00	4.091E-01	1.700E-01	4.942E-03	3.212E-03
0	1000	0	3.380E+01	8.701E+02	6.04E-02	1.194E-01	2.683E+01	6.906E+01	1.895E-03	8.678E-01	6.156E+01	9.598E-02	2.516E-02	1.046E-02	3.464E-01	2.251E-01
0	0	1000	1.478E+02	9.806E+01	6.29E+02	1.346E-02	1.173E+02	7.783E+00	2.136E-04	3.794E+00	6.938E+00	4.196E-01	2.623E+02	1.090E+02	3.904E-02	2.537E-02
50	250	700	1.394E+02	2.868E+02	4.40E+02	3.936E-02	1.106E+02	2.276E+01	6.247E-04	3.578E+00	2.029E+01	3.957E-01	1.836E+02	7.631E+01	1.142E-01	7.420E-02

Level III Distribution

Chemical name: Ethyl acetate

Distribution of mass

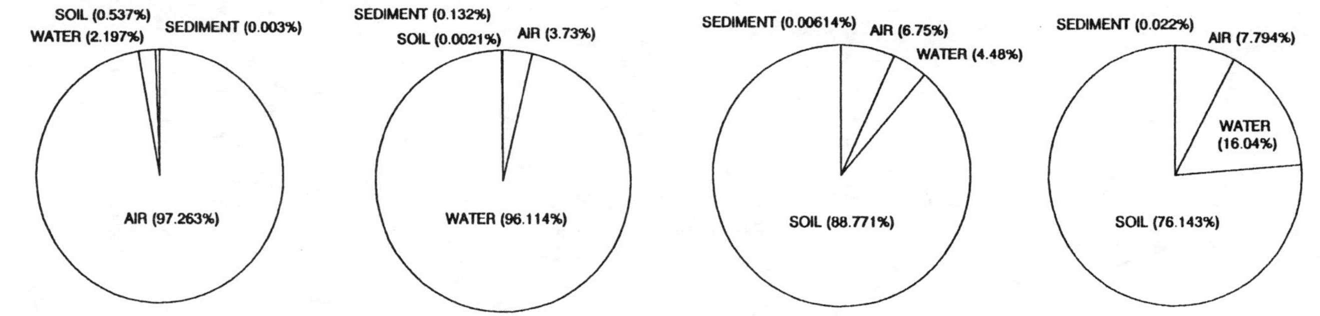

Distribution of removal rates

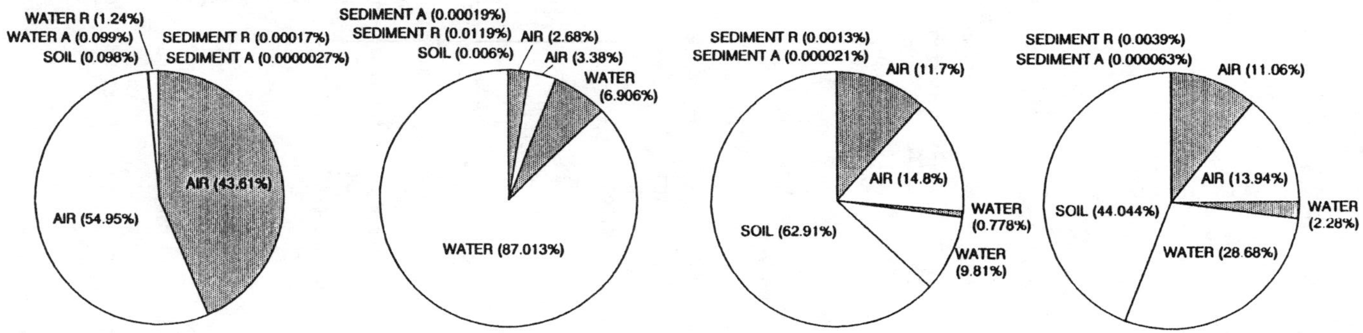

Emission rates:

E(1) = 1000 kg/h	E(1) = 0	E(1) = 0	E(1) = 50 kg/h
E(2) = 0	E(2) = 1000 kg/h	E(2) = 0	E(2) = 250 kg/h
E(3) = 0	E(3) = 0	E(3) = 1000 kg/h	E(3) = 700 kg/h

Residence time:

t = 44.84 h	t = 71.85 h	t = 174 h	t = 142 h

Reaction (white) Advection (shaded)

Chemical name: Methyl methacrylate

Level I calculation: (six-compartment model)

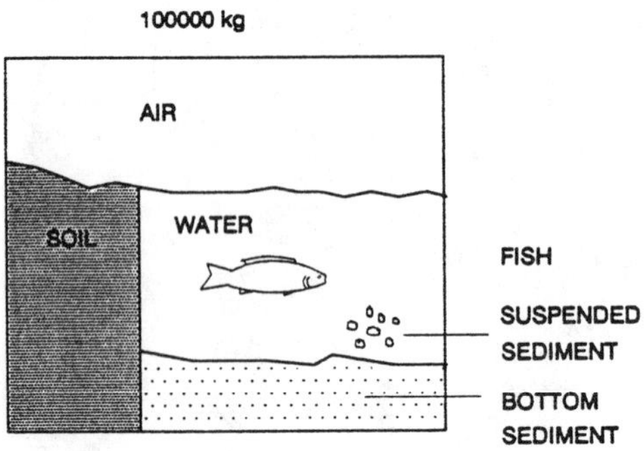

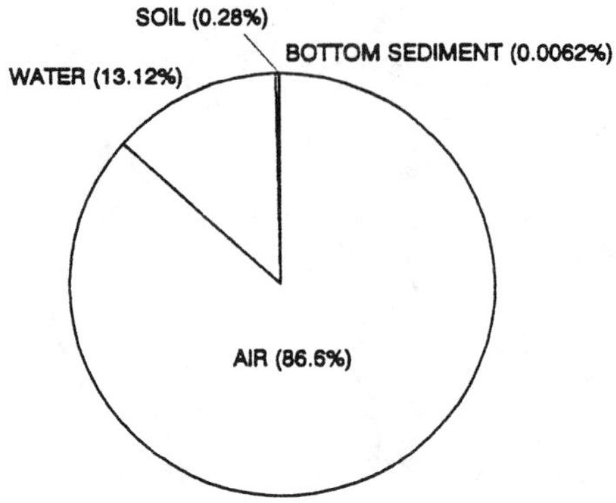

Distribution of mass

Level I calculation:

Physical-chemical properties:		Partition coeff.:	
molecular wt., g/mol	100.12	HLC	32.73154
melting point, C	-42.8	Kaw	1.32E-02
solubility, g/m3	15600	Koc	9.8352
vapor pressure, Pa	5100	BCF	1.1994
log Kow	1.38	Ksw	0.4721
fugacity ratio, F	1.0000	Ksd/w	0.9442
dissoc. const. pKa		Kssd/w	2.9506
		Kar/w	1176.471

Compartment	Z	Concentration			Amount	Amount
	mol/m3 Pa	mol/m3	mg/L (or g/m3)	ug/g	kg	%
Air	4.03E-04	8.65E-09	8.66E-07	7.31E-04	86598.44	86.5984
Water	3.06E-02	6.55E-07	6.56E-05	6.56E-05	13117	13.117
Soil	1.44E-02	3.09E-07	3.10E-05	1.29E-05	278.648	0.2786
Biota (fish)	3.66E-02	7.86E-07	7.87E-05	7.87E-05	0.01573	1.57E-05
Suspended sediment	9.01E-02	1.93E-06	1.94E-04	1.29E-04	0.1935	1.94E-04
Bottom sediment	2.88E-02	6.18E-07	6.19E-05	2.58E-05	6.1922	6.19E-03
	Total				100000	100

f = 2.144E-05 Pa

Chemical name: Methyl methacrylate

Level II calculation: (six-compartment model)

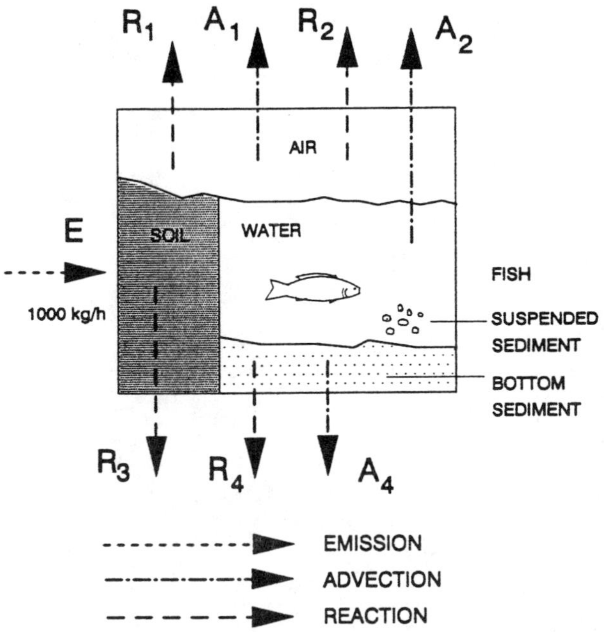

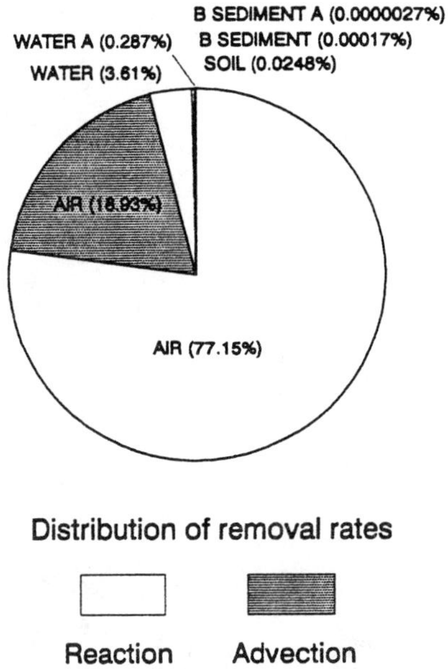

Distribution of removal rates

Level II Calculation:

Compartment	Half-life h	D Values Reaction mol/Pa.h	D Values Advection mol/Pa.h	Conc'n mol/m3	Loss Reaction kg/h	Loss Advection kg/h	Removal %
Air	17	1.64E+09	4.03E+08	1.89E-09	771.506	189.258	96.076
Water	55	7.70E+07	6.11E+06	1.43E-07	36.119	2.867	3.899
Soil	170	5.29E+05		6.76E-08	0.2482		0.0248
Biota (fish)				1.72E-07			
Suspended sediment				4.22E-07			
Bottom sediment	550	3.63E+03	5.77E+01	1.35E-07	1.71E-03	2.71E-05	1.73E-04
	Total	1.72E+09	4.10E+08		807.88	192.125	100
	R + A		2.13E+09			1000	

f = 4.686E-06 Pa

Total amt = 21855 kg

Overall residence time = 21.85 h

Reaction time = 27.05 h

Advection time = 113.75 h

Fugacity Level III calculations: (four-compartment model)

Chemical name: Methyl methacrylate

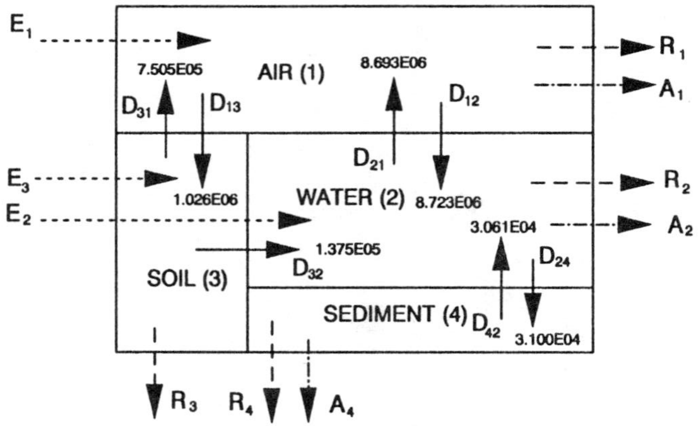

Phase Properties and Rates:

Compartment	Bulk Z mol/m3 Pa	Half-life h	D Values Reaction mol/Pa h	D Values Advection mol/Pa h
Air (1)	4.034E-04	17	1.64E+09	4.03E+08
Water (2)	3.055E-02	55	7.70E+07	6.11E+06
Soil (3)	1.646E-02	170	1.21E+06	
Sediment (4)	3.021E-02	550	1.90E+04	3.02E+02

	E(1)=1000	E(2)=1000	E(3)=1000	E(1,2,3)
Overall residence time =	19.97	68.55	153.01	125.24 h
Reaction time =	24.86	74.93	165.59	136.77 h
Advection time =	101.66	804.96	2015	1486.14 h

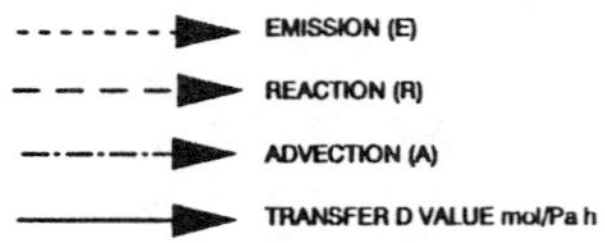

Phase Properties, Compositions, Transport and Transformation Rates:

Emission, kg/h			Fugacity, Pa				Concentration, g/m3				Amounts, kg				Total
E(1)	E(2)	E(3)	f(1)	f(2)	f(3)	f(4)	C(1)	C(2)	C(3)	C(4)	m(1)	m(2)	m(3)	m(4)	Amount, kg
1000	0	0	4.857E-06	4.650E-07	2.377E-06	2.887E-07	1.962E-07	1.423E-06	3.916E-06	8.731E-07	1.962E+04	2.845E+02	7.049E+01	4.366E-01	1.997E+04
0	1000	0	4.599E-07	1.088E-04	2.250E-07	6.756E-05	1.857E-08	3.329E-04	3.708E-07	2.044E-04	1.857E+03	6.659E+04	6.675E+00	1.022E+02	6.855E+04
0	0	1000	1.770E-06	7.307E-06	4.767E-03	4.536E-06	7.147E-08	2.235E-05	7.855E-03	1.372E-05	7.147E+03	4.471E+03	1.414E+05	6.860E+00	1.530E+05
50	250	700	1.597E-06	3.235E-05	3.337E-03	2.008E-05	6.448E-08	9.895E-05	5.499E-03	6.073E-05	6.448E+03	1.979E+04	9.898E+04	3.037E+01	1.252E+05

Emission, kg/h			Loss, Reaction, kg/h				Loss, Advection, kg/h			Intermedia Rate of Transport, kg/h T12	T21	T13	T31	T32	T24	T42
E(1)	E(2)	E(3)	R(1)	R(2)	R(3)	R(4)	A(1)	A(2)	A(4)	air-water	water-air	air-soil	soil-air	soil-water	water-sed	sed-water
1000	0	0	7.997E+02	3.585E+00	2.87E-01	5.501E-04	1.962E+02	2.845E-01	8.731E-06	4.242E+00	4.047E-01	4.987E-01	1.786E-01	3.272E-02	1.443E-03	8.847E-04
0	1000	0	7.572E+01	8.390E+02	2.72E-02	1.287E-01	1.857E+01	6.659E+01	2.044E-03	4.016E-01	9.472E+01	4.722E-02	1.691E-02	3.098E-03	3.378E-01	2.070E-01
0	0	1000	2.914E+02	5.633E+01	5.76E+02	8.644E-03	7.147E+01	4.471E+00	1.372E-04	1.546E+00	6.360E+00	1.817E-01	3.582E+02	6.562E+01	2.268E-02	1.390E-02
50	250	700	2.629E+02	2.494E+02	4.03E+02	3.826E-02	6.448E+01	1.979E+01	6.073E-04	1.394E+00	2.815E+01	1.639E-01	2.508E+02	4.594E+01	1.004E-01	6.154E-02

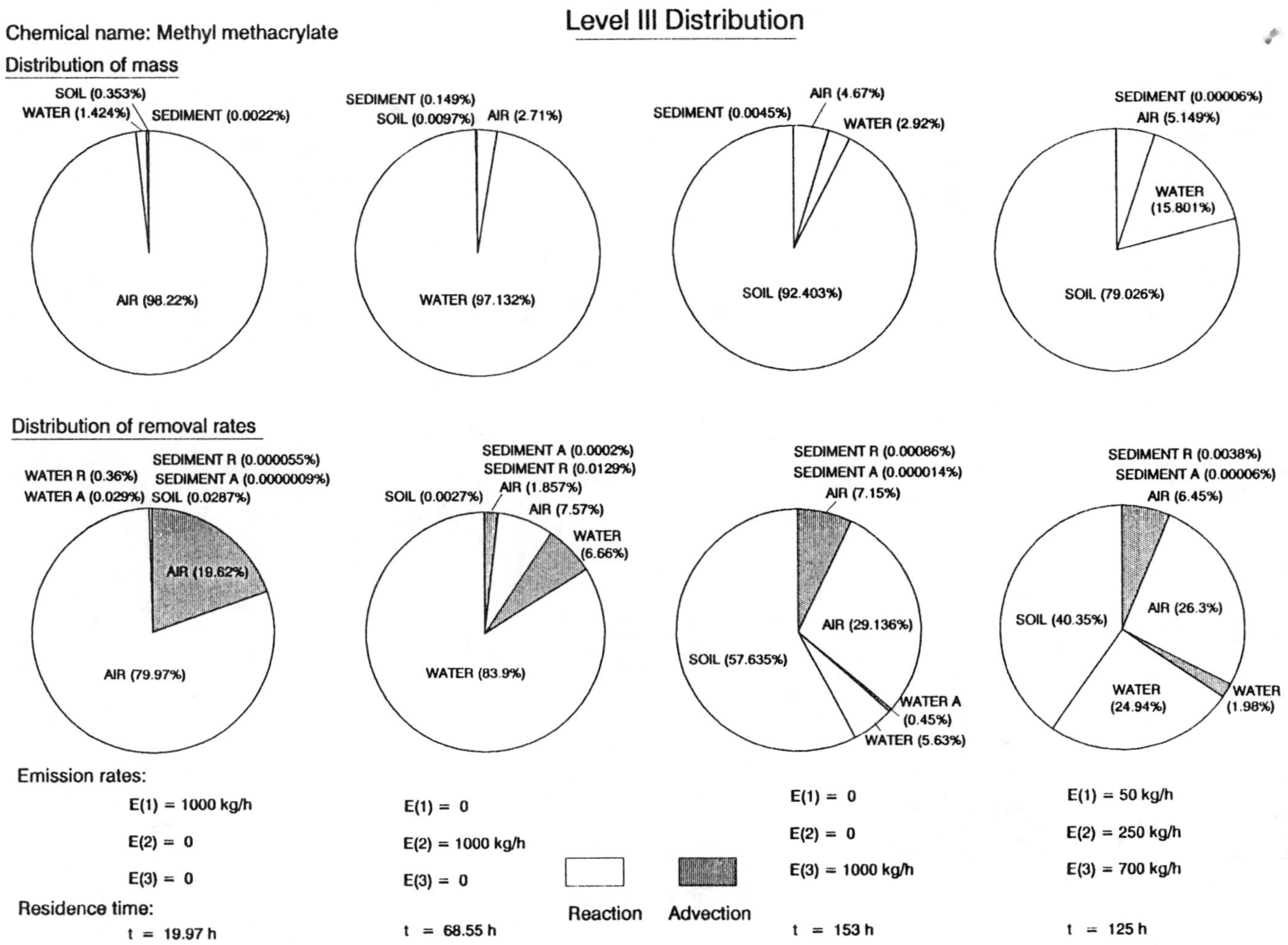

Level III Distribution
Chemical name: Methyl methacrylate
Distribution of mass
SOIL (0.353%)
WATER (1.424%)
SEDIMENT (0.0022%)
AIR (98.22%)
SEDIMENT (0.149%)
SOIL (0.0097%)
AIR (2.71%)
WATER (97.132%)
AIR (4.67%)
SEDIMENT (0.0045%)
WATER (2.92%)
SOIL (92.403%)
SEDIMENT (0.00006%)
AIR (5.149%)
WATER (15.801%)
SOIL (79.026%)
Distribution of removal rates
WATER R (0.36%)
WATER A (0.029%)
SEDIMENT R (0.000055%)
SEDIMENT A (0.0000009%)
SOIL (0.0287%)
AIR (19.62%)
AIR (79.97%)
SEDIMENT A (0.0002%)
SEDIMENT R (0.0129%)
SOIL (0.0027%)
AIR (1.857%)
AIR (7.57%)
WATER (6.66%)
WATER (83.9%)
SEDIMENT R (0.00086%)
SEDIMENT A (0.000014%)
AIR (7.15%)
AIR (29.136%)
SOIL (57.635%)
WATER A (0.45%)
WATER (5.63%)
SEDIMENT R (0.0038%)
SEDIMENT A (0.00006%)
AIR (6.45%)
SOIL (40.35%)
AIR (26.3%)
WATER (24.94%)
WATER (1.98%)
Emission rates:
E(1) = 1000 kg/h
E(2) = 0
E(3) = 0
Residence time:
t = 19.97 h
E(1) = 0
E(2) = 1000 kg/h
E(3) = 0
t = 68.55 h
Reaction
Advection
E(1) = 0
E(2) = 0
E(3) = 1000 kg/h
t = 153 h
E(1) = 50 kg/h
E(2) = 250 kg/h
E(3) = 700 kg/h
t = 125 h

Chemical name: Dimethyl phthalate (DMP)

Level I calculation: (six-compartment model)

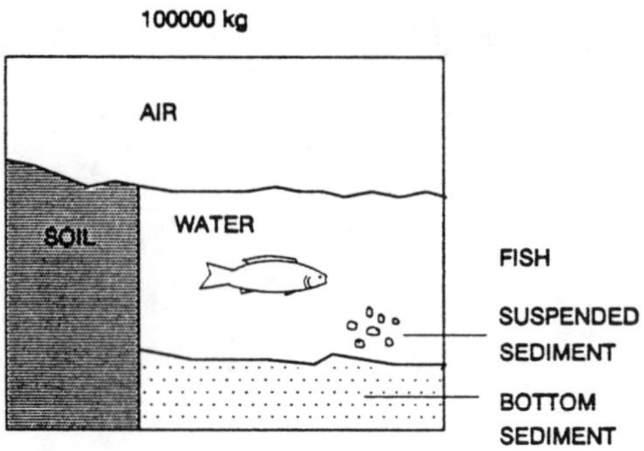

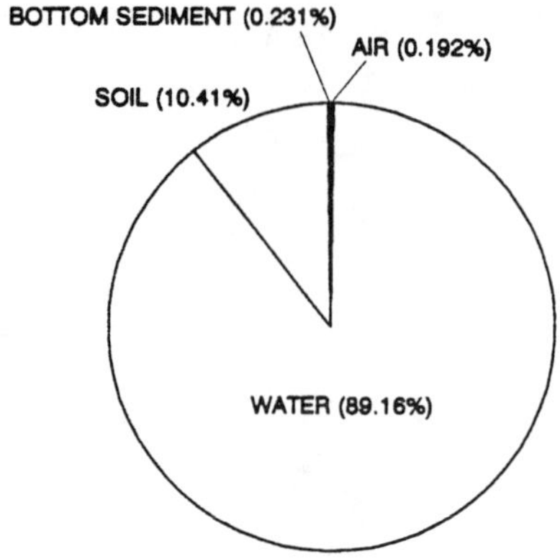

Distribution of mass

Level I calculation:

Physical-chemical properties:		Partition coeff.:	
molecular wt., g/mol	194.20	HLC	0.01068
melting point, C	5.5	Kaw	4.31E-06
solubility, g/m3	4000	Koc	54.0485
vapor pressure, Pa	0.2200	BCF	6.5913
log Kow	2.12	Ksw	2.5943
fugacity ratio, F	1.0000	Ksd/w	5.1887
dissoc. const. pKa		Kssd/w	16.2146
		Kar/w	2.73E+07

Compartment	Z	Concentration			Amount	Amount
	mol/m3 Pa	mol/m3	mg/L (or g/m3)	ug/g	kg	%
Air	4.03E-04	9.89E-12	1.92E-09	1.62E-06	192.09	0.1921
Water	9.36E+01	2.30E-06	4.46E-04	4.46E-04	89160	89.160
Soil	2.43E+02	5.96E-06	1.16E-03	4.82E-04	10409	10.409
Biota (fish)	6.17E+02	1.51E-05	2.94E-03	2.94E-03	0.5877	5.88E-04
Suspended sediment	1.52E+03	3.72E-05	7.23E-03	4.82E-03	7.2284	7.23E-03
Bottom sediment	4.86E+02	1.19E-05	2.31E-03	9.64E-04	231.31	0.2313
	Total				100000	100

f = 2.452E-08 Pa

Chemical name: Dimethyl phthalate (DMP)

Level II calculation: (six-compartment model)

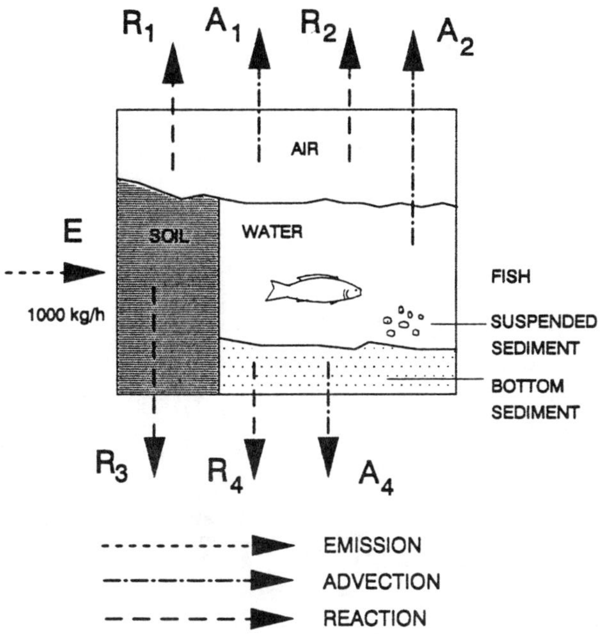

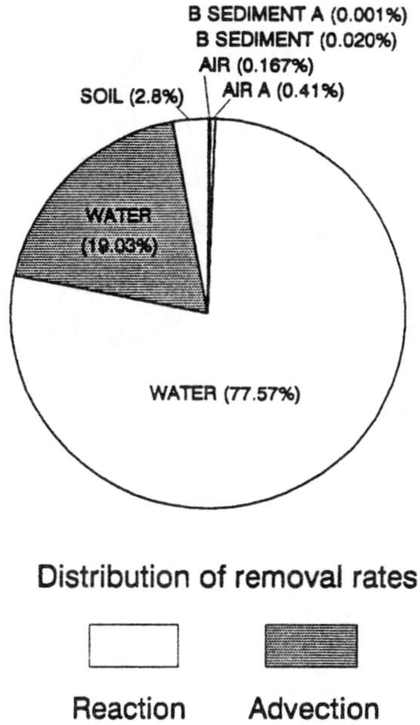

Distribution of removal rates

Level II Calculation:

Compartment	Half-life h	D Values Reaction mol/Pa.h	D Values Advection mol/Pa.h	Conc'n mol/m3	Loss Reaction kg/h	Loss Advection kg/h	Removal %
Air	170	1.64E+08	4.03E+08	2.11E-11	1.6713	4.0998	0.5771
Water	170	7.63E+10	1.87E+10	4.90E-06	775.731	190.295	96.603
Soil	550	2.75E+09		1.27E-05	27.9921		2.7992
Biota (fish)				3.23E-05			
Suspended sediment				7.94E-05			
Bottom sediment	1700	1.98E+07	9.72E+05	2.54E-05	0.2013	9.87E-03	2.11E-02
	Total	7.93E+10	1.91E+10		805.60	194.404	100
	R + A		9.84E+10			1000	

f = 5.233E-08 Pa
Total amt = 213431 kg

Overall residence time = 213.43 h
Reaction time = 264.94 h
Advection time = 1097.87 h

Fugacity Level III calculations: (four-compartment model)

Chemical name: Dimethyl phthalate

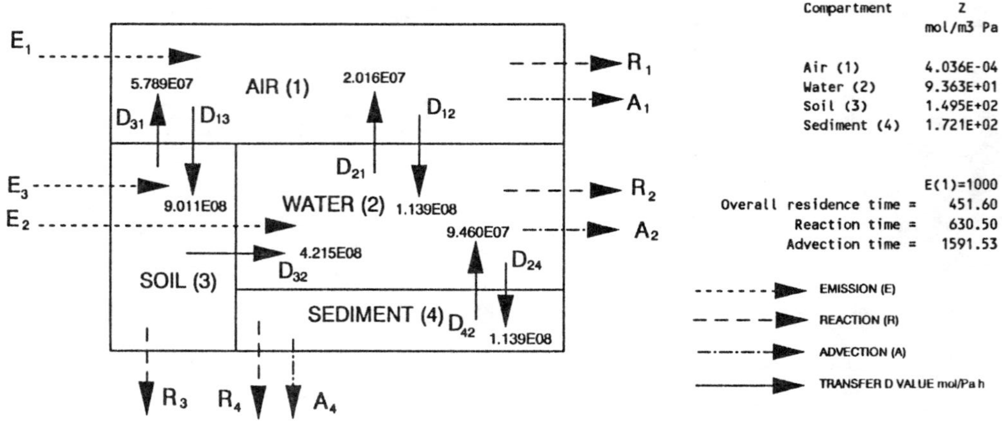

Phase Properties and Rates:

Compartment	Bulk Z mol/m3 Pa	Half-life h	D Values Reaction mol/Pa h	D Values Advection mol/Pa h
Air (1)	4.036E-04	170	1.65E+08	4.04E+08
Water (2)	9.363E+01	170	7.63E+10	1.87E+10
Soil (3)	1.495E+02	550	3.39E+09	
Sediment (4)	1.721E+02	1700	3.51E+07	1.72E+06

	E(1)=1000	E(2)=1000	E(3)=1000	E(1,2,3)	
Overall residence time =	451.60	197.68	723.64	578.54	h
Reaction time =	630.50	246.16	742.72	629.82	h
Advection time =	1591.53	1003.65	28166	7106.35	h

Phase Properties, Compositions, Transport and Transformation Rates:

Emission, kg/h			Fugacity, Pa				Concentration, g/m3				Amounts, kg				Total
E(1)	E(2)	E(3)	f(1)	f(2)	f(3)	f(4)	C(1)	C(2)	C(3)	C(4)	m(1)	m(2)	m(3)	m(4)	Amount, kg
1000	0	0	3.281E-06	7.312E-09	7.637E-07	5.633E-09	2.572E-07	1.330E-04	2.218E-02	1.882E-04	2.572E+04	2.659E+04	3.992E+05	9.410E+01	4.516E+05
0	1000	0	6.955E-10	5.414E-08	1.619E-10	4.171E-08	5.451E-11	9.845E-04	4.701E-06	1.394E-03	5.451E+00	1.969E+05	8.462E+01	6.968E+02	1.977E+05
0	0	1000	4.914E-08	6.005E-09	1.342E-06	4.626E-09	3.852E-09	1.092E-04	3.896E-02	1.546E-04	3.852E+02	2.184E+04	7.013E+05	7.729E+01	7.236E+05
50	250	700	1.986E-07	1.810E-08	9.774E-07	1.395E-08	1.557E-08	3.292E-04	2.838E-02	4.660E-04	1.557E+03	6.584E+04	5.109E+05	2.330E+02	5.785E+05

Emission, kg/h			Loss, Reaction, kg/h				Loss, Advection, kg/h			Intermedia Rate of Transport, kg/h T12	T21	T13	T31	T32	T24	T42
E(1)	E(2)	E(3)	R(1)	R(2)	R(3)	R(4)	A(1)	A(2)	A(4)	air-water	water-air	air-soil	soil-air	soil-water	water-sed	sed-water
1000	0	0	1.048E+02	1.084E+02	5.03E+02	3.836E-02	2.572E+02	2.659E+01	1.882E-03	7.254E+01	2.863E-02	5.741E+02	8.585E+00	6.252E+01	1.437E-01	1.035E-01
0	1000	0	2.222E-02	8.026E+02	1.07E-01	2.841E-01	5.451E-02	1.969E+02	1.394E-02	1.538E-02	2.120E-01	1.217E-01	1.820E-03	1.325E-02	1.064E+00	7.662E-01
0	0	1000	1.570E+00	8.903E+01	8.84E+02	3.151E-02	3.852E+00	2.184E+01	1.546E-03	1.086E+00	2.351E-02	8.599E+00	1.508E+01	1.098E+02	1.180E-01	8.499E-02
50	250	700	6.346E+00	2.684E+02	6.44E+02	9.498E-02	1.557E+01	6.584E+01	4.660E-03	4.391E+00	7.089E-02	3.475E+01	1.099E+01	8.001E+01	3.559E-01	2.562E-01

Level III Distribution

Chemical name: Dimethyl phthalate

Distribution of mass

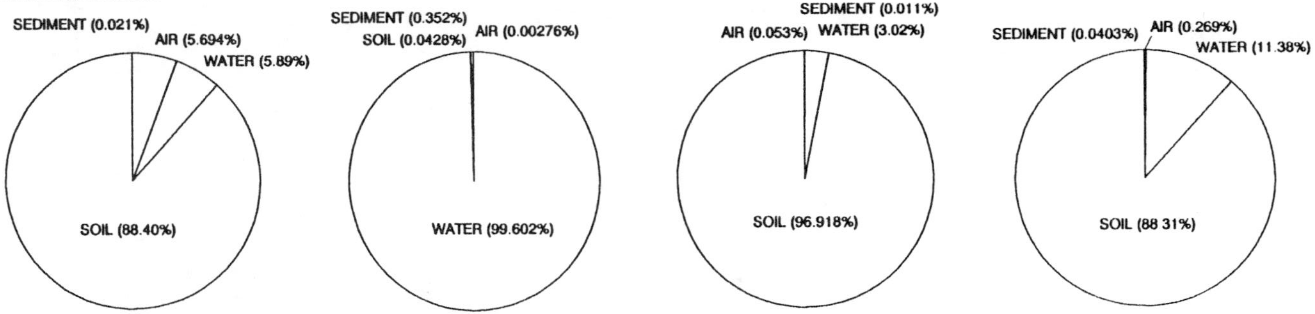

Distribution of removal rates

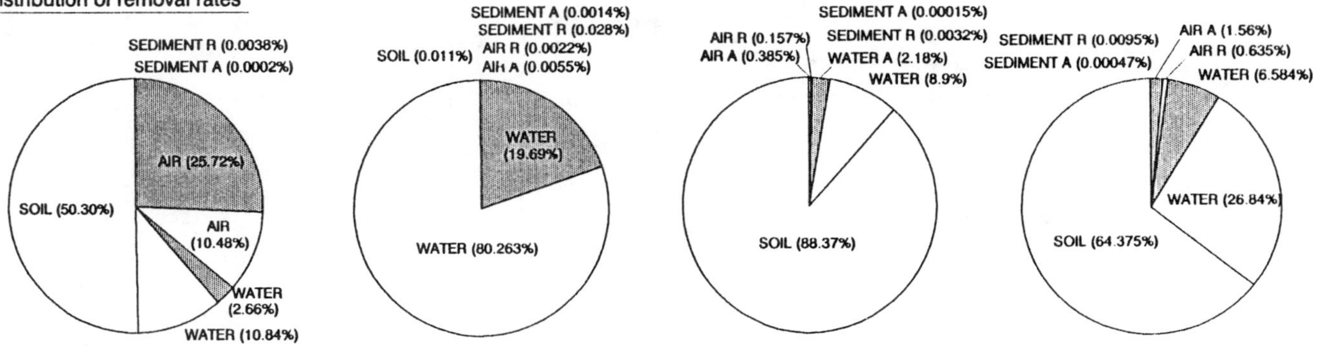

Emission rates:

E(1) = 1000 kg/h	E(1) = 0	E(1) = 0	E(1) = 50 kg/h
E(2) = 0	E(2) = 1000 kg/h	E(2) = 0	E(2) = 250 kg/h
E(3) = 0	E(3) = 0	E(3) = 1000 kg/h	E(3) = 700 kg/h

Residence time:

t = 452 h	t = 198so h	t = 724 h	t = 579 h

Reaction (unshaded) — Advection (shaded)

Chemical name: Dibutyl phthalate (DBP)

Level I calculation: (six-compartment model)

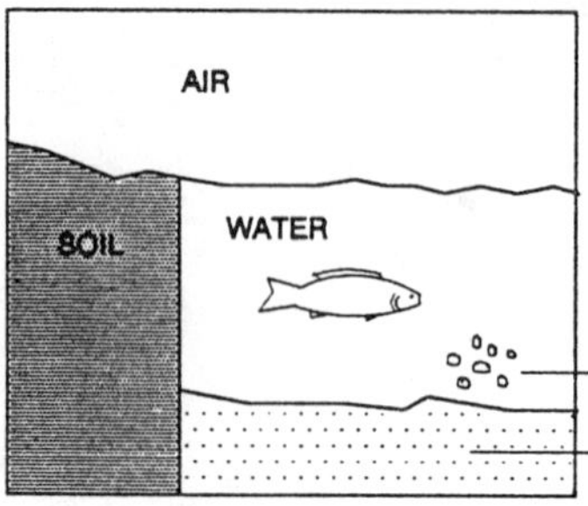

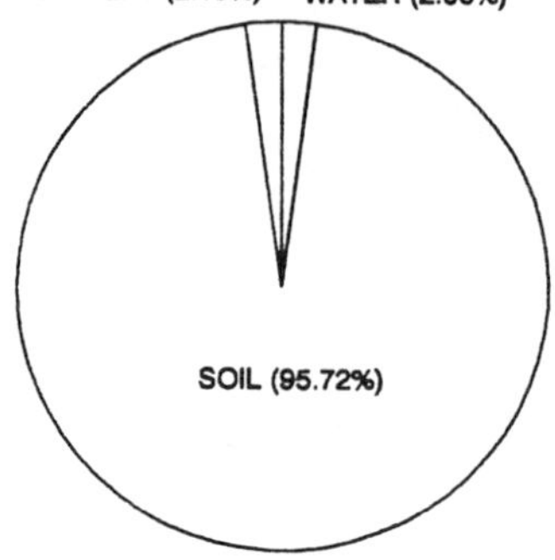

Distribution of mass

Level I calculation:

Physical-chemical properties:		Partition coeff.:	
molecular wt., g/mol	278.35	HLC	0.04647
melting point, C	-35	Kaw	1.87E-05
solubility, g/m3	11.2	Koc	21517.106
vapor pressure, Pa	0.00187	BCF	2624.0373
log Kow	4.72	Ksw	1032.8211
fugacity ratio, F	1.0000	Ksd/w	2065.6422
dissoc. const. pKa		Kssd/w	6455.132
		Kar/w	3.21E+09

Compartment	Z	Concentration			Amount	Amount
	mol/m3 Pa	mol/m3	mg/L (or g/m3)	ug/g	kg	%
Air	4.03E-04	6.94E-13	1.93E-10	1.63E-07	19.307	0.0193
Water	2.15E+01	3.70E-08	1.03E-05	1.03E-05	2059.56	2.0596
Soil	2.22E+04	3.82E-05	1.06E-02	4.43E-03	95722	95.722
Biota (fish)	5.65E+04	9.71E-05	2.70E-02	2.70E-02	5.4044	5.40E-03
Suspended sediment	1.39E+05	2.39E-04	6.65E-02	4.43E-02	66.4737	6.65E-02
Bottom sediment	4.44E+04	7.64E-05	2.13E-02	8.86E-03	2127.16	2.1272
	Total				100000	100

f = 1.719E-09 Pa

Chemical name: Dibutyl phthalate

Level II calculation: (six-compartment model)

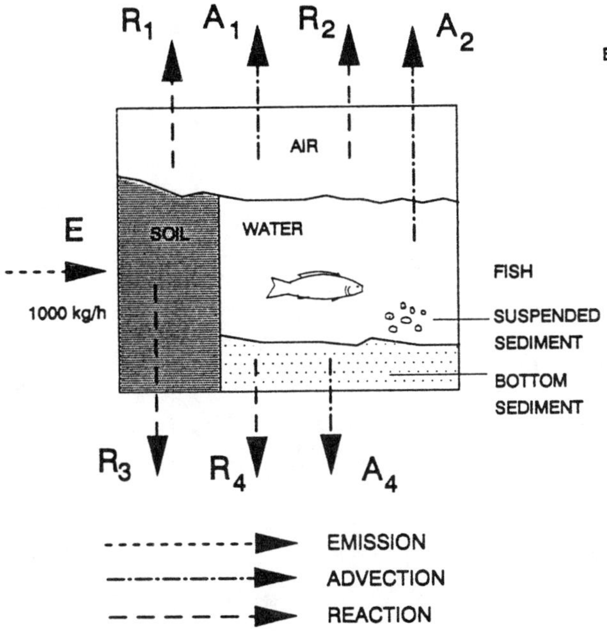

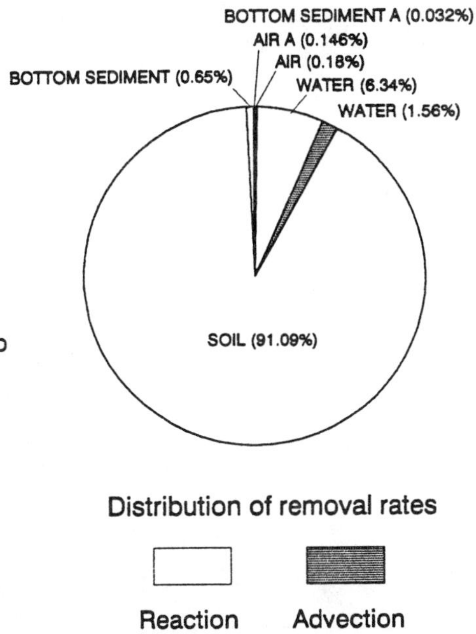

Distribution of removal rates

Level II Calculation:

Compartment	Half-life h	D Values Reaction mol/Pa.h	D Values Advection mol/Pa.h	Conc'n mol/m3	Loss Reaction kg/h	Loss Advection kg/h	Removal %
Air	55	5.08E+08	4.03E+08	5.24E-12	1.8372	1.4581	0.3295
Water	170	1.75E+10	4.30E+09	2.79E-07	63.407	15.554	7.896
Soil	550	2.52E+11		2.89E-04	910.874		91.0874
Biota (fish)				7.33E-04			
Suspended sediment				1.80E-03			
Bottom sediment	1700	1.81E+09	8.89E+07	5.77E-04	6.5488	0.3213	0.6870
	Total	2.70E+11	4.71E+09		982.67	17.334	100
	R + A		2.75E+11			1000	

f = 1.299E-08 Pa
Total amt = 755223 kg

Overall residence time = 755.22 h
Reaction time = 768.55 h
Advection time = 43569.69 h

Fugacity Level III calculations: (four-compartment model)

Chemical name: Dibutyl phthalate

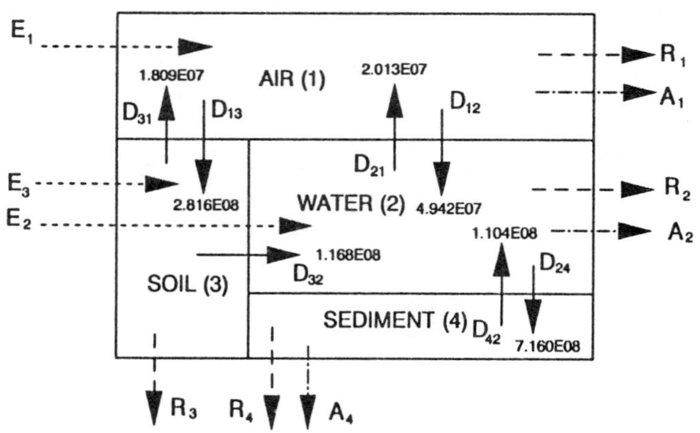

Phase Properties and Rates:

Compartment	Bulk Z mol/m3 Pa	Half-life h	D Values Reaction mol/Pa h	D Values Advection mol/Pa h
Air (1)	4.293E-04	55	5.41E+08	4.29E+08
Water (2)	2.227E+01	170	1.82E+10	4.45E+09
Soil (3)	1.112E+04	550	2.52E+11	
Sediment (4)	8.907E+03	1700	1.82E+09	8.91E+07

	E(1)=1000	E(2)=1000	E(3)=1000	E(1,2,3)	
Overall residence time =	214.54	259.19	793.36	630.88	h
Reaction time =	323.72	321.07	793.45	674.83	h
Advection time =	636.15	1344.70	6993994	9686.63	h

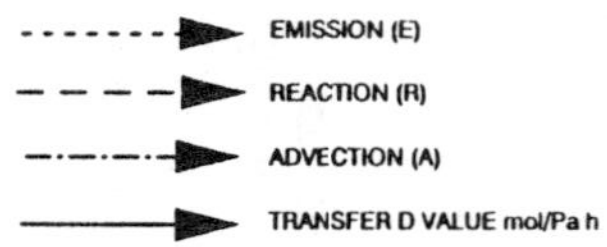

Phase Properties, Compositions, Transport and Transformation Rates:

Emission, kg/h E(1)	E(2)	E(3)	Fugacity, Pa f(1)	f(2)	f(3)	f(4)	Concentration, g/m3 C(1)	C(2)	C(3)	C(4)	Amounts, kg m(1)	m(2)	m(3)	m(4)	Total Amount, kg
1000	0	0	2.761E-06	5.870E-09	3.082E-09	2.086E-09	3.299E-07	3.638E-05	9.538E-03	5.171E-03	3.299E+04	7.276E+03	1.717E+05	2.586E+03	2.145E+05
0	1000	0	2.385E-09	1.542E-07	2.663E-12	5.478E-08	2.850E-10	9.555E-04	8.240E-06	1.358E-01	2.850E+01	1.911E+05	1.483E+02	6.791E+04	2.592E+05
0	0	1000	1.991E-10	7.181E-11	1.424E-08	2.552E-11	2.379E-11	4.451E-07	4.407E-02	6.326E-05	2.379E+00	8.901E+01	7.932E+05	3.163E+01	7.934E+05
50	250	700	1.388E-07	3.888E-08	1.012E-08	1.382E-08	1.658E-08	2.410E-04	3.133E-02	3.426E-02	1.658E+03	4.820E+04	5.639E+05	1.713E+04	6.309E+05

Emission, kg/h E(1)	E(2)	E(3)	Loss, Reaction, kg/h R(1)	R(2)	R(3)	R(4)	Loss, Advection, kg/h A(1)	A(2)	A(4)	Intermedia Rate of Transport, kg/h T12 air-water	T21 water-air	T13 air-soil	T31 soil-air	T32 soil-water	T24 water-sed	T42 sed-water
1000	0	0	4.157E+02	2.966E+01	2.16E+02	1.054E+00	3.299E+02	7.276E+00	5.171E-02	3.798E+01	3.289E-02	2.164E+02	1.552E-02	1.002E-01	1.170E+00	6.410E-02
0	1000	0	3.591E-01	7.790E+02	1.87E-01	2.768E+01	2.850E-01	1.911E+02	1.358E+00	3.281E-02	8.639E-01	1.870E-01	1.341E-05	8.658E-05	3.072E+01	1.684E+00
0	0	1000	2.997E-02	3.629E-01	9.99E+02	1.289E-02	2.379E-02	8.901E-02	6.326E-04	2.738E-03	4.024E-04	1.561E-02	7.170E-02	4.631E-01	1.431E-02	7.842E-04
50	250	700	2.090E+01	1.965E+02	7.10E+02	6.982E+00	1.658E+01	4.820E+01	3.426E-01	1.909E+00	2.179E-01	1.088E+01	5.097E-02	3.292E-01	7.749E+00	4.247E-01

Level III Distribution

Chemical name: Dibutyl phthalate

Distribution of mass

Distribution of removal rates

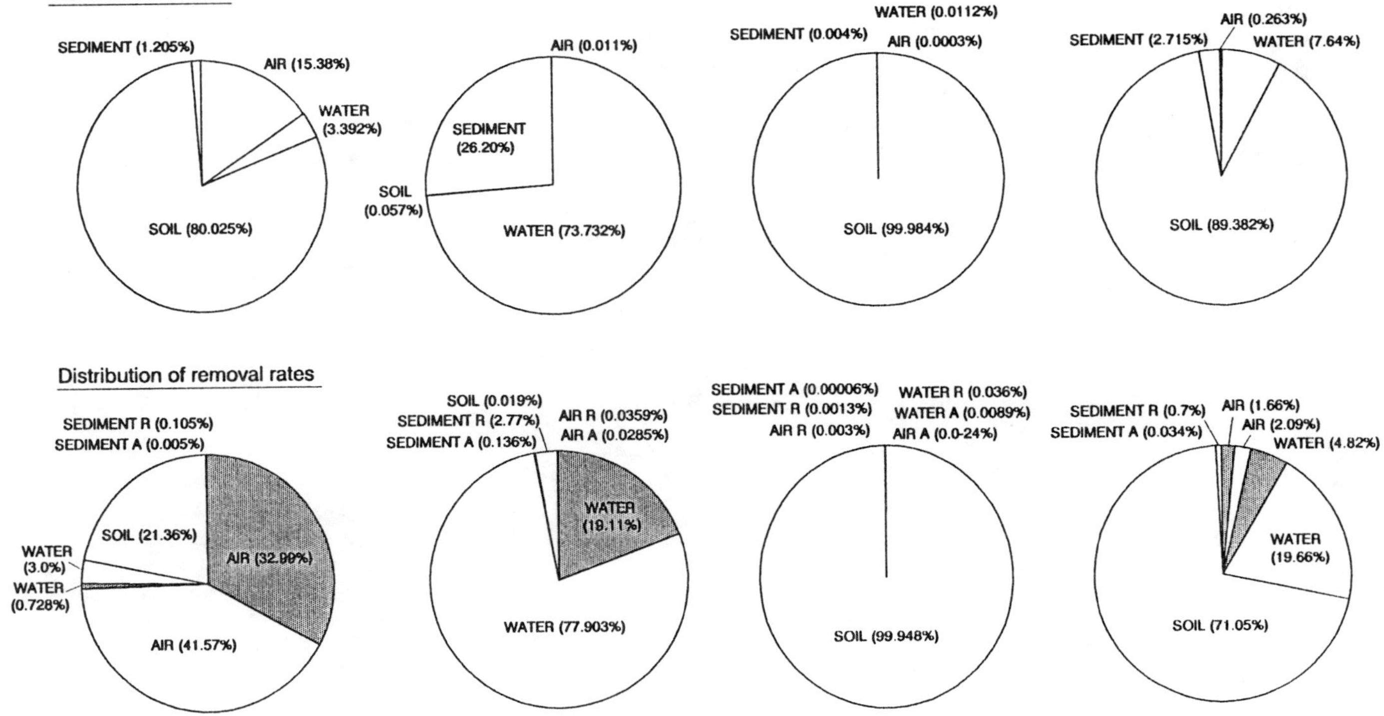

Emission rates:

E(1) = 1000 kg/h	E(1) = 0	E(1) = 0	E(1) = 50 kg/h
E(2) = 0	E(2) = 1000 kg/h	E(2) = 0	E(2) = 250 kg/h
E(3) = 0	E(3) = 0	E(3) = 1000 kg/h	E(3) = 700 kg/h

Residence time:

t = 215 h	t = 259 h	t = 793 h	t = 631 h

Reaction   Advection

Chemical name: Di(2-ethylhexyl) phthalate (DEHP)

Level I calculation: (six-compartment model)

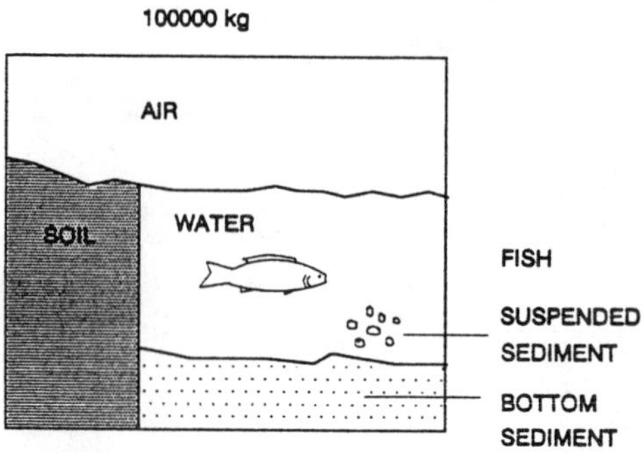

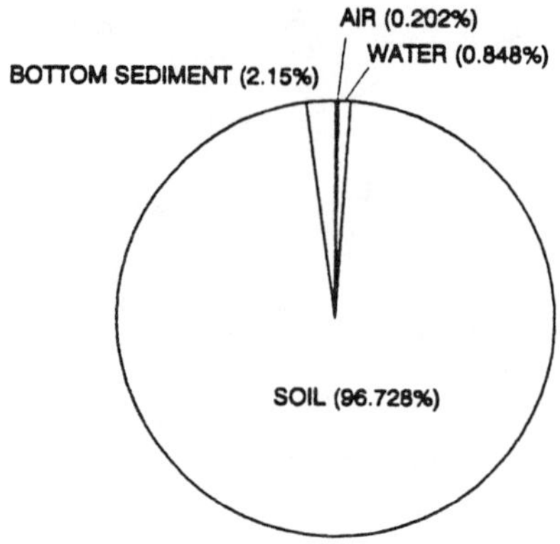

Distribution of mass

Level I calculation:

Physical-chemical properties:		Partition coeff.:	
molecular wt., g/mol	390.54	HLC	1.17847
melting point, C	-50	Kaw	4.75E-04
solubility, g/m3	0.285	Koc	52818.232
vapor pressure, Pa	0.0009	BCF	6441.2478
log Kow	5.11	Ksw	2535.2751
fugacity ratio, F	1.0000	Ksd/w	5070.5502
dissoc. const. pKa		Kssd/w	15845.469
		Kar/w	6.98E+09

Compartment	Z	Concentration			Amount	Amount
	mol/m3 Pa	mol/m3	mg/L (or g/m3)	ug/g	kg	%
Air	4.03E-04	5.16E-12	2.02E-09	1.70E-06	201.54	0.2015
Water	8.49E-01	1.09E-08	4.24E-06	4.24E-06	847.85	0.8478
Soil	2.15E+03	2.75E-05	1.07E-02	4.48E-03	96728	96.728
Biota (fish)	5.47E+03	6.99E-05	2.73E-02	2.73E-02	5.4612	5.46E-03
Suspended sediment	1.34E+04	1.72E-04	6.72E-02	4.48E-02	67.1725	6.72E-02
Bottom sediment	4.30E+03	5.50E-05	2.15E-02	8.96E-03	2149.52	2.1495
	Total				100000	100

f = 1.279E-08 Pa

Chemical name: Di(2-ethylhexyl) phthalate

Level II calculation: (six-compartment model)

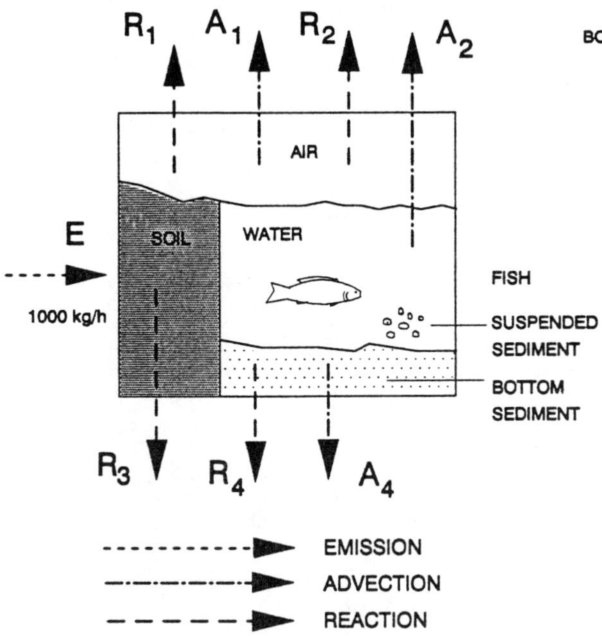

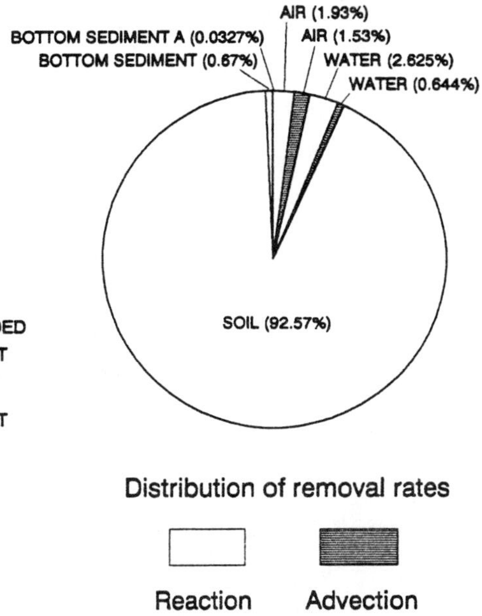

Distribution of removal rates

Level II Calculation:

Compartment	Half-life h	D Values Reaction mol/Pa.h	D Values Advection mol/Pa.h	Conc'n mol/m3	Loss Reaction kg/h	Loss Advection kg/h	Removal %
Air	55	5.08E+08	4.03E+08	3.92E-11	19.2882	15.3081	3.4596
Water	170	6.92E+08	1.70E+08	8.24E-08	26.252	6.440	3.269
Soil	550	2.44E+10		2.09E-04	925.7299		92.5730
Biota (fish)				5.31E-04			
Suspended sediment				1.31E-03			
Bottom sediment	1700	1.75E+08	8.61E+06	4.18E-04	6.6556	0.3265	0.6982
	Total	2.56E+10	5.73E+08		977.93	22.074	100
	R + A		2.62E+10			1000	

f = 9.716E-08 Pa
Total amt = 759555 kg

Overall residence time = 759.56 h
Reaction time = 776.70 h
Advection time = 34408.79 h

Fugacity Level III calculations: (four-compartment model)

Chemical name: Di(2-ethylhexyl) phthalate

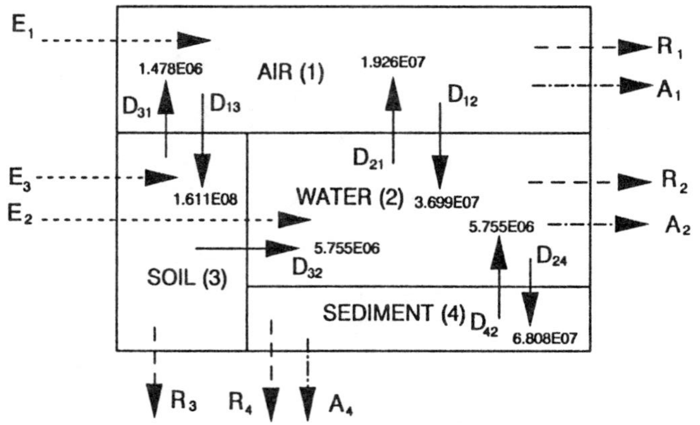

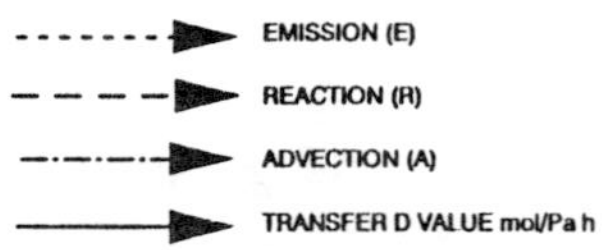

Phase Properties and Rates:

Compartment	Bulk Z mol/m3 Pa	Half-life h	D Values Reaction mol/Pa h	D Values Advection mol/Pa h
Air (1)	4.597E-04	55	5.79E+08	4.60E+08
Water (2)	9.213E-01	170	7.51E+08	1.84E+08
Soil (3)	1.076E+03	550	2.44E+10	
Sediment (4)	8.612E+02	1700	1.76E+08	8.61E+06

	E(1)=1000	E(2)=1000	E(3)=1000	E(1,2,3)	
Overall residence time =	150.43	332.14	793.50	646.01	h
Reaction time =	241.60	410.48	793.56	692.12	h
Advection time =	398.67	1740.29	11696782	9695.85	h

Phase Properties, Compositions, Transport and Transformation Rates:

Emission, kg/h			Fugacity, Pa				Concentration, g/m3				Amounts, kg				Total Amount, kg
E(1)	E(2)	E(3)	f(1)	f(2)	f(3)	f(4)	C(1)	C(2)	C(3)	C(4)	m(1)	m(2)	m(3)	m(4)	
1000	0	0	2.071E-06	7.524E-08	1.367E-08	2.646E-08	3.718E-07	2.707E-05	5.744E-03	8.898E-03	3.718E+04	5.414E+03	1.034E+05	4.449E+03	1.504E+05
0	1000	0	3.912E-08	2.513E-06	2.582E-10	8.838E-07	7.024E-09	9.043E-04	1.085E-04	2.973E-01	7.024E+02	1.809E+05	1.953E+03	1.486E+05	3.321E+05
0	0	1000	1.346E-10	5.971E-10	1.049E-07	2.100E-10	2.417E-11	2.148E-07	4.408E-02	7.062E-05	2.417E+00	4.297E+01	7.934E+05	3.531E+01	7.935E+05
50	250	700	1.134E-07	6.325E-07	7.418E-08	2.224E-07	2.036E-08	2.276E-04	3.117E-02	7.481E-02	2.036E+03	4.551E+04	5.611E+05	3.740E+04	6.460E+05

Emission, kg/h			Loss, Reaction, kg/h				Loss, Advection, kg/h			Intermedia Rate of Transport, kg/h T12	T21	T13	T31	T32	T24	T42
E(1)	E(2)	E(3)	R(1)	R(2)	R(3)	R(4)	A(1)	A(2)	A(4)	air-water	water-air	air-soil	soil-air	soil-water	water-sed	sed-water
1000	0	0	4.685E+02	2.207E+01	1.30E+02	1.814E+00	3.718E+02	5.414E+00	8.898E-02	2.992E+01	5.658E-01	1.303E+02	7.889E-03	3.072E-02	2.000E+00	9.768E-02
0	1000	0	8.850E+00	7.373E+02	2.46E+00	6.059E+01	7.024E+00	1.809E+02	2.973E+00	5.652E-01	1.890E+01	2.461E+00	1.490E-04	5.803E-04	6.682E+01	3.263E+00
0	0	1000	3.045E-02	1.752E-01	1.00E+03	1.439E-02	2.417E-02	4.297E-02	7.062E-04	1.945E-03	4.490E-03	8.469E-03	6.054E-02	2.358E-01	1.588E-02	7.752E-04
50	250	700	2.566E+01	1.855E+02	7.07E+02	1.525E+01	2.036E+01	4.551E+01	7.481E-01	1.639E+00	4.757E+00	7.137E+00	4.281E-02	1.667E-01	1.682E+01	8.212E-01

Level III Distribution

Chemical name: Di(2-ethylhexyl) phthalate

Distribution of mass

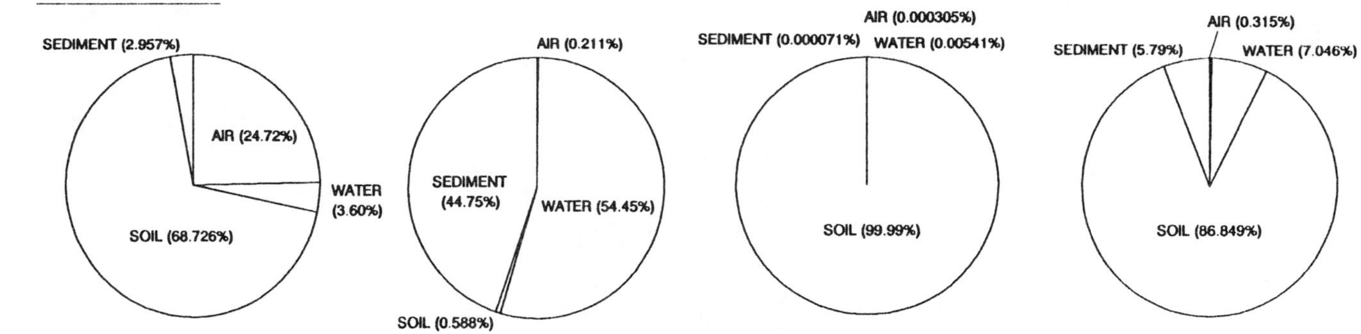

Distribution of removal rates

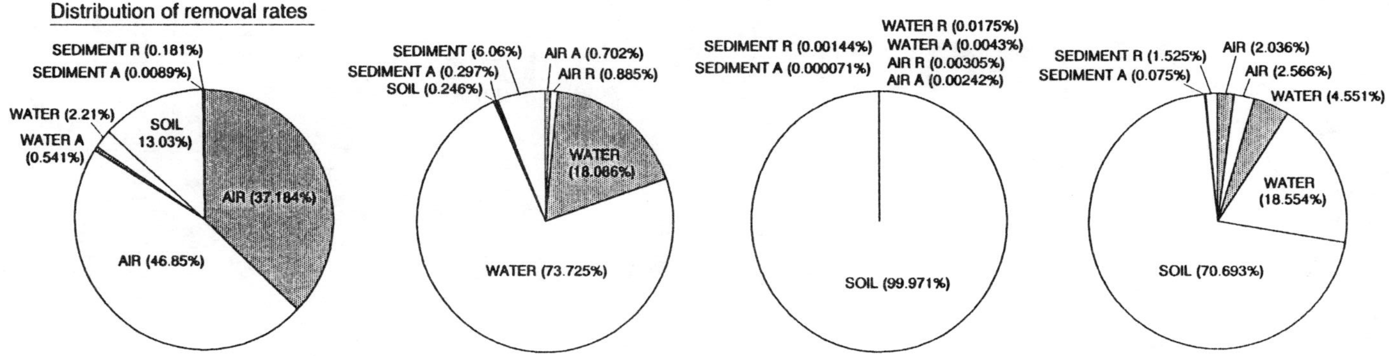

Emission rates:

E(1) = 1000 kg/h	E(1) = 0		E(1) = 0	E(1) = 50 kg/h
E(2) = 0	E(2) = 1000 kg/h		E(2) = 0	E(2) = 250 kg/h
E(3) = 0	E(3) = 0	Reaction / Advection	E(3) = 1000 kg/h	E(3) = 700 kg/h

Residence time:

t = 150 h	t = 332 h	t = 794 h	t = 646 h

Chemical name: Butyl benzyl phthalate

Level I calculation: (six-compartment model)

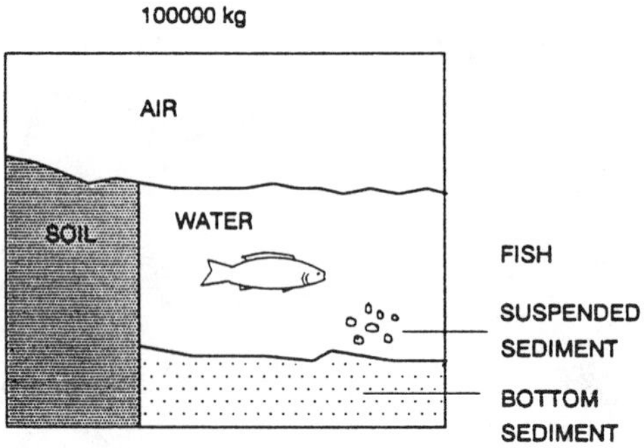

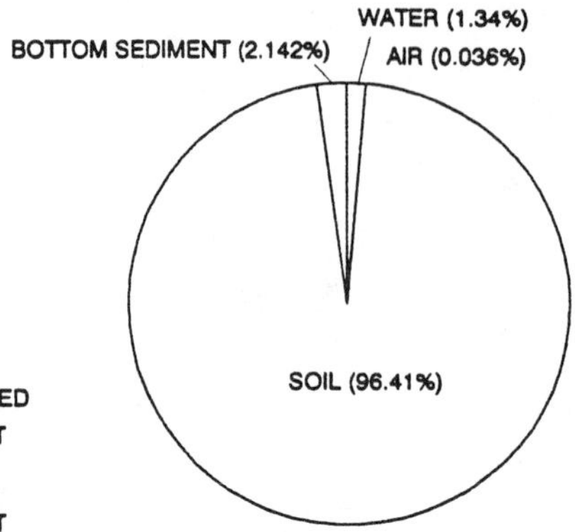

Distribution of mass

Level I calculation:

Physical-chemical properties:		Partition coeff.:	
molecular wt., g/mol	312.39	HLC	0.13355
melting point, C	-35	Kaw	5.39E-05
solubility, g/m3	2.69	Koc	33326.051
vapor pressure, Pa	0.0012	BCF	4064.1526
log Kow	4.91	Ksw	1599.6505
fugacity ratio, F	1.0000	Ksd/w	3199.3009
dissoc. const. pKa		Kssd/w	9997.815
		Kar/w	5.22E+09

Compartment	Z	Concentration			Amount	Amount
	mol/m3 Pa	mol/m3	mg/L (or g/m3)	ug/g	kg	%
Air	4.03E-04	1.15E-12	3.61E-10	3.04E-07	36.079	0.0361
Water	7.49E+00	2.14E-08	6.70E-06	6.70E-06	1339.32	1.3393
Soil	1.20E+04	3.43E-05	1.07E-02	4.46E-03	96410	96.410
Biota (fish)	3.04E+04	8.71E-05	2.72E-02	2.72E-02	5.4432	5.44E-03
Suspended sediment	7.49E+04	2.14E-04	6.70E-02	4.46E-02	66.9512	6.70E-02
Bottom sediment	2.40E+04	6.86E-05	2.14E-02	8.93E-03	2142.44	2.1424
	Total				100000	100

f = 2.863E-09 Pa

Chemical name: Butyl benzyl phthalate

Level II calculation: (six-compartment model)

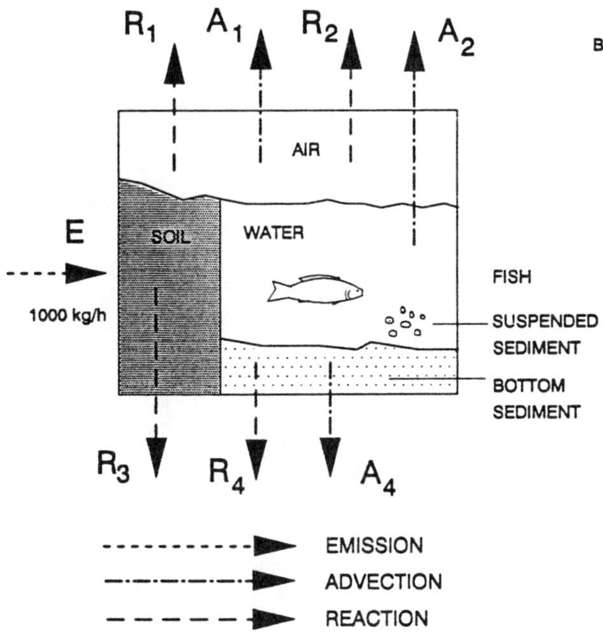

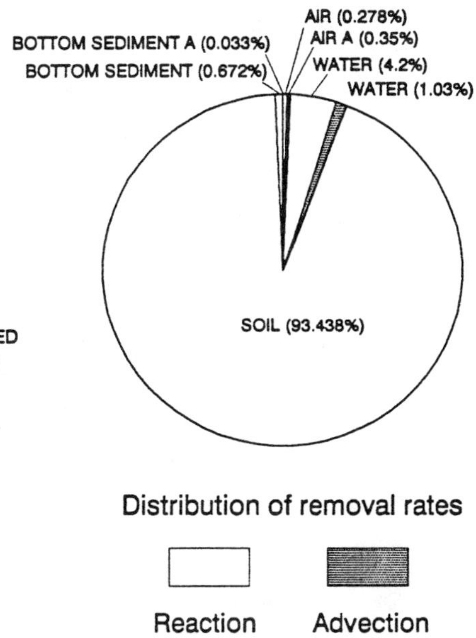

Distribution of removal rates

Level II Calculation:

Compartment	Half-life h	D Values Reaction mol/Pa.h	D Values Advection mol/Pa.h	Conc'n mol/m3	Loss Reaction kg/h	Loss Advection kg/h	Removal %
Air	55	5.08E+08	4.03E+08	8.88E-12	3.4967	2.7751	0.6272
Water	170	6.10E+09	1.50E+09	1.65E-07	41.995	10.302	5.230
Soil	550	1.36E+11		2.64E-04	934.384		93.4384
Biota (fish)				6.70E-04			
Suspended sediment				1.65E-03			
Bottom sediment	1700	9.77E+08	4.79E+07	5.28E-04	6.7178	0.3296	0.7047
	Total	1.42E+11	1.90E+09		986.59	13.407	100
	R + A		1.44E+11			1000	

f = 2.202E-08 Pa

Total amt = 769190 kg

Overall residence time = 769.19 h

Reaction time = 779.64 h

Advection time = 57373.88 h

Fugacity Level III calculations: (four-compartment model)

Chemical name: Butyl benzyl phthalate

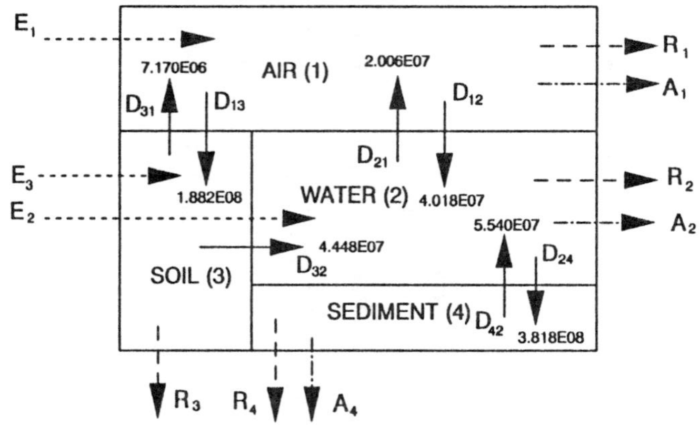

Phase Properties and Rates:

Compartment	Bulk Z mol/m3 Pa	Half-life h	D Values Reaction mol/Pa h	D Values Advection mol/Pa h
Air (1)	4.455E-04	55	5.61E+08	4.46E+08
Water (2)	7.893E+00	170	6.43E+09	1.58E+09
Soil (3)	5.991E+03	550	1.36E+11	
Sediment (4)	4.797E+03	1700	9.78E+08	4.80E+07

	E(1)=1000	E(2)=1000	E(3)=1000	E(1,2,3)
Overall residence time =	166.38	289.31	793.45	636.06 h
Reaction time =	262.79	357.57	793.52	681.10 h
Advection time =	453.48	1515.39	9697881	9618.43 h

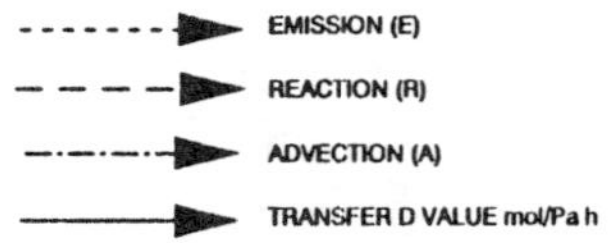

Phase Properties, Compositions, Transport and Transformation Rates:

Emission, kg/h			Fugacity, Pa				Concentration, g/m3				Amounts, kg				Total
E(1)	E(2)	E(3)	f(1)	f(2)	f(3)	f(4)	C(1)	C(2)	C(3)	C(4)	m(1)	m(2)	m(3)	m(4)	Amount, kg
1000	0	0	2.592E-06	1.242E-08	3.589E-09	4.387E-09	3.607E-07	3.063E-05	6.716E-03	6.574E-03	3.607E+04	6.126E+03	1.209E+05	3.287E+03	1.664E+05
0	1000	0	6.193E-09	3.813E-07	8.576E-12	1.347E-07	8.619E-10	9.402E-04	1.605E-05	2.018E-01	8.619E+01	1.880E+05	2.889E+02	1.009E+05	2.893E+05
0	0	1000	1.387E-10	1.254E-10	2.355E-08	4.429E-11	1.931E-11	3.092E-07	4.408E-02	6.637E-05	1.931E+00	6.184E+01	7.934E+05	3.319E+01	7.935E+05
50	250	700	1.312E-07	9.604E-08	1.667E-08	3.392E-08	1.826E-08	2.368E-04	3.119E-02	5.082E-02	1.826E+03	4.736E+04	5.615E+05	2.541E+04	6.361E+05

Emission, kg/h			Loss, Reaction, kg/h				Loss, Advection, kg/h			Intermedia Rate of Transport, kg/h T12	T21	T13	T31	T32	T24	T42
E(1)	E(2)	E(3)	R(1)	R(2)	R(3)	R(4)	A(1)	A(2)	A(4)	air-water	water-air	air-soil	soil-air	soil-water	water-sed	sed-water
1000	0	0	4.545E+02	2.497E+01	1.52E+02	1.340E+00	3.607E+02	6.126E+00	6.574E-02	3.253E+01	7.786E-02	1.524E+02	8.038E-03	4.986E-02	1.482E+00	7.592E-02
0	1000	0	1.086E+00	7.665E+02	3.64E-01	4.113E+01	8.619E-01	1.880E+02	2.018E+00	7.774E-02	2.390E+00	3.642E-01	1.921E-05	1.191E-04	4.548E+01	2.330E+00
0	0	1000	2.433E-02	2.521E-01	1.00E+03	1.353E-02	1.931E-02	6.184E-02	6.637E-04	1.741E-03	7.860E-04	8.157E-03	5.275E-02	3.272E-01	1.496E-02	7.665E-04
50	250	700	2.301E+01	1.931E+02	7.07E+02	1.036E+01	1.826E+01	4.736E+01	5.082E-01	1.647E+00	6.019E-01	7.716E+00	3.733E-02	2.316E-01	1.145E+01	5.869E-01

Level III Distribution

Chemical name: Butyl benzyl phthalate

Distribution of mass

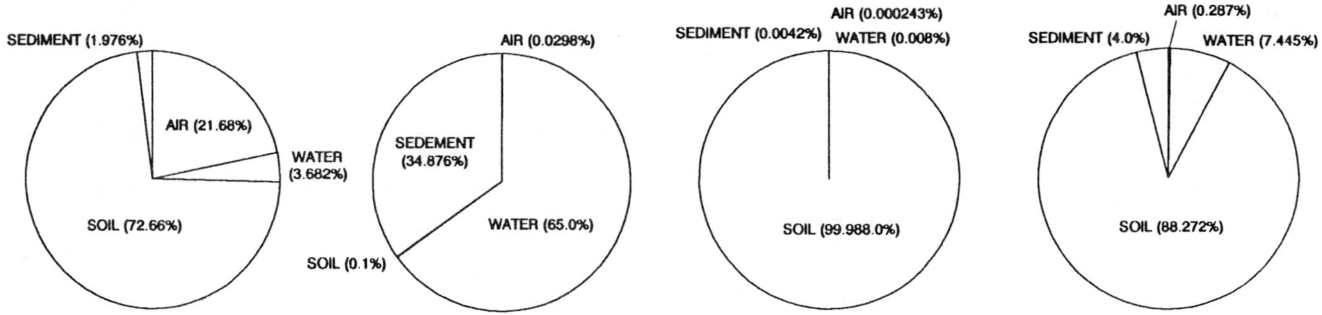

Distribution of removal rates

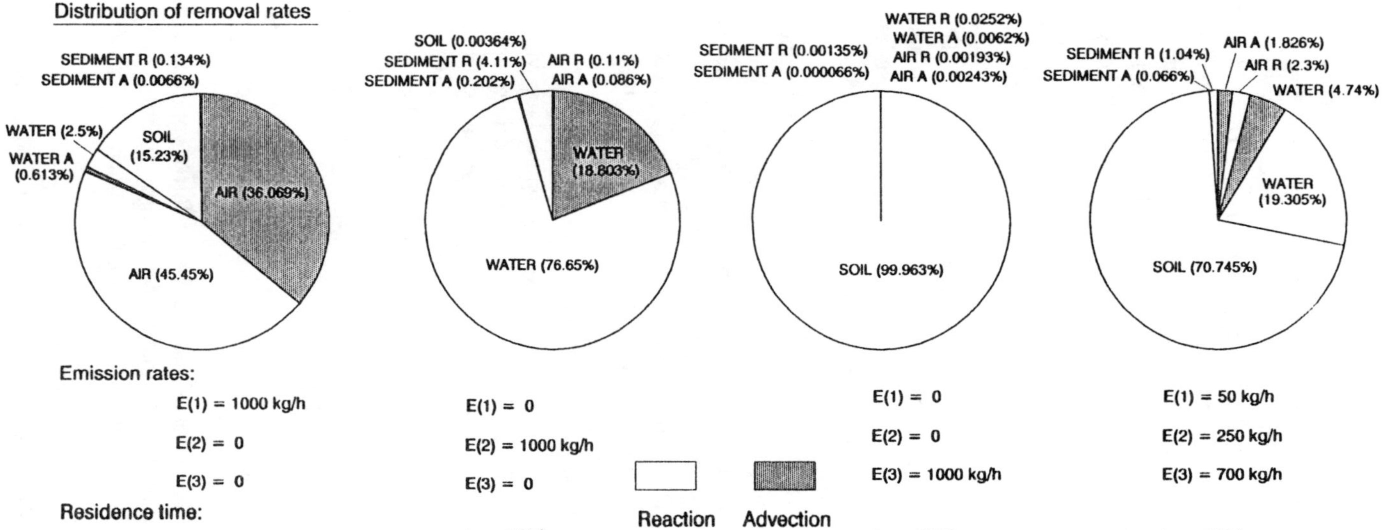

Emission rates:

E(1) = 1000 kg/h	E(1) = 0	E(1) = 0	E(1) = 50 kg/h
E(2) = 0	E(2) = 1000 kg/h	E(2) = 0	E(2) = 250 kg/h
E(3) = 0	E(3) = 0	E(3) = 1000 kg/h	E(3) = 700 kg/h

Residence time:

t = 166 h	t = 289 h	t = 793 h	t = 636 h

6.4 Commentary on the Physical-Chemical Properties and Environmental Fate

Properties and Reactivity

The selected physical-chemical properties of the esters are given in Tables 6.1 and 6.2. The aliphatic esters are fairly soluble in water and have high vapor presures and low K_{OW}s. The phthalate ester are notable because of their low solubility, high K_{OW} and low vapor pressure. Table 6.3 gives the estimated half-lives. These half-lives shoud be regarded as only tentative, average values or ranges. The literature suggests that for the aliphatic esters, air and water half-lives are typically class 3 with soil being one class higher and sediment 2 classes higher. The phthalate esters are more stable and can have half-lives of months or longer. Being more hydrophobic, they are likely to be less bioavailabe to degradation.

QSPR Plots

Figures 6.1 to 6.6 give the QSPR plots. The various molecular descriptors in Fig. 6.1 are clearly related to molar volume and to each other. The solubility data in Fig. 6.2 appear to be well behaved showing the expected decrease in solubility with increased molar volume, however, it must be noted that the data extend over 8 decades, thus a factor of 10 variation appears relatively small. It is clearly desirable to establish separate QSPRs for the these classes of esters. The same behavior is apparent in the vapor pressure data in Figure 6.3 which display a steady drop of vapor pressure with molar volume. The log K_{OW} data in Figure 6.4 suggest an increase with molar volume similar to that of other series.

The Henry's law constant data in Figure 6.5 are very scattered and the values are quite low. The K_{OW}-solubility plot in Figure 6.6 is surprisingly well behaved with a slope of about 0.9.

Because of the considerable variation in chemical structure of the esters, it is probably best to estimate the properties of a compound for which data are not available from experimentally determined data of a similar substance, making appropriate adjustments for properties such as chain length. These plots can give a guidance for these procedures. A notable feature of the esters is their very wide variation in properties with K_{OW} extending from less than 1 for methyl formate to values of over 1 million for phthalate esters.

Evaluative Calculations

Illustrated calculations are given for selected aliphatic and phthalate esters.

The aliphatic esters, being fairly volatile, partition appreciably into air as shown in the Level I diagrams. Most loss is thus by reaction or advection in air with water phase processes generally playing a more minor role as shown by Level II calculation.

The Level III calculations suggest that if discharged to air, they tend to react or advect. There will be little tendency to be absorbed in soil or in water. If discharged to water, there can be appreciable evaporation but reaction in water is most important. When discharge is to soil, there is also significant evaporation. The low K_{OW} values suggest minimal bioconcentration.

The phthalate esters, being generally more hydrophobic tend to partition more to soils and sediments, and as a result reactions in these solid, sorbing media are more important. There is a greater tendency to bioconcentrate.

The Level III calculations suggest that these aromatic esters are more susceptible to deposition from the atmosphere because of their low Henry's law constants. They are also more susceptible to sedimentation and bioconcentration. Their environmental persistence is largely controlled by their reactivity in water, soil and sediment and is generally about a month, but may be considerable longer if degradation conditions are not favorable. They are also fairly susceptible to long range atmospheric transport, especially in situations when atmospheric reactions are slow and there is little deposition. This group of chemicals is thus of more concern as being potentially more widespread and persistent in character than the alcohols and acids treated elsewhere in this volume.

6.5 REFERENCES:

ACGIH (1986) *Documentation of The Threshold Limit Values for Substances in Workroom Air.* 5th Edition, American Conference of Governmental and Industrial Hygienists (ACGIH), Inc., Cincinnati, Ohio.

Abraham, M.H. (1984) Thermodynamics of solution of homologous series of solutes in water. *J. Chem. Soc., Farad. Trans. 1* 80, 153-181.

Aichinger, G., Grady Jr., C.P.L., Tabak, H.H. (1992) Application of respirometric biodegradability testing protocol to slightly soluble organic compounds. *Water Environ. Res.* 64(7), 890-900.

Altshuller, A.P., Everson, H.E. (1953) The solubility of ethyl acetate in water. *J. Am. Chem. Soc.* 75, 1727.

Ambrose, D., Ellender, J.H., Gundry, H.A., Lee, D.A., Townsend, R. (1981) Thermodynamic properties of organic oxygen compounds. LI. The vapour pressures of some esters and fatty acids. *J. Chem. Thermodyn.* 13, 795-802.

Amidon, G.L., Williams, N.A. (1982) A solubility for non-electrolytes in water. *Int'l J. Pharm.* 11, 249-256.

Amidon, G.L., Yalkowsky, S.H., Anik, S.T., Valvani, S.C. (1975) Solubility of nonelectrolytes in polar solvents. V. Estimation of the solubility of aliphatic monofunctional compounds in water using a molecular surface area approach. *J. Phys. Chem.* 79(21), 2239-2246.

Andrews, L.J., Keefer, R.M. (1950) Cation complexes of compounds containing carbon-carbon double bonds. VI. The argentation of substituted benzenes. *J. Am. Chem. Soc.* 72, 3113-3116.

Atkinson, R. (1985) Kinetics and mechanisms of the gas-phase reactions of hydroxyl radicals with organic compounds under atmospheric conditions. *Chem. Rev.* 85, 69-201.

Atkinson, R. (1987) Structure-activity relationship for the estimation of rate constants for the gas-phase reactions of OH radicals with organic compounds. *Int'l J. Chem. Kinetics.* 19, 799-828.

Atkinson, R. (1989) Kinetics and mechanisms of the gas-phase reactions of the hydroxyl radical with organic compounds. *J. Phys. Chem. Ref. Data* Monograph 1, 1-246.

Atkinson, R., Darnall, K.R., Lloyd, A.C., Winer, A.M., Pitts, J.N., Jr. (1979) Kinetics and mechanisms for the reaction of hydroxyl radicals with organic chemicals in the gas-phase. *Adv. Photochem.* 11, 375-488.

Atlas, E., Foster, R, Giam, C.S. (1982) Air-sea exchange of high-molecular weight organic pollutants: Laboratory studies. *Environ. Sci. Technol.* 16, 283-286.

Atlas, E., Velasco, A., Sullivan, K., Giam, C.S. (1983) A radiotracer study of air-water exchange of synthetic organic compounds. *Chemosphere* 12(9/10), 1251-1258.

Balson, E.W. (1958) *The Measurement of Vapour Pressure by Torsion-Effusion Method: Surface Phenomena in Chemistry and Biology.* Pergamon Press, Oxford.

Banerjee, S. (1980) unpublished work, Syracuse Research Corporation, Syracuse, New York.

Banerjee, S. (1984) Solubility of organic mixtures in water. *Environ. Sci. Technol.* 18(8), 587-591.

Banerjee, S. (1985) Calculation of water solubility of organic compounds with UNIFAC-derived parameters. *Environ. Sci. Technol.* 19, 369-370.

Banerjee, S., Baughman, G.L. (1991) Bioconcentration factors and lipid solubility. *Environ. Sci. Technol.* 25, 536-539.

Banerjee, S., Howard, P.H. (1988) Improved estimation of solubility and partitioning through correction of UNIFAC-derived activity coefficients. *Environ. Sci. Technol.* 22, 839-841.

Banerjee, S., Howard, P.H., Lande, S.S. (1990) General structure-vapor pressure relationships for organics. *Chemosphere* 21(10-11), 1173-1180.

Barrows, M.E., Petrocelli, S.R., Macek, K.J. (1980) Bioconcentration and elimination of selected water pollutants by bluegill sunfish (*lepomis macrochiras*). In: *Dynamics, Exposure, Hazard Assessment Toxic Chemicals*. Hauqe, R., Editor, pp.379-392. Ann Arbor Science Publishers Inc., Ann Arbor, Michigan.

Blum, D.J.W., Suffet, I.H., Duguet, J.P. (1994) Quantitative structure-activity relationship using molecular connectivity for the activated carbon adsorption of organic chemicals in water. *Water Res.* 28, 687-699.

Bodor, N., Gabanyi, Z. Wong, C.K. (1989) A new method for the estimation of partition coefficient. *J. Am. Chem. Soc.* 111, 3783-3786.

Boethling, R.S., Gregg, B., Frederick, R., Gabel, N.W., Campbell, S.E., Sabljic, A. (1989) Expert systems survey on biodegradation of xenobiotic chemicals. *Ecotoxicol. Environ. Saf.* 18, 252-267.

Booth, H.S., Everson, H.E. (1948) Hydrotropic solubilities. Solubilities in 40 per cent sodium xylenesulfonate. *Ind. Eng. Chem.* 40(8), 1491-1493.

Boublik, T., Fried, V., Hala, E. (1984) *The Vapor Pressures of Pure Substances.* 2nd Edition, Elsevier, Amsterdam, The Netherlands.

Bouwer, E.J., McCarty, P.L., Lance, J.C. (1981) Trace organic behavior in soil columns during rapid infiltration of secondary wastewater. *Water Res.* 15, 151-159.

Branson, D.R. (1978) Predicting the fate of chemicals in the aquatic environment from laboratory data. In: *Estimating the Hazard of Chemical Substances to Aquatic Life. ASTM STP 657*, Cairns, Jr., J., Dickson, K.L., Maki, A.W., Eds., pp. 55-70, American Society for Testing and Materials, Philadelphia.

Brinkman, U.A., Schouten,M.J., Peereboom, J.W.C. (1979) Liquid chromatographic analysis of phthalate esters in Deutch river waters. *Intl. J. Environ. Anal. Chem.* 7, 13-23.

Brooke, D., Nielsen, I., De Bruijn, J., Hermens, J. (1990) An interlaboratory evaluation of the stir-flask method for the determination of octanol-water partition coefficients (LOG P_{OW}). *Chemosphere* 21(1/2), 119-133.

Brown, D., Thompson, R.S. (1982a) Phthalates and the aquatic enviornment: Part I. The effect of di-2-ethylhexyl phthalate (DEHP) and di-isodecyl phthalate (DIDP) on the reproduction of *dphnia magna* and observations on their bioconcentration. *Chemosphere* 4, 417-426.

Brown, D., Thompson, R.S. (1982b) Phthalates and the aquatic environment: Part II. The bioconcentration and depuration of di-2-ethylhexyl phthalate (DEHP) and diisodecyl phthalate (DIDP) in mussels (*mytilus edulis*). *Chemosphere* 11(4), 427-435.

Burrows, G. (1948) Determination of the boiling-points of liquids of low vapour pressure. *J. Soc. Chem. Ind. (London)* 65, 360-365.

Butler, J.A.V., Ramchandani, C.N. (1935) The solubility of nonelectrolytes. Part II. The influence of the polar group on the free energy of hydration of aliphatic compounds. *J. Chem. Soc.* (London) 952-955.

Buttery, R.G., Ling, J.C., Guadagni, D.G. (1969) Food volatiles. Volatility of aldehydes, ketones, and esters in dilute water solution. *J. Agric. Food & Chem.* 17, 385-389.

Buxton, G.V., Greenstock, C.L., Helman, W.P., Ross, A.B. (1988) *J. Phys Chem. Ref. Data* 17, 513-886.

Callahan, M.A., Slimak, M.W., Gabel, N.W., May, I.P., Fowler, C.F., Freed, J.R., Jennings, P., Durfee, R.L., Whitemore, F.C., Maestri, B., Mabey, W.R., Holt, B.R., Gould, C. (1979) *Water Related Environmental Fate of 129 Priority Pollutants.* EPA No.440-4-79-029a,b. Versar, Inc., Springfield, Virginia.

Campbell, I.M., Parkinson, P.E. (1978) Rate constants for reactions of hydroxyl radicals with ester vapours at 292 K. *Chem. Phys. Lett.* 53, 385-387.

Casserly, D.M., Davis, E.M., Downs, T.D., Guthrie, R.K. (1983) Sorption of organics by *selenastrum capricornutum. Water Res.* 17(11), 1591-1594.

CHEMFATE *The Environmental Fate Database*-A Computerized Database by Technical Database Services Incorporated, New York, New York.

Chen, F., Holten-Andersen, J., Tyle, H. (1993) New development of the UNIFAC model for environmental application. *Chemosphere* 26(7), 1325-1354.

Collander, R. (1951) Partition of organic compounds between higher alcohols and water. *Acta. Chem. Scand.* 5, 774-780.

Conkle, J.P., Camp, B.J., Welch, B.E. (1975) Trace composition of human respiratory gas. *Arch. Environ. Health* 30, 290-295.

Cowen, W.F., Baynes, R.K. (1980) Estimated application of gas chromatographic headspace analysis to priority pollutants. *J. Environ. Sci. Health* A15(5), 413-427.

D'Amboise, M., Hanai, T. (1982) Hydrophobicity and retention in reversed phase liquid chromatography. *J. Liq. Chromatogr.* 5(2), 229-244.

Darnall, K.R., Lloyd, A.C., Winer, A.M., Pitts, Jr., J.N. (1976) Reactivity scale for atmospheric hydrocarbons based on reaction with hydroxyl radical. *Environ. Sci. Technol.* 10(7), 692-696.

Daubert, T.E., Danner, R.P. (1989) *Physical and Thermodynamics Properties of Pure Chemicals: Data Compilation.* Hemisphere Publication Corp., New York.

Davies, R.P., Dobbs, A.J. (1984) The prediction of bioconcentration in fish. *Water Res.* 18(10), 1253-1262.

Dean, J.D., Editor (1985) *Lange's Handbook of Chemistry.* 13th Edition, McGraw-Hill, Inc., New York.

De Bruijn, J., Busser, F., Seinen, W., Hermens, J. (1989) Determination of octanol/water partition coefficients for hydrophobic organic chemicals with the "slow-stirring" method. *Environ. Toxicol. Chem.* 8, 499-512.

DeFoe, D.L., Holcombe, G.W., Hammermeister, D.E. (1990) Solubility and toxicity of eight phthalate esters to four aquatic organisms. *Environ. Toxicol. Chem.* 9, 623-636.

De Kock, A.C., Lord, D.A. (1987) A simple procedure for determining octanol-water partition coefficients using reversed phase high performance liquid chromatography (RPHPLC). *Chemosphere* 16(1), 133-142.

Deno, N.C., Berkheimer, H.E. (1960) Part I. Phase equilibria molecular transport thermodynamics. *J. Chem. Eng. Data* 5(1), 1-5.

Dobbs, A.J., Cull, M.R. (1982) Volatilization of chemicals-relative loss rates and the estimation of vapour pressures. *Environ. Pollut.* 3, 289-298.

Dobbs, A.J., Hart, G.F., Parsons, A.H. (1984) The determination of vapour pressure from relative volatilization rates. *Chemosphere* 13(5/6), 687-692.

Dojlido, J.R. (1979) *Investigations of Biodegradability and Toxicity of Organic Compounds.* Final Report 1975-79. U.S. EPA-600/2-79-163.

Donahue, D.J., Bartell, F.E. (1952) The boundary tension at water-organic liquid interfaces. *J. Phys. Chem.* 56, 480-484.

Dorfman, L.M., Adams, G.E. (1973) *Reactivity of the Hydroxyl Radical in Aqueous Solution.* NSRD-NDB-46. NTIS COM-73-50623. National Bureau of Standards, Wasington DC.

Dow Chemical (1979) Personal communication from M. Thomas of Dow Chemical Company to N.W. Gabel of Versar, Inc.

Dreisbach, R. (1952) *Pressure-Volume-Temperature Relationships of Organic Compounds.* Handbook Publishers Inc., Sandusky, Ohio.

Drossman, H., Johnson, H., Mill, T. (1988) Structure activity relationships for environmental processes 1: Hydrolysis of ester and carbamates. *Chemosphere* 17, 1509-1530.

Eadsforth, C.V. (1986) Application of reverse-phase HPLC for the determination of partition coefficients. *Pestic. Sci.* 17, 311-325.

Efroymson, R.A., Alexander, M. (1994) Biodegradation in soil of hydrophobic pollutants in nonaqueous-phase liquids (NAPLs). *Environ. Toxicol. Chem.* 13(3), 405-411.

Ellington, J.J. (1989) *Hydrolysis Rate Constants for Enhancing Property-Reactivity Relationships.* EPA/600/3-89/063. NTIS PB89-220479. U.S. Environmental Protection Agency, Environmental Research Laboratory, Athens, Georgia.

Ellington, J.J., Stancek, F.E., Payne, W.D., Trusty, C.D. (1987) *Measurement of Hydrolysis Rate Constants for Evaluation of Hazardous Waste Land Disposal. Vol. I. Data on 54 chemicals.* U.S. EPA-600/3-86-043.

Ellington, J.J., Stancil, F.E., Jr., Payne, W.D., Trusty, C.D. (1987) *Measurements of Hydrolysis Rate Constants for Evaluation of Hazardous Waste Land Disposal: Data on 54 Chemicals.* Volume II. EPA/600/53-87/019. U.S. Environmental Protection Agency, Athens, Georgia.

Elzerman, A.W., Coates, J.T. (1987) Hydrophobic organic compounds on sediments: Equilibria and kinetics of sorption. In: *Sources and Fates of Aquatic Pollutants.* Hites, R.A., Eisenreich, S.J., Editors, Advances in Chemistry Series 216, American Chemical Society, Washington, DC.

ENVIROFATE *Environmental Fate Database.* A Computerized Databse sponsored by U.S. EPA's The Office of Toxic Substances.

EPA (1985) *Environmental Fate Database.* A Computerized Databse sponsored by U.S. EPA's The Office of Toxic Substances.

Fishbein, L., Albro, P.W. (1972) Chromatographic and biological aspects of the phthalate esters. *J. Chromatography.* 70, 365-412.

Freitag, D., Ballhorn, L., Geyer, H., Körte, F. (1985) Environmental hazard profile of organic chemicals. An experimental method for the assessment of the behaviour of organic chemicals in the ecosphere by means of simple laboratory tests with ^{14}C labelled chemicals. *Chemosphere* 14(10), 1589-1616.

Freitag, D., Geyer, H., Kraus, A., Viswanathan, R., Kortzias, D., Attar, A., Klein, W., Körte, F. (1982) Ecotoxicological profile analysis. VII. Screening chemicals for their environmental behavior by comparative evaluation. *Ecotoxicol. Environ. Safety* 6, 60-81.

Fühner, H. (1924) Die wasserlöslichkeit in homologen reilen. *Chem. Ber.* 57, 510-515.

Fujisawa, S., Masuhara, E. (1980) Binding of methyl methacrylate to bovine serum albumin. *J. Den. Res.* 59, 2056-2061.

Fujisawa, S., Masuhara, E. (1981) Determination of partition coefficients of acrylates, methacrylates, and vinyl monomers using high performance liquid chromatography (HPLC). *J. Biomed. Mater. Res.* 15, 787-793.

Garten, Jr., C.T., Trabalka, J.R. (1983) Evaluation of models for predicting terrestrial food chain behavior of xenobiotics. *Environ. Sci. Technol.* 17, 590-595.

GEMS (1984) Graphical Exposure Modeling System. FAP, Fate Atmospheric Pollution. Data Base, Office of Toxic Substances, U.S. Environmental Protection Agency.

GEMS (1986) Graphical Exposure Modeling System. FAP, Fate Atmospheric Pollution. Data Base, Office of Toxic Substances, U.S. Environmental Protection Agency.

GEMS (1989) Graphical Exposure Modeling System. FAP, Fate Atmospheric Pollution. Data Base, Office of Toxic Substances, U.S. Environmental Protection Agency.

Geyer, H., Kraus, A.G., Klein, W., Richter, E., Körte, F. (1980) Relationship between water solubility and bioaccumulation potential of organic chemicals in rats. *Chemosphere* 9, 277-291.

Geyer, H., Politzki, G., Freitag, D. (1984) Prediction of ecotoxicological behavior of chemicals: Relationship between *n*-octanol/water partition coefficient and bioaccumulation of organic chemicals by alga *chlorella*. *Chemosphere* 113(2), 269-284.

Geyer, H., Sheehan, P., Kotzias, D., Freitag, D., Körte, F. (1982) Prediction of ecotoxicological behaviour of chemicals: relationship between phisico-chemical properties and bioaccumulation of organic chemicals in the mussel *mytilus edulis*. *Chemosphere* 11(11), 1121-1134.

Giam, C.S., Atlas, E., Chan, H.S., Neff, G.S. (1980) Phthalate esters, PCB and DDT residues in the Gulf of Mexico atmosphere. *Atmos. Environ.* 14, 65-69.

Gledhill, W.E., Kaley, R.G., Adams, W.J., Hicks, O., Michael, P.R., Saeger, V.W., Leblanc, G.A. (1980) An environmental safety assessment of butyl benzyl phthalate. *Environ. Sci. Technol.* 14, 301-305.

Gluck, S.J., Martin, E.J. (1990) Extended octanol-water partition coefficient determination by dual-mode centrifugal partition chromatography. *J. Liq. Chromatogr.* 13(18), 3559-3570.

Govind, R., Flaherty, P.A., Dobbs, R.A. (1991) Fate and effects of semivolatile organic pollutants during anaerobic digestion of sludge. *Water Res.* 25(5), 547-556.

Grayson, B.T., Fosbraey, L.A. (1982) Determination of the vapour pressure of pesticides. *Pestic. Sci.* 13, 269-278.

Gückel, W., Kästel, R., Lewerenz, J., Synnatschke, G. (1982) A method for determining the volatility of active ingredients used in plant protection. Part III: The temperature relationship between vapour pressure and evaporation rate. *Pestic. Sci.* 13, 161-168.

Güesten, H., Klasinc, L., Maric, D. (1984) Prediction of the abiotic degradability of organic compounds in the troposphere. *J. Atmos. Chem.* 2, 83-93.

Haag, W.R., Yao, D.C.C. (1992) Rate constants for reaction of hydroxyl radicals with several drinking water contaminants. *Environ. Sci. Technol.* 26(5), 1005-1013.

Haky, J.E., Leja, B. (1986) Evaluation of octanol-water partition coefficients using capillary gas chromatography with cold on-column injection. *Anal. Lett.* 19(1-2), 123-134.

Haky, J.E., Young, A.M. (1984) Evaluation of simple HPLC correlation method for the estimation of the octanol-water partition coefficients of organic compounds. *J. Chromatogr.* 7, 675-689.

Hamilton, D.J. (1980) Gas chromatographic measurement of volatility of herbicide esters. *J. Chromatogr.* 195, 75-83.

Hanai, T., Hubert, J. (1982) Hydrophobicity and chromatographic behaviour of aromatic acids found in urine. *J. Chromatogr.* 239, 527-536.

Hansch, C., Anderson, S.M. (1967) The effect of intramolecular hydrophobic bonding on partition coefficients. *J. Org. Chem.* 32, 2583.

Hansch, C., Leo, A. (1979) *Substituents Constants for Correlation Analysis in Chemistry and Biology*. Wiley, New York.

Hansch, C., Leo, A.J. (1982) *Medchem Project* Issue No.20, Pomona College, Claremont, California.

Hansch, C., Leo, A. (1985) *Medchem Project* Issue No. 26. Pomona College, Claremont, California.

Hansch, C., Quinlan, J.E., Lawrence, G.L. (1968) The linear free-energy relationship between partition coefficients and aqueous solubilities of organic liquids. *J. Org. Chem.* 33, 347-350.

Harnish, M., Möckel, H.J., Schulze, G. (1983) Relationship between LOG P_{OW} shake-flask values and capacity factors derived from reversed-phase HPLC for *n*-alkylbenzenes and some OECD reference substances. *J. Chromatogr.* 282, 315-332.

Hattori, Y., Kuge, Y., Nakagawa, S. (1975) Microbial decomposition of phthalate esters in environmental water. *Pollut. Control Center Osaka Perfect, Mizu Shori Gijutsu.* 16, 951-954.

Hermens, J., Leeuwangh, P. (1982) Joint toxicity of mixtures of 8 and 24 chemicals to the guppy *(poecilia reticulata). Ecotoxicol. Environ. Saf.* 6, 302-310.

Heukelekian, H., Rand, M.C. (1955) Biochemical oxygen demand of pure organic compounds. *J. Water Pollut. Control Assoc.*29, 1040-1053.

Hine, J., Mookerjee, P.K. (1975) The intrinsic hydrophilic character of organic compounds. Correlations in terms of structural contributions. *J. Org. Chem.* 40, 292-298.

Hinckley, D.A., Bidleman, T.F., Foreman, W.T. (1990) Determination of vapor pressures for nonpolar and semipolar organic compounds from gas chromatographic retention data. *J. Chem. Eng. Data* 35, 232-237.

Hinklet, D.A., Bidleman, T.F., Foreman, W.T., Tuschall, J.R. (1990) Determination of vapor pressures for nonpolar and semipolar oraganic compounds from gas chromatographic retention data. *J. Chem. Eng. Data* 35, 232-237.

Hirzy, J.W., Adams, W.J., Gledhill, W.E., Mieure, J.P. (1978) Phthalate Esters: The Environmental Issues. Unpublished seminar document, Monsanto Industrial Chemicals Company.

Hoff, J.T., Mackay, D., Gillham, R., Shiu, W.Y. (1993) Partitioning of organic chemicals at the air-water interface in environmental systems. *Environ. Sci. Technol.* 27(10), 2174-2180.

Hollifield, H.C. (1979) Rapid nephelometric estimate of water solubility of highly insoluble organic chemicals of environmental interest. *Bull. Environ. Contam. Toxicol.* 23, 579-586.

Horowitz, A., Shelton, D.R., Cornell, C.P., Tiedje, J.M. (1982) Anaerobic degradation of aromatic compounds in sediment and digested sludge. *Dev. Ind. Microbiol.* 23, 435-444.

Howard, P.H., Editor (1989) *Handbook of Fate and Exposure Data for Organic Chemicals. Vol. I-Large Production and Priority Pollutants.* Lewis Publishers, Chelsea, Michigan.

Howard, P.H., Editor (1990) *Handbook of Fate and Exposure Data for Organic Chemicals. Vol. II-Solvents.* Lewis Publishers, Chelsea, Michigan.

Howard, P.H., Editor (1993) *Handbook of Fate and Exposure Data for Organic Chemicals. Vol. IV. Sovents 2.* Lewis Publishers, Chelsea, Michigan.

Howard, P.H., Banerjee, S., Robillard, K.H. (1985) Measurement of water solubilities, octanol/water partition coefficients and vapor pressures of commercial phthalate esters. *Environ. Toxicol. Chem.* 4, 653-661.

Howard, P.H., Boethling, R.S., Jarvis, W.F., Meylan, W.M., Michalenko, E.M. (1991) *Handbook of Environmental Degradation Rates.* Lewis Publishers, Chelsea, Michigan.

Hoy, K.L. (1970) New values of the solubility parameters from vapor pressure data. *J. Paint Technol.* 42(541), 76-118.

Hoyer, Von H., Peperle, W. (1957) Dampfdruckmessungen an organischen substanzen und ihre sublimationswärmen. *Zeit. für Elcktrochemie* 62(1), 61-66.

Hulzebos, E.M., Adema, D.M.M., Breemen, E.M. D.-van, Henzen, L., van Dis, W.A., Herbold, H.A., Hoekstra, J.A., Baerselman, R., van Gestel, C.A.M. (1993) Phytotoxicity studies with *lactuca sativa* in soil and nutrient solution. *Environ. Toxicol. Chem.* 12, 1079-1094.

Inman, J.C., Strachan, S.D., Sommers, L.E., Nelson, D.W. (1984) The decomposition of phthalate esters in soil. *J. Environ. Sci. Health* B19, 245-257.

ISHOW *Information System for Hazardous Organics in Water.* A Computerized Database developed by U.S. EPA's Environmental Research Lab. and The University of Minnesota, sponsored by U.S. EPA's Office of Toxic Substances.

Isnard, P., Lambert, S. (1988) Estimating bioconcentration factors from octanol-water partition coefficient and aqueous solubility. *Chemosphere* 17(1), 21-34.

Isnard, P., Lambert, S. (1989) Aqueous solubility/*n*-octanol water partition coefficients correlations. *Chemosphere* 18, 1837-1853.

Iwasa, J., Fujita, T., Hansch, C. (1965) Substituent constants for aliphatic functions obtained from partition coefficients. *J. Med. Chem.* 8, 150-153.

Iwase, K., Komatsu, K., Hirono, S., Nakagawa, S., Moriguchi, I. (1985) Estimation of hydrophobicity based on the solvent-accessible surface area of molecules. *Chem. Pharm. Bull.*. 33(5), 2114-2121.

Jaber, H. (1982) SRI International unpublished analysis.

Johnson, B.T., Heitkamp, M.A., Jones, J.R. (1984) Environmental and chemical factors influencing the biodegradation of phthalate acid esters in freshwater sediments. *Environ. Pollut* B8, 101-108.

Johnson, B.T., Lulves, W. (1975) Biodegradation of di-*n*-butyl phthalate and di-2-ethylhexyl phthalate in freshwater hydrosoil. *J. Fish Res. Board Can.* 32, 333-339.

Johnson, B.T., Stalling, D.L., Hogan, J.W., Schoettger, R.A. (1979) Dynamics of phthalic acid esters in aquatic organisms. In: *Fate of Pollutants in the Air and Water Environments. Part II.* Suffet, I.H., Editor, Wiley-Interscience Division of John-Wiley and Sons, New York, New York.

Jones, D.C. (1929) The systems *n*-butyl alcohol-water and *n*-butyl alcohol-acetone-water. *J. Chem Soc.* (London), 799-816.

Jordan, T.E. (1954) *Vapor Pressure of Organic Compounds.* Interscience, New York.

Joshi, S.B., Dodge, M.C., Bufalini, J.J. (1982) Reactivities of selected organic compounds and contamination effects. *Atmos. Environ.* 16, 1301-1310.

Kamlet, M.J., Abraham, D.J., Doherty, R.M., Taft, R.W., Abraham, M.H. (1986) Solubility properties in polymers and biological media: 6. An equation for correlation and prediction of solubilities of liquid nonelectrolytes in blood. *J. Pharm. Sci.* 75(4), 350-354.

Kamlet, M.J., Doherty, R.M., Abboud, J.-L.M., Abraham, M.H., R.F., Taft, R.W. (1986) Linear solvation energy relationships: 36. Molecular properties governing solubilities of organic nonelectrolytes in water. *J. Pharm. Sci.* 75(4), 338-349.

Kamlet, M.J., Doherty, R.M., Abraham, M.H., Carr, P.W., Doherty, R.F., Taft, R.W. (1987) Linear solvation energy relationships. 41. Important differences between aqueous solubility relationships for aliphatic and aromatic solutes. *J. Phys. Chem.* 91, 1996-2004.

Kamlet, M.J., Doherty, R.M., Carr, P.W., Mackay, D., Abraham, M.H., Taft, R.W. (1988) Linear solvation energy relationships. 44. Parameter estimation rules that allow accurate prediction of octanol/water partition coefficients and other solubility and toxicity properties of polychlorinated biphenyls and polycyclic aromatic hydrocarbons. *Environ. Sci. Technol.* 22, 503-509.

Kamlet, M.J., Doherty, R.M., Taft, R.W., Abraham, M.H., Veith, G.D., Abraham, D.J. (1987) Solubility properties in polymers and biological media: 8. An analysis of the factors that influence toxicities of organic nonelectrolytes to the golden orfe fish (*leuciscus idus melanotus*). *Enveron. Sci. Technol.* 21(2), 149-155.

Karara, A.H., Hayton, W.L. (1984) Pharmocokinetic model for the uptake and disposition of di-2-ethylhexyl phthalate in sheephead minnows (*cyprinodon variegatus*). *Aquatic Toxicol.* 5, 181-195.

Karickhoff, S.W. (1985) Chapter 3, Pollutant sorption in environmental systems. In: *Environmental Exposure From Chemicals*. Volume I. Neely, W.B., Blau, G.E., Editors, CRC Press, Inc., Boca Raton, Florida. pp.49-64.

Kelly, T.J., Mukund, R., Spicer, C.W., Pollack, A.J. (1994) Concentrations and transormations of hazardous air polllutants. *Environ. Sci. Technol.* 28, 378A-367A.

Kenaga, E.E. (1980) Predicted bioconcentration factors and soil sorption coefficients of pesticides and other chemicals. *Ecotoxicol. Environ. Saf.* 4, 26-38.

Kenaga, E.E., Goring, C.A.I. (1980) In: *Aquatic Toxicology.* Eaton, J.G., Parrish, P.R., Hendricks, A.C., Editors, American Society for Testing and Materials, STP 707, pp.78-115.

Kerr, J.A., Stoker, D.W. (1986) The photo-oxidation of propyl acetates in synthetic air-nitrous acid-nitrogen oxide mixtures: Product formation studies and proposed kinetic mechanisms. *J. Atmos. Chem.* 4, 263-276.

Kieckbusch, T.G., King, C.J. (1979) An improved method of determining vapor-liquid equilibria for dilute organics in aqueous solution. *J. Chromatogr. Sci.* 17, 273-276.

Kincannon, D.F., Lin, Y.S. (1985) Microbial degradation of hazardous wastes by land treatment. In: *Proc. Indust. Waste Conf.* 40, 607-619.

Klečka, G.M. (1985) Chapter 6, Biodegradation. In: *Environmental Exposure From Chemicals.* Volume I. Neely, W.B., Blau, G.E., Editors, CRC Press, Inc., Boca Raton, Florida. pp.110-155.

Klein, W., Kördel, W., Weiβ, M., Poremski, H.J. (1988) Updating of the OECD test guideline 107 "partition coefficient n-octanol/water:" OECD laboratory intercomparison test on HPLC method. *Chemosphere* 17, 361-386.

Kleopfer, R.D., Fairless, B.J. (1972) Characterization of organic compounds in a municipal water supply. *Environ. Sci. Technol.* 6, 1036-1037.

Klöpffer, W., Kaufmann, G., Rippen, G., Poremski, H-J. (1982) A laboratory method for testing the volatility from aqueous solution: First results and comparison with theory. *Ecotoxicol. Environ. Saf.* 6, 545-559.

Klopman, G., Namboodiri, K., Schochet, M. (1985) Simple method of computing partition coefficient. *J. Comput. Chem.* 6, 28-38.

Kollig, H.P., Ed. (1993) *Environmental Fate Constants for Organic Chemicals Under Consideration for EPA's Hazardous Waste Identification Project.* EPA Report EPA/600/R-93/132, Environmental Research Lab., Athens, GA.

Kurane, R., Suzuki, T., Takahara, Y. (1977) Isolation of microorganisms growing on phthalate esters and degradation of phthalate ester by *pseudomonas acidovorans* 256-1. *Agric. Biol. Chem.* 41, 2119-2123.

Kurane, R., Suzuki, T., Takahara, Y. (1978) Microbial population and identification of phthalate ester-utilizing microorganisms in activated sludge inoculated with microorganisms. *Agric. Biol. Chem.* 43, 907-917.

Laurence, W.H., Bas, G.E., Purcell, W.P., Autian, J. (1972) Use of mathematical models in the study of structure-toxicity relationships of dental compounds: I. Esters of acrylic and methylacrylic acids. *J. Dent. Res.* 51, 526-535.

Leahy, D.E. (1986) Intrinsic molecular volume as a measure of the cavity term in linear solvation energy relationships: Octanol-water partition coefficients and aqueous solubilities. *J. Pharm. Sci.* 75, 629-636.

Lee, B.I., Erbar, J.H., Edmister, W.C. (1972) Thermodynamic properties at low temperatures. *Chem. Eng. Prog.* 68(9), 83.

Leggett, D.C., Parker, L.V. (1994) Modeling the equilibrium partitioning of organic contaminants between PTFE, PVC, and groundwater. *Environ. Sci. Technol.* 28, 1229-1333.

Leo, A. (1986) CLOGP-3.42 Medchem Software. Medicinal Chemistry Project, Pomona College, Claremont, California.

Leo, A., Hansch, C., Church, C. (1969) Comparison of parameters currently used for the study of structure-activity relationships. *J. Med. Chem.* 12, 766-771.

Leo, A., Hansch, C., Elkins, D. (1971) Partition coefficients and their uses. *Chem. Rev.* 71, 525-612.

Lewis, D.L., Holm, H.W. (1981) Rates of transformation of methyl parathion and diethyl phthalate by aufwuchs microorganisms. *Appl. Environ. Microbiol.* 42, 698-703.

Lewis, D.L., Holm, H.W., Kollig, H.P. (1984) Transport and fate of diethyl phthalate in aquatic ecosystems. *Environ. Toxicol. Chem.* 3, 223-231.

Leyder, F., Boulanger, P. (1983) Ultraviolet absorption, aqueous solubility, and octanol-water partition for several phthalates. *Bull. Environ. Contam. Toxicol.* 30, 152-157.

Lo, J.M., Tseng, C.L., Yang, J.Y. (1986) Radiometric method for determining solubility of organic solvents in water. *Anal. Chem.* 58, 1596-1597.

Ludzack, F.J., Ettinger, M.B. (1963) Biodegradability of organic chemicals isolated from rivers. *Purdue Univ., Eng. Bull. Ext. Ser.* 115, 278-282.

Lyman, W.J. (1985) Estimation of physical properties. In: *Environmental Exposure From Chemicals*. Volume I. Neely, W.B., Blau, G.E., Editors, Chapter 2, pp. 13-48, CRC Press, Inc., Boca Raton, Florida.

Lyman, W.J., Reehl, W.F., Rosenblatt, D.H., Editors (1982) *Handbook on Chemical Property Estimation Methods. Environmental Behavior of Organic Compounds.* McGraw-Hill, New York.

Lyman, W.J., Reehl, W.F., Rosenblatt, D.H., Editors (1990) *Handbook on Chemical Property Estimation Methods. Environmental Behavior of Organic Compounds.* American Chemical Society, Washington DC. pp.4-9, 5-4, 5-10, 7-4, 7-5, 15-15 to 15-32.

Mabey, W.R., Mill, T. (1978) Critical review of hydrolysis of organic compounds in water under environmental conditions. *J. Phys. Chem. Ref. Data* 7, 383-415.

Mabey, W.R., Smith, J.H., Podoll, R.T., Johnson, H.L., Mill, T., Chou, T.W., Gate, J., Waight-Partridge, I., Jaber, H., Vanderberg, D. (1982) *Aqueous Fate Process for Organic Priority Pollutants.* U.S. EPA Report, No. 440/4-81-14.

Mackay, D. (1982) Correlation of bioconcentration factors. *Environ. Sci. Technol.* 16, 274-278.

Mackay, D. (1985) Chapter 5, Air/water exchange coefficients. In: *Environmental Exposure From Chemicals*. Volume I. Neely, W.B., Blau, G.E., Editors, CRC Press, Inc., Boca Raton, Florida. pp.91-108.

Mackay, D., Wolkoff, A.W. (1973) Rate of evaporation of low-solubility contaminants from water bodies to atmosphere. *Environ. Sci. Technol.* 7(7), 611-614.

Marchini, S., Passerini, L., Cesareo, D., Tosato, M.L. (1988) Herbicidal triazines: Acute toxicity on daphnia, fish and plants and analysis of its relationships with structural factors. *Ecotox. Environ. Safety* 16, 148-157.

Massaldi, H.A., King, C.J. (1973) Simple technique to determine solubilities of sparingly soluble organics: Solubility and activity coefficients of *d*-limonene, *n*-butylbenzene, and *n*-hexyl acetate in water and sucrose solutions. *J. Chem. Eng. Data* 18, 393-397.

Matthiessen, P., Thain, J.E., Law, R.J., Fileman, T.W. (1993) Attempts to assess the environmental hazard posed by complex mixtures of organic chemicals in UK estuaries. *Mar. Pollut. Bull.* 26, 90-95.

Mayer, F.L. (1976) Residue dynamics of di(2-ethylhexyl) phthalate in fathead minnows (*pimephales promelas*). *J. Fish Res. Board Can.* 33, 2610-2613.

Mayer, F.L., Sanders, H.O. (1973) Toxicology of phthalic acid esters in aquatic organisms. *Environ. Health Prospect* 3, 153-157.

McBain, J.W., Richards, P.H. (1946) Solubilization of insoluble organic liquids by detergents. *Ind. Eng. Chem.* 38(6), 642-644.

McDuffie, B. (1981) Estimation of octanol/water partition coefficients for organic pollutants using reverse-phase HPLC. *Chemosphere* 10, 73-83.

McGowan, J.C. (1954) The physical toxicity of chemicals. IV. Solubilities, partition coefficients and physical toxicities. *J. Appl. Chem.* 4, 41-47.

Metcalf, R.L., Booth, G.M., Schuth, C.K., Hansen, D.J., Lu, P.Y. (1973) Uptake and fate of di-2-ethylhexyl phthalate in aquatic organisms and in a model ecosystem. *Environ. Health Perspect.* 4, 27-34.

Metcalf, R.L., Lu, P.Y. (1973) *Environmental Distribution and Metabolic Fate of Key Industrial Pollutants and Pestidcides in a Model Ecosystem.* Research Report No. 69, Water Resources Center, University of Illinois.

Meylan, W., Howard, P.H. (1991) Bond contribution method for estimating Henry's law constants. *Environ. Toxicol. Chem.* 10, 1283-1293.

Meylan, W., Howard, P.H., Boethling, R.S. (1992) Molecular topology/fragment contribution method for predicting soil sorption coefficient. *Environ. Sci. Technol.* 26(8), 1560-1567.

Mills, W.B., Dean, J.D., Porcella, D.B., Gherini, S.A., Hudson, R.J.M., Frick, W.E., Rupp, G.L., Bowie, G.L. (1982) *Water Quality Assessment: A Screening Procedure for Toxic and Conventional Pollutants.* Part I. U.S. EPA Report, No.EPA-600/6-82-004a.

Moriguchi, I. (1975) Quantitative structure activity studies. Parameters relating to hydrophobicity. *Chem. Pharm. Bull.* 23(2), 247-257.

Müller, M., Klein, W. (1992) Comparative evaluation of methods predicting water solubility for organic compounds. *Chemosphere* 25(6), 769-782.

Neely, W.B. (1979) A preliminary assessment of the environmental exposure to be expected from the addition of a chemical to a simulated ecosystem. *J. Environ. Stud.* 13, 101.

Neely, W.B. (1984) An analysis of aquatic toxicity data: Water solubility and acute LC 50 fish data. *Chemosphere* 13(7), 813-819.

Neely, W.B. (1985) Chapter 7, Hydrolysis. In: *Environmental Exposure From Chemicals.* Volume I. Neely, W.B., Blau, G.E., Editors. pp.157-174, CRC Press, Inc., Boca Raton, Florida.

Neely, W.B., Blau, G.E. (1985) Introduction to environmental exposure from chemicals. In: *Environmental Exposure From Chemicals.* Volume I. Neely, W.B., Blau, G.E., Editors, Chapter 1, pp 1-11, CRC Press, Inc., Boca Raton, Florida.

Nelson, P.E., Hoff, J.E. (1968) Food volatiles: gas chromatographic determination of partition coefficient in water-lipid systems. *J. Food Sci.* 33, 479-482.

Neuhauser, E.F., Lorhr, R.C., Malecki, M.R., Milligan, D.L., Durkin, P.R. (1985) The toxicity of selected organic chemicals to the earth worm *eisenia fetida. J. Environ. Qual.* 14(3), 383-388.

Nielson, N.M., Bundgaard, H. (1988) Glycolamide esters as biolabile prodrugs of carboxylic acid agents: synthesis, bioconversion and physicochemical properties. *J. Pharm. Sci.* 77(4), 285-298.

Nirmalakhandan, N.N., Speece, R.E. (1988) Prediction of solubility of organic chemicals based on molecular structure. *Environ. Sci. Technol.* 22(3), 328-338.

Nirmalakhandan, N.N., Speece, R.E. (1988) QSAR model for predicting Henry's constant. *Environ. Sci. Technol.* 22(11), 1349-1357.

OECD (1979) OECD, Environment Committee, Chemicals Group, OECD Chemicals Testing Programme, Expert Group, Physical Chemistry, Final Report Vol.I, Part 1 and 2, Summary of the OECD Laboratory Intercomparison, Testing Programme Part I - on the physico-chemical properties p.33, Umweltbundesamt, Bundesrepublik Deutschland, Berlin, Dec.1979.

OECD (1981) OECD Guidelines for Testing of Chemicals. Section 107, p.3. "Partition coefficient (n-octanol/water)-flask shaking method", OECD, Paris.

O'grady, D.P., Howard, P.H., Werner, A.F. (1985) Activated sludge biodegradation of 12 commercial phthalate esters. *Appl. Environ. Microbiol.* 49(2), 443-445.

Pahren, H.R., Bloodgood, D.E. (1961) Biological oxidation of several vinyl compounds. *Water Pollut. Control Fed.* 33, 233-238.

Pangrekar, J., Klopman, G., Rosekranz, H.S. (1994) Expert-system comparison of structural determinants of chemical toxicity to environmental bacteria. *Environ. Toxicol. & Chem. 13(6), 979-1001.*

Patty, F.A., Editor (1963) *Aromatic and Cyclic Dicarboxylic Acid Esters. Industrial Hygiene and Technology.* Volume II, Interscience Publishers.

Patty, F.A., Editor (1967) *Industrial Hygiene and Toxicology.* Vol. 2. Interscience, New York.

Pavlou, S.P., Weston, D.P. (1983) Initial evaluation of alternatives for development of sediment related critiria for toxic contaminants in marine waters (Puget Sound). Phase I: Development of conceptual framework. Prepared by JRB Associates for the U.S. Environmental Protection Agency under EPA Contract No. 68-01-6388.

Pavlou, S.P., Weston, D.P. (1984) Initial evaluation of alternatives for development of sediment related critiria for toxic contaminants in marine waters (Puget Sound). Phase II: Development and testing of the sediment/water equilibrium-partitioning approach. Prepared by JRB Associates for the U.S. Environmental Protection Agency under EPA Contract No. 68-01-6388.

Peakall, D.B. (1975) Phthalate esters: Occurence and biological effects. *Residue Rev.* 54, 1-41.

Peddicord, R.K. (1987) Overview of the influence of dredging and dreged material disposal on the fate and effects of sediment-associated chemicals. In: *Fate and Effects of Sediment-Bound Chemicals in Aquatic Systems* Dickson, K.L., Maki, A.W., Brungs, W.A., Editors, chapter 20, pp.317-412. Pergamon Press, New York

Perrin,D.D. (1972) *Dissociation Constants of Organic Bases in Aqueous Solutions.* IUPAC Chemical Data Series; Supplement. Butterworth, London.

Perry, E.S., Weber, W.H. (1949) Vapor pressures of phlegmatic liquids. II. High molecular weight esters and silicone oils. *J. Am. Chem. Soc.* 71, 3726-3730.

Perry, R.H., Chilton, C.H., Editors (1973) *Chemical Engineer's Handbook.* 5th Edition, McGraw-Hill Book Company, New York.

Petrasek, A.C., Kugeman, I.J., Austern, B.M., Pressley, T.A., Winslow, L.A., Wise, R.H. (1983) Fate of toxic organic compounds in wastewater treatment plants. *J. Water Pollut. Control Fed.* 55, 286-296.

Price, K.S., Waggy, G.T., Conway, R.A. (1974) Brine shrimp bioassay and seawater BOD of petrochemicals. *J. Water Pollut. Control Fed.* 46, 63-77.

Quayle, O.R. (1953) The parachors of organic compounds. *Chem. Rev.* 53, 439-589.

Radding, S.B., Liu, D.H., Johnson, H.L., Mill, T. (1977) *Review of the Environmental Fate of Selected Chemicals.* EPA Report No.560/5-77-003. U.S. Environmental Protection Agency, Office of Toxic Substances, Washington, D.C.

Ratchford, W.P., Rehbert, C.E. (1949) An improved tensimeter-still. *Anal. Chem.* 21, 1417-1419.

Reid, R.C., Prausnitz, J.M., Sherwood, T.K. (1977) *The Properties of Gases and Liquids.* 3rd Edition, McGraw-Hill Book Company, New York.

Rekker, R.F. (1977) *The Hydrophobic Fragmental Constants. Its Derivation and Application, a Means of Characterizing Membrane Systems.* Elsevier Scientific Publishing Company, Oxford, England.

Renberg, L.O., Sundström, S.G., Rosén-Olofsson, A-C. (1985) The determination of partition coefficients of organic compounds in technical products and waste waters for the estimation of their bioaccumulation potential using reversed phase thin layer chromatography. *Toxicol. Environ. Chem.* 10, 333-349.

Riddick, J., Bunger, W.B., Sakano, T.K. (1986) *Organic Solvents: Physical Properties and Method of Purification.* 4th Edition, John Wiley & Sons, New York, New York.

Rohrschneider, L. (1973) Solvent characterization by gas-liquid partition coefficients of selected solutes. *Anal. Chem.* 45(7), 1241-1247.

Roy, R.S. (1972) Spectrophotometric estimation of ester hydrolysis. *Anal. Chem.* 44, 2096-2098.

Russell, C.J., Dixon, S.L., Jurs, P.C. (1992) Computer-assisted study of the relationship between molecular structure and Henry's law constant. *Anal. Chem.* 64, 1350-1355.

Russell, D.J., McDuffie, B. (1986) Chemodynamic properties of phthalate esters: Partitioning and soil migration. *Chemosphere* 15, 1003-1021.

Russell, D.J., McDuffie, B., Fineberg, S. (1985) The effect of biodegradation on the determination of some chemodynamic properties of phthalate esters. *J. Environ. Sci. Health* A20, 927-941.

Ryan, J.A., Bell, R.M., Davidson, J.M., O'Connor, G.A. (1988) Plant uptake of non-ionic organic chemicals from soils. *Chemosphere* 17(12), 2299-2323.

Sabljic, A. (1987) Nonempirical modeling of environmental distribution and toxicity of major organic pollutants. In: *QSAR in Environmental Toxicology- II.* Kaiser, K.L.E., Editor, D. Reidel Publishing Co., Dordrecht, The Netherlands. pp.309-332.

Sabljic, A., Güsten, H., Schönherr, J., Riederer, M. (1990) Modeling plant uptake of airborne organic chemicals. 1. Plant cuticle/water partitioning and molecular connectivity. *Environ. Sci. Technol.* 24, 1321-1326.

Saeger, V.W., Tucker, E.S. (1976) Biodegradation of phthalic acid esters in river water and activated sludge. *Appl. Environ. Microbiol.* 31, 29-34.

Saito, S., Tanoue, A., Matsuo, M. (1992) Applicability of the i/o-characters to a quantitative description of bioconcentration of organic chemicals in fish. *Chemosphere* 24(1), 81-87.

Salmi, E.J., Leimu, R. (1947) The effect of structure on the rate of alkaline hydrolysis of aliphatic esters. *Suomen Kemistiehti* 20B, 43.

Sanborn, J.R. et al. (1975) Plasticizers in the environment: the fate of di-N-octylphthalate (DOP) in two model ecosystems and uptake and metabolism of DOP by aquatic organisms. *Arch. Environ. Contam. Toxicol.* 3(2), 244-255.

Sanders, H.O., Mayer, F.L., Walsh, D.F. (1973) Toxicity, residue dynamics, and reproduction effects of phthalate esters in aquatic invertebrates. *Environ. Res.* 6, 84-90.

Sangster, J. (1989) Octanol-water partition coefficients of simple organic compounds. *J. Phys. Chem. Ref. Data* 18(3), 1111-1229.

Sangster, J. (1993) LOGKOW databank, Sangster Research Labs., Montreal, Canada.

Sasaki, S. (1978) The scientific aspects of the chemical substance control law in Japan. In: *Aquatic Pollutants: Transformation and Biological Effects.* Hutzinger, O., et al. Editors, Pergamon Press, Oxford, U.K. pp.283-298.

Schouten, M.J., Peereboom, J.W.C., Brinkman, U.A.T. (1979) Liquid chromatographic analysis of phthalate esters in Dutch river water. *Int'l J. Anal. Chem.* 7, 13-23.

Schüürmann, G., Klein, W. (1988) Advances in bioconcentration prediction. *Chemosphere* 17(8), 1551-1574.

Schwarz, F.P., Miller, J. (1980) Determination of the aqueous solubilities of organic liquids at 10.0, 20.0, 30.0 °C by elution chromatography. *Anal. Chem.* 52, 2162-2164.

Scow, K,M. (1982) Chapter 9, Rate biodegradation. In: *Handbook of Chemical Property Estimation Methods, Envioronmental Behavior of Organic Compounds.* Lyman, W.J., Reehl, W.F., Rosenblatt, D.H., Eds. McGraw-Hill Book Co., New York, N.Y.

Seidell, A. (1941) *Solubilities of Organic Compounds.* Vol. 2, Van Nostrand, New York.

Shanker, R., Ramakrishna, C., Seth, P.K. (1985) Degradation of some phthalic acid esters in soil. *Environ. Pollut.* Ser. A. 39, 1-7.

Sharma, R.C., Sharma, M.M. (1970) Kinetics of alkaline hydrolysis of esters. II. Unsaturated esters and oxalic esters. *Bull. Chem. Soc. Japan* 43, 642-645.

Shelton, D.R., Boyd, S.A., Tiedje, J.M. (1984) Anaerobic biodegradation of phthalic acid esters in sludge. *Environ. Sci. Technol.* 18, 93-97.

Shimada, T., Nagasaki, Y., Yoshida, T. (1983) The acute toxicity and the accumulation of phenol and diethyl phthalate in *mugil cephalus. Kenkyu Hokuku- Kanagawa- Ken Kogai Senta* 5, 45-48.

Slave, T., Mihail, A., Burmaz, N. (1974) Degradation of some organic impurities in residual waters. *Rev. Chim.* 25, 666-670.

Small, P.A., Small, K.W., Coeley, P. (1948) The vapour pressures of some high boiling esters. *Trans. Faraday Soc.* 44, 810-816.

Steen, W.C. (1991) *Microbial Transformation Rate Constants of Structurally Diverse Man Made Chemicals*. NTIS PB91-181958. U.S. Environmental Protection Agency, Environmental Research Laboratory, Athens, Georgia.

Steen, W.C., Paris, D.F., Baughman, G.L. (1979) 177th National Meeting of American Chemical Society, Honolulu, Hawaii.

Steen, W.C., Paris, D.F., Baughman, G.L. (1979) Effects of sediment sorption on microbial degradation of toxic substances. pp. 168-173. In: *Proceedings of the Symposium on Processes Involving Contaminants and Sediments*. Baker, R., Editor, American Chemical Society, Washington, DC.

Stephen, H., Stephen, T. (1963) *Solubilities of Inorganic and Organic Compounds in Binary Systems*. Pergamon Press, New York.

Stephenson, R., Stuart, J. (1986) Mutual binary solubilities: Water-alcohols and water-esters. *J. Chem. Eng. Data* 31, 56-70.

Stephenson, R. M., Malanowski S. (1987) *Handbook of the Thermodynamics of Organic Compounds*. Elsevier Science Publishing Co., Inc., New York.

Stull, D.R. (1947) Vapor pressure of pure substances: Organic compounds. *Ind. Eng. Chem.* 39(4), 517-560.

Subba-Rao, R.V., Rubbin, H.E., Alexander, M. (1982) Kinetics and extent of mineralization of organic chemicals in trace levels in freshwater and sewage. *Appl. Environ. Microbiol.* 43, 1139-1150.

Sugatt, R.H., O'Grady, D.P., Banerjee, S., Howard, P.H., Gledhill, W.E. (1984) Shake flask biodegradation of 14 commercial phthalate esters. *Appl. Environ. Microbiol.* 47, 601-606.

Sullivan, K.F., Atlas, E.L., Giam, C.S. (1981) Loss of phthalic acid esters and polychlorinated biphenyls from seawater samples during storage. *Anal. Chem.* 53, 1718-1719.

Tabak, H.H., Quave, S.A., Mashni, C.I., Barth, E.F. (1981) Biodegradability studies with organic priority pollutant compounds. *J. Water Pollut. Control Fed.* 53, 1503-1518.

Taft, R.W., Abraham, M.H., Doherty, R.M., Kamlet, M.J. (1985) The molecular properties governing solubilities of organic nonelectrolytes in water. *Nature* 313, 384-386.

Taft, R.W., Abraham, M.H., Famini, G.R., Doherty, R.M., Abboud, J.-L.M., Kamlet, M.J. (1985) Solubility properties in polymers and biological media: 5. An analysis of the physicochemical properties which influence octanol-water partition coefficients of aliphatic and aromatic solutes. *J. Pharm. Sci.* 74(8), 807-814.

Tanii, H., Hashimoto, K. (1982) Structure-toxicity relationship of acrylates and methacrylates. *Toxicol. Lett.* 11, 125-129.

Tarr, B.D., Barron, M.G., Hayton, W.L. (1990) Effect of body size on the uptake and bioconcentration of di-2-ethylhexyl phthalate in rainbow trout. *Environ. Toxicol. & Chem.* 9, 989-995.

Tewari, Y.B., Miller, M.M., Wasik, S.P., Martire, D.E. (1982) Aqueous solubility and octanol/water partition coefficient of organic compounds at 25°C. *J. Chem. Eng. Data* 27, 451-454.

THOR (1986) THOR Database, Medicinal Chemistry Project, Pomona College, Claremont, California.

Tsonopoulos, C., Prausnitz, J.M. (1971) Activity coefficients of aromatic solutes in dilute aqueous solutions. *Ind. Eng. Chem. Fundam.* 10, 593-600.

Tucker, W.A., Lyman, W.J., Preston, A.L. (1983) Estimation of the dry deposition velocity and scavenging ratio for organic chemicals. In: *Precipitation Scavenging, Dry Deposition, and Resuspension.* Pruppacher et al., Ed., pp. 1242-1256, Elsevier Science Publishing Co., Inc., New York, N.Y.

Tute, M.S. (1971) Principle and practice of Hansch analysis: A guide structure-activity correlation for the medical chemist. *Adv. Drug Res.* 6, 1-77.

Urushigawa, Y., Yonezawa, Y. (1979) Chemico-biological interactions in biological purification systems. VI. Relation between biodegradation rate constants of di-*n*-alkyl phthalate esters and their retention times in reverse phase partition chromatography. *Chemosphere* 8(5), 317-320.

USEPA (1975) *Preliminary Asssessment of Suspected Carcinogens in Drinking Water.* Interim Report to Congress, June 1975. U.S. Environmental Protection Agency, Washington DC.

USEPA (1979) *Water Related Environmental Fate of 129 Priority Pollutants.* Volumes 1 and 2. U.S. EPA-560-/7-85-002a.

USEPA (1985) *Health and Environmental Effects Profile for Methyl methacrylate.* NTIS-PB-178785. U.S. EPA-600/x-85-364.

USEPA (1988) *PCGEMS: Personal Computer Graphical Exposure Modeling System. PCLOGP,* Office of Toxic Substances, U.S. Environmental Protection Agency, Washington DC.

Valvani, S.C., Yalkowsky, S.H., Roseman, T.J. (1981) Solubility and partitioning. IV. Aqueous solubility and octanol-water partition coefficient of liquid nonelectrolytes. *J. Pharm. Sci.* 70, 502-507.

Vaishnav, D.D., Babeu, L. (1987) Comparison of occurence and rates of chemical biodegradation in natural waters. *Bull. Environ. Contam. Toxicol.* 39, 237-244.

Veith, G.D., Austin, N.M., Morris, R.T. (1979) A rapid method for estimating LOG P for organic chemicals. *Water Res.* 13, 43-47.

Veith, G.D., Defoe, D.L., Bergstedt, B.V. (1979) Measuring and estimating the bioconcentration factor of chemicals in fish. *J. Fish Res. Board Can.* 36, 1040-1048.

Veith, G.D., Kosian, P. (1983) Chapter 15, Estimating bioconcentration potential from octanol/water partition coefficients. In: *Physical Behavior of PCBs in the Great Lakes.* Mackay, D., Paterson, S., Eisenreich, S.J., Simmons, M.S., Editors, Ann Arbor Science Publishers, Ann Arbor, Michigan. pp.269-282.

Veith, G.D., Macek, K.J., Petrocelli, S.R., Carroll, J. (1980) An evaluation of using partition coefficients and water solubility to estimate bioconcentration factors for organic chemicals in fish. In: *Aquatic Toxicology.* pp.116-129. ASTM STP 707.

Veith, G.D., Morris, R.T. (1978) *A Rapid Method for Estimating LOG P for Organic Chemicals.* EPA-600/3-78-049. U.S. Environmental Protection Agency, Ecological Research Series.

Verschueren, K. (1977) *Handbook of Environmental Data on Organic Chemicals.* Van Nostrand Reinhold, New York.

Verschueren, K. (1983) *Handbook of Environmental Data on Organic Chemicals.* 2nd Edition, Van Nostrand Reinhold, New York.

von Oepen, B.V., Kördel, W., Klein, W. (1991) Sorption of nonpolar and polar compounds to soils: Processes, measurements and experience with the applicability of the modified OECD-Guideline 106. *Chemosphere* 22(3/4), 285-304.

Wakita, K., Yoshimoto, M., Miyamoto, S. (1986) A method of calculation of the aqueous solubility of organic compounds by using new fragment solubility constants. *Chem. Pharm. Bull.* 34, 4663-4681.

Walker, W.W., Cripe, C.R., Pritchard, P.H., Bourquin, A.W. (1984) Dibutylphthalate degradation in estuarine and freshwater sites. *Chemosphere* 12, 1283-1294.

Wallington, T.J., Dagaut, P., Liu, R., Kurylo, M.J. (1988a) The gas phase reactions of hydroxyl radicals with a series of esters over the temperature range 240-440 K. *Intl. J. Chem. Kinet.* 20, 177-186.

Wallington, T.J., Dagaut, P., Kurylo, M.J. (1988b) Correlation between gas-phase and solution-phase reactivities of hydroxyl radicals toward saturated organic compounds. *J. Phys. Chem.* 92, 5024-5028.

Wasik, S.P., Tewari, Y.B., Miller, M.M., Martire, D.E. (1981) *Octanol/Water Partition Coefficients and Aqueous Solubilities of Organic Compounds.* NBSIR No. 81-2406. U.S. Dept. of Commerce, Washington DC.

Weast, R.C. (1977) *CRC Handbook of Chemistry and Physics.* 58th Edition. CRC Press, Inc., Cleveland, Ohio.

Weast, R.C. (1981-82) *CRC Handbook of Chemistry and Physics.* 62nd. Edition, CRC Press, Boca Raton, Florida.

Weast, R.C. (1982-83) *CRC Handbook of Chemistry and Physics.* 63rd. Edition, CRC Press, Boca Raton, Florida.

Weast, R., Grasselli, J.G., Editors (1989) *Handbook of Data on Organic Compounds.* Volume VI, 2nd Edition, CRC Press Inc., Boca Raton, Florida.

Werner, A.C. (1952) Vapor pressures of phthalate esters. *Ind. Eng. Chem.* 44, 2736-2740.

Winer, A.M., Lloyd, A.C., Darnall, K.R., Atkinson, R., Pitts, Jr., J.N. (1977) Rate constants for the reactions of hydroxyl radicals with *n*-propyl acetate, *sec*-butyl acetate, tetrahydofuran and peroxyacetyl nitrate. *Chem. Phys. Lett.* 51(2), 221-226.

Wofford, H.W., Wilsey, C.D., Neff, G.S., Giam, C.S., Neff, J.M. (1981) Bioaccumulation and metabolism of phthalate esters by oysters, brown shrimp and sheepshead minnows. *Ecotoxicol. Environ. Saf.* 5, 202-210.

Wolfe, N.L., Burns, L.A., Steen, W.C. (1980a) Use of linear free-energy relationships on an evaluative model to assess the fate and transport of phthalate esters in aquatic environment. *Chemosphere* 9, 393-402.

Wolfe, N.L., Steen, W.C., Burns, L.A. (1979) Unpublished report available from U.S. Environmental Protection Agency, Environmental Research Laboratory, Athens, Georgia.

Wolfe, N.L., Steen, W.C., Burns, L.A. (1980b) Phthlate ester hydrolysis: Linear-energy relationships. *Chemosphere* 9, 403-408.

Wolfe, N.L., Paris, D.F., Steen, W.C., Baughman, G.L. (1980c) Correlation of microbial degradation rates with chemical structure. *Environ. Sci. Technol.* 14, 1143-1144.

Yaguzhinskii, L.S., Smirnova, E.G., Ratnikova, L.A., Kolesova, G.M., Krasinskaya, I.P. (1973) Hydrophobic sites of mitochondrial electron transfer system. *J. Bioenerg.* 5(3), 163-174.

Yalkowsky, S.H., Morozowich, W. (1980) Chapter 3, A physical chemical basis for the design of orally active prodrugs. *Drug Design.* Vol. IX. Academic Press, Inc., New York.

Yalkowsky, S.H., Valvani, S.C., Kun, W.-Y., Dannenfelser, R.M., Editors (1987) *Arizona Database of Aqueous Solubility for Organic Compounds.* College of Pharmacy, University of Arizona, Tucson, Arizona.

Yaws, C.L., Yang, H.-C., Hopper, J.R., Hansen, K.C. (1990) Organic chemicals: water solubility data. *Chem. Eng.* July, 115-118.

Zaroogian, G.E., Heltshe, J.F., Johnson, M. (1985) Estimation of bioconcentration in marine species using structure-activity models. *Environ, Toxicol. Chem.* 4, 3-12.

Zetzsch, C. (1982) Report to Bundeminister für Forschung und Technologie, Projekttrager für unweltchemikalien.

Ziogou, K., Kirk, P.W.W., Lester, J.N. (1989) Behaviour of phthalic acid esters during batch anaerobic digestion of sludge. *Water Res.* 23(6), 743-748.

Zoeteman, B.C.J., Harmsen, K., Linders, J.B.H.J., Morra, C.F.H., Slooff, W. (1980) Persistent organic pollutants in river water and ground water of the Netherlands. *Chemosphere* 9, 231-249.

Chapter 7 Nitrogen and Sulfur Compounds

7.1 List of Chemicals and Data Compilations

Common Name: Acetonitrile
Synonym: cyanomethane, ethanenitrile, methyl cyanide
Chemical Name: acetonitrile
CAS Registry No: 75-05-8
Molecular Formula: CH_3CN
Molecular Weight: 41.05
Melting Point (°C):
-41.0 (Stull 1947)
-45.7 (Weast 1982-83; Howard 1993)
-44/-41 (Verschueren 1983)
-43.8 (Dean 1985; Riddick et al. 1986)
Boiling Point (°C):
81.8 (Stull 1947; Boublik et al. 1984)
81.6 (Mousa 1981; Weast 1982-83; Riddick et al. 1986; Howard 1993)
82.0 (Verschueren 1983; Banerjee et al. 1990)
Density (g/cm^3 at 20°C):
0.7857 (Weast 1982-83; Dean 1985)
0.790 (Verschueren 1983)
0.7822 (Riddick et al. 1986)
Molar Volume (cm^3/mol):
52.7 (calculated-density, Rohrschneider 1973)
57.4 (exptl. at normal B.P., Lee et al. 1972; quoted, Reid et al. 1977)
52.1 (Taft et al. 1985; Leahy 1986; Kamlet et al. 1986,87)
27.1 (intrinsic volume V_I, Leahy 1986; Kamlet et al. 1987)
56.3 (calculated-LeBas method)
Molecular Volume ($Å^3$):
Total Surface Area, TSA ($Å^2$):
160.1 (Iwase et al. 1985)
Dissociation Constants, pK:
29.1 (pK_a, Riddick et al. 1986; Howard 1993)
32.2 (pK_s, Riddick et al. 1986)
-10.12 (pK_{BH+}, Riddick et al. 1986)
Heat of Fusion, ΔH_{fus}, kcal/mol:
1.952 (Riddick et al. 1986)
Entropy of Fusion, ΔS_{fus}, cal/mol·K (e.u.):
Fugacity Ratio at 25°C (assuming ΔS_{fus} = 13.5 e.u.), F: 1.0

Water Solubility (g/m^3 or mg/L at 25°C):
$>3.1x10^6$ (Booth & Everson 1948)
miscible (Dean 1985)
139100, 665750 (quoted, calculated-V_I and solvatochromic p., Leahy 1986)
139100, 333670 (quoted, calculated-V_I and solvatochromic p., Kamlet et al. 1987)
infinite (Riddick et al. 1986; Howard 1993)

Vapor Pressure (Pa at 25°C):

11868 (interpolated-regression of tabulated data, Stull 1947)
13330 (27.0°C, Stull 1947)
11508 (Hoy 1970)
8306 (saturated-vapour volume, extrapolated from fitted Antoine eqn., Mousa 1981)
9864 (20°C, quoted, Verschueren 1983)
15330 (30°C, quoted, Verschueren 1983)
11919 (quoted exptl., Boublik et al. 1984)
12782 (calculated average-Antoine eqn., Boublik et al. 1984)
12310 (calculated-Antoine eqn., Dean 1985)
11840 (Howard et al. 1986; quoted, Banerjee et al. 1990)
11840 (Riddick et al. 1986; Howard 1993)
11840, 41590 (measured, calculated-solvatochromic p., Banerjee et al. 1990)

Henry's Law Constant (Pa·m^3/mol):

3.501 (exptl., Hine & Mookerjee 1975)
2.781 (calculated-bond contribution, Hine & Mookerjee 1975)
2.968 (Snider & Dawson 1985; quoted, Howard 1993)
2.033 (computed, Yaws et al. 1991)

Octanol/Water Partition Coefficient, log K_{OW}:

-0.34 (shake flask-GC, Hansch & Anderson 1967; quoted, Leo et al. 1969,71; Iwase et al. 1985; Leahy 1986; Sangster 1989)
-0.34 (quoted, Verschueren 1983)
-0.36 (calculated-TSA, Iwase et al. 1985)
-0.03 (computed, Klopman et al. 1985; quoted, Pangrekar et al. 1994)
-0.34, -0.37 (quoted, calculated-V_I and solvatochromic p., Taft et al. 1985)
-0.34, -0.45 (quoted, calculated-V_I and solvatochromic p., Leahy 1986)
-0.34, -0.394, 0.06 (quoted from Hansch & Leo 1979, calculated-CLOGP, calculated-MO indices, Bodor et al. 1989)
-0.54 (shake flask-GC, Tanii & Hashimoto 1984; quoted, Sangster 1989)
-0.34 (Hansch & Leo 1985; quoted, Howard 1993)
-0.34 (recommended, Sangster 1989)
-0.394 (calculated-QSAR, Kollig 1993)

Bioconcentration Factor, log BCF:

-0.523 (estimated-K_{OW} as per regression eqn. of Bysshe 1982, Howard 1993)

Sorption Partition Coefficient, log K_{OC}:

-0.523 (soil, estimated-K_{OW}, Lyman et al. 1982; quoted, Howard 1993)
-0.714 (calculated-K_{OW}, Kollig 1993)

Half-Lives in the Environment:

Air: photooxidation half-life of 1299-12991 hours, based on measured rate constant of $3x10^{-14}$ cm^3 $molecule^{-1}$ sec^{-1} for the vapor phase reaction with hydroxyl radicals in air (Atkinson 1985; quoted, Howard et al. 1991; Howard 1993); atmospheric transformation lifetime was estimated to be >5 days (Kelly et al. 1994).

Surface water: 168-672 hours, based on aerobic river die-away test data (Ludzack et al. 1958; quoted, Howard et al. 1991); photooxidation half-life of 314-12559 years, based on measured rate data for reaction with hydroxyl radicals in aqueous solution (Dorfman & Adams 1973; Howard et al. 1991).

Groundwater: 336-8640 hours, based on estimated aqueous aerobic biodegradation half-life (Howard et al. 1991).

Sediment:

Soil: 168-672 hours, based on estimated aqueous aerobic biodegradation half-life (Howard et al. 1991).

Biota:

Environmental Fate Rate Constants or Half-Lives:

Volatilization: based on Henry's law constant, half-life from a model river of 1 meter deep flowing at 1 m/sec with a wind velocity of 3 m/sec was estimated to be 21 hours (Lyman et al. 1982; quoted, Howard 1993).

Photolysis:

Oxidation: photooxidation half-life of 314-12559 years in water, based on measured rate data for reaction with hydroxyl radicals in aqueous solution (Dorfman & Adams 1973; Howard et al. 1991); rate constant of $(0.494 \pm 0.06)x10^{-13}$ cm^3 $molecule^{-1}$ sec^{-1} by the flash photolysis-resonance fluorescence technique for the vapor-phase reaction with OH radicals at 297.2 K in air (Harris et al. 1981; quoted, Howard 1993); rate constant of $1.90x10^{-14}$ cm^3 $molecule^{-1}$ sec^{-1} for the gas-phase reaction with hydroxyl radicals in air and $3.70x10^{-14}$ cm^3 $molecule^{-1}$ sec^{-1} for the solution-phase reaction with hydroxyl radicals in aqueous solution (Kurylo & Knable 1984; quoted, Wallington et al. 1988); measured photooxidation rate constant of $3x10^{-14}$ cm^3 $molecule^{-1}$ sec^{-1} for the vapor phase reaction with hydroxyl radicals in air (Atkinson 1985; quoted, Howard et al. 1991; Howard 1993).

Hydrolysis: $5.8x10^{-3}$ M^{-1} h^{-1} at pH 7 and 25°C with half-life longer than 150000 years (Ellington et al. 1987; quoted, Howard et al. 1991; Kollig 1993).

Biodegradation: aqueous aerobic half-life of 168-672 hours, based on aerobic river die-away test data (Ludzack et al. 1958; quoted, Howard et al. 1991); aqueous anaerobic half-life of 672-2688 hours, based on estimated aqueous aerobic biodegradation half-life (Howard et al. 1991).

Biotransformation:

Bioconcentration, Uptake (k_1) and Elimination (k_2) Rate Constants:

Common Name: Propionitrile
Synonym: propanenitrile, ethyl cyanide, cyanoethane, propyl nitrile
Chemical Name: propionitrile
CAS Registry No: 107-12-0
Molecular Formula: CH_3CH_2CN
Molecular Weight: 55.08
Melting Point (°C):
-91.9 (Stull 1947)
-92.9 (Weast 1982-83)
-92.8 (Dean 1985; Riddick et al. 1986)
-91.8 (Howard 1990)
Boiling Point (°C):
97.20 (Butler & Ramchandani 1935; Dean 1985; Howard 1990)
97.10 (Stull 1947)
97.30 (Weast 1982-83)
97.35 (Riddick et al. 1986)
97.46 (Boublik et al. 1984)
Density (g/cm^3 at 20°C):
0.7818 (Weast 1982-83; Dean 1985)
0.78182 (Riddick et al. 1986)
Molar Volume (cm^3/mol):
70.4 (Taft et al. 1985; Leahy 1986; Kamlet et al. 1986,87)
36.9 (intrinsic volume V_I, Leahy 1986; Kamlet et al. 1987)
78.5 (calculated-LeBas method)
70.5 (calculated-density)
Molecular Volume ($Å^3$):
Total Surface Area, TSA ($Å^2$):
187.5 (Iwase et al. 1985)
Dissociation Constant:
33.54 (pK_s, Riddick et al. 1986)
Heat of Fusion, ΔH_{fus}, kcal/mol:
1.206 (Riddick et al. 1986)
Entropy of Fusion, ΔS_{fus}, cal/mol·K (e.u.):
Fugacity Ratio at 25°C (assuming ΔS_{fus} = 13.5 e.u.), F: 1.0

Water Solubility (g/m^3 or mg/L at 25°C):
104950 (Seidell 1941; quoted, Hansch et al. 1968)
105200 (Hansch et al. 1968; quoted, Müller & Klein 1992)
249450 (calculated-K_{OW}, Hansch et al. 1968)
103000 (Dean 1985; Riddick et al. 1986; Howard 1990)
117760, 148250 (quoted, calculated-V_M and solvatochromic p., Kamlet et al. 1986)
117760, 214290 (quoted, calculated-V_I and solvatochromic p., Leahy 1986)
104950, 49090 (quoted, calculated-fragment const., Wakita et al. 1986)
117760, 104950 (quoted, calculated-V_I and solvatochromic p., Kamlet et al. 1987)
55000, 65000 (20°C, 30°C, shake flask-GC, Stephenson 1994)

Vapor Pressure (Pa at 25°C):

- 6005 (interpolated-regression of tabulated data, Stull 1947)
- 5332 (22.0°C, Stull 1947)
- 6140 (22.05°C, quoted exptl., Boublik et al. 1973, 1984)
- 6153 (extrapolated average-Antoine eqn., Boublik et al. 1984)
- 6140 (extrapolated-Antoine eqn., Dean 1985)
- 6133 (quoted, Kamlet et al. 1986)
- 5950 (Riddick et al. 1986; Howard 1990)
- 6306 (calculated-Antoine eqn., Stephenson & Malanowski 1987)

Henry's Law Constant (Pa·m^3/mol):

- 3.80 (partial pressure, Butler & Ramchandani 1935)
- 3.748 (Buttery et al. 1969; quoted, Hawthorne et al. 1985)
- 3.752 (quoted exptl., Hine & Mookerjee 1975)
- 3.752 (calculated-group contribution, Hine & Mookerjee 1975)
- 4.114 (calculated-bond contribution, Hine & Mookerjee 1975)
- 5.947 (quoted, Howard 1990)

Octanol/Water Partition Coefficient, log K_{OW}:

- 0.041 (shake flask-K, Collander 1951)
- 0.16 (shake flask-GC, Hansch & Anderson 1967; Hansch et al. 1968; quoted, Leo et al. 1971; Iwase et al. 1985; Sangster 1989)
- 0.16 (shake flask-UV, Hansch et al. 1968)
- -0.10 (shake flask-GC, Tanii & Hashimoto 1984; quoted, Sangster 1989)
- 0.16, 0.16 (quoted, calculated-TSA, Iwase et al. 1985)
- 0.10, 0.15 (observed, calculated-V_I and solvatochromic p., Taft et al. 1985)
- 0.10, 0.11 (observed, calculated-V_I and solvatochromic p., Leahy 1986)
- 0.16 (recommended, Sangster 1989)
- 0.16 (quoted, Howard 1990)
- 0.14 (calculated-CLOGP, Müller & Klein 1992)

Bioconcentration Factor, log BCF:

- -0.108 (estimated-K_{OW}, Lyman et al. 1982; quoted, Howard 1990)

Sorption Partition Coefficient, log K_{OC}:

- 0.079 (soil, estimated-K_{OW}, Lyman et al. 1982; quoted, Howard 1990)

Half-Lives in the Environment:

Air: half-life of 83 days, based on experimental rate constant assuming 12 hours of sunlight for the vapor-phase reaction with hydroxyl radicals in air and half-life of more than 100 days for the reaction with ozone in the atmosphere (Harris et al. 1981; quoted, Howard 1990).

Surface water:

Groundwater:

Sediment:

Soil:

Biota:

Environmental Fate Rate Constants or Half-Lives:

Volatilization: using Henry's law constant, a half-life of 13.3 hours was estimated for a model river 1 m deep flowing 1 m/sec with wind speed 3 m/sec (Lyman et al. 1982; quoted, Howard 1990).

Photolysis:

Oxidation: rate constant of $(1.94 \pm 0.20)\times10^{-13}$ cm^3 $molecule^{-1}$ sec^{-1} by flash photolysis-resonance fluorescence technique for the vapor-phase reaction with OH radicals at 298.2 K in air 1.60×10^{-13} cm^3 $molecule^{-1}$ sec^{-1} for the solution-phase reaction with hydroxyl radicals in aqueous solution (Harris et al. 1981; quoted, Wallington et al. 1988); photooxidation half-life of 83 days in air, based on experimental rate constant assuming 12 hours of sunlight for the vapor-phase reaction with hydroxyl radicals in air and half-life of more than 100 days for the reaction with ozone in the atmosphere (Harris et al. 1981; quoted, Howard 1990).

Hydrolysis:

Biodegradation:

Biotransformation:

Bioconcentration, Uptake (k_1) and Elimination (k_2) Rate Constants:

Common Name: Acrylonitrile
Synonym: cyanoethylene, propenenitrile, 2-propenenitrile, vinyl cyanide
Chemical Name: acrylonitrile, cyanoethylene
CAS Registry No: 107-13-1
Molecular Formula: $CH_2{=}CHCN$
Molecular Weight: 53.06
Melting Point (°C):
-82.0 (Stull 1947; Howard 1989)
-83.55 (Riddick et al. 1986)
Boiling Point (°C):
78.50 (Stull 1947; Weast 1977; Callahan et al. 1979)
77.30 (Riddick et al. 1986; Howard 1989)
Density (g/cm^3 at 20°C):
0.8060 (Riddick et al. 1986)
Molar Volume (cm^3/mol):
66.28 (Hoy 1970; Amidon & Williams 1982)
65.8 (Kamlet et al. 1987)
66.3 (calculated-density)
71.1 (calculated-LeBas method)
Molecular Volume (Å^3):
Total Surface Area, TSA (Å^2):
Dissociation Constant, pK_a:
Heat of Fusion ΔH_{fus}, kcal/mol:
1.489 (Riddick et al. 1986)
Entropy of Fusion ΔS_{fus}, cal/mol·K (e.u.):
Fugacity Ratio at 25°C (assuming ΔS_{fus} = 13.5 e.u.), F: 1.0

Water Solubility (g/m^3 or mg/L at 25°C):
79000 (Klein et al. 1957; quoted, Mabey et al. 1982)
75000 (quoted, Günther et al. 1968)
73500 (20°C, Windholz 1976; quoted, Callahan et al. 1979)
73240 (shake flask-LSC, Veith et al. 1980)
216160 (calculated as per Hansch et al. 1968, Veith et al. 1980)
75000 (quoted average, Valvani et al. 1981; quoted, Banerjee 1985; Howard 1989)
74950, 2784630 (quoted, calculated-K_{OW} and M.P., Amidon & Williams 1982)
73500 (20°C, Riddick et al. 1986)
80310, 175700 (quoted, calculated-fragment const., Wakita et al. 1986)
69000, 66400 (20°C, 30°C, shake flask-GC, Stephenson 1994)

Vapor Pressure (Pa at 25°C):
14340 (interpolated-regression of tabulated data, Stull 1947)
13330 (22.8°C, Stull 1947)
14100 (Hoy 1970)
13330 (22.8°C, Weast 1977; Callahan et al. 1979; Mabey et al. 1982)
11730 (20°C, quoted exptl., extrapolated-Antoine eqn., Boublik et al. 1984)
14720 (extrapolated-Antoine eqn., Boublik et al. 1984)

14370 (Daubert & Danner 1985; quoted, Howard 1989)
15240 (calculated-Antoine eqn., Dean 1985)
14530 (Howard et al. 1986; quoted, Banerjee et al. 1990)
11000 (20°C, Riddick et al. 1986)
14570 (calculated-Antoine eqn., Stephenson & Malanowski 1987)
15600 (calculated-solvatochromic p., Banerjee et al. 1990)

Henry's Law Constant (Pa·m^3/mol):
11.14 (Bocek 1976; quoted, Howard 1989)
8.918 (calculated-P/C, Mabey et al. 1982)
9.420 (quoted, WERL Treatability Database, Ryan et al. 1988)

Octanol/Water Partition Coefficient, log K_{OW}:
-0.92 (shake flask-TN, Collander 1951; quoted, Leo et al. 1971)
-0.14 (calculated per Leo et al. 1971, Callahan et al. 1979)
0.25 (shake flask-HPLC, Pratesi et al. 1979; quoted, Mabey et al. 1982)
0.0 (Fujisawa & Masuhara 1980, 1981)
1.20 (shake flask-LSC, Veith et al. 1980)
-0.92 (exptl., Valvani et al. 1981; quoted, Amidon & Williams 1982)
0.09 (shake flask-GC, Tanii & Hashimoto 1984; quoted, Sangster 1989)
0.25 (Hansch & Leo 1985; quoted, Howard 1989)
0.28 (computed, Klopman et al. 1985; quoted, Pangrekar et al. 1994)
0.12 (Lu et al. 1986; quoted, Walton et al. 1992; Bintein & Devillers 1994)
0.30 (selected, Yoshida et al. 1986)
0.25 (USEPA 1979; quoted, Ryan et al. 1988)
0.25 (quoted and recommended, Sangster 1989)
0.231 (calculated-QSAR, Kollig 1993)

Bioconcentration Factor, log BCF:
1.68 (bluegill sunfish, Barrows et al. 1978; quoted, Howard 1989)
0.00 (estimated-S, Kenaga 1980; quoted, Howard 1989)
1.68 (bluegill sunfish, Veith et al. 1980)
0.32 (calculated-K_{OW}, Veith et al. 1980)
0.017 (microorganisms-water, calculated-K_{OW}, Mabey et al. 1982)

Sorption Partition Coefficient, log K_{OC}:
0.954 (soil, calculated-S, Kenaga 1980; quoted, Howard 1989)
-0.071 (sediment-water, calculated-K_{OW}, Mabey et al. 1982)
1.101 (Captina silt loam, Walton et al. 1992)
1.006 (McLaurin sandy loam, Walton et al. 1992)
-0.0899 (calculated-K_{OW}, Walton et al. 1992)
-0.089 (calculated-K_{OW}, Kollig 1993)

Half-Lives in the Environment:
Air: 4.0 hours for photooxidation in the troposphere (Callahan et al. 1979); half-life of 3.5 days for the reaction with photochemically produced hydroxyl radicals by the sunlight (Edney et al. 1983; quoted, Howard 1989); photooxidation half-life

of 13.4-189 hours, based on measured rate constant for the reaction with hydroxyl radicals in air (Atkinson 1985; quoted, Howard et al. 1991); atmospheric transformation lifetime was estimated to be 1 - 5 to > 5 days (Kelly et al. 1994).

Surface water: 30-552 hours, based on estimated aqueous aerobic biodegradation half-life (Howard et al. 1991).

Groundwater: 60-1104 based on estimated aqueous aerobic biodegradation half-life (Howard et al. 1991).

Sediment:

Soil: half-life of less than 10 days in soil (USEPA 1979; quoted, Ryan et al. 1988); 30-552 based on estimated aqueous aerobic biodegradation half-life (Howard et al. 1991).

Biota:

Environmental Fate Rate Constants or Half-Lives:

Volatilization: coupled with the reaeration rates for oxygen in typical bodies of water, one can estimate that the half-life in a typical pond, river and lake are 6, 1.2, and 4.8 days, respectively (Lyman et al. 1982; quoted, Howard 1989).

Photolysis:

Hydrolysis: acid rate constant of 4.2×10^{-2} M^{-1} h^{-1} at pH 5.0 with half-life of 188 years and base rate constant of 6.1×10^{-1} M^{-1} h^{-1} at pH 9.0 with half-life of 13 years (Ellington et al. 1987; quoted, Howard et al. 1991, Kollig 1993); half-life of 1210 years at pH 7.0, based on measured acid and base catalyzed hydrolysis constants (Ellington et al. 1987; quoted, Howard et al. 1991).

Oxidation: aqueous oxidation rate constants for singlet oxygen of $< 1 \times 10^{8}$ $M^{-1}h^{-1}$ and for peroxy radical of 36 $M^{-1}h^{-1}$ at 25°C (Mabey et al. 1982); half-life of 4.0 hours for photooxidation in the troposphere (Callahan et al. 1979); rate constant of $(40.6 \pm 4.1) \times 10^{-13}$ cm^3 $molecule^{-1}$ sec^{-1} by flash photolysis-resonance fluorescence technique for the gas-phase reaction with OH radicals at 299 K in air (Harris et al. 1981); rate constant of $< 1 \times 10^{-19}$ cm^3 $molecule^{-1}$ sec^{-1} for the gas-phase reaction with O_3 at 296 $\pm$ 2 K (Atkinson et al. 1983; quoted, Atkinson & Carter 1984); half-life of 3.5 days for the reaction with photochemically produced hydroxyl radicals by the sunlight (Edney et al. 1983; quoted, Howard 1989); photooxidation half-life of 13.4-189 hours, based on measured rate constant for the reaction with hydroxyl radicals in air (Atkinson 1985; quoted, Howard et al. 1991).

Biodegradation: aqueous aerobic half-life of 30-552 hours, based on river die-away test data (Going et al. 1979; Ludzack et al. 1958; quoted, Howard et al. 1991); aqueous anaerobic half-life of 120-2208 hours, based on estimated aqueous aerobic biodegradation half-life (Howard et al. 1991); ^{14}C labelled acrylonitrile at concentrations up to 100 ppm was completely degraded within 2.0 days in a London soil under aerobic conditions (Donberg et al. 1992).

Biotransformation: rate constant for bacterial transformation of 3×10^{-9} $mL \cdot cell^{-1} \cdot h^{-1}$ in water (Mabey et al. 1982).

Bioconcentration, Uptake (k_1) and Elimination (k_2) Rate Constants:

Common Name: Benzonitrile
Synonym: benzenecarbonitrile, cyanobenzene, phenyl cyanide
Chemical Name: benzonitrile, benzoic acid nitrile
CAS Registry No: 100-47-0
Molecular Formula: C_6H_5CN
Molecular Weight: 103.12
Melting Point (°C):
-13.0 (Weast 1982-83; Verschueren 1983)
-12.75 (Dean 1985; Riddick et al. 1986)
Boiling Point (°C):
190.7 (Weast 1982-83; Verschueren 1983)
191.1 (Dean 1985; Riddick et al. 1986)
Density (g/cm^3 at 20°C):
1.0100 (15°C, Verschueren 1983)
1.0006 (25°C, Dean 1985; Riddick et al. 1986)
Molar Volume (cm^3/mol):
112.0 (Taft et al. 1985; Leahy 1986; Kamlet et al. 1987)
59.0 (intrinsic volume V_I, Leahy 1986; Kamlet et al. 1987)
107.9 (calculated-LeBas method)
103.1 (calculated-density)
Molecular Volume (Å^3):
Total Surface Area, TSA (Å^2):
247.5 (Iwase et al. 1985)
Dissociation Constant, pK_a::
Heat of Fusion, ΔH_{fus}, kcal/mol:
2.60 (Riddick et al. 1986)
Entropy of Fusion, ΔS_{fus}, cal/mol·K (e.u.):
Fugacity Ratio at 25°C (assuming ΔS_{fus} = 13.5 e.u.), F: 1.0

Water Solubility (g/m^3 or mg/L at 25°C):
4330 (shake flask-AS, McGowan et al. 1966)
2000 (Dean 1985; Riddick et al. 1986)
14906 (calculated-V_I and solvatochromic p., Leahy 1986)
2310, 12980 (quoted, calculated-V_I and solvatochromic p., Kamlet et al. 1987)
10000 (quoted, Isnard & Lambert 1989)
10000 (selected, Yaws et al. 1990)

Vapor Pressure (Pa at 25°C):
78.86 (calculated-Antoine eqn., Dean 1985)
100 (Riddick et al. 1986)
106 (extrapolated-Antoine eqn., Stephenson & Malanowski 1987)

Henry's Law Constant (Pa·m^3/mol):

55.32 (computed, Yaws et al. 1991)

5.311 (calculated-P/C, this work)

Octanol/Water Partition Coefficient, log K_{OW}:

1.56 (shake flask-AS, Fujita et al. 1964; quoted, Leo et al. 1969; Rekker 1977; Isnard & Lambert 1989, Sangster 1989)

1.56 (shake flask-AS, Holmes & Lough 1976; quoted, Sangster 1989)

1.67 (calculated-f const., Rekker 1977)

1.56 (shake flask at pH 7, Unger et al. 1978)

1.56 (quoted from Hansch & Leo 1979, Miyake & Terada 1982; Haky & Young 1984; Iwase et al. 1985; Bodor et al; 1989)

1.66 (RP-HPLC-k' Miyake & Terada 1982)

1.56 (quoted, Verschueren 1983)

1.56 (selected best lit. value, Garst & Wilson 1984)

1.65 ± 0.01 (exptl.-ALPM, Garst & Wilson 1984)

1.50 (HPLC-k', Haky & Young 1984; quoted, Sangster 1989)

1.56 (shake flask-GC, Tanii & Hashimoto 1984; quoted, Sangster 1989)

1.56, 1.55 (quoted, calculated-TSA, Iwase et al. 1985)

1.53 (computed, Klopman et al. 1985; quoted, Pangrekar et al. 1994)

1.56, 1.60 (quoted, calculated-solvatochromic p., Taft et al. 1985)

1.56, 1.47 (quoted, calculated-solvatochromic p., Leahy 1986)

1.69 (selected, Yoshida et al. 1986)

1.56 (RP-HPLC-capacity ratio, Minick et al. 1988)

1.56 (quoted from CLOGP, Bodor et al. 1989)

1.575 (calculated-CLOGP, Bodor et al. 1989)

2.04 (calculated-M.O. indices, Bodor et al. 1989)

1.56 (recommended, Sangster 1989)

Bioconcentration Factor, log BCF:

Sorption Partition Coefficient, log K_{OC}:

Half-Lives in the Environment:

Air:

Surface water:

Groundwater:

Sediment:

Soil:

Biota:

Environmental Fate Rate Constants or Half-Lives:

Volatilization:

Photolysis:

Oxidation: observed rate constant of 0.33×10^{-12} cm^3 $molecule^{-1}$ sec^{-1} (Zetzsch 1982) compared to calculated value of 0.42×10^{-12} cm^3 $molecule^{-1}$ sec^{-1} for the gas-phase reaction with OH radicals at room temperature in air (Atkinson 1985); observed rate constant of 0.33×10^{-12} cm^3 $molecule^{-1}$ sec^{-1} (Zetzsch 1982) compared to calculated value of 0.39×10^{-12} cm^3 $molecule^{-1}$ sec^{-1} for the gas-phase reaction with OH radicals at room temperature in air (Atkinson et al. 1985); experimental rate constant of 0.33×10^{-12} cm^3 $molecule^{-1}$ sec^{-1} (Zetzsch 1982) compared to calculated value of 0.36×10^{-12} cm^3 $molecule^{-1}$ sec^{-1} for the gas-phase reaction with OH radicals at room temperature in air (Atkinson 1987); experimental rate constant of 0.33×10^{-12} cm^3 $molecule^{-1}$ sec^{-1} (Atkinson 1989) compared to calculated value of 0.41×10^{-12} cm^3 $molecule^{-1}$ sec^{-1} by molecular orbital calculation for the gas-phase reaction with OH radicals at room temperature in air (Klamt 1993).

Hydrolysis:

Biodegradation:

Biotransformation:

Bioconcentration, Uptake (k_1) and Elimination (k_2) Rate Constants:

Common Name: Dimethylamine
Synonym: aminomethylmethane
Chemical Name: aminomethylmethane, dimethylamine
CAS Registry No: 124-40-3
Molecular Formula: CH_3NHCH_3
Molecular Weight: 45.09
Melting Point (°C):
-96.0 (Stull 1947)
-93.0 (Weast 1982-83; Howard 1990)
-92/-96 (Verschueren 1983)
-92.2 (Dean 1985; Stephenson & Malanowski 1987)
Boiling Point (°C):
7.40 (Stull 1947; Weast 1982-83; Verschueren 1983; Howard 1990)
6.89 (Boublik et al. 1984)
6.90 (Dean 1985; Stephenson & Malanowski 1987)
Density (g/cm³ at 20°C):
0.6804 (0°C, Weast 1982-83)
0.6800 (0°C, Verschueren 1983; Dean 1985)
Molar Volume (cm³/mol):
67.5 (calculated-LeBas method)
Molecular Volume (Å³):
Total Surface Area, TSA (Å²):
Dissociation Constants, pK_a:
10.732 (Perrin 1965; Weast 1982-83; Howard 1990)
10.77 (protonated cation +1, Dean 1985)
10.77 (Sangster 1989)
Heat of Fusion ΔH_{fus}, kcal/mol:
Entropy of Fusion ΔS_{fus}, cal/mol·K (e.u.):
Fugacity Ratio at 25°C (assuming ΔS_{fus} = 13.5 e.u.), F: 1.0

Water Solubility (g/m³ or mg/L at 25°C):
very soluble (Dean 1985)
620000 (selected, Yaws et al. 1990)
miscible (Stephenson 1993)

Vapor Pressure (Pa at 25°C):
236420 (extrapolated-regression of tabulated data, Stull 1947)
172220 (20°C, quoted, Verschueren 1983)
206180 (extrapolated-Antoine eqn., Boublik et al. 1984)
202620 (Daubert & Danner 1985; quoted, Howard 1990)
206000 (extrapolated-Antoine eqn., Dean 1985)
205300 (calculated-Antoine eqn., Stephenson & Malanowski 1987)

Henry's Law Constant (Pa·m^3/mol):

1.796 (quoted exptl., Hine & Mookerjee 1975; quoted, Howard 1990)
1.796 (calculated-group contribution, Hine & Mookerjee 1975)
1.030 (calculated-bond contribution, Hine & Mookerjee 1975)
1.80, 2.718 (quoted, calculated-molecular structure, Russell et al. 1992)

Octanol/Water Partition Coefficient, log K_{OW}:

-0.38 (shake flask-RC at pH 13, Wolfenden 1978; quoted, Sangster 1989)
-0.38, -0.02 (quoted, calculated, Verschueren 1983)
-0.38 (Hansch & Leo 1985; quoted, Howard 1990)
-0.38 (recommended, Sangster 1989)
-0.52 (calculated-CLOGP, Jäckel & Klein 1991)
-0.52 (calculated-CLOGP, Müller & Klein 1992)

Bioconcentration Factor, log BCF:

-0.523 (calculated-K_{OW}, Lyman et al. 1982; quoted, Howard 1990)

Sorption Partition Coefficient, log K_{OC}:

2.638 (adsorption isotherm average for five soils, Rao & Davidson 1982; quoted, Howard 1990)
0.602 (Podzol soil, von Oepen et al. 1991)
2.212 (Alfisol soil, von Oepen et al. 1991)
2.706 (sediment, von Oepen et al. 1991)

Half-Lives in the Environment:

Air: half-life of 5.9 hours was estimated for the vapor phase reaction with hydroxyl radicals of $5x10^5$ radicals/cm^3 in air (Atkinson et al. 1978; Atkinson 1985; quoted, Howard 1990); 0.892-9.20 hours, based on measured rate constant for the gas-phase reaction with hydroxyl radicals (Atkinson 1985; quoted, Howard et al. 1991) and ozone (Tuazon et al. 1978; selected, Howard et al. 1991).

Surface water: 2-79 hours, based on estimated unacclimated aqueous aerobic biodegradation half-life (Howard et al. 1991).

Groundwater: 4-158 hours, based on estimated unacclimated aqueous aerobic biodegradation half-life (Howard et al. 1991).

Sediment:

Soil: 86-336 hours, based on soil die-away test data (Tate & Alexander 1976; Greene et al. 1981; selected, Howard et al. 1991).

Biota:

Environmental Fate Rate Constants or Half-Lives:

Volatilization: using Henry's law constant, a half-life of 35.1 hours was estimated for a model river of 1 m deep flowing at 1 m/sec with a wind velocity of 3 m/sec (Lyman et al. 1982; selected, Howard 1990).

Photolysis:

Oxidation: rate constant of 6.54×10^{-11} cm^3 molecule^{-1} sec^{-1} for the gas-phase reaction with OH radicals at 299 K in the atmosphere (Atkinson et al. 1977; quoted, Carlier et al. 1986); photooxidation half-life of 5.9 hours in air was estimated for the vapor phase reaction with hydroxyl radicals of 5×10^5 radicals/cm^3 in air (Atkinson et al. 1978; Atkinson 1985; quoted, Howard 1990); rate constant of $(2.61 \pm 0.30) \times 10^{-18}$ cm^3 molecule^{-1} sec^{-1} for the gas-phase reaction with ozone at 296 ± 2 K under atmospheric conditions (Atkinson & Carter 1984; quoted, Atkinson 1985); rate constant of 6.5×10^{-11} cm^3 molecule^{-1} sec^{-1} for the gas-phase reaction with 5×10^5 OH radicals/cm^3 at room temperature having a loss rate of 2.8 day^{-1} (Atkinson 1985); photooxidation half-life of 0.892-9.20 hours, based on measured rate constant for the gas-phase reaction with hydroxyl radicals (Atkinson 1985; selected, Howard et al. 1991) and ozone in air (Tuazon et al. 1978; selected, Howard et al. 1991).

Hydrolysis:

Biodegradation: aqueous aerobic half-life of 2-79 hours, based on river die-away test data (Digeronimo et al. 1979; Dojlido 1979; selected, Howard et al. 1991); aqueous anaerobic half-life of 8-316 hours, based on estimated unacclimated aqueous aerobic biodegradation half-life (Howard et al. 1991).

Biotransformation:

Bioconcentration, Uptake (k_1) and Elimination (k_2) Rate Constants:

Common Name: Trimethylamine
Synonym: dimethylamino methane, TMA
Chemical Name: trimethylamine
CAS Registry No: 75-50-3
Molecular Formula: $CH_3N(CH_3)_2$
Molecular Weight: 59.11
Melting Point (°C):
- -117.1 (Stull 1947; Dean 1985; Stephenson & Malanowski 1987)
- -117.2 (Weast 1982-83; Howard 1990)
- -117/-124 (Verschueren 1983)

Boiling Point (°C):
- 2.9 (Stull 1947; Weast 1982-83)
- 3.5 (Verschueren 1983)
- 2.87 (Boublik et al. 1984; Howard 1990)
- 2.9 (Dean 1985; Stephenson & Malanowski 1987)

Density (g/cm^3 at 20°C):
- 0.6356 (Weast 1982-83)
- 0.6620 (-5°C, Verschueren 1983; Dean 1985)

Molar Volume (cm^3/mol):
- 95.0 (estimated-atom additivity method, Taft et al. 1985)
- 86.0 (Kamlet et al. 1986, 1987)
- 43.3 (intrinsic volume, Kamlet et al. 1987)
- 93.3 (calculated-LeBas method)

Molecular Volume ($Å^3$):
Total Surface Area, TSA ($Å^2$):
- 192.9 (Iwase et al. 1985)

Dissociation Constants, pK_a:
- 9.801, 9.987 (Perrin 1972; quoted, Howard 1990)
- 9.80 (pK_a, protonated cation +1, Dean 1985)
- 9.79 (pK_a, Sangster 1989)

Heat of Fusion ΔH_{fus}, kcal/mol:
Entropy of Fusion ΔS_{fus}, cal/mol·K (e.u.):
Fugacity Ratio at 25°C (assuming ΔS_{fus} = 13.5 e.u.), F: 1.0

Water Solubility (g/m^3 or mg/L at 25°C):
- 410000 (Dean 1985)
- 1234980, 727210 (quoted, calculated-solvatochromic p., Kamlet et al. 1987)
- 890000 (30°C, quoted, Howard 1990)
- 291000 (selected, Yaws et al. 1990)
- miscible (Stephenson 1993)

Vapor Pressure (Pa at 25°C):
- 265160 (extrapolated-regression of tabulated data, Stull 1947)
- 192485 (20°C, quoted, Verschueren 1983)

263400 (30°C, quoted, Verschueren 1983)
219306 (extrapolated-Antoine eqn., Boublik et al. 1984)
214213 (Daubert & Danner 1985; quoted, Howard 1990)
219000 (extrapolated-Antoine eqn., Dean 1985)
219895 (extrapolated-Antoine eqn., Stephenson & Malanowski 1987)

Henry's Law Constant (Pa • m^3/mol):

6.672 (quoted exptl., Hine & Mookerjee 1975)
12.713 (calculated-group contribution, Hine & Mookerjee 1975)
2.159 (calculated-bond contribution, Hine & Mookerjee 1975)
10.54 (quoted from Hine & Mookerjee 1975, Howard 1990)
10.57, 15.64 (quoted, calculated-molecular structure, Russell et al. 1992)

Octanol/Water Partition Coefficient, log K_{OW}:

0.27 (shake flask-TN, Sandell 1962; quoted, Leo et al. 1971; Rekker 1977; Iwase et al. 1985; Sangster 1989)
0.27 (calculated-f const., Rekker 1977)
0.20 (calculated-π const., Rekker 1977)
0.27 (quoted, Verschueren 1983)
0.16 (Hansch & Leo 1985; quoted, Howard 1990)
0.15 (calculated-TSA, Iwase et al. 1985)
0.16, 0.38 (quoted, calculated-solvatochromic p., Taft et al. 1985)
0.22 (quoted from Hansch & Leo 1979, Bodor et al. 1989)
0.16 (quoted from CLOGP, Bodor et al. 1989)
0.048 (calculated-CLOGP, Bodor et al. 1989)
0.08 (calculated-M.O. indices, Bodor et al. 1989)
0.16 (quoted from Pomona College unpublished data, Sangster 1989)
0.16 (recommended, Sangster 1989)

Bioconcentration Factor, log BCF:

<0.0 (estimated-K_{OW}, Lyman et al. 1982; quoted, Howard 1990)

Sorption Partition Coefficient, log K_{OC}:

1.462 (soil, estimated-K_{OW}, Lyman et al. 1982; quoted, Howard 1990)
0.602 (soil, estimated-S, Lyman et al. 1982; quoted, Howard 1990)
0.778 (Podzol soil, von Oepen et al. 1991)
2.365 (Alfisol soil, von Oepen et al. 1991)
2.831 (sediments, von Oepen et al. 1991)

Half-Lives in the Environment:

Air: half-life of 4.0 hours, based on rate constant of 6.09×10^{-11} cm^3 $molecule^{-1}$ sec^{-1} for the vapor-phase reaction with photochemically produced hydroxyl radicals of 8×10^5 radicals/cm^3 in air at 25.5°C and a half-life of 1.4 day, based on rate constant of 9.73×10^{-18} cm^3 $molecule^{-1}$ sec^{-1} for the vapor-phase reaction with ozone of 6×10^{11} molecules/cm^3 in air at 24.4°C (Atkinson 1985; GEMS 1986; quoted, Howard 1990).

Surface water: half-life of 62 days, based on rate constant of 1.3×10^{10} liter $mole^{-1}$ sec^{-1} for the reaction with photochemically produced hydroxyl radicals of 1×10^{-17} $mol \cdot L^{-1}$ in water (Mill et al. 1980; Güesten et al. 1981; quoted, Howard 1990).

Groundwater:

Sediment:

Soil:

Biota:

Environmental Fate Rate Constants or Half-Lives:

Volatilization: using Henry's law constant, a half-life of 11 hours was estimated for a model river 1 m deep flowing at 1 m/sec with a wind velocity of 3 m/sec (Lyman et al. 1982; quoted, Howard 1990).

Photolysis:

Hydrolysis:

Oxidation: rate constant of 6.09×10^{-11} cm^3 $molecule^{-1}$ sec^{-1} for the vapor-phase reaction with photochemically produced hydroxyl radicals in the atmosphere at 299 K (Atkinson et al. 1977; quoted, Carlier et al. 1986); photooxidation half-life of 62 days in water, based on rate constant of 1.3×10^{10} liter $mole^{-1}$ sec^{-1} for the reaction with photochemically produced hydroxyl radicals of 1×10^{-17} mole $liter^{-1}$ in water (Mill et al. 1980; Güesten et al. 1981; quoted, Howard 1990); rate constant of 6.10×10^{-11} cm^3 $molecule^{-1}$ sec^{-1} for the gas-phase reaction with 1×10^6 OH radicals/cm^3 with a loss rate of 5.0 day^{-1} and rate constant of 9.70×10^{-18} $cm^3 \cdot molecule^{-1}$ sec^{-1} for the gas-phase reaction with 7×10^{11} O_3 molecules/cm^3 with a loss rate of 0.6 day^{-1} both at room temperature (Atkinson et al. 1984); photooxidation half-life of 4.0 hours in air, based on rate constant of 6.09×10^{-11} cm^3 $molecule^{-1}$ sec^{-1} for the vapor-phase reaction with photochemically produced hydroxyl radicals of 8×10^5 radicals/cm^3 in air at 25.5°C and a half-life of 1.4 day, based on rate constant of 9.73×10^{-18} cm^3 $molecule^{-1}$ sec^{-1} for the vapor-phase reaction with ozone of 6×10^{11} molecules/cm^3 in air at 24.4°C (Atkinson 1985; GEMS 1986; quoted, Howard 1990).

Biodegradation:

Biotransformation:

Bioconcentration, Uptake (k_1) and Elimination (k_2) Rate Constants:

Common Name: Ethylamine
Synonym: aminoethane, ethanamine, monoethylamine
Chemical Name: aminoethane, ethylamine
CAS Registry No: 75-04-7
Molecular Formula: $CH_3CH_2NH_2$
Molecular Weight: 45.09
Melting Point (°C):
- -80.6 (Stull 1947; Verschueren 1983; Dean 1985; Weast 1982-83; Stephenson & Malanowski 1987)
- -83.0 (Verschueren 1983)
- -80.5 (Chao et al. 1990)
- -81.1 (Howard 1990)

Boiling Point (°C):
- 16.6 (Stull 1947; Dean 1985; Weast 1982-83; Howard 1990)
- 17.0 (Verschueren 1983)
- 16.73 (Boublik et al. 1984)
- 16.0 (Stephenson & Malanowski 1987)
- 16.59 (Chao et al. 1990)

Density (g/cm³ at 20°C):
- 0.6829 (Weast 1982-83)
- 0.6767 (25°C, Chao et al. 1990)

Molar Volume (cm³/mol):
- 0.66 (liquied molar volume, $V_M/100$, Leahy 1986)
- 0.335 (intrinsic volume, $V_I/100$, Kamlet et al. 1987)
- 66.0 (calculated-LeBas method)
- 66.0 (calculated-density)

Molecular Volume (Å³):
Total Surface Area, TSA (Å²):
Dissociation Constants, pK_a:
- 10.79 (Perrin 1972)
- 10.807 (20°C, Weast 1982-83)
- 10.63 (protonated cation +1, Dean 1985)
- 10.70 (Sangster 1989)

Heat of Fusion ΔH_{fus}, kcal/mol:
Entropy of Fusion ΔS_{fus}, cal/mol·K (e.u.):
Fugacity Ratio at 25°C (assuming ΔS_{fus} = 13.5 e.u.), F: 1.0

Water Solubility (g/m³ or mg/L at 25°C):
- miscible (Dean 1985; Howard 1990; Stephenson 1993)

Vapor Pressure (Pa at 25°C):
- 156200 (extrapolated-regression of tabulated data, Stull 1947)
- 121570 (20°C, quoted, Verschueren 1983)
- 172224 (30°C, quoted, Verschueren 1983)

93.325 (14.65°C, quoted exptl., Boublik et al. 1984)
139077 (extrapolated-Antoine eqn., Boublik et al. 1984)
141650 (quoted, Abraham 1984)
139698 (Daubert & Danner 1985; quoted, Howard 1990)
141000 (calculated-Antoine eqn., Dean 1985)
140087 (calculated-Antoine eqn., Stephenson & Malanowski 1987)
140900 (calculated-Cox eqn., Chao et al. 1990)

Henry's Law Constant ($Pa{\cdot}m^3/mol$):
1.012 (partial pressure, Butler & Ramchandani 1935)
0.683 (quoted exptl., Hine & Mookerjee 1975)
0.859 (calculated-group contribution, Hine & Mookerjee 1975)
0.730 (calculated-bond contribution, Hine & Mookerjee 1975)
1.033, 0.421 (quoted, calculated-molecular structure, Russell et al. 1992)

Octanol/Water Partition Coefficient, log K_{OW}:
-0.30 (shake flask-titration with ion correction, Korenman et al. 1973; quoted, Sangster 1989)
-0.13 (quoted from Pomona College unpublished data, Rekker 1977; Sangster 1989)
-0.16, -0.14 (calculated-f const., Rekker 1977)
-0.19 (calculated-π const., Rekker 1977)
-0.27, -0.08 (quoted, calculated, Verschueren 1983)
-0.13 (Hansch & Leo 1985; quoted, Howard 1990)
-0.13 (recommended, Sangster 1989)
-0.14 (calculated-CLOGP, Jäckel & Klein 1991)

Bioconcentration Factor, log BCF:
<0.0 (estimated-K_{OW}, Lyman et al. 1982; quoted, Howard 1990)

Sorption Partition Coefficient, log K_{OC}:

Half-Lives in the Environment:
Air: half-life of >9.9 days for the gas-phase reaction with hydroxyl radicals in air, based on the rate of disappearance of hydrocarbon due to reaction with hydroxyl radicals (Darnall et al. 1976); half-life of 8.6 hours, based on rate constant of 6.54×10^{-11} $cm^3{\cdot}molecules^{-1}{\cdot}sec^{-1}$ for the vapor-phase reaction with an average hydroxyl radicals of 5×10^5 $radicals/cm^3$ at 25.5°C (Atkinson 1985; quoted, Howard 1990).
Surface water: half-life of 321 days, based on a rate constant of 2.5×10^9 $L{\cdot}mol^{-1}{\cdot}sec^{-1}$ for the aqueous-phase reaction with photochemically produced hydroxyl radicals of 1×10^{-17} $mol{\cdot}L^{-1}$ (Mill et al. 1980; Güesten et al. 1981; quoted, Howard 1990).

Groundwater:
Sediment:
Soil:
Biota:

Environmental Fate Rate Constants or Half-Lives:

Volatilization: using Henry's law constant, a half-life of 2.0 days was estimated for a model river of 1 m deep flowing at 1 m/sec with a wind velocity of 3 m/sec (Lyman et al. 1982; quoted, Howard 1990).

Photolysis:

Oxidation: photooxidation half-life of >9.9 days for the gas-phase reaction with OH radicals in air, based on the rate of disappearance of hydrocarbon due to reaction with hydroxyl radicals (Darnall et al. 1976); second-order rate constant of 2.77×10^{-11} $cm^3 \cdot molecules^{-1} \cdot sec^{-1}$ for the vapor-phase reaction with hydroxyl radicals in the atmosphere at 299 K (Atkinson et al. 1977; quoted, Carlier et al. 1986); photooxidation half-life of 321 days in water, based on a rate constant of 2.5×10^9 $L \cdot mol^{-1} \cdot sec^{-1}$ for the aqueous-phase reaction with photochemically produced hydroxyl radicals of 1×10^{-17} $mol \cdot L^{-1}$ (Mill et al. 1980; Güesten et al. 1981; quoted, Howard 1990); vapor-phase rate constant of $(2.76 \pm 0.34) \times 10^{-20}$ $cm^3 \cdot molecules^{-1} \cdot sec^{-1}$ for the reaction with O_3 at 296 ± 2 K under atmospheric conditions (Atkinson & Carter 1984); photooxidation half-life of 8.6 hours in air, based on rate constant of 6.54×10^{-11} $cm^3 \cdot molecules^{-1} \cdot sec^{-1}$ for the vapor-phase reaction with an average hydroxyl radicals of 5×10^5 radicals/cm^3 at 25.5°C (Atkinson 1985; quoted, Howard 1990).

Hydrolysis:

Biodegradation:

Biotransformation:

Bioconcentration, Uptake (k_1) and Elimination (k_2) Rate Constants:

Common Name: Diethylamine
Synonym: aminoethylethane
Chemical Name: aminoethylethane, diethylamine
CAS Registry No: 109-89-7
Molecular Formula: $CH_3CH_2NHCH_2CH_3$
Molecular Weight: 73.14
Melting Point (°C):
-48.0 (Weast 1982-83; Verschueren 1983)
-50.0 (Verschueren 1983; Dean 1985; Stephenson & Malanowski 1987; Howard 1990)
-49.8 (Riddick et al. 1986)
Boiling Point (°C):
56.3 (Weast 1982-83)
57.0 (Verschueren 1983)
55.6 (Boublik et al. 1984; Riddick et al. 1986)
55.5 (Dean 1985; Stephenson & Malanowski 1987; Howard 1990)
Density (g/cm^3 at 20°C):
0.7056 (Weast 1982-83)
0.7110 (18°C, Verschueren 1983)
0.7074 (Dean 1985)
0.7070 (Riddick et al. 1986)
Molar Volume (cm^3/mol):
109.0 (exptl. at normal B.P., Lee et al. 1972; quoted, Reid et al. 1977)
111.9 (calculated-LeBas method, Reid et al. 1977)
103.7 (liquid molar volume, Kamlet et al. 1986, 1987)
53.50 (intrinsic volume V_I, Kamlet et al. 1987)
103.0 (calculated-density)
Molecular Volume ($Å^3$):
Total Surface Area, TSA ($Å^2$):
214.1 (Iwase et al. 1985)
Dissociation Constants, pK_a:
10.98 (Perrin 1965; quoted, Howard 1990)
10.797 (35°C, Perrin 1972)
10.8 (protonated cation +1, Dean 1985)
11.07 (Sangster 1989)
Heat of Fusion ΔH_{fus}, kcal/mol:
Entropy of Fusion ΔS_{fus}, cal/mol·K (e.u.):
Fugacity Ratio at 25°C (assuming ΔS_{fus} = 13.5 e.u.), F: 1.0

Water Solubility (g/m^3 or mg/L at 25°C):
815000 (14°C, quoted, Verschueren 1983)
miscible (Dean 1985)
infinite (Riddick et al. 1986; quoted, Howard 1990)
783710, 529850 (quoted, calculated-solvatochromic p., Kamlet et al. 1987)
miscible (Yaws et al. 1990)
miscible (Stephenson 1993)

Vapor Pressure (Pa at 25°C):

26660 (20°C, quoted, Verschueren 1983)
38660 (30°C, quoted, Verschueren 1983)
30110 (extrapolated-Antoine eqn., Boublik et al. 1984)
31308 (calculated-Antoine eqn., Boublik et al. 1984)
31130 (Riddick et al. 1986; quoted, Howard 1990)
31490 (extrapolated-Antoine eqn., Stephenson & Malanowski 1987)

Henry's Law Constant (Pa·m^3/mol):

2.596 (quoted exptl., Hine & Mookerjee 1975; quoted, Howard 1990)
2.537 (calculated-group contribution, Hine & Mookerjee 1975)
2.370 (calculated-bond contribution, Hine & Mookerjee 1975)

Octanol/Water Partition Coefficient, log K_{OW}:

0.43 (shake flask-K, Collander 1951; quoted, Sangster 1989)
0.57 (shake flask-TN, Sandell 1962; quoted, Leo et al. 1971; Rekker 1977; Iwase et al. 1985; Sangster 1989)
0.60, 0.61 (calculated-f const., Rekker 1977)
0.70 (calculated-π const., Rekker 1977)
0.43, 0.57 (quoted, Verschueren 1983)
0.58 (Hansch & Leo 1985; quoted, Howard 1990)
0.86 (calculated-TSA, Iwase et al. 1985)
0.81 (HPLC-k', Eadsforth 1986)
0.50, 0.59 (quoted: shake flask, calculated, Eadsforth 1986)
0.53, 0.58, 0.58, 0.74 (quoted, CLOGP, calculated-CLOGP, calculated-M.O. indices, Bodor et al. 1989)
0.58 (quoted from Pomona College unpublished data, Sangster 1989)
0.58 (recommended, Sangster 1989)
0.54 (calculated-CLOGP, Jäckel & Klein 1991)

Bioconcentration Factor, log BCF:

0.210 (calculated-K_{OW}, Lyman et al. 1982; quoted, Howard 1990)

Sorption Partition Coefficient, log K_{OC}:

1.699 (soil, calculated-K_{OW}, Lyman et al. 1982; quoted, Howard 1990)

Half-Lives in the Environment:

Air: half-life of >9.9 days for the gas-phase reaction with hydroxyl radicals in air, based on the rate of disappearance of hydrocarbon due to reaction with hydroxyl radicals (Darnall et al. 1976); half-life of 0.21 day, based on an estimated rate constant of 77.1×10^{-12} cm^3 molecule^{-1} sec^{-1} for the vapor-phase reaction with photochemically produced hydroxyl radicals of $5x10^5$ radicals/cm^3 in air (Atkinson 1987; quoted, Howard 1990).

Surface water:

Groundwater:

Sediment:

Soil:

Biota:

Environmental Fate Rate Constants or Half-Lives:

Volatilization: using Henry's law constant, a half-life of 31.6 hours was estimated for a model river 1 m deep flowing at 1 m/sec with a wind velocity of 3 m/sec (Lyman et al. 1982; quoted, Howard 1990).

Photolysis:

Oxidation: photooxidation half-life of >9.9 days for the gas-phase reaction with hydroxyl radicals in air, based on the rate of disappearance of hydrocarbon due to reaction with OH radicals (Darnall et al. 1976); photooxidation half-life of 0.21 day in air, based on an estimated second-order rate constant of $77.1x10^{-12}$ $cm^3 \cdot molecule^{-1} \cdot sec^{-1}$ for the vapor-phase reaction with photochemically produced hydroxyl radicals of $5x10^5$ radicals/cm^3 in air (Atkinson 1987; quoted, Howard 1990).

Hydrolysis:

Biodegradation:

Biotransformation:

Bioconcentration, Uptake (k_1) and Elimination (k_2) Rate Constants:

Common Name: Propylamine
Synonym: 1-aminopropane, 1-propanamine, *n*-propylamine
Chemical Name: aminopropane, *n*-propylamine
CAS Registry No: 107-10-8
Molecular Formula: $CH_3CH_2CH_2NH_2$
Molecular Weight: 59.11
Melting Point (°C):
-83.0 (Stull 1947; Weast 1982-83; Verschueren 1983; Dean 1985; Riddick et al. 1986; Stephenson & Malanowski 1987; Howard 1990)
-116.0 (Verschueren 1983)
-84.8 (Chao et al. 1990)
Boiling Point (°C):
49.25 (Butler & Ramchandani 1935)
48.50 (Stull 1947)
47.80 (Weast 1982-83; Stephenson & Malanowski 1987)
47.23 (Boublik et al. 1984; Riddick et al. 1986)
49.0 (Verschueren 1983)
46.90 (Dean 1985)
47.22 (Chao et al. 1990)
Density (g/cm³ at 20°C):
0.7173 (Weast 1982-83; Dean 1985; Riddick et al. 1986)
0.7180 (Verschueren 1983)
0.7119 (25°C, calculated, Chao et al. 1990)
Molar Volume (cm³/mol):
82.4 (liquid molar volume, Kamlet et al. 1986, 1987)
43.3 (intrinsic volume V_I, Kamlet et al. 1987)
88.2 (calculated-LeBas method)
Molecular Volume (Å³):
Total Surface Area, TSA (Å²):
220.4 (Iwase et al. 1985)
Dissociation Constants, pK:
10.74, 10.789 (20°C, Perrin 1972)
10.708 (pK_a, 20°C, Weast 1982-83)
10.568 ($pK_{BH}{}^+$, Dean 1985; Riddick et al. 1986)
10.68 (pK_a, Sangster 1989)
Heat of Fusion ΔH_{fus}, kcal/mol:
2.6229 (Riddick et al. 1986)
Entropy of Fusion ΔS_{fus}, cal/mol·K (e.u.):
Fugacity Ratio at 25°C (assuming ΔS_{fus} = 13.5 e.u.), F: 1.0

Water Solubility (g/m³ or mg/L at 25°C):
miscible (Dean 1985; Stephenson 1993)
infinite (Riddick et al. 1986; Howard 1990)
1957320, 1744460 (quoted, calculated-solvatochromic p., Kamlet et al. 1987)
miscible (Yaws et al. 1990)

Vapor Pressure (Pa at 25°C):

41800 (interpolated-regression of tabulated data, Stull 1947)
53320 (31.5°C, Stull 1947)
42102 (ebulliometry, calculated-Antoine eqn., Osborn & Douslin 1968)
32660 (20°C, quoted, Verschueren 1983)
53320 (31°C, quoted, Verschueren 1983)
38550 (22.97°C, quoted exptl., Boublik et al. 1984)
42110 (calculated-Antoine eqn., Boublik et al. 1984)
41050 (quoted, Abraham 1984)
42120 (calculated-Antoine eqn., Dean 1985)
41050 (Riddick et al. 1986; Howard 1990)
42124 (calculated-Antoine eqn., Stephenson & Malanowski 1987)
42125 (calculated-Cox eqn., Chao et al. 1990)

Henry's Law Constant (Pa m^3/mol):

1.274 (partial pressure, Butler & Ramchandani 1935)
0.784 (quoted exptl., Hine & Mookerjee 1975)
0.732 (calculated-group contribution, Hine & Mookerjee 1975)
1.330 (calculated-bond contribution, Hine & Mookerjee 1975; quoted, Howard 1990)
1.242, 0.637 (quoted, calculated-molecular structure, Russell et al. 1992)

Octanol/Water Partition Coefficient, log K_{OW}:

0.28 (shake flask-titration, Korenman et al. 1973; quoted, Sangster 1989)
0.48 (quoted from Pomona College unpublished data, Rekker 1977; Sangster 1989)
0.37, 0.39 (calculated-f const., Rekker 1977)
0.31 (calculated-π const., Rekker 1977)
0.15, 0.37 (calculated, Verschueren 1983)
0.48 (Hansch & Leo 1985; quoted, Howard 1990)
0.48 (recommended, Sangster 1989)

Bioconcentration Factor, log BCF:

-0.886 (estimated-K_{OW}, Lyman et al. 1982; quoted, Howard 1990)

Sorption Partition Coefficient, log K_{OC}:

<1.699 (soil, estimated-K_{OW}, Lyman et al. 1982; quoted, Howard 1990)

Half-Lives in the Environment:

Air: half-life of 12 hours, based on estimated seond-order rate constant of 3.21×10^{12} $cm^3 \cdot molecule^{-1} \cdot sec^{-1}$ for the vapor-phase reaction with hydroxyl radicals of $5 \times 10^5/cm^3$ at 25°C in the atmosphere (Atkinson 1987; quoted, Howard 1990).
Surface water:
Groundwater:

Sediment:
Soil:
Biota:

Environmental Fate Rate Constants or Half-Lives:

Volatilization: using Henry's law constant, a half-life of 2.44 days was estimated for a model river 1 m deep flowing at 1 m/sec with a wind speed of 3 m/sec (Lyman et al. 1982; quoted, Howard 1990).

Photolysis:

Oxidation: photooxidation half-life of 12 hours in air, based on estimated rate constant of 3.21×10^{-12} $cm^3 \cdot molecule^{-1} \cdot sec^{-1}$ for the vapor-phase reaction with hydroxyl radicals of $5 \times 10^5/cm^3$ at 25°C in the atmosphere (Atkinson 1987; quoted, Howard 1990).

Hydrolysis:

Biodegradation:

Biotransformation:

Bioconcentration, Uptake (k_1) and Elimination (k_2) Rate Constants:

Common Name: Butylamine
Synonym: 1-aminobutane, *n*-butylamine
Chemical Name: 1-aminobutane, *n*-butylamine
CAS Registry No: 109-73-9
Molecular Formula: $CH_3CH_2CH_2CH_2NH_2$
Molecular Weight: 73.14
Melting Point (°C):

- -49.1 (Weast 1982-83; Riddick et al. 1986; Stephenson & Malanowski 1987)
- -50.0 (Verschueren 1983; Howard 1990)
- -50.5 (Dean 1985)

Boiling Point (°C):

- 77.80 (Weast 1982-83; Howard 1990)
- 78.0 (Verschueren 1983; Stephenson & Malanowski 1987)
- 77.90 (Dean 1985)
- 77.07 (Riddick et al. 1986)

Density (g/cm^3 at 20°C):

- 0.7414 (Weast 1982-83)
- 0.740 (Verschueren 1983)
- 0.7392 (Riddick et al. 1986)

Molar Volume (cm^3/mol):

- 98.7 (Kamlet et al. 1986, 1987)
- 53.5 (intrinsic volume V_I, Kamlet et al. 1987)
- 110.4 (calculated-LeBas method)
- 98.84 (calculated-density)

Molecular Volume ($Å^3$):
Total Surface Area, TSA ($Å^2$):

- 220.4 (Iwase et al. 1985)

Dissociation Constants, pK:

- 10.77 (Perrin 1965; Howard 1990)
- 10.653 (Perrin 1972)
- 10.77 (pK_a, 20°C, Weast 1982-83)
- 10.64 (pK_a, protonated +1, Dean 1985)
- 10.77 (pK_{BH+}, Riddick et al. 1986)
- 10.64 (pK_a, Sangster 1989)

Heat of Fusion ΔH_{fus}, kcal/mol:
Entropy of Fusion ΔS_{fus}, cal/mol·K (e.u.):
Fugacity Ratio at 25°C (assuming ΔS_{fus} = 13.5 e.u.), F: 1.0

Water Solubility (g/m^3 or mg/L at 25°C):

- miscible (Dean 1985; Howard 1990; Stephenson 1993)
- infinite (Riddick et al. 1986)
- 667045, 549385 (quoted, calculated-solvatochromic p., Kamlet et al. 1987)
- miscible (Yaws et al. 1990)

Vapor Pressure (Pa at 25°C):

13853 (Hoy 1970)
9598 (20°C, quoted, Verschueren 1983)
13850 (quoted, Abraham 1984)
12230 (Riddick et al. 1986; quoted, Howard 1990)
12517 (extrapolated-Antoine eqn., Stephenson & Malanowski 1987)

Henry's Law Constant (Pa m^3/mol):

1.526 (partial pressure, Butler & Ramchandani 1935)
1.528 (quoted exptl., Hine & Mookerjee 1975)
1.676 (calculated-group contribution, Hine & Mookerjee 1975)
1.680 (calculated-bond contribution, Hine & Mookerjee 1975)
1.712 (quoted from Hine & Mookerjee 1975, Howard 1990)
1.528, 0.880 (quoted, calculated-molecular structure, Russell et al. 1992)

Octanol/Water Partition Coefficient, log K_{OW}:

0.68 (shake flask-K, Collander 1951; quoted, Leo et al. 1971; Rekker 1977; Sangster 1989)
0.88 (shake flask-TN, Sandell 1962; quoted, Leo et al. 1971; Rekker 1977; Iwase et al. 1985; Sangster 1989)
0.81 (quoted unpublished result, Leo et al. 1971)
0.81, 0.97 (quoted from Pomona College unpublished data, Rekker 1977)
0.90, 0.92 (calculated-f const., Rekker 1977)
0.81 (calculated-π const., Rekker 1977)
0.97, 0.88, 0.81, 0.68, 1.02 (quoted values, Yalkowsky & Morozowich 1980)
0.81, 0.91, 0.97 (calculated values, Yalkowsky & Morozowich 1980)
0.80 (inter-lab. shake flask average, Eadsforth & Moser 1983; quoted, Sangster 1989)
0.68, 0.88 (quoted, Verschueren 1983)
0.97 (Hansch & Leo 1985; quoted, Howard 1990)
0.93 (calculated-TSA, Iwase et al. 1985)
0.87 (quoted from Hansch & Leo 1979, Bodor et al. 1989)
0.97 (quoted from CLOGP, Bodor et al. 1989)
0.923 (calculated-CLOGP, Bodor et al. 1989)
0.77 (calculated-M.O. indices, Bodor et al. 1989)
0.81, 0.97, 1.02 (quoted from Pomona College unpublished data, Sangster 1989)
0.86 (recommended, Sangster 1989)
0.92 (calculated-CLOGP, Jäckel & Klein 1991)

Bioconcentration Factor, log BCF:

0.505 (calculated-K_{OW}, Lyman et al. 1982; quoted, Howard 1990)

Sorption Partition Coefficient, log K_{OC}:

1.903 (soil, calculated-K_{OW}, Lyman et al. 1982; quoted, Howard 1990)
1.176 (Podzol soil, von Oepen et al. 1991)
2.021 (Alfisol soil, von Oepen et al. 1991)
2.029 (sediment, von Oepen et al. 1991)
1.88 (soil, quoted exptl., Meylan et al. 1992)
1.78 (soil, calculated- χ & fragment contribution, Meylan et al. 1992)

Half-Lives in the Environment:

Air: estimated vapor phase half-life of 0.479 day, based on a result of reaction with photochemically produced hydroxyl radicals at a concentration of $5x10^5$ radicals/cm^3 (USEPA 1986; quoted, Howard 1990).
Surface water:
Groundwater:
Sediment:
Soil:
Biota:

Environmental Fate Rate Constants or Half-Lives:

Volatilization: using Henry's law constant, a half-life of 1.95 day was predicted for evaporation from a model river 1 m deep, flowing at 1 m/sec with a wind velocity of 3 m/sec (Lyman et al. 1982; quoted, Howard 1990).
Photolysis:
Oxidation: estimated vapor phase photooxidation half-life of 0.479 day in air, based on a result of reaction with photochemically produced hydroxyl radicals at a concentration of $5x10^5$ radicals/cm^3 (USEPA 1986; quoted, Howard 1990).
Hydrolysis:
Biodegradation:
Biotransformation:
Bioconcentration, Uptake (k_1) and Elimination (k_2) Rate Constants:

Common Name: Ethanolamine
Synonym: β-aminoethyl alcohol, ethylolamine, 2-hydroxyethylamine, β-hydroxyethylamine, monoethanolamine, MEA
Chemical Name: ethanolamine, 2-aminoethanol
CAS Registry No: 141-43-5
Molecular Formula: $HOCH_2CH_2NH_2$
Molecular Weight: 61.09
Melting Point (°C):
10.3 (Weast 1982-83; Howard 1990)
11.0 (Verschueren 1983)
10.5 (Dean 1985; Riddick et al. 1986)
Boiling Point (°C):
170.0 (Weast 1982-83)
172.0 (Verschueren 1983)
171.5 (Boublik et al. 1984)
170.8 (Dean 1985; Howard 1990)
170.95 (Riddick et al. 1986)
Density (g/cm^3 at 20°C):
1.0180 (Weast 1982-83)
1.020 (Verschueren 1983)
1.0147 (Riddick et al. 1986)
Molar Volume (cm^3/mol):
73.4 (calculated-LeBas method)
Molecular Volume (Å^3):
Total Surface Area, TSA (Å^2):
Dissociation Constant, pK:
9.48, 9.4994 (Perrin 1972; quoted, Howard 1990)
9.50 ($pK_{BH}{}^+$, Riddick et al. 1986)
Heat of Fusion ΔH_{fus}, kcal/mol:
4.90 (Riddick et al. 1986)
Entropy of Fusion ΔS_{fus}, cal/mol K (e.u.):
Fugacity Ratio at 25°C (assuming ΔS_{fus} = 13.5 e.u.), F: 1.0

Water Solubility (g/m^3 or mg/L at 25°C):
miscible (Dean 1985)
infinite (Riddick et al. 1986; quoted, Howard 1990)

Vapor Pressure (Pa at 25°C):
53.32 (20°C, quoted, Verschueren 1983)
44.15 (extrapolated average-Antoine eqn., Boublik et al. 1984)
42.51 (extrapolated-Antoine eqn., Dean 1985)
48.0 (Riddick et al. 1986)
34.66 (quoted from Dow Chemical's Handbook, Howard 1990)

Henry's Law Constant (Pa·m^3/mol):
0.0041 (calculated-bond method, Hine & Mookerjee 1975; quoted, Howard 1990)

Octanol/Water Partition Coefficient, log K_{OW}:

- -1.31 (shake flask-K, Collander 1951; quoted, Leo et al. 1971; Rekker 1977)
- -1.29 (calculated-f const., Rekker 1977)
- -1.35 (calculated-π const., Rekker 1977)
- -1.31 (quoted, Verschueren 1983)
- -1.31 (Hansch & Leo 1985; quoted, Howard 1990)

Bioconcentration Factor, log BCF:

- <0.0 (estimated-K_{OW}, Lyman et al. 1982; quoted, Howard 1990)

Sorption Partition Coefficient, log K_{OC}:

- 0.699 (soil, estimated-K_{OW}, Lyman et al. 1982; quoted, Howard 1990)

Half-Lives in the Environment:

Air: half-life of 11 hours, based on an estimated rate constant of 3.5×10^{-11} cm^3 molecule^{-1} sec^{-1} for the vapor phase reaction with photochemically produced hydroxyl radicals of 5×10^5 radicals/cm^3 in air (Atkinson 1987; quoted, Howard 1990).

Surface water:

Groundwater:

Sediment:

Soil:

Biota:

Environmental Fate Rate Constants or Half-Lives:

Volatilization:

Photolysis:

Hydrolysis:

Oxidation: photooxidation half-life of 11 hours in air, based on an estimated rate constant of 3.5×10^{-11} cm^3 molecule^{-1} sec^{-1} for the vapor phase reaction with photochemically produced hydroxyl radicals of 5×10^5 radicals/cm^3 in air (Atkinson 1987; quoted, Howard 1990).

Biodegradation:

Biotransformation:

Bioconcentration, Uptake (k_1) and Elimination (k_2) Rate Constants:

Common Name: Diethanolamine

Synonym: 2,2'-amino-diethanol, 3-aza-1,5-pentanediol, diethylolamine, bis(hydroxyethyl)amine, 2,2'-dihydroxydiethylamine, β,β'-dihydroxydiethylamine, 2,2'-iminobisethanol, 2,2'-iminodiethanol

Chemical Name: diethanolamine

CAS Registry No: 111-42-2

Molecular Formula: $HOCH_2CH_2NHCH_2CH_2OH$

Molecular Weight: 105.14

Melting Point (°C):

28.0 (Weast 1982-83; Verschueren 1983; Dean 1985; Riddick et al. 1986; Howard 1990)

Boiling Point (°C):

271.0 (Weast 1982-83)
269.1 (Verschueren 1983)
268.0 (Dean 1985)
268.39 (dec., Riddick et al. 1986)
268.8 (Howard 1990)

Density (g/cm^3 at 20°C):

1.0966 (Weast 1982-83)
1.0936 (25°C, Riddick et al. 1986)

Molar Volume (cm^3/mol):

126.7 (calculated-LeBas method)

Molecular Volume ($Å^3$):

Total Surface Area, TSA ($Å^2$):

Dissociation Constant, pK:

8.88, 8.97 (Perrin 1972; quoted, Howard 1990)
8.88 ($pK_{BH}{}^+$, Dean 1985; Riddick et al. 1986)

Heat of Fusion ΔH_{fus}, kcal/mol:

6.00 (Riddick et al. 1986)

Entropy of Fusion ΔS_{fus}, cal/mol·K (e.u.):

Fugacity Ratio at 25°C (assuming ΔS_{fus} = 13.5 e.u.), F: 0.9340

Water Solubility (g/m^3 or mg/L at 25°C):

954000 (Verschueren 1983)
964000 (Dean 1985)
954000 (20°C, Riddick et al. 1986)
miscible (quoted from Dow Chemical's Handbook, Howard 1990)

Vapor Pressure (Pa at 25°C):

<1.333 (20°C, quoted, Verschueren 1983)
0.040 (extrapolated-Antoine eqn., Dean 1985)
3×10^{-5} (Riddick et al. 1986)
0.0373 (quoted from Dow Chemical's Handbook, Howard 1990)

Henry's Law Constant ($Pa \cdot m^3/mol$):

5.42×10^{-9} (Hine & Mookerjee 1975; quoted, Howard 1990)

Octanol/Water Partition Coefficient, log K_{OW}:

-1.43 (shake flask-CR, Collander 1951; quoted, Leo et al. 1971)
-1.43, -1.51 (quoted, calculated-f const., Rekker & De Kort 1979)
-1.43 (quoted, Verschueren 1983)
-1.43 (Hansch & Leo 1985; quoted, Howard 1990)
-1.25 (computed, Klopman et al. 1985; quoted, Pangrekar et al. 1994)

Bioconcentration Factor, log BCF:

<0.0 (estimated-K_{OW}, Lyman et al. 1982; quoted, Howard 1990)

Sorption Partition Coefficient, log K_{OC}:

0.602 (soil, estimated-K_{OW}, Lyman et al. 1982; quoted, Howard 1990)

Half-Lives in the Environment:

Air: photooxidation half-life of 0.72-7.2 hours, based on estimated rate constant for the reaction with hydroxyl radicals in air (Atkinson 1987; quoted, Howard 1990; Howard et al. 1991); atmospheric transformation lifetime by reaction with water was estimated to be <1 day (Kelly et al. 1994).

Surface water: 14.4-168 hours, based on estimated unacclimated aqueous aerobic biodegradation half-life (Howard et al. 1991).

Groundwater: 28.8-336 hours, based on estimated unacclimated aqueous aerobic biodegradation half-life (Howard et al. 1991).

Sediment:

Soil: 14.4-168 hours, based on estimated unacclimated aqueous aerobic biodegradation half-life (Howard et al. 1991).

Biota:

Environmental Fate Rate Constants or Half-Lives:

Volatilization:

Photolysis:

Hydrolysis:

Oxidation: photooxidation half-life of 0.72-7.2 hours in air, based on estimated rate constant for the reaction with hydroxyl radicals in air (Atkinson 1987; quoted, Howard 1990; Howard et al. 1991).

Biodegradation: aqueous aerobic half-life of 14.4-168 hours, based on measured half-life in surface water grab sample experiment (Boethling & Alexander 1979; quoted, Howard et al. 1991) and aqueous aerobic screening test data (Gerike & Fischer 1979; Bridie et al. 1979; quoted, Howard et al. 1991); aqueous anaerobic half-life of 57.6-672 hours, based on estimated unacclimated aqueous aerobic biodegradation half-life (Howard et al. 1991).

Biotransformation:

Bioconcentration, Uptake (k_1) and Elimination (k_2) Rate Constants:

Common Name: Triethanolamine
Synonym: 2,2',2''-nitrilotriethanol, 2,2',2''-nitrilotrisethanol, triethylolamine, trihydroxytriethylamine, tris(hydroxyethyl)amine, TEA
Chemical Name: triethanolamine
CAS Registry No: 102-71-6
Molecular Formula: $(HOCH_2CH_2)_3N$
Molecular Weight: 149.19
Melting Point (°C):
21-22 (Weast 1982-83)
21.1 (Verschueren 1983)
21.6 (Dean 1985; Riddick et al. 1986)
20-21 (Howard 1990)
Boiling Point (°C):
360.0 (Verschueren 1983)
335.4 (Dean 1985; Riddick et al. 1986; Howard 1990)
Density (g/cm^3 at 20°C):
1.1242 (Weast 1982-83; Dean 1985)
1.120 (Verschueren 1983)
1.1196 (25°C, Riddick et al. 1986)
Molar Volume (cm^3/mol):
182.1 (calculated-LeBas method)
Molecular Volume ($Å^3$):
Total Surface Area, TSA ($Å^2$):
Dissociation Constant, pK:
7.92 (Perrin 1972; quoted, Howard 1990)
7.76 ($pK_{BH}{}^+$, Dean 1985; Riddick et al. 1986)
Heat of Fusion ΔH_{fus}, kcal/mol:
6.50 (Riddick et al. 1986)
Entropy of Fusion ΔS_{fus}, cal/mol·K (e.u.):
Fugacity Ratio at 25°C (assuming ΔS_{fus} = 13.5 e.u.), F: 1.0

Water Solubility (g/m^3 or mg/L at 25°C):
miscible (Dean 1985; Howard 1990)
infinite (Riddick et al. 1986)

Vapor Pressure (Pa at 25°C):
<1.33 (20°C, quoted, Verschueren 1983)
0.0131 (extrapolated-Antoine eqn., Boublik et al. 1984)
0.010 (extrapolated-Antoine eqn., Dean 1985)
<1.30 (20°C, Riddick et al. 1986)
4.79×10^{-4} (quoted from Dow Chemical's Handbook, Howard 1990)

Henry's Law Constant (Pa·m^3/mol):
3.42×10^{-14} (Hine & Mookerjee 1975; quoted, Howard 1990)

Octanol/Water Partition Coefficient, log K_{OW}:
-1.32, -1.75 (calculated, Verschueren 1983)
-1.59 (GEMS 1989; quoted, Howard 1990)

Bioconcentration Factor, log BCF:
<0.0 (estimated-K_{OW}, Lyman et al. 1982; quoted, Howard 1990)

Sorption Partition Coefficient, log K_{OC}:
0.477 (soil, estimated-K_{OW}, Lyman et al. 1982; quoted, Howard 1990)

Half-Lives in the Environment:
Air: half-life of 4.0 hours, based on an estimated rate constant of 10.4×10^{-11} cm^3 molecule^{-1} sec^{-1} for the vapor phase reaction with photochemically produced hydroxyl radicals of 5×10^5 radicals/cm^3 in air at 25°C (Atkinson 1987; quoted, Howard 1990).
Surface water:
Groundwater:
Sediment:
Soil:
Biota:

Environmental Fate Rate Constants or Half-Lives:
Volatilization:
Photolysis:
Hydrolysis:
Oxidation: photooxidation half-life of 4.0 hours in air, based on an estimated rate constant of 10.4×10^{-11} cm^3 molecule^{-1} sec^{-1} for the vapor phase reaction with photochemically produced hydroxyl radicals of 5×10^5 radicals/cm^3 in air at 25°C (Atkinson 1987; quoted, Howard 1990).
Biodegradation:
Biotransformation:
Bioconcentration, Uptake (k_1) and Elimination (k_2) Rate Constants:

Common Name: Aniline
Synonym: aminobenzene, benzeneamine, benzenamine
Chemical Name: aniline
CAS Registry No: 62-53-3
Molecular Formula: $C_6H_5NH_2$
Molecular Weight: 93.13
Melting Point (°C):
-6.20 (Stull 1947)
-6.0 (Verschueren 1983)
-6.30 (Fu & Luthy 1985; Howard 1989)
-5.98 (Riddick et al. 1986)
-6.02 (Chao et al. 1990)
Boiling Point (°C):
184.4 (Stull 1947; Riddick et al. 1986)
184.0 (Boublik et al. 1973, 1984; Verschueren 1983; Banerjee et al. 1990)
184-186 (Howard 1989)
184.17 (Chao et al. 1990)
Density (g/cm^3 at 20°C):
1.020 (Verschueren 1983)
1.0217 (Fu & Luthy 1985; Riddick et al. 1986)
1.0175 (Riddick et al. 1986; Chao et al. 1990)
Molar Volume (cm^3/mol):
91.1 (calculated-density, Chiou 1985)
91.49 (calculated-density, Fu & Luthy 1985)
88.0 (calculated-density, Jaworska & Schultz 1993)
56.2 (instrinsic V_I, Leggett & Parker 1994)
110.2 (calculated-LeBas method)
Molecular Volume ($Å^3$):
27.3 (Jäckel & Klein 1991)
Total Surface Area, TSA ($Å^2$):
226.3 (Iwase et al. 1985)
Dissociation Constant, pK_a:
4.596 (Perrin 1972; Howard 1989)
4.60 (McLeese et al. 1979; Riddick et al. 1986; Sangster 1989)
4.63 (Weast 1982-83)
4.58, 3.96 (quoted, HPLC, Miyake et al. 1987)
Enthalpy of Fusion, ΔH_{fus}, kcal/mol:
Entropy of Fusion, ΔS_{fus}, cal/mol·K (e.u.):
Fugacity Ratio at 25°C (assuming ΔS_{fus} = 13.5 e.u.), F: 1.0

Water Solubility (g/m^3 or mg/L at 25°C):
36650 (Hill & Macy 1924; quoted, Chiou 1981; Chiou & Schmedding 1981; Chiou et al. 1982; Chiou 1985)
36070 (Seidell 1941; quoted, Hansch et al. 1968; Howard 1989)
38665 (shake flask-centrifuge, Booth & Everson 1948)

36217 (shake flask-interferometry, Donahue & Bartell 1952)
34100 (Stephen & Stephen 1963; quoted, Fu & Luthy 1985,86)
36228 (shake flask-AS, Hansch et al. 1968; quoted, Müller & Klein 1992)
53339 (calculated-K_{OW}, Hansch et al. 1968)
36600 (Kenaga 1980)
33810 (quoted, Yalkowsky & Morozowich 1980)
21335 (calculated-K_{OW}, Yalkowsky & Morozowich 1980)
65930 (quoted average, Valvani et al. 1981; quoted, Banerjee 1985)
34600 (predicted-K_{OW}, Valvani et al. 1981)
37000 (30°C, quoted from SRI Research Program, Haque et al. 1980)
34000 (Verschueren 1983; quoted, Pinal et al. 1991)
56900 (calculated- γ by UNIFAC, Fu & Luthy 1985,86)
33800 (quoted, Riddick et al. 1986)
36232, 36232 (quoted, calculated-fragment const., Wakita et al. 1986)
36310 (quoted, Isnard & Lambert 1988)
34200 (selected, Yaws et al. 1990)

Vapor Pressure (Pa at 25°C):
85.71 (extrapolated-regression of tabulated data, Stull 1947)
89.52 (calculated-Antoine eqn., Dreisbach 1955)
88.64 (extrapolated-Antoine eqn., Boublik et al. 1973,84)
82.71 (calculated-Cox eqn., Chao et al. 1983)
39.99 (20°C, quoted, Verschueren 1983)
48.24 (extrapolated-Antoine eqn., Boublik et al. 1984)
65.18 (Daubert & Danner 1985; quoted, Howard 1989)
89.30 (extrapolated-Antoine eqn., Dean 1985)
65.32 (Howard et al. 1986; quoted, Banerjee et al. 1990)
89.50 (quoted, Riddick et al. 1986)
89.60 (calculated-Antoine eqn., Stephenson & Malanowski 1987)
65.32, 379.9 (quoted, calculated-solvatochromic p. , Banerjee et al. 1990)
86.70 (calculated-Cox eqn., Chao et al. 1990)

Henry's Law Constant (Pa·m³/mol):
13778 (Hakuta et al. 1977; quoted, Howard 1989)
12.16 (measured, Yoshida et al. 1983; quoted, Howard 1989)

Octanol/Water Partition Coefficient, log K_{OW}:
0.90 (shake flask-UV, Fujita et al. 1964; quoted, Leo et al. 1969,71; Mirrlees et al. 1976; Iwase et al. 1985; Sangster 1989; Kalsch et al. 1991)
0.90 (shake flask, Iwasa et al. 1965; quoted, Eadsforth & Moser 1983; Brooke et al. 1990)
0.90 (shake flask-UV, Hansch et al. 1968; quoted, Sangster 1989; Müller & Klein 1992)
0.90 (Leo et al. 1971; quoted, Carlson et al. 1975; Chiou 1981; Chiou & Schmedding 1981; Chiou et al. 1982; Chiou 1985; Brooke et al. 1990)
0.98 (quoted, Leo et al. 1971)

0.89 (shake flask-AS at pH 5.6, Umeyama et al. 1971; quoted, Sangster 1989)
0.90 (Lc-k', Carlson et al. 1975)
0.95 (calculated-π const., Carlson et al. 1975; quoted, Sangster 1989)
0.85 (shake flask, Lu & Metcalf 1975; quoted, Brooke et al. 1990)
0.94 (calculated-π const., Norrington et al. 1975; quoted, De Bruijn et al. 1989; Arnold et al. 1990; Verhaar et al. 1992)
0.90 (HPLC, Mirrlees et al. 1976; quoted, Brooke et al. 1990)
0.93 (quoted exptl., Rekker 1977; quoted, Könemann et al. 1979)
0.97 (calculated-f const., Rekker 1977; quoted, Brooke et al. 1990)
0.93 $\pm$ 0.05 (shake flask at pH 7, Unger et al. 1978)
0.90, 0.98, 0.85 (shake flask, Hansch & Leo 1979; quoted, Schultz & Moulton 1985; De Bruijn et al. 1989)
0.93, 0.91, 1.03 (quoted exptl., HPLC-k', calculated-f const., Könemann et al. 1979)
0.99 (McLeese et al. 1979)
1.00 (calculated-f const., Rekker & De Kort 1979; quoted, De Bruijn et al. 1989)
0.90 (shake flask-AS, Briggs 1981; quoted, Sangster 1989)
0.93 (quoted exptl., Valvani et al. 1981)
1.03 (RPHPLC-k', D'Amboise & Hanai 1982)
0.90 (Chiou et al. 1982; quoted, Brooke et al. 1986)
0.90 (HPLC-k', Hammers et al. 1982; quoted, De Bruijn et al. 1989; Brooke et al. 1990)
0.90 (inter-lab. shake flask average, Eadsforth & Moser 1983)
1.09 (inter-lab. HPLC average, Eadsforth & Moser 1983; quoted, Sangster 1989; Brooke et al. 1990)
0.90 (Hansch & Leo 1983; quoted, El Tayar et al. 1991)
0.90, 1.03 (quoted, calculated-f const., Harnish et al. 1983)
1.34, 1.27 (HPLC-extrapolated values, Harnish et al. 1983; quoted, Sangeter 1989)
1.08 (HPLC, Harnish et al. 1983; quoted, Sangeter 1989)
1.08 (shake flask average, OECD/EEC lab. comparison tests, Harnish et al. 1983)
0.93 (quoted from Hansch & Leo 1979, Kaiser 1983; Yoshida et al. 1986)
0.90, 0.98 (Verschueren 1983; quoted, Pinal et al. 1991)
0.79, 0.96 (exptl.-ALPM, Garst & Wilson 1984; quoted, Brooke et al. 1990)
0.89 (selected best lit. value, Garst & Wilson 1984)
0.94, 1.09 (quoted, Geyer et al. 1984)
0.90 (quoted from Hansch & Leo 1979, Haky & Young 1984; Burkhard & Kuehl 1986; Miyake et al. 1987)
0.89 (shake flask-AS at pH 7.4, El Tayar et al. 1984; quoted, Sangster 1989)
0.90, 0.99 (quoted, calculated- γ from UNIFAC, Campbell & Luthy 1985; quoted, Sangster 1989)
0.90 (Hansch & Leo 1985; quoted, Howard 1989; Arnold et al. 1990; Kalsch et al. 1991; El Tayar et al. 1991; Jaworska & Schultz 1993)
0.84 (calculated-TSA, Iwase et al. 1985)
0.81, 1.08 (HPLC-k', Eadsforth 1986; quoted, Brooke et al. 1990)
0.94, 0.91 (quoted: shake flask, calculated, Eadsforth 1986)
0.91 (estimated-HPLC-RP, Eadsforth 1986; quoted, Brooke et al. 1990)
0.98 (shake flask-AS at pH 7.5, Martin-Villodre et al. 1986; quoted, Sangster 1989)

0.93 (HPLC average, Ge et al. 1987; quoted, Sangster 1989)
0.90, 0.78 (quoted, calculated-HPLC-k' Miyake et al. 1987)
0.90, 1.18 (quoted, calculated- γ from UNIFAC, Banerjee & Howard 1988)
0.98 (quoted, Isnard & Lambert 1988,89)
0.90 (quoted from Hansch & Leo 1979, Bodor et al 1989)
0.90 (quoted from CLOGP, Bodor et al 1989)
0.915 (calculated-CLOGP, Bodor et al 1989)
1.12 (calculated-M.O. indices, Bodor et al 1989)
0.940 $\pm$ 0.006 (slow-stirring-GC, De Bruijn et al. 1989)
1.0, 0.94 (calculated-f const., π-const., De Bruijn et al. 1989)
0.90 (recommended, Sangster 1989)
1.04 (calculated-f const. as per Rekker 1977, Arnold et al. 1990)
0.94 (calculated-π const. from Norrington et al. 1975, Arnold et al. 1990)
0.942 $\pm$ 0.010 (stir-flask method by BRE, Brooke et al. 1990)
0.940 $\pm$ 0.006 (stir-flask method by RITOX, Brooke et al. 1990)
0.90 (quoted, Zok et al. 1991)
0.92 (calculated-CLOGP, Jäckel & Klein 1991; Müller & Klein 1992)
0.90, 0.6539 (quoted, calculated-original UNIFAC, Chen et al. 1993)
0.90 (quoted from EPA CLOGP data base, Hulzebos et al. 1993; Jaworska & Schultz 1993)
0.915 (calculated-QSAR, Kollig 1993)

Bioconcentration Factor, log BCF:
0.78 (fish, Lu & Metcalf 1975; quoted, Howard 1989)
0.30 (calculated-S, Kenaga 1980)
<1.0 (fish, Freitag et al. 1982; quoted, Howard 1989)
<1.0, 0.602, 3.01 (golden orfe, algae, activated sludge, Freitag et al. 1982)
0.602 (alga *chlorella fusca*, wet wt. basis, Geyer et al. 1984)
0.845 (alga *chlorella fusca*, calculated-K_{OW}, Geyer et al. 1984)
<1.0, <1.0, 2.70 (golden ide, algae, activated sludge, Freitag et al. 1985)
2.77 (*daphnia magna*, based on elimination phase, Dauble et al. 1986)
1.87 (*daphnia magna*, based on ^{14}C and exposure water, Dauble et al. 1986)
0.70 (fish, correlated-K_{OW}, Isnard & Lambert 1988)
0.78 (quoted, Isnard & Lambert 1988, 1989)
0.41 (zebrafish, Kalsch et al. 1991)
0.41 (zebrafish, Zok et al. 1991)
0.41; 1.04, -0.87, 0.03 (quoted exptl.; calculated values-K_{OW}, Bintein et al. 1993)

Sorption Partition Coefficient, log K_{OC}:
3.11 (H-montmorillonite at pH 8.35, Bailey et al. 1968; quoted, Howard 1989)
2.11 (H-montmorillonite at pH 6.80, Bailey et al. 1968; quoted, Howard 1989)
1.86 (soil average, Moreale & Van Bladel 1976; quoted, Howard 1989)
1.41 (average of seven agricultural soils, Briggs 1981; quoted, Howard 1989)
3.59 (colloidal organic carbon/ground water, Means et al.1982; quoted, Howard 1989)
2.11, 2.61 (soil, more acidic soil, Pillai et al. 1982; quoted, Howard 1989)

2.49 (nonsterile Hagerstown soil, Pillai et al. 1982)
2.11 (sterile Hagerstown soil, Pillai et al. 1982)
2.96 (nonsterile Palouse soil, Pillai et al. 1982)
2.61 (sterile Palouse soil, Pillai et al. 1982)
1.17 (soil, quoted, Sabljic 1987)
2.12, 2.05, 2.06 (calculated values: Podzol, Alfisol, sediment, von Oepen et al. 1991)
0.596 (calculated-K_{OW}, 1993)

Half-Lives in the Environment:

Air: atmospheric lifetimes of 2.3 hours in clean troposphere and 1.2 hour in moderately polluted atmosphere, based on gas-phase reaction with hydroxyl radicals at room temperature (Atkinson et al. 1987); atmospheric lifetimes of 15.0 days in clean troposphere and 5.0 days in moderately polluted atmosphere, based on gas-phase reaction with O_3 at room temperature (Atkinson et al. 1987); estimated half-life of 3.3 hours, based on reaction with photochemically produced hydroxyl radicals (Howard 1989); atmospheric transformation lifetime was estimated to be < 1 day (Kelly et al. 1994).

Surface water: half-life of 2.3 days in Rhine river (Zoeteman et al. 1980; quoted, Howard 1989); estimated half-life of 0.3-3.0 days in river waters (Zoeteman et al. 1980); half-lives of 4 to 8 hours in May sunlight with both commercial humic acids and aquatic humus as photosensitizers near-surface water and half-life of about 1 week in distilled water (Zepp et al. 1981); 6 days in eutropic pond and 21 days in an oligotrophic lake (Subba-Rao et al. 1982; quoted, Howard 1989); biodegradation half-life of 4.5 days in unpolluted and <0.5 day in polluted pond water as model environments (Lyons et al. 1984).

Groundwater: estimated half-life of 30-300 days (Zoeteman et al. 1980).

Sediment:

Soil:

Biota:

Environmental Fate Rate Constants or Half-Lives:

Volatilization: estimated half-life of 12 days from a measured Henry's law constant of 1.2×10^{-4} atm·m^3·mol^{-1} (Yoshida et al. 1983; quoted, Howard 1989) for a model river of 1 meter deep with a 1 m/sec current and a 3 m/sec wind (Lyman et al. 1982; quoted, Howard 1989).

Photolysis: first-order rate constants for photosensitized reactions in water with various humic substances as sensitizers: 0.17 h^{-1} with aquatic humus from Aucilla River, 0.12 h^{-1} with Aldrich humic acid, 0.091 h^{-1} with Fluka humic acid and 0.11 h^{-1} with Contech fulvic acid in sunlight, corresponding to half-lives of 4 to 8 hours (Zepp et al. 1981).

Oxidation: rate constant of 1×10^4 $M^{-1} \cdot sec^{-1}$ for oxidation by RO_2 radicals at 30°C in aquatic systems with half-life of 0.8 day (Howard 1972; Hendry et al. 1974; quoted, Mill 1982); rate constant of $<2 \times 10^2$ $M^{-1} \cdot sec^{-1}$ for oxidation by singlet oxygen at 25°C in aquatic systems with half-life of >100 years (Foote 1976; Mill 1979; quoted, Mill 1982); observed second-order rate constant of about

$60x10^{-12}$ $cm^3 \cdot molecule^{-1} \cdot sec^{-1}$ compared to the calculated value of 116 x 10^{-12} $cm^3 \cdot molecule^{-1} \cdot sec^{-1}$ for the gas-phase reaction with hydroxyl radicals at room temperature in air (Atkinson 1985); observed rate constant of $120x10^{-12}$ $cm^3 \cdot molecule^{-1} \cdot sec^{-1}$ (Rinke & Zetzsch 1984) compared to calculated value of $154x10^{-12}$ $cm^3 \cdot molecule^{-1} \cdot sec^{-1}$ for the gas-phase reaction with OH radicals at room temperature in air (Atkinson et al. 1985); average observed rate constant of $117x10^{-12}$ $cm^3 \cdot molecule^{-1} \cdot sec^{-1}$ (quoted from Rinke & Zetzsch 1984, Witte et al. 1986 & Atkinson et al. 1987, Atkinson 1987) versus calculated rate constant of $136x10^{-12}$ $cm^3 \cdot molecule^{-1} \cdot sec^{-1}$ for the vapor phase reaction with hydroxyl radicals in air (Atkinson 1987); rate constant of $1.10x10^{-18}$ $cm^3 \cdot molecule^{-1} \cdot sec^{-1}$ for the gas-phase reaction with O_3 at room temperature in air (Atkinson et al. 1987); rate constant of $138.5x10^{-12}$ $cm^3 \cdot molecule^{-1} \cdot sec^{-1}$ by molecular orbital calculation as compared to known experimental value of $111.0x10^{-12}$ $cm^3 \cdot molecule^{-1} \cdot sec^{-1}$ for the gas-phase reaction with OH radicals at room temperature in air (Klamt 1993).

Hydrolysis:

Biodegradation: completely degraded by a soil inoculum in 4 days (Alexander & Lustigman 1966; quoted, Verschueren 1983; Howard 1989); completly degraded in 20 days by bacteria in river mud (Calamari et al. 1980; quoted, Howard 1989); average rate of biodegradation of 19.0 mg COD g^{-1} h^{-1} for 94.5% removal (Scow 1982); biodegradation half-life of 4.5 days in unpolluted and <0.5 day in polluted pond water as model environments (Lyons et al. 1984); 0.46mM aniline solution degraded by strain Ani1 within 14 days in water (Schnell et al. 1989); average exptl. rate constant of 0.044 $hour^{-1}$ compared to the group-contribution method predicted rate constants of 0.050 $hour^{-1}$ (nonlinear) and 0.018 h^{-1} (Tabak & Govind 1993).

Biotransformation: mean bacteria transformation rate constant for all three sites of $(1.1 \pm 0.8)x10^{-11}$ $L \cdot organism^{-1} \cdot h^{-1}$ (Paris & Wolfe 1987; quoted, Steen 1991).

Bioconcentration, Uptake (k_1) and Elimination (k_2) Rate Constants:

k_1: 0.052 ± 0.0067 h^{-1} (Kalsch et al. 1991)
k_2: 7.20 ± 1.30 h^{-1} (Kalsch et al. 1991)
k_1: 11.1 ± 3.20 h^{-1} (zebrafish, Zok et al. 1991)

Common Name: 2-Chloroaniline
Synonym: 1-amino-2-chlorobenzene, *o*-aminochlorobenzene, *o*-chloroaniline, 2-chlorophenylamine
Chemical Name: 1-amino-2-chlorobenzene, *o*-chloroaniline, 2-chloroaniline
CAS Registry No: 95-51-2
Molecular Formula: $C_6H_4NH_2Cl$
Molecular Weight: 127.57
Melting Point (°C):
-1.94 (Dreisbach 1955)
-14.0 (Verschueren 1983; Howard 1989)
Boiling Point (°C):
208.8 (Kahlbaum 1898; Stull 1947; Dreisbach 1955; Verschueren 1983; Riddick et al. 1986; Howard 1989)
209.0 (Banerjee et al. 1990)
Density (g/cm³ at 20°C):
1.21253 (Kahlbaum 1898)
1.21266 (Dreisbach 1955)
1.2130 (Verschueren 1983)
1.21251 (Riddick et al. 1986)
Molar Volume (cm³/mol):
131.1 (calculated-LeBas method)
99.90 (calculated-density, Jaworska & Schultz 1993)
Molecular Volume (Å³):
Total Surface Area, TSA (Å²):
Dissociation Constant, pK_a:
2.661 (Perrin 1972; quoted, Howard 1989)
2.65 (Weast 1982-83)
2.64 (protonated cation +1, Dean 1985)
2.64 (Riddick et al. 1986)
Enthallpy of Fusion, ΔH_{fus}, kcal/mol:
Entropy of Fusion, ΔS_{fus}, cal/mol·K (e.u.):
2.84 (Riddick et al. 1986)
Fugacity Ratio at 25°C (assuming ΔS = 13.5 e.u.), F: 1.0

Water Solubility (g/m³ or mg/L at 25°C):
8760 (Dreisbach 1955; quoted, Riddick et al. 1986)
3765 (20°C, shake flask-GC, Chiou 1981; Chiou & Schmedding 1981; Chiou et al. 1982; quoted, Howard 1989; Müller & Klein 1992)
3763 (calculated-K_{OW}, Müller & Klein 1992)

Vapor Pressure (Pa at 25°C):
40.31 (extrapolated-regression of tabulated data, Kahlbaum 1898)
37.77 (extrapolated-regression of tabulated data, Stull 1947)
33.77 (Dreisbach 1955; quoted, Riddick et al. 1986)

33.88 (calculated-Antoine eqn., Dean 1985)
35.30 (torsion-weighing effusion, Piacente et al. 1985)
22.66 (Piacente et al. 1985; quoted, Howard 1989)
34.66 (Howard et al. 1986; quoted, Banerjee et al. 1990)
18.97 (extrapolated from Antoine eqn., Stephenson & Malanowski 1987)
34.66, 86.65 (quoted, calculated-solvatochromic p., Banerjee et al. 1990)

Henry's Law Constant (Pa·m^3/mol):
0.760 (calculated-P/C, Howard 1989)
0.425 (calculated-P/C, Meylan & Howard 1991)
0.143 (estimated-bond contribution, Meylan & Howard 1991)

Octanol/Water Partition Coefficient, log K_{OW}:
1.81 (shake flask, Fujita et al. 1964; quoted, Kalsch et al. 1991)
1.90 (Leo et al. 1971; quoted, Carlson et al. 1975; Chiou 1981; Chiou & Schmedding 1981; Chiou et al. 1982)
1.92 (quoted exptl., Leo et al. 1971; Rekker 1977; quoted, Könemann et al. 1979)
1.92 (HPLC-k', Carlson et al. 1975; Sangster 1993)
1.91 (calculated-π const., Carlson et al. 1975)
1.92 (quoted, McCall 1975)
1.63 (calculated-π const., Norrington et al. 1975; quoted, De Bruijn et al. 1989)
1.61 (calculated-π const., Rekker 1977)
1.91, 1.73 (quoted exptl., calculated-f const., Rekker 1977; quoted, Könemann et al. 1979)
1.90, 1.92 (shake flask, Hansch & Leo 1979; quoted, De Bruijn et al. 1989)
1.74, 1.76 (HPLC-k', calculated-f const., Könemann et al. 1979; quoted, Sangster 1993)
1.74 (calculated-f const., Rekker & De Kort 1979; quoted, De Bruijn et al. 1989)
1.91 $\pm$ 0.01 (HPLC-k', Hammers et al. 1982)
1.99 (HPLC-k'correlated, Hammers et al. 1982; quoted, De Bruijn et al. 1989)
1.90 (quoted from Hansch & Leo 1979, Kaiser 1983)
1.90 (quoted, Verschueren 1983)
1.90 (Hansch & Leo 1985; quoted, Howard 1989; Arnold et al. 1990; Kalsch et al. 1991, Jaworska & Schultz 1993; Bintein et al. 1993)
1.91 (quoted, Yoshida et al. 1986)
1.926 $\pm$ 0.021 (slow-stirring-GC, De Bruijn et al. 1989; quoted, Sangster 1993)
1.76 (calculated-f const. as per Rekker 1977, Arnold et al. 1990)
1.59 (calculated-π const. from Norrington et al. 1975, Arnold et al. 1990)
1.93 (calculated-CLOGP, Müller & Klein 1992)
1.93 (calculated-π const., Verhaar et al. 1992)
1.90, 1.4862 (quoted, calculated-original UNIFAC, Chen et al. 1993)
1.88 (recommended, Sangster 1993)

Bioconcentration Factor, log BCF:

<2.0 (Kawasaki 1980; quoted, Howard 1989)
1.30 (estimated, Canton et al. 1985; quoted, Howard 1989)
1.18 (zebrafish, Zok et al. 1991)
0.301-0.57 (carp, Tsuda et al. 1993)
1.18; 1.56, 0.73, 0.94 (quoted; calculated values-K_{OW}, Bintein et al. 1993)

Sorption Partition Coefficient, log K_{OC}:

Half-Lives in the Environment:

Air: estimated atmospheric half-life of 2 days, based on the reaction with sunlight-produced hydroxyl radicals (GEMS 1987; quoted, Howard 1989).
Surface water: volatilization half-life of 64 days from stagnant bodies of water (USEPA 1987; quoted, Howard 1989).
Groundwater:
Sediment:
Soil:
Biota: 3.6 hours in carp with excretion rate constant of 0.19 h^{-1} (Tsuda et al. 1993).

Environmental Fate Rate Constants or Half-Lives:

Volatilization: using Henry's law constant, half-life of 5.6 days was estimated for a model river of 1 meter deep (Lyman et al. 1982; quoted, Howard 1989); estimated half-life of 64 days from a representative environmental pond (stagnant) (USEPA 1987; quoted, Howard 1989).
Photolysis:
Oxidation: rate constant of 5.1×10^{-12} cm^3/molecule·sec for the reaction with hydroxyl raicals in a typical ambient atmosphere at 25°C with a half-life of about 2 days (GEMS 1987; quoted, Howard 1989).
Hydrolysis:
Biodegradation: average biodegradation rate of 25 mg COD g^{-1} h^{-1} for 95.6% removal (Scow 1982).
Biotransformation:
Bioconcentration, Uptake (k_1) and Elimination (k_2) Rate Constants:

k_1: 7.10 h^{-1} (zebrafish, Zok et al. 1991)
k_2: 0.19 h^{-1} (carp, Tsuda et al. 1993)

Common Name: 3-Chloroaniline
Synonym: 1-amino-3-chlorobenzene, *m*-chloroaniline, 3-chlorophenylamine
Chemical Name: 1-amino-3-chlorobenzene, *m*-chloroaniline, 3-chloroaniline
CAS Registry No: 108-42-9
Molecular Formula: $C_6H_4NH_2Cl$
Molecular Weight: 127.57
Melting Point (°C):
-10.4 (Stull 1947; Dean 1985; Verschueren 1983; Budavari 1989)
-10.29 (Dreisbach 1955)
Boiling Point (°C):
228.5 (Kahlbaum 1898; Stull 1947)
229.92 (Dreisbach 1955)
229.8 (Verschueren 1983)
230.5 (Dean 1985; Budavari 1989)
Density (g/cm^3 at 20°C):
1.21564 (Kahlbaum 1898)
1.21606 (Dreisbach 1955)
1.2160 (Verschueren 1983)
1.2150 (22°C, Dean 1985; Budavari 1989)
Molar Volume (cm^3/mol):
131.1 (calculated-LeBas method)
105.8 (calculated-density)
Molecular Volume (Å^3):
31.6 (Jäckel & Klein 1991)
Total Surface Area, TSA (Å^2):
Dissociation Constant, pK_a:
3.521 (Perrin 1972)
3.5 (McLeese et al. 1979)
3.46 (Weast 1982-83)
3.520 (protonated cation +1, Dean 1985)
Enthalpy of Fusion, ΔH_{fus}, kcal/mol:
2.426 (Dreisbach 1955)
Entropy of Fusion, ΔS_{fus}, cal/mol·K (e.u.):
Fugacity Ratio at 25°C (assuming ΔS_{fus} = 13.5 e.u.), F: 1.0

Water Solubility (g/m^3 or mg/L at 25°C):
5442 (20°C, shake flask-GC, Chiou 1981; Chiou & Schmedding 1981; Chiou et al. 1982; quoted, Müller & Klein 1992)
5447 (calculated-K_{OW}, Müller & Klein 1992)

Vapor Pressure (Pa at 25°C):
13.14 (extrapolated-regression of tabulated data, Kahlbaum 1898)
11.94 (extrapolated-regression of tabulated data, Stull 1947)

11.17 (Dreisbach 1955)
11.06 (calculated-Antoine eqn., Dean 1985)
15.6 (torsion-weighing effusion, Piacente et al. 1985)
9.53 (extrapolated from Antoine eqn., Stephenson & Malanowski 1987)

Henry's Law Constant ($Pa \cdot m^3/mol$):
0.223 (calculated-P/C, this work)

Octanol/Water Partition Coefficient, log K_{OW}:
1.88 (shake flask-AS, Fujita et al. 1964; quoted, Leo et al. 1971; Rekker 1977)
1.88 (Ichikawa et al. 1969; quoted, Kalsch et al. 1991)
1.88 (Leo et al. 1971; quoted, McCall 1975; Chiou 1981; Chiou & Schmedding 1981; Chiou et al. 1982)
1.90 (quoted exptl., Leo et al. 1971; Rekker 1977; quoted, Könemann et al. 1979)
1.98 (calculated-π const., Norrington et al. 1975; quoted, De Bruijn et al. 1989)
1.61 (calculated-π const., Rekker 1977)
1.73, 1.75 (calculated-f const., Rekker 1977)
1.90, 1.88 (shake flask, Hansch & Leo 1979; quoted, De Bruijn et al. 1989)
1.57, 1.76 (HPLC-k', calculated-f const., Könemann et al. 1979)
1.74 (calculated-f const., Rekker & De Kort 1979; quoted, De Bruijn et al. 1989)
1.89 $\pm$ 0.01 (HPLC-k', Hammers et al. 1982)
2.00 (HPLC-k'correlated, Hammers et al. 1982; quoted, De Bruijn et al. 1989)
1.88 (quoted from Hansch & Leo 1979, Kaiser 1983)
1.88 (quoted, Verschueren 1983)
1.910 $\pm$ 0.013 (slow-stirring-GC, De Bruijn et al. 1989; quoted, Verhaar et al. 1992)
1.88 (calculated-CLOGP 3.34, Arnold et al. 1990; quoted, Kalsch et al. 1991)
1.76 (calculated-f const. as per Rekker 1977, Arnold et al. 1990)
1.94 (calculated-π const. from Norrington et al. 1975, Arnold et al. 1990)
1.93 (calculated-CLOGP, Jäckel & Klein 1991; Müller & Klein 1992)
1.88, 1.4862 (quoted, calculated-original UNIFAC, Chen et al. 1993)

Bioconcentration Factor, log BCF:
1.06 (zebrafish, Zok et al. 1991)
-0.097 to 0.342 (average for carp, Tsuda et al. 1993)
1.06; 1.55, 0.70, 0.92 (quoted; calculated values, Bintein et al. 1993)

Sorption Partition Coefficient, log K_{OC}:

Half-Lives in the Environment:
Air:
Surface water:

Groundwater:
Sediment:
Soil:
Biota: 3.3 hours in carp with excretion rate of 0.21 h^{-1} (Tsuda et al. 1993).

Environmental Fate Rate Constants or Half-Lives:
Volatilization:
Photolysis: direct aqueous photolysis rate constant of 0.393 ± 0.006 $minute^{-1}$ with a calculated half-life of 1.76 minute (Stegeman et al. 1993).
Oxidation:
Hydrolysis: direct photohydrolysis rate constant of 0.393 ± 0.006 $minute^{-1}$ with a calculated half-life of 1.76 minute (Stegeman et al. 1993).
Biodegradation: average biodegradation rate of 6.2 mg COD g^{-1} h^{-1} for 97.2% removal (Scow 1982).
Biotransformation: mean bacteria transformation rate constant for all three sites of (2.2 ± 1.7)x10^{-12} L·$organism^{-1}$·h^{-1} (Paris & Wolfe 1987; quoted, Steen 1991).
Bioconcentration, Uptake (k_1) and Elimination (k_2) Rate Constants:
k_1: 19.1 h^{-1} (zebrafish, Zok et al. 1991)
k_2: 0.21 h^{-1} (carp, Tsuda et al. 1993)

Common Name: 4-Chloroaniline
Synonym: 1-amino-4-chlorobenzene, *p*-chloroaniline, 4-chlorophenylamine
Chemical Name: 1-amino-4-chlorobenzene, *p*-chloroaniline, 4-chloroaniline
CAS Registry No: 106-47-8
Molecular Formula: $NH_2C_6H_4Cl$
Molecular Weight: 127.57
Melting Point (°C):
70.5 (Stull 1947)
69.85 (Tsonopoulos & Prausnitz 1971)
69.9 (Schmidt-Bleek et al. 1982)
70-72 (Verschueren 1983)
Boiling Point (°C):
230.5 (Stull 1947)
231 (Schmidt-Bleek et al. 1982; Verschueren 1983)
232 (Verschueren 1983; Howard 1989; Banerjee et al. 1990)
Density (g/cm^3 at 20°C):
1.427 (19°C, Verschueren 1983)
Molar Volume (cm^3/mol):
131.1 (calculated-LeBas method)
Molecular Volume (Å^3):
31.6 (Jäckel & Klein 1991)
Total Surface Area, TSA (Å^2):
Dissociation Constant, pK_a:
3.982 (Perrin 1972; Freitag et al. 1984; quoted, Howard 1989)
4.20 (McLeese et al. 1979)
4.15 (Weast 1982-83)
3.99 (protonated +1, Dean 1985)
Heat of Fusion, ΔH_{fus}, kcal/mol:
3.75 (Tsonopoulos & Prausnitz 1971)
Entropy of Fusion, ΔS_{fus}, cal/mol K (e.u.):
13.8 (Tsonopoulos & Prausnitz 1971)
Fugacity Ratio at 25°C (assuming ΔS_{fus} = 13.5 e.u.), F: 0.3589

Water Solubility (g/m^3 or mg/L at 25°C):
3000 (Philpot et al. 1940; quoted, Tsonopoulos & Prausnitz 1971)
3900 (Kilzer et al. 1979; quoted, Howard 1989)
2620 (Schmidt-Bleek et al. 1982; Rott et al. 1982)
2896 (quoted, Müller & Klein 1992)

Vapor Pressure (Pa at 25°C):
15.19 (extrapolated-regression of tabulated data, Stull 1947)
3.33 (extrapolated, Verschueren 1977; quoted, Kilzer et al. 1979; quoted, Howard 1989)
2.50 (Schmidt-Bleek et al. 1982)

2.00 (20°C, quoted, Verschueren 1983)
6.67 (30°C, quoted, Verschueren 1983)
0.224 (extrapolated-Antoine eqn., Boublik et al. 1984)
1.636 (torsion-weighing effusion, Piacente et al. 1985)
3.33 (Howard et al. 1986; quoted, Banerjee et al. 1990)
3.66 (calculated-Antoine eqn., Stephenson & Malanowski 1987)
3.33, 32.0 (quoted, calculated-solvatochromic p., Banerjee et al. 1990)

Henry's Law Constant (Pa m^3/mol):
1.084 (calculated-P/C, Howard 1989)
0.0395 (calculated-P/C, Meylan & Howard 1991)
0.143 (estimated-bond contribution, Meylan & Howard 1991)

Octanol/Water Partition Coefficient, log K_{OW}:
1.84 (Ichikawa et al. 1969; quoted, Kalsch et al. 1991)
1.83 (quoted exptl., Leo et al. 1969,71; Rekker 1977; quoted, Könemann et al. 1979)
1.83 (LC-k', Carlson et al. 1975)
1.85 (calculated-π const., Carlson et al. 1975)
1.87 (calculated-π const., Norrington et al. 1975; quoted, De Bruijn et al. 1989)
1.61 (calculated-π const., Rekker 1977)
1.73, 1.75 (calculated-f const., Rekker 1977)
1.83 (shake flask, Hansch & Leo 1979; quoted, Kaiser 1983; Miyake et al. 1987; De Bruijn et al. 1989)
1.57, 1.76 (HPLC-k', calculated-f const., Könemann et al. 1979)
1.74 (calculated-f const., Rekker & De Kort 1979; quoted, De Bruijn et al. 1989)
1.83 (HPLC-k', Hammers et al. 1982)
2.02 (HPLC-k'correlated, Hammers et al. 1982; quoted, De Bruijn et al. 1989)
2.05 (Schmidt-Bleek et al. 1982)
1.64 (inter-lab. HPLC average, Eadsforth & Moser 1983)
1.83 (quoted, Verschueren 1983)
1.83 (selected best lit. value, Garst & Wilson 1984)
1.88 $\pm$ 0.02 (exptl.-ALPM, Garst & Wilson 1984)
2.78 (shake flask, OECD 1981 Guidelines, Geyer et al. 1984)
1.83 (Hansch & Leo 1985; quoted, Howard 1989; Arnold et al. 1990; Kalsch et al. 1991; Bintein et al. 1993)
0.41 (selected, Yoshida et al. 1986)
1.86 (calculated-solvatochromic p., Kamlet et al. 1988)
1.83 (RP-HPLC-capacity ratio, Minick et al. 1988)
1.88 $\pm$ 0.014 (slow-stirring-GC, De Brujin et al. 1989; quoted, Verhaar et al. 1992)
1.76 (calculated-f const. as per Rekker 1977, Arnold et al. 1990)
1.83 (calculated-π from Norrington et al. 1975, Arnold et al. 1990)
1.88, 2.01 (quoted, CPC-RV, El Tayar et al. 1991)
1.93 (calculated-CLOGP, Jäckel & Klein 1991; Müller & Klein 1992)
1.93 (calculated-QSAR, Kollig 1993)

Bioconcentration Factor, log BCF:

< 1.30 (golden orfe for 3 day exposure, Körte et al. 1978; quoted, Howard 1989)

3.08 (green algae for 24 hour exposure of dry wt. basis, Körte et al. 1978; quoted, Howard 1989)

2.42 (green algae for 24 hour exposure of wet wt. basis, Körte et al. 1978; quoted, Howard 1989)

< 1.0, 2.41, 3.11 (golden orfe, algae, activated sludge, Freitag et al. 1982)

2.42 (alga *chlorella fusca*, wet wt. basis, Geyer et al. 1984)

2.06 (alga *chlorella fusca*, calculated-K_{OW}, Geyer et al. 1984)

1.11, 2.42, 2.45 (golden ide, algae, activated sludge, Freitag et al. 1985)

0.91 (zebrafish, Zok et al. 1991)

-0.097-0.23 (carp, Tsuda et al. 1993)

0.91; 1.52, 0.63, 0.88 (quoted; calculated values, Bintein et al. 1993)

Sorption Partition Coefficient, log K_{OC}:

2.36-2.67 (five Belgium soils, Van Bladel & Moreale 1977; quoted, Howard 1989)

1.98-3.18 (five German soils, Rott et al. 1982; quoted, Howard 1989)

3.74 (colloidal org. matter in ground water, Means 1983; quoted, Howard 1989)

1.86 (calculated- χ , Sabljic 1987)

1.96, 1.86 (soil, quoted exptl., calculated- χ & fragment contribution, Meylan et al. 1992)

1.61 (calculated-K_{OW}, Kollig 1993)

Half-Lives in the Environment:

Air: 4.6 hours, based on estimated reaction rate with photochemically produced hydroxyl radical of 5×10^5 radicals/cm^3 in atmosphere (Wahner & Zetzsch 1983; quoted, Howard 1989).

Surface water: estimated half-life of 0.3-3.0 days in river waters (Zoeteman et al. 1980).

Groundwater: estimated half-life of 30-300 days (Zoeteman et al. 1980).

Sediment:

Soil:

Biota: 4.3 hours in carp with excretion rate of 0.16 h^{-1} (Tsuda et al. 1993).

Environmental Fate Rate Constants or Half-Lives:

Volatilization: estimated half-life of 6.4 hours from using Henry's law constant for a model river of 1 meter deep with 1 m/sec current and 3 m/sec wind (Lyman et al. 1982; quoted, Howard 1982); half-life of 3 days in an experimental pond with spiked 4-chloroaniline (Schauerte et al. 1982; quoted, Howard 1989).

Photolysis:

Hydrolysis:

Oxidation: photooxidation half-life of 4.6 hours, based on estimated reaction rate with photochemically produced hydroxyl radical of 5×10^5 radicals/cm^3 in atmosphere (Wahner & Zetzsch 1983; quoted, Howard 1989); observed rate constant of

$83x10^{-12}$ $cm^3 \cdot molecule^{-1} \cdot sec^{-1}$ (Wahner & Zetzsch 1983) compared to the calculated rate constants of $34.7x10^{-12}$ $cm^3 \cdot molecule^{-1} \cdot sec^{-1}$ and $42.8x10^{-12}$ $cm^3 \cdot molecule^{-1} \cdot sec^{-1}$ for the vapor phase reaction with hydroxyl radicals at room temperature in air (Atkinson 1985 and Atkinson et al. 1985); observed rate constant of $83x10^{-12}$ $cm^3 \cdot molecule^{-1} \cdot sec^{-1}$ (Wahner & Zetzsch 1983) compared to calculated rate constant of $54x10^{-12}$ $cm^3 \cdot molecule^{-1} \cdot sec^{-1}$ for the vapor phase reaction with hydroxyl radicals at room temperature in air (Atkinson 1987); 72.1 mg/L total organic carbon (TOC) degraded to 92% TOC after 5 hours illumination with a 250 watt tungsten lamp by photo-Fenton reaction in distilled water (Ruppert et al. 1993).

Biodegradation:

Biotransformation:

Bioconcentration, Uptake (k_1) and Elimination (k_2) Rate Constants:

k_1: 42.9 h^{-1} (zebrafish, Zok et al. 1991)

k_2: 0.16 h^{-1} (carp, Tsuda et al. 1993)

Common Name: 3,4-Dichloroaniline
Synonym:
Chemical Name: 3,4-dichloroaniline
CAS Registry No: 95-76-1
Molecular Formula: $C_6H_3NH_2Cl_2$
Molecular Weight: 162.02
Melting Point (°C):
72 (Weast 1982-83)
70-72 (Dean 1985)
Boiling Point (°C):
272.0 (Weast 1982-83)
Density (g/cm^3 at 20°C):
Molar Volume (cm^3/mol):
111.7 (calculated-density, Jaworska & Schultz 1993)
152.0 (calculated-LeBas method)
Molecular Volume ($Å^3$):
35.5 (Jäckel & Klein 1991)
Total Surface Area, TSA ($Å^2$):
Dissociation Constant, pK_a:
2.968, 3.0 (Perrin 1972)
2.0 (estimated, Wolff & Crossland 1985)
Enthalpy of Fusion, ΔH_{fus}, kcal/mol:
Entropy of Fusion, ΔS_{fus}, cal/mol·K (e.u.):
Fugacity Ratio at 25°C (assuming ΔS_{fus} = 13.5 e.u.), F: 0.3468

Water Solubility (g/m^3 or mg/L at 25°C):
92.03 (20°C, Wolff & Crossland 1985)

Vapor Pressure (Pa at 25°C):
1.30 (20°C, Wolff & Crossland 1985)
2.27 (extrapolated-Antoine eqn., Stephenson & Malanowski 1987)

Henry's Law Constant ($Pa \cdot m^3/mol$):
2.289 (calculated-P/C, this work)

Octanol/Water Partition Coefficient, log K_{OW}:
2.69 (quoted unpublished result, Leo et al. 1971, quoted, Wolff & Crossland 1985)
2.12, 2.42 (HPLC-k', calculated-f const., Könemann et al. 1979)
2.69, 2.67 (HPLC-k', HPLC-k'correlated, Hammers et al. 1982)
2.62 (inter-lab. shake flask average, Eadsforth & Moser 1983)
2.30 (inter-lab. HPLC average, Eadsforth & Moser 1983)
2.69 (quoted from Hansch & Leo 1979, Kaiser 1983; Yoshida et al. 1986)

2.14, 2.63 (HPLC-k', Eadsforth 1986)
2.69, 2.34 (quoted: shake flask, calculated, Eadsforth 1986)
2.69 (quoted, Schultz et al. 1989; Arnold et al. 1990, Zok et al. 1991)
2.79 (calculated-CLOGP, Jäckel & Klein 1991)
2.76 (calculated-π const., Verhaar et al. 1992)
2.69 (U.S. EPA CLOGP Data Base, Hulzebos et al. 1993)
2.79 (quoted, Bintein et al. 1993)

Bioconcentration Factor, log BCF:
1.48 (zebrafish, Zok et al. 1991)
1.48; 2.02, 1.88, 1.75 (fish: quoted; calculated values-K_{OW}, Bintein et al. 1993)

Sorption Partition Coefficient, log K_{OC}:
1.40 (calculated-K_{OW}, wet sediment, Wolff & Crossland 1985)
2.29 (Sabljic 1987)
2.05 (soil, quoted, Sabljic 1987)
2.29, 2.08 (soil, quoted, calculated- χ & fragment contribution, Meylan et al. 1992)

Half-Lives in the Environment:
Air:
Surface water: overall rate of loss predicted from outdoor ponds was calculated based on direct phototransformation, and indirect phototransformation to be 0.13 to 0.22 day^{-1} corresponding to a half-life of 3.2 to 5.3 days; the observed rate of loss varied from 0.11 to 0.17 day^{-1} corresponding to half-lives of 4.1 to 6.3 days (Wolff & Crossland 1985).
Groundwater:
Sediment:
Soil:
Biota:

Environmental Fate Rate Constants or Half-Lives:
Volatilization: first order rate constant was calculated to be 5.0×10^{-3} day^{-1} (Wolff & Crossland 1985).
Photolysis: rate constants for phototransformation of 0.12 to 0.20 day^{-1} (Wolff & Crossland 1985).
Oxidation:
Hydrolysis: not expected to occur (Wolff & Crossland 1985).
Biodegradation:
Biotransformation:
Bioconcentration, Uptake (k_1) and Elimination (k_2) Rate Constants:
k_1: 78.5 h^{-1} (zebrafish, Zok et al. 1991)

Common Name: *o*-Toluidine
Synonym: 2-aminotoluene, *o*-aminotoluene, 2-methylaniline, 2-methylbenzeneamine
Chemical Name: 2-aminotoluene, *o*-methylaniline, *o*-toluidine
CAS Registry No: 95-53-4
Molecular Formula: $C_6H_4(CH_3)NH_2$
Molecular Weight: 107.15
Melting Point (°C):
-16.3 (Stull 1947)
-23.68 (Dreisbach 1955)
-16.35 (Chao et al. 1990)
Boiling Point (°C):
199.7 (Stull 1947)
200.23 (Dreisbach 1955)
200.0 (Verschueren 1983; Banerjee et al. 1990)
200.32 (Boublik et al. 1984)
200.34 (Chao et al. 1990)
Density (g/cm^3 at 20°C):
0.99843 (Dreisbach 1955)
1.004 (Verschueren 1983)
0.9947 (25°C, calculated, Chao et al. 1990)
Molar Volume (cm^3/mol):
107.0 (calculated-density, Chiou 1985)
132.4 (calculated-LeBas method)
Molecular Volume ($Å^3$):
Dissociation Constants, pK_a:
4.40, 4.447, 4.46 (Perrin 1972)
4.44 (Weast 1982-83)
4.45 (protonated cation +1, Dean 1985)
4.43 (Sangster 1989)
Total Surface Area, TSA ($Å^2$):
Heat of Fusion, kcal/mol:
1.801 (Dreisbach 1955)
Entropy of Fusion, cal/mol K (e.u.):
Fugacity Ratio at 25°C (assuming ΔS = 13.5 e.u.), F: 1.0

Water Solubility (g/m^3 or mg/L at 25°C):
16330 (20°C, shake flask-GC, Chiou 1981; Chiou & Schmedding 1981; Chiou et al. 1982; quoted, Chiou 1985; Chessells et al. 1992; Müller & Klein 1992)
16287 (calculated-K_{OW}, Müller & Klein 1992)
15000 (quoted, Verschueren 1983)

Vapor Pressure (Pa at 25°C):
42.72 (extrapolated-regression of tabulated data, Stull 1947)

42.26 (Dreisbach 1955)
42.93 (extrapolated-Cox eqn., Chao et al. 1983)
13.33 (20°C, quoted, Verschueren 1983)
39.99 (30°C, quoted, Verschueren 1983)
33.96 (extrapolated-Antoine eqn., Boublik et al. 1984)
34.18 (extrapolated-Antoine eqn., Dean 1985)
42.66 (Howard et al. 1986; quoted, Banerjee et al. 1990)
36.46 (extrapolated-Antoine eqn., Stephenson & Malanowski 1987)
42.66, 113.3 (quoted, calculated-solvatochromic p., Banerjee et al. 1990)
31.10 (calculated-Cox eqn., Chao et al. 1990)

Henry's Law Constant (Pa·m^3/mol):
0.095 (calculated-P/C, this work)

Octanol/Water Partition Coefficient, log K_{OW}:
1.29 (shake flask-AS, Leo et al. 1971; quoted, Chiou 1981; Chiou et al. 1982; Chiou & Schmedding 1981; Chiou 1985)
1.32 (LC-k', Carlson et al. 1975)
1.43 (calculated-π const., Carlson et al. 1975; quoted, Sangster 1989)
1.32 (quoted, McCall 1975)
1.32, 1.63 (quoted, estimated-HPLC-RT, Veith et al. 1979)
1.29, 1.32 (quoted, Verschueren 1983)
1.42 (shake flask-AS at pH 7.5, Martin-Villodre et al. 1986; quoted, Sangster 1989)
1.34 (HPLC average, Ge et al. 1987; quoted, Sangster 1989)
1.32 (calculated-CLOGP 3.34, Arnold et al. 1990)
1.54 (calculated-f const. as per Rekker 1977, Arnold et al. 1990)
1.39 (calculated-π from Norrington et al. 1975, Arnold et al. 1990)
1.32 (recommended, Sangster 1989)
1.32 (quoted from Medchem software value, Chessells et al. 1992)
1.41 (calculated, Müller & Klein 1992)
1.43 (calculated-π const., Verhaar et al. 1992)
1.56 (calculated-QSAR, Kollig 1993)

Bioconcentration Factor, log BCF:

Sorption Partition Coefficient, log K_{OC}:
1.24 (calculated-K_{OW}, Kollig 1993)

Half-Lives in the Environment:
Air: photooxidation half-life of 0.394-3.94 hours, based on estimated rate constant for the reaction with hydroxyl radicals in air (Atkinson 1987; quoted, Howard et al.

1991); atmospheric transformation lifetime was estimated to be < 1 day (Kelly et al. 1994).

Surface water: photooxidation half-life of 62.4-3480 hours, based on estimated rate constants for reactions of representative aromatic amines with OH and RO_2 radicals (Mill & Mabey 1985; quoted, Howard et al. 1991).

Groundwater: 48-336 hours, based on estimated unacclimated aqueous aerobic biodegradation half-life (Howard et al. 1991).

Sediment:

Soil: 24-168 hours, based on estimated unacclimated aqueous aerobic biodegradation half-life (Howard et al. 1991).

Biota:

Environmental Fate Rate Constants or Half-Lives:

Volatilization:

Photolysis:

Hydrolysis:

Oxidation: rate constant of 1×10^4 M^{-1} sec^{-1} for oxidation by RO_2 radicals at 30°C in aquatic systems with half-life of 0.8 day (Howard 1972; Hendry et al. 1974; quoted, Mill 1982); rate constant of $< 2 \times 10^2$ M^{-1} sec^{-1} for oxidation by singlet oxygen at 25°C in aquatic systems with half-life of > 100 years (Foote 1976; Mill 1979; quoted, Mill 1982); photooxidation half-life of 62.4-3480 hours in water, based on estimated rate constants for reactions of representative aromatic amines with OH and RO_2 radicals (Mill & Mabey 1985; quoted, Howard et al. 1991); photooxidation half-life of 0.394-3.94 hours in air, based on estimated rate constant for the reaction with hydroxyl radicals in air (Atkinson 1987; quoted, Howard et al. 1991).

Biodegradation: decomposition by a soil microflora in > 64 days (Alexander & Lustigman 1966; quoted, Verschueren 1983); aqueous aerobic half-life of 24-168 hours, based on aqueous aerobic screening test data (Baird et al. 1977; Sasaki 1978; quoted, Howard et al. 1991); average biodegradation rate of 15.1 mg COD g^{-1} h^{-1} for 97.7% removal (Scow 1982); aqueous anaerobic half-life of 96-672 hours, based on estimated unacclimated aqueous aerobic biodegradation half-life (Howard et al. 1991).

Biotransformation:

Bioconcentration, Uptake (k_1) and Elimination (k_2) Rate Constants:

Common Name: *m*-Toluidine
Synonym: 3-aminotoluene, 3-methylbenzeneamine
Chemical Name: 3-aminotoluene, *m*-amino-methylbenzene, *m*-methylaniline, *m*-toluidine
CAS Registry No: 108-44-1
Molecular Formula: $C_6H_4(CH_3)NH_2$
Molecular Weight: 107.15
Melting Point (°C):
- -31.5 (Stull 1947)
- -30.40 (Dreisbach 1955)
- -32.0 (Verschueren 1983)
- -31.25 (Chao et al. 1990)

Boiling Point (°C):
- 203.3 (Stull 1947)
- 203.35 (Dreisbach 1955; Boublik et al. 1984)
- 203.0 (Verschueren 1983)
- 203.37 (Chao et al. 1990)

Density (g/cm³ at 20°C):
- 0.98890 (Dreisbach 1955)
- 0.990 (Verschueren 1983)
- 0.9850 (25°C, calculated, Chao et al. 1990)

Molar Volume (cm³/mol):
Molecular Volume (Å³):
Dissociation Constant, pK_a:
- 4.66, 4.712, 4.72 (Perrin 1972)
- 4.73 (Weast 1982-83)
- 4.71 (protonated cation +1, Dean 1985)
- 4.70 (Sangster 1989)

Total Surface Area, TSA (Å²): 253.4 (Iwase et al. 1985)
Heat of Fusion, kcal/mol: 1.692 (Dreisbach 1955)
Entropy of Fusion, cal/mol K (e.u.):
Fugacity Ratio at 25°C (assuming ΔS = 13.5 e.u.), F: 1.0

Water Solubility (g/m³ or mg/L at 25°C):
- 15031 (20°C, shake flask-GC, Chiou 1981; Chiou & Schmedding 1981; Chiou et al. 1982)

Vapor Pressure (Pa at 25°C):
- 49.54 (extrapolated-regression of tabulated data, Stull 1947)
- 36.64 (Dreisbach 1955)
- 33.49 (calculated-Cox eqn., Chao et al. 1983)
- 27.91 (extrapolated-Antoine eqn., Boublik et al. 1984)
- 25.66 (extrapolated-Antoine eqn., Dean 1985)
- 26.86 (extrapolated-Antoine eqn., Stephenson & Malanowski 1987)
- 25.50 (calculated-Cox eqn., Chao et al. 1990)

Henry's Law Constant (Pa·m³/mol):
- 0.257 (calculated-P/C, this work)

Octanol/Water Partition Coefficient, log K_{OW}:

1.40 (shake flask-AS, Fujita et al. 1964; quoted, Leo et al. 1971; McCall 1975; Chiou 1981; Chiou & Schmedding 1981; Chiou et al. 1982; Iwase et al. 1985; Sangster 1989)
1.43 (LC-k', Carlson et al. 1975)
1.45 (calculated-π const., Carlson et al. 1975; quoted, Sangster 1989)
1.42 (shake flask-AS, Briggs 1981; quoted, Sangster 1989)
1.35 (calculated-TSA, Iwase et al. 1985)
1.40 (calculated-CLOGP 3.34, Arnold et al. 1990)
1.54 (calculated-f const. as per Rekker 1977, Arnold et al. 1990)
1.40 (calculated-π const. from Norrington et al. 1975, Arnold et al. 1990)
1.42 (quoted from Hansch & Leo 1979, Bodor et al. 1989)
1.40, 1.43 (quoted, Verschueren 1983)
1.32 (quoted from CLOGP, Bodor et al. 1989)
1.564 (calculated-CLOGP, Bodor et al. 1989)
1.49 (calculated-M.O. indices, Bodor et al. 1989)
1.40 (recommended, Sangster 1989)
1.44 (calculated-π constant, Verhaar et al. 1992)

Bioconcentration Factor, log BCF:

Sorption Partition Coefficient, log K_{OC}:

Half-Lives in the Environment:

Air:
Surface water:
Groundwater:
Sediment:
Soil:
Biota:

Environmental Fate Rate Constants or Half-Lives:

Volatilization:
Photolysis:
Hydrolysis:
Oxidation: rate constant of 1×10^4 M^{-1} sec^{-1} for oxidation by RO_2 radicals at 30°C in aquatic systems with half-life of 0.8 day (Howard 1972; Hendry et al. 1974; quoted, Mill 1982); rate constant of $< 2 \times 10^2$ M^{-1} sec^{-1} for oxidation by singlet oxygen at 25°C in aquatic systems with half-life of >100 years (Foote 1976; Mill 1979; quoted, Mill 1982).
Biodegradation: decomposition by a soil microflora in 8 days (Alexander & Lustigman 1966; quoted, Verschueren 1983); average biodegradation rate of 30.0 mg COD g^{-1} h^{-1} for 97.7% removal (Scow 1982).
Biotransformation:
Bioconcentration, Uptake (k_1) and Elimination (k_2) Rate Constants:

Common Name: *p*-Toluidine
Synonym: 4-aminotoluene, 4-methylaniline, 4-methylbenzenamine
Chemical Name: 4-aminotoluene, *p*-amino-methylbenzene, *p*-methylaniline, *p*-toluidine
CAS Registry No: 106-49-0
Molecular Formula: $C_6H_4(CH_3)NH_2$
Molecular Weight: 107.15
Melting Point (°C):
 44.5 (Stull 1947)
 43.70 (Dreisbach 1955)
 43.75 (Tsonopoulos & Prausnitz 1971)
Boiling Point (°C):
 200.4 (Stull 1947)
 200.55 (Dreisbach 1955)
Density (g/cm^3 at 20°C): 1.043 (Verschueren 1983)
Molar Volume (cm^3/mol):
 103.60 (calculated-density, Jaworska & Schultz 1993)
Molecular Volume (Å^3):
Dissociation Constant, pK_a:
 5.02, 5.08, 5.084 (Perrin 1972)
 5.08 (Weast 1982-83; Sangster 1989)
 5.08 (protonated cation +1, Dean 1985)
 5.17 (shake flask-HPLC/UV, Johnson & Westall 1990)
Total Surface Area, TSA (Å^2): 253.8 (Iwase et al. 1985)
Heat of Fusion ΔH_{fus}, kcal/mol:
 4.14 (Tsonopoulos & Prausnitz 1971)
Entropy of Fusion ΔS_{fus}, cal/mol K (e.u.):
 13.1 (Tsonopoulos & Prausnitz 1971)
 13.77 (observed, Yalkowsky & Valvani 1980)
Fugacity Ratio at 25°C (assuming ΔS_{fus} = 13.5 e.u.), F: 0.652

Water Solubility (g/m^3 or mg/L at 25°C):
 8965 (Seidell 1941,52; quoted, Tsonopoulos & Prausnitz 1971)
 7400 (21°C, quoted, Verschueren 1983)
 6643, 5370 (20°C, shake flask-UV, calculated, Hashimoto et al. 1984)

Vapor Pressure (Pa at 25°C):
 46.27 (extrapolated-regression of tabulated data, Stull 1947)
 44.70 (Dreisbach 1955)
 38.13 (calculated-Cox eqn., Chao et al. 1983)
 40.17 (calculated-Antoine eqn., Dean 1985)
 27.03 (extrapolated-Antoine eqn., Stephenson & Malanowski 1987)

Henry's Law Constant (Pa·m^3/mol):
 0.656 (calculated-P/C, this work)

Octanol/Water Partition Coefficient, log K_{OW}:

1.39 (shake flask-AS, Fujita et al. 1964; quoted, Leo et al. 1969; McCall 1975; Verschueren 1983; Iwase et al. 1985; Sangster 1989; Schultz et al. 1987)
1.41, 1.48 (LC-k', calculated- π const., Carlson et al. 1975)
1.40 (shake flask-AS, Briggs 1981; quoted, Sangster 1989)
1.41 (selected best lit. value, Garst & Wilson 1984)
1.44 $\pm$ 0.03 (exptl.-ALPM, Garst & Wilson 1984)
1.39, 1.42 (quoted, HPLC-k', Haky & Young 1984)
1.36 (calculated-TSA, Iwase et al. 1985)
1.39 (shake flask-AS at pH 7.5, Martin-Villodre et al. 1986; quoted, Sangster 1989)
0.98 (selected, Yoshida et al. 1986)
1.41 (HPLC average, Ge et al. 1987; quoted, Sangster 1989)
1.39 (recommended, Sangster 1989)
1.40 (shake flask-HPLC/UV, Johnson & Westall 1990)
1.39 (calculated-CLOGP 3.51, Jaworska & Schultz 1993)
1.42 (calculated-π const., Verhaar et al. 1992)
1.56 (calculated-QSAR, Kollig 1993)

Bioconcentration Factor, log BCF:

Sorption Partition Coefficient, log K_{OC}:

2.51, 2.70, 2.71 (Morocco soil, Oakville soil, Milford soil, Graveel et al. 1986)
2.74, 2.22, 2.20 (Podzol soil, Alfisol soil, sediment, von Oepen et al. 1991)
1.24 (calculated-K_{OW}, Kollig 1993)

Half-Lives in the Environment:

Air:
Surface water:
Groundwater:
Sediment:
Soil:
Biota:

Environmental Fate Rate Constants or Half-Lives:

Volatilization:
Photolysis:
Hydrolysis:
Oxidation: rate constant of $1x10^4$ M^{-1} sec^{-1} for oxidation by RO_2 radicals at 30°C in aquatic systems with half-life of 0.8 day (Howard 1972; Hendry et al. 1974; quoted, Mill 1982); rate constant of $<2x10^2$ M^{-1} sec^{-1} for oxidation by singlet oxygen at 25°C in aquatic systems with half-life of >100 years (Foote 1976; Mill 1979; quoted, Mill 1982).
Biodegradation: decomposition by a microflora in 4 days (Alexander & Lustigman 1966; quoted, Verschueren 1983); average biodegradation rate of 20 mg COD g^{-1} hr^{-1} for 97.7% removal (Scow 1982).
Biotransformation:
Bioconcentration, Uptake (k_1) and Elimination (k_2) Rate Constants:

Common Name: N,N'-Dimethylaniline
Synonym: N,N'-dimethylbenzenamine
Chemical Name: N,N'-dimethylaniline
CAS Registry No: 121-69-7
Molecular Formula: $C_6H_5N(CH_3)_2$
Molecular Weight: 121.18
Melting Point (°C):
2.50 (Stull 1947)
2.40 (Tsonopoulos & Prausnitz 1971)
2.45 (Riddick et al. 1986)
2.00 (Stephenson & Malanowski 1987)
Boiling Point (°C):
193.1 (Stull 1947)
194.05 (Riddick et al. 1986)
192.0 (Stephenson & Malanowski 1987)
193.0 (Banerjee et al. 1990)
Density (g/cm^3 at 20°C): 0.95232
Molar Volume (cm^3/mol):
136.7 (Leahy 1986; Kamlet et al. 1987)
75.2 (intrinsic volume V_I, Leahy 1986; Kamlet et al. 1987)
154.6 (calculated-LeBas method)
Molecular Volume (Å^3):
Total Surface Area, TSA (Å^2):
Dissociation Constant pK:
5.15 ($pK_{BH}{}^+$, Riddick et al. 1986)
5.10 (Sangster 1993)
Heat of Fusion, ΔH_{fus}, kcal/mol:
2.73 (Tsonopoulos & Prausnitz 1971)
Entropy of Fusion, ΔS_{fus}, cal/mol K (e.u.):
9.91 (Tsonopoulos & Prausnitz 1971)
Fugacity Ratio at 25°C (assuming ΔS_{fus} = 13.5 e.u.), F: 1.0

Water Solubility (g/m^3 or mg/L at 25°C):
1105 (shake flask-GC, Chiou et al. 1982)
3576 (calculated-solvatochromic p., Leahy 1986)
1105, 1424 (quoted, calculated-V_I, solvatochromic p., Kamlet et al. 1987)
1105, 541 (quoted, calculated-V_I & M.P., Kamlet et al. 1987)
1096 (quoted, Isnard & Lambert 1989)

Vapor Pressure (Pa at 25°C):
106.3 (extrapolated-regression of tabulated data, Stull 1947)
133.3 (29.5°C, Stull 1947)
83.9 (calculated-Antoine eqn., Boublik et al. 1973)
68.95 (calculated-Cox eqn., Chao et al. 1983)
83.91 (extrapolated-Antoine eqn., Dean 1985)

69.4 (Howard et al. 1986; quoted, Banerjee et al. 1990)
670 (quoted, Riddick et al. 1986)
107 (extrapolated-Antoine eqn., Stephenson & Malanowski 1987)
55.99 (calculated-solvatochromic p., Banerjee et al. 1990)

Henry's Law Constant (Pa·m^3/mol):
11.734 (calculated-P/C, this work)

Octanol/Water Partition Coefficient, log K_{OW}:
2.31 (shake flask-AS, Fujita et al. 1964; quoted, Sangster 1993)
2.62 (shake flask-AS at pH 7.4, Rogers & Cammarata 1969; quoted, Sangster 1993)
1.66 (shake flask-AS, Leo et al. 1971)
2.29 (shake flask-AS, Yaguzhinskii et al. 1973; quoted, Sangster 1993)
2.30 (shake flask at pH 7, Unger et al. 1978)
2.28, 2.13 (quoted, calculated-V_I, solvatochromic p., Leahy 1986)
2.43 (correlated-HPLC-RT, Miyake et al. 1986; quoted, Sangster 1993)
2.62 (quoted, Yoshida et al. 1986)
2.31 (quoted, Isnard & Lambert 1989)
2.32 (CPC, El Tayar et al. 1991; quoted, Sangster 1993)
2.31 (recommended, Sangster 1993)

Bioconcentration Factor, log BCF:

Sorption Partition Coefficient, log K_{OC}:
1.99 (Alfisol soil, von Oepen et al. 1991)
2.53 (Podzol soil, von Oepen et al. 1991)
2.06 (sediment, von Oepen et al. 1991)
2.26, 1.89 (soil, quoted, calculated- χ and fragment contribution, Meylan et al. 1992)

Half-Lives in the Environment:
Air: atmospheric lifetimes of 1.9 hour in clean troposphere and 1.0 hour in moderately polluted atmosphere, based on the gas-phase reaction with hydroxyl radicals in air at room temperature (Atkinson et al. 1987); atmospheric lifetimes of 1.8 days in clean troposphere and 14 hours in moderately polluted atmosphere, based on the gas-phase reaction with O_3 in air at room temperature (Atkinson et al. 1987); half-lives of 2.7-21 hours, based on photooxidation half-life in air (Howard et al. 1991); atmospheric transformation lifetime was estimated to be <1 day (Kelly et al. 1994).
Surface water: half-lives of 19.3-1925 hours, based on photooxidation half-life in water (Howard et al. 1991).
Groundwater: half-lives of 1344-8640 hours, based on estimated unacclimated aqueous aerobic biodegradation half-life (Howard et al. 1991).

Sediment:

Soil: half-lives of 672-4320 hours, based on estimated unacclimated aqueous aerobic biodegradation half-life (Howard et al. 1991).

Biota:

Environmental Fate Rate Constants or Half-Lives:

Volatilization:

Photolysis:

Oxidation: rate constant of $1x10^4$ M^{-1} sec^{-1} for oxidation by RO_2 radicals at 30°C in aquatic systems with half-life of 0.8 day (Howard 1972; Hendry et al. 1974; quoted, Mill 1982); rate constant of $<2x10^2$ M^{-1} sec^{-1} for oxidation by singlet oxygen at 25°C in aquatic systems with half-life of >100 years (Foote 1976; Mill 1979; quoted, Mill 1982); photooxidation half-life of 2.7-21 hours, based on rate constant for reaction with hydroxyl radicals and ozone in air and half-life of 10.3-1925 hours, based on reaction with singlet oxygen in aqueous solution (Howard et al. 1991); exptl. rate constant of $1.48x10^{-10}$ cm^3 $molecule^{-1}$ sec^{-1} for the gas-phase reaction with OH radicals in the atmosphere at room temperature and calculated rate constant of $(9.10 \pm 1.00)x10^{-18}$ cm^3 $molecule^{-1}$ sec^{-1} for the gas-phase reaction with O_3 in the atmosphere at room temperature (Atkinson et al. 1987); exptl. rate constant of $148.0x10^{-12}$ cm^3 $molecule^{-1}$ sec^{-1} (Atkinson 1989) as compared to a calculated rate constant of $178x10^{-12}$ cm^3 $molecule^{-1}$ sec^{-1} by molecular orbital calculation for the gas-phase reaction with hydroxyl radicals in air at room temperature (Klamt 1993).

Hydrolysis:

Biodegradation: aerobic half-life of 672-4320 hours, based on unacclimated aqueous secreening test data and anerobic half-life of 2880-17280 hours, based on estimated unacclimated aqueous aerobic biodegradation half-life (Howard et al. 1991).

Biotransformation:

Bioconcentration, Uptake (k_1) and Elimination (k_2) Rate Constants:

Common Name: 2,6-Xylidine
Synonym: 2,6-dimethylaniline, 2,6-dimethylbenzeneamine
Chemical Name: 2,6-dimethylaniline
CAS Registry No: 87-62-7
Molecular Formula: 2,6-$(CH_3)_2C_6H_3NH_2$
Molecular Weight: 121.18
Melting Point (°C):
    11.2 (Weast 1982-83; Riddick et al. 1986)
    10-12 (Verschueren 1983; Dean 1985)
Boiling Point (°C):
    217.9 (Stull 1947; Riddick et al. 1986)
    214.0 (at 739 mm Hg, Weast 1982-83; Verschueren 1983; Banerjee et al. 1990)
    216.0 (Dean 1985)
Density (g/cm^3 at 20°C):
    0.9842 (Weast 1982-83)
    0.984 (Verschueren 1983)
Molar Volume (cm^3/mol):
    154.6 (calculated-LeBas method)
    123.2 (calculated-density)
Molecular Volume ($Å^3$):
    39.0 (Jäckel & Klein 1991)
Total Surface Area, TSA ($Å^2$):
Dissociation Constant pK:
    3.95 (pK_{BH^+}, Riddick et al. 1986)
Heat of Fusion, ΔH_{fus}, kcal/mol:
Entropy of Fusion, ΔS_{fus}, cal/mol K (e.u.):
Fugacity Ratio at 25°C (assuming ΔS_{fus} = 13.5 e.u.), F: 1.0

Water Solubility (g/m^3 or mg/L at 25°C):
    slightly soluble (Dean 1985; Budavari 1989)
    4700 (selected, this work)

Vapor Pressure (Pa at 25°C):
    43.03 (extrapolated-regression of tabulated data, Stull 1947)
    17.33 (Howard et al. 1986; quoted, Banerjee et al. 1990)
    670 (quoted, Riddick et al. 1986)
    35.99 (calculated-solvatochromic p., Banerjee et al. 1990)

Henry's Law Constant ($Pa \cdot m^3/mol$):
    17.28 (calculated-P/C, this work)

Octanol/Water Partition Coefficient, log K_{OW}:
    1.96 (calculated, Verschueren 1983)

1.91 (calculated-CLOGP, Jäckel & Klein 1991)

Bioconcentration Factor, log BCF:

Sorption Partition Coefficient, log K_{OC}:

Half-Lives in the Environment:

Air: 0.33-3.3 hours, based on estimated photooxidation half-lives in air from estimated reaction rate constant with OH radicals in air (Atkinson 1987; selected, Howard et al. 1991).

Surface water: 62.4-3480 hours, based on photooxidation half-lives in water (Howard et al. 1991).

Groundwater: 1344-8640 hours, based on estimated aqueous aerobic biodegradation half-lives (Howard et al. 1991).

Sediment:

Soil: 72-7584 hours, based on soil persistence and soil biodegradation studies (Bollag et al. 1978; Medvedev & Davidov 1981; selected, Howard et al. 1991).

Biota:

Environmental Fate Rate Constants or Half-Lives:

Volatilization:

Photolysis:

Oxidation: rate constant of 1×10^4 M^{-1} sec^{-1} for oxidation by RO_2 radicals at 30°C in aquatic systems with half-life of 0.8 day (Howard 1972; Hendry et al. 1974; quoted, Mill 1982); rate constant of $<2 \times 10^2$ M^{-1} sec^{-1} for oxidation by singlet oxygen at 25°C in aquatic systems with half-life of >100 years (Foote 1976; Mill 1979; quoted, Mill 1982); photooxidation half-life of 0.33-3.3 hours in air, based on estimated reaction rate constant with OH radicals (Atkinson 1987; selected, Howard et al. 1991) and photooxidation half-life of 62.4-3480 hours in water, based on reaction rate constants of amine class with $RO_2\bullet$ and OH radicals in water (Güesten et al. 1981; Mill & Mabey 1985; selected, Howard et al. 1991).

Hydrolysis: no hydrolyzable group (Howard et al. 1991).

Biodegradation: aqueous aerobic biodegradation half-life of 672-4320 hours, based on a biological screening study (Baird et al. 1977; selected Howard et al. 1991) and a soil degradation study (Bollag et al. 1978; selected, Howard et al. 1991); aqueous anaerobic biodegradation half-life of 2688-17280 hours, based on estimated aqueous biodegradation half-lives (Howard et al. 1991).

Biotransformation:

Bioconcentration, Uptake (k_1) and Elimination (k_2) Rate Constants:

Common Name: Diphenylamine
Synonym: N-diphenylamine, N-phenyl aniline, DPA
Chemical Name: N-diphenylamine, diphenylamine
CAS Registry No: 122-39-4
Molecular Formula: $C_6H_5NHC_6H_5$
Molecular Weight: 169.23
Melting Point (°C):
52.9 (Stull 1947)
53.0 (Tsonopoulos & Prausnitz 1971; Yalkowsky & Valvani 1980)
52.8 (Verschueren 1983)
53-54 (Dean 1985)
Boiling Point (°C):
302 (Stull 1947; Weast 1982-83; Verschueren 1983; Dean 1985)
Density (g/cm^3 at 20°C):
1.160 (Weast 1982-83; Dean 1985)
1.159 (Verschueren 1983)
Molar Volume (cm^3/mol):
200.3 (calculated-LeBas method)
145.9 (calculated-density)
Molecular Volume ($Å^3$):
Total Surface Area, TSA ($Å^2$):
334.1 (Iwase et al. 1985)
Dissociation Constant, pK_a:
0.89 (Perrin 1972)
0.90 (protonated cation +1, Dean 1985)
0.78 (Sangster 1989)
Heat of Fusion ΔH_{fus}, kcal/mol:
4.19 (Tsonopoulos & Prausnitz 1971)
Entropy of Fusion ΔS_{fus}, cal/mol K (e.u.):
12.8 (Tsonopoulos & Prausnitz 1971)
13.10 (exptl., Yalkowsky & Valvani 1980)
13.50 (estimated, Yalkowsky & Valvani 1980)
Fugacity Ratio at 25°C (assuming ΔS_{fus} = 13.5 e.u.), F: 0.5225

Water Solubility (g/m^3 or mg/L at 25°C):
174.7 (calculated-ΔS & M.P., Yalkowsky & Valvani 1980)
179.2 (calculated-M.P., Yalkowsky & Valvani 1980)
308 (quoted, Briggs 1981)
48.0 (20°C, shake flask & membrane filter-fluorophotometric, Hashimoto et al. 1982)
52, 54 (20°C, shake flask & glass fiber filters-fluorophotometric, Hashimoto et al. 1982)
300.0 (Verschueren 1983; quoted, Ma et al. 1990)
53.0 (20°C, Yalkowsky et al. 1987; quoted, Ma et al. 1990)
52.5 (Isnard & Lambert 1988; quoted, Ma et al. 1990)

Vapor Pressure (Pa at 25°C):

0.5682 (extrapolated-regression of tabulated data, Stull 1947)
0.0612 (calculated-Antoine eqn., Stephenson & Malanowski 1987)

Henry's Law Constant (Pa m^3/mol):

0.285 (calculated-P/C, Meylan & Howard 1991)
0.106 (estimated-bond contribution, Meylan & Howard 1991)
0.035 (calculated-P/C, this work)

Octanol/Water Partition Coefficient, log K_{OW}:

3.23 (shake flask-AS, Rogers & Cammarata 1969; quoted, Sangster 1989)
3.22 (calculated-M.O. indices, Rogers & Cammarata 1969; quoted, Leo et al. 1971; Rekker 1977)
3.34 (quoted unpublished result, Leo et al. 1971; quoted, Iwase et al. 1985)
3.34, 3.50, 3.72 (quoted unpublished result, Rekker 1977)
3.49 (calculated-f const., Rekker 1977)
3.50, 2.37 (quoted, estimated-HPLC-RT, Veith et al. 1979)
3.42 (calculated, Veith et al. 1979)
3.42 (shake flask as per Fujita et al. 1964, Briggs 1981)
3.22, 3.50 (quoted, Verschueren 1983)
3.33 (calculated-TSA, Iwase et al. 1985)
3.45 (Hansch & Leo 1979; quoted, Brooke et al. 1986; Bodor et al. 1989)
3.42 (shake flask-AS, Briggs 1981; quoted, Sangster 1989)
3.37 (inter-lab. shake flask average, Eadsforth & Moser 1983)
3.72 (selected best lit. value, Garst & Wilson 1984)
3.72 $\pm$ 0.03 (exptl.-ALPM, Garst & Wilson 1984)
3.62 (selected, Yoshida et al. 1986)
3.42 (quoted, Isnard & Lambert 1988)
3.50 (quoted from CLOGP, Bodor et al. 1989)
3.62 (calculated-CLOGP, Bodor et al. 1989)
2.91 (calculated-M.O. indices, Bodor et al. 1989)
3.50, 3.34 (quoted from Pomona College unpublished data, Sangster 1989)
3.50 (recommended, Sangster 1989)
3.62 (calculated-QSAR, Kollig 1993)

Bioconcentration Factor, log BCF:

1.48 (fathead minnows, Veith et al. 1979)
1.48, 2.10 (quoted, calculated-K_{OW}, Mackay 1982)
1.48 (quoted, Isnard & Lambert 1988)

Sorption Partition Coefficient, log K_{OC}:

2.54 (organic matter-water in various soils, Briggs 1981)
2.54 (soil, quoted, Sabljic 1987)

2.78 (soil, quoted exptl., Meylan et al. 1992)
2.78, 3.28 (soil, calculated- χ and fragment contribution, Meylan et al. 1992)
3.30 (calculated-K_{OW}, Kollig 1993)

Half-Lives in the Environment:

Air: photooxidation half-life of 0.247-2.47 hours, based on estimated rate data for the reaction with hydroxyl radicals in air (Atkinson 1987; selected, Howard et al. 1991).

Surface water: photooxidation half-life of 31-1740 hours in water, based on photooxidation rate constants with OH and RO_2 radicals for the amine class (Mill & Mabey 1985; Güesten et al. 1981; selected, Howard et al. 1991); 31-672 hours, based on estimated unacclimated aqueous aerobic degradation half-life and photooxidation half-life in water (Howard et al. 1991).

Groundwater: 336-1344 hours, based on estimatd aqueous aerobic biodegradation half-life (Howard et al. 1991).

Sediment:

Soil: 168-672 hours, based on estimated aqueous aerobic biodegradation half-life (Howard et al. 1991).

Biota:

Environmental Fate Rate Constants or Half-Lives:

Volatilization:

Photolysis:

Oxidation: rate constant of 1×10^4 M^{-1} sec^{-1} for oxidation by RO_2 radicals at 30°C in aquatic systems with half-life of 0.8 day (Howard 1972; Hendry et al. 1974; quoted, Mill 1982); rate constant of $< 2 \times 10^2$ M^{-1} sec^{-1} for oxidation by singlet oxygen at 25°C in aquatic systems with half-life of > 100 years (Foote 1976; Mill 1979; quoted, Mill 1982); photooxidation half-life of 31-1740 hours in water, based on photooxidation rate constants with OH and RO_2 radicals for the amine class (Mill & Mabey 1985; Güesten et al. 1981; selected, Howard et al. 1991); photooxidation half-life of 0.247-2.47 hours, based on estimated rate data for the reaction with hydroxyl radicals in air (Atkinson 1987; selected, Howard et al. 1991).

Hydrolysis:

Biodegradation: aqueous aerobic half-life of 168-672 hours, based on estimated aqueous aerobic biodegradation screening test data (Malaney 1960; quoted, Howard et al. 1991); aqueous anaerobic half-life of 672-2688 hours, based on estimated aqueous aerobic biodegradation half-life (Howard et al. 1991).

Biotransformation:

Bioconcentration, Uptake (k_1) and Elimination (k_2) Rate Constants:

Common Name: Benzidine
Synonym: *p,p'*-bianiline, 4,4'-diaminobiphenyl, 4,4'-biphenyldiamine, (1,1'-biphenyl)-4,4'-diamine
Chemical Name: benzidine
CAS Registry No: 92-87-5
Molecular Formula: $NH_2C_6H_4C_6H_4NH_2$
Molecular Weight: 184.23
Melting Point (°C):
129 (Verschueren 1977; Callahan et al. 1979; Mabey et al. 1982)
128 (Schmidt-Bleek et al. 1982)
116-129 (Verschueren 1983; Howard 1989)
Boiling Point (°C):
402 (Verschueren 1977, 1983; Callahan et al. 1979; Schmidt-Bleek et al. 1982)
~400 (Howard 1989)
Density (g/cm^3 at 20°C):
1.250 (Verschueren 1983)
Molar Volume (cm^3/mol):
213.0 (calculated-LeBas method)
Molecular Volume (Å^3):
Total Surface Area, TSA (Å^2):
352.8 (Iwase et al. 1985)
Dissociation Constants, pK_a:
4.66 (pK_1), 3.57 (pK_2) (30°C, Perrin 1965; quoted, Mabey et al. 1982; Howard 1989)
Heat of Fusion ΔH_{fus}, kcal/mol:
Entropy of Fusion ΔS_{fus}, cal/mol·K (e.u.):
Fugacity Ratio at 25°C (assuming ΔS_{fus} = 13.5 e.u.), F: 0.09578

Water Solubility (g/m^3 or mg/L at 25°C):
400 (12°C, Verschueren 1977, 1983; Callahan et al. 1979; Mabey et al. 1982)
520 (Shriner et al. 1978; quoted, Howard 1989)
360 (24°C & pH 5.9, shake flask-LSC, Means et al. 1980)
276 (20°C, Schmidt-Bleek et al. 1982)
359 (quoted, Gerstl & Helling 1987)

Vapor Pressure (Pa at 25°C):
0.724 (calculated-Trouton's rule, Mabey et al. 1982)
$1x10^{-6}$ (20°C, Schmidt-Bleek et al. 1982)

Henry's Law Constant (Pa·m^3/mol):
$3.93x10^{-6}$ (estimated, Hine & Mookerjee 1975; quoted, Howard 1989)
0.0394 (calculated-P/C at 12°C, Mabey et al. 1982)
$4.60x10^{-7}$ (calculated-P/C, this work)

Octanol/Water Partition Coefficient, log K_{OW}:

1.81 (Radding et al. 1977; quoted, Callahan et al. 1979; Ryan et al. 1988)
1.66 (shake flask-LSC at pH 5.9, Means et al. 1980)
1.40 (Schmidt-Bleek et al. 1982)
1.34 (quoted, Mabey et al. 1982)
1.34 (Hansch & Leo 1985; quoted, Howard 1989)
1.34, 1.52 (quoted, calculated-TSA, Iwase et al. 1985)
2.40 (computed, Klopman et al. 1985; quoted, Pangrekar et al. 1994)
2.01 (selected, Yoshida et al. 1986)
1.66 (quoted, Gerstl & Helling 1987)
1.58 (calculated-QSAR, Kollig 1993)

Bioconcentration Factor, log BCF:

1.74, 2.66, 2.81, 3.4 (fish, mosquitoes, snail, algae; Lu et al. 1977; quoted, Howard 1989)
1.60 (bluegills, USEPA 1980; quoted, Howard 1989)
1.00 (microorganisms-water, calculated-K_{OW}, Mabey et al. 1982)
1.90, 2.93, 3.08 (golden ide, algae, activated sludge, Freitag et al. 1985)

Sorption Partition Coefficient, log K_{OC}:

1.02 (sediment-water, calculated-K_{OW}, Mabey et al. 1982)
5.95 (Russell soil, Graveel et al. 1986)
5.68 (Chalmers soil, Graveel et al. 1986)
5.36 (Kokomo soil, Graveel et al. 1986)
5.91 (Milford soil, Graveel et al. 1986)
3.00 (calculated- χ , Gerstl & Helling 1987)
3.46, 3.44 (soil, quoted, calculated- χ & fragment contribution, Meylan et al. 1992)
1.26 (calculated-K_{OW}, Kollig 1993)

Half-Lives in the Environment:

Air: 0.312-3.12 hours, based on estimated rate constant for the reaction with hydroxyl radicals in air (Atkinson 1987; quoted, Howard et al. 1991); estimated half-life for the reaction with hydroxyl radicals and ozone was about one day (Howard 1989); atmospheric transformation lifetime was estimated to be < 1 day (Kelly et al. 1994).

Surface water: estimated half-life for the reaction with radicals and redox reactions with naturally occurring cations, etc. and perhaps with photodegradation was about one day (Howard 1989); 31.2-192 hours, based on estimated photooxidation half-life in water and estimated unacclimated aqueous aerobic biodegradation half-life (Howard et al. 1991).

Ground water: 96-484 hours, based on estimated unacclimated aqueous aerobic biodegradation half-life (Howard et al. 1991).

Sediment:

Soil: half-life of less than 10 days in soil (USEPA 1979; quoted, Ryan et al. 1988); 48-192 hours, based on aerobic soil die-away test data (Lu et al. 1977; quoted, Howard et al. 1991); overall biodegradation half-life of 76 days, when in sludge was applied to a sandy loam soil in a biological soil reactor and worked into the top 20 cm of soil (Kincannon & Yin 1985; quoted, Howard 1989).

Biota: depuration half-life from bluegills about 7 days (Lu et al. 1977; quoted, Howard 1989).

Environmental Fate Rate Constants or Half-Lives:

Volatilization:

Photolysis:

Hydrolysis:

Oxidation: aqueous oxidation rate constants for singlet oxygen of < $4\times10^{7}M^{-1}h^{-1}$ and for peroxy radical of $1.1\times10^{8}M^{-1}h^{-1}$ at 25°C (Mabey et al. 1982); photooxidation half-life of 0.312-3.12 hours, based on estimated rate constant for the reaction with hydroxyl radicals in air (Atkinson 1987; quoted, Howard et al. 1991).

Biodegradation: aqueous aerobic half-life of 48-192 hours, based on aerobic soil die-away test data (Lu et al. 1977; quoted, Howard et al. 1991); overall biodegradation half-life of 76 days, when in sludge was applied to a sandy loam soil in a biological soil reactor and worked into the top 20 cm of soil (Kincannon & Yin 1985; quoted, Howard 1989); aqueous aerobic half-life of 192-768 hours, based on estimated unacclimated aqueous aerobic biodegradation half-life (Howard et al. 1991).

Biotransformation: rate constant for bacterial transformation of 1×10^{-10} mL·cell·h^{-1} in water (Mabey et al. 1982).

Bioconcentration, Uptake (k_1) and Elimination (k_2) Rate Constants:

Common Name: 3,3'-Dichlorobenzidine
Synonym: 3,3'-dichloro-4,4'-diamino(1,1'-biphenyl), DCB
Chemical Name: 3,3'-dichlorobenzidine
CAS Registry No: 91-94-1
Molecular Formula: $NH_2C_6H_3(Cl)C_6H_3(Cl)NH_2$
Molecular Weight: 253.12
Melting Point (°C):
132 (Windholz 1976; Callahan et al. 1979; Banerjee et al. 1980)
Boiling Point (°C):
Density (g/cm³ at 20°C):
Molar Volume (cm³/mol): 254.8 (calculated-LeBas method)
Molecular Volume (Å³):
Total Surface Area, TSA (Å²):
Heat of Fusion ΔH_{fus}, kcal/mol:
Dissociation Constants, pK_b: 11.7 (Kollig 1993)
Entropy of Fusion ΔS_{fus}, cal/mol K (e.u.):
Fugacity Ratio at 25°C (assuming ΔS_{fus} = 13.5 e.u.), F: 0.08744

Water Solubility (g/m³ or mg/L at 25°C):
4.00 (22°C, as dihydrochloride, Banerjee et al. 1978; quoted, Callahan et al. 1979; Mabey et al. 1982)
3.99 (22°C, at pH 6.9 as DCB.2HCl, quoted, Verschueren 1983)
3.11 (shake flask-UV/LSC, Banerjee et al. 1980; quoted, Banerjee 1985)

Vapor Pressure (Pa at 25°C):
0.00133 (estimated, Mabey et al. 19812)

Henry's Law Constant (Pa·m³/mol):
0.0811 (calculated-P/C, Mabey et al. 1982)

Octanol/Water Partition Coefficient, log K_{OW}:
3.02 (calculated as per Leo et al.1971, Callahan et al. 1979)
3.51 (quoted, Mabey et al. 1982)
3.51 (quoted exptl., Banerjee & Howard 1988)
3.35 (calculated- γ from UNIFAC, Banerjee & Howard 1988)
3.64 (calculated-QSAR, Kollig 1993)

Bioconcentration Factor, log BCF:
2.70 (bluegill sunfish, Appleton & Sikka 1980; quoted, Verschueren 1983)
2.97 (microorganisms-water, calculated-K_{OW}, Mabey et al. 1982)
2.79, 2.97, 3.49 (fish, algae, activated sludge, Freitag et al. 1985)

Sorption Partition Coefficient, log K_{OC}:
33.19 (sediment-water, calculated-K_{OW}, Mabey et al. 1982)
4.35, 3.87 (soil: quoted, calculated- χ and fragment contribution, Meylan et al. 1992)
3.30 (calculated-K_{OW}, Kollig 1993)

Half-Lives in the Environment:

Air: 0.905-9.05 hours, based on estimated rate constant for the reaction with hydroxyl radicals in air (Atkinson 1987; quoted, Howard et al. 1991); estimated half-life for the reaction with hydroxyl radicals and ozone was about one day (Howard 1989); atmospheric transformation lifetime by photolysis was estimated to be <1 day (Kelly et al 1994).

Surface water: 0.025-0.075 hours, based on direct photolysis in distilled water in midday summer sunlight (Banerjee et al. 1978; Sikka et al. 1978; quoted, Howard et al. 1991) and approximate winter sunlight direct photolysis half-life (Banerjee et al. 1978; Sikka et al. 1978; Lyman et al. 1982; quoted, Howard et al. 1991).

Ground water: 1344-8640 hours, based on estimated unacclimated aqueous aerobic biodegradation half-life (Howard et al. 1991).

Sediment: half-life of 30 minutes by suspended microcrystalline clays may be considered the most important fate process in the aquatic environment (Callahan et al. 1979).

Soil: 672-4320 hours, based on estimated unacclimated aqueous aerobic biodegradation half-life (Howard et al. 1991).

Biota:

Environmental Fate Rate Constants or Half-Lives:

Volatilization:

Photolysis: direct aqueous photolysis rate constant of 2.1×10^{-6} h^{-1} in summer at 40° latitude (Mabey 1982); both aqueous and atmospheric photolysis half-life of 0.025-0.075 hours, based on direct photolysis in distilled water in midday summer sunlight (Banerjee et al. 1978; Sikka et al. 1978; quoted, Callahan et al. 1979; Howard et al. 1991) and approximate winter sunlight direct photolysis half-life (Banerjee et al. 1978; Sikka et al. 1978; Lyman et al. 1982; quoted, Howard et al. 1991).

Hydrolysis:

Oxidation: aqueous oxidation rate constants for singlet oxygen of $< 4 \times 10^7$ $M^{-1}h^{-1}$ and for peroxy radical of $< 4 \times 10^7$ $M^{-1}h^{-1}$ at 25°C (Mabey et al. 1982); photooxidation half-life of 31.2-1740 hours in water, based on estimated rate constants for reactions with OH and RO_2 radicals in water (Mill & Mabey 1985; quoted, Howard et al. 1991); photooxidation half-life of 0.905-9.05 hours in air, based on estimated rate constant for the reaction with hydroxyl radicals in air (Atkinson 1987; quoted, Howard et al. 1991)

Biodegradation: aqueous aerobic half-life of 672-4320 hours, based on lake die-away study test data (Appleton et al. 1978; quoted, Howard et al. 1991) and a soil die-away test (Boyd et al. 1984; quoted, Howard et al. 1991); aqueous anaerobic half-life of 2688-17280 hours, based on estimated unacclimated aqueous aerobic biodegradation half-life (Howard et al. 1991).

Biotransformation: rate constant for bacterial transformation of 3×10^{-12} $mL \cdot cell^{-1} \cdot h^{-1}$ in water (Mabey et al. 1982).

Bioconcentration, Uptake (k_1) and Elimination (k_2) Rate Constants:

Common Name: N,N'-Bianiline
Synonym: 1,2-diphenylhydrazine, hydrazobenzene
Chemical Name: 1,2-diphenylhydrazine, hydrazobenzene
CAS Registry No: 122-66-7
Molecular Formula: $C_6H_5NHNHC_6H_5$
Molecular Weight: 184.24
Melting Point (°C):
    131 (Weast 1977; Callahan et al. 1979; Mabey et al. 1982)
Boiling Point (°C):
    293 (as azobenzene, IARC 1975)
Density (g/cm^3 at 20°C):
Molar Volume (cm^3/mol):
    213.0 (calculated-LeBas method)
Molecular Volume (Å^3):
Total Surface Area, TSA (Å^2):
Dissociation constant pK_b: 13.2 (Kollig 1993)
Heat of Fusion ΔH_{fus}, kcal/mol:
Entropy of Fusion ΔS_{fus}, cal/mol·K (e.u.):
Fugacity Ratio at 25°C (assuming ΔS = 13.5 e.u.), F: 0.08945

Water Solubility (g/m^3 or mg/L at 25°C):
    0.252 (20°C, as azobenzene, Takagishi et al. 1968; quoted, Callahan et al. 1979)
    0.00184 (calculated-K_{OW}, Mabey et al. 1982)

Vapor Pressure (Pa at 25°C):
    0.00347 (quoted, Mabey et al. 1982)

Henry's Law Constant (Pa·m^3/mol):
    3.45×10^{-4} (calculated-P/C, Mabey et al. 1982)

Octanol/Water Partition Coefficient, log K_{OW}:
    3.82 (shake flask-AS as for azobenzene, Fujita et al. 1964; quoted, Callahan et al. 1979)
    3.03 (calculated as per Leo et al. 1971, Callahan et al. 1979)
    2.94 (quoted, Mabey et al. 1982)
    1.70 (calculated-QSAR, Kollig 1993)

Bioconcentration Factor, log BCF:
    2.46 (microorganisms-water, calculated-K_{OW}, Mabey et al. 1982)

Sorption Partition Coefficient, log K_{OC}:

2.62 (sediment-water, calculated-K_{OW}, Mabey et al. 1982)

1.40 (calculated-K_{OW}, Kollig 1993)

Half-Lives in the Environment:

Air: photooxidation half-life of 0.3-3.0 hours, based on estimated rate data for the reaction with hydroxyl radicals in air (Atkinson 1987; quoted, Howard et al. 1991); atmospheric transformation lifetime was estimated to be < 1 day (Kelly et al. 1994).

Surface water: photooxidation half-life of 31-1740 hours, based on photooxidation rate constants with OH and RO_2 radicals for the amine class (Güesten et al. 1981; Mill & Mabey 1985; quoted, Howard et al. 1991).

Groundwater: 1344-8640 hours, based on estimated unacclimated aqueous aerobic biodegradation half-life (Howard et al. 1991).

Sediment:

Soil: 672-4320 hours, based on estimated aqueous aerobic biodegradation half-life (Howard et al. 1991).

Biota:

Environmental Fate Rate Constants or Half-Lives:

Volatilization:

Photolysis:

Hydrolysis:

Oxidation: aqueous oxidation rate constants for singlet oxygen of $<4 \times 10^7$ $M^{-1}h^{-1}$ and for peroxy radical of $<1 \times 10^9$ $M^{-1}h^{-1}$ at 25°C (Mabey et al. 1982); photooxidation half-life of 31-1740 hours, based on photooxidation rate constants with OH and RO_2 radicals for the amine class (Güesten et al. 1981; Mill & Mabey 1985; quoted, Howard et al. 1991); photooxidation half-life of 0.3-3.0 hours in air, based on estimated rate data for the reaction with hydroxyl radicals in air (Atkinson 1987; quoted, Howard et al. 1991).

Biodegradation: aqueous aerobic half-life of 672-4320 hours, based on acclimated aerobic aqueous screening test data (Malaney 1960; quoted, Howard et al. 1991); aqueous anaerobic half-life of 2880-17280 hours, based on estimated unacclimated aqueous aerobic biodegradation half-life (Howard et al. 1991).

Biotransformation: rate constant for bacterial transformation of 1×10^{-10} $mL \cdot cell^{-1} \cdot h^{-1}$ in water (Mabey et al. 1982).

Bioconcentration, Uptake (k_1) and Elimination (k_2) Rate Constants:

Common Name: 1-Naphthylamine
Synonym: 1-naphthalenamine, α-naphthylamine, 1-NA, 1-aminonaphthalene, naphthalidine
Chemical Name: 1-naphthalenamine
CAS Registry No: 134-32-7
Molecular Formula: $C_{10}H_7NH_2$
Molecular Weight: 143.18
Melting Point (°C):
50 (Stull 1947; Verschueren 1983; Dean 1985)
48.9 (Tsonopoulos & Prausnitz 1971)
Boiling Point (°C):
300.8 (Stull 1947)
301 (Verschueren 1983; Dean 1985)
Density (g/cm³ at 20°C):
1.131 (Verschueren 1983)
1.123 (Dean 1985)
Molar Volume (cm³/mol):
161.8 (calculated-LeBas method)
Molecular Volume (Å³):
Total Surface Area, TSA (Å²): 284.3 (Iwase et al. 1985)
Dissociation Constant pK_a:
3.92 (Sangster 1993)
Heat of Fusion, ΔH_{fus}, kcal/mol:
3.40 $\pm$ 0.25 (Tsonopoulos & Prausnitz 1971)
Entropy of Fusion, ΔS_{fus}, cal/mol·K (e.u.):
10.6 $\pm$ 0.80 (Tsonopoulos & Prausnitz 1971)
Fugacity Ratio at 25°C (assuming ΔS_{fus} = 13.5 e.u.), F: 0.5659

Water Solubility (g/m³ or mg/L at 25°C):
1700 (quoted, Verschueren 1983)
590 parts in water (Budavari 1989)

Vapor Pressure (Pa at 25°C):
0.803 (extrapolated-regression of tabulated data, Stull 1947)
0.557 (extrapolated-Cox eqn., Chao et al. 1983)
0.446 (extrapolated, liquid, Antoine eqn., Stephenson & Malanowski 1987)

Henry's Law Constant (Pa·m³/mol):

Octanol/Water Partition Coefficient, log K_{OW}:
2.23 (Leo et al. 1969)
2.24 (Hansch & Leo 1979; quoted, Schultz & Moulton 1985)
2.22 (calculated, Verschueren 1983)
2.27 (shake flask-AS at pH 7.5, Martin-Villodre et al. 1986; quoted, Sangster 1993)
2.23 (quoted, Yoshida et al. 1986)
2.33 (HPLC-k', Minick et al. 1988; quoted, Sangster 1993)
2.22, 2.25 (quoted CLOGP, recommended, Sangster 1993)

Bioconcentration Factor, log BCF:

Sorption Partition Coefficient, log K_{OC}:

3.58, 3.43, 3.50 (Milford soil, Morocco soil, Oakville soil, Graveel et al. 1986)

2.63, 2.68, 3.15 (sediment, Alfisol soil, Podzol soil, von Oepen et al. 1991)

3.51 (soil, quoted exptl., Meylan et al. 1992)

3.48 (soil, calculated- χ , Meylan et al. 1992)

Half-Lives in the Environment:

Air: 0.292-2.92 hours, based on estimated photooxidation half-life in air (Howard et al. 1991).

Surface water: 0.62.4-3840 hours, based on estimated rate constants for reactions of aromatic amines with OH and RO_2 radicals in aqeous solutions (Howard et al. 1991).

Groundwater: 1344-8640 hours, based on slow biodegradation observed in an aerobic soil die-away test study and aerobic activated sludge screening tests (Howard et al. 1991).

Sediment:

Soil: 672-4320 hours, based on slow biodegradation observed in an aerobic soil die-away test study and aerobic activated sludge screening tests (Howard et al. 1991).

Biota:

Environmental Fate Rate Constants or Half-Lives:

Volatilization:

Photolysis:

Oxidation: rate constant of 1×10^{4} M^{-1} sec^{-1} for oxidation by RO_2 radicals at 30°C in aquatic systems with half-life of 0.8 day (Howard 1972; Hendry et al. 1974; quoted, Mill 1982); rate constant of $<2\times10^{2}$ M^{-1} sec^{-1} for oxidation by singlet oxygen at 25°C in aquatic systems with half-life of >100 years (Foote 1976; Mill 1979; quoted, Mill 1982); atmospheric half-life of 0.292-2.92 hours, based on estimated rate constants for the reaction with OH radicals in air and aqueous photooxidation half-life of 62.4-3480 hours, based on estimated rate constants for reaction of representative aromatic amines with OH and RO_2 radicals in aqueous solution (Howard et al. 1991); photooxidation half-lives between 0.08-0.15 hour under sunlight and 0.2-10.5 hours under UV light when adsorbed on silica and alumina on the TLC plates under simulated atmospheric conditions (Hasegawa et al. 1993).

Hydrolysis:

Biodegradation: aqueous aerobic biodegradation half-life of 672-4320 hours and aqueous anaerobic biodegradation half-life of 2688-17280 hours, based on slow biodegradation observed in aerobic soil die-away test study and aerobic activated sludge screening tests (Howard et al. 1991).

Biotransformation:

Bioconcentration, Uptake (k_1) and Elimination (k_2) Rate Constants:

Common Name: 2-Naphthylamine
Synonym: 2-naphthalenamine, β-naphthylamine, 2-NA, 2-aminonaphthalene, naphthalidine
Chemical Name: 2-naphthalenamine
CAS Registry No: 91-59-8
Molecular Formula: $C_{10}H_7NH_2$
Molecular Weight: 143.18
Melting Point (°C):
111.5 (Stull 1947)
109.9 (Tsonopoulos & Prausnitz 1971)
110.2 (Verschueren 1983)
111-113 (Dean 1985)
112.0 (Banerjee et al. 1990)
Boiling Point (°C):
306.1 (Stull 1947; Verschueren 1983; Dean 1985)
306.0 (Banerjee et al. 1990)
Density (g/cm^3):
1.061 (at 98°C, Verschueren 1983; Dean 1985)
Molar Volume (cm^3/mol):
161.8 (calculated-LeBas method)
Molecular Volume ($Å^3$):
Total Surface Area, TSA ($Å^2$):
Dissociation Constant pK_a: 4.15 (Sangster 1993)
Heat of Fusion, ΔH_{fus}, kcal/mol: 5.25 (Tsonopoulos & Prausnitz 1971)
Entropy of Fusion, ΔS_{fus}, cal/mol·K (e.u.): 13.7 (Tsonopoulos & Prausnitz 1971)
Fugacity Ratio at 25°C (assuming ΔS_{fus} = 13.5 e.u.), F: 0.141

Water Solubility (g/m^3 or mg/L at 25°C):
6.40 (18°C, Ciusa 1922; quoted, Tsonopoulos & Prausnitz 1971)
0.19 (18-20°C, Neish 1948; quoted, Tsonopoulos & Prausnitz 1971)

Vapor Pressure (Pa at 25°C):
0.634 (extrapolated-regression of tabulated data, Stull 1947)
0.369 (extrapolated-Cox eqn., Chao et al. 1983)
0.0347 (Howard et al. 1986; quoted, Banerjee et al. 1990)
0.035 (extrapolated, solid, Antoine eqn., Stephenson & Malanowski 1987)
0.362 (extrapolated, liquid, Antoine eqn., Stephenson & Malanowski 1987)
0.340 (calculated-solvatochromic p., Banerjee et al. 1990)

Henry's Law Constant ($Pa \cdot m^3/mol$):

Octanol/Water Partition Coefficient, log K_{OW}:
2.26 (shake flask, Korenman 1982)
2.25 (calculated, Verschueren 1983)
2.40 (calculated- Δ , Campbell & Luthy 1985)
2.40 (shake flask-AS, Martin-Villodre et al. 1986)
2.20, 2.34 (quoted CLOGP, recommended, Sangster 1993)

2.09 (calculated-QSAR, Kollig 1993)

Bioconcentration Factor, log BCF:

Sorption Partition Coefficient, log K_{OC}:
1.77 (calculated-K_{OW}, Kollig 1993)

Half-Lives in the Environment:

Air: photooxidation half-life of 0.30-2.90 hours, based on estimated rate constant for the vapor-phase reaction with hydroxyl radicals in air (Atkinson 1987; quoted, Howard et al. 1991).

Surface water: 62.0-3840 hours, based on estimated rate constants for reactions of aromatic amines with OH and RO_2 radicals in aqeous solutions (Güesten et al. 1981; Mill & Mabey 1985; quoted, Howard et al. 1991).

Groundwater: 1344-8640 hours, based on estimated unacclimated aqueous aerobic biodegradation half-life (Howard et al. 1991).

Sediment:

Soil: 672-4320 hours, based on unacclimated aerobic screening test data (Fochtman & Eisenberg 1979; quoted, Howard et al. 1991) and unacclimated aerobic soil grab sample data (Medvedev & Davidov 1981; quoted, Howard et al. 1981).

Biota:

Environmental Fate Rate Constants or Half-Lives:

Volatilization:

Photolysis:

Oxidation: rate constant of $1x10^4$ M^{-1} sec^{-1} for oxidation by RO_2 radicals at 30°C in aquatic systems with half-life of 0.8 day (Howard 1972; Hendry et al. 1974; quoted, Mill 1982); rate constant of $<2x10^2$ M^{-1} sec^{-1} for oxidation by singlet oxygen at 25°C in aquatic systems with half-life of >100 years (Foote 1976; Mill 1979; quoted, Mill 1982); photooxidation half-life of 0.30-2.90 hours in air, based on estimated rate constant for the vapor-phase reaction with hydroxyl radicals in air (Atkinson 1987; quoted, Howard et al. 1991); aqueous photooxidation half-life of 62-3480 hours, based on estimated rate constants for reaction of representative aromatic amines with OH and RO_2 radicals in aqueous solution (Güesten et al. 1981; Mill & Mabey 1985; quoted, Howard et al. 1991); photooxidation half-life between 0.05-0.19 hours under sunlight and 0.20-10.8 hours under UV light when adsorbed on silica and alumina TLC plates under simulated atmospheric conditions (Hasegawa et al. 1993).

Hydrolysis:

Biodegradation: aqueous aerobic biodegradation half-life of 672-4320 hours, based on unacclimated aerobic screening test data (Fochtman & Eisenberg 1979; quoted, Howard et al. 1979) and unacclimated soil grab sample data (Medvedev & Davidov 1981; quoted, Howard et al. 1991); aqueous anaerobic biodegradation half-life of 2880-17280 hours, based on estimated unacclimated aqueous aerobic bidegradation half-life (Howard et al. 1991).

Biotransformation:

Bioconcentration, Uptake (k_1) and Elimination (k_2) Rate Constants:

Common Name: 2-Nitroaniline
Synonym: 1-amino-2-nitrobenzene, *o*-aminonitrobenzene, *o*-nitroaniline, 2-nitrophenylamine
Chemical Name: 1-amino-2-nitrobenzene, *o*-nitroaniline, 2-nitroaniline
CAS Registry No: 88-74-4
Molecular Formula: $C_6H_4NH_2NO_2$
Molecular Weight: 138.12
Melting Point (°C):

71.5 (Stull 1947; Tsonopoulos & Prausnitz 1971; Weast 1982-83; Verschueren 1983)
69-70 (Dean 1985)

Boiling Point (°C):

284.5 (Stull 1947)
284.1 (Weast 1982-83; Verschueren 1983; Dean 1985)

Density (g/cm^3 at 20°C):

1.4420 (Weast 1982-83; Verschueren 1983)
1.442 (15°C, Dean 1985)

Molar Volume (cm^3/mol):

95.92 (calculated-density)
138.7 (calculated-LeBas method)

Molecular Volume (Å^3):

34.7 (Jäckel & Klein 1991)

Total Surface Area, TSA (Å^2):
Dissociation Constant, pK_a:
Heat of Fusion, ΔH_{fus}, kcal/mol:

3.85 (Tsonopoulos & Prausnitz 1971)

Entropy of Fusion, ΔS_{fus}, cal/mol K (e.u.):

11.2 (Tsonopoulos & Prausnitz 1971)

Fugacity Ratio at 25°C (assuming ΔS = 13.5 e.u.), F: 0.3589

Water Solubility (g/m^3 or mg/L at 25°C):

1212 (stirring-TN, Collet & Johnson 1926; quoted, Palit 1947; Tsonopoulos & Prausnitz 1971)
1203, 1740 (quoted, calculated-K_{OW}, Yalkowsky & Morozowich 1980)
1260 (quoted, Verschueren 1983)

Vapor Pressure (Pa at 25°C):

0.620 (extrapolated-regression of tabulated data, Stull 1947)
0.072 (Knudsen's method, calculated-Antoine eqn., Hoyer & Peperle 1957)
<13.3 (30°C, quoted, Verschueren 1983)
0.650 (extrapolated-Antoine eqn., Dean 1985)

Henry's Law Constant (Pa·m^3/mol):

Octanol/Water Partition Coefficient, log K_{OW}:

1.44 (shake flask-AS, Fujita et al. 1964; quoted, Sangster 1993)
1.83 (shake flask-AS, Hansch & Anderson 1967; quoted, Sangster 1993)
1.62 (correlated-HPLC-RT, Carlson et al. 1975; quoted, Sangster 1993)
1.81 (Hansch & Leo 1979; quoted, Deneer et al. 1987)
1.72 (shake flask, Eadsforth & Moser 1983; quoted, Sangster 1993)
1.44, 1.83 (quoted, Verschueren 1983)
1.65 (selected, Schultz & Applehans 1985)
1.67 (calculated-HPLC-k', Deneer et al. 1987; quoted, Sangster 1987)
1.502 (calculated-linear extrapolation exptl. log k at various solvent compositions, Deneer et al. 1987)
1.73 (shake flask-AS, Kramer & Henze 1990; quoted, Sangster 1993)
1.78 (calculated-CLOGP, Jäckel & Klein 1991)
1.85 (recommended, Sangster 1993)
1.78 (quoted, Bintein et al. 1993)

Bioconcentration Factor, log BCF:

0.91; 1.49, 0.55, 0.83 (quoted exptl.; calculated values-K_{OW}, Bintein et al. 1993)

Sorption Partition Coefficient, log K_{OC}:

Half-Lives in the Environment:

Air:
Surface water:
Groundwater:
Sediment:
Soil:
Biota:

Environmental Fate Rate Constants or Half-Lives:

Volatilization:
Photolysis:
Oxidation:
Hydrolysis:
Biodegradation:
Biotransformation:
Bioconcentration, Uptake (k_1) and Elimination (k_2) Rate Constants:

Common Name: 4-Nitroaniline
Synonym: 1-amino-4-nitrobenzene, *p*-aminonitrobenzene, *p*-nitroaniline, 4-nitrobenzenamine, 4-nitrophenylamine
Chemical Name: 1-amino-4-nitrobenzene, *p*-nitroaniline, 4-nitroaniline
CAS Registry No: 100-01-6
Molecular Formula: $C_6H_4NH_2NO_2$
Molecular Weight: 138.12
Melting Point (°C):
146.5 (Stull 1947)
148-149 (Weast 1982-83)
147.5 (Verschueren 1983)
146.0 (Dean 1985)
Boiling Point (°C):
336.0 (Stull 1947)
331.7 (Weast 1982-83)
260/332 (Verschueren 1983)
Density (g/cm^3 at 20°C):
1.4240 (Weast 1982-83; Verschueren 1983)
1.437 (14°C, Dean 1985)
Molar Volume (cm^3/mol):
138.7 (calculated-LeBas method)
Molecular Volume (Å^3):
34.7 (Jäckel & Klein 1991)
Total Surface Area, TSA (Å^2):
Dissociation Constant, pK_a:
Heat of Fusion, ΔH_{fus}, kcal/mol:
5.04 (Tsonopoulos & Prausnitz 1971)
Entropy of Fusion, ΔS_{fus}, cal/mol·K (e.u.):
12.0 (Tsonopoulos & Prausnitz 1971)
Fugacity Ratio at 25°C (assuming ΔS = 13.5 e.u.), F: 0.06

Water Solubility (g/m^3 or mg/L at 25°C):
568 (stirring-TN, Collet & Johnson 1926; quoted, Palit 1947; Tsonopoulos & Prausnitz 1971)
576, 603 (quoted, calculated-K_{OW}, Yalkowsky & Morozowich 1980)
380 (20°C, shake flask-membrane filter-fluorophotometric, Hashimoto et al. 1982)
390, 400 (20°C, shake flask-glass fibre filters-fluorophotometric, Hashimoto et al. 1982)
800 (19°C, quoted, Verschueren 1983)
22000 (100°C, quoted, Verschueren 1983)
800 (Dean 1985)

Vapor Pressure (Pa at 25°C):
0.035 (extrapolated-regression of tabulated data, Stull 1947)
0.200 (20°C, quoted, Verschueren 1983)
0.933 (30°C, quoted, Verschueren 1983)
0.014 (extrapolated-Antoine eqn., Dean 1985)

Henry's Law Constant (Pa·m³/mol):

Octanol/Water Partition Coefficient, log K_{OW}:

1.39 (shake flask-AS, Fujita et al. 1964; quoted, Sangster 1993)
1.39 (Hansch & Leo 1979; quoted, Schultz & Moulton 1985; Deneer et al. 1987)
0.59 (calculated- γ from UNIFAC, Campbell & Luthy 1985; quoted, Sangster 1993)
1.35 (selected, Schultz & Applehans 1985)
0.48 (selected, Yoshida et al. 1986)
1.16 (calculated-HPLC-k', Deneer et al. 1987; quoted, Sangster 1993)
1.10 (calculated-linear extrapolation exptl. values of log k at various solvent compositions, Deneer et al. 1987)
1.35 (quoted average values, Ge et al. 1987; quoted, Sangster 1993)
1.15 (correlated-HPLC-RT, Wang et al. 1989; quoted, Sangster 1993)
1.51 (shake flask-AS, Kramer & Henze 1990; quoted, Sangster 1993)
1.30 (CPC, Tsai et al. 1991; quoted, Sangster 1993)
1.31 (calculated-CLOGP, Jäckel & Klein 1991)
1.39, 1.30 (quoted, CPC-RV, El Tayar et al. 1991)
1.35 (recommended, Sangster 1993)
1.31 (quoted, Bintein et al. 19930

Bioconcentration Factor, log BCF:

Sorption Partition Coefficient, log K_{OC}:

2.26 (Alfisol soil, von Oepen et al. 1991)
2.66 (Podzol soil, von Oepen et al. 1991)
2.12 (sediment, von Oepen et al. 1991)
0.64; 1.25, -0.18, 0.41 (quoted exptl.; calculated values-K_{OW}, Bintein et al. 1993)

Half-Lives in the Environment:

Air:
Surface water:
Groundwater:
Sediment:
Soil:
Biota:

Environmental Fate Rate Constants or Half-Lives:

Volatilization:
Photolysis:
Hydrolysis:
Oxidation:
Biodegradation:
Biotransformation:
Bioconcentration, Uptake (k_1) and Elimination (k_2) Rate Constants:

Common Name: Nitrobenzene
Synonym: nitrobenzol, oil of mirbane
Chemical Name: nitrobenzene
CAS Registry No: 98-95-3
Molecular Formula: $C_6H_5NO_2$
Molecular Weight: 123.11
Melting Point (°C):
5.7 (Stull 1947; Dreisbach 1955; Howard 1989)
5.8 (Tsonopoulos & Prausnitz 1971)
5.6 (Windholz 1976; Callahan et al. 1979)
Boiling Point (°C):
210.6 (Stull 1947; Boublik et al. 1973)
210.85 (Dreisbach 1955)
211 (Windholz 1976; Callahan et al. 1979; Verschueren 1983)
Density (g/cm³ at 20°C):
1.2032 (Dreisbach 1955)
1.200 (25°C, Verschueren 1983)
Molar Volume (cm³/mol):
102 (calculated from density, Rohrschneider 1973; Chiou 1985)
112.9 (Taft et al. 1985; Leahy 1986; Kamlet et al. 1987)
63.1 (intrinsic volume V_I, Leahy 1986; Kamlet et al. 1987; Leggett & Parker 1994)
112 (calculated-LeBas method, Abernethy et al. 1988)
63.0 (intrinsic volume V_I, Abernethy et al. 1988)
Molecular Volume (Å³):
Total Surface Area, TSA (Å²):
255.6 (Iwase et al. 1985)
Dissociation Constant pK_a:
Heat of Fusion ΔH_{fus}, kcal/mol:
2.770 (Dreisbach 1955)
2.90 (Tsonopoulos & Prausnitz 1971)
Entropy of Fusion ΔS_{fus}, cal/mol·K (e.u.):
10.40 (Tsonopoulos & Prausnitz 1971)
Fugacity Ratio at 25°C (assuming ΔS_{fus} = 13.5 e.u.), F: 1.0

Water Solubility (g/m³ or mg/L at 25°C):
1204 (shake flask-centrifuge, Booth & Everson 1948)
1930 (Seidell 1941; quoted, Andrews & Keefer 1950)
2018 (shake flask-interferometry, Donahue & Bartell 1952)
1950 (quoted, Deno & Berkheimer 1960)
2060 (Hansch et al. 1968; quoted, Hine & Mookerjee 1975; Müller & Klein 1992)
4946 (calculated-K_{OW}, Hansch et al. 1968)
1900 (20°C, Verschueren 1977, 1983; quoted, Callahan et al. 1979; Mabey et al. 1982)
2093 (shake flask-LSC, Banerjee et al. 1980; quoted, Banerjee 1985)
2090 (shake flask-radioactive analysis, Veith et al. 1980)
4080 (calculated-K_{OW} as per Hansch et al. 1968, Veith et al. 1980)
1050 (quoted average observed value, Valvani et al. 1981)

5130 (predicted-K_{OW}, Valvani et al. 1981)
2043 (20-27°C, shake flask-GC, Chiou 1985)
1048, 3092 (quoted, calculated-solvatochromic p., Leahy 1986)
1900 (20°C, Riddick et al. 1986; quoted, Howard 1989)
2043, 2820 (quoted, calculated-fragment const., Wakita et al. 1986)
1950, 7590 (quoted, calculated-solvatochromic p., Kamlet et al. 1987)
1950, 1997 (quoted, calculated-V_I and M.P., Kamlet et al. 1987)
1905 (quoted, Isnard & Lambert 1988)

Vapor Pressure (Pa at 25°C):
42.06 (extrapolated-regression of tabulated data, Stull 1947)
37.86 (Dreisbach 1955; quoted, Hine & Mookerjee 1975)
34.48 (extrapolated-Antoine eqn., Boublik et al. 1973,84)
20.0 (20°C, Verschueren 1977, 1983; quoted, Callahan et al. 1979; Mabey et al. 1982; Howard 1989)
20.26, 28.37 (quoted, calculated-B.P., Mackay et al. 1982)
34.63 (extrapolated-Antoine eqn., Dean 1985)
33.33 (Howard et al. 1986; quoted, Banerjee et al. 1990)
33.33, 239.9 (quoted, calculated-solvatochromic p., Banerjee et al. 1990)

Henry's Law Constant (Pa·m³/mol):
2.367 (exptl., Hine & Mookerjee 1975)
4.511 (calculated-group contribution, Hine & Mookerjee 1975)
4.723 (calculated-bond contribution, Hine & Mookerjee 1975)
1.327 (calculated-P/C, Mabey et al. 1982)
2.472 (estimated, Lyman et al. 1982; quoted, Howard 1989)
2.367, 5.06 (quoted, calculated-molecular structure, Russell et al. 1992)

Octanol/Water Partition Coefficient, log K_{OW}:
1.85 (shake flask-AS, Fujita et al. 1964; Hansch et al. 1968; quoted, Leo et al. 1969, 1971; McCall 1975; Hansch & Leo 1979; Miyake & Terada 1982; Kaiser 1983; Miyake et al. 1987; Abernethy et al. 1988; Sangster 1989; Schultz et al. 1989)
1.85 (Leo et al. 1971; quoted, Chiou 1985; Iwase et al. 1985; Isnard & Lambert 1988; Haderlein & Schwarzenbach 1993)
1.74 (Neely 1974; quoted, Callahan et al. 1979)
1.86, 1.82 (quoted, estimated-HPLC-RT, Veith et al. 1979)
2.93 (quoted unpublished data, Veith et al. 1979)
1.83 (shake flask-LSC, Banerjee et al. 1980; quoted, Arbuckle 1983; Brooke et al. 1990; Kollig 1993)
1.83, 1.87 (measured, calculated, Veith et al. 1980)
1.85; 1.98, 1.78 (quoted, correlated-k', McDuffie 1981)
1.85 (gen. col.-HPLC, Wasik et al. 1981; Tewari et al. 1982; quoted, Sangster 1989; Brooke et al. 1990)
1.88 (shake flask-AS, Unger & Chiang 1981)
1.87 (quoted exptl., Valvani et al. 1981)
1.87 (calculated-f const., Mabey et al. 1982)
1.99 (RP-HPLC-k', Miyake & Terada 1982)

1.83 (calculated- γ from UNIFAC octanol and water solubility considered, Arbuckle 1983; quoted, Sangster 1989)
1.84 (calculated- γ from UNIFAC octanol and water solubility not considered, Arbuckle 1983; quoted, Sangster 1989)
1.85, 1.88 (quoted, Verschueren 1983)
1.83 (selected best lit. value, Garst & Wilson 1984)
1.83 $\pm$ 0.02 (exptl.-ALPM, Garst & Wilson 1984; quoted, Brooke et al. 1990)
1.84 (quoted, Geyer et al. 1984)
1.85, 1.84 (quoted, calculated- γ from UNIFAC, Campbell & Luthy 1985)
1.85 (Hansch & Leo 1985; quoted, Banerjee & Howard 1988; Howard 1989)
1.91 (calculated-TSA, Iwase et al. 1985)
1.95 (computed, Klopman et al. 1985; quoted, Pangrekar et al. 1994)
1.84 (quoted from Hansch & Leo 1979, Schultz & Moulton 1985; Bodor et al. 1989)
1.85, 1.90 (quoted, calculated-solvatochromic p., Taft et al. 1985)
1.85, 2.01 (quoted, calculated-solvatochromic p., Leahy 1986)
1.87 (Lu et al. 1986; quoted average, Walton et al. 1992)
1.85 (RP-HPLC-capacity ratio, Minick et al. 1988)
2.93 (quoted from Mackay 1982, Schüürmann & Klein 1988)
1.85 (THOR 1986; quoted, Schüürmann & Klein 1988)
1.89 (calculated-CLOGP 3.42 as per Leo 1986, Schüürmann & Klein 1988)
2.00 (selected, Yoshida et al. 1986)
1.87, 1.89 (quoted, calcd.-HPLC-k', Deneer et al. 1987; quoted, Verhaar et al. 1992)
1.85, 1.84 (quoted, calculated- γ from UNIFAC, Banerjee & Howard 1988)
1.85; 1.885, 1.60 (quoted; calculated-CLOGP, M.O., Bodor et al. 1989)
1.85, 1.88, 1.70, 1.85 (quoted values, De Bruijn et al. 1989)
1.828 $\pm$ 0.001 (slow-stirring, De Bruijn et al. 1989)
1.85 (recommended, Sangster 1989)
1.836 $\pm$ 0.051 (average, stir-flask method by BRE, Brooke et al. 1990)
1.828 $\pm$ 0.001 (average, stir-flask method by RITOX, Brooke et al. 1990)
1.79 (quoted calculated value from Rekker 1977, Brooke et al. 1990)
1.87 (quoted calculated value from Veith et al. 1979, Brooke et al. 1990)
1.70, 1.82 (quoted HPLC values from Veith et al. 1979 and 1980, Brooke et al. 1990)
1.85 (quoted from CLOGP 3.52, Gluck & Martin 1990)
1.83, 1.85, 1.88 (CPC-RV, Gluck & Martin 1990)
1.88 (calculated-CLOGP, Müller & Klein 1992)
1.87 (quoted from Walton et al. 1992, Bintein & Devillers 1994)

Bioconcentration Factor, log BCF:
1.18 (fathead minnows, Veith et al. 1979; quoted, Howard 1989)
0.06 (calculated-K_{OW}, Veith et al. 1980)
$<$1.0 (golden orfe, Freitag et al. 1982; quoted, Howard 1989)
1.38 (green algae, Freitag et al. 1982; quoted, Howard 1989)
1.38 (alga *chlorella fusca*, wet wt. basis, Geyer et al. 1984)
1.42 (alga *chlorella fusca*, calculated-K_{OW}, Geyer et al. 1984)
0.78 (*p. reticulata*, Canton et al. 1985; quoted, Howard 1989)
$<$1.0, 1.38, 1.60 (golden orfe, algae, activated sludge, Freitag et al. 1982)
$<$1.0, 1.30, 1.60 (golden ide, algae, activated sludge, Freitag et al. 1985)

1.47 ± 0.12 (guppy-fat wt. basis, Deneer et al. 1987)
1.18 (quoted, Isnard & Lambert 1988)
1.18 (quoted from Mackay 1982, Schüürmann & Klein 1988)

Sorption Partition Coefficient, log K_{OC}:
2.30 (Lincoln fine sand, calculated exptl. value, Wilson et al. 1981; quoted, Howard 1989)
2.23, 2.57 (Danish subsoils, Loekke 1984; quoted, Howard 1989)
1.63, 1.84 (two Norwegian organic soils, Seip et al. 1986; quoted, Howard 1989)
1.70 (soil, quoted, Sabljic 1987)
1.95 ± 0.84 (captina slit loam, Walton et al. 1992)
2.02 ± 1.18 (McLaurin sandy loam, Walton et al. 1992)
1.43 (predicted-K_{OW}, Walton et al. 1992)
11.51 (calculaated-K_{OW}, Kollig 1993)

Half-Lives in the Environment:

Air: atmospheric lifetimes of 180 days in clean troposphere and 90 days in moderately polluted atmosphere, based on gas-phase reaction with hydroxyl radicals in atmosphere at room temperature and atmospheric lifetimes of > 6 years in clean troposphere and > 2 years in moderately polluted atmosphere, based on gas-phase reaction with O_3 (estimated rate constant) in atmosphere at room temperature (Atkinson et al. 1987); photooxidation half-life of 0.544-5.44 hours, based on measured rate constant for reaction with hydroxyl radicals in air (Atkinson et al. 1987; quoted, Howard 1989; Howard et al. 1991); atmospheric transformation lifetime was estimated to be > 5 days (Kelly et al. 1994).

Surface water: photooxidation half-life of 125 days to 22 years, based on measured rate constant for reaction with hydroxyl radicals in water (Dorfman & Adams 1973; Anbar & Neta 1967; quoted, Howard et al. 1991); estimated half-life of 0.3-3.0 days in rivers (Zoeteman et al. 1980); half-life of 322-4728 hours, based on estimated unacclimated aqueous aerobic biodegradation half-life (Howard et al. 1991).

Groundwater: 48-9456 hours, based on estimated unacclimated aqueous anaerobic biodegradation half-life for 2,4-dinitrotoluene and estimated unacclimated aqueous aerobic biodegradation half-life (Howard et al. 1991).

Sediment:

Soil: estimated degradation half-life of 625 days in activated sludge (Freitag et al. 1985; quoted, Anderson et al. 1991); half-life of 322-4728 hours, based on aerobic soil column biodegradation study data (Kincannon & Lin 1985; quoted, Howard et al. 1991); calculated half-life of 9.1 days from first-order kinetic of degradation rate in sterilized soils (Anderson et al. 1991).

Biota:

Environmental Fate Rate Constants or Half-Lives:

Volatilization: calculated half-life from water bodies of approximately 200 hours (Mackay & Leinonen 1975; quoted, Callahan et al. 1979); using Henry's law constant one can estimate the half-life in a model river 1 m deep with a 1m/sec current and a 3 m/sec wind is 45 hours (Lyman et al. 1982; quoted, Howard 1989).

Photolysis: aqueous photolysis half-life of 67-200 days, based on measured photolysis rate constant in distilled water under midday sun at 40°N latitude (Simmons & Zepp 1986; quoted, Howard 1989; Howard et al. 1991); atmospheric photolysis half-life of 67-200 days, based on measured photolysis rate constant in distilled water under midday sun at 40°N latitude (Simmons & Zepp 1986; quoted, Howard 1989; Howard et al. 1991).

Hydrolysis:

Oxidation: photooxidation half-life of 125 days to 22 years in water, based on measured rate constant for reaction with hydroxyl radicals in water (Dorfman & Adams 1973; Anbar & Neta 1967; quoted, Howard et al. 1991); aqueous oxidation constants of $<<360\ M^{-1}\cdot h^{-1}$ for singlet oxygen and $<<1.0\ M^{-1}\cdot h^{-1}$ for peroxy radical at 25°C (Mabey et al. 1982); rate constant of $(0.09 \pm 0.02)\ M^{-1}\ sec^{-1}$ for 5/10 mM to react with ozone in water using 50-1000 mM of *t*-BuOH as scavenger at pH 2 and 20-23°C (Hoigné & Bader 1983); observed rate constant of $0.15x10^{-12}\ cm^3\ molecule^{-1}\ sec^{-1}$ for the vapor-phase reaction with OH radicals at 296K in the atmosphere (Becker et al. 1984; quoted, Carlier et al. 1986); rate constant of $0.21x10^{-12}\ cm^3\ molecule^{-1}\ sec^{-1}$ (Zetzsch 1982) compared to calculated rate constants of $0.30x10^{-12}\ cm^3\ molecule^{-1}\ sec^{-1}$ and $0.27x10^{-12}\ cm^3\cdot molecule^{-1}\cdot sec^{-1}$ for reaction with hydroxyl radicals at room temperature in air (Atkinson 1985; Atkinson et al. 1985); photooxidation half-life of 0.544-5.44 hours in air, based on measured rate constant of $0.14x10^{-12}\ cm^3\ molecule^{-1}\ sec^{-1}$ (Witte et al. 1986; quoted, Atkinson 1987) for reaction with hydroxyl radicals in air (Atkinson 1987; quoted, Howard 1989; Howard et al. 1991); observed rate constant of $0.14x10^{-12}\ cm^3\ molecule^{-1}\ sec^{-1}$ (Witte et al. 1986) compared to calculated rate constant of $0.25x10^{-12}\ cm^3\ molecule^{-1}\ sec^{-1}$ for reaction with hydroxyl radicals at room temperature in air (Atkinson 1987); exptl. rate constant of $<7.0x10^{-13}\ cm^3\ molecule^{-1}\ sec^{-1}$ for the gas phase reaction with hydroxyl radicals at room temperature and calculated rate constant of $<7.0x10^{-21}\ cm^3\ molecule^{-1}\ sec^{-1}$ for the gas phase reaction with O_3 in the atmosphere at room temperature (Atkinson et al. 1987); phototransformation decay rate constant of $0.17\ min^{-1}$ on 0.20 g/L of TiO_2, $8.8\ min^{-1}$ on 0.20 g/L of ZnO and $3.1\ min^{-1}$ on 1.0 g/L of Al_2O_3 (Minero et al. 1994) .

Biodegradation: decomposition by a soil microflora in more than 64 days (Alexander & Lustigman 1966; quoted, Verschueren 1983); aqueous anaerobic half-life of 48-300 hours, based on anaerobic natural die-away test data for 2,4-dinitrotoluene (Spanggord et al. 1980; quoted, Howard et al. 1991); average biodegradation rate of 14 mg COD $g^{-1}\ h^{-1}$ for 98% removal (Scow 1982); aqueous aerobic half-life of 322-4728 hours, based on aerobic soil column biodegradation study data (Kincannon & Lin 1985; quoted, Howard et al. 1991).

Biotransformation: first-order rate constant of $0.7\ day^{-1}$ corresponding to a half-life of 1 day in adopted activated sludge under aerobic conditions (Mills et al. 1982); rate constant for bacterial trasformation of $3x10^{-9}\ mL\cdot cell^{-1}\cdot h^{-1}$ in water (Mabey et al. 1982).

Bioconcentration, Uptake (k_1) and Elimination (k_2) Rate Constants:

Common Name: 2-Nitrotoluene
Synonym: 1-methyl-2-nitrobenzene, *o*-nitrotoluene
Chemical Name: 2-nitrotoluene, *o*-nitrotoluene
CAS Registry No: 88-72-2
Molecular Formula: $CH_3C_6H_4NO_2$
Molecular Weight: 137.13
Melting Point (°C):
- -4.1 (Stull 1947)
- -9.5 (Weast 1982-83)
- -10 (Howard 1989)

Boiling Point (°C):
- 220.4 (Kahlbaum 1898)
- 222.3 (Stull 1947)
- 221.7 (Weast 1982-83)
- 222.47 (Boublik et al. 1984)
- 222 (Howard 1989)

Density (g/cm^3 at 20°C):
- 1.1629 (Weast 1982-83)

Molar Volume (cm^3/mol):
- 127.9 (Kamlet et al. 1987)
- 153 (calculated-LeBas method, Abernethy et al. 1988)
- 73.0 (intrinsic volume V_I, Abernethy et al. 1988)
- 76.4 (intrinsic V_I, Leggett & Parker 1994)

Molecular Volume (Å^3):
Total Surface Area, TSA (Å^2):
- 269.8 (Iwase et al. 1985)

Dissociation Constant pK_a:
Heat of Fusion ΔH_{fus}, kcal/mol:
Entropy of Fusion ΔS_{fus}, cal/mol·K (e.u.):
Fugacity Ratio at 25°C (assuming ΔS_{fus} = 13.5 e.u.), F: 1.0

Water Solubility (g/m^3 or mg/L at 25°C):
- 655 (30°C, shake flask-interferometer, Gross et al. 1933)
- 656 (quoted, Deno & Berkheimer 1960)
- <233 (shake flask-centrifuge, Booth & Everson 1948)
- 324 (Hansch et al. 1968; quoted, Hine & Mookerjee 1975)
- 656, 2076 (quoted, predicted-K_{OW}, Valvani et al. 1981)
- 652 (30°C, Verschueren 1983; quoted, Howard 1989; Müller & Klein 1992)
- 656, 771 (quoted, calculated-fragment const., Wakita et al. 1986)

Vapor Pressure (Pa at 25°C):
- 23.97 (extrapolated-regression of tabulated data, Kahlbaum 1898)
- 27.69 (extrapolated-regression of tabulated data, Stull 1947)
- 27.99 (quoted, Hine & Mookerjee 1975)
- 13.33 (20°C, Verschueren 1983; quoted, Howard 1989)

1.670 (extrapolated-Antoine eqn., Boublik et al. 1984)
1.44 (extrapolated-Antoine eqn., Dean 1985)
24.80 (ebulliometry, average from extrapolated-Antoine eqn., Aim 1994)

Henry's Law Constant (Pa·m^3/mol):
5.572 (Hine & Mookerjee 1975; quoted, Howard 1985)
5.811 (quoted exptl., Hine & Mookerjee 1975)
4.723 (calculated-group contribution, Hine & Mookerjee 1975)
4.616 (calculated-bond contribution, Hine & Mookerjee 1975)

Octanol/Water Partition Coefficient, log K_{OW}:
2.30 (quoted private communication, Leo et al. 1971; quoted, Iwase et al. 1985; Haderlein & Schwarzenbach 1993)
2.30 (quoted exptl., Valvani et al. 1981)
2.30 (Hansch & Leo 1985; quoted, Howard 1989)
2.18 (calculated-TSA, Iwase et al. 1985)
2.30 (quoted, Yoshida et al. 1986)
2.30 (calculated-HPLC-k', Deneer et al. 1987; quoted, Verhaar et al. 1992)
2.40 (selected, Abernethy et al. 1988)
2.30 (unpublished data quoted from CLOGP Database, Sangster 1989)
2.30 (recommended, Sangster 1989)
2.30 (quoted from CLOGP 3.52, Gluck & Martin 1990)
2.39, 2.43, 2.58 (CPC-RV, Gluck & Martin 1990)
2.10 (calculated-CLOGP, Müller & Klein 1992)

Bioconcentration Factor, log BCF:
<2.0 (*carprinus carpio*, Sasaki 1978; Kawasaki 1980; quoted, Howard 1989)
1.52, 1.20 (calculated-K_{OW}, S, Lyman et al. 1982; quoted, Howard 1989)
2.28 $\pm$ 0.06 (guppy-fat basis, Deneer et al. 1987)

Sorption Partition Coefficient, log K_{OC}:
2.63, 2.09 (soil, calculated-K_{OW}, S, Lyman et al. 1982; quoted, Howard 1989)

Half-Lives in the Environment:
Air: half-life of 8 hours, based on a rate constant of 3.0×10^{-11} cm^3/molecules-sec for the reaction with 8×10^{-5} molecules/cm^3 photochemically produced hydroxyl radicals in air (GEMS 1986; quoted, Howard 1989).
Surface water: the midday half-life in Aucilla River water due to indirect photolysis has been calculated to be 45 minutes using an experimentally determined reaction rate constant of 0.92 hour^{-1} (Zepp et al. 1984; quoted, Howard 1989); estimated half-life of 3.2 days for a river 4 to 5 meter deep, based on monitoring data (Zoeteman et al. 1980; quoted, Howard 1989).

Groundwater:
Sediment:
Soil:
Biota:

Environmental Fate Rate Constants or Half-Lives:

Volatilization: estimated half-life of 21 hours using Henry's law constant for a model river 1 meter deep flowing 1 m/sec with a wind speed of 3 m/sec (Lyman et al. 1982; quoted, Howard 1989).

Photolysis:

Oxidation: rate constant of 3.0×10^{-11} cm^3/molecules-sec for the reaction with 8×10^{5} molecules/cm^3 photochemically produced hydroxyl radicals in air (GEMS 1986; quoted, Howard 1989).

Hydrolysis:

Biodegradation: average biodegradation rate of 32.5 mg COD g^{-1} h^{-1} for 98% removal (Scow 1982).

Biotransformation:

Bioconcentration, Uptake (k_1) and Elimination (k_2) Rate Constants:

Common Name: 4-Nitrotoluene
Synonym: 1-methyl-4-nitrobenzene, *p*-nitrotoluene
Chemical Name: 4-nitrotoluene, *p*-nitrotoluene
CAS Registry No: 99-99-0
Molecular Formula: $CH_3C_6H_4NO_2$
Molecular Weight: 137.13
Melting Point (°C):
    51.9 (Stull 1947)
    51.7 (Tsonopoulos & Prausnitz 1971)
    54.5 (Weast 1982-83)
    53-54 (Dean 1985; Howard 1989)
Boiling Point (°C):
    237.7 (Kahlbaum 1898)
    238.3 (Stull 1947; Weast 1982-83; Dean 1985; Howard 1989)
    233.26 (Boublik et al. 1984)
Density (g/cm^3 at 20°C):
    1.392 (Dean 1985)
Molar Volume (cm^3/mol):
    76.4 (intrinsic V_I, Leggett & Parker 1994)
    153 (calculated-LeBas method)
Molecular Volume ($Å^3$):
Total Surface Area, TSA ($Å^2$):
    278.5 (Iwase et al. 1985)
Dissociation Constant, pK:
    -11.27 (Perrin 1972)
Heat of Fusion ΔH_{fus}, kcal/mol:
    4.10 (Tsonopoulos & Prausnitz 1971)
Entropy of Fusion ΔS_{fus}, cal/mol·K (e.u.):
    12.6 (Tsonopoulos & Prausnitz 1971)
Fugacity Ratio at 25°C (assuming ΔS_{fus} = 13.5 e.u.), F: 0.5407

Water Solubility (g/m^3 or mg/L at 25°C):
    442 (30°C, shake flask-interferometer, Gross et al. 1933)
    <278 (shake flask-centrifuge, Booth & Everson 1948)
    302 (quoted, Tsonopoulos & Prausnitz 1971)
    442 (30°C, Verschueren 1983; quoted, Howard 1989; Müller & Klein 1992)

Vapor Pressure (Pa at 25°C):
    13.98 (extrapolated-regression of tabulated data, Kahlbaum 1898)
    22.81 (extrapolated-regression of tabulated data, Stull 1947)

5.50 (ebulliometry, fitted to Antoine eqn., Ambrose & Gundry 1980)
9.50 (extrapolated-supercooled liq., Ambrose & Gundry 1980)
13.33 (20°C, Verschueren 1983; quoted, Howard 1989)
8.347 (extrapolated-Antoine eqn., Boublik et al. 1984)
8.38 (extrapolated-Antoine eqn., Dean 1985)
15.18 (ebulliometry, average of extrapolated-Antoine eqn., Aim 1994)

Henry's Law Constant (Pa·m^3/mol):
5.065 (calculated, Hine & Mookerjee 1975; quoted, Howard 1985)

Octanol/Water Partition Coefficient, log K_{OW}:
2.37 (shake flask-AS, Fujita et al. 1964; quoted, Leo et al. 1971; Iwase et al. 1985; Sangster 1989)
2.40 (quoted unpublished result, Leo et al. 1971; quoted, Haderlein & Schwarzenbach 1993)
2.37 (Hansch & Leo 1985; quoted, Howard 1989)
2.37, 2.34 (quoted, calculated-TSA, Iwase et al. 1985)
2.40 (quoted, Yoshida et al. 1986)
2.34 (calculated-HPLC-k', Deneer et al. 1987; quoted, Verhaar et al. 1992)
2.42 (unpublished data quoted from CLOGP Database, Sangster 1989)
2.42 (recommended, Sangster 1989)
2.38 (calculated-CLOGP, Müller & Klein 1992)

Bioconcentration Factor, log BCF:
<2.0 (*carprinus carpio*, Sasaki 1978; Kawasaki 1980; quoted, Howard 1989)
1.57, 1.30 (calculated-K_{OW}, S, Lyman et al. 1982; quoted, Howard 1989)
2.37 ± 0.05 (guppy-fat basis, Deneer et al. 1987)

Sorption Partition Coefficient, log K_{OC}:
2.67, 2.18 (soil, calculated-K_{OW}, S, Lyman et al. 1982; quoted, Howard 1989)

Half-Lives in the Environment:
Air: photooxidation half-life of 8.0 hours, based on measured rate constant of 3×10^{-11} cm^3/molecules-sec at 25°C for the reaction with photochemically produced 8×10^{-5} molecules/cm^3 hydroxyl radicals (GEMS 1986; quoted, Howard 1989).
Surface water: estimated half-life of 2.7 days, based on monitoring data for a river of 4 to 5 meter deep (Zoeteman et al. 1980; quoted, Howard 1989).

Groundwater:
Sediment:
Soil:
Biota:

Environmental Fate Rate Constants or Half-Lives:

Volatilization: based on Henry's law constant an estimated half-life of 25 hours was obtained for a model river of 1 meter deep with a current of 1 m/sec and wind speed of 3 m/sec (Lyman et al. 1982; quoted, Howard 1989).

Photolysis:

Hydrolysis:

Oxidation: photooxidation half-life of 8.0 hours in air, based on measured rate constant of 3×10^{-11} cm^3/molecules-sec at 25°C for the reaction with photochemically produced 8×10^{-5} molecules/cm^3 hydroxyl radicals (GEMS 1986; quoted, Howard 1989).

Biodegradation: average biodegradation rate of 32.5 mg COD g^{-1} h^{-1} for 98% removal (Scow 1982).

Biotransformation:

Bioconcentration, Uptake (k_1) and Elimination (k_2) Rate Constants:

Common Name: 2,4-Dinitrotoluene
Synonym: dinitrotoluol, 1-methyl-2,4-dinitrobenzene, DNT
Chemical Name: 2,4-dinitrotoluene, 1-methyl-2,4-dinitrobenzene
CAS Registry No: 121-14-2
Molecular Formula: $CH_3C_6H_3(NO_2)_2$
Molecular Weight: 182.14
Melting Point (°C):
77 (Verschueren 1977; Callahan et al. 1979)
70 (Yalkowsky & Valvani 1980; Verschueren 1983)
71 (Howard 1989)
Boiling Point (°C):
300 (Verschueren 1977, 1983; Callahan et al. 1979; Howard 1989)
Density (g/cm^3 at 20°C):
1.521 (15°C, Verschueren 1983)
Molar Volume (cm^3/mol):
175.2 (calculated-LeBas method)
Molecular Volume ($Å^3$):
Total Surface Area, TSA ($Å^2$):
Dissociation Constant, pK:
-13.53 (Perrin 1972)
Heat of Fusion ΔH_{fus}, kcal/mol:
Entropy of Fusion ΔS_{fus}, cal/mol·K (e.u.):
14.01 (Yalkowsky & Valvani 1980)
Fugacity Ratio at 25°C (assuming ΔS_{fus} = 13.5 e.u.), F: 0.359

Water Solubility (g/m^3 or mg/L at 25°C):
270 (22°C, Verschueren 1977, 1983; quoted, Callahan et al. 1979; Mabey et al. 1982; Mills et al. 1982)
294 (calculated-ΔS_{fus}, Yalkowsky & Valvani 1980)
286 (calculated-M.P., Yalkowsky & Valvani 1980)
300 (22°C, Dunlap 1981; quoted, Howard 1989; Müller & Klein 1992)

Vapor Pressure (Pa at 25°C):
0.133 (59°C, Lenchitz & Velicky 1970; quoted, Callahan et al. 1979; Mills et al. 1982)
0.0177 (20°C, GC/ECD, Pella 1977)
0.029 (GC/ECD, fitted to Antoine eqn., Pella 1977)
0.0147 (20°C, Spanggord et al. 1980; quoted, Howard 1989)
0.680 (quoted, Mabey et al. 1982)
1.21×10^{-4} (extrapolated-Antoine eqn., Boublik et al. 1984)

Henry's Law Constant (Pa·m^3/mol):
0.0160 (calculated-P/C, Smith et al. 1981)
0.456 (calcuclated-P/C, Mabey et al. 1982)
0.0878 (Smith et al. 1983; quoted, Howard 1989)
32.229 (quoted from WERL Treatability database, Ryan et al. 1988)

Octanol/Water Partition Coefficient, log K_{OW}:

2.01 (calculated as per Tute 1971, Callahan et al. 1979)
2.00 (selected, Mills et al. 1982)
1.98 (Hansch & Leo 1985; quoted, Mabey et al. 1982; Howard 1989)
2.04 (calculated-HPLC-k', Deneer et al. 1987)
2.01 (quoted from USEPA 1979, Ryan et al. 1988)
2.00 (quoted, Schultz et al. 1989)
1.85 (calculated-CLOGP, Müller & Klein 1992)
2.00 (calculated-QSAR, Kollig 1993)

Bioconcentration Factor, log BCF:

1.59 (microorganisms-water, calculated-K_{OW}, Mabey et al. 1982)
1.11, 1.76 (*daphnia magna, lumbriculus variegatus*, Liu et al. 1983)
>3.30 (selanastrum capricornutum, Liu et al. 1983)
1.89 (bluegill sunfish in viscera, Liu et al. 1983)
0.602 (bluegill sunfish in muscle, Liu et al. 1983)
2.31 ± 0.03 (guppy-fat basis, Deneer et al. 1987)

Sorption Partition Coefficient, log K_{OC}:

1.65 (sediment-water, calculated-K_{OW}, Mabey et al. 1982)
1.68 (calculated-K_{OW}, Kollig 1993)

Half-Lives in the Environment:

Air: photooxidation half-life of 284-2840 hours, based on estimated rate constant for the reaction with hydroxyl radicals in air (Atkinson 1987; quoted, Howard et al. 1991); atmospheric traansformation lifetime was estimated to be 1 to 5 days (Kelly et al. 1994).

Surface water: photooxidation half-life of 3-33 hours, based on measured photooxidation rates in natural waters (Spanggord et al. 1980; Simmons & Zepp 1986; quoted, Howard et al. 1991); sunlight photolysis half-life was approximately 42 hours in pure water but ranged from 3 hours to 10 hours in three natural waters (Mabey et al. 1982).

Groundwater: 48-8640 hours, based on estimated unacclimated aqueous anaerobic and aerobic biodegradation half-life (Howard et al. 1991).

Sediment:

Soil: 672-4320 hours, based on estimated unacclimated aqueous aerobic biodegradation half-life (Howard et al. 1991).

Biota:

Environmental Fate Rate Constants or Half-Lives:

Volatilization: half-life of approximately 100 days (Mills et al. 1982).

Photolysis: diract photolysis rate constant of 1.6×10^{-2} h^{-1} for summer at 40° latitude in water (Mabey et al. 1982); aqueous photolysis half-life of 23-72 hours, based on measured photolysis rates in water (Mill & Mabey 1985; Simmons & Zepp 1986; quoted, Howard et al. 1991); atmospheric transformation lifetime was estimated to be 1 to 5 days (Kelly et al. 1994).

Hydrolysis:

Oxidation: aqueous oxidation rate constants of << 360 $M^{-1}\cdot h^{-1}$ for singlet oxygen and 144 $M^{-1}\cdot h^{-1}$ for peroxy radical at 25°C (Mabey et al. 1982); photooxidation half-life of 284-2840 hours in air, based on estimated rate constant for the reaction with hydroxyl radicals in air (Atkinson 1987; quoted, Howard et al. 1991); photooxidation half-life of 3-33 hours, based on measured photooxidation rates in natural waters (Spanggord et al. 1980; Simmons & Zepp 1986; quoted, Howard et al. 1991).

Biodegradation: aqueous anaerobic half-life of 48-240 hours, based on anaerobic natural water die-away test data (Spanggord et al. 1980; quoted, Howard et al. 1991); aqueous aerobic half-life of 672-4320 hours, based on aerobic natural water die-away test data (Spanggord et al. 1981; quoted, Howard et al. 1991).

Biotransformation: rate constant of 1×10^{-1} $mL\cdot cell^{-1}\cdot h^{-1}$ for bacterial transformation in water (Mabey 1982).

Bioconcentration, Uptake (k_1) and Elimination (k_2) Rate Constants:

Common Name: 2,6-Dinitrotoluene
Synonym: dinitrotoluol, 1-methyl-2,6-dinitrobenzene
Chemical Name: 2,6-dinitrotoluene, 1-methyl-2,6-dinitrobenzene
CAS Registry No: 606-20-2
Molecular Formula: $CH_3C_6H_3(NO_2)_2$
Molecular Weight: 182.14
Melting Point (°C):
64-66 (Verschueren 1977, 1983; Callahan et al. 1979)
66 (Weast 1982-83; Howard 1989)
Boiling Point (°C):
285 (Verschueren 1977; Callahan et al. 1979; Mabey et al. 19812; Howard 1989)
Density (g/cm^3 at 20°C):
1.2833 (111°C, Weast 1982-83; Dean 1985)
Molar Volume (cm^3/mol):
175.2 (calculated-LeBas method)
Molecular Volume (Å^3):
Total Surface Area, TSA (Å^2):
Dissociation Constant pK_a:
Heat of Fusion ΔH_{fus}, kcal/mol:
Entropy of Fusion ΔS_{fus}, cal/mol·K (e.u.):
Fugacity Ratio at 25°C (assuming ΔS_{fus} = 13.5 e.u.), F: 0.3931

Water Solubility (g/m^3 or mg/L at 25°C):
180 (20°C, estimated, Mabey et al. 1982)
~300 (selected, Mills et al. 1982)

Vapor Pressure (Pa at 25°C):
2.40 (29°C, quoted, Mabey et al. 1982)
0.0756 (GC/ECD, fitted to Antoine eqn., Pella 1977)
0.0467 (Pella 1977; quoted, Howard 1989)
0.0756 (Howard et al. 1986; quoted, Banerjee et al. 1990)
0.0756, 1.008 (quoted, calculated-solvatochromic p., Banerjee et al. 1990)

Henry's Law Constant (Pa·m^3/mol):
0.800 (calculated-P/C, Mabey et al. 1982)
32.23 (quoted from WERL Treatability database, Ryan et al. 1988)
0.022 (SOGC 1987; quoted, Howard 1989)

Octanol/Water Partition Coefficient, log K_{OW}:
2.05 (calculated as per Tute 1971, Callahan et al. 1979)
2.28 (calculated-f const., Mabey et al. 1982)
2.00 (selected, Mills et al. 1982)
1.72 (Hansch & Leo 1985; quoted, Howard 1989)
2.02 (calculated-HPLC-k', Deneer et al. 1987)
2.05 (quoted from USEPA 1979, Ryan et al. 1988)
1.72 (calculated-QSAR, Kollig 1993)

Bioconcentration Factor, log BCF:

3.72 (algal biomass, Davis et al. 1981; quoted, Howard 1989)

1.71 (microorganisms-water, calculated-K_{OW}, Mabey et al. 1982)

1.08 (calculated-K_{OW}, Lyman et al. 1982; quoted, Howard 1989)

2.44 ± 0.04 (guppy-fat basis, Deneer et al. 1987)

Sorption Partition Coefficient, log K_{OC}:

1.96 (sediment-water, calculated-K_{OW}, Mabey et al. 1982)

2.31 (soil, calculated-K_{OW}, Lyman et al. 1982; quoted, Howard 1989)

1.40 (calculated-K_{OW}, Kollig 1993)

Half-Lives in the Environment:

Air: estimated atmospheric half-life of 8 hours, based on the vapor phase reaction with hydroxyl radicals in air (GEMS 1985; quoted, Howard 1989); photooxidation half-life of 284-2840 hours, based on estimated rate constant for the reaction with hydroxyl radicals in air (Atkinson 1987; quoted, Howard et al. 1991).

Surface water: midday half-life in Aucilla river due to indirect photolysis has been estimated to be 12 minutes using experimentally determined rate constant of 3.6 h^{-1} (Zepp et al. 1984); photooxidation half-life of 2-17 hours in water, based on measured photooxidation rates in natural waters (Simmons & Zepp 1986; quoted, Howard et al. 1991).

Groundwater: 48-8640 hours, based on estimated unacclimated aqueous anaerobic biodegradation half-life 2,4-dinitrotoluene and estimated unacclimated aqueous aerobic biodegradation half-life (Howard et al. 1991).

Sediment:

Soil: 672-4320 hours, based on estimated unacclimated aqueous aerobic biodegradation half-life (Howard et al. 1991).

Environmental Fate Rate Constants or Half-Lives:

Volatilization: half-life of approximatelly 100 days (Mills et al. 1982).

Photolysis: aqueous photolysis half-life of 17-25 hours, based on measured photolysis rates in water (Simmons & Zepp 1986; Mill & Mabey 1985; quoted, Howard et al. 1991).

Hydrolysis:

Oxidation: aqueous oxidation rate constants of $<<360\ M^{-1}\cdot h^{-1}$ for singlet oxygen and $144\ M^{-1}\cdot h^{-1}$ for peroxy radical at 25°C (Mabey et al. 1982); photooxidation half-life of 2-17 hours in water, based on measured photooxidation rates in natural waters (Simmons & Zepp 1986; quoted, Howard et al. 1991); photooxidation half-life of 284-2840 hours, based on estimated rate constant for the reaction with hydroxyl radicals in air (Atkinson 1987; quoted, Howard et al. 1991).

Biodegradation: aqueous anaerobic half-life of 48-300 hours, based on anaerobic natural water die-away test data for 2,4-dinitrotoluene (Spanggord et al. 1980; quoted, Howard et al. 1991); aqueous aerobic half-life of 672-4320 hours, based on aerobic natural water die-away test data (Spanggord et al. 1981; quoted, Howard et al. 1991).

Biotransformation: rate constant of $1\times10^{-10}\ mL\cdot cell^{-1}\cdot h^{-1}$ for bacterial transformation in water (Mabey et al. 1982).

Bioconcentration, Uptake (k_1) and Elimination (k_2) Rate Constants:

Common Name: 1-Nitronaphthalene
Synonym: α-nitronaphthalene
Chemical Name: 1-nitronaphthalene, α-nitronaphthalene
CAS Registry No: 86-57-7
Molecular Formula: $C_{10}H_7NO_2$
Molecular Weight: 173.17
Melting Point (°C):
    61.5 (Weast 1982-83; Stephenson & Malanowski 1987)
Boiling Point (°C):
    304.0 (Weast 1982-83; Dean 1985; Stephenson & Malanowski 1987)
Density (g/cm^3 at 20°C):
    1.3320 (Weast 1982-83)
    1.2230 (Dean 1985)
Molar Volume (cm^3/mol):
    176.1 (calculated-LeBas method)
    135.8 (calculated-density)
Molecular Volume (Å^3):
Total Surface Area, TSA (Å^2):
Dissociation Constant pK_a:
Heat of Fusion ΔH_{fus}, kcal/mol:
    4.30 (Tsonopoulos & Prausnitz 1971)
Entropy of Fusion ΔS_{fus}, cal/mol·K (e.u.):
    13.0 (Tsonopoulos & Prausnitz 1971)
Fugacity Ratio at 25°C (assuming ΔS_{fus} = 13.5 e.u.), F: 0.4355

Water Solubility (g/m^3 or mg/L at 25°C):
    50 (Yalkowsky et al. 1987)
    9.82 (gen. col.-HPLC/UV, Yu & Xu 1992)
    9.83 (calculated-molar concn., Yu & Xu 1992)

Vapor Pressure (Pa at 25°C):
    0.202 (effusion method-fitted to Antoine eqn., Radchenko & Kitaigorodskii 1974)
    0.202 (extrapolated-Antoine eqn., Stephenson & Malanowski 1987)
    0.184 (liq., extrapolated-Antoine eqn., Stephenson & Malanowski 1987)

Henry's Law Constant (Pa m^3/mol):
    3.463 (calculated-P/C, this work)

Octanol/Water Partition Coefficient, log K_{OW}:
    3.19 (quoted from Hansch & Leo 1979, Schultz & Moulton 1985)
    3.19 (shake flask, Hansch & Leo 1987; quoted, Sangster 1993)
    3.19 (shake flask-AS, Debnath & Hansch 1992; quoted, Sangster 1993)
    3.19 (recommended, Sangster 1993)

Bioconcentration Factor, log BCF:

Sorption Partition Coefficient, log K_{OC}:

Half-Lives in the Environment:

Air:
Surface water:
Groundwater:
Sediment:
Soil:
Biota:

Environmental Fate Rate Constants or Half-Lives:

Volatilization:
Photolysis:
Hydrolysis:
Oxidation: calculated photooxidation lifetimes of about 2.0 hours, based on measured rate constants of $(5.4 \pm 1.8)\times10^{-12}$ cm^3 molecule^{-1} sec^{-1} for the gas-phase reaction with OH radicals at 298 $\pm$ 2 K in the atmosphere, $<7.2\times10^{-15}$ cm^3 molecule^{-1} sec^{-1} (quoted, Atkinson 1991) for the gas-phase reaction with NO_3 radicals at 298 $\pm$ 2 K in the atmosphere, and $<6.0\times10^{-19}$ cm^3 molecule^{-1} sec^{-1} for the gas-phase reaction with O_3 at 298 $\pm$ 2 K in the atmosphere (Atkinson et al. 1989).
Biodegradation:
Biotransformation:
Bioconcentration, Uptake (k_1) and Elimination (k_2) Rate Constants:

Common Name: Acetamide
Synonym: ethanamide
Chemical Name: acetamide, acetic acid amine
CAS Registry No: 60-35-5
Molecular Formula: CH_3CONH_2
Molecular Weight: 59.07
Melting Point (°C):
81.0 (Stull 1947)
Boiling Point (°C):
222.0 (Stull 1947)
Density (g/cm^3 at 20°C):
1.159 (Verschueren, 1983)
Molar Volume (cm^3/mol):
66.9 (calculated-LeBas method)
59.2 (calculated-density)
Molecular Volume (Å^3):
Total Surface Area, TSA (Å^2):
Dissociation Constant pK_a:
Heat of Fusion, ΔH_{fus}, kcal/mol: 3.754 (Riddick et al. 1986)
Entropy of Fusion, ΔS_{fus}, cal/mol·K (e.u.):
Fugacity Ratio at 25°C (assuming ΔS_{fus} = 13.5 e.u.), F: 0.286

Water Solubility (g/m^3 or mg/L at 25°C):
975000 (20°C, Verschueren 1983)
408000 (20°C, Riddick et al. 1986)

Vapor Pressure (Pa at 25°C):
8.61 (extrapolated-regression of tabulated data, Stull 1947)
100 (quoted, Riddick et al. 1986)
2.44 (interpolated, Antoine eqn., Stephenson & Malanowski 1987)

Henry's Law Constant (Pa·m^3/mol):
3.53×10^{-4} (calculated-P/C, this work)

Octanol/Water Partition Coefficient, log K_{OW}:
-1.58/-1.26 (calculated, Verschueren, 1983)
-1.26, -1.15, -1.09; -1.26 (quoted; recommended Sangster 1989)
-1.23 (calculated-QSAR, Kollig 1993)

Bioconcentration Factor, log BCF:

Sorption Partition Coefficient, log K_{OC}:

-1.55 (calculated-K_{OW}, Kollig 1993)

Half-Lives in the Environment:

Air: photooxidation half-life of 3.2-32 hours, based on estimated rate constant for the vapor-phase reaction with hydroxyl radicals in air (Atkinson 1987; quoted, Howard et al. 1991); atmospheric transformation lifetime was estimated to be < 1 day (Kelly et al. 1994).

Surface water: 24-168 hours, based on estimated unacclimated aqueous aerobic biodegradation half-life (Howard et al. 1991).

Groundwater: 48-336 hours, based on estimated unacclimated aqueous aerobic biodegradation half-life (Howard et al. 1991).

Sediment:

Soil: 24-168 hours, based on estimated unacclimated aqueous aerobic biodegradation half-life (Howard et al. 1991).

Biota:

Environmental Fate Rate Constants or Half-Lives:

Volatilization:

Photolysis:

Oxidation: photooxidation half-life of 3.2-32 hours in air, based on estimated rate constant for the vapor-phase reaction with hydroxyl radicals in air (Atkinson 1987; quoted, Howard et al. 1991); atmospheric transformation lifetime was estimated to be <1 day (Kelly et al. 1994).

Hydrolysis: acid rate constant of 0.03 $[M\cdot(H^+)\cdot h]^{-1}$ at pH 5 and base rate constant of 0.17 $[M\cdot(OH^-)\cdot h]^{-1}$ at pH 9 with first-order hydrolysis half-life of 3950 years at pH 7 and 25°C, (Mabey & Mill 1978; quoted, Howard et al. 1991).

Biodegradation: aqueous aerobic biodegradation half-life of 24-168 hours, based on aerobic aqueous screening test data (Malaney & Gerhold 1962,69; Urano & Kato 1986; quoted, Howard et al. 1991); aqueous anaerobic biodegradation half-life of 96-672 hours, based on estimated unacclimated aqueous aerobic biodegradation half-life (Howard et al. 1991).

Biotransformation:

Bioconcentration, Uptake (k_1) and Elimination (k_2) Rate Constants:

Common Name: Acrylamide
Synonym: propenamide
Chemical Name: acrylamide
CAS Registry No: 79-06-1
Molecular Formula: $CH_2=CHCONH_2$
Molecular Weight: 71.08
Melting Point (°C):
84.5 (Weast 1982-83; Dean 1985; Stephenson & Malanowski 1987)
84-85 (Verschueren 1983)
Boiling Point (°C):
125 (at 25 mm Hg, Dean 1985)
Density (g/cm^3 at 20°C):
1.122 (30°C, Dean 1985)
Molar Volume (cm^3/mol):
80.8 (calculated-LeBas method)
Molecular Volume ($Å^3$):
Total Surface Area, TSA ($Å^2$):
Dissociation Constant pK_a:
Heat of Fusion, ΔH_{fus}, kcal/mol:
Entropy of Fusion, ΔS_{fus}, cal/mol·K (e.u.):
Fugacity Ratio at 25°C (assuming ΔS_{fus} = 13.5 e.u.), F: 0.2579

Water Solubility (g/m^3 or mg/L at 25°C):
2050000 (quoted, Verschueren 1983)
2150000 (30°C, Dean 1985)

Vapor Pressure (Pa at 25°C):
0.616 (extrapolated-Antoine eqn., Stephenson & Malanowski 1987)

Henry's Law Constant ($Pa \cdot m^3/mol$):
1.44×10^{-4} (calculated-P/C, this work)

Octanol/Water Partition Coefficient, log K_{OW}:
-0.90 (shake flask, Fujisawa & Masuhara 1980,81; quoted, Sangster 1989)
-1.24 (calculated-HPLC-RT, Fujisawa & Masuhara 1981)
-0.67 (quoted from Pomona College unpublished data, Sangster 1989)
-0.78 (recommended, Sangster 1989)
-0.67 (EPA CLOGP Data Base, Hulzebos et al. 1993)
-0.669 (calculated-QSAR, Kollig 1993)

Bioconcentration Factor, log BCF:

Sorption Partition Coefficient, log K_{OC}:
-0.969 (calculated-K_{OW}, Kollig 1993)

Half-Lives in the Environment/Environmental Fate Rate Constants or Half-Lives: no data

Common Name: Benzamide
Synonym: benzoylamide
Chemical Name: benzamide
CAS Registry No: 55-21-0
Molecular Formula: $C_6H_5CONH_2$
Molecular Weight: 121.14
Melting Point (°C): 128-129
Boiling Point (°C): 288
Density (g/cm^3 at 20°C):
Molar Volume (cm^3/mol):
112.2 (calculated-density)
132.4 (calculated-LeBas method)
Molecular Volume ($Å^3$):
Total Surface Area, TSA ($Å^2$):
257.6 (Iwase et al. 1985)
Dissociation Constant pK_a:
Heat of Fusion, ΔH_{fus}, kcal/mol:
Entropy of Fusion, ΔS_{fus}, cal/mol·K (e.u.):
Fugacity Ratio at 25°C (assuming ΔS_{fus} = 13.5 e.u.), F: 0.0915

Water Solubility (g/m^3 or mg/L at 25°C):
134990 (quoted, Tsonopoulos & Prausnitz 1971)
14560, 30430 (quoted, calculated-K_{OW}, Yalkowsky & Morozowich 1980)
13490 (Windholz 1983; quoted, Chessells et al. 1992)
13515 (1 g in 74 mL, Merck Index, Budavari 1989)

Vapor Pressure (Pa at 25°C):
0.00522 (interpolated, Antoine eqn., Stephenson & Malanowski 1987)

Henry's Law Constant (Pa·m^3/mol):
4.52×10^{-5} (calculated-P/C, this work)

Octanol/Water Partition Coefficient, log K_{OW}:
0.64 (shake flask-AS, Fujuta et al. 1964)
0.65 (Leo et al. 1971; Hansch & Leo 1979; quoted, Iwase et al. 1985)
0.84 (HPLC-k', Hammers et al. 1982)
0.65, 0.64 (quoted, calculated-TSA, Iwase et al. 1985)
0.64, 0.65, 0.66, 0.70, 0.84;
0.64 (quoted and recommended Sangster 1989)
0.64, 0.65 (quoted, CPC-RV, El Tayar et al. 1991)
0.64 (quoted from Medchem software value, Chessells et al. 1992)
0.64 (quoted from Sangster 1989, Bintein & Devillers 1994)

Bioconcentration Factor, log BCF:

Sorption Partition Coefficient, log K_{OC}:

0.954 (sediment, von Oepen et al. 1991)
1.301 (Alfisol soil, von Oepen et al. 1991)
1.756 (Podzol soil, von Oepen et al. 1991)
1.46 (soil, quoted exptl., Meylan et al. 1992)
1.71 (soil, calculated- χ & fragment contribution, Meylan et al. 1992)
1.46 (soil, mean value, Kördel et al. 1993)

Half-Lives in the Environment:

Air: photooxidation half-life of 3.1-31 hours, based on estimated rate constant for the vapor-phase reaction with hydroxyl radicals in air (Atkinson et al. 1987; selected, Howard et al. 1991).

Surface water: 48-360 hours, based on grab sample aerobic soil column test data (Fournier & Salle 1974; selected, Howard et al. 1991).

Groundwater: 96-720 hours, based on grab sample aerobic soil column test data (Fournier & Salle 1974; selecteded, Howard et al. 1991).

Sediment:

Soil: 48-360 hours, based on grab sample aerobic soil column test data (Fournier & Salle 1974; selected, Howard et al. 1991).

Biota:

Environmental Fate Rate Constants or Half-Lives:

Volatilization:

Photolysis:

Oxidation: aqueous photooxidation half-life of 960-7.4×10^4 hours, based on measured rates for reaction with OH radicals in water (Anbar et al. 1966; Dorfman and Adams 1973; selected, Howard et al. 1991); photooxidation half-life of 3.1-31 hours in air, based on estimated rate constant for the vapor-phase reaction with hydroxyl radicals in air (Atkinson et al. 1987; selected, Howard et al. 1991).

Hydrolysis: not expected to be significant based on estimated half-lives for hydrolysis of acetamide of 261, 3950, and 46 years at pH 5,7,9, respectively, which were calculated using experimental acid and base hydrolysis rate constants for acetamide (Mabey & Mill 1978; selected, Howard et al. 1991).

Biodegradation: aqueous aerobic biodegradation half-life of 48-360 hours, and aqueous anaerobic biodegradation half-life of 192-1400 hours, both based on grab sample aerobic soil column test data (Fournier & Salle 1974; selected, Howard et al. 1991).

Biotransformation:

Bioconcentration, Uptake (k_1) and Elimination (k_2) Rate Constants:

Common Name: Urea
Synonym: carbamide, carbonyldiamide, Aquacare, Aqiadrate, Basodexam, Keratinamin, Nutraplus, Onychomal, Pastaron, Ureaphil, Ureophil, Ureapearl
Chemical Name: urea, carbamide, carbonyldiamide
CAS Registry No: 57-13-6
Molecular Formula: H_2NCONH_2
Molecular Weight: 60.06
Melting Point (°C):
135 (Weast 1982-83)
Boiling Point (°C):
Density (g/cm^3): 1.323 (Weast 1982-83)
Dissociation Constants, pKa:
Molar Volume (cm^3/mol):
58.0 (calculated-LeBas method)
Molecular Volume, TMV ($Å^3$):
Total Surface Area, TSA ($Å^2$):
Heat of Fusion, ΔH_{fus}, kcal/mol:
Entropy of Fusion, ΔS_{fus}, cal/mol·K:
Fugacity Ratio at 25°C (assuming ΔS_{fus} = 13.5 cal/mol·K), F: 0.082

Water Solubility (g/m^3 or mg/L at 25°C):
997412 (Windholz 1983; quoted, Chessells et al. 1992)
516000 (Kirk-Othmer 1983)
1000000 (quoted, Dean 1985; Budavari 1989)

Vapor Pressure (Pa at 25°C):
1.61×10^{-3} (extrapolated-Antoine eqn., Stephenson & Malanowski 1987)

Henry's Law Constant ($Pa \cdot m^3/mol$):
9.6×10^{-8} (calculated-P/C, this work)

Octanol/Water Partition Coefficient, log K_{OW}:
-1.09 (Hansch & Leo 1979)
-3.00 (Kenaga & Goring 1980)
-1.57 (HPLC-RT, Harnish et al. 1983)
-1.56 (shake flask, OECD 1981 Guidelines, Geyer et al. 1984)
-2.11 (quoted from Medchem software value, Chessells et al. 1992)
-2.11, 1.66; -2.11 (quoted CLOGP; recommended, Sangster 1993)

Bioconcentration Factor, log BCF:
4.068 (alga *chlorella fusca*, wet wt. basis, Geyer et al. 1984)
-0.100 (alga *chlorella fusca*, calculated-K_{OW}, Geyer et al. 1984)

Sorption Partition Coefficient, log K_{OC}:
0.50, 0.62 (soil, quoted, calculated- χ & fragment contribution, Meylan et al. 1992)

Half-Lives in the Environment/Environmental Fate Rate Constants and Half-Lives: no data

Common Name: Dimethylnitrosoamine
Synonym: N-nitrosodimethylamine, N-methyl-N-nitrosomethanamine, nitrous dimethylamine
Chemical Name: dimethylnitrosoamine, N-nitrosodimethylamine
CAS Registry No: 62-75-9
Molecular Formula: $CH_3N(NO)CH_3$
Molecular Weight: 74.08
Melting Point (°C):
Boiling Point (°C):

151-153 (Windholz 1976; Callahan et al. 1979)
152-153 (Verschueren 1983)

Density (g/cm³ at 20°C):

1.005 (18°C, Verschueren 1983)

Molar Volume (cm³/mol):

87.7 (calculated-LeBas method)

Molecular Volume (Å³):
Total Surface Area, TSA (Å²):
Dissociation Constant pK_a: < 1 (Kollig 1993)
Heat of Fusion ΔH_{fus}, kcal/mol:
Entropy of Fusion ΔS_{fus}, cal/mol·K (e.u.):
Fugacity Ratio at 25°C (assuming $\Delta S = 13.5$ e.u.), F:

Water Solubility (g/m³ or mg/L at 25°C):

miscible (Mirvish et al. 1976; quoted, Callahan et al. 1979; Mabey et al. 1982)

Vapor Pressure (Pa at 25°C):

1080 (quoted, Mabey et al. 1982)

Henry's Law Constant (Pa·m³/mol):

3.344 (calculated, Mabey et al. 1982)

Octanol/Water Partition Coefficient, log K_{OW}:

0.06 (Radding et al. 1977; quoted, Callahan et al. 1979; Ryan et al. 1988)
-0.68 (calculated-f const., Mabey et al. 1982)
0.76 (quoted, Kollig 1993)

Bioconcentration Factor, log BCF:

-0.96 (microorganisms-water, calculated-K_{OW}, Mabey et al. 1982)

Sorption Partition Coefficient, log K_{OC}:

-1.0 (sediment-water, calculated-K_{OW}, Mabey et al. 1982)
0.448 (calculated-K_{OW}, Kollig 1993)

Half-Lives in the Environment:

Air: 0.5-1.0 hour, based on measured rate of photolysis in the vapor phase under sunlight (Hanst et al. 1977; quoted, Howard et al. 1991); photooxidation half-life of 25.4-254 hours, based on measured rate constant for the reaction with hydroxyl radicals in air (Atkinson 1985; quoted, Howard et al. 1991).

Surface water: 0.5-1.0 hour, based on measured rate of photolysis in the vapor phase under sunlight (Hanst et al. 1977; quoted, Howard et al. 1991).

Groundwater: 1008-8640 hours, based on estimated unacclimated aqueous aerobic biodegradation half-life (Howard et al. 1991).

Sediment:

Soil: 504-4320 hours, based on aerobic soil die-away test data (Tate & Alexander 1975; Oliver et al. 1979; quoted, Howard et al. 1991).

Biota:

Environmental Fate Rate Constants or Half-Lives:

Volatilization:

Photolysis: both aqueous and atmospheric photolysis half-life of 0.5-1.0 hour, based on measured rate of photolysis in the vapor phase under sunlight (Hanst et al. 1977; quoted, Howard et al. 1991).

Hydrolysis:

Oxidation: rate constants for singlet oxygen, of $<3600\ M^{-1}h^{-1}$ and for peroxy radical of $<3600\ M^{-1}h^{-1}$ at 25°C (Mabey et al. 1982); photooxidation half-life of 25.4-254 hours in air, based on measured rate constant for the reaction with hydroxyl radicals in air (Atkinson 1985; quoted, Howard et al. 1991).

Biodegradation: aqueous aerobic half-life of 504-4320 hours, based on aerobic soil die-away test data (Tate & Alexander 1975; Oliver et al. 1979; quoted, Howard et al. 1991); aqueous anaerobic half-life of 2016-17280 hours, based on estimated unacclimated aqueous aerobic biodegradation half-life (Howard et al. 1991).

Biotransformation: rate constant for bacterial transformation of $3x10^{-12}\ mL\cdot cell^{-1}\cdot h^{-1}$ in water (Mabey et al. 1982).

Bioconcentration, Uptake (k_1) and Elimination (k_2) Rate Constants:

Common Name: Di-*n*-Propylnitrosoamine
Synonym: N-nitrosodi-*n*-propylamine, N-nitroso-N-propyl-1-propanamine
Chemical Name: di-*n*-propylnitrosoamine, N-nitrosodi-*n*-propylamine
CAS Registry No: 621-64-7
Molecular Formula: $CH_3CH_2CH_2N(NO)CH_2CH_2CH_3$
Molecular Weight: 130.19
Melting Point (°C):
Boiling Point (°C):
    205 (Windholz 1976; Callahan et al. 1979; Mabey et al. 1982)
Density (g/cm³ at 20°C):
Molar Volume (cm³/mol):
    176.5 (calculated-LeBas method)
Molecular Volume (Å³):
Total Surface Area, TSA (Å²):
Heat of Fusion ΔH_{fus}, kcal/mol:
Entropy of Fusion ΔS_{fus}, cal/mol·K (e.u.):
Fugacity Ratio at 25°C (assuming ΔS_{fus} = 13.5 e.u.), F:

Water Solubility (g/m³ or mg/L at 25°C):
    9900 (Mirvish et al. 1976; quoted, Callahan et al. 1979; Mabey et al. 1982)

Vapor Pressure (Pa at 25°C):
    53.33 (37°C, calculated-Troutin's rule, Mabey et al. 1982)

Henry's Law Constant (Pa·m³/mol):
    0.638 (calculated-P/C, Mabey et al. 1982)
    0.355 (calculated-P/C, this work)

Octanol/Water Partition Coefficient, log K_{OW}:
    1.31 (calculated as per Leo et al. 1971 from Mirvish et al. 1976 data, Callahan et al. 1979; Ryan et al. 1988)
    1.49 (calculated-f const., Mabey et al. 1982)
    1.35 (quoted, Kollig 1993)

Bioconcentration Factor, log BCF:
    0.99 (microorganisms-water, calculated-K_{OW}, Mabey et al. 1982)

Sorption Partition Coefficient, log K_{OC}:
    1.18 (sediment-water, calculated-K_{OW}, Mabey et al. 1982)
    1.09 (calculated-K_{OW}, Kollig 1993)

Half-Lives in the Environment:

Air: photooxidation half-life of 2.66-26.6 hours, based on estimated rate constant for the reaction with hydroxyl radicals in air (Atkinson 1987; quoted, Howard et al. 1991).

Surface water: 0.17-1.0 hour, based on measured rate of photolysis in the vapor phase under sunlight (Oliver 1981; quoted, Howard et al. 1991).

Groundwater: 1008-8640 hours, based on estimated unacclimated aqueous aerobic biodegradation half-life (Howard et al. 1991).

Sediment:

Soil: 504-4320 hours, based on aerobic soil die-away test data (Tate & Alexander 1975; Oliver et al. 1979; quoted, Howard et al. 1991).

Biota:

Environmental Fate Rate Constants or Half-Lives:

Volatilization:

Photolysis: both aqueous and atmospheric photolysis half-life of 0.17-1.0 hour, based on measured rate of photolysis in the vapor phase under sunlight (Oliver 1981; quoted, Howard et al. 1991).

Hydrolysis:

Oxidation: rate constants in water for singlet oxygen of $<3600\ M^{-1}h^{-1}$ and for peroxy radical of $<3600\ M^{-1}h^{-1}$ at 25°C (Mabey et al. 1982); photooxidation half-life of 2.66-26.6 hours in air, based on estimated rate constant for the reaction with hydroxyl radicals in air (Atkinson 1987; quoted, Howard et al. 1991).

Biodegradation: aqueous aerobic half-life of 504-4320 hours, based on aerobic soil die-away test data (Tate & Alexander 1975; Oliver et al. 1979; quoted, Howard et al. 1991); aqueous anaerobic half-life of 2016-17280 hours, based on estimated unacclimated aqueous aerobic biodegradation half-life (Howard et al. 1991).

Biotransformation: rate constant for bacterial biotransformation of $3\times10^{-12}\ mL\cdot cell^{-1}\cdot h^{-1}$ in water (Mabey et al. 1982).

Bioconcentration, Uptake (k_1) and Elimination (k_2) Rate Constants:

Common Name: Diphenylnitrosoamine
Synonym: N-nitrosodiphenylamine, N-nitroso-N-phenylbenzamine
Chemical Name: diphenylnitrosoamine, N-nitrosodiphenylamine
CAS Registry No: 86-30-6
Molecular Formula: $C_6H_5N(NO)C_6H_5$
Molecular Weight: 198.24
Melting Point (°C):
66.5 (Tanikaga 1969; Callahan et al. 1979; Weast 1982-83)
67.0 (Banerjee et al. 1980; Mabey et al. 1982)
Boiling Point (°C):
151-153 (Windholz 1976; Callahan et al. 1979)
Density (g/cm³ at 20°C):
Molar Volume (cm³/mol):
220.5 (calculated-LeBas method)
Molecular Volume (Å³):
Total Surface Area, TSA (Å²):
Dissociation Constant pK_a: < 0 (Kollig 1993)
Heat of Fusion ΔH_{fus}, kcal/mol:
Entropy of Fusion ΔS_{fus}, cal/mol·K (e.u.):
Fugacity Ratio at 25°C (assuming ΔS = 13.5 e.u.), F: 0.3886

Water Solubility (g/m³ or mg/L at 25°C):
35.1 (shake flask-LSC, Banerjee et al. 1980)
40.0 (calculated-S, Mabey et al. 1982)
34.7 (quoted, Isnard & Lambert 1988)

Vapor Pressure (Pa at 25°C):
13.33 (estimated, Mabey et al. 1982)

Henry's Law Constant (Pa·m³/mol):
66.87 (calculated-P/C, Mabey et al. 1982)

Octanol/Water Partition Coefficient, log K_{OW}:
2.57 (calculated as per Leo et al. 1971 from Mirvish 1976 data, Callahan et al. 1979)
3.13 (shake flask-LSC, Banerjee et al. 1980; quoted, Sangster 1993)
3.13 (quoted, Mabey et al. 1982; Isnard & Lambert 1988)
2.57 (quoted from USEPA 1979, Ryan et al. 1988)
3.13 (recommended, Sangster 1993)
3.16 (quoted, Kollig 1993)

Bioconcentration Factor, log BCF:
2.63 (microorganisms-water, calculated-K_{OW}, Mabey et al. 1982)
2.34 (quoted, Isnard & Lambert 1988)

Sorption Partition Coefficient, log K_{OC}:

2.81 (sediment-water, calculated-K_{OW}, Mabey et al. 1982)

2.84 (calculated-K_{OW}, Kollig 1993)

Half-Lives in the Environment:

Air: photooxidation half-life of 0.70-7.0 hours, based on estimated rate constant for the reaction with hydroxyl radicals in air (Atkinson 1987; quoted, Howard et al. 1991).

Surface water: 240-816 hours, based on estimated unacclimated aqueous aerobic biodegradation half-life (Howard et al. 1991).

Groundwater: 480-1632 hours, based on estimated unacclimated aqueous aerobic biodegradation half-life (Howard et al. 1991).

Sediment:

Soil: 240-816 hours, based on estimated unacclimated aqueous aerobic biodegradation half-life (Howard et al. 1991).

Biota:

Environmental Fate Rate Constants or Half-Lives:

Volatilization:

Photolysis:

Hydrolysis:

Oxidation: rate constants in water for singlet oxygen of $<3600\ M^{-1}h^{-1}$ and for peroxy radical of $<3600\ M^{-1}h^{-1}$ at 25°C (Mabey et al. 1982); photooxidation half-life of 0.70-7.0 hours in air, based on measured rate constant for the reaction with hydroxyl radicals in air (Atkinson 1987; quoted, Howard et al. 1991).

Biodegradation: aqueous aerobic half-life of 240-816 hours, based on data from one soil-die-away test; a range was bracketed around the reported half-life of 22 days (Mallik & Tesfai 1981; quoted, Howard et al. 1991); aqueous anaerobic half-life of 960-3264 hours, based on estimated unacclimated aqueous aerobic biodegradation half-life (Howard et al. 1991).

Biotransformation: rate constant for bacterial transformation of $1 \times 10^{-10}\ mL \cdot cell^{-1} \cdot h^{-1}$ in water (Mabey et al. 1982).

Bioconcentration, Uptake (k_1) and Elimination (k_2) Rate Constants:

Common Name: Pyrrole
Synonym: 1H-pyrrole
Chemical Name: pyrrole, 1H-pyrrole
CAS Registry No: 109-97-7
Molecular Formula: C_4H_4NH
Molecular Weight: 67.09
Melting Point (°C):
-23.4 (Dean 1985; Riddick et al. 1986)
-18.5 (Stephenson & Malanowski 1987)
Boiling Point (°C):
130-131 (Weast 1982-83)
129.77 (Boublik et al. 1984; Riddick et al. 1986)
130.00 (Stephenson & Malanowski 1987)
Density (g/cm³ at 20°C):
0.96910 (Weast 1982-83; Dean 1985)
0.96985 (Riddick et al. 1986)
Molar Volume (cm³/mol):
42.8 (instrinsic V_I, Hickey & Passino-Reader 1991)
78.2 (calculated-LeBas method)
69.2 (calculated-density)
Molecular Volume (Å³):
Total Surface Area, TSA (Å²):
Dissociation Constant, pK_a:
-4.40 (Perrin 1972)
-3.80 (Riddick et al. 1986)
-4.10 (Sangster 1989)
Heat of Fusion, ΔH_{fus}, kcal/mol:
1.89 (Riddick et al. 1986)
Entropy of Fusion, ΔS_{fus}, cal/mol K (e.u.):
Fugacity Ratio at 25°C (assuming ΔS_{fus} = 13.5 e.u.), F: 1.0

Water Solubility (g/m³ or mg/L at 25°C):
47000 (Dean 1985)
45000 (Riddick et al. 1986)

Vapor Pressure (Pa at 25°C):
1102 (ebulliometry, calculated-Antoine eqn., Osborn & Douslin 1968)
1136 (calculated-Cox eqn., Chao et al. 1983)
1100 (extrapolated-Antoine eqn., Boublik et al. 1984)
1100 (extrapolated-Antoine eqn., Dean 1985)
1100 (Riddick et al. 1986)
1103 (extrapolated-Antoine eqn., Stephenson & Malanowski 1987)

Henry's Law Constant (Pa m³/mol):
1.640 (calculated-P/C, this work)

Octanol/Water Partition Coefficient, log K_{OW}:

0.75 (shake flask-AS, Hansch & Anderson 1967; quoted, Leo et al. 1971; Rekker 1977; Rekker & De Kort 1979; Sangster 1989)
0.72 (calculated-f const., Rekker 1977)
0.75 (quoted best value, Garst 1984)
0.62 (HPLC-RV, Garst 1984)
0.75 (quoted from Hansch & Leo 1979, Schultz & Moulton 1985; Schultz et al. 1989)
0.75, 0.80 (quoted, calculated-f const., Rekker & De Kort 1979)
0.75 (recommended, Sangster 1989)

Bioconcentration Factor, log BCF:

Sorption Partition Coefficient, log K_{OC}:

Half-Lives in the Environment:

Air:
Surface water:
Groundwater:
Sediment:
Soil:
Biota:

Environmental Fate Rate Constants or Half-Lives:

Volatilization:
Photolysis:
Oxidation: comparison of rate constants and loss rate for the gas-phase reaction at room temperature: 1.2×10^{-10} cm^3 molecule^{-1} sec^{-1} with 1×10^6 OH radicals/cm^3 during daytime with loss rate of 10 day^{-1} having a calculated lifetime of 2.3 hours, 4.9×10^{-11} cm^3 molecule^{-1} sec^{-1} with 2.4×10^8 NO_3 radicals/cm^3 during nighttime hours with loss rate of 1000 day^{-1} having a calculated lifetime of 1.4 minute and 1.6×10^{-17} cm^3 molecule^{-1} sec^{-1} with 7×10^{11} O_3 molecules/cm^3 with loss rate of 1.0 day^{-1} having a calculated lifetime of 24 hours (Atkinson et al. 1984; quoted, Atkinson 1985; Atkinson et al. 1985); second order rate constant of 1.2×10^{-10} cm^3·molecule^{-1}·sec^{-1} for the gas-phase reaction with 5×10^5 OH radicals/cm^3 at room temperature having a loss rate of 5.2 day^{-1} (Atkinson 1985); using relative rate technique as compared to 2-methyl-2-butene, rate constant was determined to be $(4.9 \pm 1.1) \times 10^{-11}$ cm^3 molecule^{-1} sec^{-1} for the gas-phase reaction with with 2.4×10^8 NO_3 radicals/cm^3 with loss rate of 1000 day^{-1} during nighttime hours at 295 ± 1 K in the atmosphere with a calculated lifetime of 1.4 minute (Atkinson et al. 1985; quoted, Atkinson 1985); experimental rate constant of 110.0×10^{-12} cm^3 molecule^{-1} sec^{-1} as compared to a calculated value of 287.45×10^{-12} cm^3·molecule^{-1} sec^{-1} by molecular orbital calculation for the gas-phase reaction with hydroxyl radicals at room temperature in air (Klamt 1993).
Hydrolysis:
Biodegradation:
Biotransformation:
Bioconcentration, Uptake (k_1) and Elimination (k_2) Rate Constants

Common Name: Indole
Synonym: benzo[b]pyrrole, 1-benzo[b]pyrrole, 1H-indole
Chemical Name: indole
CAS Registry No: 120-72-9
Molecular Formula: C_8H_7N
Molecular Weight: 117.15
Melting Point (°C):
52.5 (Weast 1982-83; Verschueren 1983)
53.0 (Pearlman et al. 1984)
52.0 (Dean 1985; Stephenson & Malanowski 1987)
Boiling Point (°C):
254 (Weast 1982-83; Verschueren 1983; Pearlman et al. 1984; Stephenson & Malanowski 1987)
254.7 (Boublik et al. 1984)
253 (Dean 1985)
Density (g/cm^3 at 20°C):
1.220 (Weast 1982-83)
1.0643 (Dean 1985)
Molar Volume (cm^3/mol):
133.4 (calculated-density)
Molecular Volume (Å^3):
121.96 (Pearlman et al. 1984)
Total Surface Area, TSA (Å^2): 140.60 (Pearlman et al. 1984)
Dissociation Constant, pK_a:
-3.5, -3.62 (Perrin 1972)
-3.17 (Sangster 1989)
Heat of Fusion, ΔH_{fus}, kcal/mol:
Entropy of Fusion, ΔS_{fus}, cal/mol K (e.u.):
Fugacity Ratio at 25°C (assuming ΔS_{fus} = 13.5 e.u.), F: 0.5346

Water Solubility (g/m^3 or mg/L at 25°C):
3558 (shake flask-GC, Price 1976; quoted, Yalkowsky et al. 1987)
1874 (quoted, Pearlman et al. 1984)
1762 (quoted, Isnard & Lambert 1988)

Vapor Pressure (Pa at 25°C):
2.240 (extrapolated-Antoine eqn., Boublik et al. 1984)
1.565 (calculated-Antoine eqn., Stephenson & Malanowski 1987)

Henry's Law Constant (Pa·m^3/mol):
0.14 (calculated-P/C, this work)

Octanol/Water Partition Coefficient, log K_{OW}:
1.14 ± 0.01 (shake flask-AS, Iwasa et al. 1965)
2.14 (shake flask-AS, Hansch & Anderson 1967; quoted, Leo et al. 1969,71; Rekker 1977; Sangster 1989)

2.25 (shake flask-AS at pH 7.4, Rogers & Cammarata 1969; quoted, McCall 1975; Sangster 1989)
2.28 (calculated-M.O. indices, Rogers & Cammarata 1969)
2.00 (quoted unpublished result, Leo et al. 1971)
2.06 (calculated-f const., Rekker 1977)
2.00, 2.25, 2.13 (quoted, Hansch & Leo 1979)
2.25 (quoted, Veith et al. 1979)
1.66 (estimated-HPLC-RT, Veith et al. 1979; quoted, Sangster 1989)
2.06 (calculated by Rekker's method, Hanai et al. 1981)
2.00, 2.06, 2.13, 2.14, 2.25 (values quoted from Rekker 1977, Hanai & Hubert 1982)
2.17 (RP-LC, Hanai & Hubert 1982)
2.06 (calculated-f const. as per Rekker 1977, Hanai & Hubert 1982)
2.14 (inter-lab. shake flask average, Eadsforth & Moser 1983)
1.92 (inter-lab. HPLC average, Eadsforth & Moser 1983; quoted, Sangster 1989)
2.00, 2.25 (quoted, Verschueren 1983)
2.14 (selected best lit. value, Garst & Wilson 1984)
2.16 $\pm$ 0.03 (exptl.-ALPM, Garst & Wilson 1984)
2.25 (quoted from Hansch & Leo 1979, Haky & Young 1984)
2.13 (quoted from Hansch & Leo 1979, Schultz & Moulton 1985)
1.81 (HPLC-k', Eadsforth 1986; quoted, Sangster 1989)
2.13, 2.13 (quoted: shake flask, calculated, Eadsforth 1986)
2.16 (HPLC, Minick et al. 1988)
2.00, 2.12 (quoted from Pomona College unpublished data, Sangster 1989)
2.14 (recommended, Sangster 1989)
2.19 (HPLC-RT, Ritter et al. 1994)

Bioconcentration Factor, log BCF:

Sorption Partition Coefficient, log K_{OC}:

Half-Lives in the Environment:
Air:
Surface water:
Groundwater:
Sediment:
Soil:
Biota:

Environmental Fate Rate Constants or Half-Lives:
Volatilization:
Photolysis:
Oxidation:
Hydrolysis:
Biodegradation:
Biotransformation:
Bioconcentration, Uptake (k_1) and Elimination (k_2) Rate Constants:

Common Name: Pyridine
Synonym:
Chemical Name: pyridine
CAS Registry No: 110-86-1
Molecular Formula: C_5H_5N
Molecular Weight: 79.10
Melting Point (°C):
-42.0 (Stull 1947; Weast 1982-83, Jori et al. 1983; Verschueren 1983)
-41.6 (Dean 1985; Riddick et al. 1986)
Boiling Point (°C):
115.4 (Stull 1947)
115.26 (Herington & Martin 1953)
115.5 (Weast 1982-83)
115.3 (Jori et al. 1983; Boublik et al. 1984; Riddick et al. 1986)
115.0 (Verschueren 1983; Banerjee et al. 1990)
115.2 (Dean 1985)
Density (g/cm^3 at 20°C):
0.9819 (Weast 1982-83)
0.9820 (Verschueren 1983)
0.9832 (Riddick et al. 1986)
Molar Volume (cm^3/mol):
80.6 (calculated-density, Rohrschneider 1973)
80.88 (calculated-density, Yeh & Higuchi 1976)
90.8 (Taft et al. 1985)
90.5 (Kamlet et al. 1987)
47.0 (intrinsic volume V_I, Kamlet et al. 1987)
47.2 (instrinsic V_I, Hickey & Passino-Reader 1991)
93.0 (calculated-LeBas method)
Molecular Volume ($Å^3$):
Total Surface Area, TSA ($Å^2$):
Dissociation Constants, pK:
5.23 (pK_a, Leo et al. 1971; Jori et al. 1983; Zachara et al. 1987)
5.198, 5.21, 5.22, 5.229 (Perrin 1972)
5.539 (UV, Yeh & Higuchi 1976)
5.23, 5.16 (quoted, shake flask-TN, Clarke 1984)
5.17 ($pK_{BH}{}^+$, Dean 1985; Riddick et al. 1986)
5.21 (pK_a, Sangster 1989)
Heat of Fusion ΔH_{fus}, kcal/mol:
1.772 (Riddick et al. 1986)
Entropy of Fusion ΔS_{fus}, cal/mol·K (e.u.):
Fugacity Ratio at 25°C (assuming ΔS_{fus} = 13.5 e.u.), F: 1.0

Water Solubility (g/m^3 or mg/L at 25°C):
miscible (Andon & Cox 1952; Andon et al. 1954; Jori et al. 1983)
miscible (Dean 1985; Zachara et al. 1987; Stephenson 1993)
infinite (Riddick et al. 1986)

233441, 1406619 (quoted, calculated-solvatochromic p., Kamlet et al. 1987)
miscible (Yaws et al. 1990)

Vapor Pressure (Pa at 25°C):

2520 (interpolated-regression of tabulated data, Stull 1947)
2666 (24.8°C, Stull 1947)
2775 (ebulliometry, extrapolated-Antoine eqn., Herington & Martin 1953)
2774 (calculated-Antoine eqn., Andon et al. 1954)
2763 (ebulliometry, calculated-Antoine eqn., Osborn & Douslin 1968)
2789 (calculated-Antoine eqn., Cabani et al. 1971)
2767 (calculated-Cox eqn., Chao et al. 1983)
2666 (24.8°C, quoted, Jori et al. 1983)
2666 (quoted, Verschueren 1983)
2769 (extrapolated-Antoine eqn., Boublik et al. 1984)
2763 (extrapolated-Antoine eqn., Dean 1985)
2773 (Howard et al. 1986; quoted, Banerjee et al. 1990)
2700 (Riddick et al. 1986)
2773, 1653 (measured, calculated-solvatochromic p., Banerjee et al. 1990)
2478 (ebulliometry, calculated-Antoine eqn., Łencka 1990)

Henry's Law Constant (Pa·m^3/mol):

0.895 (volatility ratio-transpiration method, Andon et al. 1954)
0.900 (quoted exptl., Hine & Mookerjee 1975)
0.595 (calculated-group contribution, Hine & Mookerjee 1975)
0.766 (calculated-bond contribution, Hine & Mookerjee 1975)
1.317, 1.114 (quoted, modified gas-stripping, Hawthorne et al. 1985)
1.120 (computed, Yaws et al. 1991)
0.900, 0.305 (quoted, calculated-molecular structure, Russell et al. 1992)

Octanol/Water Partition Coefficient, log K_{OW}:

0.65 (shake flask-AS, Iwasa et al. 1965; quoted, Leo et al. 1969,71; Mirrlees et al. 1976; Jori et al. 1983; Sangster 1989)
0.64 (Gehring et al. 1967; quoted, Kaiser 1983)
0.64 (quoted unpublished result, Leo et al. 1971; quoted, Jori et al. 1983)
0.66 (HPLC, Mirrlees et al. 1976; quoted, Sangster 1989)
0.63 ± 0.02 (shake flask at pH 7, Unger et al. 1978)
0.62, 0.71 (quoted, calculated-f const., Rekker & De Kort 1979)
0.64, 1.04 (quoted, Verschueren 1983)
0.65, 0.63 (quoted, shake flask-TN, Clarke 1984; Clarke & Cahoon 1987)
0.66 (selected best lit. value, Garst & Wilson 1984)
0.63 ± 0.06 (exptl.-ALPM, Garst & Wilson 1984)
0.62 (shake flask-AS at pH 7.4, El Tayar et al. 1984; quoted, Sangster 1989)
0.65; 0.63 (quoted, shake flask-potentiometric titration, Clarke 1984)

0.65, 0.54 (quoted, calculated-γ from UNIFAC, Campbell & Luthy 1985; quoted, Sangster 1989)
0.64 (quoted from Hansch & Leo 1979, Schultz & Moulton 1985)
0.65 (quoted from Hansch & Leo 1979, Miyake et al. 1987)
0.65, 0.81 (quoted, calculated-solvatochromic p., Taft et al. 1985)
1.28 (HPLC-k', Eadsforth 1986)
0.77, 0.66 (quoted: shake flask, calculated, Eadsforth 1986)
0.67 (selected, Yoshida et al. 1986)
1.00 (selected, Zachara et al. 1987)
0.65, 0.79 (quoted, calculated-γ from UNIFAC, Banerjee & Howard 1988)
0.65 (quoted from Leo et al. 1971, Berthod et al. 1988)
0.70 (shake flask-CPC, Berthod et al. 1988)
0.63 (shake flask-HPLC at pH 7, De Voogt et al. 1988,90)
0.64, 0.66, 0.68 (quoted from Pomona College unpublished data, Sangster 1989)
0.65 (recommended, Sangster 1989)
0.64 (quoted, Sims & O'Loughlin 1989)
0.70 (RP-TLC, De Voogt et al. 1990)
0.65 (shake flask-AS, Yamagami et al. 1990)
0.65 (quoted from CLOGP 1983 of Pomona College, El Tayar et al. 1991)
0.65 (quoted from U.S. EPA ClOGP Data Base, Hulzebos et al. 1993)
0.665 (calculated-QSAR, Kollig 1993)

Bioconcentration Factor, log BCF:

Sorption Partition Coefficient, log K_{OC}:
-1.805 (estimated of Anvil Points subsurface materials, Zachara et al. 1987)
-2.541 (estimated of Loring subsurface materials, Zachara et al. 1987)
0.34 (calculated-K_{OW}, Kollig 1993)

Half-Lives in the Environment:

Air: atmospheric lifetimes of 46 days in clean troposphere and 23 days in moderately polluted atmosphere, based on the gas-phase reaction with hydroxyl radicals in air at room temperature and >4 years in clean troposphere and >1.3 year in moderately polluted atmosphere, based on the gas-phase reaction with O_3 (calculated rate constant) in air at room temperature (Atkinson et al. 1987); photooxidation half-life of 128-1284 hours, based on measured rate data for the reaction with hydroxyl radicals in air (selected, Howard et al. 1991).

Surface water: photooxidation half-life of 14.7-24.4 years, based on measured rate data for the reaction with hydroxyl radicals in aqueous solution (Dorfman & Adams 1973; selected, Howard et al. 1991); 24-168 hours, based on estimated aqueous aerobic biodegradation half-life (Howard et al. 1991).

Groundwater: 48-336 hours, based on estimated aqueous aerobic biodegradation half-life (Howard et al. 1991).

Sediment:

Soil: disappears in less than 7 days in soil suspensions (Sims & O'Loughlin 1989); half-lives of 24-168 hours, based on unacclimated grab sample of aerobic soil (Sims & Sommers 1985; selected, Howard et al. 1991).

Biota:

Environmental Fate Rate Constants or Half-Lives:

Volatilization:

Photolysis:

Oxidation: photooxidation half-life of 14.7-24.4 years in water, based on measured rate data for the reaction with hydroxyl radicals in aqueous solution (Dorfman & Adams 1973; selected, Howard et al. 1991); photooxidation half-life of 128-1284 hours in air, based on measured rate data for the reaction with hydroxyl radicals in air (Atkinson 1987; selected, Howard et al. 1991); exptl. rate constant of $(0.49 \pm 0.04)\times10^{-12}$ cm^3 $molecule^{-1}$ sec^{-1} for the gas-phase reaction with hydroxyl radicals in atmosphere at room temperature and calculated rate constant of $\geq 1.5\times10^{-16}$ cm^3 $molecule^{-1}$ sec^{-1} for the gas-phase reaction with O_3 in atmosphere at room temperature (Atkinson et al. 1987); exptl. rate constant of 0.49×10^{-12} cm^3 $molecule^{-1}$ sec^{-1} (Atkinson 1989) as compared to calculated value of 0.45×10^{-12} cm^3 $molecule^{-1}$ sec^{-1} by molecular orbital calculation for the gas-phase reaction with hydroxyl radicals at room temperature in air (Klamt 1993).

Hydrolysis:

Biodegradation: aqueous aerobic half-life of 24-168 hours, based on unacclimated grab sample of aerobic soil (Sims & Sommers 1985; quoted, Howard et al. 1991); aqueous anaerobic half-life of 168-672 hours, based on anaerobic acclimated screening test data (Naik et al. 1972; selected, Howard et al. 1991).

Biotransformation:

Bioconcentration, Uptake (k_1) and Elimination (k_2) Rate Constants:

Common Name: 2-Methylpyridine
Synonym: α-picoline, 2-picoline
Chemical Name: 2-methylpyridine, α-picoline
CAS Registry No: 109-06-8
Molecular Formula: $C_5H_4NCH_3$
Molecular Weight: 93.13
Melting Point (°C):
-70.0 (Stull 1947; Verschueren 1983; Howard 1990)
-67.0 (Weast 1982-83; Dean 1985; Riddick et al. 1986)
-64.0 (Stephenson & Malanowski 1987)
Boiling Point (°C):
128.8 (Stull 1947; Verschueren 1983; Weast 1982-83)
129.41 (Herington & Martin 1953; Boublik et al. 1984; Riddick et al. 1986)
128-129 (Dean 1985; Howard 1990)
129.50 (Stephenson & Malanowski 1987)
Density (g/cm^3 at 20°C):
0.9443 (Weast 1982-83)
0.9444 (Riddick et al. 1986)
Molar Volume (cm^3/mol):
115.2 (calculated-LeBas method)
98.62 (calculated-density)
Molecular Volume ($Å^3$):
Total Surface Area, TSA ($Å^2$):
Dissociation Constants, pK:
5.957, 6.06 (Perrin 1972)
5.97 (pK_a, 20°C, Weast 1982-83)
5.997 ($pK_{BH}{}^+$, Riddick et al. 1986; quoted, Howard 1990)
Heat of Fusion ΔH_{fus}, kcal/mol:
2.324 (Riddick et al. 1986)
Entropy of Fusion ΔS_{fus}, cal/mol·K (e.u.):
Fugacity Ratio at 25°C (assuming ΔS_{fus} = 13.5 e.u.), F: 1.0

Water Solubility (g/m^3 or mg/L at 25°C):
miscible (Andon & Cox 1952)
infinite (Riddick et al. 1986; quoted, Howard 1990)
miscible (Yaws et al. 1990)
miscible (Stephenson 1993)

Vapor Pressure (Pa at 25°C):
1370 (interpolated-regression of tabulated data, Stull 1947)
1333 (24.4°C, Stull 1947)
1277 (manometry-fitted to Antoine eqn., Hopke & Sears 1951)
1496 (ebulliometry, extrapolated-Antoine eqn., Herington & Martin 1953)
1496 (calculated-Antoine eqn., Andon et al. 1954)
1493 (ebulliometry, calculated-Antoine eqn., Osborn & Douslin 1968)
1500 (calculated-Cox eqn., Chao et al. 1983)
1067 (20°C, quoted, Verschueren 1983)

1496 (extrapolated-Antoine eqn., Boublik et al. 1984)
1494 (extrapolated-Antoine eqn., Dean 1985)
1333 (Riddick et al. 1986; quoted, Howard 1990)
1386 (extrapolated-Antoine eqn., Stephenson & Malanowski 1987)
1500 (ebulliometry, calculated-Antoine eqn., Łencka 1990)

Henry's Law Constant (Pa·m^3/mol):
1.01 (measured volatility ratio-transpiration method, Andon et al. 1954)
1.010 (quoted exptl., Hine & Mookerjee 1975; quoted, Howard 1990)
0.821 (calculated-group contribution, Hine & Mookerjee 1975)
0.749 (calculated-bond contribution, Hine & Mookerjee 1975)
2.900 (computed, Yaws et al. 1991)

Octanol/Water Partition Coefficient, log K_{OW}:
1.06 (calculated, Verschueren 1983)
0.52 (HPLC, Schultz & Moulton 1985)
1.11 (Hansch & Leo 1985; quoted, Howard 1990)
1.11 (recommended, Sangster 1989)

Bioconcentration Factor, log BCF:
0.602 (calculated-K_{OW}, Lyman et al. 1982)

Sorption Partition Coefficient, log K_{OC}:

Half-Lives in the Environment:
Air: half-life of 11.2 days, based on the gas-phase reaction with photochemically produced hydroxyl radicals in air (Atkinson 1987; quoted, Howard 1990).
Surface water:
Groundwater:
Sediment:
Soil:
Biota:

Environmental Fate Rate Constants or Half-Lives:
Volatilization: using Henry's law constant, a half-life of 88 hours was estimated for a model river 1 m deep flowing at 1 m/sec with a wind velocity of 3 m/sec (Lyman et al. 1982; quoted, Howard 1990).
Photolysis:
Oxidation: photooxidation half-life of 11.2 days in air, based on the gas-phase reaction with photochemically produced hydroxyl radicals in air (Atkinson 1987; quoted, Howard 1990).
Hydrolysis:
Biodegradation:
Biotransformation:
Bioconcentration, Uptake (k_1) and Elimination (k_2) Rate Constants:

Common Name: 3-Methylpyridine
Synonym: β-picoline, 3-picoline
Chemical Name: 3-methylpyridine, β-picoline
CAS Registry No: 108-99-6
Molecular Formula: $C_5H_4NCH_3$
Molecular Weight: 93.13
Melting Point (°C):
-18.30 (Dean 1985; Stephenson & Malanowski 1987; Howard 1993)
-18.09 (Riddick et al. 1986)
Boiling Point (°C):
144.14 (Herington & Martin 1953; Boublik et al. 1984; Riddick et al. 1986)
143.5 (Dean 1985; Verschueren 1983)
Density (g/cm³ at 20°C):
0.9443 (Weast 1982-83)
0.95658 (Riddick et al. 1986)
Molar Volume (cm³/mol):
107.4 (Kamlet et al. 1987)
57.0 (intrinsic volume V_I, Kamlet et al. 1987)
115.2 (calculated-LeBas method)
97.35 (calculated-density)
Molecular Volume (Å³):
Total Surface Area, TSA (Å²):
Dissociation Constants, pK:
5.67, 5.703 (Perrin 1972)
5.68 (pK_a, 20°C, Weast 1982-83; pK_a, protonated cation +1, Dean 1985)
5.750 ($pK_{BH}{}^+$, Riddick et al. 1986)
5.650 (pK_a, Sangster 1989)
Heat of Fusion ΔH_{fus}, kcal/mol:
3.389 (Riddick et al. 1986)
Entropy of Fusion ΔS_{fus}, cal/mol·K (e.u.):
Fugacity Ratio at 25°C (assuming ΔS_{fus} = 13.5 e.u.), F: 1.0

Water Solubility (g/m³ or mg/L at 25°C):
miscible (Andon & Cox 1952; Andon et al. 1954; Yaw et al. 1990)
infinite (Riddick et al. 1986)
102115, 601300 (quoted, calculated-solvatochromic p., Kamlet et al. 1987)
miscible (Howard 1993; Stephenson 1993)

Vapor Pressure (Pa at 25°C):
794 (ebulliometry, extrapolated-Antoine eqn., Herington & Martin 1953)
794 (calculated-Antoine eqn., Andon et al. 1954)
811 (ebulliometry, calculated-Antoine eqn., Osborn & Douslin 1968)
806 (calculated-Cox eqn., Chao et al. 1983; quoted, Howard 1993)
798 (extrapolated-Antoine eqn., Boublik et al. 1984)
800 (extrapolated-Antoine eqn., Dean 1985)
1333 (Riddick et al. 1986)
802 (extrapolated-Antoine eqn., Stephenson & Malanowski 1987)

Henry's Law Constant ($Pa \cdot m^3/mol$):

0.788 (volatility ratio-transpiration method, Andon et al. 1954)
0.784 (quoted exptl., Hine & Mookerjee 1975)
0.637 (calculated-group contribution, Hine & Mookerjee 1975)
0.749 (calculated-bond contribution, Hine & Mookerjee 1975)
1.836 (computed, Yaws et al. 1991)
0.0751 (quoted from Andon et al. 1954, Howard 1993)

Octanol/Water Partition Coefficient, log K_{OW}:

1.20, 1.20 (quoted, HPLC, Mirrlees et al. 1976)
1.20 ± 0.02 (shake flask at pH 7, Unger et al. 1978)
1.19 (HPLC, Lewis et al. 1983; quoted, Sangster 1989)
1.20 (quoted, Verschueren 1983)
1.20 (selected best lit. value, Garst 1984)
1.18 ± 0.01 (exptl.-ALPM, Garst 1984; Garst & Wilson 1984)
1.20 (Hansch & Leo 1985; quoted, Howard 1993)
1.20 (quoted and recommended, Sangster 1989)

Bioconcentration Factor, log BCF:

0.699 (calculated-K_{OW}, Lyman et al. 1982; quoted, Howard 1993)

Sorption Partition Coefficient, log K_{OC}:

2.029 (soil, calculated-K_{OW}, Lyman et al. 1982; quoted, Howard 1993)

Half-Lives in the Environment:

Air: an atmospheric half-life of 11 days from estimated photooxidation rate constant of 1.43×10^{-12} cm^3 $molecule^{-1}$ sec^{-1} for the vapor-phase reaction with 5×10^5 hydroxyl radicals/cm^3 in air at 25°C (Atkinson 1987; quoted, Howard 1993).
Surface water:
Groundwater:
Sediment:
Soil:
Biota:

Environmental Fate Rate Constants or Half-Lives:

Volatilization:
Photolysis:
Oxidation: estimated photooxidation rate constant of 1.43×10^{-12} cm^3 $molecule^{-1}$ sec^{-1} for the vapor-phase reaction with 5×10^5 hydroxyl radicals/cm^3 in air at 25°C which corresponds to an atmospheric half-life of 11 days (Atkinson 1987; quoted, Howard 1993).
Hydrolysis:
Biodegradation:
Biotransformation:
Bioconcentration, Uptake (k_1) and Elimination (k_2) Rate Constants:

Common Name: 2,3-Dimethylpyridine
Synonym: 2,3-lutidine
Chemical Name: 2,3-dimethylpyridine, 2,3-lutidine
CAS Registry No: 583-61-9
Molecular Formula: $C_5H_3N(CH_3)_2$
Molecular Weight: 107.15
Melting Point (°C): -15.5 (Stephenson & Malanowski 1987)
Boiling Point (°C):
161.157 (Coulson et al. 1959; Boublik et al. 1984)
163-164 (Weast 1982-83)
161.50 (Stephenson & Malanowski 1987)
Density (g/cm^3 at 20°C):
0.9461, 0.94212 (20, 25°C, Coulson et al. 1959)
0.9319 (25°C, Weast 1982-83)
Molar Volume (cm^3/mol):
135.9 (calculated-LeBas method)
114.98 (calculated-density)
Molecular Volume (Å^3):
Total Surface Area, TSA (Å^2):
Dissociation Constants, pK:
6.70 (20°C, Perrin 1972)
6.57 (pK_a, Weast 1982-83)
Heat of Fusion ΔH_{fus}, kcal/mol:
Entropy of Fusion ΔS_{fus}, cal/mol·K (e.u.):
Fugacity Ratio at 25°C (assuming ΔS_{fus} = 13.5 e.u.), F: 1.0

Water Solubility (g/m^3 or mg/L at 25°C):
104000 (shake flask-GC, Stephenson 1993)

Vapor Pressure (Pa at 25°C):
366.6 (calculated-Antoine eqn., Andon et al. 1954)
346.4 (ebulliometry, extrapolated-Antoine eqn., Coulson et al. 1959)
359.0 (calculated-Cox eqn., Chao et al. 1983)
425.5 (extrapolated-Antoine eqn., Boublik et al. 1984)
352.2 (extrapolated-Antoine eqn., Stephenson & Malanowski 1987)

Henry's Law Constant (Pa·m^3/mol):
0.725 (volatility ratio-transpiration method, Andon et al. 1954)
0.732 (quoted exptl., Hine & Mookerjee 1975)
0.859, 0.732 (calcd.-group contribution, bond contribution, Hine & Mookerjee 1975)

Octanol/Water Partition Coefficient, log K_{OW}:

Bioconcentration Factor, log BCF: no data
Sorption Partition Coefficient, log K_{OC}: no data
Half-Lives in the Environment/Environmental Fate Rate Constants or Half-Lives: no data

Common Name: Quinoline
Synonym: benzo[b]-pyridine, 1-benzazine
Chemical Name: quinoline
CAS Registry No: 91-22-5
Molecular Formula: C_9H_7N
Molecular Weight: 129.16
Melting Point (°C):
- -15.0 (Stull 1947)
- -15.6 (Weast 1974,1982-83; Smith et al. 1978)
- -16.15 (Mackay et al. 1982)
- -19.5 (Verschueren 1983)
- -14.9 (Dean 1985; Matzner et al. 1991)
- -14.85 (Riddick et al. 1986)

Boiling Point (°C):
- 237.70 (Stull 1947; Verschueren 1983)
- 238.05 (Weast 1974,1982-83; Smith et al. 1978; Pearlman et al. 1984; Matzner et al. 1991)
- 237.85 (Mackay et al. 1982)
- 237.63 (Boublik et al. 1984)
- 237.0 (Dean 1985; Banerjee et al. 1990)
- 237.10 (Riddick et al. 1986)

Density (g/cm^3 at 20°C):
- 1.0929 (Weast 1982-83; Fu & Luthy 1985)
- 1.095 (Verschueren 1983; Dean 1985)

Molar Volume (cm^3/mol):
- 118.1 (calculated-density, Rohrschneider 1973)
- 118.2 (calculated-density, Fu & Luthy 1985)
- 73.4 (intrinsic volume V_I, Kamlet et al. 1987)
- 144.7 (calculated-LeBas method)

Molecular Volume ($Å^3$):
- 122.0 (Pearlman et al. 1984)
- 121.2 (Zachara et al. 1990)

Total Surface Area, TSA ($Å^2$):
- 150.7 (Pearlman et al. 1984)
- 141.7 (Zachara et al. 1990)

Dissociation Constants, pK:
- 4.90 (pK_a, 20°C, Weast 1982-83; Zachara et al. 1987; Matzner et al. 1991)
- 4.80 (pK_a, protonated cation +1, Dean 1985)
- 4.94 ($pK_{BH}{}^+$, Riddick et al. 1986)
- 4.87 (pK_a, Sangster 1989)

Heat of Fusion ΔH_{fus}, kcal/mol:
- 2.58 (Riddick et al. 1986)

Entropy of Fusion ΔS_{fus}, cal/mol·K (e.u.):
Fugacity Ratio at 25°C (assuming ΔS_{fus} = 13.5 e.u.), F: 1.0

Water Solubility (g/m^3 or mg/L at 25°C):

6110 (Albersmeyer 1958; quoted, Smith et al. 1978)
60413, 3640 (observed, predicted-K_{OW}, Valvani et al. 1981)
60000 (quoted, Verschueren 1983)
6330 (quoted, Pearlman et al. 1984)
6100 (quoted, Dean 1985)
6110 (recommended, Neely & Blau 1985; quoted, Karickhoff 1985)
6840 (shake flask-HPLC/UV, Fu & Luthy 1985, 1986)
3900 (calculated-γ, Fu & Luthy 1985, 1986)
6090 (20°C, quoted, Riddick et al. 1986)
6473, 11249 (quoted, calculated-V_I & M.P., Kamlet et al. 1987)
6000 (selected, Zachara et al. 1987, 1990; quoted, Matzner et al. 1991)
5426 (centrifuge-HPLC at pH 7 & 8, Matzner et al. 1991)
8400 (20°C, shake flask-GC, Stephenson 1993)
6600 (30°C, shake flask-GC, Stephenson 1993)

Vapor Pressure (Pa at 25°C):

15.51 (extrapolated-regression of tabulated data, Stull 1947)
1.213 (extrapolated, Maczynski & Maczynska 1965; quoted, Smith et al. 1978)
1.22, 11.14 (quoted, calculated-B.P., Mackay et al. 1982)
12.83 (calculated-Cox eqn., Chao et al. 1983)
11.04 (extrapolated-Antoine eqn., Boublik et al. 1984)
11.05 (extrapolated-Antoine eqn., Dean 1985)
1.216 (recommended, Neely & Blau 1985)
12.80 (Howard et al. 1986; quoted, Banerjee et al. 1990)
11.20 (Riddick et al. 1986)
12.80, 6.145 (quoted, calculated-solvatochromic p., Banerjee et al. 1990)

Henry's Law Constant (Pa m^3/mol):

0.0253 (calculated-P/C, Smith & Bomberger 1980)
0.026 (calculated-P/C, Mackay 1985)
0.168 (calculated-P/C, Meylan & Howard 1991)
0.0697 (estimated-bond contribution, Meylan & Howard 1991)

Octanol/Water Partition Coefficient, log K_{OW}:

2.03 (shake flask-AS, Iwasa et al. 1965; quoted, Leo et al. 1969,71; McCall 1975; Mirrlees et al. 1976; Hansch & Leo 1979; Valvani et al. 1981; Haky & Young 1984; Miyake et al. 1987; Sangster 1989,93)
2.06 (shake flask-AS at pH 7.4, Rogers & Cammarata 1969; quoted, Leo et al. 1971; Hansch & Leo 1979; Sangster 1989,93)
2.09 (calculated-M.O. indices, Rogers & Cammarata 1969)
2.03 (Schultz et al. 1970; quoted, Hansch & Leo 1979)
2.04 (HPLC, Mirrlees et al. 1976; quoted, Sangster 1989,93)
2.04 ± 0.02 (shake flask at pH 7, Unger et al. 1978)

2.02, 1.96 (quoted, calculated-f const., Rekker & De Kort 1979)
2.02 (quoted from Hansch & Leo 1979, Kaiser 1983)
2.03, 2.06 (quoted, Verschueren 1983)
2.03 (quoted from Verschueren 1983, Zachara et al. 1986)
2.02 (selected best lit. value, Garst & Wilson 1984)
2.01 ± 0.02 (exptl.-ALPM, Garst & Wilson 1984)
1.88 (HPLC-k', Haky & Young 1984; quoted, Sangster 1993)
2.10 (recommended, Neely & Blau 1985)
2.04 (quoted, Karickhoff 1985)
2.04 (quoted from Hansch & Leo 1979, Schultz & Moulton 1985; Bodor et al. 1989)
2.04 (quoted, Zachara et al. 1987,90)
2.03, 2.20 (quoted, calculated- γ from UNIFAC, Banerjee & Howard 1988; quoted, Sangster 1993)
2.10 (shake flask-HPLC, De Voogt et al. 1988,90; quoted, Sangster 1993)
2.09 (28°C, shake flask-AS at pH 7.4, Go & Ngiam 1988; quoted, Sangster 1989,93)
2.03, 2.049, 2.22 (quoted from CLOGP, calculated-CLOGP, calculated-MO, Bodor et al. 1989)
2.04 (quoted, Sims & O'Laoughlin 1989)
2.03 (recommended, Sangster 1989, 1993)

Bioconcentration Factor, log BCF:

Sorption Partition Coefficient, log K_{OC}:
1.04 (Coyote Creek sediment with organic content of 1.9%, Smith et al. 1978)
1.96, 2.10, 1.67, 1.72 (estimated-K_{OW}, Karickhoff 1985)
1.42, 1.62 (estimated-S, Karickhoff 1985)
2.20 (best estimate, Karickhoff 1985)
0.251 (estimated Anvil Points subsurface materials, Zachara et al. 1987)
-1.516 (estimated Loring subsurface materials, Zachara et al. 1987)

Half-Lives in the Environment:
Air: 10-99 hours, based on an estimated rate constant for vapor phase reaction with hydroxyl radicals in air (Atkinson 1987; selected, Howard et al. 1991); atmospheric transformation lifetime was estimated to be 1 to 5 days (Kelly et al. 1994).
Surface water: half-life for all processes, except for dilution: 0.5 hour in stream, eutrophic lake and pond and 600 hours in oligotrophic lake, based on transformation and transport of quinoline predicted by the one-compartment model (Smith et al. 1978); half-life for all processes, including for dilution: 0.28 hour in stream, 0.5 hour in eutrophic lake and pond and 600 hours in oligotrophic lake , based on transformation and transport of quinoline predicted by the one-compartment model (Smith et al. 1978); half-life of 5-12 days for direct photolysis in aqueous media (Harris 1982); 72-240 hours, based on an acclimated fresh water grab sample data (Rogers et al. 1984; quoted, Howard

et al. 1991); predicted to degrade readily in sunlight in near surface lake water at 40°N latitude in summer with a half-life about 14 calendar days while its calculated half-life in winter was about 123 calendar days (Kochany & Maguire 1994).

Groundwater: 144-480 hours, based on estimated aqueous aerobic biodegradation half-life (Howard et al. 1991).

Sediment:

Soil: 72-240 hours, based on estimated aqueous aerobic biodegradation half-life (Howard et al. 1991).

Biota:

Environmental Fate Rate Constants or Half-Lives:

Volatilization: half-life of 7000 hours in stream, 35000 hours in eutrophic pond and 28000 hours in eutrophic lake and oligotrophic lake, based on transformation and transport of quinoline predicted by the one-compartment model (Smith et al. 1978).

Photolysis: 7.8×10^{-7} sec^{-1}, assuming exposed to 12 hours sunlight per day in June photolysis half-lives: 1200 hours in stream, 3000 hours in eutrophic lake and pond and 600 hours in oligotrophic lake, based on transformation and transport of quinoline predicted by the one-compartment model (Smith et al. 1978); and aqueous photolysis rate constant of 3.6×10^{-7} sec^{-1} for summer with half-life of 535 hours and 5.0×10^{-8} sec^{-1} for winter with half-life of 3851 hours both at pH 6.9 and under sunlight at 40° N (Mill et al. 1981; quoted, Howard et al. 1991); photolytic half-life of 550 hours in aquatics (Haque et al. 1980); half-life of 5-12 days for disappearance via direct photolysis in aqueous media (Harris 1982).

Oxidation: rate constant of 2.8 M^{-1} sec^{-1} for the reaction with RO_2 radical with half-life of $> 10^4$ hours in stream, eutrophic pond and lake and oligotrophic lake, based on RO_2 concentration of 10^{-9} M on transformation and transport of quinoline predicted by the one-compartment model (Smith et al. 1978); photooxidation half-life of 10-99 hours in air, based on an estimated rate constant for vapor phase reaction with hydroxyl radicals in air (Atkinson 1987; quoted, Howard et al. 1991); first-order photodegradation rate constant at 313 nm of 8.0×10^{-6} sec^{-1} in organic-free water with half-life of 24.0 hours and 8.4×10^{-6} sec^{-1} in lake water with half-life of 23 hours both saturated with air (Kochany & Maguire 1994).

Hydrolysis: no hydrolyzable groups (Howard et al. 1991).

Biodegradation: half-life of 0.5 hour in stream, eutrophic lake and pond and 10000 hours in oligotrophic lake, based on transformation and transport of quinoline predicted by the one-compartment model (Smith et al. 1978); Biodegradation rate constant of 7.4×10^{-5} $mL \cdot cell^{-1} \cdot day^{-1}$ in enrichment culture (Klečka 1985); aqueous aerobic half-life of 72-240 hours, based on an acclimated fresh water grab sample data (Rogers et al. 1984; quoted, Howard et al. 1991); aqueous anaerobic half-life of 288-960 hours, based on estimated aqueous aerobic biodegradation half-life (Howard et al. 1991).

Biotransformation:

Bioconcentration, Uptake (k_1) and Elimination (k_2) Rate Constants:

Common Name: Isoquinoline
Synonym: leucoline
Chemical Name: isoquinoline
CAS Registry No: 119-65-3
Molecular Formula: C_9H_7N
Molecular Weight: 129.16
Melting Point (°C):
- 24.6 (Stull 1947)
- 26.50 (Weast 1982-83; Dean 1985)
- 26-28 (Verschueren 1983)
- 27.0 (Pearlman et al. 1984; Kamlet et al. 1987)
- 26.48 (Riddick et al. 1986)

Boiling Point (°C):
- 240.5 (Stull 1947)
- 242.20 (at 743 mm Hg, Weast 1982-83)
- 242.0 (Verschueren 1983)
- 243.24 (Boublik et al. 1984; Riddick et al. 1986)
- 243.0 (Pearlman et al. 1984)
- 243.20 (Dean 1985)

Density (g/cm^3 at 20°C):
- 1.0986 (Weast 1982-83)
- 1.0990 (Verschueren 1983)
- 1.0910 (Dean 1985)

Molar Volume (cm^3/mol):
- 73.4 (intrinsic volume V_I, Kamlet et al. 1987)
- 144.7 (calculated-LeBas method)

Molecular Volume ($Å^3$):
- 122.0 (Pearlman et al. 1984)

Total Surface Area, TSA ($Å^2$):
- 150.5 (Pearlman et al. 1984)

Dissociation Constants, pK:
- 5.40 (Perrin 1972)
- 5.42 (pK_a, 20°C, Weast 1982-83)
- 5.40 (pK_a, protonated cation +1, Dean 1985)
- 5.38 ($pK_{BH}{}^+$, Riddick et al. 1986)
- 5.39 (pK_a, Sangster 1989)

Heat of Fusion, kcal/mol:
- 1.78 (Riddick et al. 1986)

Entropy of Fusion, cal/mol·K (e.u.):
Fugacity Ratio at 25°C (assuming ΔS = 13.5 e.u.), F: 0.9664

Water Solubility (g/m^3 or mg/L at 25°C):
- 4520 (quoted, Pearlman et al. 1984)
- 4583, 10743 (quoted, calculated-V_I & M.P., Kamlet et al. 1987)

Vapor Pressure (Pa at 25°C):

- 11.8 (extrapolated-regression of tabulated data, Stull 1947)
- 7.80 (extrapolated-Cox eqn., Chao et al. 1983)
- 6.33 (extrapolated-Antoine eqn., Boublik et al. 1984)
- 6.35 (extrapolated-Antoine eqn., Dean 1985)
- 670 (Riddick et al. 1986)

Henry's Law Constant (Pa·m³/mol):

- 19.14 (calculated-P/C, this work)

Octanol/Water Partition Coefficient, log K_{OW}:

- 2.08 (shake flask-AS, Hansch & Anderson 1967; quoted, Mirrlees et al. 1976; Sangster 1989,93)
- 2.09 (HPLC, Mirrlees et al. 1976; quoted, Sangster 1993)
- 1.82 (calc-π constant, Southworth et al. 1978)
- 2.08 (quoted, Verschueren 1983)
- 2.08 (quoted from Hansch & Leo 1979, Miyake et al. 1987)
- 2.08 (recommended, Sangster 1989,93)

Bioconcentration Factor, log BCF:

Sorption Partition Coefficient, log K_{OC}:

Half-Lives in the Environment:

- Air:
- Surface water:
- Groundwater:
- Sediment:
- Soil:
- Biota: elimination half-life of 1 minute (*daphnia pulex*, Southworth et al. 1978).

Environmental Fate Rate Constants or Half-Lives:

- Volatilization:
- Photolysis:
- Hydrolysis:
- Oxidation:
- Biodegradation:
- Biotransformation:
- Bioconcentration, Uptake (k_1) and Elimination (k_2) Rate Constants:
  - k_1: 82.0 hour^{-1} (*daphnia pulex*, 21°C, Southworth et al. 1978)
  - k_2: 34.2 hour^{-1} (*daphnia pulex*, 21°C, Southworth et al. 1978)

Common Name: Benzo(f)quinoline
Synonym: 5,6-benzoquinoline, naphthopyridine
Chemical Name: 5,6-benzoquinoline, benzo(f)quinoline
CAS Registry No: 85-02-9
Molecular Formula: $C_{13}H_9N$
Molecular Weight: 179.22
Melting Point (°C):
93.5 (Dean 1985)
Boiling Point (°C):
350 (at 721 mmHg, Dean 1985)
Density (g/cm^3 at 20°C):
Molar Volume (cm^3/mol):
196.3 (calculated-LeBas method)
Molecular Volume (Å^3):
164.575 (Pearlman et al. 1984)
Total Surface Area, TSA (Å^2):
192.89 (Pearlman et al. 1984)
Heat of Fusion, kcal/mol:
Entropy of Fusion, cal/mol·K (e.u.):
Fugacity Ratio at 25°C (assuming ΔS = 13.5 e.u.), F: 0.21

Water Solubility (g/m^3 or mg/L at 25°C):
76.1 (shake flask-GC, Smith et al. 1978)
77.1 (Mill et al. 1981)
76.0 (quoted, Steen & Karickhoff 1981)
76.2, 81.2; 78.7 (quoted values; average, Pearlman et al. 1984)

Vapor Pressure (Pa at 25°C):
0.00373 (20°C, McEachern et al. 1975; quoted, Smith et al. 1978)
0.00670 (calculated-Antoine eqn., Stephenson & Malanowski 1987)

Henry's Law Constant (Pa·m^3/mol):
0.0096 (calculated-P/C, Smith & Bomberger 1980)

Octanol/Water Partition Coefficient, log K_{OW}:
3.20 (Steen & Karickhoff 1981)

Bioconcentration Factor, log BCF:
2.18 (mixed microbial populations, Steen & Karickhoff 1981)

Sorption Partition Coefficient, log K_{OC}:

3.11 (Coyote Creek sediment, Smith et al. 1978)

4.64, 4.32 (soil, quoted, calcd.- χ and fragment contribution, Meylan et al. 1992)

Half-Lives in the Environment:

Air:

Surface water: half-lives of 0.5 hour in river water, 6.9 hours in pond water, 7.0 hours in eutrophic lake and 1.4 hours in oligotrophic lake predicted by one-compartment for all processes including dilution (Smith et al. 1978); half-life of 1 hour for disappearance via direct photolysis in aqueous media (Harris 1982).

Groundwater:

Sediment:

Soil:

Biota:

Environmental Fate Rate Constants or Half-Lives:

Volatilization: estimated half-lives of >10000 hours in river, >100000 hours in eutrophic pond, eutrophic lake and oligotrophic lake by the one compartment model (Smith et al. 1978).

Photolysis: rate constant of transformation and transport of $(1.4 \pm 0.7)\times10^{-4}$ sec^{-1} exposed to 12 hours sunlight in mid-June with estimated half-lives of 2.8 hours in river, 7.0 hours in eutrophic pond and eutrophic lake and 1.4 hour in oligotrophic lake from average photolysis rates on a summer day at 40°N latitude by the one compartment model (Smith et al. 1978); photolytic half-life of 0.52 hour in aquatics (Haque et al. 1980); half-life of 1 hour for disappearance via direct photolysis in aqueous media (Harris 1982).

Oxidation: laboratory studied rate constant of <2.8 M^{-1} sec^{-1} for the reaction with the RO_2 radicals and estimated half-lives of $>10^5$ hours in river, eutrophic pond, eutrophic lake and oligotrophic lake by the one compartment model (Smith et al. 1978).

Hydrolysis:

Biodegradation: estimated half-lives of 190 hours in river, eutrophic pond, eutrophic lake and $>10^6$ hours in oligotrophic lake by the one compartment model (Smith et al. 1978).

Biotransformation:

Bioconcentration, Uptake (k_1) and Elimination (k_2) Rate Constants:

Common Name: Carbazole
Synonym: 9H-carbazole, dibenzopyrrole
Chemical Name: carbazole
CAS Registry No: 86-74-8
Molecular Formula: $C_6H_4NHC_6H_4$
Molecular Weight: 167.21
Melting Point (°C): 247-248
248 (Pearlman et al. 1984)
Boiling Point (°C):
355 (Pearlman et al. 1984)
Density (g/cm³ at 20°C):
1.260 (25°C, Jiménez et al. 1990)
Molar Volume (cm³/mol):
192.9 (calculated-LeBas method)
132.6 (calculated-density)
Molecular Volume (Å³):
188.71 (Pearlman et al. 1984)
Total Surface Area, TSA (Å²):
155.94 (Pearlman et ak, 1984)
Heat of Fusion ΔH_{fus}, kcal/mol:
Entropy of Fusion ΔS_{fus}, cal/mol·K (e.u.):
Fugacity Ratio at 25°C (assuming ΔS_{fus} = 13.5 e.u.), F: 0.00637

Water Solubility (g/m³ or mg/L at 25°C):
1.03 ± 0.05 (20°C, shake flask-GC, Smith et al. 1978)
1.037 (Mill et al. 1981)
0.428 (20°C, shake flask-fluorophotometry, Hashimoto et al. 1982)
1.67, 1.03, 0.908; 1.204 (quoted values; lit. average, Pearlman et al. 1984)

Vapor Pressure (Pa at 25°C):
0.0933 (20°C, Smith et al. 1978)
0.0002 (effusion method, Jiménez et al. 1990)
0.00119 (extr., Antoine eqn., Stephenson & Malanowski 1987)
0.00447 (extr., liquid, Antoine eqn., Stephenson & Malanowski 1987)

Henry's Law Constant (Pa·m³/mol):
16.0 (calculated-P/C, Smith & Bomberger 1980)

Octanol/Water Partition Coefficient, log K_{OW}:
3.29 (shake flask-AS at pH 7.4, Rogers 1969)
3.01 (HPLC-k', Eadsforth 1986)
3.29, 3.50 (quoted: shake flask, calculated, Eadsforth 1986)

3.72, 3.76 (quoted from Pomona College unpublished data, Sangster 1989)
3.72 (recommended, Sangster 1989)
3.59 (HPLC, Jenke et al. 1990)
3.84 (shake flask-HPLC at pH 7, De Voogt et al. 1988)

Bioconcentration Factor, log BCF:

Sorption Partition Coefficient, log K_{OC}:
2.24 (Coyote Creek sediment, Smith et al. 1978)

Half-Lives in the Environment:
Air:
Surface water: estimated half-lives of 6.0 hours in river, 15.0 hours in eutrophic pond and eutrophic lake and 3.0 hour in oligotrophic lake from average photolysis rates on a summer day at 40°N latitude by the one compartment model (Smith et al. 1978); photolytic half-life of 1.0 hour in aquatics (Haque et al. 1980); half-life of 3 hours for disappearance via direct photolysis in aquatic media (Harris 1982).
Groundwater:
Sediment:
Soil:
Biota:

Environmental Fate Rate Constants or Half-Lives:
Volatilization: estimated half-lives of $>10^5$ hours in river, eutrophic pond, eutrophic lake and oligotrophic lake by the one compartment model (Smith et al. 1978).
Photolysis: rate constant of transformation and transport of 6.6×10^{-5} sec^{-1} exposed to midday sunlight in late January with estimated half-lives of 6.0 hours in river, 15.0 hours in eutrophic pond and eutrophic lake and 3.0 hour in oligotrophic lake from average photolysis rates on a summer day at 40°N latitude by the one compartment model (Smith et al. 1978); photolytic half-life of 1.0 hour in aquatics (Haque et al. 1980); half-life of 3 hours for disappearance via direct photolysis in aquatic media (Harris 1982).
Oxidation: laboratory investigated rate constant of 29 M^{-1} sec^{-1} for the reaction with RO_2 radicals and estimated half-lives of >240 hours in river, eutrophic pond, eutrophic lake and oligotrophic lake by the one compartment model (Smith et al. 1978).
Hydrolysis:
Biodegradation: estimated half-lives of 14 hours in river, eutrophic pond, eutrophic lake and $>10^3$ hours in oligotrophic lake by the one compartment model (Smith et al. 1978).
Biotransformation:
Bioconcentration, Uptake (k_1) and Elimination (k_2) Rate Constants:

Common Name: Benzo[*c,g*]carbazole
Synonym: 7H-dibenzo[*c,g*]carbazole
Chemical Name: 7H-dibenzo[*c,g*]carbazole
CAS Registry No:
Molecular Formula: $C_{10}H_6NHC_{10}H_6$
Molecular Weight: 267.31
Melting Point (°C):
155-159 (Smith et al. 1978)
157 (Pearlman et al. 1984)
Boiling Point (°C)
Density (g/cm^3 at 20°C):
Molar Volume (cm^3/mol):
296.1 (calculated-LeBas method)
Molecular Volume (Å^3):
237.542 (Pearlman et al. 1984)
Total Surface Area, TSA (Å^2):
267.002 (Pearlman et al. 1984)
Heat of Fusion ΔH_{fus}, kcal/mol:
Entropy of Fusion ΔS_{fus}, cal/mol·K (e.u.):
Fugacity Ratio at 25°C (assuming ΔS_{fus} = 13.5 e.u.), F: 0.0495

Water Solubility (g/m^3 or mg/L at 25°C):
0.063±0.003 (shake flask-GC, Smith et al. 1978)
0.064 (quoted, Mill et al. 1981)
0.064 (quoted, Pearlman et al. 1984)

Vapor Pressure (Pa at 25°C):
1.33×10^{-7} (estimated by comparison with benzo[a]pyrene, Smith et al. 1978)

Henry's Law Constant (Pa·m^3/mol):
0.00048 (calculated-P/C, Smith & Bomberger 1980)

Octanol/Water Partition Coefficient, log K_{OW}:
5.75 (calculated-S, Steen & Karickhoff 1981)

Bioconcentration Factor, log BCF:
4.93 (mixed microbial populations, Steen & Karickhoff 1981)

Sorption Partition Coefficient, log K_{OC}:

4.31 (Coyote Creek sediment, Smith et al. 1978)

6.03, 6.16 (soil, quoted, calculated- χ and fragment contribution, Meylan et al. 1992)

Half-Lives in the Environment:

Air:

Surface water: half-lives of 0.36 hour in river water, 1.5 hours in pond water, 1.5 hours in eutrophic lake and 0.5 hour in oligotrophic lake for all processes predicted by one-compartment model (Smith et al. 1981).

Groundwater:

Sediment:

Soil:

Biota:

Environmental Fate Rate Constants or Half-Lives:

Volatilization: estimated half-lives of 15000 hours in river, 37000 hours in eutrophic pond, 73000 hours in eutrophic lake and oligotrophic lake by the one compartment model (Smith et al. 1978).

Photolysis: rate constant of transformation and transport of 5.2×10^{-4} sec^{-1} exposed to midday sunlight in mid-January with estimated half-lives of 1.0 hour in river, 1.5 hour in eutrophic pond and eutrophic lake and 0.5 hour in oligotrophic lake assuming winter insolation by the one compartment model (Smith et al. 1978); photolytic half-life of 0.35 hour in aquatics (Haque et al. 1980).

Oxidation: laboratory studied rate constant of 830 M^{-1} sec^{-1} for the reaction with the RO_2 radicals and estimated half-lives of >700 hours in river, eutrophic pond, eutrophic lake and oligotrophic lake by the one compartment model (Smith et al. 1978).

Hydrolysis:

Biodegradation: estimated half-life to be very long in river, eutrophic pond, eutrophic lake and oligotrophic lake by the one compartment model with no acclimated cultures obtained during the screening studies (Smith et al. 1978).

Biotransformation:

Bioconcentration, Uptake (k_1) and Elimination (k_2) Rate Constants:

Common Name: Acridine
Synonym: 2,3,5,6-dibenzopyridine
Chemical Name: acridine, 2,3,5,6-dibenzopyridine
CAS Registry No: 260-94-6
Molecular Formula: $C_{13}H_9N$
Molecular Weight: 179.22
Melting Point (°C):

111 (Weast 1982-83)
108 (sublimes at 100°C, Verschueren 1983)
107-110 (Dean 1985; Matzner et al. 1991; Matzner & Bales 1994)
110 (Stephenson & Malanowski 1987)

Boiling Point (°C):

345-346 (Weast 1982-83)
346 (Verschueren 1983; Dean 1985; Stephenson & Malanowski 1987; Matzner et al. 1991; Matzner & Bales 1994)

Density (g/cm^3 at 20°C):

1.005 (Weast 1982-83)
1.10 (Verschueren 1983)

Molar Volume (cm^3/mol):

196.3 (calculated-LeBas method)

Molecular Volume (Å^3):

165.392 (Pearlman et al. 1984)

Total Surface Area, TSA (Å^2):

197.25 (Pearlman et al 1984)

Dissociation Constant, pK_a:

5.60 (Albert 1966; Matzner et al. 1991; Matzner & Bales 1994)
5.58 (20°C, Weast 1982-83)
5.60 (protonated cation +1, Dean 1985)
10.65 (Sangster 1989)

Heat of Fusion, ΔH_{fus}, kcal/mol:

4.44 $\pm$ 0.09 (McEachern et al. 1975)

Entropy of Fusion, ΔS_{fus}, cal/mol·K (e.u.):
Fugacity Ratio at 25°C (assuming ΔS_{fus} = 13.5 e.u.), F: 0.1443

Water Solubility (g/m^3 or mg/L at 25°C):

57.4 (Albert 1966; quoted, Matzner et al. 1991)
38.4 (24°C, shake flask-LSC, Means et al. 1980; quoted, Banwart et al. 1982; Zachara et al. 1986,87; Ma et al. 1990; Matzner et al. 1991)
38.5, 53.8; 46.6 (quoted values; average, Pearlman et al. 1984)
53.8 (20°C, quoted, Ma et al. 1990)
38.3 (quoted, Gerstl & Helling 1987)
54.8 (centrifuge-HPLC at pH 8, Matzner et al. 1991)
57.4, 38.4 (quoted, Matzner & Bales 1994)

Vapor Pressure (Pa at 25°C):

0.0065 (Langmuir free evaporation, McEachern et al. 1975)
0.0075 (extrapolated-Cox eqn., Chao et al. 1983)
0.2066 (static apparatus-extrapolated from Chebyshev polynomials, Sivaraman & Kobayashi 1983)
0.0065 (calculated-Antoine eqn., Stephenson & Malanowski 1987; quoted, Ma et al. 1990)

Henry's Law Constant (Pa·m^3/mol):

0.030 (calculated-P/C, Ma et al. 1990)

Octanol/Water Partition Coefficient, log K_{OW}:

3.40 (shake flask-AS, Hansch & Fujita 1964; quoted, Albert 1966; Leo et al. 1971; Mirrlees et al. 1976; Hansch & Leo 1979; Verschueren 1983; Zachara et al. 1986, 1987; Matzner et al. 1991; Matzner & Bales 1994)
3.39 (HPLC, Mirrlees et al. 1976)
3.30 (calculated-π constant, Southworth et al. 1978)
3.39 (shake flask at pH 7, Unger et al. 1978)
3.40, 3.25 (quoted, calculated-f const., Rekker & de Kort 1979)
3.62 (shake flask-LSC, Means et al. 1980; quoted, Banwart et al. 1982; Ma et al. 1990; Matzner et al. 1991; Matzner & Bales 1994)
3.29 (shake flask-AS at pH 7.4, Unger & Chiang 1981; quoted, Sangster 1989)
3.34 (calculated, Unger & Chiang 1981)
3.31 ± 0.03 (exptl.-ALPM, Garst 1984)
3.35 ± 0.02 (exptl.-ALPM, Garst & Wilson 1984)
3.40 (selected best lit. value, Garst 1984; Garst & Wilson 1984)
3.39 (quoted from Hansch & Leo 1979, Haky & Young 1984)
3.32 (shake flask-GC at pH 7.0, Haky & Leja 1986; quoted, Sangster 1989)
3.62 (quoted, Gerstl & Helling 1987)
3.40 (recommended, Sangster 1989; quoted, Bintein & Devillers 1994)
3.40 (quoted, Sims & O'Loughlin 1989)
3.40 (selected, Ma et al. 1990)

Bioconcentration Factor, log BCF:

2.40 (selected, Ma et al. 1990)

Sorption Partition Coefficient, log K_{OC}:

4.69 (average of sediments and soil samples, Means et al. 1980)
3.32 (calculated, Means et al. 1980)
-0.157 (estimated of Loring subsurface material, Zachara et al. 1987)
0.610 (estimated of Anvil Points subsurface material, Zachara et al. 1987)
3.09-3.41 (soil, calculated-K_{OW}, model of Karickhoff et al. 1979, Sabljic 1987)

3.16-3.33 (soil, calculated-K_{OW}, model of Kenaga & Goring 1980, Sabljic 1987)
2.36-2.52 (soil, calculated-K_{OW}, model of Briggs 1981, Sabljic 1987)
2.98-3.30 (soil, calculated-K_{OW}, model of Means et al. 1982, Sabljic 1987)
2.19-2.48 (soil, calculated-K_{OW}, model of Chiou et al. 1983, Sabljic 1987)
4.22, 4.26 (soil, quoted, calculated- χ , Sabljic 1987)
4.11, 3.32 (quoted, calculated- χ , Gerstl & Helling 1987)
4.11, 4.31 (soil, quoted, calculated- χ and fragment contribution, Meylan et al. 1992)

Half-Lives in the Environment:

Air:

Surface water:

Groundwater:

Sediment:

Soil:

Biota: elimination half-life of 0.188 hour (11.3 min) (*daphnia pulex*, Southworth et al. 1978).

Environmental Fate Rate Constants or Half-Lives:

Volatilization:

Photolysis:

Oxidation:

Hydrolysis:

Biodegradation:

Biotransformation:

Bioconcentration, Uptake (k_1) and Elimination (k_2) Rate Constants:

k_1: 109 h^{-1} (*daphnia pulex*, 21°C, Southworth et al. 1978)

k_2: 3.68 h^{-1} (*daphnia pulex*, 21°C, Southworth et al. 1978)

Common Name: Dimethyl sulfide
Synonym: methyl sulfide, thiobismethane, 2-thiapropane
Chemical Name: dimethyl sulfide
CAS Registry No: 75-18-3
Molecular Formula: $(CH_3)_2S$
Molecular Weight: 62.129
Melting Point (°C):
- -83.12 (Verschueren 1983)
- -98.7 (Riddick et al. 1986)

Boiling Point (°C):
- 37.33 (Riddick et al. 1986)

Density (g/cm³ at 25°C):
- 0.84228 (Riddick et al. 1986)

Molar Volume (cm³/mol):
- 73.20 (Kamlet et al. 1986)
- 77.40 (calculated-LeBas method)
- 73.76 (calculated-density)

Molecular Volume (Å³):
Total Surface Area, TSA (Å²):
Dissociation Constant pK_a:
- -6.99 (Riddick et al. 1986)

Heat of Fusion, ΔH_{fus}, kcal/mol:
- 1.9084 (Riddick et al. 1986)

Entropy of Fusion, ΔS_{fus}, cal/mol·K (e.u.):
Fugacity Ratio at 25°C (assuming ΔS_{fus} = 13.5 e.u.), F: 1.0

Water Solubility (g/m³ or mg/L at 25°C):
- 15606 (quoted, Hine & Mookerjee 1975)
- 6300 (quoted, Verschueren 1983)
- 20000 (quoted, Riddick et al. 1986)
- 19600 (selected, Yaws et al. 1990)

Vapor Pressure (Pa at 25°C):
- 66660, 64500 (25.86°C, interpolated-Antoine eqn., Zwolinski & Willhoit 1971)
- 70300 (quoted, Hine & Mookerjee 1975)
- 56000 (20°C, Verscueren 1983)
- 64470 (extrapolated-Antoine eqn., Dean 1985)
- 64650 (quoted, Riddick et al. 1986)
- 64520 (interpolated, Antoine eqn., Stephenson & Malanowski 1987)

Henry's Law Constant (Pa·m³/mol):
- 278.13 ($1/K_{AW}$, exptl., Hine & Mookerjee 1975)
- 298.02 (calculated-group contribution, Hine & Mookerjee 1975)
- 366.64 (calculated-bond contribution, Hine & Mookerjee 1975)

180.4, 184.7, 173.5 (headspace-GC, concn. of 10, 1.0, 0.1 ppm by weight, Przyjazny et al. 1983)
163.4 (quoted, Gaffney et al. 1987)
184, 1271 (quoted, calculated-molecular structure, Russell et al. 1992)

Octanol/Water Partition Coefficient, log K_{OW}:

Bioconcentration Factor, log BCF:

Sorption Partition Coefficient, log K_{OC}:

Half-Lives in the Environment:
Air:
Surface water:
Groundwater:
Sediment:
Soil:
Biota:

Environmental Fate Rate Constants or Half-Lives:
Volatilization:
Photolysis:
Oxidation: rate constant of 57×10^{-12} $cm^3 \cdot molecule^{-1} \cdot sec^{-1}$ for gas-phase reaction with O (^3P) at 296 K in air (Slagle et al. 1978); atmospheric second-order rate constants of 4.57×10^{-12} $cm^3 \cdot molecule^{-1} \cdot sec^{-1}$ for reactions with OH radicals, 9.77×10^{-13} $cm^3 \cdot molecule^{-1} \cdot sec^{-1}$ for reactions with NO_3 radicals (Sabljic & Güsten 1990; Müller & Klein 1991); rate constants for gas-phase reactions with NO_3 are, 9.92×10^{-13} $cm^3 \cdot molecule^{-1} \cdot sec^{-1}$, 7.5×10^{-13} $cm^3 \cdot molecule^{-1} \cdot sec^{-1}$, 8.1×10^{-13} $cm^3 \cdot molecule^{-1} \cdot sec^{-1}$, 1.3×10^{-12} $cm^3 \cdot molecule^{-1} \cdot sec^{-1}$ at 298 K, 1.08×10^{-12} $cm^3 \cdot molecule^{-1} \cdot sec^{-1}$, 1.04×10^{-12} $cm^3 \cdot molecule^{-1} \cdot sec^{-1}$ at 297 K (quoted, Atkinson 1991).
Hydrolysis:
Biodegradation:
Biotransformation:
Bioconcentration, Uptake (k_1) and Elimination (k_2) Rate Constants:

Common Name: Dimethyl sulfoxide
Synonym: DMSO, sulfinylbismethane, methyl sulfoxide, methylsulfinylmethane, SQ 9453, DMS-70, DMS-90, Deltan, Demasorb, Demavet, Demeso, Dermasorb, Dolicur, Domoso, Dromisol, Gamasol 90, Hyadur, Rimso-50, Sclerosol, Somipront, Somtexan, Topsym
Chemical Name: dimethyl sulfoxide
CAS Registry No: 67-68-5
Molecular Formula: $(CH_3)_2SO$
Molecular Weight: 78.129
Melting Point (°C):
18.4 (Weast 1982-83)
8, 18.5 (Verschueren 1983)
18.55 (Stephenson & Malanowski 1987)
Boiling Point (°C): 189.0
100, 189 decomposes (Verschueren 1983)
189.0 (Stephenson & Malanowski 1987)
Density (g/cm^3 at 25°C):
1.1014 (Weast 1982-83)
1.100 (Verschueren 1983)
Molar Volume (cm^3/mol):
71.0 (Taft et al. 1985; Kamlet et al. 1987)
45.5 (intrinsic volume V_I, Kamlet et al. 1987)
85.7 (calculated-LeBas method)
71.33 (calculated-density)
Molecular Volume ($Å^3$):
Total Surface Area, TSA ($Å^2$):
Dissociation Constant pK_a:
-1.54 (Riddick et al. 1986)
Heat of Fusion, ΔH_{fus}, kcal/mol:
3.434 (Riddick et al. 1986)
Entropy of Fusion, ΔS_{fus}, cal/mol·K (e.u.):
Fugacity Ratio at 25°C (assuming ΔS_{fus} = 13.5 e.u.), F: 1.0

Water Solubility (g/m^3 or mg/L at 25°C):
253000 (quoted, Riddick et a. 1986)
30395708, 31103715 (quoted, calculated-solvatochromic p., Taft et al. 1985)
27721000, 15590000 (quoted, calculated-V_I and solvatochromic p., Leahy 1986)
27721215, 22532650 (quoted, calculated-solvatochromic p., Kamlet et al. 1987)

Vapor Pressure (Pa at 25°C):
56.0 (20°C, Verschueren 1983)
80.0 (quoted, Riddick et al. 1986)
79.5 (extrapolated-Antoine eqn., Stephenson & Malanowski 1987)

Henry's Law Constant (Pa·m^3/mol):

Octanol/Water Partition Coefficient, log K_{OW}:

-1.35 (Hansch & Leo 1979; quoted, Yoshida et al. 1986)
-2.03 (quoted, Verschueren 1983)
-1.35, -1.47 (quoted, calculated-solvatochromic p., Taft et al. 1985)
-1.35, -1.23 (quoted, calcuulated-V_I and solvatochromic p., Leahy 1986)
-1.35 (quoted and recommended, Sangster 1989)
-1.35, -1.8633 (quoted, calculated-original UNIFAC, Chen et al. 1993)

Bioconcentration Factor, log BCF:

Sorption Partition Coefficient, log K_{OC}:

Half-Lives in the Environment:

Air:
Surface water:
Groundwater:
Sediment:
Soil:
Biota:

Environmental Fate Rate Constants or Half-Lives:

Volatilization:
Photolysis:
Oxidation:
Hydrolysis:
Biodegradation:
Biotransformation:
Bioconcentration, Uptake (k_1) and Elimination (k_2) Rate Constants:

Common Name: Dimethyl sulfate
Synonym: sulfluric acid dimethyl ester, DMS
Chemical Name: dimethyl sulfate
CAS Registry No: 77-78-1
Molecular Formula: $CH_3O\text{-}SO_2\text{-}OCH_3$
Molecular Weight: 126.13
Melting Point (°C):
-32.0 (Verschueren 1983)
-31.8 (Dean 1985)
Boiling Point (°C):
188 (decomposes, Verschueren 1983)
Density (g/cm^3 at 20°C):
1.33 (Verschueren 1983)
1.3322 (Dean 1985
Molar Volume (cm^3/mol):
109.7 (calculated-LeBas method)
Molecular Volume ($Å^3$):
Total Surface Area, TSA ($Å^2$):
Dissociation Constant pK_a:
Heat of Fusion, ΔH_{fus}, kcal/mol:
Entropy of Fusion, ΔS_{fus}, cal/mol·K (e.u.):
Fugacity Ratio at 25°C (assuming ΔS_{fus} = 13.5 e.u.), F: 1.0

Water Solubility (g/m^3 or mg/L at 25°C):
28000 (hydrolyzes, Verschueren 1983; Dean 1985)
28000 (18°C, Budavari 1989)

Vapor Pressure (Pa at 25°C):
<133 (20°C, Verschueren 1983)
128 (extrapolated, Antoine eqn., Stephenson & Malanowski 1987)

Henry's Law Constant ($Pa \cdot m^3/mol$):

Octanol/Water Partition Coefficient, log K_{OW}:

Bioconcentration Factor, log BCF:

Sorption Partition Coefficient, log K_{OC}:

Half-Lives in the Environment:
Air: 36.5-365 hours, based on photooxidation half-life in air from estimated rate constant for the vapor phase reaction with hydroxyl radicals in air (Atkinson 1987; quoted, Howard et al. 1991); atmospheric transformation lifetime (reaction with liquid water) was estimated to be < 1 day (Kelly et al. 1994).

Surface water: 1.2-12 hours, based on overall hydrolysis rate constant for pH 7 at 25°C (Mabey & Mill 1978; quoted, Howard et al. 1991) and for complete hydrolysis in neutral, slightly basic, and acidic aqueous solutions (Lee et al. 1980; quoted, Howard et al. 1991).

Groundwater: 1.2-12 hours, based on overall hydrolysis rate constant for pH 7 at 25°C (Mabey & Mill 1978; quoted, Howard et al. 1991) and for complete hydrolysis in neutral, slightly basic, and acidic aqueous solutions (Lee et al. 1980; quoted, Howard et al. 1991).

Sediment:

Soil: 1.2-12 hours, based on overall hydrolysis rate constant for pH 7 at 25°C (Mabey & Mill 1978; quoted, Howard et al. 1991) and for complete hydrolysis in neutral, slightly basic, and acidic aqueous solutions (Lee et al. 1980; quoted, Howard et al. 1991).

Biota:

Environmental Fate Rate Constants or Half-Lives:

Volatilization:

Photolysis:

Oxidation: atmospheric photooxidation half-life of 36.5-365 hours, based on estimated rate constant for the vapor-phase reaction with OH radicals in air (Atkinson 1987; quoted, Howard et al. 1991).

Hydrolysis: first order hydrolysis rate constant of 1.6×10^{-4} sec^{-1} at pH 7 and 25°C with a half-life of 1.2 hour (Mabey & Mill 1978; quoted, Howard et al. 1991).

Biodegradation: aqueous aerobic biodegradation half-life was estimated to be 168-672 hours and aqueous anaerobic biodegradation half-life was estimated to be 672-2688 hours (Howard et al. 1991).

Biotransformation:

Bioconcentration, Uptake (k_1) and Elimination (k_2) Rate Constants:

Common Name: Ethanethiol
Synonym: ethyl mercaptan, thioethyl alcohol, ethylhydrosulfide
Chemical Name: ethanethiol
CAS Registry No: 75-08-1
Molecular Formula: C_2H_5SH
Molecular Weight: 62.13
Melting Point (°C):
-121 (Stull 1947)
-144.4 (Weast 1982-83)
-121, -144 (Verschueren 1983)
-147.9 (Dean 1985)
Boiling Point (°C):
35.0 (Stull 1947; Weast 1982-83; Boublik et al. 1984; Dean 1985)
Density (g/cm^3 at 20°C):
0.8391 (Weast 1982-83)
0.8315 (25°C, Dean 1985)
Molar Volume (cm^3/mol):
77.4 (calculated-LeBas method)
74.72 (calculated-density)
Molecular Volume (Å^3):
Total Surface Area, TSA (Å^2):
Dissociation Constant pK_a:
Heat of Fusion, ΔH_{fus}, kcal/mol:
Entropy of Fusion, ΔS_{fus}, cal/mol·K (e.u.):
Fugacity Ratio at 25°C (assuming ΔS_{fus} = 13.5 e.u.), F: 1.0

Water Solubility (g/m^3 or mg/L at 25°C):
15606 (quoted, Hine & Mookerjee 1975)
15000 (quoted, Verschueren 1983)
6800 (quoted, Dean 1985)
6760 (20°C, Budavari 1989)
14800 (selected, Yaws et al. 1990)

Vapor Pressure (Pa at 25°C):
74630 (interpolated-regression of tabulated data, Stull 1947)
66660, 70290 (23.613°C, interpolated-Antoine eqn., Zwolinski & Wilhoit 1971)
70300 (quoted, Hine & Mookerjee 1975)
58660 (20°C, Verschueren 1983)
70110, 70290 (24.933°C, exptl., interpolated-Antoine eqn., Boublik et al. 1984)
30690 (calculated-Antoine eqn., Dean 1985)

Henry's Law Constant ($Pa \cdot m^3/mol$):

278.13 (exptl., quoted, Hine & Mookerjee 1975)
298.02 (calculated-group contribution, Hine & Mookerjee 1975)
366.64 (calculated-bond contribution, Hine & Mookerjee 1975)
360.3 (headspace-GC, Przyjazny et al. 1983)
292.37 (computed, Yaws et al. 1991)
278, 96.44 (quoted, calculated-molecular structure, Russell et al. 1992)
289.94 (calculated-P/C, this work)

Octanol/Water Partition Coefficient, log K_{OW}:

Bioconcentration Factor, log BCF:

Sorption Partition Coefficient, log K_{OC}:

Half-Lives in the Environment:

Air:
Surface water:
Groundwater:
Sediment:
Soil:
Biota:

Environmental Fate Rate Constants or Half-Lives:

Volatilization:
Photolysis:
Oxidation: rate constant of 3.0×10^{-12} $cm^3 \cdot molecule^{-1} \cdot sec^{-1}$ for gas-phase reaction with O (^3P) at 298 in air (Slagle et al. 1978); second-order rate constant of $(1.21 \pm 0.28) \times 10^{-12}$ $cm^3 \cdot molecule^{-1} \cdot sec^{-1}$ for gas-phase reactions with NO_3 radicals at 297K (Atkinson 1991).
Hydrolysis:
Biodegradation:
Biotransformation:
Bioconcentration, Uptake (k_1) and Elimination (k_2) Rate Constants:

Common Name: 1-Butanethiol
Synonym: butyl mercaptan
Chemical Name: 1-butanethiol
CAS Registry No: 109-75-5
Molecular Formula: $CH_3(CH_2)_3SH$
Molecular Weight: 90.183
Melting Point (°C): -115.67
Boiling Point (°C): 98.455
Density (g/cm^3 at 25°C): 0.83674
Molar Volume (cm^3/mol):
121.8 (calculated-LeBas method)
107.8 (calculated-density)
Molecular Volume (Å^3):
Total Surface Area, TSA (Å^2):
Dissociation Constant pK_a:
Heat of Fusion, ΔH_{fus}, kcal/mol:
2.50 (Riddick et al. 1986)
Entropy of Fusion, ΔS_{fus}, cal/mol·K (e.u.):
Fugacity Ratio at 25°C (assuming ΔS_{fus} = 13.5 e.u.), F: 1.0

Water Solubility (g/m^3 or mg/L at 25°C):
597 (quoted, Riddick et al. 1986)
600 (selected, Yaws et al. 1990)

Vapor Pressure (Pa at 25°C):
5330, 6070 (22.4°C, interpolated-Antoine eqn., Zwolinski & Wilhoit 1971)
6070 (quoted, Riddick et al. 1986)

Henry's Law Constant (Pa m^3/mol):
460.7 (headspace-GC, Przyjazny et al. 1983)
916.9 (calculated-P/C, this work)

Octanol/Water Partition Coefficient, log K_{OW}:
2.28 (quoted CLOGP, recommended, Sangster 1989)

Bioconcentration Factor, log BCF: no data

Sorption Partition Coefficient, log K_{OC}: no data

Half-Lives in the Environment/Environmental Fate Rate Constants or Half-Lives: no data

Common Name: Thiophene
Synonym: thiofuran
Chemical Name: thiophene, thiofuran
CAS Registry No: 110-02-1
Molecular Formula: C_4H_4S
Molecular Weight: 84.14
Melting Point (°C):
-38.3 (Stull 1947)
-38.2 (Weast 1982-83; Dean 1985)
-30/-38 (Verschueren 1983)
Boiling Point (°C):
84.4 (Stull 1947)
84.16 (Dreisbach 1955; Boublik et al. 1984)
84.2 (Weast 1982-83; Dean 1985)
84.0 (Verschueren 1983)
Density (g/cm^3 at 20°C):
1.05887 (25°C, Dreisbach 1955)
1.0649 (Weast 1982-83)
1.060 (Verschueren 1983)
1.0573 (25°C, Dean 1985)
1.06428, 1.05884 (20°C, 25°C, Riddick et al. 1986)
Molar Volume (cm^3/mol):
44.5 (intrinsic V_I, Hickey & Passino-Reader 1991)
88.1 (calculated-LeBas method)
79.4 (calculated from density)
Molecular Volume (Å^3):
Total Surface Area, TSA (Å^2):
Dissociation Constant pK_a:
Heat of Fusion, ΔH_{fus}, kcal/mol: 1.216 (Riddick et al. 1986)
Entropy of Fusion, ΔS_{fus}, cal/mol·K (e.u.):
Fugacity Ratio at 25°C (assuming ΔS_{fus} = 13.5 e.u.), F: 1.0

Water Solubility (g/m^3 or mg/L at 25°C):
3015 (shake flask-GC, Price 1976)
2985 (observed, Valvani et al. 1981)
3936 (predicted-K_{OW}, Valvani et al. 1981)
3600 (18°C, quoted, Verschueren 1983)
miscible (quoted, Riddick et al. 1986)
3020 (selected, Yaws et al. 1990)

Vapor Pressure (Pa at 25°C):
9670 (interpolated-regression of tabulated data, Stull 1947)
7998 (20.1°C, Stull 1947)
10620 (calculated-Antoine eqn., Dreisbach 1955)
10670, 10660 (25.09°C, interpolated-Antoine eqn., Zwolinski & Wilhoit 1971)
2450 (calculated-Cox eqn., Chao et al. 1983)

7998 (20°C, quoted, Verschueren 1983)
13330 (30°C, quoted, Verschueren 1983)
10622, 10620 (quoted exptl., calculated-Antoine eqn., Boublik et al. 1984)
10620 (calculated-Antoine eqn., Dean 1985)
10620 (quoted, Riddick et al. 1986)
10600 (extrapolated-Antoine eqn., Stephenson & Malanowski 1987)

Henry's Law Constant (Pa·m^3/mol):
224, 236, 230 (headspace-GC, concn. of 10, 1.0 and 0.1 ppm by weight, Przyjazny et al. 1983)
223.3 (calculated-P/C, this work)

Octanol/Water Partition Coefficient, log K_{OW}:
1.81 ± 0.01 (shake flask-AS, Iwasa et al. 1965; quoted, Rekker 1977; Hansch & Leo 1979; Sangster 1989)
1.79 (calculated-f const., Rekker 1977)
1.81 (quoted, Valvani et al. 1981)
1.81 (quoted, Verschueren 1983)
1.74 (HPLC, Garst 1984)
1.82 (quoted from Pomona College unpublished data, Sangster 1989)
1.81 (recommended, Sangster 1989, 1993)
1.82 (shake flask-RPHPLC-k', Yamagami & Takao 1992)
1.81 (quoted from U.S. EPA CLOGP Data Base, Hulzebos et al. 1993)

Bioconcentration Factor, log BCF:

Sorption Partition Coefficient, log K_{OC}:

Half-Lives in the Environment:
Air:
Surface water:
Groundwater:
Sediment:
Soil:
Biota:

Environmental Fate Rate Constants or Half-Lives:
Volatilization:
Photolysis:
Oxidation: comparison of rate constants and loss rates: 9.6×10^{-12} cm^3·molecule^{-1}·sec^{-1} for the gas-phase reaction with 1×10^6 OH radicals/cm^3 during daytime hours in air

at room temperature with a loss rate of 0.8 day^{-1} having a calculated lifetime of 29 hours, $3.2x10^{-14}$ $cm^3 \cdot molecule^{-1} \cdot sec^{-1}$ for the gas-phase reaction with $2.4x10^8$ NO_3 radicals/cm^3 in air at room temperature with a loss rate of 0.7 day^{-1} having a calculated lifetime of 36 hours and $<6x10^{-20}$ $cm^3 \cdot molecule^{-1} \cdot sec^{-1}$ for the gas-phase reaction with $7x10^{11}$ O_3 molecules/cm^3 in air at room temperature with a loss rate of <0.004 day^{-1} having a calculated lifetime >270 days (Atkinson et al. 1983; quoted, Atkinson et al. 1984, 1985); atmospheric rate constant of $9.70x10^{-12}$ $cm^3 \cdot molecule^{-1} \cdot sec^{-1}$ for the reaction with $5x10^5$ OH radicals/cm^3 at room temperature and having a loss rate of 0.4 day^{-1} (Atkinson 1985); using relative rate technique, an average rate constant of $(3.2 \pm 0.7)x10^{-14}$ $cm^3 \cdot molecule^{-1} \cdot sec^{-1}$ was determined for the gas-phase reaction with $2.4x10^8$ NO_3 radicals/cm^3 in nighttime air at 295 ± 1 K in the atmosphere with a calculated lifetime of 36 hours and a loss rate of 0.7 day^{-1} (Atkinson et al. 1985; quoted, Atkinson 1985); rate constants of $(3.57 \pm 0.9)x10^{-14}$ $cm^3 \cdot molecule^{-1} \cdot sec^{-1}$, and $(3.91 \pm 0.09)x10^{-14}$ $cm^3 \cdot molecule^{-1} \cdot sec^{-1}$, $(3.93 \pm 0.16)x10^{-14}$ $cm^3 \cdot molecule^{-1} \cdot sec^{-1}$ at 296 K for gas-phase reactions with NO_3 radicals (Atkinson 1991); exptl. rate constant of $9.53x10^{-12}$ $cm^3 \cdot molecule^{-1} \cdot sec^{-1}$ (Atkinson 1989) as compared a calculated value of $14.81x10^{-12}$ $cm^3 \cdot molecule^{-1} \cdot sec^{-1}$ by molecular orbital calculation for the gas-phase reaction with hydroxyl radicals at room temperature (Klamt 1993).

Hydrolysis:

Biodegradation:

Biotransformation:

Bioconcentration, Uptake (k_1) and Elimination (k_2) Rate Constants:

Common Name: Benzo(b)thiophene
Synonym: thianaphthene, thionaphthene, 1-benzothiophene
Chemical Name:
CAS Registry No: 95-15-8
Molecular Formula: C_8H_6S
Molecular Weight: 134.20
Melting Point (°C):
31.3 (Smith et al. 1978)
30.85 (Mackay et al. 1982)
29-32 (Verschueren 1983; Pearlman 1984)
Boiling Point (°C):
219.85 (Smith et al. 1978; Mackay et al. 1982)
221-222 (Verschueren 1983; Pearlman 1984)
Density (g/cm^3 at 20°C): 1.15 (Verschueren 1983)
Molar Volume (cm^3/mol):
139.7 (LeBas method)
Molecular Volume (Å^3):
118.066 (Pearlman et al. 1984)
Total Surface Area, TSA (Å^2):
146.919 (Pearlman et al. 1984)
Heat of Fusion, kcal/mol:
Entropy of Fusion ΔS_{fus}, cal/mol·K (e.u.):
Fugacity Ratio at 25°C (assuming ΔS_{fus} = 13.5 e.u.), F: 0.875

Water Solubility (g/m^3 or mg/L at 25°C):
130 (20°C, shake flask, Smith et al. 1978; quoted, Steen & Karickhoff 1981)
130.2 (Mill et al. 1981)
130.2 (quoted, Pearlman et al. 1984)

Vapor Pressure (Pa at 25°C):
26.7 (20°C, estimated from naphthalene, Smith et al. 1978; quoted, Mackay et al. 1982)
26.3, 14.80 (quoted, calculated-B.P., Mackay et al. 1982)

Henry's Law Constant (Pa·m^3/mol):
28.0 (calculated-P/C, Smith & Bomberger 1980)
24.1 (calculated-P/C, this work)

Octanol/Water Partition Coefficient, log K_{OW}:
3.09 (shake flask-AS, pH 7.5, Rogers & Cammarata 1969)
3.05 (HPLC-RT, De Voogt et al. 1988)
3.12 (quoted and recommended, Sangster 1989, 1993)
3.26 (shake flask-HPLC, De Voogt et al. 1990)

3.18 (HPLC-RT, Ritter et al. 1994)

Bioconcentration Factor, log BCF:
2.08 (mixed microbial populations, Steen & Karickhoff 1981)

Sorption Partition Coefficient, log K_{OC}:
1.77 (Coyote Creek sediment, Smith et al. 1978)
2.30 (lab. mixture of microorganisms, Smith et al. 1978)
3.49, 3.0 (soil, quoted, calc-χ and fragment contribution, Meylan et al. 1992)

Half-Lives in the Environment:
Air:
Surface water: estimated volatilization half-lives of 45 hours in river, 230 hours in eutrophic pond, 180 hours in eutrophic lake and oligotrophic lake by the one compartment model (Smith et al. 1978); photolysis rate constant of transformation and transport of $(6.9 \pm 0.7)\times10^{-7}$ sec^{-1} exposed to 12 hours sunlight per day in late May with estimated photolysis half-lives of 1200 hours in river, 2900 hours in eutrophic pond, 3500 hours in eutrophic lake and 600 hours in oligotrophic lake by the one compartment model (Smith et al. 1978).
Groundwater:
Sediment:
Soil:
Biota:

Environmental Fate Rate Constants or Half-Lives:
Volatilization: estimated half-lives of 45 hours in river, 230 hours in eutrophic pond, 180 hours in eutrophic lake and oligotrophic lake by the one compartment model (Smith et al. 1978).
Photolysis: rate constant of transformation and transport of $(6.9 \pm 0.7)\times10^{-7}$ sec^{-1} exposed to 12 hours sunlight per day in late May with estimated half-lives of 1200 hours in river, 2900 hours in eutrophic pond, 3500 hours in eutrophic lake and 600 hours in oligotrophic lake by the one compartment model (Smith et al. 1978).
Hydrolysis:
Oxidation: laboratory investigated rate constant of 83 M^{-1} sec^{-1} for the reaction with RO_2 radicals and estimated half-lives of 10^5 hours in river, eutrophic pond, eutrophic lake and oligotrophic lake by the one compartment model (Smith et al. 1978).
Biodegradation: estimated half-lives of >20 hours in river, <20 hours in eutrophic pond, 20 hours in eutrophic lake and very long half-life in oligotrophic lake, based on the biodegradation rate in the presence of alternative carbon sources will be one-half the biodegradation rate of quinoline when quinoline is the only carbon source by the one compartment model (Smith et al. 1978).
Biotransformation:
Bioconcentration, Uptake (k_1) and Elimination (k_2) Rate Constants:

Common Name: Dibenzothiophene
Synonym:
Chemical Name: dibenzothiophene
CAS Registry No: 132-65-0
Molecular Formula: $C_{12}H_8S$
Molecular Weight: 184.27
Melting Point (°C):
 99-100 (Smith et al. 1978)
 100 (Steen & Karickhoff 1981)
Boiling Point (°C):
 323.3 (Smith et al. 1978)
 331.85 (Mackay et al. 1982)
Density (g/cm^3 at 20°C):
Molar Volume (cm^3/mol): 191.3 (LeBas method)
Molecular Volume (Å^3): 163.429 (Pearlman et al. 1984)
Total Surface Area, TSA (Å^2): 193.588 (Pearlman et al. 1984)
Heat of Fusion ΔH_{fus}, kcal/mol:
Entropy of Fusion ΔS_{fus}, cal/mol·K (e.u.):
Fugacity Ratio at 25°C (assuming ΔS_{fus} = 13.5 e.u.), F: 0.185

Water Solubility (g/m^3 or mg/L at 25°C):
 1.11 ± 0.09 (28°C, measured, Smith et al. 1978)
 1.47 (24°C, shake flask-LSC, Mean et al. 1980)
 1.106 (quoted, Mill et al. 1981)
 1.50 (quoted, Steen & Karickhoff 1981)
 1.11, 1.47, 0.53; 1.032 (quoted values; average, Pearlman et al. 1984)

Vapor Pressure (Pa at 25°C):
 0.267 (20°C, estimated, Aubry et al. 1975; quoted, Smith et al. 1978)
 0.263, 0.0083 (20°C, quoted, calculated-B.P., Mackay et al. 1982)
 0.893 (static apparatus-extrapolated from Chebyshev polynomials, Sivaraman & Kobayashi 1982)
 0.586 (extrapolated-Cox eqn., Chao et al. 1983)

Henry's Law Constant (Pa·m^3/mol):
 44.3 (calculated-P/C, Smith & Bomberger 1980)

Octanol/Water Partition Coefficient, log K_{OW}:
 4.38 (shake flask-LSC, Means et al. 1980; quoted, Steen & Karickhoff 1981)
 4.33 (HPLC-RT, De Voogt et al. 1988)
 4.38 (quoted and recommended, Sangster 1989, 1993)
 4.49 (shake flask-HPLC, De Voogt et al. 1990)

Bioconcentration Factor, log BCF:
 3.20 (mixed microbial populations, Steen & Karickhoff 1981)

Sorption Partition Coefficient, log K_{OC}:

3.14 (Coyote Creek sediment, Smith et al. 1978)
4.05 (soil, Hassett et al. 1980; quoted, Sabljic 1987)
4.05; 4.17 (sediment/soil; calculated, Means et al. 1980)
4.00 (soil, calculated- χ , Sabljic 1987)
4.17 (soil, calculated-K_{OW}, model of Karickhoff et al. 1979, Sabljic 1987)
3.75 (soil, calculated-K_{OW}, model of Kenaga & Goring 1980, Sabljic 1987)
2.92 (soil, calculated-K_{OW}, model of Briggs 1981, Sabljic 1987)
4.00 (soil, calculated-K_{OW}, model of Means et al. 1982, Sabljic 1987)
3.60 (soil, calculated-K_{OW}, model of Chiou et al. 1983, Sabljic 1987)

Half-Lives in the Environment:

Air:

Surface water: half-lives of 0.5 hour from river water, 13 hours from pond water, 13 hours from eutrophic lake and 140 hours from oligotrophic lake predicted by one-compartment model for all processes; estimated volatilization half-lives of 140 hours in river, 720 hours in eutrophic pond, 580 hours in eutrophic lake and oligotrophic lake; photolysis rate constant of transformation and transport of $(2.04 \pm 0.08)\times10^{-6}$ sec^{-1} exposed to 12 hours sunlight per day in early March with estimated photolysis half-lives of 380 hours in river, 950 hours in eutrophic pond and eutrophic lake and 190 hours in oligotrophic lake; biodegradation half-lives of 13 hours in river, eutrophic pond water and 140 hours in oligotrophic lake (Smith et al. 1978); half-life of 4-8 hours for disappearance via direct photolysis in aqueous media (Harris 1982).

Environmental Fate Rate Constants or Half-Lives:

Volatilization: estimated half-lives of 140 hours in river, 720 hours in eutrophic pond, 580 hours in eutrophic lake and oligotrophic lake by the one compartment model (Smith et al. 1978).

Photolysis: solar phtolysis rate constant of 1.5×10^{-8} sec^{-1} over 24-hour day; rate constant of transformation and transport of $(2.04 \pm 0.08)\times10^{-6}$ sec^{-1} exposed to 12 hours sunlight per day in early March with estimated half-lives of 380 hours in river, 950 hours in eutrophic pond and eutrophic lake and 190 hours in oligotrophic lake from average photolysis rates on a summer day at 40°N latitude by the one compartment model (Smith et al. 1978); half-life of 4-8 hours for disappearance via direct photolysis in aquatic media (Harris 1982).

Oxidation: laboratory investigated rate constant of <7.5 $M^{-1}\cdot sec^{-1}$ for the reaction with RO_2 radicals and estimated half-lives of $>10^5$ hours in river, eutrophic pond, eutrophic lake and oligotrophic lake by the one compartment model (Smith et al. 1978).

Hydrolysis:

Biodegradation: biodegradation rate constant of 5.3×10^{-7} $mL\cdot cell^{-1}\cdot hour^{-1}$ and the estimated half-lives of 13 hours in river, eutrophic pond, and eutrophic lake and $>10^4$ hours in oligotrophic lake by the one compartment model (Smith et al. 1978).

Biotransformation:

Bioconcentration, Uptake (k_1) and Elimination (k_2) Rate Constants:

Common Name: Thiourea
Synonym: thiocarbamide
Chemical Name: thiourea
CAS Registry No: 62-56-6
Molecular Formula: H_2NCSNH_2
Molecular Weight: 76.12
Melting Point (°C):
    182 (Weast 1982-83; Verschueren 1983)
Boiling Point (°C):
    decomposes (Verschueren 1983)
Density (g/cm^3 at 20°C):
    1.405 (Weast 1982-83; Verschueren 1983)
Dissociation Constants, pK:
    2.03 (pK_1, Dean 1985)
Molar Volume (cm^3/mol):
    76.2 (calculated-LeBas method)
Molecular Volume, TMV (Å^3):
Total Surface Area, TSA (Å^2):
Heat of Fusion, ΔH_{fus}, kcal/mol:
Entropy of Fusion, ΔS_{fus}, cal/mol·K (e.u.):
Fugacity Ratio at 25°C (assuming ΔS_{fus} = 13.5 e.u.), F: 0.0280

Water Solubility (g/m^3 or mg/L at 25°C):
    91800 (13°C, quoted, Verschueren 1983)
    89822 (Windholz 1983; quoted, Chessells et al. 1992)
    110000 (Budavari 1989)
    90000 (quoted, Dean 1985)

Vapor Pressure (Pa at 25°C):

Henry's Law Constant (Pa m^3/mol):

Octanol/Water Partition Coefficient, log K_{OW}:
    -1.02 (Leo et al. 1971; quoted, Chessells et al. 1992)
    -2.38, -0.95 (calculated, Verschueren 1983)
    -1.17 (shake flask, OECD 1981 Guidelines, Geyer et al. 1984)
    -1.03 (shake flask-HPLC, Govers et al. 1986)
    -1.14, -1.05, -1.02 (quoted CLOGP, Sangster 1993)
    -0.99 (recommended, Sangster 1993)

Bioconcentration Factor, log BCF:

1.73 (alga *chlorella fusca*, wet wt. basis, Geyer et al. 1984)

-0.699 (alga *chlorella fusca*, calculated-K_{OW}, Geyer et al. 1984)

Sorption Partition Coefficient, log K_{OC}:

Half-Lives in the Environmental Compartments:

Air: 1.6-16 hours, based on estimated photooxdation half-life in air (Howard et al. 1991).

Surface water: 24-168 hours, based on estimated aqueous aerobic biodegradation half-life (Howard et al. 1991).

Ground water: 48-336 hours, based on estimated unacclimated aqueous aerobic biodegradation half-life (Howard et al. 1991).

Sediment:

Soil: 24-168 hours, based on estimated aqueous aerobic biodegradtion half-life (Howard et al. 1991)

Biota:

Environmental Fate Rate Constants and Half-Lives:

Volatilization:

Photolysis:

Oxidation: photooxidation half-life of 1.6-16 hours in air, based on rate constant for the vapor-phase reaction with OH radicals and photooxidation half-life of 2048-81927 hours in water, based on estimated rate data for reaction with OH radicals in aqueous solution (Howard et al. 1991).

Hydrolysis:

Biodegradation: aqueous aerobic biodegradation half-life of 24-168 hours, based on aqueous aerobic screening test data and aqueous anerobic biodegradation half-life of 96-672 hours, based on aqueous aerobic degradation half-life (Howard et al. 1991).

Biotransformation:

Bioconcentration Uptake (k_1) and Elimination (k_2) Rate Constants:

Common Name: Thioacetamide
Synonym: ethanethioamide, acetothioamide
Chemical Name: thioacetamide
CAS Registry No: 62-55-5
Molecular Formula: CH_3CSNH_2
Molecular Weight: 75.13
Melting Point (°C):
115-116 (Weast 1982-83)
108.5 (Verschueren 1983)
112-114 (Dean 1985)
Boiling Point (°C):
Density (g/cm³ at 20°C):
Molar Volume (cm³/mol):
84.2 (calculated-LeBas method)
Molecular Volume (Å³):
Total Surface Area, TSA (Å²):
Dissociation Constant pK_a:
Heat of Fusion, ΔH_{fus}, kcal/mol:
Entropy of Fusion, ΔS_{fus}, cal/mol·K (e.u.):
Fugacity Ratio at 25°C (assuming ΔS_{fus} = 13.5 e.u.), F: 0.1288

Water Solubility (g/m³ or mg/L at 25°C):
163000 (Dean 1985)
163000 (Budavari 1989)

Vapor Pressure (Pa at 25°C):

Henry's Law Constant (Pa·m³/mol):

Octanol/Water Partition Coefficient, log K_{OW}:
-0.46, 0 36 (quoted, Verschueren 1983)
-0.26 (quoted and recommended, Sangster 1993)

Bioconcentration Factor, log BCF:

Sorption Partition Coefficient, log K_{OC}:

Half-Lives in the Environment:

Air: half-lives of 3.2-31.7 hours, based on estimated photooxidation half-life in air (Howard et al. 1991).

Surface water: half-lives of 24-268 hours, based on estimated aqueous aerobic biodegradation half-life (Howard et al. 1991).

Groundwater: half-lives of 48-336 hours, based on estimated unacclimated aqueous aerobic biodegradation half-life (Howard et al. 1991).

Sediment:

Soil: half-lives of 24-168 hours, based on estimated aqueous aerobic biodegradation half-life (Howard et al. 1991).

Biota:

Environmental Fate Rate Constants or Half-Lives:

Volatilization:

Photolysis:

Oxidation: atmospheric half-life of 3.2-31.7 hours, based on estimated rate data for OH radicals in air (Howard et al. 1991).

Hydrolysis: first-order rate constant of 8.6×10^{-1} h^{-1} at pH 7 and 25°C (Ellington et al. 1987), corresponding to a half-life of 8064 hours (Howard et al. 1991); acid rate constant of 6.0×10^{-2} $M^{-1} \cdot h^{-1}$, corresponding to a half-life of 333 days and base rate constant of 1.4 $M^{-1} \cdot h^{-1}$, corresponging to a half-life of 289 days (Howard et al. 1991).

Biodegradation: aerobic biodegradation half-life of 24-168 h, based on aqueous aerobic screening test data and anaerobic biodegradation half-life of 96-672 h, based on estimated aqueous aerobic biodegradation half-life (Howard et al. 1991).

Biotransformation:

Bioconcentration, Uptake (k_1) and Elimination (k_2) Rate Constants:

7.2 Summary Tables

Table 7.1 Physical-chemical properties of nitrogen and sulfur containing compounds at 25 °C.

Compound	CAS no.	formula	MW	mp, °C	bp, °C	Fugacity ratio, F at 25 °C	molar volume, cm³/mol LeBas	V_I (a)	V_M (b)	TMV, Å³ (c)	total surface area TSA, Å² (c)	TSA, Å² (d)	pK_a or pK_b
Nitriles:													
acetonitrile	75-05-8	CH_3CN	41.052	-43.835	81.6	1	56.3	27.1				160.1	
propionitrile	107-12-0	C_2H_5CN	55.079	-92.78	97.35	1	78.5	36.9				187.5	
butyronitrile	109-74-0	C_3H_7CN	69.106	-111.9	117.62	1	100.7	46.5					
benzonitrile	100-47-0	C_6H_5CN	103.123		191.1	1	107.9	59				247.5	
acrylonitrile	107-13-1	C_2H_3CN	53.063	-83.55	77.3	1	71.1						
adiponitrile	111-69-3	$CN(CH_2)_4CN$	108.16	1	295	1	149.6						
Amides:													
acetamide	60-35-5	CH_3CONH_2	59.07	79-81	dec.	0.292	66.9						7.62
acrylamide	79-06-1	$H_2C{=}CHCONH_2$	71.08	84-86		0.261	80.8						
benzamide	55-21-0	$C_6H_5CONH_2$	121.14	128-129	288	0.096	132.4					257.6	
Urea:													
urea	57-13-6	H_2NCONH_2	60.06	132.7		0.0861	58.0						
Aliphatic amines:													
methylamine	74-89-5	CH_3NH_2	31.06	-92.5	-6.5	1	43.8					139.5	
dimethylamine	124-40-3	$(CH_3)_2NH$	45.09	-96	7.4	1	67.5					214.1	10.77
trimethylamine	75-50-3	$(CH_3)_3N$	59.11	-117.2	2.87	1	93.3					192.9	9.8
ethylamine	74-04-7	$CH_3CH_2NH_2$	45.09	-80.6	16.6	1	66.0						10.63
diethylamine	109-89-7	$(C_2H_5)_2NH$	73.14	-50	55.5	1	111.9					214.1	10.8
triethylamine	121-44-8	$(C_2H_5)_3N$	101.91	-114.7	88.87	1	154.8	70.4				262.9	10.78
n-propylamine	107-10-8	$C_3H_7NH_2$	59.11	-83	48.7	1	88.2					220.4	10.568
dipropylamine	142-84-7	$(C_3H_7)_2NH$	101.9			1	156.3					256.1	
diisopropylamine	108-18-9	$i(C_3H_7)_2NH$	101.9				156.3						
tripropylamine	102-69-2	$(C_3H_7)_3N$	143.27	-93.5		1	156.3	98.5				340.7	10.66

Compounds	CAS no.	formula	MW	mp, °C	bp, °C	Fugacity ratio, F at 25 °C	molar volume, cm^3/mol LeBas	V_I (a)	V_M (b)	TMV,Å^3 (c)	total surface area TSA,Å^2 (c)	TSA,Å^2 (d)	pK_a or pK_b
n-butylamine	109-73-9	$C_4H_9NH_2$	73.14	-50	77.8	1	110.4					220.4	10.64
isobutylamine	78-81-9	$iC_4H_9NH_2$	73.14	-85	68-69	1	110.4						10.41
tert-butylamine	109-73-9	$(CH_3)_3CNH_2$	73.14	-72.65	44-46	1	110.4						1.685
di-*n*-butylamine	111-92-2	$(C_4H_9)_2NH$	129.24	-60	160	1	199.2						11.25
tributylamine	102-82-9	$(C_4H_9)_3N$	185.32	-70	214	1	288						9.93
ethylenediamine	107-15-3	$H_2NCH_2CH_2NH_2$	60.10	8.5	117	1							
ethanolamine	141-43-5	$HOCH_2CH_2NH_2$	61.08	10.3	170.8	1	73.4						9.48
diethanolamine	111-42-2	$(HOCH_2CH_2)_2N$	105.14	28	268.8	0.934	126.7						8.88
triethanolamine	102-71-6	$(HOCH_2CH_2)_3N$	149.19	20	335.4	1	182.1						7.76
cyclohexylamine	108-91-8	$C_6H_{12}NH$	99.17	-17.7	135.4	1	117.4						10.66
diphenylamine	122-39-4	$(C_6H_5)_2NH$	169.23	52.8	302	0.531	200.3					334.1	0.90
α-naphthylamine	134-32-7	$C_{10}H_7NH_2$	143.19	48-50	301	0.570	161.8					284.3	3.92
β-naphthylamine	91-59-8	$C_{10}H_7NH_2$	143.19	111-113	306	0.141	161.8						4.15
4-aminobiphenyl	92-67-1	$C_6H_5C_6H_4NH_2$	169.23	52-54		0.541	168.8						4.27
Aromatic amines:													
aniline	62-53-3	$C_6H_5NH_2$	93.12	-6.3	184.4	1	110.2		88			226.3	4.596
2-chloroaniline	92-51-2	$Cl(C_6H_4)NH_2$	127.57	-14	208.84	1	131.1		99.9				2.661
3-chloroaniline	108-42-9	$Cl(C_6H_4)NH_2$	127.57	-11to-9	229	1	131.1						3.5
4-chloroaniline	106-47-8	$Cl(C_6H_4)NH_2$	127.57	70.0	232	0.359	131.1						3.982
3,4-dichloroaniline	95-76-1	$Cl_2C_6H_3NH_2$	162.06	71.5	272	0.347	152		111.7				
2,4,6-trichloroaniline	634-93-5	$C_6H_4Cl_3N$	196.46	78.5	262	0.296	172.9						
o-toluidine	95-53-4	$CH_3C_6H_4NH_2$	107.16	-24	200	1	132.4						4.45
m-toluidine	108-44-1	$CH_3C_6H_4NH_2$	107.16	-30.4	203.4	1	132.4					253.4	4.71
p-toluidine	106-49-0	$CH_3C_6H_4NH_2$	107.16	43.75	200.55	0.652	132.4		103.6			253.8	5.08
N,N'-dimethylaniline	121-69-7	$C_6H_5N(CH_3)_2$	121.18	2.45	194.05	1	154.6	75.2					5.15
2,4-xylidine	95-68-1	$(CH_3)_2C_6H_3NH_2$	121.18	-14.3	211.5	1	154.6						4.89

Compounds	CAS no.	formula	MW	mp, °C	bp, °C	Fugacity ratio, F at 25 °C	molar volume, cm^3/mol				total surface area		pK_a or pK_b
							LeBas	V_I (a)	V_M (b)	TMV,$Å^3$ (c)	TSA,$Å^2$ (c)	TSA,$Å^2$ (d)	
2,5-xylidine	95-78-3	$(CH_3)_2C_6H_3NH_2$	121.18	11.5	218	1	154.6						4.54
2,6-xylidine	87-62-7	$(CH_3)_2C_6H_3NH_2$	121.18	11.2	217.9	1	154.6						3.95
2-ethylaniline	578-54-1	$C_2H_5C_6H_4NH_2$	121.18	-44	210	1	139.6						4.42
3-ethylaniline	587-02-0	$C_2H_5C_6H_4NH_2$	121.18	-8	212	1	139.6						4.70
4-ethylaniline	589-16-2	$C_2H_5C_6H_4NH_2$	121.18	-5	216	1	139.6		120				5.00
N,N'-diethylaniline	91-66-7	$C_6H_5N(C_2H_5)_2$	149.24	-38	217	1	199.0						6.56
benzidine	92-87-5	$NH_2(C_6H_4)_2NH_2$	184.23	129	402	0.096	213					352.8	4.66
3,3'-dichlorobenzidine	91-94-1	$C_{12}H_{10}Cl_2N_2$	253.13	132-133		0.087	254.8						11.7
N,N'-bianiline	122-66-7	$(C_6H_5)_2(NH)_2$	184.24	131		0.089	213						13.2
2-nitroaniline	88-74-4	$C_6H_6N_2O_2$	138.13	73-76	284	0.335	138.7						-0.28
3-nitroaniline	99-09-2	$C_6H_6N_2O_2$	138.13	112-114		0.138	138.7						2.46
4-nitroaniline	100-01-6	$C_6H_6N_2O_2$	138.13	149-151		0.0594	138.7						1.01
2,4-dinitroaniline	97-02-9	$(O_2N)_2C_6H_3NH_2$	183.12	176-178		0.0321	167.2						-4.25
2,6-dinitroaniline	606-22-4	$(O_2N)_2C_6H_3NH_2$	183.12	134(dec)		0.0835	167.2						-5.23
3,5-dinitroaniline	618-87-1	$(O_2N)_2C_6H_3NH_2$	183.12	160-162		0.0462	167.2						0.229
Nitroaromatic:													
nitrobenzene	98-95-3	$C_6H_5NO_2$	123.11	5.6	211	1	112	63				255.6	
1,2-dinitrobenzene	528-29-0	$C_6H_4(NO_2)_2$	168.11	118		0.12	149.4						
1,3-dinitrobenzene	99-65-0	$C_6H_4(NO_2)_2$	168.11	89	297	0.233	149.4					305.2	
1,4-dinitrobenzene	99-99-0	$C_6H_4(NO_2)_2$	168.11	173	238.3	0.0344	149.4					304.3	
2-nitrotoluene	88-72-2	$CH_3C_6H_4NO_2$	137.14	-3.85	221.7	1	153					269.8	
3-nitrotoluene	99-08-1	$CH_3C_6H_4NO_2$	137.14	16.1	231.9	1	153					278.8	
4-nitrotoluene	99-99-0	$CH_3C_6H_4NO_2$	137.14	51.7	238.5	0.544	153					278.5	
2,4-dinitrotoluene	121-14-2	$CH_3C_6H_3(NO_2)_2$	182.14	70	300	0.359	175.2						
2,6-dinitrotoluene	606-20-2	$CH_3C_6H_3(NO_2)_2$	182.14	66	285	0.393	175.2						

Compounds	CAS no.	formula	MW	mp, °C	bp, °C	Fugacity ratio, F at 25 °C	molar volumem cm³/mol LeBas	V_I	V_M (a)	TMV,Å³ (b)	surface area TSA,Å² (c)	TSA,Å² (c)	pK_a or pK_b (d)
2,4,6-trinitrotoluene	118-96-7	$CH_3C_6H_2(NO_2)_3$	227.14				203.7						
1-nitronaphthalene	86-57-7	$C_{10}H_7NO_2$	173.11	62	304	0.431	176.1						
2-nitronaphthalene	58-89-5	$C_{10}H_7NO_2$	173.11	79	312.5	0.292	176.1						
4-nitrobiphenyl	92-93-3	$C_{12}H_9NO_2$	199.21	112-114		0.138	211.3						
5-nitro-acenaphthene	602-87-9	$C_{12}H_9NO_2$	199.21	101-106		0.177	211.3						
Nitrosoamines:													
N-nitrosodimethylamine	62-75-9	$(CH_3)_2NNO$	74.08		153		87.7						
N-nitrosodiethylamine	55-18-5	$(C_2H_5)_2NNO$					130.6						
di-*n*-propyl nitrosamine	621-64-7	$(C_3H_7)_2NNO$	130.19		205		176.5						
diphenyl nitrosamine	86-30-6	$(C_6H_5)_2NNO$	198.24	66.5	152	0.389	220.5						
Heterocyclic:													
1H-pyrrole	109-97-7	C_4H_5N	67.09	-23.415	129.77	1	78.2						
1-methylpyrrole	95-54-8	C_5H_7N	81.12	-57	112-113	1	104						
pyrrolidine	123-75-1	C_4H_8NH	71.12		87-88		96.6						4.453
imidazole	288-32-4	$C_3H_4N_2$	68.08	89-91	256	0.233	78.9						11.305
indazole	271-44-3	$C_7H_6N_2$	118.14	148		0.0697	130.5			105.24	134.427		
indole	120-72-9	C_8H_7N	117.14	52.5	254	0.535	133.4			121.96	140.602		
indoline	496-15-1	C_8H_9N	119.17				140.8						
pyridine	110-86-1	C_5H_5N	79.10	-41.55	115.25	1	93.0	47					5.17
2-methylpyridine	109-06-8	C_6H_7N	93.128	-66.74	129.41	1	115.2	57					5.96
3-methylpyridine	108-99-6	C_6H_7N	93.128	-18.09	144.12	1	115.2						5.68
4-methylpyridine	108-89-4	C_6H_7N	93.128	3.65	145.4	1	115.2						6.00
2,3-dimethylpyridine	583-61-9	C_7H_9N	107.155										
2,4-dimethylpyridine	108-47-4	C_6H_9N	107.155		158.4	1	135.9						
2,6-dimethylpyridine	108-48-5	C_7H_9N	107.155		144.045		135.9						6.72
2,4,6-trimethylpyridine	108-75-8	$C_8H_{11}N$	121.182		171	1	158.1						7.43

Compounds	CAS no.	formula	MW	mp, °C	bp, °C	Fugacity ratio, F at 25 °C	molar volume, cm^3/mol LeBas	V_I (a)	V_M (b)	TMV, $Å^3$ (c)	total surface area TSA, $Å^2$ (c)	TSA, $Å^2$ (d)	pK_a or pK_b
Quinolines:													
quinoline	91-22-5	C_9H_7N	129.16	-14.85	237.1	1	144.7			122.0	150.7		4.90
isoquinoline	119-65-3	C_9H_7N	129.16	26.48	243.24	0.967	144.7			121.959	150.515		5.4
3-methyl-isoquinoline		$C_{10}H_9N$	143.19	68		0.376	166.9			137.87	172.656		
2,7-dimethylquinoline	93-37-8	$C_{11}H_{11}N$	157.22	58-60		0.461	189.1						
benzo(*f*)quinoline	85-02-9	$C_{13}H_9N$	179.22	93.5	350	0.21	196.3			164.575			
benzo(*h*)quinoline	230-27-3	$C_{13}H_9N$	179.22	52	338	0.541	196.3						
9H-carbazole	86-74-8	$C_6H_4NHC_6H_4$	167.21	248	355	0.00623	192.9			188.706			
7H-dibenzo(*c,g*)carbazole		$(C_{10}H_6)_2NH$	267.31	155-156		0.052	296.1			237.542	267.002		
acridine	260-94-6	$C_{13}H_9N$	179.22	110	346	0.144	196.3			165.392	197.25		5.60
benz(*a*)acridine	225-11-6	$C_{17}H_{11}N$	229.29				244.8						
benz(*c*)acridine	225-51-4	$C_{17}H_{11}N$	229.29				244.8						
dibenz(*a,h*)acridine				228		0.00982					284.523		
Sulfur compounds:													
carbon disulfide	75-15-0	CS_2	76.131	-111.37	46.225	1	66						
dimethyl sulfate	77-78-1	$(CH_3O)_2SO_2$	126.13	-31.8	188	1	109.7						
diethyl sulfate	64-67-5	$(C_2H_5O)_2SO_2$	154.18	-24.4	208	1	138.4						
dimethyl sulfite	616-42-2	$(CH_3O)_2SO$	110.13				100.5						
dimethylsulfoxide (DMSO)	67-68-5	$(CH_3)_2SO$	78.13	18.4	189	1	85.7	45.5					
dimethylsulfone	67-71-0	$(CH_3)_2SO_2$	94.13	108-110	238	0.151	94.0						
dimethyl sulfide	75-18-3	$(CH_3)_2S$	62.13	-98.7	38	1	77.4						
dimethyl disulfide	624-92-0	$C_2H_6S_2$	94.2	-84.7	109.8	1	103						
diethyl sulfide	352-93-2	$C_4H_{10}S$	90.183	-103.93		1	128.1						
diethyl disulfide	110-81-6	$C_4H_{10}S_2$	122.24	-101.5	154	1	147.4						

Compounds	CAS no.	formula	MW	mp, °C	bp, °C	Fugacity ratio, F at 25 °C	molar volume, cm^3/mol			TMV,$Å^3$	total surface area		pK_a or pK_b
							LeBas	V_I (a)	V_M (b)	(c)	TSA,$Å^2$ (c)	TSA,$Å^2$ (d)	
Thiols:													
methanethiol	74-93-1	CH_3SH	48.1			1	55.2						10.7
ethanethiol	75-08-1	C_2H_5SH	62.13	-147.9	35	1	77.4						10.61
propanethiol	107-03-9	C_3H_7SH	76.16			1	99.6						
1-butanethiol	109-75-5	$CH_3(CH_2)_3SH$	90.183	-115.67	98.455	1	121.8						
2-butanethiol	513-53-1	C_4H_7SH	90.183	-140	85-86	1	121.8						
benzenethiol	108-98-5	C_6H_5SH	110.173		169.138		106.8						6.5
2-methylbenzenethiol	137-06-4	C_7H_8S	124.2	15		1	129						
3-methylbenzenethiol	108-40-7	C_7H_8S	124.2	-20		1	129						
4-methylbenzenethiol	106-45-6	C_7H_8S	124.2	43		0.664	129						
Thiophenes:													
thioazole	288-47-1	C_3H_3NS	85.13				85.2						
thiophene	110-02-1	C_4H_4S	84.14	-38	84	1	88.1						
2-methylthiophene	554-14-3	C_5H_6S	98.17	-63.186	112.56	1	110.3						
3-methylthiophene	616-44-4	C_5H_6S	98.17	-69	115.44	1	110.3						
benzo(*b*)thiophene	95-15-8	C_8H_6S	134.19	30.85	219.85	0.875	139.7			118.066	146.919		
dibenzothiophene	132-65-0	$C_{12}H_8S$	184.26	99	323.3	0.185	191.3			163.439			
thianthrene	92-85-3	$(C_6H_4)_2S_2$	216.32	157-159		0.049	210.9						
thioacetamide	62-55-5	CH_3CSNH_2	75.13	112-114		0.138	84.2						
thiobenzamide	2227-79-4	$C_6H_5CSNH_2$	147.2	116-118		0.126	135.8						
thiourea	62-56-6	H_2NCSNH_2	76.12	176-178		0.0321	76.2						2.03

(a) Kamlet et al. 1987, 1888; (b) Jaworska & Schultz 1993; (c) Pearlman et al. 1984; (d) Iwase et al. 1985

Table 7.2 Summary and selected physical-chemical properties of nitrogen and sulfur compounds at 25 °C.

Compound	selected properties:						H, Pa·m³/mol	
	P_S, Pa	P_L, Pa	S, g/m³	C^S, mol/m³	C_L, mol/m³	log K_{OW}	calc-P/C	exptl
Nitriles:								
acetonitrile	11840	11840	miscible			-0.34		2.75
propionitrile	5950	5950	103000	1870.0	1870.0	0.16	3.182	3.8
butyronitrile	2546	2546	33000	477.5	477.5			5.263
benzonitrile	100	100	2000	19.39	19.39	1.55	5.156	
acrylonitrile	11000	11000	75500	1423	1423	0.25	7.731	11.14
adiponitrile	0.3066	0.3066	8000	73.96	73.96	-0.32	0.0041	
Amides: RCONH2								
Acetamide (ethanamide)	2.44	8.3562	408000	6907.1	23650	-1.26	3.53×10^{-4}	
acrylamide	0.415	1.5900	2050000	2884	11050	-0.9	1.44×10^{-4}	
benzamide	0.00522	0.0544	14000	1692	17630	0.64	4.52×10^{-5}	
Urea: (NH2)2C=O								
urea	0.0016	0.0186	1000000	16650	1.93×10^{5}	-2.11	9.61×10^{-8}	
Aliphatic amines:								
methylamine	357300	357300	miscible			-0.57		1.125
dimethylamine	206200	206200	miscible			-0.38		1.8
trimethylamine	219300	219300	miscible			0.27		6.67
ethylamine	141650	141650	miscible			-0.13		1.012
diethylamine	31490	31490	miscible			0.43		2.60
triethylamine	7610	7610	55000	540	540	1.64	14.099	
n-propylamine	40740	40740	miscible			0.48		1.274
dipropylamine			53000	520	520			
diisopropylamine			12390	122				
tripropylamine			220	1.536	1.54	2.79		

Compound	selected properties: P_S, Pa	P_L, Pa	S, g/m³	C^S, mol/m³	C_L, mol/m³	log K_{OW}	H, Pa·m³/mol calc-P/C	exptl
n-butylamine	13650	13650	miscible			0.97		1.526
isobutylamine	18760	18760	miscible			0.73		
t-butylamine	48260	48260	miscible			0.4		
di-*n*-butylamine	304	304	4700	36.37	36.37	2.83	8.359	
tributylamine	5330	5330	40	0.216	0.216		2.47×10^4	
ethylenediamine								
ethanolamine	34.66	34.66	miscible			-1.31		
diethanolamine	0.0373	0.0399	miscible			-1.43		
triethanolamine	4.79×10^{-4}	4.79×10^{-4}	miscible			-1.59		
cyclohexylamine	1173	1173	miscible			1.49		
diphenylamine	0.0612	0.115	300	1.773	3.338	3.45	0.035	
α-naphthylamine	0.254	0.45				2.23		
β-naphthylamine	0.035	0.248	6.4	0.045	0.317	2.34		
4-aminobiphenyl						2.83		
Aromatic amines:								
aniline	65.19	65.190	36070	387.4	387.35	0.90	0.168	12.16
2-chloroaniline	22.66	22.660	3800	29.79	29.79	1.90	0.761	
3-chloroaniline	9.53	9.530	5440	42.64	42.64	1.88	0.223	
4-chloroaniline	2.33	6.873	3000	23.52	69.37	1.83	0.099	
3,4-dichloroaniline	1.3	3.746	92.05	0.568	1.637	2.67	2.289	
2,4,6-trichloroaniline	0.00626	0.021				3.694		
o-toluidine	13.3	13.30	15000	139.98	139.98		0.095	
m-toluidine	36	36.00	15030	140.26	140.26	1.44	0.257	
p-toluidine	45	61.48	7350	68.59	93.70	1.4	0.656	
N,N'-dimethylaniline	107	107.00	1105	9.119	9.119	2.31	11.734	
2,4-xylidine	20.5	20.50	5900	48.69	48.69		0.421	

Compound	selected properties: P_S, Pa	P_L, Pa	S, g/m^3	C^S, mol/m^3	C_L, mol/m^3	log K_{OW}	H, Pa·m^3/mol calc-P/C	exptl
2,5-xylidine			5000	41.26	41.26			
2,6-xylidine	670	670.00	4700	38.79	38.79	1.94	17.275	
2-ethylaniline			7500	61.89	61.89	1.93		
4-ethylaniline	13.5	13.50	5100	42.09	42.09	1.96	0.321	
N,N'-diethylaniline	9.7	9.70	670	4.49	4.49		2.161	
benzidine	$1.0x10^{-6}$	$1.06x10^{-5}$	400	2.17	23.10	1.81	$4.61x10^{-7}$	
3,3'-dichlorobenzidine	$5.6x10^{-5}$	$6.41x10^{-4}$	3.1	0.0122	0.140	3.51	0.005	
N,N'-bianiline	0.0035		0.252	0.0014	0.0154	3.82	$3.45x10^{-4}$	
2-nitroaniline	0.62	1.851	1200	8.687	25.93	1.78		
3-nitroaniline								
4-nitroaniline	0.035	0.589	800	5.792	97.50	1.31		
2,4-dinitroaniline						4.1		
Nitrobenzenes:								
nitrobenzene	20	20.00	1900	15.43	15.43	1.85	1.296	
1,2-dinitrobenzene	0.0052	0.0433						
1,3-dinitrobenzene	0.0081	0.0348	546	3.25	13.94	1.49	0.002	
1,4-dinitrobenzene	13.3	386.63	442	2.63	76.43	2.37	5.059	
2-nitrotoluene	17.9	17.90	651.42	4.75	4.75	2.30	3.768	
3-nitrotoluene	27.2	27.20	499.19	3.64	3.64	2.45	7.473	
4-nitrotoluene	0.653	1.2004	254.4	1.86	3.41	2.37	0.352	
2,4-dinitrotoluene	0.133	0.3705	270	1.48	4.13	2.01	0.090	
2,6-dinitrotoluene	0.0767	0.1952	200			1.72	0.070	
2,4,6-trinitrotoluene	0.00107	0.0038	210	0.925	3.29			
1-nitronaphthalene	0.702	0.072	9.82	0.057	0.132	3.19	3.50	
2-nitronaphthalene			9.24	0.053	0.183			
4-nitrobiphenyl			1.231	$6.18x10^{-3}$	0.045	3.78		

Compound	selected properties:						H, $Pa \cdot m^3/mol$	
	P_S, Pa	P_L, Pa	S, g/m^3	C^S, mol/m^3	C_L, mol/m^3	log K_{OW}	calc-P/C	exptl
5-nitro-acenaphthene			0.91	4.57×10^{-3}	0.026			
Nitrosoamines:								
N-nitrosodimethylamine			miscible			-0.57		3.343
N-nitrosodiethylamine								
di-*n*-propyl nitrosamine	27		9900	76.04		1.31	0.355	
diphenyl nitrosamine	13.33	34.27	23.05	0.116	0.299	3.13	114.6	
Heterocyclic:								
1H-pyrrole	1100	1100	45000	670.7	670.7	0.75	1.640	
1-methylpyrrole	3312							
pyrrolidine								
imidazole								
indazole			827	7.00	100.43			
indole	2.24	4.187	1874	16.00	29.90	2	0.140	
indoline			10800	90.63				
pyridine	2775	2775	miscible			0.65		0.895
2-methylpyridine	1496	1496	miscible			1.11		1.01
3-picoline	1333	1333	miscible			1.2-1.24		0.788
4-picoline	757	757	miscible			1.22		0.601
2,3-dimethylpyridine	426		104000	970.6				0.725
2,4-dimethylpyridine	456	456	miscible					0.678
2,6-dimethylpyridine	746	746	miscible					1.06
2,4,6-trimethylpyridine	5170	5170	35700	294.6	294.6		17.549	
Quinolines:								
quinoline	1.21	1.21	6110	47.31	47.31	2.06	0.026	
isoquinoline	670	693	4521	35.00	36.20		19.141	
3-methyl-isoquinoline								

Compound	selected properties: P_S, Pa	P_L, Pa	S, g/m³	C^S, mol/m³	C_L, mol/m³	log K_{OW}	H, Pa·m³/mol calc-P/C	exptl
2,7-dimethylquinoline			1795	11.42	24.77			
benzo(*f*)quinoline		0.0067	76.1	0.42	2.02	3.20	0.0096	
benzo(*h*)quinoline	0.03	0.0555						
9H-carbazole	0.0933	14.976	1.03	0.006	0.989	3.80	15.146	
7H-dibenzo[*c,g*]carbazole	$1.3x10^{-7}$	$2.5x10^{-6}$	0.063	0.236	4.532	5.75		
acridine	0.0065	0.0451	38.5	0.215	1.492	3.4	0.030	
benz(*a*)acridine						4.45		
benz(*c*)acridine								
dibenz(*a,h*)acridine								
Sulfur compounds:								
carbon disulfide	48210	48210	2100	27.584	27.584		1747.75	
dimethyl sulfate	128	128.0						
diethyl sulfate	49.1	49.10						
dimethyl sulfite								
dimethylsulfoxide	80.00	80.00	27720	354.8	354.8	-1.35	0.225	
dimethylsulfone	5.16	34.17				-1.41	200.83	
dimethyl sulfide	64650	64650	20000	321.9			7.72	
dimethyl disulfide	4000	4000	6300	66.88	66.88		59.81	
diethyl sulfide	7782	7782				1.95		
diethyl disulfide	689	689						
Thiols:								
methanethiol	201980	201980						
ethanethiol	70000		15000				289.94	
propanethiol	20635	20635				1.81		
1-butanethiol	6070	6070	597	6.620	6.620	2.28	916.94	

Compound	P_S, Pa	P_L, Pa	S, g/m³	C^S, mol/m³	C_L, mol/m³	log K_{OW}	calc-P/C	exptl
2-butanethiol	10790	10790						
benzenethiol	397	397.0				2.52		
2-methylbenzenethiol	87.4	87.40						
3-methylbenzenethiol	76.6	76.60						
4-methylbenzenethiol	85.2	128.3						
Thiophenes:								
thioazole	2287							
thiophene	10620	8000	3015	35.833	35.833	1.81	223.3	224
2-methylthiophene	3318							
3-methylthiophene	2953							
benzo(*b*)thiophene	26.66	26.7	130	0.969	1.107	3.12	24.1	
dibenzothiophene	0.267		1.11	0.0060		4.38	44.3	
thianthrene								
thioacetamide			163000	2170	15722	-0.26		
thiourea			90000	1182	36833	-0.99		

Table 7.3 Suggested half-life classes for nitrogen and sulfur containing compounds at 25 °C in various environmental compartments.

Compounds	Air class	Water class	Soil class	Sediment class
acetonitrile	6	5	5	6
propionitrile	6	5	5	6
acrylonitrile	3	4	4	5
acetamide	2	4	4	5
benzamide	2	4	4	5
dimethylamine	1	3	4	5
ethylamine	2	3	4	5
diethylamine	1	2	2	3
n-butylamine	2	3	3	4
ethanolamine	1	3	3	4
diethanolamine	1	4	4	5
cyclohexylamine	1	3	3	4
diphenylamine	1	4	5	6
α-naphthylamine	1	4	4	6
β-naphthylamine	1	4	4	6
aniline	1	4	4	6
2-chloroaniline	3	4	5	6
4-chloroaniline	1	4	4	5
o-toluidine	1	3	3	4
N,N'-dimethylaniline	1	4	5	6
2,6-xylidine	1	4	5	6
benzidine	1	4	4	5
3,3'-dichlorobenzidine	1	1	5	6
N,N'-bianiline	1	3	3	4
nitrobenzene	1	6	6	7
2-nitrotoluene	2	3	6	7
4-nitrotoluene	2	3	6	7
2,4-dinitrotoluene	2	3	6	7
2,4,6-trinitrotoluene	1	2	6	7

Compounds	Air class	Water class	Soil class	Sediment class
N-nitrosodimethylamine	1	2	6	7
N-nitrosodiethylamine	1	2	6	7
di-*n*-propyl nitrosoamine	1	2	6	7
diphenyl nitrosoamine	1	2	6	7
pyridine	5	5	6	7
3-methylpyridine	5	5	6	7
4-methylpyridine	5	5	6	7
quinoline	3	4	5	6
dimethyl sulfate	4	2	2	3
diethyl sulfate	2	2	3	4
thiophene	3	3	6	7
benzo(*b*)thiophene	4	5	6	7
thioacetamide	2	4	4	5
thiourea	1	4	4	5

where,

Class	Mean half-life (hours)	Range (hours)
1	5	< 10
2	17 (~ 1 day)	10-30
3	55 (~ 2 days)	30-100
4	170 (~ 1 week)	100-300
5	550 (~ 3 weeks)	300-1,000
6	1700 (~ 2 months)	1,000-3,000
7	5500 (~ 8 months)	3,000-10,000
8	17000 (~ 2 years)	10,000-30,000
9	~ 5 years	> 30,000

7.3 Commentary on the Physical-Chemical Properties and Environmental Fate

It has been the authors' intention to compile QSPR plots and illustrative environmental fate calculations in fugacity format for this series of chemicals. This proved to be impossible.

There is a wide diversity of nitrogen-containing organic compounds, including nitriles, amines, amides, anilines, nitro- and nitroso- compounds and various heterocyclic substances. A similar diversity applies to sulfur compounds. Trial QSPR plots showed that each chemical class tends to follow conventional structure-property relationships, but the number of chemicals in each class is limited thus it would be necessary to compile some ten sets of QSAR plots for nitrogen compounds.

There is also a considerable variability in the quality of reported physical-chemical property data for these substances, which is doubtless attributable to their less frequent appearance on lists of priority chemicals and difficulties associated with chemical analysis. In many respects the state of advancement of the environmental chemistry and ecotoxicology of these compounds is a decade or more behind that of the chlorinated aromatics. In view of these difficulties it was judged that the presentation of QSPR plots is premature.

Some difficulties were encountered when data on multimedia chemical degradation rates or half-lives were sought. With the exception of atmospheric half-lives, which are largely the result of the extensive studies by Atkinson and his coworkers (1985, 1987, 1989, 1990, 1991) few such data exist and there is a wide variation in reported values. The selected half-lives should thus be regarded as quite tentative and it is expected that there may be considerable variations from these values. Because the results of the fugacity calculations are so sensitive to these assumed half-lives it is considered unwise to present the illustrative calculations in this case.

It is suggested that the reader consult the available data for the chemical of interest, select appropriate values for the partitioning and degradation parameters and assign confidence limits to these values. Using the diskette supplied, fugacity calculations can then be completed for the selected values and a sensitivity analysis conducted varying the parameters over a perceived confidence range. This will give a range of estimates of fate which more reliably reflect the lack of certainty about parameter values in these classes.

It is hoped that as more and better data become available for these substances it will be possible to compile QSPR plots and perform more reliable evaluative calculations.

7.4 REFERENCES:

Abernethy, S.G., Mackay, D., McCarty, L.S. (1988) "Volume fraction" correlation for narcosis in aquatic organisms: The key role of partitioning. *Environ. Toxicol. & Chem.* 7, 469-481.

Abraham, M.H. (1984) Thermodynamics of solution of homologous series of solutes in water. *J. Chem. Soc., Farad. Trans. 1* 80, 153-181.

Aim, K. (1994) Saturated vapor pressure measurements on isomeric mononitrotoluenes at temperatures between 380 and 460 K. *J. Chem. Eng. Data* 39, 591-594.

Albersmeyer, W. (1958) Quantitative determination of aromatic hydrocarbons in aqueous solutions. *Gas-U. Wasserfach* 99, 269.

Albert, A. (1966) *The Acridines.* Edward Arnold, London.

Alexander, M., Lustigman, B.K. (1966) Effects of chemical structure on microbial degradation of substituted benzenes. *J. Agric. Food Chem.* 14, 410-413.

Ambrose, D. Gundry, H.A. (1980) The vapour pressure of *p*-nitrotoluene. *J. Chem. Thermodyn.* 12, 559-561.

Amidon, G.L., Williams, N.A. (1982) A solubility equation for non-electrolytes in water. *Int'l J. Pharm.* 11, 249-256.

Anbar, M., Meyerstein, D., Neta, P. (1966) The reactivity of aromatic compounds toward hydroxyl radicals. *J. Phys. Chem.* 70, 2661-2662.

Anbar, M., Neta, P. (1967) A compilation of specific bimolecular rate and hydroxyl radical with inorganic and organic compounds in aqueous solution. *Int'l J. Appl. Radiation Isotopes* 18, 493-523.

Anderson, T.A., Beauchamp, J.J., Walton, B.T. (1991) Organic chemicals in the environment. *J. Environ. Qual.* 20, 420-424.

Andon, R.J.L., Cox, J.D. (1952) Phase relationships in pyridine series. Part I. The miscibility of some pyridine homologues with water. *J. Chem. Soc.* 4601-4606.

Andon, R.J.L., Cox, J.D., Herington, E.F.G. (1954) Phase relationships in the pyridine series. Part V. The thermodynamic properties of dilute solutions of pyridine bases in water at 25°C and 40°C. *J. Chem. Soc.* 3188-3196.

Andrews, L.J., Keefer, R.M. (1950) Cation complexes of compounds containing carbon-carbon double bond. VI. The argentation of substituted benzenes. *J. Am. Chem. Soc.* 72, 3113-3116.

Appleton, H., Banerjee, S., Pack, E., Sikka, H. (1978) Fate of 3,3'-dichlorobenzidine in aquatic environment. In: *Pergamon Series in Environmental Science.* 1, 473-474.

Appleton, H., Sikka, H. (1980) Accumulation, elimination and metabolism of dichlorobenzidine in bluegill sunfish. *Environ. Sci. Technol.* 14(1), 50-54.

Arbuckle, W.B. (1983) Estimating activity coefficients for use in calculating environmental parameters. *Environ. Sci. Technol.* 17, 537-542.

Arnold, M.L., Lin, D.T., Schultz, T.W. (1990) QSAR for methyl- and/or chloro-substituted anilines and the polar narcosis mechanism of toxicity. *Chemosphere* 21(1/2), 183-191.

Atkinson, R. (1985) Kinetics and mechanisms of the gas-phase reactions of hydroxyl radicals with organic compounds under atmospheric conditions. *Chem. Rev.* 85, 69-201.

Atkinson, R. (1987) A structure-activity relationship for the estimation of rate constants for the gas-phase reactions of OH radicals with organic compounds. *Int. J. Chem. Kinetics* 19, 799-828.

Atkinson, R. (1989) Kinetics and mechanisms of the gas-phase reactions of the hydroxyl radical with organic compounds. *J. Phys. Chem. Reference Data*, Monograph No. 1, 1-246.

Atkinson, R. (1991) Kinetics and mechanisms of the gas-phase reactions of NO_3 radical with organic compounds. *J. Phys. Chem. Ref. Data* 20(3), 459-507.

Atkinson, R., Aschmann, S.M., Arey, J., Zielinska, B. (1989) Gas-phase atmospheric chemistry of 1- and 2-nitronaphthalene and 1,4-naphthoquinone. *Atmos. Environ.* 23(12), 2679-2690.

Atkinson, R., Aschmann, S.M., Carter, W.P.L. (1983) Kinetics of the reactions of ozone and hydroxyl radicals with furan and thiophene at 298 ± 2 K. *Int. J. Chem. Kinetics* 15, 51-61.

Atkinson, R., Aschmann, S.M., Pitts, Jr., J.N. (1988) Rate constants for the gas-phase reactions of the NO_3 radicals with a series of organic compounds at 296 ± 2 K. *J. Phys. Chem.* 92, 3454-3457.

Atkinson, R., Aschmann, S.M., Winer, A.M., Carter, W.P.L. (1984) Rate constants for the gas-phase reactions of hydroxyl radicals and ozone with pyrrole at 295 ± 1 K and atmospheric pressure. *Atmos. Environ.* 18, 2085-2087.

Atkinson, R., Aschmann, S.M., Winer, A.M., Carter, W.P.L. (1985) Rate constants for the gas-phase reactions of the NO_3 radicals with furan, thiophene, and pyrrole at 295 ± 1 K and atmospheric pressure. *Environ. Sci. Technol.* 19, 87-90

Atkinson, R., Aschmann, S.M., Winer, A.M., Pitts, Jr., J.N. (1985) Atmospheric gas phase loss processes for chlorobenzene, benzotrifluoride, and 4-chlorobenzofluoride, and generalization of predictive techniques for atmospheric lifetimes of aromatic compounds. *Arch. Environ. Contam. Toxicol.* 14, 417-425.

Atkinson, R., Carter, W.P.L. (1984) Kinetics and mechanisms of the gas-phase reactions of ozone with organic compounds under atmospheric conditions. *Chem. Rev.* 84, 437-470.

Atkinson, R., Perry, R.A., Pitts, J.N., Jr. (1978) Rate constants for the reactions of hydroxyl radicals with dimethylamine, trimethylamine, and ethylamine over the temperature range 298-426 K. *J. Chem. Phys.* 68, 1850-1853.

Atkinson, R., Perry, R.A., Pitts, J.N., Jr. (1977) Rate constants for the reaction of the OH radical with CH_3SH and CH_3NH_2 over the temperature range 299-426 K. *J. Chem. Phys.* 66, 1578-1581.

Atkinson, R., Tuazon, E.C., Wallington, T.J., Aschmann, S.M., Arey, J., Winer, A.M., Pitts, J.N., Jr. (1987) Atmospheric chemistry of aniline, N,N-dimethylaniline, Pyridine, 1,3,5-triazine, and nitrobenzene. *Environ. Sci. Technol.* 21, 64-72.

Aubry, M., Mayoral, M.N., Villardry, P. (1975) Determination of the vapor pressure of low volatility compounds by gas chromatography. Application to dibenzothiophene. *Boll. Soc. Chim. Fr.* 3-4 of Pt.1, 500-502.

Bailey, G.W., White, J.L., Rothberg T. (1968) Adsorption of organic herbicides by montmorillonite: Role of pH and chemical character of adsorbate. *Soil Sci. Am. Proc.* 32, 222-234.

Baird, R., Caroma, L., Jenkins, R.L. (1977) Behavior of benzidine and other aromatic amines in aerobic waste water treatment. *J. Water Pollut. Control Fed.* 49, 1606-1615.

Banerjee, S. (1985) Calculation of water solubility of organic compounds with UNIFAC-derived parameters. *Environ. Sci. Technol.* 19, 369-370.

Banerjee, S., Howard, P.H. (1988) Improved estimation of solubility and partitioning through correction of UNIFAC-derived activity coefficients. *Environ. Sci. Technol.* 22, 839-841.

Banerjee, S., Howard, P.H., Lande, S.S. (1990) General structure-vapor pressure relationships for organics. *Chemosphere 21(10/11), 1173-1180.*

Banerjee, S., Sikka, H.C., Gray, D.A., Kelly, C.M. (1978) Photodegradation of 3,3'-dichlorobenzidine. *Environ. Sci. Technol.* 12, 1425-1427.

Banerjee, S., Yalkowsky, S.H., Valvani, S.C. (1980) Water solubility and octanol/water partition coefficients of organics. Limitations of the solubility-partition coefficient correlations. *Environ. Sci. Technol.* 14, 1227-1229.

Banwart, W.L., Hassett, J.J., Wood, S.G., Means, J.C. (1982) Sorption of nitrogen-heterocyclic compounds by soils and sediments. *Soil Sci.* 133, 42-47.

Barrows, M.E. et al. (1978) *Am. Chem. Soc. Div. Environ. Chem.* 18, 345-346.

Becker, K.H., Biehl, H.M., Bruckmann, P., Fink, E.H., Führ, F., Klöpffer, W., Zellner, R., Zetzsch, C. (1984) Hydroxyl radical reaction rate constants and tropospheric lifetimes of selected environmental chemicals. *Kernforschungsanlage.* Jülich, GmbH. November 1984, ISSN 0343-7639.

Berthod, A., Han, Y.I., Armstrong, D.W. (1988) Centrifugal partition chromatography. V. Octanol-water partition coefficients, direct and indirect determination. *J. Liq. Chromatogr.* 11(7), 1441-1456.

Bintein, S., Devillers, J., Karcher, W. (1993) Nonlinear dependence of fish bioconcentration on *n*-octanol/water partition coefficient. *SAR and QSAR in Environ. Res.* 1, 29-39.

Bintein, S., Devillers, J. (1994) QSAR for organic chemical sorption in soils and sediments. *Chemosphere* 28(6), 1171-1188.

Bocek, K. (1976) Relations among activity coefficients, partition coefficients and solubilities. *Experientia Suppl.* 23, 231-240.

Bodor, N., Gabanyi, Z., Wong, C.K. (1989) A new method for the estimation of partition coefficient. *J. Am. Chem. Soc.* 111, 3783-3786.

Boethling, R.S., Alexander, M. (1979) Microbial degradation of organic compounds at trace levels. *Environ. Sci. Technol.* 13, 989-991.

Bollag, J.M., Blattmann, P., Lannio, T. (1978) Adsorption and transformation of four substituted anilines in soil. *J. Agric. & Food Chem.* 26, 1302-1306.

Booth, H.S., Everson, H.E. (1948) Hydrotropic solubilities. Solubilities in 40 per cent sodium xylenesulfonate. *Ind. & Eng. Chem.* 40(8), 1491-1493.

Boublik, T., Fried, V., Hala, E. (1973) *The Vapour Pressures of Pure Substances*. Elsevier, Amsterdam, The Netherlands.

Boublik, T., Fried, V., Hala, E. (1984) *The Vapour Pressures of Pure Substances*. Second Edition, Elsevier, Amsterdam, The Netherlands.

Bourquin, A.W. (1984) *Biodegradation in the Estuarine-Marine Environments and the Genetically Altered Microbe*. U.S. EPA 600/D-84-051. NTIS PB84-151315. Environmental Research Lab., U.S. Environmental Protection Agency, Gulf Breeze, Florida.

Boyd, S.A., Kao, C.W., Suflita, J. (1984) Fate of 3,3'-dichlorobenzidine in soil: Persistence and binding. *Environ. Tox. Chem.* 3, 201-208.

Bridie, A.L., Wolff, C.J.M., Winter, M. (1979) BOD and COD of some petrochemicals. *Water Res.* 13, 627-630.

Briggs, G.G. (1981) Theoretical and experimental relationships between soil adsorption, octanol-water partition coefficients, water solubilities, bioconcentration factors and Parachor. *J. Agric. Food & Chem.* 29, 1050-1059.

Brooke, D.N., Dobbs, A.J., Williams, N. (1986) Octanol/water partition coefficients (*P*): measurement, estimation, and interpretation, particularly for chemicals with $P > 10^5$. *Ecotoxicol. Environ. Safety* 11, 251-260.

Brooke, D.N., Nielsen, I., De Bruijn, Hermens, J. (1990) An interlaboratory evaluation of the stir-flask method for the determination of octanol-water partition coefficients (LOG P_{OW}). *Chemosphere* 21(1/2), 119-133.

Budavari, S., Editor (1989) *The Merck Index. An Encyclopedia of Chemicals, Drugs and Biologicals.* 11th Edition, Merck & Co. Inc., Rahway, New Jersey.

Burkhard, L.P., Kuehl, D.W. (1986) *n*-octanol/water partition coefficients by reverse phase liquid chromatography/mass spectrometry for eight tetrachlorinated planar molecules. *Chemosphere* 15(2), 163-167.

Butler, J.A.V., Ramchandani, C.N. (1935) The solubility of non-electrolytes. Part II. The influence of the polar group on the free energy of hydration of aliphatic compounds. *J. Chem. Soc.* 952-955.

Buttery, R.G., Ling, J.C., Guadagni, D.G. (1969) Food volatiles. Volatilities of aldehydes, ketones, and esters in dilute water solution. *J. Agric. Food & Chem.* 17, 385-389.

Bysshe, S.E. (1982) Bioconcentration factor in aquatic organisms. In: *Handbook of Chemical Property Estimation Methods: Environmental Behavior of Organic Compounds.* Lyman et al. Editors, Chapter 5, pp. 5-1 to 5-30, McGraw-Hill Book Company, New York.

Cabani, S., Conti, G., Lepori, L. (1971) Thermodynamic study on aqueous dilute solutions of organic compounds: Part 1. Cyclic amines. *Trans. Farad. Soc.* 67, 1933-1942.

Calamari, D., Da Gasso, R., Galassi, S., Provine, A., Vighi, M. (1980) Biodegradation and toxicity of selected amines on aquatic organisms. *Chemosphere* 9, 753-762.

Callahan, M.A., Slimak, M.W., Gabel, N.W., May, I.P., Fowler, C.F., Freed, J.R., Jennings, P., Durfee, R.L., Whitemore, F.C., Maestri, B., Mabey, W.R., Holt, B.R., Gould, C. (1979) *Water Related Environmental Fate of 129 Priority Pollutants.* EPA-440-4-79-029a,b. Versar, Inc., Springfield, Virginia.

Campbell, J.R., Luthy, R.G. (1985) Prediction of aromatic solute partition coefficient using the UNIFAC group contribution model. *Environ. Sci. Technol.* 19, 980-985.

Canton, J.H., Sloof, W., Kool, H.J., Struys, J., Pouw, T.J.M., Wegman, R.C.C., Piet, G.J. (1985) Toxicity, biodegradability and accumulation of a number of chlorine/nitrogen containing compounds for classification and establishing water quality criteria. *Regul. Toxicol. Pharmacol.* 5, 123-131.

Carlier, P., Hannachi, H., Mouvier, G. (1986) The chemistry of carbonyl compounds in the atmosphere - A review. *Atmos. Environ.* 20(11), 2079-2099.

Carlson, R.M., Carlson, R., Kopperman, H.L. (1975) Determination of partition coefficients by liquid chromatography. *J. Chromatogr.* 107, 219-223.

Chao, J., Lin, C.T., Chung, T.H. (1983) Vapor pressure of coal chemicals. *J. Phys. Chem. Ref. Data* 12, 1033-1063.

Chao, J., Gadalla, N.A.M., Gammon, B.E., Marsh, K.N., Rodgers, A.S., Somayajulu, G.R., Wilhoit, R.C. (1990) Thermadynamic and thermophysical properties of organic nitrogen compounds. Part I. Methanamine, ethanamine, 1-and 2-propanamine, benzenamine, 2-,3-, and 4-methylbenzenamine. *J. Phys. Chem. Ref. Data* 19(6), 1547-1615.

Chen, F., Holten-Andersen, J., Tyle, H. (1993) New development of the UNIFAC model for environmental application. *Chemosphere* 26(7), 1325-1354.

Chessells, M., Hawker, D.W., Connell, D.W. (1992) Influence of solubility in lipid on bioconcentration of hydrophobic compounds. *Ecotox. Environ. Saf.* 23, 260-273.

Chiou, C.T. (1981) Partition coefficient and water solubility in environmental chemistry. In: *Hazard Assessment of Chemicals, Current Developments.* Volume I, pp.117-153. Academic Press, New York.

Chiou, C.T. (1985) Partition coefficients of organic compounds in lipid-water systems and correlations with fish bioconcentration factors. *Environ. Sci. Technol.* 19, 57-62.

Chiou, C.T., Schmedding, D.W. (1981) Measurement and interrelation of octanol-water partition coefficient and water solubility of organic chemicals. In: *Test Protocols for Environmental Fate and Movement of Toxicants. J. Assoc. Anal. Chem.*, Arlington, Virginia.

Chiou, C.T., Schmedding, D.W., Manes, M. (1982) Partitioning of organic compounds in octanol-water system. *Environ. Sci. Technol.* 16, 4-10.

Ciusa, R. (1922) Doebner's reaction. *Gazz. Chim. Ital.* 52(II), 43-48.

Clarke, F.H. (1984) Ionization constants by curve-fitting: Application to the determination of partition coefficients. *J. Pharm. Sci.* 73(2), 226-230.

Clarke, F.H., Cahoon, N.M. (1987) Ionization constants by curve-fitting: Application to the determination of partition and distribution coefficients of acids and bases and their ions. *J. Pharm. Sci.* 76(8), 611-620.

Collander, R. (1951) Partition of organic compounds between higher alcohols and water. *Acta. Chem. Scand.* 5, 774-780.

Collet, A.R., Johnson, J. (1926) Solubility relations of isomeric organic compounds VI. Solubility of the nitroanilines in various liquids. *J. Phys. Chem.* 30, 70.

Copisarow, M. (1915) Trinitrotoluene. *Chem. News* 112, 247-248.

Coulson, E.A., Cox, J.D., Herington, E.F.G., Martin, J.F. (1959) The preparation and physical properties of the pure lutidines. *J. Chem. Soc.* (London) 1934-1940.

D'Amboise, M., Hanai, T. (1982) Hydrophobicity and retention in reversed phase liquid chromatography. *J. Liq. Chromatogr.* 5(2), 229-244.

Darnall, K.R., Lloyd, A.C., Winer, A.M., Pitts, Jr., J.N. (1976) Reactivity scale for atmospheric hydrocarbons based on reaction with hydroxyl radical. *Environ. Sci. Technol.* 10(7), 692-696.

Daubert, T.E., Danner, R.P. (1985) *Data Compilation of Properties of Pure Compounds.* pp.450. American Institute of Chemical Engineers.

Dauble, D.D., Carlile, D.W., Hanf, Jr., R.W. (1986) Bioaccumulation of fossil fuel components during single-compound and complex-mixture exposures of *daphnia magna. Bull. Environ. Contam. Toxicol.* 37, 125-132.

Davis, E.M., Murray, H.E., Leihr, J.G., Powers, E.L. (1981) Basic microbial degradation rates and chemical biproducts of selected organic compounds. *Water Res.* 15, 1125-1127.

De Bruijn, J., Busser, F., Seinen, W., Hermens, J. (1989) Determination of octanol/water partition coefficients for hydrophobic organic chemicals with the "slow-stirring" method. *Environ. Toxicol. Chem.* 8, 499-512.

De Voogt, P., Van Zijl, G.A., Govers, H., Brinkman, U.A.T. (1990) Reversed-phase TLC and structure-activity relationships of polycyclic (hetero) aromatic hydrocarbons. *J. Planar Chromatogr. - Mod. TLC,* 3(1-2), 24-33.

De Voogt, P., Wegener, J.W.M., Klamer, J.C., Van Zijl, G.A., Govers, H. (1988) Prediction of environmental fate and effects of heteroatomic polycyclic aromatics by QSARs: The position of *n*-octanol/water partition coefficients. *Biomed. Environ. Sci.* 1(2), 194-209.

De Wolf, W., Seinen, W., Hermens, J.L.M. (1993) Biotransformation and toxicokinetics of trichloroanilines in fish in relation to their hydrophobicity. *Arch. Environ. Contam. Toxicol.* 25, 110-117.

De Wolf, W., Verhaar, J.M., Hermens, J.L.M. (1991) Bioconcentration of chlorinated anilines and benzene in fish: preliminary results. *Comp. Biochem. Physiol.* 100C, 55-57.

De Wolf, W., Yedema, E.S.E., Seinen, W., Hermens, J.L.M. (1991) Bioconcentration kinetics of chlorinated anilines in guppy, *poecilia reticulata*. *Chemosphere* 28(1), 159-167.

Dean, J.D., Editor (1985) *Lange's Handbook of Chemistry.* 13th ed. McGraw-Hill, Inc., New York, N.Y..

Debnath, A.K., Hansch, C. (1992) Structure-activity relationship of genotoxic polycyclic aromatic nitro compounds. Further evidence for the importance of hydrophobicity and molecular orbital energies in genetic toxicity. *Environ. Mol. Mutagen.* 20(2), 140-144.

Deneer, J.W., Sinnige, T.L., Seinen, W., Hermens, J.L.M. (1987) Quantitative structure-activity relationships for the toxicity and bioconcentration factor of nitrobenzene derivatives towards guppy (*poecilia recticulata*). *Aqua. Toxicol.* 10, 115-129.

Deno, N.C., Berkheimer, H.E. (1960) Part I. Phase equilibria molecular transport thermodynamics. *J. Chem. Eng. Data* 5(1), 1-5.

Digeronimo, M.J., Boethling, R.S., Alexander, M. (1979) Effect of chemical structure and concentration on microbial degradation in model ecosystems. In: *Microbial Degradation of Pollutants in Marine Environments.* EPA-600/9-79-012. U.S. Environmental Protection Agency, Gulf Breeze, Florida.

Dojlido, J.R. (1979) *Investigation of Biodegradability and Toxicity of Organic Compounds.* Final Report 1975-79. U.S. EPA 600/2-79-163. Municipal Environmental Research Lab., Cincinnati, Ohio.

Donahue, D.J., Bartell, F.E. (1952) The boundary tension at water-organic liquid interfaces. *J. Phys. Chem.* 56, 480-484.

Donberg, P.A., Odelson, D.A., Klecka, G.M., Markham, D.A. (1992) Biodegradation of acrylonitrile in soil. *Environ. Toxicol. Chem.* 11, 1583-1594.

Dorfman, L.M., Adams, G.E. (1973) *Reactivity of the Hydroxyl Radical in Aqueous Solution.* NSRD-NDB-46. NTIS COM-73-50623. National Bureau of Standards, Washington, D.C.

Dreisbach, R.R. (1955) *Physical Properties of Chemical Compounds.* No. 15 of the Adv. Chem. Series. American Chemical Society, Washington D.C.

Dreisbach, R.R. (1961) *Physical Properties of Chemical Compounds-III.* No. 29 of the Adv. Chem. Series. American Chemical Society, Washington D.C.

Dunlap, K.L. (1981) In: *Kirk-Othmer Encyclopedia of Chemical Technology.* Vol.15, 3rd. Edition, John Wiley & Sons, New York. pp.925-926, 930-931.

Eadsforth, C.V. (1986) Application of reverse-phase HPLC for the determination of partition coefficients. *Pestic. Sci.* 17, 311-325.

Eadsforth, C.V., Moser, P. (1983) Assessment of reverse-phase chromatographic methods for determining partition coefficients. *Chemosphere* 12, 1459-1475.

Edney, E. et al. (1983) *Atmospheric Chemistry of Several Toxic Compounds.* US EPA-600/53-82-092.

Ellington, J.J., Stancil, F.E., Payne, W.D. (1987) *Measurement of Hydrolysis Rate Constants for Evaluation of Hazardous Waste Land Disposal. Vol. 1, Data on 32 Chemicals.* U.S. EPA-600/3-86-043. NTIS PB87-140 349/GAR.

El Tayar, N., Tsai, R.-S., Vallat, P., Altomare, C., Testa, B. (1991) Measurement of partition coefficient by various centrifugal partition chromatographic techniques. A comprehensive evaluation. *J. Chromatogr.* 556, 181-194.

El Tayar, N., van de Waterbeemd, H., Grylaki, M., Testa, B., Trager, W.F. (1984) The lipophilicity of deuterim atoms. A comparison of shake-flask and HPLC (high performance liquid chromatography) methods. *Int'l J. Pharmaceut.* 19, 271-281.

Fochtman, E.G., Eisenberg, W. (1979) Treatability of carcinogenic and other hazardous organic compounds. U.S. EPA-600/S2-79-097. U.S. Environmental Protection Agency, Cincinnati, Ohio.

Foote, C.S. (1976) *Free Radicals in Biology.* Pryor, W.A., Editor, Academic Press, Inc., New York.

Fournier, J.C., Salle, J. (1974) Microbial degradation of 2,6-dichlorobenzamide in laboratory models. I. Degradation of 2,6-dichlorobezamide in soil, comparison with the evolution of other bezamides with different substituents. Research of metabolic products. *Chemosphere* 3, 77-82.

Freitag, D., Geyer, H., Kraus, A., Viswanathan, R., Kotzias, D., Klein, W., Körte, F. (1982) Ecotoxicological profile analysis. VII. Screening chemicals for their environmental behavior by comparative evaluation. *Ecotoxicol. & Environ. Safety* 6, 60-81.

Freitag, D., Scheunert, I., Klein, W., Körte, F. (1982) Long-term fate of 4-chloroaniline-^{14}C in soil and plants under outdoor conditions. A contribution to terrestrial ecotoxicology of chemicals. *Ecotox. Environ. Saf.* 6, 60-81.

Freitag, D., Scheunert, I., Klein, W., Körte, F. (1984) Long-term fate of 4-chloroaniline ^{14}C in soil and plants under outdoor conditions. A contribution to terrestial ecotoxicology of chemicals. *J. Agric. Food Chem.* 32, 203-207.

Freitag, D., Ballhorn, L., Geyer, H., Körte, F. (1985) Environmental hazard profile of organic chemicals. An experimental method for assessment of the behavior of organic chemicals in the ecosphere by means of simple laboratory tests with ^{14}C labelled chemicals. *Chemosphere* 14, 1589-1616.

Fu, J.-K., Luthy, R.G. (1985) *Pollutant Sorption to Soils and Sediments in Organic/Aqueous Solvent Systems.* EPA/600/3-85/050. Environmental Research Laboratory, Office of Research and Development, U.S. Environmental Protection Agency, Athens, Georgia.

Fu, J.-K., Luthy, R.G. (1986) Aromatic compound solubility in solvent/water mixtures. *J. Chem. Eng.* 112, 328-345.

Fujita, T., Iwasa, J., Hansch, C. (1964) A new substituent constant derived from partition coefficients. *J. Am. Chem. Soc.* 86(23), 5175-5180.

Fujisawa, S., Masuhara, E. (1980) Binding of methyl methacrylate to bovine albumin. *J. Dent. Res.* 59, 2056-2061.

Fujisawa, S., Masuhara, E. (1981) Determination of partition coefficients of acrylates, methacrylates, and vinyl monomers using high performance liquid chromatography (HPLC). *J. Biomed. Mat. Res.* 15, 787-793.

Gaffney, J.S., Streit, W.D., Hall, J.H. (1987) Beyond acid rain. Do soluble oxidants and organic toxins interact with SO_2 and NO_2 to increase ecosystem effects? *Environ. Sci. Technol.* 21(6), 519-524.

Garst, J.E. (1984) Accurate, wide range, automated, high-performance liquid chromatographic method for the estimation of octanol/water partition coefficients. II: Equilibrium in partition coefficient measurements, additivity of substituent constants, and correlation of biological data. *J. Pharm. Sci.* 73, 1623-1629.

Garst, J.E., Wilson, W.C. (1984) Accurate, wide range, automated, high-performance liquid chromatographic method for the estimation of octanol/water partition coefficients. I: Effect of chromatographic conditions and procedure variables on accuracy and reproducibility of the method. *J. Pharm. Sci.* 73, 1616-1623.

Ge, J., Liu, W., Dong, S. (1987) Determination of partition coefficient with chemically bonded omega-hydroxysilica as HPLC column packing. *Sepu* 5(3), 182-185.

Gehring, P.J., Torkelson, T.R., Oyen, F. (1967) A comparison of the lethality of chlorinated pyridines and a study of the acute toxicity of 2-chloropyridine. *Toxicol. Appl. Pharmacol.* 11, 361-371.

GEMS (1986) *Graphical Exposure Modeling System. FAP. Fate of Atmospheric Pollutants.*

GEMS (1987) *Graphical Exposure Modeling System. FAP. Fate of Atmospheric Pollutants.*

GEMS (1989) *Graphical Exposure Modeling System. CLOGP3 Program.* Office of Toxic Substances, U.S. EPA.

Gerike, P., Fischer, W.K. (1979) A correlation study of biodegradability determinations with various chemicals in various tests. *Ecotox. Environ. Safety* 3, 159-73.

Gerstl, Z., Helling, C.S. (1987) Evaluation of molecular connectivity as a predictive method for the adsorption of pesticides in soil. *J. Environ. Sci. & Health* B22(1), 55-69.

Geyer, H., Politzki, G., Freitag, D. (1984) Prediction of ecotoxicological behaviour of chemicals: Relationship between n-octanol/water partition coefficient and bioaccumulation of organic chemicals by alga *chlorella*. *Chemosphere* 13, 269-284.

Gluck, S.J., Martin, E.J. (1990) Extended octanol-water partition coefficient determination by dual-mode centrifugal partition chromatography. *J. Liq. Chromatogr.* 13(18), 3559-3570.

Go, M.L., Ngiam, T.L. (1988) Hydrophobicity change on N-oxidation of some 4-aminoquinolines. *Chem. Pharm. Bull.* 36(4), 1393-1398.

Going, J., Kuykendahl, P., Long, S., Onstol, J., Thomas, K. (1979) *Environmental Monitoring Near Industrial Sites. Acrylonitrile.* U.S. EPA 560/6-79-003. U.S. Environmental Protection Agency, Washington, D.C.

Govers, H., Ruepert, C., Stevens, T.J., van Leeuwen, C.J. (1986) Experimental determination and prediction of partition coefficients of thioureas and their toxicity to photobacterium phosphoreum. *Chemosphere* 15(4), 383-393.

Graedel, T.E. (1978) *Chemical Compounds in the Atmosphere.* Academic Press, New York.

Graveel, J.G., Sommers, L.E., Nelson, D.W. (1986) Decomposition of benzidine, α-naphthalamine, and *p*-toluidine in soils. *J. Environ. Qual.* 15(1), 53-59.

Greene, S., Alexander, M., Leggett, D. (1981) Formation of N-nitrosodimethylamine during treatment of municipal waste water by simulated land application. *J. Environ. Qual.* 10, 416-421.

Gross, P. M., Saylor, J.H., Gorman, A. (1933) Solubilities studies. IV. The solubilities of certain slightly soluble organic compounds in water. *J. Am. Chem. Soc.* 55, 650.

Güesten, H., Filby, W.G., Schoop, S. (1981) Prediction of hydroxyl radical reaction rates with organic compounds in the gas phase. *Atmos. Environ.* 15, 1763-1765.

Günther, F.A., Westlake, W.E., Jaglan, P.S. (1968) Reported solubilities of 738 pesticide chemicals in water. *Residue Rev.* 20, 1-148.

Haderlein, S.B., Schwazenbach, R.P. (1993) Adsorption of substituted nitrobenzenes and nitrophenols to mineral surfaces. *Environ. Sci. Technol.* 27, 316-326.

Hakuta, T., Negishi, A., Goto, T., Ishizaka, S. (1977) Vapor-liquid equilibriums of some pollutants in aqueous and saline solutions. Part I. Experimental results. *Desalination* 21, 11-21.

Haky, J.E., Leja, B. (1986) Evaluation of octanol-water partition coefficients using capillary gas chromatography with cold on-column injection. *Anal. Lett.* 19, 123-134.

Haky, J.E., Young, A.M. (1984) Evaluation of a simple HPLC correlation method for the estimation of the octanol-water partition coefficients of organic compounds. *J. Chromatogr.* 7, 675-689.

Hammers, W.E., Meurs, G.J., De Ligny, C.L. (1982) Correlations between liquid chromatographic capacity ratio data on Lichrosorb RP-18 and partition coefficients in the octanol-water system. *J. Chromatogr.* 247, 1-13.

Hanai, T., Hubert, J. (1982) Hydrophobicity and chromatographic behaviour of aromatic acids found in urine. *J. Chromatogr.* 239, 527-536.

Hanai, T., Tran, C., Hubert, J. (1981) An approach to the prediction of retention times in liquid chromatography. *J. HRC & CC* 4, 454-460.

Hansch, C., Fujita, T. (1964) ρ-σ-π Analysis: Method for the correlation of biological activity and chemical structure. *J. Am. Chem. Soc.* 86, 1616-1626.

Hansch, C., Anderson, S. (1967) The effect of intermolecular hydrophobic bonding on partition coefficients. *J. Org. Chem.* 32, 2583.

Hansch, C., Leo, A. (1979) *Substituent Constants for Correlation Analysis in Chemistry and Biology.* Wiley, New York.

Hansch, C., Leo, A. (1983) *Medchem Project.* Pomona College, Claremont, California.

Hansch, C., Leo, A. (1985) *Medchem Project.* Pomona College, Claremont, California.

Hansch, C., Leo, A. (1987) *Medchem Project.* Pomona College, Claremont, California.

Hansch, C., Quinlan, J.E., Lawrence, G.L. (1968) The linear free-energy relationship between partition coefficients and the aqueous solubility of organic liquids. *J. Org. Chem.* 33, 347-350.

Hanst, P.L., Spence, J.W., Miller, M. (1977) Atmospheric chemistry of N-nitrosodimethylamine. *Environ. Sci. Technol.* 11, 403-405.

Haque, R., Falco, J., Cohen, S., Riordan, C. (1980) Role of transport and fate studies in the exposure, assessment and screening of toxic chemicals. pp.47-67. In: *Dynamics, Exposure and Hazard Assessment of Toxic Chemicals.* Haque, R., Editor, Ann Arbor Science Publishers Inc., Ann Arbor, Michigan.

Harnish, M., Möckel, H.J., Schulze, G. (1983) Relationship between log P_{OW} shake flask values and capacity factors derived from reversed phase HPLC for n-alkylbenzenes and some OECD reference substances. *J. Chromatogr.* 282, 315-332.

Harris, J.C. (1982) Rate of aqueous photolysis. In: *Handbook of Chemical Property Estimation Methods. Environmental Behavior of Organic Compounds.* Lyman, W.J., Reehl, W.F., Rosenblatt, D.H., Eds., Chapter 8, McGraw-Hill Book Co., New York.

Harris, G.W., Kleindienst, T.E., Pitts, J.N., Jr. (1981) Rate constants for the reaction of OH radicals with CH_3CN, C_2H_5CN and CH_2=CH-CN in the temperature range 298-424 K. *Chem. Phys. Lett.* 80, 479-483.

Hasegawa, K., Murase, M., Kuboshita, M., Saida, H., Shinoda, M., Miyamoto, M., Shimasaki, C., Yoshimura, T., Tsukurimichi, E., Takeuchi, S. (1993) Photooxidation of naphthaleneamines adsorbed on particles under simulated atmospheric conditions. *Environ. Sci. Technol.* 27, 1819-1825.

Hashimoto, Y., Tokura, K., Kishi, H., Strachan, W.M.J. (1984) Prediction of seawater solubility of aromatic compounds. *Chemosphere* 13(8), 881-888.

Hashimoto, Y., Tokura, K., Ozaki, K., Strachan, W.M.J. (1982) A comparison of water solubilities by the flask and micro-column methods. *Chemosphere* 11(10), 991-1001.

Hawthorne, S.B., Sievers, R.E., Barkley, R.M. (1985) Organic emissions from shale oil wastewaters and their implications for air quality. *Environ. Sci. Technol.* 19(10), 992-997.

Hein, P. (1914) Investigations on the critical state. *Z. Physik. Chem.* 86, 385-426.

Hendry, D.G., Mill, T., Piszkiewicz, L., Howard, J.A., Eigenmann, H. K. (1974) Critical review of hydrogen-atom transfer in the liquid phase. Chlorine atom, alkyltrichloromethyl, alkoxy, and alkyl peroxy radicals. *J. Phys. Chem. Ref. Data* 3, 937-978.

Herington, E.F.G., Martin, J.F. (1953) Vapour pressures of pyridine and its homologues. *Trans. Faraday Soc.* 49, 154-162.

Hermens, J., Leeuwangh, P. (1982) Joint toxicity of mixtures of 8 and 24 chemicals to the guppy *(poecilia reticulata). Ecotoxicol. & Environ. Safety* 6, 302-310.

Hickey, J.P., Passino-Reader, D.R. (1991) Linear solvation energy relationship: "Rule of Thumb" for estimation of variable values. *Environ. Sci. Technol.* 26, 1753-1760.

Hill, A.E., Macy, R.J. (1924) Ternary system. II. Silver perchlorate, aniline and water. *J. Am. Chem. Soc.* 46, 1132.

Hine, J., Mookerjee, P.K. (1975) The intrinsic hydrophilic character of organic compounds. Correlations in terms of structural contributions. *J. Org. Chem.* 40, 292-298.

Hoigné, J., Bader, H. (1983) Rate constants of reactions of ozone with organic compounds in water-I. Non-dissociating organic compounds. *Water Res.* 17, 173-183.

Holmes, H.L., Lough, C.E. (1976) *Effect of Intramolecular Hydrogen Bonding on Partition Coefficients.* Suffield Tech. Note No. DRES-TN-365. Defence Res. Establishment Suffield/Information Canada. U.S. NTIS Report No. AD-A030683.

Hopke, E.R., Sears, G.W. (1951) Vapor pressures below 1 mm Hg of several aromatic compounds. *J. Chem. Phys.* 19(11), 1345-1351.

Howard, J.A. (1972) Absolute rate constants for reactions of oxy radicals. *Adv. Free Radical Chem.* 4, 49-173.

Howard, P.H., Editor (1989) *Handbook of Environmental Fate and Exposure Data for Organic Chemicals. Vol.I, Large Production and Priority Pollutants.* Lewis Publishers, Inc., Chelsea, Michigan.

Howard, P.H., Editor (1990) *Handbook of Environmental Fate and Exposure Data for Organic Chemicals.* Vol.II, Solvents. Lewis Publishers, Inc., Chelsea, Michigan.

Howard, P.H., Editor (1993) *Handbook of Environmental Fate and Exposure Data for Organic Chemicals.* Vol.IV, Solvents 2. Lewis Publishers, Inc., Chelsea, Michigan.

Howard, P.H., Boethling, R.S., Jarvis, W.F., Meylan, W.M., Michalenko, E.M. (1991) *Handbook of Environmental Degradation Rates.* Lewis Publishers, Chelsea, Michigan.

Howard, P.H., Hueber, A.E., Mulesky, B.C., Crisman, J.S., Meylan, W., Crosby, E., Gray, D.A., Sage, G.W., Howard, K.P., LaMacchia, A., Boethling, R.S., Troast, R. (1986) BIOLOG, BIODEG, and FATE/EXPOS: New files on microbial degradation and toxicity as well as environmental fate/exposure of chemicals. *Environ. Toxicol. & Chem.* 5, 977-988.

Hoy, K.L. (1970) New values of the solubility parameters from vapor pressure data. *J. Paint Technol.* 42(541), 76-118.

Hoyer, Von H., Peperle, W. (1957) Dampfdruckmessungen an organischen substanzen und ihre sublimationswärmen. *Zeit. für Elcktrochemie* 62(1), 61-66.

Hulzebos, E.M., Adema, D.M.M., Breemen, E.M. D.-V., Henzen, L., Van Dis, W.A., Herbold, H.A., Hoekstra, J.A., Baerselman, R., Van Gestel, C.A.M. (1993) Phytotoxicity studies with *lactuca sativa* in soil and nutrient solution. *Environ. Toxicol. Chem.* 12, 1079-1094.

Hustert, K., Mansour, M., Parlar, H., Korte, F. (1981) The EPA Test-a method to determine the photochemical degradation of organic compounds in aquatic systems. *Chemosphere* 10, 995-998.

IARC (1975) *IARC Monographs on the Evaluation of Carcinogenic Risk of Chemicals to Man. Vol. 8, Some Aromatic Azo Compounds.* Internal Agency for Research on Cancer, Lyon, France. 357p.

Ichikawa, Y., Yamano, T., Fujishima, H. (1969) Relationships between the interconversion of cytochrome P540 and its activities in hydroxylations and demethylations by P450 oxidase systems. *Biochem. Biophys. Acta* 171, 32-46.

Isnard, P., Lambert, S. (1988) Estimation bioconcentration factors from octanol-water partition coefficient and aqueous solubility. *Chemosphere* 17, 21-34.

Isnard, P., Lambert, S. (1989) Aqueous solubility/*n*-octanol water partition coefficient correlations. *Chemosphere* 18, 1837-1853.

Iwasa, J., Fujita, T., Hansch, C. (1965) Substituent constants for aliphatic functions obtained from partition coefficients. *J. Med. Chem.* 8, 150-153.

Iwase, K., Komatsu, K., Hirono, S., Nakagawa, S., Moriguchi, I. (1985) Estimation of hydrophobicity based on the solvent-accessible surface area of molecules. *Chem. Pharm. Bull.* 33(5), 2114-2121.

Jäckel, H., Klein, W. (1991) Prediction of mammalian toxicity by quantitative-structure-activity relationships: Aliphatic amines and anilines. *Quant.-Strut.-Act. Relat.* 10, 198-204.

Jaworska, J.S., Schultz, T.W. (1993) Quantitative relationships of structure-activity and volume fraction for selected nonpolar and polar narcotic chemicals. *SAR and QSAR in Environ. Res.* 1, 3-19.

Jayasinghe, D.S., Brownawell, B.J., Chen, H., Westall, J.C. (1992) Determination of Henry's constants of organic compounds of low volatility: Methylanilines in methanol-water. *Environ. Sci. Technol.* 26(11), 2275-2281.

Jenke, D.R., Hayward, D.S., Kenley, R.A. (1990) Liquid chromatographic measurement of solute/solvent partition coefficients: Application to solute/container interactions. *J. Chromatogr. Sci.* 28(12), 609-612.

Jiménez, P., Moux, M.V., Turrión, C. (1990) Thermochemical properties of *N*-heterocyclic compounds. III. Enthalpies of combustion, vapour pressures and enthalpies of sublimation, and enthalpies of formation of 9*H*-carbazole, 9-methylcarbazole, and 9-ethylcarbazole. *J. Chem. Themodynamics* 22, 721-726.

Johnson, C.A., Westall, J.C. (1990) Effect of pH and KCl concentration on octanol-water distribution of methylanilines. *Environ. Sci. Technol.* 24(12), 1869-1875.

Jori, A., Calamari, D., Cattabeni, F., Domenico, A.D., Galli, C.L., Galli, E., Silano, V. (1983) Ecotoxicological profile of pyridine. *Ecotoxicol. & Environ. Safety* 7, 251-275.

Kahlbaum, G.W.A. (1898) Studien über dampfspannkraftmessungen. II. *Z. Phys. Chem.* 26, 577-658.

Kaiser, K.L.E. (1983) A non-linear function for the approximation of octanol/water partition coefficients of aromatic compounds with multiple chlorine substitution. *Chemosphere* 12, 1159-1167.

Kalsch, W.,Nagel, R., Urich, K. (1991) Uptake, elimination, and bioconcentration of ten anilines in zebrafish *(brachydanio rerio). Chemosphere* 22, 351-363.

Kamlet, M.J., Doherty, R,M., Abboud, J.-L.M., Abraham, M.H., Taft, R.W. (1986) Linear solvation energy relationships: 36. Molecular properties governing solubilities of organic nonelectrolytes in water. *J. Pharm. Sci.* 75(4), 338-349.

Kamlet, M.J., Doherty, R,M., Abraham, M.H., Carr, P.W., Doherty, R.F., Taft, R.W. (1987) Linear solvation energy relationships: 41. Important differences between aqueous solubility relationships for aliphatic and aromatic solutes. *J. Phys. Chem.* 91, 1996-2004.

Kamlet, M.J., Doherty, R,M., Taft, R.W., Abraham, M.H., Veith, G.D., Abraham, D.J. (1987) Solubility properties in polymers and biological media. 8. An analysis of the factors that influence toxicities of organic nonelectrolytes to the golden orfe fish *(leuciscus idus melanotus). Environ. Sci. Technol.* 21(2), 149-155.

Karickhoff, S.W. (1985) Chapter 3, Pollutant sorption in environmental systems. In: *Environmental Exposure from Chemicals.* Vol. I, Neely, W.B., Blau, G.E., Editors, CRC Press, Inc., Boca Raton, Florida. pp.49-64.

Karickhoff, S.W., Brown, D.S., Scott, T.A. (1979) Sorption of hydrophilic organic pollutants on natural sediments. *Water Res.* 13, 241-248.

Kawasaki, M. (1980) Experiences with test scheme under the chemical control law of Japan: An approach to structure-activity correlations. *Ecotox. Environ. Safety* 4, 444-454.

Kelly, T.J., Mukund, R., Spicer, C.W., Pollack, A.J. (1994) Concentrations and transfromations of hazardous air pollutants. *Environ. Sci. Technol.* 28, 378A-387A.

Kenaga, E.E. (1980) Predicted bioconcentration factors and soil sorption coefficients of pesticides and other chemicals. *Ecotoxicol. Environ. Safety* 4, 26-38.

Kenaga, E.E., Goring, C.A.I. (1980) In: *Aquatic Toxicology.* Eaton, J.G., Parrish, P.R., Hendricks, A.C., Editors, American Society for Testing and Materials, STP 707, pp.78-115.

Kilzer, L., Scheunert, I., Geyer, H., Klein, W., Körte, F. (1979) Laboratory screening of the volatilization rates of organic chemicals from water and soil. *Chemosphere* 8, 751-761.

Kincannon, D.F., Lin, Y.S. (1985) Microbial degradation of hazardous wastes by land treatment. *Proc. Ind. Waste Conference* 40, 607-619.

Kishida, K., Otori, T. (1980) A quantitative study on the relationship between transcorneal permeabilty of drugs and their hydrophobicity. *Jpn. J. Ophthalmol.* 24, 251-259.

Klamt, A. (1993) Estimation of gas-phase hydroxyl radical rate constants of organic compounds from molecular orbital calculations. *Chemosphere* 26(7), 1273-1289.

Klečka, G.M. (1985) Biodegradation. In: *Environmental Exposure from Chemicals.* Vol I. Neely, W.B., Blau, G.E., Eds., pp.110-155, CRC Press Inc., Boca Raton, Florida.

Klein, E., Weaver, J.W., Webre, B.G. (1957) Solubility of acrylonitrile in aqueous bases and alkali salts. *Chem. Eng. Data Ser.* 2, 72-75.

Klein, W., Kordel, W., Weiβ, M., Poremski, H.J. (1988) Updating the OECD test guideline 107 "partition coefficient n-octanol/water": OECD laboratory intercomparison test on the HPLC method. *Chemosphere* 17, 361-386.

Klopman, G., Namboodiri, K., Scochet, M. (1985) Simple method of computing partition coefficient. *J. Comput. Chem.* 6, 28-38.

Kochany, J., Maguire, R.J. (1994) Photodegradation of quinoline in water. *Chemosphere* 28(6), 1097-1110.

Kollig, H.P., Ed. (1993) *Environmental Fate Constants for Organic Chemicals under Consideration of EPA's Hazardous Waste Identification Projects.* EPA Report EPA/600/R-93/132, U.S. Environmental Research Lab., Athens GA.

Könemann, H., Zelle, R., Busser, F., Hammers, W.E. (1979) Determination of log P_{OCT} values of chloro-substituted benzenes, toluenes and anilines by high-performance liquid chromatography on ODS-silica. *J. Chromatogr.* 178, 559-565.

Kördel, W., Stutte, J., Kotthoff, G. (1993) HPLC-screening method for the determination of the adsorption-coefficient on soil-comparison of different stationary phases. *Chemosphere* 27(12), 2341-2352.

Korenman, I.M., Gurevich, Yu. N., Kulagina, T.G. (1973) Extraction of some *n*-alkylamines from aqueous solutions. *Zh. Prikl. Khim.* (Leningrad) 46(3), 683-684.

Körte, F., Freitag, D., Geyer, H., Klein, W., Kraus, A.G., Lahaniatis, E. (1978) Ecotoxicologic profile analysis-a concept for establishing ecotoxicologic priority lists for chemicals. *Chemosphere* 1, 79-102.

Kramer, C.R., Henze, U. (1990) Partitioning properties of benzene derivatives. I. Temperature dependence of the partitioning of monosubstituted benzenes and nitrobenzenes in the *n*-octanol/water system. *Z. Phys. Chem.* (Leipzig) 271(3), 503-513.

Kurylo, M.J., Knable, G.L. (1984) A kinetic investigation of the gas-phase reactions of atomic chlorine (^2P) and hydroxyl (X^2) with acetonitrile: Atmospheric significance and evidence of decreased reactivity between strong electrophiles. *J. Phys. Chem.* 88, 3305-3308.

Leahy, D.E. (1986) Intrinsic molecular volume as a measure of the cavity term in linear solvation energy relationships: Octanol-water partition coefficients and aqueous solubilities. *J. Pharm. Sci.* 75(7), 629-636.

Lee, B.I., Erbar, J.H., Edmister, W.C. (1972) Thermodynamic properties at low temperatures. *Chem. Eng. Prog.* 68(9), 83-84.

Lee, M.L., Later, D.W., Rollins, D.K., Eatough, D.J., Hansen, L.D. (1980) Dimethyl and monomethyl sulfate: Presence in coal fly ash and airborne particulate matter. *Science* 207, 186-188.

Leggett, D.C., Parker, L.V. (1994) Modeling the equilibrium partitioning of organic contaminants between PTFE, PVC, and groundwater. *Envrion. Sci. Technol.* 28, 1229-1233.

Lemaire, J. Guth, J.A., Klais, O., Leahy, J., Merz, W., Philp, J., Wilmes, R., Wolff, C.J.M. (1985) Ring test of a method for assessing the phototransformation of chemicals in water. *Chemosphere* 14, 53-77.

Lenchitz, C., Velicky, R.W. (1970) Vapor pressure and heat of sublimation of three nitrotoluenes. *J. Chem. Eng. Data* 15(3), 401-403.

Łencka, M. (1990) Measurements of vapour pressures of pyridine, 2-methyl pyridine, 2,4-dimethylpyridine, 2,6-dimethylpyridine, and 2,4,6-trimethylpyridine from 0.1 kPa to atmospheric pressure using a modified Świetosławski ebulliometer. *J. Chem. Thermodyn.* 22, 473-480.

Leo, A., Hansch, C., Elkins, D. (1971) Partition coefficients and their uses. *Chem. Rev.* 71, 525-616.

Leo, A., Hansch, C., Church, C. (1969) Comparison of parameters currently used in the study of structure-activity relationships. *J. Med. Chem.* 12, 766-771.

Lewis, S.J., Mirrlees, M.S., Taylor, P.J. (1983) Rationalizations among heterocyclic partition coefficients. Part 2. The azines. *Quant. Struct.-Act. Relat. Pharmacol. Chem. Biol.* 2, 100-111.

Lide, D.R., Milne, G.W.A., Editors (1994) *Handbook of Data on Organic Compounds*. Vol. II, 3rd. Edition, CRC Press, Boca Raton, Florida.

Liu, D.H.W. et al. (1983) *Aquatic Toxicology Hazard Assessment*. ASTM Spec. Tech. Publication 802. pp.135-150.

Loekke, H. (1985) Degradation of 4-nitrophenol in two Danish Soils. *Environ. Pollut. Ser. A.* 38, 171-181.

Lu, P.Y., Metcalf, R.L. (1975) Environmental fate and biodegradability of benzene derivatives as studies in a model aquatic ecosystem. *Environ. Health Prospect* 10, 269-284.

Lu, P.Y., Metcalf, R.L., Plummer, N., Mandel, D. (1977) The environmental fate of three carcinogens: bezo(a)pyrene, benzidine, and vinyl chloride evaluated in laboratory model ecosystems. *Arch. Environ. Contam. Toxicol.* 6, 129-142.

Lu, P.Y., Walton, B.T., Lewis, E.B., Kine, B.W., Scott, J.H., Groover, G.E. (1986) Chemical information profile of selected Resource Conservation and Recovery Act chemicals. ORNL Internal Report to Robert S. Kerr Res. Lab., U.S. Environmental Protection Agency, Oak Ridge, TN.

Ludzack, F.J., Schaffer, R.B., Bloomhuff, R.N., Ettinger, M.B. (1958) Biochemical oxidation of some commercially important organic cyanides. I. River oxidation. In: *Proc. 13th Industrial Waste Conference Eng. Bull. Purdue Univ. Eng. Ext. Ser.* pp.297-312.

Lyman, W.J., Reehl, W.F., Rosenblatt, D.H., Editors (1982) *Handbook on Chemical Property Estimation Methods. Environmental Behavior of Organic Compounds*. McGraw-Hill, New York, N.Y.

Lyons, C.D., Katz, S., Bartha, R. (1984) Mechanisms and pathways of aniline elimination from aquatic environments. *Appl. Environ. Microbiol.* 48(3), 491-496.

Ma, K.C., Shiu, W.Y., Mackay, D. (1990) *A Critically Reviewed Compilation of Physical and Chemical and Persistence Data for 110 Selected EMPPL Substances*. A report prepared for the Ontario Ministry of Environment, Water Resources Branch, Toronto, Ontario.

Mabey, W.R., Mill, T. (1978) Critical review of hydrolysis of organic compounds in water under environmental conditions. *J. Phys. Chem. Ref. Data* 7, 383-415.

Mabey, W.R., Smith, J.H., Podoll, R.T., Johnson, H.L., Mill, T., Chou, T.-W., Cates, J., Waight Partridge, I., Jaber, H., Vandenberg, D. (1982) *Aquatic Fate Process Data for Organic Priority Pollutants*. EPA Report No. 440/4-81-014, US EPA, Office of Water Regulations and Standards, Washington DC.

Mackay, D. (1982) Correlation of bioconcentration factors. *Environ. Sci. Technol.* 16, 274-278.

Mackay, D., Bobra, A., Chan, D.W., Shiu, W.Y. (1982) Vapor pressure correlations for low-volatility environmental chemicals. *Environ. Sci. Technol.* 16(10), 645-649.

Mackay, D., Leinonen, P.J. (1975) Rate of evaporation of low-solubility contaminants from water to atmosphere. *Environ. Sci. Technol.* 7, 1178-1180.

Mackay, D., Wolkoff, A.W. (1973) Rate of evaporation of low solubility contaminants from water bodies to the atmosphere. *Environ. Sci. Technol.* 7, 611-614.

Maczynski, A., Maczynska, Z. (1965) Vapor-liquid equilibrium in binary system quinoline-water. *Bull. Acad. Pol. Sci.* 13, 299-302.

Malaney, G.W. (1960) Oxidative abilities of aniline-acclimated activated sludge. *J. Water Pollut. Control Fed.* 32, 1300-1311.

Malaney, G.W., Gerhold, R.W. (1962) Structural determinants in the oxidative breakdown of aliphatic compounds by domestic activated sludge. In. *Proc. 17th Indust. Waste Conf. Purdue Univ. Ext. Ser.* 112, 249-257.

Malaney, G.W., Gerhold, R.W. (1969) Structural determinants in the oxidation of aliphatic compounds by activated sludge *J. Water Pollut. Control Fed.* 41, R18-R33.

Mallik, M.A.B., Tesfai, K. (1981) Transformation of nitrosoamines in soil and in vitro by soil organisms. *Bull. Environ. Contam. Toxicol.* 27, 115-121.

Marchini, S., Passerini, L., Cesareo, D., Tosato, M.L. (1988) Herbicidal triazines: Acute toxicity on daphnia, fish and plants and analysis of its relationships with structural factors. *Ecotox. Environ. Safety* 16, 148-157.

Martin-Villodre, A., Pla-Delfina, J.M., Moreno, J., Perez-Buendia, M.D., Miralles, J., Collado, E.F., Sanchez-Moyano, E., Del Pozo, A. (1986) Studies on the reliability of a bihyperbolic functional absorption model. I. Ring-substituted anilines. *J. Pharmacokinet. Biopharm.* 14, 615-633.

Matzner, R.A., Bales, R.C. (1994) Transport of acridine in saturated porous media. *Chemosphere* 29, 1755-1773.

Matzner, R.A., Hunter, D.R., Bales, R.C. (1991) The effects of pH and anions on the solubility and sorption behavior of acridine. In: *Organic Substances and Sediments in Water. Vol. 2, Processes and Analytical Methods*. Baker, R.A., Editor, Lewis Publishers, Chelsea, Michigan.

McCall, J.M. (1975) Liquid-liquid partition coefficients by high-performance liquid chromatography. *J. Med. Chem.* 18(6), 549-552.

McDuffie, B. (1981) Estimation of octanol/water partition coefficients for organic pollutants using reversed-phase HPLC. *Chemosphere* 10, 73-83.

McEachern, D.M., Iniguez, J.C., Ornelas (1975) Enthalpies of combustion and sublimation and vapor pressures of three benzoquinolines. *J. Chem. Eng. Data* 20, 226-228.

McEachern, D.M., Sandoval, O., Iniguez, J.C. (1975) The vapor pressures, derived enthalpies of sublimation, enthalpies of fusion, and resonance energies of acridine and phenazine. *J. Chem. Thermodyn.* 7, 299-306.

McGowan, J.C., Atkinson, P.N., Ruddle, L.H. (1966) The physical toxicity of chemicals. V. Interaction terms for solubilities and partition coefficients. *J. Appl. Chem.* 16, 99-102.

McLeese, D.W., Zitko, V., Peterson, M.R. (1979) Structure lethality relationships for phenols, anilines and other aromatic compounds in shrimp and clams. *Chemosphere* 2, 53-57.

Means, J.C. (1980) U.S. EPA Report No. 600/3-80-041.

Means, J.C. (1983) Am. Chem. Soc. 186th Nat'l. Meeting Preprints Div. Environ. Chem. 23, 250-251. Washington, DC.

Means, J.C., Hassett, J.J., Wood, S.G., Banwart, W.L., Ali, S., Khan, A. (1980) Sorption properties of polynuclear aromatic hydrocarbons and sediments: Heterocyclic and substituted compounds. In: *Polynuclear Hydrocarbons: Chemistry and Biological Effects*. Bjorseth, A., Dennis, A.J., Editors, pp.395-404, Ann Arbor Sci. Publishers, Ann Arbor, Michigan.

Means, J.C., Wood, S.G., Hassett, J.J., Banwart, W.L. (1982) Sorption of amino- and carboxy-substituted polynuclear aromatic hydrocarbons by sediments and soils. *Environ. Sci. Technol.* 16, 93-98.

Medvedev, V.A., Davidov, V.D. (1981) The transformation of various coke industry products in Chernozem soil. In: *Decomposition of Toxic and Nontoxic Compounds in Soil*. Overcash, M.R., Editor, pp.245-254., Ann Arbor Science Publishers, Ann Arbor, Michigan.

Meylan, W.M., Howard, P.H. (1991) Bond contribution method for estimating Henry's law constants. *Environ. Toxicol. Chem.* 10, 1283-1291.

Meylan, W.M., Howard, P.H., Boethling, R.S. (1992) Molecular topology/fragment contribution method for predicting soil sorption coefficients. *Environ. Sci. Technol.* 26(8), 1560-1567.

Mill, T. (1979) *Structure Reactivity Correlations for Environmental Reactions*. EPA Final Report, EPA 560/11-79-012.

Mill, T. (1982) Hydrolysis and oxidation processes in the environment. *Environ. Toxicol. & Chem.* 1, 135-141.

Mill, T., Mabey, W.R. (1985) Photochemical transformations. *Environ. Exposure Chem.* 1, 175-216.

Mill, T., Hendry, D.G., Richardson, H. (1980) Free-radical oxidants in natural waters. Science 207, 886-887.

Mill, T., Mabey, W.R., Lan, B.Y., Baraze, A. (1981) Photolysis of polycyclic aromatic hydrocarbons in water. *Chemosphere* 10, 1281-1290.

Mills, W.B., Dean, J.D., Porcella, D.B., Gherini, S.A., Hudson, R.J.M., Frick, W.E., Rupp, G.L., Bowie, G.L. (1982) *Water Quality Assessment: A Screening Procedure for Toxic and Conventional Pollutants.* Part 1, EPA Report No. EPA-600/6-82-004a, Environmental Research Lab., US EPA, Athens, GA.

Minero, C., Pelizzetti, E., Piccinini, P., Vincenti, M. (1994) Photocatalyzed transformation of nitrobenzene on TiO_2 and ZnO. *Chemosphere* 28(6), 1229-1244.

Minick, D.J., Frenz, J.H., Patrick, M.A., Brent, D.A. (1988) A comprehensive method for determining hydrophobicity constants by reversed-phase high-performance liquid chromatography. *J. Med. Chem.* 31, 1923-1933.

Mirrlees, M.S., Moulton, J.J., Murphy, C.T., Taylor, P.J. (1976) Direct measurement of octanol-water partition coefficients by high-pressure liquid chromatography. *J. Med. Chem.* 19, 615-619.

Mirvish, S.S., Issenberg, P., Sornson, H.C. (1976) Air-water and ether-water distribution of N-nitroso compounds: Implications for laboratory safety, analytic methodology, and carcinogenicity for the rat esophagus, nose, and liver. *J. Nat'l Cancer Inst.* 56(6), 1125-1129.

Miyake, K., Kitaura, F., Mizuno, N., Terada, H. (1987) Determination of partition coefficient and acid dissociation constant by high-performance liquid chromatography on porous polymer gel as stationary phase. *Chem. Pharm. Bull.* 35(1), 377-388.

Miyake, K., Mizuno, N., Terada, H. (1986) Method for determination of partition coefficient by high-performance liquid chromatography on an octadecylsilane column. Examination of its applicability. *Chem. Pharm. Bull.* 34(11), 4787-4796.

Miyake, K., Tereda, H. (1982) Determination of partition coefficients of very hydrophobic compounds by high-performance liquid chromatography on glyceryl-coated controlled-pore glass. *J. Chromatogr.* 240, 9-20.

Moreale, A., Van Bladel, R. (1976) Influence of soil properties on adsorption of pesticide-derived aniline and *p*-chloroaniline. *J. Soil Sci.* 27, 48-57.

Mousa, A.H.N. (1981) Vapour pressure and saturated-vapour volume of acetonitrile. *J. Chem. Thermodyn.* 13, 201-202.

Müller, M., Klein, W. (1991) Estimating atmospheric degradation processes by SARs. *Sci. Total Environ.* 109/110, 261-273.

Müller, M., Klein, W. (1992) Comparative evaluation of methods predicting water solubility for organic compounds. *Chemosphere* 25(6), 769-782.

Naik, M.N., Jackson, R.B., Stokes, J., Swaby, R.J. (1972) Microbial degradation and phytotoxicity of picloram and other substituted pyridines. *Soil Biol. Biochem.* 4, 13-23.

Neely, W.B., Blau, G.E. (1985) Chapter 1, Introduction to environmental exposure from chemicals. In: *Environmental Exposure from Chemicals.* Volume I., Neely, W.B., Blau, G.E., Editors, CRC Press Inc., Boca Raton, Florida. pp.1-11.

Neely, W.B., Branson, D.R., Blau, G.E. (1974) Partition coefficient to measure bioconcentration potential of chemicals in fish. *Environ. Sci. Technol.* 8, 1113-1115.

Neish, W.J.P. (1948) Solubilization of aromatic amines by purines. *Rec. Trav. Chim. Pays-Bas* 67, 361-373.

Nirmalakhandan, N.N., Speece, R.E. (1988) Prediction of aqueous solubility of organic chemicals based on molecular structure. *Environ. Sci. Technol.* 22, 328-338.

Norrington, F.E., Hyde, R.M., Williams, S.G., Wootton, R. (1975) Physicochemical-activity relations in practice. 1. A rational and self-consistent data bank. *J. Med. Chem.* 18, 604-607.

Oliver, J.E. (1981) Pesticide-derived nitrosoamines. Occurrence and environmental fate. *Agric. Symp. Ser.* 74, 349-362.

Oliver, J.E., Kearney, P.C., Kontson, A. (1979) Degradation of herbicide-related nitrosoamines in aerobic soils. *J. Agric. Food Chem.* 27, 887-891.

Osborn, A.G., Douslin, D.R. (1968) Vapor pressure relations of 13 nitrogen compounds related to petroleum. *J. Chem. Eng. Data* 13(4), 534-537.

Palit, S.R. (1947) Electronic interpretations of organic chemistry. II. Interpretation of the solubility of the organic compounds. *J. Phys. Chem.* 51, 837-857.

Pangrekar, J., Klopman, G., Rosenkranz, H.S. (1994) Expert-system comparison of structural determinants of chemical toxicity to environmental bacteria. *Environ. Toxicol. & Chem.* 13(6), 979-1001.

Paris, D.F., Wolfe, N.L. (1987) Relationship between properties of a series of anilines and their transformation by bacteria. *Appl. Environ. Microbiol.* 53, 911-916.

Paris, D.F., Wolfe, N.L., Steen, W.C., Baughman, G.L. (1983) Effect of phenol molecular structure on bacterial transformation rate constants in pond and river samples. *Appl. Environ. Microbiol.* 45, 1153-1155.

Pearlman, R.S., Yalkowsky, S.H., Banerjee, S. (1984) Water solubilities of polynuclear aromatic and heteroaromatic compounds. *J. Phys. Chem. Ref. Data* 13, 555-562.

Peijnendburg, W.J.G.M., de Beer, K.G.M., den Hollander, H.A., Stegeman, M.H.L., Verboom, H. (1993) Kinetics, products, mechanisms and QSARs for the hydrolytic transformation of aromatic nitriles in anaerobic sediment slurries. *Environ. Toxicol. Chem.* 12, 1149-1161.

Pella, P.A. (1977) Measurement of the vapor pressures of TNT, 2,4-DNT, 2,6-DNT, and EGDN. *J. Chem. Thermodyn.* 9, 301-305.

Perrin, D.D. (1965) *Dissociation Constants of Organic Bases in Aqueous Solutions*. Butterworth, London.

Perrin, D.D. (1972) *Dissociation Constants of Organic Bases in Aqueous Solutions*. IUPAC Chemical Data Series; Supplement. Butterworth, London.

Philpot, J.P., Rhodes, E.C., Davies, C.W. (1940) The determination of mobilities and dissociation constants by means of conductivity titrations. *J. Chem. Soc.* 84-87.

Piacente, V., Scardala, P., Gigli, R. (1985) Vaporization study of *o*-, *m*-, and *p*-chloroaniline by torsion-weighing effusion vapor pressure measurements. *J. Chem. Eng. Data* 30, 372-376.

Pillai, P., Helling, C.S., Dragun, J. (1982) Soil catalyzed oxidation of aniline. *Chemosphere* 11, 299, 317.

Pinal, R., Lee, L.S., Rao, P.S.C. (1991) Prediction of the solubility of hydrophobic compounds in nonideal solvent mixtures. *Chemosphere* 22(9/10), 939-951.

Politzki, G.R., Bieniek, D., Lahaniatis, E.S., Scheunert, I., Klein, W., Körte, F. (1982) Determination of vapour pressures of some organic chemicals adsorbed on silicagel. *Chemosphere* 11(12), 1217-1229.

Pratesi, P., Villa, L., Ferri, V., de Micheli, C., Grana, E., Grieco, C., Silipo, C., Victtoria, A. (1979) On the additive-constitutive properties of substituent hydrophobic parameters in a set of muscarinic agents. *Il Farmaco Ed. Sci.* 34(7), 579-587.

Price, L.C. (1976) Aqueous solubility of petroleum as applied to its origin and primary migration. *Am. Assoc. Petrol. Geol. Bull.* 60, 213-244.

Przyjazny, A., Janicki, W., Chrzanowski, W., Staszewski, R. (1983) Headspace gas chromatographic determination of distribution coefficients of selected organosulphur compounds and their dependence on some parameters. *J. Chromatogr.* 280, 249-260.

Radchenko, L.G., Kitaigorodskii, A.I. (1974) The vapour pressures and heats of sublimation of naphthalene, biphenyl, octafluoronaphthalene, decafluorobiphenyl, acenaphthene and α-nitronaphthalene. *Russian J. Phys. Chem.* 48(11), 1595-1597.

Radding, S.B., Liu, D.H., Johnson, H.L., Mill, T. (1977) *Review of the Environmental Fate of Selected Chemicals. EPA 560/5-77-003.* U.S. Environmental Protection Agency, Office of Toxic Substance, Washington, D.C. pp.147.

Rao, P.S.C., Davidson, J.M. (1982) *Retention and Transformation of Selected Pesticides and Phosphorus in Soil-Water System. A Critical Review.* EPA-600/S3-82-060.

Reid, R.C., Prausnitz, J.M., Sherwood, T.K. (1977) *The Properties of Gases and Liquids.* 3rd ed., McGraw-Hill Book Company, New York.

Rekker, R.F. (1977) *The Hydrophobic Fragmental Constant.* Its Derivation and Application. A Means of Characterizing Membrane Systems. Vol. 1, Pharmacochemistry Library, Nauta, W.T., Rekker, R.F., Editors. Elsevier Scientific Publishing Company, Oxford.

Rekker, R.F., De Kort, H.M. (1979) The hydrophobic fragment constant: An extension to a 1000 data point set. *Eur. J. Med. Chem.* 14, 479-488.

Riddick, J.A., Bunger, W.B., Sakano, T.K. (1986) *Organic Solvents.* 4th Edition. John Wiley and Sons, New York.

Rinke, M., Zetzsch, C. (1984) Rate constants for the reactions of hydroxyl radicals with aromatics: benzene, phenol, aniline, and 1,2,4-trichlorobenzene. *Ber. Bunsen-Ges. Phys. Chem.* 88, 55-62.

Rippen, G. (1987) *Handbuch Umweltchemikalien.* Ecomed, Landsberg.

Ritter, S., Hauthal, W.H., Maurer, G. (1994) Partition coefficients of some environmentally important organic compounds between 1-octanol and water from reversed-phase high-performance liquid chromatography. *J. Chem. Eng. Data* 39, 414-417.

Roberson, R.E., Sugamori, S.E. (1966) The hydrolysis of dimethyl sulfate and diethyl sulfate in water. *Can. J. Chem.* 44, 1728-1730.

Rogers, J.E., Li, S.W., Felice, L.J. (1984) *Microbial Transformation Kinetics of Xenobiotics in Aquatic Environment.* EPA-600/3-84-043. (NTIS-PB84-162866). Battelle Pacific West Labs., Richland, WA.

Rogers, K.S. (1969) Rabbit erythrocyte hemolysis by lipophile aryl molecules. *Proc. Soc. Exp. Biol. Med.* 130(4), 1140-1142.

Rogers, K.S., Cammarata, A. (1969) Superdelocalizability and charge density. A correlation with partition coefficients. *J. Med. Chem.* 12, 692-693.

Rohrschneider, L. (1973) Solvent characterization by gas-liquid partition coefficients of selected solutes. *Anal. Chem.* 45(7), 1241-1247.

Rott, B., Viswanathan, R., Freitag, D., Korte, F. (1982) Vergleichende untersuchung der anwendbrakeit von umweltchemikalien. *Chemosphere* 11, 531-538.

Ruppert, G., Bauer, R., Heisler, G., Novalic, S. (1993) Mineralization of cyclic organic water contaminants by the photo-Fenton reaction-Influence of structure and substituents. *Chemosphere* 27(8), 1339-1347.

Russell, C.J., Dixon, S.L., Jurs, P.C. (1992) Computer-assisted study of the relationship between molecular structure and Henry's law constant. *Anal. Chem.* 64, 1350-1355.

Ryan, J.A., Bell, R.M., Davidson, J.M., O'Connor, G.A. (1988) Plant uptake of non-ionic organic chemicals from soils. *Chemosphere* 17(12), 2299-2323.

Sabljic, A. (1987) On the prediction of soil sorption coefficients of organic pollutants from molecular structure: Application of molecular topology model. *Environ. Sci. Technol.* 21, 358-366.

Sabljic, A., Güsten, H. (1990) Predicting the night-time NO_3 radical reactivity in the troposphere. *Atmos. Environ.* 24A(1), 73-78.

Sandell, K.B. (1962) Distribution of aliphatic amines between water and mixture of organic solvents. *Naturwissenschaften* 49, 12-13.

Sangster, J. (1989) Octanol-water partition coefficients of simple organic compounds. *J. Phys. Chem. Ref. Data* 18(3), 1111-1230.

Sangster, J. (1993) LOGKOW Data Bank, Sangster Research Lab., Montreal, Canada.

Sasaki, S. (1978) The scientific aspects of the chemical substance control law in Japan. In: *Aquatic Pollutants: Transformation and Biological Effects.* Hutzinger, O. et al., Editors, Pergamon Press, Oxford, U.K. pp.283-298.

Schauerte, W., Lay, J.P., Klein, W., Korte, F. (1982) Long-term fate xenobiotics in aquatic ecosystems. *Ecotoxicol. Environ. Safety* 6, 560-569.

Schmidt-Bleek, F., Haberland, W., Klein, A.W., Caroli, S. (1982) Steps towards environmental hazard assessment of new chemicals. *Chemosphere* 11(4), 383-415.

Schnell, S., Bak, F., Pfennig, N. (1989) Anaerobic degradation of aniline and dihydroxybenzenes by newly isolated sulfate-reducing bacteria and description of *Desulfobacterium anilini. Arch. Microbiol.* 152, 556-563.

Schultz, Q.E., Jung, C., Moller, K.E. (1970) Schätzung des verteilungkoeffizienten mit hilfe quantenchemischer molekülgröβen. *Z. Naturforsch.*, 25B, 1024-1026.

Schultz, T.W., Applehans, F.M. (1985) Correlation for the acute toxicity of multiple N-substituted aromatic molecules. *Ecotoxicol. Environ. Saf.* 10, 75.

Schultz, T.W., Dawson, D.A., Lin, D.T. (1989) Compare toxicity of selected nitrogen-containing aromatic compounds in the *tetrahymena pyriformis* and *pimephales promelas* test systems. *Chemosphere* 18(11/12), 2283-2291.

Schultz, T.W., Moulton, B.A. (1985) Structure-activity relationships for nitrogen-containing aromatic molecules. *Environ. Toxicol. Chem.* 4, 353-359.

Schüürmann, G., Klein, W. (1988) Advances in bioconcentration prediction. *Chemosphere* 17(8), 1551-1574.

Scow, K.M. (1982) Chapter 9, Rate of biodegradation. In: *Handbook of Chemical Property Estimation Methods. Environmental Behavior of Organic Compounds.* Lyman, W.J., Reehl, W.F., Rosenblatt, D.H., Editors, McGraw-Hill Book Co., Ney York, New York.

Scully, F.E., Jr., Hoigne, J. (1987) Rate constants for reactions of singlet oxygen with phenols and other compounds in water. *Chemosphere* 16, 681-694.

Seidell, A. (1941) *Solubilities of Organic Compounds.* Van Nostrand Co. Inc., New-York.

Seidell, A. (1952) *Solubilities of Organic Compounds.* Van Nostrand Co. Inc., New-York.

Seip, H.M., Alstad, J., Carlberg, G.E., Martinsen, K., Skaane, P. (1986) Measurements of mobility of organic compounds in soils. *Sci. Total Environ.* 50, 87-101.

Shriner, C.D. et al. (1978) *Reviews of the Environmental Effects of Pollutants. II. Benzidine.* USEPA-600/1-78-024.

Sikka, H.C., Appleton, H.T., Banerjee, S. (1978) *Fate of 3,3'-Dichlorobenzidine in Aquatic Environment.* EPA-600/3-78-068. Syracuse Research Corp., Syracuse, New York.

Simmons, M.S., Zepp, R.G. (1986) Influence of humic substances on photolysis of nitroaromatic compounds in aquatic systems. *Water Res.* 20, 899-904.

Sims, G.K., O'Loughlin, E.J. (1989) Degradation of pyridines in the environment. *Critical Rev. Environ. Control.* 19, 309-340.

Sims, G.K., Sommers, L.E. (1985) Degradation of pyridine derivatives in soil. *J. Environ. Qual.* 14, 580-584.

Sivaraman, A., Kobayashi, R. (1982) Investigation of vapor pressures and heats of vaporization of condensed aromatic compounds at elevated temperatures. *J. Chem. Eng. Data* 27, 264-269.

Sivaraman, A., Kobayashi, R. (1983) Vapor pressures and enthalpies of vaporization of thianthrene, acridine, and 9-methylanthracene at elevated temperatures. *J. Chem. Thermodynamics* 15, 1127-1135.

Slagle, I.R., Balocchi, F., Gutman, D. (1978) Study of the reactions of oxygen atoms with hydrogen sulfide, methanethiol, ethanethiol and methyl sulfide. *J. Phys. Chem.* 82(12), 1333-1336.

Smith, H.W. (1921) The nature of secondary valence. *J. Phys. Chem.* 25, 204-263; 605-627.

Smith, J.H., Bomberger, D.C. (1980) Prediction of volatilization rate of chemicals in water. In: *Hydrocarbons and Halogenated Hydrocarbons in the Environment*, Afghan, B.K., Mackay, D., Eds., pp. 445-451, Plenum Press, New York, N.Y.

Smith, J.H., Bomberger, D.C., Jr., Haynes, D.L. (1981) Volatilization rates of intermediate and low volatility chemicals from water. *Chemosphere* 10(5), 281-289.

Smith, J.H., Mabey, W.R., Bahonos, N. Holt, B.R., Lee, S.S., Chou, T.W., Venberger, D.C., Mill, T. (1978) *Envirnmental Pathways of Selected Chemicals in Fresh Water Systems: Part II. Laboratory Studies.* Interagency Energy-Environment Research Program Report. EPA-600/7-78-074. Environmental Reasearch Laboratory Office of Research and Development. U.S. Environment Protection Agency, Athens, Georgia.

Smith, J.H., Mackay, D., Ng, C.W.K. (1983) Volatilization of pesticides from water. *Residue Rev.* 85, 73-88.

Snider, J.R., Dawson, G.A. (1985) Tropospheric light alcohols, carbonyls, and acetonitrile: concentrations in the southwestern United States and Henry's law data. *J. Geophys. Res.* 90(D2), 3797-3805.

SOGC (1987) *Dinitrotoluenes: BUA Substance Report 12.* The Advisory Board for Environmentally Significant Hazardous Waste (BUA) of the Society of German Chemists, Editors. Weinheim, Fed. Republic of Germany: VCH Verlagsgesellschaft p.2.

Southworth, G.R., Beauchamp, J.J., Schmieder, P.K. (1978) Bioaccumulation potential and acute toxicity of synthetic fuels in freshwater biota: azzarenes. *Environ. Sci. Technol.* 12, 1062-1066.

Spanggord, R.J., Mabey, W.R., Mill, T., Chou, T.W., Smith, J.H. (1981) *Environmental Fate Studies on Certain Munitions Wastewater Constituents. Part 1. Model Validation*. NTIS AD-A129 373/7. SRI International, Menlo Park, California. 79pp.

Spanggord, R.J., Mill, T., Chou, T.W., Mabey, W.R., Smith, J.H., Lee, S. (1980) *Environmental Fate Studies on Certain Munitions Wastewater Constituents. Final Report. Phase I-Literature Review*. SRI Project No. LSU-7934. Contract No. DAMD 17-78-C-8081. U.S. Army Medical Research and Development Command, Fort Detrick, MD.

Steen, W.C. (1991) *Microbial Transformation Rate Constants of Structually Diverse Man-made Chemicals*. EPA 600/3-91/016. PB91-181958. Environmental Research Lab., Office of Research and Development, U.S. Environmental Protection Agency, Athens, Georgia.

Steen, W.C., Karickhoff, S.W. (1981) Biosorption of hydrophobic organic pollutants by mixed microbial populations. *Chemosphere* 10, 27-32.

Stegeman, M.H.L., Peijnenburg, W.J.G.M., Verboom, H. (1993) A quantitative structure-activity relationship for the direct photohydrolysis of meta-substituted halobenzene directives in water. *Chemosphere* 26(5), 837-849.

Stephen, H., Stephen, Y. (1963) *Solubilities of Inorganic and Organic Compounds*. Vol. 1 and 2, Pergamon Press, Oxford.

Stephenson, R.M. (1993) Mutual solubility of water and pyridine derivatives. *J. Chem. Eng. Data* 38, 428-431.

Stephenson, R.M. (1993) Mutual solubility of water and aliphatic amines. *J. Chem. Eng. Data* 38, 625-629.

Stephenson, R.M. (1993) Mutual solubilities: water + cyclic amines, water + alkanolamines, and water + polyamines. *J. Chem. Eng. Data* 38, 634-637.

Stephenson, R.M. (1994) Mutual solubility of water and nitriles. *J. Chem. Eng. Data* 39, 255-227.

Stephenson, R.M., Malanowski, S. (1987) *Handbook of Thermodynamics of Organic Compounds*. Elsevier Science Publishing Co. Inc., New York.

Stull, D.R. (1947) Vapor pressure of pure substances: Organic compounds. *Ind. Eng. Chem.* 39(4), 517-560.

Subba-Rao, R.V., Rubbin, H.E., Alexander, M. (1982) Kinetics and extent of mineralization of organic chemicals in trace levels in freshwater and sewage. *Appl. Environ. Microbiol.* 43, 1139-1150.

Sudhakar-Barik, Sethunathan, N. (1978) Metabolism of nitrophenols in flooded soils. *J. Environ. Qual.* 7, 349-352.

Tabak, H.H., Govind, R. (1993) Prediction of biodegradation kinetics using a nonlinear group contribution method. *Environ. Toxicol. Chem.* 12, 251-260.

Taft, R.W., Abraham, M.H., Doherty, R.M., Kamlet, M.J. (1985) The molecular properties governing solubilities of organic nonelectrolytes in water. *Nature* 313, 384-386.

Taft, R.W., Abraham, M.H., Famini, G.R., Doherty, R.M., Abboud, J.-L.M., Kamlet, M.J. (1985) Solubility properties in polymers and biological media. 5: An analysis of the phsicochemical properties which influence octanol-water partition coefficients of aliphatic and aromatic solutes. *J. Pharm. Sci.* 74(8), 807-814.

Takagishi, T., Katayama, A., Konishi, K., Kuroki, N. (1968) The solubilities of azobenzene derivatives in water. *Kolloid-Zeitschrift Zeit. Polym.* 232(1), 693-699.

Tanii, H., Hashimoto, K. (1984) Studies on the mechanism of acute toxicity of nitriles in mice. *Arch. Toxicol.* 55, 47-54.

Tanikaga, R. (1969) Photolysis of nitrosobenzene. *Bull. Chem. Soc. Japan* 42(1), 210-214.

Tate, R.L., III, Alexander, M. (1975) Stability of nitrosoamines in samples of lake water, soil and sewage. *J. Nat'l Cancer Inst.* 54(2), 327-330.

Tate, R.L., III, Alexander, M. (1976) Microbial formation and degradation of dimethylamine. *Appl. Environ. Microbiol.* 31, 399-403.

Taylor, C.A., Rinkenbach, W.H. (1923) The solubility of trinitrotoluene in organic solvents. *J. Am. Chem. Soc.* 45, 44.

Tewari, Y.B., Miller, M.M., Wasik, S.P., Martire, D.E. (1982) Aqueous solubility and octanol/water partition coefficient of organic compounds at 25°C. *J. Chem. Eng. Data* 27, 451-454.

Timmermans, J. (1950) *Physical-Chemical Constants of Pure Organic Compounds.* Vol.1, Elsevier Publishing Company, Inc., New York.

Tsai, R.S., El Tayar, N., Testa, B. (1991) Toroidal coil centrifugal partition chromatography, a method for measuring partition coefficients. *J. Chromatogr.* 538(1), 119-123.

Tsonopoulos, C., Prausnitz, J.M. (1971) Activity coefficients of aromatic solutes in dilute aqueous solutions. *Ind. Eng. Chem. Fundam.* 10, 593-600. Supplement Materials.

Tsuda, T., Aoki, M., Kojima, M., Fujita, T. (1993) Accumulation and excretion of chloroanilines by carp. *Chemosphere* 26, 2301-2306.

Tuazon, E.C., Winer, A.M., Pitts, J.N., Jr. (1978) Fourier transform infrared detection of nitroamines in irradiated amine-NO_x systems. *Environ. Sci. Technol.* 12, 954-958.

Tute, M.S. (1971) Principles and practice of Hansch analysis: a guide to structure-activity correlation for the medicinal chemist. *Adv. Drug Res.* 6, 1-77.

Umeyama, H., Nagai, T., Nogami, H. (1971) Mechanism of adsorption of phenols by carbon black from aqueous solution. *Chem. Pharm. Bull.* 19(8), 1714-1721.

Unger, S.H., Chiang, G.H. (1981) Octanol-physiological buffer distribution coefficients of lipophilic amines by reversed-phase high-performance liquid chromatography and their correlation with biological activity. *J. Med. Chem.* 24(3), 262-270.

Unger, S.H., Cook, J.R., Hollenberg, J.S. (1978) Simple procedure for determining octanol-aqueous partition, distribution, and ionization coefficients by reversed-phase high-pressure liquid chromatography. *J. Pharm. Sci.* 67(10), 1664-1667.

Urano, K., Kato, Z. (1986) Evaluation of biodegradation ranks of priority organic compounds. *J. Haz. Mat.* 13, 135-145.

USEPA (1979) *Status Assessment of Toxic Chemicals: Benzidine.* U.S. EPA-600/2-79-210.

USEPA (1980) *Ambient Water Quality Criteria for Benzidine.* pp. B1 to B6.

USEPA (1986) EXAMS II Computer Modeling System.

USEPA (1987) EXAMS II Computer Modeling System.

Valvani, S.C., Yalkowsky, S.H., Roseman, T.J. (1981) Solubility and partitioning. IV. Aqueous solubility and octanol-water partition coefficient of liquid nonelectrolytes. *J. Pharm. Sci.* 70, 502-507.

Van Bladel, R., Moreale, A. (1977) Adsorption of herbicide-derived *p*-chloroaniline residues in soils: A predictive equation. *J. Soil Sci.* 28, 93-102.

Veith, G.D., Austin, N.M., Morris, R.T. (1979) A rapid method for estimating log *p* for organic chemicals. *Water Res.* 13, 43-47.

Veith, G.D., Broderius, S.J. (1987) Structure-toxicity relationships for industrial chemicals causing tyre (II) narcosis syndrome. In: *QSAR in Environmental Toxicology - II.* Kaiser, K.L.E., Ed., pp. 385-391, D. Reidel Publishing Co., Dordrecht, The Netherlands.

Veith, G.D., DeFoe, D.L., Bergstedt, B.V. (1979) Measuring and estimating the bioconcentration factor of chemicals in fish. *J. Fish Res. Board Can.* 36, 1040-1048.

Veith, G.D., Macek, K.J., Petrocelli, S.R., Carroll, J. (1980) An evaluation of using partition coefficients and water solubility to estimate bioconcentration factors for organic chemicals in fish. In: *Aquatic Toxicology.* ASTM STP707, American Society for Testing and Materials. Eaton, J.G., Parrish, P., Hendricks, A.C., Editors. pp.116-129.

Verhaar, H.M., van Leeuwen, C.J., Hermens, J.L.M. (1992) Classifying environmental pollutants. 1. Structure-activity relationships for prediction of aquatic toxicity. *Chemosphere* 25, 471-491.

Verschueren, K. (1977) *Handbook of Environmental Data on Organic Chemicals.* Van Nostrand Reinhold, New York.

Verschueren, K. (1983) *Handbook of Environmental Data on Organic Chemicals.* 2nd. Edition, Van Nostrand Reinhold, New York.

von Oepen, B., Kördel, W., Klein, W. (1991) Sorption of nonpolar and polar compounds in soils: Processes, measurements and experience with the applicability of the modified OECD-guideline 106. *Chemosphere* 22, 285-304.

Wahner, A., Zetzsch, C. (1983) Rate constants for the addition of hydroxyl raicals to aromatics (benzene, *p*-chloroaniline, and *o-, m- and p*-dichlorobenzene) and the unimolecular decay of the adduct. Kinetics into a quasi-equilibrium. *J. Phys. Chem.* 87, 4945-4951.

Wakita, K., Yoshimoto, M., Miyamoto, S. (1986) A method for calculation of the aqueous solubility of organic compounds by using new fragment solubility constants. *Chem. Pharm. Bull.* 34, 4663-4681.

Wallington, T.J., Dagaut, P., Kurylo, M.J. (1988) Correlation between gas-phase and solution-phase reactivities of hydroxyl radicals toward saturated organic compounds. *J. Phys. Chem.* 92, 5024-5028.

Walton, B.T., Hendricks, M.S., Anderson, T.A., Griest, W.H., Merriweather, R., Beauchamp, J.J., Francis, C.W. (1992) Soil sorption of volatile and semivolatile organic compounds in a mixture. *J. Environ. Quality* 21, 552-558.

Wang, L., Xu, L., Xu, O., Tian, L., Zhang, Z. (1989) Determination of partition coefficients of organic acids and bases and the correlation of partition coefficients in different systems. *Huanjing Kexue Xuebao* 9(4), 418-424.

Wasik, S.P., Tewari, Y.B., Miller, M.M., Martire, D.E. (1981) *Octaol/Water Partition Coefficients and Water Solubilities of Organic Compounds.* NBSIR No.81-2406. U.S. Dept. of Commerce, Washington, D.C.

Weast, R.C., Editor (1982-83) *Handbook of Chemistry and Physics.* 63rd Edition, CRC Press, Boca Raton, Florida.

Wilson, J.T., Enfield, C.G., Dunlap, W.J. (1981) Transport and fate of selected organic pollutants in a sandy soil. *J. Environ. Qual.* 10, 501-506.

Windholz, M., Editor (1976) *The Merck Index.* Volume 9, Merck & Co. Inc., Rahway, New Jersey.

Windholz, M., Editor (1983) *The Merck Index.* Volume 10, Merck & Co. Inc., Rahway, New Jersey.

Wiśniewska, B., Łencka, M., Rogalski, M. (1986) Vapour pressures of 2,4-, 2,6-, and 3,5-dimethylpyridine at temperatures from 267-360 K. *J. Chem. Thermodyn.* 18, 703-708.

Witte, F., Urbanik, E., Zetzsch, C. (1986) Temperature dependence of the rate constants for the addition of hydroxyl radical to benzene and to some monosubstituted aromatics (aniline, bromobenzene, and nitrobenzene) and the unimolecular decay of the adducts. Part 2. Kinetics into a quasi-equilibrium. *J. Phys. Chem.* 90, 3251-3259.

Wolfenden, R.V. (1978) Interaction of the peptide bond with solvent water: A vapor phase analysis. *Biochemistry* 17(1), 201-204.

Wolff, C.J.M., Crossland, N.O. (1985) Fate and effects of 3,4-dichloroaniline in the laboratory and in outdoor ponds. 1. Fate. *Environ. Toxicol. Chem.* 4, 481-487.

Yaguzhinskii, L.S., Smirnova, E.G., Ratnikova, L.A., Kolesova, G.M., Krasinskaya, I.P. (1973) Hydrophobic sites of the mitochondrial electron transfer system. *J. Bioenerg.* 5(3), 163-174.

Yalkowsky, S.H., Morozowich, W. (1980) Chapter 3, A physical chemical basis for the design of orally active prodrugs. In: *Drug Design.* Vol. IX. Academic Press, Inc., New York.

Yalkowsky, S.H., Valvani, S.C. (1980) Solubility and partitioning I: Solubility of nonelectrolytes in water. *J. Pharm. Sci.* 69(8), 912-922.

Yalkowsky, S.H., Valvani, S.C., Kun, W.-Y., Dannenfelser, R.M., Editors (1987) *Arizona Database of Aqueous Solubility for Organic Compounds.* College of Pharmacy, University of Arizona, Tucson, Arizona.

Yamagami, C., Takao, N. (1992) Hydrophobicity parameters determined by reversed-phase liquid chromatography. V. Relationship between capacity factor and the octanol-water partition coefficient for simple heteroaromatic compounds and their ester derivatives. *Chem. Pharm. Bull.* 40(4), 925-929.

Yamagami, C., Takao, N., Fujita, T. (1990) Hydrophobicity parameter of diazines. 1. Analysis and prediction of partition coefficients of monosubstituted diazines. *Quant. Struct.-Act. Relat.* 9(4), 313-320.

Yaws, C.L., Yabg, H.-C., Hopper, J.R., Hansen, K.C. (1990) Organic chemicals: water solubility data. *Chem. Eng.* July, 115-118.

Yaws, C., Yang, H.C., Pan, X. (1991) Henry's law constants for 362 organic compounds in water. *Chem. Eng.* 98(11), 179-185.

Yeh, K.C., Higuchi, W.I. (1976) Oil-water distribution of *p*-alkylpyridines. *J. Pharm. Sci.* 65, 80-86.

Yoshida, K., Mizuno, T., Ose, Y., Sato, T. (1986) The estimation for toxicity of chemicals on fish by physico-chemical properties. *Chemosphere* 15(2), 195-203.

Yoshida, K., Shigeoka, T., Yamauchi, F. (1983) Non-steady state equilibrium model for the preliminary prediction of the fate of chemicals in the environment. *Ecotox. Environ. Saf.* 7, 179-190.

Yu, G., Xu, X. (1992) Investigation of aqueous solubilities of nitro-PAH by dynamic couple-column HPLC. *Chemosphere* 24(12), 1699-1705.

Zachara, J.M., Ainsworth, C.C., Cowan, C.E., Schmidt, R.L. (1990) The effect of nitrogen on the sorption of ionizable N-containing aromatic compounds by amorphous silica. *Environ. Sci. Technol.* 24, 118-126.

Zachara, J.M., Ainsworth, C.C., Cowan, C.E., Thomas, B.L. (1987) Sorption of binary mixtures of aromatic nitrogen heterocyclic compounds on subsurface materials. *Environ. Sci. Technol.* 21, 397-402.

Zachara, J.M., Ainsworth, C.C., Felice, L.J., Resch, C.T. (1986) Quinoline sorption to subsurface materials: Role of pH and retention of the organic cation. *Environ. Sci. Technol.* 20, 620-627.

Zepp, R.G., Schlotzhauer, P.F., Simmons, M.F., Miller, G.C., Baughman, G.L., Wolfe, N.L. (1984) Dynamics of pollutant photoreactions in the hydrosphere. *Fresenius Z. Anal. Chem.* 319, 119-125.

Zepp, R.G., Gaughman, G.L., Schlotzhauer, P.F. (1981) Comparison of photochemical behavior of various humic substances in water: I. Sunlight induced reactions of aquatic pollutants photosensitized by humic substances. *Chemosphere* 109-118.

Zetzsch, C. (1982) Predicting the rate of OH-addition to aromatics using σ^+-electrophilic substituent constants for mono- and polysubstituted benzene. 15th Informal Conference on Photochemistry, June 27-July 1, 1982, Stanford, California.

Zoeteman, B.C.F., Harmsen, K.M., Linders, J.B.H.J., Morra, C.F.H., Slooff, W. (1980) Persistent organic pollutants in river water and groundwater of the Netherlands. *Chemosphere 9, 231-249.*

Zok, S., Görge, G., Kalsch, W., Nagel, R. (1991) Bioconcentration, metabolism and toxicity of substituted anilines in the zebrafish (*brachydanio rerio). Sci. Total Environ.* 109/110, 411-421.

Zwolinski, B.J., Wilhoit, R.C. (1971) *Handbook of Vapor Pressures and Heats of Vaporization of Hydrocarbons and Related Compounds.* API 44, TRC Publication No. 101, Texas A & M University, College Station, Texas.

List of Symbols and Abbreviations:

A_i	area of phase i, m^2
ALPM	Automated Log-P Measurement
BCF	bioconcentration factor
C	molar concentration, mol/L or mmol/m^3
C^S	saturated aqueous concentration, mol/L or mmol/m^3
C_L	subcooled liquid concentration, mol/L or mmol/m^3
C_S	solid molar concentration, mol/L or mmol/m^3
C_A	concentration in air phase, mol/L or mmol/m^3
C_W	concentration in water phase, mol/L or mmol/m^3
^{14}C	radioactive labelled carbon-14 compound
CC	Countercurrent Chromatography
COD	chemical oxygen demand
CPC	Centrifugal Partition Chromatography
D	D values, mol/Pa·h
D_A	D values for advection, mol/Pa·h
D_{Ai}	D values for advective loss in phase i, mol/Pa·h
D_R	D value for reaction, mol/Pa·h
D_{Ri}	D value for reaction loss in phase i, mol/Pa·h
D_{ij}	intermedia D values, mol/Pa·h
D_{VW}	intermedia D value for air-water diffusion (absorption), mol/Pa·h
D_{RW}	intermedia D value for air-water dissolution, mol/Pa·h
D_{QW}	D value for total particle transport (dry and wet), mol/Pa·h
D_{RS}	D value for rain dissolution (air-soil), mol/Pa·h
D_{QS}	D value for wet and dry deposition (air-soil), mol/Pa·h
D_{VS}	D value for total soil-air transport, mol/Pa·h
D_S	D value for air-soil boundary layer diffusion, mol/Pa·h
D_{SW}	D value for water transport in soil, mol/Pa·h
D_{SA}	D value for air transport in soil, mol/Pa·h
D_{Ti}	total transport D value in bulk phase i, mol/Pa·h
DOC	dissolved organic carbon
DOM	dissolved organic matter
DTA	Differential Thermal Analyzer
E	emission rate, mol/h or kg/h
EPICS	Equilibrium Partitioning In Closed System
F	Fugacity ratio
f	fugacity, Pa
f_i	fugacity in pure phase i, Pa
f-const.	fragmental constants

fluo.	fluorescence method
G	advective inflow, m^3/h
G_B	advective inflow to bottom sediment m^3/h
ΔG_v	Gibbs's free energy of vaporization kJ/mol or kcal/mol
GC	gas chromatography
GC/FID	GC analysis with flame ionization detector
GC/ECD	GC analysis with electron capture detector
GC-RT	GC retention time
gen. col.	generator-column
H, HLC	Henry's law constant, $Pa \cdot m^3/mol$
ΔH_{fus}	enthalpy of fusion, kcal/mol
ΔH_v	enthalpy of vaporization, kJ/mol or kcal/mol
HPLC	high pressure liquid chromatography
HPLC/UV	HPLC analysis with UV detector
HPLC/fluo.	HPLC analysis with fluorescence detector
HPLC-k'	HPLC-capacity factor correlation
HPLC-RI	HPLC-retention index correlation
HPLC-RT	HPLC-retention time correlation
HPLC-RV	HPLC-retention volume correlation
IP	ionization potential
J	intermediate quantities for fugacity calculation
K	Kjeldahl method
k	first-order rate constant, h^{-1} ($hour^{-1}$)
k_i	first-order rate constant in phase i, h^{-1}
k_A	air-water mass transfer coefficient, air-side, m/h
k_W	air-water mass transfer coefficient, water-side, m/h
K_{AW}	dimensionless air/water partition coefficient
K_B	bioconcentration factor
K_h	association coefficient
K_{OC}	organic-carbon sorption partition coefficient
K_{OM}	organic-matter sorption partition coefficient
K_{OW}	octanol-water partition coefficient
K_p	sorption coefficient
k_1	uptake/accumulation rate constant, d^{-1} (day^{-1})
k_2	elimination/clearance/depuration rate constant, d^{-1}
k_b	biodegradation rate constant, d^{-1}
k_h	hydrolysis rate constant, d^{-1}
k_p	photolysis rate constant, d^{-1}
L	lipid content of fish

LSC	Liquid Scintillation Counting
m_i	amount of chemical in phase i, mol or kg
M	total amount of chemical, mol or kg
MCI	molecular connectivity indices
MO	molecular orbital calculation
M.P.	melting point, °C
MR	molar refraction
MS	Mass Spectrometry
MW	molecular weight, g/mol
n_C	number of carbon atoms
n_{Cl}	number of chlorine atoms
P	vapor pressure, Pa (Pascal)
P_L	liquid or subcooled liquid vapor pressure, Pa
P_S	solid vapor pressure, Pa
Q	scavenging ratio
QSAR	Quantitative Structural-Activity Relationship
QSPR	Quantitative Structure-Property Relationship
RC	Radiochemical method
RP-HPLC	Reversed Phase High Pressure Liquid Chromatography
RP-TLC	Reversed Phase Thin Layer Chromatography
S	water solubility, mg/L or g/m^3
ΔS_{fus}	entropy of fusion, J/mol·K or cal/mol·K (e.u.)
$S_{octanol}$	solubility in octanol
SD	standard deviation
t	residence time, h (hour)
t_o	overall residence time, h
t_A	advection persistence time, h
t_B	sediment burial residence time, h
t_R	reaction persistence time, h
$t_{1/2}$	half-life, h
T_{ij}	intermedia transport rate, mol/h or kg/h
T	system temperature, K
T_B	boiling point, K
T_M	melting point, K
TLC	thin-layer chromatography
TMV	total molecular volume per molecule, $Å^3$ (Angstrom^3)
TN	titration method
TSA	total surface area per molecule, $Å^2$
U_1	air side, air-water MTC (same as k_A), m/h

U_2	water side, air-water MTC (same as k_W), m/h
U_3	rain rate (same as U_R), m/h
U_4	aerosol deposition rate, m/h
U_5	soil-air phase diffusion MTC, m/h
U_6	soil-water phase diffusion MTC, m/h
U_7	soil-air boundary layer MTC, m/h
U_8	sediment-water MTC, m/h
U_9	sediment deposition rate, m/h
U_{10}	sediment resuspension rate, m/h
U_{11}	soil-water run-off rate, m/h
U_{12}	soil-solids run-off rate, m/h
U_R	rain rate, m/h
U_Q	dry deposition velocity, m/h
U_B	sediment burial rate, m/h
UV	UV spectrometry
UNIFAC	UNIQUAC Functional Group Activity Coefficients
V_i	volume of pure phase i, m^3
V_S	volume of bottom sediment, m^3
V_{Bi}	volume of bulk phase i, m^3
V_I	intrinsic molar volume, cm^3/mol
V_M	molar volume, cm^3/mol
v_i	volume fraction of phase i
v_Q	volume fraction of aerosol
VOC	volatile organic chemicals
W	molecular mass, g/mol
Z_i	fugacity capacity of phase i, mol/m^3 Pa
Z_{Bi}	fugacity capacity of bulk phase i, mol/m^3 Pa

Greek characters:

π-const.	substituent constants
γ	solute activity coefficient
γ_o	solute activity coefficient in octanol phase
γ_W	solute activity coefficient in water phase
ρ_i	density of pure phase i, kg/m^3
ρ_{Bi}	density of bulk phase i, kg/m^3
χ	molecular connectivity indices
ϕ_{OC}	organic carbon fraction
ϕ_i	organic carbon fraction in phase i

APPENDICES:

A.1 BASIC COMPUTER PROGRAM FOR FUGACITY CALCULATIONS

```
10 REM Fugacity Level I,II and III program, 6 compartments,(GENDISS)
20 REM Select condensed print
30 WIDTH "lpt1:",250
40 LPRINT CHR$(15)
50 DIM N$(9),V(9),Z(9),C(9),F(9),M(9),P(9),CG(9),CU(9),DEN(9),ORG(9),VZ(9),DR(9),
DA(9),CB(9),A(9),PA(9),PR(9),RK(9),GA(9)
60 DIM NR(9),NA(9),I(9),GD(9,9),D(9,9),N(9,9),GRA(9),TD(9,9),HL(9),U(20),UY(20),
U$(20)
70 REM N$  = six phases : air, water, soil, sediment, susp sedt and fish
80 REM V   = volume of the six phases  (m3)
90 REM DEN = density of the six phases (kg/m3)
100 REM HT  = depth of air, water, soil and sediment (m)
110 REM AR  = area of air, water, soil and sediment  (m2)
120 REM ORG = the fraction of organic carbons in sediment and susp sedt
130 REM Z   = Z values for each phase (mol/m3.Pa)
140 REM VZ  = VZ values for each phase (mol/Pa)
150 REM F   = fugacity values for each phase (Pa)
160 REM C   = concentration of chemical in each phase (mol/m3)
170 REM CG  = concentration of chemical in each phase (g/m3)
180 REM CU  = concentration of chemical in each phase (ug/g)
190 REM M   = the total amount of chemical in each phase (mol)
200 REM MK  = the total amount of chemical in each phase (kg)
210 REM P   = the mole percent of chemical in each phase (%)
220 N$(1)="Air      "
230 N$(2)="Water    "
240 N$(3)="Soil     "
250 N$(4)="Sediment "
260 N$(5)="Susp sedt"
270 N$(6)="Fish     "
280 ART=100000!*1000000!:FAR(2)=.1:FAR(3)=1-FAR(2)
290 AR(2)=FAR(2)*ART:AR(3)=FAR(3)*ART:AR(1)=ART:AR(4)=AR(2)'areas m2
300 HT(1)=1000:HT(2)=20:HT(3)=.1:HT(4)=.01
310 V(1)=AR(1)*HT(1):V(2)=AR(2)*HT(2):V(3)=AR(3)*HT(3):V(4)=AR(4)*HT(4)
320 V(5)=.000005*V(2):V(6)=.000001*V(2)
330 REM input properties
340 PRINT "Select chemical, chemical to be user-specified =1, benzene= 2, HCB= 3, Hexa-CB= 4,
BAP=5,TriCB= 6, TCE= 7, p-cresol =8, test=9, to exit type 10 "
350 INPUT QC
360 ON QC GOTO 370,510,520,530,540,550,560,570,580,4360
370 INPUT "Name of chemical ",CHEM$
380 INPUT "Temperature eg 25 deg C";TC
390 INPUT "Melting point temperature or data temperature if chemical is liquid eg 80 deg C";TM
400 INPUT "Molecular mass eg 200 g/mol";WM
401 PRINT "For a non-dissociating chemical input a pKa of zero"
402 INPUT "Dissociation constant pKa eg 5.0";PKA
403 IF PKA=0 GOTO 410
404 INPUT "pH of data determination eg 7.0";PHD
406 INPUT "pH of environment eg 6.0";PHE
410 INPUT "Vapor pressure eg 2 Pa ";P
420 INPUT "Water solubility eg 50 g/m3 ";SG
430 INPUT "Log octanol water coefficient eg 4.0 ";LKOW
440 PRINT "Input overall reaction rate half-lives eg 100 h "
450 PRINT "For zero reaction rate enter a fictitiously long half life eg 1E11 h"
460 INPUT "Half life in air        ";HL(1)
470 INPUT "Half life in water      ";HL(2)
```

```
0 INPUT "Half life in soil        ";HL(3)
0 INPUT "Half life in sediment   ";HL(4)
0 GOTO 590
0
EM$ = "Benzene":TC = 25!:WM = 78.11:P = 12700:SG = 1780:LKOW = 2.13:TM = 5.53:HL(1) = 17:HL(2) = 170:H
) = 550:HL(4) = 1700:GOTO 590
0
EM$ = "Hexachlorobenzene(HCB)":TC = 25!:WM = 284.8:P = .0023:SG = .005:LKOW = 5.5:TM = 230!:HL(1) = 17
0:
(2) = 55000!:HL(3) = 55000!:HL(4) = 55000!:GOTO 590
0 CHEM$ = "Hexachlorobiphenyl Hexa-CB)":TC = 25!:WM = 350:P = .0005:SG = .0035:LKOW = 6.8:TM = 106.85:
(1) = 5500:HL(2) = 55000!:HL(3) = 55000!:HL(4) = 55000!:GOTO 590
0
EM$ = "Benzo[a]pyrene(BAP)":TC = 25!:WM = 252.3:P = 7.3E-07:SG = .0038:LKOW = 5.98:TM = 175!:HL(1) = 5:
(2) = 5:HL(3) = 5500:HL(4) = 17000:GOTO 590
0 CHEM$ = "1,2,3-Trichlorobenzene":TC = 25!:WM = 181.45:P = 28!:SG = 21!:LKOW = 4.1:TM = 53!:HL(1) = 550:
(2) = 1700:HL(3) = 5500:HL(4) = 17000!:GOTO 590
0
EM$ = "Trichloroethylene(TCE)":TC = 25!:WM = 131.5:P = 9870:SG = 1100:LKOW = 2.29:TM = -73!:HL(1) = 170:
(2) = 5500:HL(3) = 5500:HL(4) = 17000:GOTO 590
0 CHEM$ = "p-Cresol":TC = 25:WM = 108.14:P = 14.4:SG = 1.8:LKOW = 1.92:TM = 34.8:HL(1) = 5:
(2) = 17:HL(3) = 55:HL(4) = 170:GOTO 590
0 CHEM$ = "Test phenolic":TC = 27.5:WM = 100!:P = .00001:SG = .001:LKOW = 6!:TM = 127.5:HL(1) = 100:
(2) = 1000:HL(3) = 5000:HL(4) = 50000!:PKA = 5:PHD = 6:PHE = 7:GOTO 590
0 MTK = 100000!
0 MT = MTK*1000/WM
0 REM   Input for Fugacity Level II program
0 EK = 1000
0 E = EK*1000/WM
5 PRINT "The emission rates for Level III are 1000 kg/h to each of air, water, and soil in turn "
0 PRINT "Input desired emission rates of chemical for fourth Level III calculation kg/h"
0 INPUT "Emission into air       ";IK4(1)
0 INPUT "Emission into water    ";IK4(2)
0 INPUT "Emission into soil      ";IK4(3)
0 GRA(1) = 100
0 GRA(2) = 1000
0 GRA(4) = 50000!
0 S = SG/WM 'solubility mol/m3
0 H = P/S   'Henry's law constant Pa.m3/mol
0 KOW = 10^LKOW  'Octanol-water partition coefficient
0 KOC = .41*KOW  'Organic carbon-water partition coefficient
0 KFW = .05*KOW  'Fish-water bioconcentration factor
0 TK = TC + 273.15 'Temperature K
0 RG = 8.314     'Gas constant
0 IF TM>TC GOTO 790 ELSE GOTO 810
0 FR = EXP(6.79*(1-(TM + 273.15)/TK))
0 GOTO 820
0 FR = 1
0 PL = P/FR
0 ORG(3) = .02:ORG(4) = .04:ORG(5) = .2'Organic carbon contents g/g
0 DEN(1) = .029*101325!/RG/TK:DEN(2) = 1000:DEN(3) = 2400 'Densities kg/m3
0 DEN(4) = 2400:DEN(5) = 1500:DEN(6) = 1000:DEN(7) = 2000 'Densities kg/m3
1 IF PKA = 0 GOTO 860
2 ID = 10^(PHD-PKA)  'Ratio ionic to non-ionic species at data pH
4 IE = 10^(PHE-PKA)  'Ratio ionic to non ionic species at environmental pH
0 REM calculate Z values, E and D refer to data and environment
1 REM N, I, and T refer to non-ionic , ionic and total
```

```
870 Z(1)=1/RG/TK
880 ZTD(2)=1/H: ZID(2)=ZTD(2)*ID/(1+ID):ZND(2)=ZTD(2)/(1+ID)
882 ZNE(2)=ZND(2):ZIE(2)=ZNE(2)*IE:ZTE(2)=ZNE(2)+ZIE(2)
885 Z(2)=ZTE(2)
890 Z(3)=ZTD(2)*DEN(3)*ORG(3)*KOC/1000
900 Z(4)=ZTD(2)*DEN(4)*ORG(4)*KOC/1000
910 Z(5)=ZTD(2)*DEN(5)*ORG(5)*KOC/1000
920 Z(6)=ZTD(2)*DEN(6)*KFW/1000
930 K71=6000000!/PL
940 Z(7)=Z(1)*K71
950 K12=Z(1)/Z(2) 'Partition coefficients
960 K32=Z(3)/Z(2)
970 K42=Z(4)/Z(2)
980 K52=Z(5)/Z(2)
990 K62=Z(6)/Z(2)
1000 REM calculate distribution
1010 VZT=0
1020 FOR N= 1 TO 6
1030 VZ(N)=V(N)*Z(N)
1040 VZT=VZT+VZ(N)
1050 NEXT N
1060 F1=MT/VZT 'fugacity
1070 FOR N=1 TO 6
1080 F(N)=F1
1090 C(N)=F(N)*Z(N) 'concentration mol/m3
1100 M(N)=C(N)*V(N) 'amount mol
1110 MK(N)=M(N)*WM/1000
1120 P(N)=100*M(N)/MT 'percentages
1130 CG(N)=C(N)*WM  'concentration g/m3
1140 CU(N)=CG(N)*1000/DEN(N)'concentration ug/g
1150 NEXT N
1160 REM print out results
1170 LPRINT " PROGRAM 'GENDISS':SIX COMPARTMENT FUGACITY LEVEL I CALCULATION "
1180 LPRINT " "
1190 LPRINT "Properties of "CHEM$
1200 LPRINT " "
1210 LPRINT "Temperature deg C                ";TC
1220 LPRINT "Molecular mass g/mol             ";WM
1230 LPRINT "Melting point deg C              ";TM
1240 LPRINT "Fugacity ratio                   ";FR
1241 IF PKA=0 GOTO 1250
1242 LPRINT "Dissociation constant pKa        ";PKA
1244 LPRINT "pH of data determination         ";PHD
1246 LPRINT "pH of environment                ";PHE
1247 LPRINT "Ratio ionic to non-ionic (data)  ";ID
1248 LPRINT "Ratio ionic to non-ionic (envir) ";IE
1250 LPRINT "Vapor pressure Pa                ";P
1260 LPRINT "Sub-cooled liquid vapor press Pa ";PL
1270 LPRINT "Solubility g/m3                  ";SG
1280 LPRINT "Solubility mol/m3                ";S
1290 LPRINT "Henry's law constant Pa.m3/mol   ";H
1300 LPRINT "Log octanol-water p-coefficient  ";LKOW
1310 LPRINT "Octanol-water partn-coefficient  ";KOW
1320 LPRINT "Organic C-water ptn-coefficient  ";KOC
1330 LPRINT "Fish-water partition coefficient ";KFW
1340 LPRINT "Air-water partition coefficient  ";K12
1350 LPRINT "Soil-water partition coefficient ";K32
```

```
1360 LPRINT "Sedt-water partition coefficient ";K42
1370 LPRINT "Susp sedt-water partn coeffnt    ";K52
1380 LPRINT "Aerosol-air partition coeff    ";K71
1390 LPRINT "Aerosol Z value                ";Z(7)
1400 LPRINT "Aerosol density kg/m3          ";DEN(7)
1401 LPRINT " "
1402 IF PKA=0 GOTO 1410
1403 LPRINT " Z values in water "
1404 LPRINT "        At data pH                  At environmental pH"
1405 LPRINT "   Neutral     Ionic      Total       Neutral      Ionic       Total"
1406 LPRINT USING "  ##.###^^^^  ##.###^^^^  ##.###^^^^  ##.###^^^^
ZND(2),ZID(2),ZTD(2),ZNE(2),ZIE(2),ZTE(2)
1408 LPRINT " "
1410 LPRINT " "
1420 LPRINT "Amount of chemical moles         ";MT
1430 LPRINT "Amount of chemical kilograms     ";MTK
1440 LPRINT "Fugacity Pa                  ";F1
1450 LPRINT "Total of VZ products             ";VZT
1460 LPRINT " "
1470 LPRINT "Phase properties and compositions"
1480 LPRINT " "
1490 LPRINT "Phase      "TAB(15) N$(1) TAB(30) N$(2) TAB(45) N$(3) TAB(60) N$(4) TAB(75) N$(5) TAB(90)
$(6)
1500 LPRINT "Volume m3   "TAB(15) V(1) TAB(30) V(2) TAB(45) V(3) TAB(60) V(4) TAB(75) V(5) TAB(90) V(6)
1510 LPRINT "Density kg/m3"TAB(15) DEN(1) TAB(30) DEN(2) TAB(45) DEN(3) TAB(60) DEN(4) TAB(75) DEN(5)
AB(90) DEN(6)
1520 LPRINT "Depth m     "TAB(15) HT(1) TAB(30) HT(2) TAB(45) HT(3) TAB(60) HT(4)
1530 LPRINT "Area  m2    "TAB(15) AR(1) TAB(30) AR(2) TAB(45) AR(3) TAB(60) AR(4)
1540 LPRINT "Frn org carb " TAB(45) ORG(3) TAB(60) ORG(4) TAB(75) ORG(5)
1550 LPRINT "Z mol/m3.Pa "TAB(15) Z(1) TAB(30) Z(2) TAB(45) Z(3) TAB(60) Z(4) TAB(75) Z(5) TAB(90) Z(6)
1560 LPRINT "VZ mol/Pa   "TAB(15) VZ(1) TAB(30) VZ(2) TAB(45) VZ(3) TAB(60) VZ(4 ) TAB(75) VZ(5) TAB(90)
Z(6)
1570 LPRINT "Fugacity Pa "TAB(15) F(1) TAB(30) F(2) TAB(45) F(3) TAB(60) F(4) TAB(75) F(5) TAB(90) F(6)
1580 LPRINT "Conc mol/m3 "TAB(15) C(1) TAB(30) C(2) TAB(45) C(3) TAB(60) C(4) TAB(75) C(5) TAB(90) C(6)
1590 LPRINT "Conc g/m3   "TAB(15) CG(1) TAB(30) CG(2) TAB(45) CG(3) TAB(60) CG(4 ) TAB(75) CG(5)
AB(90) CG(6)
1600 LPRINT "Conc ug/g   "TAB(15) CU(1) TAB(30) CU(2) TAB(45) CU(3) TAB(60) CU(4) TAB(75) CU(5) TAB(90)
J(6)
1610 LPRINT "Amount mol  "TAB(15) M(1) TAB(30) M(2) TAB(45) M(3) TAB(60) M(4) TAB(75) M(5) TAB(90)
(6)
1620 LPRINT "Amount kg   "TAB(15) MK(1) TAB(30) MK(2) TAB(45) MK(3) TAB(60) MK(4) TAB(75) MK(5)
AB(90) MK(6)
1630 LPRINT "Amount %    "TAB(15) P(1) TAB(30) P(2) TAB(45) P(3) TAB(60) P(4) TAB(75) P(5) TAB(90) P(6)
1640 LPRINT CHR$(12)
1650 REM Fugacity Level II program, 6 compartments
1660 REM calculate total inflows
1670 GA(1)=V(1)/GRA(1)
1680 GA(2)=V(2)/GRA(2)
1690 GA(4)=V(4)/GRA(4)
1700 REM calculate D values
1710 NRT=0:NRTK=0:NAT=0:NATK=0:VZT=0:MT=0:DT=0:DTA=0:DTR=0 'set totals to zero
1720 FOR N= 1 TO 4
1730 RK(N)=.693/HL(N) 'rate constants from half lives
1740 DR(N)=V(N)*Z(N)*RK(N):DA(N)=GA(N)*Z(N) 'reaction and advection D values
1750 DTR=DTR+DR(N):DTA=DTA+DA(N) 'total D values
1760 NEXT N
1770 DT=DTR+DTA 'total D value
```

```
1780 F2=E/DT 'fugacity
1790 FOR N=1 TO 6
1800 F(N)=F2
1810 C(N)=F(N)*Z(N) 'concentration mol/m3
1820 M(N)=C(N)*V(N) 'amount mol
1830 MK(N)=M(N)*WM/1000
1840 MT=MT+M(N) 'total amount
1850 CG(N)=C(N)*WM 'concentration g/m3
1860 CU(N)=CG(N)*1000/DEN(N) 'concentration ug/g
1870 NR(N)=V(N)*C(N)*RK(N):NRK(N)=NR(N)*WM/1000 'reaction rates mol/h and kg/h
1880 NA(N)=GA(N)*C(N):NAK(N)=NA(N)*WM/1000 'advection rates mol/h and kg/h
1890 NRT=NRT+NR(N):NAT=NAT+NA(N) 'total rates mol/h
1900 NRTK=NRTK+NRK(N):NATK=NATK+NAK(N) 'total rates kg/h
1910 NEXT N
1920 NT=NRT+NAT:NTK=NRTK+NATK
1930 MTK=MT*WM/1000
1940 FOR N=1 TO 6
1950 P(N)=100*M(N)/MT 'percentages of amount
1960 PR(N)=100*NR(N)/NT 'percentages of reaction rate
1970 PA(N)=100*NA(N)/NT 'percentages of advection rate
1980 NEXT N
1990 IF NRT=0 THEN TR=0 ELSE TR=MT/NRT
2000 IF NAT=0 THEN TA=0 ELSE TA=MT/NAT
2010 TOV=MT/NT 'overall residence time h
2020 REM print out results
2030 LPRINT
2040 LPRINT  "SIX COMPARTMENT FUGACITY LEVEL II CALCULATION  ";CHEM$
2050 LPRINT " "
2060 LPRINT "Emission rate of chemical mol/h  ";E
2070 LPRINT "Emission rate of chemical  kg/h  ";EK
2080 LPRINT "Fugacity Pa                      ";F2
2090 LPRINT "Total amount of chemical mol     ";MT
2100 LPRINT "Total amount of chemical kg      ";MTK
2110 LPRINT " "
2120 LPRINT "Phase properties,compositions and rates"
2130 LPRINT " "
2140 LPRINT "Phase       "TAB(15) N$(1) TAB(30) N$(2) TAB(45) N$(3) TAB(60) N$(4) TAB(75) N$(5) TAB(90)
N$(6)
2150 LPRINT "Adv.flow m3/h"TAB(15) GA(1) TAB(30) GA(2) TAB(45) GA(3) TAB(60) GA(4)
2160 LPRINT "Adv.restime h"TAB(15) GRA(1) TAB(30) GRA(2) TAB(45) GRA(3) TAB(60) GRA(4)
2170 LPRINT "Rct halflife h"TAB(15) HL(1) TAB(30) HL(2) TAB(45) HL(3) TAB(60) HL(4)
2180 LPRINT "Rct rate c.h-1"TAB(15) RK(1) TAB(30) RK(2) TAB(45) RK(3) TAB(60) RK(4)
2190 LPRINT "Fugacity Pa "TAB(15) F(1) TAB(30) F(2) TAB(45) F(3) TAB(60) F(4) TAB(75) F(5) TAB(90) F(6)
2200 LPRINT "Conc mol/m3 "TAB(15) C(1) TAB(30) C(2) TAB(45) C(3) TAB(60) C(4) TAB(75) C(5) TAB(90) C(6)
2210 LPRINT "Conc g/m3   "TAB(15) CG(1) TAB(30) CG(2) TAB(45) CG(3) TAB(60) CG(4 ) TAB(75) CG(5)
TAB(90) CG(6)
2220 LPRINT "Conc ug/g   "TAB(15) CU(1) TAB(30) CU(2) TAB(45) CU(3) TAB(60) CU(4) TAB(75) CU(5) TAB(90
CU(6)
2230 LPRINT "Amount mol  "TAB(15) M(1) TAB(30) M(2) TAB(45) M(3) TAB(60) M(4) TAB(75) M(5) TAB(90)
M(6)
2240 LPRINT "Amount kg   "TAB(15) MK(1) TAB(30) MK(2) TAB(45) MK(3) TAB(60) MK(4) TAB(75) MK(5)
TAB(90) MK(6)
2250 LPRINT "Amount %    "TAB(15) P(1) TAB(30) P(2) TAB(45) P(3) TAB(60) P(4) TAB(75) P(5) TAB(90) P(6)
2260 LPRINT "D rct mol/Pa.h"TAB(15) DR(1) TAB(30) DR(2) TAB(45) DR(3) TAB(60) DR(4)
2270 LPRINT "D adv mol/Pa.h"TAB(15) DA(1) TAB(30) DA(2) TAB(45) DA(3) TAB(60) DA(4)
2280 LPRINT "Rct rate mol/h"TAB(15) NR(1) TAB(30) NR(2) TAB(45) NR(3) TAB(60) NR(4)
2290 LPRINT "Adv rate mol/h"TAB(15) NA(1) TAB(30) NA(2) TAB(45) NA(3) TAB(60) NA(4)
```

```
00 LPRINT "Rct rate kg/h "TAB(15) NRK(1) TAB(30) NRK(2) TAB(45) NRK(3) TAB(60) NRK(4)
10 LPRINT "Adv rate kg/h "TAB(15) NAK(1) TAB(30) NAK(2) TAB(45) NAK(3) TAB(60) NAK(4)
20 LPRINT "Reaction %    "TAB(15) PR(1) TAB(30) PR(2) TAB(45) PR(3) TAB(60) PR(4)
30 LPRINT "Advection %   "TAB(15) PA(1) TAB(30) PA(2) TAB(45) PA(3) TAB(60) PA(4)
40 LPRINT " "
50 LPRINT "Total advection D value     ";DTA
60 LPRINT "Total reaction D value      ";DTR
70 LPRINT "Total D value               ";DT
80 LPRINT " "
90 LPRINT "Output by reaction    mol/h ";NRT
00 LPRINT "Output by advection   mol/h ";NAT
10 LPRINT "Total output by reaction and advection mol/h ";NT
20 LPRINT" "
30 LPRINT "Output by reaction    kg/h  ";NRTK
40 LPRINT "Output by advection   kg/h  ";NATK
50 LPRINT "Total output by reaction and advection kg/h  ";NTK
60 LPRINT" "
70 LPRINT "Overall residence time   h  ";TOV
80 LPRINT "Reaction residence time  h  ";TR
90 LPRINT "Advection residence time h  ";TA
00 LPRINT CHR$(12)
10 LPRINT
20 REM Fugacity Level III Program
30 REM Set bulk phase volumes, densities and Z values
40 VB(1)=V(1):VB(2)=V(2):VB(3)=1.8E+10:VB(4)=5E+08
50 VA(1)=1:VQ(1)=2E-11 'volume fractions
60 VW(2)=1:VP(2)=.000005:VF(2)=.000001
70 VA(3)=.2:VW(3)=.3:VE(3)=.5
80 VW(4)=.8:VS(4)=.2
90 DENB(1)=VA(1)*DEN(1)+VQ(1)*DEN(7)
00 DENB(2)=VW(2)*DEN(2)+VP(2)*DEN(5)+VF(2)*DEN(6)
10 DENB(3)=VW(3)*DEN(2)+VA(3)*DEN(1)+VE(3)*DEN(3)
20 DENB(4)=VW(4)*DEN(2)+VS(4)*DEN(4)
30 ZB(1)=VA(1)*Z(1)+VQ(1)*Z(7)
40 ZB(2)=VW(2)*Z(2)+VP(2)*Z(5)+VF(2)*Z(6)
50 ZB(3)=VA(3)*Z(1)+VW(3)*Z(2)+VE(3)*Z(3)
60 ZB(4)=VW(4)*Z(2)+VS(4)*Z(4)
70 KB12=ZB(1)/ZB(2)
80 KB32=ZB(3)/ZB(2)
90 KB42=ZB(4)/ZB(2)
00 REM Parameters
10 U(1)=5          :U$(1)="air side air-water MTC             "
20 U(2)=.05        :U$(2)="water side air-water MTC           "
30 U(3)=.0001      :U$(3)="rain rate                          "
40 U(4)=6E-10      :U$(4)="aerosol deposition velocity        "
50 U(5)=.02        :U$(5)="soil air phase diffusion MTC       "
60 U(6)=.00001     :U$(6)="soil water phase diffusion MTC     "
70 U(7)=5          :U$(7)="soil air boundary layer MTC        "
80 U(8)=.0001      :U$(8)="sediment-water diffusion MTC       "
90 U(9)=.0000005   :U$(9)="sediment deposition velocity       "
00 U(10)=.0000002:U$(10)="sediment resuspension velocity     "
10 U(11)=.00005   :U$(11)="soil water runoff rate             "
20 U(12)=1E-08    :U$(12)="soil solids runoff rate            "
30 'Calculate D values
40 DRW=AR(2)*U(3)*Z(2)
50 DQW=AR(2)*U(4)*Z(7)
60 DVWA=AR(2)*U(1)*Z(1)
```

```
2870 DVWW=AR(2)*U(2)*Z(2)
2880 DVW=1/(1/DVWA+1/DVWW)
2890 D(2,1)=DVW
2900 D(1,2)=DVW+DQW+DRW
2910 DVSB=AR(3)*U(7)*Z(1)
2920 DVSA=AR(3)*U(5)*Z(1)
2930 DVSW=AR(3)*U(6)*Z(2)
2940 DRS=AR(3)*U(3)*Z(2)
2950 DQS=AR(3)*U(4)*Z(7)
2960 DVS=1/(1/DVSB+1/(DVSW+DVSA))
2970 D(3,1)=DVS
2980 D(1,3)=DVS+DRS+DQS
2990 DSWD=AR(2)*U(8)*Z(2)
3000 DSD=AR(2)*U(9)*Z(5)
3010 DSR=AR(4)*U(10)*Z(4)
3020 D(2,4)=DSWD+DSD
3030 D(4,2)=DSWD+DSR
3040 DSWW=AR(3)*U(11)*Z(2)
3050 DSWS=AR(3)*U(12)*Z(3)
3060 D(3,2)=DSWW+DSWS
3070 D(2,3)=0
3075 IK(1)=1000:IK(2)=0:IK(3)=0:QQ=1 'first Level III emissions
3080 REM calculate total chemical inflows
3090 IN=0:INK=0
3100 FOR N=1 TO 4
3110 I(N)=IK(N)*1000/WM
3120 IN=IN+I(N):INK=INK+IK(N)
3130 NEXT N
3140 REM calculate reaction and advection D values for bulk phases
3150 GAB(1)=VB(1)/GRA(1)
3160 GAB(2)=VB(2)/GRA(2)
3170 GAB(4)=VB(4)/GRA(4)
3180 VZBT=0
3190 FOR N= 1 TO 4
3200 RK(N)=.693/HL(N)
3210 DR(N)=VB(N)*ZB(N)*RK(N):DA(N)=GAB(N)*ZB(N)
3220 VZB(N)=VB(N)*ZB(N)
3230 VZBT=VZBT+VZB(N)
3240 NEXT N
3250 FOR N=1 TO 4
3260 FOR NN=1 TO 4
3270 GD(N,NN)=D(N,NN)/ZB(N)
3280 IF GD(N,NN)=0 GOTO 3300 ELSE GOTO 3290
3290 TD(N,NN)=.693*VB(N)/GD(N,NN)
3300 NEXT NN
3310 NEXT N
3320 DT(1)=DR(1)+DA(1)+D(1,2)+D(1,3)
3330 DT(2)=DR(2)+DA(2)+D(2,1)+D(2,3)+D(2,4)
3340 DT(3)=DR(3)+DA(3)+D(3,1)+D(3,2)
3350 DT(4)=DR(4)+DA(4)+D(4,2)
3360 J1=I(1)/DT(1)+I(3)*D(3,1)/DT(3)/DT(1)
3370 J2=D(2,1)/DT(1)
3380 J3=1-D(3,1)*D(1,3)/DT(1)/DT(3)
3390 J4=D(1,2)+D(3,2)*D(1,3)/DT(3)
3400 F(2)=(I(2)+J1*J4/J3+I(3)*D(3,2)/DT(3)+I(4)*D(4,2)/DT(4))/(DT(2)-J2*J4/J3-D(2,4)*D(4,2)/DT(
3410 F(1)=(J1+F(2)*J2)/J3
3420 F(3)=(I(3)+F(1)*D(1,3))/DT(3)
```

```
3430 F(4)= (I(4)+F(2)*D(2,4))/DT(4)
3440 NRT=0:NAT=0:MT=0
3450 FOR N=1 TO 4
3460 C(N)=F(N)*ZB(N)
3470 M(N)=C(N)*VB(N)
3480 MK(N)=M(N)*WM/1000
3490 MT=MT+M(N)
3500 CG(N)=C(N)*WM
3510 CU(N)=CG(N)*1000/DENB(N)
3520 NR(N)=F(N)*DR(N):NRK(N)=NR(N)*WM/1000
3530 NA(N)=F(N)*DA(N):NAK(N)=NA(N)*WM/1000
3540 NRT=NRT+NR(N):NAT=NAT+NA(N)
3550 NEXT N
3555 F(6)=F(2):C(6)=F(6)*Z(6):CG(6)=C(6)*WM:CU(6)=CG(6)*1000/DEN(6)
3560 MTK=MT*WM/1000
3570 NRTK=NRT*WM/1000:NATK=NAT*WM/1000
3580 NT=NRT+NAT:NTK=NT*WM/1000
3590 FOR N=1 TO 4
3600 P(N)=100*M(N)/MT
3610 PR(N)=100*NR(N)/NT
3620 PA(N)=100*NA(N)/NT
3630 NEXT N
3640 N(1,2)=D(1,2)*F(1):NK(1,2)=N(1,2)*WM/1000
3650 N(1,3)=D(1,3)*F(1):NK(1,3)=N(1,3)*WM/1000
3660 N(2,1)=D(2,1)*F(2):NK(2,1)=N(2,1)*WM/1000
3670 N(2,4)=D(2,4)*F(2):NK(2,4)=N(2,4)*WM/1000
3680 N(3,1)=D(3,1)*F(3):NK(3,1)=N(3,1)*WM/1000
3690 N(3,2)=D(3,2)*F(3):NK(3,2)=N(3,2)*WM/1000
3700 N(4,2)=D(4,2)*F(4):NK(4,2)=N(4,2)*WM/1000
3710 TR=MT/(NRT+.0000001)
3720 TA=MT/(NAT+.0000001)
3730 TOV=MT/NT
3740 TOVD=TOV/24
3750 REM print out results
3760 LPRINT " FOUR COMPARTMENT FUGACITY LEVEL III CALCULATION",CHEM$
3770 LPRINT
3780 LPRINT "Bulk phase properties,compositions and rates"
3790 LPRINT " "
3800 LPRINT "Phase       "TAB(15) N$(1) TAB(30) N$(2) TAB(45) N$(3) TAB(60) N$(4) TAB(75) "Total" TAB(90)
N$(6)
3810 LPRINT "Bulk vol m3 "TAB(15) VB(1) TAB(30) VB(2) TAB(45) VB(3) TAB(60) VB(4)
3820 LPRINT "Density kg/m3"TAB(15) DENB(1) TAB(30) DENB(2) TAB(45) DENB(3) TAB(60) DENB(4)
3830 LPRINT "Bulk Z value"TAB(15) ZB(1) TAB(30) ZB(2) TAB(45) ZB(3) TAB(60) ZB(4 ) TAB(90) Z(6)
3840 LPRINT "Bulk VZ     "TAB(15) VZB(1) TAB(30) VZB(2) TAB(45) VZB(3) TAB(60) VZB(4) TAB(75) VZBT
3850 LPRINT "Emission mol/h"TAB(15) I(1) TAB(30) I(2) TAB(45) I(3) TAB(60) I(4) TAB(75) IN
3860 LPRINT "Emission kg/h "TAB(15) IK(1) TAB(30) IK(2) TAB(45) IK(3) TAB(60) IK(4) TAB(75) INK
3870 LPRINT "Fugacity Pa "TAB(15) F(1) TAB(30) F(2) TAB(45) F(3) TAB(60) F(4) TAB(90) F(6)
3880 LPRINT "Conc mol/m3 "TAB(15) C(1) TAB(30) C(2) TAB(45) C(3) TAB(60) C(4) TAB(90) C(6)
3890 LPRINT "Conc g/m3   "TAB(15) CG(1) TAB(30) CG(2) TAB(45) CG(3) TAB(60) CG(4 ) TAB(90) CG(6)
3900 LPRINT "Conc ug/g   "TAB(15) CU(1) TAB(30) CU(2) TAB(45) CU(3) TAB(60) CU(4) TAB(90) CU(6)
3910 LPRINT "Amount mol  "TAB(15) M(1) TAB(30) M(2) TAB(45) M(3) TAB(60) M(4) TAB(75) MT
3920 LPRINT "Amount kg   "TAB(15) MK(1) TAB(30) MK(2) TAB(45) MK(3) TAB(60) MK(4) TAB(75) MTK
3930 LPRINT "Amount %    "TAB(15) P(1) TAB(30) P(2) TAB(45) P(3) TAB(60) P(4)
3940 LPRINT "Adv.flow m3/h"TAB(15) GAB(1) TAB(30) GAB(2) TAB(45) GAB(3) TAB(60) GAB(4)
3950 LPRINT "D rct mol/Pa.h"TAB(15) DR(1) TAB(30) DR(2) TAB(45) DR(3) TAB(60) DR(4 )
3960 LPRINT "D adv mol/Pa.h"TAB(15) DA(1) TAB(30) DA(2) TAB(45) DA(3) TAB(60) DA(4 )
3970 LPRINT "Rct rate mol/h"TAB(15) NR(1) TAB(30) NR(2) TAB(45) NR(3) TAB(60) NR(4 ) TAB(75) NRT
```

```
3980 LPRINT "Rct rate kg/h "TAB(15) NRK(1) TAB(30) NRK(2) TAB(45) NRK(3) TAB(60) NRK(4 ) TAB(75) NRTK
3990 LPRINT "Adv rate mol/h"TAB(15) NA(1) TAB(30) NA(2) TAB(45) NA(3) TAB(60) NA(4 ) TAB(75) NAT
4000 LPRINT "Adv rate kg/h "TAB(15) NAK(1) TAB(30) NAK(2) TAB(45) NAK(3) TAB(60) NAK(4 ) TAB(75) NATK
4010 LPRINT "Reaction %    "TAB(15) PR(1) TAB(30) PR(2) TAB(45) PR(3) TAB(60) PR(4 )
4020 LPRINT "Advection %   "TAB(15) PA(1) TAB(30) PA(2) TAB(45) PA(3) TAB(60) PA(4 )
4030 LPRINT " "
4040 LPRINT "Overall residence time   h  ";TOV
4050 LPRINT "Reaction residence time  h  ";TR;
4060 LPRINT "    Advection residence time h  ";TA
4070 LPRINT
4080 LPRINT "Intermedia Data.  Half times   Equiv flows   D values    Rates  of transport "
4090 LPRINT "                          h          m3/h        mol/Pa.h    mol/h      kg/h"
4100 LPRINT "Air to water     ";:LPRINT USING " ##.####^^^^ ";TD(1,2);GD(1,2);D(1,2) ;N(1,2);NK(1,2)
4110 LPRINT "Air to soil      ";:LPRINT USING " ##.####^^^^ ";TD(1,3);GD(1,3);D(1,3) ;N(1,3);NK(1,3)
4120 LPRINT "Water to air     ";:LPRINT USING " ##.####^^^^ ";TD(2,1);GD(2,1);D(2,1) ;N(2,1);NK(2,1)
4130 LPRINT "Water to sediment";:LPRINT USING " ##.####^^^^ ";TD(2,4);GD(2,4);D(2,4) ;N(2,4);NK(2,4)
4140 LPRINT "Soil to air      ";:LPRINT USING " ##.####^^^^ ";TD(3,1);GD(3,1),D(3,1) ,N(3,1);NK(3,1)
4150 LPRINT "Soil to water    ";:LPRINT USING " ##.####^^^^ ";TD(3,2);GD(3,2);D(3,2) ;N(3,2);NK(3,2)
4160 LPRINT "Sediment to water";:LPRINT USING " ##.####^^^^ ";TD(4,2);GD(4,2),D(4,2) ,N(4,2);NK(4,2)
4170 LPRINT "   Transport velocity parameters              m/h              m/year  "
4180 FOR I=1 TO 12
4190 UY(I)=U(I)*8760
4200 LPRINT TAB(5) I TAB(10) U$(I) TAB(45) U(I) TAB(60) UY(I)
4210 NEXT I
4220 LPRINT "Individual process D values "
4230 LPRINT "Air-water diffusion (air-side)      ";DVWA TAB(50);
4240 LPRINT "Air-water diffusion (water-side)  ";DVWW
4250 LPRINT "Air-water diffusion (overall)       ";DVW
4260 LPRINT "Rain dissolution to water           ";DRW  TAB(50);
4270 LPRINT "Aerosol deposition to water         ";DQW
4280 LPRINT "Rain dissolution to soil            ";DRS  TAB(50);
4290 LPRINT "Aerosol deposition to soil          ";DQS
4300 LPRINT "Soil-air diffusion (air-phase)    ";DVSA TAB(50);
4310 LPRINT "Soil-air diffusion (water-phase) ";DVSW
4320 LPRINT "Soil-air diffusion (bndry layer) ";DVSB TAB(50);
4330 LPRINT "Soil-air diffusion (overall)       ";DVS
4340 LPRINT "Water-sediment diffusion            ";DSWD
4350 LPRINT "Water-sediment deposition           ";DSD TAB(50);
4360 LPRINT "Sediment-water resuspension         ";DSR
4370 LPRINT "Soil-water runoff (water)           ";DSWW TAB(50);
4380 LPRINT "Soil-water runoff (solids)          ";DSWS
4385 LPRINT CHR$(12)
4390 ON QQ GOTO 4412,4414,4416,4418
4412 IK(1)=0:IK(2)=1000:IK(3)=0:QQ=QQ+1 :GOTO 3080
4414 IK(1)=0:IK(2)=0:IK(3)=1000:QQ=QQ+1 :GOTO 3080
4416 IK(1)=IK4(1):IK(2)=IK4(2):IK(3)=IK4(3):QQ=QQ+1 :GOTO 3080
4418 PRINT "If you wish to run another Level III calculation input 1 ,otherwise input 2."
4419 INPUT QQQ
4420 ON QQQ GOTO 4422,4500
4422 PRINT "Input emission rates of chemical for Level III calculation kg/h"
4430 INPUT "Emission into air       ";IK(1)
4440 INPUT "Emission into water     ";IK(2)
4450 INPUT "Emission into soil      ";IK(3)
4460 GOTO 3080
4500 END
```

PROGRAM 'GENDISS':SIX COMPARTMENT FUGACITY LEVEL I CALCULATION

roperties of pentachlorophenol

emperature deg C	25
olecular mass g/mol	266.34
elting point deg C	174
ugacity ratio	.0335979
issociation constant pKa	4.74
H of data determination	5.1
H of environment	5.1
atio ionic to non-ionic (data)	2.290869
atio ionic to non-ionic (envir)	2.290869
apor pressure Pa	.00415
ub-cooled liquid vapor press Pa	.1235197
olubility g/m3	14
olubility mol/m3	.0525644
enry's law constant Pa.m3/mol	7.895078E-02
og octanol-water p-coefficient	5.05
ctanol-water partn-coefficient	112201.9
rganic C-water ptn-coefficient	46002.79
ish-water partition coefficient	5610.096
ir-water partition coefficient	3.185016E-05
oil-water partition coefficient	2208.134
edt-water partition coefficient	4416.268
usp sedt-water partn coeffnt	13800.84
erosol-air partition coeff	4.857527E+07
erosol Z value	19596.13
erosol density kg/m3	2000

Z values in water

At data pH			At environmental pH		
Neutral	Ionic	Total	Neutral	Ionic	Total
3.849E+00	8.817E+00	1.267E+01	3.849E+00	8.817E+00	1.267E+01

mount of chemical moles	375460
mount of chemical kilograms	100000
ugacity Pa	1.443674E-09
otal of VZ products	2.600726E+14

hase properties and compositions

hase	Air	Water	Soil	Sediment	Susp sedt	Fish
olume m3	1E+14	2E+11	8.999999E+09	1E+08	999999.9	200000
ensity kg/m3	1.185413	1000	2400	2400	1500	1000
epth m	1000	20	.1	.01		
rea m2	1E+11	1E+10	9E+10	1E+10		
rn org carb			.02	.04	.2	
mol/m3.Pa	4.034179E-04	12.66612	27968.49	55936.98	174803.1	71058.15
Z mol/Pa	4.034179E+10	2.533224E+12	2.517164E+14	5.593698E+12	1.74803E+11	1.421163E+10
ugacity Pa	1.443674E-09	1.443674E-09	1.443674E-09	1.443674E-09	1.443674E-09	1.443674E-09
onc mol/m3	5.824037E-13	1.828574E-08	4.037736E-05	8.075473E-05	2.523585E-04	1.025848E-04
onc g/m3	1.551174E-10	4.870224E-06	1.075411E-02	2.150821E-02	6.721317E-02	2.732242E-02
onc ug/g	1.308551E-07	4.870224E-06	4.480878E-03	8.961756E-03	4.480879E-02	2.732242E-02
mount mol	58.24037	3657.148	363396.3	8075.473	252.3585	20.51695
mount kg	15.51174	974.0448	96786.96	2150.821	67.21316	5.464484
mount %	1.551174E-02	.9740449	96.78696	2.150822	6.721316E-02	5.464485E-03

SIX COMPARTMENT FUGACITY LEVEL II CALCULATION pentachlorophenol

Emission rate of chemical mol/h	3754.599
Emission rate of chemical kg/h	1000
Fugacity Pa	3.425497E-08
Total amount of chemical mol	8908780
Total amount of chemical kg	2372765

Phase properties,compositions and rates

Phase	Air	Water	Soil	Sediment	Susp sedt	Fish
Adv.flow m3/h	1E+12	2E+08	0	2000		
Adv.restime h	100	1000	0	50000		
Rct halflife h	550	550	1700	5500		
Rct rate c.h-1	.00126	.00126	4.076471E-04	.000126		
Fugacity Pa	3.425497E-08	3.425497E-08	3.425497E-08	3.425497E-08	3.425497E-08	3.425497E-[illegible]
Conc mol/m3	1.381907E-11	4.338775E-07	9.580596E-04	1.916119E-03	5.987873E-03	2.434094E-[illegible]
Conc g/m3	3.68057E-09	1.155589E-04	.2551696	.5103392	1.59481	.6482966
Conc ug/g	3.104884E-06	1.155589E-04	.1063207	.2126413	1.063207	.6482966
Amount mol	1381.907	86775.5	8622536	191611.9	5987.872	486.8189
Amount kg	368.057	23111.79	2296526	51033.91	1594.81	129.6593
Amount %	1.551174E-02	.9740447	96.78694	2.150821	6.721316E-02	5.464484E-[illegible]
D rct mol/Pa.h	5.083066E+07	3.191862E+09	1.026114E+11	7.048059E+08		
D adv mol/Pa.h	4.034179E+08	2.533224E+09	0	1.11874E+08		
Rct rate mol/h	1.741202	109.3371	3514.952	24.1431		
Adv rate mol/h	13.81907	86.77551	0	3.832238		
Rct rate kg/h	.4637518	29.12085	936.1722	6.430274		
Adv rate kg/h	3.68057	23.11179	0	1.020678		
Reaction %	4.637518E-02	2.912085	93.61721	.6430273		
Advection %	.368057	2.311179	0	.1020678		

Total advection D value	3.048516E+09
Total reaction D value	1.065589E+11
Total D value	1.096075E+11

Output by reaction mol/h 3650.173
Output by advection mol/h 104.4268
Total output by reaction and advection mol/h 3754.6

Output by reaction kg/h 972.1871
Output by advection kg/h 27.81304
Total output by reaction and advection kg/h 1000

Overall residence time h 2372.764
Reaction residence time h 2440.646
Advection residence time h 85311.24

FOUR COMPARTMENT FUGACITY LEVEL III CALCULATION pentachlorophenol

Bulk phase properties,compositions and rates

Phase	Air	Water	Soil	Sediment	Total	Fish
Bulk vol m3	1E+14	2E+11	1.8E+10	5E+08		
Density kg/m3	1.185413	1000.009	1500.237	1280		
Bulk Z value	4.038098E-04	13.61119	13988.04	11197.53		71058.15
Bulk VZ	4.038098E+10	2.722239E+12	2.517848E+14	5.598764E+12	2.601462E+14	
Emission mol/h	3754.599	0	0	0	3754.599	
Emission kg/h	1000	0	0	0	1000	
Fugacity Pa	6.116072E-06	2.906724E-08	7.526238E-09	2.736145E-08		2.906724E-08
Conc mol/m3	2.46973E-09	3.956398E-07	1.052774E-04	3.063805E-04		2.065464E-03
Conc g/m3	6.577879E-07	1.053747E-04	2.803957E-02	8.160139E-02		.5501156
Conc ug/g	5.549017E-04	1.053738E-04	1.869009E-02	6.375109E-02		.5501156
Amount mol	246973	79127.96	1894992	153190.3	2374284	
Amount kg	65778.78	21074.94	504712.2	40800.7	632366.6	
Amount %	10.402	3.332709	79.81323	6.452063		
Adv.flow m3/h	1E+12	2E+08	0	10000		
D rct mol/Pa.h	5.088004E+07	3.430021E+09	1.026393E+11	7.054443E+08		
D adv mol/Pa.h	4.038098E+08	2.722239E+09	0	1.119753E+08		
Rct rate mol/h	311.186	99.70122	772.4881	19.30198	1202.677	
Rct rate kg/h	82.88128	26.55443	205.7445	5.140888	320.3211	
Adv rate mol/h	2469.73	79.12795	0	3.063805	2551.922	
Adv rate kg/h	657.7878	21.07494	0	.816014	679.6788	
Reaction %	8.28813	2.655443	20.57445	.514089		
Advection %	65.7788	2.107494	0	8.160141E-02		

Overall residence time h 632.3668

Reaction residence time h 1974.165 Advection residence time h 930.3905

Intermedia Data.	Half times h	Equiv flows m3/h	D values mol/Pa.h	Rates of transport mol/h	kg/h
Air to water	8.5082E+02	8.1451E+10	3.2891E+07	2.0116E+02	5.3577E+01
Air to soil	2.2136E+02	3.1307E+11	1.2642E+08	7.7319E+02	2.0593E+02
Water to air	9.3824E+04	1.4772E+06	2.0107E+07	5.8445E-01	1.5566E-01
Water to sediment	2.1276E+03	6.5144E+07	8.8668E+08	2.5773E+01	6.8645E+00
Soil to air	1.5351E+07	8.1258E+02	1.1366E+07	8.5547E-02	2.2784E-02
Soil to water	2.1235E+06	5.8742E+03	8.2169E+07	6.1842E-01	1.6471E-01
Sediment to water	3.1154E+04	1.1122E+04	1.2454E+08	3.4076E+00	9.0758E-01

	Transport velocity parameters	m/h	m/year
1	air side air-water MTC	5	43800
2	water side air-water MTC	.05	438
3	rain rate	.0001	.876
4	aerosol deposition velocity	6E-10	5.256E-06
5	soil air phase diffusion MTC	.02	175.2
6	soil water phase diffusion MTC	.00001	.0876
7	soil air boundary layer MTC	5	43800
8	sediment-water diffusion MTC	.0001	.876
9	sediment deposition velocity	.0000005	.00438
10	sediment resuspension velocity	.0000002	.001752
11	soil water runoff rate	.00005	.438
12	soil solids runoff rate	1E-08	.0000876

Individual process D values

Air-water diffusion (air-side)	2.017089E+07	Air-water diffusion (water-side)	6.33306E+09
Air-water diffusion (overall)	2.010685E+07		
Rain dissolution to water	1.266612E+07	Aerosol deposition to water	117576.8
Rain dissolution to soil	1.139951E+08	Aerosol deposition to soil	1058191
Soil-air diffusion (air-phase)	726152.2	Soil-air diffusion (water-phase)	1.139951E+07
Soil-air diffusion (bndry layer)	1.815381E+08	Soil-air diffusion (overall)	1.136645E+07
Water-sediment diffusion	1.266612E+07		
Water-sediment deposition	8.740152E+08	Sediment-water resuspension	1.11874E+08
Soil-water runoff (water)	5.699754E+07	Soil-water runoff (solids)	2.517164E+07

FOUR COMPARTMENT FUGACITY LEVEL III CALCULATION pentachlorophenol

Bulk phase properties,compositions and rates

Phase	Air	Water	Soil	Sediment	Total	Fish
Bulk vol m3	1E+14	2E+11	1.8E+10	5E+08		
Density kg/m3	1.185413	1000.009	1500.237	1280		
Bulk Z value	4.038098E-04	13.61119	13988.04	11197.53		71058.15
Bulk VZ	4.038098E+10	2.722239E+12	2.517848E+14	5.598764E+12	2.601462E+14	
Emission mol/h	0	3754.599	0	0	3754.599	
Emission kg/h	0	1000	0	0	1000	
Fugacity Pa	1.77151E-08	5.409513E-07	2.179963E-11	5.092059E-07		5.409513E-0
Conc mol/m3	7.153533E-12	7.362992E-06	3.049341E-07	5.701847E-03		3.843899E-0
Conc g/m3	1.905272E-09	1.961059E-03	8.121615E-05	1.51863		10.23784
Conc ug/g	1.607264E-06	1.961043E-03	5.413555E-05	1.18643		10.23784
Amount mol	715.3533	1472598	5488.815	2850924	4329726	
Amount kg	190.5272	392211.8	1461.891	759315	1153179	
Amount %	1.652191E-02	34.01135	.1267705	65.84536		
Adv.flow m3/h	1E+12	2E+08	0	10000		
D rct mol/Pa.h	5.088004E+07	3.430021E+09	1.026393E+11	7.054443E+08		
D adv mol/Pa.h	4.038098E+08	2.722239E+09	0	1.119753E+08		
Rct rate mol/h	.9013452	1855.474	2.237499	359.2164	2217.829	
Rct rate kg/h	.2400643	494.1869	.5959355	95.67369	590.6966	
Adv rate mol/h	7.153533	1472.598	0	57.01847	1536.77	
Adv rate kg/h	1.905272	392.2118	0	15.1863	409.3034	
Reaction %	2.400643E-02	49.4187	5.959356E-02	9.567369		
Advection %	.1905272	39.22119	0	1.51863		

Overall residence time h 1153.179
Reaction residence time h 1952.236 Advection residence time h 2817.419

Intermedia Data.	Half times h	Equiv flows m3/h	D values mol/Pa.h	Rates of transport mol/h	kg/h
Air to water	8.5082E+02	8.1451E+10	3.2891E+07	5.8266E-01	1.5519E-01
Air to soil	2.2136E+02	3.1307E+11	1.2642E+08	2.2395E+00	5.9648E-01
Water to air	9.3824E+04	1.4772E+06	2.0107E+07	1.0877E+01	2.8969E+00
Water to sediment	2.1276E+03	6.5144E+07	8.8668E+08	4.7965E+02	1.2775E+02
Soil to air	1.5351E+07	8.1258E+02	1.1366E+07	2.4778E-04	6.5995E-05
Soil to water	2.1235E+06	5.8742E+03	8.2169E+07	1.7913E-03	4.7708E-04
Sediment to water	3.1154E+04	1.1122E+04	1.2454E+08	6.3417E+01	1.6890E+01

	Transport velocity parameters	m/h	m/year
1	air side air-water MTC	5	43800
2	water side air-water MTC	.05	438
3	rain rate	.0001	.876
4	aerosol deposition velocity	6E-10	5.256E-06
5	soil air phase diffusion MTC	.02	175.2
6	soil water phase diffusion MTC	.00001	.0876
7	soil air boundary layer MTC	5	43800
8	sediment-water diffusion MTC	.0001	.876
9	sediment deposition velocity	.0000005	.00438
10	sediment resuspension velocity	.0000002	.001752
11	soil water runoff rate	.00005	.438
12	soil solids runoff rate	1E-08	.0000876

Individual process D values

Air-water diffusion (air-side)	2.017089E+07	Air-water diffusion (water-side)	6.33306E+09
Air-water diffusion (overall)	2.010685E+07		
Rain dissolution to water	1.266612E+07	Aerosol deposition to water	117576.8
Rain dissolution to soil	1.139951E+08	Aerosol deposition to soil	1058191
Soil-air diffusion (air-phase)	726152.2	Soil-air diffusion (water-phase)	1.139951E+07
Soil-air diffusion (bndry layer)	1.815381E+08	Soil-air diffusion (overall)	1.136645E+07
Water-sediment diffusion	1.266612E+07		
Water-sediment deposition	8.740152E+08	Sediment-water resuspension	1.11874E+08
Soil-water runoff (water)	5.699754E+07	Soil-water runoff (solids)	2.517164E+07

FOUR COMPARTMENT FUGACITY LEVEL III CALCULATION pentachlorophenol

Bulk phase properties,compositions and rates

Phase	Air	Water	Soil	Sediment	Total	Fish
Bulk vol m3	1E+14	2E+11	1.8E+10	5E+08		
Density kg/m3	1.185413	1000.009	1500.237	1280		
Bulk Z value	4.038098E-04	13.61119	13988.04	11197.53		71058.15
Bulk VZ	4.038098E+10	2.722239E+12	2.517848E+14	5.598764E+12	2.601462E+14	
Emission mol/h	0	0	3754.599	0	3754.599	
Emission kg/h	0	0	1000	0	1000	
Fugacity Pa	6.908563E-10	4.358869E-10	3.654806E-08	4.103072E-10		4.358869E-10
Conc mol/m3	2.789745E-13	5.93294E-09	5.112358E-04	4.594426E-06		3.097331E-05
Conc g/m3	7.430207E-11	1.580179E-06	.1361625	1.223679E-03		8.249431E-03
Conc ug/g	6.268031E-08	1.580166E-06	9.076068E-02	9.559994E-04		8.249431E-03
Amount mol	27.89745	1186.588	9202245	2297.213	9205756	
Amount kg	7.430208	316.0358	2450926	611.8396	2451861	
Amount %	3.030436E-04	1.288963E-02	99.96186	2.495409E-02		
Adv.flow m3/h	1E+12	2E+08	0	10000		
D rct mol/Pa.h	5.088004E+07	3.430021E+09	1.026393E+11	7.054443E+08		
D adv mol/Pa.h	4.038098E+08	2.722239E+09	0	1.119753E+08		
Rct rate mol/h	3.515079E-02	1.495101	3751.268	.2894488	3753.088	
Rct rate kg/h	9.362061E-03	.3982051	999.1128	.0770918	999.5974	
Adv rate mol/h	.2789745	1.186588	0	4.594426E-02	1.511507	
Adv rate kg/h	7.430208E-02	.3160358	0	1.223679E-02	.4025747	
Reaction %	9.362062E-04	3.982051E-02	99.91128	7.709179E-03		
Advection %	7.430208E-03	3.160358E-02	0	1.223679E-03		

Overall residence time h 2451.861
Reaction residence time h 2452.849 Advection residence time h 6090450

Intermedia Data.	Half times h	Equiv flows m3/h	D values mol/Pa.h	Rates of transport mol/h	kg/h
Air to water	8.5082E+02	8.1451E+10	3.2891E+07	2.2723E-02	6.0519E-03
Air to soil	2.2136E+02	3.1307E+11	1.2642E+08	8.7338E-02	2.3262E-02
Water to air	9.3824E+04	1.4772E+06	2.0107E+07	8.7643E-03	2.3343E-03
Water to sediment	2.1276E+03	6.5144E+07	8.8668E+08	3.8649E-01	1.0294E-01
Soil to air	1.5351E+07	8.1258E+02	1.1366E+07	4.1542E-01	1.1064E-01
Soil to water	2.1235E+06	5.8742E+03	8.2169E+07	3.0031E+00	7.9985E-01
Sediment to water	3.1154E+04	1.1122E+04	1.2454E+08	5.1100E-02	1.3610E-02

	Transport velocity parameters	m/h	m/year
1	air side air-water MTC	5	43800
2	water side air-water MTC	.05	438
3	rain rate	.0001	.876
4	aerosol deposition velocity	6E-10	5.256E-06
5	soil air phase diffusion MTC	.02	175.2
6	soil water phase diffusion MTC	.00001	.0876
7	soil air boundary layer MTC	5	43800
8	sediment-water diffusion MTC	.0001	.876
9	sediment deposition velocity	.0000005	.00438
10	sediment resuspension velocity	.0000002	.001752
11	soil water runoff rate	.00005	.438
12	soil solids runoff rate	1E-08	.0000876

Individual process D values

Air-water diffusion (air-side)	2.017089E+07	Air-water diffusion (water-side)	6.33306E+09
Air-water diffusion (overall)	2.010685E+07		
Rain dissolution to water	1.266612E+07	Aerosol deposition to water	117576.8
Rain dissolution to soil	1.139951E+08	Aerosol deposition to soil	1058191
Soil-air diffusion (air-phase)	726152.2	Soil-air diffusion (water-phase)	1.139951E+07
Soil-air diffusion (bndry layer)	1.815381E+08	Soil-air diffusion (overall)	1.136645E+07
Water-sediment diffusion	1.266612E+07		
Water-sediment deposition	8.740152E+08	Sediment-water resuspension	1.11874E+08
Soil-water runoff (water)	5.699754E+07	Soil-water runoff (solids)	2.517164E+07

FOUR COMPARTMENT FUGACITY LEVEL III CALCULATION pentachlorophenol

Bulk phase properties,compositions and rates

Phase	Air	Water	Soil	Sediment	Total	Fish
Bulk vol m3	1E+14	2E+11	1.8E+10	5E+08		
Density kg/m3	1.185413	1000.009	1500.237	1280		
Bulk Z value	4.038098E-04	13.61119	13988.04	11197.53		71058.15
Bulk VZ	4.038098E+10	2.722239E+12	2.517848E+14	5.598764E+12	2.601462E+14	
Emission mol/h	187.73	938.6498	2628.22	0	3754.6	
Emission kg/h	50	250	700	0	1000	
Fugacity Pa	3.10716E-07	1.369963E-07	2.59654E-08	1.289568E-07		1.369963E-07
Conc mol/m3	1.254702E-10	1.864683E-06	3.632052E-04	1.443997E-03		9.734701E-03
Conc g/m3	3.341772E-08	4.966397E-04	9.673606E-02	.3845941		2.59274
Conc ug/g	2.819078E-05	4.966355E-04	6.448051E-02	.3004641		2.59274
Amount mol	12547.02	372936.6	6537693	721998.4	7645175	
Amount kg	3341.772	99327.92	1741249	192297.1	2036216	
Amount %	.1641168	4.878065	85.51397	9.443844		
Adv.flow m3/h	1E+12	2E+08	0	10000		
D rct mol/Pa.h	5.088004E+07	3.430021E+09	1.026393E+11	7.054443E+08		
D adv mol/Pa.h	4.038098E+08	2.722239E+09	0	1.119753E+08		
Rct rate mol/h	15.80924	469.9001	2665.072	90.9718	3241.753	
Rct rate kg/h	4.210634	125.1532	709.8151	24.22943	863.4083	
Adv rate mol/h	125.4702	372.9366	0	14.43997	512.8467	
Adv rate kg/h	33.41772	99.32792	0	3.845941	136.5916	
Reaction %	.4210634	12.51532	70.98152	2.422943		
Advection %	3.341772	9.932792	0	.3845942		

Overall residence time h 2036.216
Reaction residence time h 2358.347 Advection residence time h 14907.33

Intermedia Data.	Half times h	Equiv flows m3/h	D values mol/Pa.h	Rates of transport mol/h	kg/h
Air to water	8.5082E+02	8.1451E+10	3.2891E+07	1.0220E+01	2.7219E+00
Air to soil	2.2136E+02	3.1307E+11	1.2642E+08	3.9281E+01	1.0462E+01
Water to air	9.3824E+04	1.4772E+06	2.0107E+07	2.7546E+00	7.3365E-01
Water to sediment	2.1276E+03	6.5144E+07	8.8668E+08	1.2147E+02	3.2353E+01
Soil to air	1.5351E+07	8.1258E+02	1.1366E+07	2.9513E-01	7.8606E-02
Soil to water	2.1235E+06	5.8742E+03	8.2169E+07	2.1336E+00	5.6825E-01
Sediment to water	3.1154E+04	1.1122E+04	1.2454E+08	1.6060E+01	4.2775E+00

	Transport velocity parameters	m/h	m/year
1	air side air-water MTC	5	43800
2	water side air-water MTC	.05	438
3	rain rate	.0001	.876
4	aerosol deposition velocity	6E-10	5.256E-06
5	soil air phase diffusion MTC	.02	175.2
6	soil water phase diffusion MTC	.00001	.0876
7	soil air boundary layer MTC	5	43800
8	sediment-water diffusion MTC	.0001	.876
9	sediment deposition velocity	.0000005	.00438
10	sediment resuspension velocity	.0000002	.001752
11	soil water runoff rate	.00005	.438
12	soil solids runoff rate	1E-08	.0000876

Individual process D values

Air-water diffusion (air-side)	2.017089E+07	Air-water diffusion (water-side)	6.33306E+09
Air-water diffusion (overall)	2.010685E+07		
Rain dissolution to water	1.266612E+07	Aerosol deposition to water	117576.8
Rain dissolution to soil	1.139951E+08	Aerosol deposition to soil	1058191
Soil-air diffusion (air-phase)	726152.2	Soil-air diffusion (water-phase)	1.139951E+07
Soil-air diffusion (bndry layer)	1.815381E+08	Soil-air diffusion (overall)	1.136645E+07
Water-sediment diffusion	1.266612E+07		
Water-sediment deposition	8.740152E+08	Sediment-water resuspension	1.11874E+08
Soil-water runoff (water)	5.699754E+07	Soil-water runoff (solids)	2.517164E+07

Level III Calculation: pentachlorophenol

Phase Properties and Rates:

Compartment	Bulk Z mol/m3 Pa	Half-life h	D Values Reaction mol/Pa h	D Values Advection mol/Pa h
Air (1)	4.038E−04	550	5.09E+07	4.04E+08
Water (2)	1.361E+01	550	3.43E+09	2.72E+09
Soil (3)	1.399E+04	1700	1.03E+11	
Bottom sediment (4)	1.120E+04	5500	7.05E+08	1.12E+08

Intermedia D values:

	mol/Pa h
air/water D12	3.289E+07
air/soil, D13	1.264E+08
water/air, D21	2.011E+07
water/sed., D24	8.867E+08
soil/air, D31	1.137E+07
soil/water, D32	8.217E+07
sed./water, D42	1.245E+08

	E(1)=1000	E(2)=1000	E(3)=1000	E(1,2,3)	
Overall residence time =	632.37	1153.18	2451.86	2036.22	h
Reaction time =	1974.17	1952.24	2452.85	2358.35	h
Advection time =	930.39	2817.42	6090446	14907.33	h

Phase Properties, Compositions, Transport and Transformation Rates:

Emission, kg/h			Fugacity, Pa				Concentration, g/m3				Amounts, kg				Total amount, kg
E(1)	E(2)	E(3)	f(1)	f(2)	f(3)	f(4)	C(1)	C(2)	C(3)	C(4)	m(1)	m(2)	m(3)	m(4)	
1000	0	0	6.116E−06	2.907E−08	7.526E−09	2.736E−08	6.578E−07	1.054E−04	2.804E−02	8.160E−02	6.578E+04	2.107E+04	5.047E+05	4.080E+04	6.324E+05
0	1000	0	1.772E−08	5.410E−07	2.180E−11	5.092E−07	1.905E−09	1.961E−03	8.122E−05	1.519E+00	1.905E+02	3.922E+05	1.462E+03	7.593E+05	1.153E+06
0	0	1000	6.909E−10	4.359E−10	3.655E−08	4.103E−10	7.430E−11	1.580E−06	1.362E−01	1.224E−03	7.430E+00	3.160E+02	2.451E+06	6.118E+02	2.452E+06
50	250	700	3.107E−07	1.370E−07	2.597E−08	1.290E−07	3.342E−08	4.966E−04	9.674E−02	3.846E−01	3.342E+03	9.933E+04	1.741E+06	1.923E+05	2.036E+06

Emission, kg/h			Loss, Reaction, kg/h				Loss, Advection, kg/h			Intermedia Rate of Transport, kg/h T12	T21	T13	T31	T32	T24	T42
E(1)	E(2)	E(3)	R(1)	R(2)	R(3)	R(4)	A(1)	A(2)	A(4)	air-water	water-air	air-soil	soil-air	soil-water	water-sed	sed-water
1000	0	0	8.288E+01	2.655E+01	2.06E+02	5.141E+00	6.578E+02	2.107E+01	8.160E−01	5.358E+01	1.557E−01	2.059E+02	2.278E−02	1.647E−01	6.864E+00	9.076E−01
0	1000	0	2.401E−01	4.942E+02	5.96E−01	9.567E+01	1.905E+00	3.922E+02	1.519E+01	1.552E−01	2.897E+00	5.965E−01	6.599E−05	4.771E−04	1.278E+02	1.689E+01
0	0	1000	9.362E−03	3.982E−01	9.99E+02	7.709E−02	7.430E−02	3.160E−01	1.224E−02	6.052E−03	2.334E−03	2.326E−02	1.106E−01	7.999E−01	1.029E−01	1.361E−02
50	250	700	4.211E+00	1.252E+02	7.10E+02	2.423E+01	3.342E+01	9.933E+01	3.846E+00	2.722E+00	7.337E−01	1.046E+01	7.861E−02	5.683E−01	3.235E+01	4.277E+00

Output:

pentachlorophenol

Level I calculation:

Physical–chemical properties:		Partition coeff.:		Z values in water:	
molecular wt., g/mol	266.34	HLC	0.07895	at data pH	5.1
melting point, C	174	Kaw	3.185E–05	neutral	3.849
solubility, g/m3	14	Koc	46002.76	ionic	8.817
vapor pressure, Pa	0.00415	BCF	5610.09	total	12.666
log Kow	5.05	Ksw	2208.13	at environ. pH	5.1
fugacity ratio, F	0.0336	Ksd/w	4416.26	neutral	3.849
dissoc. const. pKa	4.74	Kssd/w	13800.83	ionic	8.817
		Kar/w	4.858E+07	total	12.666

Compartment	Z	Concentration			Amount	Amount
	mol/m3 Pa	mol/m3	mg/L (or g/m3)	ug/g	kg	%
Air	4.034E–04	5.824E–13	1.551E–10	1.309E–07	15.512	0.0155
Water	1.267E+01	1.829E–08	4.870E–06	4.870E–06	974.05	0.9740
Soil	2.797E+04	4.038E–05	1.075E–02	4.481E–03	96787	96.787
Biota (fish)	7.106E+04	1.026E–04	2.732E–02	2.732E–02	5.4645	5.46E–03
Suspended sediment	1.748E+05	2.524E–04	6.721E–02	4.481E–02	67.213	6.72E–02
Bottom sediment	5.594E+04	8.075E–05	2.151E–02	8.962E–03	2150.82	2.1508
	Total				100000	100
		f =	1.444E–09	Pa		

Level II Calculation:

Compartment	Half-life	D Values D(reaction)	D(advec'n)	Conc'n	Loss Reaction	Loss Advection	Removal	
	h			mol/m3	kg/h	kg/h	%	
Air	550	5.08E+07	4.03E+08	1.38E–11	0.464	3.681	0.414	
Water	550	3.19E+09	2.53E+09	4.34E–07	29.121	23.112	5.223	
Soil	1700	1.03E+11		9.58E–04	936.172		93.617	
Biota (fish)				2.43E–03				
Suspended sediment				5.99E–03				
Bottom sediment	5500	7.05E+08	1.12E+08	1.92E–03	6.430	1.0207	0.7451	
	Total	1.06E+11	2.94E+09		972.19	27.81	100	
	R + A		1.09E+11			1000		
	f =	3.425E–08	Pa	Overall residence time =			2372.76	h
Total amount =		2372764	kg	Reaction time =			2440.65	h
				Advection time =			85311.18	h

For E(A,B,C), i.e., E(1)=50, E(2)=250 and E(3)=700 kg/h

Compartment	Fugacity f's Pa	Concentration C mol/m3	S mg/L	ug/g	amount m, mol C*VBi	amount %	Loss Reaction mol/h (Ci*DRi)	Loss Advection mol/h (Ci*DAi)	Loss rate mol/h (Ci*DL)	Reaction %	Advection %	Removal %
Air (1)	3.107E-07	1.255E-10	3.342E-08	2.819E-05	1.25E+04	0.16412	1.58E+01	125.4702	0.00E+00	0.4211	3.3418	3.76
Water (2)	1.370E-07	1.865E-06	4.966E-04	4.966E-04	3.73E+05	4.87807	4.70E+02	372.9366	0	12.5153	9.9328	22.45
Soil (3)	2.597E-08	3.632E-04	9.674E-02	6.448E-02	6.54E+06	85.51398	2.67E+03	0.0000	0	70.9815	0.0000	70.98
Bottom sediment (4)	1.290E-07	1.444E-03	3.846E-01	3.005E-01	7.22E+05	9.44384	9.10E+01	14.4400	0.00E+00	2.4229	0.3846	2.81E+00
Total					7.65E+06	100	3.24E+03	512.8468	0.0000			100
								3.75E+03				

For E(A), i.e., E(1) = 1000 kg/h conditions:

Compartment	Fugacity f, Pa	Concentration mol/m3	mg/L	ug/g	amount m, mol	amount %	Loss Reaction mol/h	Loss Advection mol/h	Loss rate mol/h	Reaction %	Advection %	Removal %
Air (1)	6.116E-06	2.470E-09	6.578E-07	5.549E-04	2.47E+05	10.40200	311.19	2469.7303	0.00E+00	8.2881	65.7788	74.07
Water (2)	2.907E-08	3.956E-07	1.054E-04	1.054E-04	7.91E+04	3.33271	99.7012	79.1280	0	2.6554	2.1075	4.76
Soil (3)	7.526E-09	1.053E-04	2.804E-02	1.869E-02	1.89E+06	79.81323	772.4881	0.0000	0	20.5744	0.0000	20.5744
Bottom sediment (4)	2.736E-08	3.064E-04	8.160E-02	6.375E-02	1.53E+05	6.452E+00	19.3020	3.064E+00	0.00E+00	5.14E-01	8.16E-02	5.96E-01
Total					2.37E+06		1.20E+03	2551.9220	0.0000			100
								3755				

For E(B), i.e., E(2) = 1000 kg/h conditions:

Compartment	Fugacity f, Pa	Concentration mol/m3	mg/L	ug/g	amount m, mol	amount %	Reaction mol/h	Advection mol/h	Loss rate mol/h	Reaction %	Advection %	Removal %
Air (1)	1.772E-08	7.154E-12	1.905E-09	1.607E-06	715	0.01652	0.9013	7.1535	0.00E+00	0.0240	0.1905	0.21
Water (2)	5.410E-07	7.363E-06	1.961E-03	1.961E-03	1.47E+06	34.01137	1855.47	1472.5987	0	49.4187	39.2212	88.64
Soil (3)	2.180E-11	3.049E-07	8.122E-05	5.414E-05	5488.82	0.12677	2.2375	0.0000	0	0.0596	0.0000	0.0596
Bottom sediment (4)	5.092E-07	5.702E-03	1.519E+00	1.186E+00	2850922	65.84534	359.2162	57.0184	0.00E+00	9.5674	1.52E+00	11.0860
Total					4329725.3		2217.83	1536.7707	0.0000			100
								3755				

For E(C) i.e., E(3) = 1000 kg/h conditions:

Compartment	Fugacity f, Pa	Concentration mol/m3	mg/L	ug/g	amount m, mol	amount %	Reaction mol/h	Advection mol/h	Loss rate mol/h	Reaction %	Advection %	Removal %
Air (1)	6.909E-10	2.790E-13	7.430E-11	6.268E-08	2.79E+01	0.00030	0.04	0.2790	0.00E+00	0.0009	0.0074	0.01
Water (2)	4.359E-10	5.933E-09	1.580E-06	1.580E-06	1.19E+03	0.01289	1.4951	1.1866	0	0.0398	0.0316	0.07
Soil (3)	3.655E-08	5.112E-04	1.36E-01	9.076E-02	9.20E+06	99.96185	3751.27	0.0000	0	99.9113	0.0000	99.91
Bottom sediment (4)	4.103E-10	4.594E-06	1.224E-03	9.560E-04	2.30E+03	0.02495	0.2894	4.594E-02	0.00E+00	7.71E-03	1.22E-03	8.93E-03
Total					9205756.8		3753.09	1.5115	0.0000			100
								3755				

Level III summary

--	E	E(A)	E(B)	E(C)
Total emission rate mol/h	3754.5993	3754.5993	3754.5993	3754.5993
Total VZ products	2.601E+14	2.601E+14	2.601E+14	2.601E+14
Total amount of chemicals	7645175.0	2374283.7	4329725.3	9205756.8
Total advection D value	1.068E+11	1.068E+11	1.068E+11	1.068E+11
Total reaction D value	3.24E+09	3.24E+09	3.24E+09	3.24E+09
Total D value	1.101E+11	1.101E+11	1.101E+11	1.101E+11
Output by reaction, mol/h	3241.7526	1202.6773	2217.8294	3753.0878
Output by advection, mol/h	512.84675	2551.9220	1536.7707	1.5115078
Output by losses	0	0	0	0
Overall residence time, h	2036.2159	632.36673	1153.1788	2451.8612
Reaction residence time, h	2358.3462	1974.1651	1952.2354	2452.8487
Advection residence time, h	14907.328	930.39039	2817.4179	6090445.8

Compartment	Subcomp't	volume fraction vi	Fugacity capacity Zi	Bulk vol. VB, m3	Bulk ZB(i) sum(viZi)	partition coeff. (Zi/Zw)	Bulk VZ=VB*Zi	Bulk den. sum(vidi) kg/m3	Bulk Adv. flow GAB=VB/Gi
Air (1)	Air	1	4.034E-04	1.000E+14	0.0004038	2.97E-05	4.038E+10	1.18541328	1.00E+12
	Aerosol	2.00000E-11	1.960E+04						
Water (2)	Water	1	1.267E+01	2.000E+11	13.611191	2208.1323	2.722E+12	1000.0085	2.00E+08
	Particulate	0.000005	1.748E+05						
	Biota(fish)	0.000001	7.106E+04						
Soil (3)	Air	0.2	4.034E-04	1.800E+10	1.40E+04	4416.2646	2.518E+14	1500.23708	
	Water	0.3	1.267E+01						
	Solids	0.5	27968.465						
Bottom sediment (4)	Water	0.8	12.666118	5.00E+08	11197.519	5610.0922	5.599E+12	1280	10000
	Solids	0.2	55936.930						
					25199.163		2.601E+14		

Level III Intermedia Data:

----------------	D values Dij mol/Pa h	Eq. flows Dij/Z(i) GDij, m3/h	half-life .693Vi/G t(1/2), h	Rate of transport Dij*f(i) N, mol/h	N*MW/1000 Nk, kg/h
Air to water (D12)	3.2891E+07	8.145E+10	8.508E+02	1.022E+01	2.72E+00
Air to soil (D13)	1.2642E+08	3.131E+11	2.214E+02	3.928E+01	1.05E+01
Water to air (D21)	2.0107E+07	1.477E+06	9.382E+04	2.755E+00	7.34E-01
Water to sed. (D24)	8.8668E+08	6.514E+07	2.128E+03	1.215E+02	3.24E+01
Soil to air (D31)	1.1366E+07	8.126E+02	7.676E+06	2.951E-01	7.86E-02
Soil to water (D32)	8.2169E+07	5.874E+03	1.062E+06	2.134E+00	5.68E-01
Sed. to water (D42)	1.2454E+08	1.112E+04	6.231E+03	1.606E+01	4.28E+00

Equations for Dij values:

D12 = A(2)*(1/(1/U(1)*Z(1)+1/(U(2)*Z(2)+U(4)*Z(7))

D13 = A(3)*(1/(1/(U(5)*Z(1)+U(6)*Z(2)))+U(3)*Z(2)+U(4)*Z(7))

D21 = A(2)*(1/(1/(U(1)*Z(1)+1/(U(2)*Z(2)))

D24 = A(2)*(U(8)*Z(2)+U(9)*Z(5))

D31 = A(3)*(1/(1/(U(5)*Z(1)+U(6)*Z(2)+1/(U(7)*Z(1)))+U(3)*Z(2)+U(4)*Z(7))

D32 = A(3)*(U(11)*Z(2)+U(12)*Z(3))

D42 = A(4)*(U(8)*Z(2)+U(10)*Z(4))

				E(A)		E(B)		E(C)	
----------------	D values Dij mol/Pa h	Eq. flows Dij/Z(i) GDij, m3/h	half-life .693Vi/G t(1/2), h	Rate of transport Dij*f(i) N, mol/h	N*MW/100 Nk, kg/h	Rate of transport Dij*f(i) N, mol/h	N*MW/100 Nk, kg/h	Rate of transport Dij*f(i) N, mol/h	N*MW/1000 Nk, kg/h
Air to water (D12)	3.2891E+07	8.145E+10	8.508E+02	2.012E+02	5.36E+01	5.827E-01	1.55E-01	2.272E-02	6.052E-03
Air to soil (D13)	1.2642E+08	3.131E+11	2.214E+02	7.732E+02	2.06E+02	2.240E+00	5.96E-01	8.734E-02	2.326E-02
Water to air (D21)	2.0107E+07	1.477E+06	9.382E+04	5.845E-01	1.56E-01	1.088E+01	2.90E+00	8.764E-03	2.334E-03
Water to sed. (D24)	8.8668E+08	6.514E+07	2.128E+03	2.577E+01	6.86E+00	4.797E+02	1.28E+02	3.865E-01	1.029E-01
Soil to air (D31)	1.1366E+07	8.126E+02	7.676E+06	8.555E-02	2.28E-02	2.478E-04	6.60E-05	4.154E-01	1.106E-01
Soil to water (D32)	8.2169E+07	5.874E+03	1.06E+06	6.184E-01	1.65E-01	1.791E-03	4.77E-04	3.003E+00	7.999E-01
Sed. to water (D42)	1.2454E+08	1.112E+04	6.231E+03	3.408E+00	9.08E-01	6.342E+01	1.69E+01	5.110E-02	1.361E-02

Phase properties and rates:

Compartment	Rate const. k, 1/hr	D Values Reaction VB*ZB*k mol/Pa h	Advection GAB*ZB mol/Pa h	DTs	Js	Js, E(A)	Js, E(B)	Js, E(C)
Air (1)	0.00126	5.088E+07	4.038E+08	6.140E+08	3.062E-07	6.115E-06	0.000E+00	6.766E-10
Water (2)	0.00126	3.430E+09	2.722E+09	7.059E+09	3.275E-02	3.275E-02	3.275E-02	3.275E-02
Soil (3)	0.000407647	1.026E+11	0	1.027E+11	1.000E+00	1.000E+00	1.000E+00	1.000E+00
Bottom sediment (4)	0.000126	7.054E+08	1.120E+08	9.420E+08	3.299E+07	3.299E+07	3.299E+07	3.299E+07
Total	Total	1.068E+11	3.238E+09					
	R + A		1.101E+11					

Define dimension of the generic environment (unit world):

Compartment	Volume, Vi m3	Depth, h m	Area, A m2	Density d, kg/m3	Fugacity Cap., Zi mol/Pa m3	ViZi	*input Advective flow, G mol/h	Residence time, V/G t(R),h	*input Reaction half-life t(1/2),h	*input Rate const. 0.693/t k, 1/hr	*input Emission rate, E mol/h	E(A) mol/h	E(B) mol/h	E(C) mol/h
Air (1)	1.00E+14	1000	1.000E+11	1.1854132	4.034E-04	4.03E+10	1.000E+12	100	550	1.260E-03	187.73	3754.60	0	0
Water (2)	2.00E+11	20	1.000E+10	1000	1.267E+01	2.53E+12	2.000E+08	1000	550	1.260E-03	938.65	0	3754.60	0
Soil (3)	9.00E+09	0.1	9.000E+10	2400	2.797E+04	2.52E+14	0		1700	4.076E-04	2628.22	0	0	3754.60
bottom sediment (4)	1.00E+08	0.01	1.000E+10	2400	5.594E+04	5.59E+12	2000	50000	5500	1.260E-04	0.00	0	0	0
Sus. sediment (5)	1.00E+06			1500	1.748E+05	1.75E+11			1.00E+11	6.9300E-12	3754.60			
Biota (fish) (6)	2.00E+05			1000	7.106E+04	1.42E+10			1.00E+11	6.9300E-12				
Aerosol (7)	2000			2000	1.960E+04	2.60E+14								

Level I calculation:

Compartment	Volume, Vi, m3	Fugacity capacity Zi	ViZi	Conc'n, c c = f*Zi mol/m3	Amount m= ciVi mol	Amount =m*MW/1E kg	Amount %	Conc'n,S mg/L (or g/m3)	Conc'n (S/d)*1000 ug/g
Air (1)	1.00E+14	4.034E-04	4.034E+10	5.824E-13	5.82E+01	1.551E+01	1.55E-02	1.55E-10	1.31E-07
Water (2)	2.00E+11	1.267E+01	2.533E+12	1.829E-08	3.66E+03	9.740E+02	9.74E-01	4.87E-06	4.87E-06
Soil (3)	9.00E+09	2.797E+04	2.517E+14	4.038E-05	3.63E+05	9.679E+04	9.68E+01	1.08E-02	4.48E-03
Bottom sediment (4)	1.00E+08	5.594E+04	5.594E+12	8.075E-05	8.08E+03	2.151E+03	2.15E+00	2.15E-02	8.96E-03
Sus. sediment (5)	1.00E+06	1.748E+05	1.748E+11	2.524E-04	2.52E+02	6.721E+01	6.72E-02	6.72E-02	4.48E-02
Biota (fish) (6)	2.00E+05	7.106E+04	1.421E+10	1.026E-04	2.05E+01	5.464E+00	5.46E-03	2.73E-02	2.73E-02
			2.601E+14		375459.93	100000	100		

Level II phase properties and rates:

Compartment	Rate const. k, 1/hr	D(reaction) VZk	D(advec'n) GZ	conc'n c =f*Z mol/m3	Amount m= ciVi mol	Amount m*MW/1000 kg	conc'n mg/L (or g/m3)	Conc'n ug/g (S/d)*1000	Loss Reaction mol/h Vck	Loss Advection mol/h Gc	% Loss reaction	% Loss advection	Removal %
Air (1)	0.00126	5.083E+07	4.034E+08	1.382E-11	1.38E+03	3.68E+02	3.681E-09	3.10E-06	1.741E+00	1.382E+01	4.64E-02	0.37	0.41
Water (2)	0.00126	3.192E+09	2.533E+09	4.339E-07	8.68E+04	2.31E+04	1.156E-04	1.16E-04	1.093E+02	8.678E+01	2.91E+00	2.311	5.223
Soil (3)	0.000407647	1.026E+11		9.581E-04	8.62E+06	2.30E+06	2.552E-01	1.06E-01	3.515E+03		9.36E+01		9.36E+01
Bottom sediment (4)	0.000126	7.048E+08	1.119E+08	1.916E-03	1.92E+05	5.10E+04	5.103E-01	2.13E-01	2.414E+01	3.832E+00	6.43E-01	1.02E-01	7.45E-01
Sus. sediment (5)	6.93000E-12	1.211E+00		5.988E-03	5.99E+03	1.59E+03	1.595E+00	1.06E+00	4.150E-08		1.11E-09		1.11E-09
Biota (fish) (6)	6.93000E-12	9.849E-02		2.434E-03	4.87E+02	1.30E+02	6.483E-01	6.48E-01	3.374E-09		8.99E-11		8.99E-11
	Total	1.066E+11	3.049E+09		8908779.0	2372764.2			3650.17	104.43	97.22	2.78	100
	R + A		1.096E+11							3754.59938			

Level II summary:

Total amount of chemicals,	8908779.0	moles	2372764	kg
Total reaction D value	1.066E+11			
Total advection D value	3.05E+09			
Total D value	1.096E+11			
Fugacity, E/sum D values	3.425E-08			
Output by reaction, mol/h	3650.1725		972.18695	kg/h
Output by advection, mol/h	104.42685		27.813049	kg/h
Total output, mol/h	3754.5993		1000	kg/h
Overall resistence time, h	2372.7642			
Reaction resistence time, h	2440.6460			
Advection resistence time, h	85311.183			

Kp(3) = 0.41*Kow*y(3)	920	
Kp(4) = 0.41*Kow*y(4)	9201	
Kp(5) = 0.41*Kow*y(5)	1840	
Bioconcentraion factor, BCF or Kb, K(6) = 0.050*Kow	5610	
If ionization ratio ID<=0.001, Z(2)=ZTD(2) otherwise Z(2)=ZTE(2)		
Air/water partition coeff., Kaw, Z(1)/Z(2)	3.185E-05	
Soil/water partition coeff., Ksw, Z(3)/Z(2)	2.208E+03	
Sediment/water partition coeff., Ksd/w, Z(4)/Z(2)	4.416E+03	
Sus. sediment/water partition coeff., Kssd/w, Z(5)/Z(2)	1.380E+04	
Aerosol/water partition coeff., Kar/w, Z(aerosol)/Z(air)	4.858E+07	
Densities, g/cm3 or kg/L		
air, d(1) = (0.029*101325/RT)	0.0011854	
water, d(2)	1	
soil, d(3) =	2.4	
bottom sediment, d(4) =	2.4	
suspended sediment, d(5) =	1.5	
biota, d(6) =	1	
aerosol density, g/m3, d(7) =	2	
Fugacity capacities, Z:		
Z(1) or Z(air) = 1/RT	4.034E-04	
Z(2) = ZTD(2) or Z(water) = 1/H = c/p WHEN pKa >> pH	1.267E+01	
ZID(2) = ZTD(2)*ID/(1+ID)	8.8172494	
ZND(2) = ZTD(2)/(1+ID)	3.8488689	
ZNE(2) = ZND(2), Z for neutral forms	3.8488689	
ZIE(2) = ZNE(2)*IE	8.8172494	
ZTE(2) = ZNE(2)+ZIE(2)	12.666118	
Z(2) = ZTE(2) WHEN dissociation is appreciable, see TEXT	12.666118	
Z(3) or Z(soil) = Kp(s)*ZTD(2)*d(s)	2.797E+04	
Z(4) or Z(bottom sediment) = Kp(bs)*ZTD(2)*d(bs)	5.594E+04	
Z(5) or Z(suspended sediment) = Kp(ss)*ZTD(2)*d(ss)	1.748E+05	
Z(6) or Z(biota) = K *ZTD(2)*d(B)	7.106E+04	
Z(7) or Z(aerosol) = Z(1)*6*E6/p(L)RT	1.960E+04	
Fugacity (Level I), f = total no. of moles/sum(ViZi)	1.444E-09	
Fugacity (Level II), f = emission/sum(D values)	3.425E-08	
Transport parameters:	m/h	m/yr
	k	k/8760
air-water MTC, air side, U(1)	5	43800
air-water MTC, water side, U(2)	0.05	438
rain rate, U(3) = 0.85/8760	0.0001	0.876
aerosol deposition velocity, U(4)	6.00E-10	0.0000052
soil air phase diffusion MTC, U(5)	0.02	175.2
soil water phase diffusion MTC, U(6)	1.00E-05	0.0876
soil air boundary layer MTC, U(7)	5	43800
sediment-water MTC, U(8)	0.0001	0.876
sediment deposition rate, U(9)	5.00E-07	0.00438
sediment resuspended rate, U(10)	2.00E-07	0.001752
soil water runoff rate, U(11)	5.00E-05	0.438
soil solids runoff rate, U(12)	1.00E-08	0.0000876

A2. Fugacity calculations using Lotus 123 spreadsheet program.

Input Parameters in Column C:	Column C		
Chemical name:	pentachlorophenol		
system, C	25		
molecular weight, g/mol	266.34		
melting point, C	174		
dissociation constant, log pKa	4.74	WHEN none available enter 12	
pH of data determination	5.1	WHEN none available enter 6	
pH of environment	5.1		
solubility, g/m3	14		
vapor pressure, Pa	0.00415		
log Kow	5.05		
degradation half-lives, h			
– in air compartment	550		
– in water compartment	550		
– in soil compartment	1700		
– in sediment compartment	5500		
emission rate, kg/h			
– to air compartment only	1000	0	0
– to water compartment only	0	1000	0
– to soil compartment only	0	0	1000
– to all compartments	5	25	700

Go To A288 to check both input and output results.
/GRAPH Name Use (L1, L2, L3...) – to view distribution pie charts

Fugacity Level I, II and III Calculations:

* Amount of chemicals, moles		375460	moles	100000	kg	
* Emission rate of chemicals, E =		3754.60	mol/h	1000	kg/h	
Gas constant, Pa m3/mol K, R=		8.314				
System temperature, in C		25	* input data			
System temperature in K, T = (t + 273.15)		298.15				
* Molecular weight, g/mol	MW =	266.34	* input data			*
* Melting point, t in C	M.P.	174	* input data			*
M.P. or Tm = (mp+273.15) in degree Kelvin		447.15				*
Fugacity ratio = exp(6.79(1–Tm/T)) for solid comp'ds, F		0.0336				*
* Dissociation constant pKa, WHEN none available, enter 12		4.74	* input data			
* pH of data determination, pHd, WHEN none available, enter 6		5.1	* input data			
* pH of environment, pHe		5.1	* input data			
Ratio ionic to non–ionic (data deter'n), ID = 10^(pHd–pKa)		2.2908676				
Ratio ionic to non–ionic (environ.), IE = 10^(pHe–pKa)		2.2908676				
* Solublitiy, g/m3 or mg/L	S =	14	* input data			
molar solubility, mol/m3, c=S/MW		5.26E–02				
* Vapor pressure, Pa	P =	0.00415	* input data			
Vapor pressure, subcooled liquid value, Pa		1.235E–01				
Henry's law constant, Pa m3/mol, H=p/c		0.0790				
* Octanol/water partition coefficient,	log Kow	5.05	* input data			
Kow =		112202				
Partition coefficient, organic C, Koc = 0.41*Kow*y		46002.76				
for soil (mole fraction organic C), y(3) =		0.02				
suspended sediment, y(5) =		0.2				
bottom sediment, y(6) =		0.04				

* input

Adv. flow, G(air)	1.0000E+12	m3/h
Adv. flow G(water)	5.00E+08	m3/h
Adv. flow, G(sed.)	2000	m3/h

Half–lives, hours		
t(air)	550	h
t(water)	550	h
t(soil)	1700	h
t(sediment)	5500	h

Residence time: h		
air	100	h
water	1000	h
sediment	50000	h

Emission Rates, kg/h :

	air	water	soil
E(A,B,C)	50	250	700
E(A)	1000	0	0
E(B)	0	1000	0
E(C)	0	0	1000

Index